Sex Differences in Cardiac Disease
Pathophysiology, Presentation, Diagnosis and Management

Sex Differences in Cardiac Disease

Pathophysiology, Presentation, Diagnosis and Management

Edited by

Niti R. Aggarwal, MD, FACC, FASNC
Senior Associate Consultant
Mayo Clinic
Assistant Professor of Medicine
Mayo Clinic College of Medicine and Science
Rochester, MN, United States

Malissa J. Wood, MD, FACC
Co-director Corrigan Women's Heart
Health Program
Massachusetts General Hospital
Boston, MA
Medical Director
MGH Danvers Ambulatory Cardiology Center
Danvers, MA
Associate Professor of Medicine
Harvard Medical School
Boston, MA, United States

ELSEVIER

Elsevier
125 London Wall, London EC2Y 5AS, United Kingdom
525 B Street, Suite 1650, San Diego, CA 92101, United States
50 Hampshire Street, 5th Floor, Cambridge, MA 02139, United States
The Boulevard, Langford Lane, Kidlington, Oxford OX5 1GB, United Kingdom

Notices
Knowledge and best practice in this field are constantly changing. As new research and experience broaden our understanding, changes in research methods, professional practices, or medical treatment may become necessary.

Practitioners and researchers must always rely on their own experience and knowledge in evaluating and using any information, methods, compounds, or experiments described herein. In using such information or methods they should be mindful of their own safety and the safety of others, including parties for whom they have a professional responsibility.

To the fullest extent of the law, neither the Publisher nor the authors, contributors, or editors, assume any liability for any injury and/or damage to persons or property as a matter of products liability, negligence or otherwise, or from any use or operation of any methods, products, instructions, or ideas contained in the material herein.

Library of Congress Cataloging-in-Publication Data
A catalog record for this book is available from the Library of Congress

British Library Cataloguing-in-Publication Data
A catalogue record for this book is available from the British Library

ISBN: 978-0-12-819369-3

For information on all Elsevier publications
visit our website at https://www.elsevier.com/books-and-journals

Publisher: Stacy Masucci
Acquisitions Editor: Katie Chan
Editorial Project Manager: Tracy I. Tufaga
Production Project Manager: Kiruthika Govindaraju
Cover Designer: Christian J. Bilbow

Typeset by SPi Global, India

Niti R. Aggarwal, MD

To my parents, Monica and Ranjit, for their unconditional love, steadfast belief in me, and for teaching me the value of high expectations

Malissa J. Wood, MD

To my late parents, Betty and Bob Wood, for their love and constant support, and for showing me each day how grit, determination, and hard work would be instrumental in helping me reach my goals

Contents

Contributors

Beth L. Abramson, MD, MSc, FRCPC
University of Toronto; Division of Cardiology, St. Michael's
    Hospital, Toronto, ON, Canada
Peripheral Arterial Disease

Niti R. Aggarwal, MD
Department of Cardiovascular Disease, Mayo Clinic College
    of Medicine and Science, Rochester, MN, United States
Heart Failure With Preserved Ejection Fraction
Pulmonary Arterial Hypertension
Ventricular Arrhythmias
CardioOncology
Cardiovascular Medications
Women's Heart Programs

Jack Aguilar, MD
Barbra Streisand Women's Heart Center, Smidt Heart
    Institute, Cedars-Sinai Medical Center, Los Angeles, CA,
    United States
Coronary Microvascular Dysfunction

Christina K. Anderson, MD
Division of Cardiology, Department of Internal Medicine,
    Rush Medical College, Chicago, IL, United States
Atrial Fibrillation

Zoltan Arany, MD, PhD
Division of Cardiology, University of Pennsylvania
    Perelman School of Medicine, Philadelphia, PA, United
    States
Peripartum Cardiomyopathy

C. Noel Bairey Merz, MD
Barbra Streisand Women's Heart Center, Smidt Heart
    Institute, Cedars-Sinai Medical Center, Los Angeles, CA,
    United States
Coronary Microvascular Dysfunction

Ami B. Bhatt
Division of Cardiology, Department of Medicine,
    Massachusetts General Hospital, Boston, MA, United
    States
Congenital Heart Disease

Laurie Bossory, MD
Division of Cardiology, Department of Internal Medicine,
    The Ohio State University Wexner Medical Center,
    Columbus, OH, United States
Acute Coronary Syndrome

Konstantinos Dean Boudoulas, MD
Division of Cardiology, Department of Internal Medicine,
    The Ohio State University Wexner Medical Center,
    Columbus, OH, United States
Acute Coronary Syndrome

Renee P. Bullock-Palmer, MD
Department of Cardiology, Deborah Heart and Lung Center,
    Browns Mills, NJ, United States
Stable Ischemic Heart Disease

Sarah Chuzi, MD
Department of Medicine, Northwestern University Feinberg
    School of Medicine, Chicago, IL, United States
Heart Failure With Reduced Ejection Fraction

Daniela Crousillat, MD
Cardiac Ultrasound Laboratory, Massachusetts General
    Hospital, Harvard Medical School, Boston, MA, United
    States
Valvular Heart Disease

Anne B. Curtis, MD
Department of Medicine, Jacobs School of Medicine and
    Biomedical Sciences, University at Buffalo, Buffalo,
    New York, United States
Atrial Fibrillation

Esther Davis, MBBS, DPhil
Cardiology Division, Massachusetts General Hospital;
    Harvard Medical School, Boston, MA, United States
Myocardial Infarction With Nonobstructive Coronary
    Disease

Anita Deswal, MD, MPH
Cardiology, University of Texas MD Anderson Cancer
    Center, Houston, TX, United States
Sex Differences in Heart Failure With Preserved Ejection
    Fraction
Pulmonary Arterial Hypertension

Mariana Garcia, MD
Emory University, School of Medicine, Division of
    Cardiology, Atlanta, GA, United States
Epidemiology and Prevalence

Eugenia Gianos, MD
Northwell Health, Department of Cardiology, Donald and Barbara Zucker School of Medicine at Hofstra/Northwell, New York, NY, United States
Sex Hormones and Their Impact on Cardiovascular Health

Ridhima Goel, MD
Interventional Cardiovascular Research and Clinical Trials, The Zena and Michael A. Wiener Cardiovascular Institute, Icahn School of Medicine at Mount Sinai Hospital, New York, NY, United States
Interventions in Ischemic Heart Disease

Rajiv Gulati, MD, PhD
Department of Cardiovascular Diseases, Mayo Clinic College of Medicine and Science, Rochester, MN, United States
Spontaneous Coronary Artery Dissection

Sonia A. Henry, MD
Northwell Health, Department of Cardiology, Northshore University Hospital, Zucker School of Medicine at Hoftsra Northwell, Manhasset, NY, United States
Disparity in Care Across the CVD Spectrum

Jeff C. Huffman, MD
Department of Psychiatry, Massachusetts General Hospital/ Harvard Medical School, Boston, MA, United States
Psychosocial Issues in Cardiovascular Disease

Sasha De Jesus, MD
Northwell Health, Department of Cardiology, Donald and Barbara Zucker School of Medicine at Hofstra/Northwell, New York, NY, United States
Sex Hormones and Their Impact on Cardiovascular Health

Deborah N. Kalkman, MD, PhD
Interventional Cardiovascular Research and Clinical Trials, The Zena and Michael A. Wiener Cardiovascular Institute, Icahn School of Medicine at Mount Sinai Hospital, New York, NY, United States; Amsterdam UMC, University of Amsterdam, Heart Center, Department of Clinical and Experimental Cardiology, Amsterdam Cardiovascular Sciences, Amsterdam, The Netherlands
Interventions in Ischemic Heart Disease

Cynthia Kos, DO
Department of Cardiology, Deborah Heart and Lung Center, Browns Mills, NJ, United States
Stable Ischemic Heart Disease

Yamini Krishnamurthy, MD
Division of Cardiology, Department of Medicine, Columbia University Irving Medical Center, New York, NY, United States
Congenital Heart Disease

Gautam Kumar, MD
Division of Cardiology, Emory University School of Medicine; Atlanta VA Medical Center, Atlanta, GA, United States
Takotsubo Syndrome

Sonali Kumar, MD
Emory Clinical Cardiovascular Research Institute and Emory Women's Heart Center, Emory University School of Medicine, Atlanta, GA, United States
Takotsubo Syndrome

Benjamin Laliberte, PharmD, BCCP
Department of Pharmacy, Massachusetts General Hospital, Boston, MA, United States
Unique Features of Cardiovascular Pharmacology in Pregnancy and Lactation

Emily Lau, MD
Department of Medicine, Division of Cardiology, Massachusetts General Hospital, Boston, MA, United States
Pregnancy and Cardiovascular Disease

Ana Micaela León, BA
Emory University School of Medicine, Atlanta, GA, United States
Takotsubo Syndrome

Jennifer Lewey, MD, MPH
Division of Cardiology, University of Pennsylvania Perelman School of Medicine, Philadelphia, PA, United States
Peripartum Cardiomyopathy

Christina M. Luberto, PhD
Department of Psychiatry, Massachusetts General Hospital/ Harvard Medical School, Boston, MA, United States
Psychosocial Issues in Cardiovascular Disease

Rekha Mankad, MD
Department of Cardiovascular Diseases, Mayo Clinic, Rochester, MN, United States
CardioRheumatology

JoAnn E. Manson, MD, DrPH
Brigham and Women's Hospital, Harvard Medical School, Boston, MA, United States
Epidemiology and Prevalence

Stephanie Trentacoste McNally, MD
Northwell Health, Department of Obstetrics and Gynecology, Donald and Barbara Zucker School of Medicine at Hofstra/Northwell, New York, NY, United States
Sex Hormones and Their Impact on Cardiovascular Health

Roxana Mehran, MD
Interventional Cardiovascular Research and Clinical Trials, The Zena and Michael A. Wiener Cardiovascular Institute, Icahn School of Medicine at Mount Sinai Hospital, New York, NY, United States
Interventions in Ischemic Heart Disease

Laxmi S. Mehta, MD
Division of Cardiology, Department of Internal Medicine, The Ohio State University Wexner Medical Center, Columbus, OH, United States
Acute Coronary Syndrome

Puja K. Mehta, MD
Emory Clinical Cardiovascular Research Institute and Emory Women's Heart Center, Emory University School of Medicine, Atlanta, GA, United States
Takotsubo Syndrome

Theofanie Mela, MD
Cardiac Arrhythmia Service, Massachusetts General Hospital, Boston, MA, United States
Role of ICD and CRT

Erin D. Michos, MD, MHS
Division of Cardiology, The Johns Hopkins School of Medicine, Baltimore, MD, United States
Prevention of Cardiovascular Disease

Jennifer H. Mieres, MD
Zucker School of Medicine at Hoftsra Northwell, Northwell Health, Lake Success, NY, United States
Disparity in Care Across the CVD Spectrum

Iva Minga, MD
Division of Cardiovascular Medicine, Department of Internal Medicine, Northshore University Health System, Evanston, IL, United States
CardioOncology

Anum Minhas, MD
Division of Cardiology, The Johns Hopkins School of Medicine, Baltimore, MD, United States
Prevention of Cardiovascular Disease

Selma F. Mohammed, MD, PhD
Creighton University; Catholic Health Initiative Heart and Vascular Institute, Omaha, NE, United States
Heart Failure With Preserved Ejection Fraction
Pulmonary Arterial Hypertension

Sharon L. Mulvagh, MD, FRCP (C)
Dalhousie University, Halifax, NS, Canada
Women's Heart Programs

Ajith P. Nair, MD
Section of Cardiology, Department of Medicine, Baylor College of Medicine, Houston, TX, United States
Heart Failure With Preserved Ejection Fraction
Pulmonary Arterial Hypertension

Anna O'Kelly, MD
Department of Medicine, Massachusetts General Hospital, Boston, MA, United States
Pregnancy and Cardiovascular Disease

Tochi M. Okwuosa, DO
Rush University Medical Center, Division of Cardiovascular Disease, Chicago, IL, United States
CardioOncology

Elyse R. Park, PhD, MPH
Department of Psychiatry, Massachusetts General Hospital/Harvard Medical School, Boston, MA, United States
Psychosocial Issues in Cardiovascular Disease

Hena Patel, MD
University of Chicago, Division of Cardiovascular Disease, Chicago, IL, United States
CardioOncology

Odayme Quesada, MD
Barbra Streisand Women's Heart Center, Smidt Heart Institute, Cedars-Sinai Medical Center, Los Angeles, CA, United States
Coronary Microvascular Dysfunction

Stacey E. Rosen, MD
Northwell Health, Department of Cardiology, Donald and Barbara Zucker School of Medicine at Hofstra/Northwell, New York, NY, United States
Sex Hormones and Their Impact on Cardiovascular Health

Andrea M. Russo, MD
Cooper Medical School of Rowan University, Camden, NJ, United States
Ventricular Arrhythmias

Amy Sarma, MD
Cardiology Division, Massachusetts General Hospital; Harvard Medical School, Boston, MA, United States
Myocardial Infarction With Nonobstructive Coronary Disease

Dawn C. Scantlebury, MBBS
Faculty of Medical Sciences, University of the West Indies, Cave Hill, St. Michael, Barbados
Sex Hormones and Their Impact on Cardiovascular Health

Nandita S. Scott, MD
Department of Medicine, Division of Cardiology, Massachusetts General Hospital, Boston, MA, United States
Pregnancy and Cardiovascular Disease

Ashish Sharma, MD
Internal Medicine Residency Program, Mercer University School of Medicine Coliseum Medical Center, Macon, GA, United States
Takotsubo Syndrome

Garima Sharma, MD
Division of Cardiology, The Johns Hopkins School of Medicine, Baltimore, MD, United States
Prevention of Cardiovascular Disease

Chrisandra Shufelt, MD, MS
Barbra Streisand Women's Heart Center, Smidt Heart Institute, Cedars-Sinai Medical Center, Los Angeles, CA, United States
Coronary Microvascular Dysfunction

Kajenny Srivaratharajah, MD, MSc, FRCPC
Department of Medicine, McMaster University, Hamilton Health Sciences, Hamilton, ONT, Canada
Peripheral Arterial Disease

Juan Tamargo, MD, PhD
Department of Pharmacology, School of Medicine, Universidad Complutense, Instituto de Investigación Sanitaria Gregorio Marañón, Madrid, Spain
Cardiovascular Medications

María Tamargo, MD
Department of Cardiology, Hospital General Universitario Gregorio Marañón, Instituto de Investigación Sanitaria Gregorio Marañón, Madrid, Spain
Cardiovascular Medications

Pamela Telisky, DO
Department of Cardiology, Deborah Heart and Lung Center, Browns Mills, NJ, United States
Stable Ischemic Heart Disease

Marysia S. Tweet, MD
Department of Cardiovascular Diseases, Mayo Clinic College of Medicine and Science, Rochester, MN, United States
Spontaneous Coronary Artery Dissection

Birgit Vogel, MD
Interventional Cardiovascular Research and Clinical Trials, The Zena and Michael A. Wiener Cardiovascular Institute, Icahn School of Medicine at Mount Sinai Hospital, New York, NY, United States
Interventions in Ischemic Heart Disease

Annabelle Santos Volgman, MD
Division of Cardiology, Department of Internal Medicine, Rush Medical College, Chicago, IL, United States
Atrial Fibrillation

Esther Vorovich, MD
Department of Medicine, Northwestern University Feinberg School of Medicine, Chicago, IL, United States
Heart Failure With Reduced Ejection Fraction

Janet Wei, MD
Barbra Streisand Women's Heart Center, Smidt Heart Institute, Cedars-Sinai Medical Center, Los Angeles, CA, United States
Coronary Microvascular Dysfunction

Nanette K. Wenger, MD
Emory University School of Medicine; Emory Heart and Vascular Center; Emory Women's Heart Center, Atlanta, GA, United States
Introduction: Past, Present, and Future of Heart Disease in Men and Women

Clyde W. Yancy, MD, MS
Department of Medicine, Northwestern University Feinberg School of Medicine, Chicago, IL, United States
Heart Failure With Reduced Ejection Fraction

Gloria Y. Yeh, MD, MPH
Division of General Medicine and Primary Care, Beth Israel Deaconess, Medical Center/Harvard Medical School, Boston, MA, United States
Psychosocial Issues in Cardiovascular Disease

Debbie C. Yen, PharmD, BCCP
Internal Medicine Clinic, 10th Medical Group, United States Air Force Academy, Colorado Springs, CO, United States
Unique Features of Cardiovascular Pharmacology in Pregnancy and Lactation

Evin Yucel, MD
Cardiac Ultrasound Laboratory, Massachusetts General Hospital, Harvard Medical School, Boston, MA, United States
Valvular Heart Disease

Preface

Sex-based differences in cardiovascular disease (CVD) have traditionally been poorly appreciated and understood, with women being underrepresented in research, clinical trials, and publications. Previously, research studies were largely conducted on men, and results were extrapolated to women. Over the past decade there has been an improved appreciation of sex differences and a dramatic increase in data specifically addressing cardiovascular conditions in women and men. Commensurate with this abundant body of literature and research, CVD deaths in both men and women have declined. Despite this progress, disparities in CVD care persist with women being less likely to receive evidence-based medicine and interventions, and often experiencing worse morbidity and mortality. There is a growing recognition that there are biological differences between men and women that contribute to sex-specific differences in risk factors, pathophysiology, presentation, diagnostic algorithms, and response to therapy. There is a clinical need for a comprehensive book consolidating the vast literature addressing sex-specific differences in CVD, and highlighting the areas of unmet need. This book fills that void, and provides a systematic review of current literature addressing these differences in a cohesive and accessible manner. The chapters are written by well-recognized experts in their respective fields. Each chapter includes figures and tables created to provide concise summaries which cover the specific concepts presented in each chapter. Where appropriate, a clinical case is presented at the beginning of each chapter to better highlight the key clinical features of the topic. The authors also provide clear, concise summary points at the conclusion of each chapter. Most chapters conclude with an "Editor's summary," an infographic designed by the Editors, highlighting key features of the chapter.

Dr. Nanette K. Wenger graciously authored the Introduction which beautifully sets the stage for the sections and chapters that follow. The book includes 13 sections and a total of 29 chapters.

The second and third sections of the book focus on epidemiology of heart disease (including discussion of risk factors unique to women) and sex differences in ischemic heart disease. CVD continues to serve as the leading cause of death in men and women, and there is an ongoing need to increase awareness of CVD in women (Chapter 2). There exist several barriers to women seeking CVD care, including the desire to put others before self, caretaker responsibilities, inadequate self-confidence, social stigma, and inadequate financial resources. While there has been a decline in CVD-related mortality among older women, there has been a relative stagnation in mortality among younger women, especially in maternal mortality (Chapter 21). It is now recognized that the traditional risk factors confer a differential risk in women, compared to men (Chapters 2 and 3). For instance, diabetes confers a 45% higher risk of ischemic heart disease in women compared to men. Furthermore, there is increasing recognition of the presence and impact of novel risk factors unique or more common in women, including gestational diabetes, preeclampsia, and early menopause and menarche, and incorporated in risk scores (Chapter 3). The presence of chronic autoimmune disorders such as systemic lupus erythematosus and rheumatoid arthritis also significantly increases the risk of developing premature coronary and systemic atherosclerosis (Chapter 23). Despite the high burden of disease and presence of these risk factors, women often do not seek care even during an acute coronary syndrome (ACS) (Chapter 4). Compared to men, women often experience delays in presentation, longer door to needle times, less pharmacotherapy, and fewer revascularizations at the time of hospital discharge after an ACS. Physiologic and anatomic differences in women such as smaller coronary artery size, more coronary tortuosity, less coronary calcification, and presence of higher fractional flow reserve volumes for any given stenosis might contribute to the higher prevalence of adverse events in women undergoing revascularization compared to men (Chapter 6). While obstructive, atherosclerotic CAD remains the most common etiology of acute coronary syndromes overall, less common etiologies of ACS including spontaneous coronary artery dissection (SCAD), myocardial infarction with nonobstructive coronary arteries (MINOCA), microvascular coronary disease, vasospasm, and coronary embolism can be seen in both sexes, but are more common in women (Chapters 5, 8, and 9). Recognition of these less common forms of coronary artery disease and evaluation for the full spectrum of ischemic heart disease is imperative due to long-term prognostic significance (Chapter 7).

The fourth section of the book is devoted to sex differences in heart failure. Sex differences persist in the form of heart failure, with women having a lower lifetime risk of experiencing heart failure with reduced ejection fraction (HFrEF) compared to men (Chapter 10). In contrast, at any given age, prevalence of HFpEF is higher in women compared with men (Chapter 11). HFrEF is associated with higher mortality and higher rates of ventricular assist device usage and heart transplantation in men. Despite these data, women living with heart failure experience worse quality of life compared to men (Chapter 10). Takotsubo syndrome is one form of left ventricular systolic dysfunction with 90% of cases occurring in women (Chapter 13). Although less frequent, men with this cardiomyopathy have a higher rate of out-of-hospital arrest and sudden cardiac death. While the overall risk of death is low, peripartum cardiomyopathy (PPCM) is a significant contributor to the increasing maternal mortality rate in the United States. PPCM disproportionately affects black women, and timely diagnosis and aggressive treatment are essential to reduce the morbidity and mortality of affected women (Chapter 14). Pulmonary arterial hypertension, while more common in women, is associated with a higher overall mortality rate in men (Chapter 12).

Successful strategies utilized in the management of ventricular and atrial arrhythmias include both pharmacologic and procedural interventions. A nuanced approach to management is required given the significant sex- and gender-related differences in responses to these therapies. Atrial fibrillation, the most common sustained arrhythmia, is associated with greater risk of disabling stroke in women. Despite this, women are less likely to be prescribed systemic anticoagulation (Chapter 16). Women have a lower lifetime risk of experiencing sudden cardiac arrest at any age (Chapter 17). Men with sudden cardiac arrest are more likely to present with ventricular tachycardia or fibrillation. In contrast, women are more likely to present with pulseless electric activity or asystole. Women are underrepresented in trials of antiarrhythmic drugs and implantable cardioverter defibrillator therapy (Chapters 17 and 18).

In addition to being more likely to experience atypical symptoms in the presence of ischemic heart disease and arrhythmias, women with peripheral vascular disease are also less likely to experience symptoms of claudication than are men. Disparities in the care of women with vascular disease include less frequent revascularization and less aggressive medical therapy (Chapter 19).

Given advances in treatment of congenital heart disease, today more adults than children are living with congenital heart disease. These patients require comprehensive team-based care given the cardiac and extra-cardiac complications associated with congenital heart disease. Pregnancy poses a particular risk to women with congenital heart disease (Chapter 20) and it provides a window into a woman's overall cardiovascular health. Complications of pregnancy, including hypertensive disorders of pregnancy, portend increased risk of future cardiovascular events and should be taken into consideration in overall cardiovascular risk assessment. Women with preexisting cardiac disease should undergo a thorough preconception risk assessment and should be managed by a multidisciplinary cardio-obstetric team throughout pregnancy. Clinicians caring for women of reproductive age with cardiovascular disease must familiarize themselves with the safety profiles of medications frequently used in the management of heart disease (Chapters 21 and 22).

Over the past decade, in addition to cardio-obstetrics described above, there has been an emergence and growth of several additional novel cardiovascular disciplines: cardio-rheumatology and cardiooncology (Chapters 23 and 24). There has been a rapid increase in the numbers and types of therapies available for treatment of inflammatory conditions and cancer. While many of these therapies are associated with dramatically improved outcomes, many have adverse cardiovascular side effects and impact on overall cardiovascular health. Furthermore, there is also a growing recognition of the common risk factors between coronary artery disease and breast cancer.

In Chapter 25, the authors provide a detailed review of the impact of reproductive hormones on the cardiovascular system. This comprehensive chapter covers reproductive health (contraception and treatment of infertility), menopausal replacement therapy, transgender hormone use, and therapeutic use of hormones in treatment of cancer.

Mounting evidence supports the link between psychological health and cardiovascular disease (Chapter 26). The impact of psychosocial stressors on cardiovascular risk appears to be stronger in women than in men, and has been associated with increased inflammation, platelet activation, endothelial dysfunction, activated hypothalamus-pituitary-adrenal axis, and unhealthy behaviors. Clinicians providing care to patients with or at risk for cardiovascular disease must familiarize themselves with the impact of psychosocial stressors on their patients' cardiovascular health.

The numbers and classes of medications used in the prevention and treatment of cardiovascular disease increase on nearly a daily basis. Many of these medications have differing pharmacodynamics, pharmacokinetics, efficacy, and side effect profiles in men and women (Chapter 27).

Throughout the book, disparities in cardiovascular care are emphasized in their respective chapters. In Chapter 28, the authors assess the current landscape of cardiovascular care and directly address these disparities. Racial disparities persist with regard to the presence and treatment of hypertension, stroke, atrial fibrillation, and heart failure risk factors. Black women often experience worse outcomes compared to age-matched black or white men. These differences are related to both biological differences and socioeconomic determinants of health. We must recognize gaps in care in order to effectively develop strategies to minimize them so that we may provide more equitable care to all our patients. The book closes with Chapter 29. This chapter contains a detailed discussion of future directions including sex-specific pathways of cardiovascular care. The emergence of women's heart programs has not only advanced care for women with cardiovascular disease but also raised awareness about sex differences in prevention, presentation, and treatment of cardiovascular disease. Shining a light on and addressing these differences through innovation and modification of our care platforms will hopefully lead to improved outcomes for all of our patients. We were in the midst of editing and reviewing the chapters when we received the news of the impending COVID-19 pandemic. We recognize that the impact of this unprecedented health concern has been felt across the globe. We are so appreciative of all healthcare workers and in particular the authors of chapters in this book. They all continued to take care of patients and their families while writing their contributions to this book. We recognize the increasing evidence that there are sex differences in risk for and responses to the COVID-19 virus; however, much of this information came to light as this book was going to press. In an effort to keep this book current and to publish it on time, COVID-19-related topics will not be covered in this edition, but will most certainly provide historic perspective of this pandemic in future editions.

We hope the book will help clinicians, researchers, and public health experts to recognize the unique aspects of sex-specific CVD and form a substantial knowledge base and foundation for improved and equitable CVD care in both men and women. We believe this book will be a useful resource for a wide variety of practitioners, cardiologists, oncologists, gynecologists, family practitioners, internists, pharmacists, physician assistants, and respective trainees, who treat men and women with heart disease.

Niti R. Aggarwal, MD
Malissa J. Wood, MD

Acknowledgments

We, the editors, are grateful to friends, colleagues, and mentors for their constant guidance, tremendous support, and invaluable collaboration. We also extend our deepest gratitude to the authors who have thoroughly researched the literature and contributed considerable time and effort to maintaining the high standard of this book. We are appreciative of the collaboration of coauthors in the various chapters that resulted from this publication.

Lastly, we are indebted to our patients for constant inspiration and entrusting us with their care.

Niti R. Aggarwal, MD
Malissa J. Wood, MD

Section I

Introduction

Chapter 1

Introduction: Past, Present, and Future of Heart Disease in Men and Women

Nanette K. Wenger

For most of the 20th century, cardiovascular disease was addressed as a problem for men, with coronary disease and myocardial infarction the foremost morbid and mortal problems. Women were widely considered protected from heart disease (absent evidence-based data) by their hormones in the premenopausal years and by the subsequent widespread application of menopausal hormone therapy. Although women in the Framingham Heart Study had a higher incidence of angina, this was overwhelmed by the predominance of myocardial infarction in men, with its attendant 40% hospital mortality. Angina was not viewed as a serious problem. Hypertension had yet to be accepted as a lethal condition, and heart failure prevalence would only increase in subsequent years consequent to the increased survival of both women and men from more acute cardiovascular problems. Not surprisingly, the emergence of clinical research studies and in particular randomized controlled trials of cardiovascular prevention and therapies involved exclusively or predominantly men and typically middle-aged white men. Stroke was considered untreatable and not widely accepted as a common consequence of uncontrolled hypertension. Rheumatic heart disease with its surgically amenable mitral stenosis predominated in women, and the problem of cardiac arrhythmias, and in particular atrial fibrillation, had yet to receive widespread acknowledgment. Ignored was the fact that the more women than men died annually from cardiovascular disease, a statistic that persisted until 2013–2014 (**Figure 1**).

The 1992 NHLBI Conference: Cardiovascular Health and Disease in Women highlighted the flawed assumption that women did not experience heart disease until elderly age and were not as seriously at risk as men. It presented new information appropriate for clinical application but identified knowledge gaps that impeded quality cardiovascular care for women, displaying a research agenda for the next decades [1]. The importance of sex and gender differences in cardiovascular disease was promulgated by the 2001 IOM report "Exploring the Biological Contributions to Human Health. Does Sex Matter?" [2], which advocated the need for the evaluation of sex-based differences in human disease and in medical research, with translation of these differences into clinical practice.

Not until the late 1990s and early 2000s did the randomized controlled trials of menopausal hormone therapy including the Heart and Estrogen Progestin Replacement Study (HERS) in women with heart disease and the hormone trials of the Women's Health Initiative (WHI) in healthy women [3–5] identify that menopausal hormone therapy did not prevent incident or recurrent cardiovascular disease and thus was not indicated for primary or secondary prevention. The importance of these trials was the refocusing of attention on established cardiovascular preventive therapies for women.

Yet, as recently as 2003, the AHRQ "Report on the Diagnosis and Treatment of CHD in Women" [6, 7] displayed that most contemporary recommendations for the prevention, diagnostic testing, and medical and surgical treatment of CHD in women were extrapolated from studies conducted predominantly in middle-aged men and that there remained fundamental knowledge gaps regarding the biology, clinical manifestations, and optimal management strategies for women.

Advocacy also increased the awareness of cardiovascular disease in women, beginning with the NHLBI Heart Truth Campaign in 2004 and the American Heart Association's Go Red for Women Initiative in the same year, as well as the work of WomenHeart, the National Coalition for Women with Heart Disease.

Subsequent clinical research studies provided data specifically for women, such as the Women's Health Study, showing that aspirin provided stroke protection but not protection from MI [8], contrary to data for men in the Physician's Health Study.

Gender-specific data derived from the CRUSADE Quality Improvement Registry of women with non-ST elevation acute coronary syndrome [9] demonstrated that the

Sex Differences in Cardiac Disease. https://doi.org/10.1016/B978-0-12-819369-3.00010-1

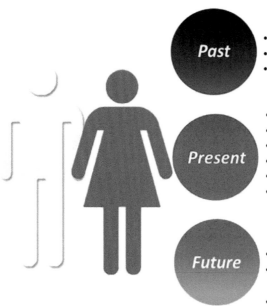

- CVD data from men extrapolated to women
- Women largely excluded from research studies
- CVD in women not recognized and ignored

- CVD continues to be #1 killer of men and women
- Increased recognition of sex differences in CVD
- Emergence of sex-specific guidelines in CVD
- Increased advocacy and awareness of CVD in women
- Women with worse prognosis after CVD event
- Improved female inclusion in clinical, cellular, and FDA drug trials

- Individualized sex-specific care
- Recognition of sex-specific CV risk factors: obstetrics, rheumatology, oncology
- Expanded landscape for women's CV health research

FIGURE 1   **The Past, Present, and Future of Heart Disease in Men and Women.** CV, cardiovascular; CVD, cardiovascular disease; FDA, Food and Drug Administration. *Image courtesy of Niti R. Aggarwal.*

prognosis with an acute coronary syndrome was worse in women who incurred an increase in hospital death, myocardial infarction, heart failure, stroke, and the need for transfusion. Yet women were less likely to receive coronary interventions and guideline-based medical based therapies despite their high-risk status. The question was raised as to whether the worse prognosis for women was related to their raised baseline risk or to suboptimal admission and discharge therapies: was this biology, bias, or both?

A concept-changing paradigm derived from the NHLBI Women's Ischemia Syndrome Evaluation (WISE) study. At the time, women with abnormal noninvasive diagnostic studies, in the absence of obstructive disease of the epicardial coronary arteries at angiography, were considered to represent a "false-positive" noninvasive test, based on the male model of disease [10, 11]. Subsequent data from the WISE cohorts identified that myocardial ischemia was the villain, associated with adverse clinical outcomes in women in the absence of obstructive coronary disease, today termed INOCA [12]. This highlighted the importance of microvascular disease and nonobstructive coronary disease in women, with current research actively targeting the optimal diagnostic procedures to identify this complex pathophysiologic spectrum. Clinical studies are currently underway to attempt to improve the outcome of women with microvascular disease and nonobstructive coronary artery disease, with unimpressive data yet forthcoming [13, 14].

The Women's Antioxidant Cardiovascular Study (WACS) and the Women's Antioxidant with Folic Acid Study (WAFACS) [15, 16] identified that vitamin C and beta carotene as well as folic acid and vitamin D supplements did not prevent incident or recurrent cardiovascular disease in women and removed these ineffective therapies from the recommended regimens. Shortly thereafter, the AHA Women's CVD Prevention Guideline 2011 Update [17] highlighted that pregnancy complications, specifically preeclampsia, gestational diabetes, pregnancy-induced hypertension, preterm delivery, and small for gestational age infants were all early indicators of an increase in cardiovascular risk, placing a detailed history of pregnancy complications as a routine component of risk assessment for women. At the same time, this guideline identified an increased risk of cardiovascular disease with systemic autoimmune collagen vascular disease, warranting screening for conventional coronary risk factors and interventions as appropriate. The 2011 Update of the Cardiovascular Prevention Guideline also addressed stroke prevention in women with atrial fibrillation, stating that atrial fibrillation increased the stroke risk 4–5 fold and that undertreatment with anticoagulants doubled the risk of recurrent stroke.

More recently, a report from the Get With the Guidelines CAD database [18] showed women to have a doubled STEMI mortality compared with their male peers, 10.2% vs. 5.5%. This was predominantly an initial 24-h increase in mortality and was associated with a decrease in the application of early aspirin, beta blockers, reperfusion therapy, and timely reperfusion. It was not that physicians chose to treat their women patients differently, but rather this represented a lack of recognition of ST-elevation myocardial infarction; remediation of this problem offered opportunities to lessen gender disparities in care and improve clinical outcomes for women.

The Institute of Medicine (IOM) 2010 Report on Women's Health Research: Progress, Pitfalls, and Promise [19] further highlighted that medical research historically neglected the health needs of women, even though there was major progress in reducing cardiovascular mortality. It identified that women were not a homogeneous group, with disparities in disease burden among subgroups of women, particularly those socially disadvantaged because of race, ethnicity, income level, and educational attainment. The IOM highlighted that the lack of analysis and reporting of sex-stratified analyses limited the ability to identify potentially important sex and gender differences, including differences in care, and advocated for translation of women's health research findings into both clinical practice and public health policies, with effective communication of research-based health messages to women, the public, providers, and policy makers.

The result of the emerging sex-specific research and recommendations was stunning. Until the year 2000, the decline in cardiovascular mortality in the United States occurred predominantly in men, but beginning in 2000, the cardiovascular mortality decline was more prominent in women and, as previously noted, for the first time in 2003–2014 fewer women than men died annually from cardiovascular disease.

But all this may not bode well for the future—since that time there has been a leveling or increase of cardiovascular mortality for both women and men, predominantly in the younger age groups 35–50 years of age, likely reflecting the US epidemic of obesity and sedentary lifestyle.

Women's participation and data in clinical trials remains a challenge, with a Cochrane review of 258 cardiovascular clinical trials showing that only one-third examined outcome by sex, although among those with sex-based analyses, 20% reported significant differences in outcomes between women and men [20]. Worthy of mention is that the exclusion of elderly patients from clinical trials doubly disadvantages women, who have their predominance of coronary and other cardiovascular events at older age. Promise for the future is offered by HR2101: The Research for All Act of 2015, 114th Congress (2015–2016). In addition to the mandate by the Government Accountability Office to update reports on women and minorities in medical research at both the NIH and the FDA, the National Institutes of Health were legislatively ordered to update guidelines on the inclusion of women and minorities in clinical research. Of equal importance, they were mandated to ensure that both male and female cells and tissues, as well as animals, be included in basic research, with the results disaggregated according to sex and sex differences examined. Similarly, the FDA was mandated to ensure that clinical drug trials for expedited drug approval were sufficient to determine the safety and effectiveness for both women and men, with the outcomes supported by the results of clinical trials that separately examined outcomes for women and men.

The past decade has seen Scientific Statements on women and peripheral artery disease [21]. Guidelines for the Prevention of Stroke in Women [22] define sex-specific stroke risk factors including pregnancy, preeclampsia, gestational diabetes, oral contraceptive use, menopausal hormone use and changes in hormonal status, as well as risk factors that were stronger or had an increased prevalence in women. The role of noninvasive testing in women for the clinical evaluation of suspected ischemic heart disease received repeated attention [23, 24], as did the sex differences in the cardiovascular consequences of diabetes mellitus showing that the gender advantage of a decrease in cardiovascular events in women compared with comparably aged men was lost in the context of type 2 diabetes [25].

Also highlighted was the intersection of cardiovascular disease and breast cancer, targeting their overlapping risk factors and the risk that current breast cancer treatments may accelerate cardiovascular disease and resultant left ventricular dysfunction, and identifying the need for surveillance, prevention, and secondary management of cardiotoxicity during breast cancer treatment [26]. Most recently, a cooperation between the American Heart Association and the American College of Obstetricians and Gynecologists [27] cited that 90% of US women have at least one risk factor for cardiovascular disease, with women less likely to receive guideline-recommended therapies and advocating that healthy lifestyles and behaviors should be discussed at each OB/GYN visit with enhanced screening for cardiovascular disease and cardiovascular risk factors. Shared information should be used to assess risk, initiate interventions, and facilitate significant lifestyle changes.

The ideal vision for women's cardiovascular health research in the next decade is that the landscape must be expanded to include beliefs and behaviors; local, national, and global community issues; economic and environment issues; ethical aspects; legislative and political issues; public policy; and societal/sociocultural variables. All these can only be ascertained by examining gender differences in cardiovascular disease, with the application of personalized or individualized medicine beginning with sex-based examination of differences.

References

[1] Wenger NK, Speroff L, Packard B. Cardiovascular health and disease in women. N Engl J Med 1993;329:247–56.

[2] Wizeman TM, Pardue M-L, editors, Committee on Understanding the Biology of Sex and Gender Differences, Board on Health Sciences Policy, Institute of Medicine. Exploring the biological contributions to human health. Does sex matter? National Academy Press; 2001.

[3] Hulley S, Grady D, Bush T, Furberg C, Herrington D, Riggs B, et al. Randomized trial of estrogen plus progestin for secondary prevention of coronary heart disease in postmenopausal women. Heart and Estrogen/progestin Replacement Study (HERS) Research Group. JAMA 1998;280:605–13.

[4] Rossouw JE, Anderson GL, Prentice RL, LaCroix AZ, Kooperberg C, Stefanick ML, et al. Risks and benefits of estrogen plus progestin in healthy postmenopausal women: principal results From the Women's Health Initiative randomized controlled trial. JAMA 2002;288:321–33.

[5] Anderson GL, Limacher M, Assaf AR, Bassford T, Beresford SAA, Black H, et al. Effects of conjugated equine estrogen in postmenopausal women with hysterectomy: the Women's Health Initiative randomized controlled trial. JAMA 2004;291:1701–12.

[6] Agency for Healthcare Research and Quality. Results of systematic review of research on diagnosis and treatment of coronary heart disease in women. Evidence report/technology assessment number 80. AHRQ Publication. No. 03-E034, DHHS. Washington, DC: U.S. Department of Health and Human Services; 2003.

[7] Agency for Healthcare Research and Quality. Diagnosis and treatment of coronary heart disease in women: systematic review of evidence on selected topics. AHRQ, No. 03-E036. Washington, DC: U.S. Department of Health and Human Services; 2003.

[8] Ridker PM, Cook NR, Lee I-M, Gordon D, Gaziano JM, Manson JE, et al. A randomized trial of low-dose aspirin in the primary prevention of cardiovascular disease in women. N Engl J Med 2005;352:1293–304.

[9] Blomkalns AL, Chen AY, Hochman JS, Peterson ED, Trynosky K, Diercks DB, et al. Gender disparities in the diagnosis and treatment of non-ST-segment elevation acute coronary syndromes: large-scale observations from the CRUSADE (Can Rapid Risk Stratification of Unstable Angina Pectoris Suppress Adverse Outcomes With Early Implementation of the American College of Cardiology/American Heart Association Guidelines) National Quality Improvement Initiative. J Am Coll Cardiol 2005;45:832–7.

[10] Bairey Merz CN, Shaw LJ, Reis SE, Bittner V, Kelsey SF, Olson M, et al. Insights from the NHLBI-Sponsored Women's Ischemia Syndrome Evaluation (WISE) Study. Part II. Gender differences in presentation, diagnosis, and outcome with regard to gender-based pathophysiology of atherosclerosis and macrovascular and microvascular coronary disease. J Am Coll Cardiol 2006;47(3 Suppl):S21–9.

[11] Pepine CJ, Ferdinand KC, Shaw LJ, Light-McGroary KA, Shah RU, Gulati M, et al. Emergence of nonobstructive coronary artery disease: a women's problem and need for change in definition on angiography. Am J Cardiol 2015;66:1918–33.

[12] Herscovici R, Sedlak T, Wei J, Pepine CJ, Handberg E, Bairey Merz CN. Ischemia and no obstructive coronary artery disease (INOCA): what is the risk? J Am Heart Assoc 2018;7(17). https://doi.org/10.1161/JAHA.118.008868.

[13] AlBadri A, Bairey Merz CN, Johnson BD, Wei J, Mehta PK, Cook-Wiens G, et al. The impact of abnormal coronary reactivity on long-term clinical outcomes in women. J Am Coll Cardiol 2019;73:684–93.

[14] Wenger NK. The feminine face of heart disease. Challenges and opportunities. J Am Coll Cardiol 2019;73:694–7.

[15] Cook NR, Albert CM, Gaziano JM, Zaharris E, MacFadyen J, Danielson E, et al. A randomized factorial trial of vitamins C and E and beta carotene in the secondary prevention of cardiovascular events in women: results from the Women's Antioxidant Cardiovascular Study. Arch Intern Med 2007;167:1610–8.

[16] Albert CM, Cook NR, Gaziano JM, Zaharris E, MacFadyen J, Danielson E, et al. Effect of folic acid and B vitamins on risk of cardiovascular events and total mortality among women at high risk for cardiovascular disease: a randomized trial. JAMA 2008;299:2027–36.

[17] Mosca L, Benjamin EJ, Berra K, Bezanson JL, Dolor RJ, Lloyd-Jones DM, et al. Effectiveness-based guidelines for the prevention of cardiovascular disease in women—2011 update: a guideline from the American Heart Association. Circulation 2011;123:1243–62.

[18] Jneid H, Fonarow GC, Cannon CP, Hernandez AF, Palacios IF, Maree AO, et al. Sex differences in medical care and early death after acute myocardial infarction. Circulation 2008;118:2803–10.

[19] Committee on Women's Health Research, Institute of Medicine. Women's health research: progress, pitfalls, and promise. Washington, DC: National Academies Press; 2010.

[20] Wenger NK, Ouyang P, Miller VM, Bairey Merz CN. Strategies and methods for clinical scientists to study sex-specific cardiovascular health and disease in women. J Am Coll Cardiol 2016;67:2186–8.

[21] Hirsch AT, Allison MA, Gomes AS, Corriere MA, Duval S, Ershow AG, et al. A call to action: women and peripheral artery disease: a scientific statement from the American Heart Association. Circulation 2012;125:1449–72.

[22] Bushnell C, McCullough LD, Awad IA, Chireau MV, Fedder WN, Furie KL, et al. Guidelines for the prevention of stroke in women: a statement for healthcare professionals from the American Heart Association/American Stroke Association. Stroke 2014;45:1545–88.

[23] Mieres JH, Gulati M, Bariey Merz N, Berman DS, Gerber TC, Hayes SN, et al. Role of noninvasive testing in the clinical evaluation of women with suspected ischemic heart disease: a consensus statement from the American Heart Association. Circulation 2014;130:350–79.

[24] Regensteiner JG, Golden S, Huebschmann AG, Barrett-Connor E, Chang AY, Chyun D, et al. Sex differences in the cardiovascular consequences of diabetes mellitus: a scientific statement from the American Heart Association. Circulation 2015;132:2424–47.

[25] Baldassarre LA, Raman SV, Min JK, Mieres JH, Gulati M, Wenger NK, et al. Noninvasive imaging to evaluate women with stable ischemic heart disease. JACC Cardiovasc Imaging 2016;9:421–35.

[26] Mehta LS, Watson KE, Barac A, Beckie TM, Bittner V, Cruz-Flores S, et al. Cardiovascular disease and breast cancer: where these entities intersect: a scientific statement from the American Heart Association. Circulation 2018;137:e30–66.

[27] Brown HL, Warner JJ, Gianos E, Gulati M, Hill AJ, Hollier LM, et al. Promoting risk identification and reduction of cardiovascular disease in women trough collaboration with obstetricians and gynecologists: a presidential advisory from the American Heart Association and the American College of Obstetricians and Gynecologists. Circulation 2018;137:e843–52.

Section II

Epidemiology and Prevalence

Chapter 2

Epidemiology and Prevalence

Mariana Garcia and JoAnn E. Manson

Abstract

Cardiovascular disease (CVD) is the leading cause of death in both men and women in the United States and most other developed countries. In 2016, CVD claimed the lives of 412,244 women in the United States. Growing knowledge of sex differences in symptoms/clinical presentation, pathophysiology of CVD, reliability of diagnostic tests, and responses to treatments, as well as an improved recognition of nontraditional risk factors specific to women, have resulted in improved clinical outcomes in women. However, despite these advances, CVD continues to be the leading cause of morbidity and mortality in both women and men in the United States. Women face disparities in diagnosis, treatment, and research related to heart disease. Persistent differences in risk factor prevalence, healthcare access, low levels of awareness by both the public and medical communities leading to underdiagnosis, delayed adoption of evidence-based guidelines in women, and major knowledge gaps are some of the factors contributing to CVD disparities. The objective of this chapter is to address the known sex differences in CVD based on epidemiologic and clinical data, characterize current sex-specific health disparities, discuss challenges, and identify strategic pathways to improve the cardiovascular health of women.

Definitions

The terms coronary heart disease (CHD), coronary artery disease (CAD), and cardiovascular disease (CVD), as used in this topic, are defined as follows:

- Coronary artery disease: general disease process affecting the coronary arteries (usually caused by atherosclerosis).

Sex Differences in Cardiac Disease. https://doi.org/10.1016/B978-0-12-819369-3.00018-6

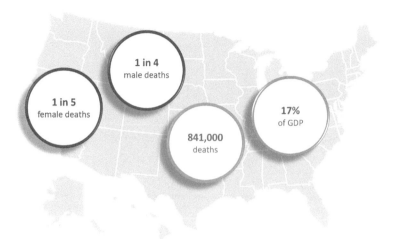

FIGURE 1   Epidemiology of Cardiovascular Disease in the United States. GDP, gross domestic product.

- Coronary heart disease: includes specifically the diagnosis of angina, myocardial infarction, silent myocardial infarction, and coronary death. It may also include coronary revascularization.
- Cardiovascular disease: represents several heart and blood vessel diseases, including CHD (as above), hypertension, atrial fibrillation, heart failure, cerebrovascular disease (including stroke and transient ischemic attack), peripheral arterial disease (PAD), and aortic atherosclerosis (including thoracic and abdominal aortic aneurysm).

CVD is the leading cause of death for both men and women in the United States (US), resulting in one of every four male deaths and one of every five female deaths, and accounting for 840,678 deaths in 2016 (**Figure 1**) [1, 2]. Moreover, nearly half (48%) of US adults aged 20 and older have some form of CVD, which includes CHD, heart failure (HF), stroke and/or hypertension, and other vascular conditions. This percentage is projected to continue to rise [1, 3].

There is a need to increase the awareness of CVD in women, as nearly half of women in the United States do not recognize that heart disease is the leading cause of death in women [1].

Furthermore, CVD mortality in young women (< 55 years of age) remains high and even greater than that in men [1]. The latest heart disease and stroke statistics (2019 update) indicates that while there has been a decline in CVD related mortality among older women, there has been relative stagnation in CHD mortality among young women over the last decade [1]. Moreover, the CVD risk burden is greater in young women compared to men of similar age, although hospitalization rates are lower [4]. The CVD burden is increasing

due in part to the increasing prevalence of obesity, poor diet, sedentary lifestyle, and the rise in type 2 diabetes—all major risk factors for heart disease (**Table 1**).

Between 2000 and 2011, the annual rate of decline for all CVD mortality in the United States averaged 3.8%. The annual rates (percent [95% confidence interval (CI)]) of decline of all CVD was 4.0% (3.8–4.1%) for females and 3.7% (3.5–3.9%) for males between 2000 and 2011 [6].

However, since 2011, this progress has plateaued with the overall decline in CVD mortality rates flattening to less than 1% per year [7]. More alarmingly, in 2015, the death rate from heart disease increased by 1% for the first time since 1969, according to the Centers for Disease Control and Prevention's National Center for Health Statistics [8].

According to data from the National Health and Nutrition Examination Survey (NHANES) 2013–2016 cycle, among US women age 20 and older (mean age 48 years), 44.7% had some form of CVD. This prevalence constitutes ~60 million American women living with CVD. In addition, CVD accounted for 412,244 female deaths in 2016, which is more than the total combined female deaths from cancer, accidents, and diabetes combined [1]. CVD in women is still underrecognized and undertreated, and women face disparities in diagnosis, treatment, and research. The risk in women is often underestimated due to the misperception that females are "protected" against heart disease. This underrecognition of heart disease and differences in clinical presentation in women leads to less aggressive treatment strategies and a lower representation of women in clinical trials. Women generally make up only about 20% of enrolled patients, even though women represent 40–50% of participants in longitudinal studies and registries of CVD [9, 10].

TABLE 1 Sex Differences in Traditional CVD Risk Factors

Risk Factor	Sex-Based Differences
Diabetes mellitus	Women with DM have a 3-fold excess risk of fatal CAD compared with nondiabetic women.
	MI: earlier occurrence and higher mortality in diabetic women compared with diabetic men. Lower revascularization rates in diabetic women compared with diabetic men.
	HF: diabetic women have a higher risk of developing HF compared with diabetic men.
	Stroke: DM is a stronger risk factor for stroke in women compared with men.
	PAD: DM is a stronger risk factor for the development of claudication in women compared with men. Decreased long-term survival in women undergoing revascularization and increased postsurgical mortality are seen in diabetic women with PAD compared with diabetic men with PAD.
Hypertension	Higher prevalence of HTN in women over age 60 than in men of the same age.
	Less well controlled in women than men.
	Similar associations with CHD and stroke risk in women and men.
Dyslipidemia	Among women, dyslipidemia has the highest population attributable risk at 47.1%, compared with all other known risk factors for CVD.
	Dyslipidemia is a strong risk factor for CHD in both men and women, with similar benefits from cholesterol-lowering therapy.
Obesity	The impact of obesity on the development of CAD appears to be greater in women than in men. In the Framingham Heart Study, obesity was linked to a 64% higher risk of CAD in women and a 46% higher risk in men, compared to normal body mass index.
Physical inactivity	The prevalence of inactivity and sedentary behaviors is higher among women than men, but moderate-to-vigorous exercise is associated with major reductions in CVD risk in both.
Smoking	In a recent meta-analysis, smoking was a stronger risk factor for CAD (by ~25%) in women than in men in most age gropus.

BP, blood pressure; CAD, coronary artery disease; CVD, cardiovascular disease; DM, diabetes mellitus; HF, heart failure; HTN, hypertension; LDL, low-density lipoprotein; MI, myocardial infarction; PAD, peripheral arterial disease. *Reproduced with permission from [5].*

Despite recent increases in awareness, many women are still uninformed about their CVD risk. A 2012 survey conducted by the American Heart Association (AHA) found that only 36% of black women and 34% of Hispanic women knew that heart disease is their leading cause of death, compared to 65% of white women (**Figure 2**). Less than 25% of women could name hypertension and high cholesterol as risk factors for heart disease, and less than 50% knew the major symptoms of heart disease. Black and Hispanic women were less likely than white women to be aware of heart attack symptoms [12].

Major strides aimed at improving the understanding of sex and gender differences in CVD have been implemented over the past 20 years (**Figure 3**). Scientists, healthcare professionals, the public, and policy makers have initiated educational campaigns focusing on increased awareness of sex and gender differences in the presentation of disease and on recognizing the impact of heart disease in women. The rate of awareness of heart disease as the leading cause of death in the United States nearly doubled between 1997 (when the AHA launched its first campaign for women) and 2009 [14–16]. The extent to which efforts to close gaps and

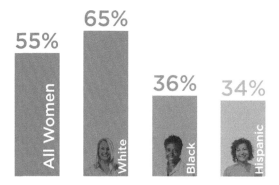

FIGURE 2   Awareness That Cardiovascular Disease Is the Leading Cause of Death in Women. According to 2012 survey conducted by the American Heart Association, only 55% of women were aware that CVD was the leading cause of death in women, with even further decreased awareness in black and Hispanic women. *Data from [11].*

heighten awareness of heart disease in women are causally linked to lower CVD mortality or improved clinical outcomes for women is not established.

The goal of this chapter is to review the trends in awareness of heart disease in women, to evaluate the distribution and determinants of CVD in women, characterize contemporary sex differences in the burden of CVD, and discuss health disparities and current challenges.

Sex-Specific Awareness of Heart Disease in the Public and Medical Community

Public Awareness

In 1994, an AHA statement reported that 45.2% of deaths in women were from CVD [17, 18]. In 1995, the AHA published the brochure *Silent Epidemic* and started encouraging greater awareness of CVD in women. Three years later,

FIGURE 3   Professional and Patient-Lead Advances Regarding Cardiovascular Disease in Women. ACC, American College of Cardiology; AHA, American Heart Association; ASA, American Stroke Association; CAD, coronary artery disease; CVD, cardiovascular disease; IOM, Institute of Medicine; JACC, Journal of the American College of Cardiology; NEJM, New England Journal of Medicine; NHLBI, National Heart, Lung, and Blood Institute; WISE, Women's Ischemic Syndrome Evaluation. *Reproduced with permission from [13].*

the AHA commissioned a national survey to assess awareness and knowledge of CVD in women. In this poll of 1000 women, only 8% of those surveyed knew that CVD was a woman's greatest health threat and only one in three women correctly identified heart disease as their leading cause of death [19]. The results of the survey were presented in Washington, DC, against a background of the Capitol building and 500,000 red carnations, representing the number of women who died annually of CVD. Consequently, the AHA has conducted triennial surveys to monitor national trends in awareness of heart disease among women.

The remarkable decrease in CVD mortality starting in 2000 among women in the United States coincided with the combined efforts to raise awareness by many organizations and the results of the landmark trials on primary and secondary prevention of CHD in women [20, 21]. In 2002, the National Heart, Lung, and Blood Institute (NHLBI) launched The Heart Truth, the first federally sponsored national campaign to raise awareness and to reduce mortality in women in the United States. The Red Dress symbol was released as part of this campaign and has now become a national symbol for women and heart disease. Subsequently, the AHA partnered with NHLBI and other organizations to launch the Go Red for Women initiative in February 2004. The campaign incorporated awareness of heart disease in women through celebrity speakers, National Wear Red Day, Go Red Luncheons, and events that prompted and empowered women to take action. Since the launch of the Red Dress Campaign, dedicated to increasing recognition of CVD in women, awareness of heart disease as a leading cause of death among US women has almost doubled at 54% in 2012, in comparison with 30% in 1997 [22]. These activities coincided with the release of the first-ever evidence-based guidelines focused on the prevention of heart disease in women [23]. The AHA Go Red for Women campaign has expanded its scope to encompass the creation of risk-assessment tools, disease-management guidelines and their implementation, and sex-specific research to decrease the high morbidity of CVD in women.

Despite recent advances, public awareness of CVD remains suboptimal. This is particularly true among women. In 2014, the Women's Heart Alliance conducted a nationwide survey of women ages 25–60 years. Overall, 45% of women were unaware that heart disease is the number one killer of women in the United States, and this knowledge gap is worse among younger women (58% of women between the ages of 25 and 29 vs. 35% of women between 50 and 60). The results also showed low awareness in women with lower levels of education: 55% of women with some college or less vs. 28% of women with college degrees and a similar gradient by income levels. Hispanic women (73%) and African American women (55%) were more likely to be

unaware of heart disease being the leading cause of death than non-Hispanic (NH) white women (34%) [11].

Public awareness among women is also low regarding linkages between heart disease and risk factors such as diabetes (43%), autoimmune disease (19%), pregnancy complications (21%), early menopause (10%), and irregular periods (5%) [11].

Even though the awareness of atypical signs of a heart attack has risen due to national awareness initiatives, with 18% in 2012 [12] up from 10% in 1997 [22], the overall recognition levels remain quite low. In the Women's Heart Alliance survey, less than half the women knew that jaw pain, cold sweats, nausea, or anxiety could be symptoms of heart disease and/or heart attack [11].

Even when women correctly recognized the symptoms of heart disease, not all women reported they would seek medical attention or call 9-1-1 if they had these symptoms. According to a poll conducted by the AHA in 2012 among US women > 25 years of age, when asked what they would do first if they thought they were experiencing signs of a heart attack, 65% of women in 2012 reported that they would call 9-1-1 compared with 53% in 2009. When asked what they would do first if they thought someone else were experiencing signs of a heart attack, 81% of women reported that they would call 9-1-1 [12].

The 2014 Women's Heart Alliance survey also showed that nearly 71% of women rarely raised the issue of heart health with their physicians even if they had risk factors for CVD [11]. Reasons for this low prioritization were most frequently lack of knowledge and competing demands. However, the underlying reasons for the misperception of CVD risk, suboptimal prioritization of cardiovascular health, and avoiding medical care are poorly understood. Women reported that factors influencing these behaviors included prioritizing others over self, caretaker responsibilities, tendency to minimize personal health concerns to avoid placing a burden on others, inadequate financial resources, and lack of personal confidence to make a lifestyle change [11]. Social stigma, in particular regarding body weight, constituted a significant barrier to women discussing heart health and taking action to reduce risk. Nearly half the women surveyed canceled or postponed healthcare visits because of weight issues [11] (**Figure 4**).

Patients' perceptions of discussions with their healthcare providers regarding heart disease issues have changed over time. In 2003, 38% of women reported discussing heart disease with their doctor, and, after the AHA Go Red for Women Initiative, this number rose to 54% in 2005 [22]. Alarmingly and for unclear reasons, the percentage of women reporting discussions with their providers about

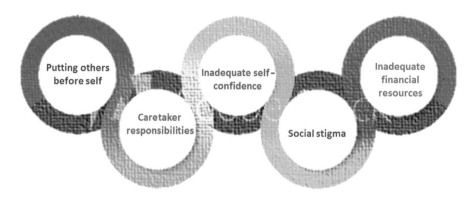

FIGURE 4   Potential Barriers for Women Seeking Care. Women often do not prioritize their cardiovascular health and reported several barriers that account for this behavioral trend. *Data from [11].*

heart disease risks has markedly declined, from 48% in 2009 [15] to 21% in 2012 [12]. These rates are even lower among Hispanic women than among white or black women [11].

In a study of young and mid-life women (ages 18–55) hospitalized with acute myocardial infarction (MI), ~50% perceived themselves to be at low risk for heart disease despite nearly all patients (98%) having at least one established risk factor, and 64% having three or more risk factors [24]. Furthermore, in those patients with acute MI and no previous history of CHD, women were 21% less likely to report prior provider discussions about heart disease and ways to modify CVD risk than men who were hospitalized with acute MI [24].

Medical Awareness

In February 2004, AHA released "Evidence-Based Guidelines for Cardiovascular Disease Prevention in Women" to assist healthcare providers in determining appropriate preventive CVD care based on a woman's future risk. Successful adoption of practice guidelines is related to physician awareness and agreement, self-efficacy, outcome expectancy, and practice habits, in addition to patient- and system-related factors [25]. That same year, a national on-line survey of physicians demonstrated that fewer than one in five physicians were aware that more women died each year of CVD than men [26]. Ten years later, in 2014, the Women's Heart Alliance surveyed 200 primary care providers (PCPs) and 100 cardiologists to determine their self-reported preparedness to address CVD risks in their female patients [11]. PCPs reported CVD as a top health concern in women, but less important than weight-related health concerns and breast health. Most physicians reported suboptimal training in assessing CVD risks in women (22% PCPs and 42% cardiologists) [11]. A single-center survey of 80 postgraduate trainees demonstrated that although 60% recognized the importance of and need for the implementation of gender-based concepts in their formal curriculum, nearly

70% of respondents reported no or only minimal formal training regarding sex- and gender-specific medicine concepts in their education programs or didactic lectures [27].

The perception of lower CVD risk in women than in men has been implicated as the primary factor in underutilization of preventive recommendations [26, 28]. The 2004 online survey assessed knowledge and incorporation of national CVD prevention guidelines in 300 PCPs, 100 obstetricians/gynecologists (Ob/Gyns), and 100 cardiologists [26]. Cardiologists and PCPs had a high level of awareness of contemporary hypertension and lipid guidelines (>90% for both), but a lower awareness of the AHA's CVD prevention guidelines for women (80% and 60%, respectively) [26]. Ob/Gyns reported 60% awareness of hypertension guidelines, 45% awareness of lipid guidelines, and 60% awareness of CVD prevention guidelines for women. Incorporation of CVD prevention guidelines for women into practice was <42% in all three groups [26].

Similarly, the Women's Heart Alliance survey demonstrated that 16% of PCPs and 22% of cardiologists were implementing the AHA's guidelines for CVD risk assessment in women. Specifically, they reported that they: (1) discussed personal and family medical history and pregnancy complications that further increase heart disease risk; (2) asked about any heart disease symptoms; (3) asked about smoking, diet, and physical activity habits; (4) screened for depression among women with heart disease; (5) conducted a physical examination that included blood pressure, body mass index (BMI), and waist circumference; (6) measured cholesterol, triglycerides, and glucose levels; (7) calculated 10-year and lifetime heart disease risk; and (8) talked with women about what each of these means for their heart health [11]. Only one in four PCPs and one in five cardiologists reported implementing at least five of the eight recommended CVD risk assessments in women [11]. Physician utilization of the American College of Cardiologists/AHA Atherosclerotic Cardiovascular Disease pooled cohort equation risk estimator was better (44% PCPs and 53% cardiologists) [11]. Despite medical society-endorsed guidelines for

CVD prevention, major gaps remain in physicians' self-perception of preparedness to assess women's CVD risks and in the application of guidelines in clinical practice [11].

The Burden of Heart Disease in Men and Women

Despite enormous declines in the burden of CVD in the past decades, mainly due to improvements in primary and secondary prevention, CVD disease remains the main cause of premature death and disability among men and women worldwide. Statistics from the Global Burden of Disease Study show that CVD in 2013 accounted for 35% of all deaths in women and 32% in men throughout the world [29]. In the United States, CVD is not only the leading cause of death, but also one of the costliest chronic conditions, constituting 17% of overall national health expenditures [30, 31].

Based on NHANES 2013–2016 data, the prevalence of CVD (comprising CHD, HF, stroke, and hypertension) in adult women ≥ 20 years of age was 44.7% (60 million) vs. 51.2% (61.5 million) in adult men [1]. In 2011, the AHA commissioned a report that showed that by 2035, ~40% of the US population would suffer from CVD. Unfortunately, that prediction came true in 2015. That same year, the death rate from heart disease rose by 1% for the first time since 1969 [8]. The AHA repeated its projections of the prevalence and economic burden of hypertension, CHD, HF,

stroke, atrial fibrillation, and other heart diseases from 2015 to 2035 [3]. This latest study projects that by 2035, in the US, there will be:

- 123.2 million with hypertension
- 24 million with CHD
- 11.2 million suffering from stroke
- 7.2 million with atrial fibrillation
- 8.8 million with HF

The prevalence of total CVD is higher among men than women; however, the prevalence of hypertension, HF, and stroke is higher among women than among men. The prevalence is expected to increase for all conditions among both sexes [3]. Total costs of total CVD in the United States in 2015 were higher among males ($325 billion) than females ($252 billion) and are projected to increase to $591 billion and $525 billion, respectively, by 2035. However, for congestive heart failure, stroke, and atrial fibrillation, the costs are higher among women than among men.

Most of the burden of CVD can be explained by a set of traditional risk factors that affect both men and women, including smoking, overweight and obesity, hypertension, diabetes, and elevated cholesterol. Increased recognition of the prevalence of traditional CVD risk factors, and their differential impact on women, as well as emerging, nontraditional risk factors unique to or more common in women, contribute to a new understanding of mechanisms leading to worsening outcomes for women (**Figure 5**).

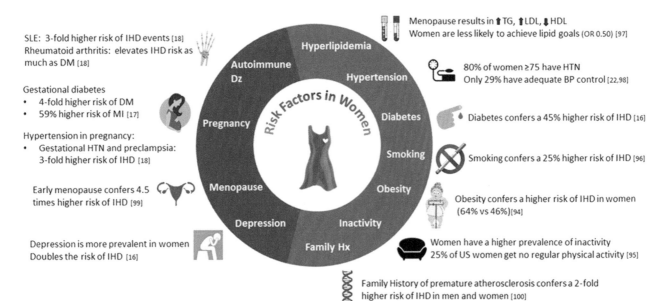

FIGURE 5   Traditional and Nontraditional Cardiovascular Risk Factors in Women. BP, blood pressure; DM, diabetes; HDL, high-density lipoprotein; HTN, hypertension; IHD, ischemic heart disease; LDL, low-density lipoprotein; MI, myocardial infarction; SLE, systemic lupus erythematosus; TG, triglycerides. *Reproduced with permission from [32].*

Prevalence and Epidemiology of Risk Factors in Women

Diabetes

More than 11.7 million US women have a diagnosis of diabetes mellitus (DM) and ~95% of these women have type 2 diabetes mellitus (T2DM) [33]. The increasing prevalence of T2DM is concerning because it is a potent risk factor for CVD and has long been recognized to confer greater risk for CVD death in women compared with men [34]. A growing body of literature shows that there are appreciable and clinically relevant differences in how diabetes affects the risk of CVD in men and women.

A recent pooled analysis summarizing data from 64 cohorts, including nearly 900,000 individuals and more than 28,000 incident CHD events, showed that the presence of diabetes nearly tripled the risk of incident CHD in women (RR 2.82 [95% CI 2.35; 3.38]), whereas it little more than doubled the risk in men (RR 2.16 [95% CI 1.82; 2.56]) [35]. Therefore, diabetes conferred a 44% (95% CI 27–63%) greater excess risk for incident CHD in women compared with men.

Moreover, there is a threefold excess fatal CAD risk in women with T2DM compared with nondiabetic women (95% CI, 1.9–4.8) [36]. In addition, a pooled analysis on data from 750,000 individuals and more than 12,000 incident stroke events provided strong evidence that women with DM have a 27% (95% CI 10, 46%) greater increased risk of stroke compared with their male counterparts; the pooled relative risk of stroke associated with diabetes was 2.28 (95% CI 1.93, 2.69) in women and 1.83 (95% CI 1.60, 2.08) in men, independent of sex differences in other major cardiovascular risk factors [37].

Women who progress from normoglycemia to prediabetes have higher levels of endothelial dysfunction, higher blood pressure, and more abnormalities in their fibrinolysis and thrombosis pathways compared with those who do not, and such differences are more pronounced than in men [38].

Early diagnosis of diabetes is essential, particularly in racial/ethnic groups at high risk for diabetes, such as African Americans, Hispanics, American Indians, and Pacific Islander Americans. The burden of diabetes varies greatly by race/ethnicity, with blacks having the highest age-adjusted prevalence, followed by Hispanics, Asians, and then NH whites [39]. Compared with white postmenopausal women in the United States, there is a more than twofold higher risk of diabetes in blacks and approximately twofold higher risk in Hispanics and Asians [40].

Hypertension

Hypertension significantly increases the risk of MI, HF, atrial fibrillation, and stroke. Premenopausal women are at a higher risk of hypertensive end-organ damage than age-matched men, including microalbuminuria (13.7% vs. 6.2%, $P = 0.002$) and left ventricular hypertrophy (26.4% vs. 8.8%, $P < 0.0001$) [41]. According to NHANES 2013–2016 data, the prevalence of hypertension is higher in women than men (77.8% in women vs. 70.8% in men in the 65–74 age group, 85.6% in women vs. 80% in men in the > 74 age group), but less than half receive adequate treatment [42]. Hypertension is often poorly controlled in older women; only 23% of women vs. 38% of men > 80 years have a BP < 140/90 mmHg [43]. The prevalence of hypertension is 25.3% among Hispanic women, and only 37.5% have controlled BP [44].

Dyslipidemia

Elevated serum lipid levels are the greatest contributor to development of ischemic heart disease worldwide, and clinical trials have shown that low-density lipoprotein cholesterol reduction with statins leads to improved CVD outcomes [45]. Statin therapy has similar proportional benefits for women and men in CVD event reduction [46].

Nevertheless, sex differences in statin treatment and adherence to guideline-recommended lipid management are well documented [47–49]. Female patients have historically received less aggressive lipid management than male patients, and the reasons underlying this remain poorly understood. A recent study comprising a large sample of US adults seen in community practice found that women were less likely than men to receive guideline-recommended statin therapy (67.0% vs. 78.4%; $P < 0.001$) or to receive a statin at the guideline-recommended intensity (36.7% vs. 45.2%; $P < 0.001$). Potential causes of these disparities include the following: women were less likely to report having been offered statin therapy, more likely to decline statin therapy when offered, and more likely to discontinue statin therapy after starting [50].

Obesity

More than two in three adults in the United States are classified as overweight or obese, and the prevalence of obesity is higher among women than among men [51]. While elevated BMI is associated with increased fatal and nonfatal ischemic heart disease in both women and men, sex differences in fat distribution have been implicated in ischemic heart disease. Women predominantly accumulate subcutaneous fat, whereas men accumulate significantly more visceral fat.

It is established that abdominal or central adiposity is an important predictor of chronic disease risk independent of total adiposity [52]. For example, individuals with a higher proportion of abdominal fat have a greater risk of developing CHD [53, 54], T2DM [55, 56], and cancer [57].

The trunk-to-leg fat ratio can be considered a marker of body shape. Studies have reported heterogeneity of body fat distribution across racial groups. Abdominal visceral adiposity has been reported to be significantly greater in white men and women compared with black men and women, and white women have lower measures of subcutaneous adipose tissue than black women [58]. Black women tend to have more "pear-shaped" bodies, i.e., they tend to have more subcutaneous fat deposited in the hips and thighs vs. in the abdominal areas [59].

BMI, waist circumference, and waist-to-hip ratio are common surrogate measures of adiposity in clinical and public health practice. Body fat distribution changes according to menopausal status, with central obesity more pronounced in postmenopausal women [60]. Postmenopausal women have increased visceral fat accrual, which has implications for development of insulin resistance, inflammatory responses, lipolysis, and CVD [61]. A recent study of postmenopausal women aged 50–79 years participating in the Women's Health Initiative found that in this age group across all racial/ethnic groups, waist circumference was a better predictor of diabetes risk (compared with waist to hip ratio or BMI), especially for Asian women [62].

The prevalence of obesity among US women varies markedly by race/ethnicity. According to the Centers for Disease Control and Prevention, the prevalence of obesity among adults and youth in the United States in 2015–2016 was 38.0% in NH white, 54.8% in NH black, 14.8% in NH Asian, and 50.6% in Hispanic women [63]. Based on a recent US Department of Health and Human Services Office of Minority Health report, American Indian or Alaska Native adults are 50% more likely to be obese than NH whites (43.7% vs. 28.5%) [64].

Physical Inactivity

According to data from a 2011 National Health Interview Survey in adults, inactivity was higher among women than men (33.2% vs. 29.9%, age-adjusted) and increased with age from 26.1% to 33.4%, 40.0%, and 52.4% among adults 18–44, 45–64, 65–74, and ≥ 75 years of age, respectively [65].

Smoking

Tobacco use increases CVD risk, including progression of atherosclerosis, MI, and sudden cardiac death. Importantly, in women, the combination of smoking with oral contraceptive use has a synergistic effect on risk of acute MI, stroke, and venous thromboembolism.

A meta-analysis conducted in 2011 reported that in all age groups, with the exception of the youngest (30–44 years), women had a 25% increased risk for CAD conferred by cigarette smoking compared with men [66].

A meta-analysis to estimate the effect of smoking on stroke according to sex showed evidence of a more harmful effect of smoking in women than in men in Western (relative risk ratio [RRR], 1.10 [1.02–1.18]) but not in Asian (RRR, 0.97 [0.87–1.09]) populations [67]. A large population-based case-control study found that women who were current smokers and used oral contraceptives had an 8.8-fold higher risk (odds ratio [OR] 8.79, 95% CI 5.73–13.49) of venous thrombosis than nonsmoking women who did not use oral contraceptives [68].

Psychological Stress

Psychological stress remains an understudied, underappreciated and poorly managed cardiovascular risk factor. The INTERHEART study provided the first substantive data supporting the relationship between stress (OR 1.45), depression (OR 1.55) and first MI [69]. After an MI, depression, trauma, and perceived stress are disproportionately common in younger women compared with their male counterparts or older patients [70, 71] and are powerful predictors of cardiovascular risk in young women [72–74]. Moreover, young women after MI have a twofold likelihood of developing mental stress-induced MI compared with men (22% vs. 11%, $P = 0.009$) [75]. Future research will be vital to better address this important association and develop strategies to mitigate this major risk factor.

Risk Factors Unique to Women

Early Menopause

After menopause the incidence of CVD increases substantially. One systematic review found an increased risk of CVD, CVD mortality, and all-cause mortality in women who experienced early menopause [76].

Adverse Pregnancy Outcomes

Pregnancy is a metabolic stress test and provides a unique opportunity in women to analyze their future risk of developing CVD. The 2011 guidelines for CVD prevention in women incorporated adverse pregnancy outcomes as cardiac risk factors [77]. These include gestational diabetes, preeclampsia, eclampsia, hypertensive disorders of pregnancy, and preterm delivery, all associated with increased future heart disease risk [78]. A diagnosis of preeclampsia doubles the risk of future diabetes and stroke [79] and is associated with a nearly fourfold elevated risk for developing hypertension within 14 years of pregnancy [80]. Gestational DM increases the risk of developing T2DM by sevenfold, which is a major risk factor for subsequent atherosclerotic cardiovascular disease, but also raises CVD risk (twofold for stroke and fourfold for MI) independently of the interim development of overt T2DM [81].

CH
2

Multiparity

According to a recent study, multiparity is associated with poorer cardiovascular health, especially among women with five or more live births. The Life's Simple 7 cardiovascular health score, defined according to AHA criteria, was recently examined in relation to parity (a score of 0–8 was considered inadequate, 9–10 average, and 11–14 optimal). Among women with a mean (SD) age of 62 (10) years, the mean (SD) cardiovascular health score was lower with higher parity (8.9 [2.3], 8.7 [2.3], 8.5 [2.2], and 7.8 [2.0] for 0, 1–2, 3–4, and ≥ 5 live births, respectively) [82].

Incidence and Prevalence of Coronary Heart Disease

According to the Heart Disease and Stroke Statistics 2019 update from the AHA, using data from 2013 to 2016, an estimated 18.2 million Americans > 20 years of age have CHD, with a prevalence of 9.4 million (7.4%) in men and 8.8 million (6.2%) in women [1]. Based on data from 1995 to 2012, 23% of women age 45 and older who have an initial recognized MI die within a year compared with 18% of men. However, within 5 years after a first MI, 47% of women and 36% of men will die [1] (**Figure 6**).

Observed sex differences in the incidence and presentation of CHD in population might be partially explained by distinct pathophysiological processes leading to MI [1]. Women have less obstructive and extensive epicardial artery disease than men but are more prone to have impaired coronary vasomotor function and microcirculatory dysfunction. Population-based studies reported a lower calcium score [83] and atheroma volume [84] in women when compared with men.

A study, performed within the Rotterdam cohort, investigated sex-based differences of the vascular tree. Median coronary calcium scores, mean carotid intima-media thickness, and carotid plaque scores were higher in men than in

women in all age groups. The findings of this study showed that sex differences in the coronary vessels were larger than in the other vascular beds; moreover, the observed sex differences in atherosclerosis of the coronary arteries were particularly pronounced in younger participants but still present in the older age groups. The age-adjusted male:female odds ratios of having a calcium score above 1000 in the two lowest age tertiles were 6.9 (95% CI: 3.4, 13.9) and 7.4 (95% CI: 4.3, 12.7), showing that men have a substantially higher burden of coronary calcification than women [83].

A systematic analysis demonstrated that women have less plaque in terms of percent atheroma volume ($33.9 \pm 10.2\%$ vs. $37.8 \pm 10.3\%$, $P < 0.001$) and total atheroma volume (148.7 ± 66.6 mm^3 vs. 194.7 ± 84.3 mm^3, $P < 0.001$). With medical therapy, the rate of change in both measures did not differ between sexes [84].

On the other hand, based on available experimental and clinical data, microvascular dysfunction is considered as a major etiological factor for CHD in the absence of significant coronary obstruction, particularly in women [85]. Further clarification of the pathophysiological processes underlying CHD may help with tailoring sex-specific strategies for the prevention, detection, and management of CHD. Due to sex differences in pathophysiology and clinical manifestation of CHD (discussed in Chapter 3), clinical symptoms of myocardial ischemia in women are often regarded as "atypical" and likely to be ignored or misdiagnosed [86]. As a consequence, women with overt CHD may have delays in diagnosis and treatment, contributing to worse prognosis and outcomes [87].

Data from 44 years of follow-up in the original Framingham Study cohort and 20 years of surveillance of their offspring have allowed ascertainment of the incidence of initial coronary events including both recognized and clinically unrecognized MI, angina pectoris, unstable angina, and sudden and nonsudden coronary deaths [88–90]. The following observations were noted:

- For adults aged 40 years, the lifetime risk of developing CHD is 49% in men and 32% in women. At age 70 years, this risk increases to 35% in men and 24% in women.
- For total coronary events, the incidence rises steeply with age, with women lagging behind men by 10 years. For the more serious manifestations of coronary disease, such as MI and sudden death, women lag behind men in incidence by 20 years, but the sex ratio for incidence narrows progressively with advancing age [91]. The incidence at ages 65–94 compared to ages 35–64 more than doubles in men and triples in women.
- In premenopausal women, serious manifestations of coronary disease, such as MI and sudden death, are

| Prevalence : | 9.4 million (7.4%) | 8.8 million (6.2%) |
| 5 yr mortality after MI: | 36% | 47% |

FIGURE 6   Burden of Coronary Artery Disease in the United States. *Data from [1].*

relatively uncommon. Beyond menopause, the incidence and severity of coronary disease increases abruptly, with rates three times those of women the same age who remain premenopausal [88].

Incidence and Prevalence of Heart Failure

According to data from NHANES 2013–2016, an estimated 6.2 million Americans ≥ 20 years of age had HF [1]. By 2030, the incidence of HF is projected to rise by 46%, affecting more than 8 million individuals. HF affects both sexes equally and is a leading cause of morbidity and mortality.

By the age of 40, men and women have equal lifetime risks of developing HF. At 40 years of age, the lifetime risk of developing HF for both men and women is one in five [1, 92]. Occurrence of HF increases with advancing age, and women at older ages are at greater risk than men [93].

Incidence rates of HF in men approximately double with each 10-year increase in age from 65 to 85 years; however, the HF incidence rate triples for women between ages 65–74 and 75–84 years [1].

In the Atherosclerosis Risk in Communities Study (ARIC), the age-adjusted incidence rate per 1000 person-years was lowest for NH white women (3.4) compared with all other groups, including NH white men (6.0), NH black women (8.1), and NH black men (9.1) [94].

Incidence rates in black women were more similar to those of NH black men than of white women [94].

The lifetime risk of developing HF differs by sex and race. Data from the NHLBI-sponsored Chicago Heart Association Detection Project in Industry, ARIC, and the Cardiovascular Health Study cohorts indicate that lifetime risks for HF were 30–42% in white males, 20–29% in black males, 32–39% in white females, and 24–46% in black females [95].

Patients with HF and preserved ejection fraction are more often female and older compared to those with HF and reduced systolic function [96, 97]. According to the 2019 AHA Heart Disease and Stroke Statistics Update, white women had the highest proportion of hospitalized HF with preserved ejection fraction (59%), whereas black men had the highest proportion of hospitalized HF with reduced ejection fraction (70%) [1].

Over the past 50 years, the incidence of HF has declined among women but not among men [98]. Survival after the onset of HF has improved in both sexes, however (~ 12% per decade; $P = 0.01$ for men and $P = 0.02$ for women) [98]. Men and younger adults have experienced larger survival gains compared to women and the elderly [99].

Incidence and Prevalence of Atrial Fibrillation

Estimates of the prevalence of atrial fibrillation in the United States ranged from ~ 2.7 million to 6.1 million in 2010 [100, 101].

In the Framingham Heart Study, atrial fibrillation incidence (per 1000 person-years) was 3.8 in men and 1.6 in women [102]. The Olmsted County Minnesota Study reported the atrial fibrillation incidence (per 1000 person-years) in men to be 4.7 compared with 2.7 in women [101]. Despite higher incidence of atrial fibrillation in men, lifetime risks of atrial fibrillation in women and men are similar owing to longer life expectancy for women. In the Framingham Heart Study [103], the lifetime risks of atrial fibrillation in women and men at age 60 years were 23.4% and 25.8%, respectively.

Atrial fibrillation incidence has been shown to increase markedly with increasing age in both women and men, reaching as high as 30.4 per 1000 person-years in women and 32.9 per 1000 person-years in men by age 85–89 years [104]. In a Medicare sample, per 1000 person-years, the age- and sex-standardized incidence of atrial fibrillation was 27.3 in 1993 and 28.3 in 2007, representing a 0.2% mean annual change ($P = 0.02$). Of individuals with incident atrial fibrillation in 2007, ≈ 55% were females, 91% were white, 84% had hypertension, 36% had HF, and 30% had cerebrovascular disease [105].

Atrial fibrillation is associated with worse symptoms and quality of life [106], and increased risk of complications such as stroke [107, 108] and mortality [109], in women compared to men.

Incidence and Prevalence of Peripheral Arterial Disease

According to data from several US cohorts during the 1970s to 2000s and the 2000 US Census, 6.5 million Americans ≥ 40 years (5.5%) are estimated to have low ankle-brachial index (ABI) (< 0.9) [110]. Of these, one-fourth have severe PAD (ABI < 0.7) [110]. The population-based prevalence of PAD in women has been incompletely evaluated. In contrast to the abundant data defining the sex-specific prevalence of CHD and stroke, few population surveys of PAD have been performed, and ongoing PAD surveillance is not currently conducted in any state or nation. Thus, in calculating the relative prevalence of PAD, it has been noted that many population-based studies of PAD do not report prevalence for women separately. In the published literature, the results are mixed with respect to differences in prevalence by sex. In a systematic review, the prevalence of PAD in women 45–93 years of age was reported to range from 3% to 29% [111].

In-hospital mortality was higher in females than males, regardless of disease severity or procedure performed, even after adjustment for age and baseline comorbidities: 0.5% vs. 0.2% after percutaneous transluminal angioplasty or stenting for intermittent claudication; 1.0% vs. 0.7% after open surgery for intermittent claudication; 2.3% vs. 1.6% after percutaneous transluminal angioplasty or stenting for critical limb ischemia; and 2.7% vs. 2.2% after open surgery for critical limb ischemia ($P < 0.01$ for all comparisons) [112].

Temporal Trends

Over the past few decades, the incidence of CHD has decreased in developed nations. An analysis of the NHANES 1 Epidemiologic Follow-up Study examined the incidence of CHD in two cohorts: one comprised 10,869 patients between 1971 and 1982, and a second included 9774 patients between 1982 and 1992 [43]. A decrease was found in the incidence of CHD from 133 to 114 cases per 10,000 people per year of follow-up, as well as an overall decline in CVD from 294 to 225 cases per 10,000 patients. The Mayo Clinic, as part of the Rochester Epidemiology Project, utilized a population-based data resource including the complete medical records of all of the residents of Olmsted County, Minnesota to analyze the incidence of newly diagnosed CHD in the local population, over a 10-year period from 1988 to 1998 [44]. The study reported a decline, from 57 to 50 cases per 10,000 people, in the age-adjusted incidence of new CHD, including CHD diagnosed by angiography, unstable angina, MI, and sudden death.

Temporal trends in sudden cardiac death and nonsudden CHD death in the Framingham Heart Study original and offspring cohorts, from 1950 to 1999, were examined. From 1950–1969 to 1990–1999, overall CHD death rates decreased by 59% (95% CI 47–68%, P(trend) < 0.001). Nonsudden CHD death decreased by 64% (95% CI 50–74%, P(trend) < 0.001), and sudden cardiac death rates decreased by 49% (95% CI 28–64%, P(trend) < 0.001). These trends were seen in both men and women, in those with and without a prior history of CHD, and in smokers and nonsmokers [113].

Although the numbers are remarkable—mortality rates due to CHD between 1980 and 2002 fell almost 52% for men and 49% for women [45]—CHD remains the number one cause of death in the United States, and these trends are not experienced equally among different demographic groups. The overall decline in CHD mortality rates have flattened to less than 1% per year since 2011, and rates have even worsened for the most at-risk populations. While decline in mortality rates have slowed in men, from −6.2% between 1980 and 1989 to −0.5% in 2000–2002, the plateauing of the decline in mortality rates is even more marked in women,

from −5.4% to 1.5% during the same period. Recent data suggest a worrisome reduction of the rate of decline in CHD mortality among younger people between the ages of 35 and 54 years, showing no decrease, or even a slight increase, in recent years [46–49].

Young adults, especially women, continue to show much slower reductions in CHD mortality [114]. Between 2001 and 2010 in the United States, there was no significant reduction in MI hospitalization rates among young people (< 55 years of age) [4], a stark contrast to Medicare population studies, which demonstrated > 20% reduction in hospitalization rates in the same time period [115].

In the Atherosclerosis Risk in Communities Surveillance study, which conducts hospital surveillance of MI in four US communities, the proportion of MI hospitalizations attributable to young patients increased from 1995 to 2014 and was especially pronounced among women. The prevalence of hypertension and diabetes among young patients admitted with MI increased over time as well. Compared with young men, young women presenting with MI had a lower likelihood of receiving guideline-based MI therapies [116].

Cardiovascular Health Disparities Persist

The reasons as to why CVD mortality differs among different demographic groups are multidimensional and complex in origin. Components that contribute to sex and gender differences in CV medicine include not only sex-specific pathophysiology and risk factors, but also racial/ethnic differences in CVD, disparities in care, and diverse research methodologies (**Figure 7**). CVD mortality is influenced not only by sex and race/ethnicity but also by geographic location, economic status, and education. There are disparities in diagnosis, treatment, research, and outcomes across groups. The disparities can also be attributed to poorer preventive care related to economic or educational factors and poorer lifestyle behaviors among some groups, such as a higher prevalence of smoking, obesity, and lack of physical activity [53].

The burden of CVD in the United States differs geographically, with the southeast having particularly high mortality rates [117]. This pattern is consistent across sex and age strata and has been observed in both white and black populations [118].

Women have traditionally been considered as one subgroup, but the vast differences among women of varying races and ethnic groups warrants attention [54]. African American and Native American [55] women with a history of CVD, for example, have poorer overall health, more complications, and higher mortality than other female groups in the United States [56]. MI prevalence by race and ethnicity interacts with age. Among women < 55 years of

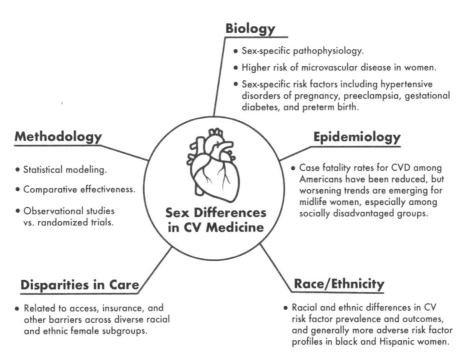

Biology

- Sex-specific pathophysiology.
- Higher risk of microvascular disease in women.
- Sex-specific risk factors including hypertensive disorders of pregnancy, preeclampsia, gestational diabetes, and preterm birth.

Methodology

- Statistical modeling.
- Comparative effectiveness.
- Observational studies vs. randomized trials.

Sex Differences in CV Medicine

Epidemiology

- Case fatality rates for CVD among Americans have been reduced, but worsening trends are emerging for midlife women, especially among socially disadvantaged groups.

Disparities in Care

- Related to access, insurance, and other barriers across diverse racial and ethnic female subgroups.

Race/Ethnicity

- Racial and ethnic differences in CV risk factor prevalence and outcomes, and generally more adverse risk factor profiles in black and Hispanic women.

FIGURE 7 **Components That Contribute to Sex and Gender Differences in Cardiovascular Medicine.** CV, cardiovascular; CVD, cardiovascular disease.

age with MI, black women have higher mortality rates than white women even after adjustment for chronic renal failure, time to presentation, insurance, and treatment in the first 24 h (each 5-year decrement in age was associated with a 4.3% increase in mortality [95% CI, 2.4–6.3%]) [119].

In the United States, African Americans, Hispanics, Asians, and Native Americans have less access to health care than their white counterparts and are therefore less likely to receive preventive care or counseling on lifestyle risk modification [57]. Almost 50% of the uninsured nonelderly population in the United States is part of an ethnic minority group, with African American and Hispanic women having the highest uninsured rates in the country [58].

Members of minority groups have not had the same decline in CVD mortality rates as whites in recent years, and these differences are particularly prominent among women. African American women who are younger than 55 years and develop acute MI are more likely to have contributing factors such as hypertension, diabetes, kidney disease, or obesity [50]. For reasons that are not yet clear, racial disparities exist even when there is good access to health care and medical insurance coverage [60]. African Americans with normal renal function have a 42% higher mortality rate, according to data from the NHANES [61]. Irrespective of race/ethnicity, women tend to have a higher incidence of undiagnosed MI than men, and MI rates are increasing among women while decreasing among men [59].

Race/Ethnicity

Racial and ethnic differences in cardiovascular health have persisted over the past several decades and there is little evidence of narrowing disparities. Moreover, racial and ethnic disparities are generally larger among women than men [120]. NH blacks have a higher incidence of CHD, HF, stroke, and overall CVD mortality as compared with NH whites [121, 122]. The CVD burden in Hispanics as compared with NH whites is mixed, whereas Hispanics have lower overall CVD mortality, and incidence rates for CHD and stroke appear to be higher [123].

Contemporary Challenges

Enduring disparities in overall cardiovascular health, defined by seven health factors and behaviors—diet, physical activity, smoking status, BMI, blood pressure, blood glucose, and total cholesterol—exist for NH black and Mexican-American women as compared with NH white women. Of these factors, among women, NH whites had significantly lower BMI and glucose scores, and higher physical activity scores, as compared with NH blacks and Mexican-Americans. NH black women have significantly higher blood pressure levels, compared with NH whites and Mexican-Americans [120]. In a study using NHANES data on adults >20 years, in cycles 1999/2000 through 2011/2012, black women had significantly lower mean CV health scores as compared with NH white women at each survey cycle (age adjusted means score difference=0.93; P=0.001 in 2011/2012) and Mexican-American women had significantly lower mean scores as compared with NH white women at almost all survey cycles (age adjusted means score difference=0.71; P=0.02 in 2011/2012) [120].

The racial differences among disorders of pregnancy and their contributions to long-term CV risk have not been clearly established and warrant further study. Prior studies have had limited power to estimate ethnic and racial differences as most studies have included predominantly North American and European populations, even though there are known differences in the prevalence of these diseases between populations. As an example, preeclampsia is more common among women of African-Caribbean origin [124], but these women are underrepresented in current CVD follow-up cohorts.

Opportunities for Equitable Outcomes

Improving CVD outcomes in women is a complex issue which requires a multifaceted approach including education to improve awareness, recognition of sex differences in prevention, diagnosis, treatment, and outcomes, treatment centers that specialize in sex- and gender-specific care, and legislation and policy that specifically address women's cardiac health.

Public awareness campaigns need to focus on younger women and minorities in a manner that is culturally appropriate to each group; social media can be used to reach women in specific ethnic and racial minority groups and can be used to encourage heart-healthy behavior and CVD awareness. Community organizations, churches, schools, and workplaces should collaborate in neighborhoods to promote CVD knowledge and adoption of healthy habits for CVD prevention [125–127]. Community groups are better able to understand the challenges of their communities and can offer education that is affordable and accessible to women. This is of utmost importance for women at lower socioeconomic levels.

Evidence suggests that CVD mortality rates may be improved if barriers to health care are reduced or removed for at-risk populations. Within the Department of Veteran Affairs, a healthcare system in the United States with fewer access barriers for its eligible patients, African American patients had a 24% lower all-cause mortality and 37% lower incidence of CHD than in other healthcare settings, after multivariable adjustments. Women in the Veteran's Affairs system showed a 45% lower mortality and 32% lower incidence of CHD [61].

Sex-specific research would fill critical knowledge gaps, as underrepresentation of women in research has

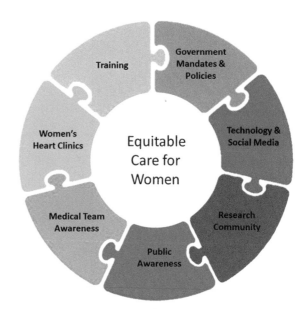

FIGURE 8   Overarching Framework for Equitable CV Health in Women. *Reproduced with permission from [32].*

led to routine extrapolations of findings in men to women, whether appropriate or not. Analysis of the 2007 women's CVD prevention guidelines revealed that women comprised only 25% of the participants in 156 CHD trials [128]. The 2010 report issued by the Institute of Medicine (now the National Academy of Medicine) calls for ongoing efforts to include more women in clinical trials and the National Institute of Health requires consideration of sex in research grant applications [129]. In 2010 a report published by the Institute of Medicine called for ongoing efforts to enhance the inclusion of women in clinical trials [130]. Novel strategies such as oversampling, focused recruitment of women, better incentives for researchers to include women, and social networking-enabled recruitment may be beneficial [131].

Methodologic Limitations of CVD Research in Women

Hypotheses and research frameworks regarding biological differences between women and men are dependent on observational data. Observational findings are influenced by social and cultural determinants of health, variable healthcare-seeking behaviors, access to care, and other factors. The result is that cardiovascular health in women may be influenced by many gendered determinants of health. Moreover, an observation of sex differences may also be influenced by socioeconomic and other nonclinical factors. Often, clinical trial sample sizes for women are much smaller than those for men. Similarly, matched cohorts are problematic, because they fail to reflect the "real-world" phenotype and experiences of women [132, 133]. Matching women to men, regardless of the statistical rigor used, represents a selected subset of women. The resulting failure to document sex differences in clinical outcomes may reflect such limitations.

New methodologies that can help unravel the complex interactions between biology and environment and lead to differential patterns of CVD outcomes among women should be developed. This could ultimately lead to new insights regarding sex- and gender-based differences in vulnerabilities and exposures, thus informing efforts to improve CVD outcomes for both women and men.

Barriers to enrollment and continued participation in clinical trials by women must be identified and addressed including transportation, childcare, stipend for time missed from work, and better education about the goals of research. Implementation of these strategies will likely improve inclusion of women in clinical trials [134].

Women's Heart Centers

Dedicated Women's Heart Centers are specialized clinics that are uniquely capable of identifying, characterizing, treating, and preventing heart disease in women, while also addressing important research gaps and developing new diagnostic tools and treatments [135]. These centers offer a personalized approach with a sex- and gender-specific care model and address risks that predominantly affect women as well as educate women on how to recognize CV disease symptoms and entities either unique to or more common in women. With a multidisciplinary team of clinicians versed in fields like cardio-oncology, cardio-rheumatology, and pregnancy disorders, these centers are well positioned to identify at-risk women and to advance the goal of sex-specific medicine [136] (**Figure 8**).

Government Mandates, Policies, and Advocacy

Significant strides in women's cardiovascular health have occurred due to unique partnerships between advocacy groups, policy makers, and researchers. Landmark actions include the 2014 FDA Action Plan of 27 points to enhance the collection and availability of sex-specific data and development of a new women's health research plan to answer specific sex-specific concerns and promote enrollment of diverse subpopulations [137].

Despite recommendations and legislations addressing the gaps in sex- and gender-based disparities in CVD, these are inadequately implemented [138]. Transforming research into policy and development of equitable care for women is a key component of improving CVD care of women. Policies supporting the health of women must be expanded and adopted to specifically address access to high-quality care, as well as aspects of financing and payment models [139].

Focusing on these issues is important and will require policy changes, investments, and collaboration among many stakeholders [139]. Federal agencies, researchers, clinicians, patients, women advocates, and policy makers must commit to work together to fully implement and expand existing policies to attain these goals.

Key Points

 In the United States, CVD remains the leading cause of death in women. Sex- and gender-specific disparities in outcomes persist, particularly in subsets of women disadvantaged by race, ethnicity, income level, or educational attainment.

 Sex-specific impact of traditional and novel risk factors for heart disease should be considered for improved risk stratification of at-risk women.

 Increased awareness of heart disease in women, attention to social determinants of health, health literacy, improved adherence to sex-specific guidelines, and adequate inclusion of women in research trials are necessary to address the existing disparities in research and clinical care.

 Optimal CV care for women should entail authentic partnerships among women, their communities, and academic- and community-serving health systems that engage in strategic planning to redesign care to meet the needs of diverse subgroups of women.

Acknowledgment

We graciously acknowledge the graphical assistance of Eduardo Garcia and Dr. Niti Aggarwal in preparation of the figures.

Sources of Funding

Dr. JoAnn Manson receives funding from the National Institutes of Health (HL34594, CA138962, and U01 HL145386).

References

[1] Benjamin EJ, Muntner P, Alonso A, Bittencourt MS, Callaway CW, Carson AP, et al. Heart disease and stroke statistics—2019 update: a report from the American Heart Association. Circulation 2019;139(10):e56–66. https://doi.org/10.1161/CIR.0000000000000659.

[2] Centers for Disease Control and Prevention NC for HS. Underlying Cause of Death 1999–2017 on CDC WONDER Online Database. Data are from the Multiple Cause of Death Files, 1999–2017, as compiled from data provided by the 57 vital statistics jurisdictions through the Vital Statistics Cooperative Program, Hyattsville, Maryland; 2017.

[3] Khavjou O, Phelps D, Leib A. Projections of cardiovascular disease prevalence and costs: 2015–2035. RTI Int; 2016.

[4] Gupta A, Wang Y, Spertus JA, Geda M, Lorenze N, Nkonde-Price C, et al. Trends in acute myocardial infarction in young patients and differences by sex and race, 2001 to 2010. J Am Coll Cardiol 2014;64(4):337–45. https://doi.org/10.1016/j.jacc.2014.04.054.

[5] Garcia M, Mulvagh SL, Bairey Merz CN, Buring JE, Manson JE. Cardiovascular disease in women. Circ Res 2016;118(8):1273–93. https://doi.org/10.1161/CIRCRESAHA.116.307547.

[6] Sidney S, Quesenberry CP, Jaffe MG, Sorel M, Nguyen-Huynh MN, Kushi LH, et al. Recent trends in cardiovascular mortality in the United States and public health goals. JAMA Cardiol 2016;1(5):594. https://doi.org/10.1001/jamacardio.2016.1326.

[7] Mensah GA, Wei GS, Sorlie PD, Fine LJ, Rosenberg Y, Kaufmann PG, et al. Decline in cardiovascular mortality: possible causes and implications. Circ Res 2017;120(2):366–80. https://doi.org/10.1161/CIRCRESAHA.116.309115.

[8] Xu J, Murphy SL, Kochanek KD, Arias E. Mortality in the United States, 2015. NCHS Data Brief 2016;267:1–8.

[9] Mehta LS, Beckie TM, DeVon HA, Grines CL, Krumholz HM, Johnson MN, et al. Acute myocardial infarction in women. Circulation 2016;133(9):916–47. https://doi.org/10.1161/CIR.0000000000000351.

[10] Wenger NK. Are we there yet? Closing the gender gap in coronary heart disease recognition, management and outcomes. Expert Rev Cardiovasc Ther 2013;11(11):1447–50. https://doi.org/10.1586/14779072.2013.845526.

[11] Bairey Merz CN, Andersen H, Sprague E, Burns A, Keida M, Walsh MN, et al. Knowledge, attitudes, and beliefs regarding cardiovascular disease in women: the Women's Heart Alliance. J Am Coll Cardiol 2017;70(2):123–32. https://doi.org/10.1016/j.jacc.2017.05.024.

[12] Mosca L, Hammond G, Mochari-Greenberger H, Towfighi A, Albert MA. Fifteen-year trends in awareness of heart disease in women. Circulation 2013;127(11):1254–63. https://doi.org/10.1161/CIR.0b013e318287cf2f.

[13] Lundberg GP, Mehta LS, Volgman AS. Specialized care for women: the impact of Women's Heart Centers. Curr Treat Options Cardiovasc Med 2018;20(9):76. https://doi.org/10.1007/s11936-018-0656-5.

[14] Roger VL, Go AS, Lloyd-Jones DM, Adams RJ, Berry JD, Brown TM, et al. Heart disease and stroke statistics—2011 update. Circulation 2011;123(4):e18–209. https://doi.org/10.1161/CIR.0b013e3182009701.

[15] Mosca L, Mochari-Greenberger H, Dolor RJ, Newby LK, Robb KJ. Twelve-year follow-up of American women's awareness of cardiovascular disease risk and barriers to heart health. Circ Cardiovasc Qual Outcomes 2010;3(2):120–7. https://doi.org/10.1161/CIRCOUTCOMES.109.915538.

[16] Ford ES, Ajani UA, Croft JB, Critchley JA, Labarthe DR, Kottke TE, et al. Explaining the decrease in U.S. deaths from coronary disease, 1980–2000. N Engl J Med 2007;356(23):2388–98. https://doi.org/10.1056/NEJMsa053935.

[17] American Heart Association. 1997 heart stroke facts: statistical update. Dallas:TX, American Heart Association; 1996.

[18] Mosca L, Manson JE, Sutherland SE, Langer RD, Manolio T, Barrett-Connor E. Cardiovascular disease in women. Circulation 1997;96(7):2468–82. https://doi.org/10.1161/01.CIR.96.7.2468.

[19] Mosca L, Jones WK, King KB, Ouyang P, Redberg RF, Hill MN. Awareness, perception, and knowledge of heart disease risk and prevention among women in the United States. American Heart Association Women's Heart Disease and Stroke Campaign Task Force. Arch Fam Med 2000;9(6):506–15.

[20] Hulley S, Grady D, Bush T, Furberg C, Herrington D, Riggs B, et al. Randomized trial of estrogen plus progestin for secondary prevention of coronary heart disease in postmenopausal women. Heart and Estrogen/progestin Replacement Study (HERS) Research Group. JAMA 1998;280(7):605–13.

[21] Rossouw JE, Anderson GL, Prentice RL, LaCroix AZ, Kooperberg C, Stefanick ML, et al. Risks and benefits of estrogen plus progestin in healthy postmenopausal women: principal results from the Women's Health Initiative randomized controlled trial. JAMA 2002;288(3):321–33.

[22] Mosca L, Ferris A, Fabunmi R, Robertson RM, American Heart Association. Tracking women's awareness of heart disease: an American Heart Association national study. Circulation 2004;109(5):573–9. https://doi.org/10.1161/01.CIR.0000115222.69428.C9.

[23] Mosca L, Banka CL, Benjamin EJ, Berra K, Bushnell C, Dolor RJ, et al. Evidence-based guidelines for cardiovascular disease prevention in women. Circulation 2004;109(5):672–93. https://doi.org/10.1161/01.CIR.0000114834.85476.81.

[24] Leifheit-Limson EC, D'Onofrio G, Daneshvar M, Geda M, Bueno H, Spertus JA, et al. Sex differences in cardiac risk factors, perceived risk, and health care provider discussion of risk and risk modification among young patients with acute myocardial infarction: the VIRGO study. J Am Coll Cardiol 2015;66(18):1949–57. https://doi.org/10.1016/j.jacc.2015.08.859.

[25] Cabana MD, Rand CS, Powe NR, Wu AW, Wilson MH, Abboud PA, et al. Why don't physicians follow clinical practice guidelines? JAMA 1999;282(15):1458. https://doi.org/10.1001/jama.282.15.1458.

[26] Mosca L, Linfante AH, Benjamin EJ, Berra K, Hayes SN, Walsh BW, et al. National study of physician awareness and adherence to cardiovascular disease prevention guidelines. Circulation 2005;111(4):499–510. https://doi.org/10.1161/01.CIR.0000154568.43333.82.

[27] Dhawan S, Bakir M, Jones E, Kilpatrick S, Bairey Merz CN. Sex and gender medicine in physician clinical training: results of a large, single-center survey. Biol Sex Differ 2016;7(Suppl 1):37. https://doi.org/10.1186/s13293-016-0096-4.

[28] Barnhart J, Lewis V, Houghton JL, Charney P. Physician knowledge levels and barriers to coronary risk prevention in women: survey results from the Women and Heart Disease Physician Education Initiative. Womens Health Issues 2007;17(2):93–100. https://doi.org/10.1016/j.whi.2006.11.003.

[29] GBD 2013 Mortality and Causes of Death Collaborators. Global, regional, and national age-sex specific all-cause and cause-specific mortality for 240 causes of death, 1990–2013: a systematic analysis for the Global Burden of Disease Study 2013. Lancet 2015;385(9963):117–71. https://doi.org/10.1016/S0140-6736(14)61682-2.

[30] Cohen JW, Krauss NA. Spending and service use among people with the fifteen most costly medical conditions, 1997. Health Aff 2003;22(2):129–38. https://doi.org/10.1377/hlthaff.22.2.129.

[31] Trogdon JG, Finkelstein EA, Nwaise IA, Tangka FK, Orenstein D. The economic burden of chronic cardiovascular disease for major insurers. Health Promot Pract 2007;8(3):234–42. https://doi.org/10.1177/1524839907303794.

[32] Aggarwal NR, Patel HN, Mehta LS, Sanghani RM, Lundberg GP, Lewis SJ, et al. Sex differences in ischemic heart disease: advances, obstacles, and next steps. Circ Cardiovasc Qual Outcomes 2018;11(2):e004437. https://doi.org/10.1161/CIRCOUTCOMES.117.004437.

[33] Anon. National diabetes statistics report. Centers for Disease Control and Prevention; 2020. Available from: https://www.cdc.gov/diabetes/data/statistics/statistics-report.html. [Accessed 3 October 2019].

[34] Barrett-Connor EL, Cohn BA, Wingard DL, Edelstein SL. Why is diabetes mellitus a stronger risk factor for fatal ischemic heart disease in women than in men? The Rancho Bernardo Study. JAMA 1991;265(5):627–31.

[35] Peters SAE, Huxley RR, Woodward M. Diabetes as risk factor for incident coronary heart disease in women compared with men: a systematic review and meta-analysis of 64 cohorts including 858,507 individuals and 28,203 coronary events. Diabetologia 2014;1542–51. https://doi.org/10.1007/s00125-014-3260-6.

[36] Manson JE, Colditz GA, Stampfer MJ, Willett WC, Krolewski AS, Rosner B, et al. A prospective study of maturity-onset diabetes mellitus and risk of coronary heart disease and stroke in women. Arch Intern Med 1991;151(6):1141–7.

[37] Peters SAE, Huxley RR, Woodward M. Diabetes as a risk factor for stroke in women compared with men: a systematic review and meta-analysis of 64 cohorts, including 775,385 individuals and 12,539 strokes. Lancet 2014;383(9933):1973–80. https://doi.org/10.1016/S0140-6736(14)60040-4.

[38] Donahue RP, Rejman K, Rafalson LB, Dmochowski J, Stranges S, Trevisan M. Sex differences in endothelial function markers before conversion to pre-diabetes: does the clock start ticking earlier among women? The Western New York Study. Diabetes Care 2007;30(2):354–9. https://doi.org/10.2337/dc06-1772.

[39] Anon. Estimates of diabetes and its burden in the United States. National diabetes statistics report. Available from: https://www-cdc-gov.proxy.library.emory.edu/diabetes/pdfs/data/statistics/national-diabetes-statistics-report.pdf; 2020. [Accessed 20 December 2019].

[40] Ma Y, Hébert JR, Manson JE, Balasubramanian R, Liu S, Lamonte MJ, et al. Determinants of racial/ethnic disparities in incidence of diabetes in postmenopausal women in the U.S.: the Women's Health Initiative 1993–2009. Diabetes Care 2012;35(11):2226–34. https://doi.org/10.2337/dc12-0412.

[41] Palatini P, Mos L, Santonastaso M, Saladini F, Benetti E, Mormino P, et al. Premenopausal women have increased risk of hypertensive target organ damage compared with men of similar age. J Women's Health 2011;20(8):1175–81. https://doi.org/10.1089/jwh.2011.2771.

[42] Gu Q, Burt VL, Paulose-Ram R, Dillon CF. Gender differences in hypertension treatment, drug utilization patterns, and blood pressure control among US adults with hypertension: data from the national health and nutrition examination survey 1999-2004. Am J Hypertens 2008;21(7):789–98. https://doi.org/10.1038/ajh.2008.185.

[43] Lloyd-Jones DM, Evans JC, Levy D. Hypertension in adults across the age spectrum: current outcomes and control in the community. JAMA 2005;294(4):466–72. https://doi.org/10.1001/jama.294.4.466.

[44] Sorlie PD, Allison MA, Aviles-Santa ML, Cai J, Daviglus ML, Howard AG, et al. Prevalence of hypertension, awareness, treatment, and control in the Hispanic Community Health Study/Study of Latinos. Am J Hypertens 2014;27(6):793–800. https://doi.org/10.1093/ajh/hpu003.

[45] Grundy SM, Stone NJ, Guideline Writing Committee for the 2018 Cholesterol Guidelines. 2018 cholesterol clinical practice guidelines: synopsis of the 2018 American Heart Association/American College of Cardiology/Multisociety Cholesterol Guideline. Ann Intern Med 2019;170(11):779–83. https://doi.org/10.7326/M19-0365.

[46] Kostis WJ, Cheng JQ, Dobrzynski JM, Cabrera J, Kostis JB. Meta-analysis of statin effects in women versus men. J Am Coll Cardiol 2012;59(6):572–82. https://doi.org/10.1016/j.jacc.2011.09.067.

[47] Virani SS, Woodard LD, Ramsey DJ, Urech TH, Akeroyd JM, Shah T, et al. Gender disparities in evidence-based statin therapy in patients with cardiovascular disease. Am J Cardiol 2015;115(1):21–6. https://doi.org/10.1016/j.amjcard.2014.09.041.

[48] Peterson ED, DeLong ER, Masoudi FA, O'Brien SM, Peterson PN, Rumsfeld JS, et al. Position statement on composite measures for healthcare performance assessment. American College of Cardiology Foundation/American Heart Association task force on performance measures (Writing Committee to develop a position statement on composite measures). J Am Coll Cardiol 2010;2010:1755–66. https://doi.org/10.1016/j.jacc.2010.02.016.

[49] Salami JA, Warraich HJ, Valero-Elizondo J, Spatz ES, Desai NR, Rana JS, et al. National trends in nonstatin use and expenditures among the US adult population from 2002 to 2013: insights from medical expenditure panel survey. J Am Heart Assoc 2018;7(2). https://doi.org/10.1161/JAHA.117.007132.

[50] Nanna MG, Wang TY, Xiang Q, Goldberg AC, Robinson JG, Roger VL, et al. Sex differences in the use of statins in community practice. Circ Cardiovasc Qual Outcomes 2019;12(8). https://doi.org/10.1161/CIRCOUTCOMES.118.005562.

[51] Flegal KM, Carroll MD, Kit BK, Ogden CL. Prevalence of obesity and trends in the distribution of body mass index among US adults, 1999-2010. JAMA 2012;307(5):491. https://doi.org/10.1001/jama.2012.39.

[52] Fox CS, Massaro JM, Hoffmann U, Pou KM, Maurovich-Horvat P, Liu C-Y, et al. Abdominal visceral and subcutaneous adipose tissue compartments. Circulation 2007;116(1):39–48. https://doi.org/10.1161/circulationaha.106.675355.

[53] de Koning L, Merchant AT, Pogue J, Anand SS. Waist circumference and waist-to-hip ratio as predictors of cardiovascular events: meta-regression analysis of prospective studies. Eur Heart J 2007;28(7):850–6. https://doi.org/10.1093/eurheartj/ehm026.

[54] Folsom AR, Kushi LH, Anderson KE, Mink PJ, Olson JE, Hong CP, et al. Associations of general and abdominal obesity with multiple health outcomes in older women: the Iowa Women's Health Study. Arch Intern Med 2000;160(14):2117–28. https://doi.org/10.1001/archinte.160.14.2117.

[55] MacKay MF, Haffner SM, Wagenknecht LE, D'Agostino RB, Hanley AJG. Prediction of type 2 diabetes using alternate anthropometric measures in a multi-ethnic cohort: the insulin resistance atherosclerosis study. Diabetes Care 2009;32(5):956–8. https://doi.org/10.2337/dc08-1663.

[56] Marcadenti A, Fuchs SC, Moreira LB, Wiehe M, Gus M, Fuchs FD. Accuracy of anthropometric indexes of obesity to predict diabetes mellitus type 2 among men and women with hypertension. Am J Hypertens 2011;24(2):175–80. https://doi.org/10.1038/ajh.2010.212.

[57] Barberio AM, Alareeki A, Viner B, Pader J, Vena JE, Arora P, et al. Central body fatness is a stronger predictor of cancer risk than overall body size. Nat Commun 2019;10(1). https://doi.org/10.1038/s41467-018-08159-w.

[58] Katzmarzyk PT, Bray GA, Greenway FL, Johnson WD, Newton RL, Ravussin E, et al. Racial differences in abdominal depot-specific adiposity in white and African American adults. Am J Clin Nutr 2010;91(1):7–15. https://doi.org/10.3945/ajcn.2009.28136.

[59] Tchernof A, Després J-P. Pathophysiology of human visceral obesity: an update. Physiol Rev 2013;93(1):359–404. https://doi.org/10.1152/physrev.00033.2011.

[60] Garaulet M, Pérez-Llamas F, Baraza JC, Garcia-Prieto MD, Fardy PS, Tébar FJ, et al. Body fat distribution in pre- and post-menopausal women: metabolic and anthropometric variables. J Nutr Heal Aging 2002;6(2):123–6.

[61] Fuente-Martín E, Argente-Arizón P, Ros P, Argente J, Chowen JA. Sex differences in adipose tissue. Adipocyte 2013;2(3):128–34. https://doi.org/10.4161/adip.24075.

[62] Luo J, Hendryx M, Laddu D, Phillips LS, Chlebowski R, LeBlanc ES, et al. Racial and ethnic differences in anthropometric measures as risk factors for diabetes. Diabetes Care 2019;42(1):126–33. https://doi.org/10.2337/dc18-1413.

[63] Anon. Prevalence of obesity among adults and youth: United States, 2015–2016. Available from: https://www.cdc.gov/nchs/products/databriefs/db288.htm. [Accessed 7 December 2019].

[64] Anon. Content—the office of minority health. Available from: https://minorityhealth.hhs.gov/omh/content.aspx?lvl=3%26lvlID=62%26ID=6457. [Accessed 7 January 2020].

[65] Schiller JS, Lucas JW, Ward BW, Peregoy JA. Summary health statistics for US adults: National Health Interview Survey, 2011; 2012.

[66] Huxley RR, Woodward M. Cigarette smoking as a risk factor for coronary heart disease in women compared with men: a systematic review and meta-analysis of prospective cohort studies. Lancet 2011;378(9799):1297–305. https://doi.org/10.1016/S0140-6736(11)60781-2.

[67] Peters SAE, Huxley RR, Woodward M. Smoking as a risk factor for stroke in women compared with men: a systematic review and meta-analysis of 81 cohorts, including 3 980 359 individuals and 42 401 strokes. Stroke 2013;44(10):2821–8. https://doi.org/10.1161/STROKEAHA.113.002342.

[68] Pomp ER, Rosendaal FR, Doggen CJM. Smoking increases the risk of venous thrombosis and acts synergistically with oral contraceptive use. Am J Hematol 2008;83(2):97–102. https://doi.org/10.1002/ajh.21059.

[69] Rosengren A, Hawken S, Ôunpuu S, Sliwa K, Zubaid M, Almahmeed WA, et al. Association of psychosocial risk factors with risk of acute myocardial infarction in 11 119 cases and 13 648 controls from 52 countries (the INTERHEART study): case-control study. Lancet 2004;364(9438):953–62. https://doi.org/10.1016/S0140-6736(04)17019-0.

[70] Mallik S, Spertus JA, Reid KJ, Krumholz HM, Rumsfeld JS, Weintraub WS, et al. Depressive symptoms after acute myocardial infarction. Arch Intern Med 2006;166(8):876. https://doi.org/10.1001/archinte.166.8.876.

[71] Smolderen KG, Spertus JA, Gosch K, Dreyer RP, D'Onofrio G, Lichtman JH, et al. Depression treatment and health status outcomes in young patients with acute myocardial infarction. Circulation 2017;135(18):1762–4. https://doi.org/10.1161/CIRCULATIONAHA.116.027042.

[72] Korkeila J, Vahtera J, Korkeila K, Kivimäki M, Sumanen M, Koskenvuo K, et al. Childhood adversities as predictors of incident coronary heart disease and cerebrovascular disease. Heart 2010;96(4):298–303. https://doi.org/10.1136/hrt.2009.188250.

[73] Rich-Edwards JW, Mason S, Rexrode K, Spiegelman D, Hibert E, Kawachi I, et al. Physical and sexual abuse in childhood as predictors of early-onset cardiovascular events in women. Circulation 2012;126(8):920–7. https://doi.org/10.1161/CIRCULATIONAHA.111.076877.

[74] Shah AJ, Ghasemzadeh N, Zaragoza-Macias E, Patel R, Eapen DJ, Neeland IJ, et al. Sex and age differences in the association of depression with obstructive coronary artery disease and adverse cardiovascular events. J Am Heart Assoc 2014;3(3). https://doi.org/10.1161/JAHA.113.000741.

[75] Vaccarino V, Sullivan S, Hammadah M, Wilmot K, Al Mheid I, Ramadan R, et al. Mental stress-induced-myocardial ischemia in young patients with recent myocardial infarction. Circulation 2018;137(8):794–805. https://doi.org/10.1161/CIRCULATIONAHA.117.030849.

[76] Muka T, Oliver-Williams C, Kunutsor S, Laven JSE, Fauser BCJM, Chowdhury R, et al. Association of age at onset of menopause and time since onset of menopause with cardiovascular outcomes, intermediate vascular traits, and all-cause mortality: a systematic review and meta-analysis. JAMA Cardiol 2016;1(7):767–76. https://doi.org/10.1001/jamacardio.2016.2415.

[77] Mosca L, Benjamin EJ, Berra K, Bezanson JL, Dolor RJ, Lloyd-Jones DM, et al. Effectiveness-based guidelines for the prevention of cardiovascular disease in women—2011 update. Circulation 2011;123(11):1243–62. https://doi.org/10.1161/CIR.0b013e31820faaf8.

[78] Wenger NK. Recognizing pregnancy-associated cardiovascular risk factors. Am J Cardiol 2014;113(2):406–9. https://doi.org/10.1016/j.amjcard.2013.08.054.

[79] Carr DB, Newton KM, Utzschneider KM, Tong J, Gerchman F, Kahn SE, et al. Preeclampsia and risk of developing subsequent diabetes. Hypertens Pregnancy 2009;28(4):435–47. https://doi.org/10.3109/10641950802629675.

[80] Bellamy L, Casas J-P, Hingorani AD, Williams DJ. Pre-eclampsia and risk of cardiovascular disease and cancer in later life: systematic review and meta-analysis. BMJ 2007;335(7627):974. https://doi.org/10.1136/bmj.39335.385301.BE.

[81] Vrachnis N, Augoulea A, Iliodromiti Z, Lambrinoudaki I, Sifakis S, Creatsas G. Previous gestational diabetes mellitus and markers of cardiovascular risk. Int J Endocrinol 2012;2012:1–6. https://doi.org/10.1155/2012/458610.

[82] Ogunmoroti O, Osibogun O, Kolade OB, Ying W, Sharma G, Vaidya D, et al. Multiparity is associated with poorer cardiovascular health among women from the Multi-Ethnic Study of Atherosclerosis. Am J Obstet Gynecol 2019;221(6):631.e1–631.e16. https://doi.org/10.1016/j.ajog.2019.07.001.

[83] Kardys I, Vliegenthart R, Oudkerk M, Hofman A, Witteman JCM. The female advantage in cardiovascular disease: do vascular beds contribute equally? Am J Epidemiol 2007;166(4):403–12. https://doi.org/10.1093/aje/kwm115.

[84] Nicholls SJ, Wolski K, Sipahi I, Schoenhagen P, Crowe T, Kapadia SR, et al. Rate of progression of coronary atherosclerotic plaque in women. J Am Coll Cardiol 2007;49(14):1546–51. https://doi.org/10.1016/j.jacc.2006.12.039.

[85] Collins LM, Lanza ST. Latent class and latent transition analysis: with applications in the social, behavioral, and health sciences. John Wiley & Sons; 2013.

[86] Anderson RD, Pepine CJ. Gender differences in the treatment for acute myocardial infarction. Circulation 2007;115(7):823–6. https://doi.org/10.1161/CIRCULATIONAHA.106.685859.

[87] Manteuffel M, Williams S, Chen W, Verbrugge RR, Pittman DG, Steinkellner A. Influence of patient sex and gender on medication use, adherence, and prescribing alignment with guidelines. J Womens Health (Larchmt) 2014;23(2):112–9. https://doi.org/10.1089/jwh.2012.3972.

[88] Gordon T, Kannel WB, Hjortland MC, McNamara PM. Menopause and coronary heart disease. The Framingham Study. Ann Intern Med 1978;89(2):157–61.

[89] Kannel WB, Cupples LA, D'Agostino RB. Sudden death risk in overt coronary heart disease: the Framingham Study. Am Heart J 1987;113(3):799–804.

[90] Lerner DJ, Kannel WB. Patterns of coronary heart disease morbidity and mortality in the sexes: a 26-year follow-up of the Framingham population. Am Heart J 1986;111(2):383–90.

[91] Benjamin EJ, Virani SS, Callaway CW, Chamberlain AM, Chang AR, Cheng S, et al. Heart disease and stroke statistics—2018 update: a report from the American Heart Association. Circulation 2018;137(12):e67–492. https://doi.org/10.1161/CIR.0000000000000558.

[92] Lloyd-Jones DM, Larson MG, Leip EP, Beiser A, D'Agostino RB, Kannel WB, et al. Lifetime risk for developing congestive heart failure: the Framingham Heart Study. Circulation 2002;106(24):3068–72. https://doi.org/10.1161/01.CIR.0000039105.49749.6F.

[93] Goldberg RJ, Spencer FA, Farmer C, Meyer TE, Pezzella S. Incidence and hospital death rates associated with heart failure: a community-wide perspective. Am J Med 2005;118(7):728–34. https://doi.org/10.1016/j.amjmed.2005.04.013.

[94] Loehr LR, Rosamond WD, Chang PP, Folsom AR, Chambless LE. Heart failure incidence and survival (from the Atherosclerosis Risk in Communities study). Am J Cardiol 2008;101(7):1016–22. https://doi.org/10.1016/j.amjcard.2007.11.061.

[95] Huffman MD, Berry JD, Ning H, Dyer AR, Garside DB, Cai X, et al. Lifetime risk for heart failure among white and black americans: cardiovascular lifetime risk pooling project. J Am Coll Cardiol 2013;61(14):1510–7. https://doi.org/10.1016/j.jacc.2013.01.022.

[96] Hogg K, Swedberg K, McMurray J. Heart failure with preserved left ventricular systolic function: epidemiology, clinical characteristics, and prognosis. J Am Coll Cardiol 2004;317–27. https://doi.org/10.1016/j.jacc.2003.07.046.

[97] Masoudi FA, Havranek EP, Smith G, Fish RH, Steiner JF, Ordin DL, et al. Gender, age, and heart failure with preserved left ventricular systolic function. J Am Coll Cardiol 2003;41(2):217–23. https://doi.org/10.1016/S0735-1097(02)02696-7.

[98] Levy D, Kenchaiah S, Larson MG, Benjamin EJ, Kupka MJ, Ho KKL, et al. Long-term trends in the incidence of and survival with heart failure. N Engl J Med 2002;347(18):1397–402. https://doi.org/10.1056/NEJMoa020265.

[99] Roger VL, Weston SA, Redfield MM, Hellermann-Homan JP, Killian J, Yawn BP, et al. Trends in heart failure incidence and survival in a community-based population. JAMA 2004;292(3):344–50. https://doi.org/10.1001/jama.292.3.344.

[100] Go AS, Hylek EM, Phillips KA, Chang Y, Henault LE, Selby JV, et al. Prevalence of diagnosed atrial fibrillation in adults: national implications for rhythm management and stroke prevention: the AnTicoagulation and Risk Factors in Atrial Fibrillation (ATRIA) Study. JAMA 2001;285(18):2370–5. https://doi.org/10.1001/jama.285.18.2370.

[101] Miyasaka Y, Barnes ME, Gersh BJ, Cha SS, Bailey KR, Abhayaratna WP, et al. Secular trends in incidence of atrial fibrillation in Olmsted County, Minnesota, 1980 to 2000, and implications on the projections for future prevalence. Circulation 2006;114(2):119–25. https://doi.org/10.1161/CIRCULATIONAHA.105.595140.

[102] Schnabel RB, Yin X, Gona P, Larson MG, Beiser AS, McManus DD, et al. 50 year trends in atrial fibrillation prevalence, incidence, risk factors, and mortality in the Framingham Heart Study: a cohort study. Lancet 2015;386(9989):154–62. https://doi.org/10.1016/S0140-6736(14)61774-8.

[103] Lloyd-Jones DM, Wang TJ, Leip EP, Larson MG, Levy D, Vasan RS, et al. Lifetime risk for development of atrial fibrillation: the Framingham Heart Study. Circulation 2004;110(9):1042–6. https://doi.org/10.1161/01.CIR.0000140263.20897.42.

[104] Wilke T, Groth A, Mueller S, Pfannkuche M, Verheyen F, Linder R, et al. Incidence and prevalence of atrial fibrillation: an analysis based on 8.3 million patients. Europace 2013;15(4):486–93. https://doi.org/10.1093/europace/eus333.

[105] Piccini JP, Hammill BG, Sinner MF, Jensen PN, Hernandez AF, Heckbert SR, et al. Incidence and prevalence of atrial fibrillation and associated mortality among Medicare beneficiaries, 1993-2007. Circ Cardiovasc Qual Outcomes 2012;5(1):85–93. https://doi.org/10.1161/CIRCOUTCOMES.111.962688.

[106] Humphries KH, Kerr CR, Connolly SJ, Klein G, Boone JA, Green M, et al. New-onset atrial fibrillation: sex differences in presentation, treatment, and outcome. Circulation 2001;103(19):2365–70. https://doi.org/10.1161/01.CIR.103.19.2365.

[107] Emdin CA, Wong CX, Hsiao AJ, Altman DG, Peters SAE, Woodward M, et al. Atrial fibrillation as risk factor for cardiovascular disease and death in women compared with men: systematic review and meta-analysis of cohort studies. BMJ 2016;352. https://doi.org/10.1136/bmj.h7013.

[108] Bushnell C, McCullough LD, Awad IA, Chireau MV, Fedder WN, Furie KL, et al. Guidelines for the prevention of stroke in women: a statement for healthcare professionals from the American Heart Association/American Stroke Association. Stroke 2014;45(5):1545–88. https://doi.org/10.1161/01.str.0000442009.06663.48.

[109] Benjamin EJ, Wolf PA, D'Agostino RB, Silbershatz H, Kannel WB, Levy D. Impact of atrial fibrillation on the risk of death: the Framingham Heart Study. Circulation 1998;98(10):946–52. https://doi.org/10.1161/01.CIR.98.10.946.

[110] Centers for Disease Control and Prevention (CDC). Lower extremity disease among persons aged ≥40 years with and without diabetes—United States, 1999–2002. Morb Mortal Wkly Rep 2005;54(45):1158–60.

[111] Higgins P, Higgins A. Epidemiology of peripheral arterial disease in women. J Epidemiol 2003;13(1):1–14. https://doi.org/10.2188/jea.13.1.

[112] Lo RC, Bensley RP, Dahlberg SE, Matyal R, Hamdan AD, Wyers M, et al. Presentation, treatment, and outcome differences between men and women undergoing revascularization or amputation for lower extremity peripheral arterial disease. J Vasc Surg 2014;59(2). https://doi.org/10.1016/j.jvs.2013.07.114.

[113] Fox CS, Evans JC, Larson MG, Kannel WB, Levy D. Temporal trends in coronary heart disease mortality and sudden cardiac death from 1950 to 1999: the Framingham heart study. Circulation 2004;110(5):522–7. https://doi.org/10.1161/01.CIR.0000136993.34344.41.

[114] Wilmot KA, O'Flaherty M, Capewell S, Ford ES, Vaccarino V. Coronary heart disease mortality declines in the United States from 1979 through 2011 clinical perspective. Circulation 2015;132(11):997–1002. https://doi.org/10.1161/CIRCULATIONAHA.115.015293.

[115] Chen J, Normand S-LT, Wang Y, Drye EE, Schreiner GC, Krumholz HM. Recent declines in hospitalizations for acute myocardial infarction for medicare fee-for-service beneficiaries. Circulation 2010;121(11):1322–8. https://doi.org/10.1161/CIRCULATIONAHA.109.862094.

[116] Arora S, Stouffer GA, Kucharska-Newton AM, Qamar A, Vaduganathan M, Pandey A, et al. Twenty year trends and sex differences in young adults hospitalized with acute myocardial infarction. Circulation 2019;139(8):1047–56. https://doi.org/10.1161/CIRCULATIONAHA.118.037137.

[117] Interactive atlas of heart disease and stroke | cdc.gov. 2020. Accessed 8 December 2019.

[118] Anon. Women and heart disease | cdc.gov. Available from: https://www.cdc.gov/heartdisease/women.htm; 2020. [Accessed 8 December 2019].

[119] Manhapra A, Canto JG, Vaccarino V, Parsons L, Kiefe CI, Barron HV, et al. Relation of age and race with hospital death after acute myocardial infarction. Am Heart J 2004;148(1):92–8. https://doi.org/10.1016/j.ahj.2004.02.010.

[120] Pool LR, Ning H, Lloyd-Jones DM, Allen NB. Trends in racial/ethnic disparities in cardiovascular health among US adults from 1999–2012. J Am Heart Assoc 2017;6(9). https://doi.org/10.1161/JAHA.117.006027.

[121] Mozaffarian D, Benjamin EJ, Go AS, Arnett DK, Blaha MJ, Cushman M, et al. Heart disease and stroke statistics—2016 update. Circulation 2016;133(4):e38–360. https://doi.org/10.1161/CIR.0000000000000350.

[122] Pearson-Stuttard J, Guzman-Castillo M, Penalvo JL, Rehm CD, Afshin A, Danaei G, et al. Modeling future cardiovascular disease mortality in the United States. Circulation 2016;133(10):967–78. https://doi.org/10.1161/CIRCULATIONAHA.115.019904.

[123] Rodriguez CJ, Allison M, Daviglus ML, Isasi CR, Keller C, Leira EC, et al. Status of cardiovascular disease and stroke in Hispanics/Latinos in the United States: a science advisory from the American Heart Association. Circulation 2014;130(7):593–625. https://doi.org/10.1161/CIR.0000000000000071.

[124] Caughey AB, Stotland NE, Washington AE, Escobar GJ. Maternal ethnicity, paternal ethnicity, and parental ethnic discordance. Obstet Gynecol 2005;106(1):156–61. https://doi.org/10.1097/01.AOG.0000164478.91731.06.

[125] Victor RG, Lynch K, Li N, Blyler C, Muhammad E, Handler J, et al. A cluster-randomized trial of blood-pressure reduction in black barbershops. N Engl J Med 2018;378(14):1291–301. https://doi.org/10.1056/NEJMoa1717250.

[126] Pronk NP, Katz AS, Lowry M, Payfer JR. Reducing occupational sitting time and improving worker health: the take-a-stand project, 2011. Prev Chronic Dis 2012;9. https://doi.org/10.5888/pcd9.110323.

[127] Gittner LS, Hassanein SE, Murphy PJ. Church-based heart health project: health status of Urban African Americans. Perm J 2007;11(3). https://doi.org/10.7812/tpp/06-126.

[128] Melloni C, Berger JS, Wang TY, Gunes F, Stebbins A, Pieper KS, et al. Representation of women in randomized clinical trials of cardiovascular disease prevention. Circ Cardiovasc Qual Outcomes 2010;3(2):135–42. https://doi.org/10.1161/CIRCOUTCOMES.110.868307.

[129] Clayton JA, Collins FS. Policy: NIH to balance sex in cell and animal studies. Nature 2014;509(7500):282–3. https://doi.org/10.1038/509282a.

[130] Institute of Medicine (US) Committee on Women's Health Research. Women's health research: progress, pitfalls, and promise. Washington, DC: National Academies Press (US); 2010.

[131] Tweet M, Gulati R, Aase LA, Hayes SN. Spontaneous coronary artery dissection: a disease-specific, social networking community-initiated study. Elsevier; 2011.

[132] Nicolini F, Vezzani A, Fortuna D, Contini GA, Pacini D, Gabbieri D, et al. Gender differences in outcomes following isolated coronary artery bypass grafting: long-term results. J Cardiothorac Surg 2016;11(1):144. https://doi.org/10.1186/s13019-016-0538-4.

[133] Leipsic J, Taylor CM, Gransar H, Shaw LJ, Ahmadi A, Thompson A, et al. Sex-based prognostic implications of nonobstructive coronary artery disease: results from the international multicenter CONFIRM study. Radiology 2014;273(2):393–400. https://doi.org/10.1148/radiol.14140269.

[134] Infanti JJ, O'Dea A, Gibson I, McGuire BE, Newell J, Glynn LG, et al. Reasons for participation and non-participation in a diabetes prevention trial among women with prior gestational diabetes mellitus (GDM). BMC Med Res Methodol 2014;14(1):13. https://doi.org/10.1186/1471-2288-14-13.

[135] Garcia M, Miller VM, Gulati M, Hayes SN, Manson JE, Wenger NK, et al. Focused cardiovascular care for women. Mayo Clin Proc 2016;91(2):226–40. https://doi.org/10.1016/j.mayocp.2015.11.001.

[136] AlBadri A, Wei J, Mehta PK, Shah R, Herscovici R, Gulati M, et al. Sex differences in coronary heart disease risk factors: rename it ischaemic heart disease! Heart 2017;103(20):1567–8. https://doi.org/10.1136/heartjnl-2017-311921.

[137] Anon. FDA action plan to enhance the collection and availability of demographic subgroup data. Available from: https://www.fda.gov/media/89307/download. [Accessed 8 December 2019].

[138] Wood SF, Mieres JH, Campbell SM, Wenger NK, Hayes SN. Advancing Women's heart health through policy and science: highlights from the first National Policy and science summit on Women's cardiovascular health. Women's Heal Issues 2016;26(3):251–5. https://doi.org/10.1016/j.whi.2016.03.001.

[139] Havranek EP, Mujahid MS, Barr DA, Blair IV, Cohen MS, Cruz-Flores S, et al. Social determinants of risk and outcomes for cardiovascular disease. Circulation 2015;132(9):873–98. https://doi.org/10.1161/CIR.0000000000000228.

Section III

Ischemic Heart Disease

Chapter 3

Prevention of Cardiovascular Disease

Garima Sharma, Anum Minhas, and Erin D. Michos

Clinical Case

Ms. M is a 52-year-old black woman with no known cardiovascular disease (CVD) who is interested in starting postmenopausal hormone replacement therapy (HRT) because she continues to have significant bothersome vasomotor symptoms following her final menstrual period approximately 18 months ago. Prior to starting HRT, she was referred by her gynecologist for a cardiovascular risk assessment. Notably, the patient's father had a myocardial infarction at age 48. The patient experienced preeclampsia with her third pregnancy at age 36 and is currently taking a thiazide diuretic for antihypertensive medication. She has never smoked and does not have diabetes. Her blood pressure is 136/84 mmHg, total cholesterol 220 mg/dL, high-density lipoprotein (HDL) cholesterol 56 mg/dL, triglycerides 210 mg/dL (nonfasting), and low-density lipoprotein (LDL) cholesterol 122 mg/dL. How would you counsel her regarding her cardiovascular (CV) risk profile?

Sex Differences in Cardiac Disease. https://doi.org/10.1016/B978-0-12-819369-3.00013-7

Abstract

Cardiovascular disease (CVD) remains the leading cause of death for both men and women globally, as well as being a leading contributor to morbidity and healthcare expenditures. This disease burden can be reduced by prioritizing a preventive approach to cardiovascular health. Both sex and gender differences contribute to cardiovascular health and the implementation of CVD prevention strategies. Healthcare professionals should be cognizant of sex-differential effects of the traditional risk factors for CVD and of emerging sex-specific CVD risk factors. Female-specific "risk-enhancing" factors include those related to pregnancy (hypertensive disorders of pregnancy, preterm delivery, small for gestational age infant, gestational diabetes, parity) and hormonal risk (early menarche, polycystic ovarian syndrome, early menopause, use of hormone replacement therapy). In men, vascular erectile dysfunction, stemming from endothelial dysfunction and subclinical atherosclerosis, is a red flag that portends CVD risk. For both men and women, if decisions for the initiation of preventive pharmacotherapies (i.e., statins) still remain uncertain after considering these risk-enhancing factors, the use of coronary artery calcium scores can refine risk assessment and guide shared decision making. Understanding sex and gender unique risk factors may improve current risk assessment strategies for primary prevention of CVD and facilitate comprehensive preventive care to both female and male patients.

Introduction

Cardiovascular disease (CVD) is the leading cause of death in women and accounted for 412,244 deaths in women in 2016 in the United States [1]. Yet, in 1997, only 30% of American women surveyed were aware that CVD was the leading cause of death in women; this did increase to 54% in 2009 but had subsequently plateaued when last surveyed in 2012 [2]. In addition, while CVD-related mortality declined in the overall US population prior to 2000, this decline was largely observed in men, with rates of CVD death remaining relatively stable or even increasing among women. In response to this, the American Heart Association (AHA) launched the Go Red for Women campaign and published the first women-specific guidelines for CVD prevention in 1999, which were last updated in 2011 [3]. With these efforts, since 2000, CVD death rates fortunately started to decline in women, but total CVD deaths remained higher in women compared to men until 2013 when they became approximately equal or even slightly lower in women. However, since 2013, tracking along with the developing obesity and diabetes epidemics, there has been a concerning plateauing or even slight rise in CVD mortality in both men

and women [1]. This is a cause of concern for both men and women and an indication that more intensive preventive efforts are still sorely needed.

Furthermore, despite the decline in CVD mortality among women since 2000, age and racial disparities still exist. Most of the decline in CVD mortality has been noted in older women, whereas for younger women aged < 55 years, there has been stagnation in the decline in CVD mortality with minimal improvement between 1999 and 2011 [4]. The proportion of acute myocardial infarction has actually increased in young adults (aged 35–54 years) between 1995 and 2014 and become more pronounced for women [5]. Of particular concern, statistics show that deaths from heart disease have increased by 7% between 2011 and 2017 among middle-aged women 45–64 years old, a greater relative change than any other sex or age group [6]. Furthermore, in the United States, the rate of CVD in black females is higher at 286/100,000 compared to 206/100,000 for white females, with lower documented rates of awareness of coronary artery disease (CAD) and stroke among black women compared to white women [1].

Women are less likely to receive preventive treatment or guidance, such as lipid-lowering therapy, aspirin [7], and therapeutic lifestyle changes compared to men at similar CVD risk [8, 9]. There is also a disparity in antihypertensive management, and women with hypertension are less likely to have their blood pressure (BP) at goal [10]. There are also gender disparities in patient-reported outcomes with women being more likely to report poor communication with their healthcare provider and less satisfaction with their healthcare experience, which may also be contributing to suboptimal implementation of preventive efforts [7].

Both sex and gender differences contribute to women's health and the implementation of CVD prevention strategies. Sex (or biological) differences in the cardiovascular system occur secondary to differences in gene expression from the sex chromosomes, which may be further modified by sex differences in hormones and result in sex-specific gene expression and function. These differences cause variations in the prevalence and presentation of cardiovascular pathology, including autonomic dysregulation, hypertension, diabetes, and vascular and cardiac remodeling. In contrast, gender refers to social construct, including expected roles and behaviors, and varies across societies and over historical time. Gender differences are unique to humans and arise from sociocultural practices (behaviors, environment, lifestyle, nutrition) [11]. Sex and gender both influence women's cardiovascular health, as health is determined by both biology and the expression of gender.

Hormonal changes, pregnancy-related disorders, autoimmune diseases, menopause, and cancer therapies can all have an impact on endothelial function, vascular anatomy, and myocardial contractility. Women with heart disease often present later, receive care not consistent with

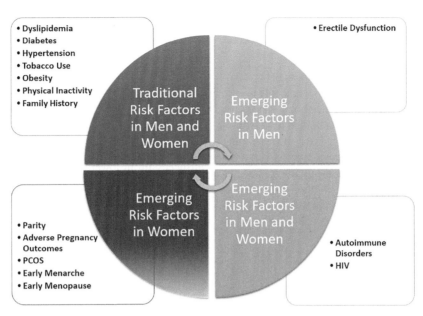

FIGURE 1 Traditional and Emerging Risk Factors for CVD in Men and Women. HIV, human immunodeficiency virus; PCOS, polycystic ovary syndrome.

accepted guidelines, and have less access to diagnostic and therapeutic resources. Understanding the differences and challenges of treating CVD in women is essential to improving population health [12].

This chapter will highlight the important differences in the traditional and nontraditional risk factors for CVD in women and men (**Figure 1**) and address sex-specific considerations in the prevention of CVD.

Traditional (Modifiable) Risk Factors and Sex Differences

The traditional risk factors for atherosclerotic CVD (ASCVD) are pertinent in both men and women. The primary traditional ASCVD risk factors are diabetes mellitus, hypertension, dyslipidemia, tobacco use, obesity, and physical inactivity [13], which are discussed further in the sections below. However, the prevalence and magnitude of risk for these factors is variable between the sexes and may impact risk factor optimization among men and women [12]. For instance, hypertension, diabetes mellitus, and smoking are more potent risk factors for myocardial infarction in women than in men [14].

The 2013 and 2019 American College of Cardiology (ACC)/AHA Primary Prevention Guidelines on the assessment of CVD risk therefore recommend use of the sex-specific Pooled Cohort Equations, also called the ASCVD risk calculator, which incorporates sex and race differences in addition to the traditional risk factors [13, 15]. After 10-year ASCVD risk estimation, patients can be categorized as <5% (low risk), 5–7.5% (borderline risk), 7.5–<20% (intermediate risk), and ≥20% (high risk). Lifestyle

intervention is generally enough for most low-risk individuals, while high-risk individuals are recommended for both lifestyle and preventive pharmacotherapy (i.e., statins). Individuals at borderline or intermediate risk often need additional information to help refine their risk further, as discussed below. Using the Pooled Cohort Equations, our case patient Ms. M was estimated to have a 10-year ASCVD risk of 5.4%, which would be considered only borderline risk. However, prior studies have suggested that risk calculators based on traditional risk factors alone may underestimate ASCVD risk for women [16–18]. In later sections, we will discuss how consideration of "risk-enhancing factors" and the selection use of coronary artery calcium (CAC) measurement can help better refine risk estimation in women.

Diabetes Mellitus

Diabetes mellitus is a metabolic disorder characterized by elevated blood glucose levels, most commonly defined as a hemoglobin A1c ≥6.5%, with type 2 diabetes mellitus (T2DM) stemming from insulin resistance. There is rising incidence of T2DM among US adults, with 9.4% of adults carrying the diagnosis [1]. An even greater percentage, roughly a third, have prediabetes and are at risk for diabetes [1]. The relative risk for CVD is 44% greater in women than in men with diabetes [19]. Furthermore, diabetes is a stronger risk factor for CAD mortality among women than men, with a more than threefold excess fatal CAD risk in women with T2DM compared to women without diabetes [19–21]. Women with T2DM also have a much higher adjusted hazard ratio (HR) for fatal CAD [HR 14.7 (95% confidence interval (CI) 6.2–35.3)] compared to men with T2DM [HR 3.8 (2.5–5.7)] [22]. Beyond CAD, women with diabetes also have a higher risk of developing heart failure, of having a stroke, and of developing peripheral arterial disease compared to men [11].

This increase in ASCVD in women with diabetes, compared to men, may be related to an overall worse risk profile in women, including an increase in metabolic syndrome, worse atherogenic dyslipidemia, endothelial dysfunction, and hypercoagulability [23–25]. Prior to developing frank diabetes, women with prediabetes may also have a worse risk factor profile compared to men with prediabetes [26, 27]. Despite the increased risk of ASCVD with diabetes, the control of diabetes among both men and women remains suboptimal at 30% in women vs. 20% in men between 2013 and 2016 [28].

Hypertension

Hypertension, or elevated BP, is defined as systolic BP (SBP) ≥ 130 mmHg and diastolic BP (DBP) ≥ 80 mmHg [29]. The prevalence of hypertension is slightly lower in the United States among women than men, with roughly 42% of women and 49% of men carrying the diagnosis [28]. Ambulatory BP monitoring has shown that women tend to have lower SBP than men until menopause, after which SBP is actually higher in women compared to men. Animal studies have suggested that this may be related to androgen effects on the renin-angiotensin system, possibly leading to increased vasoconstrictor substance production and reduction in nitric oxide availability [30].

Blood pressure levels, though, are associated with comparable risk of myocardial infarction in both sexes [3, 31–33]. More importantly, across a spectrum of ages (40–80 years), a log-linear relationship has been demonstrated between SBP 115 to 180 mmHg and DBP 75 to 105 mmHg and risk of ASCVD, based on a large meta-analysis of 61 studies [34]. Both SBP and DBP are associated with increased ASCVD risk [35]. Each 20 mmHg rise in SBP and 10 mmHg rise in DBP is associated with doubling in mortality from heart disease, stroke, and other vascular diseases [34].

Fortunately, there has been an increase in recent years in the number of those taking antihypertensive medications, with 64% of women and 53% of men taking medications [28]. Similarly, BP control has improved as well, with 30% of women and 22% of men achieving control [28]. Nevertheless, more ASCVD deaths are attributable to hypertension than any other modifiable ASCVD risk factor, and hypertension remains a challenge to control in both sexes [13]. Treatment of hypertension is discussed in the management section below.

Dyslipidemia

Evidence from genetic studies, epidemiologic studies, and clinical interventional studies has confirmed that the serum cholesterol levels reflective of the apolipoprotein-B containing lipoproteins, namely low-density lipoprotein cholesterol (LDL-C) but also the triglyceride-rich lipoprotein particles, are causally related to ASCVD [36–38]. High-density lipoprotein cholesterol (HDL-C), while generally higher in women than in men, has a U-shaped epidemiologic relationship with mortality risk [39] and is not thought to be causally related to ASCVD [38]. To date, therapeutic strategies designed to raise HDL-C have been unsuccessful in lowering ASCVD risk [40]. Thus, HDL-C should no longer be thought of as "good" cholesterol as its levels at the high end of spectrum are not necessarily protective against ASCVD. Current AHA/ACC Cholesterol Guidelines still predominantly focus on LDL-C as the primary therapeutic target [41].

While elevated cholesterol conveys risk in both sexes, it has been shown that dyslipidemia may carry the greatest population-adjusted risk among women (47.1%), compared to all other traditional ASCVD risk factors [14]. Sex differences in total cholesterol levels differ by age with total cholesterol levels being similar for women and men < 35 years, lower in women compared to men for ages 35–49 years, and then higher in women compared to men after age 50 [28]. This correlates with known increases in total cholesterol and LDL-C after menopause [42]. Compared to men, women are less likely to have lipids treated and controlled [28].

Several prior studies have demonstrated that adults with lower cholesterol levels have lower risk of ASCVD, and that even a 1 mmol/L reduction in total cholesterol is associated with about a half [HR 0.44 (95% CI 0.42–0.48)], a third [0.66 (0.65–0.68)], and a sixth [0.83 (0.81–0.85)] lower CAD mortality in both sexes at ages 40–49, 50–69, and 70–89 years, respectively [43]. For high-risk patients, clinical interventional studies have reaffirmed the principle of "lower is better" for LDL-C [36].

All patients benefit from therapeutic lifestyle changes including a heart-healthy diet and regular physical activity. The decision of when to initiate and how intensively to implement pharmacologic therapy depends on the absolute risk of the patient. Statins remain the first-line pharmacological therapy for secondary prevention of ASCVD and for high-risk primary prevention [41]. For more details on statin and other lipid-targeted therapeutics, see the preventive pharmacologic section below.

However, it is concerning that despite comparable recommendations for treatment for both sexes, men are more likely to be treated for and obtain control of dyslipidemia than women [44]. Among individuals eligible for statin therapy, women are less likely to be offered any statins (67% vs. 78%) or guideline-directed statin intensity than men (37% vs. 45%), and women also declined (4% vs. 2%) or discontinued (11% vs. 6%) statin therapy more frequently than men (all $P < 0.001$) [45]. Only 48% of women compared to 67% of men are treated for dyslipidemia, and only 40% of women compared to 63% of men achieve control [28]. This difference in treatment is independent of underlying CVD status, and may suggest disparity in appropriately identifying dyslipidemia among women compared to men. Even among those with established ASCVD, for which statin treatment is a class I indication, women are less likely to be prescribed or be adherent to statin treatment [7, 46, 47]. More work needs to be done to reduce these disparities in implementation of these guideline-endorsed prevention recommendations.

Tobacco Use

Tobacco use, such as smoking (including second-hand smoke) and smokeless tobacco, is a risk factor for ASCVD and increases all-cause mortality. This was initially recognized in 1964 in the US Surgeon General's report on the health consequences of smoking, which concluded that smokers had a higher death rate from CAD than nonsmokers, and since then, multiple studies have confirmed this [48–50]. More recently, there is evidence that Electronic Nicotine Delivery Systems (ENDS), or e-cigarettes, which emit aerosol with nicotine and toxic gases, likely also increase CVD [51, 52].

Smoking confers risk for both sexes but is a more potent risk factor for women than men for myocardial infarction, resulting in a relative risk of 2–5 for women, which increases with dose and is greater than the twofold risk seen in men [21, 53]. First-time myocardial infarction also occurs prematurely more often in women than men who smoke [54]. Additional risk factors such as use of oral contraceptives further synergistically increase risk of acute myocardial infarction in female smokers compared to male [55, 56]. Overall, though, smoking has decreased in the United States, with lower use among women (at 18%) compared to men (at 22%) [28]. Clinicians should ask women about current and past smoking as well as secondary tobacco smoke exposures, use of e-cigarettes, and assess readiness to quit at each visit. They should provide counseling and offer nicotine replacement or other pharmacotherapy, combined with a behavioral modification program [13].

Obesity

The rising incidence of obesity in the United States and worldwide has affected both men and women. Both obesity, defined as body mass index (BMI) $\geq 30\,\text{kg/m}^2$, and overweight, defined as BMI = 25–29.9 kg/m^2, are associated with increased risk of ASCVD [57, 58]. Weight gain during adult years imparts a significantly elevated risk for heart failure (particularly heart failure with preserved ejection fraction) and also ASCVD, as observed in prospective cohort studies [59, 60].

The effect of obesity on development of CVD is greater in women than men, with the relative risk of CAD at 64% in women compared to 46% in men in the Framingham Heart Study [57]. Recently, the independent association between high BMI and large waist circumference and ASCVD in women has also been demonstrated [58, 61, 62]. A recent large cohort study showed that within each BMI category, CAD risk increased with increasing waist circumference, and within each waist category, CAD risk increased with rising BMI [58]. The lowest risk for cumulative CAD incidence is in women with BMI $< 25\,\text{kg/m}^2$ and waist circumference $< 70\,\text{cm}$, with the highest incidence in women with BMI $\geq 30\,\text{kg/m}^2$ and waist circumference $\geq 80\,\text{cm}$ [58].

In addition, obesity is associated with hypertension, diabetes, and dyslipidemia, all of which are also independent risk factors for ASCVD, thereby increasing overall risk of ASCVD even more, especially for women. The Framingham Offspring Study demonstrated that clustering of three or more risk factors was associated with a 2.4 (95% CI 1.6–3.4) and 5.9 (2.6–13.7) times greater risk for CAD in men and women, respectively (both $P < 0.001$) [63].

In the United States, average BMI is greater among women (29.6 kg/m^2) compared to men (29.0 kg/m^2) [28]. This is explained by a greater percentage of obese women (42%) compared to men (37%) [28]. Given the easy access to affordable food and sedentary lifestyles of today's society, it is likely that BMI will continue to rise among the population unless strategic interventions are made to counter this epidemic.

The 2019 ACC/AHA Primary Prevention Guidelines recommend that, for adults who are overweight or obese, counseling and comprehensive lifestyle interventions should be implemented, which include calorie restriction and achieving and maintaining weight loss to improve the cardiovascular risk profile [13]. While the guidelines recommend calculating a BMI at least annually, they also state that it may be reasonable to additionally measure waist circumference to identify those at higher cardiometabolic risk.

Physical Inactivity

Sedentary lifestyle is associated with multiple adverse effects. A large meta-analysis demonstrated that sedentary behavior, independent of physical activity, was associated with all-cause mortality [HR 1.2 (95% CI 1.1–1.4)], cardiovascular mortality [1.2 (1.1–1.3)], and incident CVD [1.1 (1.0–1.7)] [64].

Conversely, physical activity is associated with lower risk of CAD/CVD in a dose-response fashion, as demonstrated by several large observational studies [65–69]. Specifically, individuals who perform 150 min/week of moderate-intensity leisure-time physical activity have a 14% lower risk of CAD [relative risk (RR) 0.86 (95% CI 0.77–0.96)], compared to those with no activity, and those who perform 300 min/week of moderate-intensity physical activity have an even greater 20% reduction in risk of CAD [0.80 (0.74–0.88)] [66]. Women seem to derive even greater benefit from physical activity than men [66].

While the 2019 ACC/AHA Primary Prevention Guidelines recommend that all adults should engage in at least 150 min/week of moderate intensity or 75 min/week of vigorous-intensity aerobic physical activity, there is a startlingly large proportion of adults that remain inactive [13]. The 2011 National Health Interview Survey demonstrated that more women than men were inactive (33% vs. 30%), and physical inactivity increased with age, with more than 50% of adults ≥ 75 years reporting inactivity [70]. These data highlight the importance of implementing strategies to increase physical activity and reduce sedentary behavior among all adults.

Even brisk walking (a moderate intensity activity) can have substantial CAD risk reduction in women [69]. Women with ASCVD who do not meet physical activity goals incur substantially higher medical expenditures [71]. There are sociodemographic disparities among women who do not achieve goals, with older women, minorities, and those with lower education and income being less likely to achieve recommended physical activity levels [71]. Given the substantial benefits on ASCVD reduction and other health outcomes, all women should be encouraged to move more [72]. The guidelines state that any activity, even if less than the recommended amounts, is better than no or little activity [13, 73]. Obtainment of recommended levels of moderate to vigorous activity, as well as reduction in sitting time, remain two separate goals [73].

Family History of Premature CAD

A family history of premature CAD, defined as the presence of a first-degree relative with clinical CAD or sudden death (men < 55, women < 65 years old) has been recognized as a potent risk factor for incident ASCVD events, with an approximate doubling of risk [74]. The strength of a family history of CAD as a cardiovascular risk factor increases with increasing prematurity of parental CAD, with the highest risk estimated in patients with a maternal history of CAD < 50 years old [odds ratio 3.15 (95% CI 2.18–2.44)] [75]. A positive sibling history is predictive as well, even after considering other ASCVD risk factors and parental history [76].

Among women estimated to be at low 10-year ASCVD risk, women with a family history of premature ASCVD are more likely to have significant subclinical atherosclerosis as measured by CAC scores [17]. Furthermore, a family history of premature CAD significantly magnifies the risk of any CAC, CAC > 100, and CAC > 75th age/sex percentile, when added to other cardiovascular risk factors [77].

A prior study from the Multi-Ethnic Study of Atherosclerosis (MESA) demonstrated that CAC effectively risk-stratifies asymptomatic persons with a family history of premature CAD [78]. This suggests that CAC may help determine which patients will likely benefit from aspirin and lipid-modifying therapy in this high-risk subset. The 2019 ACC/AHA Primary Prevention Guidelines consider family history of premature ASCVD as a "risk enhancing" factor that may favor the initiation or intensification of statin therapy; however, when risk-based decisions are uncertain, a CAC score may be useful to guide shared decision making [13].

Emerging Risk Factors for CVD in Women

The above traditional risk factors are important for both women and men; however, there has been recent recognition of factors unique to women related to pregnancy and sex hormones that elevate a woman's risk for future ASCVD [79] (**Figure 2**), as summarized below.

Parity

During pregnancy, there are normal physiologic changes that occur in lipids, glucose, and weight; pregnancy also confers additional stressors such as endothelial dysfunction, inflammation, and hemostatic processes. These pregnancy related factors may influence a woman's subsequent cardiovascular risk. Women may gain weight with each subsequent pregnancy, with multiparous women being more likely to have an elevated BMI later in life [80]. Women with multiparity, particularly five or more live pregnancies, are less likely to be in ideal cardiovascular health when they reach middle to older age, particularly driven by elevated BMI [80]. Furthermore, a meta-analysis of 10 cohort studies found parity to be related to subsequent CVD risk in a dose-response relationship [81]. Thus, after delivery, more attention needs to be placed on the mother's health and well-being, as some of this risk could be potentially mediated by more intensive lifestyle interventions after each pregnancy.

Adverse Pregnancy Outcomes Overview

Adverse pregnancy outcomes, such as hypertensive disorders of pregnancy, gestational diabetes, preterm birth, and delivery of small for gestational age infant, are now being recognized as emerging risk factors for premature CVD in women [82–85]. Current guidelines for management of CVD disease in pregnancy largely concentrate on valvular heart disease, congenital heart disease, arrhythmias, and dilated cardiomyopathies. However, many women are at risk for developing future CVD from pregnancy related complications such as preeclampsia, gestational hypertension, gestational diabetes, and preterm birth, and are often lost to follow-up [86]. Recently, the American

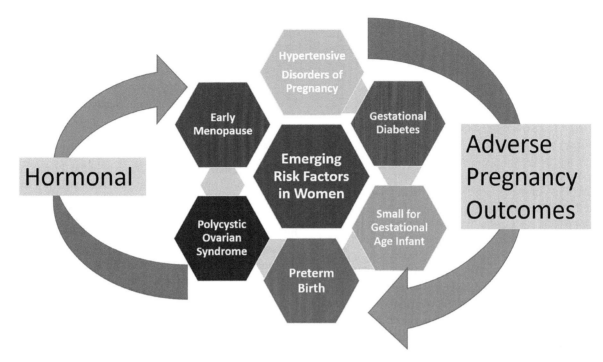

FIGURE 2   Emerging Risk Factors of Cardiovascular Disease in Women.

College of Obstetricians and Gynecologists (ACOG) released new guidance on the management of hypertension during pregnancy, one of the leading causes of maternal death [87]. Similarly, new information recognizing other adverse pregnancy outcomes including gestational diabetes and postpartum diabetes is now available on ways to improve prevention and early intervention in these high-risk women [88].

Recently, there has been an alarming trend in adverse pregnancy outcomes, as suggested by a rise in the incidence of hypertensive disorders of pregnancy, preterm birth, and percentage of low birthweight infants in the United States [89]. Adverse pregnancy outcomes are associated with increased risk of developing chronic hypertension, left ventricular dysfunction, vascular dysfunction, and renal dysfunction [90]. Even after adjusting for sociodemographic factors, preexisting conditions, and clinical and behavioral conditions associated with the current pregnancy, women with any adverse pregnancy outcomes experience a 19% increased risk of CVD [91]. Understanding these unique risk factors specific to women and the pathogenesis of adverse pregnancy outcomes-related cardiovascular dysfunction is an unmet need, and preparing the cardiovascular workforce to be better equipped to risk-stratify these women is a critical area for future research and advocacy in this area.

Although separate diagnoses, adverse pregnancy outcomes share key features that suggest a shared pathogenesis [90]. There is suggestion of placental insufficiency and development of pro-inflammatory and antiangiogenic milieu. Abnormal placental spiral artery remodeling, placental oxidative stress causing maternal endothelial dysfunction, and inflammation likely lead to placental ischemia and incomplete placentation [90]. Antiangiogenic proteins are formed due to placental ischemia and can be detected in maternal circulation. These factors can cause vascular, cardiac, and renal damage during pregnancy and are associated with CVD postpartum.

Cardiologists and primary care physicians should partner to develop a comprehensive pregnancy history tool for cardiovascular risk assessment and take a detailed pregnancy history which incorporates direct questions highlighting adverse pregnancy outcomes [92]. Pregnancy provides a unique opportunity for both patients and clinicians to engage in improving overall health. Many studies have shown that women are motivated to change their lifestyle habits during pregnancy and the postpartum period. Obstetricians and cardiologists can improve the cardiovascular health of women through enhanced collaboration [93].

Preeclampsia and Hypertensive Disorders of Pregnancy

Preeclampsia is a multisystem disease that occurs after 20 weeks of gestation, mediated by abnormalities in the placental vasculature leading to both short-term and long-term endothelial dysfunction and inappropriate vasoconstriction in multiple vascular beds [82]. Although several factors are involved in the pathophysiology of preeclampsia, placental insufficiency may be the most important, leading to the induction of an antiangiogenic state in the mother and development of endothelial dysfunction in several organs, resulting in the clinical manifestations of the disease [94].

A large meta-analysis showed that women with a history of preeclampsia have a 71% increased risk of cardiovascular mortality, a 2.5-fold increase in risk of CAD, and a fourfold increase in heart failure when compared to normal cohorts [83]. Another study of more than 60,000 women reported that women with gestational hypertension and preeclampsia had almost a threefold and sixfold increased rate of chronic hypertension in later life, respectively [95]. It also showed that women who were hypertensive during their first pregnancy had a 70% increased risk of T2DM and 30% increased prevalence of hypercholesterolemia later in life [95]. New data have shown that preeclampsia is associated with complications such as preterm birth or a small for gestational age infant. A Norwegian study with a mean follow-up of 17.2 years found that women with preeclampsia alone had a twofold increased risk of a major cardiovascular event, whereas women with preeclampsia plus preterm birth or small for gestational age infant were four times as likely to have a major cardiac event compared to women with uncomplicated pregnancies [96].

Women with chronic hypertension with superimposed preeclampsia also have high incidence of preterm birth, small for gestational age infants, and perinatal death [97]. Indeed, as mentioned in the 2019 ACC/AHA Primary Prevention Guidelines, a history of preeclampsia is considered a "risk enhancing" factor that would move a woman to a higher predicted ASCVD risk category and favor consideration of statin therapy for prevention [13].

Preterm Birth

Preterm birth, defined as gestation length < 37 weeks, occurs in approximately 12.5% of births in the United States each year, affecting roughly one in eight infants [82]. One study examined the associations of adverse pregnancy outcomes with a range of cardiovascular risk factors, and found that preterm birth corresponded with higher SBP 18 years after delivery [98]. In the decade following pregnancy, women who delivered before 34 weeks had higher blood pressure, atherogenic lipids, and carotid intimal media thickness compared to women with term births [99]. A meta-analysis found that women with preterm birth were 63% more likely to experience a CVD event for up to 50 years than women with no preterm birth, and 93% more likely to experience CVD death compared to women with no previous

history [100]. The highest risks for future maternal CVD occur when preterm birth is before 32 weeks [84]. These findings highlight the potential burden of CVD in women with preterm birth and the need for careful consideration of short- and long-term follow-up and management of women with a history of pregnancy complications.

Small for Gestational Age Infant

Small for gestational age is variously defined using growth curves and/or observed clinical characteristics, including fundal height, various body proportions, soft tissue measurements, and a birth weight of < 10% percentile or birth weight deviation of 1 to 2 standard deviations [101]. Having a small for gestational age infant is an independent marker of CVD-related mortality and morbidity in the mother [102]. Delivery of a small for gestational age infant has been associated with increased maternal CVD risk and is dose dependent with an increase in the risk of maternal CVD according to both the severity of small for gestational age and number of previous deliveries of small for gestational age infants [103]. As a result of the shared mechanisms of actions of many adverse pregnancy outcomes, it is likely that their cooccurrence may afford a greater risk for maternal CVD than independently [100]. There is a need for better research to inform evidence on CVD prevention in these women, but certainly the presence of an adverse pregnancy outcome should be a "red flag" that suggests a woman should be followed more closely for CVD risk with more intensive implementation of lifestyle measures and screening and treatment of traditional CVD risk factors.

Gestational Diabetes Mellitus

Gestational diabetes mellitus is linked with several acute maternal health risks such as preeclampsia, preterm birth, excess growth of the fetus causing birth lacerations and delivery by Caesarean section, as well as long-term development of T2DM, metabolic syndrome, and CVD [82, 85, 104, 105]. Although maternal glucose tolerance often normalizes shortly after pregnancy, women with gestational diabetes have a substantially increased risk of developing T2DM later in life [104]. Studies have reported that women are more than seven times as likely to develop diabetes after gestational diabetes and that approximately 50% of mothers with gestational diabetes will develop diabetes within 10 years, making gestational diabetes one of the strongest predictors of T2DM [104, 106]. Lifestyle interventions that target diet, activity, and behavioral strategies can effectively modify body weight. In randomized controlled trials, lifestyle intervention and medical treatment decreased the number who progressed to diabetes by approximately 50%

in women with previous gestational diabetes [107]. Even in the absence of conventional risk factors of ASCVD, gestational diabetes is associated with higher CVD risk in young women [108]. Most guidelines recommend breastfeeding (if able), a lifelong healthy lifestyle including weight loss, as well as an oral glucose tolerance test 2–6 months post-delivery and every 1–3 years [82, 88]. These high-risk women would benefit from systematic follow-up, programs facilitating early diagnosis and intervention, and increased awareness among patients and clinicians. Future clinical trials should test and use dietary, lifestyle, and pharmacologic interventions that might prevent or delay the onset of T2DM in affected women.

Polycystic Ovary Syndrome

Polycystic ovary syndrome (PCOS) is a complex disorder comprising both hormonal and metabolic abnormalities that include impaired glucose tolerance, T2DM, vascular disease, dyslipidemia, and obstructive sleep apnea [109, 110]. Insulin resistance and the consequent development of hyperinsulinemia contribute to the constellation of cardio-metabolic abnormalities noted above. There is an increased prevalence of cardiovascular risk factors in PCOS even though there are conflicting data about cardiovascular event risk directly attributed to PCOS [110]. Some prior studies and meta-analyses have shown that women with PCOS have greater CAD and stroke risk compared to women without PCOS [111–115]. While some studies have shown this association to be independent of BMI [114], other studies have suggested that this excess CVD risk may be explained by traditional CVD risk factors and that the absolute risk conferred by PCOS may be small [116]. In addition, these studies suggest that there is heterogeneity of CVD risk across certain PCOS subtypes [110, 116].

There is evidence of early endothelial dysfunction in patients with PCOS and normal weight suggesting early vascular damage [117]. Lifelong exposure to an adverse cardiovascular risk profile in women with PCOS may lead to premature atherosclerosis [117, 118]. If ASCVD risk is uncertain for the treatment of dyslipidemia in PCOS, a CAC score for women older than 40 years might help identify those with evidence of subclinical atherosclerosis who warrant more intensified preventive treatment [13, 118]. Future research should include prospective longitudinal studies in peri- and postmenopausal women with PCOS to better estimate the longer-term risk of cardiac morbidity and mortality in this population. In the meantime, regular screening for risk factors such as dyslipidemia, impaired glucose tolerance, sleep apnea, and hypertension and timely early interventions are critical to reduce the overall risk burden [119].

Menarche

Early onset of menarche has been shown to be associated with increased CVD risk in women [120]. For example, among women participating in the Women's Ischemic Syndrome Evaluation (WISE) study, compared to women who reported menarche at age 12, those women who reported onset of menarche at or before age 10 had a 4.5-fold increased risk of subsequent CVD [121]. This highlights the importance of taking a detailed reproductive/hormonal history in women.

Early Menopause

The prevalence of most of the conventional CV risk factors, especially T2DM, hypertension, dyslipidemia, and obesity, and other risk factors in postmenopausal women is high [122]. Early menopause (before age 45 years) compounds many traditional CVD risk factors, including changes in body fat distribution, reduced glucose tolerance, abnormal plasma lipids, increased blood pressure, increased sympathetic tone, endothelial dysfunction, and vascular inflammation [123]. Early menopause is an independent risk factor for CVD, increasing risk by approximately 1.3-fold [120, 124]. Indeed the 2019 ACC/AHA Guidelines have indicated that early menopause before the age of 40 is a "risk enhancing" factor that would elevate one's estimated ASCVD risk into a higher category and potentially favor statin therapy [13].

Menopause influences risk factors for CVD because it increases total serum cholesterol by 2–20% and triglycerides by 7–35%, antithrombin III, factor VII coagulant activity VIIc, and plasma fibrinogen [125]. Postmenopausal women with higher endogenous testosterone levels relative to estrogen levels are at higher cardiovascular risk [126]. In premenopausal women, endogenous estrogen has beneficial effects in preventing atherosclerosis by improving plasma lipids, maintaining endothelial cell integrity and promoting nitric oxide production. Conversely, in the setting of established atherosclerosis, exogenous estrogen can increase matrix metalloproteinase expression, which could lead to instability of the fibrous cap and rupture of the atheromatous plaque [125, 127].

Hormone Replacement Therapy

Hormone replacement therapy (HRT) with estrogens in older postmenopausal women without established ASCVD has not been shown to be associated with a reduction in CVD events [128, 129]. Studies have shown that there is a lower incidence of CAD in women taking HRT within 10 years of their menopause, but stroke risk remained elevated [130]. Data in women under the age of 60 years with symptoms or other indications have shown that initiating HRT near menopause provided a favorable benefit-risk ratio. The Danish Osteoporosis Prevention Study undertaken to investigate the long-term effects of HRT on cardiovascular outcomes in recently postmenopausal women found that after 10 years of randomized treatments, women receiving HRT that started early after the menopause had a significant reduction in risk of cardiovascular mortality, heart failure, or myocardial infarction [131]. This "timing hypothesis" is reflected in current National Institute for Health and Care Excellence (NICE) and International Menopause Society guideline recommendations, indicating the beneficial effects of estrogen therapy in women when started in a time-dependent manner, i.e., closer to menopause [132].

Thus, while HRT is still not recommended for the sole purpose of ASCVD prevention, it may be reasonably safe to initiate for the management of menopausal women aged <60 years or within 10 years of menopause. For women who initiate HRT more than 10 or 20 years from menopause onset or are aged 60 years or older, the benefit-risk ratio appears less favorable because of the greater absolute risks of CAD, stroke, venous thromboembolism, and dementia [133]. HRT initiated within 10 years of menopause is also associated with less CAC, a marker of subclinical atherosclerosis ($P = 0.02$ by rank test) [134].

As mentioned above, HRT use purely for reduction of CVD risk is not indicated, but is primarily used for relief of genitourinary and vasomotor menopausal symptoms. Thus, HRT can be considered in women in an individualized manner, closer to menopause, with careful considerations of their overall CVD risks. If ASCVD risk remains uncertain after calculation of 10-year risk estimation and consideration of risk-enhancing factors, a CAC score can be helpful to further refine risk and guide risk-based decisions [13]. The 2019 ACC/AHA Guidelines for Primary Prevention of CVD stated that measurement of CAC is reasonable for intermediate or select borderline-risk individuals as a risk decision aid to guide shared decision making about preventive therapies in cases where there is otherwise uncertainty of risk (Class of Recommendation IIa) [13].

Emerging Risk Factors for CVD in Men

Erectile Dysfunction

Erectile dysfunction (ED), defined as the inability to achieve or maintain erection for satisfactory sexual intercourse, is very common, affecting 40% of men older than 40% and 70% of those over 70 years of age [135]. ED can be subcategorized as organic or psychogenic. Vascular ED, stemming from atherosclerosis and/or endothelial dysfunction, is the most common organic cause of ED and is associated with significantly increased ASCVD risk [136, 137].

Vascular ED is fueled by the same traditional risk factors for ASCVD such as diabetes, smoking, insufficient physical activity, obesity, high BP, and high cholesterol. However, ED appears to be a risk marker for future ASCVD events even independently of these shared risk factors [136, 137]. Penile arteries are smaller in caliber compared to coronary arteries and thus perhaps more susceptible to the hemodynamic effects of atherosclerosis and endothelial dysfunction [138]. As a result, ED symptoms can predate CAD symptoms by 2–5 years. This is why the presence of vascular ED can signal the presence of processes that also lead to CVD events and may offer an opportunity to implement intensified preventive efforts.

Men should be routinely screened for ED, and if it is present, should have screening of traditional risk factors and management of those risk factors as appropriate. Intensified lifestyle measures should be implemented. If the decision to treat with statin medication is uncertain, CAC scoring may be useful to further refine risk in men with ED who were otherwise deemed at low or intermediate 10-year risk by the Pooled Cohort Equations [13]. An analysis from the MESA study determined that a CAC score > 100 at baseline was a significant predictor of the presence of ED 9 years later [HR 1.43 (95% CI 1.09–1.88)] after adjusting for cardiovascular risk factors [139]. This study suggests that CAC > 100 is a useful threshold in accurately risk-stratifying which ED patients would benefit from preventive therapy, including statins and possibly aspirin.

Novel Risk Markers in Women and Men

The 2019 ACC/AHA Primary Prevention Guidelines have indicated that the presence of autoimmune disease or HIV infection are "risk enhancing" factors that would favor the initiation or intensification of statin therapy [13]. A CAC score may again be beneficial to further risk-stratify this population. These factors are briefly discussed below.

Inflammatory Disorders

Autoimmune disorders affect approximately 8% of the population, of which the majority are women (78%) [140]. The presence of autoimmune and other inflammatory disorders has been linked to accelerated atherosclerosis and increased risk of ASCVD events. Rheumatoid arthritis portends a 50% increased incidence of cardiovascular mortality [141]. Among patients with systemic lupus erythematosus (SLE), the relative risk of myocardial infarction is between 2- and 10-fold higher than the general population [142]. Psoriatic disorders are also associated with increased risk of myocardial infarction and cardiovascular mortality [143]. As a result, the need to improve ASCVD risk assessment and treatment in patients with rheumatologic disorders is

FIGURE 3 **Nonpharmacologic and Pharmacologic Approaches to Addressing CVD in Men and Women.** ASCVD, atherosclerotic cardiovascular disease.

paramount. Prior studies have suggested that individuals with these inflammatory disorders have increased subclinical atherosclerosis compared to controls [144–146], and thus CAC scoring may also be of use in this population to further refine risk.

Human Immunodeficiency Virus (HIV) Infection

HIV infection is associated with an almost twofold increased incidence of myocardial infarction after adjusting for traditional cardiovascular risk factors [147]. Individuals with HIV also have increased coronary plaque, independent of ASCVD risk factors [148, 149]. While HIV infection has traditionally been considered a disease of younger persons, antiretroviral therapy has improved the survival of HIV-infected individuals, leading to an increased prevalence of HIV-infected adults aged 60 and older and the subsequent emergence of chronic diseases such as ASCVD in this population. This changing demographic highlights the need for more awareness of cardiovascular risk stratification and treatment in patients living with HIV.

Pharmacologic Prevention Treatments

A healthy lifestyle remains the foundation of all preventive efforts. The recommendations for physical activity, healthy diet [150], smoking cessation, and weight maintenance are discussed above in the respective risk factor sections. In this section, we will discuss pharmacologic interventions.

Nonpharmacologic and pharmacologic approaches to CVD prevention are shown in **Figure 3**.

Antihypertensive Management

The 2017 ACC AHA Hypertension Guidelines changed the definition of hypertension to ≥ 130/80 mmHg, with BP range 130–139/80–89 mmHg now classified as stage 1 hypertension [29]. Stage 2 hypertension is ≥ 140 SBP or ≥ 90 mmHg DBP. Nonpharmacologic lifestyle interventions should be implemented for all individuals with hypertension, including those who also require antihypertensive medications. These nonpharmacologic measures include weight loss, heart-healthy diet pattern, reduction in dietary sodium, increased dietary potassium, reduction in alcohol intake, and increased physical activity.

For those with stage 1 hypertension and < 10% year estimated ASCVD risk, only nonpharmacologic therapy and continued monitoring are needed. However, for stage 1 hypertension with 10-year risk ≥ 10% or established ASCVD or chronic kidney disease and for those with stage 2 hypertension, both pharmacologic therapy and non-pharmacologic intervention are recommended [13, 29]. The first-line antihypertensive medications include diuretics, calcium channel blockers, or angiotensin-converting enzyme inhibitors (ACE-Is) or angiotensin receptor blockers (ARBs). In adults with confirmed hypertension, a BP target of < 130/80 mmHg is recommended.

There are no sex differences in the general recommendations for hypertension management. However, ACE-Is and ARBs are contraindicated in pregnancy and should be used with caution in women who may become pregnant. For women who develop hypertension while taking oral contraceptives, the first treatment is to stop the oral contraceptives and switch to another form of birth control. ACE-I-induced cough and peripheral edema associated with calcium channel blockers are more common in women than in men [151]. The ACOG recommends treatment of preeclampsia with severe features; hypertension management during pregnancy is beyond the scope of this review and one should refer to ACOG guidelines [87].

Statins and Other Lipid-Directed Pharmacotherapy

Statins remain the first-line pharmacological therapy for secondary prevention among those with established ASCVD and for higher-risk primary prevention for ASCVD risk reduction [41]. They should be thought of as "prevention medications" rather than simply cholesterol lowering medications. Statin therapy is associated with significant decreases in cardiovascular events and in all-cause mortality in both women and men. In the largest primary prevention trial to date, JUPITER, which enrolled 6801 women, statins were shown to significantly reduce ASCVD events in women [RR 0.63 (0.49–0.82)] with no difference when compared with men and a trend toward reduced mortality [RR 0.78 (0.53–1.15)] [152]. Statin therapy should be used in appropriate patients without regard to sex [153], and indeed the 2018 AHA/ACC Cholesterol Guidelines do not have sex-specific recommendations [41].

For primary prevention, the 2019 ACC/AHA Guidelines recommend that in all adults, both men and women, between ages 40 and 75 years, without ASCVD or diabetes mellitus, the 10-year ASCVD risk using the Pooled Cohort Equations should be calculated as the starting point to guide risk-based treatment decisions such as statins [13]. While moderate-to-high intensity statin therapy is generally recommended for those at intermediate risk (7.5–19.9%) and high intensity statin for those with ≥20% 10-year risk, this should be in the context of a clinician-patient risk discussion given concerns that the Pooled Cohort Equations may actually overestimate ASCVD risk [154]. Risk discussion should include a review of major risk factors, the potential benefits of lifestyle and statin therapies, the potential for adverse effects and drug-drug interactions, consideration of costs of therapy, and patient preferences and values in shared decision making.

For those at borderline (5–7.4%) and intermediate risk (7.5–19.9%), the presence of the "risk-enhancing factors" may strengthen decisions to initiate or intensify statin therapy. Risk-enhancing factors, important for both men and women, include: family history of premature ASCVD; persistently elevated LDL-C levels ≥ 160 mg/dL (≥ 4.1 mmol/L); metabolic syndrome; chronic kidney disease; chronic inflammatory disorders (e.g., rheumatoid arthritis, psoriasis, or chronic HIV); high-risk ethnicity (e.g., South Asian); and if measured, having elevated levels of apolipoprotein B, lipoprotein (a), high-sensitivity C-reactive protein (hsCRP), or persistently elevated triglycerides. In addition to the above, there are also risk-enhancing factors unique to women such as early menopause and adverse pregnancy outcomes such as preeclampsia, which were discussed above. However, after considering these risk-enhancing factors, if risk-based decisions for statin therapy still remain uncertain, then measurement of CAC score can help further refine risk and guide the statin treatment decision [13].

Statins are also recommended in primary prevention for those aged 20 and older with severe primary hypercholesterolemia with LDL-C ≥ 190 mg/dL (high-intensity statin) and for patients age 40–75 years with diabetes (at least moderate-intensity statin, with higher intensity statin for patients with diabetes with multiple ASCVD risk factors). A 10-year risk estimation is not needed for these individuals before initiating statin therapy. For primary prevention patients with heterozygous familial hypercholesterolemia with an LDL-C threshold ≥ 100 mg/dL despite maximally tolerated statin, add-on ezetimibe, and/or PCSK9 inhibition are recommended for further LDL-C lowering, given the substantial lifetime ASCVD risk of these patients.

For patients with established clinical ASCVD (secondary prevention) aged < 75 years, a high intensity statin is recommended. At least a moderate intensity statin is recommended for those over age 75, but with considerations of patient comorbidities, frailty, drug-drug interactions, and anticipated life-expectancy. For secondary prevention patients treated with a maximally tolerated statin, a threshold LDL-C of ≥ 70 mg/dL may prompt consideration of the further addition of a nonstatin therapy with proven efficacy (such as ezetimibe or PCSK9 inhibitors) for further LDL-C lowering [41]. In addition, for statin-treated secondary prevention patients or patients with diabetes with multiple risk factors, who have residual triglyceride elevation, icosapent ethyl [purified eicosapentaenoic acid (EPA)] at a dose of 4 g/day may also further reduce ASCVD risk [155].

Despite the proven benefit of statin therapy among both sexes, as mentioned above, women are less likely to be prescribed or be adherent to statin treatment [7, 45–47]. In terms of patient beliefs, women compared to men were more likely to believe that statins caused muscle aches and pains [45]. However, statin therapy has been shown to have excellent safety in both men and women, with risk of serious muscle injury being less than 0.1% and risk of serious liver toxicity around 0.001% [156]. There is a small risk of newly diagnosed diabetes mellitus, around 0.2% per year of treatment; however, this modest increased risk occurs predominantly among individuals who already have underlying insulin resistance [156]. Furthermore, the benefits of the reduction of major vascular events conferred by statin therapy outweigh the small risk of new onset of diabetes. Increased lifestyle changes should be recommended to help decrease the risk of progression to diabetes.

Aspirin Therapy

For decades, low-dose aspirin (75–100 mg) has been widely administered for ASCVD prevention. By irreversibly inhibiting platelet function, aspirin reduces risk of atherothrombosis but with the trade-off of increased risk of bleeding, particularly in the gastrointestinal tract. Aspirin is well established for secondary prevention of ASCVD and is widely recommended for this indication. However, in primary prevention, the absolute risks of vascular events are lower than in secondary prevention, but the complication rates (i.e., bleeding) are comparable, making the role of aspirin much less certain in this population.

The largest primary prevention study of aspirin in women was the 2005 Women's Health Study (WHS), which randomized nearly 40,000 initially healthy women > 45 years to 100 mg alternate-day dosing of aspirin or placebo. The WHS found that low-dose aspirin reduced the risk of stroke over a 10-year follow-up without reducing the risk of myocardial infarction. Subgroup analyses showed that aspirin significantly reduced the risk of major cardiovascular events, ischemic stroke, and myocardial infarction among women ≥ 65 years old [157]. However, more recent studies published in 2018 have shown that in the modern era on a background of contemporary therapy, aspirin should not be used in the routine primary prevention of ASCVD due to lack of net benefit [158–160]. In fact, for healthy older adults aged ≥ 70 years, there was even an increased risk of mortality attributed to aspirin noted, without heterogeneity by sex [161]. The Aspirin for Reducing Events in the Elderly (ASPREE) trial failed to demonstrate a cardiovascular benefit with aspirin for primary prevention; however, aspirin did confer an increased risk for major hemorrhage, which was not significantly different between men and women [158]. Furthermore, an updated 2019 meta-analysis, which included these new trials, found that the number needed to treat to cause a major bleeding was actually lower than the number needed to treat to prevent an ASCVD event (210 vs. 265), suggesting more harm than benefit [162].

Clinicians should qualitatively evaluate for bleeding risk and withhold aspirin in primary prevention patients with increased risk such as prior gastrointestinal bleeding or ulceration, known bleeding disorder, severe liver disease, thrombocytopenia, concurrent anticoagulation or nonsteroidal antiinflammatory drug (NSAID) use, or uncontrolled hypertension. The 2019 ACC/AHA Prevention Guidelines [13] made the following recommendations: (1) low-dose aspirin might be considered for primary prevention of ASCVD in select higher ASCVD adults aged 40–70 years who are not at increased bleeding risk; (2) low-dose aspirin should not be administered on a routine basis for primary prevention of ASCVD among adults > 70 years; and (3) low-dose aspirin should not be administered for primary prevention among adults at any age who are at increased bleeding risk. There were no sex-specific recommendations for aspirin. There may still be select primary prevention patients where aspirin may be reasonable. One might still consider low-dose aspirin (75–100 mg/day) among current smokers, those with a strong family history of premature ASCVD, those with very elevated cholesterol suboptimally treated with statins, those with subclinical atherosclerosis such as a CAC score > 100, and select patients with diabetes at high ASCVD risk, if these individuals are at low risk for bleeding.

Novel Diabetes Medications for ASCVD Outcome Reduction

As stated above, a healthy lifestyle including a heart-healthy diet, regular physical activity, and weight management remain the mainstay of diabetes prevention and management. For those adults with T2DM who require additional glucose lowering, it is reasonable to initiate metformin as first-line therapy along with lifestyle therapies at the time of diagnosis. For adults with T2DM and additional ASCVD risk factors who require glucose-lowering therapy despite initial lifestyle modifications and metformin, there should be consideration for initiation of a sodium-glucose cotransporter 2 (SGLT-2) inhibitor or a glucagon-like peptide-1 (GLP-1) receptor agonist to improve glycemic control and reduce CVD risk.

Notably SGLT-2 inhibitors have been shown to significantly reduce heart failure events and reduce risk of progression of chronic kidney disease, and these benefits appear independent of A1c lowering, as demonstrated in large randomized clinical trials (EMPA-REG, CANVAS, DECLARE, CREDENCE) [163–166]. GLP-1 receptor agonists have also been shown to reduce major cardiovascular events in both the secondary [167–169] and primary [170] prevention populations, as demonstrated in a meta-analysis of GLP-1 receptor agonist trials [171]. So while discussion of pharmacologic management of diabetes is beyond the scope of this chapter, generally these therapies should be thought of as "cardiovascular preventive medications" rather than simply as diabetes drugs. Of note, since these clinical trials of SGLT-2 inhibitors and GLP-1 receptor agonists largely enrolled patients with established CVD or at high risk for CVD, trial enrollment was predominantly male (> 60%); however, no heterogeneity of benefit by sex was noted. In light of this new trial data, both the ACC and the American Diabetes Association have put forward consensus statements regarding the use of these novel therapies for CVD prevention in T2DM, generally as second-line therapy following metformin and lifestyle modifications [172, 173]. No sex-specific recommendations were made.

Enrollment of Women in Cardiovascular Trials

Randomized clinical trials generate the largest evidence base for shaping guideline recommendations. However, despite recommendations by the National Institutes of Health and the Food and Drug Administration (FDA), female patients continued to be underrepresented in cardiovascular clinical trials, relative to their disease burden in the population [174, 175]. This leads to limited data regarding the efficacy and safety of cardiovascular therapeutics in women.

Conclusion

Despite substantial progress over the past two decades, more work remains to be done to further improve cardiovascular health in women. For many decades, CVD research has focused primarily on men, thus leading to an under-appreciation of sex differences from an etiologic, diagnostic, and therapeutic perspective. The first step needs to be toward increasing the participation of women in clinical trials so that gender-specific data can be studied. The scientific community needs to advocate for more research on gender and sex differences in CVD. More focus on the special populations, including veterans and individuals who are lesbian, gay, bisexual, transsexual, or questioning (LGBTQ), is also an important area of future research. Healthcare professionals should be cognizant of sex-differential effects of the conventional risk factors for CVD, and of emerging risk factors in men (erectile dysfunction) and women (hypertensive disorders of pregnancy, gestational diabetes) and hormonal risk (e.g., PCOS, early menopause) that can be identified during reproductive life and that may improve current risk assessment strategies for primary prevention of CVD.

A patient-centered approach should focus on shared decision-making tools and incorporate the current ABCDEF model of prevention [13]: **A**ssessment of Risk, **A**spirin use (selectively), **B**lood pressure management, **C**holesterol assessment and statin therapy, **C**igarette smoking cessation, **D**iet, **D**iabetes management and prevention, **E**xercise prescription, and assessing **F**amily history. This will help provide comprehensive preventive care to both female and male patients.

Key Points

1. Women are at increased CVD risk, most notably at older ages; however, young women < 50 years are more likely to incur fatality after myocardial infarction compared to younger men.

2. CVD risk can be different by sex: disparity exists in the risk conferred by traditional CVD risk factors and there are unique risks posted by pregnancy and hormones for women.

3. Despite the increased risk of ASCVD with diabetes, diabetes control remains suboptimal with only 30% of women and 20% of men achieving control of their diabetes.

4. Women are less likely to be offered any statins (67% vs. 78%, $P < 0.001$) or guideline-directed statin intensity than men (37% vs. 45%, $P < 0.001$), and women are more likely to decline or discontinue therapy.

5. Women with a history of preeclampsia have a 71% increased risk of CV mortality, a 2.5-fold increase in risk of CAD, and a fourfold increase in heart failure when compared to women with uncomplicated pregnancies.

6. Women are more than seven times as likely to develop diabetes after gestational diabetes and approximately 50% of mothers with gestational diabetes will develop diabetes within 10 years, making gestational diabetes one of the strongest predictors of T2DM.

7. Early menopause and a history of adverse pregnancy outcome like preeclampsia are "risk enhancers" that would favor more intensive preventive attention in women such as consideration for statin therapy.

8. Women continue to be underrepresented in cardiovascular clinical trials, which limits the evidence base regarding safety and efficacy of these therapies in women.

Back to Clinical Case

The clinical case describes a 52-year-old black woman with a positive family history of myocardial infarction, history of preeclampsia and hypertension, who seeks counseling prior to starting HRT for vasomotor symptoms. Her estimated 10-year ASCVD risk is calculated to be 5.4%, which would be considered borderline risk. However, she has several notable "risk-enhancing" factors including her family history of premature CAD and her personal history of preeclampsia. With preeclampsia she has a 2.5-fold increased risk of developing CAD, a 2-fold increased risk of experiencing a major cardiovascular event, and a 4-fold increased risk of diabetes compared to age-matched controls. Undoubtedly these risk factors place her in a higher risk category. Promotion of a healthy lifestyle is paramount to reduce her lifetime risk of ASCVD. A CAC score could also be considered to risk-stratify her further, and guide risk-based decisions.

Funding

Dr. Michos is supported by the Amato Fund in Women's Cardiovascular Health at Johns Hopkins University. Dr. Minhas is supported by National Heart, Lung, and Blood Institute training grant T32HL007024.

References

[1] Benjamin EJ, Muntner P, Alonso A, Bittencourt MS, Callaway CW, Carson AP, et al. Heart disease and stroke statistics—2019 update: a report from the American Heart Association. Circulation 2019;139(10):e56–e528.

[2] Mosca L, Mochari-Greenberger H, Dolor RJ, Kristin Newby L, Robb KJ. Twelve-year follow-up of American women's awareness of cardiovascular disease risk and barriers to heart health. Circ Cardiovasc Qual Outcomes 2010;3(2):120–7.

[3] Mosca L, Benjamin EJ, Berra K, Bezanson JL, Dolor RJ, Lloyd-Jones DM, et al. Effectiveness-based guidelines for the prevention of cardiovascular disease in women—2011 update: a guideline from the American Heart Association. Circulation 2011;123(11):1243–62.

[4] Wilmot KA, O'Flaherty M, Capewell S, Ford ES, Vaccarino V. Coronary heart disease mortality declines in the United States from 1979 through 2011: evidence for stagnation in young adults, especially women. Circulation 2015;132(11):997–1002.

[5] Arora S, Stouffer GA, Kucharska-Newton AM, Qamar A, Vaduganathan M, Pandey A, et al. Twenty year trends and sex differences in young adults hospitalized with acute myocardial infarction. Circulation 2019;139(8):1047–56.

[6] Curtin SC. Trends in cancer and heart disease death rates among adults aged 45–64: United States, 1999–2017. Natl Vital Stat Rep 2019;68(6):1–8.

[7] Okunrintemi V, Valero-Elizondo J, Patrick B, Salami J, Tibuakuu M, Ahmad S, et al. Gender differences in patient-reported outcomes among adults with atherosclerotic cardiovascular disease. J Am Heart Assoc 2018;7(24):e010498.

[8] Mosca L, Linfante AH, Benjamin EJ, Berra K, Hayes SN, Walsh BW, et al. National study of physician awareness and adherence to cardiovascular disease prevention guidelines. Circulation 2005;111(4):499–510.

[9] Abuful A, Gidron Y, Henkin Y. Physicians' attitudes toward preventive therapy for coronary artery disease: is there a gender bias? Clin Cardiol 2005;28(8):389–93.

[10] Gu Q, Burt VL, Paulose-Ram R, Dillon CF. Gender differences in hypertension treatment, drug utilization patterns, and blood pressure control among US adults with hypertension: data from the National Health and Nutrition Examination Survey 1999-2004. Am J Hypertens 2008;21(7):789–98.

[11] Garcia M, Mulvagh SL, Bairey Merz CN, Buring JE, Manson JE. Cardiovascular disease in women: clinical perspectives. Circ Res 2016;118(8):1273–93.

[12] McKibben RA, Al Rifai M, Mathews LM, Michos ED. Primary prevention of atherosclerotic cardiovascular disease in women. Curr Cardiovasc Risk Rep 2016;10:1. https://doi.org/10.1007/s12170-015-0480-3.

[13] Arnett DK, Blumenthal RS, Albert MA, Buroker AB, Goldberger ZD, Hahn EJ, et al. 2019 ACC/AHA guideline on the primary prevention of cardiovascular disease: executive summary: a report of the American College of Cardiology/American Heart Association Task Force on Clinical Practice Guidelines. J Am Coll Cardiol 2019.

[14] Yusuf S, Hawken S, Ounpuu S, Dans T, Avezum A, Lanas F, et al. Effect of potentially modifiable risk factors associated with myocardial infarction in 52 countries (the INTERHEART study): case-control study. Lancet 2004;364(9438):937–52.

[15] Goff Jr DC, Lloyd-Jones DM, Bennett G, Coady S, D'Agostino RB, Gibbons R, et al. 2013 ACC/AHA guideline on the assessment of cardiovascular risk: a report of the American College of Cardiology/American Heart Association Task Force on Practice Guidelines. Circulation 2014;129(25 Suppl. 2):S49–73.

[16] Michos ED, Nasir K, Braunstein JB, Rumberger JA, Budoff MJ, Post WS, et al. Framingham risk equation underestimates subclinical atherosclerosis risk in asymptomatic women. Atherosclerosis 2006;184(1):201–6.

[17] Michos ED, Vasamreddy CR, Becker DM, Yanek LR, Moy TF, Fishman EK, et al. Women with a low Framingham risk score and a family history of premature coronary heart disease have a high prevalence of subclinical coronary atherosclerosis. Am Heart J 2005;150(6):1276–81.

[18] Michos ED, Blumenthal RS. How accurate are 3 risk prediction models in US women? Circulation 2012;125(14):1723–6.

[19] Huxley R, Barzi F, Woodward M. Excess risk of fatal coronary heart disease associated with diabetes in men and women: meta-analysis of 37 prospective cohort studies. BMJ 2006;332(7533):73–8.

[20] Natarajan S, Liao Y, Cao G, Lipsitz SR, McGee DL. Sex differences in risk for coronary heart disease mortality associated with diabetes and established coronary heart disease. Arch Intern Med 2003;163(14):1735–40.

[21] Jonsdottir LS, Sigfússon N, Gudnason V, Sigvaldason H, Thorgeirsson G. Do lipids, blood pressure, diabetes, and smoking confer equal risk of myocardial infarction in women as in men? The Reykjavik Study. J Cardiovasc Risk 2002;9(2):67–76.

[22] Juutilainen A, Kortelainen S, Lehto S, Rönnemaa T, Pyörälä K, Laakso M. Gender difference in the impact of type 2 diabetes on coronary heart disease risk. Diabetes Care 2004;27(12):2898–904.

[23] Steinberg HO, Paradisi G, Cronin J, Crowde K, Hempfling A, Hook G, et al. Type II diabetes abrogates sex differences in endothelial function in premenopausal women. Circulation 2000;101(17):2040–6.

[24] Carr ME. Diabetes mellitus: a hypercoagulable state. J Diabetes Complicat 2001;15(1):44–54.

[25] Pradhan AD. Sex differences in the metabolic syndrome: implications for cardiovascular health in women. Clin Chem 2014;60(1):44–52.

[26] Haffner SM, Miettinen H, Stern MP. Relatively more atherogenic coronary heart disease risk factors in prediabetic women than in prediabetic men. Diabetologia 1997;40(6):711–7.

[27] Donahue RP, Rejman K, Rafalson LB, Dmochowski J, Stranges S, Trevisan M. Sex differences in endothelial function markers before conversion to pre-diabetes: does the clock start ticking earlier among women? The Western New York Study. Diabetes Care 2007;30(2):354–9.

[28] Peters SAE, Muntner P, Woodward M. Sex differences in the prevalence of, and trends in, cardiovascular risk factors, treatment, and control in the United States, 2001 to 2016. Circulation 2019;139(8):1025–35.

[29] Whelton PK, Carey RM, Aronow WS, Casey DE, Collins KJ, Himmelfarb CD, et al. 2017 ACC/AHA/AAPA/ABC/ACPM/AGS/APhA/ASH/ASPC/NMA/PCNA guideline for the prevention, detection, evaluation, and management of high blood pressure in adults: executive summary: a report of the American College of Cardiology/American Heart Association Task Force on Clinical Practice Guidelines. Circulation 2018;138(17):e426–83.

[30] Reckelhoff JF. Gender differences in the regulation of blood pressure. Hypertension 2001;37(5):1199–208.

[31] van den Hoogen PC, van Popele NM, Feskens EJ, van der Kuip DA, Grobbee DE, Hofman A, et al. Blood pressure and risk of myocardial infarction in elderly men and women: the Rotterdam study. J Hypertens 1999;17(10):1373–8.

[32] Psaty BM, Furberg CD, Kuller LH, Cushman M, Savage PJ, Levine D, et al. Association between blood pressure level and the risk of myocardial infarction, stroke, and total mortality: the cardiovascular health study. Arch Intern Med 2001;161(9):1183–92.

[33] Miura K, Nakagawa H, Ohashi Y, Harada A, Taguri M, Kushiro T, et al. Four blood pressure indexes and the risk of stroke and myocardial infarction in Japanese men and women: a meta-analysis of 16 cohort studies. Circulation 2009;119(14):1892–8.

[34] Lewington S, Clarke R, Qizilbash N, Peto R, Collins R, Prospective Studies Collaboration. Age-specific relevance of usual blood pressure to vascular mortality: a meta-analysis of individual data for one million adults in 61 prospective studies. Lancet 2002;360(9349):1903–13.

[35] Flint AC, Conell C, Ren X, Banki NM, Chan SL, Rao VA, et al. Effect of systolic and diastolic blood pressure on cardiovascular outcomes. N Engl J Med 2019;381(3):243–51.

[36] Ference BA, Ginsberg HN, Graham I, Ray KK, Packard CJ, Bruckert E, et al. Low-density lipoproteins cause atherosclerotic cardiovascular disease. 1. Evidence from genetic, epidemiologic, and clinical studies. A consensus statement from the European Atherosclerosis Society Consensus Panel. Eur Heart J 2017;38(32):2459–72.

[37] Ference BA, Kastelein JJP, Ray KK, Ginsberg HN, John Chapman M, Packard CJ, et al. Association of triglyceride-lowering LPL variants and LDL-C-lowering LDLR variants with risk of coronary heart disease. JAMA 2019;321(4):364–73.

[38] Holmes MV, Asselbergs FW, Palmer TM, Drenos F, Lanktree MB, Nelson CP, et al. Mendelian randomization of blood lipids for coronary heart disease. Eur Heart J 2015;36(9):539–50.

[39] Hamer M, O'Donovan G, Stamatakis E. High-density lipoprotein cholesterol and mortality: too much of a good thing? Arterioscler Thromb Vasc Biol 2018;38(3):669–72.

[40] Allard-Ratick MP, Kindya BR, Khambhati J, Engels MC, Sandesara PB, Rosenson RS, et al. HDL: fact, fiction, or function? HDL cholesterol and cardiovascular risk. Eur J Prev Cardiol 2019. 2047487319848214.

[41] Grundy SM, Stone NJ, Bailey AL, Beam C, Birtcher KK, Blumenthal RS, et al. 2018 AHA/ACC/AACVPR/AAPA/ABC/ACPM/ADA/AGS/APhA/ASPC/NLA/PCNA guideline on the management of blood cholesterol: executive summary: a report of the American College of Cardiology/American Heart Association Task Force on Clinical Practice Guidelines. Circulation 2019;139(25):e1046–81.

[42] Jensen J, Nilas L, Christiansen C. Influence of menopause on serum lipids and lipoproteins. Maturitas 1990;12(4):321–31.

[43] Prospective Studies Collaboration, Lewington S, Whitlock G, Clarke R, Sherliker P, Emberson J, et al. Blood cholesterol and vascular mortality by age, sex, and blood pressure: a meta-analysis of individual data from 61 prospective studies with 55,000 vascular deaths. Lancet 2007;370(9602):1829–39.

[44] Zhao M, Vaartjes I, Graham I, Grobbee D, Spiering W, Klipstein-Grobusch K, et al. Sex differences in risk factor management of coronary heart disease across three regions. Heart 2017;103(20):1587–94.

[45] Nanna MG, Wang TY, Xiang Q, Goldberg AC, Robinson JG, Roger VL, et al. Sex differences in the use of statins in community practice. Circ Cardiovasc Qual Outcomes 2019;12(8):e005562.

[46] Buja A, Boemo DG, Furlan P, Bertoncello C, Casale P, Baldovin T, et al. Tackling inequalities: are secondary prevention therapies for reducing post-infarction mortality used without disparities? Eur J Prev Cardiol 2014;21(2):222–30.

[47] Smolina K, Ball L, Humphries KH, Khan N, Morgan SG. Sex disparities in post-acute myocardial infarction pharmacologic treatment initiation and adherence: problem for young women. Circ Cardiovasc Qual Outcomes 2015;8(6):586–92.

[48] National Center for Chronic Disease Prevention and Health Promotion (US) Office on Smoking and Health. The health consequences of smoking—50 years of progress: a report of the surgeon general. Atlanta, GA: Centers for Disease Control and Prevention (US); 2014. Available from: https://www.ncbi.nlm.nih.gov/books/NBK179276/.

[49] Banks E, Joshy G, Korda RJ, Stavreski B, Soga K, Egger S, et al. Tobacco smoking and risk of 36 cardiovascular disease subtypes: fatal and non-fatal outcomes in a large prospective Australian study. BMC Med 2019;17(1):128.

[50] Mons U, Müezzinler A, Gellert C, Schöttker B, Abnet CC, Bobak M, et al. Impact of smoking and smoking cessation on cardiovascular events and mortality among older adults: meta-analysis of individual participant data from prospective cohort studies of the CHANCES consortium. BMJ 2015;350:h1551.

[51] Qasim H, Karim ZA, Rivera JO, Khasawneh FT, Alshbool FZ. Impact of electronic cigarettes on the cardiovascular system. J Am Heart Assoc 2017;6(9):e006353.

[52] Bhatnagar A, Whitsel LP, Ribisl KM, Bullen C, Chaloupka F, Piano MR, et al. Electronic cigarettes: a policy statement from the American Heart Association. Circulation 2014;130(16):1418–36.

[53] Njolstad I, Arnesen E, Lund-Larsen PG. Smoking, serum lipids, blood pressure, and sex differences in myocardial infarction. A 12-year follow-up of the Finnmark Study. Circulation 1996;93(3):450–6.

[54] Grundtvig M, Hagen TP, German M, Reikvam A. Sex-based differences in premature first myocardial infarction caused by smoking: twice as many years lost by women as by men. Eur J Cardiovasc Prev Rehabil 2009;16(2):174–9.

[55] Lidegaard O. Smoking and use of oral contraceptives: impact on thrombotic diseases. Am J Obstet Gynecol 1999;180(6 Pt 2):S357–63.

[56] Pomp ER, Rosendaal FR, Doggen CJM. Smoking increases the risk of venous thrombosis and acts synergistically with oral contraceptive use. Am J Hematol 2008;83(2):97–102.

[57] Hubert HB, Feinleib M, McNamara PM, Castelli WP. Obesity as an independent risk factor for cardiovascular disease: a 26-year follow-up of participants in the Framingham Heart Study. Circulation 1983;67(5):968–77.

[58] Canoy D, Cairns BJ, Balkwill A, Wright FL, Green J, Reeves G, et al. Coronary heart disease incidence in women by waist circumference within categories of body mass index. Eur J Prev Cardiol 2013;20(5):759–62.

[59] Fliotsos M, Zhao D, Rao VN, Ndumele CE, Guallar E, Burke GL, et al. Body mass index from early-, mid-, and older-adulthood and risk of heart failure and atherosclerotic cardiovascular disease: MESA. J Am Heart Assoc 2018;7(22):e009599.

[60] Ndumele CE, Matsushita K, Lazo M, Bello N, Blumenthal RS, Gerstenblith G, et al. Obesity and subtypes of incident cardiovascular disease. J Am Heart Assoc 2016;5(8):e003921.

[61] Czernichow S, Kengne A-P, Stamatakis E, Hamer M, Batty GD. Body mass index, waist circumference and waist-hip ratio: which is the better discriminator of cardiovascular disease mortality risk?: evidence from an individual-participant meta-analysis of 82 864 participants from nine cohort studies. Obes Rev 2011;12(9):680–7.

[62] Flint AJ, Rexrode KM, Hu FB, Glynn RJ, Caspard H, Manson JE, et al. Body mass index, waist circumference, and risk of coronary heart disease: a prospective study among men and women. Obes Res Clin Pract 2010;4(3):e171–81.

[63] Wilson PW, Kannel WB, Silbershatz H, D'Agostino RB. Clustering of metabolic factors and coronary heart disease. Arch Intern Med 1999;159(10):1104–9.

[64] Biswas A, Oh PI, Faulkner GE, Bajaj RR, Silver MA, Mitchell MS, et al. Sedentary time and its association with risk for disease incidence, mortality, and hospitalization in adults: a systematic review and meta-analysis. Ann Intern Med 2015;162(2):123–32.

[65] Lear SA, Hu W, Rangarajan S, Gasevic D, Leong D, Iqbal R, et al. The effect of physical activity on mortality and cardiovascular disease in 130 000 people from 17 high-income, middle-income, and low-income countries: the PURE study. Lancet 2017;390(10113):2643–54.

[66] Sattelmair J, Pertman J, Ding EL, Kohl HW, Haskell W, Lee I-M. Dose response between physical activity and risk of coronary heart disease: a meta-analysis. Circulation 2011;124(7):789–95.

[67] Kyu HH, Bachman VF, Alexander LT, Mumford JE, Afshin A, Estep K, et al. Physical activity and risk of breast cancer, colon cancer, diabetes, ischemic heart disease, and ischemic stroke events: systematic review and dose-response meta-analysis for the Global Burden of Disease Study 2013. BMJ 2016;354:i3857.

[68] Florido R, Zhao D, Ndumele CE, Lutsey PL, McEvoy JW, Windham BG, et al. Physical activity, parental history of premature coronary heart disease, and incident atherosclerotic cardiovascular disease in the atherosclerosis risk in communities (ARIC) study. J Am Heart Assoc 2016;5(9):e003505.

[69] Chomistek AK, Henschel B, Eliassen AH, Mukamal KJ, Rimm EB. Frequency, type, and volume of leisure-time physical activity and risk of coronary heart disease in young women. Circulation 2016;134(4):290–9.

[70] Schiller JS, Lucas JW, Peregoy JA. Summary health statistics for U.S. adults: National Health Interview Survey, 2011. Vital Health Stat 2012;10(256):1–218.

[71] Okunrintemi V, Benson EA, Tibuakuu M, Zhao D, Ogunmoroti O, Valero-Elizondo J, et al. Trends and costs associated with suboptimal physical activity among US women with cardiovascular disease. JAMA Netw Open 2019;2(4):e191977.

[72] Michos ED, Blaha MJ. Encouraging young women to move more: linking physical activity in young adulthood to coronary risk in women. Circulation 2016;134(4):300–3.

[73] Piercy KL, Troiano RP, Ballard RM, Carlson SA, Fulton JE, Galuska DA, et al. The physical activity guidelines for Americans. JAMA 2018;320(19):2020–8.

[74] Lloyd-Jones DM, Nam BH, D'Agostino RB, Levy D, Murabito JM, Wang TJ, et al. Parental cardiovascular disease as a risk factor for cardiovascular disease in middle-aged adults: a prospective study of parents and offspring. JAMA 2004;291(18):2204–11.

[75] Weijmans M, van der Graaf Y, Reitsma JB, Visseren FL. Paternal or maternal history of cardiovascular disease and the risk of cardiovascular disease in offspring. A systematic review and meta-analysis. Int J Cardiol 2015;179:409–16.

[76] Murabito JM, Pencina MJ, Nam BH, D'Agostino RB, Wang TJ, Lloyd-Jones D, et al. Sibling cardiovascular disease as a risk factor for cardiovascular disease in middle-aged adults. JAMA 2005;294(24):3117–23.

[77] Michos ED, Nasir K, Rumberger JA, Vasamreddy C, Braunstein JB, Budoff MJ, et al. Relation of family history of premature coronary heart disease and metabolic risk factors to risk of coronary arterial calcium in asymptomatic subjects. Am J Cardiol 2005;95(5):655–7.

[78] Patel J, Al Rifai M, Blaha MJ, Budoff MJ, Post WS, Polak JF, et al. Coronary artery calcium improves risk assessment in adults with a family history of premature coronary heart disease: results from multiethnic study of atherosclerosis. Circ Cardiovasc Imaging 2015;8(6):e003186.

[79] Agarwala A, Michos ED, Samad Z, Ballantyne CM, Virani SS. The use of sex-specific factors in the assessment of women's cardiovascular risk. Circulation 2020;141(7):592–9.

[80] Ogunmoroti O, Osibogun O, Kolade OB, Ying W, Sharma G, Vaidya D, et al. Multiparity is associated with poorer cardiovascular health among women from the Multi-Ethnic Study of Atherosclerosis. Am J Obstet Gynecol 2019;221(6):631 e1–631 e16.

[81] Li W, Ruan W, Lu Z, Wang D. Parity and risk of maternal cardiovascular disease: a dose-response meta-analysis of cohort studies. Eur J Prev Cardiol 2019;26(6):592–602.

[82] Hauspurg A, Ying W, Hubel CA, Michos ED, Ouyang P. Adverse pregnancy outcomes and future maternal cardiovascular disease. Clin Cardiol 2018;41(2):239–46.

[83] Wu P, Haththotuwa R, Kwok CS, Babu A, Kotronias RA, Rushton C, et al. Preeclampsia and future cardiovascular health: a systematic review and meta-analysis. Circ Cardiovasc Qual Outcomes 2017;10(2):e003497.

[84] Wu P, Gulati M, Kwok CS, Wong CW, Narain A, O'Brien S, et al. Preterm delivery and future risk of maternal cardiovascular disease: a systematic review and meta-analysis. J Am Heart Assoc 2018;7(2):e007809.

[85] Kramer CK, Campbell S, Retnakaran R. Gestational diabetes and the risk of cardiovascular disease in women: a systematic review and meta-analysis. Diabetologia 2019;62(6):905–14.

[86] Regitz-Zagrosek V, Roos-Hesselink JW, Bauersachs J, Blomström-Lundqvist C, Cífková R, De Bonis M, et al. 2018 ESC guidelines for the management of cardiovascular diseases during pregnancy. Eur Heart J 2018;39(34):3165–241.

[87] ACOG Writing Group. Gestational Hypertension and Preeclampsia: ACOG Practice Bulletin Summary, No. 222. Obstet Gynecol 2020;135(6):1492–5. https://doi.org/10.1097/aog.0000000000003892.

[88] Committee on Practice Bulletins—Obstetrics. ACOG Practice Bulletin No. 190: gestational diabetes mellitus. Obstet Gynecol 2018;131(2):e49–64.

[89] Martin JA, Hamilton BE, Osterman MJK, Driscoll AK, Drake P. Births: final data for 2017. Natl Vital Stat Rep 2018;67(8):1–50.

[90] Lane-Cordova AD, Khan SS, Grobman WA, Greenland P, Shah SJ. Long-term cardiovascular risks associated with adverse pregnancy outcomes: JACC review topic of the week. J Am Coll Cardiol 2019;73(16):2106–16.

[91] Cain MA, Salemi JL, Tanner JP, Kirby RS, Salihu HM, Louis JM. Pregnancy as a window to future health: maternal placental syndromes and short-term cardiovascular outcomes. Am J Obstet Gynecol 2016;215(4):484 e1–484 e14.

[92] Roberts JM, Catov JM. Pregnancy is a screening test for later life cardiovascular disease: now what? Research recommendations. Womens Health Issues 2012;22(2):e123–8.

[93] Davis MB, Walsh MN. Cardio-obstetrics. Circ Cardiovasc Qual Outcomes 2019;12(2):e005417.

[94] Rana S, Lemoine E, Granger JP, Karumanchi SA. Preeclampsia. Circ Res 2019;124(7):1094–112.

[95] Stuart JJ, Tanz LJ, Missmer SA, Rimm EB, Spiegelman D, James-Todd TM, et al. Hypertensive disorders of pregnancy and maternal cardiovascular disease risk factor development: an observational cohort study. Ann Intern Med 2018;169(4):224–32.

[96] Riise HK, Sulo G, Tell GS, Igland J, Nygård O, Vollset SE, et al. Incident coronary heart disease after preeclampsia: role of reduced fetal growth, preterm delivery, and parity. J Am Heart Assoc 2017;6(3):e004158.

[97] Bramham K, Parnell B, Nelson-Piercy C, Seed PT, Poston L, Chappell LC. Chronic hypertension and pregnancy outcomes: systematic review and meta-analysis. BMJ 2014;348:g2301.

[98] Fraser A, Nelson SM, Macdonald-Wallis C, Cherry L, Butler E, Sattar N, et al. Associations of pregnancy complications with calculated cardiovascular disease risk and cardiovascular risk factors in middle age: the Avon Longitudinal Study of parents and children. Circulation 2012;125(11):1367–80.

[99] Catov JM, Dodge R, Barinas-Mitchell E, Sutton-Tyrrell K, Yamal JM, Piller LB, et al. Prior preterm birth and maternal subclinical cardiovascular disease 4 to 12 years after pregnancy. J Women's Health (Larchmt) 2013;22(10):835–43.

[100] Grandi SM, Filion KB, Yoon S, Ayele HT, Doyle CM, Hutcheon JA, et al. Cardiovascular disease-related morbidity and mortality in women with a history of pregnancy complications. Circulation 2019;139(8):1069–79.

[101] Ewing AC, Ellington SR, Shapiro-Mendoza CK, Barfield WD, Kourtis AP. Full-term small-for-gestational-age newborns in the U.S.: characteristics, trends, and morbidity. Matern Child Health J 2017;21(4):786–96.

[102] Pariente G, Sheiner E, Kessous R, Michael S, Shoham-Vardi I. Association between delivery of a small-for-gestational-age neonate and long-term maternal cardiovascular morbidity. Int J Gynaecol Obstet 2013;123(1):68–71.

[103] Ngo AD, Roberts CL, Chen JS, Figtree G. Delivery of a small-for-gestational-age infant and risk of maternal cardiovascular disease—a population-based record linkage study. Heart Lung Circ 2015;24(7):696–704.

[104] Damm P, Houshmand-Oeregaard A, Kelstrup L, Lauenborg J, Mathiesen ER, Clausen TD. Gestational diabetes mellitus and long-term consequences for mother and offspring: a view from Denmark. Diabetologia 2016;59(7):1396–9.

[105] Lowe LP, Metzger BE, Dyer AR, Lowe J, McCance DR, Lappin TR, et al. Hyperglycemia and Adverse Pregnancy Outcome (HAPO) Study: associations of maternal A1C and glucose with pregnancy outcomes. Diabetes Care 2012;35(3):574–80.

[106] Bellamy L, Casas JP, Hingorani AD, Williams D. Type 2 diabetes mellitus after gestational diabetes: a systematic review and meta-analysis. Lancet 2009;373(9677):1773–9.

[107] Ratner RE, Christophi CA, Metzger BE, Dabelea D, Bennett PH, Pi-Sunyer X, et al. Prevention of diabetes in women with a history of gestational diabetes: effects of metformin and lifestyle interventions. J Clin Endocrinol Metab 2008;93(12):4774–9.

[108] Fadl H, Magnuson A, Östlund I, Montgomery S, Hanson U, Schwarcz E. Gestational diabetes mellitus and later cardiovascular disease: a Swedish population based case-control study. BJOG 2014;121(12):1530–6.

[109] Hoffman LK, Ehrmann DA. Cardiometabolic features of polycystic ovary syndrome. Nat Clin Pract Endocrinol Metab 2008;4(4):215–22.

[110] Osibogun O, Ogunmoroti O, Michos ED. Polycystic ovary syndrome and cardiometabolic risk: opportunities for cardiovascular disease prevention. Trends Cardiovasc Med 2020;30(7):399–404.

[111] Zhao L, Zhu Z, Lou H, Zhu G, Huang W, Zhang S, et al. Polycystic ovary syndrome (PCOS) and the risk of coronary heart disease (CHD): a meta-analysis. Oncotarget 2016;7(23):33715–21.

[112] Zhou Y, Wang X, Jiang Y, Ma H, Chen L, Lai C, et al. Association between polycystic ovary syndrome and the risk of stroke and all-cause mortality: insights from a meta-analysis. Gynecol Endocrinol 2017;33(12):904–10.

[113] Glintborg D, Rubin KH, Nybo M, Abrahamsen B, Andersen M. Cardiovascular disease in a nationwide population of Danish women with polycystic ovary syndrome. Cardiovasc Diabetol 2018;17(1):37.

[114] de Groot PC, Dekkers OM, Romijn JA, Dieben SW, Helmerhorst FM. PCOS, coronary heart disease, stroke and the influence of obesity: a systematic review and meta-analysis. Hum Reprod Update 2011;17(4):495–500.

[115] Hart R, Doherty DA. The potential implications of a PCOS diagnosis on a woman's long-term health using data linkage. J Clin Endocrinol Metab 2015;100(3):911–9.

[116] Carmina E, Lobo RA. Is there really increased cardiovascular morbidity in women with polycystic ovary syndrome? J Women's Health (Larchmt) 2018;27(11):1385–8.

[117] Orio Jr F, Palomba S, Cascella T, De Simone B, Di Biase S, Russo T, et al. Early impairment of endothelial structure and function in young normal-weight women with polycystic ovary syndrome. J Clin Endocrinol Metab 2004;89(9):4588–93.

[118] Talbott EO, Zborowski JV, Rager JR, Boudreaux MY, Edmundowicz DA, Guzick DS. Evidence for an association between metabolic cardiovascular syndrome and coronary and aortic calcification among women with polycystic ovary syndrome. J Clin Endocrinol Metab 2004;89(11):5454–61.

[119] Dokras A. Cardiovascular disease risk in women with PCOS. Steroids 2013;78(8):773–6.

[120] Peters SA, Woodward M. Women's reproductive factors and incident cardiovascular disease in the UK Biobank. Heart 2018;104(13):1069–75.

[121] Lee JJ, Cook-Wiens G, Johnson BD, Braunstein GD, Berga SL, Stanczyk FZ, et al. Age at menarche and risk of cardiovascular disease outcomes: findings from the National Heart Lung and Blood Institute-sponsored women's ischemia syndrome evaluation. J Am Heart Assoc 2019;8(12):e012406.

[122] Tandon VR, Mahajan A, Sharma S, Sharma A. Prevalence of cardiovascular risk factors in postmenopausal women: a rural study. J Midlife Health 2010;1(1):26–9.

[123] Rosano GM, Vitale C, Marazzi G, Volterrani M. Menopause and cardiovascular disease: the evidence. Climacteric 2007;10(Suppl. 1):19–24.

[124] Atsma F, Bartelink ML, Grobbee DE, van der Schouw YT. Postmenopausal status and early menopause as independent risk factors for cardiovascular disease: a meta-analysis. Menopause 2006;13(2):265–79.

[125] Newson L. Menopause and cardiovascular disease. Post Reprod Health 2018;24(1):44–9.

[126] Zhao D, Guallar E, Ouyang P, Subramanya V, Vaidya D, Ndumele CE, et al. Endogenous sex hormones and incident cardiovascular disease in post-menopausal women. J Am Coll Cardiol 2018;71(22):2555–66.

[127] Ouyang P, Michos ED, Karas RH. Hormone replacement therapy and the cardiovascular system lessons learned and unanswered questions. J Am Coll Cardiol 2006;47(9):1741–53.

[128] Rossouw JE, Anderson GL, Prentice RL, LaCroix AZ, Kooperberg C, Stefanick ML, et al. Risks and benefits of estrogen plus progestin in healthy postmenopausal women: principal results from the Women's Health Initiative randomized controlled trial. JAMA 2002;288(3):321–33.

[129] Hulley S, Grady D, Bush T, Furberg C, Herrington D, Riggs B, et al. Randomized trial of estrogen plus progestin for secondary prevention of coronary heart disease in postmenopausal women. Heart and Estrogen/progestin Replacement Study (HERS) Research Group. JAMA 1998;280(7):605–13.

[130] Rossouw JE, Prentice RL, Manson JE, Wu L, Barad D, Barnabei VM, et al. Postmenopausal hormone therapy and risk of cardiovascular disease by age and years since menopause. JAMA 2007;297(13):1465–77.

[131] Schierbeck LL, Rejnmark L, Tofteng CL, Stilgren L, Eiken P, Mosekilde L, et al. Effect of hormone replacement therapy on cardiovascular events in recently postmenopausal women: randomised trial. BMJ 2012;345:e6409.

[132] Lumsden MA, Davies M, Sarri G, Guideline Development Group for Menopause: Diagnosis and Management (NICE Clinical Guideline No. 23). Diagnosis and management of menopause: the National Institute of Health and Care Excellence (NICE) guideline. JAMA Intern Med 2016;176(8):1205–6.

[133] The NAMS 2017 Hormone Therapy Position Statement Advisory Panel. The 2017 hormone therapy position statement of The North American Menopause Society. Menopause 2017;24(7):728–53.

[134] Manson JE, Allison MA, Rossouw JE, Carr JJ, Langer RD, Hsia J, et al. Estrogen therapy and coronary-artery calcification. N Engl J Med 2007;356(25):2591–602.

[135] Selvin E, Burnett AL, Platz EA. Prevalence and risk factors for erectile dysfunction in the US. Am J Med 2007;120(2):151–7.

[136] Dong J-Y, Zhang Y-H, Qin L-Q. Erectile dysfunction and risk of cardiovascular disease: meta-analysis of prospective cohort studies. J Am Coll Cardiol 2011;58(13):1378–85.

[137] Uddin SMI, Mirbolouk M, Dardari Z, Feldman DI, Cainzos-Achirica M, DeFilippis AP, et al. Erectile dysfunction as an independent predictor of future cardiovascular events. Circulation 2018;138(5):540–2.

[138] Montorsi P, Ravagnani PM, Galli S, Rotatori F, Briganti A, Salonia A, et al. The artery size hypothesis: a macrovascular link between erectile dysfunction and coronary artery disease. Am J Cardiol 2005;96(12B):19M–23M.

[139] Feldman DI, Cainzos-Achirica M, Billups KL, DeFilippis AP, Chitaley K, Greenland P, et al. Subclinical vascular disease and subsequent erectile dysfunction: the multiethnic study of atherosclerosis (MESA). Clin Cardiol 2016;39(5):291–8.

[140] Fairweather D, Rose NR. Women and autoimmune diseases. Emerg Infect Dis 2004;10(11):2005–11.

[141] Avina-Zubieta JA, Choi HK, Sadatsafavi M, Etminan M, Esdaile JM, Lacaille D. Risk of cardiovascular mortality in patients with rheumatoid arthritis: a meta-analysis of observational studies. Arthritis Rheum 2008;59(12):1690–7.

[142] Manzi S, Meilahn EN, Rairie JE, Conte CG, Medsger TA, Jansen-McWilliams L, et al. Age-specific incidence rates of myocardial infarction and angina in women with systemic lupus erythematosus: comparison with the Framingham Study. Am J Epidemiol 1997;145(5):408–15.

[143] Armstrong EJ, Harskamp CT, Armstrong AW. Psoriasis and major adverse cardiovascular events: a systematic review and meta-analysis of observational studies. J Am Heart Assoc 2013;2(2):e000062.

[144] Asanuma Y, Oeser A, Shintani AK, Turner E, Olsen N, Fazio S, et al. Premature coronary-artery atherosclerosis in systemic lupus erythematosus. N Engl J Med 2003;349(25):2407–15.

[145] Chung CP, Oeser A, Raggi P, Gebretsadik T, Shintani AK, Sokka T, et al. Increased coronary-artery atherosclerosis in rheumatoid arthritis: relationship to disease duration and cardiovascular risk factors. Arthritis Rheum 2005;52(10):3045–53.

[146] Shaharyar S, Warraich H, McEvoy JW, Oni E, Ali SS, Karim A, et al. Subclinical cardiovascular disease in plaque psoriasis: association or causal link? Atherosclerosis 2014;232(1):72–8.

[147] Triant VA, Lee H, Hadigan C, Grinspoon SK. Increased acute myocardial infarction rates and cardiovascular risk factors among patients with human immunodeficiency virus disease. J Clin Endocrinol Metab 2007;92(7):2506–12.

[148] Post WS, Budoff M, Kingsley L, Palella FJ, Witt MD, Li X, et al. Associations between HIV infection and subclinical coronary atherosclerosis. Ann Intern Med 2014;160(7):458–67.

[149] Chow D, Young R, Valcour N, Kronmal RA, Lum CJ, Parikh NI, et al. HIV and coronary artery calcium score: comparison of the Hawaii Aging with HIV Cardiovascular Study and Multi-Ethnic Study of Atherosclerosis (MESA) cohorts. HIV Clin Trials 2015;16(4):130–8.

[150] Fischer NM, Pallazola VA, Xun H, Cainzos-Achirica M, Michos ED. The evolution of the heart-healthy diet for vascular health: a walk through time. Vasc Med 2020. 1358863X19901287.

[151] Engberding N, Wenger NK. Management of hypertension in women. Hypertens Res 2012;35(3):251–60.

[152] Mora S, Glynn RJ, Hsia J, MacFadyen JG, Genest J, Ridker PM. Statins for the primary prevention of cardiovascular events in women with elevated high-sensitivity C-reactive protein or dyslipidemia: results from the Justification for the Use of Statins in Prevention: an Intervention Trial Evaluating Rosuvastatin (JUPITER) and meta-analysis of women from primary prevention trials. Circulation 2010;121(9):1069–77.

[153] Kostis WJ, Cheng JQ, Dobrzynski JM, Cabrera J, Kostis JB. Meta-analysis of statin effects in women versus men. J Am Coll Cardiol 2012;59(6):572–82.

[154] DeFilippis AP, Young R, McEvoy JW, Michos ED, Sandfort V, Kronmal RA, et al. Risk score overestimation: the impact of individual cardiovascular risk factors and preventive therapies on the performance of the American Heart Association-American College of Cardiology-Atherosclerotic Cardiovascular Disease risk score in a modern multi-ethnic cohort. Eur Heart J 2017;38(8):598–608.

[155] Bhatt DL, Steg PG, Miller M, Brinton EA, Jacobson TA, Ketchum SB, et al. Cardiovascular risk reduction with icosapent ethyl for hypertriglyceridemia. N Engl J Med 2019;380(1):11–22.

[156] Newman CB, Preiss D, Tobert JA, Jacobson TA, Page RL, Goldstein LB, et al. Statin safety and associated adverse events: a scientific statement from the American Heart Association. Arterioscler Thromb Vasc Biol 2019;39(2):e38–81.

[157] Ridker PM, Cook NR, Lee IM, Gordon D, Gaziano JM, Manson JE, et al. A randomized trial of low-dose aspirin in the primary prevention of cardiovascular disease in women. N Engl J Med 2005;352(13):1293–304.

[158] McNeil JJ, Wolfe R, Woods RL, Tonkin AM, Donnan GA, Nelson MR, et al. Effect of aspirin on cardiovascular events and bleeding in the healthy elderly. N Engl J Med 2018;379(16):1509–18.

[159] ASCEND Study Collaborative Group, Bowman L, Mafham M, Wallendszus K, Stevens W, Buck G, et al. Effects of aspirin for primary prevention in persons with diabetes mellitus. N Engl J Med 2018;379(16):1529–39.

[160] Gaziano JM, Brotons C, Coppolecchia R, Cricelli C, Darius H, Gorelick PB, et al. Use of aspirin to reduce risk of initial vascular events in patients at moderate risk of cardiovascular disease (ARRIVE): a randomised, double-blind, placebo-controlled trial. Lancet 2018;392(10152):1036–46.

[161] McNeil JJ, Nelson MR, Woods RL, Lockery JE, Wolfe R, Reid CM, et al. Effect of aspirin on all-cause mortality in the healthy elderly. N Engl J Med 2018;379(16):1519–28.

[162] Zheng SL, Roddick AJ. Association of aspirin use for primary prevention with cardiovascular events and bleeding events: a systematic review and meta-analysis. JAMA 2019;321(3):277–87.

[163] Zinman B, Wanner C, Lachin JM, Fitchett D, Bluhmki E, Hantel S, et al. Empagliflozin, cardiovascular outcomes, and mortality in type 2 diabetes. N Engl J Med 2015;373(22):2117–28.

[164] Neal B, Perkovic V, Mahaffey KW, de Zeeuw D, Fulcher G, Erondu N, et al. Canagliflozin and cardiovascular and renal events in type 2 diabetes. N Engl J Med 2017;377(7):644–57.

[165] Wiviott SD, Raz I, Bonaca MP, Mosenzon O, Kato ET, Cahn A, et al. Dapagliflozin and cardiovascular outcomes in type 2 diabetes. N Engl J Med 2019;380(4):347–57.

[166] Perkovic V, Jardine MJ, Neal B, Bompoint S, Heerspink HJL, Charytan DM, et al. Canagliflozin and renal outcomes in type 2 diabetes and nephropathy. N Engl J Med 2019;380(24):2295–306.

[167]   Marso SP, Daniels GH, Brown-Frandsen K, Kristensen P, Mann JF, Nauck MA, et al. Liraglutide and cardiovascular outcomes in type 2 diabetes. N Engl J Med 2016;375(4):311–22.

[168]   Marso SP, Bain SC, Consoli A, Eliaschewitz FG, Jódar E, Leiter LA, et al. Semaglutide and cardiovascular outcomes in patients with type 2 diabetes. N Engl J Med 2016;375(19):1834–44.

[169]   Hernandez AF, Green JB, Janmohamed S, D'Agostino RB, Granger CB, Jones NP, et al. Albiglutide and cardiovascular outcomes in patients with type 2 diabetes and cardiovascular disease (Harmony Outcomes): a double-blind, randomised placebo-controlled trial. Lancet 2018;392(10157):1519–29.

[170]   Gerstein HC, Colhoun HM, Dagenais GR, Diaz R, Lakshmanan M, Pais P, et al. Dulaglutide and cardiovascular outcomes in type 2 diabetes (REWIND): a double-blind, randomised placebo-controlled trial. Lancet 2019;394:121–30.

[171]   Zelniker TA, Wiviott SD, Raz I, Im K, Goodrich EL, Furtado RHM, et al. Comparison of the effects of glucagon-like peptide receptor agonists and sodium-glucose cotransporter 2 inhibitors for prevention of major adverse cardiovascular and renal outcomes in type 2 diabetes mellitus. Circulation 2019;139(17):2022–31.

[172]   Das SR, Everett BM, Birtcher KK, Brown JM, Cefalu WT, Januzzi JL, et al. 2018 ACC expert consensus decision pathway on novel therapies for cardiovascular risk reduction in patients with type 2 diabetes and atherosclerotic cardiovascular disease. A report of the American College of Cardiology task force on expert consensus decision pathways. J Am Coll Cardiol 2018;72:3200–23.

[173]   American Diabetes Association. 10. Cardiovascular disease and risk management: Standards of Medical Care in Diabetes-2019. Diabetes Care 2019;42(Suppl. 1):S103–23.

[174]   Khan SU, Khan MZ, Riaz H, Subramanian CR, Khan MU, Lone AN, et al. Abstract 13723: representation of women and older patients in trials of lipid lowering therapy: a systematic review. Circulation 2019;140(Suppl. 1):A13723.

[175]   Jin X, Chandramouli C, Allocco B, Gong E, Lam CSP, Yan LL. Women's participation in cardiovascular clinical trials from 2010 to 2017. Circulation 2020;141(7):540–8.

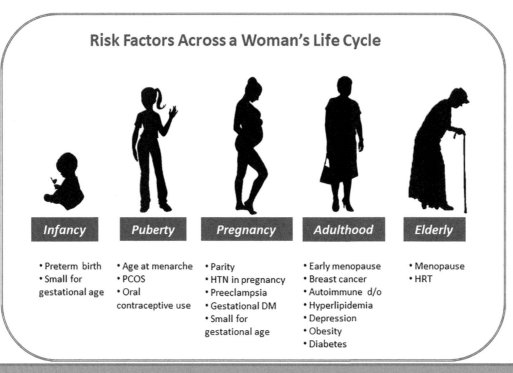

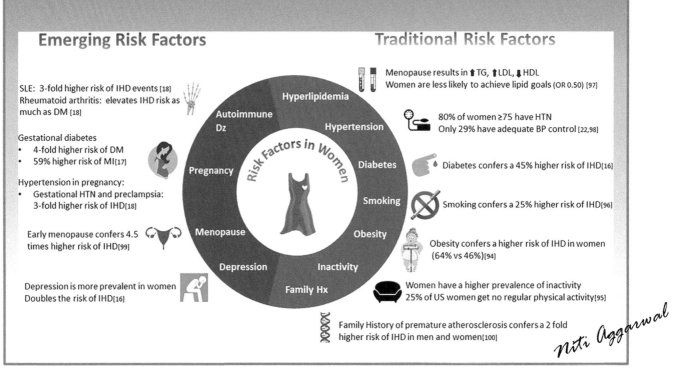

Traditional and Emerging Risk Factors Across a Woman's Life Cycle. Many traditional risk factors for ischemic heart disease impart a differential risk for women compared with men. The role of emerging nontraditional cardiac risks unique to or predominant in women are also being increasingly recognized across the entire lifespan of women. BP, blood pressure; DM, diabetes mellitus; d/o, disorder; HDL, high-density lipoprotein; HRT, hormone replacement therapy; HTN, hypertension; IHD, ischemic heart disease; LDL, low-density lipoprotein; MI, myocardial infarction; PCOS, polycystic ovarian syndrome; SLE, systemic lupus erythematosus; TG, triglycerides. *Reproduced with permission from [1]. Image courtesy of Niti R. Aggarwal.*

[1] Aggarwal NR, Patel HN, Mehta LS, Sanghani RM, Lundberg GP, Lewis SJ, et al. Sex differences in ischemic heart disease: advances, obstacles, and next steps. Circ Cardiovasc Qual Outcomes 2018;11(2):e004437. https://doi.org/10.1161/CIRCOUTCOMES.117.004437.

Chapter 4

Acute Coronary Syndrome

Laurie Bossory, Konstantinos Dean Boudoulas,
and Laxmi S. Mehta

Clinical Case

A 50-year-old Asian female with a history of Hashimoto's thyroiditis, depression, and Raynaud's phenomenon presented to the emergency room with stuttering chest pain. Days prior to admission she had chest pain at rest, and on the date of admission she had chest pain that awoke her from her sleep. She had dyspnea and upper back pain. Her blood pressure and heart rates were controlled. On examination, she was alert and oriented. Lungs were clear to auscultation. Abdomen was soft and nontender. Cardiac exam revealed regular rate and rhythm with normal S1 S2, no murmurs, no jugular vein distention (JVD), and no peripheral edema. The patient's electrocardiogram (ECG) on presentation demonstrated ST segment elevation in the anterior leads. Given her ongoing symptoms, she was emergently taken to the catheterization laboratory and found to have 99% blockage of the mid-left anterior descending (LAD) and the remaining coronary arteries and branches were normal. What other medications would you consider administering to this patient prior to proceeding with cardiac catheterization?

Abstract

Acute coronary syndrome (ACS) remains the leading cause of death in both men and women, with a rising mortality rate seen in young women. Sex-related differences exist in baseline risk factors, coronary anatomy and physiology, clinical presentation, management, and outcomes of ACS patients. Women who present with ACS are more frequently older and present with more comorbidities compared to men. Chest pain is the most frequent symptom of ACS; however, atypical symptoms are more frequently reported in women. Sex differences also exist in the pathophysiologic mechanisms of myocardial infarction with plaque rupture being the most common cause of coronary thrombosis; however, in younger women, plaque erosion is more common. On coronary angiography, women are more likely to have normal coronary arteries compared to men. They are less likely than men to be hospitalized for acute myocardial infarction (MI), but when hospitalized they have longer length of stay and higher in-hospital mortality. Women are less likely to be treated with guideline-directed medical therapy, to undergo invasive coronary revascularization procedures, or to be referred to cardiac rehabilitation. Despite recognition of sex differences in risk factors, presenting symptoms, coronary anatomy and physiology, and outcomes following ACS, the underlying etiologies for these differences are largely unknown. This challenge is further compounded by inadequate representation of women in ACS trials. Adequate inclusion of women in clinical trials, sex-specific a priori analysis, and reporting of results are necessary to better understand the plethora of unknowns and improve sex-specific disparities in ACS.

Introduction

According to the American Heart Association 2019 heart disease statistics update, the prevalence of cardiovascular disease is 61.5 million in males and 60.0 million in females, representing 51.2% and 44.7% of the US population, respectively [1]. Non-Hispanic black males have the highest prevalence of cardiovascular disease at 60.1% in those above age 20, followed by non-Hispanic black females at 57.1% and non-Hispanic white males at 50.6%. The prevalence of myocardial infarction is 5.1 million (4.0%) in males and 3.3 million (2.3%) in females. Mortality is more than 428,000 for males and more than 412,000 for females each year, with coronary heart disease accounting for 44% of deaths attributable to cardiovascular disease. Cardiovascular disease remains the number one cause of death for men and women each year, more than both cancer and lung disease combined. Although nearly as many women die from coronary artery disease (CAD) as men, they are consistently underrepresented in clinical trials and cohort studies. Sex differences exist in the presentation, pathophysiology, and diagnosis of acute coronary syndromes (ACSs). Women are less likely to receive guideline-derived medical therapy or referral to cardiac rehab.

Clinical Presentation

In general, women who present with an ACS are older and have more comorbidities than men (**Figure 1**). Women presenting with non-ST-segment elevation myocardial infarction (NSTEMI) are often 4–5 years older than men and more frequently have diabetes, hypertension, and heart failure [2]. Similar findings are present for women who present with ST elevation myocardial infarction (STEMI); they are consistently older than men and more often have a history of diabetes, hypertension, and cardiogenic shock at time of presentation [3].

The majority of patients with ACS present with classic symptoms of central chest pain or discomfort. However, atypical symptoms are more frequent in women than men, but can occur in either both sexes. These atypical symptoms include pleuritic chest pain, back pain, dyspnea, jaw/shoulder pain, dizziness, indigestion, and palpitations, among others [4]. A multicenter prospective study that examined the influence of sex on symptoms in those presenting to the emergency room with possible ACS demonstrated shoulder pain [odds ratio (OR) 2.53, 95% confidence interval (CI) 1.29–4.96] and arm pain (OR 2.15, 95% CI 1.10–4.20) were predictive of ACS in women, but not in men (OR 1.11, 95% CI 0.67–1.85; OR 1.21, 95% CI 0.74–1.99, respectively). In addition, in men shortness of breath was predictive of a non-ACS diagnosis (OR 0.49, 95% CI 0.30–0.79), but it was not predictive in women (OR 1.36, 95% CI 0.68–2.70) [5].

Highlighting how difficult it can be to discern symptoms predictive of ACS among race and sex, a prospective

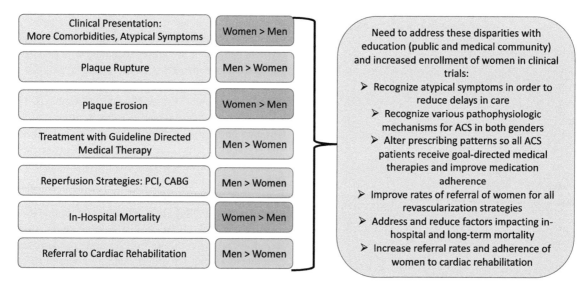

FIGURE 1 Sex Differences in Acute Coronary Syndrome and Disparities That Need to be Addressed. ACS, acute coronary syndrome; CABG, coronary artery bypass grafting; PCI, percutaneous coronary intervention.

emergency room cohort study analyzed differences in chest pain symptoms and 30-day ACS outcomes. It found that in black men diaphoresis was associated with ACS, and in white men left arm radiation and substernal pain were associated with ACS. In black women diaphoresis, palpitations and left arm radiation were associated with ACS, and no symptoms were predictive of ACS in white women [6]. In the Gender and Sex Determinants of Cardiovascular Disease: From Bench to Beyond Premature Acute Coronary Syndrome (GENESIS PRAXY) study, among young patients (55 years or younger) with ACS, chest pain was the most prevalent symptom regardless of sex. Young women reported more symptoms and the lack of chest pain was more likely to be present in young women than men (19.0% vs. 13.7%, $P=0.03$) [7]. The inconsistency of symptoms suggestive of ACS in the different studies points to the lack of reliability of any specific symptoms aside from chest pain. The atypical symptoms experienced by women may contribute to delays in presentation [3, 8], diagnosis, and ultimately receiving treatment. Symptoms also vary in women based on age (**Table 1**) [9].

Prodromal cardiac symptoms are symptoms occurring in the days to weeks prior to a cardiac event. Prodromal symptoms are more likely to be present in women than men under the age of 55 [10]. Symptoms included unusual fatigue, sleep disturbances, anxiety, and arm weakness or discomfort. Chest pain was not common in either sex, rep-

resenting a prodromal symptom in only 24% of patients in the prodromal stage. This is an important finding that young patients, both men and women, presented with atypical symptoms and rarely chest pain. Recognition of prodromal symptoms is important given their ability to predict future cardiovascular events in women [11, 12], and yet rarely results in initiation of risk-reduction therapies [4, 13-16].

Pathophysiologic Mechanisms

The basis of atherosclerotic disease within the vascular endothelium is thought to be related to inflammation. It is hypothesized that there are sex differences in hemostasis and inflammatory markers in ACS. Men with STEMI were found to have significantly higher levels of fibrinogen, C-reactive protein, and interleukin-6 as compared to women, reflecting a strong inflammatory component of ACS in men. The highest level of D-dimer concentration was found in women presenting with NSTEMIs [17]. Some studies claim that elevated D-dimer was thought to be predictive of complications with NSTEMI such as congestive heart failure and restenosis following percutaneous coronary intervention (PCI) [18, 19].

Compared with men, women are more likely to have unusual pathophysiologic mechanisms (**Figure 2**) underlying their ACS such as coronary artery spasm (CAS), Takotsubo cardiomyopathy, or spontaneous coronary artery dissection (SCAD). Overall, the most common cause of coronary thrombosis regardless of presentation in both men and women remains plaque rupture. This is responsible for 76% of deaths in myocardial infarctions in men and 55% in women [21]. Plaque rupture occurs when the thin fibrous cap of an atherosclerotic lesion is disrupted, exposing the highly thrombogenic core, resulting in partial to complete occlusion of the coronary artery [22]. When no plaque rupture is identified, the term plaque erosion is often used. Plaque erosion is more common among younger women and is defined as an acute thrombus with an underlying intima rich in smooth muscle cells and proteoglycan matrix; there is no endothelium beneath the thrombus on microscopic imaging (**Figure 3**) [24]. In previous studies evaluating etiology of ACS between males and females, plaque rupture was found to be particularly infrequent among younger females, possibly reflecting the protective effects of estrogen in premenopausal women [25]. Differences in plaque characteristics among men and women have been inconsistently observed. In a recent study, nonculprit plaque lesions in men evaluated with optical coherence tomography (OCT) were shown to contain larger lipid cores than in female patients [26]. However, there are several studies that found no sex differences among plaque characteristics using OCT in acute and nonacute presentations of coronary disease [27, 28].

The prevalence of ischemic heart disease (IHD) with no obstructive coronary arteries (INOCA) is increasing and is

TABLE 1 Symptom Differences in Younger Women vs. Older Women		
Symptom Differences	Younger Women	Older Women
Acute	↑ Atypical chest symptoms	↓ Atypical chest pain
	↑ Chest pain	↓ Chest pain
	↑ Chest pressure	↓ Typical angina pain
	↑ Nausea	
	↓ Diaphoresis	↑ Dyspnea
Prodromal	↑ Prodromal symptoms	↑ Sleep disturbance

Note. The *Younger Women* group included women younger than 65 years with acute symptoms and those younger than 50 years with prodromal symptoms. The *Older Women* group included women 65 years or older with acute symptoms and those 50 years or older with prodromal symptoms. ↑ = increased; ↓ = decreased. *Reproduced with permission from [9].*

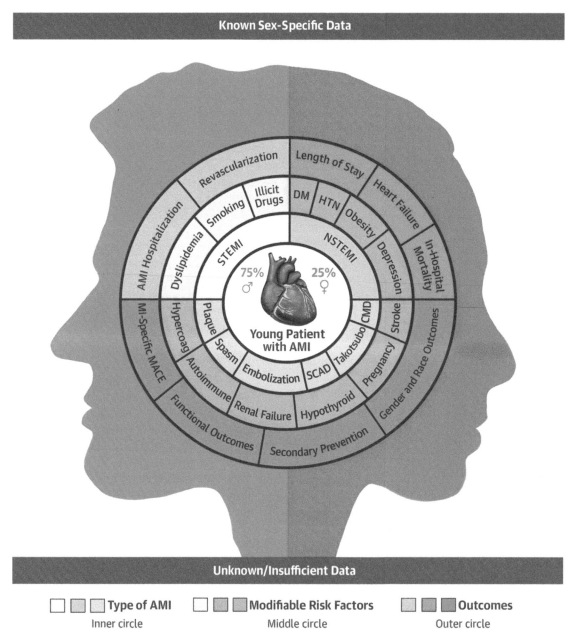

FIGURE 2   Clues for Diagnosis of AMI in the Young. Visual summarizing established (top half) and lesser known associations (bottom half) for AMI in the young by type of AMI, modifiable traditional and nontraditional risk factors, and outcomes. Factors attributed to male sex are shown in *blue* and female sex are in *orange*. AMI, acute myocardial infarction; CMD, coronary microvascular dysfunction; DM, diabetes mellitus; HTN, hypertension; hypercoag, hypercoagulability; MACE, major adverse cardiac event(s); MI, myocardial infarction; NSTEMI, non-ST-segment elevation myocardial infarction; SCAD, spontaneous coronary artery dissection; STEMI, ST-segment elevation myocardial infarction. *Reproduced with permission from [20].*

associated with higher costs due to increased association with major adverse cardiac events, angina, and heart failure hospitalizations. INOCA is defined as a clinical condition with stable symptoms suggestive of IHD along with objective evidence of myocardial ischemia with noninvasive testing and the absence of obstructive CAD. Epicardial stenosis of ≥50% or fractional flow reserve <0.8 are considered to be flow-limiting obstructive CAD. Coronary

microvascular dysfunction is one proposed mechanism for INOCA and is associated with higher cardiac events, but there is a gap in the appropriate risk stratification and treatment of these patients [29].

Women presenting with ACS are also more likely to have normal coronary arteries on coronary angiogram compared with men [23]. A subset of these ACS patients develop

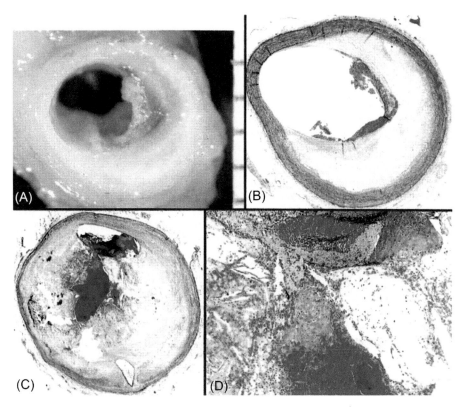

FIGURE 3 **Plaque Erosion in a Young Female Smoker Who Presented With Sudden Cardiac Death.** The images in (A–D) show an eccentric plaque and associated subocclusive thrombus. *Reproduced with permission from [23].*

myocardial infarction with nonobstructive coronary arteries (MINOCA), which is seen in ~5–6% of patients who undergo coronary angiography due to an acute myocardial infarction (AMI) [30]. CAS is one mechanism responsible for MINOCA and it is an intense contraction of the coronary smooth muscle caused by an increase in sensitivity to intracellular calcium [31]. The intense vasoconstriction of the coronary arteries can cause total or subtotal occlusion of the vessel, which can present as stable angina, unstable angina, AMI, and sudden cardiac death. The prevalence of CAS varies among different countries, but appears to be higher in the Japanese population [32]. Previously, it has been thought to be associated with SCAD through a series of case reports; however, a recent retrospective study from Cleveland Clinic showed that CAS has a low prevalence in patients with a prior history of SCAD, though the study was limited by its small sample size [33]. Diagnosis of CAS remains important as treatment varies from typical ACS and is often treated with calcium channel blockers or nitrates. Another mechanism responsible for these presentations of MINOCA is microvascular dysfunction. This also involves a constriction response seen mainly in the coronary microcirculation, and can cause a significant reduction in coronary blood flow [34]. See Chapter 9 for further discussion on MINOCA.

As discussed in Chapter 5, SCAD is increasingly being recognized as an important cause of MINOCA in young to middle-aged women. Believed to occur due to weakening of the coronary arterial wall, it is often associated with fibromuscular dysplasia. It is often triggered by intense emotional and physical stressors, resulting in catecholamine surges, thereby increasing arterial shear stress.

Role of Biomarkers

Biomarkers play an integral role in clinical management of ACS and risk stratification for cardiovascular disease. High sensitivity cardiac troponin (hs-cTn) is used throughout the world to aid in the early diagnosis of myocardial infarction. The universal definition of AMI requires a rise and fall in troponin and evidence of elevated cardiac troponin with at least one value above the 99th percentile, as well as one of the following: symptoms of myocardial ischemia, new ischemic ECG changes, development of pathologic Q waves, imaging with new regional wall motion abnormality of new loss of viable myocardium, or coronary thrombus on angiography [35]. There are several factors that influence the results of troponin; these include sex, age, renal function, and duration of chest pain [36]. Sex is often thought to be the least influential of these factors.

Sex-specific cutoff levels for hs-cTn have been explored with the thought that female sex is associated with a lower hs-cTn level. In 2015, a prospective cohort study suggested use of sex-specific hs-cTn in patients presenting with suspected ACS could double the diagnosis of myocardial infarctions in women, potentially improving mortality and increasing evidence-based treatment [37]. Women in this study were on average older, equally as likely to present with chest pain, and on admission had lower serum troponin concentration than men. In contrast, one of the largest studies published in 2016 included more than 2700 patients presenting with ACS, and did not demonstrate any benefit to sex-specific cutoffs [38]. In this study, the use of sex-specific hs-cTn resulted in the reclassification of only three patients: the diagnosis of two women was upgraded from unstable angina to AMI and the diagnosis of only one man was downgraded from AMI to unstable angina. Given the minuscule incremental benefit of using sex-specific troponin values, the study recommended that sex-neutral uniform 99th percentile hs-cTn should remain the standard of care in the diagnosis of AMI. Similar results were reported in a subsequent study of 1200 patients, which showed that use of sex-specific troponins reclassified only 4.5% of ACS patients, with increasing rates of AMI in women and decreasing rates of AMI in the male population [39], and without any meaningful benefit on outcomes. Similarly, sex-specific hs-cTn was unable to predict adverse outcomes in patients admitted to the coronary care unit [40].

Although initial studies suggest the use of sex-specific hs-cTn may be beneficial to increase sensitivity of AMI/NSTEMI diagnosis in women [41], subsequent studies have refuted this, demonstrating reclassification of only a small percentage of patients and a negligible impact on outcomes or predictive power for future adverse cardiac events in patients.

Sex-Based Disparities in Management

Delay in Presentation

Significant delay in seeking treatment for ACS symptoms is more common in women [4, 42, 43]. Young women in the Variation in Recovery: Role of Gender on Outcomes of Young AMI Patients (VIRGO) trial who received reperfusion therapy were more likely to present with no symptoms or atypical symptoms (16% vs. 10%; $P=0.008$) and more likely to have a delayed presentation of >6h after symptom onset (35% vs. 23%; $P=0.002$) compared to young men [44]. There are several factors contributing to delay in seeking treatment for ACS, including inaccurate symptom attribution, lack of awareness of self-risk, and barriers to self-care [45, 46]. It is plausible that the delays in seeking acute treatment for AMI may contribute to the poorer outcomes in women.

Guideline Recommendations

Although many of the references from the guidelines for management of STEMI and NSTEMI are based on stud-

ies underpowered to specifically address sex differences, the indications for reperfusion and pharmacologic therapy are outlined in American College of Cardiology/American Heart Association (ACC/AHA) guidelines for management of STEMI and NSTEMI [47, 48], with no overt sex-specific recommendations.

Pharmacotherapy

The ACC/AHA guidelines for management of NSTEMI should be treated with the same pharmacologic agents as men. Pharmacotherapy with antiplatelet agents, beta-blockers, statin, angiotensin-converting enzyme inhibitors (ACEIs), and angiotensin receptor blockers (ARBs) have been shown to be equally efficacious in both sexes, resulting in reduced morbidity and mortality postmyocardial infarction [48]. For instance in the Pravastatin or Atorvastatin Evaluation and Infection Therapy-Thrombolysis in Myocardial Infarction 22 (PROVE IT-TIMI 22) trial, both women and men had reduction in the primary end point of death, myocardial infarction, revascularization after 30 days, and stroke when treated with intensive vs. standard statin therapy. In this study, women treated with intensive statin therapy had a 25% relative reduction [hazard ratio (HR) 0.75, 95% CI 0.57–0.99, $P=0.04$] compared to men who had a 14% relative reduction (HR 0.86, 95% CI 0.75–0.99, $P=0.04$; P-interaction, 0.38) for the primary end point [49]. Despite the evidence of equal efficacy, women are often underprescribed guideline-recommended therapies. Data of patients undergoing PCI from the American College of Cardiology-National Cardiovascular Data Registry (ACC-NCDR) showed that women were less likely to be treated with aspirin in the acute setting (90.2% vs. 91.7%, OR 1.16, 95% CI 1.13–1.20, $P<0.01$) and at discharge (94.7% vs. 95.6%, OR 1.17, 95% CI 1.12–1.21, $P<0.01$). Also, women were less likely to be discharged on statin (81.0% vs. 84.5%, OR 1.10, 95% CI 1.07–1.13, $P<0.01$) therapy [50]. Subsequent data from the Get With The Guidelines—Coronary Artery Disease (GWTG-CAD) registry showed that at discharge following an AMI, women compared to age-matched men were less likely to be prescribed an ACEI/ARB (72.1% vs. 78.5%, OR 0.73, $P<0.002$ in <45 years of age, 74.6% vs. 77.4%, OR 0.90, $P<0.0005$ in those >45 years of age), and lipid-lowering medication (84.6% vs. 90.7%, OR 0.63, $P<0.001$ in those <45 years; 82.8% vs. 89.3%, OR 0.70, $P<0.0001$) [51].

Sex differences have also been observed in medication adherence following ACS with female sex associated with worse adherence [52]. A 2015 study reviewing six major lipid trials found similar reductions in lipid profiles in men and women, and reduced risk of major cardiac events occurring in 10.3% of men and 9% of women. Women had slightly higher discontinuation rates of statin therapy due to adverse events and slightly higher rates of myalgias (11.3%

vs.9.4% with atorvastatin and 10.8% vs. 7.7% with simvastatin) compared to men [53].

It is important to note that ACEIs and ARB therapy are pregnancy category C for the first trimester and category D for the second and third trimester. Statins are considered pregnancy category X. These medications may not be prescribed in young women of reproductive age who undergo ACS.

In the most recent ACC/AHA clinical practice guidelines on management of cholesterol, sex-specific risk-enhancing factors, including history of premature menopause (before age 40) and preeclampsia, which impact the timing of initiating statin therapy for primary prevention of cardiovascular disease, are emphasized [54]. Such sex-specific guidelines can help with improved sex-specific risk reduction and improved outcomes.

Potential Bias With Revascularization Therapies

Women have historically had a higher risk of operative mortality following coronary artery bypass grafting (CABG). In 2016, a retrospective analysis of more than 2.2 million patients through the Nationwide Inpatient Sample database from 2003 to 2012 reviewed the temporal trends in sex-specific mortality in patients undergoing surgical coronary revascularization [55]. The investigators found that the annual rate of bypass surgery had decreased by 54% in men and 58% in women over the 10-year period, but in-hospital mortality was still greater in women (3.2% vs. 1.8%, $P < 0.001$). Female sex was an independent predictor of mortality after multivariate analysis across all age groups (OR 1.4, 95% CI 1.36–1.43, $P < 0.001$). The in-hospital mortality declined at a faster rate in women compared to men, albeit sex differences in mortality still persisted. Similar findings were reported from another group that reviewed patient data from 1972 through 2011 including more than 57,000 patients, of whom only 19% were women [56]. Women had a lower survival than men following CABG (65% and 31% at 10 and 20 years, respectively, vs. 74% and 41%, $P < 0.001$). Worse outcomes were attributed to a multitude of factors, including old age, delayed diagnosis and treatment, greater prevalence of comorbidities (diabetes, hypertension, and heart failure), and underuse of arterial grafts. A single-center study of nearly 18,000 patients in the Netherlands also confirmed women undergoing CABG were older and had a greater number of preoperative comorbidities, as well as fewer number of bypass grafts, and were less likely to receive an internal mammary artery graft [57], all of which contribute to worse prognosis in women post-CABG.

A study out of Iceland and Sweden looked at the effects of age and gender on patients with severe coronary disease referred for PCI or CABG [58]. Women with one vessel CAD were more likely to be treated medically and less likely to undergo PCI as compared to men. This may be explained by an attempt to avoid complications with the knowledge that women have more in-hospital complications following PCI than men. Further, women with left main disease, or two- or three-vessel CAD, were less likely to undergo CABG as compared to men. Overall, serious bleeding complications were infrequent, but were four times more frequent in women than men. Despite higher in-hospital complication rates in women than men, there were no significant gender differences in 30-day mortality in ACS patients undergoing PCI or CABG in this study.

In addition to decreased likelihood for undergoing CABG, women with ACS are also less likely to undergo percutaneous interventions. Despite sex-neutral recommendations for pursuing aggressive early intervention in patients with STEMI and NSTEMI [47, 48], several studies demonstrate women were less likely to undergo reperfusion therapy by either fibrinolytic therapy or PCI [3, 59]. This disparity likely reflects the older age, atypical presentation, more comorbidities, and higher risk of bleeding seen in women on presentation.

Furthermore, there have been many studies that propose revascularization strategies in patients with multivessel disease presenting with STEMI. The 2015 ACC/AHA update modified their recommendations of PCI of a noninfarct artery from a class III (harm) to a class IIb recommendation for patients with STEMI and multivessel disease who are hemodynamically stable [60]. The majority of studies influencing this update incorporated less than 30% women in their study population, some as low as 12% women [60–63]. The 2017 Culprit Lesion Only PCI Versus Multivessel PCI in Cardiogenic Shock (CULPRIT-SHOCK) study evaluated PCI strategies in AMI in patients with cardiogenic shock and concluded that PCI of the culprit lesion only, rather than immediate multivessel PCI, resulted in lower 30-day composite risk of death or renal failure [64]. The percentages of women in the two study groups were 25% and 22%, for culprit lesion and multivessel revascularization, respectively. As with prior major studies, these results translate more readily to the male patients and may be considered underpowered for treatment among female patients with multivessel disease presenting with a STEMI.

Cardiac Rehabilitation After ACS

Cardiac rehabilitation is an accepted and internationally recommended intervention following ACS that has been shown to reduce mortality, reduce hospitalizations, and prevent subsequent myocardial infarctions [65, 66]. Comprehensive cardiac rehabilitation programs enhance exercise in a monitored environment, as well as provide education regarding risk factors and address lifestyle modifications. Referral and completion of cardiac rehabilitation following an AMI is a class I recommendation for both men and women in evidence-based guidelines [47, 48]. The recommendation and benefit of cardiac rehabilitation are also incorporated into the effectiveness-based cardiovascular disease prevention guidelines for women [67].

Women are commonly underreferred for cardiac rehabilitation and have reduced rates of cardiac rehabilitation completion, despite the knowledge that women completing cardiac rehab may experience a greater reduction in mortality compared with men [68, 69]. In a meta-analysis of sex differences in cardiac rehabilitation with studies mostly out of Canada and the United States, mean adherence to cardiac rehabilitation was 69% in men and 64% in women [70]. In a similar meta-analysis, on average men were less likely to be referred for cardiac rehabilitation following ACS compared to women (49% vs. 40%) [71]. Lack of referrals and completion have previously been attributed to several patient-related factors including socioeconomic status, depression, obesity, discomfort of exercise, and family obligations; however, equally important is the lack of support from physicians [4, 70]. Home-based or women-only rehabilitation programs have been proposed as an approach to increase compliance and completion of cardiac rehabilitation. Other options include more flexible hours so that women with childcare responsibilities have more options for participation, screening, and counseling for depression since mood can be a cause of poor compliance, and offering alternative exercise formats such as dance classes [72]. Unfortunately, many alternative models of cardiac rehabilitation were unsuccessful in improving adherence [73].

For men and women who complete standardized cardiac rehabilitation programs, women had poorer psychosocial health at baseline and less improvement in healthcare quality of life compared to men [74]. Notably, osteoporosis and menopause are two conditions affecting physiological and psychological conditions in older women [75]. Structured research programs addressing these aspects are lacking.

Disparities in Outcomes

Despite a nearly equal prevalence of cardiovascular disease in men and women, and representing the leading cause of death, there are still dramatic sex and racial disparities in prevention, treatment, and outcomes (**Table 2**). Multiple

TABLE 2 Sex-Based Disparities in Outcomes and Quality of Care

Factors	Setting	Sex-Specific Outcomes
Diagnostic testing	Stable IHD	1. Less than 1 in 10 women with angina and abnormal stress test had any change in pharmacotherapy or referral to diagnostic angiography
Delay in reperfusion	STEMI	1. Women had longer median first medical C2D times compared with men (80 vs. 75 min) 2. Women experienced a 30-min prehospital delay from symptom onset to hospital presentation compared with men 3. Young women were more likely to exceed door-to-needle time guidelines for PCI during STEMI compared with age-matched men (67% vs. 32%) (OR, 2.62)
Fewer revascularizations	ACS	1. Women were less likely to undergo revascularization after STEMI and NSTEMI 2. Women with documented one-vessel disease were less likely to undergo PCI compared with men (OR, 0.78) 3. Women were less likely to be referred for surgical revascularization (OR, 0.81) 4. Young women were particularly less likely to have revascularization compared with men (28% vs. 13%) 5. Despite known survival benefits of arterial grafts over vein grafts as conduits, women undergoing CABG surgery were less likely to receive arterial grafts compared with men
	Stable IHD	1. Despite higher angina class, women were less likely to undergo coronary angiography (31% vs. 49% men) 2. Women with stable angina and confirmed CAD were less likely to undergo PCI compared with men (OR, 0.70)

TABLE 2 Sex-Based Disparities in Outcomes and Quality of Care—cont'd		
Factors	Setting	Sex-Specific Outcomes
Less pharmacotherapy	Primary prevention	1. Women were less likely to have IHD risk factors measured, and young women (aged 35–54 years) were 37% less likely to be prescribed guideline-recommended medications 2. Women were 65% less likely to have assessment of their smoking status, body habitus, blood pressure, and lipid profile
	ACS	1. In the CRUSADE study, women were less likely to receive aspirin, ACE inhibitors, and statins on hospital discharge, even after adjustment for higher comorbidities in women 2. Black women were significantly less likely to receive appropriate secondary prevention measures compared with age-matched white patients after MI, despite having ≥ 3 risk factors 3. Medicare claims data demonstrated similar prescriptions patterns at hospital discharge but reported a 30–35% lower 12-month medication adherence after MI among black and Hispanic women compared with white men
	Stable IHD	1. Women report a significantly lower use of statin and aspirin therapy compared with men 2. Women were less likely to achieve guideline-directed secondary prevention targets for lipid (OR, 0.5), glucose (OR, 0.78), physical activity (OR, 0.74), or body mass index (OR, 0.82). Similar findings were confirmed by the EUROASPIRE III and IV surveys in Europe
Cardiac rehabilitation	ACS	1. Despite higher event rates and worse outcomes after MI, women were less likely to access cardiac rehabilitation 2. Women were 32% less likely to be referred to cardiac rehabilitation (39.6% vs. 49.4%) 3. Women were 36% less likely to enroll in cardiac rehabilitation (38.5% vs. 45.0%) 4. Women adhered to median of 71.9% cardiac rehabilitation sessions compared with 75.6% of men
Morbidity after MI	ACS	1. All women with acute MI had a 26% higher 1-year rate of rehospitalization even after adjusting for comorbidities, with even higher rehospitalizations in black women 2. Women with NSTEMIs had a higher in-hospital risk of recurrent MI (OR, 1.1) and heart failure (OR, 1.4). Similar outcome disparities were noted in STEMI 3. Black women were more likely to have angina at 1 year after MI treated with PCI (49% vs. 31% in white men) 4. Young women with acute MI experienced more angina and depression, had worse quality of life, and were less likely to return to work within 12 months after their MI
	Stable IHD	1. Women with suspected angina but angiographically normal vessels had more hospitalizations and repeat catheterizations for chest pain or ACS (OR, 4.1) 2. These women exhibited a decreased quality of life 3. 57% of women with confirmed CAD reported recurrent angina (compared with 47% of men)

Continued

TABLE 2 Sex-Based Disparities in Outcomes and Quality of Care—cont'd		
Factors	Setting	Sex-Specific Outcomes
Mortality	ACS	1. Women with STEMI had higher in-hospital mortality compared with men (10.2% vs. 5.5%)
		2. Women with ACS had higher in-hospital mortality after coronary angiography
		3. Men and women have similar 30-day post-MI mortality after adjusting for comorbidities and angiographic severity of disease
		4. Younger women (<55 years) had a twofold higher postinfarct in-hospital and 1-year mortality
	Stable IHD	1. Women with stable chest pain had higher in-hospital mortality at time of angiography (OR, 1.25) and higher 1-year mortality compared with men
		2. Women with angiographic CAD had a twofold higher risk of 1-year death, even after adjusting for severity of disease and comorbidities

ACE, angiotensin-converting enzyme; ACS, acute coronary syndrome; C2D, contact-to-device time; CABG, coronary artery bypass graft; CAD, coronary artery disease; CRUSADE, Can Rapid Risk Stratification of Unstable Angina Patients Suppress Adverse Outcomes With Early Implementation of the American College of Cardiology/American Heart Association Guidelines; EUROASPIRE, European Action on Secondary Prevention Through Intervention to Reduce Events; IHD, ischemic heart disease; MI, myocardial infarction; NSTEMI, non-ST-segment elevation MI; OR, odds ratio; PCI, percutaneous coronary intervention; STEMI, ST-segment elevation MI. *Reproduced with permission from [76].*

prior studies have shown that women presenting with ACS are less likely to be treated with guideline-directed medical therapy and less likely to undergo coronary angiogram and receive timely revascularization (**Figure 1**) [4]. The sex disparities in outcome after ACS are more prevalent in certain subsets of population particularly disadvantaged by age, race, and ethnicity.

Older women have also been shown to have lower revascularization rates than men [77] with corresponding increased mortality following an ACS event [78]. Invasive therapies are likely underutilized in elderly or frail patients due to increased risk of complications, particularly the associated high risk of bleeding with elderly female patients [79].

Young women with ACS represent an understudied population with an excess mortality risk compared to men. In a 2017 study out of Israel, when evaluated based upon age, women less than 55 years of age were found to have increased hospital mortality and increased 30-day major adverse cardiac and cerebral events and mortality [8]. Women were found to have a 4.1-fold higher risk of in-hospital mortality compared to men. Younger women were also found to have less invasive management and less aggressive secondary prevention medication management. In addition to increased mortality risk compared with young men, young women are also at a higher risk for readmission following ACS [80]. Young women have increased comorbidities, increased length of stay, and increased in-hospital mortality compared to young men, and temporal trends

show no change in AMI hospitalization rates and increased comorbidities over time (**Figure 4**) [81].

Modifiable risk factors (such as diabetes, hypertension, obesity, dyslipidemia, and smoking) were recently shown to be highly prevalent at the time of first AMI in young adults less than 60 years of age, with more than 90% having at least one modifiable risk factor (**Figure 5**). Furthermore, these risk factor progressively increased in prevalence over time, raising concern that even more robust preventative strategies need to be incorporated in these patients. In addition, data from the VIRGO study demonstrated than in young AMI patients (age ≤ 55 years), women were more likely not only to have traditional cardiac risk factors (heart failure, diabetes, smoking, and morbid obesity), but also to have increased mental health issues (depression and stress) [83]. More young women (39%) experienced depressive symptoms at the time of their AMI compared to men (22%, $P < 0.0001$) and almost half of women (48%) compared to a quarter of men (24%, $P < 0.0001$) reported a lifetime history of depression [84]. At the time of the AMI and during the first 12 months of recovery, women had higher levels of perceived stress compared to men; however, stress levels decreased in a similar pattern over time in both genders [85]. Similarly, in GENESIS PRAXY, the nontraditional factors of depression, anxiety, and stress at home were more frequent in young women compared to young men (**Figure 3**) [86]. In the Translational Research Investigating Underlying Disparities in Acute Myocardial Infarction Patients' Health Status (TRIUMPH) study of AMI patients

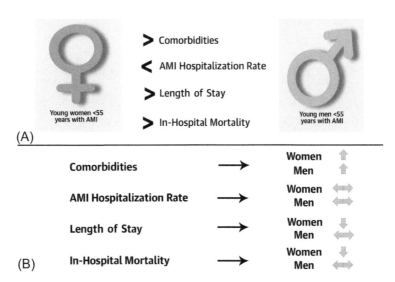

FIGURE 4 **AMI Trends From 2001 to 2010 in Young Women and Men.** (A) Overall comparisons of comorbidities, hospitalization rates, and short-term outcomes in young women with AMIs compared with young men. (B) Trends in young women and men with AMIs, 2001 to 2010. AMI, acute myocardial infarction. *Reproduced with permission from [81].*

≥ 18 years of age, increased long-term mortality was associated with depression; however, this may be in large part due to untreated depression [87].

Significant disparities continue to exist among women of racial and ethnic minorities. Black women are known to have the highest prevalence of myocardial infarction [88–91]. Black and Hispanic women with AMI typically present later to the hospital and have more comorbidities at time of presentation, such as diabetes, hypertension, obesity, and heart failure, compared with non-Hispanic white women [92–95]. Black women are less likely to be referred for coronary angiography and reperfusion therapy compared with white women and black men [93, 96, 97].

Among the 2699 women with established CAD from the Heart and Estrogen/progestin Replacement Study (HERS), the mortality rate among black women was higher than white women (16% vs. 8%, $P < 0.001$). There was a 60% greater risk of CHD events in black women compared to white women even after adjustment for demographic factors and comorbid conditions (HR 1.60, 95% CI 1.11–2.32, $P = 0.01$). Despite having higher rates of cardiac risk factors, black women were less likely to be treated with statins or aspirin and also less likely to reach optimal blood pressure (BP) or low-density lipoprotein (LDL) control compared to white women (56% vs. 63%, $P = 0.01$; 30% vs. 38%, $P = 0.04$) [94].

In a study out of Vienna evaluating mortality differences from a STEMI registry, unadjusted in-hospital mortality and long-term mortality were higher in women; however, following adjustment for confounders, multivariate analysis did not reveal differences in mortalities between men and women [3]. This is similar to the French multicenter registry, which also showed no difference for in-hospital mortality [98]. The STEMI study out of Australia also did not show differences for in-hospital mortality among men and women, but did show more major adverse cardiac events at 6 months for women [99], possibly reflecting delayed negative outcomes of disparities in treatment.

A recent analysis of ACS in China reviewed more than 82,000 cases from 2014 to 2018 and found less evidence-based acute treatments and less secondary prevention for women than men [100]. Women were on average older with more comorbidities, higher prevalence of diabetes, hypertension, dyslipidemia, and chronic kidney disease. Notably, statistically significant were several disparities in acute management of ACS: eligible female patients received less dual antiplatelet therapy (DAPT), statin therapy, and heparin during hospitalization compared to male patients. Women presenting with STEMI were less likely to receive acute reperfusion than men (50.2% vs. 59.5%, $P < 0.001$). Women were also less likely to receive DAPT, statins, ACEIs/ARBs, smoking cessation counseling, and cardiac rehabilitation counseling at discharge. These findings were similar to other recent studies looking at ACS in China [101, 102], as well as a large cohort study of Australian STEMI patients [99]. These studies did not present any medical reasoning for decreased intervention or less prescriptions for standard goal-directed medical therapy. Differences in management may be confounded by the often-atypical presentation of ACS in women or by intrinsic bias of clinicians.

Overall, factors contributing to increased mortality in women presenting with ACS likely include their increased

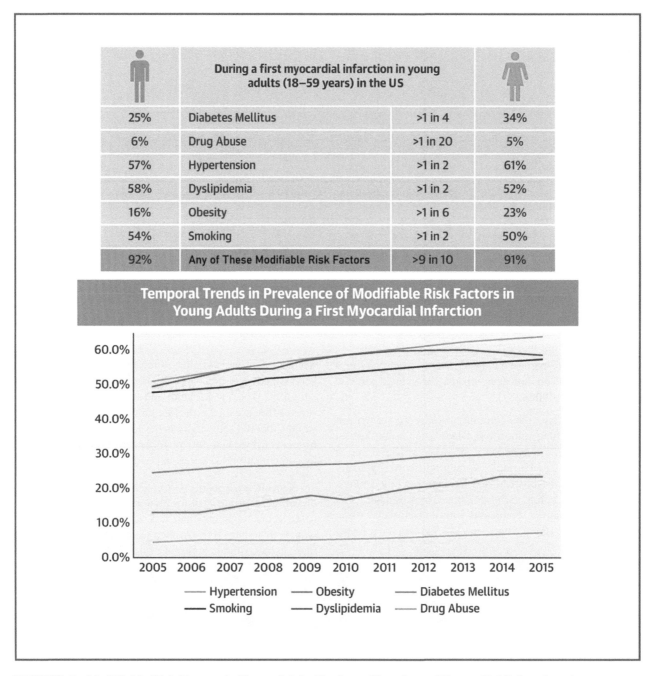

FIGURE 5   Modifiable Risk Factors in Young Adults During a First Acute Myocardial Infarction. Among young adults hospitalized for a first AMI in the United States, > 90% had at least one modifiable RF, the rates of which increased temporally between 2005 and 2015 ($P_{trend} < 0.001$ for all). Dyslipidemia, smoking, and drug abuse were more prevalent in men, and diabetes mellitus, hypertension, and obesity were more prevalent in women. AMI, acute myocardial infarction; RF, risk factor. *Reproduced with permission from [82].*

Mutually Beneficial Partnership

- **Patient-Family Council**
  - Diverse Community Representation
  - Creation of an Office of Healthcare Equity and Community Partnership
- **Generalist-Specialist Leadership/Partnership**
  - Enhanced Continuity of Care
- **Advanced Practitioners**
  - Diverse Clinicians
- **Nursing Allied Health Professionals**
  - Technicians/Technologists
- **Administration**
  - New Care Models
  - Marketing
  - Health Plan Representation

FIGURE 6 Sex and Gender Team Science: Redefining Patient-Centeredness for Women. *Reproduced with permission from [106].*

age at presentation, increased comorbidities including diabetes, hypertension, and shock, and their often-atypical presenting features which can result in time delays. Although there have been improvements in recognizing sex disparities over the past several decades, recent studies have shown that disparities still exist and there is much to learn regarding the individualization of treatment between men and women.

Clinical pathways are often used in quality improvement as a means to help eliminate disparities. They provide clinicians with a set of guidelines and checklists to ensure equality of treatment and prevent innate bias. The clinical pathway for ACS implemented in China, a country known for its gender disparities, showed a significantly greater proportion of male and female patients discharged on optimal medical therapy and improvement in the number of eligible STEMI patients receiving reperfusion therapy [103].

The first step toward eliminating sex disparities begins with adequate representation in clinical trials. Despite nearly equal prevalence of disease among men and women, women often represent a small fraction of participants in major studies, with even worse representation for minorities. This change will likely need to stem from policy changes that incorporate more women in cardiovascular trials. Dedicated centers for women's health were developed to help improve outcomes for women with cardiovascular disease. They have in part been successful over the last decade as awareness and sex-specific treatment have increased and include a patient centeredness approach with shared decision making while fully engaging the entire patient care team (**Figure 6**) [104–106].

Conclusion

Despite cardiovascular disease being the leading cause of mortality of men and women, there are sex-specific differences in the management and outcomes of ACS patients. Sex differences in clinical presentation and pathophysiologic mechanisms of ACS may in part explain the treatment delays seen in women. In addition, women are less likely to be treated with guideline-recommended medical therapies or undergo coronary revascularization procedures compared to men presenting with ACS, affecting clinical outcomes. Further, women remain underrepresented in clinical trials. Educating the public and medical communities, and the development of dedicated women's health centers, will assist in addressing these sex-specific disparities in care.

CH
4

Key Points

1. Women who present with acute coronary syndrome (ACS) are more frequently older and present with more comorbidities compared to men.

2. Chest pain is the most frequent symptom of ACS in both men and women; however, atypical symptoms are more frequently reported in women.

3. Plaque rupture is the most common cause of coronary thrombosis; however, plaque erosion is more common in younger women.

4. Women presenting with ACS are more likely to have normal coronary arteries on coronary angiography compared to men.

5. Women presenting with ACS are less likely to be treated with guideline-directed medical therapy or undergo revascularization with percutaneous coronary intervention or coronary artery bypass grafting.

6. Despite proven cardiovascular benefits of cardiac rehabilitation, women are commonly underreferred and have reduced completion rates of cardiac rehabilitation.

Back to Clinical Case

The clinical case depicted a 50-year-old woman presenting with an ST elevation myocardial infarction (STEMI). She was emergently taken to the catheterization laboratory and found to have 99% blockage of the mid-LAD and the remaining coronary arteries and branches were normal. Intracoronary nitroglycerin was administered with dilation of the mid-LAD and only minimal luminal irregularities present in the vessel. These findings were most consistent with coronary spasm. Left ventriculogram revealed mild hypokinesis of the anterior wall. Upon further discussion with the patient, she admitted to increased work-related stress and depression over the prior 3 months. The patient was treated with aspirin, long-acting nitroglycerin formulation, and calcium channel blocker. Following discharge, she successfully completed cardiac rehab. She has undergone counseling for depression and high levels of work-related stress. Follow-up echocardiogram at 6 months showed normal global and regional wall motion. She has remained angina-free on this medication regimen. With regard to the question posed with the introduction to the case, vasospasm could be considered in the differential in a young woman presenting with STEMI and sublingual nitroglycerin should be promptly administered. Immediate resolution of pain and ECG changes would support the possible diagnosis of coronary vasospasm.

References

[1] Benjamin EJ, Muntner P, Alonso A, Bittencourt MS, Callaway CW, Carson AP, et al. Heart disease and stroke statistics—2019 update: a report from the American Heart Association. Circulation 2019;139(10):e56–e528.

[2] Kragholm K, Halim SA, Yang Q, Schulte PJ, Hochman JS, Melloni C, et al. Sex-stratified trends in enrollment, patient characteristics, treatment, and outcomes among non-ST-segment elevation acute coronary syndrome patients: insights from clinical trials over 17 years. Circ Cardiovasc Qual Outcomes 2015;8(4):357–67.

[3] Piackova E, Jäger B, Farhan S, Christ G, Schreiber W, Weidinger F, et al. Gender differences in short- and long-term mortality in the Vienna STEMI registry. Int J Cardiol 2017;244:303–8.

[4] Mehta LS, Beckie TM, DeVon HA, Grines CL, Krumholz HM, Johnson MN, et al. Acute myocardial infarction in women: a scientific statement from the American Heart Association. Circulation 2016;133(9):916–47.

[5] Devon HA, Rosenfeld A, Steffen AD, Daya M. Sensitivity, specificity, and sex differences in symptoms reported on the 13-item acute coronary syndrome checklist. J Am Heart Assoc 2014;3(2):e000586.

[6] Allabban A, Hollander JE, Pines JM. Gender, race and the presentation of acute coronary syndrome and serious cardiopulmonary diagnoses in ED patients with chest pain. Emerg Med J 2017;34(10):653–8.

[7] Khan NA, Daskalopoulou SS, Karp I, Eisenberg MJ, Pelletier R, Tsadok MA, et al. Sex differences in acute coronary syndrome symptom presentation in young patients. JAMA Intern Med 2013;173(20):1863–71.

[8] Sabbag A, Matetzky S, Porter A, Iakobishvili Z, Moriel M, Zwas D, et al. Sex differences in the management and 5-year outcome of young patients (< 55 years) with acute coronary syndromes. Am J Med 2017;130(11):1324 e15–22.

[9] DeVon HA, Pettey CM, Vuckovic KM, Koenig MD, McSweeney JC. A review of the literature on cardiac symptoms in older and younger women. J Obstet Gynecol Neonatal Nurs 2016;45(3):426–37.

[10] Khan NA, Daskalopoulou SS, Karp I, Eisenberg MJ, Pelletier R, Tsadok MA, et al. Sex differences in prodromal symptoms in acute coronary syndrome in patients aged 55 years or younger. Heart 2017;103(11):863–9.

[11] McSweeney J, Cleves MA, Fischer EP, Moser DK, Wei J, Pettey C, et al. Predicting coronary heart disease events in women: a longitudinal cohort study. J Cardiovasc Nurs 2014;29(6):482–92.

[12] O'Keefe-McCarthy S, Ready L. Impact of prodromal symptoms on future adverse cardiac-related events: a systematic review. J Cardiovasc Nurs 2016;31(1):E1–10.

[13] Blomkalns AL, Chen AY, Hochman JS, Peterson ED, Trynosky K, Diercks DB, et al. Gender disparities in the diagnosis and treatment of non-ST-segment elevation acute coronary syndromes: large-scale observations from the CRUSADE (Can Rapid Risk Stratification of Unstable Angina Patients Suppress Adverse Outcomes With Early Implementation of the American College of Cardiology/American Heart Association Guidelines) National Quality Improvement Initiative. J Am Coll Cardiol 2005;45(6):832–7.

[14] Koopman C, Vaartjes I, Heintjes EM, Spiering W, van Dis I, Herings RM, et al. Persisting gender differences and attenuating age differences in cardiovascular drug use for prevention and treatment of coronary heart disease, 1998-2010. Eur Heart J 2013;34(41):3198–205.

[15] Maddox TM, Ho PM, Roe M, Dai D, Tsai TT, Rumsfeld JS. Utilization of secondary prevention therapies in patients with nonobstructive coronary artery disease identified during cardiac catheterization: insights from the National Cardiovascular Data Registry Cath-PCI Registry. Circ Cardiovasc Qual Outcomes 2010;3(6):632–41.

[16] Patel MR, Chen AY, Peterson ED, Newby LK, Pollack CV, Brindis RG, et al. Prevalence, predictors, and outcomes of patients with non-ST-segment elevation myocardial infarction and insignificant coronary artery disease: results from the can rapid risk stratification of unstable angina patients suppress ADverse outcomes with early implementation of the ACC/AHA guidelines (CRUSADE) initiative. Am Heart J 2006;152(4):641–7.

[17] Siennicka A, Jastrzebska M, Smialkowska K, Oledzki S, Chelstowski K, Klysz M, et al. Gender differences in hemostatic and inflammatory factors in patients with acute coronary syndromes: a pilot study. J Physiol Pharmacol 2018;69(1):91–8.

[18] Turker Y, Dogan A, Ozaydin M, Kaya S, Onal S, Akkaya M, et al. Association of thrombotic and fibrinolytic factors with severity of culprit lesion in patients with acute coronary syndromes without ST elevation. South Med J 2010;103(4):289–94.

[19] Charoensri N, Pornratanarangsi S. D-dimer plasma levels in NSTE-ACS patient. J Med Assoc Thail 2011;94(Suppl. 1):S39–45.

[20] Safdar B. Clues to diagnose myocardial infarction in the young: no longer a needle in the haystack. J Am Coll Cardiol 2019;73(5):585–8.

[21] Falk E, Nakano M, Bentzon JF, Finn AV, Virmani R. Update on acute coronary syndromes: the pathologists' view. Eur Heart J 2013;34(10):719–28.

[22] Bentzon JF, Otsuka F, Virmani R, Falk E. Mechanisms of plaque formation and rupture. Circ Res 2014;114(12):1852–66.

[23] Bairey Merz CN, Shaw LJ, Reis SE, Bittner V, Kelsey SF, Olson M, et al. Insights from the NHLBI-sponsored Women's Ischemia Syndrome Evaluation (WISE) study: part II: gender differences in presentation, diagnosis, and outcome with regard to gender-based pathophysiology of atherosclerosis and macrovascular and microvascular coronary disease. J Am Coll Cardiol 2006;47(3 Suppl):S21–9.

[24] Yahagi K, Davis HR, Arbustini E, Virmani R. Sex differences in coronary artery disease: pathological observations. Atherosclerosis 2015;239(1):260–7.

[25] Burke AP, Farb A, Malcom GT, Liang Y, Smialek J, Virmani R. Effect of risk factors on the mechanism of acute thrombosis and sudden coronary death in women. Circulation 1998;97(21):2110–6.

[26] Tian J, Wang X, Tian J, Yu B. Gender differences in plaque characteristics of nonculprit lesions in patients with coronary artery disease. BMC Cardiovasc Disord 2019;19(1):45.

[27] Sun R, Sun L, Fu Y, Liu H, Xu M, Ren X, et al. Culprit plaque characteristics in women vs men with a first ST-segment elevation myocardial infarction: in vivo optical coherence tomography insights. Clin Cardiol 2017;40(12):1285–90.

[28] Bharadwaj AS, Vengrenyuk Y, Yoshimura T, Baber U, Hasan C, Narula J, et al. Multimodality intravascular imaging to evaluate sex differences in plaque morphology in stable CAD. JACC Cardiovasc Imaging 2016;9(4):400–7.

[29] Bairey Merz CN, Pepine CJ, Walsh MN, Fleg JL. Ischemia and no obstructive coronary artery disease (INOCA). Circulation 2017;135(11):1075–92.

[30] Tamis-Holland JE, Jneid H, Reynolds HR, Agewall S, Brilakis ES, Brown TM, et al. Contemporary diagnosis and management of patients with myocardial infarction in the absence of obstructive coronary artery disease: a scientific statement from the American Heart Association. Circulation 2019;139(18):e891–908.

[31] Yasue H, Mizuno Y, Harada E. Coronary artery spasm—clinical features, pathogenesis and treatment. Proc Jpn Acad Ser B Phys Biol Sci 2019;95(2):53–66.

[32] Hung M-J, Hu P, Hung M-Y. Coronary artery spasm: review and update. Int J Med Sci 2014;11(11):1161–71.

[33] White Solaru K, Heupler F, Cho L, Kim ESH. Prevalence of coronary vasospasm using coronary reactivity testing in patients with spontaneous coronary artery dissection. Am J Cardiol 2019;123(11):1812–5.

[34] De Vita A, Manfredonia L, Lamendola P, Villano A, Ravenna SE, Bisignani A, et al. Coronary microvascular dysfunction in patients with acute coronary syndrome and no obstructive coronary artery disease. Clin Res Cardiol 2019;108(12):1364–70.

[35] Thygesen K, Alpert JS, Jaffe AS, Chaitman BR, Bax JJ, Morrow DA, et al. Fourth universal definition of myocardial infarction (2018). J Am Coll Cardiol 2018;72(18):2231–64.

[36] Twerenbold R, Boeddinghaus J, Nestelberger T, Wildi K, Rubini Gimenez M, Badertscher P, et al. Clinical use of high-sensitivity cardiac troponin in patients with suspected myocardial infarction. J Am Coll Cardiol 2017;70(8):996–1012.

[37] Shah ASV, Griffiths M, Lee KK, McAllister DA, Hunter AL, Ferry AV, et al. High sensitivity cardiac troponin and the under-diagnosis of myocardial infarction in women: prospective cohort study. BMJ 2015;350:g7873.

[38] Rubini Gimenez M, Twerenbold R, Boeddinghaus J, Nestelberger T, Puelacher C, Hillinger P, et al. Clinical effect of sex-specific cutoff values of high-sensitivity cardiac troponin T in suspected myocardial infarction. JAMA Cardiol 2016;1(8):912–20.

[39] Mueller-Hennessen M, Lindahl B, Giannitsis E, Biener M, Vafaie M, deFilippi CR, et al. Diagnostic and prognostic implications using age- and gender-specific cut-offs for high-sensitivity cardiac troponin T—sub-analysis from the TRAPID-AMI study. Int J Cardiol 2016;209:26–33.

[40] Eggers KM, Jernberg T, Lindahl B. Prognostic importance of sex-specific cardiac troponin T 99(th) percentiles in suspected acute coronary syndrome. Am J Med 2016;129(8):880 e1–880 e12.

[41] Slagman A, Searle J, Vollert JO, Storchmann H, Büschenfelde DMZ, von Recum J, et al. Sex differences of troponin test performance in chest pain patients. Int J Cardiol 2015;187:246–51.

[42] Moser DK, Kimble LP, Alberts MJ, Alonzo A, Croft JB, Dracup K, et al. Reducing delay in seeking treatment by patients with acute coronary syndrome and stroke: a scientific statement from the American Heart Association Council on cardiovascular nursing and stroke council. Circulation 2006;114(2):168–82.

[43] Rosenfeld AG, Lindauer A, Darney BG. Understanding treatment-seeking delay in women with acute myocardial infarction: descriptions of decision-making patterns. Am J Crit Care 2005;14(4):285–93.

[44] D'Onofrio G, Safdar B, Lichtman JH, Strait KM, Dreyer RP, Geda M, et al. Sex differences in reperfusion in young patients with ST-segment-elevation myocardial infarction: results from the VIRGO study. Circulation 2015;131(15):1324–32.

[45] Lichtman JH, Leifheit-Limson EC, Watanabe E, Allen NB, Garavalia B, Garavalia LS, et al. Symptom recognition and healthcare experiences of young women with acute myocardial infarction. Circ Cardiovasc Qual Outcomes 2015;8(2 Suppl. 1):S31–8.

[46] DeVon HA. Promoting cardiovascular health in women across the life span. J Obstet Gynecol Neonatal Nurs 2011;40(3):335–6.

[47] O'Gara PT, Kushner FG, Ascheim DD, Casey DE, Chung MK, de Lemos JA, et al. 2013 ACCF/AHA guideline for the management of ST-elevation myocardial infarction: a report of the American College of Cardiology Foundation/American Heart Association task force on practice guidelines. J Am Coll Cardiol 2013;61(4):e78–e140.

[48] Amsterdam EA, Wenger NK, Brindis RG, Casey DE, Ganiats TG, Holmes DR, et al. 2014 AHA/ACC guideline for the management of patients with non-ST-elevation acute coronary syndromes: a report of the American College of Cardiology/American Heart Association task force on practice guidelines. Circulation 2014;130(25):e344–426.

[49] Truong QA, Murphy SA, McCabe CH, Armani A, Cannon CP, TIMI Study Group. Benefit of intensive statin therapy in women: results from PROVE IT-TIMI 22. Circ Cardiovasc Qual Outcomes 2011;4(3):328–36.

[50] Akhter N, Milford-Beland S, Roe MT, Piana RN, Kao J, Shroff A. Gender differences among patients with acute coronary syndromes undergoing percutaneous coronary intervention in the American College of Cardiology-National Cardiovascular Data Registry (ACC-NCDR). Am Heart J 2009;157(1):141–8.

[51] Bangalore S, Fonarow GC, Peterson ED, Hellkamp AS, Hernandez AF, Laskey W, et al. Age and gender differences in quality of care and outcomes for patients with ST-segment elevation myocardial infarction. Am J Med 2012;125(10):1000–9.

[52] Kumbhani DJ, Fonarow GC, Cannon CP, Hernandez AF, Peterson ED, Peacock WF, et al. Predictors of adherence to performance measures in patients with acute myocardial infarction. Am J Med 2013;126(1):74 e1–9.

[53] Hsue PY, Bittner VA, Betteridge J, Fayyad R, Laskey R, Wenger NK, et al. Impact of female sex on lipid lowering, clinical outcomes, and adverse effects in atorvastatin trials. Am J Cardiol 2015;115(4):447–53.

[54] Grundy SM, Stone NJ, Bailey AL, Beam C, Birtcher KK, Blumenthal RS, et al. 2018 AHA/ACC/AACVPR/AAPA/ABC/ACPM/ADA/AGS/APhA/ASPC/NLA/PCNA guideline on the management of blood cholesterol: executive summary: a report of the American College of Cardiology/American Heart Association task force on clinical practice guidelines. J Am Coll Cardiol 2019;73(24):3168–209.

[55] Swaminathan RV, Feldman DN, Pashun RA, Patil RK, Shah T, Geleris JD, et al. Gender differences in in-hospital outcomes after coronary artery bypass grafting. Am J Cardiol 2016;118(3):362–8.

[56] Attia T, Koch CG, Houghtaling PL, Blackstone EH, Sabik EM, Sabik JF. Does a similar procedure result in similar survival for women and men undergoing isolated coronary artery bypass grafting? J Thorac Cardiovasc Surg 2017;153(3):571–579, e9.

[57] Ter Woorst JF, van Straten AHM, Houterman S, Soliman-Hamad MA. Sex difference in coronary artery bypass grafting: preoperative profile and early outcome. J Cardiothorac Vasc Anesth 2019;33(10):2679–84.

[58] Gudnadottir GS, Andersen K, Thrainsdottir IS, James SK, Lagerqvist B, Gudnason T. Gender differences in coronary angiography, subsequent interventions, and outcomes among patients with acute coronary syndromes. Am Heart J 2017;191:65–74.

[59] Pagidipati NJ, Peterson ED. Acute coronary syndromes in women and men. Nat Rev Cardiol 2016;13(8):471–80.

[60] Levine GN, Bates ER, Blankenship JC, Bailey SR, Bittl JA, Cercek B, et al. 2015 ACC/AHA/SCAI focused update on primary percutaneous coronary intervention for patients with ST-elevation myocardial infarction: an update of the 2011 ACCF/AHA/SCAI guideline for percutaneous coronary intervention and the 2013 ACCF/AHA guideline for the management of ST-elevation myocardial infarction. J Am Coll Cardiol 2016;67(10):1235–50.

[61] Wald DS, Morris JK, Wald NJ, Chase AJ, Edwards RJ, Hughes LO, et al. Randomized trial of preventive angioplasty in myocardial infarction. N Engl J Med 2013;369(12):1115–23.

[62] Engstrom T, Kelbæk H, Helqvist S, Høfsten DE, Kløvgaard L, Holmvang L, et al. Complete revascularisation versus treatment of the culprit lesion only in patients with ST-segment elevation myocardial infarction and multivessel disease (DANAMI-3-PRIMULTI): an open-label, randomised controlled trial. Lancet 2015;386(9994):665–71.

[63] Gershlick AH, Khan JN, Kelly DJ, Greenwood JP, Sasikaran T, Curzen N, et al. Randomized trial of complete versus lesion-only revascularization in patients undergoing primary percutaneous coronary intervention for STEMI and multivessel disease: the CvLPRIT trial. J Am Coll Cardiol 2015;65(10):963–72.

[64] Thiele H, Akin I, Sandri M, Fuernau G, de Waha S, Meyer-Saraei R, et al. PCI strategies in patients with acute myocardial infarction and cardiogenic shock. N Engl J Med 2017;377(25):2419–32.

[65] Martin B-J, Hauer T, Arena R, Austford LD, Galbraith PD, Lewin AM, et al. Cardiac rehabilitation attendance and outcomes in coronary artery disease patients. Circulation 2012;126(6):677–87.

[66] Anderson L, Thompson DR, Oldridge N, Zwisler AD, Rees K, Martin N, et al. Exercise-based cardiac rehabilitation for coronary heart disease: cochrane systematic review and meta-analysis. J Am Coll Cardiol 2016;67(1):1–12.

[67] Mosca L, Benjamin EJ, Berra K, Bezanson JL, Dolor RJ, Lloyd-Jones DM, et al. Effectiveness-based guidelines for the prevention of cardiovascular disease in women—2011 update: a guideline from the American Heart Association. J Am Coll Cardiol 2011;57(12):1404–23.

[68] Colbert JD, Martin BJ, Haykowsky MJ, Hauer TL, Austford LD, Arena RA, et al. Cardiac rehabilitation referral, attendance and mortality in women. Eur J Prev Cardiol 2015;22(8):979–86.

[69] Minges KE, Strait KM, Owen N, Dunstan DW, Camhi SM, Lichtman J, et al. Gender differences in physical activity following acute myocardial infarction in adults: a prospective, observational study. Eur J Prev Cardiol 2017;24(2):192–203.

[70] Oosenbrug E, Marinho RP, Zhang J, Marzolini S, Colella TJ, Pakosh M, et al. Sex differences in cardiac rehabilitation adherence: a meta-analysis. Can J Cardiol 2016;32(11):1316–24.

[71] Colella TJF, Gravely S, Marzolini S, Grace SL, Francis JA, Oh P, et al. Sex bias in referral of women to outpatient cardiac rehabilitation? A meta-analysis. Eur J Prev Cardiol 2015;22(4):423–41.

[72] Galati A, Piccoli M, Tourkmani N, Sgorbini L, Rossetti A, Cugusi L, et al. Cardiac rehabilitation in women: state of the art and strategies to overcome the current barriers. J Cardiovasc Med (Hagerstown) 2018;19(12):689–97.

[73] Grace SL, Midence L, Oh P, Brister S, Chessex C, Stewart DE, et al. Cardiac rehabilitation program adherence and functional capacity among women: a randomized controlled trial. Mayo Clin Proc 2016;91(2):140–8.

[74] Terada T, Chirico D, Tulloch HE, Scott K, Pipe AL, Reed JL. Sex differences in psychosocial and cardiometabolic health among patients completing cardiac rehabilitation. Appl Physiol Nutr Metab 2019;44(11):1237–45.

[75] Witvrouwen I, Van Craenenbroeck EM, Abreu A, Moholdt T, Kränkel N. Exercise training in women with cardiovascular disease: differential response and barriers—review and perspective. Eur J Prev Cardiol 2019. 2047487319838221.

[76] Aggarwal NR, Patel HN, Mehta LS, Sanghani RM, Lundberg GP, Lewis SJ, et al. Sex differences in ischemic heart disease. Circ Cardiovasc Qual Outcomes 2018;11(2):e004437.

[77] De Carlo M, Morici N, Savonitto S, Grassia V, Sbarzaglia P, Tamburrini P, et al. Sex-related outcomes in elderly patients presenting with non-ST-segment elevation acute coronary syndrome: insights from the Italian elderly ACS study. JACC Cardiovasc Interv 2015;8(6):791–6.

[78] Patel A, Goodman SG, Yan AT, Alexander KP, Wong CL, Cheema AN, et al. Frailty and outcomes after myocardial infarction: insights from the CONCORDANCE registry. J Am Heart Assoc 2018;7(18):e009859.

[79] Vicent L, Martinez-Selles M. Frailty and acute coronary syndrome: does gender matter? J Geriatr Cardiol 2019;16(2):138–44.

[80] Dreyer RP, Ranasinghe I, Wang Y, Dharmarajan K, Murugiah K, Nuti SV, et al. Sex differences in the rate, timing, and principal diagnoses of 30-day readmissions in younger patients with acute myocardial infarction. Circulation 2015;132(3):158–66.

[81] Gupta A, Wang Y, Spertus JA, Geda M, Lorenze N, Nkonde-Price C, et al. Trends in acute myocardial infarction in young patients and differences by sex and race, 2001 to 2010. J Am Coll Cardiol 2014;64(4):337–45.

[82] Yandrapalli S, Nabors C, Goyal A, Aronow WS, Frishman WH. Modifiable risk factors in young adults with first myocardial infarction. J Am Coll Cardiol 2019;73(5):573–84.

[83] Bucholz EM, Strait KM, Dreyer RP, Lindau ST, D'Onofrio G, Geda M, et al. Editor's choice-sex differences in young patients with acute myocardial infarction: a VIRGO study analysis. Eur Heart J Acute Cardiovasc Care 2017;6(7):610–22.

[84] Smolderen KG, Strait KM, Dreyer RP, D'Onofrio G, Zhou S, Lichtman JH, et al. Depressive symptoms in younger women and men with acute myocardial infarction: insights from the VIRGO study. J Am Heart Assoc 2015;4(4).

[85] Xu X, Bao H, Strait KM, Edmondson DE, Davidson KW, Beltrame JF, et al. Perceived stress after acute myocardial infarction: a comparison between young and middle-aged women versus men. Psychosom Med 2017;79(1):50–8.

[86] Choi J, Daskalopoulou SS, Thanassoulis G, Karp I, Pelletier R, Behlouli H, et al. Sex- and gender-related risk factor burden in patients with premature acute coronary syndrome. Can J Cardiol 2014;30(1):109–17.

[87] Smolderen KG, Buchanan DM, Gosch K, Whooley M, Chan PS, Vaccarino V, et al. Depression treatment and 1-year mortality after acute myocardial infarction: insights from the TRIUMPH registry (translational research investigating underlying disparities in acute myocardial infarction patients' health status). Circulation 2017;135(18):1681–9.

[88] Shaw LJ, Shaw RE, Bairey Merz CN, Brindis RG, Klein LW, Nallamothu B, et al. Impact of ethnicity and gender differences on angiographic coronary artery disease prevalence and in-hospital mortality in the American College of Cardiology-National Cardiovascular Data Registry. Circulation 2008;117(14):1787–801.

[89] Hozawa A, Folsom AR, Sharrett AR, Chambless LE. Absolute and attributable risks of cardiovascular disease incidence in relation to optimal and borderline risk factors: comparison of African American with white subjects—atherosclerosis risk in communities study. Arch Intern Med 2007;167(6):573–9.

[90] Clark LT, Ferdinand KC, Flack JM, Gavin JR, Hall WD, Kumanyika SK, et al. Coronary heart disease in African Americans. Heart Dis 2001;3(2):97–108.

[91] Safford MM, Brown TM, Muntner PM, Durant RW, Glasser S, Halanych JH, et al. Association of race and sex with risk of incident acute coronary heart disease events. JAMA 2012;308(17):1768–74.

[92] Canto JG, Taylor HA, Rogers WJ, Sanderson B, Hilbe J, Barron HV. Presenting characteristics, treatment patterns, and clinical outcomes of non-black minorities in the National Registry of myocardial infarction 2. Am J Cardiol 1998;82(9):1013–8.

[93] Mehta RH, Marks D, Califf RM, Sohn S, Pieper KS, Van de Werf F, et al. Differences in the clinical features and outcomes in African Americans and whites with myocardial infarction. Am J Med 2006;119(1):70, e1–8.

[94] Jha AK, Varosy PD, Kanaya AK, Hunninghake DB, Hlatky MA, Waters DD, et al. Differences in medical care and disease outcomes among black and white women with heart disease. Circulation 2003;108(9):1089–94.

[95] Rodriguez CJ, Allison M, Daviglus ML, Isasi CR, Keller C, Leira EC, et al. Status of cardiovascular disease and stroke in Hispanics/Latinos in the United States: a science advisory from the American Heart Association. Circulation 2014;130(7):593–625.

[96] Vaccarino V, Rathore SS, Wenger NK, Frederick PD, Abramson JL, Barron HV, et al. Sex and racial differences in the management of acute myocardial infarction, 1994 through 2002. N Engl J Med 2005;353(7):671–82.

[97] Chen J, Rathore SS, Radford MJ, Wang Y, Krumholz HM. Racial differences in the use of cardiac catheterization after acute myocardial infarction. N Engl J Med 2001;344(19):1443–9.

[98] Donataccio MP, Puymirat E, Parapid B, Steg PG, Eltchaninoff H, Weber S, et al. In-hospital outcomes and long-term mortality according to sex and management strategy in acute myocardial infarction. Insights from the French ST-elevation and non-ST-elevation myocardial infarction (FAST-MI) 2005 registry. Int J Cardiol 2015;201:265–70.

[99] Khan E, Brieger D, Amerena J, Atherton JJ, Chew DP, Farshid A, et al. Differences in management and outcomes for men and women with ST-elevation myocardial infarction. Med J Aust 2018;209(3):118–23.

[100] Hao Y, Liu J, Yang N, Smith SC, Huo Y, et al. Sex differences in in-hospital management and outcomes of patients with acute coronary syndrome. Circulation 2019;139(15):1776–85.

[101] Du X, Spatz ES, Dreyer RP, Hu S, Wu C, Li X, et al. Sex differences in clinical profiles and quality of care among patients with ST-segment elevation myocardial infarction from 2001 to 2011: insights from the China Patient-Centered Evaluative Assessment of Cardiac Events (PEACE)-retrospective study. J Am Heart Assoc 2016;5(2).

[102] Zheng X, Dreyer RP, Hu S, Spatz ES, Masoudi FA, Spertus JA, et al. Age-specific gender differences in early mortality following ST-segment elevation myocardial infarction in China. Heart 2015;101(5):349–55.

[103] Du X, Patel A, Li X, Wu Y, Turnbull F, Gao R. Treatment and outcomes of acute coronary syndromes in women: an analysis of a multicenter quality improvement Chinese study. Int J Cardiol 2017;241:19–24.

[104] Lundberg GP, Mehta LS, Sanghani RM, Patel HN, Aggarwal NR, Aggarwal NT, et al. Heart centers for women. Circulation 2018;138(11):1155–65.

[105] Lundberg GP, Mehta LS, Volgman AS. Specialized care for women: the impact of women's heart centers. Curr Treat Options Cardiovasc Med 2018;20(9):76.

[106] Shaw LJ, Pepine CJ, Xie J, Mehta PK, Morris AA, Dickert NW, et al. Quality and equitable health care gaps for women: attributions to sex differences in cardiovascular medicine. J Am Coll Cardiol 2017;70(3):373–88.

Clinical Presentation

Older
More comorbidities
Atypical symptoms more often
than men
Present with lower troponins

Pathophysiology

Plaque rupture more common in men
Plaque erosion more common in women
More likely to have normal coronary
arteries & MINOCA
Often due to Takotsubo, spasm & SCAD

Sex Specific Disparities in Management

Delay

- Delay in presentation

- Longer door to needle
time

Less Pharmacotherapy

- Less likely to get
Aspirin during ACS
- Less likely to get
Aspirin, ACEI/ARB and
statin on hospital
discharge
- Less adherent to meds
due to higher side
effects

Less Revascularization

- More likely to be
treated medically
- Less likely to have
angiography or PCI
- Less likely to be
referred for CABG
- Less likely to receive
arterial grafts during
CABG

Less Rehabilitation

- Less likely to be
referred or enrolled in
cardiac rehabilitation

- Less likely to complete
cardiac rehabilitation

Sex Specific Disparities in Outcomes

More events & death

Higher risk of
rehospitalization
Young women with ACS had
4-fold higher risk of recurrent
MI, HF & in-hospital mortality
compared to men

More chest pain

Black women more likely to
have angina 1 year after MI

Worse QOL

Women have more
depressive symptoms &
worse QOL after MI

Niti Aggarwal

Presentation, Management and Outcomes of Acute Coronary Syndrome in Women Compared to Men. Women with ACS are often older, have more comorbidities, and often experience atypical symptoms. This results in delayed diagnosis. Additionally women are often prescribed less pharmacotherapy, revascularization and rehabilitation. This contributes to disparities in outcome with women experiencing more morbidity and mortality. ACEI, angiotensin converting enzyme inhibitor; ACS, acute coronary syndrome; ARB, angiotensin receptor blocker; CABG, coronary artery bypass grafting; HF, heart failure; MI, myocardial infarction; MINOCA, myo-cardial infarction with normal coronary arteries; PCI, percutaneous coronary intervention; QOL, quality of life; SCAD, spontaneous coronary artery dissection. *Image courtesy of Niti R. Aggarwal.*

Chapter 5

Spontaneous Coronary Artery Dissection

Marysia S. Tweet* and Rajiv Gulati

Clinical Case

A 47-year-old female with history of migraines noted chest discomfort shortly following a run. Initially a stuttering discomfort, it soon was accompanied with a severe chest pressure, nausea, and generalized lethargy. She called for the emergency medical services; initial physical exam was normal, and on-site electrocardiogram showed nonspecific T-wave changes. She was brought to the emergency department (Figure 1A). Her first cardiac troponin-T was normal at <0.02 ng/L, but the following two levels were abnormal with a rising pattern concerning for acute coronary syndrome. Coronary angiography showed a smooth stenosis of the obtuse marginal coronary artery with normal distal coronary blood flow (Figure 1B and C). Initially thought as spasm, intracoronary nitroglycerin was delivered without change in stenosis. Intracoronary imaging showed intramural hematoma and the absence of atherosclerosis, indicating a diagnosis of spontaneous coronary artery dissection (SCAD). What additional testing is recommended, and how would you manage her?

Continued

---

* Dr. Tweet is supported by the Office of Research on Women's Health (NIH HD65987).

Clinical Case—cont'd

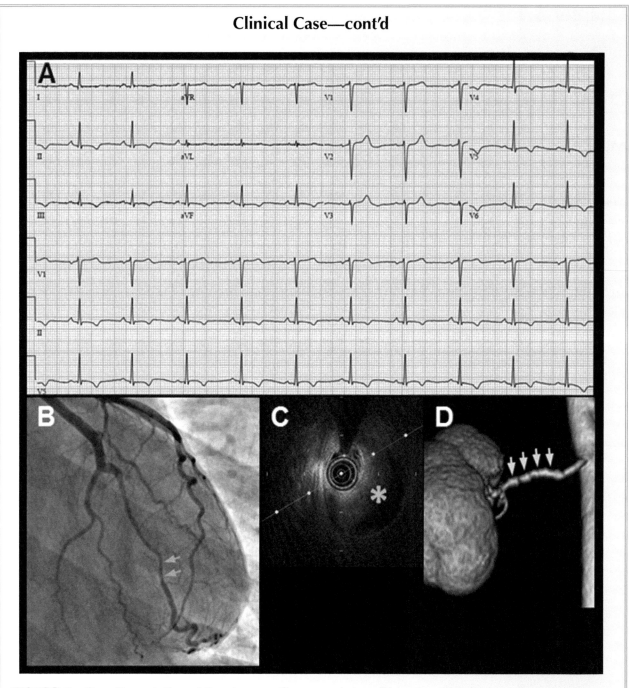

FIGURE 1   Case Presentation of Spontaneous Coronary Artery Dissection (SCAD). (A) Electrocardiogram with nonspecific T-wave changes in the inferior and lateral precordial leads. (B) SCAD of the obtuse marginal artery *(arrows)*. (C) Intramural hematoma on optical coherence tomography *(asterisk)*. (D) Renal fibromuscular dysplasia (FMD) *(arrows)*.

Abstract

Awareness and diagnosis of spontaneous coronary artery dissection (SCAD) are increasing, as this has been recently realized as an important cause of myocardial infarction in women. The mechanism of SCAD is different than atherosclerosis, with a relatively younger patient demographic and natural history. These differences influence the approach to management. Herein, current knowledge regarding SCAD is reviewed with a focus on presentation, diagnosis, and management.

Definition and Epidemiology

Spontaneous coronary artery dissection (SCAD) is a nonatherosclerotic etiology of acute coronary syndrome (ACS) [1, 2] as demonstrated in clinical case (Figure 1). SCAD causes myocardial ischemia or infarction due to a non-iatrogenic dissection or intramural hematoma of the coronary artery with luminal narrowing and subsequent reduction in coronary blood flow (**Figure 2**). Prevalence of SCAD in angiographic series of ACS cases is 0.16–4%, although this estimate is much higher (8.7–35%) among young women presenting with ACS (**Figure 3**) [1–7]. Although incidence is poorly understood, an Olmsted county study found the incidence of SCAD from 1979 to 2009 to be 0.26 per 100,000 persons [8]. With heightened awareness and recognition amongst patients and healthcare teams, identification of hematoma with intravascular imaging, and the development of

SCAD-specific angiographic classification, SCAD is now considered to be more common that previously reported [1].

Most who have SCAD are women (81–95%) with a mean age of 44–53 years, although it can affect an age span from the teenage years to 80s [1, 2]. SCAD less commonly occurs in men but when it does, it has been often associated with extreme exertion [8, 9]. Concomitant conditions and provoking factors include extracoronary vascular abnormalities (EVA) such as fibromuscular dysplasia (FMD), extreme stress, pregnancy or postpartum state, connective diseases, migraines, and familial predisposition. Typical atherosclerotic risk factors are not as frequent among SCAD patients compared to patients with atherothrombotic myocardial infarction (MI) [1, 2].

Distribution of SCAD

SCAD can occur in any coronary artery, including branch vessels such as septal perforators and right ventricular branches. However, SCAD is observed to occur most often in the left anterior descending coronary and/or its diagonal and septal branches (**Figure 4**). The left circumflex, obtuse marginal, and ramus branches are involved in 15–45% of cases and the right coronary artery and its branches are involved in 10–39% of cases. SCAD usually involves the mid and distal segments of the coronary arteries. Only up to 4% of cases involve the left main coronary artery. Multivessel SCAD, or SCAD involving more than one coronary artery, occurs in 9–23% of reported cases [1, 10].

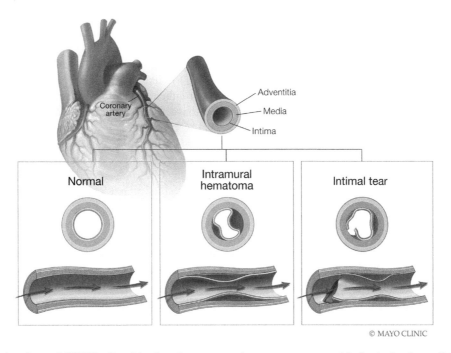

FIGURE 2 Mechanism of SCAD. Graphic showing a normal coronary artery with the intimal, medial, and adventitial layers of a normal coronary, intramural hematoma, and an intimal tear. *Reproduced with permission from Mayo Foundation for Medical Education and Research. All rights reserved.*

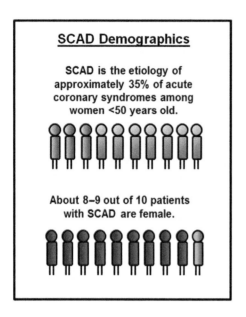

FIGURE 3 SCAD Demographics. Nonatherosclerotic spontaneous coronary artery dissection (SCAD) predominantly affects women.

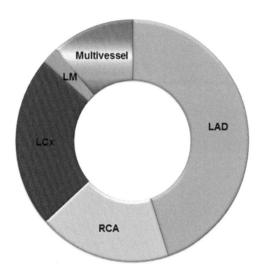

FIGURE 4 Coronary Distribution of Spontaneous Coronary Artery Dissection. LAD, left anterior descending artery; LCx, left circumflex; LM, left main; Multivessel, multiple coronary arteries; RCA, right coronary artery.

Pathophysiology

Much remains unclear regarding the pathophysiology of SCAD. One hypothesis is that SCAD may initiate as dissection of the coronary artery intima and/or media with subsequent luminal compressive occlusion from blood accumulation within the media (**Figure 2**). Alternatively, hematoma within the media/adventitia may occur without an associated intimal dissection. In some cases, an intimal tear (rupture) may occur secondary to pressure from medial he-

matoma accumulation [11]. Early pathologic studies have suggested that hypereosinophilia may be associated with vessel dissection [12, 13], but this association has been refuted [14]. Other studies assessing histopathology of SCAD have identified: relative proliferation of the coronary adventitial vasa vasorum [15]; elastic fiber and collagen fragmentation and loss of muscle cells consistent with cystic medial necrosis [14]; and intimal fibroplasia suggestive of coronary FMD [16].

Genetic Predisposition Including Heritable Connective Tissue Diseases

Approximately 5.1–8.2% of patients with SCAD have a heritable connective tissue disease, the most common of which are vascular Ehlers-Danlos IV (vascular subtype), Marfan syndrome, and Loeys-Dietz syndrome [17, 18]. Since these specific diseases may affect frequency and type of future care along with familial counseling, it is reasonable for SCAD patients to consider assessment by a geneticist and discussion regarding genetic testing. While genetic panels vary, genes commonly tested for mutations on aortopathy comprehensive panels are as follows: COL3A1, SMAD3, FBN1, FBN2, SMAD3, TGFβ, TGFβR1, TGFβR2, ACTA2, CBS, and NOTCH1. Other genetic mutations have been found in the context of SCAD including PKD1 in a patient with autosomal dominant polycystic kidney disease and LMX1B in a patient with Nail-patella syndrome [18].

Registry data suggest that only 1% of patients with history of SCAD have a relative with SCAD [19]. The degree of these relationships varies—examples of pairs include mother-daughter, sister-sister, niece-aunt—and not all family members have SCAD. These findings are consistent with complex inheritance patterns, and the occurrence of SCAD is likely a complex process with both genetic and environmental inciting factors [19]. Recent genetic studies have found a common variant in a risk allele of the PHACTR1/EDN1 gene which is associated with SCAD and FMD (**Figure 5**) [20]. In addition, whole exome sequencing has revealed an association of both familial and sporadic SCAD with variants in the TLN1 gene, which is responsible for encoding a large cytoplasmic protein involved in arterial structure and integrity [21]. A whole exome sequencing study of the Chinese Han population reported variants in TSR1 among SCAD patients which is expressed in the coronary artery tissues on histochemistry [22]. This finding has not been replicated; however, 82% of the 85 SCAD patients in that study were male with comorbidities of diabetes, hypertension, and dyslipidemia in 28%, 52%, and 34%, respectively, indicating different patient demographics in that SCAD cohort compared to other studies. A recent genome-wide association study identified five replicated risk loci and candidate genes for SCAD; these risk loci were of large effect size and most are also associated with arteriopathies

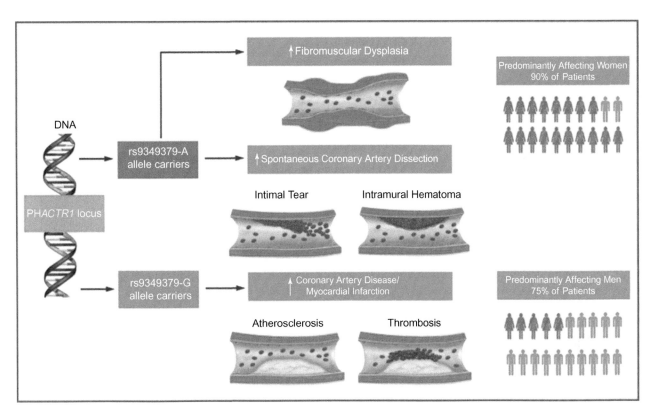

FIGURE 5   Genetic Insights Into Spontaneous Coronary Artery Dissection DNA, deoxyribonucleic acid. *Reproduced with permission from [20].*

[22, 23]. These findings are hypothesis generating, particularly with regard to better understanding the mechanism of SCAD although at this time, SCAD-specific genetic testing is not clinically tested or available. Because of the low likelihood that SCAD will affect a relative, family members do not undergo routine screening. Regardless, family members are encouraged to be aware of their relative's medical history.

Precipitating Factors

Several other factors are associated with SCAD (**Figure 6**). The role of both endogenous and exogenous hormones in the predilection and evolution of SCAD is poorly understood [24]. The female sex hormones are implicated in SCAD as it predominantly affects women and occurs during and after pregnancy. While case reports have associated SCAD with exogenous hormone use, causation is uncertain particularly since exogenous hormones are used frequently in the general population [1]. Some have suggested that perhaps the variation of hormones, rather than the mere presence, may contribute to SCAD, although further research is necessary to further explore this hypothesis [24]. Emotional and physical stressors also have been associated with the precipitation of SCAD, although they are challenging to quantify. A Canadian cohort of 750 patients reported the presence of

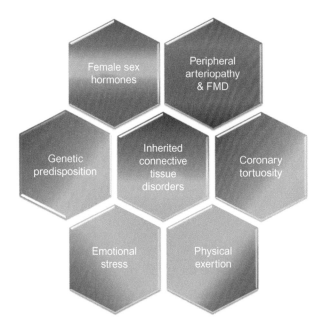

FIGURE 6   Contributing Factors to Spontaneous Coronary Artery Dissection. FMD, fibromuscular dysplasia.

emotional and physical stress in 50% and 29% of patients, respectively [10]. In a cohort of 585 patients, 41% described having enough stress to affect health at time of SCAD [25]. Similarly, in a cohort of 113 patients it was found that 27%

associated SCAD symptom onset during exercise and 31% during emotional stress [26]. In addition, studies have observed that the association of physical exertion with SCAD is more common among men, although women have also reported this association [8, 9, 26].

Clinical Presentation

With rare exception, SCAD presents as an ACS: non-ST segment elevation myocardial infarction (NSTEMI), ST segment elevation myocardial infarction (STEMI), and cardiac arrest [1, 2]. Most patients present with typical ischemic symptoms including chest pressure, nausea, diaphoresis, and referred sensations or pain to the arm, neck, jaw, or back. Some present with atypical symptoms such as dizziness, fatigue, headache, and syncope [1]. The electrocardiogram can be normal or show changes such as T-wave inversion, ST segment elevation, q-waves, and occasionally ventricular arrhythmias. There are no known electrocardiographic features specific to SCAD. Cardiac enzymes levels are almost universally elevated, although initial troponin levels may be within normal range, especially among those patients who seek evaluation early [1, 27].

Diagnosis

SCAD commonly occurs in young women. In the absence of conventional atherosclerotic risk factors, there is risk that these women may be misdiagnosed and dismissed home from the emergency department. Therefore, the diagnosis of SCAD requires a high level of clinical suspicion with a corresponding work-up for ACS.

Invasive coronary angiography is the gold standard, diagnostic study for SCAD. To help facilitate the description of SCAD, it has been categorized into the following subtypes on coronary angiography:

Type 1 with a visible disruption of the coronary artery wall on angiography,
Type 2A with a smooth stenosis consistent with intramural hematoma with eventual improvement in distal coronary caliber,
Type 2B with a smooth stenosis consistent with intramural hematoma with tapering and no distal flow,
Type 3 which resembles focal atherosclerosis, and
Type 4 which is a total occlusion due to SCAD and may resemble atherosclerosis or thrombus (**Figure 7**) [2, 28].

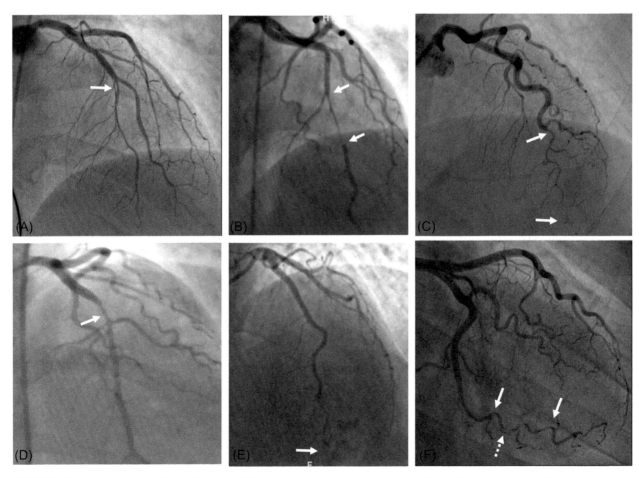

FIGURE 7   Angiographic Classification of Spontaneous Coronary Artery Dissection. Examples of the angiographic classification including type 1 (A), type 2A (B), type 2B (C), type 3 (D), and type 4 (E). Panel F shows an angiogram with features of both types 1 and 2A. *Reproduced with permission from [2].*

In cases of diagnostic uncertainty and suspected type 3 dissection, intracoronary imaging with optical coherence tomography (OCT) or intravascular ultrasound (IVUS) should be considered to visualize the coronary artery wall and assess for atherosclerosis, coronary wall disruption, and/or intramural hemorrhage [2]. However, these diagnostic tools are not available in all catheterization laboratories, and intravascular imaging is associated with risk of propagation of coronary dissection or iatrogenic dissection [29–32]. While both IVUS and OCT can visualize the coronary wall, they each have strengths and limitations. As compared to the high 10–15 μm resolution of OCT, the resolution of IVUS is only 100–150 μm. The OCT catheters have a small profile with a faster rotational and pullback speed. However, OCT requires a contrast bolus to displace the intracoronary blood, whereas IVUS does not [33, 34].

Coronary tortuosity is highly prevalent among patients with SCAD, when compared to age and sex-matched controls [35]. While this has implications for possible pathophysiologic mechanisms and predisposition for SCAD, the presence of coronary tortuosity should prompt the team to further scrutinize the coronary vessels, particularly small branches and distal vessels, for subtle SCAD. This is particularly relevant as SCAD due to a LAD hematoma may be misdiagnosed as Takotsubo cardiomyopathy, although the possibility of concurrent SCAD and Takotsubo cardiomyopathy has been proposed for some patients especially if there was stress preceding the onset of symptoms [36, 37].

Noninvasive coronary computed tomography angiography (CCTA) can visualize SCAD, especially in the proximal vessels. However, it is not the recommended first diagnostic approach in those presenting with ACS, even if SCAD is high on the differential. An obvious dissection plane usually is not visible on CCTA, contributing to the possibility of misdiagnosis; other limitations include motion artifact and potentially misleading collateral supply of contrast to obstructed vessels [35, 38]. CCTA cannot always exclude SCAD given typical involvement of small to distal vessels that may be poorly visualized by CCTA. The most common features of acute SCAD on CCTA include an abrupt luminal stenosis, intramural hematoma, and tapered luminal stenosis, with appearances which can be similar to noncalcified plaque [38]. However, CCTA can be helpful as a noninvasive approach to further assess progression vs. healing of the coronaries in those with persistent symptoms [39]. Subendocardial contrast delay suggesting recent ischemia and cardiac structure can also be identified with CCTA. Subsequent healing visualized on CCTA may also facilitate diagnosis on SCAD in retrospect [40].

In a series of 277 SCAD patients, the left ventricular ejection fraction was >50% in the majority of patients (74%), as might be expected among patients who have SCAD in a distal or branch vessel [41]. Cardiac magnetic resonance (CMR) imaging may be used to help discern the diagnosis among patients in whom the occurrence of SCAD is uncertain with the aim of finding abnormalities in the distribution of a suspicious coronary artery. However, it is important to note that some patients with SCAD can have normal findings on CMR despite recent SCAD [42]. A recent CMR series of 158 SCAD patients found that the majority of patients had small or no infarctions on follow-up imaging with 39% of patients without delayed gadolinium enhancement [43]. A series of 43 SCAD patients with ventriculograms showed wall motion abnormalities out of proportion to the coronary involvement in 24 (55.8%) patients [37]. Whether this may have represented stunning or injury from unappreciated coronary hematoma remains to be further studied. Therefore, in the acute setting, CMR may help to distinguish concomitant Takotsubo in those with left ventricular dysfunction that is out of proportion to the coronary territory affected by SCAD.

Acute Inpatient Management

With the absence of SCAD-specific trials and incomplete understanding of pathophysiology of SCAD, initial treatment upon arrival to the hospital relies on the American College of Cardiology/American Heart Association (ACC/AHA) expert consensus guidelines for acute ACS [44, 45]. However, it is imperative to remember that extrapolation of therapy and management of ACS guidelines focused on atherosclerotic plaque do not necessarily apply to SCAD vessels due to the differing mechanism of myocardial ischemia. Presently, initiation of dual antiplatelet therapy (DAPT), intravenous heparin, beta blockade, and statins are all advised until the diagnosis of SCAD is made, at which time an individualized approach to ongoing medical therapy should be taken. Advanced therapies such as ventricular support devices, extracorporeal membrane oxygenation, and vasopressors are used as indicated for ACS and associated complications [1].

Unlike atherosclerosis, SCAD frequently heals, with evidence even within the first week of presentation [1, 2, 11]. More than 90% of SCAD lesions heal by 1 month after SCAD [10]. Therefore, if a SCAD patient is hemodynamically stable without significant obstruction of coronary artery blood flow [e.g., thombolysis in myocardial infarction (TIMI) flow of 2–3], management with medications alone is recommended (**Figure 8**). DAPT is continued during the acute period in conservatively managed patients, although timing of its use is controversial for SCAD due to the lack on intracoronary thrombus. Intravenous heparin is often stopped after the diagnostic angiogram in conservatively managed patients due to concerns for hematoma expansion, although practice may vary dependent on individual factors as data are lacking regarding acute SCAD management. It is important to note that in a study of 240 participants with SCAD managed conservatively, 17.5% had early (usually within 1 week)

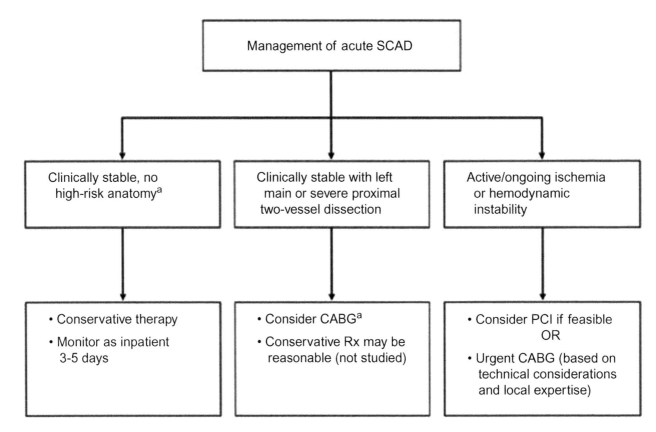

FIGURE 8   Algorithm for Management of Acute Spontaneous Coronary Artery Dissection. CABG, coronary artery bypass grafting; PCI, percutaneous coronary intervention; Rx, management; SCAD, spontaneous coronary artery dissection. [a]Left main or proximal 2-vessel coronary artery dissection. *Reproduced with permission from [1].*

worsening of ischemia due to extension of the SCAD lesion. A recent study of 750 SCAD patients observed that 2.3% of the 648 patients who were conservatively managed went on to revascularization during the admission. In that same study, 18 (17%) of the 106 patients underwent unplanned percutaneous intervention [10].

Revascularization is recommended in the setting of major vessel occlusion or flow reduction, with clinical concern for ongoing ischemia/infarction including pain, arrhythmia, or hemodynamic disturbance. No single percutaneous interventional strategy is recommended over others, but it should be recognized that the goal of revascularization is acute restoration of flow. Observational studies and case series have demonstrated natural healing of SCAD vessels treated conservatively, and this likely applies to vessels treated with balloon angioplasty. Stent placement may be less important in SCAD given the favorable natural history together with the risk of stent-induced worsening of dissection, risk of undersizing of stents in the setting of diffuse intramural hematoma, and longer-term risks of restenosis. CABG as an acute strategy should be considered in selected cases, recognizing good early clinical outcomes but late occurrence of graft occlusion due to spontaneous healing of native vessels [1, 2, 32].

Percutaneous coronary intervention (PCI) is associated with higher than expected complications than for atherosclerotic ACS [32]. One study conducted prior to the recommendation of conservative treatment for SCAD found a procedural failure rate of 53% among 89 SCAD patients undergoing PCI with 13% requiring urgent coronary artery bypass grafting for bailout [32]. This study used SCAD-specific definitions for procedural success. Similarly, another study of 103 SCAD patients who underwent PCI found that only 29% of the procedures were successful, with 41% deemed as "partial success" and 30% as unsuccessful [10]. Challenges unique to PCI for SCAD include an inability to cross the lesion with a wire, unintentional wire entry into the false lumen, and extension of hematoma or worsening of intimal dissection with balloon dilation [1, 2, 32]. There is also increased risk of proximal coronary catheter-induced dissection during both diagnostic angiography and interventions [29]. However, improving coronary blood flow can be critical and, for some, lifesaving. Several approaches to address the challenges of PCI in those with SCAD have been described. One approach includes stenting proximal and distal to the culprit stenosis to reduce propagation of intramural hematoma, direct stenting without balloon predilation to reduce propagation, or plain old balloon angioplasty to improve coronary flow [1]. Intracoronary imaging can also be used to assist with stent placement, including confirming that a stent is adequately deployed. Cautious use of a cutting balloon to decompress an intramural hematoma may help to improve coronary blood flow, although risks include propagation of a coronary tear or coronary perforation [46, 47]. A technique used for treatment of chronic total occlusion led to a successful result for a patient in whom the intracoronary wire was noted to be in the false lumen [48].

If a patient undergoes percutaneous intervention, DAPT is continued. However, for medically managed patients, level and duration of antiplatelet therapy are debatable when considering the presence of hematoma and lack of informative data. Some in practice will therefore dismiss on low-dose aspirin alone, although this is an area to be further studied [1, 2]. Due to risk of interval progression of SCAD requiring revascularization within the first 6 days, close inpatient monitoring is warranted for those

patients conservatively managed [11]. Therefore, 3–5 days of monitoring are recommended for conservatively treated patients by consensus guidelines. Other medications such as beta blockers, angiotensin-converting enzyme inhibitors, and antianginal therapies are to be started and titrated according to other factors such as left ventricular function, blood pressure, heart rate, and persistent symptoms.

Outpatient Medications

At dismissal and outpatient follow-up, medications are tailored to the individual (**Figure 9**) [1, 2, 49]. DAPT is continued for those with history of percutaneous intervention. Contrary to guidelines for atherosclerotic ACS, the consensus for SCAD patients who are conservatively managed without angioplasty or stenting is that DAPT can be de-escalated to aspirin prior to 1 year. This recommendation

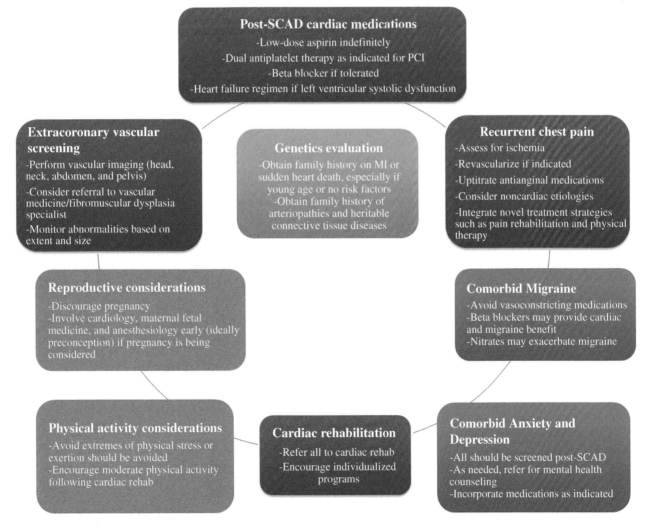

FIGURE 9   Outpatient Care for Patients With Spontaneous Coronary Artery Dissection. MI, myocardial infarction; PCI, percutaneous coronary intervention; SCAD, spontaneous coronary artery dissection. *Reproduced with permission from [49].*

CH
5

is based upon expert consensus, the observation that intra-coronary thrombus is uncommon among those with SCAD, and theoretical concern that DAPT may expand intramural hematoma. However, there are currently no data to indicate benefit or harm of DAPT or aspirin. Chronic anticoagulation such as warfarin is warranted only if there is an indication beyond SCAD such as left ventricular thrombus or atrial fibrillation. Optimal medical therapy including beta blockers and angiotensin-converting enzyme inhibitors is pertinent, with most benefit expected in those with reduced ventricular systolic function. Although further studies are needed, an observational study, comprising 327 patients, found that SCAD patients taking beta blockers had decreased risk of subsequent recurrence with a hazard ratio of 0.36, $P=0.004$ [50, 51]. As SCAD patients often have migraine history, beta blockers may occasionally improve migraine-related symptoms [25]. Hypertension has been associated with increased risk of recurrent SCAD, highlighting the pertinence of managing hypertension if present [50]. For those patients with continued ischemic symptoms, antianginal therapies such as short- and long-acting nitrates, calcium channel blockers, and ranolazine may be beneficial. Occasionally, patients will continue to have persistent pain for which work-up of other etiologies and noncardiac treatment regiments may need to be considered. Statin therapy has been used if coexisting hyperlipidemia is present; otherwise its utility remains in question and is not routinely prescribed in those with history of SCAD with a normal lipid profile [1, 2, 49, 50].

Cardiac Rehabilitation

Referral and participation in cardiac rehabilitation is the standard of care for all patients post-ACS, including those with SCAD. Unfortunately, in a study of 354 patients, the most common reason (67%) for the 85 patients who did not complete cardiac rehabilitation was lack of referral by the healthcare provider [52]. Reluctance to refer may be related to clinician concern that exercise could incite additional SCAD due to its association with extreme exertion. Lack of risk factors or baseline physical activity may also misguide clinicians into thinking that cardiac rehabilitation is not necessary. However, studies thus far have shown that cardiac rehabilitation is both safe and beneficial for patients with SCAD in regards to both physical and mental well-being [53, 54]. As some with SCAD may already have a higher than average exercise routine, the cardiac rehabilitation program can be tailored to the needs of that individual [55] along with considerations suggested by the AHA SCAD consensus (**Figure 10**) [1].

Symptom Recurrence

Approximately half of patients will experience recurrent chest discomfort, dyspnea, or other symptoms after SCAD. Recurrent symptoms do require consideration of additional work-up, with a strategy as suggested by the 2018 AHA Scientific Statement on SCAD (**Figure 11**) [1]. Ischemic causes of the symptoms might include interval worsening of SCAD occlusion, endothelial, and/or microvascular

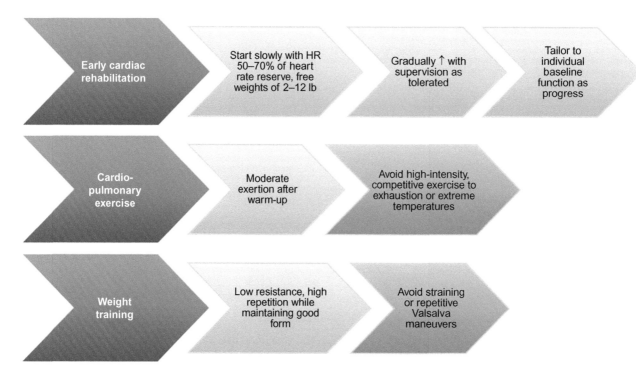

FIGURE 10 Exercise Considerations After Acute Coronary Syndrome Due to Spontaneous Coronary Artery Dissection. HR, heart rate.

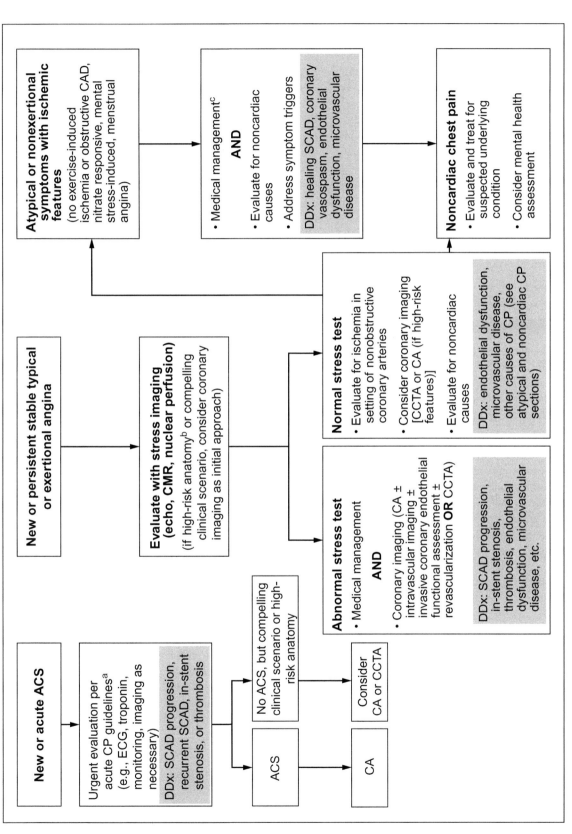

FIGURE 11 **Evaluation and Management of Chest Pain After SCAD.** ACS, acute coronary syndrome; CA, coronary angiography; CAD, coronary artery disease; CCTA, coronary computed tomography angiography; CMR, cardiac magnetic resonance; CP, chest pain; DDx, differential diagnosis; ECG, electrocardiogram; Echo, echocardiography; SCAD, Spontaneous Coronary Artery Dissection. [a]Current guidelines. [b]High-risk anatomy indicates SCAD affecting the left main or two proximal coronary arteries. [c]Medical management for post-SCAD CP without obstructive disease: long-acting nitrates, calcium channel blockers, or ranolazine. *Reproduced with permission from [1].*

dysfunction, in-stent thrombosis, or stenosis. Pericarditis and/or costochondritis may also need to be considered. However, patients can often have a reassuring evaluation despite ongoing symptoms. One may speculate that the symptoms of pain may be related to underlying adventitial healing as the majority of SCAD vessels heal on interval follow-up. However, noncardiac etiologies should also be considered. These may include medication side effects, gastrointestinal reflux, musculoskeletal pain, and neurologic pain [1]. A subgroup of patients self-reported catamenial chest pain symptoms, the etiology of which is poorly understood, and is speculated as possibly being the result of a relative estrogen withdrawal with associated coronary vasospasm or microvascular dysfunction. Current management strategies for those patients include uptitration of antianginals prior to menstruation. Whether hormonal supplementation would be beneficial or harmful is not well understood [24, 56, 57].

Symptom management may include the addition and/or uptitration of antianginals which can require an ongoing process of trial and error. Since many patients with SCAD also have history of migraine headaches, beta blockers or calcium channel blockers are common medications to manage angina. For similar reasons, if long-acting nitrates are used, starting at the smallest possible dose can sometimes minimize symptoms without causing severe migraines or hypotension. If nitrates are not tolerated, some patients find relief with ranolazine.

Occasionally, patients will experience symptom relief with medications commonly used for neuropathic or chronic pain syndromes such as gabapentin, pregabalin, or antidepressants such as the serotonin and norepinephrine reuptake inhibitors and tricyclic antidepressants. Certainly, any deleterious side effects of such medications should be accounted for based on the patient history with counseling as deemed appropriate. Use of proton pump inhibitors or other medications for reflux may also reduce symptoms, especially if patients are experiencing aspirin-related gastritis. Finally, as ACS is a significant life event, antidepressants, anxiolytics, and dedicated therapy may benefit the health of some patients with the potential to improve symptom management and coping strategies. This may be particularly important for those patients struggling with depression, anxiety, or posttraumatic stress disorder (PTSD) following SCAD.

Fibromuscular Dysplasia and Other Extracoronary Vascular Abnormalities

FMD and other EVA such as aneurysms, dissections, and dilatations are common among those with history of SCAD [1]. FMD is a noninflammatory and nonatherosclerotic arterial disease predominantly diagnosed on vascular imaging by either invasive angiography, noninvasive computed tomography angiography (CTA), or magnetic resonance angiography [58]. Prevalence of FMD and other arteriopathies has ranged from 25% to 86% in published series [1, 26]. The wide variation may be in part due to differences in the sensitivity of various screening modalities and lack of comprehensive imaging in all SCAD patients. The most common arteries involved include the renal artery (29–74%), the carotid artery (22–52%), and the iliac artery (19–82%) [26, 59–61], although FMD can affect nearly any mid-sized artery [58, 59]. Because the discovery of an asymptomatic aneurysm or dissection may affect future monitoring and care recommendations, all patients with history of SCAD are advised to undergo at least one time head to pelvis comprehensive imaging for FMD and other EVA. In those with abnormalities, input by vascular medicine colleagues may help guide further monitoring and care based on the imaging findings [1].

Pregnancy-Associated SCAD and Reproductive Counseling

SCAD predominantly affects women, many of whom are of childbearing age, and approximately 4.7–18% of women with SCAD experience it during or after a pregnancy [2, 8, 10]. Furthermore, among women who have MI during or after pregnancy, SCAD is a common etiology. For instance, a case series of 150 published and compiled cases of MI associated with pregnancy found SCAD as the etiology in 43% [62]. However, in a cohort study of 55 million pregnancy-related hospitalizations of which 4471 were due to MI, only 14.5% were attributed to coronary dissection [63]. A Canadian cohort study of 4.4 million pregnancies from 2008 to 2012 found an incidence of 1.8 SCADs per 100,000 pregnancies [64]. Several series have demonstrated that pregnancy-associated SCAD is associated with a deleterious presentation including proximal and left main vessel involvement, large infarctions, cardiogenic shock, and low ventricular systolic function with presenting features that are more severe when compared to SCAD not associated with pregnancy (**Figure 12**). Pregnancy-associated SCAD (P-SCAD) can occur anytime during or after pregnancy, with most cases reported during the early postpartum period. Compared to the US data, women with P-SCAD were more often multiparous, with history of infertility therapies and preeclampsia. Whether these observations are risk markers vs. confounders is yet to be better understood [64–67].

Since the exact risk incurred by future pregnancy is unknown, avoidance of future pregnancies is generally advised [1]. The literature is limited to a small series in which one out of eight SCAD women with pregnancy after SCAD experienced another SCAD in the postpartum period. Unfortunately, her SCAD was life-threatening, extended to

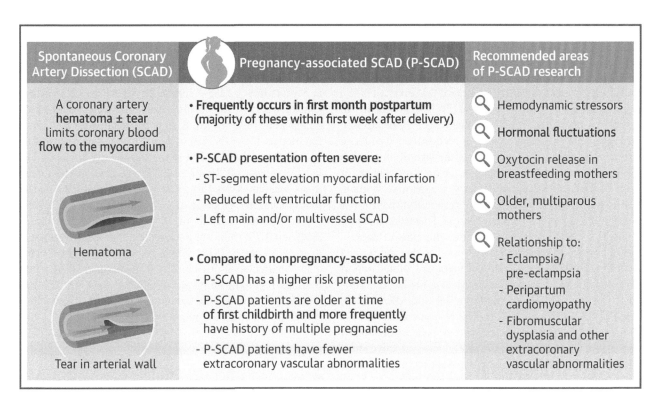

FIGURE 12 **Features Observed in Women With P-SCAD.** P-SCAD, pregnancy-associated spontaneous coronary artery dissection; SCAD, spontaneous coronary artery dissection. *Reproduced with permission from [65].*

the left main coronary artery, and required emergency coronary artery bypass grafting [68]. Importantly, three (38%) of these women were initially misdiagnosed as having MI from coronary vasospasm [68]. Another series observed that 2 of 11 women with pregnancy after SCAD experienced SCAD [69]. A recent cohort study incorporating more women with history of pregnancy after SCAD found that two of 22 women with 31 pregnancies experienced recurrent SCAD [70]. Therefore, pregnancy after SCAD is without risk. However, the desire for biological children may be such that some women pursue pregnancy, women may conceive unintentionally, or they may not even be fully aware that SCAD was the cause of a prior MI. In such situations, a tertiary team approach integrating cardiology, maternal fetal medicine, and obstetric anesthesiology expertise is appropriate. Further care includes review of medications with adjustment as indicated based on safety, cessation of teratogenic medications such as statins and angiotensin-converting enzyme inhibitors (ACEIs)/angiotensin receptor blockers (ARBs), and careful planning regarding delivery (vaginal preferred unless there are obstetric indications) based upon current consensus for the cardiovascular care of pregnant women [57, 71–73].

Given the unknown risk and potential association of estrogen and progesterone and variations of these hor-

mones with SCAD, nonhormonal forms of contraception such as partner vasectomy, tubal ligation, or intrauterine devices (including those with local hormone release) are preferred over hormonal options when possible among SCAD patients [1]. Patients who are on hormone replacement therapy at the time of the SCAD are generally advised to discontinue these medications unless there is a compelling clinical indication to continue [1, 74]. Postmenopausal hormone therapy, which involves less potent formulations than contraceptives, is limited to the smallest dose for the shortest time period [1].

Psychological Health Considerations

For most, SCAD is a life-changing event. As such, post-SCAD symptoms, fears about what is known and unknown and the event itself and risk of recurrence may provoke, contribute to, or exacerbate anxiety, depression, and post-traumatic stress disorder. One study of 158 SCAD patients found that 33% of survey respondents were receiving treatment for depression and 38% for anxiety, many cases of which were predating the SCAD event. In that study, younger women and those with postpartum SCAD had higher measures of both depression and anxiety [75]. In another study of 512 SCAD patients, 82% screened positive for trauma with "serious, life-threatening illness (heart attack, etc.)" as the most significant contributor in 53%. Patients indicated at least mild symptoms of PTSD (28%), depression (32%), and anxiety (41%) [76, 77]. Another study of 14 female SCAD patients invited to a psychosocial group found high prevalence of stress (93%), insomnia (57%), anxiety (71%), depression (36%), and PTSD (43%) [76, 78]. Frequently in the context of limited data, SCAD patients may be advised to pursue lifestyle modifications as recommended to those with atherosclerotic CAD. However, the lack of modifiable risk factors and uncertainty may be anxiety-provoking for these patients. Women, particularly postpartum women, have a high burden of caretaking responsibilities further contributing to their psychosocial stress. Therefore, it is pertinent that patients are evaluated for symptoms of depression, anxiety, and posttraumatic stress disorder and treated accordingly with medications and therapy. Online support groups and nonprofit organizations such as the SCAD survivor Facebook page, SCAD Research, Inc., SCAD Alliance, Beat SCAD, and WomenHeart: The National Coalition for Women with Heart Disease may also be sources of support for patients with history of SCAD. The most effective strategies for therapy and behavioral intervention in this unique patient cohort require further attention and study.

Long-Term Outcomes

Overall, mortality is low after the diagnosis of SCAD in multiple series, although these retrospective data are generated from SCAD survivors and cannot, for example, account for those who may have died from undiagnosed SCAD. When compared to those with ACS due to atherosclerosis, those diagnosed with SCAD have better survival in two series. This is likely related to lack of comorbidities; however, these series are limited by a small number of patients [8, 79]. The risk of major adverse cardiovascular events remains present, much of which is related to recurrent SCAD [1, 2]. Other adverse events include recurrent MI (unrelated to SCAD, e.g., in-stent thrombosis), heart failure and associated complications, stroke or transient ischemic attack, and recurrent chest pains [1, 2]. Overall rates of major adverse cardiac events (MACE) in published series vary (in part because of the heterogeneous definitions of MACE): 2.5% at 30 days in a prospective Canadian cohort [10]; 14.6% at 6 years in an Italian series [80]; 37% at 5 years in a Japanese series [4]; 19.9% at 3.1 years in a Canadian series [50]; and 10-year Kaplan-Meir estimates of 47.4% in a US series [8]. Risk of recurrent SCAD has been reported as 15% (29/189) of patients over a median follow-up of 2.3 years with a 5-year Kaplan-Meir estimate of 27% in a US series [32]; 10.4% (34/327) at a median of 3.1 years in a Canadian series [50]; 11% (7/63) of patients after the first month after a median of 34 months in a Japanese series [4]; 4.7% after a median of 22 months in the Italian series [80]; and 4.7% (3/63) of patients at a median of 34 months in a Swiss series [81]. SCAD recurrence involves previously unaffected arteries in 77–100% of cases [1, 8]. These findings highlight ongoing outpatient follow-up for those with SCAD and provide the impetus for further research as preventative strategies are not well determined.

Conclusion

SCAD is increasingly recognized as an important etiology of MI predominantly in women with minimal atherosclerotic risk. Advancements in the catheterization laboratory, including the use of intracoronary imaging, have facilitated recognition of intracoronary hematoma. Patient engagement and advocacy have enhanced awareness and facilitated research endeavors. Importantly, the etiology, natural history, and management of SCAD are different than that for atherosclerotic disease. While much has been learned about SCAD, ongoing and future research endeavors will likely reveal further insights. Challenges with future research include the need for consistent terminology used among researchers, collaboration and recruitment of a sufficient number of patients into studies, and the transition from predominantly retrospective to prospective study designs [82]. Ongoing efforts aimed toward these goals including collaboration across institutions ultimately may facilitate such research endeavors.

Key Points

1 Prevalence of SCAD in angiographic series of ACS cases is 0.16–4%, although this estimate is higher (8.7–35%) among young women presenting with ACS.

2 SCAD often occurs in young women without conventional risk factors, many of whom have a predisposing arteriopathy. Possible precipitants include extreme physical or emotional distress.

3 SCAD is associated with fibromuscular dysplasia, coronary tortuosity, and other arteriopathies such as aneurysm, dissection, and dilatation.

4 Pregnancy-associated SCAD is associated with a severe presentation including left main dissection, left ventricular systolic failure, and ST-elevation myocardial infarction. It can occur anytime during or after pregnancy, although it most often occurs in the early postpartum.

5 Invasive coronary angiography is the first-line strategy for the diagnosis of SCAD, with adjunctive intracoronary imaging when necessary.

6 The angiographic appearance of SCAD is described by four subtypes; coronary tortuosity is a common feature among those with SCAD.

7 Given spontaneous healing of the SCAD in most patients and risk for catheter-induced complications, medical management approach is favored unless there is concern for hemodynamic instability, ongoing ischemia, or compromise of a large vessel with a significant flow limiting lesion. Medical management is poorly studied and extrapolated from the atherosclerotic ACS and expert consensus.

8 Even though SCAD-related mortality is low, high rates of adverse events including 10–20% risk of recurrent SCAD merit continued attention.

Back to Clinical Case

The case describes a 47-year-old hemodynamically stable female with SCAD and normal distal coronary flow. Ideally it is recommended that she be managed medically conservatively. Emergent coronary revascularization is typically reserved for patients with hemodynamic instability or ongoing ischemia/infarction together with flow reduction in a major vessel. She was discharged home on a medication regimen of daily low-dose aspirin, beta blocker, and low-dose angiotensin-converting enzyme inhibitor. She enrolled in cardiac rehabilitation shortly after dismissal. Outpatient computed tomography angiography (CTA) imaging demonstrated bilateral cervical, vertebral, and renal fibromuscular dysplasia (**Figure 1D**). Her follow-up evaluation included discussion about the knowns and unknowns about SCAD, including the observation of recurrent SCAD in 10–20% of patients.

References

[1] Hayes SN, Kim ESH, Saw J, Adlam D, Arslanian-Engoren C, Economy KE, et al. Spontaneous coronary artery dissection: current state of the science: a scientific statement from the American Heart Association. Circulation 2018;137:e523–57.

[2] Adlam D, Alfonso F, Maas A, Vrints C, Writing Committee. European Society of Cardiology, acute cardiovascular care association, SCAD study group: a position paper on spontaneous coronary artery dissection. Eur Heart J 2018;39:3353–68.

[3] Kim Y, Han X, Ahn Y, Kim MC, Sim DS, Hong YJ, et al. Clinical characteristics of spontaneous coronary artery dissection in young female patients with acute myocardial infarction in Korea. Korean J Intern Med 2019. https://doi.org/10.3904/kjim.2019.118.

[4] Nakashima T, Noguchi T, Haruta S, Yamamoto Y, Oshima S, Nakao K, et al. Prognostic impact of spontaneous coronary artery dissection in young female patients with acute myocardial infarction: a report from the Angina Pectoris–Myocardial Infarction Multicenter Investigators in Japan. Int J Cardiol 2016;207:341–8.

[5] Rashid HNZ, Wong DTL, Wijesekera H, Gutman SJ, Shanmugam VB, Gulati R, et al. Incidence and characterisation of spontaneous coronary artery dissection as a cause of acute coronary syndrome—a single-centre Australian experience. Int J Cardiol 2016;202:336–8.

[6] Nishiguchi T, Tanaka A, Ozaki Y, Taruya A, Fukuda S, Taguchi H, et al. Prevalence of spontaneous coronary artery dissection in patients with acute coronary syndrome. Eur Heart J Acute Cardiovasc Care 2013;5(3):263–70.

[7] Saw J, Aymong E, Mancini GBJ, Sedlak T, Starovoytov A, Ricci D. Nonatherosclerotic coronary artery disease in young women. Can J Cardiol 2014;30:814–9.

[8] Tweet MS, Hayes SN, Pitta SR, Simari RD, Lerman A, Lennon RJ, et al. Clinical features, management, and prognosis of spontaneous coronary artery dissection. Circulation 2012;126:579–88.

[9] Fahmy P, Prakash R, Starovoytov A, Boone R, Saw J. Pre-disposing and precipitating factors in men with spontaneous coronary artery dissection. J Am Coll Cardiol Intv 2016;9:866–8.

[10] Saw J, Starovoytov A, Humphries K, Sheth T, So D, Minhas K, et al. Canadian spontaneous coronary artery dissection cohort study: in-hospital and 30-day outcomes. Eur Heart J 2019;40:1188–97.

[11] Waterbury TM, Tweet MS, Hayes SN, Eleid MF, Bell MR, Lerman A, et al. Early natural history of spontaneous coronary artery dissection. Circ Cardiovasc Interv 2018;11, e006772.

[12] Robinowitz M, Virmani R, McAllister HA. Spontaneous coronary artery dissection and eosinophilic inflammation: a cause and effect relationship? Am J Med 1982;72:923–8.

[13] Day C, Ansford A, Boris T, Milne N, Ong B. Spontaneous coronary artery dissection a post-mortem case series. Pathology 2013;45:S90.

[14] Briguori C, Bellevicine C, Visconti G, Focaccio A, Aprile V, Troncone G. In vivo histological assessment of a spontaneous coronary artery dissection. Circulation 2010;122:1044–6.

[15] Kwon T-G, Gulati R, Matsuzawa Y, Aoki T, Guddeti RR, Herrmann J, et al. Proliferation of coronary adventitial vasa vasorum in patients with spontaneous coronary artery dissection. JACC Cardiovasc Imaging 2016;9:891–2.

[16] Moulson N, Kelly J, Iqbal MB, Saw J. Histopathology of coronary fibromuscular dysplasia causing spontaneous coronary artery dissection. J Am Coll Cardiol Intv 2018;11:909–10.

[17] Henkin S, Negrotto SM, Tweet MS, Kirmani S, Deyle DR, Gulati R, et al. Spontaneous coronary artery dissection and its association with heritable connective tissue disorders. Heart 2016;102(11):876–81.

[18] Kaadan MI, MacDonald C, Ponzini F, Duran J, Newell K, Pitler L, et al. Prospective cardiovascular genetics evaluation in spontaneous coronary artery dissection. Circ Genom Precis Med 2018;11, e001933.

[19] Goel K, Tweet M, Olson TM, Maleszewski JJ, Gulati R, Hayes SN. Familial spontaneous coronary artery dissection: evidence for genetic susceptibility. JAMA Intern Med 2015;175:821–6.

[20] Adlam D, Olson TM, Combaret N, Kovacic JC, Iismaa SE, Al-Hussaini A, et al. Association of the PHACTR1/EDN1 genetic locus with spontaneous coronary artery dissection. J Am Coll Cardiol 2019;73:58–66.

[21] Turley TN, Theis JL, Sundsbak RS, Evans JM, O'Byrne MM, Gulati R, et al. Rare missense variants in TLN1 are associated with familial and sporadic spontaneous coronary artery dissection. Circ Genom Precis Med 2019;12, e002437.

[22] Sun Y, Chen Y, Li Y, Li Z, Li C, Yu T, et al. Association of TSR1 variants and spontaneous coronary artery dissection. J Am Coll Cardiol 2019;74:167–76.

[23] Turley TN, O'Byrne MM, Kosel ML, de Andrade M, Gulati R, Hayes SN, et al. Identification of susceptibility loci for spontaneous coronary artery dissection. JAMA Cardiol 2020;5(8):1–10.

[24] Tweet MS, Miller VM, Hayes SN. The evidence on estrogen, progesterone, and spontaneous coronary artery dissection. JAMA Cardiol 2019;4:403–4.

[25] Kok SN, Hayes SN, Cutrer FM, Raphael CE, Gulati R, Best PJM, et al. Prevalence and clinical factors of migraine in patients with spontaneous coronary artery dissection. J Am Heart Assoc 2018;7, e010140.

[26] Sharma S, Kaadan MI, Duran JM, Ponzini F, Mishra S, Tsiaras SV, et al. Risk factors, imaging findings, and sex differences in spontaneous coronary artery dissection. Am J Cardiol 2019;123:1783–7.

[27] Lindor RA, Tweet MS, Goyal KA, Lohse CM, Gulati R, Hayes SN, et al. Emergency department presentation of patients with spontaneous coronary artery dissection. J Emerg Med 2017;52:286–91.

[28] Saw J. Coronary angiogram classification of spontaneous coronary artery dissection. Catheter Cardiovasc Interv 2014;84:1115–22.

[29] Prakash R, Starovoytov A, Heydari M, Mancini GBJ, Saw J. Catheter-induced iatrogenic coronary artery dissection in patients with spontaneous coronary artery dissection. J Am Coll Cardiol Intv 2016;9:1851.

[30] Macaya F, Salinas P, Gonzalo N, Escaned J. Repeated intracoronary imaging in spontaneous coronary artery dissection. Weigh Benefits Risks 2017;10:2342.

[31] Saw JWL, Alsulaimi A, Fung A, Parsa A, Starovoytov A, Kiani G, et al. Use of intracoronary imaging for diagnosis of spontaneous coronary artery dissection. J Am Coll Cardiol 2019;73:1055.

[32] Tweet MS, Eleid MF, Best PJM, Lennon RJ, Lerman A, Rihal CS, et al. Spontaneous coronary artery dissection: revascularization versus conservative therapy. Circ Cardiovasc Interv 2014;7:777–86.

[33] Waksman R, Kitabata H, Prati F, Albertucci M, Mintz GS. Intravascular ultrasound versus optical coherence tomography guidance. J Am Coll Cardiol 2013;62:S32–40.

[34] Bezerra HG, Attizzani GF, Sirbu V, Musumeci G, Lortkipanidze N, Fujino Y, et al. Optical coherence tomography versus intravascular ultrasound to evaluate coronary artery disease and percutaneous coronary intervention. J Am Coll Cardiol Intv 2013;6:228–36.

[35] Eleid MF, Guddeti RR, Tweet MS, Lerman A, Singh M, Best PJ, et al. Coronary artery tortuosity in spontaneous coronary artery dissection: angiographic characteristics and clinical implications. Circ Cardiovasc Interv 2014.

[36] Hausvater A, Smilowitz NR, Saw J, Sherrid M, Ali T, Espinosa D, et al. Spontaneous coronary artery dissection in patients with a provisional diagnosis of takotsubo syndrome. J Am Heart Assoc 2019;8, e013581.

[37] Duran JM, Naderi S, Vidula M, Michalak N, Chi G, Lindsay M, et al. Spontaneous coronary artery dissection and its association with takotsubo syndrome: novel insights from a tertiary center registry. Catheter Cardiovasc Interv 2019;95(3):485–91.

[38] Tweet MS, Akhtar NJ, Hayes SN, Best PJM, Gulati R, Araoz PA. Spontaneous coronary artery dissection: acute findings on coronary computed tomography angiography. Eur Heart J Acute Cardiovasc Care 2018. https://doi.org/10.1177/2048872617753799.

[39] Tweet MS, Gulati R, Williamson EE, Vrtiska TJ, Hayes SN. Multimodality imaging for spontaneous coronary artery dissection in women. JACC Cardiovasc Imaging 2016;9:436–50.

[40] Guo LQL, Wasfy MM, Hedgire S, Kalra M, Wood M, Prabhakar AM, et al. Multimodality imaging of spontaneous coronary artery dissection: case studies of the Massachusetts General Hospital. Coron Artery Dis 2016;27:70–1.

[41] Franco C, Starovoytov A, Heydari M, Mancini GBJ, Aymong E, Saw J. Changes in left ventricular function after spontaneous coronary artery dissection. Clin Cardiol 2017;40:149–54.

[42] Tan NY, Hayes SN, Young PM, Gulati R, Tweet MS. Usefulness of cardiac magnetic resonance imaging in patients with acute spontaneous coronary artery dissection. Am J Cardiol 2018;122:1624–9.

[43] Al-Hussaini A, Abdelaty AMSEK, Gulsin GS, Arnold JR, Garcia-Guimaraes M, Premawardhana D, et al. Chronic infarct size after spontaneous coronary artery dissection: implications for pathophysiology and clinical management. Eur Heart J 2020;41(23):2197–205.

[44] Amsterdam EA, Wenger NK, Brindis RG, Casey DE, Ganiats TG, Holmes DR, et al. 2014 AHA/ACC guideline for the management of patients with non-ST-elevation acute coronary syndromes: a report of the American College of Cardiology/American Heart Association Task Force on Practice Guidelines. J Am Coll Cardiol 2014;64:e139–228.

[45] O'Gara PT, Kushner FG, Ascheim DD, Casey DE, Chung MK, de Lemos JA, et al. 2013 ACCF/AHA guideline for the management of ST-elevation myocardial infarction: a report of the American College of Cardiology Foundation/American Heart Association Task Force on Practice Guidelines. Circulation 2013;127:e362–425.

[46] Main A, Lombardi WL, Saw J. Cutting balloon angioplasty for treatment of spontaneous coronary artery dissection: case report, literature review, and recommended technical approaches. Cardiovasc Diagn Ther 2019;9:50–4.

[47] Yumoto K, Sasaki H, Aoki H, Kato K. Successful treatment of spontaneous coronary artery dissection with cutting balloon angioplasty as evaluated with optical coherence tomography. JACC Cardiovasc Interv 2014;7:817–9.

[48] Daniels DV, Pargaonkar VS, Nishi T, Tremmel JA. Antegrade dissection re-entry after subintimal wiring of an occluded vessel from spontaneous coronary artery dissection. JACC Case Rep 2020;2:72–6.

[49] Tweet MS, Kok SN, Hayes SN. Spontaneous coronary artery dissection in women: what is known and what is yet to be understood. Clin Cardiol 2018;41:203–10.

[50] Saw J, Humphries K, Aymong E, Sedlak T, Prakash R, Starovoytov A, et al. Spontaneous coronary artery dissection: clinical outcomes and risk of recurrence. J Am Coll Cardiol 2017;70:1148–58.

[51] Tweet MS, Olin JW. Insights into spontaneous coronary artery dissection: can recurrence be prevented? J Am Coll Cardiol 2017;70:1159–61.

[52] Krittanawong C, Tweet MS, Hayes SE, Bowman MJ, Gulati R, Squires RW, et al. Usefulness of cardiac rehabilitation after spontaneous coronary artery dissection. Am J Cardiol 2016;117:1604–9.

[53] Chou AY, Prakash R, Rajala J, Birnie T, Isserow S, Taylor CM, et al. The first dedicated cardiac rehabilitation program for patients with spontaneous coronary artery dissection: description and initial results. Can J Cardiol 2016;32(4):554–60.

[54] Silber TC, Tweet MS, Bowman MJ, Hayes SN, Squires RW. Cardiac rehabilitation after spontaneous coronary artery dissection. J Cardiopulm Rehabil Prev 2015;35:328–33.

[55] Patterson M, Hayes SN, Squires RW, Tweet MS. Home-based cardiac rehabilitation in a young athletic woman following spontaneous coronary artery dissection. J Clin Exerc Physiol 2016;5:6–11.

[56] Tweet MS, Codsi E, Best PJM, Gulati R, Rose CH, Hayes SN. Menstrual chest pain in women with history of spontaneous coronary artery dissection. J Am Coll Cardiol 2017;70:2308–9.

[57] Regitz-Zagrosek V, Roos-Hesselink JW, Bauersachs J, Blomström-Lundqvist C, Cífková R, De Bonis M, et al. 2018 ESC guidelines for the management of cardiovascular diseases during pregnancy. Eur Heart J 2018;39:3165–241.

[58] Gornik HL, Persu A, Adlam D, Aparicio LS, Azizi M, Boulanger M, et al. First International Consensus on the diagnosis and management of fibromuscular dysplasia. Vasc Med 2019;24:164–89.

[59] Prasad M, Tweet MS, Hayes SN, Leng S, Liang JJ, Eleid MF, et al. Prevalence of extracoronary vascular abnormalities and fibromuscular dysplasia in patients with spontaneous coronary artery dissection. Am J Cardiol 2015;115:1672–7.

[60] Saw J, Aymong E, Sedlak T, Buller CE, Starovoytov A, Ricci D, et al. Spontaneous coronary artery dissection: association with predisposing arteriopathies and precipitating stressors and cardiovascular outcomes. Circ Cardiovasc Interv 2014;7:645–55.

[61] Macaya F, Moreu M, Ruiz-Pizarro V, Salazar CH, Pozo E, Aldazábal A, et al. Screening of extra-coronary arteriopathy with magnetic resonance angiography in patients with spontaneous coronary artery dissection: a single-centre experience. Cardiovasc Diagn Ther 2019;9:229–38.

[62] Elkayam U, Jalnapurkar S, Barakkat MN, Khatri N, Kealey AJ, Mehra A, et al. Pregnancy-associated acute myocardial infarction: a review of contemporary experience in 150 cases between 2006 and 2011. Circulation 2014;129:1695–702.

[63] Smilowitz NR, Gupta N, Guo Y, Zhong J, Weinberg CR, Reynolds HR, et al. Acute myocardial infarction during pregnancy and the puerperium in the united states. Mayo Clin Proc 2018;93:1404–14.

[64] Faden MS, Bottega N, Benjamin A, Brown RN. A nationwide evaluation of spontaneous coronary artery dissection in pregnancy and the puerperium. Heart 2016;102:1974.

[65] Tweet MS, Hayes SN, Codsi E, Gulati R, Rose CH, Best PJM. Spontaneous coronary artery dissection associated with pregnancy. J Am Coll Cardiol 2017;70:426–35.

[66] Havakuk O, Goland S, Mehra A, Elkayam U. Pregnancy and the risk of spontaneous coronary artery dissection. Circ Cardiovasc Interv 2017;10, e004941.

[67] Ito H, Taylor L, Bowman M, Fry ETA, Hermiller JB, Van Tassel JW. Presentation and therapy of spontaneous coronary artery dissection and comparisons of postpartum versus nonpostpartum cases. Am J Cardiol 2011;107:1590–6.

[68] Tweet MS, Hayes SN, Gulati R, Rose CH, Best PJM. Pregnancy after spontaneous coronary artery dissection: a case series. Ann Intern Med 2015;162:598–600.

[69] Cauldwell M, Steer PJ, von Klemperer K, Kaler M, Grixti S, Hale J, et al. Maternal and neonatal outcomes in women with history of coronary artery disease. Heart 2020;106:380.

[70] Tweet MS, Young KA, Rose CH, Best PJ, Gulati R, Hayes SN. Abstract 13043: pregnancy after spontaneous coronary artery dissection: a case series of 22 women and 31 pregnancies. Circ Am Heart Assoc 2019;140.

[71] Codsi E, Tweet MS, Rose CH, Arendt KW, Best PJM, Hayes SN. Spontaneous coronary artery dissection in pregnancy: what every obstetrician should know. Obstet Gynecol 2016;128:731–8.

[72] Halpern DG, Weinberg CR, Pinnelas R, Mehta-Lee S, Economy KE, Valente AM. Use of medication for cardiovascular disease during pregnancy: JACC state-of-the-art review. J Am Coll Cardiol 2019;73:457–76.

[73] Mehta LS, Warnes CA, Bradley E, Burton T, Economy K, Mehran R, et al. Cardiovascular considerations in caring for pregnant patients: a scientific statement from the American Heart Association. Circulation 2020. https://doi.org/10.1161/CIR.0000000000000772.

[74] Lundberg G, Wenger NK. Menopause hormone therapy: what a cardiologist needs to know. https://www.acc.org/latest-in-cardiology/articles/2019/07/17/11/56/menopause-hormone-therapy; 2019.

[75] Liang JJ, Tweet MS, Hayes SE, Gulati R, Hayes SN. Prevalence and predictors of depression and anxiety among survivors of myocardial infarction due to spontaneous coronary artery dissection. J Cardiopulm Rehabil Prev 2014;34:138–42.

[76] Johnson AK, Hayes SN, Sawchuk C, Johnson MP, Best PJ, Gulati R, et al. Analysis of posttraumatic stress disorder, depression, anxiety, and resiliency within the unique population of spontaneous coronary artery dissection survivors. J Am Heart Assoc 2020;9, e014372.

[77] Johnson A, Tweet M, Best P, Gulati R, Hayes S. Prevalence of post-traumatic stress disorder and other psychological sequelae in patients who have experienced spontaneous coronary artery dissection. J Am Coll Cardiol 2019;73:252.

[78] Edwards KS, Vaca KC, Naderi S, Tremmel JA. Patient-reported psychological distress after spontaneous coronary artery dissection: evidence for post-traumatic stress. J Cardiopulm Rehabil Prev 2019;39:E20–e23.

[79] Lobo AS, Cantu SM, Sharkey SW, Grey EZ, Storey K, Witt D, et al. Revascularization in patients with spontaneous coronary artery dissection and ST-segment elevation myocardial infarction. J Am Coll Cardiol 2019;74:1290–300.

[80] Lettieri C, Zavalloni D, Rossini R, Morici N, Ettori F, Leonzi O, et al. Management and long-term prognosis of spontaneous coronary artery dissection. Am J Cardiol 2015;116:66–73.

[81] Rogowski S, Maeder MT, Weilenmann D, Haager PK, Ammann P, Rohner F, et al. Spontaneous coronary artery dissection. Catheter Cardiovasc Interv 2017;89:59–68.

[82] Kim ESH, Tweet MS. Spontaneous coronary artery dissection: a call for consensus and research advancement. J Am Coll Cardiol 2019;74:1301–3.

Chapter 6

Interventions in Ischemic Heart Disease

Deborah N. Kalkman, Birgit Vogel, Ridhima Goel,
and Roxana Mehran

Clinical Case

A 69-year-old female patient with known hypertension underwent coronary angiography due to anginal symptoms 3 years ago. The coronary angiogram back then showed a right dominant system with intermediate lesions in a small-sized posterior descending artery and ramus intermedius. Treatment was conservative, with optimization of medical therapy. However, the patient's blood pressure had not been well controlled over the last 2 years.

The patient now experiences recurrent severe angina and presents with a troponin rise. An electrocardiogram (ECG) did not show any ST elevation. The coronary angiogram now shows a 60% stenosis of the left anterior descending artery (LAD) and complete occlusion of the small- sized second obtuse marginal branch (OM2). The latter was considered the culprit lesion. How would you manage this patient?

Abstract

The field of percutaneous coronary interventions (PCI) in ischemic heart disease is in continuous evolution. New insights on coronary stenting, dual antiplatelet therapy (DAPT), and mechanical support are generated daily. However, most data report on best therapies in the overall PCI population, which may not be directly applicable to women. Women are a minority in the PCI population, accounting for approximately one-third of the patient population treated with PCI.

Women undergoing PCI have more comorbidities than their male counterparts. They have smaller coronary vessels and are more likely to experience radial artery spasm. As a result, trans radial access is less commonly used in women than in men. Women also have a higher chance of bleeding after PCI, which affects decision making regarding stent type and DAPT regimen. Many operators prefer stents that allow for shorter DAPT duration. Clinical outcomes after PCI in women remain less favorable compared to men, with higher rates of target lesion failure and higher all-cause death. These sex-related differences are driven by variation in comorbidities, clinical presentation, and lesion

Sex Differences in Cardiac Disease. https://doi.org/10.1016/B978-0-12-819369-3.00014-9

characteristics. While the field of mechanical assist devices is evolving rapidly, only very limited sex-specific data are available on this topic. Further research is needed to improve outcomes in women undergoing interventions for ischemic heart disease.

Introduction

Percutaneous coronary intervention (PCI) celebrated its 40th anniversary in 2017. Much has changed since the introduction of the first coronary balloon dilatation by Doctor Andreas Gruntzig [1]. Innovations have led to first-, second-, and third-generation drug-eluting stents (DES), a transition from transfemoral to transradial procedures and newer more potent antithrombotic therapy. In addition, multiple options for mechanical support are now available [2–4].

While much is known about coronary interventions for ischemic heart disease in the general PCI population, it is important to note that the general PCI population consists of twice as many men as women [5–7]. Therefore, sex-specific data are needed to understand the unique needs of women undergoing interventions for ischemic heart disease.

This chapter provides an overview of sex-specific data on interventions for ischemic heart disease such as procedural characteristics, including radial vs. femoral approach and stent choice, postintervention dual antiplatelet therapy (DAPT) strategies, mechanical support during interventions, as well as clinical outcomes.

Sex Differences in Pathophysiology of Coronary Artery Disease

Coronary artery disease (CAD) manifestations vary by sex, and are impacted by the anatomical, physiological, and hormonal differences between women and men (**Figure 1**). Female patients have smaller hearts, with lower ventricular wall thickness, lower left ventricular end-diastolic volumes, lower levels of circulating hemoglobin per unit plasma leading to lower oxygen-carrying capacity, higher pulse rate, and higher ejection fraction, in addition to narrower coronary artery dimensions and higher coronary blood flow, which all are factors considered to impact CAD development [8–10]. Female sex hormones have been suggested to have vascular antiinflammatory properties and to impact arterial wall remodeling, the coagulation pathway, and the lipid metabolism, resulting in a cardioprotective effect [8, 11–14]. Studies have found that at the onset of CAD, women are older than men with higher burden of comorbidities such as hypertension, diabetes mellitus, dyslipidemia, anemia, obesity, kidney disease, and heart failure [15]. In addition, certain CAD risk factors predominantly or exclusively affect women and include depression, hormonal disturbances, autoimmune disorders or chronic inflammatory conditions, and radio/chemotherapy for breast and gynecological cancers [12, 15, 16].

Female CAD symptoms are more often deemed atypical, which has been linked to a delay in development of a correct diagnosis and administration of appropriate treatment [17–20]. In addition, it has been documented that

Anatomic
- Smaller Heart
- Lower LV thickness
- Lower LV diastolic volume
- Smaller radial artery
- More coronary tortuosity

Physiologic
- Higher EF
- Higher pulse rate
- Less O_2 carrying capacity
- Higher coronary artery flow
- Higher bleeding tendency

Hormonal
- Estrogen affects lipid metabolism
- Estrogen impact arterial wall remodeling
- Estrogen has anti-inflammatory role on vasculature
- Estrogen & Progesterone alter coagulation pathway

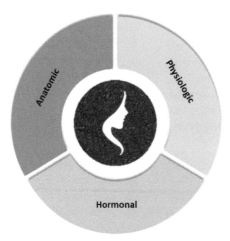

FIGURE 1 Sex Differences in Coronary Artery Pathophysiology. Anatomic, physiologic, and hormonal differences persist between men and women. EF, ejection fraction; LV, left ventricle; O_2, oxygen. *Image courtesy of Niti R. Aggarwal.*

a greater proportion of female compared to male patients do not have significant coronary artery obstruction when presenting with angina symptoms. A pilot report from the Women's Ischemia Syndrome Evaluation study showed that a majority proportion of female subjects undergoing invasive evaluation have no or nonobstructive CAD as assessed by coronary angiography [21]. These findings were further confirmed in patients with both stable and acute symptoms [22, 23]. Female patients also have higher prevalence of severe coronary tortuosity on coronary angiography [24].

Symptomatic nonobstructive disease can present as MINOCA (myocardial infarction in the absence of obstructive coronary artery disease) or INOCA (ischemia and no obstructive coronary artery disease), and is more prevalent in female patients [25, 26]. MINOCA represents a condition that can be caused by coronary mechanisms (e.g., coronary artery dissection, coronary spasm, and coronary emboli), but can also be mimicked by myocardial disorders (e.g., myocarditis, Takotsubo syndrome, and other cardiomyopathies) or noncardiac conditions (e.g., pulmonary embolism) [27]. The underlying pathophysiological mechanisms of INOCA are not well understood but appear to include factors such as diffuse atherosclerotic plaques and coronary microvascular dysfunction (CMD) [25]. CMD has been categorized as structural and functional alterations of the microcirculation leading to inappropriate response to vasodilator stimuli and pose significant diagnostic challenge due to lack of imaging techniques [25]. Studies have suggested a role of increased sympathetic pathway activities, arterial vessel remodeling and stiffness, and decreased vascular density leading to regional ischemia in nonobstructive disease. In addition, CMD may also have an impact on symptoms in patients with significant epicardial stenosis, mainly due to inward remodeling of the coronary resistance and capillary rarefaction distal to the coronary stenosis [25]. A report from the Women's Ischemia Syndrome Evaluation-Coronary Vascular Dysfunction study showed that women with INOCA and impaired coronary flow reserve as an indicator for microvascular dysfunction are more likely to have adverse outcomes if accompanied by a low baseline resting coronary flow velocity [28].

In a substudy from the CONFIRM registry, the investigators found that the presence of nonobstructive plaques in the left main vessel had higher 5-year risk of adverse events in female but not male patients [29]. When compared with male patients with nonobstructive left main plaque, female patients had an 80% higher risk of events [29].

In addition to having different CAD presentation and underlying pathophysiology, female patients also have different plaque morphology. The intravascular ultrasound (IVUS) subanalysis from the PROSPECT study showed that female patients with acute coronary syndrome (ACS) have less extensive coronary lesions, and less frequently had plaque rupture [30]. Another substudy on plaque characteristics from the PROSPECT trial showed that women <65 years had fewer nonculprit lesions, with smaller plaque volume and higher percentage of fibrotic content [31]. Another study on ACS patients showed that although female subjects have overall lower plaque volumes in culprit lesions, they have higher dense calcium and necrotic material in these lesions with thinner fibrotic caps, making them more vulnerable [32]. The investigators from the DanRisk study also reported that asymptomatic female patients who developed plaques during the course of their study have reduced fibrin clot lysability compared to female patients without plaques and compared to male patients with plaques [33] (**Table 1**).

In a subgroup analysis from the Fractional Flow Reserve Versus Angiography for Multivessel Evaluation (FAME) trial, sex differences were investigated in patients

TABLE 1 Plaque Morphology in Female Patients				
Study Author (Reference)	Imaging	Female Subjects/ Sample Size	Population	Findings
Dickerson et al. [34]	MDCT	50/94	Patients without CAD	Female subjects have smaller LAD and RCA
Hong et al. [32]	Virtual histology-intravascular ultrasound	108/362	Patients with ACS	Females have smaller coronary cross-sectional area, smaller plaque mass in culprit lesions, but greater dense calcium and necrotic core volumes
Qureshi et al. [35]	MDCT	498/916	Symptomatic patients referred for CCTA	Female subjects more often had no or minimal plaque, which was significant even after adjustment for risk factors

Continued

TABLE 1 Plaque Morphology in Female Patients—cont'd				
Study Author (Reference)	Imaging	Female Subjects/ Sample Size	Population	Findings
Bharadwaj et al. [36]	OCT/IVUS	115/383	Patients with stable CAD	Females had lower plaque burden, but no difference in plaque characteristics
Ramanathan et al. [33]	CCTA	71/138	Asymptomatic patients without CAD	Asymptomatic women with coronary plaques assessed by CCTA have reduced fibrin clot lysability compared to both women without coronary plaques and men
Lansky et al. [30]	IVUS	167/697	Patients with ACS	Female subjects have less frequent plaque rupture and lower necrotic material or calcium content
Chiha et al. [24]	ICA	281/870	Symptomatic patients without prior CAD	Female subjects have higher severe coronary tortuosity
Nasir et al. [37]	MDCTA	148/416	Symptomatic patients	Female subjects have lower calcium content

ACS, acute coronary syndrome; CAD, coronary artery disease; CCTA, cardiac-computed tomographic angiography; ICA, invasive coronary angiography; IVUS, intravascular ultrasound; LAD, left anterior descending artery; MDCT, multidetector computed tomography; MDCTA, multidetector computed tomographic angiography; OCT, optical coherence tomography; RCA, right coronary artery.

undergoing fractional flow reserve (FFR)-guided PCI. The FAME trial randomized 744 men and 261 women with multivessel disease to angiography-guided PCI or FFR-guided PCI. The sex-specific analysis found that women undergoing PCI were older with more hypertension, but no sex-related differences in terms of major adverse cardiac events (MACE) at 2-year follow-up. Higher FFR values and lower numbers of significant lesions defined as a FFR value of 0.80 or below were found in women compared with men. An FFR-guided PCI strategy was equally beneficial in both sexes [38].

Procedural Characteristics of Coronary Interventions in Women

Radial Approach

PCI by femoral approach has been largely replaced by radial approach, due to the reduced bleeding risk and less complications at the access site [39–41]. In the MATRIX randomized clinical trial evaluating transfemoral and transradial PCI in ACS, a reduction in all-cause mortality was observed with radial PCI [42].

The SAFE-PCI for women trial evaluated whether a radial approach would reduce bleeding or vascular complications in women undergoing PCI [43]. A total of 1787 women (691 undergoing PCI) were randomized at 60 sites. There was no significant difference in the primary efficacy end point, defined as any Bleeding Academic Research Consortium [44] type 2, 3, or 5 bleeding or vascular complications requiring intervention, between radial or femoral access among women undergoing PCI. However, among the total study population including women undergoing cardiac catheterization or PCI, radial access significantly reduced bleeding and vascular complications.

In routine clinical practice, women receive less radial PCI [45]. This might be due to smaller radial arteries, or because their radial arteries are more prone to spasm [46, 47]. As a result, higher transradial approach failure is observed in women undergoing radial PCI [47].

Stent Choice

Sex-specific outcomes have been described for many different DES types, such as paclitaxel-eluting stent, zotarolimus-eluting stents, sirolimus-eluting stents, and everolimus-eluting stents [48–53]. A recent study by Bourantas et al. showed a consistent higher target lesion failure in women compared to men (**Figure 2**). It is hypothesized that this difference is driven by baseline and

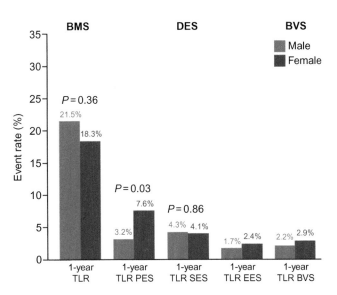

FIGURE 2 Clinical Outcomes After Percutaneous Coronary Intervention With Different Devices. BMS, bare-metal stent; BVS, bioresorbable vascular scaffold; DES, drug-eluting stent; EES, everolimus-eluting stent; PES, paclitaxel-eluting stent; SES, sirolimus-eluting stent; TLR, target lesion revascularization. *Reproduced with permission from [54].*

lesion characteristics, such as higher calcific burden in female patients with stable angina, determining higher chance of device failure [55]. Interestingly stent thrombosis (ST) rates seem to be lower in women, compared to men, specifically in patients treated with bioresorbable vascular scaffold (BVS) [56–58].

The WIN-DES Collaborative Patient-Level Pooled Analysis, including all female patients from 26 randomized trials of DES, reported on the safety and efficacy of women undergoing PCI with bare-metal stents (BMS), early-generation DES, and newer-generation DES. Great improvements in clinical outcomes are observed with the use of early-generation DES compared to BMS, and further improvements with newer-generation DES are seen [59]. Further analyses show that new-generation DES are safe and efficient in female patients presenting with acute myocardial infarction and complex PCI [60, 61].

While there are no female-specific stents currently on the market, there are stents that may be preferred by PCI operators in female patients. Stents that allow for shorter DAPT duration might be beneficial in women undergoing PCI due to their higher risk of postintervention bleeding. Stents with small stent struts, a pro-healing layer, or biodegradable polymers are examples of stents that may specifically work well in women [62–64].

Thrombotic Events, Antiplatelet Strategies, and Bleeding

Intraprocedural thrombotic events, such as new or worsened thrombus, abrupt vessel closure, distal embolization, and no reflow or slow reflow with thrombus, occur at similar frequency in men and women, and are associated

in both sexes with similar degree of increased MACE [65]. Therefore, higher thrombotic burden does not offer an explanation for the worse short- and long-term outcomes observed in women vs. men after presentation with ACS and myocardial infarction [65]. More information on PCI outcomes in ST-elevated myocardial infarction (STEMI), non-STEMI (NSTEMI), and elective procedures can be found in the section clinical outcomes after PCI.

A meta-analysis compared the effectiveness of bivalirudin vs. heparin plus routine glycoprotein IIb/IIIa inhibitors (GPI) in women. Major bleeding was decreased in both men and women with bivalirudin compared to heparin plus GPI. Bivalirudin compared to heparin alone showed a lower bleeding benefit in women [66].

Bleeding after PCI in women has been evaluated in a subgroup analysis of the Harmonizing Outcomes with Revascularization and Stents in Acute Myocardial Infarction (HORIZONS-AMI) study. The investigators found that women with STEMI have a higher risk of bleeding as compared to men, which most likely is caused by higher number of comorbidities in women undergoing primary PCI. Registry data has confirmed these findings [67, 68]. In a subgroup analysis of the GLOBAL LEADERS Randomized Clinical Trial, bleeding and hemorrhagic stroke were observed more frequently in women undergoing PCI compared to men [69].

On the other side, high platelet reactivity was seen more frequently in women vs. men from the ADAPT-DES study (Assessment of Dual Antiplatelet Therapy with Drug-Eluting Stents). Interestingly, no differences were observed in association between high platelet reactivity and ST between men and women, and high platelet reactivity was associated with significantly reduced bleeding only among women [70]. A study from the Novara Atherosclerosis Study Group evaluated the prevalence of high on-treatment residual platelet reactivity in patients receiving aspirin, clopidogrel, or ticagrelor. They found that sex did not impact high on-treatment residual platelet reactivity in patients receiving DAPT [71].

DAPT Choice

In the Cangrelor Versus Standard Therapy to Achieve Optimal Management of Platelet Inhibition (CHAMPION PHOENIX) trial, investigators randomized elective or urgent PCI patients, who were receiving guideline-recommended therapy to intravenous platelet adenosine diphosphate receptor antagonist cangrelor or a loading dose of 600 or 300 mg of clopidogrel. In a subgroup analysis on sex differences, cangrelor reduced the rate of major adverse cardiovascular events, including death, myocardial infarction, ischemia-driven revascularization, or ST, and second-

ary endpoint ST in women and men [72]. While severe bleeding was similar in men and women, the odds of moderate bleeding associated with cangrelor were significantly increased in women but not in men. The net clinical benefit favored cangrelor in both women and men.

In a large meta-analysis potent $P2Y_{12}$ inhibitors, including prasugrel, ticagrelor, and intravenous cangrelor, showed similar efficacy and safety in men and women, reducing the risk of MACE by 14% in women (**Table 2**). This MACE reduction comes at the cost of more bleeding, however, with similar risk between men and women in this meta-analysis. The authors conclude that sex should not influence patient selection for the treatment with potent $P2Y_{12}$ [73].

DAPT Duration

It has been repeatedly shown that women receive less guideline-recommended DAPT [74]. Expert consensus states that women should not be treated differently from

TABLE 2 Efficacy and Safety of the Potent $P2Y_{12}$ Inhibitors in Men vs. Women

	Women		Men		
	Potent $P2Y_{12}$ Inhibitor	Comparator Arm	Potent $P2Y_{12}$ Inhibitor	Comparator Arm	*P* Interaction (Sex)
MACE	8.30	9.80	7.30	8.60	0.93
CV death, MI, or stroke	8.80	10.20	7.60	9.04	0.60
CV death	3.48	4.10	2.66	3.19	0.86
MI	5.24	6.22	4.71	5.86	0.65
Stroke	1.07	1.15	0.91	0.81	0.72
Stent thrombosis (definite)	0.58	1.30	0.82	1.38	0.85
Stent thrombosis (definite or probable)	1.38	2.26	1.60	2.56	0.94
TIMI non-CABG major bleeding	1.60	1.27	1.43	0.95	0.62
TIMI non-CABG minor bleeding	1.66	1.02	0.80	0.58	0.055
TIMI non-CABG major or minor bleeding	3.30	2.53	2.30	1.59	0.76
ICH	0.33	0.34	0.37	0.24	0.24
All-cause death	3.93	4.30	3.22	3.44	0.99

Values are % unless otherwise indicated. MACE is defined as the primary end point in each of the individual trials. In the CHAMPION PCI-PLATFORM trials, outcome provided was death, MI, or stroke. CABG, coronary artery bypass grafting; CV, cardiovascular; ICH, intracranial hemorrhage; MACE, major adverse cardiovascular event; MI, myocardial infarction; TIMI, thrombolysis in myocardial infarction. *Adapted with permission from [73].*

men with regard to guideline-recommended therapies [75]. In the 2017 *European Society of Cardiology (ESC)* focused update on DAPT in CAD, the section on sex consideration also states that there is no convincing evidence for a sex-related difference in the efficacy and safety of currently available DAPT type or duration across studies [76]. On the other side, the 2016 *American College of Cardiology/American Heart Association (ACC/AHA)* focused update on duration of DAPT in patients with CAD guideline does identify female sex as a risk for increased bleeding, which may favor shorter-duration DAPT. However, no sex-specific recommendations for the duration of DAPT after PCI are provided [77].

In the PARIS (Patterns of Nonadherence to Antiplatelet Regimens in Stented Patients) study, DAPT cessation was higher in women, with increased rates of discontinuation, disruption for bleeding, and disruption due to noncompliance. However, impact of DAPT cessation was similar in men and women [78]. In this analysis, female sex remained a predictor of bleeding after adjustment for baseline characteristics, but not a predictor for ischemic events. The GLOBAL LEADERS Randomized Clinical Trial, evaluating 1 month of DAPT followed by 23 months of ticagrelor monotherapy or 12 months of DAPT followed by 12 months of aspirin monotherapy, showed that ticagrelor monotherapy was associated with lower risk of 1-year bleeding events in men, but not in women, with significant interaction between sex and treatment allocation [69]. At 2 years, there were no significant sex-related differences in efficacy and safety between the two treatment strategies.

Clinical Outcomes After PCI

It is now well recognized that higher rates of major adverse events after PCI in women are observed than in men [79–84]. This increased risk is driven by differences in baseline characteristics, such as the prevalence of diabetes mellitus or chronic kidney disease, the presentation with ACS, and the complexity of coronary lesions [55, 60, 61, 85–88]. Higher all-cause mortality rate has been attributed to more chronic disease and heart failure in women [89, 90]. Differences in baseline characteristics may also contribute to the differences observed in coronary plaque characteristics and the increased rates of hospital readmission after PCI in women vs. men [55, 91].

Primary PCI in ACS

Numerous studies have reported on outcomes according to sex after primary PCI in ACS patients [79, 92–98]. The Treatment with Adenosine Diphosphate Receptor Inhibitors: Longitudinal Assessment of Treatment Patterns and Events after Acute Coronary Syndrome (TRANSLATE-ACS) observational study showed large differences in demographic, clinical, and treatment profiles between men and women undergoing primary PCI, which explains the difference in outcomes [79]. This was confirmed in several other studies and a large meta-analysis [93, 94, 97]. In another analysis evaluating clinical outcomes in the setting of universal access to primary PCI, women and especially older women had significantly higher adjusted risk of 1-year rehospitalization for myocardial infarction [92]. An analysis by Kosmidou et al. in patients with STEMI undergoing primary PCI also showed that women had a higher 1-year rate of death or heart failure (HF) hospitalization compared to men. However, the authors concluded that this finding could not be explained by sex-specific differences in the magnitude or prognostic impact of infarct size or by differences in postinfarct cardiac function [96].

Very early aggressive revascularization in women with NSTEMI has been reported to improve long-term outcome and may be considered as the preferred treatment strategy [98].

Spontaneous Coronary Artery Dissection

Spontaneous coronary artery dissection (SCAD) is a non-traumatic, noniatrogenic, and nonatherosclerotic spontaneous separation of the coronary artery wall [99]. SCAD is seen in approximately 1–4% of patients presenting with ACS [100, 101]. However, SCAD is observed more frequently in women [102–104]. In women aged < 50, reports suggest SCAD to be the cause of MI in up to 35% of cases [105, 106]. Intravascular imaging with intravascular ultrasound or OCT is advised when SCAD is suspected. In stable patients conservative treatment is recommended, as in the majority of cases the coronary artery heals spontaneously [99]. However, PCI or coronary artery bypass grafting (CABG) should be considered in patients with ongoing ischemia or scenarios such as left main dissection, sustained ventricular arrhythmia, or hemodynamic instability [99, 107].

When stenting is needed to treat SCAD, stent lengths tend to be long in order to cover both edges of the dissection by > 5 mm, and are, therefore, at increased risk of in-stent restenosis and thrombosis [99, 108].

Complex PCI Including Left Main Stenting

The EXCEL (Evaluation of XIENCE Versus Coronary Artery Bypass Surgery for Effectiveness of Left Main Revascularization) trial randomized patients with unprotected left main disease to either PCI with everolimus-eluting stents or CABG. The primary end point was the composite of all-cause death, myocardial infarction, or stroke at 3 years. In a subgroup analysis, they found that sex was not an independent predictor of adverse outcomes. As reported in many other studies, it was shown that women undergoing PCI had a trend toward worse outcomes due to their higher number of comorbidities and increased periprocedural complications [109].

In the Interventional Cardiology Research In-cooperation Society-left MAIN revascularization (IRIS-MAIN) registry, women had similar long-term outcomes after PCI with DES for left main CAD despite differences in clinical and lesion characteristics [110].

Stent Thrombosis

Subgroup analyses of different studies investigating stent types have shown no difference in the occurrence of ST between the sexes. The TAXUS-IV trial demonstrated no sex differences between ST rate in Taxus-treated patients or BMS-treated women [49]. A pooled analysis of several sirolimus-eluting stent trials also did not show a difference in occurrence of ST between men and women [48]. In the Patient Related Outcomes with Endeavor Versus Cypher stenting Trial (PROTECT) similar cumulative incidence of ST was observed in men and women at 5-year follow-up. The favorable effect of the zotarolimus-eluting stent vs. the sirolimus-eluting stent was not modified by sex [111]. Similarly, ST rates were similar between men and women at 12 months after everolimus-eluting stent placement [52]. However, it should be noted that overall ST event rates are very low, and none of the studies were statistically powered to detect meaningful difference for this end point between women and men.

In a subgroup analysis of the WIN-DES collaboration, predictors of definite ST in women were assessed. Young age, diabetes mellitus, NSTEMI, and smaller stent diameter were predictors of ST (**Table 3**). Although the occurrence of ST was uncommon, ST was associated with increased mortality in women undergoing PCI with stent placement [112].

Residual Angina After PCI

Residual angina is defined as persistence of angina pectoris due to myocardial ischemia in a patient treated with either surgical coronary revascularization or PCI, with no period of complete dissolution of symptoms [113, 114]. Prevalence of residual angina is common and some studies identified female sex as a predictor of residual angina [115, 116]. There are multiple factors that impact residual angina; the most important one is incomplete revascularization. In women undergoing PCI, incomplete revascularization can occur due to small vessel diameter, diffuse lesions, and concomitant microvascular disease, resulting in increased incidence of residual angina after PCI. Moreover, women have higher rates of no flow-limiting stenosis, caused by coronary stenosis or microvascular disease, coronary spasm, or SCAD [117, 118]. In the latter case PCI is not the

TABLE 3 Predictors of Definite Stent Thrombosis in Women

	Hazard Ratio	95% Confidence Interval	P-Value
Stent diameter (per 1 mm decrease)	3.76	1.66–8.53	<0.01
DM without insulin vs. no DM	2.25	1.27–3.99	<0.01
NSTEMI vs. stable angina	1.97	1.04–3.75	0.04
Multivessel disease	1.57	0.86–2.87	0.14
Age (per 1 year decrease)	1.03	1.00–1.05	0.04
New-generation DES vs. BMS	0.60	0.33–1.07	0.09

BMS, bare-metal stents; DES, drug-eluting stents; DM, diabetes mellitus; NSTEMI, non-ST-segment elevation myocardial infarction. *Adapted with permission from [112].*

recommended treatment and will not cause symptom relief. It is therefore of great importance that PCI operators should focus on the underlying mechanism of angina symptoms in their female patients and should treat accordingly.

Management After Elective PCI in Women

There are no sex-specific management guidelines for female patients who underwent PCI for stable angina. The American College of Cardiology 2014 Update of the Guideline for the Diagnosis and Management of Patients with Stable Ischemic Heart Disease states: "As is the case for many chronic conditions, studies specifically geared toward answering clinical questions about the management of stable ischemic heart disease (SIHD) in women, older adults, and persons with chronic kidney disease are lacking" [119, 120]. The European Society of Cardiology 2014 Guideline on myocardial revascularization does not address sex differences in management after elective PCI [119]. Therefore, current guidelines for the general population should be followed.

Treatment is focused on risk factor modification and permanent lifestyle changes, to prevent progression of atherosclerotic disease and minimize the risk of restenosis and

ST. Secondary prevention targeting low low-density lio-protein (LDL) levels as well as optimal low blood pressure and adequate antithrombotic therapy should be started in all patients after PCI. Currently there are no female-specific targets for blood pressure or lipid levels. Future research should explore if different targets for women compared to men will be beneficial for the intermediate and long-term outcomes after PCI.

Mechanical Support

Only limited sex-specific data exist on the role of mechanical support in high-risk percutaneous interventions in women. One study by the percutaneous ventricular assist device Working Group evaluated sex differences in clinical outcomes in patients with cardiogenic shock (CS) undergoing PCI with simultaneous use of Impella (Abiomed Inc., Danvers, Massachusetts). In this study, there were only a small number of female patients (62 men vs. 19 women); however, primary endpoint all-cause in-hospital mortality did not significantly differ between sexes. In addition, all secondary outcomes (myocardial infarction, stroke, CS, heart failure, dialysis requirement, bleeding within 72h, blood transfusion, dysrhythmia, MACE, and survival at 30days) were similar in men and women [121]. A publication from the Global cVAD (catheter-based ventricular assist device) Registry investigators analyzed sex-specific outcomes in patients with CS undergoing PCI for acute myocardial infarction and Impella 2.5 support. Also in this analysis, women comprised only 27% of patients. They were older, smaller, and had a higher Society of Thoracic Surgeons mortality risk score compared to men. Overall survival to discharge was not statistically different between sexes after propensity score matching [122]. A small series of female patients with SCAD underwent placement of Impella for management of CS or for support of high-risk PCI. All four of these patients survived to discharge and the Impella provided transient hemodynamic support until recovery of ventricular function occurred [123].

Mechanical circulatory support in heart failure patients can bridge the time to transplant and extend life as destination therapy. Data from the prospective Interagency Registry for Mechanically Assisted Circulatory Support (INTERMACS) and data from other sites using mechanical support were combined for a sex group analysis evaluating clinical outcomes after simultaneous biventricular external assist device, temporary ventricular assist device (VAD), single ventricle extracorporeal VAD, or implantable VAD. An important finding was that, compared to men, only a small percentage of women receive MCS. In the INTERMACS database, only 21% of patients receiving MCS were female. Second, women were younger and less likely to be white. Third, women with MCS had significantly more major adverse events than men, mainly driven by increased mortality and incident stroke [124]. A review article by Hsich summarized that survival has improved for both sexes implanted with VADs with no sex differences in mortality [125]. Moreover, continuous flow devices are shown to be superior to pulsatile devices. Unfortunately, neurological adverse events, hemorrhagic, and ischemic stroke are observed more frequently in women after left VAD (LVAD). However, Hsich et al. conclude that the risk of neurological events is outweighed by the mortality benefits and that VADs should be implanted in eligible patients regardless of sex who have failed medical therapy [125–128].

Conclusion

A great deal of progress has been made in the field of PCI since the first coronary intervention was performed in 1977. The introduction of smaller catheters, balloons, and higher-quality stents has reduced risks and complications in both sexes. Despite this progress, clinical outcomes after PCI remain worse in women compared to men. Women still experience not only higher all-cause mortality but also higher rates of repeat revascularization. This may be in part due to a higher burden of comorbidities, smaller vessel size, and a greater burden of coronary calcification in women compared to men. Women are more likely than men to experience post-PCI bleeding and this should be factored into stent choice. Optimal duration of DAPT after PCI remains uncertain.

CH
6

Key Points

 1 Women undergoing PCI have different baseline characteristics compared to men, with a higher number of comorbidities (such as diabetes mellitus and chronic kidney disease).

 2 Coronary lesions in women tend to have a smaller vessel diameter and have more calcifications.

 3 Radial approach for PCI is performed less in women compared to men.

 4 Clinical outcomes after PCI are worse in women compared to men. Not only is higher all-cause mortality observed, but also higher repeat revascularization is seen in women.

 5 Stent choice for PCI in women should be driven by options for shorter duration of DAPT after stent placement due to higher risk of post-PCI bleeding in women.

 6 Duration of dual antiplatelet regimen after PCI in women is an important issue, because of increased risk of bleeding in women. However, debate on optimal duration is still ongoing as no sex-specific data are available.

 7 Limited sex-specific data are available for mechanical support, because the majority of mechanical support devices are implanted in men. However, survival in women and men with pVAD or CVAD is similar, though higher rates of neurological events are reported in women treated with LVAD.

 8 Technical challenges associated with PCI in women compared with men include more common coronary tortuosity, smaller vessel size, higher bleeding risk at access site, and higher likelihood of SCAD as underlying etiology of acute myocardial ischemia.

Back to Clinical Case

The case describes a 69-year-old female with known coronary artery disease, treated medically, now presenting with NSTEMI. Coronary angiography demonstrated complete occlusion of a small OM2. PCI with stent placement was indicated. Wiring of the OM2 was performed and predilatation with a 2.0×20 mm balloon was applied at 8 atm, followed by predilatation with a 2.5×15 mm balloon at 8 atm. The patient was asked to participate in a clinical trial evaluating two different stent types. The patient provided informed consent and was randomized to a new-generation DES with a biodegradable polymer. A 2.5×18 mm stent was placed at 10 atm with a good final result.

No adverse outcomes were observed during hospital admission. Her medication was optimized prior to discharge. Antihypertensive treatment and statin therapy were increased and DAPT was prescribed for 12 months, followed by lifelong aspirin monotherapy.

The patient has planned yearly visits at her cardiologist in the outpatient clinic. All is going well, with no residual angina.

References

[1] Stefanini GG, Byrne RA, Windecker S, Kastrati A. State of the art: coronary artery stents—past, present and future. EuroIntervention 2017;13:706–16.

[2] McKavanagh P, Zawadowski G, Ahmed N, Kutryk M. The evolution of coronary stents. Expert Rev Cardiovasc Ther 2018;16:219–28.

[3] Kolkailah AA, Alreshq RS, Muhammed AM, Zahran ME, Anas El-Wegoud M, Nabhan AF. Transradial versus transfemoral approach for diagnostic coronary angiography and percutaneous coronary intervention in people with coronary artery disease. Cochrane Database Syst Rev 2018;4(4):CD012318. https://doi.org/10.1002/14651858.CD012318.pub2. PMID: 29665617; PMCID: PMC6494633.

[4] Atkinson TM, Ohman EM, O'Neill WW, Rab T, Cigarroa JE, Interventional Scientific Council of the American College of Cardiology. A practical approach to mechanical circulatory support in patients undergoing percutaneous coronary intervention: an interventional perspective. JACC Cardiovasc Interv 2016;9:871–83.

[5] Benjamin EJ, Virani SS, Callaway CW, Chamberlain AM, Chang AR, Cheng S, et al. Heart disease and stroke statistics—2018 update: a report from the American Heart Association. Circulation 2018;137:e67–e492.

[6] Epps KC, Holper EM, Selzer F, Vlachos HA, Gualano SK, Abbott JD, et al. Sex differences in outcomes following percutaneous coronary intervention according to age. Circ Cardiovasc Qual Outcomes 2016;9:S16–25.

[7] Potts J, Sirker A, Martinez SC, Gulati M, Alasnag M, Rashid M, et al. Persistent sex disparities in clinical outcomes with percutaneous coronary intervention: insights from 6.6 million PCI procedures in the United States. PLoS One 2018;13:e0203325.

[8] McSweeney JC, Rosenfeld AG, Abel WM, Braun LT, Burke LE, Daugherty SL, et al. Preventing and experiencing ischemic heart disease as a woman: state of the science. Circulation 2016;133:1302–31.

[9] Hiteshi AK, Li D, Gao Y, Chen A, Flores F, Mao SS, et al. Gender differences in coronary artery diameter are not related to body habitus or left ventricular mass. Clin Cardiol 2014;37:605–9.

[10] Patel MB, Bui LP, Kirkeeide RL, Gould KL. Imaging microvascular dysfunction and mechanisms for female-male differences in CAD. JACC Cardiovasc Imaging 2016;9:465–82.

[11] Novella S, Heras M, Hermenegildo C, Dantas AP. Effects of estrogen on vascular inflammation. Arterioscler Thromb Vasc Biol 2012;32:2035–42.

[12] Meyer MR, Barton M. Estrogens and coronary artery disease: new clinical perspectives. Adv Pharmacol 2016;77:307–60. https://doi.org/10.1016/bs.apha.2016.05.003. 27451102.

[13] Boese AC, Kim SC, Yin K-J, Lee J-P, Hamblin MH. Sex differences in vascular physiology and pathophysiology: estrogen and androgen signaling in health and disease. Am J Phys Heart Circ Phys 2017;313:H524–45.

[14] Wingate S. Cardiovascular anatomy and physiology in the female. Crit Care Nurs Clin North Am 1997;9:447–52.

[15] Garcia M, Mulvagh SL, Merz CNB, Buring JE, Manson JE. Cardiovascular disease in women. Circ Res 2016;118:1273–93.

[16] Mehta PK, Bess C, Elias-Smale S, Vaccarino V, Quyyumi A, Pepine CJ, et al. Gender in cardiovascular medicine: chest pain and coronary artery disease. Eur Heart J 2019;40:3819–26.

[17] Canto JG, Canto EA, Goldberg RJ. Time to standardize and broaden the criteria of acute coronary syndrome symptom presentations in women. Can J Cardiol 2014;30:721–8.

[18] Khan NA, Daskalopoulou SS, Karp I, Eisenberg MJ, Pelletier R, Tsadok MA, et al. Sex differences in acute coronary syndrome symptom presentation in young patients. JAMA Intern Med 2013;173:1863–71.

[19] Bugiardini R, Ricci B, Cenko E, Vasiljevic Z, Kedev S, Davidovic G, et al. Delayed care and mortality among women and men with myocardial infarction. J Am Heart Assoc 2017;6:e005968.

[20] Maas AHEM. Characteristic symptoms in women with ischemic heart disease. Curr Cardiovasc Risk Rep 2019;13:17.

[21] Sharaf BL, Pepine CJ, Kerensky RA, Reis SE, Reichek N, Rogers WJ, et al. Detailed angiographic analysis of women with suspected ischemic chest pain (pilot phase data from the NHLBI-sponsored Women's Ischemia Syndrome Evaluation [WISE] Study Angiographic Core Laboratory). Am J Cardiol 2001;87:937–41.

[22] Shaw LJ, Shaw RE, Merz CN, Brindis RG, Klein LW, Nallamothu B, et al. Impact of ethnicity and gender differences on angiographic coronary artery disease prevalence and in-hospital mortality in the American College of Cardiology—National Cardiovascular Data Registry. Circulation 2008;117:1787–801.

[23] Jespersen L, Hvelplund A, Abildstrøm SZ, Pedersen F, Galatius S, Madsen JK, et al. Stable angina pectoris with no obstructive coronary artery disease is associated with increased risks of major adverse cardiovascular events. Eur Heart J 2011;33:734–44.

[24] Chiha J, Mitchell P, Gopinath B, Burlutsky G, Kovoor P, Thiagalingam A. Gender differences in the prevalence of coronary artery tortuosity and its association with coronary artery disease. IJC Heart Vasc 2017;14:23–7.

[25] Padro T, Manfrini O, Bugiardini R, Canty J, Cenko E, De Luca G, et al. ESC Working Group on coronary pathophysiology and microcirculation position paper on 'coronary microvascular dysfunction in cardiovascular disease'. Cardiovasc Res 2020;116:741–55.

[26] Smilowitz NR, Mahajan AM, Roe MT, Hellkamp AS, Chiswell K, Gulati M, et al. Mortality of myocardial infarction by sex, age, and obstructive coronary artery disease status in the ACTION registry-GWTG (acute coronary treatment and intervention outcomes network registry-get with the guidelines). Circ Cardiovasc Qual Outcomes 2017;10:e003443.

[27] Ibanez B, James S, Agewall S, Antunes MJ, Bucciarelli-Ducci C, Bueno H, et al. 2017 ESC guidelines for the management of acute myocardial infarction in patients presenting with ST-segment elevation: the task force for the management of acute myocardial infarction in patients presenting with ST-segment elevation of the European Society of Cardiology (ESC). Eur Heart J 2018;39:119–77.

[28] Suppogu N, Wei J, Nelson MD, Cook-Wiens G, Cheng S, Shufelt CL, et al. Resting coronary velocity and myocardial performance in women with impaired coronary flow reserve: results from the Women's Ischemia Syndrome Evaluation-Coronary Vascular Dysfunction (WISE-CVD) study. Int J Cardiol 2020;309:19–22. https://doi.org/10.1016/j.ijcard.2020.01.053. PMID: 32037132; PMCID: PMC7195998.

[29] Xie JX, Eshtehardi P, Varghese T, Goyal A, Mehta PK, Kang W, et al. Prognostic significance of nonobstructive left main coronary artery disease in women versus men. Circ Cardiovasc Imaging 2017;10:e006246.

[30] Lansky AJ, Ng VG, Maehara A, Weisz G, Lerman A, Mintz GS, et al. Gender and the extent of coronary atherosclerosis, plaque composition, and clinical outcomes in acute coronary syndromes. JACC Cardiovasc Imaging 2012;5:S62–72.

[31] Ruiz-García J, Lerman A, Weisz G, Maehara A, Mintz GS, Fahy M, et al. Age- and gender-related changes in plaque composition in patients with acute coronary syndrome: the PROSPECT study. EuroIntervention 2012;8:929–38.

[32] Hong YJ, Jeong MH, Choi YH, Ma EH, Cho SH, Ko JS, et al. Gender differences in coronary plaque components in patients with acute coronary syndrome: virtual histology-intravascular ultrasound analysis. J Cardiol 2010;56:211–9.

[33] Ramanathan R, Gram JB, Sidelmann JJ, Dey D, Kusk MW, Nørgaard BL, et al. Sex difference in fibrin clot lysability: association with coronary plaque composition. Thromb Res 2019;174:129–36.

[34] Dickerson JA, Nagaraja HN, Raman SV. Gender-related differences in coronary artery dimensions: a volumetric analysis. Clin Cardiol 2010;33:E44–9.

[35] Qureshi W, Blaha MJ, Nasir K, Al-Mallah MH. Gender differences in coronary plaque composition and burden detected in symptomatic patients referred for coronary computed tomographic angiography. Int J Cardiovasc Imaging 2013;29:463–9.

[36] Bharadwaj AS, Vengrenyuk Y, Yoshimura T, Baber U, Hasan C, Narula J, et al. Multimodality intravascular imaging to evaluate sex differences in plaque morphology in stable CAD. JACC Cardiovasc Imaging 2016;9:400–7.

[37] Nasir K, Gopal A, Blankstein R, Ahmadi N, Pal R, Khosa F, et al. Noninvasive assessment of gender differences in coronary plaque composition with multidetector computed tomographic angiography. Am J Cardiol 2010;105:453–8.

[38] Kim HS, Tonino PAL, De Bruyne B, Yong ASC, Tremmel JA, Pijls NHJ, et al. The impact of sex differences on fractional flow reserve-guided percutaneous coronary intervention: a FAME (fractional flow reserve versus angiography for multivessel evaluation) substudy. JACC Cardiovasc Interv 2012;5:1037–42.

[39] Bertrand OF, Bélisle P, Joyal D, Costerousse O, Rao SV, Jolly SS, et al. Comparison of transradial and femoral approaches for percutaneous coronary interventions: a systematic review and hierarchical Bayesian meta-analysis. Am Heart J 2012;163:632–48.

[40] Ratib K, Mamas MA, Anderson SG, Bhatia G, Routledge H, De Belder M, et al. Access site practice and procedural outcomes in relation to clinical presentation in 439,947 patients undergoing percutaneous coronary intervention in the United Kingdom. JACC Cardiovasc Interv 2015;8:20–9.

[41] Ferrante G, Rao SV, Jüni P, Da Costa BR, Reimers B, Condorelli G, et al. Radial versus femoral access for coronary interventions across the entire spectrum of patients with coronary artery disease a meta-analysis of randomized trials. J Am Coll Cardiol Intv 2016;9:1419–34.

[42] Valgimigli M, Gagnor A, Calabró P, Frigoli E, Leonardi S, Zaro T, et al. Radial versus femoral access in patients with acute coronary syndromes undergoing invasive management: a randomised multicentre trial. Lancet 2015;385:2465–76.

[43] Rao SV, Hess CN, Barham B, Aberle LH, Anstrom KJ, Patel TB, et al. A registry-based randomized trial comparing radial and femoral approaches in women undergoing percutaneous coronary intervention the SAFE-PCI for women (Study of Access Site for Enhancement of PCI for Women) trial. J Am Coll Cardiol Intv 2014;7:857–67.

[44] Mehran R, Rao SV, Bhatt DL, Gibson CM, Caixeta A, Eikelboom J, et al. Standardized bleeding definitions for cardiovascular clinical trials a consensus report from the Bleeding Academic Research Consortium. Circulation 2011;123:2736–U144.

[45] Feldman DN, Swaminathan RV, Kaltenbach LA, Baklanov DV, Kim LK, Wong SC, et al. Adoption of radial access and comparison of outcomes to femoral access in percutaneous coronary intervention an updated report from the National Cardiovascular Data Registry (2007–2012). Circulation 2013;127:2295–306.

[46] Saito S, Ikei H, Hosokawa G, Tanaka S. Influence of the ratio between radial artery inner diameter and sheath outer diameter on radial artery flow after transradial coronary intervention. Catheter Cardiovasc Interv 1999;46:173–8.

[47] Dehghani P, Mohammad A, Bajaj R, Hong T, Suen CM, Sharieff W, et al. Mechanism and predictors of failed transradial approach for percutaneous coronary interventions. JACC Cardiovasc Interv 2009;2:1057–64.

[48] Solinas E, Nikolsky E, Lansky AJ, Kirtane AJ, Morice MC, Popma JJ, et al. Gender-specific outcomes after sirolimus-eluting stent implantation. J Am Coll Cardiol 2007;50:2111–6.

[49] Lansky AJ, Costa RA, Mooney M, Midei MG, Lui HK, Strickland W, et al. Gender-based outcomes after paclitaxel-eluting stent implantation in patients with coronary artery disease. J Am Coll Cardiol 2005;45(8):1180–5. https://doi.org/10.1016/j.jacc.2004.10.076. PMID: 15837246.

[50] Russ MA, Wackerl C, Zeymer U, Hochadel M, Kerber S, Zahn R, et al. Gender based differences in drug eluting stent implantation—data from the German ALKK registry suggest underuse of DES in elderly women. BMC Cardiovasc Disord 2017;17:68.

[51] Mikhail GW, Gerber RT, Cox DA, Ellis SG, Lasala JM, Ormiston JA, et al. Influence of sex on long-term outcomes after percutaneous coronary intervention with the paclitaxel-eluting coronary stent: results of the "TAXUS Woman" analysis. JACC Cardiovasc Interv 2010;3(12):1250–9. https://doi.org/10.1016/j.jcin.2010.08.020. PMID: 21232718.

[52] Batchelor W, Kandzari DE, Davis S, Tami L, Wang JC, Othman I, et al. Outcomes in women and minorities compared with white men 1 year after everolimus-eluting stent implantation: insights and results from the PLATINUM diversity and PROMUS element plus post-approval study pooled analysis. JAMA Cardiol 2017;2:1303–13.

[53] Nakatani D, Ako J, Tremmel JA, Waseda K, Otake H, Koo B-K, et al. Sex differences in neointimal hyperplasia following endeavor zotarolimus-eluting stent implantation. Am J Cardiol 2011;108:912–7.

[54] Bourantas CV, Bajaj R, Tufaro V, Kilic Y, Serruys PW. Sex differences in clinical outcomes following bioresorbable scaffold implantation: a paradigm shift? EuroIntervention 2019;15(7):574–6. https://doi.org/10.4244/EIJV15I7A105. PMID: 31538628.

[55] Inoue F, Yamaguchi S, Ueshima K, Fujimoto T, Kagoshima T, Uemura S, et al. Gender differences in coronary plaque characteristics in patients with stable angina: a virtual histology intravascular ultrasound study. Cardiovasc Interv Ther 2010;25:40–5.

[56] Kerkmeijer LSM, Tijssen RYG, Hofma SH, Pinxterhuis TH, Kraak RP, Kalkman DN, et al. A paradox in sex-specific clinical outcomes after bioresorbable scaffold implantation: 2-year results from the AIDA trial. Int J Cardiol 2020;300:93–8. https://doi.org/10.1016/j.ijcard.2019.08.045. PMID: 31511193.

[57] Shreenivas S, Kereiakes DJ, Ellis SG, Gao R, Kimura T, Onuma Y, et al. Efficacy and safety of the absorb bioresorbable vascular scaffold in females and males: results of an individual patient-level pooled meta-analysis of randomized controlled trials. JACC Cardiovasc Interv 2017;10:1881–90.

[58] Baquet M, Hoppmann P, Grundmann D, Schmidt W, Kufner S, Theiss HD, et al. Sex and long-term outcomes after implantation of the absorb bioresorbable vascular scaffold for treatment of coronary artery disease. EuroIntervention 2019;15:615–22.

[59] Stefanini GG, Baber U, Windecker S, Morice MC, Sartori S, Leon MB, et al. Safety and efficacy of drug-eluting stents in women: a patient-level pooled analysis of randomised trials. Lancet 2013;382:1879–88.

[60] Giustino G, Baber U, Aquino M, Sartori S, Stone GW, Leon MB, et al. Safety and efficacy of new-generation drug-eluting stents in women undergoing complex percutaneous coronary artery revascularization: from the WIN-DES collaborative patient-level pooled analysis. JACC Cardiovasc Interv 2016;9:674–84.

[61] Giustino G, Harari R, Baber U, Sartori S, Stone GW, Leon MB, et al. Long-term safety and efficacy of new-generation drug-eluting stents in women with acute myocardial infarction: from the women in innovation and drug-eluting stents (WIN-DES) collaboration. JAMA Cardiol 2017;2:855–62.

[62] de Winter RJ, Katagiri Y, Asano T, Milewski KP, Lurz P, Buszman P, et al. A sirolimus-eluting bioabsorbable polymer-coated stent (MiStent) versus an everolimus-eluting durable polymer stent (Xience) after percutaneous coronary intervention (DESSOLVE III): a randomised, single-blind, multicentre, non-inferiority, phase 3 trial. Lancet 2018;391:431–40.

[63] von Birgelen C, Zocca P, Buiten RA, Jessurun GAJ, Schotborgh CE, Roguin A, et al. Thin composite wire strut, durable polymer-coated (Resolute Onyx) versus ultrathin cobalt–chromium strut, bioresorbable polymer-coated (Orsiro) drug-eluting stents in allcomers with coronary artery disease (BIONYX): an international, single-blind, randomised non-inferiority trial. Lancet 2018;392:1235–45.

[64] Woudstra P, Kalkman DN, den Heijer P, Menown IBA, Erglis A, Suryapranata H, et al. 1-year results of the REMEDEE registry: clinical outcomes after deployment of the abluminal sirolimus-coated bioengineered (combo) stent in a multicenter, prospective all-comers registry. JACC Cardiovasc Interv 2016;9:1127–34.

[65] Schoos MM, Mehran R, Dangas GD, Yu J, Baber U, Clemmensen P, et al. Gender differences in associations between intraprocedural thrombotic events during percutaneous coronary intervention and adverse outcomes. Am J Cardiol 2016;118:1661–8.

[66] Mina GS, Firouzbakht T, Modi K, Dominic P. Gender-based outcomes of bivalirudin versus heparin in patients undergoing percutaneous coronary interventions: meta-analysis of randomized controlled trials. Catheter Cardiovasc Interv 2017;90:735–42.

[67] Yu J, Mehran R, Grinfeld L, Xu K, Nikolsky E, Brodie BR, et al. Sex-based differences in bleeding and long term adverse events after percutaneous coronary intervention for acute myocardial infarction: three year results from the HORIZONS-AMI trial. Catheter Cardiovasc Interv 2015;85:359–68.

[68] Othman H, Khambatta S, Seth M, Lalonde TA, Rosman HS, Gurm HS, et al. Differences in sex-related bleeding and outcomes after percutaneous coronary intervention: insights from the blue cross blue shield of Michigan cardiovascular consortium (BMC2) registry. Am Heart J 2014;168:552–9.

[69] Chichareon P, Modolo R, Kerkmeijer L, Tomaniak M, Kogame N, Takahashi K, et al. Association of sex with outcomes in patients undergoing percutaneous coronary intervention: a subgroup analysis of the GLOBAL LEADERS randomized clinical trial. JAMA Cardiol 2020;5(1):21–9. https://doi.org/10.1001/jamacardio.2019.4296. PMID: 31693078; PMCID: PMC7029729.

[70] Yu J, Mehran R, Baber U, Ooi S-Y, Witzenbichler B, Weisz G, et al. Sex differences in the clinical impact of high platelet reactivity after percutaneous coronary intervention with drug-eluting stents: results from the ADAPT-DES Study (Assessment of Dual Antiplatelet Therapy With Drug-Eluting Stents). Circ Cardiovasc Interv 2017;10(2):e003577. https://doi.org/10.1161/CIRCINTERVENTIONS.116.003577. PMID: 28193677.

[71] Verdoia M, Pergolini P, Rolla R, Nardin M, Barbieri L, Daffara V, et al. Gender differences in platelet reactivity in patients receiving dual antiplatelet therapy. Cardiovasc Drugs Ther 2016;30:143–50.

[72] O'Donoghue ML, Bhatt DL, Stone GW, Steg PG, Gibson CM, Hamm CW, et al. Efficacy and safety of cangrelor in women versus men during percutaneous coronary intervention: insights from the cangrelor versus standard therapy to achieve optimal management of platelet inhibition (CHAMPION PHOENIX) trial. Circulation 2016;133:248–55.

[73] Lau ES, Braunwald E, Murphy SA, Wiviott SD, Bonaca MP, Husted S, et al. Potent P2Y12 inhibitors in men versus women: a collaborative meta-analysis of randomized trials. J Am Coll Cardiol 2017;69:1549–59.

[74] Blomkalns AL, Chen AY, Hochman JS, Peterson ED, Trynosky K, Diercks DB, et al. Gender disparities in the diagnosis and treatment of non-ST-segment elevation acute coronary syndromes. J Am Coll Cardiol 2005;45:832–7.

[75] Gutierrez-Chico JL, Mehilli J. Gender differences in cardiovascular therapy: focus on antithrombotic therapy and percutaneous coronary intervention. Drugs 2013;73:1921–33.

[76] Valgimigli M, Bueno H, Byrne RA, Collet JP, Costa F, Jeppsson A, et al. 2017 ESC focused update on dual antiplatelet therapy in coronary artery disease developed in collaboration with EACTS. Eur Heart J 2018;39:213–54.

[77] Levine GN, Bates ER, Bittl JA, Brindis RG, Fihn SD, Fleisher LA, et al. 2016 ACC/AHA guideline focused update on duration of dual antiplatelet therapy in patients with coronary artery disease. J Am Coll Cardiol 2016;68:1082–115.

[78] Yu J, Baber U, Mastoris I, Dangas G, Sartori S, Steg PG, et al. Sex-based differences in cessation of dual-antiplatelet therapy following percutaneous coronary intervention with stents. JACC Cardiovasc Interv 2016;9:1461–9.

[79] Hess CN, McCoy LA, Duggirala HJ, Tavris DR, O'Callaghan K, Douglas PS, et al. Sex-based differences in outcomes after percutaneous coronary intervention for acute myocardial infarction: a report from TRANSLATE-ACS. J Am Heart Assoc 2014;3:e000523.

[80] Mehilli J, Kastrati A, Dirschinger J, Bollwein H, Neumann FJ, Schomig A. Differences in prognostic factors and outcomes between women and men undergoing coronary artery stenting. JAMA 2000;284:1799–805.

[81] Cowley MJ, Mullin SM, Kelsey SF, Kent KM, Gruentzig AR, Detre KM, et al. Sex differences in early and long-term results of coronary angioplasty in the NHLBI PTCA registry. Circulation 1985;71:90–7.

[82] Heer T, Hochadel M, Schmidt K, Mehilli J, Zahn R, Kuck KH, et al. Sex differences in percutaneous coronary intervention-insights from the coronary angiography and PCI registry of the German Society of Cardiology. J Am Heart Assoc 2017;6(3):e004972. https://doi.org/10.1161/JAHA.116.004972. Erratum in: J Am Heart Assoc. 2017;6(9): PMID: 28320749; PMCID: PMC5524024.

[83] Kunadian V, Qiu W, Lagerqvist B, Johnston N, Sinclair H, Tan Y, et al. Gender differences in outcomes and predictors of all-cause mortality after percutaneous coronary intervention (data from United Kingdom and Sweden). Am J Cardiol 2017;119:210–6.

[84] Aggarwal NR, Patel HN, Mehta LS, Sanghani RM, Lundberg GP, Lewis SJ, et al. Sex differences in ischemic heart disease: advances, obstacles, and next steps. Circ Cardiovasc Qual Outcomes 2018;11:e004437.

[85] Farhan S, Baber U, Vogel B, Aquino M, Chandrasekhar J, Faggioni M, et al. Impact of diabetes mellitus on ischemic events in men and women after percutaneous coronary intervention. Am J Cardiol 2017;119:1166–72.

[86] Baber U, Giustino G, Sartori S, Aquino M, Stefanini GG, Steg PG, et al. Effect of chronic kidney disease in women undergoing percutaneous coronary intervention with drug-eluting stents: a patient-level pooled analysis of randomized controlled trials. JACC Cardiovasc Interv 2016;9:28–38.

[87] Guedeney P, Claessen BE, Mehran R, Kandzari DE, Aquino M, Davis S, et al. Small-vessel PCI outcomes in men, women, and minorities following platinum chromium everolimus-eluting stents: insights from the pooled PLATINUM diversity and PROMUS element plus post-approval studies. Catheter Cardiovasc Interv 2019;94:82–90.

[88] Fath-Ordoubadi F, Barac Y, Abergel E, Danzi GB, Kerner A, Nikolsky E, et al. Gender impact on prognosis of acute coronary syndrome patients treated with drug-eluting stents. Am J Cardiol 2012;110:636–42.

[89] Raphael CE, Singh M, Bell M, Crusan D, Lennon RJ, Lerman A, et al. Sex differences in long-term cause-specific mortality after percutaneous coronary intervention: temporal trends and mechanisms. Circ Cardiovasc Interv 2018;11:e006062.

[90] Bucholz EM, Butala NM, Rathore SS, Dreyer RP, Lansky AJ, Krumholz HM. Sex differences in long-term mortality after myocardial infarction: a systematic review. Circulation 2014;130:757–67.

[91] Kwok CS, Potts J, Gulati M, Alasnag M, Rashid M, Shoaib A, et al. Effect of gender on unplanned readmissions after percutaneous coronary intervention (from the nationwide readmissions database). Am J Cardiol 2018;121:810–7.

[92] Zheng H, Foo LL, Tan HC, Richards AM, Chan SP, Lee C-H, et al. Sex differences in 1-year rehospitalization for heart failure and myocardial infarction after primary percutaneous coronary intervention. Am J Cardiol 2019;123:1935–40.

[93] Wada H, Ogita M, Miyauchi K, Tsuboi S, Konishi H, Shitara J, et al. Contemporary sex differences among patients with acute coronary syndrome treated by emergency percutaneous coronary intervention. Cardiovasc Interv Ther 2017;32:333–40.

[94] Pancholy SB, Shantha GPS, Patel T, Cheskin LJ. Sex differences in short-term and long-term all-cause mortality among patients with ST-segment elevation myocardial infarction treated by primary percutaneous intervention: a meta-analysis. JAMA Intern Med 2014;174:1822–30.

[95] Udell JA, Fonarow GC, Maddox TM, Cannon CP, Frank Peacock W, Laskey WK, et al. Sustained sex-based treatment differences in acute coronary syndrome care: insights from the American Heart Association get with the guidelines coronary artery disease registry. Clin Cardiol 2018;41:758–68.

[96] Kosmidou I, Redfors B, Selker HP, Thiele H, Patel MR, Udelson JE, et al. Infarct size, left ventricular function, and prognosis in women compared to men after primary percutaneous coronary intervention in ST-segment elevation myocardial infarction: results from an individual patient-level pooled analysis of 10 randomized trials. Eur Heart J 2017;38:1656–63.

[97] Jakobsen L, Niemann T, Thorsgaard N, Nielsen TT, Thuesen L, Lassen JF, et al. Sex- and age-related differences in clinical outcome after primary percutaneous coronary intervention. EuroIntervention 2012;8:904–11.

[98] Mueller C, Neumann FJ, Roskamm H, Buser P, Hodgson JM, Perruchoud AP, et al. Women do have an improved long-term outcome after non-ST-elevation acute coronary syndromes treated very early and predominantly with percutaneous coronary intervention a prospective study in 1,450 consecutive patients. J Am Coll Cardiol 2002;40(2):245–50.

[99] Saw J, Mancini GBJ, Humphries KH. Contemporary review on spontaneous coronary artery dissection. J Am Coll Cardiol 2016;68:297–312.

[100] Tweet MS, Hayes SN, Pitta SR, Simari RD, Lerman A, Lennon RJ, et al. Clinical features, management, and prognosis of spontaneous coronary artery dissection. Circulation 2012;126:579–88.

[101] Nishiguchi T, Tanaka A, Ozaki Y, Taruya A, Fukuda S, Taguchi H, et al. Prevalence of spontaneous coronary artery dissection in patients with acute coronary syndrome. Eur Heart J Acute Cardiovasc Care 2016;5:263–70.

[102] Saw J. Spontaneous coronary artery dissection. Interv Cardiol 2015;10:142–3.

[103] Poon K, Bell B, Raffel OC, Walters DL, Jang I-K. Spontaneous coronary artery dissection: utility of intravascular ultrasound and optical coherence tomography during percutaneous coronary intervention. Circ Cardiovasc Interv 2011;4:e5–7.

[104] Vrints CJM. Spontaneous coronary artery dissection. Heart 2010;96:801–8.

[105] Nakashima T, Noguchi T, Haruta S, Yamamoto Y, Oshima S, Nakao K, et al. Prognostic impact of spontaneous coronary artery dissection in young female patients with acute myocardial infarction: a report from the angina pectoris-myocardial infarction multicenter investigators in Japan. Int J Cardiol 2016;207:341–8.

[106] Rashid HNZ, Wong DTL, Wijesekera H, Gutman SJ, Shanmugam VB, Gulati R, et al. Incidence and characterisation of spontaneous coronary artery dissection as a cause of acute coronary syndrome—a single-Centre Australian experience. Int J Cardiol 2016;202:336–8.

[107] Hassan S, Prakash R, Starovoytov A, Saw J. Natural history of spontaneous coronary artery dissection with spontaneous angiographic healing. JACC Cardiovasc Interv 2019;12:518–27.

[108] Lempereur M, Fung A, Saw J. Stent mal-apposition with resorption of intramural hematoma with spontaneous coronary artery dissection. Cardiovasc Diagn Ther 2015;5:323–9.

[109] Serruys PW, Cavalcante R, Collet C, Kappetein AP, Sabik JF, Banning AP, et al. Outcomes after coronary stenting or bypass surgery for men and women with unprotected left main disease the EXCEL trial. JACC Cardiovasc Interv 2018;11:1234–43.

[110] Shin E-S, Lee CW, Ahn J-M, Lee PH, Chang M, Kim M-J, et al. Sex differences in left main coronary artery stenting: different characteristics but similar outcomes for women compared with men. Int J Cardiol 2018;253:50–4.

[111] Ten Haaf M, Appelman Y, Wijns W, Steg G, Mauri L, Rademaker-Havinga T, et al. Frequency of stent thrombosis risk at 5 years in women versus men with zotarolimus-eluting compared with sirolimus-eluting stent. Am J Cardiol 2016;118:1178–86.

[112] Kerkmeijer LS, Claessen BE, Baber U, Sartori S, Chandrasekhar J, Stefanini GG, et al. Incidence, determinants and clinical impact of definite stent thrombosis on mortality in women: from the WIN-DES collaborative patient-level pooled analysis. Int J Cardiol 2018;263:24–8.

[113] Abbate A, Biondi-Zoccai GGL, Agostoni P, Lipinski MJ, Vetrovec GW. Recurrent angina after coronary revascularization: a clinical challenge. Eur Heart J 2007;28:1057–65.

[114] Lemos PA, Hoye A, Serruys PW. Recurrent angina after revascularization: an emerging problem for the clinician. Coron Artery Dis 2004;15(Suppl. 1):S11–5.

[115] Vogel B, Goel R, Kunadian V, Kalkman DN, Chieffo A, Appelman Y, et al. Residual angina in female patients after coronary revascularization. Int J Cardiol 2019;286:208–13.

[116] Kok MM, van der Heijden LC, Sen H, Danse PW, Löwik MM, Anthonio RL, et al. Sex difference in chest pain after implantation of newer generation coronary drug-eluting stents: a patient-level pooled analysis from the TWENTE and DUTCH PEERS trials. JACC Cardiovasc Interv 2016;9:553–61.

[117] Corcoran D, Young R, Adlam D, McConnachie A, Mangion K, Ripley D, et al. Coronary microvascular dysfunction in patients with stable coronary artery disease: the CE-MARC 2 coronary physiology sub-study. Int J Cardiol 2018;266:7–14.

[118] Yip A, Saw J. Spontaneous coronary artery dissection—a review. Cardiovasc Diagn Ther 2015;5:37–48.

[119] Authors/Task Force members, Windecker S, Kolh P, Alfonso F, Collet J-P, Cremer J, et al. 2014 ESC/EACTS guidelines on myocardial revascularization: the task force on myocardial revascularization of the European Society of Cardiology (ESC) and the European Association for Cardio-Thoracic Surgery (EACTS) developed with the special contribution of the European Association of Percutaneous Cardiovascular Interventions (EAPCI). Eur Heart J 2014;35:2541–619.

[120] Fihn SD, Blankenship JC, Alexander KP, Bittl JA, Byrne JG, Fletcher BJ, et al. 2014 ACC/AHA/AATS/PCNA/SCAI/STS focused update of the guideline for the diagnosis and management of patients with stable ischemic heart disease: a report of the American College of Cardiology/American Heart Association Task Force on Practice Guidelines, and the American Association for Thoracic Surgery, Preventive Cardiovascular Nurses Association, Society for Cardiovascular Angiography and Interventions, and Society of Thoracic Surgeons. J Am Coll Cardiol 2014;64:1929–49.

[121] Doshi R, Patel K, Decter D, Jauhar R, Meraj P. Gender disparities with the use of percutaneous left ventricular assist device in patients undergoing percutaneous coronary intervention complicated by cardiogenic shock: from pVAD working group. Indian Heart J 2018;70:S90–5.

[122] Joseph SM, Brisco MA, Colvin M, Grady KL, Walsh MN, Cook JL, et al. Women with cardiogenic shock derive greater benefit from early mechanical circulatory support: an update from the cVAD registry. J Interv Cardiol 2016;29:248–56.

[123] Sharma S, Polak S, George Z, LeDoux J, Sohn R, Stys A, et al. Management of spontaneous coronary artery dissection complicated by cardiogenic shock using mechanical circulatory support with the Impella device. Catheter Cardiovasc Interv 2019. https://doi.org/10.1002/ccd.28677. Epub ahead of print. PMID: 31876350.

[124] McIlvennan CK, Lindenfeld J, Kao DP. Sex differences and in-hospital outcomes in patients undergoing mechanical circulatory support implantation. J Heart Lung Transplant 2017;36:82–90.

[125] Hsich EM. Sex differences in advanced heart failure therapies. Circulation 2019;139:1080–93.

[126] Bogaev RC, Pamboukian SV, Moore SA, Chen L, John R, Boyle AJ, et al. Comparison of outcomes in women versus men using a continuous-flow left ventricular assist device as a bridge to transplantation. J Heart Lung Transplant 2011;30:515–22.

[127] Sherazi S, Kutyifa V, McNitt S, Papernov A, Hallinan W, Chen L, et al. Effect of gender on the risk of neurologic events and subsequent outcomes in patients with left ventricular assist devices. Am J Cardiol 2017;119:297–301.

[128] Hsich EM, Naftel DC, Myers SL, Gorodeski EZ, Grady KL, Schmuhl D, et al. Should women receive left ventricular assist device support?: findings from INTERMACS. Circ Heart Fail 2012;5:234–40.

Editor's Summary: Interventions in Ischemic Heart Disease

Niti Aggarwal

Increased response to treatment
- Greater benefit from DES due to greater reduction in intimal hyperplasia
- Exhibit greater reduction in plaque regression with statins

Higher bleeding risk
- Increased risk of post-PCI bleeding
- Favor shorter duration of DAPT

Less radial access site
- Smaller radial arteries
- More prone to spasm

Worse outcomes
- Higher adverse events after PCI
- Greater burden of residual symptoms

Smaller vessel size
- Less plaque burden & thrombus load needed for obstructive disease
- Greater risk of restenosis
- Incomplete revascularization resulting in more residual symptoms

Plaque burden & composition
- Less plaque volume
- Plaque erosion >> plaque rupture
- Less coronary calcifications
- Less fibrous tissue
- More lipid rich plaque

Higher FFR values
- Higher FFR for any given stenosis due to small vessel size, BSA and supplied myocardium

Coronary tortuosity more common
Associated with increased calcifications, more ischemia and increased procedural complexity

Coronary Artery Disease Features in Women Compared to Men. Women have less plaque burden, lower coronary artery calcium scores, and higher FFR scores for the same degree of stenosis. They also have more coronary tortuosity, higher bleeding risk and smaller more spasmodic radial arteries which may explain the worse outcomes in women after PCI compared to men. BSA, body surface area; DAPT, dual antiplatelet therapy; DES, drug eluding stent; FFR, fractional flow reserve; PCI, percutaneous coronary intervention. *Image courtesy of Niti R. Aggarwal.*

Chapter 7

Stable Ischemic Heart Disease

Renee P. Bullock-Palmer, Pamela Telisky, and Cynthia Kos

Clinical Case

A 52-year-old female with a 15-year history of benign essential hypertension, type 2 diabetes mellitus, hyperlipidemia, and tobacco smoking presented with a 10-month history of exertional dyspnea and exertional chest pain. The chest pain was described as a burning substernal sensation that was moderate in intensity and relieved with rest. She also has noted a reduction in her functional capacity to less than two blocks over the last year. Her blood pressure on examination was 128/80 and heart rate was 70 beats/min. Her cardiovascular examination was unremarkable with normal heart sounds, regular rate, and rhythm with no audible murmurs and no gallops as well as no rubs. Chest was clear to auscultation bilaterally. There was no jugular venous distension. No pedal edema.

Continued

Sex Differences in Cardiac Disease. https://doi.org/10.1016/B978-0-12-819369-3.00027-7

Clinical Case—cont'd

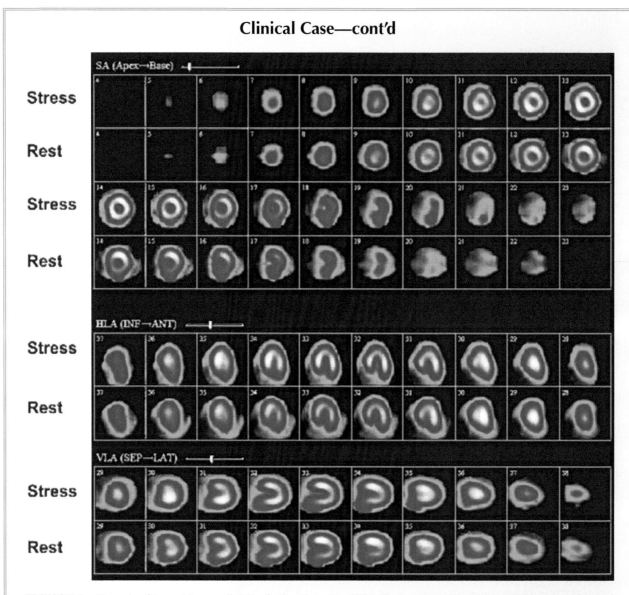

FIGURE 1   Exercise Stress Myocardial Perfusion Imaging With Technetium 99m-SPECT Showing Normal Myocardial Perfusion on Stress and Rest Images. Left ventricular ejection fraction was normal at 65%. ANT, anterior; HLA, horizontal long axis; INF, inferior; LAT, lateral; SA, short axis; SEP, septal; SPECT, single-photon emission computed tomography; VLA, vertical long axis.

Electrocardiogram (ECG) showed normal sinus rhythm with secondary ST-T changes. She gave a history of having a prior exercise nuclear stress test that was performed 15 months ago which was reported as being normal but it was noted that her exercise capacity was reduced for age at 6.4 metabolic equivalents (METs). She walked for less than 5 min on the treadmill; the test was stopped due to intolerable dyspnea after achieving 85% of maximum predicted heart rate for age. No ischemic ECG noted. Heart rate and blood pressure response were normal. Myocardial perfusion was described as being normal (**Figure 1**) with normal systolic wall motion of the left ventricle, calculated left ventricular ejection fraction was 65%, normal. Patient presented for a second opinion for her symptoms but was reluctant to have any further tests that exposed her to radiation unless it was deemed to be necessary. How would you manage her going forward?

Abstract

Stable ischemic heart disease is defined as a condition in which patients with established or suspected ischemic heart disease have a myocardial fixed supply and demand mismatch resulting in chronic angina or chronic anginal equivalents such as exertional dyspnea occurring with a predictable workload with no suspected unstable or acute symptoms as to suggest an acute coronary syndrome. There are sex-specific differences in the pathophysiology and presentation of coronary artery disease (CAD). These differences determine the diagnostic and management strategy in women. Appreciation of these differences determines treatment outcomes. The prevalence of nonobstructive CAD is greater in women compared to men. However, the clinical outcomes are worse in these women with nonobstructive CAD compared to men. Therefore, the diagnostic strategy that emphasizes anatomic assessment with luminogram with coronary angiography rather than physiologic assessment to determine endothelial dysfunction will often result in missed diagnosis of women with nonobstructive CAD.

Introduction

Accurate and timely diagnosis of coronary artery disease (CAD) is critical in appropriately and adequately treating stable ischemic heart disease (SIHD) in women. Unfortunately, according to the National Health and Nutrition Examination Survey (NHANES) data from 1992 to 2002, CAD is prevalent in more than half of the female population aged >55 years [1]. Despite this fact, when compared with men, fewer women (78% vs. 73%, respectively) are referred for exercise treadmill stress testing and despite a higher angina class, even fewer women (48% vs. 31%, respectively) are referred for diagnostic coronary angiogram [2]. Differences between men and women also exist with regard to presentation and manifestation of CAD. In fact, 58% of women who die suddenly from an acute myocardial

Key Points

1. Timely diagnosis of CAD is critical in appropriately treating SIHD in women.

2. CAD is prevalent in more than half of the female population older than 55 years of age.

3. Fewer women compared to men are referred for exercise treadmill stress testing and diagnostic coronary angiogram.

4. Gender differences exist with regard to presentation and manifestation of CAD.

infarction (MI) had no classic warning symptoms [3]. In addition, 38% women vs. 25% of men will die within 1 year of having their first MI [3]. Therefore, early detection of symptoms and accurate and timely diagnosis of CAD is important to decrease CAD-related mortality.

Definition and Prevalence

Myocardial ischemia is defined as a decreased supply of blood flow to the myocardium relative to the workload demands of the myocardium. This supply and demand mismatch may be due to either an obstruction of the macrovascular or microvascular circulation or increased demands of the myocardium relative to the supply of blood flow to the myocardium due to anemia, infection, or high-output state such as thyrotoxicosis. Obstruction of the macrovascular circulation may be due to coronary atherosclerosis, vasospasm of the coronary arteries, or coronary anomalies resulting in compression of the ostium of the epicardial vessel. Microvascular obstructive CAD is often secondary to endothelial dysfunction and has a greater prevalence in females compared to men [4]. Microvascular CAD is an often underappreciated and underrecognized cause of myocardial ischemia.

SIHD is defined as a condition in which patients with established or suspected ischemic heart disease (IHD) have a myocardial fixed supply and demand mismatch resulting in chronic angina or chronic anginal equivalents such as exertional dyspnea occurring with a predictable workload with no suspected unstable or acute symptoms as to suggest an acute coronary syndrome (ACS). At least 50% of patients with SIHD have symptoms of angina pectoris [5]. As of 2016, over 5 million females in the United States have angina pectoris with a greater prevalence in females compared to males [5.1 million (3.7%) females vs. 4.3 million (3.5%) males] [6]. This gender disparity is greater in the non-Hispanic (NH) black female (3.8% of NH black females vs. 3.6% of NH black males) as well as the Hispanic female population (3.6% of Hispanic females vs. 2.6% of Hispanic males) when compared to non-Hispanic white females (3.8% of NH white females vs. 3.8% of NH white males) [6]. In non-Hispanic Asian or Pacific Islander females, the prevalence is less compared with their male counterparts (1.6% of females vs. 2.0% of males) [6].

While the annual mortality related to CAD has decreased among females in the United States and based on epidemiologic data in 2017 is less than that in men [152,619 (41.7%) in females vs. 213,295 (58.3%) in males], this decrease in mortality has plateaued in the last 4 years. In addition, CAD is still the leading cause of mortality for women in the United States and there is an increase in the mortality rates in younger women who are less than 50 years of age compared to older women and compared to their male counterparts [6].

Key Points

① Myocardial ischemia is defined as a supply and demand mismatch that may be secondary to obstructive macrovascular or microvascular disease.

② Microvascular obstructive CAD has a greater prevalence in females compared to males and is often an underrecognized cause of myocardial ischemia.

③ There is a greater prevalence of angina in females compared to males, with sex disparity being greatest in NH black females and Hispanic females.

④ CAD is still the leading cause of mortality for women in the United States with an increase in the mortality rates in younger women aged <50 years compared to older women and compared to their male counterparts.

Sex Differences in Pathophysiology

Despite the fact that angina is more prevalent in females compared to males, females are more likely to have nonobstructive CAD compared to males and are more likely to have coronary microvascular dysfunction (CMD) when compared to males [7]. Adults who underwent coronary angiography for acute ST-segment elevation and non-ST-segment-elevation myocardial infarction (MI) in the National Cardiovascular Data Registry ACTION Registry-GWTG (Acute Coronary Treatment and Intervention Outcomes Network Registry-Get With the Guidelines) from 2007 to 2014 showed that among 322,523 patients with MI, myocardial infarction with nonobstructive coronary arteries (MINOCA) occurred in 18,918 (5.9%) [7]. MINOCA was more common in women than men (10.5% vs. 3.4%; $P < 0.0001$), and women had higher mortality than men overall (3.6% vs. 2.4%; $P < 0.0001$) [7]. This is particularly so when there is the presence of CMD with a coronary flow reserve (CFR) of <2 [7].

Sex differences in pathophysiology of CAD is due to a number of sex-related factors due to differences in coronary artery anatomy and function. There are also sex differences in risk factor burden and impact on CAD. Females also have several unique nontraditional CAD risk factors such as systemic inflammatory diseases. Presentation of ischemic heart disease also varies between the sexes [7].

Coronary Artery Anatomy and Pathology

The myocardium is supplied by a complex network of coronary arteries and prearterioles. The epicardial coronary arteries are a low-resistance arterial system that supplies a complex high-resistance network of prearterioles which then supply the coronary microcirculation capillary bed. The prearteriolar circulatory system contributes to most of the resistance circuit of the heart. This periarteriolar circulatory system contributes to the autoregulatory system which serves to increase myocardial blood supply to match the myocardial demand of the heart [7].

Females have been noted to have smaller-sized coronary arteries [8] when compared to males. This sex difference is independent of left ventricular (LV) mass, body mass index (BMI), body surface area, and age. The smaller-sized coronary arteries are thought to result in higher shear stress in the coronary arteries in females compared to males. This higher shear stress is also believed to result in the sex differences in the distribution of CAD as lower shear stress is believed to result in a more focal area of coronary plaque formation leading to stenosis compared with higher shear stress which is believed to result in more diffuse coronary plaque deposition. This diffuse coronary plaque distribution is thought to contribute to the higher prevalence of nonobstructive CAD in females compared to males [10.5% (F) vs. 3.4% (M), $P < 0.0001$] [8, 9].

Another additional sex difference in CAD involves the fact that the occurrence of coronary plaque erosion is more common in females compared to males, while coronary plaque rupture is more common in males [10]. Culprit lesions in men that result in plaque rupture usually have a high plaque burden with a thin fibrous cap and a large atherosclerotic lipid-rich core with positive remodeling of the vessel. However, in women the characteristic plaque features leading to plaque erosion include more fibrous plaque ($P < 0.001$), less thin-cap fibroatheroma ($P < 0.001$), a lower plaque burden ($P = 0.003$), and a reduced remodeling index ($P = 0.003$) [11].

Coronary Function

Coronary artery endothelial dysfunction is more common in females compared to males despite the higher incidence of nonobstructive CAD in females [10]. This coronary artery endothelial dysfunction occurs in the microvascular myocardial circulation and is therefore usually missed on routine invasive coronary angiography as well as coronary cardiac computed tomography. Although endothelial dysfunction in the presence of nonobstructive CAD (INOCA) is more common in females compared with males, this finding is not benign. INOCA in females is associated with a higher symptom burden as well as a higher utilization of medical services due to their higher symptom burden [6]. There was a statistically significant difference in angina hospitalization between women with no CAD and women with INOCA ($P = 0.03$) [12]. When anginal characteristics were added into the multivariate model, including the chest pain descriptors of typical angina, number of symptoms checked, pain frequency, and usual severity of pain (severity at its worst), "chest pain severity at its worst" had a statistically significant hazard ratio (HR) of 2.16 [95% confidence interval (CI) 1.29–3.64, $P = 0.004$] [12]. INOCA in females was associated with an almost fourfold greater risk for recurrent hospitalization because of angina (stable and unstable), heart failure, stroke, and MI (HR 3.9, 95% CI 3.3–4.6, $P < 0.001$) vs. asymptomatic reference individuals [12]. Within the first year, women with INOCA had a higher risk of major adverse cardiovascular events (MACE) than men with INOCA (adjusted HR 2.43, 95% CI 1.08–5.49). Furthermore, women with INOCA had a 2.55-fold higher risk of MACE than women with no CAD (95% CI 1.33–4.889) [13].

Risk Factor Burden and Emerging Risk Factors in Women

The incidence of CAD in females generally occurs 10 years later when compared with men. This later occurrence of this pathology is thought to be contributed by the onset of menopause with decreased estrogen levels and the loss of the cardioprotective effects of estrogen particularly as it relates to endothelial function [11].

Traditional Risk Factors

The traditional risk factors related to CAD include hypertension, diabetes mellitus (DM), hyperlipidemia, tobacco smoking, obesity, physical inactivity, increased age, and family history of premature CAD. At least 50% of age-related increases in the prevalence of IHD in women is due to increased prevalence of IHD risk factors associated with increasing age [11]. Females have a greater risk factor burden compared to males; over 80% of middle-aged women have at least one traditional risk factor [11]. Hypertension, DM, and tobacco smoking are more potent risk factors for MI in females compared with males with odds ratios of 1.5, 1.6, and 1.3, respectively [10]. In addition, females with DM have a sixfold higher rate of CAD-related mortality compared to males [11].

Physical inactivity in women is associated with greater rates of traditional risk factors such as obesity, DM, hypertension, and dyslipidemia [11]. Physical inactivity is also more prevalent in older females compared to similarly aged males [11]. High-risk minority females have higher rates of physical inactivity.

Menopause is associated with increased triglyceride levels as well as increased low-density lipoprotein (LDL) levels and decreased high-density lipoprotein (HDL) levels in women [11]. Increased triglyceride levels have been shown to be a stronger predictor of IHD risk in women compared to men. However, it is still being debated whether this relationship is related to the ratio of triglycerides to HDL [11].

Emerging Risk Factors

There have been several emerging risk factors for IHD in females; these include systemic inflammatory disorders such as systemic lupus erythematosus (SLE) and rheumatoid arthritis (RA). SLE has been associated with a threefold higher risk of IHD while the presence of RA results in a similar risk of IHD as seen in females with DM. There are also other emerging risk factors such as early natural or surgical-induced menopause, which confers a 4.5-fold high risk for IHD [10]. Mental stress is also an emerging risk factor with a greater prevalence of depression in females which is associated with a twofold higher risk for IHD [10]. There are also pregnancy-related emerging risk factors such as gestational diabetes, which is associated with a fourfold higher risk of developing DM and is also associated with a 59% higher risk of MI [10]. Gestational hypertension and preeclampsia are also associated with a threefold higher risk of IHD [10].

Key Points

1. Although angina is more prevalent in females compared to males, females are more likely to have nonobstructive CAD compared to males and more likely to have coronary microvascular disease compared to males.

2. There are sex differences in pathophysiology of CAD due to differences in coronary anatomy and function.

3. Coronary artery endothelial dysfunction is more common in females compared to males despite the higher incidence of nonobstructive CAD in females.

4. Women with INOCA had a higher risk of MACE than men with INOCA.

5. In addition to the traditional risk factors for CAD, there are emerging risk factors that are unique to women.

Clinical Presentation

Definition of Angina

Three features of chest pain that are typically thought to be anginal include exertional exacerbation, a substernal location, and relief with rest or with nitroglycerin within 10 min. Chest pain is classified as being typical angina if all three of these features are present; if two out of the three features are present, it is classified as atypical angina; and if one or none of the three features are present, it is classified as nonanginal.

CH
7

Women with IHD often present with chest pain similarly to men [73.2% of women vs. 72.3% of men (*P* = 0.30)] [14], and women were more likely to characterize the chest pain as "crushing/pressure/squeezing/tightness" [52.5% of women vs. 46.2% of men (*P* < 0.001)] [14]. Women were more likely to present with three or more associated symptoms than men (e.g., epigastric symptoms, palpitations, and pain or discomfort in the jaw, neck, arms, or between the shoulder blades; 61.9% of women vs. 54.8% of men, *P* < 0.001) [15]. Despite the fact that females with IHD often present with chest pain, the utilization of several global risk score assessment tools such as Framingham, Diamond, and Forrester, modified Diamond and Forrester and combined Diamond-Forrester, and Coronary Artery Surgery Study (CASS) often confer a lower risk of cardiac events and a lower pretest likelihood for coronary disease for females compared to males [14]. A higher percentage of women were characterized by their providers as having low risk (< 30%) pretest probability for obstructive CAD compared to men, while a higher percentage of men were characterized as having high risk (> 70%) [14]. This underappreciation and underrecognition of the "at risk" female usually results in misdiagnosis or delayed diagnosis of IHD in women [11]. This missed diagnosis and delayed diagnosis often result in worse outcomes for females compared to males.

Clinical Recognition

The prompt diagnosis of IHD is challenged by underrecognition of risk factors and pertinent symptoms by women themselves as well as healthcare professionals. Average age at first MI is 65.6 years for males and 72.0 years for females [6]. Cardiac risk factors such as older age, hyperlipidemia, smoking, insulin resistance, obesity, and family history affect both genders; however, prolonged smoking and insulin resistance are proving to be more harmful to women than men [7]. Despite educational efforts such as the Red Dress Campaign, a recent survey found that 45% of women are unaware of their risk factors and the deadly impact of delayed recognition and treatment of IHD [16]. Women have experienced on average 30-min delays from symptom onset to hospital evaluation compared to their male counterparts [17]. Cardiovascular risk factors are often comorbid and confounding in women at the time of presentation. Healthcare professionals have access to point of care calculators to characterize risk; however, these algorithms have limitations. For example, nearly 75% of asymptomatic middle-aged to octogenarian women are considered low risk by the Framingham risk score [18], whereas the Reynold risk score reclassified 15% of intermediate-risk women to high risk in the Women's Health Study. In addition, these risk scores often do not incorporate the novel risk factors affecting women such as early menopause and disorders of pregnancy. If pretest probability tools such as the Diamond-Forrester approach are utilized, women who present with atypical but common symptoms of dyspnea,

nausea, fatigue, jaw pain, or lightheadedness often experience a delay in the recognition of ischemia [19]. Delayed recognition leads to delayed revascularization, which may account for similar adverse outcome rates with men despite a lower atherosclerotic burden.

Assessment of Pretest Probability

The assessment of a patient's pretest probability for having obstructive CAD is crucial in determining the best diagnostic path. The clinically assessed pretest probability of CAD determines the type of test that should be ordered for the patient to determine the presence of obstructive CAD in the patient presenting with angina or anginal equivalent. The pretest probability is based on sex, age, and symptoms [20]. For example, a 39-year-old female presenting with atypical angina would have a very low (< 10%) pretest probability of having obstructive CAD. However, a female aged > 50 years presenting with similar atypical angina will have an intermediate (10–90%) pretest probability of having obstructive CAD. Females aged > 60 years presenting with atypical angina or nonanginal chest pain will have an intermediate pretest probability of having obstructive CAD; those presenting with typical angina will have a high (> 90%) pretest probability of having obstructive CAD. Pretest probability in females should also take into account the presence of clinical comorbidities and the presence of cardiac risk factors, particularly for patients with DM as well as females with emerging cardiac risk factors [10].

Key Points

 Incidence of IHD occurs a decade later in women than men.

 Smoking and insulin resistance are among the most harmful risk factors to women.

 On average, women presenting with classic angina present to the hospital later, wait longer for evaluation, and have longer door-to-balloon times than men.

 The assessment of a patient's pretest probability for having obstructive CAD is crucial in determining the best diagnostic path.

Role of Noninvasive Imaging

Noninvasive cardiac imaging has a very important role in diagnosing IHD as well as following females with IHD to determine disease progression in appropriately selected cases. Women have a higher symptom burden with regards to angina despite a lower incidence of obstructive CAD [10].

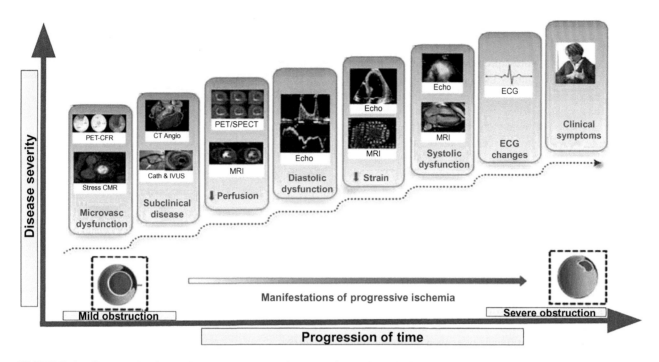

FIGURE 2 **Schematic Outlining the Ischemic Cascade in Patients With Coronary Artery Disease (CAD).** Angio, angiogram; cath, catheterization; CMR, cardiac magnetic resonance; CT, computed tomography; ECG, electrocardiogram; echo, echocardiogram; IVUS, intravascular ultrasound; MRI, magnetic resonance imaging; PET, positron emission tomography; PET-CFR, positron emission tomography-coronary flow reserve; SPECT, single-photon emission tomography. *Reproduced with permission from [21].*

As CAD progresses over time, there are various stages that occur during the disease process. The first stage usually is subclinical and begins with the deposition of cholesterol plaque along the intima of the coronary vessels resulting in fatty streaks. These plaques are usually nonobstructive and therefore nonischemic and can usually be detected by coronary computed tomography (CT) angiogram and coronary CT without contrast for coronary artery calcium scoring. The next stage of this disease process is the development of endothelial dysfunction, which can be detected by coronary malperfusion with impaired CFR that may be depicted on cardiac positron emission tomography (cPET) myocardial blood flow assessment or on stress cardiac magnetic resonance (CMR). As CAD progresses to further layering of plaque and the development of intermediate obstructive disease, this can be detected with perfusion abnormalities on nuclear myocardial perfusion single-photon emission computed tomography (SPECT) and cPET imaging suggesting the presence of myocardial ischemia. The further stages of this disease process usually results in significant obstruction of coronary blood flow (CBF) resulting in a greater degree of ischemia that then causes myocardial wall motion abnormalities that can be detected by stress echocardiography or cardiac stress magnetic resonance imaging. At the extreme end of this disease spectrum, as the severity of the obstruction progresses to near-occlusion of blood flow, more severe ischemia develops resulting in ST-T segment changes on electrocardiography (ECG), which can be detected on exer-

cise stress ECG. At the extreme end of the progression, the patient may develop angina at rest. The progression of CAD along these stages is known as the "ischemic cascade," which should be considered when determining the best diagnostic strategy for patients who may present at different stages of this disease spectrum (**Figure 2**) [7, 20, 21].

Key Points

1   **Women have a higher symptom burden with regards to angina despite a lower incidence of obstructive CAD.**

2   **The "ischemic cascade" should be considered when determining the best diagnostic strategy for patients who may present at different stages of this disease spectrum.**

Noninvasive Functional and Physiologic Assessment

The type of test ordered to diagnose obstructive CAD is determined by the patient's functional status, ECG, and risk factors. The indications, strengths, and limitations of these tests will be discussed in the next few sections of this chapter.

Exercise Treadmill Stress Testing

Patients in whom the presence of SIHD is suspected with a low-to-intermediate pretest probability should be referred for exercise treadmill stress ECG (exercise tolerance test, ETT) if they have good functional status, Duke Activity Status Index (DASI) score >5 metabolic equivalents (METs), are not on digoxin with an interpretable ECG with no baseline ST-T segment abnormalities, no left bundle branch block (LBBB), no ventricular paced rhythm, and no ECG [preexcitation (Wolff-Parkinson-White) pattern]. The WOMEN trial indicated no advantage of ETT with myocardial perfusion imaging (MPI) over ETT alone in low- to intermediate-risk females who are able to exercise, as there was no difference after 2 years of follow-up in MACE (98.0% for ETT and 97.7% for MPI; $P=0.59$) [22].

Functional status provided with an ETT is an important clinical assessment that determines patient cardiovascular outcomes. The National Institutes of Health (NIH) sponsored the Women's Ischemia Syndrome Evaluation (WISE) study, which showed that female patients with poorer functional status as determined by the DASI had a shorter time to death or MI than did women with greater functional status and a higher DASI score [23]. There are also prognostic markers on ETT such as the Duke treadmill score, heart rate recovery, blood pressure response to exercise, and the presence of ventricular ectopy with exercise. Patients with a high-risk Duke treadmill score (<−10) even with normal myocardial perfusion have been shown to have a high annual cardiac event rate of 3.6% [24].

TABLE 1 Sex-Specific Sensitivity and Specificity of Several Types of Cardiac Stress Tests in Women Compared to Coronary Angiography

Test	Sensitivity (%)	Specificity (%)
Exercise ECG	61	70
Stress echocardiography	81	86
Stress myocardial perfusion SPECT	82	81
Pharmacological myocardial perfusion SPECT	89	65
Pharmacological myocardial perfusion PET	90	89
Pharmacological stress CMRI	89	84

CMRI, cardiac magnetic resonance imaging; ECG, electrocardiogram; PET, positron emission tomography; SPECT, single-photon emission computed tomography. *Refs. [25–30].*

ETT has decreased sensitivity (Sn) and specificity (Sp) for the detection of obstructive CAD (Sn of 61% and Sp of 70%) [25] (**Table 1**). This decreased sensitivity may be attributed to the fact that female patients are less able to exercise to the maximal limit due to the presence of obesity, orthopedic limitations, and/or diabetes, and they are less likely to achieve their target heart rate, leading to a submaximal stress test with decreased sensitivity [31]. Exercise-induced ST-T segment depression on ECG in the absence of CAD has been described in women in relation to fluctuating estrogen levels during the menstrual cycle or from hormonal replacement therapy in menopause, thus resulting in decreased diagnostic accuracy compared to men [32].

Stress Imaging Tests With Echocardiography

Exercise is the preferred mode of stress testing as this provides important clinical and physiologic hemodynamic data that pharmacologic testing cannot provide. In patients who are able to exercise but have an uninterpretable ECG, exercise stress testing with adjunctive imaging with echocardiography or SPECT MPI is recommended [33, 34]. In females who are unable to exercise with an uninterpretable ECG, pharmacologic stress testing with dobutamine stress echocardiography or pharmacologic MPI with SPECT or PET is recommended [35].

Key Points

1. **Women with low-to-intermediate pretest probability who are able to exercise can undergo ETT without imaging with reliable results.**

2. **Functional status provided with an ETT is an important clinical assessment that determines patient cardiovascular outcomes.**

3. **Patients with a high-risk Duke treadmill score (<−10) even with normal myocardial perfusion have been shown to have a high annual cardiac event rate of 3.6%.**

4. **Exercise-induced ST-T segment depression on ECG in the absence of CAD has been described in women in relation to fluctuating estrogen levels during the menstrual cycle or from hormonal replacement therapy in menopause, thus resulting in decreased diagnostic accuracy compared to men.**

Using aggregated data available in studies including approximately 1000 women with suspected CAD, stress echocardiography has demonstrated good diagnostic accuracy for diagnosing CAD with a sensitivity of 81% (89% in women with multivessel disease) and specificity of 86%; overall accuracy was 84% [25] (**Table 1**).

Stress echocardiography provides the ability to assess myocardial systolic wall motion abnormalities immediately postexercise and myocardial risk may be further refined with the calculation of the wall motion score index (WMSI). A high WMSI of >1.7 has been associated with a greater likelihood of the presence of severe multivessel CAD and/or severe left main CAD [36]. The use of ultrasound enhancing agents (UEA) provides improved visualization of the endocardium and therefore improved test accuracy [37]. This is particularly important in obese females and female patients with large breasts or breast implants. UEA is also useful to assess myocardial perfusion with stress echocardiography and has been shown to improve test accuracy when combined with wall motion analysis [38, 39]. Advanced techniques such as myocardial strain imaging can also be used to assess the effect of exercise on myocardial mechanical function and is an emerging area in myocardial ischemia detection [40]. Dobutamine echocardiography (DSE) is also useful in determining the presence of viable hibernating myocardium that will likely improve with coronary artery revascularization (**Figure 3**) [26]. Fluorodeoxyglucose 18 (FDG-18) cardiac PET, when compared with other viability studies such as DSE, thallium-201, technetium-99m (Tc-99m), and CMR, has the highest sensitivity (92%) and negative predictive value (87%) in the detection of viable hibernating myocardium. However, DSE has the greatest specificity (78%) and positive predictive value (75%) in this detection when compared with these other imaging modalities [26].

Key Points

1. **Exercise is the preferred mode of stress testing as this provides important clinical and physiologic hemodynamic data that pharmacologic testing cannot provide.**

2. **A high WMSI of > 1.7 has been associated with a greater likelihood of the presence of severe multivessel CAD and/or severe left main CAD.**

3. **The use of ultrasound enhancing agents (UEA) provides improved visualization of the endocardium and therefore improved test accuracy.**

4. **Advanced techniques such as myocardial strain imaging can also be used to assess the effect of exercise on myocardial mechanical function and is an emerging area in myocardial ischemia detection.**

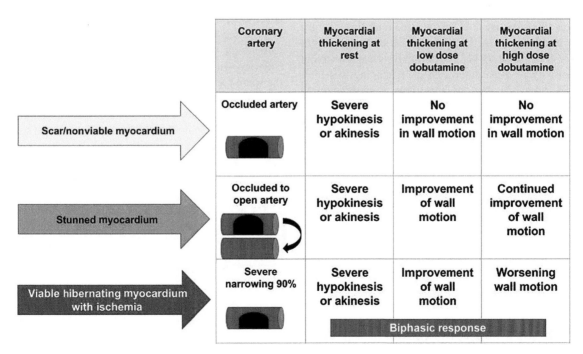

	Coronary artery	Myocardial thickening at rest	Myocardial thickening at low dose dobutamine	Myocardial thickening at high dose dobutamine
Scar/nonviable myocardium	Occluded artery	Severe hypokinesis or akinesis	No improvement in wall motion	No improvement in wall motion
Stunned myocardium	Occluded to open artery	Severe hypokinesis or akinesis	Improvement of wall motion	Continued improvement of wall motion
Viable hibernating myocardium with ischemia	Severe narrowing 90%	Severe hypokinesis or akinesis	Improvement of wall motion	Worsening wall motion
			Biphasic response	

FIGURE 3 **Diagram Outlining the Myocardial Response to Intravenous Dobutamine in the Presence of Myocardial Scar With Nonviable Myocardium, Stunned Myocardium, and Viable Hibernating Myocardium.** Differentiating infarcted nonviable (scar) myocardium from viable hibernating myocardium and stunned myocardium.

Stress Imaging With MPI

Stress imaging with SPECT MPI is recommended in symptomatic patients with suspected or established SIHD with an uninterpretable ECG. Although the ideal stress agent is exercise, for patients who are unable to exercise, pharmacologic stress testing with more often vasodilatory stress agents such as regadenoson, adenosine, or dipyridamole may be used. Less frequently inotropic and chronotropic stress agents such as dobutamine may also be used [41]. Dypyridamole, adenosine, and dobutamine dosing is weight based, which may pose a challenge for obese females especially when the limits of the dosing of these drugs are reached. However, regadenoson does not require weight-based dosing with a rapid onset of action, and therefore has a much shorter infusion time, and due to this agent being a more selective A2A adenosine receptor agonist there are fewer side effects compared to the other agents [41]. The ADVANCE MPI trial in which a third of the study population was female showed that when compared to adenosine, regadenoson had a good safety profile, was tolerated well, and was equally effective in females as in males [42].

An analysis of over 30 studies evaluating women with suspected or known CAD had shown that SPECT MPI had a sensitivity and specificity of 82% and 81%, respectively, for diagnosing obstructive CAD [27] (**Table 1**). With regards to coronary territory, it has been shown that vasodilator stress SPECT MPI with Tc-99m had sensitivity, specificity, and normalcy rates of 93%, 78%, and 88%, respectively [30]. However, similar to men, in women the sensitivity of the detection of obstructive CAD in the left circumflex territory was lower relative to the left anterior descending and right coronary artery territory [43]. Meta-analysis has shown that the sensitivity of dipyridamole SPECT imaging was 89% (95% CI, 84–93%), but the specificity of dipyridamole SPECT imaging was lower at 65% (95% CI, 54–74%) [44] (**Table 1**). Use of gated SPECT MPI with Tc-99m has improved the accuracy of SPECT MPI with specificity (Sp) of 92.2% when compared with nongated SPECT MPI with Tc-99m (Sp of 84.4%) or with thallium-201 (Tl-201) (Sp of 67.2%); this is due to better image quality with Tc-99m and the utility of wall motion assessment with gated SPECT to rule out attenuation artifact [45].

SPECT MPI is an established widely available technique that has been well studied with regards to diagnosing and risk-stratifying patients with an intermediate pretest probability for the presence of SIHD [32]. Besides myocardial perfusion, there are several other useful prognostic data that can be determined from stress testing with SPECT MPI such as LV ejection fraction and systolic wall motion post-stress as well as transient ischemic dilation and ischemic ECG changes with stress testing, particularly vasodilatory stress testing [46–50]; these can be used to further risk-stratify patients as low, intermediate, and high risk [32].

Sex-specific limitations of SPECT MPI in females, which have been summarized in **Table 2**, include potentially lower sensitivity for the detection of obstructive CAD in women due to lower myocardial mass in females and depth-dependent blur on SPECT MPI [32]. In addition, due to smaller hearts and higher LV ejection fraction in females, LV end-systolic volumes may be underestimated, resulting in higher normal limits for transient ischemic dilation (TID). Utilization of sex-based normal limits of LVEF and LV volumes is recommended to improve the diagnostic accuracy of SPECT MPI [32].

There are a few limitations of SPECT MPI that are especially relevant for women, which include attenuation from soft tissue such as large and/or dense breasts. This attenuation can also occur in obese patients as well, and reduces the specificity of the test. While gated SPECT MPI can assist with determining soft tissue attenuation vs. a real perfusion defect, this is useful for primarily fixed defects on stress and rest SPECT MPI images. However, there are several technological advances that have occurred in SPECT MPI with regards to attenuation correction with line source or low-dose CT emission imaging. These techniques have improved the diagnostic accuracy of SPECT MPI. In addition, the use of combination of supine and prone imaging on the standard Anger camera has been shown to improve the specificity and normality of SPECT MPI without decreasing the sensitivity of the test [52]. More recently, newer technology with wide beam iterative reconstruction (IR) software with the standard Anger camera have improved the spatial resolution of SPECT MPI due to noise reduction and correcting for depth-dependent blur [33] with resolution recovery. This improved spatial resolution allows for faster acquisition times, and the use of combined supine and prone SPECT MPI stress imaging has become more practical and has also been shown to improve the diagnostic accuracy of SPECT MPI [34]. In addition, newer solid-state cadmium zinc telluride (CZT) cameras with improved radiotracer count statistics in addition to IR software as outlined previously using combined supine and upright stress imaging acquisition have improved the accuracy of SPECT MPI especially for females, reclassifying 77% of equivocal scans in females to either normal or abnormal compared to 69% of equivocal scans for men [53]. The diagnostic accuracy of SPECT MPI with this solid-state CZT camera system with IR and supine and upright stress imatint has been shown to be high in females [area under receiver operating characteristic (ROC) curve, 0.822 (95% CI, 0.685–0.959)]

TABLE 2 An Outline of Sex-Specific Limitations of Single-Photon Emission Computed Tomography Myocardial Perfusion Imaging (SPECT MPI) and Potential Solutions

Sex-Specific Limitations for SPECT MPI	Potential Solutions
Lower sensitivity for the detection of obstructive CAD in women due to lower myocardial mass in females and depth-dependent blur on SPECT MPI [32]	• Wide beam iterative reconstruction software improves the spatial resolution of SPECT MPI due to noise reduction and correcting for depth-dependent blur with resolution recovery [33]
Due to smaller hearts and higher LVEF in females, LV end-systolic volumes may be underestimated in females resulting in higher normal limits for TID [32]	• Utilization of sex-based normal limits of LVEF and LV volumes is recommended to improve diagnostic accuracy of SPECT MPI [32]
There may be reduced specificity in females due to attenuation from soft tissue such as large and/or dense breasts; this attenuation can also occur in obese patients	• Gated SPECT MPI can assist with determining soft tissue attenuation vs. a real perfusion defect for primarily fixed defects on stress and rest SPECT MPI images [32] • Attenuation correction or supine and prone/upright imaging may be helpful [32]
Diagnosing multivessel CAD or severe LM disease with balanced ischemia remains a challenge in SPECT MPI stress imaging due to relative perfusion rather than absolute myocardial blood flow being assessed	• PET MPI with absolute myocardial blood flow assessment is useful in detecting multivessel ischemia
Microvascular CAD cannot be reliably diagnosed on SPECT MPI especially without the assessment of absolute myocardial blood flow	• PET MPI with assessment of coronary flow reserve can be useful in detecting microvascular CAD [32]
Radiation exposure in medical imaging	• Ensure appropriate use of imaging tests for the correct clinical indications • Nuclear SPECT imaging with iterative reconstruction, depth resolution recovery and noise reduction software as well as the use of solid-state CZT cameras have enabled improved spatial resolution and increased camera sensitivity that has allowed for a lower required dose of the radiotracer and, therefore, lower radiation exposure [33] • In addition, cardiac PET is associated with decreased radiation exposure as low as 2–3 mSv particularly with new technological advancement such as 3D-reconstructed PET imaging [51]

3D, three-dimensional; CAD, coronary artery disease; CZT, cadmium zinc telluride; LM, left main; LV, left ventricular; LVEF, left ventricular ejection fraction; MPI, myocardial perfusion imaging; PET, positron emission tomography; SPECT, single-photon emission computed tomography; TID, transient ischemic dilation.

and is similar to that of males (ROC area 0.884; 95% CI, 0.836–0.933) [32].

The issue of diagnosing multivessel CAD or severe LM disease with balanced ischemia remains a challenge in SPECT MPI stress imaging due to relative perfusion rather than absolute myocardial blood flow being assessed. In addition, microvascular CAD cannot be reliably diagnosed on SPECT MPI especially without the assessment of absolute myocardial blood flow.

There is an increasing awareness of radiation exposure in medical imaging. Several technical advances have led to imaging with lower radiation dose exposure, which is relevant to nuclear imaging and is further discussed later in this chapter.

Hybrid imaging with SPECT MPI and CT imaging have been an emerging area in the field. This has been found to be useful in not only defining stress-induced ischemia but also determining the presence of calcified and noncalcified coronary plaque. This additional incorporation of coronary artery calcium (CAC) score with SPECT MPI has been shown to be useful in further stratifying patients' cardiovascular risk. It has been shown that CAC of 0 in symptomatic patients is associated with low rates of myocardial ischemia (<3% in patients with CAC <100) and is useful in determining the need for downstream testing [32]. In symptomatic patients, a CAC > 1000 was associated with myocardial ischemia in > 25% of the cases [32]. A large study cohort of asymptomatic low- to intermediate-risk patients of whom 45.4% of the study population were female had shown that women were older, had a greater prevalence of CAC, and had a higher 15-year mortality when compared to men [54]. This indicates that CAC scoring is a useful tool to improve cardiac risk stratification beyond traditional risk factors. While sequential SPECT MPI with coronary computed tomography angiography (CTA) is not the routine standard of care, it is important to note that patients with an abnormal SPECT MPI along with an abnormal CTA are associated with an annual death rate of 6% and are independently associated with a high risk of death and MI ($P < 0.005$) [55].

Key Points

1. **Stress imaging with SPECT MPI is recommended in symptomatic patients with suspected or established SIHD with an uninterpretable ECG.**

2. **Use of gated SPECT MPI with Tc-99m has improved the accuracy of SPECT MPI.**

3. **Besides myocardial perfusion, there are several other useful prognostic data that can be determined from stress testing with SPECT MPI.**

4. **Limitations of SPECT MPI, especially relevant for women, include attenuation artifacts due to large breasts, and radiation exposure, both of which are improved with technological advances in hardware and software reconstruction.**

5. **Hybrid imaging with SPECT MPI and CT imaging has been found to be useful to also determine the presence of calcified and noncalcified coronary plaque for further risk factor stratification.**

Stress PET MPI

Stress PET MPI has significantly greater accuracy over SPECT MPI, with ability to detect smaller perfusion defects owing to its improved spatial and temporal resolution [32]. This is very valuable for obese females and females with large breast tissue that leads to increased attenuation artifacts, as this decreases the rate of false-positive tests. Due to the smaller heart size in women, there is a greater chance of false-negative tests due to partial volume effects [32]. However, the improved spatial resolution of PET MPI minimizes the occurrence of false-negative tests [32]. The sensitivity and specificity of stress PET MPI are 90% and 89%, respectively, for the detection of angiographically significant CAD [28] (**Table 1**). PET MPI has significantly greater accuracy (area under the ROC curve of 0.95) compared with SPECT MPI (area under the ROC curve of 0.90; $P < 0.0001$) in the detection of obstructive CAD [32].

The significant strengths of stress PET MPI are not only the improved accuracy in females but also the low radiation dose exposure due to the short-lived radiotracers and the ability to assess absolute myocardial flow and determine CFR, which is useful in diagnosing microvascular CAD. The presence of microvascular CAD with impaired CFR is associated with a twofold higher mortality rate in females compared to males [10]. Microvascular disease will be discussed in more detail in Chapter 8. Assessment of absolute blood flow is also useful to determine if the patient was adequately stressed or may have been a nonresponder to regadenoson due to caffeine or theophylline intake within 12 h of the stress test.

There are several functional risk markers on PET MPI that can be used to further risk-stratify patients such as defect size and severity. A normal PET MPI is associated with a low risk (<1% annual cardiac event rate), while an abnormal scan is associated with a 4.2% annual cardiac event rate [32]. In addition, the predicted CAD death rate increased by 33% for every 10% increase in the percentage of myocardial ischemia and increased by > 50% for every 10% increase in the percentage of myocardial scar [32]. The assessment of LVEF and LV volumes with PET MPI adds important prognostic information for females similar to males. LVEF was inversely related to CAD mortality. The induction of TID and/or new wall motion abnormalities and/or decreased LVEF on peak vasodilator stress PET MPI images relative to the resting images is associated with worse cardiovascular outcomes and is a useful prognostic marker [32]. The prognostic value of PET MPI is similar in females and males [32].

Limitations of PET MPI include the lack of wide availability of the test as well as the inability to use exercise stress due to the short half-lives of the PET radiotracers. However, it is encouraging that the use of PET MPI is growing both in the United States and worldwide with over 200 centers performing PET MPI in the United States [32].

1 Stress PET MPI has significantly greater accuracy over SPECT MPI due to improved spatial and temporal resolution.

2 PET MPI offers a high prognostic value and ability to predict CV outcomes in both men and women. The significant strengths of stress PET MPI include the low radiation dose exposure and the ability to diagnose microvascular CAD.

3 Limitations of PET MPI include the lack of wide availability of the test as well as the inability to use exercise stress due to the short half-lives of the PET radiotracers.

Cardiac Magnetic Resonance Imaging

Cardiac magnetic resonance imaging (CMRI) is an advanced imaging modality with regards to the detection of myocardial ischemia and infarction. CMRI has several advantages over echocardiography and nuclear MPI. These advantages include the lack of ionizing radiation exposure, the lack of being limited by the patient's body habitus and poor acoustic window, high spatial resolution, and very good diagnostic accuracy with regards to the detection of CAD. Stress CMRI has been used more routinely with dobutamine stress to assess for the presence of stress-induced regional wall motion abnormalities, and is particularly useful in patients with poor acoustic windows [31]. More recently stress CMRI has been performed with vasodilatory agents and this is now considered a first-line stress CMRI modality. The Clinical Evaluation of Magnetic Resonance Imaging in Coronary Heart Disease (CE-MARC) trial was a single-center study including 752 patients and had shown that stress CMRI had greater accuracy in the detection of myocardial ischemia when compared to SPECT [29]. Sensitivity was high for both sexes with no difference (89% vs. 86%, $P=0.57$) [29]. Specificity was 83.5% with no sex difference (**Table 1**). In this study, it was shown that SPECT had a significantly lower sensitivity in women compared to men (51% vs. 71%, $P=0.007$) [29]. Other studies such as the Study of Gadodiamide Injection in Myocardial Perfusion Magnetic Resonance Imaging (MR-IMPACT II) have shown similar findings with greater diagnostic accuracy for stress CMRI vs. SPECT (area under the curve, 0.76 ± 0.04 vs. 0.63 ± 0.05; $P=0.033$) [56]. Stress CMRI for women has high diagnostic accuracy for detecting ischemia (87%, $n=147$) and is a great prognostic tool (hazard ratio: ~50, $n=168$) [31]. Similar to PET imaging, CMRI is able to diagnose both macrovascular and microvascular disease.

Late gadolinium enhancement (LGE) is another routine diagnostic CMR tool in the assessment of CAD. LGE is useful in the detection of subendocardial and transmural infarction (**Figure 4**). Often LGE on CMRI has detected previously unrecognized MI which is more often missed in women compared to men. Women had a greater prevalence

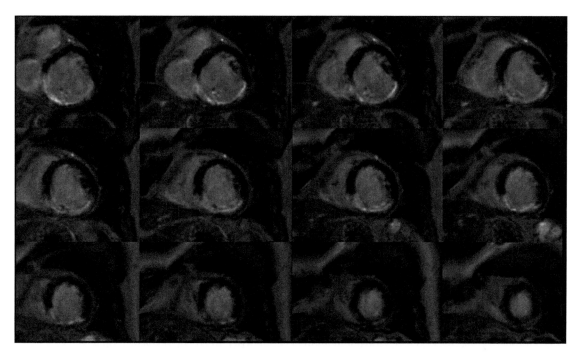

FIGURE 4 Cardiac Magnetic Resonance Imaging With Late Gadolinium Enhancement Seen in the Inferior and Inferoseptal Walls Indicating the Presence of a Myocardial Infarct in These Wall Segments.

Key Points

1. CMRI has several advantages over echocardiography and nuclear myocardial perfusion imaging which include the lack of ionizing radiation exposure, the lack of being limited by the patient's body habitus and poor acoustic window, high spatial resolution, and very good diagnostic accuracy with regards to the detection of CAD.

2. Stress CMRI for women has a high diagnostic accuracy for detecting ischemia and is a great prognostic tool.

3. Often LGE on CMRI detects previously unrecognized myocardial infarction which is more often missed in women compared to men.

4. Emerging areas in CMRI show a lot of promise in the detection of CAD in women.

of clinically unrecognized MI (45%) compared with recognized MI (18%) [57]. This has important clinical implications as patients with clinically unrecognized ischemic heart disease (IHD) are most likely to not be on guideline- directed medical therapy and therefore more likely to have worse outcomes [31, 58]. In addition, CMRI-detected clinically unrecognized MI was an independent predictor of CAD mortality with a hazard ratio of 17.4 [59].

Emerging areas in CMRI such as myocardial perfusion and metabolic imaging and CMRI with blood oxygenation level-dependent (BOLD) imaging to detect microvascular disease, CMRI spectroscopy, and exercise CMRI show a lot of promise in the detection of CAD in women [31].

Multimodality Imaging Features of Myocardial Infarction

There are several imaging features of MI that may be seen on echocardiography, MPI, and CMRI as outlined in **Table 3** and shown in **Figures 4–6**.

TABLE 3 Multimodality Imaging Features of Myocardial Infarction

Imaging Modality	Imaging Features Suggestive of Infarction
Echocardiography	1. Thinning of the myocardium 2. Nontransmural infarcts show varying degrees of hypokinesis dependent on the amount of infarcted myocardium present 3. Transmural infarcts usually show akinesis or dyskinesis 4. Increased echogenicity with old infarcts due to increased deposition of calcium related to a fibrosis 5. Myocardial infarcted segments are associated with abnormal GLS
SPECT myocardial perfusion imaging	1. Decreased myocardial perfusion radiotracer uptake of technetium-99m or thallium-201 in any myocardial segment on rest images associated with hypokinesis in the affected myocardial segment
PET myocardial perfusion imaging	1. Decreased myocardial perfusion radiotracer uptake of rubidium (Rb-82) chloride or ammonia (N-13) in any myocardial segment on rest images associated with hypokinesis in the affected myocardial segment 2. Myocardial infarcts with viable hibernating myocardium are associated with the uptake of FDG-18
Cardiac MRI	1. Thinning of the myocardium 2. Nontransmural infarcts show varying degrees of hypokinesis dependent on the amount of infarcted myocardium present 3. Transmural infarcts usually show akinesis or dyskinesis 4. Late gadolinium enhancement occurs in the infarcted myocardial wall segments

FDG-18, fluorodeoxyglucose 18; GLS, global longitudinal strain; MRI, magnetic resonance imaging; PET, positron emission tomography; SPECT, single photon emission CT.

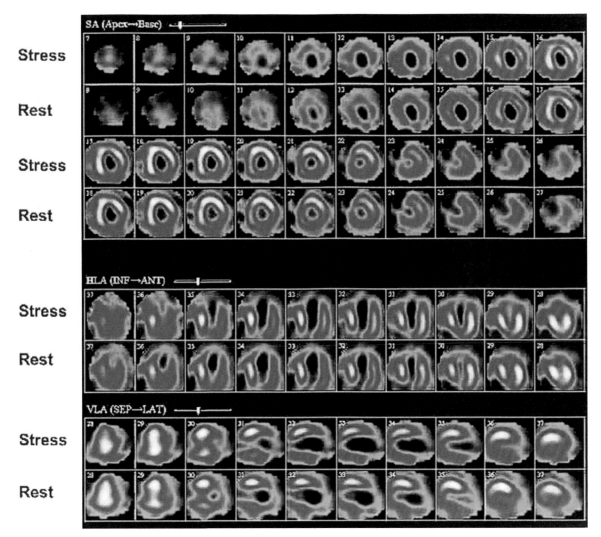

FIGURE 5   Regadenoson Stress Myocardial Perfusion Imaging With Technetium 99m-SPECT Showing Dilated Left Ventricle With a Large Fixed Perfusion Defect on Stress and Rest Involving the Entire Apex and Extending to the Mid-Anterior Wall and the Entire Inferior and Inferolateral Wall With Severe Photon Reduction. Wall motion showed severely reduced left ventricular ejection fraction of 15% with apical dyskinesis, severe hypokinesis of the inferior, inferolateral and mid-anterior walls, and moderate hypokinesis of the remaining wall segments. This indicates the presence of a large myocardial infarct in these aforementioned walls. On coronary angiography, the patient had severe (95%) stenosis in the distal left main coronary artery and 90% mid-right coronary artery stenosis. ANT, anterior; HLA, horizontal long axis; INF, inferior; LAT, lateral; SA, short axis; SEP, septal; SPECT, single-photon emission computed tomography; VLA, vertical long axis.

Noninvasive Anatomical Assessment

Coronary computed tomography angiography (CTA) has emerged as a very useful tool to diagnose the full spectrum of CAD which may extend from minimal nonobstructive CAD with small noncalcified plaque to severe obstructive CAD with occluded and heavily calcified coronary vessels (**Figures 7–9**). Invasive coronary angiography has been traditionally viewed as the gold standard in di-

agnosing the presence of CAD despite the fact that invasive coronary angiogram provides a luminogram of the coronary artery vessels and often misses the presence of eccentric coronary plaque with coronary artery positive vessel remodeling. The presence of nonobstructive CAD plaques seen on coronary CTA is associated with an elevated mortality risk [31]. The number of coronary vessels with nonobstructive plaque is associated with a twofold to

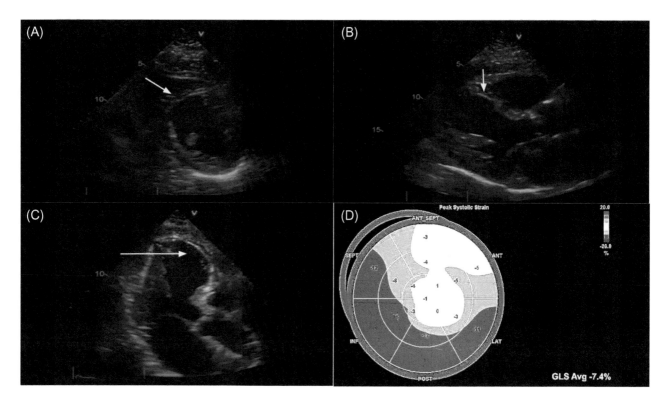

FIGURE 6 **Transthoracic Echocardiogram Showing Thinning and Akinesis of the Anteroseptal Wall on the Parasternal Short Axis** *(Arrow)*. **(A), Parasternal Long Axis** *(Arrow)* **(B), and Apical Three-Chamber View** *(Arrow)* **(C). (D) Abnormal GLS of −7.4% Involving the Anterior and Anteroseptal Walls.** These echocardiographic findings indicate the presence of myocardial infarction in the anteroseptal wall representing the left anterior descending (LAD) coronary artery territory. GLS, global longitudinal strain.

sixfold elevated risk of death [31]. There was a graded increase in this risk based on the number of vessels or segments with identifiable nonobstructive plaque observed with coronary CTA or invasive coronary angiography [31]. Coronary CTA is effective as a second-line procedure for risk stratification of women following an indeterminate or intermediate-risk stress test. Coronary CTA is also effective as a second-line procedure for risk stratification of women with persistent chest pain (or equivalent) symptoms following a negative stress test [60].

Plaque characteristics determined on coronary CTA have implications with regards to mortality risks. Noncalcified plaque was associated with the highest 6-year death rate of 9.6% when compared to mixed plaque (3.3%) and calcified plaque (1.4%) [31]. Similar to invasive coronary angiography, women have less obstructive CAD seen on coronary CTA compared with men, despite the fact that they had higher mortality risk with multivessel CAD. Moreover, the number of nonobstructive plaques predicted death in women even when adjusted for obstruc-

tive CAD; this has not been found to be true for men [31]. The CONFIRM registry had also shown that nonobstructive CAD was associated with an approximately twofold increased major adverse risk in women [61]. The 5-year outcomes were higher for women than men with three vessels associated with coronary plaque with the relative hazard for three-vessel CAD being higher for women as compared with men (HR in women: 4.21; 95% CI: 2.47–7.18; $P < 0.0001$ vs. HR in men: 3.27; 95% CI: 1.96–5.45; $P < 0.0001$) [61]. The coronary artery calcification (CAC) score in symptomatic patients is independently predictive of all-cause mortality as well as MI in both men and women. In patients with nonobstructive CAD, mortality increases proportionally with the CAC score from a mortality rate of 0.8% in patients with a CAC score of 0 to a mortality rate of 9.8% for CAC of ≥ 400 [31]. **Figure 10** outlines a proposed multimodality diagnostic strategy for the assessment of women with SIHD [32], while **Figure 11** outlines the various high-risk, intermediate-risk, and low-risk features on multimodality noninvasive cardiac imaging assessment in SIHD [5].

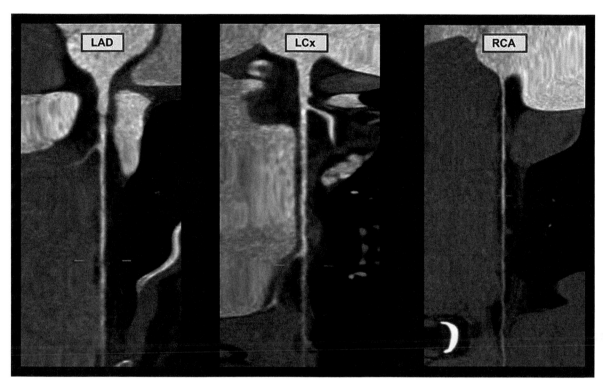

FIGURE 7   Coronary Computed Tomography Showing Straightened Curved Multiplanar Reformatting (Curved MPR) of the LAD, LCx, and RCA With No Evidence of Any Coronary Plaque or Stenosis. LAD, left anterior descending; LCx, left circumflex; RCA, right coronary artery.

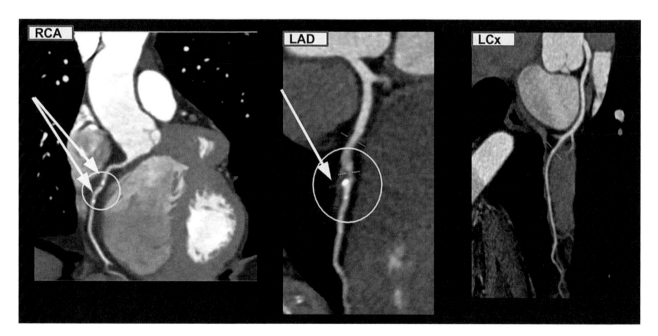

FIGURE 8   Coronary Computed Tomography Showing Curved Multiplanar Reformatting (Curved MPR) of the LAD, LCx, and RCA With Evidence of a Partially Calcified Plaque in the Mid-LAD Resulting in Severe (70–99%) Stenosis. There is also a partially calcified plaque in the mid-RCA resulting in mild (25–49%) stenosis. LAD, left anterior descending; LCx, left circumflex; RCA, right coronary artery.

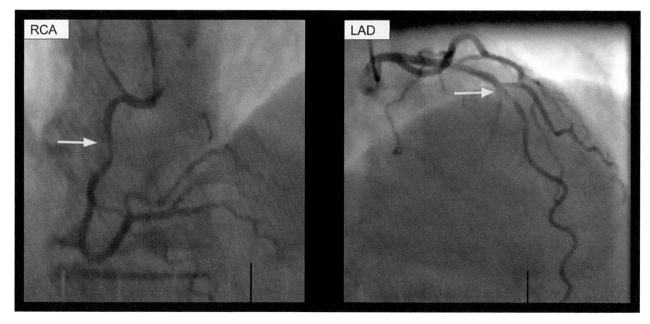

FIGURE 9 Fractional Flow Reserve (FFR) of the Mid-RCA Was 0.98 and Was Therefore Thought Not to Be Physiologically Significant, Therefore No Percutaneous Coronary Intervention (PCI) of the RCA Was Performed. Cardiac Catheterization Showing Severe (>90%) Stenosis in the Mid-LAD That Was Treated With a Drug Eluting Stent (DES). LAD, left anterior descending; RCA, right coronary artery.

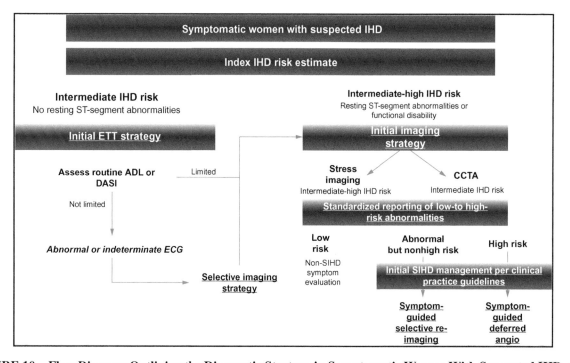

FIGURE 10   Flow Diagram Outlining the Diagnostic Strategy in Symptomatic Women With Suspected IHD. ADL, activities of daily living; angio, angiogram; CCTA, coronary computed tomography angiography; DASI, Duke Activity Status Index; ECG, electrocardiogram; ETT, exercise tolerance treadmill test; IHD, ischemic heart disease. *Reproduced with permission from [32].*

Noninvasive Risk Stratification

High risk (3% annual death or MI)

1. Severe resting LV dysfunction (LVEF 35%) not readily explained by noncoronary causes

2. Resting perfusion abnormalities 10% of the myocardium in patients without prior history or evidence of MI

3. Stress ECG findings including 2 mm of ST-segment depression at low workload or persisting into recovery, exercise-induced ST-segment elevation, or exercise-induced VT/VF

4. Severe stress-induced LV dysfunction (peak exercise LVEF 45% or drop in LVEF with stress 10%)

5. Stress-induced perfusion abnormalities encumbering 10% myocardium or stress segmental scores indicating multiple vascular territories with abnormalities

6. Stress-induced LV dilation

7. Inducible wall motion abnormality (involving two segments or two coronary beds)

8. Wall motion abnormality developing at low dose of dobutamine (10 mg/kg/min) or at a low heart rate (120 beats/min)

9. CAC score 400 Agatston units

10. Multivessel obstructive CAD (70% stenosis) or left main stenosis (50% stenosis) on CCTA

Intermediate risk (1–3% annual death or MI)

1. Mild/moderate resting LV dysfunction (LVEF 35–49%) not readily explained by noncoronary causes

2. Resting perfusion abnormalities in 5–9.9% of the myocardium in patients without a history or prior evidence of MI

3. 1 mm of ST-segment depression occurring with exertional symptoms

4. Stress-induced perfusion abnormalities encumbering 5–9.9% of the myocardium or stress segmental scores (in multiple segments) indicating one vascular territory with abnormalities but without LV dilation

5. Small wall motion abnormality involving one to two segments and only one coronary bed

6. CAC score 100–399 Agatston units

7. One vessel CAD with 70% stenosis or moderate CAD stenosis (50–69% stenosis) in two arteries on CCTA

Low risk (1% annual death or MI)

1. Low-risk treadmill score (score 5) or no new ST-segment changes or exercise-induced chest pain symptoms when achieving maximal levels of exercise

2. Normal or small myocardial perfusion defect at rest or with stress encumbering 5% of the myocardium

3. Normal stress or no change of limited resting wall motion abnormalities during stress

4. CAC score 100 Agatston units

5. No coronary stenosis 50% on CCTA

FIGURE 11 An Outline of the Low-Risk, Intermediate-Risk, and High-Risk Noninvasive Imaging Features That Are Useful in the Risk Stratification of Patients With Stable Ischemic Heart Disease. CAC, coronary artery calcium; CAD, coronary artery disease; CCTA, coronary computed tomography angiography; ECG, electrocardiogram; LV, left ventricular; LVEF, left ventricular ejection fraction; MI, myocardial infarction; VF, ventricular fibrillation; VT, ventricular tachycardia. *Data adapted from [5].*

Key Points

1. The presence of nonobstructive coronary artery disease plaques seen on coronary computed tomography angiography (coronary CTA) is associated with an elevated mortality risk.

2. Coronary CTA is effective as a second-line procedure for risk stratification of women following an indeterminate or intermediate-risk stress test and for women with persistent chest pain (or equivalent) symptoms following a negative stress test.

3. Plaque characteristics determined on coronary CTA have implications with regards to mortality risks.

4. Women have less obstructive CAD seen on coronary CTA compared with men, despite the fact that they had higher mortality risk with multivessel CAD.

5. In patients with nonobstructive CAD, mortality increases proportionally with the CAC score.

Relevance of Radiation Exposure Regarding Diagnostic Testing

There has been an increased awareness both within the medical community and among the greater public with regards to radiation exposure during various medical imaging tests. This is important for females as the radiation exposure standard limits set for females have been extrapolated from data based on a standard 70 kg male. Therefore, it is likely that the radiation dose thresholds that have been extrapolated to females are above the true dose limitations for females, especially those with a smaller body habitus with weight less than 50 kg. It is therefore imperative that there is judicial selection of appropriate noninvasive testing to diagnose IHD with the least risk of radiation dose exposure. Fortunately, technical advances in coronary CTA with prospective-gated acquisition with IR have allowed lower radiation exposure from coronary CTA to as low as 5 mSv compared to previous levels of 12 mSv a decade ago [30]. Similarly, advances in nuclear SPECT imaging with IR, depth resolution recovery and noise reduction software, and the use of solid-state CZT cameras have enabled improved spatial resolution and increased camera sensitivity that has allowed for a lower required dose of the radiotracer and therefore lower radiation exposure [33]. In addition, cardiac PET is associated with decreased radiation exposure as low as 2–3 mSv particularly with new technological advancement such as three-dimensional (3D) reconstructed PET imaging [51].

The best way of minimizing radiation dose exposure for females is ensuring that the most appropriate noninvasive test is ordered for the right patient at the right time. This is imperative, as not every female patient will meet indications for cardiac CT or nuclear SPECT or PET imaging. Many of these patients may be appropriately referred for treadmill exercise ECG stress or stress echocardiography for diagnosis of IHD. Therefore, it is important to assess the patient's pretest probability for the presence of obstructive CAD to then determine the best diagnostic strategy. It is also important to determine the patient's functional capacity as well as interpretability of the electrocardiogram to determine if the patient is better suited for a treadmill exercise ECG or exercise stress echo which have no associated radiation exposure compared to other tests such as coronary CTA or nuclear PET or SPECT imaging, which are associated with varying degrees of radiation exposure. The average radiation doses for several cardiac imaging studies is shown in **Table 4** [62, 63].

TABLE 4 Radiation Dose Associated With Several Cardiac Tests	
Test	Radiation Dose Exposure (mSv)
Annual background radiation in the United States	3
Diagnostic cardiac catheterization	7
Cardiac CT	
Retrospective-gated helical cardiac CT	19
Retrospective-gated helical cardiac CT with dose modulation	12
ECG-triggered prospective axial cardiac CT	3
Cardiac CT for coronary artery calcium score only	2
Myocardial Perfusion Imaging	
Dual isotope SPECT with Tl-201 and 99m-Tc	22
99m-Tc rest stress SPECT	11
99m-Tc low-dose rest stress SPECT	6–9

TABLE 4 Radiation Dose Associated With Several Cardiac Tests—Cont'd	
Test	Radiation Dose Exposure (mSv)
99m-Tc low-dose stress only SPECT	3
82-Rb-rest stress PET	3
13-NH3-rest stress PET	2

CT, computed tomography; ECG, electrocardiogram; PET, positron emission tomography; SPECT, single-photon emission computed tomography; Tl-201, thallium-201; 13-NH3, 13-ammonia; 82-Rb, 82-rubidium; 99m-Tc, 99m-technetium. *Refs. [62, 63].*

Key Points

1. It is likely that the radiation dose thresholds that have been extrapolated to females are above the true dose limitations for females, especially those with a smaller body habitus with weight less than 50 kg.

2. It is imperative that there is judicial selection of appropriate noninvasive testing to diagnose IHD with the least risk of radiation dose exposure.

3. Technical advances in coronary CTA and MPI with SPECT and use of PET imaging have allowed for lower radiation exposure.

4. The best way of minimizing radiation dose exposure for females is ensuring that the most appropriate noninvasive test is ordered.

Management

Lifestyle Modification and Cardiac Rehabilitation

Management of cardiovascular risk factors includes a dedication to lifestyle modifications; however, the true impact of intervention such as physical activity and cardiac rehabilitation is not well known, as outcome studies focusing on women do not make up the breadth of research. Women completing cardiac rehabilitation experience the greatest reduction in mortality (HR 0.36, 95% CI 0.28–0.45) with a relative greater benefit for women compared to men

(HR 0.51, 95% CI 0.46–0.56) [64]. However, despite this, women were less likely to be referred for cardiac rehabilitation (31.1% of women vs. 42.2% of men, P<0.0001) and were less likely to complete cardiac rehabilitation (50.1% of women vs. 60.4% of men, P<0.0001) [64]. This sex disparity persisted even after adjusting for demographic and clinical characteristics. In comparison to men, referral to cardiac rehabilitation was significantly lower in women [adjusted odds ratio (OR) 0.74, 95% CI 0.69–0.79] as was completion of the program (adjusted OR 0.73, 95% CI 0.66–0.81) [64]. In a recent statement for healthcare professionals from the American Heart Association, a call was made to not only include women in studies but also create cardiac rehabilitation programs designed to increase women's continued participation [11].

> ## Key Points
>
> **1** Women are less likely to be referred to, enrolled in, and complete cardiac rehab.
>
> **2** Women are underrepresented in outcome studies involving lifestyle modification and cardiac rehabilitation.

Pharmacotherapy

Despite published efficacy and guideline endorsement, women are less likely to receive guideline-recommended medications for primary and secondary prevention of IHD. The CRUSADE study showed that women are less likely to receive aspirin [87.5% (F) vs. 90.4% (M), adjusted OR 0.91, 95% CI (0.85–0.98)], angiotensin-converting enzyme (ACE) inhibitors [55.3% (F) vs. 55.5% (M), adjusted OR 0.93, 95% CI (0.88–0.98)], and statin [55.9% (F) vs. 63.4% (M), adjusted OR 0.92, 95% CI (0.88–0.98)] on hospital discharge despite having higher comorbidities [65]. Young women (35–54 years old) in particular are less likely to be prescribed guideline-recommended medications compared to age-matched men [OR 0.63, 95% CI (0.52–0.77)] [66]. Less than 1/10 with angina and abnormal stress tests had any change in medications or referral to diagnostic angiography [22]. A cross-sectional study of ambulatory female veterans (40–85 years old) of Caucasian, black, and Hispanic/Latino ethnicity showed that only 31.1% of women with previous diagnosis of heart disease received antiplatelet agents, and 53% of postmenopausal women continued hormone replacement therapy despite its known cardiovascular adverse effects [67]. In the same study, only 53.4% of current/former female smokers were advised to start smoking cessation therapy, and only 30% of those with HDL levels <40mg/mL were advised to start an exercise program [67].

In addition, although females have a greater prevalence of nonobstructive CAD, there is a twofold increase in cardiac mortality in these patients when there is associated coronary artery endothelial dysfunction [10]. In view of this misperception, females who are diagnosed as having myocardial ischemia after noninvasive stress testing are often diagnosed as having a "false-positive" stress test when their invasive coronary angiogram shows no evidence of coronary artery stenoses. This leads to downstream mismanagement of these patients as they are often not placed on appropriate guideline-directed medical therapy (GDMT) such as antianginal therapy and in cases when patients have undiagnosed minimal or nonobstructive CAD, these patients are then not appropriately treated with GDMT such as aspirin and statin therapy [10].

> ## Key Points
>
> **1** Women are less likely to receive guideline-directed medications for primary and secondary IHD prevention.
>
> **2** Studies have shown that despite increased risk of IHD, women are less likely to be cautioned against hormonal therapy or the cardiovascular impact of smoking, or encouraged to exercise.

Percutaneous or Surgical Revascularization

Coronary angiography is considered the gold standard of diagnosing obstructive epicardial CAD. To date, females remain significantly underrepresented in guideline-changing heart disease research and trials. This has created important limitations in the evidence base for cardiovascular medicine for women [7]. The focus of IHD management has historically been on revascularization of obstructive CAD rather than the goal of relieving symptoms and improving survival [7].

Guideline updates have a class I indication for coronary angiography in patients with SIHD with unacceptable ischemic symptoms despite GDMT and who are candidates for coronary revascularization [68]. Revascularization is considered when there is an angiographically significant stenosis of >70%. This angiographic assessment of stenosis severity relies on comparisons to an adjacent, nondiseased reference segment, which may lead to an underestimation of the lesion severity in a diffusely diseased coronary artery. Despite this class I indication, women are 40–50% less likely to have an angiogram after adjustment for clinical factors, including results of noninvasive testing [2]. This disparity was similar regardless of region and center, and persisted even after evaluation by a cardiologist [2].

Both American and European guidelines recommend coronary angiography within 48 h for all high-risk non-ST elevation-acute coronary syndrome (NSTE-ACS) patients regardless of sex. However, when considering the current trials, there are differences in baseline risk, as well as timing and type of revascularization, compounded by the persistently low representation of women [69].

There have been more investigations on the use of percutaneous interventions with advancements in drug eluting stents (DES). Guidelines for coronary angiography, pertaining to PCI or coronary artery bypass grafting (CABG), are evolving. When considering a revascularization method, multiple studies have analyzed percutaneous DES placement, CABG, and optimum medical therapy. The SYNTAX (Synergy Between Percutaneous Coronary Intervention with TAXUS and Cardiac Surgery) trial was one of the first randomized controlled studies which evaluated CABG vs. DES implantation [68, 70–72]. The SYNTAX trial had a small female population with only 24.4% of the study group in the CABG arm ($n = 348$) and 28% of the study group in the DES arm ($n = 357$) being female; although a sex-specific analysis was not performed, this study found a lower rate of death, stroke, and MI in the CABG-treated patients [73]. A SYNTAX score is a point-based scoring system determined by location, severity, and extent of coronary stenosis. Those who were intermediate with SYNTAX scores of 23–32, and high with SYNTAX score > 33, major adverse cardiac and cerebrovascular events (MACCE) occurred more often in those receiving DES compared to those who received CABG [70]. Current guidelines recommend a Heart Team approach to determine the most appropriate revascularization strategy for revascularization in individual patients with DM and complex multivessel CAD. In this population, CABG is ideally preferred over PCI to improve survival [68].

Despite guidelines, females were less likely to receive guideline-based revascularization procedures even with confirmed CAD. Adjusted for age, diabetes, and symptom severity, women remained significantly less likely to undergo revascularization than their male counterparts [2]. Among patients with proven CAD, women remained significantly less likely to have a revascularization procedure (OR, 0.70; 95% CI, 0.52–0.94; $P = 0.019$) [2]. Reasons influencing this bias include increased complication rates, poorer outcomes, and possibly an unwillingness to undergo invasive procedures or surgery [2]. Women have smaller epicardial coronary arteries, even after accounting for body habitus and LV mass, which may not allow them to be candidates for revascularization based on vessel size and accommodation for stent placement [7]. In very small vessels, less than 2 mm, the use of stents is prohibited [74].

CABG is the treatment of choice with significant obstruction of the left main coronary artery, or with triple vessel epicardial disease and LV systolic dysfunction. Female sex is conventionally considered a risk factor for CABG and has been included in many cardiac operative risk evaluation scores as a poor prognostic factor [75]. Studies have shown that in some select patients with mild-to-moderate angina and obstructive disease that is angiographically suitable for revascularization, there was no difference in death or MI rates between the medical treatment group and the invasive revascularization group [32, 69, 76, 77]. Females comprised only 15–30% of patients in these studies [69, 76, 77]. The recent ISCHEMIA trial has shown that in patients with stable CAD and moderate or severe myocardial ischemia, there was no evidence that an initial invasive strategy, as compared with an initial conservative strategy, reduced the risk of ischemic cardiovascular events or death from any cause over a median of 3.2 years [13.3% of the invasive group compared with 15.5% of the conservative arm ($P = 0.34$)] [77]. It is important to note that a limited number of women were enrolled in the study (23% of the total enrolled 5179 patients were female) as many were excluded from randomization when compared to men due to less ischemia and more nonobstructive CAD [77].

As with any invasive procedure, there are known risks involved. According to the American College of Cardiology (ACC) National Cardiovascular Data Registry CathPCI Registry, coronary angiography carried a 1.5% incidence of procedural complications with diagnostic angiograms [68]. This included death, stroke, MI, bleeding, infection, contrast allergic or anaphylactoid reaction, arrhythmias, and need for emergency revascularization [68]. Women have a relatively poorer outcome and higher rates of com-

---

Key Points

 The limited enrollment of women in coronary revascularization trials has created limitations in female-specific evidence-based practice.

 Women are less likely to be recommended to have invasive testing, and less likely to undergo revascularization.

 CABG is preferred to PCI in both men and women with diabetes and multivessel CAD.

 Women have smaller epicardial vessels with less incidence of complete revascularization.

 Women have a higher rate of complications and higher in-hospital mortality rates with either PCI or CABG.

plications with either CABG or PCI compared to men [68]. Those with confirmed CAD were also significantly more likely to suffer death or nonfatal MI, even after multivariable adjustment for baseline risk factors, giving the female gender a predictor of complications and having twofold increased risk for death or MI [2, 7]. Women undergoing PCI demonstrated an increased risk of bleeding, requiring blood transfusions (adjusted HR, 1.32; 95% CI, 1.04–1.67) and hemorrhagic stroke at 2 years (adjusted HR, 4.76; 95% CI, 1.92–11.81) [78]. Women undergoing revascularization in the setting of ACS also had a higher risk of mortality and recurrent ACS compared to men; the 1-year rate of death or recurrent ACS was 10.6% for men compared with 13.1% for women (HR, 1.24; 95% CI 1.16–1.33) [79].

INOCA and Microvascular Disease

Nonobstructive epicardial coronary arteries is stenosis less than 50% of the vessel lumen at angiography, or fractional flow reserve (FFR) of greater than 0.8 which is not considered to be physiologically significant. Approximately 5–25% of ACS patients exhibit no obstructive atherosclerotic disease [69]. However, all-cause mortality is significantly higher among those without obstructive atherosclerosis [69].

Nonobstructive CAD is often present in combination with coronary vasomotor disorders, including spasm, endothelial dysfunction of coronary arteries, and microvascular disease [80]. The result of this high burden of functional CAD, as well as incomplete revascularization even with anatomical disease, may be a predisposition to higher burden of symptoms [79]. Women with anginal symptoms have twice as often ischemia with nonobstructive coronary arteries (INOCA) [80]. INOCA is not a benign condition and carries a higher 5-year event rate in symptomatic women, particularly when signs of ischemia are present [80].

Invasive testing is important to discriminate between vasospasm of the epicardial coronary arteries and the coronary microvasculature. Diagnosis of vasomotor disorders is best with invasive measures of the index of microvascular resistance (IMR), CFR, and performance of vasoreactivity testing, typically with acetylcholine or adenosine. Nonendothelial-dependent microvascular dysfunction is defined as a CFR in response to adenosine < 2.5, endothelial-dependent microvascular dysfunction is defined as change in CBF in response to acetylcholine < 50%, and coronary spasm is defined as change in coronary artery diameter in response to acetylcholine < 90% in addition to chest pain and ECG changes [81]. Abnormal results from such invasive tests are associated with adverse outcomes [80, 82]. These adverse outcomes include death, nonfatal MI, nonfatal stroke, and hospital stay for congestive heart failure, angina, and other vascular events. Low CFR, defined as

the ratio of average peak velocity after adenosine to average baseline velocity just before adenosine, significantly predicted increased risk for a major adverse event as described in the WISE trial [82, 83]. This continues to be an area of interest to researchers as we acknowledge the evolving differences with INOCA and the importance of microvascular dysfunction that affects innumerable people, and to a greater extent, women.

Treatment for these vasomotor disorders is focused on lifestyle modifications and adequate control of traditional risk factors. Blood pressure control should be strict, more so than what is recommended by current guidelines. Short-acting nitrates may be helpful to relieve symptoms; however, long-acting nitrate may exacerbate symptoms. Calcium antagonists, mainly diltiazem, at high doses, and even combined with low-dose selective beta-blockers, can provide some relief. In the event of refractory angina, xanthine derivatives, ivabradine, nicorandil, and ranolazine may be of benefit. No standard treatment regimen is suited for every patient and it should be tailored to each individual [80, 84–86].

Key Points	
	INOCA is not a benign condition.
	Diagnosis of vasomotor disorders is established with index of microvascular resistance (IMR) and coronary flow reserve (CFR).
	Treatment is individualized, with a focus on risk factor modification.

Conclusion

SIHD in women encompasses varying sex-specific features with regards to the coronary artery anatomy, clinical presentation, and diagnosis. Women have smaller coronary arteries compared to men. They also have a lower prevalence of obstructive CAD despite greater clustering of risk factors including older age, obesity, hypertension, and diabetes. In addition, they may have novel risk factors such as gestational diabetes and systemic inflammation which are not included in most contemporary cardiovascular risk assessment scores. Women often present with atypical symptoms compared to men, resulting in increased diagnostic challenges. It is important that these sex differences are considered when determining the best diagnostic strategy for women. It is also important to consider and minimize the radiation dose exposure when considering diagnostic testing in women.

Despite increased awareness of heart disease in women, and documented comparable treatment efficacies in both sexes, women continue to be under prescribed medications and are less likely to be referred for revascularization procedures and cardiac rehabilitation when compared to their male counterparts. Overall, women experience higher cardiovascular morbidity and mortality compared to men. These disparities in outcome underscore the need for sex-specific management algorithms. Therefore, ongoing efforts for public and clinician awareness are pivotal and improved understanding of the pathophysiology behind these sex differences is important. In addition, there remains a need for inclusion of more women in study trials as well as national registries to assist in designing more sex-specific guidelines for diagnosing and managing women with SIHD.

Back to Clinical Case

The case described a 52-year-old female with hypertension, diabetes type 2, hyperlipidemia, and nicotine dependence, with a 10-month history of typical angina, with a previously normal exercise SPECT study, and a normal rest echocardiogram. She was referred for exercise stress echocardiography. The patient exercised for three and a half minutes on the treadmill with Bruce protocol and she achieved 4.6 METs. The test was terminated due to limiting dyspnea at 88% maximum predicted heart rate. She had nondiagnostic changes on the ECG due to baseline ST-T segment abnormalities at baseline. Echocardiographic images (**Figures 12–15**) obtained within 1 min postexercise showed akinesis in the entire anterior, anteroseptal, and anterolateral walls, and dyskinetic apex which suggested the presence of a large

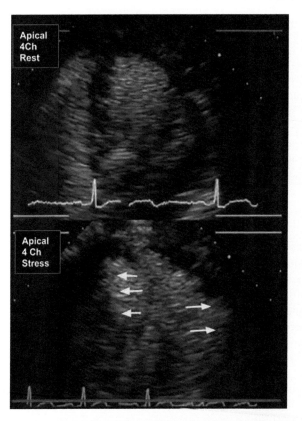

FIGURE 12 **Stress Echo Images Showing Apical Four-Chamber Views at Rest Showing Normal Wall Motion, the Apical Four-Chamber View Post-Stress Showed Ischemic Dilation of the Left Ventricle With Akinesis of the Anterolateral and Inferoseptal Walls Representing Multivessel Severe Ischemia in the Left Circumflex and Likely the Left Anterior Descending Coronary Artery Territories.**

Back to Clinical Case—cont'd

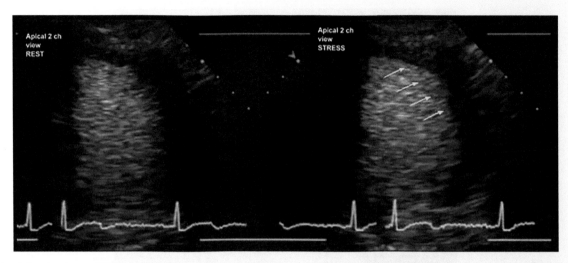

FIGURE 13   Stress Echo Images Showing the Apical Two-Chamber Views With Normal Systolic Wall Motion of the Anterior and Inferior Walls. Poststress apical two-chamber view showed akinesis of the anterior wall post-stress indicating severe ischemia in the left anterior descending coronary artery territory.

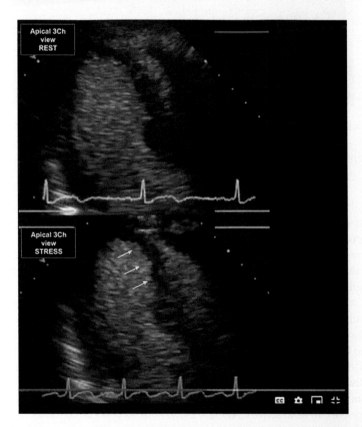

FIGURE 14   Stress Echo Images Showing the Apical Three-Chamber Views at Rest With Normal Wall Motion of the Inferolateral and Anteroseptal Walls. Postexercise the post-stress images showed severe hypokinesis of the apicoseptal and mid-anteroseptal walls indicating the presence of severe myocardial ischemia in the left anterior descending artery territory.

Continued

Back to Clinical Case—cont'd

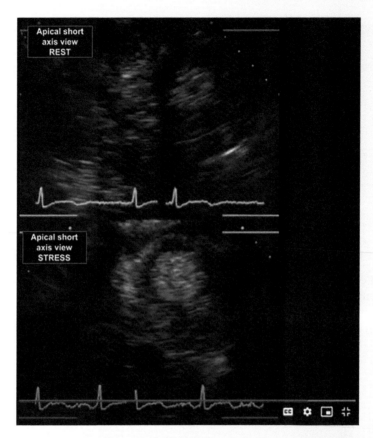

FIGURE 15 Stress Echo Images Showing the Apical Short Axis Views at Rest With Normal Systolic Wall Motion. Postexercise images demonstrate dilation of the left ventricle with severe hypokinesis of the anteroapical, apicolateral, and apicoseptal wall, and akinesis of the apicoinferior wall indicating the presence of severe myocardial ischemia.

area of severe multivessel ischemia. She was also noted to have dilation of the left ventricle postexercise. Wall motion score index was calculated as 1.8. Cardiac catheterization (**Figure 16**) revealed 99% ostial subtotal occlusion of the left anterior descending (LAD) coronary artery with thrombus and 90% ostial first obtuse marginal (OM1) disease. Patient proceeded to have CABG with left internal mammary artery (LIMA) to LAD, saphenous vein graft (SVG) to first diagonal (D1), and the obtuse marginal (OM1) arteries. The patient recovered and her symptoms were resolved. She was referred for cardiac rehabilitation and did well.

This case highlights several challenges in the diagnosis of CAD in women and highlighted several teaching points which were reviewed in this chapter. It is important to note that the patient had left ventricular hypertrophy (LVH); this is relevant with regards to SPECT MPI as the presence of LVH and the smaller cavity size has been associated with decreased sensitivity due to the limited spatial resolution of SPECT MPI resulting in the underappreciation of perfusion defects [32]. In addition, the patient had barely reached her target heart rate at 85% maximum predicted heart rate for age, which may also decrease the sensitivity of the test in detecting myocardial ischemia due to flow-limiting coronary stenosis. The patient's limited functional capacity on the nuclear stress test was also another indication that she may be at a higher risk for the presence of obstructive CAD [23, 24, 32]. Stress echocardiography is not associated with ionizing radiation and was a good consideration for this patient, who was concerned about unnecessary radiation exposure. In addition, stress echo had shown even greater markers for ischemia with poor functional capacity of 4.6 METs which was a poor prognostic feature, especially as she exhibited severe ischemia at this reduced workload. This patient had a high wall motion score index of 1.8 and this indicated a high likelihood of a large coronary territory involvement with severe CAD and poor prognosis [36]. The patient proceeded to have CABG [left internal mammary artery (LIMA) to

Back to Clinical Case—cont'd

left anterior descending (LAD) bypass graft and saphenous vein graft (SVG) to the first obtuse marginal bypass graft] which has been associated with reduced risk for revascularization and better cardiovascular outcomes compared with PCI especially in the presence of diabetes mellitus and ostial LAD/distal left main disease and multivessel CAD [68]. She also benefited from cardiac rehabilitation, which has been associated with improved cardiovascular outcomes despite the fact that females are less likely to be referred for cardiac rehab compared to males [64].

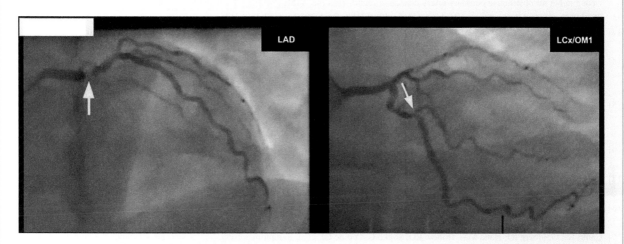

FIGURE 16 Invasive Coronary Angiogram Showing Severe 99% Stenosis of the Proximal (Ostial) Portion of the LAD and Severe >95% Stenosis of the Ostium of the First OM1 of the LCx. Patient proceeded to have coronary artery bypass graft (CABG) surgery with left internal mammary artery (LIMA) to LAD bypass graft and saphenous vein (SVG) to the first obtuse marginal bypass graft. LAD, left anterior descending; LCx, left circumflex; OM1, obtuse marginal branch.

References

[1] Anon. National Health and Nutrition Examination Survey (NHANES) data. Centers for Disease Control and Prevention; 1992-2002. https://wwwc-dcgov/nchs/nhanes/indexhtm. [Accessed 16 February 2020].

[2] Daly C, Clemens F, Lopez Sendon JL, Tavazzi L, Boersma E, Danchin N, et al. Gender differences in the management and clinical outcomes of stable angina. Circulation 2006;113:490–8.

[3] Thom T, Haase N, Rosamond W, Howard VJ, Rumsfeld J, Manolio T, et al. Heart disease and stroke statistics—2006 update: a report from the American Heart Association Statistics Committee and Stroke Statistics Subcommittee. Circulation 2006;113(6):e85–e151. https://doi.org/10.1161/CIRCULATIONAHA.105.171600.

[4] Mygind ND, Michelsen MM, Pena A, Frestad D, Dose N, Aziz A, et al. Coronary microvascular function and cardiovascular risk factors in women with angina pectoris and no obstructive coronary artery disease: the iPOWER study. J Am Heart Assoc 2016;5(3):e003064. https://doi.org/10.1161/JAHA.115.003064.

[5] Fihn S, Gardin J, Abrams J, Berra K, Blankenship J, Dallas A, et al. 2012 ACCF/AHA/ACP/AATS/PCNA/SCAI/STS guideline for the diagnosis and management of patients with stable ischemic heart disease. Circulation 2012;126(25). https://doi.org/10.1161/CIR.0b013e318277d6a0.

[6] Virani SS, Alonso A, Benjamin EJ, Bittencourt MS, Callaway CW, Carson AP, et al. American Heart Association Council on Epidemiology and Prevention Statistics Committee and Stroke Statistics Subcommittee. Heart disease and stroke statistics—2020 update: a report from the American Heart Association. Circulation 2020;141(9):e139–596. https://doi.org/10.1161/CIR.0000000000000757. Epub 2020 Jan 29 31992061.

[7] Taqueti VR. Sex differences in the coronary system. Adv Exp Med Biol 2018;1065:257–78. https://doi.org/10.1007/978-3-319-77932-4_17. Review 30051390.

[8] Gould KL, Johnson NP, Bateman TM, Beanlands RS, Bengel FM, Bober R, et al. Anatomic versus physiologic assessment of coronary artery disease. Role of coronary flow reserve, fractional flow reserve, and positron emission tomography imaging in revascularization decision-making. J Am Coll Cardiol 2013;62:1639–53. PubMed:23954338.

[9] Smilowitz NR, Mahajan AM, Roe MT, Hellkamp AS, Chiswell K, Gulati M, et al. Mortality of myocardial infarction by sex, age, and obstructive coronary artery disease status in the ACTION Registry-GWTG (Acute Coronary Treatment and Intervention Outcomes Network Registry-get with the guidelines). Circ Cardiovasc Qual Outcomes 2017;10(12):e003443. https://doi.org/10.1161/CIRCOUTCOMES.116.003443. 29246884.

[10] Aggarwal NR, Patel HN, Mehta LS, Sanghani RM, Lundberg GP, Lewis SJ, et al. Sex differences in ischemic heart disease: advances, obstacles, and next steps. Circ Cardiovasc Qual Outcomes 2018;11(2):e004437. https://doi.org/10.1161/CIRCOUTCOMES.117.004437. Review 29449443.

[11] McSweeney JC, Rosenfeld AG, Abel WM, Braun LT, Burke LE, Daugherty SL, et al. Preventing and experiencing ischemic heart disease as a woman: state of the science: a scientific statement from the American Heart Association. Circulation 2016;133(13):1302–31. https://doi.org/10.1161/CIR.0000000000000381. Epub 2016 Feb 29. Review. No abstract available 26927362.

[12] Aldiwani H, Zaya M, Suppogu N, Quesada O, Johnson BD, Mehta PK, et al. Angina hospitalization rates in women with signs and symptoms of ischemia but no obstructive coronary artery disease: a report from the WISE (Women's Ischemia Syndrome Evaluation) study. J Am Heart Assoc 2020;9(4):e013168. https://doi.org/10.1161/JAHA.119.013168. Epub 2020 Feb 17 32063125.

[13] Sedlak TL, Lee M, Izadnegahdar M, Merz CN, Gao M, Humphries KH. Sex differences in clinical outcomes in patients with stable angina and no obstructive coronary artery disease. Am Heart J 2013;166:38–44.

[14] Hemal K, Pagidipati NJ, Coles A, Dolor RJ, Mark DB, Pellikka PA, et al. Sex differences in demographics, risk factors, presentation, and noninvasive testing in stable outpatients with suspected coronary artery disease: insights from the PROMISE trial. JACC Cardiovasc Imaging 2016;9(4):337–46. 27017234.

[15] Lichtman JH, Leifheit EC, Safdar B, Bao H, Krumholz HM, Lorenze NP, et al. Sex differences in the presentation and perception of symptoms among young patients with myocardial infarction: evidence from the VIRGO study (variation in recovery: role of gender on outcomes of young AMI patients). Circulation 2018;137(8):781–90. https://doi.org/10.1161/CIRCULATIONAHA.117.031650. 29459463.

[16] Bairey Merz CN, Andersen H, Sprague E, Burns A, Keida M, Walsh MN, et al. Knowledge, attitudes, and beliefs regarding cardiovascular disease in women: the Women's Heart Alliance. J AM Coll Cardiol 2017;70(2):123–32.

[17] Cushman M, Arnold PM, Pasty BM, Manolio TA, Kuller LH, Burke GL, et al. C-reactive protein and the 10-year incidence of coronary heart disease in older men and women: the cardiovascular health study. Circulation 2005;112:25–31.

[18] Pasternak RC, Abrams J, Greenland P, Smaha LA, Wilson PW, Houston-Miller N. Task force #1—identification of coronary artery risk: is there a detection gap? J Am Coll Cardiol 2003;41:1863–74.

[19] Canto JG, Goldberg RJ, Hand MM, Bonow RO, Sopko G, Pepine CJ, et al. Symptom presentation of women with acute coronary syndromes: myth vs reality. Arch Intern Med 2007;167:2405–13. Khan NA, Daskalopoulou SS, Karp I, Eisenberg MJ, Pellitier R, Tsaok MA, Dasgupta K, Norris CM, Pilote L. Sex differences in acute coronary syndrome presentation in young patients. JAMA Intern Med 2013;173:1863–71.

[20] Bullock-Palmer RP. Prevention, detection and management of coronary artery disease in minority females. Ethn Dis 2015;25(4):499–506. https://doi.org/10.18865/ed.25.4.499. Review 26674268.

[21] Aggarwal NR, Bond RM, Mieres JH. The role of imaging in women with ischemic heart disease. Clin Cardiol 2018;41:194–202. https://doi.org/10.1002/clc.22913.

[22] Shaw LJ, Mieres JH, Hendel RH, Boden WE, Gulati M, Veledar E, et al. Comparative effectiveness of exercise electrocardiography with or without myocardial perfusion single photon emission computed tomography in women with suspected coronary artery disease: results from the What Is the Optimal Method for Ischemia Evaluation in Women (WOMEN) trial. Circulation 2011;124(11):1239–49. https://doi.org/10.1161/CIRCULATIONAHA.111.029660. Epub 2011 Aug 15.

[23] Shaw LJ, Olson MB, Kip K, Kelsey SF, Johnson BD, Mark DB, et al. The value of estimated functional capacity in estimating outcome: results from the NHBLI-Sponsored Women's Ischemia Syndrome Evaluation (WISE) Study. J Am Coll Cardiol 2006;47(3 Suppl):S36–43. https://doi.org/10.1016/j.jacc.2005.03.080.

[24] Hachamovitch R, Berman DS, Kiat H, Cohen I, Cabico JA, Friedman J, et al. Exercise myocardial perfusion SPECT in patients without known coronary artery disease: incremental prognostic value and use in risk stratification. Circulation 1996;93(5):905–14. 8598081.

[25] Mieres JH, Shaw LJ, Arai A, Budoff MJ, Flamm SD, Hundley WG, et al. Role of noninvasive testing in the clinical evaluation of women with suspected coronary artery disease: consensus statement from the Cardiac Imaging Committee, Council on Clinical Cardiology, and the Cardiovascular Imaging and Intervention Committee, Council on Cardiovascular Radiology and Intervention, American Heart Association. Circulation 2005;111(5):682–96. https://doi.org/10.1161/01.CIR.0000155233.67287.60.

[26] Anavekar NS, Chareonthaitawee P, Narula J, Gersh BJ. Revascularization in patients with severe left ventricular dysfunction: is the assessment of viability still viable? J Am Coll Cardiol 2016;67(24):2874–87. https://doi.org/10.1016/j.jacc.2016.03.571. Review 27311527.

[27] Sanders GD, Patel MR, Chatterjee R, Ross AK, Bastian LA, Coeytaux RR, et al. Noninvasive technologies for the diagnosis of coronary artery disease in women: identification of future research needs from comparative effectiveness review no. 58. Rockville, MD: AHRQ Comparative Effectiveness Reviews; 2013. Report No.: 13-EHC072-EF.

[28] Di Carli MF, Dorbala S, Meserve J, El Fakhri G, Sitck A, Moore SC. Clinical myocardial perfusion PET/CT. J Nucl Med 2007;48:783–93.

[29] Greenwood JP, Motwani M, Maredia N, Brown JM, Everett CC, Nixon J, et al. Comparison of cardiovascular magnetic resonance and single-photon emission computed tomography in women with suspected coronary artery disease from the Clinical Evaluation of Magnetic Resonance Imaging in Coronary Heart Disease (CE-MARC) Trial. Circulation 2014;129:1129–38. PubMed:24357404.

[30] Stocker TJ, Deseive S, Leipsic J, Hadamitzky M, Chen MY, Rubinshtein R, et al. Reduction in radiation exposure in cardiovascular computed tomography imaging: results from the prospective multicenter registry on radiation dose estimates of cardiac CT angiography in daily practice in 2017 (PROTECTION VI). Eur Heart J 2018;39(41):3715–23. https://doi.org/10.1093/eurheartj/ehy546.

[31] Baldassarre LA, Raman SV, Min JK, Mieres JH, Gulati M, Wenger NK, et al. Noninvasive imaging to evaluate women with stable ischemic heart disease. JACC Cardiovasc Imaging 2016;9(4):421–35. https://doi.org/10.1016/j.jcmg.2016.01.004. Review 27056162.

[32] Taqueti VR, Dorbala S, Wolinsky D, Abbott B, Heller GV, Bateman TM, et al. Myocardial perfusion imaging in women for the evaluation of stable ischemic heart disease-state-of-the-evidence and clinical recommendations. J Nucl Cardiol 2017;24(4):1402–26. https://doi.org/10.1007/s12350-017-0926-8. Epub 2017 Jun 5 28585034.

[33] Abbott BG, Case JA, Dorbala S, Einstein AJ, Galt JR, Pagnanelli R, et al. Contemporary cardiac SPECT imaging-innovations and best practices: an information statement from the American Society of Nuclear Cardiology. J Nucl Cardiol 2018;25(5):1847–60. https://doi.org/10.1007/s12350-018-1348-y. 30143954.

[34] Bloom SA, Meyers K. Reducing radiation to patients and improving image quality in a real-world nuclear cardiology laboratory. J Nucl Cardiol 2017;24(6):1871–7. https://doi.org/10.1007/s12350-017-0851-x. Epub 2017 Mar 22. Review 28332179.

[35] Mieres JH, Gulati M, Bairey Merz N, Berman DS, Gerber TC, Hayes SN, et al. Role of noninvasive testing in the clinical evaluation of women with suspected ischemic heart disease: a consensus statement from the American Heart Association. Circulation 2014;130:350–79.

[36] Yao SS, Bangalore S, Chaudhry FA. Prognostic implications of stress echocardiography and impact on patient outcomes: an effective gatekeeper for coronary angiography and revascularization. J Am Soc Echocardiogr 2010;23(8):832–9. https://doi.org/10.1016/j.echo.2010.05.004. 20554154.

[37] Plana JC, Mikati IA, Dokainish H, Lakkis N, Abukhalil J, Davis R, et al. A randomized cross-over study for evaluation of the effect of image optimization with contrast on the diagnostic accuracy of dobutamine echocardiography in coronary artery disease The OPTIMIZE Trial. JACC Cardiovasc Imaging 2008;1(2):145–52. https://doi.org/10.1016/j.jcmg.2007.10.014. 19356420.

[38] Elhendy A, O'Leary EL, Xie F, McGrain AC, Anderson JR, Porter TR. Comparative accuracy of real-time myocardial contrast perfusion imaging and wall motion analysis during dobutamine stress echocardiography for the diagnosis of coronary artery disease. J Am Coll Cardiol 2004;44(11):2185–91. 15582317.

[39] Porter TR, Xie F. Myocardial perfusion imaging with contrast ultrasound. JACC Cardiovasc Imaging 2010;3(2):176–87. https://doi.org/10.1016/j.jcmg.2009.09.024. Review 20159645.

[40] Leitman M, Tyomkin V, Peleg E, Zyssman I, Rosenblatt S, Sucher E, et al. Speckle tracking imaging in normal stress echocardiography. J Ultrasound Med 2017;36:717–24. https://doi.org/10.7863/ultra.16.04010.

[41] Henzlova MJ, Duvall WL, Einstein AJ, Travin MI, Verberne HJ. ASNC imaging guidelines for SPECT nuclear cardiology procedures: stress, protocols, and tracers. J Nucl Cardiol 2016;23(3):606–39. https://doi.org/10.1007/s12350-015-0387-x. No abstract available. Erratum in: J Nucl Cardiol 2016 Jun;23(3):640–2. 26914678.

[42] Cerqueira MD, Nguyen P, Staehr P, Underwood SR, Iskandrian AE, Investigators A-MT. Effects of age, gender, obesity, and diabetes on the efficacy and safety of the selective A2A agonist regadenoson versus adenosine in myocardial perfusion imaging integrated ADVANCE-MPI trial results. JACC Cardiovasc Imaging 2008;1:307–16.

[43] Amanullah AM, Kiat H, Friedman JD, Berman DS. Adenosine technetium-99m sestamibi myocardial perfusion SPECT in women: diagnostic efficacy in detection of coronary artery disease. J Am Coll Cardiol 1996;27:803–9.

[44] Kim C, Kwok YS, Heagerty P, Redberg R. Pharmacologic stress testing for coronary disease diagnosis: a meta-analysis. Am Heart J 2001;142(6):934–44. 11717594.

[45] Taillefer R, DePuey EG, Udelson JE, Beller GA, Latour Y, Reeves F. Comparative diagnostic accuracy of Tl-201 and Tc-99m sestamibi SPECT imaging (perfusion and ECG-gated SPECT) in detecting coronary artery disease in women. J Am Coll Cardiol 1997;29:69–77.

[46] Doukky R, Olusanya A, Vashistha R, Saini A, Fughhi I, Mansour K, et al. Diagnostic and prognostic significance of ischemic electrocardiographic changes with regadenoson-stress myocardial perfusion imaging. J Nucl Cardiol 2015;22(4):700–13. https://doi.org/10.1007/s12350-014-0047-6. Epub 2015 Apr 24 25907352.

[47] Doukky R, Frogge N, Bayissa YA, Balakrishnan G, Skelton JM, Confer K, et al. The prognostic value of transient ischemic dilatation with otherwise normal SPECT myocardial perfusion imaging: a cautionary note in patients with diabetes and coronary artery disease. J Nucl Cardiol 2013;20(5):774–84. https://doi.org/10.1007/s12350-013-9765-4. Epub 2013 Aug 9 23929206.

[48] Lester D, El-Hajj S, Farag AA, Bhambhvani P, Tauxe L, Heo J, et al. Prognostic value of transient ischemic dilation with regadenoson myocardial perfusion imaging. J Nucl Cardiol 2016;23(5):1147–55. https://doi.org/10.1007/s12350-015-0272-7. Epub 2015 Oct 21 26490267.

[49] Nakanishi R, Gransar H, Slomka P, Arsanjani R, Shalev A, Otaki Y, et al. Predictors of high-risk coronary artery disease in subjects with normal SPECT myocardial perfusion imaging. J Nucl Cardiol 2016;23(3):530–41. https://doi.org/10.1007/s12350-015-0150-3. Epub 2015 May 14 25971987.

[50] Sharir T, Kang X, Germano G, Bax JJ, Shaw LJ, Gransar H, et al. Prognostic value of poststress left ventricular volume and ejection fraction by gated myocardial perfusion SPECT in women and men: gender-related differences in normal limits and outcomes. J Nucl Cardiol 2006;13(4):495–506. 16919573.

[51] Dilsizian V, Bacharach SL, Beanlands RS, Bergmann SR, Delbeke D, Dorbala S, et al. ASNC imaging guidelines/SNMMI procedure standard for positron emission tomography (PET) nuclear cardiology procedures. J Nucl Cardiol 2016;23(5):1187–226. https://doi.org/10.1007/s12350-016-0522-3. Epub 2016 Jul 8. No abstract available 27392702.

[52] Slomka PJ, Nishina H, Abidov A, Hayes SW, Friedman JD, Berman DS, et al. Combined quantitative supine-prone myocardial perfusion SPECT improves detection of coronary artery disease and normalcy rates in women. J Nucl Cardiol 2007;14:44–52.

[53] Ben-Haim S, Almukhailed O, Neill J, Slomka P, Allie R, Shiti D, et al. Clinical value of supine and upright myocardial perfusion imaging in obese patients using the D-SPECT camera. J Nucl Cardiol 2014;21:478–85.

[54] Kelkar AA, Schultz WM, Khosa F, Schulman-Marcus J, O'Hartaigh BW, Gransar H, et al. Long-term prognosis after coronary artery calcium scoring among low-intermediate risk women and men. Circ Cardiovasc Imaging 2016;9:e003742.

[55] Pazhenkottil AP, Nkoulou RN, Ghadri JR, Herzog BA, Buechel RR, Kuest SM, et al. Prognostic value of cardiac hybrid imaging integrating single-photon emission computed tomography with coronary computed tomography angiography. Eur Heart J 2011;32:1465–71.

[56] Schwitter J, Wacker CM, Wilke N, Al-Saadi N, Sauer E, Huettle K, et al. Superior diagnostic performance of perfusion-cardiovascular magnetic resonance versus SPECT to detect coronary artery disease: the secondary endpoints of the multicenter multivendor MR-IMPACT II (Magnetic Resonance Imaging for Myocardial Perfusion Assessment in Coronary Artery Disease Trial). J Cardiovasc Magn Reson 2012;14:61. PubMed:22938651.

[57] Barbier CE, Bjerner T, Johansson L, Lind L, Ahlstrom H. Myocardial scars more frequent than expected: magnetic resonance imaging detects potential risk group. J Am Coll Cardiol 2006;48:765–71.

[58] Schelbert EB, Cao JJ, Sigurdsson S, Wu E, Parker MA, Lee DC, et al. Prevalence and prognosis of unrecognized myocardial infarction determined by cardiac magnetic resonance in older adults. JAMA 2012;308:890–6.

[59] Kim HW, Klem I, Shah DJ, Wu E, Meyers SN, Parker MA, et al. Unrecognized non-Q-wave myocardial infarction: prevalence and prognostic significance in patients with suspected coronary disease. PLoS Med 2009;6:e1000057.

[60] Truong QA, Rinehart S, Abbara S, Achenbach S, Berman DS, Bullock-Palmer R, et al. Coronary computed tomographic imaging in women: an expert consensus statement from the Society of Cardiovascular Computed Tomography. J Cardiovasc Comput Tomogr 2018;12(6):451–66. https://doi.org/10.1016/j.jcct.2018.10.019. Epub 2018 Oct 23. Review 30392926.

[61] Min JK, Dunning A, Lin FY, Achenbach S, Al-Mallah M, Budoff MJ, et al. Age- and sex-related differences in all-cause mortality risk based on coronary computed tomography angiography findings results from the International Multicenter CONFIRM (Coronary CT Angiography Evaluation for Clinical Outcomes: An International Multicenter Registry) of 23,854 patients without known coronary artery disease. J Am Coll Cardiol 2011;58(8):849–60. https://doi.org/10.1016/j.jacc.2011.02.074. 21835321.

[62] Meinel FG, Nance JW, Harris BS, De Cecco CN, Costello P, Schoepf UJ. Radiation risks from cardiovascular imaging tests. Circulation 2014;130(5):442–5. https://doi.org/10.1161/CIRCULATIONAHA.113.005340.

[63] Einstein AJ. Effects of radiation exposure from cardiac imaging: how good are the data? J Am Coll Cardiol 2012;59(6):553–65. https://doi.org/10.1016/j.jacc.2011.08.079. Review 22300689.

[64] Colbert JD, Martin BJ, Haykowsky MJ, Hauer TL, Austford LD, Arena RA, et al. Cardiac rehabilitation referral, attendance and mortality in women. Eur J Prev Cardiol 2015;22(8):979–86. https://doi.org/10.1177/2047487314545279. Epub 2014 Oct 2 25278001.

[65] Blomkalns AL, Chen AY, Hochman JS, Peterson ED, Trynosky K, Diercks DB, et al. Gender disparities in the diagnosis and treatment of non-ST-segment elevation acute coronary syndromes: large-scale observations from the CRUSADE (Can Rapid Risk Stratification of Unstable Angina Patients Suppress Adverse Outcomes With Early Implementation of the American College of Cardiology/American Heart Association Guidelines) National Quality Improvement Initiative. J Am Coll Cardiol 2005;45:832–7.

[66] Hyun KK, Redfern J, Patel A, Peiris D, Brieger D, Sullivan D, et al. Gender inequalities in cardiovascular risk factor assessment and management in primary healthcare. Heart 2017;103:492–8.

[67] Canter DL, Atkins MD, McNeal CJ, Bush RL. Risk factor treatment in veteran women at risk for cardiovascular disease. J Surg Res 2009;157(2):175–80. https://doi.org/10.1016/j.jss.2008.07.014. Epub 2008 Aug 24 19482299.

[68] Fihn SD, Blankenship JC, Alexander KP, Bittl JA, Byrne JG, Fletcher BJ, et al. ACC/AHA/AATS/PCNA/SCAI/STS focused update of the guideline for the diagnosis and management of patients with stable ischemic heart disease. J Am Coll Cardiol 2014;64:1929–49.

[69] Crea F, Battipaglia I, Andreotti F. Sex differences in mechanisms, presentation and management of ischaemic heart disease. Atherosclerosis 2015;241:157–68.

[70] Serruys PW, Morice MC, Kappetein AP, Colombo A, Holmes DR, Mack MJ, et al. Percutaneous coronary intervention versus coronary artery bypass grafting for severe coronary artery disease. N Engl J Med 2009;360:961–72.

[71] Morice MC, Serruys PW, Kappetein AP, Feldman TE, Ståhle E, Colombo A, et al. Outcomes in patients with de novo left main disease treated with either percutaneous coronary intervention using paxlitaxel-eluting stents or coronary artery bypass graft treatment in the SYNTAX trial. Circulation 2010;121:2645–53.

[72] Kappetein AP, Feldman TE, Mack MJ, Morice MC, Holmes DR, Ståhle E, et al. Comparison of coronary bypass surgery with drug-eluting stenting for the treatment of left main and/or three-vessel disease; 3-year follow-up of the SYNTAX trial. Eur Heart J 2011;32:2125–34.

[73] Kappetein AP, Head SJ, Morice MC, Banning AP, Serruys PW, Mohr FW, et al. Treatment of complex coronary artery disease in patients with diabetes; 5-year results comparing outcomes of bypass surgery and percutaneous coronary artery intervention in the SYNTAX trial. Eur J Cardiothorac Surg 2013;43:1006–13.

[74] Gudnadottir G, Anderson K, Thrainsdottir IS, James SK, Lagerqvist B, Gudnason T. Gender differences in coronary angiography, subsequent interventions, and outcomes among patients with acute coronary syndromes. Am Heart J 2017;191:65–74.

[75] Pina I, Zheng Q, She L, Szwed H, Lang IM, Farsky PS, et al. Sex difference in patients with ischemic heart failure undergoing surgical revascularization, results from the STITCH trial (Surgical Treatment for Ischemic Heart Failure). Circulation 2018;137:771–80.

[76] Boden WE, O'Rourke RA, Teo KK, Hartigan PK, Maron DJ, Kostuk WJ, et al. Optimal medical therapy with or without PCI for stable coronary disease. N Engl J Med 2007;356:1503–16.

[77] Maron DJ, Hochman JS, Reynolds HR, Bangalore S, O'Brien SM, Boden WE, et al. Initial invasive or conservative strategy for stable coronary disease. N Engl J Med 2020;382:1395–407. https://doi.org/10.1056/NEJMoa1915922.

[78] Chichareon P, Modolo R, Kerkmeijer L, Tomaniak M, Kogame N, Takahashi K, et al. Association of sex with outcomes in patients undergoing percutaneous coronary intervention, a subgroup analysis of the GLOBAL LEADERS randomized clinical trial. JAMA Cardiol 2020;5(1):21–9.

[79] Udell JA, Koh M, Qui F, Austin P, Wijeysundera HC, Bagai A, et al. Outcomes of women and men with acute coronary syndrome treated with and without percutaneous coronary revascularization. J Am Heart Assoc 2017;6:E004319.

[80] Maas A. Characteristic symptoms in women with ischemic heart disease. Curr Cardiovasc Risk Rep 2019;13–7.

[81] Bairey Merz CN, Pepine CJ, Walsh MN, Fleg JL. Ischemia and no obstructive coronary artery disease (INOCA): developing evidence-based therapies and research agenda for the next decade. Circulation 2017;135(11):1075–92. https://doi.org/10.1161/CIRCULATIONAHA.116.024534. Review 28289007.

[82] Pepine CJ, Anderson RD, Sharaf BL, Reis SE, Smith KM, Handberg EM, et al. Coronary microvascular reactivity to adenosine predicts adverse outcome in women evaluated for suspected ischemia results from the National Heart, Lung and Blood Institute WISE (Women's Ischemia Syndrome Evaluation) study. J Am Coll Cardiol 2010;55(25):2825–32. PubMed:20579538.

[83] Shufelt C, Pacheco C, Tweet M, Miller V. Sex-specific physiology and cardiovascular disease. Adv Exp Med Biol 2019;1065:433–54.

[84] Task Force Members, Montalescot G, Sechtem U, Achenbach S, Andreotti F, Arden C, et al. ESC guidelines on the management of stable coronary artery disease: the task force on the management of stable coronary artery disease of the European Society of Cardiology. Eur Heart J 2013;34(38):2949–3003. https://doi.org/10.1093/eurheartj/eht296. Epub 2013 Aug 30. No abstract available. Erratum in: Eur Heart J. 2014 Sep 1;35(33):2260–1 239962862013.

[85] Ong P, Athanasiadis A, Sechtum U. Pharmacotherapy for coronary microvascular dysfunction. Eur Heart J Cardiovasc Pharmacother 2015;1(1):65–71.

[86] Rambarat CA, Elgendy IY, Handberg EM, Bairey Merz CN, Wei J, Minissian MB, et al. Late sodium channel blockade improves angina and myocardial perfusion in patients with severe coronary microvascular dysfunction: Women's Ischemia Syndrome Evaluation-Coronary Vascular Dysfunction ancillary study. Int J Cardiol 2019;276:8–13.

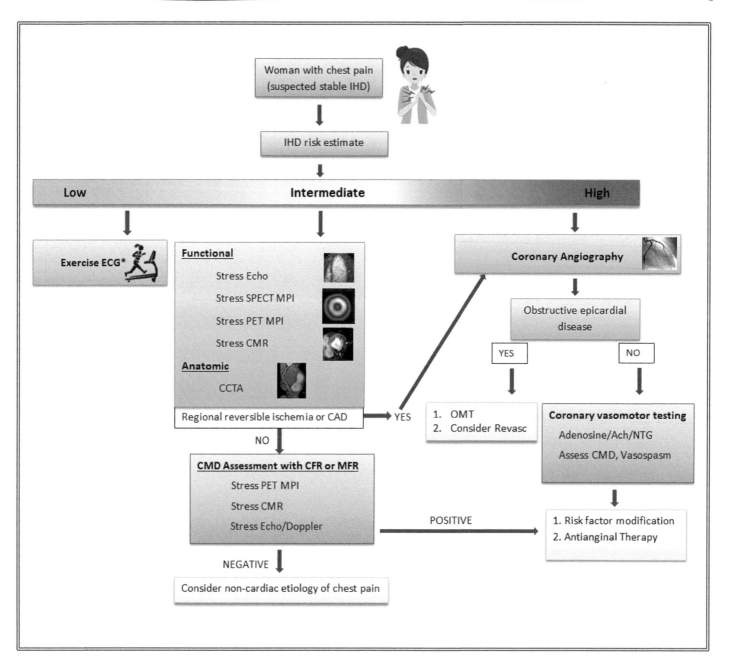

Diagnostic Work-Up for Woman With Suspected Stable IHD. The image encompasses a comprehensive diagnostic algorithm for a woman with suspected stable IHD. Imaging for obstructive epicardial and nonobstructive microvascular disease are depicted in orange and purple boxes respectively; management options are depicted in yellow boxes. *If resting ECG normal, and woman able to perform ≥ 5 METS; otherwise, proceed to "Intermediate" pathway and if a functional option is chosen, perform pharmacologic rather than exercise stress testing. Ach, acetylcholine; CCTA, coronary computed tomographic angiography; CFR, coronary flow reserve; CMD, coronary microvascular dysfunction; CMR, cardiac magnetic resonance imaging; ECG, electrocardiogram; Echo, echocardiogram; IHD, ischemic heart disease; MFR, myocardial flow reserve; MPI, myocardial perfusion imaging; NTG, nitroglycerin; OMT, optimal medical therapy; Revasc, coronary revascularization; SPECT, single-photon emission tomography. *Reproduced with permission from [1].*

Reference

[1] Koilpillai P, Aggarwal NR, Mulvagh SL. State of the art in noninvasive imaging of ischemic heart disease and coronary microvascular dysfunction in women: indications, performance, and limitations. Curr Atheroscler Rep 2020;22(12):73. https://doi.org/10.1007/s11883-020-00894-0.

Chapter 8

Coronary Microvascular Dysfunction

Jack Aguilar, Janet Wei, Odayme Quesada, Chrisandra Shufelt, and C. Noel Bairey Merz

Clinical Case

A 54-year-old female with hypertension and dyslipidemia presented to the emergency department after experiencing chest pain at rest for 12 h. Her only medication was hydrochlorothiazide that she has taken daily. On presentation, her physical exam was unremarkable and her vital signs were within normal limits. Her electrocardiogram (ECG) demonstrated normal sinus rhythm and normal duration of complexes without evidence of T-wave inversions or ST elevations. Her troponin was initially elevated at 2.98 ng/mL (ref <0.05 ng/mL) which continued to rise on repeat testing and peaked at 4.29 ng/mL. An echocardiogram demonstrated normal left ventricular function with ejection fraction of 64%, with no evidence of wall motion abnormalities. Coronary angiography showed no obstructive coronary artery disease (CAD) and no coronary vasospasm was reported. She was diagnosed with a non-ST-segment elevation myocardial infarction (NSTEMI) with no obstructive CAD and was referred for further evaluation and work-up in the outpatient setting. What further testing would you do next? What risk factors of coronary microvascular dysfunction are highlighted in this situation?

Abstract

Coronary microvascular dysfunction (CMD) is an increasingly recognized, novel entity of ischemic heart disease, defined as the impaired function of coronary microcirculation. Patients with CMD present with signs and symptoms of ischemic heart disease without angiographic evidence of obstructive coronary artery disease (CAD) (defined as >50% diameter stenosis). CMD is more prevalent in women and is associated with an increased risk of major adverse cardiovascular events (MACE). Diagnosis can be made by both invasive and noninvasive modalities in the setting of objective findings of myocardial ischemia and the absence of obstructive CAD. There has been growing literature in the therapeutic options for this patient population, though studies remain limited and many are still in the early stages of investigation. Current treatment options have been based on traditional therapies for obstructive coronary heart disease, including lipid-lowering medications, beta-blockers, angiotensin-converting enzyme inhibitors or angiotensin receptor blockers, calcium channel blockers, nitrates, and novel antianginal and antiischemic medications. Current trials are ongoing and investigating the effects of intensive medical therapy as compared to usual care for women with evidence of myocardial ischemia and absence of obstructive coronary arterial disease. This chapter covers the clinical features, risk factors, prognosis, pathophysiology, diagnostic modalities, and potential treatment options for patients with CMD.

Introduction

Cardiovascular disease remains a leading cause of death worldwide. Though myocardial ischemia is classically studied in patients with obstructive CAD, currently up to half of patients undergoing elective angiography for known or suspected angina have no obstructive epicardial CAD [1], now termed ischemia and no obstructive coronary artery disease (INOCA). Women who present with signs and symptoms of ischemia are more likely than men to have no evidence of obstructive CAD [2, 3]. About 50–65% of patients who present with angina and have nonobstructive CAD are thought to have CMD, which is defined as the impaired function of the coronary microvessels [4].

The Women's Ischemia Syndrome Evaluation (WISE) identified that up to 50% of women undergoing clinically indicated coronary angiography have no or nonobstructive CAD defined as coronary artery stenosis <50%, yet have a 2.5% yearly risk of major cardiovascular events (death, nonfatal myocardial infarction (MI), nonfatal stroke, and heart failure hospitalization) [5]. These women are often reassured despite growing evidence of the adverse prognosis associated with INOCA [6–8].

While historically considered a benign prognosis, after 10 years follow-up, cardiovascular death or MI occurred in 6.7% of the WISE women with no evidence of CAD and 12.8% of women with nonobstructive CAD [9]. Other studies demonstrate that women with ischemia and no CAD are four times more likely than men to be readmitted to the hospital within 180 days of acute coronary syndrome or chest pain [5, 10]. In the American College of Cardiology (ACC) national cardiovascular data registry, there was a higher prevalence of nonobstructive CAD in women (51%) than men (32%) [11]. Men with nonobstructive CAD have elevated rates of adverse events in the first year after angiography, though women are more than three times as likely as men to experience a cardiac event within the first-year postangiography [12]. These differences in outcomes between men and women with nonobstructive CAD may be related to differences in medical management, as men are more likely to be treated with angiotensin receptor blockers, while women are more likely to receive beta-blockers and diuretics [13].

Treatment of ischemic heart disease has largely been focused on treating obstructive CAD, and few studies have looked at treatment for patients with INOCA or CMD. There has been a growing body of literature detailing the pathophysiology, risk factors, diagnosis, and treatment of patients with CMD. For these patients, more studies and clinical trials are needed to understand benefit from medications classically used for patients with obstructive CAD, as well as novel treatments that target microvascular function.

Pathophysiology of Coronary Microvasculature Disease

Coronary arteries can be thought of as three different compartments: large epicardial arteries, prearterioles, and arterioles. The large epicardial coronary arteries function largely in a capacitance role and offer little resistance to coronary blood flow (CBF) [14]. The coronary microvasculature (vessels <300 µm in diameter) is comprised of prearterioles (diameter ~100–300 µm) and arterioles (diameter <100 µm) [15]. Prearterioles have a characteristic measurable drop in pressure along their path and represent the intermediate compartment of the coronary circulation [15]. Arterioles are the site of metabolic regulation of myocardial blood flow [14], matching myocardial blood supply to oxygen demand based on their response to substances produced during myocardial metabolism [15]. This function is critical, as the normal functioning heart is largely dependent on oxidative phosphorylation; thus any increase in cardiac activity requires an almost instantaneous match in oxygen availability [16, 17]. Both prearterioles and arterioles are below resolution limits of current angiographic systems and cannot be visualized by angiography [14].

CMD can result from impaired vasodilation, increased vasoconstriction, or both of the coronary microvessels. Causes of impaired vasodilation can be endothelium dependent [flow-mediated, acetylcholine (Ach), histamine, bradykinin, etc.] or endothelial independent (adenosine and catecholamines), which can be used to assess the function of the coronary microvasculature [14].

Arteries maintain a constant shear stress despite changes in blood flow velocity and viscosity; thus increases in coronary diameters are proportional to increase in blood flow, maintaining a constant shear stress [18]. Proximal pre-arterioles are most responsive to changes in flow, with dilatation largely determined by vasodilators released by endothelial cells in response to shear stress changes. These vasodilators include nitric oxide (NO), endothelium-derived hyperpolarizing factor (EDHF), and prostacyclin [16]. Endothelial production of NO can be initiated by specific receptors, such as bradykinin, histamine, and muscarinic receptors, or by mechanical deformation from shear forces or pulsatile strain caused by blood flow [14]. With endothelial dysfunction, there is an impaired endothelial response to vasodilatory mechanisms or paradoxical vasoconstriction in response to administered endothelial-dependent agonist [17]. This inappropriate response is thought to be mediated by a decrease in either production or availability of NO [17, 19]. There may also be a potential impairment in response of vascular smooth muscle to NO [20]. This imbalance either in response to or in availability of NO leads to a state that favors vasoconstriction despite the mismatch of oxygen to metabolic demand.

Under normal physiologic conditions, CBF is regulated by metabolites that are produced in proportion to the oxygen consumed, such as carbon dioxide and reactive oxygen species [17]. Hypoxia-driven vasodilation is mediated by adenosine [21], which is formed by degradation of adenine nucleotides and diffuses from myocytes to stimulate A2 receptors on smooth muscle cells, leading to arterial dilatation [16]. Rise in interstitial concentration of adenosine parallels increase in CBF, though inhibition of adenosine does not reduce the magnitude of hyperemia, suggesting that multiple substances may play roles in regulation of CBF [16, 22].

There are both sympathetic and parasympathetic contributions to CBF. The net effect of sympathetic stimulation on coronary resistance vessels is vasodilation, as epinephrine and norepinephrine lead to coronary vessel dilation even without alpha-receptor blockade [16, 23]. Alpha-receptor stimulation leads to coronary vasoconstriction and has been repeatedly demonstrated when there is adrenergic activation of the heart [24, 25]. Parasympathetic stimulation leads to vasodilation mediated by NO [16].

Intracoronary infusion of Ach in humans results in increased CBF and epicardial coronary artery dilation except in atherosclerotic epicardial vessels, which show paradoxical vasoconstriction [14]. Normally, Ach causes vasodilation through release of NO in sufficient quantity to oppose Ach's direct muscarinic receptor activation of smooth muscles [19]. However, when endothelium is damaged, Ach causes vasoconstriction via activation of smooth muscle muscarinic receptors [19, 26].

Risk Factors for CMD

Risk factors for CMD include hypertension, age, diabetes mellitus, dyslipidemia, elevated C-reactive protein (CRP), and estrogen deficiency. Though CAD and CMD share risk factors, these risk factors accounted for less than 20% of CMD among patients in WISE [27]. Aging leads to increased arterial wall stiffness and thickening of media, while hypertension leads to remodeling of small arteries and increased peripheral resistance from arteriolar vasoconstriction and reduction in density of microvessels [28]. Diabetes mellitus can lead to systemic microvascular dysfunction [29] and may be associated with CMD [30]. Chronic hyperglycemia has been associated with significantly reduced endothelial-dependent and -independent coronary vasodilator capacity [31]. In a study of 54 hypertensive patients without CAD, hypercholesterolemia was associated with reduced coronary flow reserve (CFR) [32]. In addition, a correlation between CFR and low-density lipoprotein (LDL) was seen, but not with total cholesterol, suggesting a pathologic role of LDL in CMD [33].

Chronic inflammatory rheumatoid diseases, such as systemic lupus erythematosus, systemic sclerosis, and rheumatoid arthritis, have been shown to be associated with increased risk of premature CAD and stroke [34], with studies suggesting the degree of inflammation affecting the risk of cardiovascular events [35, 36]. The proposed mechanism by which chronic inflammatory disease leads to endothelial dysfunction is through the reduction in synthesis and bioavailability of NO through tumor necrosis factor (TNF)-alpha-induced inhibition of endothelial NO synthase [36]. However, it is difficult to identify a single mechanism to explain the relationship between CMD and chronic inflammation, given the different molecular pathways involved in different forms of chronic inflammatory diseases. In a study of individuals with suspected CMD due to the presence of angina, abnormal stress testing, and normal coronary angiography, the high sensitivity CRP (hsCRP) level was higher than in the healthy control group [37]. Amongst 18 patients with systemic lupus erythematous and no obstructive CAD, 44% had CMD based on myocardial perfusion reserve index (MPRI) compared to control individuals [38]. In patients with ankylosing spondylitis, CFR was shown to be significantly lower than in the control group; these patients also had elevated levels of hsCRP and TNF-alpha, which were independently correlated to the CFR level [39].

There are some studies suggesting that low estrogen levels may contribute to endothelial dysfunction [40], though further studies are needed to thoroughly establish the relationship. Postmenopausal hormone therapy remains contraindicated for the prevention of cardiovascular disease in women [41], and a clinical trial conducted in WISE patients showed no beneficial change in markers of ischemia [40].

Clinical Features and Prognosis

Patients with CMD can present with signs and symptoms of ischemic heart disease and are found to have no obstructive CAD (defined as >50% diameter stenosis) [42]. CMD is relatively more prevalent in women than men, with women comprising 70% of patients with CMD [43]. Women have worse outcomes after infarction compared to age-matched men [44], potentially due to CMD. Patients with CMD can present with both typical angina, defined as substernal chest pain precipitated by physical exertion or emotional stress and relieved by rest or nitroglycerin, and nontypical anginal symptoms. In a study on patients with CMD and no CAD, patients with typical anginal symptoms had worse symptom burden and coronary endothelial dysfunction [45].

Women with INOCA and thought to have CMD are at elevated risk for MACE, predominantly heart failure hospitalizations and diastolic dysfunction [46]. CMD has been found to have a 2.5% annual rate of MACE including stroke, heart failure, or MI [47]. Along with increased risk of MACE, CMD has also been associated with diastolic dysfunction and hospitalizations for heart failure with preserved ejection fraction (HFpEF). In a study on patients with impaired CFR and without flow-limiting CAD, a CFR <2, reflecting CMD, was associated with an adjusted haz-ard ratio of 2.2 ($P=0.01$) for MACE and an adjusted hazard ratio of 2.43 ($P=0.03$) for HFpEF hospitalization [48]. In another study on 157 patients, 14% of those with CMD had adverse cardiovascular events by the 28-month follow-up, compared with none of the control individuals [49].

Women with nonobstructive CAD were threefold more likely than men without obstructive CAD to have adverse cardiovascular event within the first year of cardiac catheterization [12]. Sedlak et al. found that women with nonobstructive CAD had a higher risk of MACE than men with nonobstructive CAD with an adjusted hazard ratio of 2.43, and an adjusted hazard ratio of 2.55 for MACE compared to women with no CAD [12]. Coronary reactivity testing can be used to diagnose CMD, with measurement of CFR used as both a marker for diagnosis [5] and predictor of adverse events [42]. Data from a study done on men and women evaluated for suspected CAD with no visual evidence of CAD on positron emission tomography (PET) myocardial perfusion imaging demonstrated a higher rate of MACE in patients with CFR <2 at 3 years as compared to those with CFR ≥2 (**Figure 1**) [42, 50]. The authors suggest that CFR is a predictor of MACE in patients with CMD, which is associated with adverse outcomes regardless of sex [50].

Diagnosis of CMD

CMD can be diagnosed using invasive angiography techniques or noninvasive techniques [PET, transthoracic Doppler echocardiography (TTDE), and cardiac magnetic resonance imaging (CMRI)]. The Coronary Vasomotion Disorders International Study Group (COVADIS) recommends the following criteria for the diagnosis of CMD:

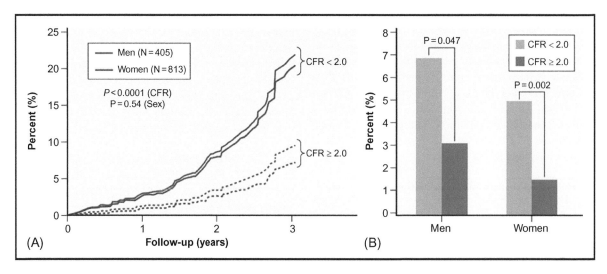

FIGURE 1   Sex Effects on Coronary Microvascular Dysfunction Measured by Positron Emission Tomography and Adverse Outcomes in Symptomatic Patients Without Obstructive Coronary Artery Disease. CFR, coronary flow reserve. *Adapted with permission from [50].*

(1) presence of symptoms suggestive of myocardial ischemia; (2) objective documentation of myocardial ischemia, as assessed by currently available techniques; (3) absence of obstructive CAD [<50% coronary diameter reduction and/or fractional flow reserve (FFR) > 0.80]; and (4) confirmation of a reduced CBF reserve and/or inducible microvascular spasm [51].

Invasive Modalities

Functional coronary testing involves using a stimulus administered at doses to achieve maximal vasodilation, allowing for the measurement of CBF at maximal vasodilation and at rest. The most common vasodilatory agents used include adenosine, Ach, nitroglycerin, ergotamine, and dipyridamole [26]. CBF is measured as the amount of flow through a given coronary artery per unit time, and the ratio of maximal to basal CBF gives the CFR. CFR can be calculated using a Doppler wire or thermodilution method [26, 52].

Intracoronary Doppler wire recording is at present the invasive method of choice to assess CMD. This allows for the direct measurement of CBF velocity in single epicardial arteries using the Doppler effect, which establishes the velocity and direction of blood flow derived by determining the frequency of shift resulting from the emitted and returning ultrasound waves [53]. **Figure 2** shows an example of CFR measurement using the intracoronary Doppler flow wire by measuring velocities both at rest and after adenosine stimulus. Other methods used include thermodilution and gas washout, both based on Fick's principle, though they require accurate measurement of temperature or gas concentration in the coronary sinus. In a study done on 40 patients with suspected CAD looking at direct comparison between Doppler flow velocity and thermodilution measurements of CFR as compared to PET measurement of CFR, a moderate correlation was found between thermodilution- and Doppler-derived CFR, with a stronger correlation between Doppler-derived CFR and PET-derived CFR [54].

CFR is measured invasively and calculated by taking the time-averaged peak hyperemic coronary flow velocity divided by the resting flow velocity, as measured on invasive Doppler [26]. In WISE, CFR less than 2.32 best predicted adverse outcomes in women with ischemia with nonobstructive coronary arteries, with a 5-year major adverse coronary event rate of 27% vs. 9.3% in women with a CFR greater than 2.32 [47]. Traditionally, CFR <2.5 is used as threshold for diagnosis of CMD [5]. However, CFR has been shown to vary according to age and sex and thus values obtained should be compared to appropriate age- and sex-matched individuals [15].

Adenosine is the most commonly used substance to evaluate CFR [15, 55]. Adenosine at a dose of 140 μg/kg/min for 1.5–6 min was found to achieve maximal microvascular dilation [56]. The side effects of adenosine include bradycardia, due to atrioventricular or sinoatrial nodal blockade and bronchoconstriction, both mediated by purinergic A1 receptor stimulation. The advantages of adenosine include its very short half-life of 10 s, which allows for rapid regression of side effects and repetition of test during the same session. Another commonly used substance to assess endothelium-independent coronary microvascular dilatation is dipyridamole, which acts by inhibiting adenosine degradation by adenosine deaminase [57].

Ach is used to assess endothelial-dependent dilation. Ach-administered intracoronary can be used to diagnose endothelial dysfunction if the epicardial coronary artery diameter decreased by >20% compared to baseline quantitative coronary angiography [49, 58]. Under normal conditions,

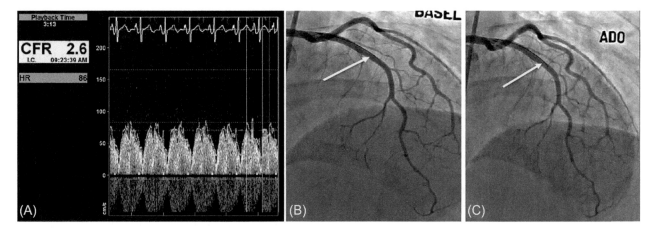

FIGURE 2   Example of CFR Measurement (A) Using Intracoronary Doppler Wire Measurement of Coronary Blood Flow. Stable positioning of the Doppler flow wire is shown (*yellow arrow*) at baseline (B), as well as after adenosine administration (C). ADO, adenosine; CFR, coronary flow reserve; HR, heart rate.

Ach causes vasodilation. In contrast with impaired endothelial response, Ach leads to vasoconstriction due to muscarinic receptor activation [19].

Normal coronary endothelial function is defined as a threefold to fourfold increase in CBF in response to vasodilators, and a reduction in CBF is defined as endothelial dysfunction [59]. Nonendothelial-dependent microvascular dysfunction is defined as CFR <2.5 with intracoronary infusion of vasodilators [47]. The change in epicardial coronary artery diameter and measurement of blood flow velocity using Doppler flow wires allows for the response of coronary vasculature to vasodilator substances to be measured and quantified [26]. The definitions of test findings are summarized in **Table 1** [42].

An additional method of assessing the coronary microvasculature is via Doppler-derived hyperemic microvascular resistance (hMR) and thermodilution-derived index of microcirculatory resistance (IMR) [60]. The IMR estimates flow with thermodilution, while hMR incorporates Doppler flow velocity. A recent study by Williams et al. comparing both hMR and IMR found both measures to be predictors of microvascular disease as determined invasively by CFR measurement and CMRI-derived MPRI, though hMR had superior sensitivity over IMR [60].

Noninvasive Modalities

There are various noninvasive methods to assess coronary microvasculature, which measure CBF and myocardial perfusion. CBF can be assessed by Doppler ultrasound, cardiac PET, and CMRI [61–63]. Noninvasive absolute myocardial blood flow and myocardial flow reserve (MFR) are currently uniquely measured by PET [64], while CMRI is most commonly used to measure MPRI [65], which can also be used to assess the coronary microvasculature.

TTDE is a noninvasive method that allows the measurement of CBF velocity taken as an indirect measure of CBF, mainly in the left anterior descending coronary artery (LAD), which is easier to visualize [66, 67]. However, in several patients, CBF in the left circumflex and in the posterior interventricular descending artery can also be assessed [67]. TTDE of the posterior descending artery is unsatisfactory in most patients [15]. TTDE of the LAD during intravenous dipyramidole-induced stress has prognostic capabilities [68, 69]. CFR is measured as the ratio of hyperemic diastolic peak flow velocity during maximal vasodilator stimulation over the basal flow velocity. Reproducibility studies have shown low interobserver and intraobserver variability [67]. Several studies have investigated the reliability of TTDE by comparing results with those obtained with the gold standard method of intracoronary Doppler flow wire recordings [70–72] or with PET [72]. Advantages of TTDE are that it is completely noninvasive, easily available at bedside, not time consuming, cheap, and suitable for serial assessments to follow coronary microvascular function over time or to check effects of therapeutic interventions. Limitations are that there is only evidence to support reliability when coronary microvascular function is assessed in the LAD, an adequate echocardiographic window cannot be obtained in all patients, and the test requires appropriate experience by the operator.

Myocardial contrast echocardiography (MCE) can also image myocardial blood flow by echocardiography. Microbubbles are injected intravenously, acting as a contrast agent, which allows for visualization of the microcirculation by detecting acoustic emissions [15]. The signal from the microbubbles is proportional to their concentration in the blood, and can thus be used to relate to myocardial blood volume and myocardial blood flow (MBF) [73]. Myocardial blood flow can be estimated by examining microbubble intensity over time and correlates well with MBF measurements obtained from PET [74]. MCE has been validated against single photon emission CT (SPECT) and invasive angiography [75]. MCE is easy to perform and relatively inexpensive, though there have been reports of major adverse events in critically ill patients after contrast administration [76].

CMRI first-pass perfusion studies allow the evaluation of the changes in myocardial signal intensity of gadolinium; quantitative measures of MBF at rest and during hyperemia can then be determined based on intensity curves of regions

TABLE 1 Definitions of Coronary Microvascular and Macrovascular Dysfunction

CMD Pathways	Microvascular Dysfunction	Macrovascular Dysfunction
Non-endothelium dependent	CFR in response to adenosine <2.5	Change in coronary artery diameter in response to nitroglycerin <20%
Endothelium dependent	Change in CBF in response to acetylcholine <50%	No change in coronary artery diameter in response to acetylcholine
Coronary spasm	Chest pain + ECG changes Change in coronary artery diameter in response to acetylcholine <90%	

CBF, coronary blood flow; CFR, coronary flow reserve; CMD, coronary microvascular dysfunction; ECG, electrocardiogram. *From [42].*

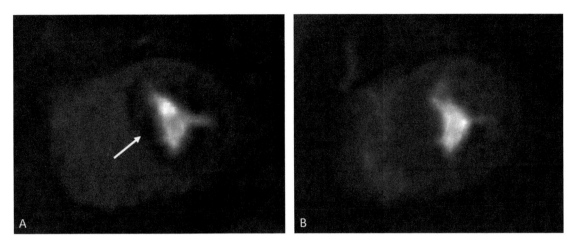

FIGURE 3 Cardiac MRI in Patient With Stable Angina and No Evidence of Coronary Atherosclerosis Who Underwent Coronary Reactivity Testing With Doppler Flow Wire and Was Found to Have a CFR of 2.1 Consistent With Coronary Microvascular Dysfunction. This patient's cardiac MRI showed a circumferential subendocardial perfusion defect (*white arrow*) at stress with adenosine (A) but not at rest (B). Calculated MPRI was 1.19. CFR, coronary flow reserve; MPRI, myocardial perfusion reserve index; MRI, magnetic resonance imaging.

of interest [77]. In a study by Panting et al., patients with cardiac syndrome X were shown to have impairment of subendocardial CBF in response to adenosine [78]. In a study of 118 women with INOCA in the WISE cohort, CMRI-derived MPRI predicted CMD status with moderate sensitivity and specificity, 73% and 74%, respectively [65]. The use of stress perfusion CMRI has been used in women with symptoms of myocardial ischemia and no obstructive CAD to assess myocardial blood flow and was found to be predictive of adverse events [79]. There are several advantages to CMRI, namely, it has high spatial resolution and is radiation free. In addition, it allows for the assessment of subendocardial and subepicardial perfusion [80, 81]. A novel, noninvasive diagnostic method is using CMRI to assess MPRI in response to adenosine (**Figure 3**). In a study of women with previously confirmed CMD, there were lower MPRI values globally and in the subendocardial and subepicardial regions in comparison to controls [82].

PET is a well-validated and accurate noninvasive approach to assess coronary vasomotor function [83]. Imaging with PET involves rest and vasodilator-stress perfusion studies allowing for quantification of regional and global myocardial blood flow and calculation of CFR. PET offers an accurate measure of MBF both at rest and during hyperemia, which helps establish the presence of CMD if reduction in MFR is caused by impaired increase in MBF or by an increase in basal MBF. When combined with computed tomography (CT), it can be used to calculate the coronary artery calcium score simultaneously [84, 85]. The accuracy of PET for quantitative noninvasive measurement of CFR and MBF has been validated in experimental animals and humans with good reproducibility [86].

Treatment

Management of patients with CMD remains difficult due to limited studies done in this patient population. As such, there is limited guidance and wide variability in how clinicians treat these patients. Clinicians currently employ strategies similar to patients with nonobstructive CAD, with recommendation of lifestyle modifications such as encouraging diet, exercise, and smoking cessation, in addition to treatment of modifiable risk factors such as hyperlipidemia, hypertension, and diabetes (**Figure 4**).

Statins

Statins have been shown to improve endothelium-independent coronary microvascular function in patients with high cholesterol levels [87]. Statins may be beneficial if endothelial dysfunction is suspected as the cause of ischemic heart disease [88], though there are no large randomized control studies on statin use in endothelial dysfunction.

Beta-Blockers

Beta-blockers reduce myocardial oxygen consumption and improve coronary perfusion by prolonging diastolic time. There are few studies that have evaluated the effects of beta-blockers on symptoms in microvascular angina. In early studies, atenolol showed favorable effects regarding symptoms and stress test results [89, 90]. Beta-blockers have been shown to improve endothelial function; in a small study of 18 patients with cardiac syndrome X, nebivolol increased flow-mediated dilation [91]. In a more recent small study comparing nebivolol to metoprolol in 30 patients with

Pharmacologic	Nonpharmacologic
• Nitrates	• Exercise
• Statins	• Cognitive behavioral therapy
• ACE-I	• Transcendental meditation
• ACE-I + aldosterone blockade	• Transcutaneous electrical nerve stimulation
• Calcium antagonists	
• Low-dose tricyclic antidepressants	
• Estrogens	
• PDE-5 inhibitors	
• Exercise	
• L-arginine	
• Ranolazine	
• Ivabradine	
• Ranolazine + ivabradine	
• Metformin	
• Rho-kinase inhibitors	
• Endothelin receptor blockers	

FIGURE 4   **Potential Therapies for Coronary Microvascular Dysfunction.** ACE-I, angiotensin-converting enzyme inhibitor; PDE-5, phosphodiesterase-5. *From [42] with permission.*

cardiac syndrome X, patients treated with nebivolol had improved exercise stress test results [92].

ACE Inhibitors/ARBs

Angiotensin-converting enzyme (ACE) inhibitors have been proposed as therapeutic agents in CMD, thought to be due to their mechanism in lowering serum and tissue angiotensin II, and have been found to increase the availability of NO and reduce oxidative stress in cardiac syndrome X patients [93]. In a study done on 63 normotensive women and CMD, treatment with ramipril had no significant effect on coronary flow velocity reserve or symptoms compared to placebo, leading the authors to suggest that benefits from angiotensin-converting enzyme inhibitors may be mediated by blood pressure reduction [94].

Calcium Channel Blockers

Calcium channel blockers are effective vasodilators and used to improve reduced vasodilatory capacity of coronary microvasculature. They function through blockade of transmembrane influx of calcium ions into arterial smooth muscles. In one randomized controlled trial, patients treated with verapamil or nifedipine had improved symptoms and exercise stress test parameters as compared to placebo [95]. In a recent study by Ford et al., patients with angina and no obstructive CAD were randomized to stratified medical therapy or standard of care and assessed at 6 months for anginal symptoms [96]. In this study, the intervention group had a significant reduction in anginal symptoms, and more patients in the intervention arm were likely to be prescribed calcium channel blockers.

Nitrates

Nitrates have the capacity to reduce cardiac work through reduction of preload, although they also have coronary vasodilator effects. In an observational study of 99 cardiac syndrome X patients, both sublingual nitrates and oral nitrates relieved episodes of chest pain in 42% of patients, along with calcium antagonists [97]. Nitrates have been shown to significantly improve exercise stress test results in patients with CAD, though patients with microvascular angina did not have the same beneficial results, suggesting that the dilator effect of nitrates on small coronary vessels is poor [98]. The reason for these results is not known, though there may be a component attributable to the myocardial hypoperfusion caused by hypotension and reflex adrenergic activity.

Xanthine Derivatives

Aminophylline, a xanthine derivative, is a nonselective adenosine-receptor antagonist used as antianginal and antiischemic treatment. The benefits of xanthine derivatives may be related to xanthine's inhibition of arteriolar dilator effects of adenosine, through antagonism of the A2 receptors on smooth muscle cells. In nondysfunctional coronary microvessels, this effect may favor CBF redistribution toward areas with CMD, where release of adenosine is increased, attenuating excess dilation of microvasculature in a well-perfused area and shunting blood to a poorly perfused area. There has been some improvement in symptoms and exercise capacity with the use of short-term oral aminophylline [99, 100] in patients with signs and symptoms of ischemia and normal coronary angiography.

Ranolazine

Ranolazine is an antiischemic drug with a not completely known mechanism of action, though it is thought to act by inhibiting the inward late Na current, thus reducing intracellular calcium overload in cardiomyocytes during ischemia and improving myocardial relaxation and left ventricular diastolic function [101]. It is increasingly becoming an adjunctive antianginal agent in patients with refractory angina when traditional antianginal agents are ineffective or contraindicated. In a small randomized controlled crossover trial of 20 women with microvascular angina and evidence of ischemic perfusion defects on adenosine CMRI, ranolazine resulted in significantly improved MPRI as compared to placebo [102]. In a study of 45 patients with microvascular angina, ranolazine was shown to improve both angina status and exercise stress test results as compared to both ivabradine and placebo at 4 weeks [103]. In one study, ranolazine was shown to significantly increase CFR after 8 weeks of treatment compared to placebo [104]. In another randomized placebo-controlled crossover trial on patients with evidence of CMD, no differences in angina, stress MPRI, diastolic filling, or quality of life were seen in patients who received short-term ranolazine for 2 weeks [105]. In a more recent study, symptomatic patients with no obstructive CAD and limited MPRI were randomized in a double-blind crossover trial to ranolazine or placebo; this study looked at patients subgrouped by CFR <2.5, showing improved angina and quality of life in patients with reduced CFR as by the Seattle Angina Questionnaire (SAQ) (**Figure 5**) [106]. In a meta-analysis of randomized controlled trials, ranolazine was found to improve MPRI and reduce angina in patients with CMD [107].

Ivabradine

Ivabradine selectively reduces the activity of the sinus node through inhibition of the pacemaker I_f current, which involves both Na^+ and K^+ flows. This leads to a decrease in the rate of diastolic depolarization and heart rate; however, unlike beta-blockers, ivabradine does not have negative inotropic effects or cause vasoconstriction [108, 109]. In patients with stable CAD, use of ivabradine showed improved coronary flow velocity and CFR, and this improvement in coronary microvascular function remained even after correction for heart rate [110]. Ivabradine was also shown to improve angina and exercise stress test results in patients with microvascular angina in a small placebo-controlled trial [103]. In addition, in a meta-analysis of randomized controlled trials, ivabradine and ranolazine were shown to reduce angina in patients with CMD [107].

Estrogens

Menopause hormone therapy in women is not recommended for prevention or secondary prevention of cardiovascular disease and in some women, based on age and time since menopause, is associated with increased risks of cardiovascular disease. In a randomized placebo-controlled trial, transdermal 17-beta-estradiol reduced spontaneous episodes of chest pain in women with microvascular angina, though did not significantly impact exercise-induced myocardial ischemia and nonischemic episodes during ECG Holter monitoring [111]. In the WISE study, women were randomized to oral estradiol with norethindrone or placebo for 12 weeks and it was found that while the hormone therapy improved chest pain symptoms, it did not improve ischemia

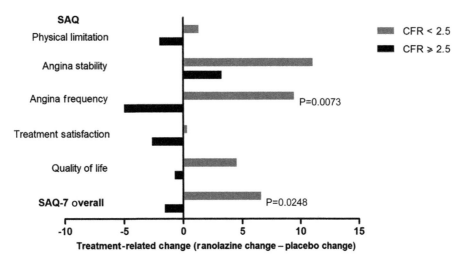

FIGURE 5 **Mean 2-Week Change in Seattle Angina Questionnaire (SAQ and SAQ7) Score Showing Improvement in Angina Stability, Physical Limitation, Angina Frequency, Treatment Satisfaction, and Quality of Life as Compared to Placebo in Patients With CFR <2.5.** CFR, coronary flow reserve. *Reproduced with permission from [106].*

or endothelial dysfunction [40]. Other studies have also found that menopause hormone therapy has not shown substantial improvement in electrocardiographic indicators of ischemia [112].

Other Pharmacotherapies

Atrasentan, an endothelin-A receptor antagonist, was shown to increase CBF in patients with microvascular angina after 6 months of treatment compared to placebo [113]. Fasudil, a rho kinase inhibitor that reduces smooth muscle cell hypercontraction, was shown to improve pacing-induced myocardial ischemic parameters such as anginal symptoms and maximum ST-segment depression [114]. Nicorandil, an adenosine triphosphate-sensitive nitrate-potassium-channel agonist, has been shown to improve peak exercise capacity in patients with microvascular angina [115].

Nonpharmacologic Therapies

There may be a role for patient symptom control by non-pharmacologic therapies. Cognitive behavioral therapy was shown to improve symptom frequency in women with cardiac syndrome X [116]. Similarly, cardiac rehabilitation for women with cardiac syndrome X was shown to improve symptom severity, quality of life, and exercise tolerance [117, 118].

Future Directions

There is currently an ongoing randomized controlled trial that will assess the prognostic and symptomatic benefit of treatment in women with suspected microvascular disease with either statins, angiotensin convertase enzyme inhibitors or angiotensin-receptor blockers, and aspirin compared to usual care [96]. There is also a current multicenter, prospective, randomized controlled trial being done studying the effects intensive medical therapy [treatment with intensive statin, ACE inhibitor or angiotensin receptor blocker (ARB), and aspirin] as compared to usual care for women who have signs or symptoms of ischemia but no obstructive disease [Women's Ischemia Trial to Reduce Events in Non-obstructive CAD (WARRIOR) Trial, NCT03417388].

Summary and Conclusion

Obstructive CAD is well understood with abundant treatment strategies, and dysfunction of the coronary microvasculature, termed CMD, has emerged as a novel entity of ischemic heart disease with newly understood prevalence and prognostic implications. The coronary microvasculature cannot be directly imaged in vivo; however, a number of invasive and noninvasive techniques are available for measurement and diagnosis. CMD is associated with poor outcomes and long-term cardiovascular risk. There is a lack of evidence-based data to currently guide treatment of this patient population; however, clinical trials are ongoing. Further studies should be focused on refining and standardizing the assessment of CMD and should examine long-term outcomes of patients with this increasingly recognized condition.

Key Points

1. Coronary microvascular dysfunction is an increasingly recognized novel disease process of ischemic heart disease, resulting from impaired vasodilation or increased vasoconstriction, or both, of coronary microvessels.

2. Patients with CMD present with signs and symptoms of ischemic heart disease and are found to have no obstructive coronary artery disease (defined as >50% diameter stenosis). More than half of patients who present with angina and nonobstructive CAD are thought to have CMD.

3. Women make up the majority of patients with CMD and have largely gone untreated. Studies have shown that untreated CMD is associated with an increased risk of major adverse cardiovascular events.

4. There are both invasive and noninvasive methods to diagnose CMD. The invasive method of choice to assess for CMD is intracoronary Doppler wire recording, while noninvasive techniques include cardiac positron emission tomography, transthoracic Doppler echocardiography, and cardiac magnetic resonance imaging.

5. Risk factors for CMD are similar to those for patients with obstructive CAD. These include hypertension, age, diabetes mellitus, dyslipidemia, elevated CRP, and estrogen deficiency.

6. Clinicians currently employ treatment strategies for patients with CMD similar to those used for patients with nonobstructive CAD, as current evidence for treatment in this patient population is limited to small sample-sized studies.

7. Current pharmacological treatment options for CMD include lipid-lowering medications, beta-blockers, angiotensin-converting enzyme inhibitors or angiotensin receptor blockers, calcium channel blockers, and novel antianginal and antiischemic medications. Nonpharmacologic treatment options include cardiac rehabilitation and cognitive behavioral therapy.

8. Evidence regarding efficacy of current treatments options in CMD is still needed, with promising new trials currently underway to investigate the different medication modalities and their effect on long-term outcomes.

CH
8

Back to Clinical Case

The clinical case describes a 54-year-old female who presented with prolonged chest pain at rest and who was diagnosed with NSTEMI with nonobstructive CAD after coronary angiography. Women, like this patient, are more likely than men to present with signs and symptoms of ischemia and to have no evidence of CAD on angiography. More than half of these patients are thought to have coronary microvascular dysfunction (CMD) requiring further testing for diagnosis. Risk factors for CMD include hypertension, diabetes mellitus, advanced age, dyslipidemia, chronic inflammatory disease, and estrogen deficiency. This patient had both hypertension and dyslipidemia as risk factors for CMD. Untreated patients with CMD have an elevated annual rate of major adverse cardiovascular events. This patient underwent coronary reactivity testing that demonstrated a coronary flow reserve (CFR) of 1.8 after adenosine-induced vasoconstriction, with resolution of vasoconstriction after nitroglycerin administration (**Figure 6**). She was started on aspirin, carvedilol, and simvastatin, and referred to cardiac rehab. Her antihypertensive regimen was changed from hydrochlorothiazide to a calcium channel blocker. She was given sublingual nitroglycerin as needed for angina. One year after starting treatment, she had an improvement in her symptoms and had no recurrent myocardial infarctions or hospitalizations for angina.

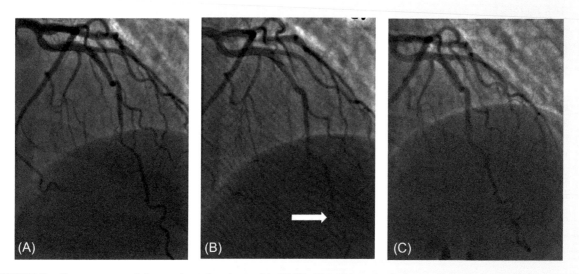

FIGURE 6 Coronary reactivity testing of patient with CMD. This figure includes the baseline coronary angiogram (A) and the angiogram showing adenosine-induced vasoconstriction (B) with a corresponding CFR of 1.8. Resolution of vasoconstriction with nitroglycerin is also shown (C). CFR, coronary flow reserve; CMD, coronary microvascular dysfunction.

Source of Funding

This work was supported by contracts from the National Heart, Lung, and Blood Institutes Nos. N01-HV-68161, N01-HV-68162, N01-HV-68163, N01-HV-68164, Grants U0164829, U01 HL649141, U01 HL649241, K23HL125941, K23HL127262, T32HL69751, R01 HL090957, 1R03AG032631 from the National Institute on Aging, GCRC Grant M01-RR00425 from the National Center for Research Resources, the National Center for Advancing Translational Sciences Grant UL1TR000124 and UL1TR001427, and the Edythe L. Broad and the Constance Austin Women's Heart Research Fellowships, the Barbra Streisand Women's Cardiovascular Research and Education Program, the Linda Joy Pollin Women's Heart Health Program, and the Erika Glazer Women's Heart Health Project, all in the Barbra Streisand Women's Heart Center at the Cedars-Sinai Smidt Heart Institute, Los Angeles, California.

This work is solely the responsibility of the authors and does not necessarily represent the official views of the National Heart, Lung, and Blood Institute or National Institutes of Health.

References

[1] Patel MR, Peterson ED, Dai D, Brennan JM, Redberg RF, Anderson HV, et al. Low diagnostic yield of elective coronary angiography. N Engl J Med 2010;362(10):886–95.

[2] Banks K, Lo M, Khera A. Angina in women without obstructive coronary artery disease. Curr Cardiol Rev 2010;6(1):71–81.

[3] Sullivan AK, Holdright DR, Wright CA, Sparrow JL, Cunningham D, Fox KM. Chest pain in women: clinical, investigative, and prognostic features. BMJ 1994;308(6933):883–6.

[4] Marinescu MA, Löffler AI, Ouellette M, Smith L, Kramer CM, Bourque JM. Coronary microvascular dysfunction, microvascular angina, and treatment strategies. JACC Cardiovasc Imaging 2015;8(2):210–20.

[5] Kothawade K, Bairey Merz CN. Microvascular coronary dysfunction in women: pathophysiology, diagnosis, and management. Curr Probl Cardiol 2011;36(8):291–318.

[6] Bugiardini R, Bairey MC. Angina with "normal" coronary arteries: a changing philosophy. JAMA 2005;293:477–84.

[7] Gulati M, Cooper-DeHoff RM, McClure C, Johnson BD, Shaw LJ, Handberg EM, et al. Adverse cardiovascular outcomes in women with nonobstructive coronary artery disease: a report from the Women's Ischemia Syndrome Evaluation Study and the St James Women Take Heart Project. Arch Intern Med 2009;169(9):843–50.

[8] Eastwood JA, Johnson BD, Rutledge T, Bittner V, Whittaker KS, Krantz DS, et al. Anginal symptoms, coronary artery disease, and adverse outcomes in Black and White women: the NHLBI-sponsored Women's Ischemia Syndrome Evaluation (WISE) study. J Womens Health (Larchmt) 2013;22(9):724–32.

[9] Sharaf B, Wood T, Shaw L, Johnson BD, Kelsey S, Anderson RD, et al. Adverse outcomes among women presenting with signs and symptoms of ischemia and no obstructive coronary artery disease: findings from the National Heart, Lung, and Blood Institute-sponsored Women's Ischemia Syndrome Evaluation (WISE) angiographic core laboratory. Am Heart J 2013;166(1):134–41.

[10] Humphries KH, Pu A, Gao M, Carere RG, Pilote L. Angina with "normal" coronary arteries: sex differences in outcomes. Am Heart J 2008;155(2):375–81.

[11] Shaw LJ, Shaw RE, Merz CN, Brindis RG, Klein LW, Nallamothu B, et al. Impact of ethnicity and gender differences on angiographic coronary artery disease prevalence and in-hospital mortality in the American College of Cardiology-National Cardiovascular Data Registry. Circulation 2008;117(14):1787–801.

[12] Sedlak TL, Lee M, Izadnegahdar M, Merz CN, Gao M, Humphries KH. Sex differences in clinical outcomes in patients with stable angina and no obstructive coronary artery disease. Am Heart J 2013;166(1):38–44.

[13] Johnston N, Schenck-Gustafsson K, Lagerqvist B. Are we using cardiovascular medications and coronary angiography appropriately in men and women with chest pain? Eur Heart J 2011;32(11):1331–6.

[14] Crea F, Camici PG, Bairey Merz CN. Coronary microvascular dysfunction: an update. Eur Heart J 2014;35(17):1101–11.

[15] Camici PG, d'Amati G, Rimoldi O. Coronary microvascular dysfunction: mechanisms and functional assessment. Nat Rev Cardiol 2015;12(1):48–62.

[16] Duncker DJ, Bache RJ. Regulation of coronary blood flow during exercise. Physiol Rev 2008;88(3):1009–86.

[17] Goodwill AG, Dick GM, Kiel AM, Tune JD. Regulation of coronary blood flow. Compr Physiol 2017;7(2):321–82.

[18] Lupi A, Buffon A, Finocchiaro ML, Conti E, Maseri A, Crea F. Mechanisms of adenosine-induced epicardial coronary artery dilatation. Eur Heart J 1997;18(4):614–7.

[19] AlBadri A, Wei J, Mehta PK, Landes S, Petersen JW, Anderson RD, et al. Acetylcholine versus cold pressor testing for evaluation of coronary endothelial function. PLoS One 2017;12(2), e0172538.

[20] Durand MJ, Gutterman DD. Diversity in mechanisms of endothelium-dependent vasodilation in health and disease. Microcirculation 2013;20(3):239–47.

[21] Deussen A, Ohanyan V, Jannasch A, Yin L, Chilian W. Mechanisms of metabolic coronary flow regulation. J Mol Cell Cardiol 2012;52(4):794–801.

[22] Yada T, Richmond KN, Van Bibber R, Kroll K, Feigl EO. Role of adenosine in local metabolic coronary vasodilation. Am J Physiol 1999;276(5):H1425–33.

[23] Sun D, Huang A, Mital S, Kichuk MR, Marboe CC, Addonizio LJ, et al. Norepinephrine elicits beta2-receptor-mediated dilation of isolated human coronary arterioles. Circulation 2002;106(5):550–5.

[24] Chilian WM. Functional distribution of alpha 1- and alpha 2-adrenergic receptors in the coronary microcirculation. Circulation 1991;84(5):2108–22.

[25] Heusch G, Baumgart D, Camici P, Chilian W, Gregorini L, Hess O, et al. Alpha-adrenergic coronary vasoconstriction and myocardial ischemia in humans. Circulation 2000;101(6):689–94.

[26] Wei J, Mehta PK, Johnson BD, Samuels B, Kar S, Anderson RD, et al. Safety of coronary reactivity testing in women with no obstructive coronary artery disease: results from the NHLBI-sponsored WISE (Women's Ischemia Syndrome Evaluation) study. JACC Cardiovasc Interv 2012;5(6):646–53.

[27] Wessel TR, Arant CB, McGorray SP, Sharaf BL, Reis SE, Kerensky RA, et al. Coronary microvascular reactivity is only partially predicted by atherosclerosis risk factors or coronary artery disease in women evaluated for suspected ischemia: results from the NHLBI Women's Ischemia Syndrome Evaluation (WISE). Clin Cardiol 2007;30(2):69–74.

[28] Tanaka M, Fujiwara H, Onodera T, Wu DJ, Matsuda M, Hamashima Y, et al. Quantitative analysis of narrowings of intramyocardial small arteries in normal hearts, hypertensive hearts, and hearts with hypertrophic cardiomyopathy. Circulation 1987;75(6):1130–9.

[29] Girach A, Vignati L. Diabetic microvascular complications—can the presence of one predict the development of another? J Diabetes Complications 2006;20(4):228–37.

[30] Sucato V, Evola S, Novo G, Novo S. Diagnosis of coronary microvascualar dysfunction in diabetic patients with cardiac syndrome X: comparison by current methods. Recenti Prog Med 2013;104(2):63–8.

[31] Di Carli MF, Janisse J, Grunberger G, Ager J. Role of chronic hyperglycemia in the pathogenesis of coronary microvascular dysfunction in diabetes. J Am Coll Cardiol 2003;41(8):1387–93.

[32] Galderisi M, de Simone G, Cicala S, Parisi M, D'Errico A, Innelli P, et al. Coronary flow reserve in hypertensive patients with hypercholesterolemia and without coronary heart disease. Am J Hypertens 2007;20(2):177–83.

[33] Kaufmann PA, Gnecchi-Ruscone T, Schäfers KP, Lüscher TF, Camici PG. Low density lipoprotein cholesterol and coronary microvascular dysfunction in hypercholesterolemia. J Am Coll Cardiol 2000;36(1):103–9.

[34] Dregan A, Charlton J, Chowienczyk P, Gulliford MC. Chronic inflammatory disorders and risk of type 2 diabetes mellitus, coronary heart disease, and stroke: a population-based cohort study. Circulation 2014;130(10):837–44.

[35] Gargiulo P, Marsico F, Parente A, Paolillo S, Cecere M, Casaretti L, et al. Ischemic heart disease in systemic inflammatory diseases. An appraisal. Int J Cardiol 2014;170(3):286–90.

[36] Faccini A, Kaski JC, Camici PG. Coronary microvascular dysfunction in chronic inflammatory rheumatoid diseases. Eur Heart J 2016;37(23):1799–806.

[37] Lanza GA, Sestito A, Cammarota G, Grillo RL, Vecile E, Cianci R, et al. Assessment of systemic inflammation and infective pathogen burden in patients with cardiac syndrome X. Am J Cardiol 2004;94(1):40–4.

[38] Ishimori ML, Martin R, Berman DS, Goykhman P, Shaw LJ, Shufelt C, et al. Myocardial ischemia in the absence of obstructive coronary artery disease in systemic lupus erythematosus. JACC Cardiovasc Imaging 2011;4(1):27–33.

[39] Caliskan M, Erdogan D, Gullu H, Yilmaz S, Gursoy Y, Yildirir A, et al. Impaired coronary microvascular and left ventricular diastolic functions in patients with ankylosing spondylitis. Atherosclerosis 2008;196(1):306–12.

[40] Merz CN, Olson MB, McClure C, Yang YC, Symons J, Sopko G, et al. A randomized controlled trial of low-dose hormone therapy on myocardial ischemia in postmenopausal women with no obstructive coronary artery disease: results from the National Institutes of Health/National Heart, Lung, and Blood Institute-sponsored Women's Ischemia Syndrome Evaluation (WISE). Am Heart J 2010;159(6):987. e981–987.

[41] Mosca L, Benjamin EJ, Berra K, Bezanson JL, Dolor RJ, Lloyd-Jones DM, et al. Effectiveness-based guidelines for the prevention of cardiovascular disease in women—2011 update: a guideline from the American Heart Association. J Am Coll Cardiol 2011;57(12):1404–23.

[42] Bairey Merz CN, Pepine CJ, Walsh MN, Fleg JL. Ischemia and no obstructive coronary artery disease (INOCA): developing evidence-based therapies and research agenda for the next decade. Circulation 2017;135(11):1075–92.

[43] Jones E, Eteiba W, Merz NB. Cardiac syndrome X and microvascular coronary dysfunction. Trends Cardiovasc Med 2012;22(6):161–8.

[44] Safdar B, Nagurney JT, Anise A, DeVon HA, D'Onofrio G, Hess EP, et al. Gender-specific research for emergency diagnosis and management of ischemic heart disease: proceedings from the 2014 Academic Emergency Medicine Consensus Conference Cardiovascular Research Workgroup. Acad Emerg Med 2014;21(12):1350–60.

[45] AlBadri A, Leong D, Bairey Merz CN, Wei J, Handberg EM, Shufelt CL, et al. Typical angina is associated with greater coronary endothelial dysfunction but not abnormal vasodilatory reserve. Clin Cardiol 2017;40(10):886–91.

[46] AlBadri A, Lai K, Wei J, Landes S, Mehta PK, Li Q, et al. Inflammatory biomarkers as predictors of heart failure in women without obstructive coronary artery disease: a report from the NHLBI-sponsored Women's Ischemia Syndrome Evaluation (WISE). PLoS One 2017;12(5), e0177684.

[47] Pepine CJ, Anderson RD, Sharaf BL, Reis SE, Smith KM, Handberg EM, et al. Coronary microvascular reactivity to adenosine predicts adverse outcome in women evaluated for suspected ischemia results from the National Heart, Lung and Blood Institute WISE (Women's Ischemia Syndrome Evaluation) study. J Am Coll Cardiol 2010;55(25):2825–32.

[48] Taqueti VR, Solomon SD, Shah AM, Desai AS, Groarke JD, Osborne MT, et al. Coronary microvascular dysfunction and future risk of heart failure with preserved ejection fraction. Eur Heart J 2018;39(10):840–9.

[49] Suwaidi JA, Hamasaki S, Higano ST, Nishimura RA, Holmes DR, Lerman A. Long-term follow-up of patients with mild coronary artery disease and endothelial dysfunction. Circulation 2000;101(9):948–54.

[50] Murthy VL, Naya M, Taqueti VR, Foster CR, Gaber M, Hainer J, et al. Effects of sex on coronary microvascular dysfunction and cardiac outcomes. Circulation 2014;129(24):2518–27.

[51] Ong P, Camici PG, Beltrame JF, Crea F, Shimokawa H, Sechtem U, et al. International standardization of diagnostic criteria for microvascular angina. Int J Cardiol 2018;250:16–20.

[52] Pijls NH, De Bruyne B, Smith L, Aarnoudse W, Barbato E, Bartunek J, et al. Coronary thermodilution to assess flow reserve: validation in humans. Circulation 2002;105(21):2482–6.

[53] Joye JD, Schulman DS. Clinical application of coronary flow reserve using an intracoronary Doppler guide wire. Cardiol Clin 1997;15(1):101–29.

[54] Everaars H, de Waard GA, Driessen RS, Danad I, van de Ven PM, Raijmakers PG, et al. Doppler flow velocity and thermodilution to assess coronary flow reserve: a head-to-head comparison with [^{15}O]H$_2$O PET. JACC Cardiovasc Interv 2018;11(20):2044–54.

[55] Gould KL, Johnson NP, Bateman TM, Beanlands RS, Bengel FM, Bober R, et al. Anatomic versus physiologic assessment of coronary artery disease. Role of coronary flow reserve, fractional flow reserve, and positron emission tomography imaging in revascularization decision-making. J Am Coll Cardiol 2013;62(18):1639–53.

[56] Webb CM, Collins P, Di Mario C. Normal coronary physiology assessed by intracoronary Doppler ultrasound. Herz 2005;30(1):8–16.

[57] McGuinness ME, Talbert RL. Pharmacologic stress testing: experience with dipyridamole, adenosine, and dobutamine. Am J Hosp Pharm 1994;51(3):328–46 [quiz 404-325].

[58] Lee BK, Lim HS, Fearon WF, Yong AS, Yamada R, Tanaka S, et al. Invasive evaluation of patients with angina in the absence of obstructive coronary artery disease. Circulation 2015;131(12):1054–60.

[59] Kuruvilla S, Kramer CM. Coronary microvascular dysfunction in women: an overview of diagnostic strategies. Expert Rev Cardiovasc Ther 2013;11(11):1515–25.

[60] Williams RP, de Waard GA, De Silva K, Lumley M, Asrress K, Arri S, et al. Doppler versus thermodilution-derived coronary microvascular resistance to predict coronary microvascular dysfunction in patients with acute myocardial infarction or stable angina pectoris. Am J Cardiol 2018;121(1):1–8.

[61] Cole JS, Hartley CJ. The pulsed Doppler coronary artery catheter preliminary report of a new technique for measuring rapid changes in coronary artery flow velocity in man. Circulation 1977;56(1):18–25.

[62] Hartley CJ, Cole JS. An ultrasonic pulsed Doppler system for measuring blood flow in small vessels. J Appl Physiol 1974;37(4):626–9.

[63] Ganz W, Tamura K, Marcus HS, Donoso R, Yoshida S, Swan HJ. Measurement of coronary sinus blood flow by continuous thermodilution in man. Circulation 1971;44(2):181–95.

[64] Rimoldi OE, Camici PG. Positron emission tomography for quantitation of myocardial perfusion. J Nucl Cardiol 2004;11(4):482–90.

[65] Thomson LE, Wei J, Agarwal M, Haft-Baradaran A, Shufelt C, Mehta PK, et al. Cardiac magnetic resonance myocardial perfusion reserve index is reduced in women with coronary microvascular dysfunction. A National Heart, Lung, and Blood Institute-sponsored study from the Women's Ischemia Syndrome Evaluation. Circ Cardiovasc Imaging 2015;8(4).

[66] Kiviniemi T. Assessment of coronary blood flow and the reactivity of the microcirculation non-invasively with transthoracic echocardiography. Clin Physiol Funct Imaging 2008;28(3):145–55.

[67] Meimoun P, Tribouilloy C. Non-invasive assessment of coronary flow and coronary flow reserve by transthoracic Doppler echocardiography: a magic tool for the real world. Eur J Echocardiogr 2008;9(4):449–57.

[68] Cortigiani L, Rigo F, Gherardi S, Bovenzi F, Picano E, Sicari R. Implication of the continuous prognostic spectrum of Doppler echocardiographic derived coronary flow reserve on left anterior descending artery. Am J Cardiol 2010;105(2):158–62.

[69] Sicari R, Rigo F, Cortigiani L, Gherardi S, Galderisi M, Picano E. Additive prognostic value of coronary flow reserve in patients with chest pain syndrome and normal or near-normal coronary arteries. Am J Cardiol 2009;103(5):626–31.

[70] Hozumi T, Yoshida K, Akasaka T, Asami Y, Ogata Y, Takagi T, et al. Noninvasive assessment of coronary flow velocity and coronary flow velocity reserve in the left anterior descending coronary artery by Doppler echocardiography: comparison with invasive technique. J Am Coll Cardiol 1998;32(5):1251–9.

[71] Hildick-Smith DJ, Maryan R, Shapiro LM. Assessment of coronary flow reserve by adenosine transthoracic echocardiography: validation with intracoronary Doppler. J Am Soc Echocardiogr 2002;15(9):984–90.

[72] Lethen H, Tries HP, Brechtken J, Kersting S, Lambertz H. Comparison of transthoracic Doppler echocardiography to intracoronary Doppler guidewire measurements for assessment of coronary flow reserve in the left anterior descending artery for detection of restenosis after coronary angioplasty. Am J Cardiol 2003;91(4):412–7.

[73] Wei K, Jayaweera AR, Firoozan S, Linka A, Skyba DM, Kaul S. Quantification of myocardial blood flow with ultrasound-induced destruction of microbubbles administered as a constant venous infusion. Circulation 1998;97(5):473–83.

[74] Muro T, Hozumi T, Watanabe H, Yamagishi H, Yoshiyama M, Takeuchi K, et al. Assessment of myocardial perfusion abnormalities by intra-venous myocardial contrast echocardiography with harmonic power Doppler imaging: comparison with positron emission tomography. Heart 2003;89(2):145–9.

[75] Porter TR, Xie F. Myocardial perfusion imaging with contrast ultrasound. JACC Cardiovasc Imaging 2010;3(2):176–87.

[76] Grayburn PA. Product safety compromises patient safety (an unjustified black box warning on ultrasound contrast agents by the Food and Drug Administration). Am J Cardiol 2008;101(6):892–3.

[77] Jerosch-Herold M, Seethamraju RT, Swingen CM, Wilke NM, Stillman AE. Analysis of myocardial perfusion MRI. J Magn Reson Imaging 2004;19(6):758–70.

[78] Panting JR, Gatehouse PD, Yang GZ, Grothues F, Firmin DN, Collins P, et al. Abnormal subendocardial perfusion in cardiac syndrome X detected by cardiovascular magnetic resonance imaging. N Engl J Med 2002;346(25):1948–53.

[79] Doyle M, Weinberg N, Pohost GM, Bairey Merz CN, Shaw LJ, Sopko G, et al. Prognostic value of global MR myocardial perfusion imaging in women with suspected myocardial ischemia and no obstructive coronary disease: results from the NHLBI-sponsored WISE (Women's Ischemia Syndrome Evaluation) study. JACC Cardiovasc Imaging 2010;3(10):1030–6.

[80] Lanza GA, Crea F. Primary coronary microvascular dysfunction: clinical presentation, pathophysiology, and management. Circulation 2010;121(21):2317–25.

[81] Patel AR, Epstein FH, Kramer CM. Evaluation of the microcirculation: advances in cardiac magnetic resonance perfusion imaging. J Nucl Cardiol 2008;15(5):698–708.

[82] Shufelt CL, Thomson LE, Goykhman P, Agarwal M, Mehta PK, Sedlak T, et al. Cardiac magnetic resonance imaging myocardial perfusion reserve index assessment in women with microvascular coronary dysfunction and reference controls. Cardiovasc Diagn Ther 2013;3(3):153–60.

[83] Taqueti VR, Di Carli MF. Coronary microvascular disease pathogenic mechanisms and therapeutic options: JACC state-of-the-art review. J Am Coll Cardiol 2018;72(21):2625–41.

[84] Plank F, Friedrich G, Dichtl W, Klauser A, Jaschke W, Franz WM, et al. The diagnostic and prognostic value of coronary CT angiography in as-ymptomatic high-risk patients: a cohort study. Open Heart 2014;1(1), e000096.

[85] Hou ZH, Lu B, Gao Y, Jiang SL, Wang Y, Li W, et al. Prognostic value of coronary CT angiography and calcium score for major adverse cardiac events in outpatients. JACC Cardiovasc Imaging 2012;5(10):990–9.

[86] Feher A, Sinusas AJ. Quantitative assessment of coronary microvascular function: dynamic single-photon emission computed tomography, posi-tron emission tomography, ultrasound, computed tomography, and magnetic resonance imaging. Circ Cardiovasc Imaging 2017;10(8).

[87] Caliskan M, Erdogan D, Gullu H, Topcu S, Ciftci O, Yildirir A, et al. Effects of atorvastatin on coronary flow reserve in patients with slow coronary flow. Clin Cardiol 2007;30(9):475–9.

[88] Reriani MK, Dunlay SM, Gupta B, West CP, Rihal CS, Lerman LO, et al. Effects of statins on coronary and peripheral endothelial function in humans: a systematic review and meta-analysis of randomized controlled trials. Eur J Cardiovasc Prev Rehabil 2011;18(5):704–16.

[89] Romeo F, Gaspardone A, Ciavolella M, Gioffrè P, Reale A. Verapamil versus acebutolol for syndrome X. Am J Cardiol 1988;62(4):312–3.

[90] Ferrini D, Bugiardini R, Galvani M, Gridelli C, Tollemeto D, Puddu P, et al. Opposing effects of propranolol and diltiazem on the angina threshold during an exercise test in patients with syndrome X. G Ital Cardiol 1986;16(3):224–31.

[91] Kayaalti F, Kalay N, Basar E, Mavili E, Duran M, Ozdogru I, et al. Effects of nebivolol therapy on endothelial functions in cardiac syndrome X. Heart Vessels 2010;25(2):92–6.

[92] Erdamar H, Sen N, Tavil Y, Yazici HU, Turfan M, Poyraz F, et al. The effect of nebivolol treatment on oxidative stress and antioxidant status in patients with cardiac syndrome-X. Coron Artery Dis 2009;20(3):234–8.

[93] Camici PG, Marraccini P, Gistri R, Salvadori PA, Sorace O, L'Abbate A. Adrenergically mediated coronary vasoconstriction in patients with syndrome X. Cardiovasc Drugs Ther 1994;8(2):221–6.

[94] Michelsen MM, Rask AB, Suhrs E, Raft KF, Høst N, Prescott E. Effect of ACE-inhibition on coronary microvascular function and symptoms in normotensive women with microvascular angina: a randomized placebo-controlled trial. PLoS One 2018;13(6), e0196962.

[95] Cannon RO, Watson RM, Rosing DR, Epstein SE. Efficacy of calcium channel blocker therapy for angina pectoris resulting from small-vessel coronary artery disease and abnormal vasodilator reserve. Am J Cardiol 1985;56(4):242–6.

[96] Ford TJ, Stanley B, Good R, Rocchiccioli P, McEntegart M, Watkins S, et al. Stratified medical therapy using invasive coronary function testing in angina: the CorMicA trial. J Am Coll Cardiol 2018;72(23 Pt A):2841–55.

[97] Kaski JC, Rosano GM, Collins P, Nihoyannopoulos P, Maseri A, Poole-Wilson PA. Cardiac syndrome X: clinical characteristics and left ventricular function. Long-term follow-up study. J Am Coll Cardiol 1995;25(4):807–14.

[98] Russo G, Di Franco A, Lamendola P, Tarzia P, Nerla R, Stazi A, et al. Lack of effect of nitrates on exercise stress test results in patients with microvascular angina. Cardiovasc Drugs Ther 2013;27(3):229–34.

[99] Yoshio H, Shimizu M, Kita Y, Ino H, Kaku B, Taki J, et al. Effects of short-term aminophylline administration on cardiac functional reserve in patients with syndrome X. J Am Coll Cardiol 1995;25(7):1547–51.

[100] Elliott PM, Krzyzowska-Dickinson K, Calvino R, Hann C, Kaski JC. Effect of oral aminophylline in patients with angina and normal coronary arteriograms (cardiac syndrome X). Heart 1997;77(6):523–6.

[101] Hasenfuss G, Maier LS. Mechanism of action of the new anti-ischemia drug ranolazine. Clin Res Cardiol 2008;97(4):222–6.

[102] Mehta PK, Goykhman P, Thomson LE, Shufelt C, Wei J, Yang Y, et al. Ranolazine improves angina in women with evidence of myocardial ischemia but no obstructive coronary artery disease. JACC Cardiovasc Imaging 2011;4(5):514–22.

[103] Villano A, Di Franco A, Nerla R, Sestito A, Tarzia P, Lamendola P, et al. Effects of ivabradine and ranolazine in patients with microvascular angina pectoris. Am J Cardiol 2013;112(1):8–13.

[104] Tagliamonte E, Rigo F, Cirillo T, Astarita C, Quaranta G, Marinelli U, et al. Effects of ranolazine on noninvasive coronary flow reserve in patients with myocardial ischemia but without obstructive coronary artery disease. Echocardiography 2015;32(3):516–21.

[105] Bairey Merz CN, Handberg EM, Shufelt CL, Mehta PK, Minissian MB, Wei J, et al. A randomized, placebo-controlled trial of late Na current inhibition (ranolazine) in coronary microvascular dysfunction (CMD): impact on angina and myocardial perfusion reserve. Eur Heart J 2016;37(19):1504–13.

[106] Rambarat CA, Elgendy IY, Handberg EM, Bairey Merz CN, Wei J, Minissian MB, et al. Late sodium channel blockade improves angina and myocardial perfusion in patients with severe coronary microvascular dysfunction: Women's Ischemia Syndrome Evaluation-Coronary Vascular Dysfunction ancillary study. Int J Cardiol 2019;276:8–13.

[107] Zhu H, Xu X, Fang X, Zheng J, Zhao Q, Chen T, et al. Effects of the antianginal drugs ranolazine, nicorandil, and ivabradine on coronary microvascular function in patients with nonobstructive coronary artery disease: a meta-analysis of randomized controlled trials. Clin Ther 2019;41(10):2137–52. e2112.

[108] Bairey Merz CN, Pepine CJ, Shimokawa H, Berry C. Treatment of coronary microvascular dysfunction. Cardiovasc Res 2020;116(4):856–70.

[109] Koruth JS, Lala A, Pinney S, Reddy VY, Dukkipati SR. The clinical use of ivabradine. J Am Coll Cardiol 2017;70(14):1777–84.

[110] Skalidis EI, Hamilos MI, Chlouverakis G, Zacharis EA, Vardas PE. Ivabradine improves coronary flow reserve in patients with stable coronary artery disease. Atherosclerosis 2011;215(1):160–5.

[111] Rosano GMC. Symptomatic response to 17-beta-estradiol in women with syndrome-X [Lefroy DC, Peters NS, Lindsay DC, Sarrel PM, Collins P, eds.]. J Am Coll Cardiol 1994;A6.

[112] Kaski JC. Cardiac syndrome X in women: the role of oestrogen deficiency. Heart 2006;92(Suppl. 3):iii5–9.

[113] Reriani M, Raichlin E, Prasad A, Mathew V, Pumper GM, Nelson RE, et al. Long-term administration of endothelin receptor antagonist improves coronary endothelial function in patients with early atherosclerosis. Circulation 2010;122(10):958–66.

[114] Fukumoto Y, Mohri M, Inokuchi K, Ito A, Hirakawa Y, Masumoto A, et al. Anti-ischemic effects of fasudil, a specific Rho-kinase inhibitor, in patients with stable effort angina. J Cardiovasc Pharmacol 2007;49(3):117–21.

[115] Chen JW, Lee WL, Hsu NW, Lin SJ, Ting CT, Wang SP, et al. Effects of short-term treatment of nicorandil on exercise-induced myocardial ischemia and abnormal cardiac autonomic activity in microvascular angina. Am J Cardiol 1997;80(1):32–8.

[116] Asbury EA, Kanji N, Ernst E, Barbir M, Collins P. Autogenic training to manage symptomology in women with chest pain and normal coronary arteries. Menopause 2009;16(1):60–5.

[117] Asbury EA, Slattery C, Grant A, Evans L, Barbir M, Collins P. Cardiac rehabilitation for the treatment of women with chest pain and normal coronary arteries. Menopause 2008;15(3):454–60.

[118] Laksanakorn W, Laprattanagul T, Wei J, Shufelt C, Minissian M, Mehta PK, et al. Cardiac rehabilitation for cardiac syndrome X and microvascular angina: a case report. Int J Case Rep Images 2015;6(4):239–44.

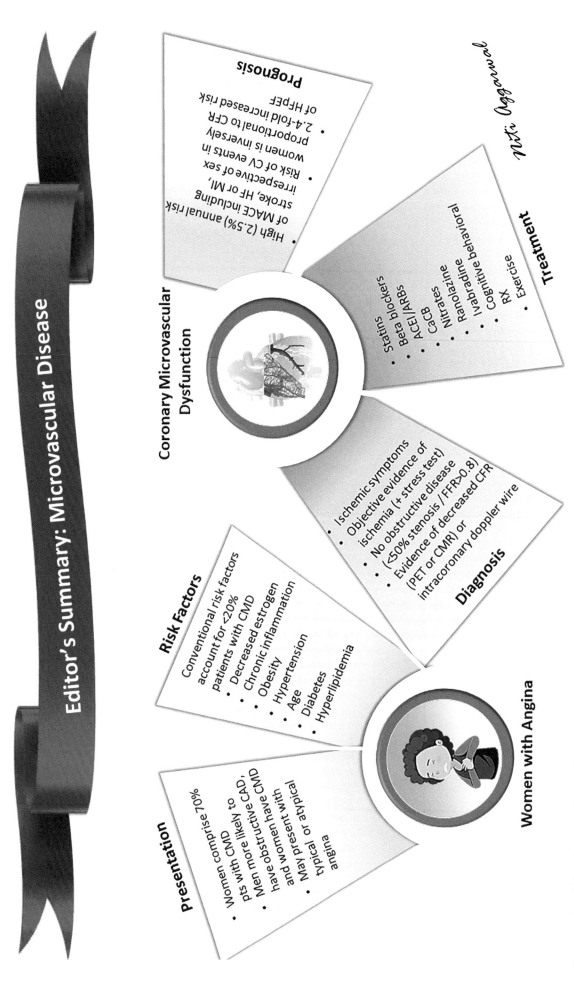

Editor's Summary: Microvascular Disease

Niti Aggarwal

Coronary Microvascular Dysfunction

Prognosis
- High (2.5%) annual risk of MACE including stroke, HF or MI, irrespective of sex
- Risk of CV events in women is inversely proportional to CFR
- 2.4-fold increased risk of HFpEF

Treatment
- Statins
- Beta blockers
- ACEI/ARBs
- CaCB
- Nitrates
- Ranolazine
- Ivabradine
- Cognitive behavioral
- RX
- Exercise

Diagnosis
- Ischemic symptoms
- Objective evidence of ischemia (+ stress test)
- No obstructive disease (<50% stenosis / FFR>0.8)
- Evidence of decreased CFR (PET or CMR) or intracoronary doppler wire

Risk Factors
- Conventional risk factors account for <20% patients with CMD
- Decreased estrogen
- Chronic inflammation
- Obesity
- Hypertension
- Age
- Diabetes
- Hyperlipidemia

Women with Angina

Presentation
- Women comprise 70% pts with CMD
- Men more likely to have obstructive CAD, and women with CMD
- May present with typical or atypical angina

Presentation, Risk Factors, Diagnosis and Management of Coronary Microvascular Dysfunction. Commonly seen in women, CMD may present with typical or atypical symptoms, and often without conventional risk factors. Early diagnosis with PET or CMR may be pivotal to improve symptoms and decrease the risk of adverse events. ACEI, angiotensin converting enzyme inhibitor; ARB, angiotensin receptor blocker; CaCB, calcium channel blocker; CAD, coronary artery disease; CFR, coronary flow reserve; CMD, coronary microvascular dysfunction; CMR, cardiac magnetic resonance; CV, cardiovascular; FFR, fractional flow reserve; HF, heart failure; HFpEF, heart failure with preserved ejection fraction; MACE, major adverse cardiovascular event; MI, myocardial infarction; PET, positron emission tomography; RX, therapy. *Image courtesy of Niti R. Aggarwal.*

Chapter 9

Myocardial Infarction With Nonobstructive Coronary Disease

Esther Davis and Amy Sarma

Clinical Case

A 60-year-old woman presented with nausea and chest discomfort following an argument with her son. Her past medical history was significant for asthma, hyperlipidemia, and an episode of chest pain 2 weeks prior in the context of strenuous exercise for which she did not seek medical attention. An electrocardiogram (ECG) demonstrated normal sinus rhythm without ST-segment or T-wave changes. High-sensitivity troponin-T was 58 ng/L (normal < 14 ng/L). An echocardiogram revealed regional hypokinesis of the inferior and inferolateral walls. In this context, she was taken for coronary angiography, which did not demonstrate coronary plaque, luminal narrowing, or evidence of coronary thrombus in the epicardial vessels. The vessels were highly tortuous, but there was no evidence of coronary dissection. What is the patient's diagnosis and how should one manage her going forward?

Abstract

Acute coronary syndrome (ACS) in the absence of obstructive coronary disease has been recognized for more than 80 years. However, recognition that these patients represent a unique group with differing treatment needs is more recent. Myocardial infarction with nonobstructed coronary arteries (MINOCA) accounts for approximately 6% of ACS presentations. MINOCA is a working diagnosis, which should prompt further etiologic investigation to identify underlying pathophysiology and institute disease-specific therapies. Multiple atherosclerotic and nonatherosclerotic mechanisms may lead to a presentation with MINOCA and nonischemic processes may mimic MINOCA. Women, particularly younger women with fewer traditional cardiac risk factors, are overrepresented in MINOCA presentations and women younger than 55 are five times more likely than men to experience MINOCA. MINOCA is associated with excess mortality and recurrent adverse cardiovascular events, and patients are less likely to receive evidence-based treatment. Further, there is a paucity of data to guide treatment strategies. Additional research is needed to understand optimal management of these patients.

Sex Differences in Cardiac Disease. https://doi.org/10.1016/B978-0-12-819369-3.00007-1

Definition

The clinical phenomenon of an acute coronary syndrome (ACS) in the absence of obstructive coronary disease has been recognized for more than 80 years. In 1939, the *Archives of Internal Medicine* published a report of 15 patients (7 of whom were women) with autopsy findings of extensive myocardial necrosis, but no significant luminal narrowing of the coronary vasculature [1]. Despite this, the recognition that these patients represent a clinically unique group with differing treatment needs is a relatively recent development with the term "MINOCA" coined in 2012 [2] and the first major position statement on this condition published as recently as 2017 [3]. Since then, there has been further evolution of the concept of MINOCA, increasing the specificity of the diagnosis [4].

The first working group position paper published by the European Society of Cardiology (ESC) on MINOCA proposed that diagnostic criteria includes primary acute myocardial infarction (AMI) based on the third universal definition of myocardial infarction (MI) without obstructive coronary disease on coronary angiography (**Table 1**) [3]. This statement emphasized the critical concept that MINOCA should be considered a "working diagnosis" that prompts further etiologic investigation rather than a final diagnosis. Given that MINOCA carries with it an increased risk of recurrent major adverse cardiovascular events (MACE) and mortality, elucidating the underlying pathophysiology enables implementation of disease-specific therapies (where available) that may reduce long-term risk [5–7]. Subsequently, the fourth universal definition of MI was published in 2018 with a strong emphasis on distinguishing myocardial injury (as reflected by troponin release from nonischemic etiologies) from MI due to coronary ischemia [8]. In this context, the American Heart Association (AHA) revised the concept of MINOCA, advocating for its use solely in the context of MI in the absence of obstructive coronary arteries and excluding nonischemic cardiomyopathies from the definition [4]. From a practical standpoint, Pasupathy et al. have suggested the use of "troponin-positive nonobstructive coronary arteries" (**Figure 1**) to reflect the concept of MINOCA as a working diagnosis (prior to elucidation of true myocardial infarction from nonischemic causes of myocardial injury) in order to distinguish MINOCA as a working diagnosis from MINOCA as a designation of myocardial infarction with nonobstructive coronary disease [9]. While the evolving concept proposed by the AHA will serve to make MINOCA a more specific diagnosis reflective of true coronary events, it is important to note that many studies to date investigating outcomes of MINOCA have included nonischemic etiologies when evaluating the current data on the topic discussed in this chapter. Further specificity in the diagnosis may enable a better understanding of optimal diagnostic and therapeutic approaches for MINOCA in future analyses.

TABLE 1 ESC Diagnostic Criteria for Myocardial Infarction With Nonobstructed Coronary Arteries

The diagnosis of MINOCA is made immediately upon coronary angiography in a patient presenting with features consistent with an acute myocardial infarct as detailed by the following criteria:

(1) AMI criteria[a]
  (a) Positive cardiac biomarker (preferably cardiac troponin) defined as a rise and/or fall in serial levels, with at least one value above the 99th percentile upper reference limit
  (b) Corroborative clinical evidence of infarction evidenced by at least one of the following:
    (i) Symptoms of ischemia
    (ii) New or presumed new significant ST-T changes or new LBBB
    (iii) Development of pathological Q waves
    (iv) Imaging evidence of new loss of viable myocardium or new RWMA
    (v) Intracoronary thrombus evident on angiography or at autopsy
(2) Nonobstructive coronary arteries on angiography
  • Defined as the absence of obstructive CAD on angiography (i.e., no coronary artery stenosis ≥ 50%) in any potential infarct-related artery
  • This includes both patients with
    ○ Normal coronary arteries (no stenosis > 30%)
    ○ Mild coronary atheromatosis (stenosis > 30% but < 50%)
(3) No clinically overt specific cause for the acute presentation
  • At the time of angiography, the cause and thus a specific diagnosis for the clinical presentation is not apparent
  • Accordingly, there is a necessity to further evaluate the patient for the underlying cause of the MINOCA presentation

---

[a] *AMI definition as per third universal definition of myocardial infarction (2012).*
AMI, acute myocardial infarction; CAD, coronary artery disease; ESC, European Society of Cardiology; LBBB, left bundle branch block; MINOCA, myocardial infarction with nonobstructed coronary arteries; RWMA, regional wall motion abnormalities. *Reproduced with permission from [3].*

Epidemiology

Meta-analysis data suggest an overall prevalence of MINOCA of 6% among patients presenting with ACS who undergo coronary angiography [10]. However, significant heterogeneity in prevalence has been described between

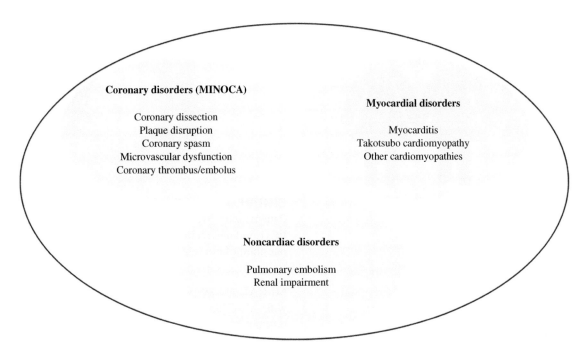

FIGURE 1 **Troponin-Positive Nonobstructive Coronary Arteries (TP-NOCA).** MINOCA, myocardial infarction with nonobstructed coronary arteries. *Reproduced with permission from [9].*

studies (with a range of 1%–14%, I^2 statistic 99%) potentially due to differences in extent of investigation with respect to underlying pathophysiology as well as study definitions of MINOCA [10]. Indeed, study heterogeneity remains a limitation of the MINOCA literature to date.

Patients who experience MINOCA exhibit unique clinical characteristics as compared to those who experience myocardial infarction in the context of obstructive coro-

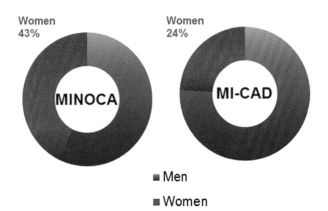

FIGURE 2 **Gender Distribution of MINOCA and MI-CAD.** Women comprise a larger portion of patients who experience myocardial infarction with nonobstructed coronary arteries (MINOCA) as compared to those who experience MI with obstructive coronary artery disease (MI-CAD). *Data from [10]. Image courtesy of Niti R. Aggarwal.*

nary artery disease (MI-CAD). Women comprise a larger proportion of patients who experience MINOCA (43%) as compared to those with MI-CAD (24%, $P < 0.001$) [10] (**Figure 2**). Further, MINOCA tends to occur at a younger age (mean 58.8 years) than MI-CAD (61.3 years, $P < 0.001$) [10]. Among patients under the age of 55, women are nearly five times more likely to experience MINOCA as compared with men [14.9% vs. 3.5%, unadjusted odds ratio (OR) 4.84; 95% confidence interval (CI) 3.29–7.13] [6] (**Figure 3**). Further, MINOCA appears more prevalent in nonwhite populations as compared with white populations (14.9% vs. 10.0%, unadjusted OR 1.57; 95% CI 1.21–2.04) [6]. In another analysis, a higher proportion of MINOCA as compared with MI-CAD patients were of black race (23% vs. 9% for men, 16% vs. 13% for women, $P < 0.0001$) [11].

Risk factor profiles differ among patients with MINOCA as compared with MI-CAD. In a large pooled meta-analysis, the prevalence of hyperlipidemia was lower among those with MINOCA (21%) vs. MI-CAD (32%, $P < 0.001$) but comorbid diabetes, hypertension, smoking, and family history did not otherwise differ [10]. More recent, large registry data have suggested a more distinct baseline phenotype of patients with MINOCA as compared with MI-CAD, with a lower prevalence of smoking (men 36% vs. 41%, $P < 0.0001$; women 23% vs. 35%, $P < 0.0001$) or diabetes (men 19% vs. 24%, $P < 0.001$; women 21% vs. 32%, $P < 0.001$), and higher prevalence of atrial fibrillation or flutter (men 9% vs. 4%, $P < 0.0001$; women 8% vs. 6%, $P < 0.0001$) [11]. Women, but not men, presenting with MINOCA also had lower rates

FIGURE 3 Incidence of MINOCA in Patients Aged, <55 Years, Women Are Five Times More Likely to Experience MINOCA Than Men. MINOCA, myocardial infarction with nonobstructive coronary arteries. *Data from [6]. Image courtesy of Niti R. Aggarwal.*

of hypertension (69% vs. 73%, *P* < 0.0001) and peripheral artery disease (4% vs. 7%, *P* < 0.0001) [11]. Similar results have been reported among a younger female population enrolled in the Variation in Recovery: Role of Gender on Outcomes of Young AMI Patient (VIRGO) study, in which at least one traditional atherosclerotic risk factor (i.e., smoking, diabetes, obesity, hypertension, dyslipidemia, or family history) was found in 99% of patients with MI-CAD versus 91% of those with MINOCA (*P* < 0.001) [6]. Conversely, those with MINOCA exhibited a higher prevalence of hypercoagulable syndromes (3.0% vs. 1.3%, *P*=0.036) as compared with the MI-CAD population.

The impact of hormonal exposure, menopausal state, and reproductive history on risk of MINOCA versus MI-CAD remains an area of investigation. In the VIRGO study of AMI in young adults, no differences emerged with respect to rates of polycystic ovarian disease, oral contraceptive pill use, or parity between women who had MINOCA versus MI-CAD [6]. Among those who had been pregnant, there did not appear to be an effect of pregnancy-associated complications (including preeclampsia, stillbirth, and miscarriage) on the risk of MINOCA versus MI-CAD and those who experienced gestational diabetes were at higher risk for

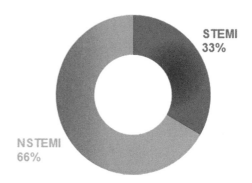

FIGURE 4 ECG Findings in Patients With MINOCA. Patients with MINOCA can present with a wide range of ECG findings including ST-segment elevation MI in up to one-third of cases. *Data from [10].*

MI-CAD (16.8% vs. 10.9%, *P*=0.028). However, women with MI-CAD (vs. MINOCA) were more likely to be postmenopausal (55.2% vs. 41.2%, *P* < 0.001) and to have experienced menarche at an earlier age (12.0 vs. 13.0 years, *P*=0.003) [6].

Clinical Presentation of Patient With MINOCA

The initial clinical presentation of MINOCA is indistinguishable from MI-CAD, with chest pain being the most common presenting symptom in both (~85%) [6]. ST-elevation myocardial infarction (STEMI) occurs in approximately a third of MINOCA cases, which can present with a wide variety of electrocardiogram (ECG) manifestations [10, 11] (**Figure 4**). While potentially influenced by the heterogeneity of underlying mechanisms, those presenting with MINOCA tend to have lower peak troponin levels as compared to MI-CAD [6, 11].

Causes of MINOCA

There are several coronary and noncoronary etiologies that encompass the diagnosis of MINOCA, and these are summarized in **Table 2**.

Coronary Etiologies

Plaque Disruption

Disruption of atherosclerotic plaque without occlusion or significant narrowing of the epicardial coronary arteries is an important cause of MINOCA and responsible for 5–20% of type 1 MIs [13]. These plaques are commonly missed on coronary angiography and often require intracoronary imaging with the use of high-resolution optical coherence

Continued

TABLE 2 Diagnosis and Management of Patients With a Working Diagnosis of MINOCA

Cause	Definition	Prevalence of MINOCA	Sex Differences	Key Diagnostic Test	Treatment
Ischemic MINOCA					
Plaque disruption (erosion, rupture, and calcific nodules)*	Disruption of atherosclerotic plaque without occlusion or significant narrowing of the epicardial coronary arteries	Up to 40% when intracoronary imaging is undertaken	Plaque erosion more common in women under the age of 50 Plaque rupture more common in men	Detailed evaluation of coronary angiogram IVUS OCT	Similar to MI-CAD Aspirin Dual antiplatelet therapy with clopidogrel/ticagrelor High-intensity statins β-blockers ACE inhibitors/ARBs
Coronary embolism	Ischemia due to embolism into the coronary arteries	Unclear medical management for heart failure β-blockers ACE inhibitors/ARBs	Unknown	Detailed evaluation of coronary angiogram IVUS OCT	Antiplatelet or anticoagulant therapy Other targeted therapies for hypercoagulable condition
Coronary artery spasm	Transient total or subtotal coronary artery occlusion (>90% constriction) with angina and ischemic ECG changes either spontaneously or in response to a provocative stimulus	Up to 46% of patients tested with early vasoprovocation testing	Vasoreactivity testing is more commonly positive in women	Vasoreactivity testing during coronary angiography	Calcium channel blocker Other antispastic agents (nitrates, nicorandil, and cilostazol) Statin

TABLE 2 Diagnosis and Management of Patients With a Working Diagnosis of MINOCA—cont'd

Cause	Definition	Prevalence of MINOCA	Sex Differences	Key Diagnostic Test	Treatment
Microvascular disease*	Abnormal coronary microvascular resistance that is clinically evident as an inappropriate coronary blood flow response, impaired myocardial perfusion, and/or myocardial ischemia	31.5% of those with positive vasoreactivity testing	Microvascular disease is generally more common in women	Vasoreactivity testing during coronary angiography Coronary flow reserve by positron emission tomography or stress MRI Thermodilution	Conventional antianginal therapies (e.g., calcium channel blocker, β-blocker) Unconventional antianginal therapies (e.g., L-arginine, ranolazine, dipyridamole, aminophylline, imipramine, and α-blockers)
SCAD	Epicardial coronary artery dissection that is not associated with atherosclerosis or trauma and not iatrogenic	Unknown (up to 4% of ACS)	Up to 25% of ACS events in women under 50	Detailed evaluation of coronary angiogram IVUS	Aspirin β-blocker Clopidogrel Avoid statin unless hyperlipidemia
Type 2 myocardial infarction	Ischemia due to supply-demand mismatch	Unknown	Unknown	Clinical and laboratory assessment for causes of supply-demand mismatch	Treat underlying supply-demand mismatch

Nonischemic MINOCA Mimics

Myocarditis		Inflammatory disorder of the myocardium, idiopathic or due to infectious agents, systemic diseases, drugs, or toxins	Up to 33% of MINOCA presentations	1.5–1.7:1 female to male	Clinical features CMR Echocardiogram	Medical management for heart failure β-blockers ACE inhibitors/ARBs Exercise avoidance Immunosuppression in select cases
Takotsubo cardiomyopathy		Syndrome of acute ECG changes and acute reversible heart failure in a characteristic pattern in the absence of occlusive CAD	18%–27% of MINOCA presentations	Up to 90% of cases in women	Left ventriculogram Echocardiogram CMR	Medical management for heart failure β-blockers ACE inhibitors/ARBs
Cardiomyopathies		Disorder of the myocardium with structural and functional abnormality in the absence of an ischemic, valvular, hypertensive, or congenital cause	Up to 25% of patients undergoing CMR (includes Takotsubo patients)	1.3–1.5:1 female to male	Left ventriculogram Echocardiogram CMR	Medical management for heart failure β-blockers ACE inhibitors/ARBs

ACE, angiotensin-converting enzyme; ACS, acute coronary syndrome; ARBs, angiotensin receptor blockers; CMR, cardiac magnetic resonance; ECG, electrocardiogram; IVUS, intravascular ultrasound; MI-CAD, MI with obstructive coronary artery disease; MINOCA, myocardial infarction with nonobstructed coronary arteries; MRI, magnetic resonance imaging; OCT, optical coherence tomography; SCAD, spontaneous coronary artery dissection.
*Figures are reproduced with permission from [14, 36].

tomography (OCT) or intravascular ultrasound (IVUS) for visualization (**Figure 5**). The term encompasses the imaging and pathological findings of plaque rupture or ulceration, plaque erosion, and calcified nodules. Ruptured plaques are those that are collagen poor and lipid rich, with a thin fibrous cap that fractures, creating continuity between the coronary lumen and plaque cavity and resulting in a robust inflammatory response (**Figure 6**) [4, 15]. The mechanisms of plaque rupture have not been completely defined; however, the release of matrix metalloprotein, high shear stress points, macrophage calcification, and iron deposition have been implicated in the pathway leading to plaque rupture [16]. Plaque rupture has been noted to be more common in men compared to women in a sudden cardiac death registry although prevalence increases above the age of 50 [17]. In contrast, plaque erosion—characterized by overlying thrombus formation and endothelial cell apoptosis without plaque discontinuity—occurs with lipid-poor plaques, rich in proteoglycans and glycosaminoglycans. Plaque erosion is believed to underlie 30–35% of sudden cardiac deaths and 25–40% of ACS [16, 18]. Data from a sudden cardiac death registry suggest that plaque erosion is more common in women younger than 50 and is found in up 82% of cases (in contrast to a decreased prevalence over the age of 50). Overall plaque erosion was more common in women (58%)

compared to men (24%) [17]. Plaque rupture or ulceration may underlie close to 40% of cases of MINOCA where intracoronary imaging is undertaken [19, 20]. Calcified nodules—in which fibrous cap disruption occurs in the context of a calcified plaque—likely account for a smaller subset of events and are most readily identified by OCT. To date, there have been no identified sex differences in patients with evidence of calcified nodules [16].

Coronary Embolism

Coronary embolism is a rare cause of MI, though potentially underdiagnosed in the setting of MINOCA, especially in cases of spontaneous thrombolysis where detection is challenging. The possibility of a hypercoagulable state (either due to inherited or acquired disorders) warrants investigation in cases of potential embolism, though they may occur spontaneously as well. Among patients with MINOCA, hereditary thrombophilias (including factor V leiden, and protein C and S deficiencies) have been reported in as many as 14–24% of patients [10, 21]. In a study comprised of 84 MINOCA patients, 15.5% were found to have antiphospholipid antibody syndrome [21]. There are additional data suggesting that prothrombotic gene variants are more common in young women presenting with AMI, though it is unclear what percentage of this study population had MI-CAD

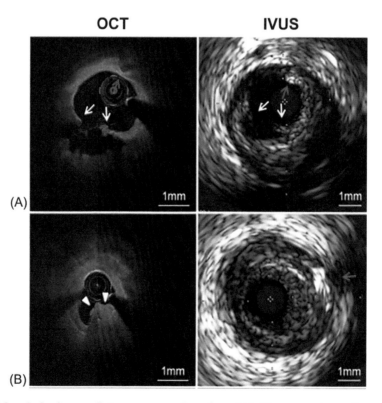

OCT   **IVUS**

FIGURE 5 Plaque Morphologies on Intracoronary Imaging. (A) Plaque rupture: disrupted fibrous cap *(white arrows)*, large lipid core (1–10 o'clock), and spotty calcification *(red arrow)* were observed on both OCT and IVUS. (B) Plaque erosion: OCT demonstrated white thrombus *(arrowheads)* on the luminal surface. There was no evidence of fibrous cap disruption. IVUS revealed deep spotty calcification *(red arrow)*. IVUS, intravascular ultrasound; OCT, optical coherence tomography. *Adapted with permission from [14].*

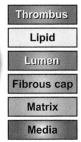

Plaque erosion	**Plaque rupture**
Lipid poor	Lipid rich
Proteoglycan and glycosaminoglycan rich	Collagen poor, thin fibrous cap
Nonfibrillar collagen breakdown	Interstitial collagen breakdown
Few inflammatory cells	Abundant inflammation
Endothelial cell apoptosis	Smooth muscle cell apoptosis
Secondary neutrophil involvement	Macrophage predominance
Female predominance	Male predominance
High triglycerides	High LDL

FIGURE 6   Contrast Between Superficial Erosion and Fibrous Cap Rupture as Causes of Atrial Thrombosis. LDL, low-density lipoprotein. *Reproduced with permission from [15].*

[22]. Other causes of embolic MINOCA may include atrial fibrillation, infective endocarditis, cardiac tumor, or exogenous hormone use.

Coronary Artery Spasm

Coronary artery spasm is defined as "transient total or subtotal coronary artery occlusion (>90% constriction) with angina and ischemic ECG changes [occurring] either spontaneously or in response to a provocative stimulus" [23]. Spontaneous forms result from disorders of vasomotor tone. Further, coronary spasm often occurs at the site of atherosclerotic disease, but this is not always the case, with 6% of patients experiencing spasm at sites without any evidence (by OCT in a small study) of underlying atherosclerotic disease [24]. In addition, spasm may occur in response to certain drugs including fluorouracil, cocaine, and methamphetamines. Among MINOCA patients formally tested, the rate of inducible spasm is high, ranging from 24% to 58% [5, 9]. There are multiple predictors of coronary spasm, including younger age, male sex, and history of prior angina, but traditional risk factors for atherosclerotic disease (including hypertension, diabetes, hyperlipidemia, family history of coronary disease, smoking, and peak troponin levels) may less commonly track with risk of spasm [25]. Given current hesitation to pursue provocative testing in the acute setting, the prevalence of causative coronary artery spasm in MINOCA may be underestimated [3]. However, given that vasomotor dysfunction may also result from myocardial injury (whether ischemic or nonischemic in etiology), it can simultaneously be challenging to determine after an MI whether coronary artery spasm was initially causative or simply consequential [4]. In contrast to MI-CAD, MINOCA-related spasm is less likely to exhibit time dependence from the initial event: in a meta-

analysis of 402 MINOCA patients, the incidence of provocable spasm did not differ among those tested during versus 6 weeks after an MI. In contrast, those within 6 weeks of an obstructive atherosclerotic infarct are more likely to exhibit inducible spasm as compared to those tested after 6 weeks [9]. In addition, coronary artery spasm may be more prevalent among certain racial groups, with positive testing more prevalent after MINOCA in Japanese (81%) and Korean (61%) cohorts [10].

Sex differences with respect to coronary artery spasm remain underinvestigated. The incidence of positive vasoreactivity and ability to induce such reactivity at lower doses of stimuli may be higher in women as compared to men [26]. In addition, among a Japanese cohort, young women (aged less than 50) with vasospasm experienced a higher incidence of adverse outcomes with a rate of MACE of 18% at 60-month follow-up, an incidence significantly higher than that seen in older women (MACE 8% in women age 50–64 years and 4% in women aged >65, $P < 0.01$). Younger age was significantly associated with MACE events in women but not men. Overall, however, sex differences in coronary artery vasospasm require further investigation [27].

Microvascular Dysfunction

Coronary microvascular dysfunction (CMD) is defined as abnormal coronary microvascular resistance (either arteriolar or prearteriolar) that results in inappropriate coronary blood flow response, impaired myocardial perfusion, and/or myocardial ischemia that cannot be accounted for by epicardial coronary disease [28]. It is increasingly recognized among patients with nonobstructive epicardial disease and ischemic chest pain (with a prevalence ranging from 30% to 50%) and occurs more commonly in women as compared with men, particularly those with traditional risk factors for atherosclerotic disease [29]. While CMD is an important etiology of stable ischemic disease, its prevalence in the context of MINOCA is not as well understood. As with inducible epicardial coronary artery spasm, given that CMD can be a consequence of MI, it remains challenging to determine the proportion of CMD found after MINOCA that may have been truly causative in the initial event rather than a consequence [4]. CMD is discussed in more detail in Chapter 8.

Spontaneous Coronary Artery Dissection

Spontaneous coronary artery dissection (SCAD) is defined as "an epicardial coronary artery dissection that is not associated with atherosclerosis or trauma and not iatrogenic" [30]. While recognition of SCAD is increasing, its true clinical prevalence is at present difficult to estimate in light of continued underdiagnosis of this condition [30]. Often, SCAD results in luminal occlusion and, therefore, would not result in MINOCA classification. However, SCAD can often be missed on conventional lumenography, which does not specifically image the arterial wall, and thus in the context of a high clinical suspicion, intracoronary imaging methods (e.g., OCT or IVUS) should be considered in order to better clarify the diagnosis, though these methods must be balanced against the potential risk of dissection propagation. SCAD is discussed in greater detail in Chapter 5.

Noncoronary Etiologies

In light of the fourth universal definition of MI as a troponin elevation resulting from MI rather than injury (from diverse nonischemic mechanisms), the most recent AHA scientific statement moves the diagnosis of MINOCA away from inclusion of nonischemic cardiomyopathies [4]. While such cardiomyopathies still remain within MINOCA as a "working diagnosis," they should no longer be included in the overall concept of MINOCA as specific clinical entity reflecting infarction without epicardial coronary obstruction. Nonetheless, these nonischemic mimics remain important to understand both in order to aid in distinguishing them from true MINOCA and because to date these nonischemic diagnoses had been included in investigations of MINOCA.

Takotsubo Cardiomyopathy

Takotsubo cardiomyopathy is a syndrome of acute chest pain or heart failure, troponin elevation, and systolic dysfunction (often with regional wall motion abnormalities, most classically affecting the apex with basal hyperkinesis) that often presents similarly to MI-CAD. Often coronary angiography without obstructive disease distinguishes Takotsubo cardiomyopathy from MI-CAD and, therefore, this clinical entity has previously been classified as MINOCA [3]. Thus, it has been included in MINOCA cohort studies, where it has been found to account for between 18% and 27% of cases [31, 32].

Myocarditis

Myocarditis is an inflammatory disorder of the myocardium that can present with acute onset chest pain and troponin elevation, but which can be distinguished from MI-CAD based on imaging, immunological, histological, and immunohistochemical criteria [12]. The etiology of this condition is often unexplained, but a variety of infectious agents, systemic diseases, drugs, and toxins have been implicated. Central to distinguishing this nonischemic mimic from true MINOCA is the use of cardiovascular magnetic resonance imaging (MRI). Myocarditis has been estimated to be responsible for up to 33% of suspected MINOCA presentations, though it may account for a greater proportion if timely cardiac MRI is pursued (see the section Cardiac MRI) [10, 31].

Cardiac MRI findings can support the diagnosis of myocarditis based on the published "Lake Louise Criteria" [33]. Suggestive MRI features include myocardial edema manifested by regional or global signal increase in T2-weighted images, increased early gadolinium enhancement, and/or at least one area of nonischemic late gadolinium enhancement (**Figure 7**) [33]. When cardiac MRI fails to provide a diagnosis or is clinically inappropriate, 2007 guidelines

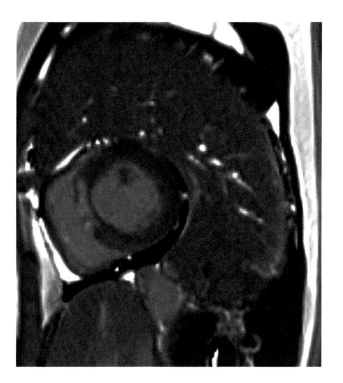

FIGURE 7 Cardiac MRI Image of Myocarditis. A mid-short axis late gadolinium enhancement image showing near transmural late gadolinium enhancement in the anterior and inferior septum. The endocardium is spared suggesting a nonischemic cause. There is a small circumferential pericardial effusion. Together these features are most suggestive of myopericarditis. MRI, magnetic resonance imaging.

recommend (class I indication) endomyocardial biopsy in patients with unexplained fulminant heart failure or complicated unexplained heart failure (of 2 weeks to 3 months' duration) that fails to respond to usual care [34]. The yield of endomyocardial biopsy, however, is variable and in a series of 109 cases of unexplained fulminant heart failure, biopsy was diagnostic in 38.5% and resulted in a change in clinical management in 27.5%, with myocarditis being the most common diagnosis [35]. In light of modest sensitivity and nontrivial procedural complication risk, a more recent position statement has proposed more specific criteria for patients with clinically suspected myocarditis in whom biopsy is recommended, which are summarized in **Table 3** [12].

Distinguishing myocarditis from true ischemic MINOCA events is vital as there are substantial differences in treatment and prognosis. Myocarditis is generally associated with a favorable prognosis with more than 50% of cases experiencing resolution within the first month. There is, however, a subgroup of patients who will experience persistent cardiac dysfunction (25%) or deteriorate acutely [12]. Patients are managed for ventricular dysfunction using standard evidence-based guidelines with additional specific recommendations to avoid exercise for a period of up to 6 months. Depending on the etiology, further management with immunosuppression or steroid therapy may be indicated [12].

TABLE 3 Diagnostic Criteria for Clinically Suspected Myocarditis

Clinical Presentations[a]

Acute chest pain, pericarditic, or pseudo-ischemic

New-onset (days up to 3 months) or worsening of dyspnoea at rest or exercise, and/or fatigue, with or without left and/or right heart failure signs

Subacute/chronic (> 3 months) or worsening of dyspnoea at rest or exercise, and/or fatigue, with or without left and/or right heart failure signs

Palpitation, and/or unexplained arrhythmia symptoms and/or syncope, and/or aborted sudden cardiac death
Unexplained cardiogenic shock

Diagnostic Criteria
I. ECG/Holter/stress test features Newly abnormal 12 lead ECG and/or Holter and/or stress testing, any of the following: I to III degree atrioventricular block, or bundle branch block, ST/T-wave change (ST elevation or non-ST elevation, T-wave inversion), sinus arrest, ventricular tachycardia or fibrillation and asystole, atrial fibrillation, reduced R-wave height, intraventricular conduction delay (widened QRS complex), abnormal Q waves, low voltage, frequent premature beats, and supraventricular tachycardia
II. Myocardiocytolysis markers Elevated TnT/TnI
III. Functional and structural abnormalities on cardiac imaging (echo/angio/CMR) New, otherwise unexplained LV and/or RV structure and function abnormality (including incidental finding in apparently asymptomatic subjects): regional wall motion or global systolic or diastolic function abnormality, with or without ventricular dilatation, with or without increased wall thickness, with or without pericardial effusion, and with or without endocavitary thrombi
IV. Tissue characterization by CMR Edema and/or LGE of classical myocarditic pattern (see text)

[a] *If the patient is asymptomatic ≥ 2 diagnostic criteria should be met.*
Clinically suspected myocarditis if ≥ 1 clinical presentation and ≥ 1 diagnostic criteria from different categories, in the absence of (1) angiographically detectable coronary artery disease (coronary stenosis ≥ 50%); (2) known preexisting cardiovascular disease or extra-cardiac causes that could explain the syndrome (e.g., valve disease, congenital heart disease, hyperthyroidism, etc.) (see text). Suspicion is higher with higher number of fulfilled criteria. CMR, cardiac magnetic resonance; ECG, electrocardiogram; echo, echocardiogram; LGE, late gadolinium enhancement; LV, left ventricle; RV, right ventricle; TnI, troponin I; TnT, troponin T. *Reproduced with permission from [12].*

Supply-Demand Mismatch

Both the ESC and AHA consensus statements emphasize the importance of identification of noncardiac causes of chest pain and the consideration of demand ischemia (type 2 MI) as an initial diagnostic step (**Figure 8**) [3, 4].

Noncardiac causes of chest pain that may be associated with troponin elevation include pulmonary embolism, sepsis, stroke, or end-stage renal failure. Demand ischemia, which may be seen in a variety of clinical settings including profound anemia, hypotension, respiratory failure, shock,

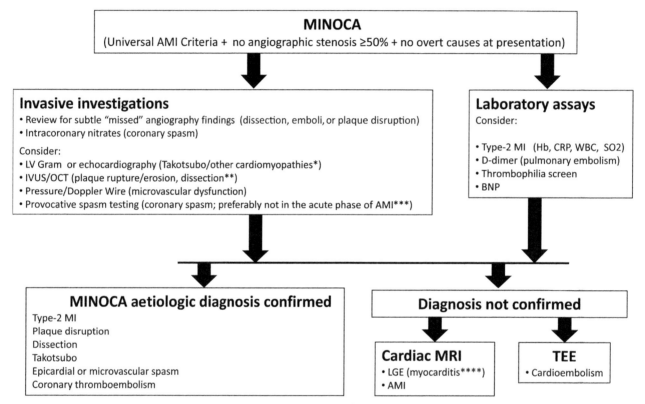

FIGURE 8 **European Society of Cardiology Recommended Diagnostic and Therapeutic Algorithm for Myocardial Infarction With Nonobstructive Coronary Arteries.** *Takotsubo cardiomyopathy cannot be diagnosed with certainty in the acute phase as the definition requires follow-up imaging to document recovery of left ventricular function. In the authors' experience, some patients with apparent Takotsubo have unrecognized ischemic injury or myocarditis. We therefore recommend cardiac MRI when Takotsubo cardiomyopathy is suspected. **Plaque disruption (rupture, or erosion) should be suspected and intracoronary imaging considered whenever an alternate etiology of the clinical presentation such as myocarditis or vasospasm has not been clearly identified, particularly among those patients with evidence of atherosclerosis on the coronary angiogram. Intravascular ultrasound and intracoronary OCT frequently show more atherosclerotic plaque than may be appreciated on angiography. They also increase sensitivity for dissection. If intracoronary imaging is to be performed, it is appropriate to carry out this imaging at the time of the acute cardiac catheterization, after diagnostic angiography. Patients should be made aware of the additional information the test can provide and the small increase in risk associated with intracoronary imaging. ***Provocative testing for coronary artery spasm has been safely performed by experienced clinical researchers in selected patients with a recent acute myocardial infarction. However, death cases have been reported (Per Tornvall Tornberg, personal communication) and this should not be a standard procedure among the patients, particularly in the acute phase. ****Clinically suspected myocarditis (no angiographic stenosis ≥ 50% plus nonischemic pattern on cardiac magnetic resonance imaging) by ESC Task Force criteria. Diagnosis of certainty and aetiological diagnosis of myocarditis requires EMB (histology, immunohistology, and infectious agents by PCR). AMI, acute myocardial infarction; BNP, B-type natriuretic peptide; CRP, C-reactive protein; EMB, endomyocardial biopsy; ESC, European Society of Cardiology; Hb, hemoglobin; IVUS, intravascular ultrasound; LGE, late gadolinium enhancement; LV, left ventricle; MI, myocardial infarction; MINOCA, myocardial infarction with nonobstructed coronary arteries; MRI, magnetic resonance imaging; OCT, optical coherence tomography; SO2, oxygen saturation; TEE, transesophageal echocardiography; WBC, white blood cell count. *Reproduced with permission from [3].*

heart failure, and tachyarrhythmias, may also lead to myocardial necrosis and troponin elevation even in the absence of obstructive coronary disease [2, 37]. These causes of type II MI are distinct from MINOCA, a term reserved for the characterization of a type I event. The history, physical examination, and judicious use of laboratory testing would ideally lead to identification of these diagnoses prior to invasive coronary angiogram.

Truly Unexplained MINOCA

A meticulous diagnostic workup will reveal an underlying cause in ~90% of cases [3, 31]. However, a few patients will have truly unexplained MINOCA despite testing. These patients present a challenge in term of ongoing management. As a population, they are poorly defined and no currently available literature has described their characteristics. It is likely that as diagnostic techniques improve and are utilized to a greater degree in cases of MINOCA, the proportion of truly unexplained cases will diminish.

Diagnostic Evaluation

Identification of the mechanistic etiology in cases of MINOCA is important to allow for evidence-based treatment (where available) and to provide an accurate prognosis. In addition, many cases may be multifactorial in etiology, further highlighting the need for thorough investigation in all patients.

Invasive Imaging

A diagnosis of MINOCA requires the absence of obstructive coronary artery disease on angiogram and current consensus statements recommend careful review for subtle and potentially missed findings such as distal obstructive disease, small vessel obstruction, dissection, or plaque disruption [3, 4]. An arbitrary threshold of less than 50% luminal obstruction has been used to define nonobstructive disease [38]. Within this classification, "normal coronaries" typically imply a complete lack of disease, minimal luminal irregularities signify disease severity <30%, and mild to moderate nonobstructive atherosclerosis includes lesions ≥30% but <50% [4]. Classification of lesion severity is typically by visual estimate and, therefore, highly subjective. When lesion severity is in question, fractional flow reserve (FFR) can be utilized to ensure that hemodynamically significant obstructive lesions are not missed as the cause of MI [4].

Invasive coronary imaging techniques are available to clarify the presence of plaque (and its characteristics), coronary dissection, thrombus, or other vascular patholo-

gies. Both IVUS and OCT provide detailed information on the structure of the coronary arterial wall including the capacity to identify plaque characteristics, dissection flaps, and intramural hematomas, all of which may clarify the underlying diagnosis in cases of MINOCA [39]. Of these techniques, OCT offers the greatest imaging resolution. A single study of 38 patients with MINOCA who underwent OCT demonstrated plaque disruption in 24%; 11% had plaque erosion, and 18% demonstrated coronary thrombus [39]. Interestingly in this group, there was no relationship between plaque disruption and late gadolinium enhancement on cardiac MRI, suggesting that patients with a coronary lesion may not necessarily demonstrate evidence of this on cardiac MRI. In patients investigated with IVUS, the prevalence of plaque rupture has been demonstrated to be between 35% and 40% [19, 20].

A potential limitation of invasive coronary imaging is that the decision to proceed to this technique must be made during the index angiogram, without the benefit of other noninvasive imaging strategies (e.g., cardiac MRI), which may suggest a specific coronary etiology or indicate a nonischemic etiology. Furthermore, although the reported complication rate of intracoronary imaging is low, potential risks may be higher in certain subsets of patients (e.g., SCAD) [40]. However, OCT has been shown to impact immediate management decisions in up to 16% of MINOCA patients (in a single small study), with percutaneous coronary interventions (PCI) performed in the majority [39] and, therefore, should be strongly considered at the time of initial coronary angiography.

Despite the potential yield, there are no data available on what proportion of patients with MINOCA are assessed using IVUS or OCT or what best practices for implementation are, especially when the potential culprit vessel is unclear. A pragmatic approach may be to utilize OCT or IVUS in patients who have some degree of angiographic disease, as studies suggest that these patients are more likely to have plaque erosion or rupture and, therefore, are more likely to have lesions reclassified as those that might benefit from PCI in the immediate setting [39, 41]. There is, however, evidence that plaque rupture occurs in areas of vessels without visible plaque in up to 43% of patients [41] and, therefore, in the absence of clear clinical indication of the culprit region (i.e., by regional ECG changes or wall motion abnormalities) a strategy for intracoronary imaging may not be clear. Studies to date have typically investigated imaging of one to two vessels, though further data are required to better clarify both patient and vessel selection [39, 41].

Beyond the identification of potentially obstructive plaque, understanding whether nonobstructive plaque may be present (which can result in ischemia from downstream showering of thrombus or atheroemboli) and what its characteristics may be through intracoronary imaging may alter medical management decisions (e.g., administration of statin or antiplatelet therapies) even if PCI is not indicated in real time.

Functional Coronary Testing

Coronary Spasm Provocation Testing

Even when ultimately thought to be the causative etiology, spontaneous spasm may not be evident at the time of coronary angiography. Thus, when suspected, spasm should be induced at the time of coronary angiography by administration of intracoronary acetylcholine, which can demonstrate both epicardial coronary spasm as well as microvascular dysfunction [23]; microvascular dysfunction becomes apparent when clinical symptoms and ECG changes occur in the absence of epicardial artery spasm.

There is major variation in the prevalence of vasospasm in MINOCA cohorts, with suggestion of ethnic and geographic variation [42]. Overall, meta-analysis data suggest a prevalence of 27% among MINOCA patients with higher incidence described in studies conducted among Japanese and Korean—as compared to the European and American—cohorts [10]. However, there has also been limited clinical utilization of provocation testing in European and United States centers, particularly in the acute setting, and the 2017 ESC guidelines recommend avoiding such testing during the acute admission. This recommendation, however, is likely influenced by mortality associated with ergotamine-based testing, which was generally tested outside of the coronary angiography setting with less monitoring than catheterization-based testing [4, 41]. In a more recent Italian study, however, coronary artery spasm was found to be inducible in 46% of MINOCA patients [5]. Patients with inducible coronary spasm had higher mortality, cardiac death, and readmission for ACS [5]. In this study, testing was conducted within 48 h of index admission and procedure-related arrhythmias occurred in 5% with no other serious adverse events. On this basis, the AHA have included testing for coronary artery spasm (using modern protocols and invasive monitoring) as part of their diagnostic algorithm for MINOCA without specific recommendation to delay testing from the acute admission [4].

Microvascular Testing

CMD is common in patients with stable ischemic chest pain and no obstructive disease on angiogram, though its prevalence in MINOCA is not entirely clear [4]. Montone and colleagues reported that of patients with positive acetylcholine provocation tests, 35.1% had microvascular spasm defined by the epicardial arterial spasm of ≤90% with classic symptoms [5]. Diagnostic evaluation for microvascular dysfunction can be performed invasively through measurement of coronary flow reserve or noninvasively with the utilization of positron emission tomography (the most validated and accurate noninvasive modality), cardiac MRI, or dynamic myocardial perfusion computed tomography [44]. CMD is discussed in more detail in Chapter 8.

Left Ventricular Assessment

When coronary angiography does not independently elucidate the etiology of MINOCA, ventriculography may provide immediate data regarding overall left ventricular systolic function and regional wall motion abnormalities that may render nonischemic etiologies (e.g., dilated cardiomyopathies and Takotsubo cardiomyopathy) higher on the differential or suggest a causative coronary artery (e.g., through a focal wall motion abnormality) that may warrant further intracoronary imaging. Given that patients with MINOCA should undergo timely coronary angiography in line with current ACS guidelines, the coronary assessment may precede an assessment of ventricular function, which can otherwise be evaluated through echocardiography or cardiac MRI.

Cardiac MRI

Cardiac MRI is central to the diagnostic evaluation in MINOCA and is included in both AHA and ESC guidelines [3, 4]. Cardiac MRI gives detailed information on cardiac structure and function as well as providing a noninvasive mechanism of cardiac tissue characterization. In the context of MINOCA, cardiac MRI has the capacity to provide a diagnosis or reclassify a clinical diagnosis and may have significant impact on clinical management. Many MINOCA etiologies are associated with classic cardiac MRI findings. Cardiac MRI can distinguish between ischemic and nonischemic insults and may provide evidence of otherwise unsuspected diagnoses such as preexisting dilated, infiltrative cardiomyopathy or hypertrophic cardiomyopathy [10, 32]. Ischemic events are characterized by subendocardial or transmural late gadolinium enhancement (LGE), which typically follows a coronary territorial distribution [9]. In contrast, nonischemic myocardial injury—particularly that seen in myocarditis—is classically associated with subepicardial or mid-wall enhancement. Myocardial edema imaging using T2-weighted sequences can be suggestive of Takotsubo cardiomyopathy, characterized by edema and characteristic wall motion abnormalities without evidence of LGE.

An ideal cardiac MRI protocol for MINOCA has not been definitively established and variability in imaging protocols used to date likely accounts for some of the variability in diagnostic yield. To allow for a complete diagnostic evaluation, studies should include full anatomical evaluation of both the left and right ventricle with cine imaging including a full left ventricular volume stack, T2-weighted edema imaging, and LGE. In addition, variability in diagnostic yield may be influenced by the timing of scanning relative to the patient's presentation. A systemic review of 26 cardiac MRI studies in patients with troponin-positive nonobstructive coronary arteries suggested that a diagnosis was attained in 74% of patients, with the remaining patients having normal cardiac MRI studies. In this review, myocarditis was the most common diagnosis (33%), followed by ischemic injury (24%) and Takotsubo cardiomyopathy (18%) (**Figure 9**) [10]. In a more recent retrospective study of 388 patients with cardiac MRI obtained an average of 37 days after presentation, cardiac MRI identified a cause for troponin rise in 74% of patients [32]. A diagnosis of infarction was identified in 25%, myocarditis and cardiomyopathy in 24.7% each, and the remaining 25.5% had a normal cardiac MRI study [32]. In cohorts scanned as early as 6 days after presentation, the yield of cardiac MRI was higher; up to 87% had a relatively higher detection rate of Takotsubo cardiomyopathy (27%) and myocarditis (37%) [31]. On this basis, it has been suggested that cardiac MRI be performed as close to the acute presentation as feasible, ideally within the first week [31, 45, 46]. Definitive diagnosis based on

cardiac MRI may lead to changes in length of hospital stay, medical therapy, and decisions to avoid or undertake invasive procedures. Moreover, cardiac MRI findings have been shown to predict the intermediate-term prognosis of patients with troponin-positive nonobstructive coronary arteries. The finding of cardiomyopathy made by cardiac MRI was associated with the highest mortality of 15%, compared to 4%, 2%, and 2% in patients with MI, myocarditis, and normal cardiac MRI, respectively (**Figure 10**) [32].

Between 13% and 30% of patients have no abnormalities detected on cardiac MRI [31, 32, 45]. Though cardiac MRI is highly sensitive for the detection of myocardial ischemia and injury, it is hypothesized that a subset of MINOCA patients either have too little myonecrosis or necrotic myocytes that are too diffusely distributed (without a sufficient contiguous area of concentration) for detection by LGE [3]. It is also likely that yield of cardiac MRI is affected by sequences obtained and timing of examination, with studies obtained too far out from the acute event missing healed or transient changes. A negative cardiac MRI can lead to uncertainty regarding the underlying diagnosis and management, though it does not rule out a coronary event in the context of a high level of suspicion for true MINOCA. For example, MINOCA studies that have utilized IVUS have demonstrated that up to 25% of patients who experience plaque disruption or rupture have normal cardiac MRI [20].

Assessment for Coronary Thromboembolism

All coronary angiograms should be examined for the presence of coronary thromboembolism as the etiology for MINOCA, though such events can be missed by luminal angiography. When suspicion remains, intracoronary imaging can potentially aid in diagnosis. However, a causative thrombus may still never be detected on examination of coronary arteries in some cases. Nonetheless, 14% of MINOCA patients (vs. ~5% of the general population) are found to have hereditary thrombophilias on testing, most commonly activated protein C and factor V Leiden [10].

The ESC consensus statement suggests a clinical assessment including assessment of a history of risk factors for venous or systemic thromboembolism, laboratory screening for inherited thrombophilias and acquired thrombotic conditions and echocardiography to exclude shunt, cardiac tumor, or valvular source of emboli [3]. The AHA statement additionally posits that it is reasonable to consider inherited hypercoagulable states in MINOCA, especially in younger women [4].

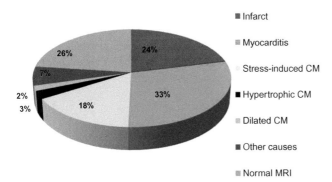

- ■ Infarct
- ■ Myocarditis
- ▢ Stress-induced CM
- ■ Hypertrophic CM
- ▢ Dilated CM
- ■ Other causes
- ■ Normal MRI

24%
26%
7%
2%
3%
18%
33%

FIGURE 9   Etiology of MINOCA Based on Cardiac MRI Publications. A systematic review of 26 cardiac MRI studies in troponin-positive nonobstructive disease patients revealed cause of event in 74% patients. CM, cardiomyopathy; MRI, magnetic resonance imaging. *Data from [10].*

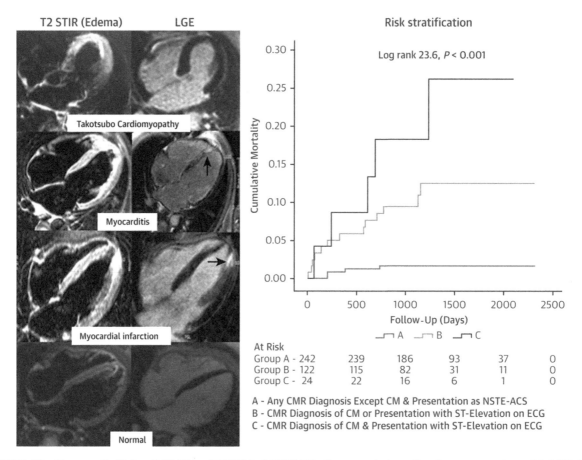

FIGURE 10   Prognostic Role of CMR and ECG in MINOCA. In a population of patients presenting with MINOCA, CMR identified four main categories: cardiomyopathy (Takotsubo), myocarditis, myocardial infarction, and normal. In the Takotsubo cardiomyopathy case, there is transmural edema of the mid-apical segments but no LGE; in the myocarditis case, there is diffuse edema with diffuse patchy epicardial and mid-wall LGE *(black arrow)*, subsequent endomyocardial biopsy confirmed giant cells myocarditis; in the myocardial infarction case, there is focal transmural edema of the apical anterolateral wall with corresponding focal transmural LGE *(black arrow)*. No abnormalities were noted in the normal case. (Right) Kaplan-Meier curves showing the risk of mortality according to the risk group (group A: any CMR diagnosis except cardiomyopathy and presentation as NSTE-ACS; group B: CMR diagnosis of cardiomyopathy or presentation as STEMI; and group C: CMR diagnosis of cardiomyopathy and presentation as STEMI). CM, cardiomyopathy; CMR, cardiac magnetic resonance; ECG, electrocardiogram; LGE, late gadolinium enhancement; MINOCA, myocardial infarction with nonobstructed coronary arteries; NSTE-ACS, non-ST elevation-acute coronary syndrome; STEMI, ST-elevation myocardial infarction. *Reproduced with permission from [32].*

Role of Psychological Stress

Many of the underlying causes of MINOCA have been associated with psychological stress including SCAD [47], coronary microvascular disease [48], coronary artery spasm [49], and acute plaque disruption [50]. MINOCA mimic Takotsubo cardiomyopathy is also intimately linked to psychological stressors [51]. Whether the relationship of MINOCA to stress varies with underlying etiology and what if any impact gender has on these factors remain unknown. Among women enrolled in the VIRGO study, no differences emerged with respect to depression and stress using standardized questionnaires between patients with

obstructive CAD versus MINOCA, though this requires further study [6].

MINOCA also demonstrates circadian variability, as has been described with MI-CAD. The relationship of MINOCA presentations to time of day, day of the week, and major holidays has been investigated in 9092 patients in the Swedish Web-system for Enhancement and Development of Evidence-based care in Heart disease Evaluated According to Recommended Therapies (SWEDEHEART) study. Presentations were most common in the morning (peaking at 8 am 2.25, 95% CI 1.96–2.59) and on Mondays [1.28 (1.18–1.38) [52]. There was, however, no evidence of seasonal variation or relationship of MINOCA presentations to major seasonal holidays or events, in contrast to such findings in the MI-CAD population [52, 53].

Management

General Management Principles

In contrast to MI-CAD, best practices for secondary prevention in MINOCA are unclear. At present, MINOCA patients are less likely to be discharged on guideline-directed therapies established for coronary artery disease (CAD) as compared to those who experience MI-CAD [25, 52, 53]. There is some evidence for the use of secondary prevention therapies across the heterogeneous spectrum of MINOCA, though the current heterogeneity in diagnoses among available cohorts has limited a more nuanced understanding of secondary prevention strategies for specific diagnoses. The SWEDEHEART study undertook a retrospective evaluation of the use of secondary prevention medication in 9136 patients with MINOCA who survived at least 30 days after the index event [7]. In this cohort, the composite outcome of all-cause mortality and hospitalization for MI/stroke/heart failure was reduced by 18% in patients who received ACE inhibitors/ARB treatment [hazard ratio (HR) 0.82; 95% CI 0.73–0.93] and by 23% in those treated with statins (HR 0.77; 95% CI 0.68–0.87) [7]. Similarly, there was a 14% reduction in major adverse cardiac events in patients on beta-blocker therapy (HR 0.86; 95% CI 0.74–1.101) [7]. In the SWEDEHEART cohort, the efficacy of these medications was similar across both sexes. Although MINOCA is associated with lower rates of conventional cardiovascular risk factors, it is important to note that more than 90% of patients will have at least one traditional risk factor including hypertension (54.9%), diabetes (17.4%), dyslipidemia (54.9%), smoking (34.5%), obesity (42.1%), and family history of CAD (61%), and aggressive risk factor reduction remains important for this population [6]. However, prospective, randomized studies are required to better understand effective treatment strategies for MINOCA. Further, given the heterogeneity in potential underlying causes for MINOCA, optimal treatment differs and needs to be tailored to the underlying causative pathophysiology.

Treatment of Underlying Cause

Plaque Disruption

Based on the available evidence, patients found to have causative plaque should be treated with evidence-based secondary prevention strategies derived from the MI-CAD population [3, 4]. ESC guidelines suggest the use of dual antiplatelet therapy (DAPT) for 1 year and lifelong single antiplatelet therapy thereafter [3]. Although the SWEDEHEART study did not show benefit of DAPT, this may reflect heterogeneity in underlying MINOCA diagnoses rather than a lack of effect of antiplatelet agents in the context of plaque disruption, though further prospective studies of this are required in the context of MINOCA [7]. Both guidelines for ACS management and the effects of the SWEDEHEART study suggest that treatment with ACE inhibitors/ARBs and statins are likely to be beneficial [7]. Patients with MINOCA are also less likely to be referred to cardiac rehabilitation services posthospitalization [25]. Given that they are likely to benefit as MI-CAD patients do, all patients should be referred to cardiac rehabilitation following MI, regardless of whether the mechanism was MI-CAD or MINOCA; however, prospective studies are again warranted in this setting.

There are limited data on coronary intervention in the context of nonobstructive plaque disruption. The recent AHA position statement does not suggest routine stenting in patients with nonobstructive plaque disruption as an etiology of MI [4]. Indeed, the mechanism of MI in this context is unknown, and may be due to downstream microemboli or superimposed coronary spasm. Resultantly, it remains unclear whether these lesions will respond to stenting in the same manner as clearly obstructive disease. Pending further study, pharmacotherapy rather than an interventional strategy is generally recommended.

Coronary Vasospasm

Avoidance of precipitating factors, where identified, is of primary importance in patients with vasospasm. Smoking cessation has also been shown to be effective in symptom reduction and avoidance of cardiovascular events in this population [56]. Calcium channel blockers are the mainstay of medical treatment for patients with vasospastic angina and have been shown to predict MI-free survival in a Japanese study of 245 patients followed for an average of 80 months [57]. Patients with MINOCA are substantially more likely to be discharged from hospital on a calcium channel blocker than those with obstructive CAD [25], possibly representing clinical suspicion of vasospasm as an underlying cause. When calcium channel blockers do not offer sufficient symptom control, long-acting nitrates may be beneficial, although there are some concerns about the impact of nitrate tolerance and that there may be some risk associated with the combination of nitrates and nicorandil [58]. Other reported although less commonly used therapies include cilostazol, pioglitazone, and magnesium [56].

Statin therapy has been shown to reduce the incidence of acetylcholine-induced vasospasm in patients treated with calcium channel blockers [59]. It is possible, therefore, that some of the benefit associated with statins in the SWEDEHEART study reflected a positive effect in the vasospasm population. It therefore seems reasonable to give statins in this population, particularly if there is concomitant evidence of atherosclerotic disease [7].

Coronary Microvascular Dysfunction

Understanding of management of CMD is limited compared to our understanding of macrovascular disease and therapies, which have been developed for epicardial coronary disease may not be as effective for treatment of CMD [28]. Calcium channel blockers and beta-blockers are generally beneficial in alleviating symptoms with limited benefit to nitrates. Other drugs that may relieve symptoms by improving endothelial function include statins, L-arginine, and enalapril. Dipyridamole and ranolazine may help by improving microvascular vasodilation. For refractory symptoms, visceral analgesic agents such as imipramine and aminophylline may be considered. Therapies for CMD are further discussed in Chapter 8.

Coronary Emboli

Few management guidelines exist for the treatment of coronary emboli. When a reversible cause is identified, this cause should be treated. However, in the absence of a reversible cause, one opinion suggests that patients with reversible risk factors for thrombosis should be anticoagulated for 3 months and that those with persistent risk factors or unexplained emboli may benefit from longer-term anticoagulation [60]. However, this remains a question that requires further investigation.

Prognosis

The prognosis of patients with MINOCA has traditionally been considered better than that of MI with obstructive CAD. Nonetheless, the diagnosis carries with it an increased risk of MACE and mortality. A 2015 meta-analysis demonstrated an overall in-hospital mortality of 0.9% and a 1-year mortality of 4.7% for MINOCA [10]. This may not completely reflect medium-term prognosis, with the SWEDEHEART study reporting a major adverse cardiac event in 24% patients at 4 years, including mortality in 13.4%, recurrent MI in 7.1%, and stroke and heart failure hospitalization in 4.3% and 6.4%, respectively [7]. The assumption that MINOCA carries an improved prognosis compared to obstructive CAD has recently been brought into doubt with recent studies showing no difference in prognosis between those with MINOCA and those with obstructive CAD [25].

The presence or absence of nonobstructive CAD is potentially relevant to prognosis. The outcomes of women with MINOCA have been specifically considered in the Women's Ischemia Syndrome Evaluation (WISE) co-

hort, which followed up 917 women who presented with symptoms consistent with AMI and divided them based on angiography into those with no CAD, those with non-obstructive CAD, and those with obstructive CAD [61]. In this group, the cumulative 10-year rate of mortality or recurrent MI of women with no CAD was 6.7%, and in those with nonocclusive CAD it was 12.8%, although in both cases this was lower than the event rate in women with occlusive CAD (25.9%). The cause of the MINOCA presentation was not reported in this group and, therefore, may have included those with nonischemic pathophysiology. The inclusion of cardiomyopathic processes is likely to artificially elevate the mortality as cardiac MRI-based studies have shown that the medium-term prognosis is worst in those with cardiomyopathy (with 15% 3.5-year mortality), while those with ischemic or no identified cause for MINOCA had a mortality of 2–4% [32].

MINOCA patients often experience high rates of recurrent symptoms, with consequent impact on quality of life and hospital readmissions. In the WISE cohort, 15.9% of women with no CAD were readmitted for worsening or refractory chest pain within 5 years of the index event; the readmission rate was higher for those with nonobstructive CAD (22.3%, $P < 0.001$) [61]. Other studies have suggested that as many as a quarter of patients experience recurrent anginal symptoms within 1 year of a MINOCA presentation, a level that does not significantly differ from patients with obstructive CAD [62]. In addition, 7.1% of patients experienced recurrent MI, 4.3% stroke, and 6.4% were hospitalized for heart failure [7]. Patients with MINOCA may also experience progression of nonocclusive coronary artery disease to the point where it becomes obstructive. Among patients who experienced repeat infarction in the SWEDEHEART MINOCA registry, 47% had evidence of occlusive disease on repeat angiogram [63].

The risk of mortality in MINOCA appears to be in part predicated on the presence of traditional cardiovascular risks with age (HR mortality 1.07, HR MACE 1.05, $P < 0.001$), smoking (HR mortality 1.53, HR MACE 1.38, $P = 0.001$), diabetes (HR mortality 1.36, $P = 0.018$, HR MACE 1.44, $P < 0.001$), hypertension (HR MACE 1.25, $P = 0.001$), previous MI (HR MACE 1.38, $P = 0.025$), chronic obstructive pulmonary disease (COPD) (HR mortality 2.33, HR MACE 1.63, $P < 0.001$), previous cerebrovascular accident (CVA) (HR mortality 1.53, $P = 0.007$, HR MACE 1.69, $P < 0.001$), and renal impairment (HR MACE 1.01, $P = 0.027$, HR mortality 1.02, $P = 0.03$) predicting MACE and/or mortality in the SWEDEHEART cohort [64]. Other factors that have been associated with adverse outcome include ST elevation at presentation, atypical presenting symptoms, worsening heart failure or heart failure symptoms, and nonuse of ACE inhibitors/ARBs and statin medication [25, 64].

Although women experience higher event rates after MI than men, this may not be the case in MINOCA. In both Korean registry data and the SWEDEHEART study, sex was not an independent predictor of mortality or MACE [25, 64]. This replicates the findings of an earlier study looking at the impact of sex on outcomes in MI, which reported that excess mortality after MI was limited to those with obstructive CAD, with no gender difference emerging with respect to outcomes following MINOCA [11, 65]. However, in a study of MINOCA patients younger than 55 years of age (VIRGO), all reported deaths were in women (as compared with men) [6]. Further study will be required to better understand the risk factors for adverse events after MINOCA and means of reducing this risk.

Interesting, many recurrent MIs following MINOCA are found to be due to MI-CAD rather than recurrent MINOCA. In the SWEDEHEART registry, 340 patients with prior MINOCA underwent coronary angiography for recurrent events, of whom 47% were found to have obstructive lesions as the cause of the recurrent MI. Thus, a history of MINOCA should not dissuade providers from pursuing repeat coronary angiography for recurrent events as many of these recurrent events will be due to obstructive disease. Of further interest, mortality at 38 months did not differ between patients who experienced recurrent MI-CAD (11.9%) vs. recurrent MINOCA (13.9%, $P=0.54$) [63].

Conclusion

The diagnostic workup for a patient with potential MINOCA can be complex and multifaceted. As emphasized by both the AHA and ESC consensus statements, MINOCA is a working diagnosis that should prompt thorough investigation into the underlying causative pathophysiology in order to offer patients a definitive diagnosis (when possible) and available targeted therapy. Diagnostic algorithms have been published by both groups (**Figure 10**). The AHA consensus statement emphasizes the importance of reserving the term MINOCA for patients with a type I coronary event (in the absence of obstructive disease) and, therefore, excluding patients with nonischemic etiologies for troponin rise, including myocarditis, cardiomyopathy, or Takotsubo cardiomyopathy [4].

The potential diagnostic pathway of a patient presenting with ACS symptoms who may subsequently be found to have MINOCA is outlined in **Figure 8**. An overt alternative clinical diagnosis may be evident on early history or examination and the diagnosis of MINOCA is no longer appropriate in this context (i.e., type II MI or nonischemic cardiomyopathy). Patients felt to be experiencing a type I MI should proceed to coronary angiography, which can help with clarification of angiographic disease severity and etiology, with the use of FFR, instantaneous wave-free ratio (iFR), or intracoronary imaging when appropriate. When plaque is suspected—even in the context of nonobstructive disease—intracoronary imaging may be helpful to further clarify whether characteristics may merit intervention. If no culprit lesion is identified, provocative testing for vasospasm or microvascular function testing can be considered. If invasive testing does not clarify a coronary diagnosis, patients should undergo cardiac MRI, which can aid in distinguishing between nonischemic mimics or confirm whether a true ischemic event may have occurred. The yield of cardiac MRI is greatest when performed soon after the index event and, therefore, should be considered prior to hospital discharge or within 1 week of presenting event whenever feasible. Finally, if a coronary embolus is suspect after above testing, hypercoagulability testing and potential investigation for an intraatrial shunt or intracardiac source by echocardiography should be considered.

Prognosis after MINOCA is not benign, with many women experiencing recurrent chest pain. Further, there is a 4.7% 1-year risk of mortality and risk of recurrent MIs [7, 10], which can represent both recurrent MINOCA as well as MI-CAD. Providers and patients should be aware of these risks and understand that MINOCA should, therefore, be aggressively investigated and treated. A challenge, however, remains that optimal treatment following MINOCA remains uncertain given a lack of available data. Current studies include heterogeneous etiologies of MINOCA and, therefore, conclusions are difficult to generalize. It is likely that treatment of MINOCA is best if targeted to the underlying pathophysiology, though prospective studies of this are greatly needed in order to potentially improve the prognosis following MINOCA.

CH
9

Key Points

1 Approximately 6% of ACS patients will have nonobstructive coronary artery disease on angiography and receive a working diagnosis of MINOCA while the cause of infarction is established.

2 MINOCA indicates a type 1 myocardial infarction (with nonobstructive coronary arteries) with atherosclerotic and nonatherosclerotic causes including coronary plaque disruption, coronary embolism, spontaneous coronary artery disease, coronary spasm, and microvascular dysfunction.

3 Mimickers of MINOCA should be excluded, such as:

- other causes of troponin elevation (pulmonary emboli, sepsis, and renal failure);
- overlooked distal or branch vessel stenosis; and
- nonischemic myopathic processes.

4 Women, particularly younger women with fewer cardiac risk factors, are disproportionally represented in the MINOCA population.

5 The diagnosis of MINOCA should prompt a detailed workup to establish the cause. This should at minimum include coronary angiography and assessment of cardiac anatomy (preferably with cardiac MRI), but may also include invasive coronary imaging, provocation testing, and thrombophilia screening.

6 Coronary spasm is a common cause of MINOCA, with first-line treatment being calcium channel blockers.

7 A meticulous workup will reveal the cause in as many as 90% of patients and allow disease-specific therapies (where available) to be initiated.

8 Patients diagnosed with MINOCA are less likely to receive guideline-based ACS therapies. Treatment with beta-blockers, statins, and ACE/ARB inhibition has demonstrated reduction in major adverse cardiovascular events.

9 However, several knowledge gaps persist. Prospective trials and large multicenter registers are needed to guide future management of these patients.

Back to Clinical Case

The case represented a 60-year-old female with positive troponins, but with angiographically normal coronary arteries. In other words, her presentation was consistent with MINOCA. A closer review of her angiogram still did not demonstrate any branch stenosis. A cardiac MRI was subsequently obtained and revealed an ischemic pattern of late gadolinium enhancement (LGE) consistent with an infarct in the basal to mid-inferior and inferolateral walls (**Figure 11**). A hypercoagulability panel was without evidence of inherited thrombophilia. There was no evidence of patent foramen ovale with the use of agitated saline contrast on transthoracic echocardiogram. Computed tomography angiography (CTA) did not reveal fibromuscular dysplasia or other vascular arteriopathy. Thus, while an infarct was ultimately identified, its etiology was not fully elucidated, leaving the patient and her providers with lingering questions regarding optimal management strategies and long-term prognosis. In the presence of recurrent symptoms, management could include microvascular functional testing with stress cardiac MRI or quantitative positron emission tomography to rule out microvascular disease. In addition, one could consider provocative testing during repeat angiography with acetyl choline.

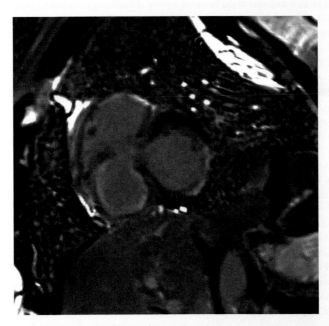

FIGURE 11   Cardiac MRI of 60-Year-Old Woman Presenting With MINOCA. A basal short-axis cardiac MRI image demonstrating subendocardial late gadolinium enhancement in the inferior and inferolateral walls consistent with myocardial infarction. MINOCA, myocardial infarction with nonobstructed coronary arteries; MRI, magnetic resonance imaging.

Acknowledgment

The authors gratefully acknowledge the assistance of Dr. Sandeep Hedgire (MD) and Dr. Jena Depetris (MD) for their assistance in the preparation of **Figures 7 and 11**.

References

[1] Gross H, Sternberg WH. Myocardial infarction without significant lesions of coronary arteries. Arch Intern Med 1939;64:249–67.

[2] Beltrame JF. Assessing patients with myocardial infarction and nonobstructed coronary arteries (MINOCA). J Intern Med 2013;273:182–5.

[3] Agewall S, Beltrame JF, Reynolds HR, Niessner A, Rosano G, Caforio ALP, et al. ESC working group position paper on myocardial infarction with non-obstructive coronary arteries. Eur Heart J 2017;38(3):143–53.

[4] Tamis-Holland JE, Jneid H, Reynolds HR, Agewall S, Brilakis ES, Brown TM, et al. Contemporary diagnosis and management of patients with myocardial infarction in the absence of obstructive coronary artery disease: a scientific statement from the American Heart Association. Circulation 2019;139:e891–908.

[5] Montone RA, Niccoli G, Fracassi F, Russo M, Gurgoglione F, Cammà G, et al. Patients with acute myocardial infarction and non-obstructive coronary arteries: safety and prognostic relevance of invasive coronary provocative tests. Eur Heart J 2018;39:91–8.

[6] Safdar B, Spatz ES, Dreyer RP, Beltrame JF, Lichtman JH, Spertus JA, et al. Presentation, clinical profile, and prognosis of young patients with myocardial infarction with nonobstructive coronary arteries (MINOCA): results from the VIRGO study. J Am Heart Assoc 2018;7(13):e009174.

[7] Lindahl B, Baron T, Erlinge D, Hadziosmanovic N, Nordenskjöld A, Gard A, et al. Medical therapy for secondary prevention and long-term outcome in patients with myocardial infarction with nonobstructive coronary artery disease. Circulation 2017;135:1481–9.

[8] Thygesen K, Alpert JS, Jaffe AS, Chaitman BR, Bax JJ, Morrow DA, et al. Fourth universal definition of myocardial infarction (2018). Circulation 2018;138:e618–51.

[9] Pasupathy S, Tavella R, Beltrame JF. Myocardial infarction with nonobstructive coronary arteries (MINOCA): the past, present, and future management. Circulation 2017;135(16):1490–3.

[10] Pasupathy S, Air T, Dreyer RP, Tavella R, Beltrame JF. Systematic review of patients presenting with suspected myocardial infarction and nonobstructive coronary arteries. Circulation 2015;131:861–70.

[11] Smilowitz NR, Mahajan AM, Roe MT, Hellkamp AS, Chiswell K, Gulati M, et al. Mortality of myocardial infarction by sex, age, and obstructive coronary artery disease status in the ACTION registry-GWTG (Acute Coronary Treatment and Intervention Outcomes Network registry-get with the guidelines). Circ Cardiovasc Qual Outcomes 2017;10:e003443.

[12] Caforio ALP, Pankuweit S, Arbustini E, Basso C, Gimeno-Blanes J, Felix SB, et al. Current state of knowledge on aetiology, diagnosis, management, and therapy of myocarditis: a position statement of the European Society of Cardiology Working Group on Myocardial and Pericardial Diseases. Eur Heart J 2013;34(26):2636–48. 2648a–d.

[13] Thygesen K, Alpert JS, Jaffe AS, Simoons ML, Chaitman BR, White HD, et al. Third universal definition of myocardial infarction. J Am Coll Cardiol 2012;60:1581–98.

[14] Higuma T, Soeda T, Abe N, Yamada M, Yokoyama H, Shibutani S, et al. A combined optical coherence tomography and intravascular ultrasound study on plaque rupture, plaque erosion, and calcified nodule in patients with ST-segment elevation myocardial infarction incidence, morphologic characteristics, and outcomes after percutaneous coronary intervention. JACC Cardiovasc Interv 2015;8(9):116.

[15] Libby P, Pasterkamp G. Requiem for the 'vulnerable plaque'. Eur Heart J 2015;36(43):2984–7.

[16] Yahagi K, Davis HR, Arbustini E, Virmani R. Sex differences in coronary artery disease: pathological observations. Atherosclerosis 2015;239:260–7.

[17] Arbustini E, Dal Bello B, Morbini P, Burke AP, Bocciarelli M, Specchia G, et al. Plaque erosion is a major substrate for coronary thrombosis in acute myocardial infarction. Heart 1999;82:269–72.

[18] Yamamoto E, Yonetsu T, Kakuta T, Soeda T, Saito Y, Yan BP, et al. Clinical and laboratory predictors for plaque erosion in patients with acute coronary syndromes. J Am Heart Assoc 2019;8:e012322.

[19] Ouldzein H, Elbaz M, Roncalli J, Cagnac R, Carrié D, Puel J, et al. Plaque rupture and morphological characteristics of the culprit lesion in acute coronary syndromes without significant angiographic lesion: analysis by intravascular ultrasound. Ann Cardiol Angeiol 2012;61:20–6.

[20] Reynolds HR, Srichai MB, Iqbal SN, Slater JN, Mancini GBJ, Feit F, et al. Mechanisms of myocardial infarction in women without angiographically obstructive coronary artery disease. Circulation 2011;124:1414–25.

[21] Stepien K, Nowak K, Wypasek E, Zalewski J, Undas A. High prevalence of inherited thrombophilia and antiphospholipid syndrome in myocardial infarction with non-obstructive coronary arteries: comparison with cryptogenic stroke. Int J Cardiol 2019;290:1–6.

[22] Tomaiuolo R, Bellia C, Caruso A, Di Fiore R, Quaranta S, Noto D, et al. Prothrombotic gene variants as risk factors of acute myocardial infarction in young women. J Transl Med 2012;10:235.

[23] Beltrame JF, Crea F, Kaski JC, Ogawa H, Ong P, Sechtem U, et al. International standardization of diagnostic criteria for vasospastic angina. Eur Heart J 2017;38:2565–8.

[24] Shin E-S, Ann SH, Singh GB, Lim KH, Yoon H-J, Hur S-H, et al. OCT-defined morphological characteristics of coronary artery spasm sites in vasospastic angina. JACC Cardiovasc Imaging 2015;8:1059–67.

[25] Choo EH, Chang K, Lee KY, Lee D, Kim JG, Ahn Y, et al. Prognosis and predictors of mortality in patients suffering myocardial infarction with non-obstructive coronary arteries. J Am Heart Assoc 2019;8:e011990.

[26] Aziz A, Hansen HS, Sechtem U, Prescott E, Ong P. Sex-related differences in vasomotor function in patients with angina and unobstructed coronary arteries. J Am Coll Cardiol 2017;70:2349–58.

[27] Kawana A, Takahashi J, Takagi Y, Yasuda S, Sakata Y, Tsunoda R, et al. Gender differences in the clinical characteristics and outcomes of patients with vasospastic angina—a report from the Japanese Coronary Spasm Association. Circ J 2013;77:1267–74.

[28] Beltrame JF, Crea F, Camici P. Advances in coronary microvascular dysfunction. Heart Lung Circ 2009;18:19–27.

[29] Ong P, Camici PG, Beltrame JF, Crea F, Shimokawa H, Sechtem U, et al. International standardization of diagnostic criteria for microvascular angina. Int J Cardiol 2018;250:16–20.

[30] Hayes SN, Kim ESH, Saw J, Adlam D, Arslanian-Engoren C, Economy KE, et al. Spontaneous coronary artery dissection: current state of the science: a scientific statement from the American Heart Association. Circulation 2018;137:e523–57.

[31] Pathik B, Raman B, Mohd Amin NH, Mahadavan D, Rajendran S, McGavigan AD, et al. Troponin-positive chest pain with unobstructed coronary arteries: incremental diagnostic value of cardiovascular magnetic resonance imaging. Eur Heart J Cardiovasc Imaging 2016;17:1146–52.

[32] Dastidar AG, Baritussio A, De Garate E, Drobni Z, Biglino G, Singhal P, et al. Prognostic role of CMR and conventional risk factors in myocardial infarction with nonobstructed coronary arteries. JACC Cardiovasc Imaging 2019;12(10):1973–82.

[33] Friedrich MG, Sechtem U, Schulz-Menger J, Holmvang G, Alakija P, Cooper LT, et al. Cardiovascular magnetic resonance in myocarditis: a JACC white paper. J Am Coll Cardiol 2009;53:1475–87.

[34] Cooper LT, Baughman KL, Feldman AM, Frustaci A, Jessup M, Kuhl U, et al. The role of endomyocardial biopsy in the management of cardio-vascular disease: a scientific statement from the American Heart Association, the American College of Cardiology, and the European Society of Cardiology. Endorsed by the Heart Failure Society of America and the Heart Failure Association of the European Society of Cardiology. J Am Coll Cardiol 2007;50:1914–31.

[35] Bennett MK, Gilotra NA, Harrington C, Rao S, Dunn JM, Freitag TB, et al. Evaluation of the role of endomyocardial biopsy in 851 patients with unexplained heart failure from 2000-2009. Circ Heart Fail 2013;6:676–84.

[36] Taqueti VR. Coronary microvascular dysfunction in vasospastic angina provocative role for the microcirculation in macrovessel disease prognosis. J Am Coll Cardiol 2019;74(19):2361.

[37] Sandoval Y, Smith SW, Thordsen SE, Apple FS. Supply/demand type 2 myocardial infarction: should we be paying more attention? J Am Coll Cordiol 2014;63:2079–87.

[38] Patel MR, Calhoon JH, Dehmer GJ, Grantham JA, Maddox TM, Maron DJ, et al. ACC/AATS/AHA/ASE/ASNC/SCAI/SCCT/STS 2017 appropriate use criteria for coronary revascularization in patients with stable ischemic heart disease: a report of the American College of Cardiology appropriate use criteria task force, American Association for Thoracic Surgery, American Heart Association, American Society of Echocardiography, American Society of Nuclear Cardiology, Society for Cardiovascular Angiography and Interventions, Society of Cardiovascular Computed Tomography, and Society of Thoracic Surgeons. J Am Coll Cardiol 2017;69:2212–41.

[39] Opolski MP, Spiewak M, Marczak M, Debski A, Knaapen P, Schumacher SP, et al. Mechanisms of myocardial infarction in patients with nonobstructive coronary artery disease: results from the optical coherence tomography study. JACC Cardiovasc Imaging 2019;12(11 Pt 1):2210–21.

[40] Saw J, Mancini GBJ, Humphries K, Fung A, Boone R, Starovoytov A, et al. Angiographic appearance of spontaneous coronary artery dissection with intramural hematoma proven on intracoronary imaging. Catheter Cardiovasc Interv 2016;87:E54–61.

[41] Iqbal SN, Feit F, Mancini GBJ, Wood D, Patel R, Pena-Sing I, et al. Characteristics of plaque disruption by intravascular ultrasound in women presenting with myocardial infarction without obstructive coronary artery disease. Am Heart J 2014;167:715–22.

[42] Pristipino C, Beltrame JF, Finocchiaro ML, Hattori R, Fujita M, Mongiardo R, et al. Major racial differences in coronary constrictor response between Japanese and Caucasians with recent myocardial infarction. Circulation 2000;101:1102–8.

[43] Buxton A, Goldberg S, Hirshfeld JW, Wilson J, Mann T, Williams DO, et al. Refractory ergonovine-induced coronary vasospasm: importance of intracoronary nitroglycerin. Am J Cardiol 1980;46:329–34.

[44] Taqueti VR, Di Carli MF. Coronary microvascular disease pathogenic mechanisms and therapeutic options: JACC state-of-the-art review. J Am Coll Cardiol 2018;72:2625–41.

[45] Dastidar AG, Rodrigues JCL, Johnson TW, De Garate E, Singhal P, Baritussio A, et al. Myocardial infarction with nonobstructed coronary arteries: impact of CMR early after presentation. JACC Cardiovasc Imaging 2017;10:1204–6.

[46] Pasupathy S, Tavella R, McRae S, Beltrame JF. Myocardial infarction with non-obstructive coronary arteries—diagnosis and management. Eur Cardiol 2015;10:79–82.

[47] Saw J, Aymong E, Sedlak T, Buller CE, Starovoytov A, Ricci D, et al. Spontaneous coronary artery dissection: association with predisposing arteriopathies and precipitating stressors and cardiovascular outcomes. Circ Cardiovasc Interv 2014;7:645–55.

[48] Asbury EA, Creed F, Collins P. Distinct psychosocial differences between women with coronary heart disease and cardiac syndrome X. Eur Heart J 2004;25:1695–701.

[49] Hung M-Y, Mao C-T, Hung M-J, Wang J-K, Lee H-C, Yeh C-T, et al. Coronary artery spasm as related to anxiety and depression: a nationwide population-based study. Psychosom Med 2019;81:237–45.

[50] Lagraauw HM, Kuiper J, Bot I. Acute and chronic psychological stress as risk factors for cardiovascular disease: insights gained from epidemiological, clinical and experimental studies. Brain Behav Immun 2015;50:18–30.

[51] Dawson DK. Acute stress-induced (takotsubo) cardiomyopathy. Heart 2018;104:96–102.

[52] Nordenskjold AM, Eggers KM, Jernberg T, Mohammad MA, Erlinge D, Lindahl B. Circadian onset and prognosis of myocardial infarction with non-obstructive coronary arteries (MINOCA). PLoS One 2019;14:e0216073.

[53] Mohammad MA, Karlsson S, Haddad J, Cederberg B, Jernberg T, Lindahl B, et al. Christmas, national holidays, sport events, and time factors as triggers of acute myocardial infarction: SWEDEHEART observational study 1998-2013. BMJ 2018;363:k4811.

[54] Bainey KR, Welsh RC, Alemayehu W, Westerhout CM, Traboulsi D, Anderson T, et al. Population-level incidence and outcomes of myocardial infarction with non-obstructive coronary arteries (MINOCA): insights from the Alberta contemporary acute coronary syndrome patients invasive treatment strategies (COAPT) study. Int J Cardiol 2018;264:12–7.

[55] Maddox TM, Ho PM, Roe M, Dai D, Tsai TT, Rumsfeld JS. Utilization of secondary prevention therapies in patients with nonobstructive coronary artery disease identified during cardiac catheterization. Circ Cardiovasc Qual Outcomes 2010;3:632–41.

[56] Beltrame JF, Crea F, Kaski JC, Ogawa H, Ong P, Sechtem U, et al. The who, what, why, when, how and where of vasospastic angina. Circ J 2016;80:289–98.

[57] Yasue H, Takizawa A, Nagao M, Nishida S, Horie M, Kubota J, et al. Long-term prognosis for patients with variant angina and influential factors. Circulation 1988;78:1–9.

[58] Takahashi J, Nihei T, Takagi Y, Miyata S, Odaka Y, Tsunoda R, et al. Prognostic impact of chronic nitrate therapy in patients with vasospastic angina: multicentre registry study of the Japanese Coronary Spasm Association. Eur Heart J 2015;36:228–37.

[59] Yasue H, Mizuno Y, Harada E, Itoh T, Nakagawa H, Nakayama M, et al. Effects of a 3-hydroxy-3-methylglutaryl coenzyme A reductase inhibitor, fluvastatin, on coronary spasm after withdrawal of calcium-channel blockers. J Am Coll Cardiol 2008;51:1742–8.

[60] Raphael CE, Heit JA, Reeder GS, Bois MC, Maleszewski JJ, Tilbury RT, et al. Coronary embolus: an underappreciated cause of acute coronary syndromes. JACC Cardiovasc Interv 2018;11:172–80.

CH
9

[61] Sharaf B, Wood T, Shaw L, Johnson BD, Kelsey S, Anderson RD, et al. Adverse outcomes among women presenting with signs and symptoms of ischemia and no obstructive coronary artery disease: findings from the National Heart, Lung, and Blood Institute-sponsored Women's Ischemia Syndrome Evaluation (WISE) angiographic core laboratory. Am Heart J 2013;166:134–41.

[62] Grodzinsky A, Arnold SV, Gosch K, Spertus JA, Foody JM, Beltrame J, et al. Angina frequency after acute myocardial infarction in patients without obstructive coronary artery disease. Eur Heart J Qual Care Clin Outcomes 2015;1:92–9.

[63] Nordenskjöld AM, Lagerqvist B, Baron T, Jernberg T, Hadziosmanovic N, Reynolds HR, et al. Reinfarction in patients with myocardial infarction with nonobstructive coronary arteries (MINOCA): coronary findings and prognosis. Am J Med 2019;132:335–46.

[64] Nordenskjold AM, Baron T, Eggers KM, Jernberg T, Lindahl B. Predictors of adverse outcome in patients with myocardial infarction with non-obstructive coronary artery (MINOCA) disease. Int J Cardiol 2018;261:18–23.

[65] Gehrie ER, Reynolds HR, Chen AY, Neelon BH, Roe MT, Gibler WB, et al. Characterization and outcomes of women and men with non-ST-segment elevation myocardial infarction and nonobstructive coronary artery disease: results from the Can Rapid Risk Stratification of Unstable Angina Patients Suppress Adverse Outcomes with Early Implementation of the ACC/AHA Guidelines (CRUSADE) quality improvement initiative. Am Heart J 2009;158:688–94.

Heart Failure

Chapter 10

Heart Failure With Reduced Ejection Fraction

Clyde W. Yancy, Esther Vorovich, and Sarah Chuzi

Clinical Case

A 72-year-old female is referred for further management of heart failure with reduced ejection fraction (25%) secondary to nonischemic cardiomyopathy, which was diagnosed 2 years prior. The patient has a history of hypertension and obesity. Her blood pressure is 120/80 and her heart rate is 75 bpm. She has edema on her lower extremities but an otherwise normal cardiovascular exam. She complains of fatigue and dyspnea with activities of daily living and with ambulating around her home. Review of systems is also positive for depressive symptoms, including lack of energy, insomnia, and decreased interest in activities. Her medications include carvedilol 25 mg twice daily, lisinopril 40 mg daily, and spironolactone 12.5 mg daily. Laboratory studies including electrolyte panel are within normal limits. Her electrocardiogram shows normal sinus rhythm at a rate of 75 bpm with a left bundle branch block and QRS interval of 150 ms. A prior ischemic evaluation was negative, and a cardiac magnetic resonance imaging study confirmed dilated cardiomyopathy. How would you manage this patient? Would your management strategy be different if this was a 72-year-old man?

Abstract

Heart failure (HF) with reduced ejection fraction (HFrEF), defined as clinical signs and symptoms of volume overload and/or low cardiac output with an ejection fraction (EF) of ≤40%, affects an estimated 3.5 million patients in the United States. Women account for 36% of this population. Studies in the last decade have highlighted differences between women and men with HFrEF involving multiple aspects of the syndrome including epidemiology, pathophysiology, outcomes, and treatment. Women with HFrEF have fewer comorbidities, lower rates of hospitalization, and superior survival compared to men, but they also experience more symptoms, poorer functional status, and worse health-related quality of life (HRQOL). While disparities in the prescription of guideline-directed medical therapy

Sex Differences in Cardiac Disease. https://doi.org/10.1016/B978-0-12-819369-3.00022-8

(GDMT) have narrowed somewhat, women are still undertreated with diuretics and devices, and are less likely to be referred to disease management programs. Further, women remain significantly underrepresented in clinical HF trials, limiting available data guiding potential sex-specific clinical approaches. These differences in presentation, outcomes, and treatment are likely secondary to both biological and psychosocial factors. This chapter will outline important sex-specific differences in HFrEF while also addressing gaps in the literature and future directions.

Epidemiology

Prevalence

HF is a global epidemic affecting more than 37 million individuals worldwide and an estimated 6.5 million Americans [1, 2]. Characterized by a constellation of signs and symptoms of fluid retention and/or low tissue perfusion, HF is often categorized according to baseline left ventricular EF, which is used to both phenotype patients and guide therapy. Currently, patients are clinically categorized by EF into three categories: reduced EF ($\leq$40%), preserved EF ($\geq$50%), or borderline/mid-range EF (41–49%) [1, 3]. HF with reduced ejection fraction (HFrEF) accounts for 50–60% of HF cases [4, 5], affecting an estimated 2.4–5.8% of the overall population [6]. Recent data from an American Heart Association Get With The Guidelines® (GWTG) study indicated that 36% of HFrEF cases occur in women, compared to 63% of HF with preserved ejection fraction (HFpEF) cases and 47% of HF with borderline EF [4].

Incidence and Lifetime Risk

An estimated 1 million new cases of HF (all phenotypes) are diagnosed annually, with nearly half occurring in women [1]. In addition, HFrEF is reported to account for about 52% of incident HF cases in a study by Ho and colleagues using data from multiple longitudinal cohorts [7]. In this study, women had a lower incidence of HFrEF than men over a 12-year median follow-up period, with male sex identified as a significant independent predictor of incident HF [hazard ratio (HR) 1.84, 95% confidence interval (CI) 1.55–2.19]. Further, the incidence of HF is declining over time with a greater reduction in women with HFrEF compared to men [8].

The lifetime risk of HFrEF is also lower in women. Using data from the Cardiovascular Health Study (CHS) and the Multi-Ethnic Study of Atherosclerosis (MESA) cohorts, Pandey and colleagues found that women had a lower lifetime risk of HFrEF compared to men at 45 years (5.8%, 95% CI 5.0–6.6 vs. 10.6%, 95% CI 9.4–11.8), 55 years (5.8%, 95% CI 5.0–6.6 vs. 10.5%, 95% CI 9.3–11.8), 65 years (5.8%, 95% CI 5.0–6.6 vs. 10.5%, 95% CI 9.2–11.7), and 75 years (5.2%, 95% CI 4.4–6.6 vs. 9.5%, 95% CI 8.2–10.8) [9]. In contrast, the lifetime risk of developing HFpEF was comparable for both sexes across all age groups [9].

Risk Factors

Nonmodifiable Risk Factors

Age and Race

Age is a nonmodifiable risk factor for HF (all phenotypes) [10]. For HFrEF specifically, increasing age (per decade > age 60) was associated with greater incident risk of HFrEF (HR 2.5, 95% CI 1.7–3.6) in a recent prospective community-based cohort study of 3800 patients without HF at enrollment [11]. An analysis from the Women's Health Initiative (WHI) cohort of 42,170 postmenopausal women showed more HFrEF hospitalizations for increasing age 60–69 years (HR 1.48, 95% CI 1.11–1.97) and 70–79 years (HR 2.76, 95% CI 2.01–3.79) [12]. Furthermore, the lifetime risk for incident HFrEF in women is 5.8% at age 45 years vs. 5.2% at age 75 years, suggesting a balance between increasing incidence of HFrEF and competing risk of mortality in older women [9].

The role of race in women at risk for HFrEF is less well studied. Black individuals have a lifetime risk of incident HFrEF of 7.7% compared to 7.9% for nonblack individuals at age 45; however, this analysis was not further stratified based on sex [9]. In the WHI cohort, hospitalization due to HFrEF was lower for Hispanic (HR 0.54, 95% CI 0.33–0.90) women but not black women compared to white women [12]. More data are needed to draw firm conclusions about the role of race and its interaction with gender in HFrEF.

Reproductive Factors

Endogenous sex hormone exposure and reproductive factors may affect the risk of HFrEF. The number of live births has been associated with increased left ventricle (LV) mass, end-diastolic volume, end-systolic volume, and decreased left ventricular ejection fraction (LVEF) [13]. Shorter total reproductive duration (time from menarche to menopause) has been associated with increased risk of HFrEF in age-adjusted models (HR 0.98, 95% CI 0.97–0.99) although these associations did not persist after multivariable adjustment [13, 14]. There was no relationship between age at first pregnancy or number of live births and incident HFrEF [14]. Data are emerging to suggest that women who experience early menopause (<45 years) are at increased risk for developing HF (HR 1.66, 95% CI 1.01–2.73) [15]; however, there are limited data regarding early menopause and prediction of HFrEF specifically.

Modifiable Risk Factors

Several comorbidities are associated with HFrEF, but sex-specific differences have been identified. Women with HFrEF have less coronary artery disease (CAD) and lower rates of coronary revascularization [16] and myocardial infarction (MI) [1], but have more hypertension (HTN) [16, 17] than men. Women who suffer an MI have increased risk of developing HFrEF compared to those without MI (HR 2.5, 95% CI 1.6–3.90) [12], as do those with HTN (HR 1.99, 95% CI 1.59–2.51) [12], which comprises nearly 50% of the population attributable risk for HFrEF in women [12]. In contrast, history of CAD (other than MI) and atrial fibrillation are not associated with the risk of incident HFrEF. The prevalence of various traditional cardiovascular and noncardiovascular risk factors and their associated risk of incident HFrEF development in women are detailed in **Table 1**.

Etiologies of HFrEF

HF can be caused by a myriad of cardiac and noncardiac conditions, hereditary defects, toxic exposures, and systemic diseases [1]. Ischemic heart disease is the most common etiology of HFrEF, accounting for approximately two-thirds of cases [18]. Men experience higher rates of CAD and MI, higher total incidence of HF attributable to ischemic heart disease, and earlier and greater mortality than women with ischemic heart disease [1, 19]. There are a number of biological explanations for these phenomena. On the cellular level, women appear to be more resistant to cardiomyocyte loss in the context of MI and may be relatively protected from apoptosis and cell death as compared to men [20]. In subjects who died of fatal MI, peri-infarct apoptosis was greater in men than women [21] and women had a greater amount of myocardium at risk salvaged after acute MI with percutaneous coronary intervention [22]. In mouse models, female mice demonstrated a lesser degree of ventricular dilation and hypertrophy following acute MI than male mice [23]. While still under investigation, estrogens may mediate a large portion of the observed sex differences in response to acute coronary ischemia [24]. Acute myocarditis and dilated or idiopathic cardiomyopathy also exhibit a male predominance, demonstrating a male to female ratio of 1.3–1.5:1 and 1.5–1.7:1, respectively [25]. Sex-specific differences in gene expression, immune system functioning, remodeling, and inflammation have been proposed as potential etiologic explanations [25]. Hypertrophic cardiomyopathy and arrhythmogenic right ventricular cardiomyopathy also demonstrate a male predominance; however, these are outside the scope of this chapter.

Certain etiologies of HFrEF demonstrate a female sex predominance. Chemotherapy-related cardiotoxicity, primarily due to anthracyclines, disproportionately

TABLE 1 Sex-Specific Data on Prevalence of Risk Factors in HFrEF and Risk of Incident HFrEF

Risk Factor	Prevalence in Women (vs. Men) With HFrEF	Risk of Incident HFrEF in Women
Traditional Cardiovascular Disease Risk Factors		
HTN	↑	↑
Coronary heart disease	↓	≈
Myocardial infarction	↓	↑
Valvular heart disease	↑	↑
Atrial fibrillation/flutter	↓	≈
Other Risk Factors		
Older age	↑	↑
Diabetes	≈	↑
Obesity	↑	≈
COPD	↓	≈
Cigarette smoking (current or former)	↓	↑

≈ No difference in prevalence in women vs. men; risk factor not associated with increased risk of incident HFrEF vs. women without the risk factor.
COPD, chronic obstructive pulmonary disease; HFrEF, heart failure with reduced ejection fraction; HTN, hypertension.

affects women due to the prominent (although declining) role of anthracyclines in breast cancer [26], although sex may also be an additive risk factor independent of dose [27]. Takotsubo's (stress cardiomyopathy) is also more common in women, who represent 90% of patients in the International Takotsubo Registry [28]. The reasons for the uneven sex distribution in Takotsubo's may be related to excessive vasoconstriction, impaired endothelium-dependent vasodilation, and augmented sympathetic activation in response to mental stress [29].

Peripartum cardiomyopathy (PPCM) represents the one true sex-specific etiology of HF, affecting an estimated 1 in 3000 deliveries in the United States, with a higher incidence in African Americans, women > 30 years of age, and women with a history of pregnancy-associated HTN [30]. Although the etiology of PPCM remains uncertain, autoimmune inflammatory processes triggered by fetal or placental antigens are suspected to play a role [31], as are genetic and vascular factors. In a prospective cohort study of 100 women with PPCM followed through 1 year postpartum (the Investigations of Pregnancy-Associated Cardiomyopathy, or IPAC cohort), EF had recovered in 72% of PPCM patients by 1 year; however, 13% of patients had experienced major events or had persistent severe cardiomyopathy with EF < 35% at the end of follow-up [32].

Outcomes

Symptoms and Quality of Life

HFrEF poses significant physical and emotional burdens. Women consistently report more signs and symptoms of HF than men despite having more recently diagnosed HF, higher mean LVEF, and similar N-terminal-pro hormone B-type natriuretic peptide than men [16, 17]. Women with HFrEF also demonstrate worse HRQOL and increased disease severity as assessed by both the physician and the patient with large differences between women and men in their reported mobility, ability to undertake usual activities, and ability to participate in self-care [16, 17].

The reasons for these differences in symptoms and HRQOL are not clear, nor are they easily explained by major differences in physiologic markers of HF severity or comorbidities. Differences in hemodynamics may play a role, as women have been shown to exhibit higher end-diastolic pressures despite lower chamber volumes [33]. Studies have suggested that women may experience more perceived impairment due to less social support or more depression [34]. Excess social stressors, decreased access to care, and other socioeconomic factors may also be contributory, but more data are needed.

Hospitalization

HF hospitalization is a poor prognostic marker [35], and recent studies show that women with HFrEF may have a lower risk of HF hospitalization than men. Dewan and colleagues examined 3357 women enrolled in two large HFrEF trials: the Prospective Comparison of Angiotensin Receptor-Neprilysin Inhibitor with Angiotensin-converting Enzyme Inhibitor to Determine Impact on Global Mortality and Morbidity in Heart Failure (PARADIGM-HF) trial and

the Aliskiren Trial to Minimize Outcomes in Patients with Heart Failure (ATMOSPHERE). In this study, the risks of first hospitalization for HF (adjusted HR 0.80, 95% CI 0.72–0.89) and recurrent HF hospitalization (adjusted incidence rate ratio 0.69, 95% CI 0.61–0.79) were lower in women than in men [16]. Similarly, in a large cohort of 40,000 patients with HFrEF from the Swedish Heart Failure Registry, the risks of both cardiovascular and HF hospitalization were significantly lower in women [17]. The reasons for the disconnect between greater symptom burden and lower rates of hospitalization for women highlighted in these studies are unclear. As previously discussed, differences in access to health care, less caregiver support, less healthcare-seeking behaviors, and differential perceptions of disease may play roles. Alternatively, differential treatment in the clinic or emergency room (with fewer resultant admissions) may be contributory and requires fuller exploration (**Figure 1**).

Mortality

HFrEF mortality remains high, approaching 40% within 5 years of onset in a large community-based sample study [36]. In contrast to patients with non-HF forms of cardiovascular disease, women with HFrEF exhibit an independent survival benefit when compared to men [16, 37, 38]. In a pooled analysis of five randomized HFrEF trials, female sex was independently associated with significantly longer survival (adjusted HR 0.77, 95% CI 0.81–0.85), as was nonischemic etiology (HR 0.80, 95% CI 0.72–0.89) [37]. Dewan and colleagues also reported lower all-cause (adjusted HR 0.68, 95% CI 0.62–0.74) and cardiovascular (adjusted HR 0.70, 95% CI 0.63–0.77) mortality in women with HFrEF than men [16] (**Figure 2**).

There is no clear pathophysiologic explanation for the better survival observed among women with HFrEF, which persists even after adjustment for EF [37, 39]. Myocardial remodeling and response to stress may be more favorable in women [40]. In a registry of 927 patients with HFrEF, women experienced increased rates of reverse remodeling (defined as ≥15% reduction in LV end-systolic volume index) compared to men, irrespective of disease severity and

More ischemic cardiomyopathy

Greater risk for HF hospitalization

Higher all-cause and cardiovascular mortality

Represent >75% VAD and transplant patients

More signs/symptoms of HF

Worse quality of life

Less likely to be prescribed diuretics

More likely to be prescribed digoxin

Less likely to receive ICD or CRT

FIGURE 1 Sex-Specific Differences in HFrEF. CRT, cardiac resynchronization therapy; HF, heart failure; ICD, implantable cardioverter device; VAD, ventricular assist device.

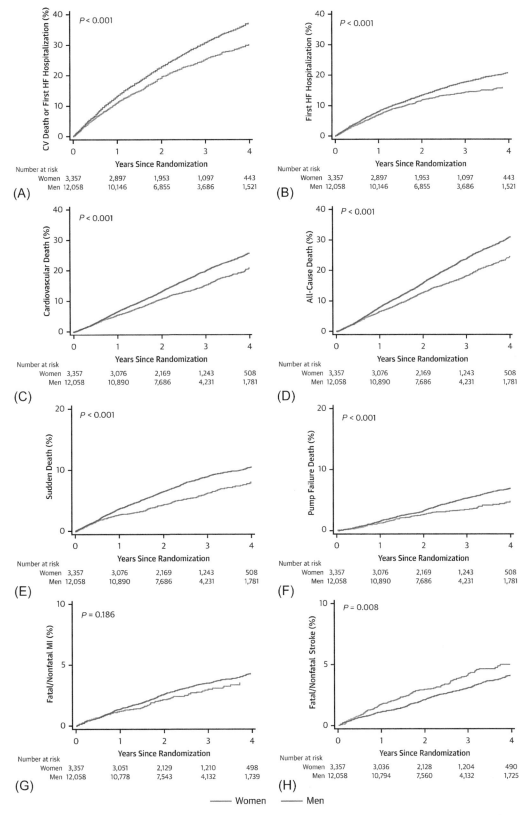

FIGURE 2 **Clinical Outcomes in Men and Women With Heart Failure With Reduced Ejection Fraction.** Cumulative event curves for (A) primary composite outcome, (B) hospitalization for HF, (C) cardiovascular death, (D) all-cause death (Kaplan-Meier), (E) sudden death, (F) pump failure or death, (G) fatal or nonfatal myocardial infarction, and (H) fatal or nonfatal stroke. The risk tables below the graphs show the numbers at risk of the event of interest. Women were at reduced risk for all outcomes except stroke when compared with men. All *P* values were unadjusted. CV, cardiovascular; HF, heart failure; MI, myocardial infarction. *Reproduced with permission from [16].*

etiology [41]. Women also have lower circulating plasma norepinephrine levels which are known to negatively impact ventricular remodeling [42]. Other theories under investigation include differential sex associations between biomarkers of inflammation (such as C-reactive protein and interleukin 6) and mortality [43], sex differences in gene expression profiles involving metabolism, fibrosis, and inflammation pathways [44], and sex differences in mitochondrial activity [44]. Finally, there appears to be an interaction between sex and outcomes in specific genetic cardiomyopathies with both titin (TTN) [45] and lamin A/C (LMNA) [46] mutations showing decreased disease severity and mortality for women.

Treatment

Women remain conspicuously under-enrolled in trials of HF (pharmacologic and device) therapies, many of which were underpowered to examine sex-specific differences and/or did not include prespecified sex-specific analyses (**Table 2**). An analysis of publicly available Food and Drug Administration (FDA) reviews evaluating 36 cardiovascular drug approvals from 2005 to 2015 demonstrated that women accounted for only 34% of trial participants [65]. Further, three of the included HF trials showed that the participation prevalence for women (the percentage of women in the trial divided by the percentage of women in the disease popula-

TABLE 2 Sex-Specific Analyses of Pharmacologic and Device Therapies for HFrEF

Therapy or Drug	Trial (Year)	Ejection Fraction (%)	Proportion of Females	Effects in Women
Beta-Blocker				
Carvedilol	Copernicus (2002) [47]	<25	20%	– ↓ Death or hospitalization for CV reason and death or hospitalization for HF
Bisoprolol	CIBIS-II (1999) [48]	≤35	19%	– ↓ All-cause mortality
Metoprolol succinate	MERIT-HF (1999) [49]	≤40	23%	– 21% ↓ in primary combined end point of all-cause mortality/all cause hospitalizations (164 vs. 137 pts.; *P*=0.044) – ↓ Number of CV hospitalizations by 29% – ↓ Number of HF hospitalizations by 42%
Carvedilol	US Carvedilol (1996) [50]	≤35	23%	– ↓ Mortality (HR 0.23; 95% CI 0.07–0.69)
ACE Inhibitor				
Enalapril	CONSENSUS (1987) [51]	NS	30%	– No sex-specific analyses reported
Enalapril	SOLVD (1991) [52]	≤35	20%	– No sex-specific analyses reported
Angiotensin Receptor Blocker				
Valsartan	Val-heft (2001) [53]	<40	20%	– No reduction in mortality – ↓ HF hospitalizations (HR 0.74; 95% CI 0.55–0.98)
Candesartan	CHARM Trials Pooled Analysis (2004) [54]	≤40	25%	– ↓ Combined end point of CV death or HF hospitalization
Mineralocorticoid Receptor Antagonists				
Spironolactone	RALES (1999) [55]	≤35	27%	– ↓ Mortality
Eplerenone	EMPHASIS-HF (2010) [56]	≤35	22%	– ↓ Combined end point of mortality and HF hospitalization

TABLE 2 Sex-Specific Analyses of Pharmacologic and Device Therapies for HFrEF—cont'd

Therapy or Drug	Trial (Year)	Ejection Fraction (%)	Proportion of Females	Effects in Women
Eplerenone	EPHESUS (2003) [57]	≤40	30%	– ↓ Mortality from any cause – No ↓ death from CV causes or hospitalization for CV events
Angiotensin-Receptor Neprilysin Inhibitor				
Sacubitril/ valsartan	PARADIGM-HF (2014)	≤40 ≤35	22%	– ↓ Primary end point of CV mortality and HF hospitalization, but driven by ↓ hospitalizations
Implantable Cardioverter-Defibrillators				
Primary and secondary prevention ICDs	Curtis et al. (2007) [58]	–	52% (primary prevention) 48% (secondary prevention)	– Women 3× less likely to receive ICD for primary prevention, 2.5× less likely for secondary prevention compared to men
Primary prevention ICDs	Ghanbari et al. (2009), meta-analysis [59]	–	20%	– No reduction in all-cause mortality with ICD in women
Primary prevention ICDs	Zeitler et al. (2015) [60]	≤35	26%	– Survival benefit seen in women age > 65 with primary prevention ICD (HR 0.79, 95% CI 0.66–0.95) – No interaction between sex and presence of ICD with respect to survival
Cardiac Resynchronization Therapy				
CRT	MADEIT-CRT [61]	≤30	25%	– ↓ Probability of VT/VF death (HR 0.62, $P < 0.001$) – Female sex independent predictor of CRT response
Ventricular Assist Devices				
LVAD	Bogaev et al. (2011) [62]	–	22%	– No survival differences on LVAD support between men and women at 18 months, but fewer women underwent OHT than men (40% vs. 55%, $P = 0.001$)
LVAD	Hsich et al. (2012) [63]	–	21%	– No mortality differences between men and women – Women with ↑ risk of neurologic event (HR 1.44, $P = 0.020$)
LVAD	Meteeren et al. (2017) [64]	–	20%	– No survival differences at 1, 3, or 5 years by sex – No differences in complications between sexes

CI, confidence interval; CRT, cardiac resynchronization therapy; CV, cardiovascular; HF, heart failure; HFrEF, HF with reduced ejection fraction; HR, hazard ratio; ICD, implantable cardioverter device; LVAD, left ventricular assist device; OHT, orthotopic heart transplantation; VF, ventricular fibrillation; VT, ventricular tachycardia.

tion based on a gender-stratified estimate in the literature) was 0.5 to 0.6, indicating underrepresentation relative to the disease population. Thus, the underenrollment of women in HF trials is not solely explained by the differences in disease prevalence.

Pharmacologic Treatments

Pharmacotherapy is the cornerstone of treatment for HFrEF and includes beta-blockers (BBs), renin-angiotensin-aldosterone system (RAAS) antagonists such as angiotensin-converting enzyme inhibitors (ACEIs) and angiotensin II receptor blockers (ARBs), mineralocorticoid receptor antagonists (MRAs), hydralazine and isordil in African Americans, and, recently, angiotensin receptor-neprilysin inhibitors (ARNIs). The goals of GDMT are to reduce morbidity and mortality. Due to the aforementioned lack of sex-specific data, GDMT should be applied uniformly for men and women until more studies are performed. It should be noted, however, that women have a higher risk of experiencing adverse drug reactions and more severe adverse drug reactions than men [66]. Drug responses in women may be affected by endogenous estrogen levels, which vary throughout a woman's life [67]. Further, a recent post-hoc analysis of a large prospective study in 11 European countries of male ($n = 1308$) and female ($n = 402$) patients with HFrEF demonstrated sex differences in dosing requirements in order to achieve lower hazard of death or hospitalization. In men, the lowest hazards of death or hospitalization for HF occurred at 100% of the recommended dose of ACEIs or ARBs and BBs, but women showed approximately 30% lower risk at only 50% of the recommended doses, with no further decrease in risk at higher dose levels [68]. Thus, an understanding that there are likely underinvestigated sex-differences in pharmacokinetics and pharmacodynamics when it comes to GDMT in HFrEF is necessary.

A summary of the sex-stratified analyses of major trials underlying the recommendations for BBs, RAAS antagonists, and MRAs in HFrEF is presented in **Table 2**. Sex-stratified analyses of early randomized controlled trials in BBs [47–50], ARBs [53, 54], and MRAs [55–57] showed that the benefits of these medications extend to women with HF, resulting in reductions in all-cause and cardiovascular hospitalizations and mortality in most trials. In contrast, early landmark trials in ACEIs (CONSENSUS [51] and SOLVD [52]) did not include sex-stratified analyses, limiting conclusions about the benefits of ACEIs in women. Two meta-analyses of ACEIs in HF demonstrated only a trend toward benefit from ACEIs; however, the cumulative sum of women in these analyses (~1500) was significantly lower than the number of patients included in the SOLVD trial alone, highlighting the lack of power to discern sex-specific treatment differences with currently available data [69, 70].

The African-American Heart Failure Trial (A-HEFT), enrolled 420 women (41%) and demonstrated that the combination of hydralazine and isordil improved survival and quality of life (QOL) and reduced hospitalizations for African American women who were already taking ACEIs, ARBs, or BBs [71]. The PARADIGM-HF trial [72] included 22% women ($n = 1832$) and demonstrated a reduction in the primary combined end point of cardiovascular (CV) mortality and HF hospitalization for women taking an ARNI as compared to an ACEI. Of note, this benefit in women was driven by a reduction in hospitalizations rather than mortality, the interpretation of this again being limited by small sample size.

Loop diuretics and digoxin have not been shown to improve mortality but do improve symptoms and morbidity, respectively. However, trial analyses show that women are less likely to be prescribed diuretics despite more evidence of congestion [16]. A post-hoc analysis of data from the Digitalis Investigation Group (DIG) study showed that digoxin therapy was associated with an increased risk of death among women, but not among men [73], likely due to higher serum digoxin concentrations in women [74]. Recent analyses including patients from the PARADIGM-HF and ATMOSPHERE trials [16], as well as from a Swedish HF registry [17], showed that women with HFrEF were more likely to be prescribed digoxin than men, highlighting the potential clinical relevance of the aforementioned understudied sex differences in pharmacokinetics and dynamics.

Device Therapies

Women are also underrepresented in device-based trials (**Figure 3**). Despite the life-saving capabilities of implantable cardioverter devices (ICDs), women make up less than a quarter of the population receiving these devices in randomized controlled trials and national registries [76]. In an analysis of Medicare claims data, women were three times less likely to receive an ICD for primary prevention and 2.5 times less likely to receive an ICD for secondary prevention compared to men [58]. In their analysis of PARADIGM-HF and ATMOSPHERE, Dewan and colleagues found that 8.6% of women received an ICD vs. 16.6% of men ($P < 0.0001$) [16]. Disparities in ICD implantation persisted even after adjustment for relevant covariates. The cause of these disparities is not known but may be related to the increased rates of device complications observed in women [76, 77] or disparities in counseling [78] whereby women are less likely to receive appropriate education about ICDs. Alternatively, although most individual trials and registries have not shown any sex differences in survival after ICD implantation, a 2009 meta-analysis of pooled data from five primary prevention trials showed no reduction in all-cause mortality among women who received ICDs (HR 1.01, 95%

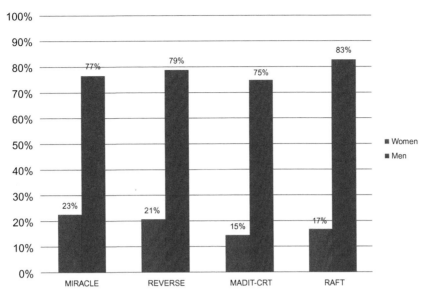

FIGURE 3 **Proportion of Women Compared With Men in Clinical Trials of CRT.** *Reproduced with permission from [75].*

CI 0.76–1.33) [59], although this finding has not been consistent across studies [60]. More studies are needed to better understand the risks and benefits of primary and secondary prevention ICD for women.

In contrast to ICDs, cardiac resynchronization therapy appears to provide women with a greater benefit than men [61], which may be due to the higher prevalence of left bundle branch block [16, 75, 79] and greater amounts of dyssynchrony seen for any given QRS length in women [75]. Despite what appears to be increased benefit with cardiac resynchronization therapy (CRT), women may undergo CRT implantation at lower rates than men. In Dewan and colleagues' study, 4.1% of women received CRT vs. 6.9% of men ($P < 0.0001$). Similarly, in the SwedeHF study, 3.6% of women received CRT (and/or ICD) vs. 7.3% of men ($P < 0.05$) [17]. In an analysis from the National Inpatient Sample Database of 311,009 patients undergoing CRT implantation in the United States between 2006 and 2012, men were significantly more likely to undergo cardiac resynchronization therapy-defibrillator (CRT-D) implant compared to women, and these differences increased significantly over the study period [80].

Nonpharmacologic/Nondevice-Based Treatments

Nonpharmacologic interventions, including lifestyle modifications, disease management programs, and cardiac rehabilitation, are all important components of HF treatment, and data suggest that women are underutilizing these aspects of HF care. Women with HFrEF are less likely to be enrolled in a disease management program than men [16]. Cardiac rehabilitation utilization remains low among women with heart disease in general, who are substantially less likely to be referred to a rehabilitation program [odds ratio (OR)=0.68] [81], to enroll in rehabilitation once referred (OR=0.73) [82], or to complete a full course of rehabilitation (OR=0.73) [83] as compared with men. Data from PARADIGM-HF demonstrated that women remain less likely to be prescribed an exercise regimen (15.0% vs. 18.1%; $P = 0.002$ [16]) in the modern era, despite prior studies showing that women derive a larger morbidity and mortality benefit [84]. Little is known about disparities in HF lifestyle counseling; however, a small study of 41 men and 27 women demonstrated that women had a better understanding of and were more adherent to a sodium-restricted diet (as measured by 24-h urine excretion) and were more educated on the signs and symptoms of fluid overload [85]. This study is limited by its small sample size, and more data are needed regarding gender and lifestyle modifications in HF.

VAD and Transplant

Ventricular assist device (VAD) implantation and orthotopic heart transplantation (OHT) are the only two therapies that prolong survival and improve QOL in eligible patients with advanced American College of Cardiology (ACC)/American Heart Association (AHA) stage D HF. Utilization of early generation pulsatile VADs in women was significantly limited due to the VADs' large size. The advent of smaller, continuous flow VAD, introduced in 2008, changed the landscape for all advanced HF patients and especially for women. However, despite the availability of smaller pumps, women continue to represent a vast minority of VAD patients in clinical trials (19–25%) as well as in real-world practice (21.1% from 2008 to 2011 and 21.3% from 2012 to 2017) [86, 87]. Further, important sex differences in regard to acuity and complications exist. At the time of implantation, women are more likely to be in cardiogenic shock, more likely to be supported by temporary mechanical circulatory support, and have worse hemodynamics [88]. After implantation, women require longer ventilatory and inotropic support resulting in prolonged intensive care stays [88] and have a higher incidence of neurologic complications (both hemorrhagic and ischemic) in multiple studies [89]. Women also report significantly more problems related to pain/discomfort and anxiety/depression at 3 and 6 months after implant [90]. It is unclear whether these gender differences are related to referral strategies, implant timing, or gender-specific outpatient care postimplantation. Despite these differences, multiple studies have demonstrated comparable survival between men and women in the contemporary era of continuous flow pumps [88, 89, 91].

Sex-specific differences in transplant exist as well. Women undergo OHT less frequently than men, representing only 23.6% of heart transplant recipients in the United States and 21% internationally [92]. In an analysis of 33,069 (25% women) patients on the heart transplant active wait-list from 2004 to 2015, rates of OHT were lower in women than in men listed as United Network for Organ Sharing (UNOS) status 1A but higher in women than in men listed as UNOS status 2 [93]. Women are not only more sensitized at baseline but also appear to experience a differential increase in sensitization with VAD implantation compared to men, creating an additional hurdle to OHT [94]. Not surprisingly, wait-list mortality showed an inverse relation with women demonstrating higher mortality when listed status 1A and lower mortality when listed status 2 than men [95, 96], although this higher risk may be improving over time [88, 97]. The higher wait-list mortality risk

for women listed 1A was present despite shorter waiting times and may be due to higher acuity at time of listing [96]. The improved wait-list survival for women listed status 2 was likely due to sex-based differences in peak oxygen consumption termed V02 (an often-used metric to decide on initiating transplant listing) as women demonstrate improved survival as compared to men at any given V02 [98]. Posttransplant, women have improved survival: median survival for women is 12.2 years as compared to 11.4 years for men ($P < 0.0001$) [99]. Women tend to have more rejection and exhibit more functional limitations and symptoms, whereas men develop more coronary allograft vasculopathy and malignancy [100].

Knowledge Gaps and Future Directions

Significant advances have been made in the understanding and treatment of HFrEF. However, sex-specific differences have emerged, which highlight a number of gaps in our understanding of how sex affects the epidemiology, pathophysiology, outcomes, and treatment of HFrEF in women. The role of social stressors, social support, and gender bias in the management of HF patients must also be further explored, with findings helping to guide medical education bias training. Mechanisms for increasing female enrollment in HF trials and a priori planning for sex-specific analyses should be implemented across future HFrEF investigations.

Conclusion

HFrEF is prevalent in women and, while associated with lower rates of hospitalization and better survival as compared to men, is also associated with more symptoms, worse functional status, and worse HRQOL. Disparities in treatment have narrowed but remain. Further investigation into the mechanisms underlying sex differences in presentation and outcome is warranted.

Key Points

1 Women account for 36% of patients with HFrEF, compared to 63% of HFpEF cases. They have lower incidence and lifetime risk of HFrEF than men.

2 Important risk factors for incident HFrEF in women include: age, history of myocardial infarction, and hypertension. When compared to men with HFrEF, women are older, more often obese, and are more likely to have preexisting hypertension than men, but less likely to have a previous myocardial infarction, coronary artery disease, or history of coronary revascularization.

3 Importantly, women with HFrEF have lower risk of HF hospitalization and mortality than men with HFrEF, but experience more signs and symptoms of HF and worse health-related quality of life. The reasons for these discrepancies are not entirely clear but may be due to a combination of biologic and psychosocial factors.

4 Women are underenrolled in trials of HF pharmacologic and device therapies. While most agents are proven to be effective in women, the insufficient number of women in many pivotal trials limits conclusions about drug effect. Nonetheless, due to the lack of sex-specific data, guideline-directed medical therapy should be prescribed uniformly for men and women.

5 Gender disparities in the prescription of most guideline-directed medical therapies have narrowed; however, women are still underprescribed diuretics, ICDs, and CRT.

6 Sex-specific differences in ventricular assist devices and heart transplantation therapy and outcomes are notable. Further research is warranted to understand how to improve gender disparities.

Back to Clinical Case

The clinical case describes a 72-year-old woman with symptomatic HFrEF. When compared to men with HFrEF, women are older and are more likely to be obese or hypertensive but are less likely to have coronary artery disease. Importantly, women experience more signs and symptoms of HF, more depressive symptoms, and worse health-related quality of life than men despite having more recently diagnosed HF and similar N-terminal-pro hormone B-type natriuretic peptide. Women are also less likely to be prescribed diuretics or device therapy. This patient has New York Heart Association (NYHA) class III heart failure, evidence of volume overload, and a class I indication for cardiac resynchronization therapy. Even though women are underenrolled in HF trials, pharmacologic and device therapies should be prescribed equally for both sexes. Thus, this patient was initiated on diuretic therapy and counseled about the risks and benefits of cardiac resynchronization therapy. She was referred to cardiac behavioral medicine for further management of her depressive symptoms and psychosocial functioning with regard to her heart failure. This case highlights the sex-specific differences in the presentation of HFrEF, including the increased prevalence of psychosocial disturbances among women with HFrEF, and in the prescription of guideline-directed treatments.

References

[1] Benjamin EJ, Virani SS, Callaway CW, Chamberlain AM, Chang AR, Cheng S, et al. Heart disease and stroke statistics—2018 update: a report from the American Heart Association. Circulation 2018;137(12):E67–E492.

[2] Bui AL, Horwich TB, Fonarow GC. Epidemiology and risk profile of heart failure. Nat Rev Cardiol 2010;8:30.

[3] Yancy CW, Jessup M, Bozkurt B, Butler J, Casey DE, Colvin MM, et al. 2017 ACC/AHA/HFSA focused update of the 2013 ACCF/AHA guideline for the management of heart failure: a report of the American College of Cardiology/American Heart Association Task Force on clinical practice guidelines and the Heart Failure Society of America. J Am Coll Cardiol 2017;70(6):776–803.

[4] Steinberg BA, Zhao X, Heidenreich PA, Peterson ED, Bhatt DL, Cannon CP, et al. Trends in patients hospitalized with heart failure and preserved left ventricular ejection fraction: prevalence, therapies, and outcomes. Circulation 2012;126(1):65–75.

[5] Chioncel O, Lainscak M, Seferovic PM, Anker SD, Crespo-Leiro MG, Harjola V-P, et al. Epidemiology and one-year outcomes in patients with chronic heart failure and preserved, mid-range and reduced ejection fraction: an analysis of the ESC Heart Failure Long-Term Registry. Eur J Heart Fail 2017;19(12):1574–85.

[6] van Riet EES, Hoes AW, Wagenaar KP, Limburg A, Landman MAJ, Rutten FH. Epidemiology of heart failure: the prevalence of heart failure and ventricular dysfunction in older adults over time. A systematic review. Eur J Heart Fail 2016;18(3):242–52.

[7] Ho JE, Enserro D, Brouwers FP, Kizer JR, Shah SJ, Psaty BM, et al. Predicting heart failure with preserved and reduced ejection fraction: the International Collaboration on Heart Failure Subtypes. Circ Heart Fail 2016;9(6).

[8] Gerber Y, Weston SA, Redfield MM, Chamberlain AM, Manemann SM, Jiang R, et al. A contemporary appraisal of the heart failure epidemic in Olmsted County, Minnesota, 2000 to 2010. JAMA Intern Med 2015;175(6):996–1004.

[9] Pandey A, Omar W, Ayers C, LaMonte M, Klein L, Allen NB, et al. Sex and race differences in lifetime risk of heart failure with preserved ejection fraction and heart failure with reduced ejection fraction. Circulation 2018;137(17):1814–23.

[10] Bleumink GS, Knetsch AM, Sturkenboom MCJM, Straus SMJM, Hofman A, Deckers JW, et al. Quantifying the heart failure epidemic: prevalence, incidence rate, lifetime risk and prognosis of heart failure: the Rotterdam study. Eur Heart J 2004;25(18):1614–9.

[11] Gong FF, Jelinek MV, Castro JM, Coller JM, McGrady M, Boffa U, et al. Risk factors for incident heart failure with preserved or reduced ejection fraction, and valvular heart failure, in a community-based cohort. Open Heart 2018;5(2):e000782.

[12] Eaton CB, Pettinger M, Rossouw J, Martin LW, Foraker R, Quddus A, et al. Risk factors for incident hospitalized heart failure with preserved versus reduced ejection fraction in a multiracial cohort of postmenopausal women. Circ Heart Fail 2016;9(10):e002883.

[13] Parikh NI, Lloyd-Jones DM, Ning H, Ouyang P, Polak JF, Lima JA, et al. Association of number of live births with left ventricular structure and function. The multi-ethnic study of atherosclerosis (MESA). Am Heart J 2012;163(3):470–6.

[14] Hall PS, Nah G, Howard BV, Lewis CE, Allison MA, Sarto GE, et al. Reproductive factors and incidence of heart failure hospitalization in the Women's Health Initiative. J Am Coll Cardiol 2017;69(20):2517–26.

[15] Rahman I, Akesson A, Wolk A. Relationship between age at natural menopause and risk of heart failure. Menopause 2015;22(1):12–6.

[16] Dewan P, Rørth R, Jhund PS, Shen L, Raparelli V, Petrie MC, et al. Differential impact of heart failure with reduced ejection fraction on men and women. J Am Coll Cardiol 2019;73(1):29–40.

[17] Stolfo D, Uijl A, Vedin O, Strömberg A, Faxén UL, Rosano GMC, et al. Sex-based differences in heart failure across the ejection fraction spectrum: phenotyping, and prognostic and therapeutic implications. JACC Heart Fail 2019;7(6):505–15.

[18] Ponikowski P, Voors AA, Anker SD, Bueno H, Cleland JGF, Coats AJS, et al. 2016 ESC guidelines for the diagnosis and treatment of acute and chronic heart failure: the task force for the diagnosis and treatment of acute and chronic heart failure of the European Society of Cardiology (ESC) developed with the special contribution of of the Heart Failure Association (HFA) of the ESC. Eur Heart J 2016;37(27):2129–200.

[19] Franke J, Lindmark A, Hochadel M, Zugck C, Koerner E, Keppler J, et al. Gender aspects in clinical presentation and prognostication of chronic heart failure according to NT-proBNP and the Heart Failure Survival Score. Clin Res Cardiol 2015;104(4):334–41.

[20] Guerra S, Leri A, Wang X, Finato N, Di Loreto C, Beltrami CA, et al. Myocyte death in the failing human heart is gender dependent. Circ Res 1999;85(9):856–66.

[21] Biondi-Zoccai GGL, Abate A, Bussani R, Camilot D, Giorgio FD, Marino M-PD, et al. Reduced post-infarction myocardial apoptosis in women: a clue to their different clinical course? Heart 2005;91(1):99–101.

[22] Mehilli J, Ndrepepa G, Kastrati A, Nekolla SG, Markwardt C, Bollwein H, et al. Gender and myocardial salvage after reperfusion treatment in acute myocardial infarction. J Am Coll Cardiol 2005;45(6):828–31.

[23] Cavasin MA, Tao Z, Menon S, Yang X-P. Gender differences in cardiac function during early remodeling after acute myocardial infarction in mice. Life Sci 2004;75(18):2181–92.

[24] Bouma W, Noma M, Kanemoto S, Matsubara M, Leshnower BG, Hinmon R, et al. Sex-related resistance to myocardial ischemia-reperfusion injury is associated with high constitutive ARC expression. Am J Physiol Circ Physiol 2010;298(5):H1510–7.

[25] Fairweather D, Cooper LT, Blauwet LA. Sex and gender differences in myocarditis and dilated cardiomyopathy. Curr Probl Cardiol 2013;38(1):7–46.

[26] Giordano SH, Lin Y-L, Kuo YF, Hortobagyi GN, Goodwin JS. Decline in the use of anthracyclines for breast cancer. J Clin Oncol 2012;30(18):2232–9.

[27] Dobbs NA, Twelves CJ, Gillies H, James CA, Harper PG, Rubens RD. Gender affects doxorubicin pharmacokinetics in patients with normal liver biochemistry. Cancer Chemother Pharmacol 1995;36(6):473–6.

[28] Templin C, Ghadri JR, Diekmann J, Napp LC, Bataiosu DR, Jaguszewski M, et al. Clinical features and outcomes of takotsubo (stress) cardiomyopathy. N Engl J Med 2015;373(10):929–38.

[29] Martin EA, Prasad A, Rihal CS, Lerman LO, Lerman A. Endothelial function and vascular response to mental stress are impaired in patients with apical ballooning syndrome. J Am Coll Cardiol 2010;56(22):1840–6.

[30] Elkayam U. Clinical characteristics of peripartum cardiomyopathy in the United States: diagnosis, prognosis, and management. J Am Coll Cardiol 2011;58(7):659–70.

[31] Gleicher N, Elkayam U. Peripartum cardiomyopathy, an autoimmune manifestation of allograft rejection? Autoimmun Rev 2009;8(5):384–7.

[32] McNamara DM, Elkayam U, Alharethi R, Damp J, Hsich E, Ewald G, et al. Clinical outcomes for peripartum cardiomyopathy in North America: results of the IPAC study (investigations of pregnancy-associated cardiomyopathy). J Am Coll Cardiol 2015;66(8):905–14.

[33] Cioffi G, Stefenelli C, Tarantini L, Opasich C. Prevalence, predictors, and prognostic implications of improvement in left ventricular systolic function and clinical status in patients > 70 years of age with recently diagnosed systolic heart failure. Am J Cardiol 2003;92(2):166–72.

[34] Bennett SJ, Perkins SM, Lane KA, Deer M, Brater DC, Murray MD. Social support and health-related quality of life in chronic heart failure patients. Qual Life Res 2001;10(8):671–82.

[35] Lin AH, Chin JC, Sicignano NM, Evans AM. Repeat hospitalizations predict mortality in patients with heart failure. Mil Med 2017;182(9–10):e1932–7.

[36] Tsao CW, Lyass A, Enserro D, Larson MG, Ho JE, Kizer JR, et al. Temporal trends in the incidence of and mortality associated with heart failure with preserved and reduced ejection fraction. JACC Heart Fail 2018;6(8):678–85.

[37] Frazier CG, Alexander KP, Newby LK, Anderson S, Iverson E, Packer M, et al. Associations of gender and etiology with outcomes in heart failure with systolic dysfunction: a pooled analysis of 5 randomized control trials. J Am Coll Cardiol 2007;49(13):1450–8.

[38] Tabassome S, Murielle M-K, Christian F-B, Patrice J. Sex differences in the prognosis of congestive heart failure. Circulation 2001;103(3):375–80.

[39] O'Meara E, Clayton T, McEntegart MB, McMurray JJV, Piña IL, Granger CB, et al. Sex differences in clinical characteristics and prognosis in a broad spectrum of patients with heart failure: results of the Candesartan in Heart failure: Assessment of Reduction in Mortality and morbidity (CHARM) program. Circulation 2007;115(24):3111–20.

[40] Piro M, Della Bona R, Abbate A, Biasucci LM, Crea F. Sex-related differences in myocardial remodeling. J Am Coll Cardiol 2010;55(11):1057–65.

[41] Aimo A, Vergaro G, Castiglione V, Barison A, Pasanisi E, Petersen C, et al. Effect of sex on reverse remodeling in chronic systolic heart failure. JACC Heart Fail 2017;5(10):735–42.

[42] Ghali JK, Krause-Steinrauf HJ, Adams KF, Khan SS, Rosenberg YD, Yancy CW, et al. Gender differences in advanced heart failure: insights from the BEST study. J Am Coll Cardiol 2003;42(12):2128–34.

[43] Meyer S, van der Meer P, van Deursen VM, Jaarsma T, van Veldhuisen DJ, van der Wal MH, et al. Neurohormonal and clinical sex differences in heart failure. Eur Heart J 2013;34(32):2538–47.

[44] De Bellis A, De Angelis G, Fabris E, Cannatà A, Merlo M, Sinagra G. Gender-related differences in heart failure: beyond the 'one-size-fits-all' paradigm. Heart Fail Rev 2019;25(2):245–55.

[45] Herman DS, Lam L, Taylor MRG, Wang L, Teekakirikul P, Christodoulou D, et al. Truncations of titin causing dilated cardiomyopathy. N Engl J Med 2012;366(7):619–28.

[46] van Rijsingen IAW, Nannenberg EA, Arbustini E, Elliott PM, Mogensen J, Hermans-van Ast JF, et al. Gender-specific differences in major cardiac events and mortality in lamin A/C mutation carriers. Eur J Heart Fail 2013;15(4):376–84.

[47] Packer M, Fowler MB, Roecker EB, Coats AJ, Katus HA, Krum H, et al. Effect of carvedilol on the morbidity of patients with severe chronic heart failure. Circulation 2002;106(17):2194–9.

[48] CIBIS-II Investigators and Committees. The cardiac insufficiency bisoprolol study II (CIBIS-II): a randomised trial. Lancet 1999;353(9146):9–13.

[49] Hjalmarson A, Goldstein S, Fagerberg B, Wedel H, Waagstein F, Kjekshus J, et al. Effect of metoprolol CR/XL in chronic heart failure: metoprolol CR/XL randomised intervention trial in-congestive heart failure (MERIT-HF). Lancet 1999;353(9169):2001–7.

[50] Packer M, Bristow MR, Cohn JN, Colucci WS, Fowler MB, Gilbert EM, et al. The effect of carvedilol on morbidity and mortality in patients with chronic heart failure. N Engl J Med 1996;334(21):1349–55.

[51] CONSENSUS Trial Study Group. Effects of enalapril on mortality in severe congestive heart failure. N Engl J Med 1987;316(23):1429–35.

[52] The SOLVD Investigators, Yusuf S, Pitt B, Davis CE, Hood WB, Cohn JN. Effect of enalapril on survival in patients with reduced left ventricular ejection fractions and congestive heart failure. N Engl J Med 1991;325(5):293–302.

[53] Cohn JN, Tognoni G, Valsartan Heart Failure Trial Investigators. A randomized trial of the angiotensin-receptor blocker valsartan in chronic heart failure. N Engl J Med 2001;345(23):1667–75.

[54] Young JB, Dunlap ME, Pfeffer MA, Probstfield JL, Cohen-Solal A, Dietz R, et al. Mortality and morbidity reduction with candesartan in patients with chronic heart failure and left ventricular systolic dysfunction. Circulation 2004;110(17):2618–26.

[55] Pitt B, Zannad F, Remme WJ, Cody R, Castaigne A, Perez A, et al. The effect of spironolactone on morbidity and mortality in patients with severe heart failure. N Engl J Med 1999;341(10):709–17.

[56] Zannad F, McMurray JJV, Krum H, van Veldhuisen DJ, Swedberg K, Shi H, et al. Eplerenone in patients with systolic heart failure and mild symptoms. N Engl J Med 2010;364(1):11–21.

[57] Pitt B, Remme W, Zannad F, Neaton J, Martinez F, Roniker B, et al. Eplerenone, a selective aldosterone blocker, in patients with left ventricular dysfunction after myocardial infarction. N Engl J Med 2003;348(14):1309–21.

[58] Curtis LH, Al-Khatib SM, Shea AM, Hammill BG, Hernandez AF, Schulman KA. Sex differences in the use of implantable cardioverter-defibrillators for primary and secondary prevention of sudden cardiac death. JAMA 2007;298(13):1517–24.

[59] Ghanbari H, Dalloul G, Hasan R, Daccarett M, Saba S, David S, et al. Effectiveness of implantable cardioverter-defibrillators for the primary prevention of sudden cardiac death in women with advanced heart failure: a meta-analysis of randomized controlled trials. JAMA Intern Med 2009;169(16):1500–6.

[60] Zeitler EP, Hellkamp AS, Fonarow GC, Hammill SC, Curtis LH, Hernandez AF, et al. Primary prevention implantable cardioverter-defibrillators and survival in older women. JACC Heart Fail 2015;3(2):159–67.

[61] Tompkins CM, Kutyifa V, Arshad A, McNitt S, Polonsky B, Wang PJ, et al. Sex differences in device therapies for ventricular arrhythmias or death in the multicenter automatic defibrillator implantation trial with cardiac resynchronization therapy (MADIT-CRT) trial. J Cardiovasc Electrophysiol 2015;26(8):862–71.

[62] Bogaev RC, Pamboukian SV, Moore SA, Chen L, John R, Boyle AJ, et al. Comparison of outcomes in women versus men using a continuous-flow left ventricular assist device as a bridge to transplantation. J Heart Lung Transplant 2011;30(5):515–22.

[63] Hsich EM, Naftel DC, Myers SL, Gorodeski EZ, Grady KL, Schmuhl D, et al. Should women receive left ventricular assist device support? Findings from INTERMACS. Circ Heart Fail 2012;5(2):234–40.

[64] van Meeteren J, Maltais S, Dunlay SM, Haglund NA, Beth Davis M, Cowger J, et al. A multi-institutional outcome analysis of patients undergoing left ventricular assist device implantation stratified by sex and race. J Heart Lung Transplant 2017;36(1):64–70.

[65] Scott PE, Unger EF, Jenkins MR, Southworth MR, McDowell T-Y, Geller RJ, et al. Participation of women in clinical trials supporting FDA approval of cardiovascular drugs. J Am Coll Cardiol 2018;71(18):1960–9.

[66] Tamargo J, Rosano G, Walther T, Duarte J, Niessner A, Kaski JC, et al. Gender differences in the effects of cardiovascular drugs. Eur Heart J Cardiovasc Pharmacother 2017;3(3):163–82.

[67] Taylor AL. Heart failure in women. Curr Heart Fail Rep 2015;12(2):187–95.

[68] Santema BT, Ouwerkerk W, Tromp J, Sama IE, Ravera A, Regitz-Zagrosek V, et al. Identifying optimal doses of heart failure medications in men compared with women: a prospective, observational, cohort study. Lancet 2019;394(10205):1254–63.

[69] Shekelle PG, Rich MW, Morton SC, Atkinson CSW, Tu W, Maglione M, et al. Efficacy of angiotensin-converting enzyme inhibitors and beta-blockers in the management of left ventricular systolic dysfunction according to race, gender, and diabetic status: a meta-analysis of major clinical trials. J Am Coll Cardiol 2003;41(9):1529–38.

[70] Garg R, Yusuf S, Collaborative Group on ACE Inhibitor Trials. Overview of randomized trials of angiotensin-converting enzyme inhibitors on mortality and morbidity in patients with heart failure. JAMA 1995;273(18):1450–6.

[71] Taylor AL, Lindenfeld J, Ziesche S, Walsh MN, Mitchell JE, Adams K, et al. Outcomes by gender in the African-American Heart Failure Trial. J Am Coll Cardiol 2006;48(11):2263–7.

[72] McMurray JJV, Packer M, Desai AS, Gong J, Lefkowitz MP, Rizkala AR, et al. Angiotensin–neprilysin inhibition versus enalapril in heart failure. N Engl J Med 2014;371(11):993–1004.

[73] Rathore SS, Wang Y, Krumholz HM. Sex-based differences in the effect of digoxin for the treatment of heart failure. N Engl J Med 2002;347(18):1403–11.

[74] Adams KF, Patterson JH, Gattis WA, O'Connor CM, Lee CR, Schwartz TA, et al. Relationship of serum digoxin concentration to mortality and morbidity in women in the digitalis investigation group trial: a retrospective analysis. J Am Coll Cardiol 2005;46(3):497–504.

[75] Nishimura M, Birgersdotter-Green U. Gender-based differences in cardiac resynchronization therapy response. Card Electrophysiol Clin 2019;11(1):115–22.

[76] Peterson PN, Daugherty SL, Wang Y, Vidaillet HJ, Heidenreich PA, Curtis JP, et al. Gender differences in procedure-related adverse events in patients receiving implantable cardioverter-defibrillator therapy. Circulation 2009;119(8):1078–84.

[77] Masoudi FA, Go AS, Magid DJ, Cassidy-Bushrow AE, Gurwitz JH, Liu TI, et al. Age and sex differences in long-term outcomes following implantable cardioverter-defibrillator placement in contemporary clinical practice: findings from the Cardiovascular Research Network. J Am Heart Assoc 2015;4(6):e002005.

[78] Hess PL, Hernandez AF, Bhatt DL, Hellkamp AS, Yancy CW, Schwamm LH, et al. Sex and race/ethnicity differences in implantable cardioverter-defibrillator counseling and use among patients hospitalized with heart failure. Circulation 2016;134(7):517–26.

[79] Bogaev RC. Gender disparities across the spectrum of advanced cardiac therapies: real or imagined? Curr Cardiol Rep 2016;18(11):108.

[80] Chatterjee NA, Borgquist R, Chang Y, Lewey J, Jackson VA, Singh JP, et al. Increasing sex differences in the use of cardiac resynchronization therapy with or without implantable cardioverter-defibrillator. Eur Heart J 2017;38(19):1485–94.

[81] Colella TJF, Gravely S, Marzolini S, Grace SL, Francis JA, Oh P, et al. Sex bias in referral of women to outpatient cardiac rehabilitation? A meta-analysis. Eur J Prev Cardiol 2014;22(4):423–41.

[82] Samayoa L, Grace SL, Gravely S, Scott LB, Marzolini S, Colella TJF. Sex differences in cardiac rehabilitation enrollment: a meta-analysis. Can J Cardiol 2014;30(7):793–800.

[83] Feola M, Garnero S, Daniele B, Mento C, Dell'Aira F, Chizzolini G, et al. Gender differences in the efficacy of cardiovascular rehabilitation in patients after cardiac surgery procedures. J Geriatr Cardiol 2015;12(5):575–9.

[84] Piña IL, Bittner V, Clare RM, Swank A, Kao A, Safford R, et al. Effects of exercise training on outcomes in women with heart failure: analysis of HF-ACTION (heart failure—a controlled trial investigating outcomes of exercise TraiNing) by sex. JACC Heart Fail 2014;2(2):180–6.

[85] Chung ML, Moser DK, Lennie TA, Worrall-Carter L, Bentley B, Trupp R, et al. Gender differences in adherence to the sodium-restricted diet in patients with heart failure. J Card Fail 2006;12(8):628–34.

[86] Mehra MR, Uriel N, Naka Y, Cleveland JC, Yuzefpolskaya M, Salerno CT, et al. A fully magnetically levitated left ventricular assist device—final report. N Engl J Med 2019;380(17):1618–27.

[87] Kormos RL, Cowger J, Pagani FD, Teuteberg JJ, Goldstein DJ, Jacobs JP, et al. The Society of Thoracic Surgeons Intermacs database annual report: evolving indications, outcomes, and scientific partnerships. Ann Thorac Surg 2019;107(2):341–53.

[88] Weymann A, Patil NP, Sabashnikov A, Mohite PN, García Sáez D, Amrani M, et al. Gender differences in continuous-flow left ventricular assist device therapy as a bridge to transplantation: a risk-adjusted comparison using a propensity score-matching analysis. Artif Organs 2015;39(3):212–9.

[89] Boyle AJ, Jorde UP, Sun B, Park SJ, Milano CA, Frazier OH, et al. Pre-operative risk factors of bleeding and stroke during left ventricular assist device support: an analysis of more than 900 heartmate II outpatients. J Am Coll Cardiol 2014;63(9):880–8.

[90] Grady KL, Wissman S, Naftel DC, Myers S, Gelijins A, Moskowitz A, et al. Age and gender differences and factors related to change in health-related quality of life from before to 6 months after left ventricular assist device implantation: findings from interagency registry for mechanically assisted circulatory support. J Heart Lung Transplant 2016;35(6):777–88.

[91] Goldstein DJ, Mehra MR, Naka Y, Salerno C, Uriel N, Dean D, et al. Impact of age, sex, therapeutic intent, race and severity of advanced heart failure on short-term principal outcomes in the MOMENTUM 3 trial. J Heart Lung Transplant 2018;37(1):7–14.

[92] OPTN/SRTR 2016 annual data report: preface. Am J Transplant 2018;18:1–9.

[93] Hsich EM, Thuita L, McNamara DM, Rogers JG, Valapour M, Goldberg LR, et al. Variables of importance in the Scientific Registry of Transplant Recipients database predictive of heart transplant waitlist mortality. Am J Transplant 2019;19(7):2067–76.

[94] Alba AC, Tinckam K, Foroutan F, Nelson LM, Gustafsson F, Sander K, et al. Factors associated with anti-human leukocyte antigen antibodies in patients supported with continuous-flow devices and effect on probability of transplant and post-transplant outcomes. J Heart Lung Transplant 2015;34(5):685–92.

[95] Hsich EM, Blackstone EH, Thuita L, McNamara DM, Rogers JG, Ishwaran H, et al. Sex differences in mortality based on united network for organ sharing status while awaiting heart transplantation. Circ Heart Fail 2017;10(6):e003635.

[96] Hsich EM, Starling RC, Blackstone EH, Singh TP, Young JB, Gorodeski EZ, et al. Does the UNOS heart transplant allocation system favor men over women? JACC Heart Fail 2014;2(4):347–55.

[97] Morris AA, Cole RT, Laskar SR, Kalogeropoulos A, Vega JD, Smith A, et al. Improved outcomes for women on the heart transplant wait list in the modern era. J Card Fail 2015;21(7):555–60.

[98] Hsich E, Chadalavada S, Krishnaswamy G, Starling RC, Pothier CE, Blackstone EH, et al. Long-term prognostic value of peak oxygen consumption in women versus men with heart failure and severely impaired left ventricular systolic function. Am J Cardiol 2007;100(2):291–5.

[99] Khush KK, Cherikh WS, Chambers DC, Harhay MO, Hayes D, Hsich E, et al. The International Thoracic Organ Transplant Registry of the International Society for Heart and Lung Transplantation: thirty-sixth adult heart transplantation report—2019; focus theme: donor and recipient size match. J Heart Lung Transplant 2019;38(10):1056–66.

[100] Hickey KT, Doering LV, Chen B, Carter EV, Sciacca RR, Pickham D, et al. Clinical and gender differences in heart transplant recipients in the NEW HEART study. Eur J Cardiovasc Nurs 2017;16(3):222–9.

Prevalence

■ Women
■ Men

HFrEF

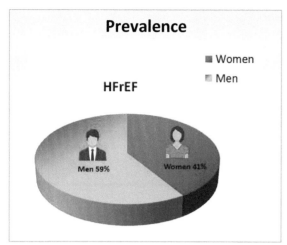

Men 59%   Women 41%

Lifetime risk of HFrEF by age and sex

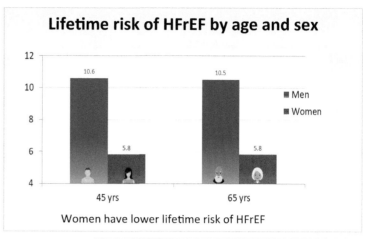

Women have lower lifetime risk of HFrEF

Etiology & Pathophysiology

Men
Ischemic cardiomyopathy
Myocarditis
ARVD

Women
Peripartum cardiomyopathy
Takotsubo cardiomyopathy
Chemo-induced cardiomyopathy
Duchenne muscular dystrophy

Women:
- Have less CAD & less likely to have ischemic CM
- More likely to have concentric rather than eccentric remodeling

Comorbidities

Diabetes: A more potent risk factor in women, results in LV remodeling.

Hypertension: Poorly controlled BP in women. Risk of HF with hypertension is 3X in women, & 2X in men.

Obesity: More prevalent in women

Smoking: Fewer women smoke. But 88% higher risk of HF in female smokers, vs 45% in males.

Socioeconomic status: Lower in women

Symptoms & Quality of Life

Women have:
- Increased symptom burden
- Decreased QOL scores
- Increased anxiety & depression
- Have higher NYHA class than men, regardless of EF

Class III-IV: Men 45%, women 48%

Trials & Prognosis

- Women are largely underrepresented in pharmacologic and device trials
- Women have fewer hospitalizations & lower mortality even after adjusting for EF

Management

- Women are less likely to be prescribed diuretics despite more congestion on exam
- More likely to be prescribed digoxin
- Less likely to receive ICD or CRT, and less likely to get appropriate ICD shocks
- Less likely to receive LV assist device or cardiac transplant

Niti Aggarwal

Sex differences in HFrEF. Women have a lower problems of HFrEF, more likely to have comorbidities less likely to have ischemia, have fewer hospitalizations and less likely to be prescribed CRT, ICD and mechanical devices compared to men. ARVD, arrhythmogenic right ventricular dysplasia; BP, blood pressure; CAD, coronary artery disease; CM, cardiomyopathy; CRT, cardiac resynchronization therapy; EF, ejection fraction; HF, heart failure; HFrEF, heart failure with reduced EF; ICD, implantable cardioverter defibrillator; LV, left ventricle; NYHA, New York Heart Association; QOL, quality of life. *Image courtesy of Niti R. Aggarwal.*

Chapter 11

Heart Failure With Preserved Ejection Fraction

Selma F. Mohammed, Niti R. Aggarwal, Ajith P. Nair, and Anita Deswal

Clinical Case

A 77-year-old woman with hypertension, persistent atrial fibrillation, on warfarin, with atherosclerotic peripheral arterial disease and carotid artery atherosclerosis, stage 3 chronic kidney disease, anemia, and mild obstructive sleep apnea, on continuous positive airway pressure (CPAP) presented with dyspnea on exertion and progressive effort intolerance over 2 months. Her cardiovascular risk factors include a 40-pack-year tobacco history. Her body mass index (BMI) was normal. She was normotensive and euvolemic on exam. Her cardiac examination revealed irregularly irregular heart rhythm and a holosystolic murmur in the left lower sternal border.

Her 12-lead electrocardiogram (ECG) showed atrial fibrillation and her echocardiogram revealed a normal left ventricular size and wall thickness, left ventricular ejection fraction of 65%, indeterminate diastolic function, severe biatrial dilation, normal right ventricular size with mild systolic dysfunction, and moderate tricuspid valve regurgitation. What is the differential diagnosis of this presentation, and how would you manage the patient next?

Sex Differences in Cardiac Disease. https://doi.org/10.1016/B978-0-12-819369-3.00019-8

CH
11

Abstract

Heart failure with preserved ejection fraction (HFpEF) is the most common type of heart failure (HF) in women and increases in prevalence with age. HFpEF is associated with multiple cardiovascular and noncardiovascular comorbidities. Diagnosis of HFpEF is challenging and effective disease-modifying therapies are lacking. HFpEF is associated with poor outcomes and is an area of unmet needs. In this chapter, we will review the epidemiology, pathophysiology, clinical presentation, diagnostic approach, prognosis, and management of patients with HFpEF with emphasis on differences by sex. We will also summarize key points and future directions for practicing clinicians and researchers in the field of HFpEF.

Epidemiology of HFpEF

Prevalence

Heart failure (HF) constitutes a substantial public-health burden with a prevalence rate of approximately 2.2% [1]. Approximately 6.2 million adult Americans are affected by HF and it is projected that by the year 2030, over 8 million people in the United States will have HF [1–3]. The rise in the burden of HF is attributable to the changes in population demographics and increased prevalence of HF risk factors [1–3].

Heart failure with preserved ejection fraction (HFpEF) accounts for approximately half of the prevalent HF (i.e., about 3.1 million adults in the United States). Estimates of HFpEF prevalence vary according to the population studied

and diagnostic criteria used to define HFpEF [4, 5]. The EPICA (Epidemiology of Heart Failure and Learning) observational cross-sectional study from Portugal reported point prevalence of HFpEF by age and sex [6]. In this Southwestern European community study, the prevalence of HFpEF was slightly higher than heart failure with reduced ejection fraction (HFrEF) (1.7% vs. 1.3%). The prevalence of HFpEF and HFrEF increased steadily with age in both men and women. At any given age, HFpEF was higher in women compared with men, whereas the opposite was true for HFrEF [7]. Importantly, the increase in HFpEF with age was steeper in women as compared with men [6]. Similarly, in patients hospitalized with HF, as demonstrated in the Get With The Guidelines-Heart Failure (GWTG-HF) registry, women comprised 41% and 68% of the HFrEF and HFpEF subjects, respectively [8] (**Figure 1**).

Incidence

Specific epidemiologic data on the incidence of HFpEF in the population are scarce due to limited availability of community data and/or inaccessibility of information on left ventricular ejection fraction (LVEF) at the time of HF diagnosis. Further, incidence studies are largely from North America. It is well established the incidence of HF and HF subtype varies depending on sex, age, race, and risk factor profile. Overall, roughly half of incident HF cases are accounted for by HFrEF and the other half by HFpEF [9, 10].

A community surveillance study from Olmsted County, Minnesota, USA reported the incidence of HFpEF by age and sex and trends in HF incidence in Olmsted County, Minnesota between the years 2000 and 2010 [9]. This study showed a substantial decline in the age- and sex-adjusted incidence rates of HF over time from the year 2000

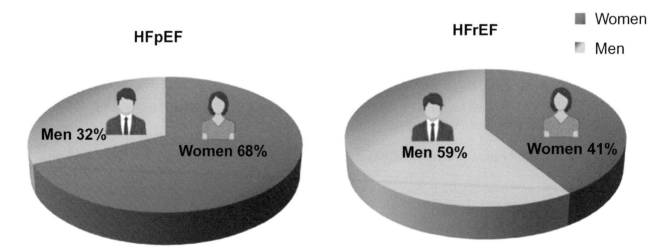

FIGURE 1 **Sex-Specific Prevalence of Heart Failure.** The figure demonstrates the sex-specific prevalence of HFpEF and HFrEF. Women are more likely to have HFpEF and men more likely to have HFrEF. HFpEF, heart failure with preserved ejection fraction; HFrEF, heart failure with reduced ejection fraction. *Data from [8].*

(315.8 per 100,000) to the year 2010 (219.3 per 100,000). Intriguingly, the decline in the incidence of HF over time was greater in women than men and for HFrEF than HFpEF [9]. Accordingly, the proportional incident HFpEF vs. HFrEF increased over time, signaling a shift toward HFpEF as the predominant incident HF [9]. A similar trend is also described by Owan et al. [4]. The reasons for these sex-time and HF-type interactions with time are unclear [9]. However, given the relatively homogenous population of Olmsted County with predominantly white race, older population, and higher education, these studies may not be completely generalizable to other communities, which may have younger individuals with more diversity.

Cumulative incidence of HFpEF was also studied in a community setting using three other prospective US cohorts: the Framingham Heart Study (FHS), the Cardiovascular Health Study (CHS), and the Prevention of Renal and Vascular Endstage Disease (PREVEND) [10]. The cumulative incidence of HF overall and relative incidence of HFpEF vs. HFrEF varied by cohort. Sex-specific incidence data were not provided; however, male sex was associated with a higher risk of HFrEF, but not HFpEF [10]. This study underscores the heterogeneity of HF incidence across different populations [10].

Lifetime Risk

Pandey et al. defined lifetime risk of HFpEF and HFrEF by sex and race in two prospective cohort studies [the CHS and the Multiethnic Study of Atherosclerosis (MESA)]. Among 12,417 individuals over 45 years of age, the lifetime risk of development of HFpEF was similar in men and women (10.4 and 10.7 at index age 45), whereas lifetime risk of HFrEF was higher in men than women (10.6 and 5.8 at index age 45) [11]. When analysis was stratified by sex, women had expectedly a higher lifetime risk of HFpEF compared with HFrEF (**Figure 2**) [11]. The lifetime risk of HFpEF is unchanged in women aged 45 and 65 years at 10.7, suggesting a trade-off between HFpEF and competing risk of mortality in older patients. Interestingly, blacks had a lower lifetime risk for HFpEF compared with nonblacks (7.7% vs. 11.2%); racial breakdown of risk by sex was not included in this analysis.

Collectively, these studies supported prior suggestions by others that after accounting for age and risk factors, the incidence and lifetime risk of development of HFpEF are similar in men and women [9, 10, 12].

Risk Factors and Comorbidities in HFpEF

Traditional and nontraditional cardiovascular disease risk factors predispose to HFpEF and influence its pathophysiology.

Risk Factors

Nonmodifiable risk factors for HFpEF include age, sex, race, and reproductive characteristics.

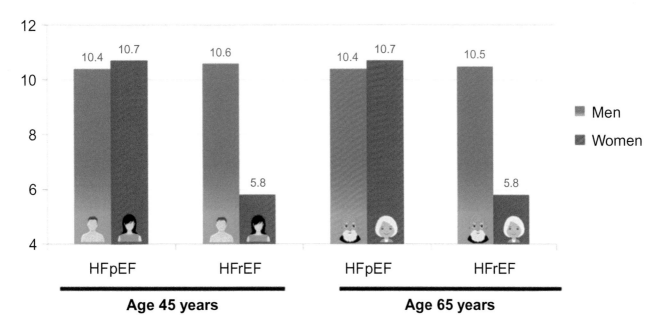

FIGURE 2   Lifetime Risk of Heart Failure by Age and Sex. Age and sex differences in lifetime risk of HFpEF and HFrEF from a multiethnic cohort. HFpEF, heart failure with preserved ejection fraction; HFrEF, heart failure with reduced ejection fraction. *Data from [11].*

Age

Age is a consistent potent risk factor for HFpEF in both men and women [10, 11]. Age-related effects on left ventricular geometry and remodeling, particularly hypertrophy, are more prominent in women compared with men [13].

Furthermore, increasing age is associated with increase in arterial and ventricular systolic and diastolic stiffness. Women have a steeper increase in arterial stiffness and left ventricular mass with age coupled with increased left ventricular systolic and diastolic stiffness [13–18]. The increase in left ventricular mass in women is not paralleled by increase in left ventricular volume, hence unfavorable geometric remodeling with concentric hypertrophy occurs in women [19]. This remodeling predisposes older women to development of HFpEF. Indeed, among a cohort of patients with HFpEF, women were older than men (71.4 ± 8.7 vs. 68.9 ± 9.6 years, $P < 0.001$) [20].

Race

Interaction of race and sex in patients with HFpEF is an area that is understudied. The lifetime risk of HFpEF is approximately 1.5-fold higher in nonblacks (11.2%) than blacks (7.7%); however, breakdown by race and sex is not reported [21]. In a subanalysis of the multiracial HF cohort from the Women's Health Initiative study (42,170 postmenopausal women), white (vs. African American or Hispanic) race was associated with higher incidence of HFpEF as well as HFrEF in postmenopausal women [22].

Reproductive Characteristics

Unique to women are reproductive characteristics and changes in sex hormone exposure during the physiological states of menstrual cycles, pregnancy, peripartum, and menopause.

Estrogen has beneficial cardiovascular effects that are exerted through enhancing the cyclic guanosine monophosphate (GMP) pathway (both the natriuretic peptide and the nitric oxide systems), counteracting the renin-angiotensin-aldosterone system, modulation of the calcium channel influx and activity in the sarcoplasmic reticulum, titin isoform switches and mechanical properties, inhibition of adverse extracellular matrix remodeling, and attenuation of oxidative stress (**Figure 3A and B**) [23–25]. Lack of estrogen has been shown to increase left ventricular (LV) hypertrophy, while hormone replacement therapy prevents development of LV hypertrophy post menopause [26]. In addition, estrogen supplementation or lower levels of circulating androgens have also been linked to higher levels of B-type natriuretic peptide (BNP) and atrial natriuretic peptide (ANP), which in turn may be cardioprotective [27] (**Figure 3A and B**). Declining estrogen levels in postmenopausal women may result in lower levels of natriuretic peptides, and may explain the higher sex-specific prevalence of HFpEF seen in this age group (**Figure 3A and B**).

Because of attenuation of the cardiovascular protective effects of estrogen, postmenopausal women are more prone to development of HFpEF. Among a cohort of 28,516 postmenopausal women enrolled in the Women's Health Initiative, a shorter total reproductive duration was associated with a higher incidence of HF [28], an association that was more apparent in women who experienced a natural menopause [28]. Intriguingly, nulliparity was also associated with incident HFpEF, although this may be because of residual confounding [28].

Comorbidities

Cardiovascular and noncardiovascular comorbidities are common in women at risk of or with HFpEF. The burden of comorbidities in HFpEF was studied in a community cohort from Olmsted County, Minnesota between 2000 and 2010 [29]. In this study, the authors found that nearly 86% of HF patients suffer from at least two extra chronic conditions. The most common comorbidity in HF regardless of LVEF or sex was hypertension [29]. Importantly, patients with HFpEF had one extra comorbidity compared with HFrEF; however, the number of comorbidities was similar among men and women with HFpEF (men: 4.5 and women: 4.4)

FIGURE 3 **Summary of the Likely Protective Mechanisms of Estrogen Against Cardiovascular Disease.** (A) The protective effect of estrogen in cardiovascular disease is associated with reduced fibrosis, stimulation of angiogenesis, and vasodilation, improved mitochondrial function, and reduced oxidative stress. (B) This schematic represents the regulatory mechanisms of estrogen (17b-estradiol [E2]) on biological processes contributing to the development of HFpEF. The decline of E2 at menopause might contribute to the pathogenesis of HFpEF. ACE, angiotensin-converting enzyme; ANG, angiotensin; AngII, angiotensin II; ANP, atrial natriuretic peptide; AT1, angiotensin II receptor type 1; BH4, tetrahydrobiopterin; BNP, B-type natriuretic peptide; CKD, chronic kidney disease; ECM, extracellular matrix; eNOS, endothelial nitric oxide synthase; ET-1, endothelin-1; FAO, fatty acid oxidation; HFpEF, heart failure with preserved ejection fraction; MMP2, matrix metalloproteinase 2; NFκB, nuclear factor kappa-light-chain-enhancer of activated B cells; NO, nitric oxide; RAAS, renin-angiotensin-aldosterone system; ROS, reactive oxygen species; TNF-α, tumor necrosis factor-alpha; VEGF, vascular endothelial growth factor. *Reproduced with permission from [23, 24].*

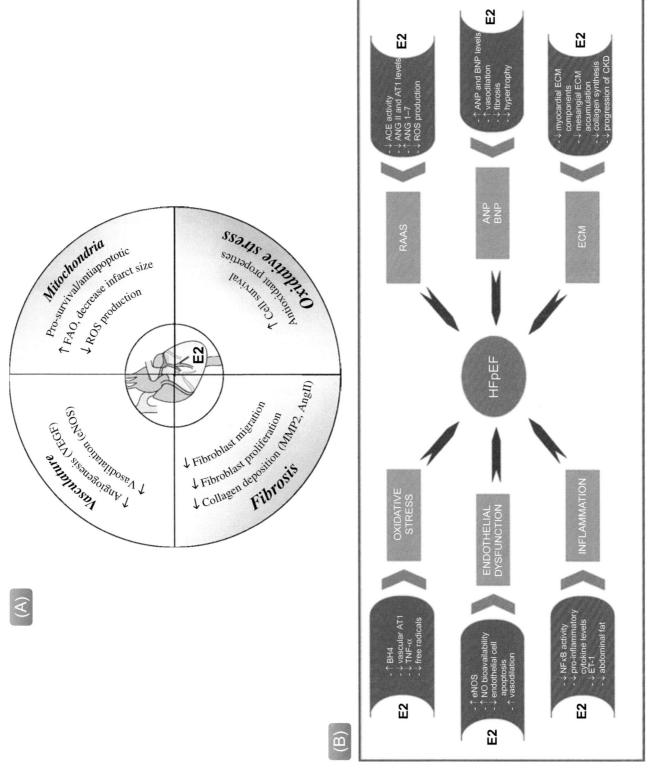

FIGURE 3 See legend on opposite page

CH
11

(**Figure 4**) [29]. The most common comorbidities were also similar in men and women: hypertension, hyperlipidemia, cardiac arrhythmias, coronary artery disease (CAD), arthritis, diabetes mellitus, and cancer [29]. The burden and type of comorbidities may differ in ethnically diverse populations (**Figure 5**).

Hypertension

Approximately a third of the adult US population is hypertensive [30]. At a population level, hypertension is less common in premenopausal women compared with men, but more common in postmenopausal women compared with men [30]. Hypertension is the most common comorbidity in HFpEF and is a major risk factor for HF irrespective of ejection fraction [1]. In patients with HFpEF in the Swedish HF registry, women were more likely to have hypertension compared to men (72% vs. 69%, $P < 0.005$) [31]. Hypertension portends a threefold increase in risk of HFpEF in women vs. a twofold increase in men. Sex differences exist in left ventricular adaptation to pressure overload with women with hypertension developing more concentric left ventricular hypertrophy (LVH) compared with men [32].

Obesity

Nearly 40% of the US population is obese, with equal prevalence in men and women [33]. Analysis from the Strong Heart Study, which included nearly 3000 individuals from three communities, showed that for any given BMI, women have greater adiposity compared with men [34]. Obesity influences cardiac remodeling and function; De Simone et al. found that left ventricular mass (unindexed) was lower in women compared with men and in nonobese women and

men compared with their sex-matched counterparts [34]. However, height- or fat-free mass indexed left ventricular mass was higher in women compared with men [34]. In response to obesity, women demonstrate a more pronounced increase in LV mass and wall thickness [35].

Obesity is also associated with diastolic dysfunction, with a greater impact on diastolic function in women [34, 36]. A large pooled cohort of about 22,000 participants from four community-based cohorts showed that among women, obesity is a strong risk factor for incident HFpEF [hazard ratio (HR): 1.38 per 1 standard deviation (SD) of BMI], but not for HFrEF [37]. Further, data from the Women's Health Initiative demonstrated a sex-race interaction, wherein obesity was a more prominent risk factor for HFpEF in African American women than white women [22]. Patients with HFpEF are more obese compared with patients with HFrEF [10], and among patients with HFpEF, obesity is more prevalent in women compared with men [38]; for example, the mean BMI was $29.8 \, kg/m^2$ in women vs. $28.7 \, kg/m^2$ in men, $P < 0.001$ in one study [20].

Diabetes Mellitus

Diabetes mellitus is prevalent in 10% of the US population (90–95% type II), with equal incidence and prevalence in women and men [39]. Diabetes mellitus preferentially predisposes women to development of HF. Data from the Framingham study showed that diabetes mellitus increases the risk of incident HF by fivefold in women and 2.4-fold in men [40], and that diabetes mellitus was strongly associated with adverse LV remodeling (LV mass and wall thickness) in women [41]. Approximately 45% of patients

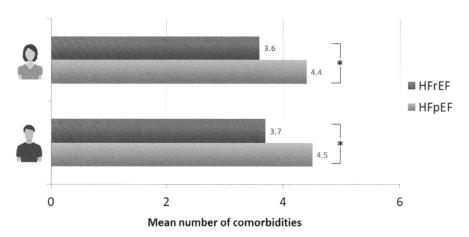

FIGURE 4   Multimorbidity Burden in Men and Women With Heart Failure. The figure illustrates the number of comorbidities developed within 5 years preceding the onset of heart failure. Men and women had similar numbers of chronic conditions. However, patients with HFpEF had on average a higher number of comorbidities compared to those with HFrEF. HFpEF, heart failure with preserved ejection fraction; HFrEF, heart failure with reduced ejection fraction. *$P < 0.001$. *Data from [29].*

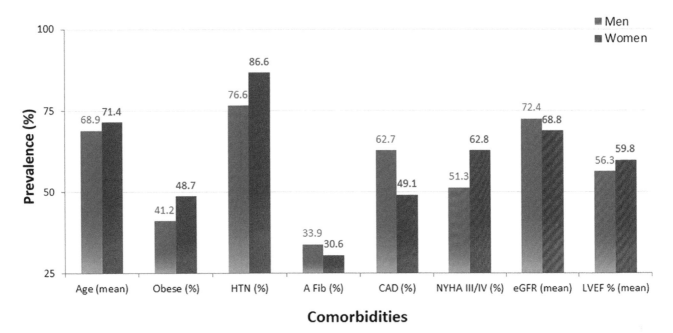

FIGURE 5 Prevalence of Sex-Specific Characteristics in Patients With HFpEF. Women who have HFpEF are older, more obese, more likely to have hypertension, and more likely to have preserved ejection fraction compared to age-matched men. In contrast, women are less likely to have atrial fibrillation, CAD and renal dysfunction. A Fib, atrial fibrillation; CAD, coronary artery disease; eGFR, estimated glomerular filtration rate; HFpEF, heart failure with preserved ejection fraction; HTN, hypertension; LVEF, left ventricular ejection fraction; NYHA, New York Heart Association class. All $P < 0.001$. *Data from [20].*

with HFpEF have diabetes mellitus [42]. However, it has also been observed that women with HFpEF are less likely to have diabetes mellitus [31]. More research is needed to understand differential sex-specific effects of diabetes mellitus in HFpEF.

Coronary Artery Disease

Compared with men, women are less likely to develop epicardial CAD, but are more likely to have microvascular disease. This was seen in the Women's Ischemia Syndrome Evaluation (WISE) study, wherein more than half of the women referred for angiography for suspected myocardial ischemia on stress testing had normal or insignificant CAD, consistent with microvascular coronary disease [43].

Epicardial CAD is present in approximately 35–60% of the HFpEF patients depending on the study [44–47]. Expectedly, women with HFpEF are less likely to have epicardial CAD (9.1% vs. 62.9%, $P < 0.001$) and less angina (41.1% vs. 48.7, $P < 0.001$) compared with men [20, 38, 48, 49]. Although less prevalent in women than in men, CAD remains a potent risk factor for HFpEF in women. In the MESA, an elevated coronary calcium score (a surrogate for CAD) correlated directly with increased prevalence of HFpEF in women [50]. In addition, coronary microvascular

rarefaction is common in HFpEF as seen in an autopsy study of patients with HFpEF [51]. Mechanisms by which CAD (epicardial or microvascular) may contribute to HFpEF include ischemia causing diastolic dysfunction and myocardial fibrosis.

Atrial Fibrillation

The risk of HFpEF in patients with atrial fibrillation did not differ by sex [52]. Data from the community showed that atrial fibrillation is present in 50–65% of the patients with HFpEF [53, 54], its prevalence is higher in men compared with women with HFpEF [54], it impairs exercise capacity, and it confers a poor prognosis [53, 55].

Risk Factors Specific to Women

Women with history of preeclampsia are predisposed to LV diastolic dysfunction and may have increased risk of HFpEF [56, 57]. Additionally, women who received radiation therapy for a history of breast cancer are at increased risk for HFpEF [58]. Although the exact mechanism for this is unknown, it is hypothesized that radiation contributes to development of coronary microvascular compromise, which in turn results in myocyte interstitial fibrosis, resulting in increased myocardial stiffness and diastolic function.

Pathophysiology of HFpEF

The pathophysiology of HFpEF is complex and intersects many entities (**Figure 6**). Some key pathophysiologic perturbations in HFpEF involve abnormalities in cardiac and vascular properties at rest and impaired reserve with exercise (**Figure 7**). These result in decreased cardiac output and/or increased LV filling pressures (**Figure 7**). Many of these pathophysiologic changes are more prominent in women and may explain the sex-specific predilection to HFpEF.

LV Diastolic Dysfunction

LV diastolic dysfunction is a cardinal feature of HFpEF and causes elevation of LV filling pressure and left atrial hypertension. Women have worse LV remodeling with concentric remodeling and hypertrophy compared to men who have eccentric remodeling [60]. Women with HFpEF have more LV diastolic dysfunction (both impaired relaxation and increased ventricular stiffness) and higher LV filling pressure, which appear to be intrinsic sex-related differences [38, 61]. Diastolic reserve with exercise [measured as pulmonary capillary wedge pressure (PCWP) to peak exercise workload or cardiac output] is also lower in women with HFpEF compared with men (0.8 vs. 0.6 mmHg/W, $P=0.001$) [59]. In fact, the odds of low diastolic function reserve (PCWP/cardiac output slope >2 mmHg/L/min) in women with HFpEF are approximately twofold higher compared with men [61]. Long-standing diastolic dysfunction results in left atrial hypertension and adverse remodeling, which further

contribute to the pathophysiology of HFpEF. This is particularly prominent in patients with atrial fibrillation [55].

LV Systolic Dysfunction

Although LVEF is normal in patients with HFpEF, these patients have impaired systolic function as assessed by global longitudinal strain [62]. Women with HFpEF have higher LVEF and more LV systolic stiffness (end systolic elastane) than men [48]. These mild abnormalities are exaggerated during exercise, whereby stroke volume and cardiac output reserve are diminished [59, 61, 63]. Lau et al. reported that peak exercise cardiac output is nearly 2 L lower in women compared with men [61].

Arterial Stiffness

HFpEF is characterized by increased arterial stiffness. Arterial stiffness (assessed by net afterload or effective arterial elastance) is higher in women with HFpEF compared with men with a steeper increase in arterial stiffness with age [38, 49, 64]. This is primarily due to increase in the pulsatile load [15]. Hence, ventricular-vascular coupling [arterial elastance (Ea)/end-systolic elastance (Ees) ratio] is more abnormal in women than men [65].

Pulmonary Hypertension and Right Ventricular Dysfunction

Pulmonary hypertension (PH) is present in approximately three-quarters of patients with HFpEF, with a higher

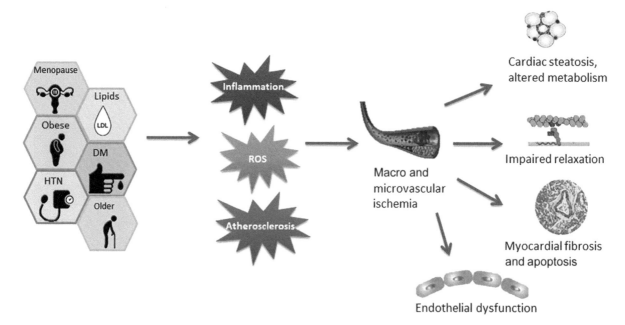

FIGURE 6   Pathophysiology of Heart Failure With Preserved Ejection Fraction (HFpEF) in Women. DM, diabetes mellitus; HTN, hypertension; ROS, reactive oxygen species.

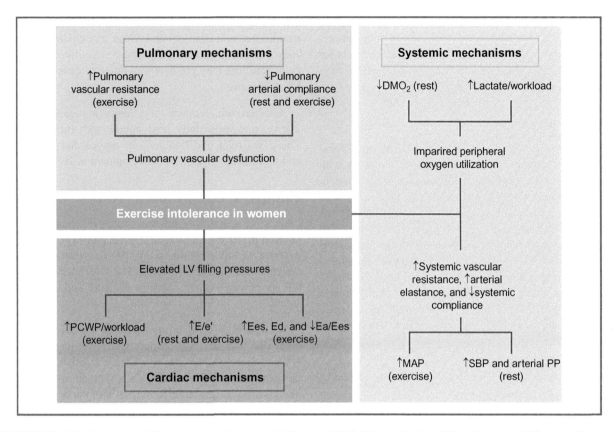

FIGURE 7 **Mechanisms of Exercise Intolerance in Women With Heart Failure With Preserved Ejection Fraction (HFpEF).** DMO_2, peripheral oxygen diffusion; Ea, arterial elastance; Ees, end-systolic elastance; LV, left ventricular; MAP, mean arterial pressure; PCWP, pulmonary capillary wedge pressure; PP, pulse pressure; SBP, systolic blood pressure. *Reproduced with permission from [59].*

prevalence in women (82%) vs. men (58%) [66]. PH manifests initially as postcapillary with later progression to combined precapillary and postcapillary PH in many patients. Women with HFpEF have lower pulmonary artery compliance than men [59].

Right ventricular (RV) dysfunction is present in about a third of HFpEF patients [67]. Interestingly, male sex is a predictor of RV dysfunction in HFpEF independent of PH [68]. This sex-specific association was also described in HFrEF, and animal studies suggested a role for testosterone on RV response to stress. Conversely, women have a lower RV reserve with exercise as measured by change in RV ejection fraction with exercise, despite similar pulmonary vascular reserve [pulmonary artery pressure/cardiac output (CO) slope] [61].

Coronary Microvascular Dysfunction

Coronary microvascular dysfunction is more common in women, whereas coronary epicardial disease is more common in men [43]. Both can contribute to diastolic dysfunction and HF via ischemia. In patients with HFpEF, coronary

flow reserve measured noninvasively by echo (echocardiogram) was present in 75% of patients without significant epicardial CAD and coronary microvascular dysfunction was associated with systemic endothelial dysfunction (reactive hyperemia index) [69]. Indeed, the severity of coronary microvascular rarefaction correlated with the amount of myocardial fibrosis in HFpEF [51]. Additionally, patients with microvascular dysfunction were found to have a five-fold increased risk of needing HFpEF hospitalization during a median follow-up of 4 years, highlighting the mechanistic importance of microvascular dysfunction [70].

Peripheral and Skeletal Musculature Abnormalities

Alterations in skeletal muscle structure (loss of muscle mass and/or increase in the interstitial space) and function (fiber type, mitochondrial dysfunction, and inefficient energy utilization) are common in HFpEF and lead to limited peripheral oxygen extraction with exercise [71]. Peripheral oxygen extraction is lower in women compared with men. The odds of having abnormal peripheral oxygen extraction are 2.5-fold higher in women compared with men, indicating more peripheral abnormalities in women [61]. Furthermore, women had a greater rise in mixed venous lactate levels for a given workload, reflecting impairments in peripheral oxygen consumption in patients with HFpEF.

Systemic Inflammation

Several years ago, Paulus and Tschöpe proposed a paradigm of comorbidity-driven systemic inflammation with resultant coronary endothelial microvascular inflammation, decreased nitric oxide bioavailability, and hence impaired protein kinase G signaling, leading to myocardial fibrosis and hypertrophy and cardiomyocyte stiffness [72]. Indeed, women have more microvascular CAD than men and cardiometabolic diseases in women are strongly associated with the development of HFpEF [37, 43]. Further, estrogen is considered immune enhancing, whereas testosterone is considered immunosuppressive [73]. These sex associations are modified by environmental factors, lifestyle, and stage of life; however, these data lend support to this paradigm of HFpEF and the sex differences described.

Chronotropic Incompetence

Chronotropic incompetence is prevalent in nearly 57% of patients with HFpEF [74]. This along with the reduced stroke volume reserve impairs the heart's ability to augment its cardiac output with exercise in patients with HFpEF. Women have a greater reliance on heart rate, given their lower LV stroke volumes compared to men. As a result, women are more sensitive to chronotropic reserve compared to men.

Atrial Fibrillation or Left Atrial Function

Atrial fibrillation is frequently seen in patients with HFpEF, and atrial dysfunction is associated with worse exercise capacity, worse pulmonary hypertension (PH), and higher all-cause mortality [53, 55]. Women have a larger left atrial volume index and lower left atrial ejection fraction than men in the setting of atrial fibrillation [75]. Although the risk of HFpEF in patients with atrial fibrillation did not differ by sex [52], atrial fibrillation was associated with a higher risk of hospitalization in women compared to age-matched men with HFpEF [76], suggesting a stronger implication of atrial fibrillation in women with HFpEF.

Molecular and Cellular Changes

In addition to organ-level changes, alterations at the cellular level may also drive the development of HFpEF. Intracellular calcium plays a key role in left ventricular relaxation.

According to murine models, female myocytes have smaller calcium currents and lower excitation-contraction coupling gain than men [77]. These sex differences in calcium handling may predispose women to developing more LV dysfunction.

Furthermore, women have lower myocardial glucose uptake, and more fatty acid uptake and utilization [78]. These sex-related differences in myocardial metabolism and inefficiency may affect development of HFpEF.

Clinical Presentation of HFpEF

Like HF in general, symptoms of HFpEF range from dyspnea on exertion or at rest, orthopnea, paroxysmal nocturnal dyspnea, or ankle edema to nonspecific symptoms such as cough, anorexia, confusion, etc. The most specific signs for HF are elevated jugular venous pressure and a third heart sound.

Compared to men, women usually are diagnosed at an older age, and are more symptomatic with more advanced NYHA class [31, 49, 79, 80]. This is consistent with cardiopulmonary exercise testing data, wherein women exhibited a lower exercise tolerance [61]. Women enrolled in the Phosphodiesterase-5 Inhibition to Improve Clinical Status and Exercise Capacity in Heart Failure with Preserved Ejection Fraction (RELAX) trial had a lower peak VO$_2$ than men (-1.88 mL/kg/min), even after adjustment for the latter's heavier weight [81]. Specifically, women had lower peripheral O$_2$ extraction with exercise as well as worse systolic and diastolic reserve in response to exercise compared with men [49, 61].

Women with acute decompensated HFpEF had a higher systolic blood pressure and were less likely to have peripheral edema. Women with HFpEF and acute HF also had a lower blood urea nitrogen [24 (17–35) vs. 26 (19–39) mg/dL, $P < 0.0001$], creatinine [1.2 (0.9–1.6) vs. 1.5 (1.1–2.0) mg/dL; $P < 0.0001$], and troponin T [0.02 (0.01–0.06) vs. 0.04 (0.02–0.07) µg/L; $P = 0.0002$] at presentation [82]. Women also had lower BNP on presentation [20].

Phenotypes of HFpEF

An analysis of the TOPCAT (Treatment of Preserved Cardiac Function Heart Failure with an Aldosterone Antagonist Trial) identified three HFpEF phenotypes. Phenotype 2 was older in age, had more women, had more concentric LV remodeling, more atrial fibrillation and left atrial enlargement, and less obesity and diabetes. Further, this phenotype was characterized by more arterial stiffness and elevated biomarkers of innate immunity and vascular calcification and portended an intermediate risk of mortality [83].

Diagnosis of HFpEF

HF is a clinical syndrome that is diagnosed based on a constellation of symptoms and signs compatible with decreased cardiac output and/or elevated filling pressures at rest or during exercise. This encompasses a wide spectrum of presentations ranging from exertional dyspnea to acute decompensated HF; making a diagnosis in the former setting is more challenging than the latter. In addition to clinical signs/symptoms, an LVEF ≥ 50% and cardiac structural and functional abnormalities (left atrial enlargement, diastolic dysfunction, and left ventricular remodeling) are suggested to make the diagnosis of HFpEF. Echocardiography is a widely available and accessible imaging modality that provides comprehensive assessment of cardiac structure and function. Cardiac magnetic resonance imaging (MRI) is the gold standard for quantitative assessment of mass and volumes. Further cardiac magnetic resonance (CMR) provides better myocardial tissue characterization with late gadolinium enhancement and T1 mapping. Additionally, CMR-derived assessment of diastolic functional parameters including measurement of transmitral velocities, myocardial tagging, left atrial volume, and LV mass is feasible [84] (**Figure 8**). This is particularly helpful in assessing ischemia and fibrosis as well as discerning infiltrative and inflammatory cardiomyopathies [84].

In addition to imaging, diagnosis of HFpEF is supported by elevated cardiac biomarkers [N-terminal pro B-type natriuretic peptide (NT-proBNP)]. In cases of diagnostic uncertainty, cardiopulmonary exercise testing, diastolic stress echocardiography or MRI testing, or invasive hemodynamic assessment should be pursued to confirm the diagnosis [85, 86].

The European Society of Cardiology (ESC) diagnostic criteria for HFpEF require: (1) symptoms/signs of HF,

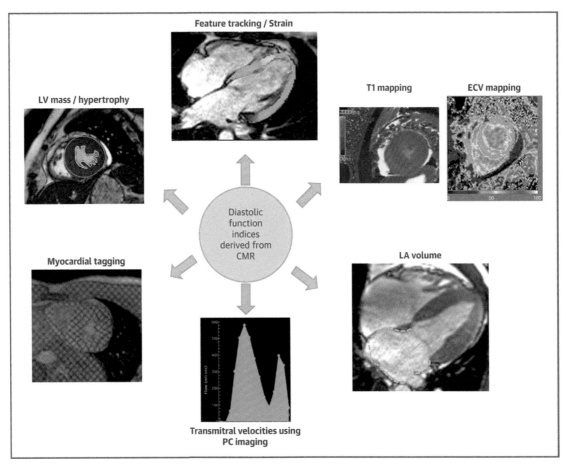

FIGURE 8   Assessment of Diastolic Function Using Cardiac MRI. Several left ventricular diastolic function indices can be derived from cardiac MRI including left ventricular mass, myocardial tracking, transmitral velocities, strain assessment, T1 mapping, and ECV mapping. CMR, cardiac magnetic resonance; ECV, extracellular volume; LA, left atrial; LV, left ventricular; PC, phase contrast. *Reproduced with permission from [84].*

(2) an LVEF $\geq$ 50%, (3) elevated levels of natriuretic peptides (NPs) (BNP > 35 pg/mL and/or NT-proBNP > 0.125 pg/mL), and (4) evidence of cardiac functional and structural alterations [86]. Further, the ESC 2016 guidelines recommend invasive measurement of LV filling pressure or stress test when there is diagnostic uncertainty [86]. The 2013 American College of Cardiology (ACC)/American Heart Association (AHA) guidelines do not include elevated biomarkers or cardiac structural abnormalities in the definition of HFpEF.

Neither the ESC nor the ACC/AHA guidelines recommend any sex-specific diagnostic criteria. However, it should be noted that women with HFpEF have more diastolic dysfunction, higher LVEF, and lower natriuretic peptide levels than men [7, 80]. Recently, the H2FPEF score (obesity, hypertension, atrial fibrillation, PH, age, and E/e' ratio) was developed to define the probability of HFpEF [87] (**Figure 9**). The odds of having HFpEF increased by a factor of 2 for every 1-unit increase in the score. Given the homogeneity of the population studied, external validation is required before wide application of this precise score.

Prognosis

Symptoms and Quality of Life

Exercise intolerance is the hallmark of the disease. The reduction in exercise capacity and impairment of quality of life (QoL) in the HFpEF population are similar to that of HFrEF, despite fundamental differences in ventricular and arterial structure and the extent of cardiac dysfunction [88].

Data from the Digitalis Investigation Group (DIG), the TOPCAT (Treatment of Preserved Cardiac Function Heart Failure With an Aldosterone Antagonist) trial, and the Swedish Heart Failure Registry showed that women with HFpEF are more symptomatic with more advanced New York Heart Association (NYHA) functional class than men [31, 79, 89]. In a sex-specific analysis of 378 patients with HFpEF from a prospective European, multicenter study of HFpEF (Karolinska-Rennes), Faxén et al. administered a battery of QoL assessments and found that women had worse scores in general, but disease-specific QoL in women was similar to men independent of age and disease severity [90]. A recent analysis of the Efficacy and Safety of LCZ696 Compared to Valsartan, on Morbidity and Mortality in Heart Failure Patients With Preserved Ejection Fraction (PARAGON-HF) trial, showed that women with HFpEF had worse QoL scores than men and female sex was one of the strongest predictors of lower QoL in HFpEF [91]. These findings were further confirmed in a cohort of HF patients, where women with HFpEF also had worse QoL scores [higher Minnesota Living With Heart Failure (MLWHF) questionnaire scores as compared to their male counterparts (median 44.0 in women compared to 37.0 in men, $P < 0.001$)]. Further, in HFpEF, the relationship between exercise capacity and QoL differs between men and women. An ancillary analysis of the RELAX and NEAT-HFpEF trials showed an association between 6-min walk test distance and QoL in men but not in women with HFpEF [92].

Hospitalizations

HF contributes to over a million annual primary hospitalizations in the United States (equally shared by HFpEF and HFrEF) and 1–2% of all hospitalizations [1, 93]. Throughout

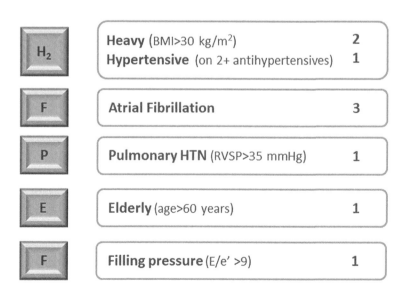

FIGURE 9 **H₂FPEF Score for Diagnosis of Heart Failure With Preserved Ejection Fraction (HFpEF).** BMI, body mass index; HTN, hypertension; RVSP, right ventricular systolic pressure. *Data from [87].*

TABLE 1 Hospitalizations in Heart Failure With Preserved Ejection Fraction (HFpEF) in Men and Women			
Outcome	Men (%)	Women (%)	HR (95% CI) Women vs. Men
Hospitalization	31.0	33.1	0.98 (0.91–1.05)

CI, confidence interval; HR, hazard ratio. *Data from [31].*

the course of HFpEF, approximately 34% of the subjects with HFpEF are hospitalized for acute HF exacerbation [94]. HF also prominently contributes to hospitalization burden in those hospitalized for non-HF indications [93].

Sex-specific data on HF hospitalization in patients with HFpEF have been inconsistent. In the GWTG-HF registry, patients with HFpEF had higher rates of all-cause readmission, whereas patients with HFrEF had higher rates of HF-specific readmission [95]. In contrast, the Irbesartan in Heart Failure with Preserved Ejection Fraction (I-PRESERVE) trial, women with HFpEF had a lower incidence of first all-cause hospitalization compared to men (184.3 in women vs. 231.4 in men per 1000 person-years) [38]. Similar results for women experiencing higher rates for HFpEF hospitalizations were confirmed in other studies [31, 48, 49]. These results were however not consistent, and other studies reported similar rates of HF hospitalizations among men and women with HFpEF (**Table 1**) [38, 79, 96].

Mortality

Mortality in patients with HFpEF is reported to be similar to or lower than mortality in HFrEF, depending on population demographics, LVEF cutoffs, and study setting [4, 5, 9, 95, 97–100]. Overall, HF portends a poor prognosis with approximately 20–25% mortality at 1 year, 50–75% mortality at 5 years, and 90% mortality at 10 years after HF diagnosis [4, 5, 9, 99].

Population-based studies from the FHS and Olmsted County, MN described mortality of HFpEF in the community [9, 99]. In a more contemporary analysis from Olmsted County, MN, Gerber et al. studied 2644 incident HF cases between 2000 and 2010 and described mortality of 20.2% at 1 year and 52.6% at 5 years after diagnosis. The risk of cardiovascular death was lower for HFpEF than for HFrEF [adjusted hazard ratio 0.79; 95% confidence interval (CI), 0.67–0.93]; however, sex-specific data were not provided [9]. An individual patient meta-analysis of 41,972 HF patients [of whom 10,347 had preserved ejection fraction (EF)] showed a 32% lower mortality for HFpEF compared with HFrEF at 3 years (23.4% vs. 26.3%). These results were consistent even after adjustment for age, sex, etiology, and comorbidities (adjusted hazard ratio 0.68; 95% CI, 0.64, 0.71) and were similar in men and women [100].

Notably, overall mortality rate in the Meta-Analysis Global Group in Chronic Heart Failure (MAGGIC) population was substantially lower than the community cohorts. This difference may be explained by inclusion of selective, relatively healthier patients in clinical trials [4, 5, 9, 95, 97–100]. In contrast, some reports have described a lower mortality for women compared to men with HFpEF (**Figure 10**) [31]. Stolfo et al. included 9957 patients with HFpEF enrolled in the Swedish Heart Failure Registry between 2000 and 2012 and found a lower risk of death in women with HFpEF compared to age-matched men (adjusted hazard ratio of 0.81) [31] (**Figure 10**). Similar findings were reported from the Candesartan in Heart failure: Assessment of Reduction in Mortality and morbidity (CHARM), and the I-PRESERVE and the DIG trials [38, 79, 96].

Cause of Death With HFpEF and HFrEF

The frequency of cardiovascular vs. noncardiovascular causes of death in HFpEF vary as expected given the heterogeneity of the population under study (**Figure 11**) [102]. Cardiovascular causes account for at least half of the deaths in HFpEF in the community, and a higher percentage (around 70–80%) in patients enrolled in clinical trials [38, 100–107]. In a sex-specific analysis of the Framingham cohort, 39% of deaths in men and 49% in women were due to a cardiovascular cause [101].

In the I-PRESERVE trial, cardiovascular deaths were more common than noncardiovascular deaths among both women and men (70.9% and 68.2% of all deaths, respectively), with sudden death being the most common cause of cardiovascular death [38]. Sudden cardiac death accounted for approximately one-fifth of the total deaths in HFpEF patients enrolled in the Americas region of the TOPCAT trial and followed over a median of 3 years. Interestingly, male sex was an independent predictor of sudden death or aborted cardiac arrest [108].

In summary, cardiovascular causes are the major causes of death in HFpEF, accounting for 60% of overall deaths in HFpEF. Approximately 40% of cardiovascular deaths in HFpEF are due to sudden death and a quarter to a third of deaths are due to progressive HF. Sex-specific data are needed to better understand and risk-stratify patients with HFpEF.

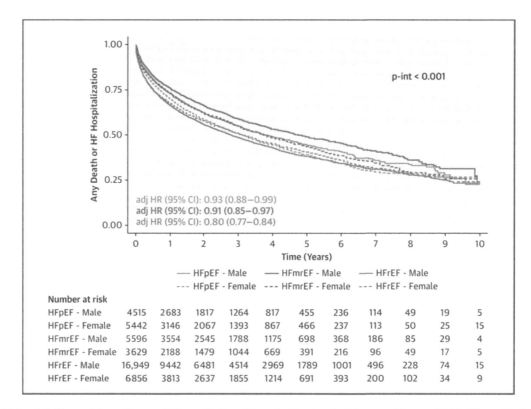

FIGURE 10   All-Cause Mortality and Heart Failure Hospitalization in Men vs. Women With Heart Failure. Figure depicts sex-specific Kaplan-Meier curves for time to all-cause mortality or heart failure hospitalization by heart failure phenotype. The unadjusted risk of mortality/HF hospitalization was higher in females than in males in HFpEF and HFmrEF patients, but lower in HFrEF patients. Adjusted risk of all-cause mortality/HF hospitalization was lower in females in all HF phenotypes, with increasing differences at lower EF. CI, confidence interval; HF, heart failure; HFmrEF, heart failure with midrange ejection fraction; HFpEF, heart failure with preserved ejection fraction; HFrEF, heart failure with reduced ejection fraction; HR, hazard ratio. *Reproduced with permission from [31].*

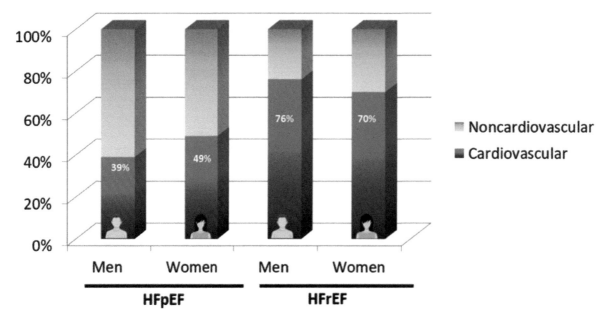

FIGURE 11   Cause of Death in Men and Women With Heart Failure. Men with HFrEF are more likely to die of cardiovascular causes compared to phenotype-matched women. In contrast, women with HFpEF are more likely to die of cardiovascular etiologies compared to men. HFpEF, heart failure with preserved ejection fraction; HFrEF, heart failure with reduced ejection fraction. *Data from [101].*

Management of HFpEF

In contrast to HFrEF, there are no proven disease-modifying therapies for HFpEF.

The HF guidelines mainly recommend use of diuretics to achieve euvolemia, blood pressure control, treatment of arrhythmias and ischemia, and management of comorbidities [86, 109]. Sex-based differences exist in access to therapy, pharmacokinetics/pharmacodynamics, and response to and outcomes after pharmacological therapy.

Management of Acute HFpEF

Data from the Acute Decompensated HF National Registry Emergency Module (ADHERE-EM), which included 83 hospitals in the United States (2004–2005), showed no difference in emergency department management of women with HFpEF (including diuretics, vasodilators, and vasoactive medications) [82].

Management of Chronic HFpEF

Beta-Blockers

Although beta-blockers are a mainstay for HFrEF, evidence for their use in HFpEF is limited. A prospective observational study found that women with HFpEF discharged on beta-blockers after hospitalization for decompensated HF experienced a higher rate of 6-month HF rehospitalization. Beta-blocker use remained an independent predictor of subsequent HF hospitalization in women, even after adjustment for covariates [OR 11 (2–62), $P=0.006$]. In contrast, no relationship between beta-blocker use and rehospitalization was found in men [110]. This difference may be explained on the basis of increased wave reflection and peripheral vascular resistance as a result of beta-blockade with resultant increase in arterial afterload in women more than men or the pharmacokinetics of β-blockers leading to increased drug exposure and thus more effects on heart rate and afterload in women compared with men [111]. In contrast, data from the Swedish HF Registry showed reduction of all-cause mortality with beta-blockers but not HF hospitalizations in HFpEF, without any sex differences [112]. To summarize, the evidence is controversial and does not endorse broad use of beta-blockers in women or men with HFpEF.

Renin-Angiotensin Blockers/Neprilysin Inhibitors

Large studies of angiotensin-converting enzyme (ACE) inhibitors and angiotensin receptor blockers (ARBs) in HFpEF have not shown any sex difference in response to therapy. In the CHARM program, there was no sex difference in response to candesartan [96]. Prespecified subgroup analysis of the largest HFpEF trial to date, the Efficacy and Safety of LCZ696 Compared to Valsartan, on Morbidity and Mortality in Heart Failure Patients With Preserved Ejection Fraction (PARAGON-HF) trial, patients with HFpEF were randomized to receive the angiotensin receptor neprilysin inhibitor, sacubitril-valsartan, or the ARB, valsartan, with a primary composite end point of total (first and recurrent) HF hospitalizations and cardiovascular death [80, 113]. Sex was one of the 12 prespecified subgroup analyses performed. Among the 4796 participants analyzed (52% women), the primary end point was not significantly different between the two groups; however, there was a sex interaction with a suggestion of benefit of sacubitril-valsartan over valsartan in women [113]. Subsequent ancillary analysis showed that the effect was predominantly related to reduction in HF hospitalizations and not in mortality. Further, the effect of sacubitril-valsartan was greater at the lower range of LVEF, but the threshold of benefit started at a higher LVEF value in women than men [80]. In response to sacubitril-valsartan, there was no sex difference in NYHA improvement; however, the self-reported physical function, symptoms, and QoL score (measured by the Kansas City Cardiomyopathy Questionnaire—Clinical Summary Score) were less in women vs. men [80]. The current AHA/ACC HF guidelines have a sex-neutral weak recommendation for use of ARBs to reduce hospitalization in HFpEF (Class IIb, level of evidence B) [109].

Mineralocorticoid Receptor Antagonists

Mineralocorticoid antagonists (MRAs) decrease myocardial fibrosis and prevent cardiac remodeling in patients with HFpEF. Sex differences in the effects of aldosterone on cardiac remodeling and clinical response to MRAs were assessed in a post-hoc analysis of the TOPCAT (Treatment of Preserved Cardiac Function Heart Failure With an Aldosterone Antagonist) trial, in the subset of 1767 patients enrolled from the Americas, which included 882 women. All-cause mortality was lower in women randomized to spironolactone vs. those randomized to placebo (hazard ratio: 0.66; $P=0.01$); however, no difference was noted in mortality in men randomized to spironolactone vs. placebo (P-value for sex interaction$=0.02$) [89]. These results suggest possible sex-specific responses to mineralocorticoid antagonism. In another sex-specific analysis of TOPCAT of the effect of spironolactone across the range of LVEF ($\geq 45\%$), women appeared to benefit across the whole LVEF spectrum of HFpEF, whereas men only benefited at the lower end of LVEF [114].

The current 2017 AHA/ACC HF guidelines again provide only a weak sex-neutral recommendation to consider use of mineralocorticoid receptor antagonists in selected patients (Class IIb, Level B-R) [86, 109].

Diuretics

The current AHA/ACC HF guidelines have a class I recommendation for use of diuretics for symptomatic relief of patients with HF and serve as the mainstay of therapy for patients with HFpEF [109]. Data regarding sex differences in diuretic utilization and management in chronic HFpEF are lacking.

Conclusion

HFpEF is the predominant form of HF in women. Sex differences and interactions in HFpEF are observed in epidemiology, pathophysiology, clinical characteristics, and morbidity outcomes, along with possible differential responses to certain therapies. Further, sex hormones may contribute to the pathogenesis of HFpEF. However, further research is needed to better understand if differential treatment by sex is needed to improve outcomes in these patients.

Existing Gaps in Knowledge and Future Directions

Several challenges exist and limit our understanding of sex differences in HFpEF. First, most of the studies that defined sex differences in HFpEF were retrospective. Prospective sex-specific studies on the epidemiology, pathophysiology, and response to therapy in HFpEF across various racial and ethnic groups are needed to advance our understanding of this disease. Longitudinal studies designed to evaluate the sex-specific outcomes and treatment of HFpEF are paramount. Second, it is controversial whether HFpEF is pathophysiologically heterogeneous (in contrast to HFrEF). Issues of defining whether the heterogeneity of HFpEF indicates different diseases or if there a unifying pathophysiologic process in all HFpEF patients and whether sex plays a role in this heterogeneity are key to identifying therapies for HFpEF. Third, women are underrepresented in HF trials. Despite better representation of women in more recent HFpEF trials including I-PRESERVE and TOPCAT (with >50% women), adequate representation of women in clinical trials of HFpEF represents a major hurdle to the field, wherein women are disproportionately affected. Finally, global data on HFpEF are lacking. More data from developing countries and women across various racial and ethnic groups are needed to inform and advance the field.

Key Points

1 Heart failure with preserved ejection fraction (HFpEF) is the most common subtype of heart failure in women and accounts for half of the cases of HF for men and women combined.

2 Multimorbidity is common in women with HFpEF (average of 4.4 comorbidities per patient), with hypertension representing the most common comorbidity.

3 Women with HFpEF are older, more obese, more likely to have hypertension, and less likely to have atrial fibrillation and coronary artery disease compared to men.

4 Women are have more systolic dysfunction (strain) abnormalities, despite higher LVEF compared to men.

5 Women with HFpEF demonstrate more substantial features of impaired diastolic reserve (E/e') with a greater rise in PCWP indexed to cardiac output and workload during exercise compared to men.

6 The pathophysiology of HFpEF is complex and involves an interplay of cardiac dysfunction, increased arterial stiffness, coronary microvascular dysfunction, and systemic inflammation.

7 In patients with undifferentiated dyspnea (particularly older women), cardiopulmonary exercise testing, stress echocardiography, cardiac MRI, and right heart catheterization with/without exercise are helpful diagnostic tools to define the etiology of dyspnea and rule in/rule out HFpEF.

8 HFpEF portends a grim prognosis with severely diminished quality of life, frequent hospitalizations, and high mortality (approximately 50% within 5 years of diagnosis depending on disease stage and comorbidities).

9 Women with HFpEF have similar rates of hospitalization, and slightly lower similar all-cause mortality, compared to age-matched men.

10 Emerging evidence suggests sex-specific responses to pharmacological therapies for HFpEF.

Back to Clinical Case

The patient presented in the clinical case was a typical patient with HFpEF (elderly woman with several cardiovascular comorbidities, with symptoms of progressive dyspnea on exertion, normal left ventricular ejection fraction, and nonobstructive coronary artery disease on coronary angiography). The differential diagnosis for this case is broad including CAD, HFpEF, specific cardiomyopathies (restrictive, infiltrative, and inflammatory), pericardial disease, pulmonary vascular disease, and pulmonary parenchymal disease. Coronary angiography demonstrated moderate nonobstructive CAD. Additional diagnostic workup for this patient included a cardiopulmonary exercise test that showed a low peak VO_2 with a normal pulmonary response to exercise and evidence of a cardiac output limitation. Diastolic function assessment was challenging on echocardiogram due to atrial fibrillation. She had no echocardiographic features of an infiltrative process or pericardial disease. Symptoms were likely secondary to HFpEF. In the absence of any trials demonstrating proven pharmacologic benefit, current recommendations include symptomatic management and optimization of comorbidity status. Lifestyle modifications were recommended, with emphasis on exercise and cardiorespiratory fitness to improve her functional capacity.

References

[1] Benjamin EJ, Muntner P, Alonso A, Bittencourt MS, Callaway CW, Carson AP, et al. Heart disease and stroke statistics—2019 update: a report from the American Heart Association. Circulation 2019;139:e56–e528.

[2] Heidenreich PA, Trogdon JG, Khavjou OA, Butler J, Dracup K, Ezekowitz MD, et al. Forecasting the future of cardiovascular disease in the United States: a policy statement from the American Heart Association. Circulation 2011;123:933–44.

[3] Heidenreich PA, Albert NM, Allen LA, Bluemke DA, Butler J, Fonarow GC, et al. Forecasting the impact of heart failure in the United States: a policy statement from the American Heart Association. Circ Heart Fail 2013;6:606–19.

[4] Owan TE, Hodge DO, Herges RM, Jacobsen SJ, Roger VL, Redfield MM. Trends in prevalence and outcome of heart failure with preserved ejection fraction. N Engl J Med 2006;355:251–9.

[5] Bhatia RS, Tu JV, Lee DS, Austin PC, Fang J, Haouzi A, et al. Outcome of heart failure with preserved ejection fraction in a population-based study. N Engl J Med 2006;355:260–9.

[6] Ceia F, Fonseca C, Mota T, Morais H, Matias F, de Sousa A, et al. Prevalence of chronic heart failure in Southwestern Europe: the EPICA study. Eur J Heart Fail 2002;4:531–9.

[7] Harada E, Mizuno Y, Kugimiya F, Shono M, Maeda H, Yano N, et al. Sex differences in heart failure with preserved ejection fraction reflected by B-type natriuretic peptide level. Am J Med Sci 2018;356:335–43.

[8] Shah KS, Xu H, Matsouaka RA, Bhatt DL, Heidenreich PA, Hernandez AF, et al. Heart failure with preserved, borderline, and reduced ejection fraction: 5-year outcomes. J Am Coll Cardiol 2017;70:2476–86.

[9] Gerber Y, Weston SA, Redfield MM, Chamberlain AM, Manemann SM, Jiang R, et al. A contemporary appraisal of the heart failure epidemic in Olmsted County, Minnesota, 2000 to 2010. JAMA Intern Med 2015;175:996–1004.

[10] Ho JE, Enserro D, Brouwers FP, Kizer JR, Shah SJ, Psaty BM, et al. Predicting heart failure with preserved and reduced ejection fraction: the international collaboration on heart failure subtypes. Circ Heart Fail 2016;9.

[11] Pandey A, Omar W, Ayers C, LaMonte M, Klein L, Allen NB, et al. Sex and race differences in lifetime risk of heart failure with preserved ejection fraction and heart failure with reduced ejection fraction. Circulation 2018;137:1814–23.

[12] Ho JE, Lyass A, Lee DS, Vasan RS, Kannel WB, Larson MG, et al. Predictors of new-onset heart failure: differences in preserved versus reduced ejection fraction. Circ Heart Fail 2013;6:279–86.

[13] Lieb W, Xanthakis V, Sullivan LM, Aragam J, Pencina MJ, Larson MG, et al. Longitudinal tracking of left ventricular mass over the adult life course: clinical correlates of short- and long-term change in the Framingham offspring study. Circulation 2009;119:3085–92.

[14] Hayward CS, Kelly RP. Gender-related differences in the central arterial pressure waveform. J Am Coll Cardiol 1997;30:1863–71.

[15] Redfield MM, Jacobsen SJ, Borlaug BA, Rodeheffer RJ, Kass DA. Age- and gender-related ventricular-vascular stiffening: a community-based study. Circulation 2005;112:2254–62.

[16] Cheng S, Xanthakis V, Sullivan LM, Lieb W, Massaro J, Aragam J, et al. Correlates of echocardiographic indices of cardiac remodeling over the adult life course: longitudinal observations from the Framingham Heart Study. Circulation 2010;122:570–8.

[17] Coutinho T, Borlaug BA, Pellikka PA, Turner ST, Kullo IJ. Sex differences in arterial stiffness and ventricular-arterial interactions. J Am Coll Cardiol 2013;61:96–103.

[18] Okura H, Takada Y, Yamabe A, Kubo T, Asawa K, Ozaki T, et al. Age- and gender-specific changes in the left ventricular relaxation: a Doppler echocardiographic study in healthy individuals. Circ Cardiovasc Imaging 2009;2:41–6.

[19] Subramanya V, Zhao D, Ouyang P, Lima JA, Vaidya D, Ndumele CE, et al. Sex hormone levels and change in left ventricular structure among men and post-menopausal women: the Multi-Ethnic Study of Atherosclerosis (MESA). Maturitas 2018;108:37–44.

[20] Dewan P, Rorth R, Raparelli V, Campbell RT, Shen L, Jhund PS, et al. Sex-related differences in heart failure with preserved ejection fraction. Circ Heart Fail 2019;12, e006539.

[21] Pandey A, Parashar A, Kumbhani D, Agarwal S, Garg J, Kitzman D, et al. Exercise training in patients with heart failure and preserved ejection fraction: meta-analysis of randomized control trials. Circ Heart Fail 2015;8:33–40.

[22] Eaton CB, Pettinger M, Rossouw J, Martin LW, Foraker R, Quddus A, et al. Risk factors for incident hospitalized heart failure with preserved versus reduced ejection fraction in a multiracial cohort of postmenopausal women. Circ Heart Fail 2016;9.

[23] Iorga A, Cunningham CM, Moazeni S, Ruffenach G, Umar S, Eghbali M. The protective role of estrogen and estrogen receptors in cardiovascular disease and the controversial use of estrogen therapy. Biol Sex Differ 2017;8:33.

[24] Sabbatini AR, Kararigas G. Menopause-related estrogen decrease and the pathogenesis of HFpEF: JACC review topic of the week. J Am Coll Cardiol 2020;75:1074–82.

[25] Li S, Gupte AA. The role of estrogen in cardiac metabolism and diastolic function. Methodist Debakey Cardiovasc J 2017;13:4–8.

[26] Pedram A, Razandi M, Lubahn D, Liu J, Vannan M, Levin ER. Estrogen inhibits cardiac hypertrophy: role of estrogen receptor-beta to inhibit calcineurin. Endocrinology 2008;149:3361–9.

[27] Lam CS, Cheng S, Choong K, Larson MG, Murabito JM, Newton-Cheh C, et al. Influence of sex and hormone status on circulating natriuretic peptides. J Am Coll Cardiol 2011;58:618–26.

[28] Hall PS, Nah G, Howard BV, Lewis CE, Allison MA, Sarto GE, et al. Reproductive factors and incidence of heart failure hospitalization in the Women's Health Initiative. J Am Coll Cardiol 2017;69:2517–26.

[29] Chamberlain AM, St Sauver JL, Gerber Y, Manemann SM, Boyd CM, Dunlay SM, et al. Multimorbidity in heart failure: a community perspective. Am J Med 2015;128:38–45.

[30] Ramirez LA, Sullivan JC. Sex differences in hypertension: where we have been and where we are going. Am J Hypertens 2018;31:1247–54.

[31] Stolfo D, Uijl A, Vedin O, Stromberg A, Faxen UL, Rosano GMC, et al. Sex-based differences in heart failure across the ejection fraction spectrum: phenotyping, and prognostic and therapeutic implications. JACC Heart Fail 2019;7:505–15.

[32] Levy D, Larson MG, Vasan RS, Kannel WB, Ho KK. The progression from hypertension to congestive heart failure. JAMA 1996;275:1557–62.

[33] Hales CM, Carroll MD, Fryar CD, Ogden CL. Prevalence of obesity among adults and youth: United States, 2015–2016. NCHS Data Brief 2017;(288):1–8.

[34] De Simone G, Devereux RB, Chinali M, Roman MJ, Barac A, Panza JA, et al. Sex differences in obesity-related changes in left ventricular morphology: the Strong Heart Study. J Hypertens 2011;29:1431–8.

[35] Kuch B, Muscholl M, Luchner A, Doring A, Riegger GA, Schunkert H, et al. Gender specific differences in left ventricular adaptation to obesity and hypertension. J Hum Hypertens 1998;12:685–91.

[36] Kim HL, Kim MA, Oh S, Kim M, Park SM, Yoon HJ, et al. Sex difference in the association between metabolic syndrome and left ventricular diastolic dysfunction. Metab Syndr Relat Disord 2016;14:507–12.

[37] Savji N, Meijers WC, Bartz TM, Bhambhani V, Cushman M, Nayor M, et al. The association of obesity and cardiometabolic traits with incident HFpEF and HFrEF. JACC Heart Fail 2018;6:701–9.

[38] Lam CS, Carson PE, Anand IS, Rector TS, Kuskowski M, Komajda M, et al. Sex differences in clinical characteristics and outcomes in elderly patients with heart failure and preserved ejection fraction: the Irbesartan in Heart Failure with Preserved Ejection Fraction (I-PRESERVE) trial. Circ Heart Fail 2012;5:571–8.

[39] Hales CM, Carroll MD, Fryar CD, Ogden CL. National Diabetes Statistics Report, 2017: estimates of diabetes and its burden in the United States; 2017.

[40] Kannel WB, Hjortland M, Castelli WP. Role of diabetes in congestive heart failure: the Framingham study. Am J Cardiol 1974;34:29–34.

[41] Galderisi M, Anderson KM, Wilson PW, Levy D. Echocardiographic evidence for the existence of a distinct diabetic cardiomyopathy (the Framingham Heart Study). Am J Cardiol 1991;68:85–9.

[42] McHugh K, DeVore AD, Wu J, Matsouaka RA, Fonarow GC, Heidenreich PA, et al. Heart failure with preserved ejection fraction and diabetes: JACC state-of-the-art review. J Am Coll Cardiol 2019;73:602–11.

[43] Sharaf BL, Pepine CJ, Kerensky RA, Reis SE, Reichek N, Rogers WJ, et al. Detailed angiographic analysis of women with suspected ischemic chest pain (pilot phase data from the NHLBI-sponsored Women's Ischemia Syndrome Evaluation [WISE] Study Angiographic Core Laboratory). Am J Cardiol 2001;87:937–41. A3.

[44] Shah SJ. Evolving approaches to the management of heart failure with preserved ejection fraction in patients with coronary artery disease. Curr Treat Options Cardiovasc Med 2010;12:58–75.

[45] Hwang SJ, Melenovsky V, Borlaug BA. Implications of coronary artery disease in heart failure with preserved ejection fraction. J Am Coll Cardiol 2014;63:2817–27.

[46] Mentz RJ, Broderick S, Shaw LK, Fiuzat M, O'Connor CM. Heart failure with preserved ejection fraction: comparison of patients with and without angina pectoris (from the Duke Databank for Cardiovascular Disease). J Am Coll Cardiol 2014;63:251–8.

[47] Trevisan L, Cautela J, Resseguier N, Laine M, Arques S, Pinto J, et al. Prevalence and characteristics of coronary artery disease in heart failure with preserved and mid-range ejection fractions: a systematic angiography approach. Arch Cardiovasc Dis 2018;111:109–18.

[48] Gori M, Lam CS, Gupta DK, Santos AB, Cheng S, Shah AM, et al. Sex-specific cardiovascular structure and function in heart failure with preserved ejection fraction. Eur J Heart Fail 2014;16:535–42.

[49] Beale AL, Nanayakkara S, Kaye DM. Impact of sex on ventricular-vascular stiffness and Long-term outcomes in heart failure with preserved ejection fraction: TOPCAT trial substudy. J Am Heart Assoc 2019;8, e012190.

[50] Sharma K, Al Rifai M, Ahmed HM, Dardari Z, Silverman MG, Yeboah J, et al. Usefulness of coronary artery calcium to predict heart failure with preserved ejection fraction in men versus women (from the multi-ethnic study of atherosclerosis). Am J Cardiol 2017;120:1847–53.

[51] Mohammed SF, Hussain S, Mirzoyev SA, Edwards WD, Maleszewski JJ, Redfield MM. Coronary microvascular rarefaction and myocardial fibrosis in heart failure with preserved ejection fraction. Circulation 2015;131:550–9.

[52] Chamberlain AM, Gersh BJ, Alonso A, Kopecky SL, Killian JM, Weston SA, et al. No decline in the risk of heart failure after incident atrial fibrillation: a community study assessing trends overall and by ejection fraction. Heart Rhythm 2017;14:791–8.

[53] Zakeri R, Chamberlain AM, Roger VL, Redfield MM. Temporal relationship and prognostic significance of atrial fibrillation in heart failure patients with preserved ejection fraction: a community-based study. Circulation 2013;128:1085–93.

[54] Sartipy U, Dahlstrom U, Fu M, Lund LH. Atrial fibrillation in heart failure with preserved, mid-range, and reduced ejection fraction. JACC Heart Fail 2017;5:565–74.

[55] Zakeri R, Borlaug BA, McNulty SE, Mohammed SF, Lewis GD, Semigran MJ, et al. Impact of atrial fibrillation on exercise capacity in heart failure with preserved ejection fraction: a RELAX trial ancillary study. Circ Heart Fail 2014;7:123–30.

[56] Bokslag A, Franssen C, Alma LJ, Kovacevic I, Kesteren FV, Teunissen PW, et al. Early-onset preeclampsia predisposes to preclinical diastolic left ventricular dysfunction in the fifth decade of life: an observational study. PLoS One 2018;13, e0198908.

[57] Alma LJ, Bokslag A, Maas A, Franx A, Paulus WJ, de Groot CJM. Shared biomarkers between female diastolic heart failure and pre-eclampsia: a systematic review and meta-analysis. ESC Heart Fail 2017;4:88–98.

[58] Saiki H, Petersen IA, Scott CG, Bailey KR, Dunlay SM, Finley RR, et al. Risk of heart failure with preserved ejection fraction in older women after contemporary radiotherapy for breast cancer. Circulation 2017;135:1388–96.

[59] Beale AL, Nanayakkara S, Segan L, Mariani JA, Maeder MT, van Empel V, et al. Sex differences in heart failure with preserved ejection fraction pathophysiology: a detailed invasive hemodynamic and echocardiographic analysis. JACC Heart Fail 2019;7:239–49.

[60] Lundorff IJ, Sengelov M, Godsk Jorgensen P, Pedersen S, Modin D, Eske Bruun N, et al. Echocardiographic predictors of mortality in women with heart failure with reduced ejection fraction. Circ Cardiovasc Imaging 2018;11, e008031.

[61] Lau ES, Cunningham T, Hardin KM, Liu E, Malhotra R, Nayor M, et al. Sex differences in cardiometabolic traits and determinants of exercise capacity in heart failure with preserved ejection fraction. JAMA Cardiol 2019;5(1):30–7.

[62] Kraigher-Krainer E, Shah AM, Gupta DK, Santos A, Claggett B, Pieske B, et al. Impaired systolic function by strain imaging in heart failure with preserved ejection fraction. J Am Coll Cardiol 2014;63:447–56.

[63] Nanayakkara S, Telles F, Beale AL, Evans S, Vizi D, Marwick TH, et al. Relationship of degree of systolic dysfunction to variations in exercise capacity and hemodynamic status in HFpEF. JACC Cardiovasc Imaging 2020;13(2 Pt 1):528–30.

[64] Mohammed SF, Borlaug BA, Roger VL, Mirzoyev SA, Rodeheffer RJ, Chirinos JA, et al. Comorbidity and ventricular and vascular structure and function in heart failure with preserved ejection fraction: a community-based study. Circ Heart Fail 2012;5:710–9.

[65] Borlaug BA, Redfield MM. Diastolic and systolic heart failure are distinct phenotypes within the heart failure spectrum. Circulation 2011;123:2006–13 [discussion 2014].

[66] Thenappan T, Shah SJ, Gomberg-Maitland M, Collander B, Vallakati A, Shroff P, et al. Clinical characteristics of pulmonary hypertension in patients with heart failure and preserved ejection fraction. Circ Heart Fail 2011;4:257–65.

[67] Mohammed SF, Hussain I, AbouEzzeddine OF, Takahama H, Kwon SH, Forfia P, et al. Right ventricular function in heart failure with preserved ejection fraction: a community-based study. Circulation 2014;130:2310–20.

[68] Melenovsky V, Hwang SJ, Lin G, Redfield MM, Borlaug BA. Right heart dysfunction in heart failure with preserved ejection fraction. Eur Heart J 2014;35:3452–62.

[69] Shah SJ, Lam CSP, Svedlund S, Saraste A, Hage C, Tan RS, et al. Prevalence and correlates of coronary microvascular dysfunction in heart failure with preserved ejection fraction: PROMIS-HFpEF. Eur Heart J 2018;39:3439–50.

[70] Taqueti VR, Solomon SD, Shah AM, Desai AS, Groarke JD, Osborne MT, et al. Coronary microvascular dysfunction and future risk of heart failure with preserved ejection fraction. Eur Heart J 2018;39:840–9.

[71] Del Buono MG, Arena R, Borlaug BA, Carbone S, Canada JM, Kirkman DL, et al. Exercise intolerance in patients with heart failure: JACC state-of-the-art review. J Am Coll Cardiol 2019;73:2209–25.

[72] Paulus WJ, Tschope C. A novel paradigm for heart failure with preserved ejection fraction: comorbidities drive myocardial dysfunction and remodeling through coronary microvascular endothelial inflammation. J Am Coll Cardiol 2013;62:263–71.

[73] Klein SL, Flanagan KL. Sex differences in immune responses. Nat Rev Immunol 2016;16:626–38.

[74] Borlaug BA, Olson TP, Lam CS, Flood KS, Lerman A, Johnson BD, et al. Global cardiovascular reserve dysfunction in heart failure with preserved ejection fraction. J Am Coll Cardiol 2010;56:845–54.

[75] Yoshida K, Obokata M, Kurosawa K, Sorimachi H, Kurabayashi M, Negishi K. Effect of sex differences on the association between stroke risk and left atrial anatomy or mechanics in patients with atrial fibrillation. Circ Cardiovasc Imaging 2016;9:e004999. https://doi.org/10.1161/CIRCIMAGING.116.004999.

[76] O'Neal WT, Sandesara P, Hammadah M, Venkatesh S, Samman-Tahhan A, Kelli HM, et al. Gender differences in the risk of adverse outcomes in patients with atrial fibrillation and heart failure with preserved ejection fraction. Am J Cardiol 2017;119:1785–90. https://doi.org/10.1016/j.amjcard.2017.02.045.

[77] Parks RJ, Ray G, Bienvenu LA, Rose RA, Howlett SE. Sex differences in SR Ca(2 +) release in murine ventricular myocytes are regulated by the cAMP/PKA pathway. J Mol Cell Cardiol 2014;75:162–73.

[78] Peterson LR, Soto PF, Herrero P, Mohammed BS, Avidan MS, Schechtman KB, et al. Impact of gender on the myocardial metabolic response to obesity. JACC Cardiovasc Imaging 2008;1:424–33.

[79] Deswal A, Bozkurt B. Comparison of morbidity in women versus men with heart failure and preserved ejection fraction. Am J Cardiol 2006;97:1228–31.

[80] McMurray JJV, Jackson AM, Lam CSP, Redfield MM, Anand IS, Ge J, et al. Effects of sacubitril-valsartan, versus valsartan, in women compared to men with heart failure and preserved ejection fraction: insights from PARAGON-HF. Circulation 2019;141(5):338–51.

[81] Mohammed SF, Borlaug BA, McNulty S, Lewis GD, Lin G, Zakeri R, et al. Resting ventricular-vascular function and exercise capacity in heart failure with preserved ejection fraction: a RELAX trial ancillary study. Circ Heart Fail 2014;7:580–9.

[82] Zsilinszka R, Shrader P, DeVore AD, Hardy NC, Mentz RJ, Pang PS, et al. Sex differences in the management and outcomes of heart failure with preserved ejection fraction in patients presenting to the emergency department with acute heart failure. J Card Fail 2016;22:781–8.

[83] Cohen JB, Schrauben SJ, Zhao L, Basso MD, Cvijic ME, Li Z, et al. Clinical phenogroups in heart failure with preserved ejection fraction: detailed phenotypes, prognosis, and response to spironolactone. JACC Heart Fail 2020;8:172–84.

[84] Chamsi-Pasha MA, Zhan Y, Debs D, Shah DJ. CMR in the evaluation of diastolic dysfunction and phenotyping of HFpEF: current role and future perspectives. JACC Cardiovasc Imaging 2020;13:283–96.

[85] Obokata M, Kane GC, Reddy YN, Olson TP, Melenovsky V, Borlaug BA. Role of diastolic stress testing in the evaluation for heart failure with preserved ejection fraction: a simultaneous invasive-echocardiographic study. Circulation 2017;135:825–38.

[86] Ponikowski P, Voors AA, Anker SD, Bueno H, Cleland JGF, Coats AJS, et al. 2016 ESC guidelines for the diagnosis and treatment of acute and chronic heart failure: The Task Force for the diagnosis and treatment of acute and chronic heart failure of the European Society of Cardiology (ESC) developed with the special contribution of the Heart Failure Association (HFA) of the ESC. Eur Heart J 2016;37:2129–200.

[87] Reddy YNV, Carter RE, Obokata M, Redfield MM, Borlaug BA. A simple, evidence-based approach to help guide diagnosis of heart failure with preserved ejection fraction. Circulation 2018;138:861–70.

[88] Lewis EF, Lamas GA, O'Meara E, Granger CB, Dunlap ME, McKelvie RS, et al. Characterization of health-related quality of life in heart failure patients with preserved versus low ejection fraction in CHARM. Eur J Heart Fail 2007;9:83–91.

[89] Merrill M, Sweitzer NK, Lindenfeld J, Kao DP. Sex differences in outcomes and responses to spironolactone in heart failure with preserved ejection fraction: a secondary analysis of TOPCAT trial. JACC Heart Fail 2019;7:228–38.

[90] Faxen UL, Hage C, Donal E, Daubert JC, Linde C, Lund LH. Patient reported outcome in HFpEF: sex-specific differences in quality of life and association with outcome. Int J Cardiol 2018;267:128–32.

[91] Chandra A, Vaduganathan M, Lewis EF, Claggett BL, Rizkala AR, Wang W, et al. Health-related quality of life in heart failure with preserved ejection fraction: the PARAGON-HF trial. JACC Heart Fail 2019.

[92] Honigberg MC, Lau ES, Jones AD, Coles A, Redfield MM, Lewis GD, et al. Sex differences in exercise capacity and quality of life in heart failure with preserved ejection fraction: a secondary analysis of the RELAX and NEAT-HFpEF trials. J Card Fail 2020;26:276–80.

[93] Blecker S, Paul M, Taksler G, Ogedegbe G, Katz S. Heart failure-associated hospitalizations in the United States. J Am Coll Cardiol 2013;61:1259–67.

[94] Dunlay SM, Redfield MM, Weston SA, Therneau TM, Hall Long K, Shah ND, et al. Hospitalizations after heart failure diagnosis a community perspective. J Am Coll Cardiol 2009;54:1695–702.

[95] Cheng RK, Cox M, Neely ML, Heidenreich PA, Bhatt DL, Eapen ZJ, et al. Outcomes in patients with heart failure with preserved, borderline, and reduced ejection fraction in the Medicare population. Am Heart J 2014;168:721–30.

[96] O'Meara E, Clayton T, McEntegart MB, McMurray JJ, Pina IL, Granger CB, et al. Sex differences in clinical characteristics and prognosis in a broad spectrum of patients with heart failure: results of the Candesartan in Heart failure: Assessment of Reduction in Mortality and morbidity (CHARM) program. Circulation 2007;115:3111–20.

[97] Roger VL, Weston SA, Redfield MM, Hellermann-Homan JP, Killian J, Yawn BP, et al. Trends in heart failure incidence and survival in a community-based population. JAMA 2004;292:344–50.

[98] Fonarow GC, Stough WG, Abraham WT, Albert NM, Gheorghiade M, Greenberg BH, et al. Characteristics, treatments, and outcomes of patients with preserved systolic function hospitalized for heart failure: a report from the OPTIMIZE-HF Registry. J Am Coll Cardiol 2007;50:768–77.

[99] Lee DS, Gona P, Vasan RS, Larson MG, Benjamin EJ, Wang TJ, et al. Relation of disease pathogenesis and risk factors to heart failure with preserved or reduced ejection fraction: insights from the Framingham heart study of the national heart, lung, and blood institute. Circulation 2009;119:3070–7.

[100] Mcta-analysis Global Group in Chronic Heart Failure. The survival of patients with heart failure with preserved or reduced left ventricular ejection fraction: an individual patient data meta-analysis. Eur Heart J 2012;33:1750–7.

[101] Lee DS, Gona P, Albano I, Larson MG, Benjamin EJ, Levy D, et al. A systematic assessment of causes of death after heart failure onset in the community: impact of age at death, time period, and left ventricular systolic dysfunction. Circ Heart Fail 2011;4:36–43.

[102] Vaduganathan M, Patel RB, Michel A, Shah SJ, Senni M, Gheorghiade M, et al. Mode of death in heart failure with preserved ejection fraction. J Am Coll Cardiol 2017;69:556–69.

[103] Henkel DM, Redfield MM, Weston SA, Gerber Y, Roger VL. Death in heart failure: a community perspective. Circ Heart Fail 2008;1:91–7.

[104] Aschauer S, Zotter-Tufaro C, Duca F, Kammerlander A, Dalos D, Mascherbauer J, et al. Modes of death in patients with heart failure and preserved ejection fraction. Int J Cardiol 2017;228:422–6.

[105] Yusuf S, Pfeffer MA, Swedberg K, Granger CB, Held P, McMurray JJ, et al. Effects of candesartan in patients with chronic heart failure and preserved left-ventricular ejection fraction: the CHARM-Preserved Trial. Lancet 2003;362:777–81.

[106] Ahmed A, Rich MW, Fleg JL, Zile MR, Young JB, Kitzman DW, et al. Effects of digoxin on morbidity and mortality in diastolic heart failure: the ancillary digitalis investigation group trial. Circulation 2006;114:397–403.

[107] Zile MR, Gaasch WH, Anand IS, Haass M, Little WC, Miller AB, et al. Mode of death in patients with heart failure and a preserved ejection fraction: results from the Irbesartan in Heart Failure With Preserved Ejection Fraction Study (I-Preserve) trial. Circulation 2010;121:1393–405.

[108] Vaduganathan M, Claggett BL, Chatterjee NA, Anand IS, Sweitzer NK, Fang JC, et al. Sudden death in heart failure with preserved ejection fraction: a competing risks analysis from the TOPCAT trial. JACC Heart Fail 2018;6:653–61.

[109] Yancy CW, Jessup M, Bozkurt B, Butler J, Casey Jr DE, Colvin MM, et al. 2017 ACC/AHA/HFSA focused update of the 2013 ACCF/AHA guideline for the management of heart failure: a report of the American College of Cardiology/American Heart Association Task Force on Clinical Practice Guidelines and the Heart Failure Society of America. J Card Fail 2017;23:628–51.

[110] Farasat SM, Bolger DT, Shetty V, Menachery EP, Gerstenblith G, Kasper EK, et al. Effect of beta-blocker therapy on rehospitalization rates in women versus men with heart failure and preserved ejection fraction. Am J Cardiol 2010;105:229–34.

[111] Luzier AB, Killian A, Wilton JH, Wilson MF, Forrest A, Kazierad DJ. Gender-related effects on metoprolol pharmacokinetics and pharmacodynamics in healthy volunteers. Clin Pharmacol Ther 1999;66:594–601.

[112] Lund LH, Benson L, Dahlstrom U, Edner M, Friberg L. Association between use of beta-blockers and outcomes in patients with heart failure and preserved ejection fraction. JAMA 2014;312:2008–18.

[113] Solomon SD, McMurray JJV, Anand IS, Ge J, Lam CSP, Maggioni AP, et al. Angiotensin-neprilysin inhibition in heart failure with preserved ejection fraction. N Engl J Med 2019;381:1609–20.

[114] Solomon SD, Claggett B, Lewis EF, Desai A, Anand I, Sweitzer NK, et al. Influence of ejection fraction on outcomes and efficacy of spironolactone in patients with heart failure with preserved ejection fraction. Eur Heart J 2016;37:455–62.

Sex-Specific Characteristics of Patients with HFpEF

Prevalence

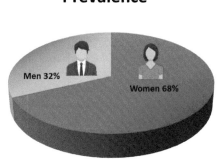

Men 32% Women 68%

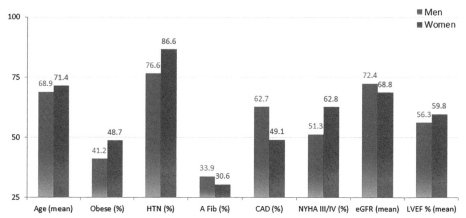

Legend: ■ Men ■ Women

Category	Men	Women
Age (mean)	68.9	71.4
Obese (%)	41.2	48.7
HTN (%)	76.6	86.6
A Fib (%)	33.9	30.6
CAD (%)	62.7	49.1
NYHA III/IV (%)	51.3	62.8
eGFR (mean)	72.4	68.8
LVEF % (mean)	56.3	59.8

Pathophysiology of HFpEF Specific to Women

Inflammation
More comorbidities & autoimmune d/o contribute to increased inflammation

Chronotropic Incompetence
Increased reliance on HR given lower SV

LV Structure
- Smaller LV stroke volume
- More concentric remodeling
- Greater LV stiffness in women, worse with exercise
- Higher LVEF, but impaired LV strain

Peripheral Vasculature
- Arterial-ventricular coupling declines with age, especially with exercise
- Increased vascular stiffness, lower compliance
- Higher pulse pressure
- More coronary microvascular dysfunction
- Increased lactate/workload
- Decreased peripheral O2 handling

Atrial Function
Worse outcome with AFib

Pulmonary Function
- Increased pulm vascular resistance
- Decreased pulm artery compliance
- Increased PCWP/workload w/ exercise

Metabolic & Cellular Level
- Decreased glucose uptake and greater metabolic inefficiency
- Less cardiomyocyte apoptosis, and less myocyte hypertrophy

Niti Aggarwal

Diagnosis

H₂	**Heavy** (BMI>30 kg/m2)	2
	Hypertensive (on 2+ antihypertensives)	1
F	**Atrial Fibrillation**	3
P	**Pulmonary HTN** (RVSP>35 mm Hg)	1
E	**Elderly** (age>60 years)	1
F	**Filling pressure** (E/e' >9)	1

H₂FpEF score defines the probability of HFpEF

Outcomes

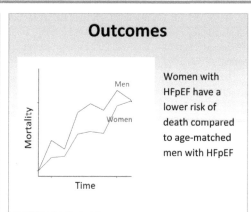

Women with HFpEF have a lower risk of death compared to age-matched men with HFpEF

Management

Beta blockers- broad usage not recommended in men & women. CHARM- no sex-difference to ARB PARAGON-HF: Sacubitril-valsartan is better than valsartan to reduce hospitalization in women. Diuretics: Sex-neutral class I recommendation in men & women.

Sex-Differences in HFpEF. Afib, atrial fibrillation; BMI, body mass index; CAD, coronary artery disease; d/o, disorder; HFpEF, heart failure with preserved ejection fraction; HTN, hypertension; LV, left ventricle; LVEF, left ventricular ejection fraction; NYHA, New York heart Association; O₂, oxygen; PCWP, pulmonary capillary wedge pressure; pulm, pulmonary. *Image courtesy of Niti R. Aggarwal.*

Chapter 12

Pulmonary Arterial Hypertension

Ajith P. Nair, Selma F. Mohammed, Niti R. Aggarwal, and Anita Deswal

Clinical Case

A 27-year-old female presented with progressive shortness of breath over 1 year. She was initially treated with inhaled bronchodilators after she was diagnosed with asthma. However, she continued to have progressive dyspnea and eventually experienced chest pain and syncope with moderate exertion. She had no past medical or surgical history, and her family history was unremarkable. There was no history of substance abuse and she had no travels outside the United States. On examination her blood pressure was 105/65 mmHg and her body mass index (BMI) was normal. Her jugular venous pressure was elevated but she had no peripheral edema. Cardiac examination demonstrated a regular rhythm with a prominent P2 and a right ventricular heave. Her lungs were clear. Noninvasive testing included an electrocardiogram which showed sinus rhythm with a right axis deviation and ST depressions in V1–V3 suggestive of right ventricular strain. An echocardiogram revealed a small left ventricular cavity size with septal shift consistent with right ventricular pressure and volume overload. There was right atrial enlargement, severe right ventricular enlargement with moderate dysfunction, moderate tricuspid regurgitation, and an estimated pulmonary artery pressure of 90 mmHg. A 6-min walk test was performed, and it was reduced at 330 m. How would you manage her next?

Abstract

Pulmonary arterial hypertension (PAH), a form of pulmonary hypertension (PH), is a complex disorder with sexual dimorphism. The disease preferentially affects women, and female sex influences both the development of PAH and right ventricular remodeling. Large registries have demonstrated that while women are more likely to develop PAH, older men with PAH have increased mortality. Sex hormones may impact right ventricular function as well as response to PAH-specific medications. Despite advancements in therapeutics for PAH, mortality remains high for both men and women. Further investigation is necessary to fully elucidate the impact of sex on PAH pathogenesis and help guide novel therapeutics.

Sex Differences in Cardiac Disease. https://doi.org/10.1016/B978-0-12-819369-3.00011-3

Introduction

PH is a heterogenous group of disorders with a broad spectrum of histological changes, where mean pulmonary arterial pressure (mPAP) at rest is greater than 20 mmHg [1]. This mean pulmonary arterial (PA) pressure is two standard deviations above the mean PA pressure in normal subjects, and this is lower than the previously defined value of 25 mmHg. There are no criteria for what defines exercise-induced PH as there is a broad variation in pulmonary pressures during exercise with advancing age, whereas resting values are preserved across age groups. The 6th World Symposium on Pulmonary Hypertension classifies PH into five groups (see **Table 1**) [1]. PAH, a subset of PH, is a proliferative vascular disorder characterized by pulmonary arteriolar obstruction, resulting in right heart failure and death. It includes idiopathic, heritable, drug-induced, connective tissue, and autoimmune related PH. It is defined as a mPAP > 20 mmHg with a normal wedge pressure (or left atrial or left ventricular end diastolic pressure) < 15 mmHg and a pulmonary vascular resistance (PVR) > 3 Woods Units (see **Table 2**). While PAH occurs more frequently in women, the role of sex and sex hormones in development and pathophysiology of this complex disorder is still poorly defined. This chapter discusses the epidemiology, pathophysiology, therapeutic targets, and clinical outcomes of PAH with emphasis on sex differences.

TABLE 1 Sixth World Symposium Classification of Pulmonary Hypertension

Group 1: Pulmonary Arterial Hypertension

1.1 Idiopathic (IPAH)
1.2 Heritable (HPAH)
1.3 Drug and toxin induce PAH

1.4 Associated with (APAH)
    1.4.1 Connective tissue disorder
    1.4.2 HIV infection
    1.4.3 Portal hypertension
    1.4.4 Congenital heart disease
    1.4.5 Schistosomiasis

1.5 PAH long-term responders to CCB
1.6 PAH with venous/capillary involvement (PVOD/ PCH)
1.7 Persistent pulmonary hypertension of the newborn syndrome

Group 2: Left Heart Disease

2.1 PH due to HFpEF
2.2 PH with HFrEF
2.3 Valvular heart disease
2.4 Congenital/acquired cardiovascular conditions leading to postcapillary PH

TABLE 1 Sixth World Symposium Classification of Pulmonary Hypertension—cont'd

Group 3: Hypoxia or Lung Disease

3.1 Obstructive lung disease
3.2 Restrictive lung disease
3.3 Other lung disease with mixed restrictive/ obstructive pattern
3.4 Hypoxia without lung disease
3.5 Developmental lung disorders

Group 4: PH Due to Pulmonary Artery Obstruction

4.1 CTEPH
4.2 Other pulmonary artery obstructions

Group 5: Miscellaneous

5.1 Hematological disorders
5.2 Systemic and metabolic disorders
5.3 Others
5.4 Complex congenital heart disease

CCB, calcium channel blockers; CTEPH, chronic thromboembolic pulmonary hypertension; HFpEF, heart failure with preserved ejection fraction; HFrEF, heart failure with reduced ejection fraction; HPAH, heritable pulmonary arterial hypertension; IPAH, idopathic pulmonary arterial hypertension; PAH, pulmonary arterial hypertension; PCH, pulmonary capillary hemangiomatosis; PH, pulmonary hypertension; PVOD, pulmonary veno-occlusive disease.

TABLE 2 Hemodynamic Definitions of Pulmonary Hypertension

Definitions	Characteristics	Clinical Groups
Precapillary PH	mPAP > 20 mmHg PCWP < 15 PVR > 3 WU	1, 3, 4, and 5
Isolated postcapillary PH	mPAP > 20 mmHg PCWP > 15 mmHg PVR < 3 WU	2 and 5
Combined precapillary and postcapillary PH	mPAP > 20 mmHg PCWP > 15 mmHg PVR > 3 WU	2 and 5

mPAP, mean pulmonary arterial pressure; PH, pulmonary hypertension; PCWP, pulmonary capillary wedge pressure; PVR, pulmonary vascular resistance.

PAH Epidemiology and Sex-Based Differences

The incidence of PAH is 2.0–7.6 cases per million adults per year and the prevalence is 11–26 cases per million adults [2–4]. Multiple registries have demonstrated an increased prevalence of PAH among women (see **Table 3**) [2–13]. A National Institutes of Health multicenter registry from the 1980s demonstrated a 1.7:1 female to male ratio, and COMPERA (Comparative, Prospective Registry of Newly Initiated Therapies for Pulmonary Hypertension), which enrolled patients in Europe from 2007 to 2011, demonstrated a 1.8:1 female/male ratio [5, 12]. However, this sex predilection is more prevalent in younger patients with a

TABLE 3 Pulmonary Arterial Hypertension Registries							
Registry	PAH Cohort	Enrollment Dates	Subjects	Mean Age	Sex Ratio (Female/ Male)	1-Year Survival, %	5-Year Survival, %
US NIH [5]	IPAH, HPAH	1981–88 (prospective)	187	36	1.7:1	68	34
US PHC [6]	WHO Group 1	1982–2006 (retrospective)	578	46 ± 14	3.1:1	86	61
Scottish Morbidity Record [2]	IPAH, CTD, CHD	1986–2001 (retrospective)	374	50 ± 13	2.3:1	NA	NA
French [3]	WHO Group 1	2002–03 (prospective)	674	50 ± 15	1.9:1	87	NA
Mayo [7]	WHO Group 1	1995–2004 (prospective)	484	52 ± 15	3.1:1	81.1	47.9
Chinese [8]	IPAH, HPAH	1999–2004 (prospective)	72	36 ± 12	2.4:1	68	20.8
Spanish [9]	WHO Group 1	1998–2008 (retrospective, prospective)	1028	45 ± 17	2.4:1	87[a]	65[a]
REVEAL [4]	WHO Group 1	2006–07 (prospective)	2525	53 ± 14	3.9:1	86.3 incident 90.4 prevalent	61.2 incident 65.4 prevalent
United Kingdom/ Ireland [10]	IPAH, HPAH, anorexigen	2001–09	482	50 ± 17	1.4:1	93	60
New Chinese [11]	IPAH, CTD	2008–11	276	IPAH 33 ± 15 CTD 41 ± 14	3.1:1	92.1 IPAH 85.4 CTD	NA
COMPERA [12]	WHO Group 1	2007–11 (prospective)	1283	68 (55–75)	1.8:1	92	NA

[a] *Includes 162 patients with chronic thromboembolic pulmonary hypertension.*
CHD, coronary heart disease; COMPERA, Comparative, Prospective Registry of Newly Initiated Therapies for Pulmonary Hypertension; CTD, connective tissue disease; HPAH, heritable pulmonary arterial hypertension; IPAH, idiopathic pulmonary arterial hypertension; NA, not applicable; PAH, pulmonary arterial hypertension; REVEAL, registry to evaluate early and long-term PAH disease management; US NIH, United States National Institutes of Health; US PHC, United States Pulmonary Hypertension Connection; WHO, World Health Organization.

female/male ratio of 2.3:1 among 18–65-year-old patients vs. 1.2:1 in those older than 65 years [12]. These findings suggest that changes in hormonal levels with age play an important role in the pathogenesis of PAH. REVEAL (Registry to Evaluate Early and Long-Term PAH Disease Management) was a US registry that included 54 US centers and enrolled 2967 patients between 2006 and 2007. Among the 2525 adults meeting established hemodynamic criteria for PAH, 79.5% were women [4, 14]. There was ethnic variation in disease prevalence with a female to male ratio of 3.2:1 in Caucasians, 4.7:1 in Hispanics, 5.5:1 in blacks, and 3.9:1 in other races [14]. In Chinese registries, women accounted for 71–76% of the cohort, and the ratio of female to male patients was 3.1:1 [8, 11].

Heritable PAH (HPAH) demonstrates a similar epidemiological bias with women being twice as likely to be affected than men. Mutations in bone morphogenetic protein type II (BMPR2), which involve transforming growth factor beta (TGF-beta) cell signaling, are the most common genetic abnormality observed in patients with HPAH with considerable variability in expression and onset of disease [15, 16]. In a pooled analysis of eight cohorts of patients with idiopathic, heritable, and drug-associated PAH tested for BMPR2 mutation, 448 (29%) of the 1550 patients had a BMPR2 mutation and 69% of carriers were female [17].

PAH associated with connective tissue disorders demonstrates an even more robust bias toward females. Connective tissue disorders occur more frequently in women, and women with systemic sclerosis are eight times more likely to develop PAH than men and women with systemic lupus erythematosus (SLE) women are 17 times more likely to develop PAH [18, 19]. In an analysis of the REVEAL registry, 90.2% of patients with connective tissue associated PAH were women and among those patients, 88.5% with

systemic sclerosis-associated PAH and 94.5% of SLE-associated PAH were women [20]. Patients with systemic sclerosis-associated PAH had higher B-type natriuretic peptide (BNP) levels and the poorest survival of all connective tissue-associated PAH subgroups.

Although portal hypertension is more common in males, female sex is a risk factor for the development of PH. Up to 6% of patients with advanced liver disease may develop PH. In a multicenter case-control study of patients with portal hypertension, female sex was associated with an increased risk of PH [adjusted odds ratio=2.90, 95% confidence interval (CI) 1.20–7.01, P=0.018] in addition to autoimmune hepatitis, which implicated both hormonal and immunological factors in the development of PH [21]. The REVEAL registry enrolled 174 patients with PH and 52% were female [22]. Female risk is also a major factor in the development of methamphetamine-associated PAH [23], whereas it does not appear to be a major determinant in human immunodeficiency virus infection or congenital heart disease.

PAH Mortality

Despite the increased prevalence of PAH among females, survival is worse in men vs. women. In the REVEAL registry, 1-year survival was 91% and male sex increased the risk for mortality by twofold in patients older than 60 years of age [hazard ratio (HR) 2.2, 95% CI 1.6–3.0] [4]. Likewise, in a French registry of 354 patients with idiopathic, familial, and anorexigen-associated PAH, mortality was independently associated with male sex, right ventricular (RV) hemodynamic dysfunction, and exercise impairment. Males were at 2.6-fold increased risk for mortality compared to females [24]. Further data from REVEAL reinforced that survival was improved in women for subjects >60 years (see **Figure 1**). Among 2318 women and 651 men ana-

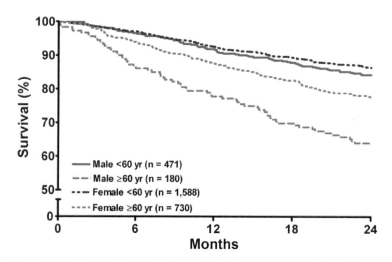

FIGURE 1 Two-Year Survival From Time of Enrollment in the REVEAL Registry by Age > 60 and Sex. Females had better survival estimates for 2 years than males in patients with pulmonary arterial hypertension. Stratifying by age demonstrated similar survival between men and women aged < 60 years at enrollment. In contrast, men aged ≥ 60 years had lower survival compared to age-matched women. *Reproduced with permission from [25].*

lyzed, men had higher mean pulmonary artery pressures and mean right atrial pressures. Survival at 2 years was similar between the sexes below age < 60 years at enrollment, whereas in those older than 60 years there was a difference in survival that was notable after 18 months (64.0%±3.6% vs. 77.5%±1.6%, respectively; $P < 0.001$) [25]. At 5 years, the overall survival rates among previously diagnosed patients were 57.4%±2.4% vs. 67.5%±1.2% in men vs. women, respectively, and in newly diagnosed patients the survival rates were 53.0%±4.0% vs. 62.9%±2.1% in men vs. women, respectively [26]. In patients with systemic sclerosis, male sex was associated with an increased mortality risk at 1, 2, and 3 years (HR 3.9, 95% CI 1.1–13.9) [27].

Right Ventricular Function in PAH

Changes in sex hormones with aging appear to impact PAH epidemiology and disease progression. Right ventricular failure is the ultimate cause of death in PAH, and sex hormones have an impact on right ventricular function and remodeling. This may account for the dichotomy between men and women with relation to disease prevalence and outcomes. The multiethnic study of atherosclerosis (MESA) study included 5098 participants, and RV volumes and mass were available for 4204 participants. Cardiac magnetic resonance imaging (MRI) was used to assess RV indices. Older age was associated with lower RV mass with larger age-related decline in males vs. females. Males also had lower RV ejection fractions compared to women [28, 29]. Among 1957 men and 1738 postmenopausal women on hormonal therapy, high estradiol levels were associated with higher RV ejection fractions and lower RV end-systolic volumes. Higher dehydroepiandrosterone (DHEA) levels were associated with greater mass and volumes and higher RV stroke volumes in women. In men, high testosterone levels were associated with greater RV mass and volumes and higher RV stroke volumes [29]. Higher estradiol levels were associated with improved RV systolic function in women on hormone therapy, whereas higher androgen levels were associated with greater RV mass and volumes in both men and women. These results were also corroborated by findings from the Framingham Heart Study Offspring cohort in which 1794 cohort members who underwent cardiac MRI were compared to a reference group of 1336 adults without cardiovascular and pulmonary disease. RV ejection fraction was greater in women, whereas men had greater RV volumes, but these differences were attenuated with advancing age [30]. Using subjects from MESA-RV, Ventetuolo and colleagues demonstrated that genetic variations in cytochrome P4501B1 (CYP1B1), the primary enzyme involved in estradiol (E2) metabolism, were associated with higher right ventricular ejection fraction (RVEF) in postmenopausal women [31]. Genetic polymorphisms in CYP1B1 have also been implicated in disease penetrance in BMPR2

carriers, suggesting a key role in estrogen and estrogen metabolism in modifying PAH pathogenesis [32]. In contrast, circulating testosterone levels may worsen RV function and stimulate hypertrophy and fibrosis in animal models of RV load stress [33, 34]. This relationship has been demonstrated in humans where androgen receptor genotype may impact RV function [35].

Right ventricular function has been demonstrated to be the strongest predictor of outcomes in PAH irrespective of changes in PVR with therapy [36]. Progressive right ventricular-pulmonary artery decoupling is a predictor of outcomes in PAH [37]. A cohort of 63 patients with idiopathic, familial, and anorexigen-associated PAH who underwent radionuclide angiography and right heart catheterization, older age and male sex were associated with lower RVEF [38]. In a retrospective cohort study of 101 patients with idiopathic, heritable, and drug-associated PAH, both male and female patients had comparable PVR and RVEF at baseline and had similar reductions in PVR 1 year after treatment. However, women demonstrated improvements in RVEF, whereas it deteriorated in males. Transplant-free survival was worse in males and 39% of this difference was mediated through changes in RVEF [39]. Maladaptive changes in RV function have also been demonstrated by cardiac MRI among men vs. women with idiopathic PAH despite similar afterload [40].

PAH Pathology and Impact of Sex Hormones

Sex hormones have demonstrated a role in the pathogenesis of PAH in animal models, but these results have often been contradictory in human observations. The contradictory findings can be dubbed the "estrogen paradox" [41, 42]. PAH and right ventricular failure demonstrate a sexual dimorphism, which may be driven by estrogen and the conversion of androgens into estrogen (aromatization), receptor signaling and their interaction with an individual's genetic substrate, the BMPR2 pathway, and the RV (see **Table 4**). Mutations in BMPR2 are present in 70–80% of families with HPAH and in 25% of patients with idiopathic PAH [43, 44]. Despite the fact that pathogenic BMPR2 mutation causes PAH, the penetrance of the disease phenotype is incomplete. The estimated penetrance in male carriers is roughly 14%, whereas in female carriers it is approximately 42% [45]. Female sex is thus the single most important factor influencing penetrance of BMPR2, likely driven by estrogen metabolism [32].

Sex hormones are key modulators of PAH pathogenesis. In addition, steroidogenic hormones such as aldosterone also play a role in pulmonary vascular remodeling and myocardial fibrosis. Testosterone can act as a pulmonary vasodilator through calcium antagonism [46].

TABLE 4 Sex Hormone Effects on the Pulmonary Vasculature and Right Ventricle in PAH

	Pulmonary Vasculature	Right Ventricle
Testosterone	Unclear	↑Fibrosis
DHEA	↑Nitric oxide ↓Inflammation	↑Ischemic resistance ↓Fibrosis
Estradiol	↑Proliferation ↓BMPR2	Unclear
16-α-hydroxyestrone	↑Proliferation ↓BMPR2	Unclear

BMPR2, bone morphogenetic protein type II; DHEA, dehydroepiandrosterone; PAH, pulmonary arterial hypertension.

Dehydroepiandrosterone can abrogate hypoxic vasocontriction [47]. Estrogen can have dichotomous effects on the vasculature and PAH development. Estrogen can mediate vasodilation via estrogen alpha and beta receptors through nitric oxide (NO)-dependent mechanisms [48]. Estradiol is metabolized by the cytochrome P450 enzymes into metabolites, which can have different actions on inflammation [32]. Austin and colleagues demonstrated that BMPR2 gene expression is reduced in humans and mice through direct estrogen receptor alpha binding to the BMPR2 promotor [49]. This reduced BMPR2 expression may contribute to the increased predominance of women with idiopathic and heritable PAH.

Genetic variations in estrogen signaling are associated with increased risk of PH. Single-nucleotide polymorphisms in estrogen receptor 1 and aromatase, the enzyme that converts testosterone to E2, were associated with increased risk of PH [50]. The polymorphism in aromatase was associated with increased plasma E2 levels.

In an effort to evaluate the contribution of sex hormones to PAH in men, Ventetuolo and colleagues assessed estradiol (the most common circulating form estrogen, E2), testosterone, and dehydroepiandrosterone-sulfate (DHEA-S) levels in men with idiopathic, heritable, or connective tissue disease-associated PAH. Higher E2 conferred an elevated risk for PAH in men with 50-fold increased risk for every 1-unit increased in E2. High DHEA-S levels were associated with reduced risk of PAH and a better hemodynamic profile with lower right atrial pressures and PVR [35]. Differences in sex hormone processing and signaling contributed to the evolution of pulmonary vascular disease in men.

In postmenopausal females with idiopathic PAH, connective tissue disease, and congenital heart disease-associated PAH, high E2 and low DHEA-S levels were associated with increased risk and severity of PAH in a study by Baird and colleagues [51]. Every unit decrease in DHEA-S, total testosterone, bioavailable testosterone, and sex hormone binding globulin increased the risk of PAH by threefold to fourfold. Reductions in DHEA-S were associated with lower 6-min walk distance, higher right atrial pressures, increased PVR, and increased risk of death over 2 years.

Increased expression of aromatase has been detected in female human pulmonary artery smooth muscle cells from women and the use of the aromatase inhibitor anastrozole has been shown to attenuate PH in animal models [52]. In a study of 18 patients with PAH, anastrozole significantly reduced E2 levels and improved 6-min walk distance compared with placebo although it had no effect on circulating biomarkers, functional class, health-related quality of life, or tricuspid annular plane systolic excursion (TAPSE) [53].

PAH Therapies

Currently approved therapies for PAH target three primary pathways that are key in the pathogenesis of the disease: the NO, prostacyclin, and endothelin pathways (see **Figure 2**) [54]. The imbalance of vasoactive mediators plays a primary role in the development and progression in the proliferative pathological changes in PAH. These pathways are influenced by sex hormones.

Phosphodiesterase-5 Inhibitors

NO binds to soluble guanylate cyclase (sGC), which leads to the production of cyclic guanosine monophosphate (cGMP), which promotes arteriole vasodilation and suppresses cell proliferation. Whole-body NO production is greater in premenopausal women than in men [55].

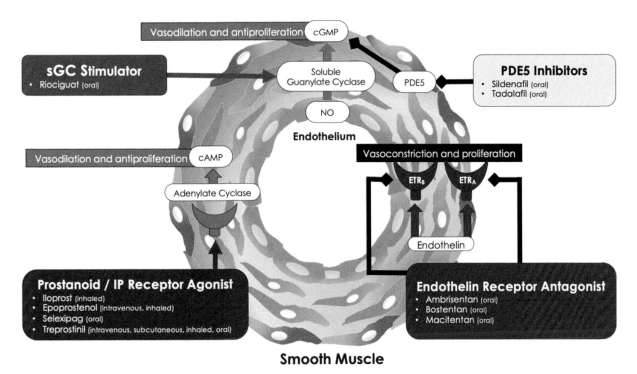

FIGURE 2 Molecular Targets of Pulmonary Arterial Hypertension Therapy. sGC stimulators promote increased cGMP via soluble guanylate cycle. This has vasodilatory and antiproliferative effects. Prostacyclins increase cAMP, which also has vasodilatory and antiproliferative effects. PDE5 inhibitors act on phosphodiesterase-5 to prevent the breakdown of cGMP. Endothelin acts upon ETR_A and ETR_B, which promote vasoconstriction and smooth muscle cell proliferation. Endothelial receptor antagonists block ETa and ETb. cAMP, cyclic adenosine monophosphate; cGMP, cyclic guanosine monophosphate; ETR_A, endothelin receptor A; ETR_B, endothelin receptor B; IP, prostacyclin; NO, nitric oxide; PDE5, phosphodiesterase-5; sGC, soluble guanylate cyclase. *Adapted with permission from [54].*

Phosphodiesterase-5 (PDE5) inhibition prevents the degradation of cGMP. Sildenafil and tadalafil are the two PDE5 inhibitors that have demonstrated short- and long-term clinical benefit in PAH [56, 57]. In a retrospective analysis of the major trial evaluating tadalafil for PAH, men and premenopausal women had greater functional improvement when treated with tadalafil than older women, although no impact on time to clinical worsening was noted [58].

Soluble Guanylate Cyclase

Riociguat is an sGC stimulator and acts independent of NO to promote positive vascular remodeling and pulmonary vasodilation. Riociguat has demonstrated benefits in both PAH and chronic thromboembolic pulmonary hypertension (CTEPH) to increase exercise tolerance and decrease time to clinical worsening [59, 60]. The maximum concentrations of riociguat and its metabolite M1 were significantly higher in women than in men (35% and 50% higher, respectively) in a pharmacokinetic assessment of the drug, although to-

tal exposure was similar [61]. In animal models, DHEA increased sGC levels in the pulmonary vasculature and reversed hypoxic pulmonary vascular constriction [47].

Endothelin Receptor Antagonist

Endothelin-1 (ET1) binds to endothelin receptors A and B. Endothelin A activation leads to pulmonary vasoconstriction and smooth muscles cell proliferation while endothelin B acts to clear ET1 and mediate endothelial cell vasodilation and NO and prostacyclin release. ET1 levels are elevated in PAH, and endothelin receptor antagonists (ERAs) improve functional status and reduce morbidity and mortality in patients with PAH [13, 62, 63]. Men have greater levels of ET1 and greater levels of endothelin-mediated vasoconstriction than women [64, 65]. Women with PAH have greater responses to ERAs than men in a pooled analysis of six randomized placebo-controlled ERA trials. Placebo-adjusted 6-min walk distance improved by 44.1 m in women vs. 16.7 m in men after 12 weeks of therapy [66].

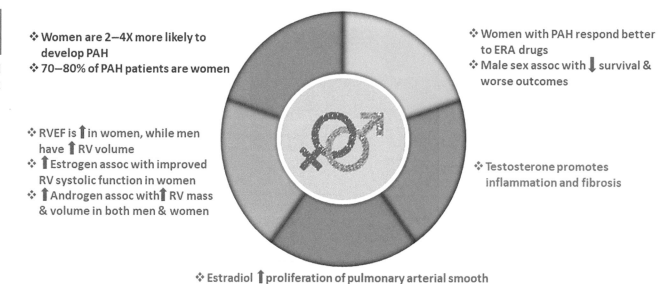

❖ Women are 2–4X more likely to develop PAH
❖ 70–80% of PAH patients are women

❖ RVEF is ⬆ in women, while men have ⬆ RV volume
❖ ⬆ Estrogen assoc with improved RV systolic function in women
❖ ⬆ Androgen assoc with ⬆ RV mass & volume in both men & women

❖ Women with PAH respond better to ERA drugs
❖ Male sex assoc with ⬇ survival & worse outcomes

❖ Testosterone promotes inflammation and fibrosis

❖ Estradiol ⬆ proliferation of pulmonary arterial smooth muscle cells, leading to PAH
❖ Estradiol ⬇ BMPR2 expression resulting in ⬆ lung fibrosis

FIGURE 3 **Sex Differences in Pulmonary Arterial Hypertension.** The figure summarizes the key sex differences in epidemiology, pathophysiology, clinical impact of sex hormones, response to treatment, and outcomes of patients with pulmonary arterial hypertension. Assoc, associated; BMPR2, bone morphogenetic protein type II; ERA, endothelin receptor antagonist; PAH, pulmonary arterial hypertension; RV, right ventricle; RVEF, right ventricular ejection fraction.

Prostacyclins

Prostacyclins are released by endothelial cells and promote pulmonary vasodilation and have antithrombotic and antiproliferative properties. Prostacyclins can be administered in oral, inhaled, subcutaneous, and intravenous forms. Despite significant side effects, epoprostenol is an intravenous prostacyclin that reduced mortality in advanced PAH patients with class IV symptoms [67]. The PROSPECT registry included 336 US patients treated with temperature-stable epoprostenol (Veletri). Freedom from hospitalization at 1 year was lower in males than females ($38.3\% \pm 5.9\%$ vs. $54.6\% \pm 3.2\%$, respectively; $P < 0.015$) and the overall 1-year survival estimate was $84.0\% \pm 2.1\%$ [68].

A multiparametric risk stratification approach is necessary to define initial therapy and to follow response to therapy in order to maximize clinical outcomes. The 2015 European Society of Cardiology (ESC) guidelines suggest risk stratification using clinical, imaging, biomarker, functional capacity, and hemodynamic measures to define patients at low, medium, and high risk for mortality. These variables can be used for the periodic assessment of therapeutic response to medications and prognosis [69]. The REVEAL 2.0 risk calculator can estimate 12-month mortality, and male age > 60 years is among the variables used to discriminate risk [70].

PAH and Pregnancy

Pregnancy carries a high risk in women with PH. The third trimester of pregnancy and first month postpartum carry the greatest risk of mortality due to right ventricular failure, pulmonary hypertensive crisis, and possibly pulmonary embolism. General anesthesia poses an increased mortality risk and should generally be avoided. During the third trimester, there is up to a 30% increase in plasma volume, which can contribute to right ventricular failure. PVR can increase immediately postdelivery and precipitate refractory right ventricular collapse. Epoprostenol has been used since 2001, and there have been successful outcomes in pregnancies due to the use of PAH-specific medications and through multidisciplinary teams at experienced PAH centers [71–73]. Compared to pretreatment era, mortality has improved from 38% to 25% with 78% deaths occurring in the first month after delivery [74]. Monitoring with Swan-Ganz catheter is debatable, although central venous pressure goals of 10–12 mmHg may be reasonable during delivery [75]. ERA and riociguat are contraindicated due to teratogenic effects (category X). Prostacyclins and phosphodiesterase inhibitors are pregnancy risk category B and are utilized through pregnancy. Epoprostenol is the most effective medication for use and should be utilized in the third trimester and continued for up to several months after delivery to ensure hemodynamic stability [76]. Deliveries can be vaginal or cesarean and appropriately planned with utilizing epidural anesthesia as opposed to general anesthesia [74].

Conclusion

The incidence and prevalence of nearly all forms of PAH demonstrate a strong female predominance, and are likely influenced by genetics, epigenetics, and the complex role of sex hormones (**Figure 3**). While women are more likely to develop idiopathic, heritable, drug-induced, connective tissue disease (CTD)-associated PAH and PH, male patients with PAH are at increased mortality risk. Women with PAH appear to tolerate increased RV load, have better right ventricular-pulmonary artery coupling, and live longer than men. Estrogen signaling plays a distinct role in the development of PAH, and maladaptive change in right ventricular function may be partially due to the influence of androgens on right ventricular fibrosis and remodeling. Age and temporal changes in sex hormone milieu may also influence changes in the pulmonary vasculature and the right ventricle. Key sex differences in PAH are summarized in **Figure 3**. Despite adequate inclusion of women in PAH trials, sex-specific analysis of data is often not reported. This may be necessary to elucidate the impact of sex on PAH therapeutics.

Key Points

1. The incidence and prevalence of nearly all forms of PAH demonstrate a strong female predominance.

2. While women are more likely to develop idiopathic, heritable, drug-induced, connective tissue disease-associated PAH and portopulmonary hypertension, male patients with PAH are at twofold to fourfold increased risk for mortality.

3. Women with PAH appear to better tolerate increased RV load, have better right ventricular-pulmonary artery coupling, and live longer than men.

4. Estrogen signaling plays a distinct role in the development of PAH, and maladaptive changes in right ventricular function may be partially due to the influence of androgens on right ventricular fibrosis and remodeling.

5. Age and temporal change in the sex hormonal milieu may also influence changes in the pulmonary vasculature and the right ventricle.

6. Despite adequate inclusion of women in PAH trials, sex-specific analysis of data is often not reported.

Back to Clinical Case

The clinical case described a 27-year-old female with progressive 1-year history of dyspnea, and subsequent development of chest pain and syncope with moderate exertion, with reduced exercise capacity, found to have severe right ventricular enlargement, moderated dysfunction, and elevated pulmonary artery pressure on echocardiography. She underwent cardiac catheterization which demonstrated a mean pulmonary artery pressure of 50 mmHg, a pulmonary capillary wedge pressure of 5 mmHg, a right atrial pressure of 15 mmHg, and a cardiac index of 1.8 L/min/m^2. An evaluation for associated causes of pulmonary arterial hypertension was unremarkable and she was diagnosed with idiopathic pulmonary arterial hypertension. She was started on intravenous epoprostenol followed by tadalafil and macitentan. This therapy resulted in a hemodynamic and clinical response, and her 1-year follow-up appointment echocardiography revealed improvement in the features of pulmonary hypertension and her 6 min walk distance increased to 440 m.

References

[1] Simonneau G, Montani D, Celermajer DS, Denton CP, Gatzoulis MA, Krowka M, et al. Haemodynamic definitions and updated clinical classification of pulmonary hypertension. Eur Respir J 2019;53(1):1801913.

[2] Peacock AJ, Murphy NF, McMurray JJV, Caballero L, Stewart S. An epidemiological study of pulmonary arterial hypertension. Eur Respir J 2007;30(1):104–9.

[3] Humbert M, Sitbon O, Chaouat A, Bertocchi M, Habib G, Gressin V, et al. Pulmonary arterial hypertension in France: results from a national registry. Am J Respir Crit Care Med 2006;173(9):1023–30.

[4] Badesch DB, Raskob GE, Elliott CG, Krichman AM, Farber HW, Frost AE, et al. Pulmonary arterial hypertension: baseline characteristics from the REVEAL registry. Chest 2010;137(2):376–87.

[5] Rich S, Dantzker DR, Ayres SM, Bergofsky EH, Brundage BH, Detre KM, et al. Primary pulmonary hypertension. A national prospective study. Ann Intern Med 1987;107(2):216–23.

[6] Thenappan T, Shah SJ, Rich S, Gomberg-Maitland M. A USA-based registry for pulmonary arterial hypertension: 1982-2006. Eur Respir J 2007;30(6):1103–10.

[7] Kane GC, Maradit-Kremers H, Slusser JP, Scott CG, Frantz RP, McGoon MD. Integration of clinical and hemodynamic parameters in the prediction of long-term survival in patients with pulmonary arterial hypertension. Chest 2011;139(6):1285–93.

[8] Jing Z-C, Xu X-Q, Han Z-Y, Wu Y, Deng K-W, Wang H, et al. Registry and survival study in Chinese patients with idiopathic and familial pulmonary arterial hypertension. Chest 2007;132(2):373–9.

[9] Escribano-Subias P, Blanco I, Lopez-Meseguer M, Lopez-Guarch CJ, Roman A, Morales P, et al. Survival in pulmonary hypertension in Spain: insights from the Spanish registry. Eur Respir J 2012;40(3):596–603.

[10] Ling Y, Johnson MK, Kiely DG, Condliffe R, Elliot CA, Gibbs JS, et al. Changing demographics, epidemiology, and survival of incident pulmonary arterial hypertension: results from the pulmonary hypertension registry of the United Kingdom and Ireland. Am J Respir Crit Care Med 2012;186(8):790–6.

[11] Zhang R, Dai L-Z, Xie W-P, Yu Z-X, Wu B-X, Pan L, et al. Survival of Chinese patients with pulmonary arterial hypertension in the modern treatment era. Chest 2011;140(2):301–9.

[12] Hoeper MM, Huscher D, Ghofrani HA, Delcroix M, Distler O, Schweiger C, et al. Elderly patients diagnosed with idiopathic pulmonary arterial hypertension: results from the COMPERA registry. Int J Cardiol 2013;168(2):871–80.

[13] Pulido T, Adzerikho I, Channick RN, Delcroix M, Galiè N, Ghofrani HA, et al. Macitentan and morbidity and mortality in pulmonary arterial hypertension. N Engl J Med 2013;369(9):809–18.

[14] Frost AE, Badesch DB, Barst RJ, Benza RL, Elliott CG, Farber HW, et al. The changing picture of patients with pulmonary arterial hypertension in the United States: how REVEAL differs from historic and non-US contemporary registries. Chest 2011;139(1):128–37.

[15] International PPH Consortium, Lane KB, Machado RD, Pauciulo MW, Thomson JR, Phillips JA, et al. Heterozygous germline mutations in BMPR2, encoding a TGF-beta receptor, cause familial primary pulmonary hypertension. Nat Genet 2000;26(1):81–4.

[16] Machado RD, Pauciulo MW, Thomson JR, Lane KB, Morgan NV, Wheeler L, et al. BMPR2 haploinsufficiency as the inherited molecular mechanism for primary pulmonary hypertension. Am J Hum Genet 2001;68(1):92–102.

[17] Evans JDW, Girerd B, Montani D, Wang X-J, Galiè N, Austin ED, et al. BMPR2 mutations and survival in pulmonary arterial hypertension: an individual participant data meta-analysis. Lancet Respir Med 2016;4(2):129–37.

[18] Chung L, Farber HW, Benza R, Miller DP, Parsons L, Hassoun PM, et al. Unique predictors of mortality in patients with pulmonary arterial hypertension associated with systemic sclerosis in the REVEAL registry. Chest 2014;146(6):1494–504.

[19] Foderaro A, Ventetuolo CE. Pulmonary arterial hypertension and the sex hormone paradox. Curr Hypertens Rep 2016;18(11):84.

[20] Chung L, Liu J, Parsons L, Hassoun PM, McGoon M, Badesch DB, et al. Characterization of connective tissue disease-associated pulmonary arterial hypertension from REVEAL: identifying systemic sclerosis as a unique phenotype. Chest 2010;138(6):1383–94.

[21] Kawut SM, Krowka MJ, Trotter JF, Roberts KE, Benza RL, Badesch DB, et al. Clinical risk factors for portopulmonary hypertension. Hepatology 2008;48(1):196–203.

[22] Krowka MJ, Miller DP, Barst RJ, Taichman D, Dweik RA, Badesch DB, et al. Portopulmonary hypertension: a report from the US-based REVEAL registry. Chest 2012;141(4):906–15.

[23] Zhao SX, Kwong C, Swaminathan A, Gohil A, Crawford MH. Clinical characteristics and outcome of methamphetamine-associated pulmonary arterial hypertension and dilated cardiomyopathy. JACC Heart Fail 2018;6(3):209–18.

[24] Humbert M, Sitbon O, Chaouat A, Bertocchi M, Habib G, Gressin V, et al. Survival in patients with idiopathic, familial, and anorexigen-associated pulmonary arterial hypertension in the modern management era. Circulation 2010;122(2):156–63.

[25] Shapiro S, Traiger GL, Turner M, McGoon MD, Wason P, Barst RJ. Sex differences in the diagnosis, treatment, and outcome of patients with pulmonary arterial hypertension enrolled in the registry to evaluate early and long-term pulmonary arterial hypertension disease management. Chest 2012;141(2):363–73.

[26] Farber HW, Miller DP, Poms AD, Badesch DB, Frost AE, Muros-Le Rouzic E, et al. Five-year outcomes of patients enrolled in the REVEAL registry. Chest 2015;148(4):1043–54.

[27] Chung L, Domsic RT, Lingala B, Alkassab F, Bolster M, Csuka ME, et al. Survival and predictors of mortality in systemic sclerosis-associated pulmonary arterial hypertension: outcomes from the pulmonary hypertension assessment and recognition of outcomes in scleroderma registry. Arthritis Care Res 2014;66(3):489–95.

[28] Kawut SM, Lima JAC, Barr RG, Chahal H, Jain A, Tandri H, et al. Sex and race differences in right ventricular structure and function: the multiethnic study of atherosclerosis-right ventricle study. Circulation 2011;123(22):2542–51.

[29] Ventetuolo CE, Ouyang P, Bluemke DA, Tandri H, Barr RG, Bagiella E, et al. Sex hormones are associated with right ventricular structure and function: the MESA-right ventricle study. Am J Respir Crit Care Med 2011;183(5):659–67.

[30] Foppa M, Arora G, Gona P, Ashrafi A, Salton CJ, Yeon SB, et al. Right ventricular volumes and systolic function by cardiac magnetic resonance and the impact of sex, age, and obesity in a longitudinally followed cohort free of pulmonary and cardiovascular disease: the Framingham heart study. Circ Cardiovasc Imaging 2016;9(3):e003810.

[31] Ventetuolo CE, Mitra N, Wan F, Manichaikul A, Barr RG, Johnson C, et al. Oestradiol metabolism and androgen receptor genotypes are associated with right ventricular function. Eur Respir J 2016;47(2):553–63.

[32] Austin ED, Cogan JD, West JD, Hedges LK, Hamid R, Dawson EP, et al. Alterations in oestrogen metabolism: implications for higher penetrance of familial pulmonary arterial hypertension in females. Eur Respir J 2009;34(5):1093–9.

[33] Marsh JD, Lehmann MH, Ritchie RH, Gwathmey JK, Green GE, Schiebinger RJ. Androgen receptors mediate hypertrophy in cardiac myocytes. Circulation 1998;98(3):256–61.

[34] Hemnes AR, Maynard KB, Champion HC, Gleaves L, Penner N, West J, et al. Testosterone negatively regulates right ventricular load stress responses in mice. Pulm Circ 2012;2(3):352–8.

[35] Ventetuolo CE, Baird GL, Barr RG, Bluemke DA, Fritz JS, Hill NS, et al. Higher estradiol and lower dehydroepiandrosterone-sulfate levels are associated with pulmonary arterial hypertension in men. Am J Respir Crit Care Med 2016;193(10):1168–75.

[36] van de Veerdonk MC, Kind T, Marcus JT, Mauritz G-J, Heymans MW, Bogaard HJ, et al. Progressive right ventricular dysfunction in patients with pulmonary arterial hypertension responding to therapy. J Am Coll Cardiol 2011;58(24):2511–9.

[37] Vanderpool RR, Pinsky MR, Naeije R, Deible C, Kosaraju V, Bunner C, et al. RV-pulmonary arterial coupling predicts outcome in patients referred for pulmonary hypertension. Heart 2015;101(1):37–43.

[38] Kawut SM, Al-Naamani N, Agerstrand C, Berman Rosenzweig E, Rowan C, Barst RJ, et al. Determinants of right ventricular ejection fraction in pulmonary arterial hypertension. Chest 2009;135(3):752–9.

[39] Jacobs W, van de Veerdonk MC, Trip P, de Man F, Heymans MW, Marcus JT, et al. The right ventricle explains sex differences in survival in idiopathic pulmonary arterial hypertension. Chest 2014;145(6):1230–6.

[40] Swift AJ, Capener D, Hammerton C, Thomas SM, Elliot C, Condliffe R, et al. Right ventricular sex differences in patients with idiopathic pulmonary arterial hypertension characterised by magnetic resonance imaging: pair-matched case controlled study. PLoS One 2015;10(5):e0127415.

[41] Docherty CK, Harvey KY, Mair KM, Griffin S, Denver N, MacLean MR. The role of sex in the pathophysiology of pulmonary hypertension. Adv Exp Med Biol 2018;1065:511–28.

[42] Lahm T, Tuder RM, Petrache I. Progress in solving the sex hormone paradox in pulmonary hypertension. Am J Physiol Lung Cell Mol Physiol 2014;307(1):L7–26.

[43] Deng Z, Morse JH, Slager SL, Cuervo N, Moore KJ, Venetos G, et al. Familial primary pulmonary hypertension (gene PPH1) is caused by mutations in the bone morphogenetic protein receptor-II gene. Am J Hum Genet 2000;67(3):737–44.

[44] Morrell NW, Aldred MA, Chung WK, Elliott CG, Nichols WC, Soubrier F, et al. Genetics and genomics of pulmonary arterial hypertension. Eur Respir J 2019;53(1):1801899.

[45] Larkin EK, Newman JH, Austin ED, Hemnes AR, Wheeler L, Robbins IM, et al. Longitudinal analysis casts doubt on the presence of genetic anticipation in heritable pulmonary arterial hypertension. Am J Respir Crit Care Med 2012;186(9):892–6.

[46] Jones RD, English KM, Pugh PJ, Morice AH, Jones TH, Channer KS. Pulmonary vasodilatory action of testosterone: evidence of a calcium antagonistic action. J Cardiovasc Pharmacol 2002;39(6):814–23.

[47] Oka M, Karoor V, Homma N, Nagaoka T, Sakao E, Golembeski SM, et al. Dehydroepiandrosterone upregulates soluble guanylate cyclase and inhibits hypoxic pulmonary hypertension. Cardiovasc Res 2007;74(3):377–87.

[48] Lahm T, Crisostomo PR, Markel TA, Wang M, Wang Y, Tan J, et al. Selective estrogen receptor-alpha and estrogen receptor-beta agonists rapidly decrease pulmonary artery vasoconstriction by a nitric oxide-dependent mechanism. Am J Physiol Regul Integr Comp Physiol 2008;295(5):R1486–93.

[49] Austin ED, Hamid R, Hemnes AR, Loyd JE, Blackwell T, Yu C, et al. BMPR2 expression is suppressed by signaling through the estrogen receptor. Biol Sex Differ 2012;3(1):6.

[50] Roberts KE, Fallon MB, Krowka MJ, Brown RS, Trotter JF, Peter I, et al. Genetic risk factors for portopulmonary hypertension in patients with advanced liver disease. Am J Respir Crit Care Med 2009;179(9):835–42.

[51] Baird GL, Archer-Chicko C, Barr RG, Bluemke DA, Foderaro AE, Fritz JS, et al. Lower DHEA-S levels predict disease and worse outcomes in post-menopausal women with idiopathic, connective tissue disease- and congenital heart disease-associated pulmonary arterial hypertension. Eur Respir J 2018;51(6):1800467.

[52] Mair KM, Wright AF, Duggan N, Rowlands DJ, Hussey MJ, Roberts S, et al. Sex-dependent influence of endogenous estrogen in pulmonary hypertension. Am J Respir Crit Care Med 2014;190(4):456–67.

[53] Kawut SM, Archer-Chicko CL, DeMichele A, Fritz JS, Klinger JR, Ky B, et al. Anastrozole in pulmonary arterial hypertension. A randomized, double-blind, placebo-controlled trial. Am J Respir Crit Care Med 2017;195(3):360–8.

[54] Parikh V, Bhardwaj A, Nair A. Pharmacotherapy for pulmonary arterial hypertension. J Thorac Dis 2019;11(Suppl. 14):S1767–81.

[55] Forte P, Kneale BJ, Milne E, Chowienczyk PJ, Johnston A, Benjamin N, et al. Evidence for a difference in nitric oxide biosynthesis between healthy women and men. Hypertension 1998;32(4):730–4.

[56] Galie N, Ghofrani HA, Torbicki A, Barst RJ, Rubin LJ, Badesch D, et al. Sildenafil citrate therapy for pulmonary arterial hypertension. N Engl J Med 2005;353(20):2148–57.

[57] Galie N, Brundage BH, Ghofrani HA, Oudiz RJ, Simonneau G, Safdar Z, et al. Tadalafil therapy for pulmonary arterial hypertension. Circulation 2009;119(22):2894–903.

[58] Rusiecki J, Rao Y, Cleveland J, Rhinehart Z, Champion HC, Mathier MA. Sex and menopause differences in response to tadalafil: 6-minute walk distance and time to clinical worsening. Pulm Circ 2015;5(4):701–6.

[59] Ghofrani HA, D'Armini AM, Grimminger F, Hoeper MM, Jansa P, Kim NH, et al. Riociguat for the treatment of chronic thromboembolic pulmonary hypertension. N Engl J Med 2013;369(4):319–29.

[60] Ghofrani HA, Galie N, Grimminger F, Grünig E, Humbert M, Jing ZC, et al. Riociguat for the treatment of pulmonary arterial hypertension. N Engl J Med 2013;369(4):330–40.

[61] Frey R, Saleh S, Becker C, Muck W. Effects of age and sex on the pharmacokinetics of the soluble guanylate cyclase stimulator riociguat (BAY 63-2521). Pulm Circ 2016;6(Suppl. 1):S58–65.

[62] Rubin LJ, Badesch DB, Barst RJ, Galie N, Black CM, Keogh A, et al. Bosentan therapy for pulmonary arterial hypertension. N Engl J Med 2002;346(12):896–903.

[63] Galie N, Olschewski H, Oudiz RJ, Torres F, Frost A, Ghofrani HA, et al. Ambrisentan for the treatment of pulmonary arterial hypertension: results of the ambrisentan in pulmonary arterial hypertension, randomized, double-blind, placebo-controlled, multicenter, efficacy (ARIES) study 1 and 2. Circulation 2008;117(23):3010–9.

[64] Miyauchi T, Yanagisawa M, Iida K, Ajisaka R, Suzuki N, Fujino M, et al. Age- and sex-related variation of plasma endothelin-1 concentration in normal and hypertensive subjects. Am Heart J 1992;123(4 Pt 1):1092–3.

[65] Stauffer BL, Westby CM, Greiner JJ, Van Guilder GP, Desouza CA. Sex differences in endothelin-1-mediated vasoconstrictor tone in middle-aged and older adults. Am J Physiol Regul Integr Comp Physiol 2010;298(2):R261–5.

[66] Gabler NB, French B, Strom BL, Liu Z, Palevsky HI, Taichman DB, et al. Race and sex differences in response to endothelin receptor antagonists for pulmonary arterial hypertension. Chest 2012;141(1):20–6.

[67] Barst RJ, Rubin LJ, Long WA, McGoon MD, Rich S, Badesch DB, et al. A comparison of continuous intravenous epoprostenol (prostacyclin) with conventional therapy for primary pulmonary hypertension. N Engl J Med 1996;334(5):296–301.

[68] Frantz RP, Schilz RJ, Chakinala MM, Badesch DB, Frost AE, McLaughlin VV, et al. Hospitalization and survival in patients using epoprostenol for injection in the PROSPECT observational study. Chest 2015;147(2):484–94.

[69] Galie N, Humbert M, Vachiery JL, Gibbs S, Lang I, Torbicki A, et al. 2015 ESC/ERS guidelines for the diagnosis and treatment of pulmonary hypertension: the joint task force for the diagnosis and treatment of pulmonary hypertension of the European Society of Cardiology (ESC) and the European Respiratory Society (ERS): endorsed by: Association for European Paediatric and Congenital Cardiology (AEPC), International Society for Heart and Lung Transplantation (ISHLT). Eur Respir J 2015;46(4):903–75.

[70] Benza RL, Gomberg-Maitland M, Elliott CG, Farber HW, Foreman AJ, Frost AE, et al. Predicting survival in patients with pulmonary arterial hypertension: the REVEAL risk score calculator 2.0 and comparison with ESC/ERS-based risk assessment strategies. Chest 2019;156(2):323–37.

[71] Duarte AG, Thomas S, Safdar Z, Torres F, Pacheco LD, Feldman J, et al. Management of pulmonary arterial hypertension during pregnancy: a retrospective, multicenter experience. Chest 2013;143(5):1330–6.

[72] Jais X, Olsson KM, Barbera JA, Blanco I, Torbicki A, Peacock A, et al. Pregnancy outcomes in pulmonary arterial hypertension in the modern management era. Eur Respir J 2012;40(4):881–5.

[73] Smith JS, Mueller J, Daniels CJ. Pulmonary arterial hypertension in the setting of pregnancy: a case series and standard treatment approach. Lung 2012;190(2):155–60.

[74] Bedard E, Dimopoulos K, Gatzoulis MA. Has there been any progress made on pregnancy outcomes among women with pulmonary arterial hypertension? Eur Heart J 2009;30(3):256–65.

[75] Olsson KM, Channick R. Pregnancy in pulmonary arterial hypertension. Eur Respir Rev 2016;25(142):431–7.

[76] Franco V, Ryan JJ, McLaughlin VV. Pulmonary hypertension in women. Heart Fail Clin 2019;15(1):137–45.

Chapter 13

Takotsubo Syndrome

Ashish Sharma, Sonali Kumar, Ana Micaela León, Gautam Kumar, and Puja K. Mehta

Clinical Case

A 68-year-old female presented to the Emergency Department with substernal chest pain and shortness of breath after an episode of syncope. She had a past medical history of hypertension, diabetes, and anxiety. She reported significant stress and anxiety surrounding her family's recent move to a new city away from friends of many years. Physical exam was pertinent for blood pressure 80/60 mmHg, pulse 130 beats per minute (bpm), respiratory rate 30 breaths per minute, and oxygen saturation 92% on room air. Other findings included jugular venous pressure (JVP) of 20 mmHg, bibasilar crackles, 2 + pitting edema bilaterally in the lower extremities, and cold extremities.

Laboratory evaluation revealed a troponin 4.5 ng/mL, BNP (brain natriuretic peptide) of 500 ng/L, potassium 4.5 meQ/L, and creatinine of 2.0 mg/dL. Electrocardiogram (ECG) demonstrated ST segment elevation in leads V2–V6 with deep T-wave inversions and QTc of 500 ms. Chest X-ray was pertinent for cephalization of pulmonary vessels and bilateral Kerley B lines and was consistent with pulmonary edema. Given her presentation, she was suspected of having an acute coronary syndrome (ACS) and was taken to the cardiac catheterization laboratory. Coronary angiography revealed no obstruction of the coronary arteries and left ventricular end diastolic pressure of 30 mmHg with left ventriculography showing apical hypokinesis. Transthoracic echocardiogram revealed a dilated left ventricle with a depressed function of 30% with apical akinesis, and basal segments were hypercontractile. Moderate-to-severe mitral regurgitation was noted due to systolic anterior motion of the anterior leaflet of the mitral valve, with mild-to-moderate tricuspid regurgitation. Aortic valve was trileaflet with mild aortic regurgitation and a normal aortic root measurement. Right ventricular size and function were normal. What is the etiology of her presentation, and how would you manage her going forward?

Abstract

Over the past several decades, there has been an increasing awareness of sex differences in cardiovascular disease presentation and pathophysiology. Stress-induced cardiomyopathy or Takotsubo syndrome (TTS) is one such condition that overwhelmingly affects postmenopausal women who have recently experienced a physical or an emotional stressor. The stressor activates a neurohormonal cascade causing a catecholamine storm that provokes symptoms similar to those described in acute coronary syndrome and heart failure. There are characteristic echocardiographic features that can indicate that a Takotsubo event has occurred. Treatment focuses on management of acute heart failure and monitoring for arrhythmias, because these patients can be at high risk of adverse outcomes and complications during the acute phase. This review focuses on our contemporary understanding of pathophysiology, diagnosis, and management of this condition that differentially impacts women.

Background and Diagnostic Criteria

TTS, also known as Takotsubo cardiomyopathy, stress cardiomyopathy, or broken heart syndrome, is a condition that is precipitated by an emotional or a physical stressor that mimics ACS and myocardial infarction with heart failure symptoms (**Figure 1**). Even positive emotional stressors (such as a surprise birthday party) can trigger TTS, described as happy heart syndrome. Reported cases date back to 1986, including a woman who had just lost her child to suicide [2]. In 1990, Dr. Hikaru Sato described it as a "Takotsubo-like left ventricular dysfunction" due to the similarities in shape between the ballooning left ventricular (LV) apex and the octopus-trapping pot used by Japanese fishermen [3]. Over the past two decades there has been an increased recognition of this condition, and studies have noted an increased prevalence among postmenopausal women. Signs and symptoms of the clinical presentation, including ECG abnormalities and different anatomical variants, have become more established over time. Several al-

gorithms and criteria have been proposed and revised to aid the clinician in diagnosing TTS (**Table 1**).

Women are more likely to present with ACS in the setting of no obstructive epicardial stenosis on coronary angiography. This presentation of myocardial infarction with no obstructive coronary arteries (MINOCA) includes coronary and noncoronary causes [7, 8]. During angiography, MINOCA is considered to be a working diagnosis that comprises patients who have had acute myocardial infarctions without any obstruction as well as patients with atherosclerosis that is insufficient to compromise myocardial perfusion. Coronary causes include emboli/thrombi, coronary microvascular and endothelial dysfunction, spontaneous coronary dissection, as well as coronary vasospasm. Noncoronary causes include myocarditis. When a patient presents with MINOCA, TTS should be on the differential diagnosis.

The diagnostic criteria published by the Mayo Clinic in 2004 include transient hypokinesis, akinesis, or dyskinesis of the LV mid-segments, with or without involvement of the apex (usually but not necessarily) in the presence of a stressful trigger [9]. Regional wall motion abnormalities must extend beyond a single epicardial vascular distribution; however, there are rare exceptions where the wall motion abnormalities are limited to a territory supplied by a single epicardial vessel. The Mayo Clinic criteria include the need for new electrocardiographic (ECG) abnormalities such as ST-segment elevation and/or T-wave inversion as well as the absence of pheochromocytoma and myocardi-

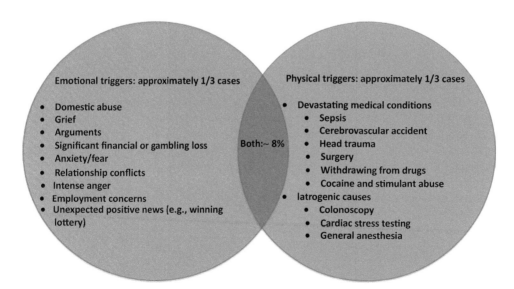

FIGURE 1   Emotional and Physical Triggers for TTS. In a study of the 1759 patients in the International Takotsubo Registry Study, 36% were found to have a physical trigger, 27.7% were found to have an emotional trigger, 7.8% were found to have both an emotional and physical trigger, and 28.5% were found to have no trigger. TTS, Takotsubo syndrome. *Data from [1].*

TABLE 1 Major Criteria Currently Used to Diagnose Takotsubo Syndrome

Revised Mayo Clinic Criteria [4]	Heart Failure Association of the European Society of Cardiology Criteria [5]	International Takotsubo Diagnostic Criteria [6]
Transient wall motion abnormalities (hypokinesia, dyskinesia, akinesia) of the mid and apical portion of the LV that extend beyond the territory of a single vessel	Transient wall motion abnormalities (hypokinesia, dyskinesia, akinesia) of the mid and apical portion of the LV that extend beyond the territory of a single vessel; recovery of ventricular dysfunction on follow-up cardiac imaging at 3–6 months	Transient wall motion abnormalities (hypokinesia, dyskinesia, akinesia) of the mid and apical portion of the LV that extend beyond the territory of a single vessel; however, there may be rare cases in which a regional wall motion abnormality is in the path of a single coronary artery. Having significant CAD is not a contraindication to TTS
Presence of a stressful trigger (not necessary)	Presence of a stressful trigger (not necessary)	Emotional, physical, or a combination of triggers (not necessary); classic patient is a postmenopausal woman
Obstructive CAD and rupture of an acute plaque should be excluded	Obstructive CAD and rupture of an acute plaque should be excluded	Accounting for neurological disorders, such as subarachnoid hemorrhage, cerebrovascular accident, that can be potential triggers for TTS
ECG abnormalities such as ST segment elevation or inversion of T wave; elevation of troponin that is discordant with the degree of LV systolic dysfunction	During the acute phase (3 months), ECG abnormalities including ST changes, left bundle branch block, and QTc prolongation must be new and reversible; degree of troponin elevation should be discordant with the amount of myocardium that is dysfunctional. High levels of serum natriuretic peptide (BNP or NT-proBNP) present during the acute phase	ECG abnormalities such as ST-segment elevation, ST segment depression, T-wave inversion, QTc prolongation; degree of troponin or creatinine kinase elevation is underwhelming compared to the degree of ventricular dysfunction. High levels of serum natriuretic peptide (BNP or NT-proBNP) during the acute phase
Exclusion of myocarditis and pheochromocytoma	Exclusion of other pathological conditions that could lead to transient LV abnormalities, such as hypertrophic cardiomyopathy and myocarditis	Exclusion of myocarditis

BNP, brain natriuretic peptide; CAD, coronary artery disease; ECG, electrocardiogram; LV, left ventricle; NT-proBNP, N-terminal pro-brain natriuretic peptide; TTS, Takotsubo syndrome.

tis [9]. The 2015 Heart Failure Association (HFA) of the European Society of Cardiology task force criteria include transient wall motion abnormalities of the left ventricle (LV) and right ventricle that may be preceded by an emotional or physical stressful trigger [10]. The regional wall motion abnormalities usually extend beyond a single vascular distribution, resulting in circumferential dysfunction of the ventricular segments. Atherosclerosis-mediated coronary artery disease (CAD), including acute plaque rupture, thrombus formation, and coronary dissection or other pathological conditions to explain LV dysfunction such as hypertrophic cardiomyopathy and viral myocarditis, must be excluded. During the first 3 months (acute phase), new and reversible ECG abnormalities such as ST segment elevation, ST depression, left bundle branch block (LBBB), T-wave inversion, and/or QTc prolongation should be present.

Serum natriuretic peptides (BNP or NT-proBNP) are significantly elevated during the acute phase. The rise in troponin is usually out of proportion (smaller) than the amount of myocardium affected. By 3–6 months, there should be recovery of ventricular systolic function on cardiac imaging [10]. The International Takotsubo (InterTAK) Diagnostic Criteria expand further to state that postmenopausal women are predominantly affected and triggers may include neurological disorders [such as subarachnoid hemorrhage (SAH), stroke/transient ischemic attack (TIA), seizures] as well as pheochromocytoma [11]. The InterTAK Diagnostic Criteria score can help determine the pretest probability of TTS; it includes five historical factors and two ECG-related factors, ranging up to 100 points. A female with an emotional and physical stressor, absence of ST-segment depression, concomitant anxiety or depression, and QTc prolongation would receive a score of 100.

Epidemiology

The precise incidence of TTS remains unknown, but it is estimated that there are 15–30 cases per 100,000 per year in the United States, with similar estimates in Europe [6]. Approximately 2% of patients with suspected ACS may actually have TTS. Similarities in presentation between ACS and TTS present a diagnostic challenge that may blur true incidence and prevalence rates, although better access to emergent angiography has led to an increase in recognition [10, 12, 13].

A majority of patients with TTS present with chest pain (67%) and dyspnea (17%), and some can be critically ill at presentation, with cardiogenic shock (4.2% of cases) and ventricular fibrillation (1.5%) [5, 12, 13]. Data from the Nationwide Inpatient Sample showed men had mortality rates more than two times higher compared to women (8.4% vs. 3.6%), most likely due to higher comorbid conditions of ventricular arrhythmias and sudden cardiac arrests [1, 14]. Acute respiratory failure, psychiatric or neurological disorders, systolic dysfunction, elevated troponin levels, and physical stressors on admissions were independent factors that predicted hospital course complications [1, 13, 15]. Rates of inpatient mortality in TTS have historically ranged from 1% to 5%, but newer data suggest 30-day mortality to be in the 4–6% range, which would be comparable to ST-elevation myocardial infarction (STEMI) and non-ST-elevation myocardial elevation (NSTEMI) [13].

Nearly 90% of patients with TTS are postmenopausal women, with no differences in prevalence rates seen among various ethnic groups (**Table 2**) [13]. Up to 80% of women who develop TTS are older than 50 years, and those above 55 years had a 10.7 times higher chance of developing TTS than similar age men, and 4.8 times higher chance than women below 55 years old. In contrast, there is no

TABLE 2 Sex Differences in Takotsubo Syndrome

Variables	Men	Women
Prevalence	~10–15%	~85–90%
Age	80% are >50 years	No association
Inpatient complications[a]	More common	Less common
All-cause mortality (first 30 days and beyond)	Higher	Lower

[a] *Complications include arrhythmias, cardiac arrest, ventilatory support, and intracranial hemorrhage.*

association between age and being diagnosed with TTS in men, who tend to be younger than women at age of diagnosis [5]. One study showed the median age at presentation to be 76 years in women and 72 years in men. Emotional triggers are more common in women whereas physical triggers had a higher prevalence for men [13, 16, 17]. Complications during the hospital course, such as arrhythmias, cardiac arrest, ventilatory support, and intracranial hemorrhage, are more common among men [5]. Rates of major adverse cardiac events (MACE) and inpatient mortality, as well as all-cause mortality during the first 30 days of admission, were higher in men compared to women [12, 13, 15]. The increased risk of all-cause mortality as well as cardiac and neurological adverse events extends beyond the first 30 days in men as well [5].

Research on TTS initially focused on the relative prevalence in Asians compared to Caucasians. However, recent studies have shifted to establishing differences in prevalence, presenting symptoms and complications in Caucasians, African Americans, and Hispanics. Compared to African Americans and Hispanics, Caucasians have shown an overall higher prevalence of TTS, emotional triggers and history of anxiety, depression, and/or selective serotonin reuptake inhibitor (SSRI) and benzodiazepine use [18]. Compared to Hispanics, African Americans are less likely to present with chest pain but often have longer hospital stays that are more likely complicated by acute heart failure and acute respiratory failure [18]. Caucasians typically report chest pain during initial presentation as compared to other ethnic groups [5]. African Americans (45.2%) have higher rates of in-hospital complications such as respiratory failure requiring ventilator support compared to Caucasians (18.3%) [13, 18]. Caucasians were more likely to be younger and have elevated troponin levels at presentation compared to African Americans (1.76 ± 4.2

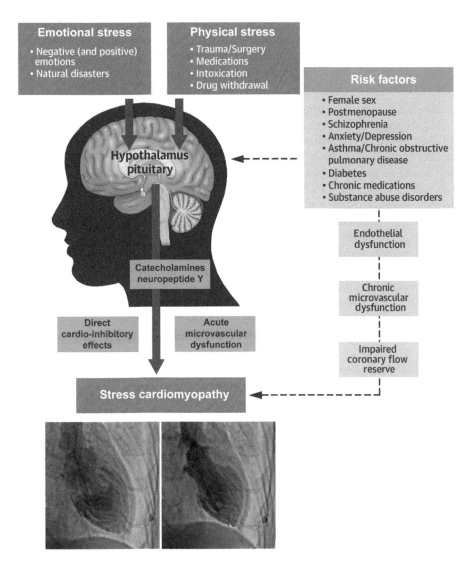

FIGURE 2   Pathophysiology of Takotsubo Syndrome. In a predisposed person with risk factors for endothelial dysfunction, in response to either emotional or physical stress, an intense stimulation of adrenergic stimuli may trigger Takotsubo cardiomyopathy, resulting in neurogenic-mediated cardiac stunning. *Reproduced with permission from [20].*

vs. 0.65±1.3) [18, 19]. African American patients show a higher prevalence of QT prolongation and T-wave inversion, whereas ST-segment elevation is more common in Asians, and ST-segment depression more common in non-African Americans [13, 18, 19].

Pathophysiology

Although the exact pathogenesis and underlying mechanisms contributing to TTS are not fully understood, sympathetic and neurohormonal activation and the resultant increase in catecholamines in response to an acute stressor form the basis of TTS (**Figure 2**) [21, 22]. Studies conducted on a small number of patients experiencing the acute phase of TTS have demonstrated significant in-

creases in cerebral blood flow to the hippocampus (limbic system), brainstem, and basal ganglia [23], which form in part the neurological structures involved in the human stress response (along with the neocortex, the rest of the limbic system, the reticular formation, and the spinal cord) [24]. Once the inciting stressor is perceived as sufficiently threatening, complex reciprocal interactions between limbic and neocortical structures [25] trigger a stress response, which in turn stimulates the hypothalamic-pituitary-adrenal axis [24]. Chromaffin cells of the adrenal medulla synthesize and release both norepinephrine and epinephrine into the circulation. The nerve impulses also descend through the spinal cord to the stellate ganglia, the origin point of sympathetic cardiac innervation [21]. Norepinephrine activates postsynaptic alpha and beta adrenoceptors [26]. Therefore, the totality of catecholamines present at postsynaptic adrenoceptors is a combination of circulating norepinephrine and epinephrine from the adrenal medulla and local release of norepinephrine by the sympathetic nerve [27].

Studies have demonstrated that circulating catecholamine levels are increased during the acute phases of TTS in nearly 75% of patients [28]. These circulating catecholamine and neuropeptide levels in TTS are several times higher than the levels found in patients with STEMI and continue to remain elevated a week after TTS begins [29]. Increased norepinephrine spillover in the coronary sinus has been reported, suggesting high catecholamine release from nerve endings in the myocardium [30]. Contraction band necrosis, caused by a calcium overload in the cardiomyocytes, is one of the histopathologic hallmarks of TTS and is characterized by contracted sarcomeres, eosinophilic bands, and a mononuclear inflammatory response [31]. Similarly, increased local norepinephrine has also been seen in stunned myocardium after a neurogenic cause such as SAH [32].

The mechanisms of how stress response leads to LV dysfunction patterns seen in TTS remain unclear [33]. An oxygen supply to demand mismatch is created from the increased heart rate and contractility caused by the catecholamine surge [34]. The resultant areas of myocellular hypoxia are further aggravated by metabolic changes which cause uncoupling of oxidative phosphorylation in mitochondria [35, 36]. Change in membrane permeability results in electrolyte abnormalities such as hypokalemia, hypocalcemia, and hypomagnesemia, all of which can contribute to myocardial toxicity [37]. Catecholamines can also be free radicals that interfere with calcium and sodium transporters which could cause further cardiomyocyte injury [38]. Differences in the regional densities of adrenergic receptors could explain the typical presentation of LV dysfunction seen in TTS, with beta 2 adrenergic receptors more heavily expressed in the apex than in the basal segments of the LV [10]. On the other hand, there is a greater density of β1 adrenergic receptors and sympathetic nerve terminals at the base compared to the apex of the LV [39]. Both catecholamines elicit positive inotropic responses via Gs-coupling protein; however, surges in catecholamines cause beta 2 adrenoceptors to switch from Gs (stimulatory) to Gi (inhibitory) signaling, which may cause a negative inotropic effect [40]. The resultant apical ballooning may confer cardioprotection and limit the degree of acute injury [20].

Invasive and noninvasive diagnostic tools have shown abnormal endothelial dysfunction and coronary microvascular responses in patients with TTS [41, 42]. Coronary microvascular dysfunction in TTS is further supported by reduced thrombolysis in myocardial infarction (TIMI) frame count during emergent angiography, perfusion defects in the LV decreasing contractility on contrast echocardiography, and decreased tracer uptake during acute phase with return to normal at follow-up on nuclear imaging [20, 43].

Endothelial dysfunction may explain the higher prevalence of TTS in postmenopausal women [43]. Decreased estrogen levels are associated with increased sympathetic drive and endothelial dysfunction [44]. Animal studies have shown that estrogen supplementation along with alpha and beta adrenoceptor blockers prevented stress-induced LV apical ballooning [45]. Estrogen treatment upregulated levels of atrial natriuretic peptide and heat shock protein 70, both substances with known cardioprotective benefits [21]. One study found that TTS women more frequently reported irregular menses, higher parity, and use of postmenopausal hormone therapy compared to myocardial infarction and healthy female controls [46]. Another study found no difference in prevalence of menopause, years since menopause, and rates of oophorectomy in females with TTS, females with ST-elevation myocardial infarction, and women with neither diagnosis.

Clinical Presentation

TTS presents similarly to ACS (ST-elevation or non-ST-elevation myocardial infarctions), with symptoms, electrocardiographic changes, and biomarker elevation indicative of myocardial injury (i.e., troponin elevation) [14, 19]. In approximately two-thirds of cases, a significant emotional or a physical stressor can be identified within the prior 1–5 days, and a majority of TTS cases are seen in postmenopausal women [21]. While most postmenopausal women (> 75%) are classically known to present with chest pain, they may also present with shortness of breath (nearly 50%), dizziness (> 25%), and/or syncope (5–10%) [13]. It is also important to note that the healthcare provider should ask about an emotional/physical stress since patients may not initially volunteer that information. The classic patient tends to be in respiratory distress, hypotensive, with signs of hypoperfusion. Cardiopulmonary physical exam can reveal signs of heart failure, including jugular venous distention, crackles often at the bases, sinus tachycardia with a gallop, and a systolic murmur [due to LV outflow tract obstruction (LVOTO) or mitral regurgitation]. When a patient with such signs or symptoms presents in the acute care setting, it is important to maintain a broad differential diagnosis as there are many conditions that share a similar clinical presentation, including but not limited to ACS, aortic dissection, spontaneous coronary artery dissection, endogenous catecholamine overload (e.g., pheochromocytoma), exogenous catecholamine overload (e.g., cocaine abuse), pulmonary embolism, myocarditis, and peripartum cardiomyopathy [47]. Distinguishing between primary TTS and secondary

TTS such as that precipitated by an underlying medical condition is important to do since it has management implications [10].

Diagnostic Testing

In the patient with TTS, ECG may show ST changes (elevations or depressions), deep T-wave inversions, and QT prolongation (**Figure 3**). The presence of ST-segment changes, negative T waves, bundle branch blocks, atrioventricular block, low voltage, and/or ventricular arrhythmias are more suggestive of acute myocarditis [10]. The most specific criteria for TTS includes ST segment elevation in lead aVR combined with ST segment elevation in anteroseptal leads (more than two of three, V1–V2–V3) [20, 48]. In cases of non-ST segment elevation, ST segment elevation in aVR combined with concomitant T-wave inversion in any lead was also found to be 100% specific for TTS (**Figure 4**) [48]. The lack of reciprocal changes, abnormal Q waves, and the sum of ST segment elevation in leads V4–V6 being greater than the sum of ST segment elevation in leads V1–V3 is reported to be highly specific and sensitive for TTS [49]. Due to the shared and dynamic ECG findings in ACS, coronary angiography and ventriculography are usually performed to determine the diagnosis of TTS [20].

Patients with TTS tend to have lower troponin levels compared to the extent of myocardium involved but tend to have higher BNP levels as compared to patients with ACS. In a study of 1750 patients with TTS, troponin levels were elevated in 87.0% of patients, with mean levels similar to those with ACS; however, the peak troponin significantly

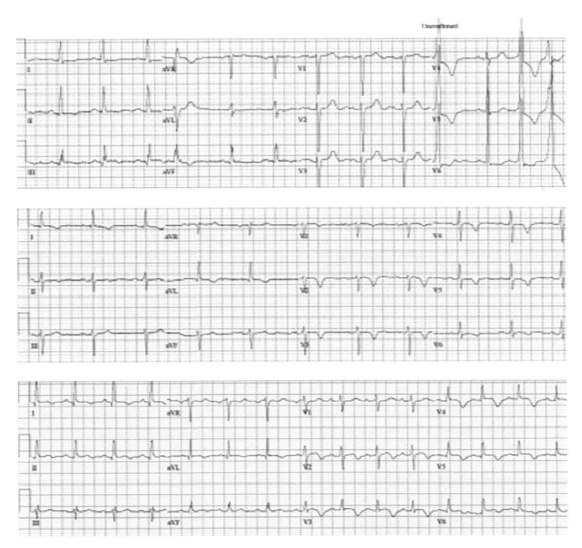

FIGURE 3   Example Electrocardiographic Patterns in Takotsubo Syndrome. ECG patterns can be dynamic and vary, with ST-T-wave changes. T-wave inversions may develop, with or without QTc prolongation, and therefore serial ECGs should be obtained in the acute setting. ECG, electrocardiogram.

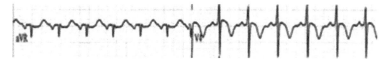

FIGURE 4 **Non-ST Segment Elevation and Takotsubo Syndrome.** In cases of non-ST segment elevation, ST elevation in aVR combined with concomitant T-wave inversion in any lead was also found to be 100% specific for TTS.

differed between the two conditions. While patients with TTS had a 1.8 factor increase in troponin levels compared to admission value, patients with ACS increased by a factor of 6 compared to admission value ($p < 0.001$). BNP levels were found to be 5.9 times higher than the normal limit in patients with TTS, far exceeding levels in patients with ACS [13].

The classical pattern on echocardiogram (in approximately 50–80% of cases) is a LV regional wall motion abnormality with hypokinesis of the mid- and apical regions and hypercontractility of the basal region. Apical ballooning is often associated with other features suggestive of a diagnosis of TTS such as LVOTO due to basal hypercontractility and mitral regurgitation due to systolic anterior motion of the anterior leaflet of the mitral valve. There can be diagnostic challenges in differentiating apical ballooning from anteroapical stunning due to myocardial ischemia in the case of obstruction of the left anterior descending artery (LAD) and patients with multivessel CAD may have numerous and extensive wall motion abnormalities mimicking TTS [20]. In the inverted Takotsubo or basal variant, circumferential basal hypokinesia and apical hypercontractility are present, also known as the "nutmeg" or "artichoke" heart. In the mid-left ventricular variant (MLV), there is circumferential mid-ventricular hypokinesis in addition to basal and apical hypercontractility. There are rarer variants including biventricular apical dysfunction, dysfunction sparing the apical tip, and isolated right ventricular Takotsubo syndrome. More than one subtype can occur in the same patient [10]. Similar patterns of TTS can be seen on ventriculography and are highlighted in **Figure 5**. Young women with TTS often have atypical echocardiographic findings of midventricular ballooning vs. the classic apical ballooning [50]. A recent study demonstrated that nearly 40% of patients may have a "mid-ventricular" variant, in which the apex is spared and there is akinesis of the mid-ventricular wall. An atypical variant in which there is akinesis of the basal wall with sparing of the apex and mid-ventricular wall has also been reported. Despite there being typical and atypical forms, there have been no studies demonstrating differences in patient characteristics or prognosis. Variants involving hyperkinesis of the basal region and akinesis of the mid-apical region can lead to LVOTO or functional mitral valve regurgitation.

Most patients with TTS ultimately undergo coronary angiography due to a presentation that mimics ACS. In a study of 2399 adult patients with TTS, the 261 patients who were managed without coronary angiography were found

to be slightly older (66.3 years vs. 65.7 years, $p = 0.009$), had a higher prevalence of chronic kidney disease (11.5% vs. 5% $p < 0.001$), and atrial fibrillation (15.3% vs. 10.2%, $p = 0.012$) compared to those managed with coronary angiography [51]. Age, sex, race, prevalence of diabetes, and hypertension did not differ between the two groups. The incidence of shock (3.4% vs. 3.5%, $p = ns$), and rate of major operations during the hospital stay [4.6% vs. 2.4%, $p = $ not significant (ns)] were similar between the groups; however, patients who did not undergo angiography had a higher in-hospital mortality rate compared to those who did (2.7% vs. 0.9%, $p = 0.012$). Even after adjusting for clinical and procedural variables, not undergoing coronary angiography was found to be an independent predictor of higher in-hospital mortality in patients with TTS [$p = 0.033$, odds ratio (OR) 2.70 (1.08–6.75)]. Patients with TTS who do not undergo angiography have a higher in-hospital mortality compared to those who do undergo coronary angiography, possibly due to misdiagnosis and the incorrect inclusion of patients with ACS in the former group.

Speckle tracking echocardiography (STE) is another modality that has been utilized in the detection of stress cardiomyopathy. One study assessed LV global longitudinal strain (GLS) alterations using STE in patients that were recovering from aneurysmal SAH on days 1, 3, and 7 after admission. On day 1 impaired GLS in SAH patients compared to controls was clearly observed [$-16.7(-18.7/-13.7)\%$ vs. $-20(-22/-19)\%$, $p < 0.0001$]. GLS improved on day 7 as is expected with stress cardiomyopathy.

When the diagnosis of TTS is being considered, a ventriculography is usually performed unless contraindicated (e.g., in the case of an apical thrombus) as it is diagnostic for the condition and is able to diagnose certain forms (e.g., mid-ventricular) not well appreciated by echocardiography [51]. Coronary computed tomography angiography is a noninvasive alternative to invasive coronary angiography to exclude significant obstructive CAD in patients who are stable and have a high likelihood of having TTS (e.g., transthoracic echocardiogram consistent with the diagnosis) [20]. Coronary angiograms should be carefully reviewed by experienced operators to definitively exclude the presence of spontaneous coronary artery dissection, which has been observed to overlap with TTS [52].

The pattern of the late gadolinium enhancement (LGE) detected by cardiac magnetic resonance (CMR) imaging also differs between TTS, myocardial infarction, and

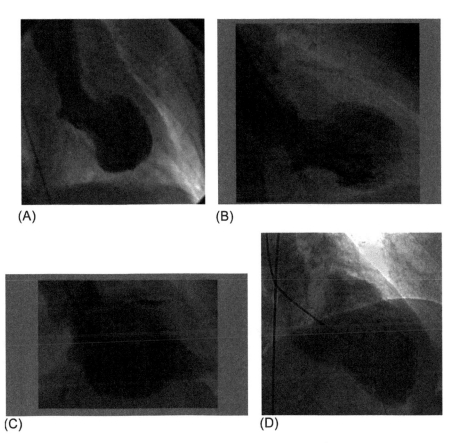

FIGURE 5 Variations in Ventricular Contraction Patterns in Takotsubo Syndrome. Illustration demonstrating the pathophysiology of stress cardiomyopathy in addition to the role of emotional stress, physical stress, and risk factors. Panel (A) demonstrates the apical (classical) pattern (end-systolic image from left ventriculography). Panel (B) demonstrates the mid-ventricular form (end-systolic image from left ventriculography). Panel (C) demonstrates the inverted or basal type (end-systolic image from left ventriculography). Panel (D) demonstrates the focal type (end-systolic image from left ventriculography) [5, 25].

myocarditis and aids in diagnosis in challenging cases. In the initial phase of TTS, edema appears as high signal intensity on T2-weighted CMR, and typically aligns with the region of wall motion abnormality in TTS. The acute phase of TTS is characterized by myocardial edema with no evidence of LGE on CMR [53]. The TTS-induced ventricular dysfunction is transient and without significant tissue fibrosis as observed during acute phase and 3-month follow-up CMR images [54] (**Figure 6**). The classic hallmark of TTS is the absence of LGE. Even when present, LGE is much lower intensity than that seen with myocardial infarction (signal intensity < 5 standard deviation) [20]. In contrast to TTS, both myocarditis and myocardial infarction typically have macroscopic fibrosis. A patient with apical infarction may mimic the apical ballooning pattern on end-systolic functional imaging, but exhibits subendocardial and often transmural delayed enhancement on CMR (**Figure 7**). In contrast, a patient with myocarditis often presents with patchy mid-myocardial or subepicardial delayed enhancement.

Treatment and Clinical Course

The objective of management in TTS patients is to prevent any further complications such as arrhythmias while providing supportive care to maximize recovery, which usually happens in a couple of weeks. In the acute phase, close monitoring for QT interval prolongation, conduction abnormalities, and arrhythmias is needed, along with treatment of heart failure [5, 57].

In patients presenting with LVOTO and shock, use of inotropes such as levosimendan, dobutamine, dopamine, and milrinone should be avoided due to their ability to increase ventricular contractility, which in turn would exacerbate the outflow tract obstruction. Low-dose beta-blockers may be used to decrease the LVOTO and increase cardiac output [20]. Esmolol, a short-acting beta-adrenergic receptor blocker, has been shown to be an effective option to promote hemodynamic stability [58]. Vasopressors such as phenylephrine or vasopressin may be used to augment blood pressure levels without exacerbating the outflow tract obstruction

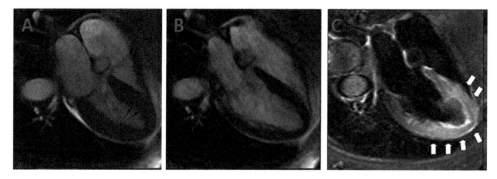

FIGURE 6 Cardiac MRI of Patient With Takotsubo Pattern. End-diastolic (A) and end-systolic (B) four-chamber view cine imaging demonstrating wall motion abnormalities (*black arrow*), 2-days after onset of symptoms. T2-weighted images in long-axis views (C) demonstrating normal signal intensity in the basal segments of the left myocardium and global edema in the mid- and apical segments (*white arrows*). MRI, magnetic resonance imaging. *Adapted with permission from [55].*

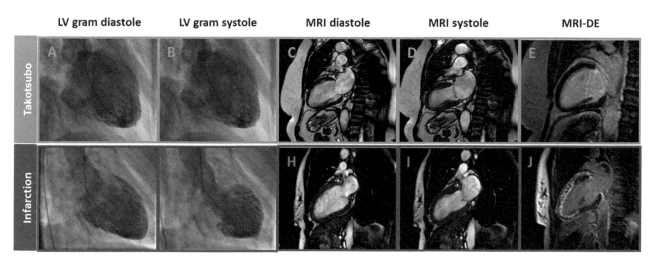

FIGURE 7 Comparison of Takotsubo Pattern vs. Myocardial Infarction on Cardiac MRI. Top blue panel represents a patient with Takotsubo cardiomyopathy, and bottom red panel a patient with apical myocardial infarction. End-systolic images of the left ventriculogram (B, G) cine images of cardiac MRI (D, I) demonstrate the typical apical ballooning shape in both patients. However, the late gadolinium enhancement images of MRI demonstrate no scar in patient with Takotsubo (panel E). In contrast, the patient with apical infarct has a dense transmural apical scar with microvascular obstruction (panel F). DE, delayed enhancement; LV gram, left ventriculogram; MRI, magnetic resonance imaging. *Adapted with permission from [56].*

[20]; however, catecholamines such as norepinephrine should be avoided due to their inciting role in TTS [59]. Low-dose vasopressors may also be utilized as a bridging therapy [20]. Patients who develop severe cardiogenic shock should be evaluated for further means of circulatory support [10]. Intra-aortic balloon pumps augment blood flow to the coronary arteries during diastole and may potentially be harmful to patients with TTS due to increased transaortic pressures during systole, which could exacerbate LVOTO [60]. However, some studies have also proven that the intra-aortic balloon pump is an effective temporary measure [61]. Studies have also shown both benefits and drawbacks to using extracorporeal membrane oxygenation

(ECMO) in TTS. Patients with TTS experiencing right ventricular failure or those that have achieved return of spontaneous circulation after cardiac arrest and have been put on ECMO have shown positive results [62]. However, because the shunt created by ECMO enhances afterload, care should be taken to avoid ECMO use in patients with TTS and LVOTO [60].

Angiotensin-converting enzyme inhibitors (ACEi) are a mainstay treatment for systolic heart failure with reduced ejection fraction, and studies have shown their therapeutic benefits in TTS. One study shows the mortality rate of TTS patients treated with ACEi to be one-fourth that of those that did not receive ACEi at the 1-year point [13]. Another study reports that patients treated with ACEi have lower rates of TTS recurrence than those who did not receive ACEi medications [62, 64]. Although no randomized control trials have been conducted thus far, some studies have shown no difference in recurrence or severity of TTS with administration of beta-blockers [65, 66].

Further research is needed to determine the use of aspirin or dual antiplatelet therapy in TTS patients [60]. Prevalence rates of anxiety disorder in TTS patient is estimated to be 13% and mood disorders to be 9% [67]. Compared to matched myocardial infarction controls, patients with TTS had 10% prevalence of anxiety compared to 1% in myocardial infarction (MI), and 20% prevalence of depression compared to 8% in those with MI [13]. Although there is no data on the use of antidepressant or anxiolytic medications in the prevention or delay of a TTS event, the mechanism by which these medicines work (lowered catecholamine reuptake) may actually prove to be a contraindication to use in TTS [68–70].

Complications and Prognosis

While the majority of patients with TTS have a favorable prognosis with resolution of ventricular dysfunction, the condition is not benign. Several complications of TTS have been described including LVOTO, severe mitral regurgitation, LV thrombus, cardiogenic shock, arrhythmias including complete heart block, and rarely death. In a study of 101 patients with TTS, compared to a control group with an anterior STEMI, TTS group had lower rates of sudden cardiac arrest, pulmonary edema, and cardiogenic shock (14.7% vs. 30.7%, $p=0.0078$) [71]. Three patients in the TTS group and one patient in the anterior STEMI group had cardiac rupture. The TTS group was also found to have a significantly lower rate of in-hospital mortality, mortality at 30 days, 3 months, 1 year, and 2.5 years ($p=0.035$, $p=0.0226$, $p=0.0075$, $p=0.009$, respectively). Annual rates of mortality among patients with TTS are 3.5% with significant predisposition to those with older age, those whose condition was precipitated by a physical stressor, and the atypical ballooning form of the condition. The annual recurrence rate was found to be 1%.

Recurrence

Takotsubo syndrome is known to carry a significant risk of recurrence, which is defined as another distinct TTS episode occurring after the total resolution of wall motion abnormality caused by the preceding TTS event [72]. TTS recurrence was seen in 4.7% of patients translating to an overall incidence rate of 18.7 cases per 1000 patient-years. Of patients with recurrence, there was no significant difference noted in sex nor were there differences in triggers, ballooning patterns, left ventricular ejection fraction, and

initial symptoms. The prevalence of comorbid conditions such as hypercholesterolemia (23.1% vs. 35.2%; $p=0.045$) and diabetes mellitus (4.6% vs. 15.7%; $p=0.012$) was decreased in patients with recurrence. Meanwhile, psychiatric [anxiety, adjustment, and affective disorders (42.6% vs. 32.5%; $p=0.028$)] and neurologic [migraine, cerebrovascular disease, seizure (37.1% vs. 24.7%; $p=0.028$)] disorders were seen more frequently in recurrent TTS patients. Nearly 35% of patients had a different ballooning pattern among their TTS episodes and 47% had a different trigger type. Definitive therapy for the prevention of recurrence has not been established thus far [72].

Knowledge Gaps

A number of questions remain unanswered about TTS. In patients with no identifiable physical or psychological stressor, what triggers the sympathetic stimulation resulting in a catecholamine storm? Does estrogen truly play a protective role in the pathogenesis of TTS? Studies pointing toward a genetic component of TTS have sought to explain the reasons for recurrence among patients as well as reported familial cases of the syndrome. Is there a genetic predisposition among people who experience TTS? What strategies can prevent recurrent events? Should beta-blockers and ACEi be continued indefinitely for years after the recovery of LV? Are there novel neurohormonal and inflammatory biomarkers that are prognostic in the TTS population?

Conclusion

TTS should be on the differential for both men and women who present with signs and symptoms of acute myocardial infarction and/or heart failure, with symptoms, ECG changes, and characteristic LV wall motion abnormalities. Classically, echocardiography reveals apical ballooning and LV basal hypercontractility, with a reduced ejection fraction. While no obstructive CAD on angiography is usually noted in TTS, obstructive CAD does not exclude a TTS diagnosis. In approximately two-thirds of cases of TTS, a significant emotional or a physical stressor can be identified within the prior 1–5 days, and a majority of cases are seen in postmenopausal women. A catecholamine storm in response to acute stressor, in the setting of underlying coronary endothelial dysfunction and abnormal coronary vasoreactivity, is implicated. While LV dysfunction recovers within a few weeks in a majority of cases, TTS is no longer believed to be benign. Management of TTS involves close ECG monitoring for QT prolongation and arrhythmias, and the treatment of heart failure. Mechanistic research in brain-heart connection, i.e., autonomic reactivity and connections to vascular reactivity and myocardial stress, is needed to further our understanding of why TTS differentially impacts women.

CH
13

Key Points

1. Ninety percent of cases of Takotsubo syndrome are in women with mean ages 66–68 years.

2. Male sex is associated with 2.46 times higher inpatient mortality.

3. TTS presents with transient regional wall motion abnormalities, identified by LV gram, TTE, or cardiac MRI.

4. Emotional and physical triggers are identified in one-third of the cases, resulting in sympathetic and neurohormonal response.

5. ECG abnormalities include ST elevation especially in lead aVR and T-wave inversions.

6. Patients with TTS have a much smaller rise in troponin levels from admission value, compared to patients with ACS.

7. In patients with TTS, BNP levels are significantly higher than the normal limit, far exceeding levels in patients with ACS.

8. Absence of scar using MRI-based delayed enhancement imaging can synch the diagnosis of TTS and rule out myocardial infarction and myocarditis.

9. While no obstructive CAD on angiography is usually noted in TTS, obstructive CAD does not exclude a TTS diagnosis. It is important to rule out ACS with the presence of obstructive CAD, usually with invasive angiography.

Back to Clinical Case

The case presented earlier in the chapter depicts a 68-year-old female with substernal chest pain, life stress, heart failure on exam, found to be having a ST-elevation myocardial infarction, but with no obstructive coronary artery disease. Echocardiogram demonstrated an ejection fraction of 30%, with apical akinesis. The patient was ultimately deemed to have Takotsubo syndrome (TTS) and was treated based on American College of Cardiology (ACC)/American Heart Association (AHA) guidelines for management of acute heart failure [55]. She was treated with phenylepherine and a percutaneous left ventricular assist device (Impella CP) to provide temporary hemodynamic support. On day 2 of her hospital course, her blood pressure improved to 110/80 mmHg, pulse was 78 bpm, and extremities were warm and well perfused. On day 4, she was started on metoprolol and lisinopril. Her clinical status continued to improve. On day 7, she was discharged to a subacute rehabilitation facility and agreed to participating in cognitive behavioral therapy to better learn how to cope with the stressors involving her recent move.

References

[1] Sharkey SW, Maron BJ. Epidemiology and clinical profile of Takotsubo cardiomyopathy. Circ J 2014;78(9):2119–28.

[2] Ryan T, Fallon JT. Case records of the Massachusetts General Hospital. Weekly clinicopathological exercises. Case 18-1986. A 44-year-old woman with substernal pain and pulmonary edema after severe emotional stress. N Engl J Med 1986;314(19):1240–7.

[3] Dote K, Sato H, Tateishi H, Uchida T, Ishihara M. Myocardial stunning due to simultaneous multivessel coronary spasms: a review of 5 cases. J Cardiol 1991;21(2):203–14.

[4] Stawiarski K, Ramakrishna H. Redefining Takotsubo syndrome and its implications. J Cardiothorac Vasc Anesth 2019;1–5.

[5] Dias A, Núñez Gil IJ, Santoro F, Madias JE, Pelliccia F, Brunetti ND, et al. Takotsubo syndrome: state-of-the-art review by an expert panel – part 1. Cardiovasc Revasc Med 2019;20(1):70–9.

[6] Brinjikji W, El-Sayed AM, Salka S. In-hospital mortality among patients with takotsubo cardiomyopathy: a study of the National Inpatient Sample 2008 to 2009. Am Heart J 2012;164:215–21.

[7] Mehta PK, Beltrame JF. Myocardial infarction with non-obstructive coronary arteries: a humbling diagnosis in 2018. Heart 2019;105(7):506–7.

[8] Pasupathy S, Air T, Dreyer RP, Tavella R, Beltrame JF. Systematic review of patients presenting with suspected myocardial infarction and non-obstructive coronary arteries. Circulation 2015;131(10):861–70.

[9] Scantlebury D, Prasad A. Diagnosis of Takotsubo cardiomyopathy. Circulation 2014;78:2129–39.

[10] Lyon A, Bossone E, Schneider B, Sechtem U, Citro R, Underwood SR, et al. Current state of knowledge on Takotsubo syndrome: a position statement from the taskforce on Takotsubo syndrome of the heart failure Association of the European Society of cardiology. Eur J Heart Fail 2016;18(1):8–27.

[11] Ghadri J-R, Wittstein IS, Prasad A, Sharkey S, Dote K, Akashi YJ, et al. International expert consensus document on Takotsubo syndrome. Part I. Clinical characteristics, diagnostic criteria, and pathophysiology. Eur Heart J 2018;39(22):2032–46.

[12] Gil IJN, Andres M, Delia MA, Sionis A, Martín A, Bastante T, et al. Characterization of Tako-tsubo cardiomyopathy in Spain: results from the RETAKO National Registry. Rev Esp Cardiol 2015;68(6):505–12.

[13] Templin C, Ghadri JR, Diekmann J, Napp LC, Bataiosu DR, Jaguszewski M, et al. Clinical features and outcomes of Takotsubo (stress) cardiomyopathy. N Engl J Med 2015;373(10):929–38.

[14] Prasad A, Lerman A, Rihal CS. Apical ballooning syndrome (Tako-Tsubo or stress cardiomyopathy): a mimic of acute myocardial infarction. Am Heart J 2008;155(3):408–17.

[15] Sharkey SW, Windenburg DC, Lesser JR, Maron MS, Hauser RG, Lesser JN, et al. Natural history and expansive clinical profile of stress (tako-tsubo) cardiomyopathy. J Am Coll Cardiol 2010;55(4):333–41.

[16] Dias A, Franco E, Figueredo VM, Hebert K. Can previous oophorectomy worsen the clinical course of takotsubo cardiomyopathy females? Age and gender-related outcome analysis. Int J Cardiol 2014;177(3):1134–6.

[17] Dias A, Franco E, Figueredo VM, Hebert K, Quevedo HC. Occurrence of Takotsubo cardiomyopathy and use of antidepressants. Int J Cardiol 2014;174(2):433–6.

[18] Dias A, Franco E, Koshkelashvili N, Pressman GS, Hebert K, Figueredo VM. Racial and ethnic differences in Takotsubo cardiomyopathy presentation and outcomes. Int J Cardiol 2015;194:100–3.

[19] Franco E, Dias A, Koshkelashvili N, Pressman GS, Herbert K, Figueredo VM. Distinctive electrocardiographic features in African Americans diagnosed with Takotsubo cardiomyopathy. Ann Noninvasive Electrocardiol 2016;21(5):486–92.

[20] Medina de Chazal H, Del Buono MG, Keyser-Marcus L, Ma L, Moeller FG, Berrocal D, et al. Stress cardiomyopathy diagnosis and treatment JACC state-of-the-art review. J Am Coll Cardiol 2018;72(16):1955–71.

[21] Pelliccia F, Kaski JC, Crea F, Camici PG. Pathophysiology of Takotsubo syndrome. Circulation 2017;135(24):2426–41.

[22] Steptoe A, Kivimäki M. Stress and cardiovascular disease. Nat Rev Cardiol 2012;9:360–70.

[23] Suzuki H, Matsumoto Y, Kaneta T, Sugimura K, Takahashi J, Fukumoto Y, et al. Evidence for brain activation in patients with Takotsubo cardiomyopathy. Circ J 2014;78(1):256–8.

[24] Crossman A, Neary D. Neuroanatomy. London: Churchill Livingstone; 2000.

[25] LeDoux JE. Emotion circuits in the brain. Annu Rev Neurosci 2000;23:155–84.

[26] Lymperopoulos A, Rengo G, Koch WJ. Adrenal adrenoceptors in heart failure: fine-tuning cardiac stimulation. Trends Mol Med 2007;13:503–11.

[27] Florea VG, Cohn JN. The autonomic nervous system and heart failure. Circ Res 2014;114:1815–26.

[28] Akashi YJ, Nef HM, Lyon AR. Epidemiology and pathophysiology of Takotsubo syndrome. Nat Rev Cardiol 2015;12(7):387–97.

[29] Wittstein IS, Thiemann DR, Lima JAC, Baughman KL, Schulman SP, Gerstenblith G, et al. Neurohumoral features of myocardial stunning due to sudden emotional stress. N Engl J Med 2005;352:539–48.

[30] Kume T, Akasaka T, Kawamoto T, Yoshitani H, Watanabe N, Neishi Y, et al. Assessment of coronary microcirculation in patients with Takotsubo-like left ventricular dysfunction. Circ J 2005;69:934–9.

[31] Basso C, Thiene G. The pathophysiology of myocardial reperfusion: a pathologist's perspective. Heart 2006;92(11):1559–62.

[32] Ohtsuka T, Hamada M, Kodama K, Sasaki O, Suzuki M, Hara Y, et al. Images in cardiovascular medicine: neurogenic stunned myocardium. Circulation 2000;101:2122–4.

[33] Templin C, Napp L, Ghadri JR. Takotsubo syndrome: underdiagnosed, underestimated, but understood? J Am Coll Cardiol 2016;67:1937–40.

[34] Zhang X, Szeto C, Gao E, Tang M, Jin J, Fu Q, et al. Cardiotoxic and cardioprotective features of chronic β-adrenergic signaling. Circ Res 2013;112:498–509.

[35] Okonko DO, Shah AM. Heart failure: mitochondrial dysfunction and oxidative stress in CHF. Nat Rev Cardiol 2015;12:6–8.

[36] Behonick GS, Novak MJ, Nealley EW, Baskin SI. Toxicology update: the cardiotoxicity of the oxidative stress metabolites of catecholamines (aminochromes). J Appl Toxicol 2001;21:S15–22.

[37] Borkowski BJ, Cheema Y, Shahbaz AU, Bhattacharya SK, Weber KT. Cation dyshomeostasis and cardiomyocyte necrosis: the Fleckenstein hypothesis revisited. Eur Heart J 2011;32:1846–53.

[38] Y-Hassan S. Acute cardiac sympathetic disruption in the pathogenesis of the Takotsubo syndrome: a systematic review of the literature to date. Cardiovasc Revasc Med 2014;159:35–42.

[39] Ancona F, Bertoldi L, Ruggieri F, Cerri M, Magnoni M, Beretta L, et al. Takotsubo cardiomyopathy and neurogenic stunned myocardium: similar albeit different. Eur Heart J 2016;37:2830–2.

[40] Heubach JF, Ravens U, Kaumann AJ. Epinephrine activates both Gs and Gi pathways, but norepinephrine activates only the Gs pathway through human beta2-adrenoceptors overexpressed in mouse heart. Mol Pharmacol 2004;65:1313–22.

[41] Pelliccia F, Greco C, Vitale C, Rosano G, Gaudio C, Kaski JC. Takotsubo syndrome (stress cardiomyopathy): an intriguing clinical condition in search of its identity. Am J Med 2014;127:699–704.

[42] Naegele M, Flammer AJ, Enseleit F, Roas S, Frank M, Hirt A, et al. Endothelial function and sympathetic nervous system activity in patients with Takotsubo syndrome. Int J Cardiol 2016;224:226–30.

[43] Kaski JC. Cardiac syndrome X in women: the role of oestrogen deficiency. Heart 2006;92:iii5–9.

[44] Vitale C, Mendelsohn ME, Rosano GMC. Gender differences in the cardiovascular effect of sex hormones. Nat Rev Cardiol 2009;6:532–42.

[45] Ueyama T, Ishikura F, Matsuda A, Asanuma T, Ueda K, Ichinose M, et al. Chronic estrogen supplementation following ovariectomy improves the emotional stress-induced cardiovascular responses by indirect action on the nervous system and by direct action on the heart. Circ J 2007;71:565–73.

[46] Salmoirago-Blotcher E, Dunsiger S, Swales HH, Aurigemma GP, Ockene I, Rosman L, et al. Reproductive history of women with Takotsubo cardiomyopathy. Am J Cardiol 2016;118:1922–8.

[47] Gopalakrishnan P, Zaidi R, Sardar MR. Takotsubo cardiomyopathy: pathophysiology and role of cardiac biomarkers in differential diagnosis. World J Cardiol 2017;9(9):723–30.

[48] Frangieh AH, Obeid S, Ghadri J-R, Imori Y, D'Ascenzo F, Kovac M, et al. ECG criteria to differentiate between Takotsubo (stress) cardiomyopathy and myocardial infarction. J Am Heart Assoc 2016.

[49] Ogura R, Hiasa Y, Takahashi T, Yamaguchi K, Fujiwara K, Ohara Y, et al. Specific findings of the standard 12-lead ECG in patients with `Takotsubo' cardiomyopathy. Circulation 2003;67(8):687–90.

[50] Bertin N, Brosolo G, Antonini-Canterin F, Citro R, Minisini R, Alassas K, et al. Takotsubo syndrome in young fertile women. Acta Cardiol 2019;1–9.

[51] Hajali RH, Schuett AB, Suero G, Susco B. Importance of coronary angiography in the diagnosis of Takotsubo cardiomyopathy. J Am Coll Cardiol 2013;61(10).

[52] Duran JM, Naderi S, Vidula M, Michalak N, Chi G, Lindsay M, et al. Spontaneous coronary artery dissection and its association with takotsubo syndrome: novel insights from a tertiary center registry. Catheter Cardiovasc Interv 2020;95(3):485–91.

[53] Eitel I, Lücke C, Grothoff M, Sareban M, Schuler G, Thiele H, et al. Inflammation in Takotsubo cardiomyopathy: insights from cardiovascular magnetic resonance imaging. Eur Radiol 2010;20:422–31.

[54] Testa M, Feola M. Usefulness of myocardial positron emission tomography/nuclear imaging in Takotsubo cardiomyopathy. World J Radiol 2014;6:502–6.

[55] Athanasiadis A, Schneider B, Sechtem U. Role of cardiovascular magnetic resonance in takotsubo cardiomyopathy. Heart Fail Clin 2013;9(2):167–76 [viii].

[56] Eitel I, Behrendt F, Schindler K, Gutberlet M, Schuler G, Thiele H. Takotsubo cardiomyopathy or myocardial infarction? Answers from delayed enhancement magnetic resonance imaging. Int J Cardiol 2009;135(1):e9–12.

[57] Yancy CW, Jessup M, Bozkurt B, Butler J, Casey DE, Colvin MM, et al. 2017 ACC/AHA/HFSA focused update of the 2013 ACCF/AHA guideline for the Management of Heart Failure: a report of the American College of Cardiology/American Heart Association task force on clinical practice guidelines and the Heart Failure Society of America. J Card Fail 2017;23(8):628–51.

[58] Santoro F, Ieva R, Ferraretti A, Fanelli M, Musaico F, Tarantino N, et al. Hemodynamic effects, safety, and feasibility of intravenous esmolol infusion during Takotsubo cardiomyopathy with left ventricular outflow tract obstruction: results from a multicenter registry. Cardiovasc Ther 2016;34(3):161–6.

[59] Citro R, Rigo F, D'Andrea A, Ciampi Q, Parodi G, Provenza G, et al. Echocardiographic correlates of acute heart failure, cardiogenic shock, and in-hospital mortality in tako-tsubo cardiomyopathy. JACC Cardiovasc Imaging 2014;7(2):119–29.

[60] Dias A, Gil IJN, Santoro F, Madias JE, Pelliccia F, Brunetti ND, et al. Takotsubo syndrome: state-of-the-art review by an expert panel. Part 2. Cardiovasc Revasc Med 2019;20(2):153–66.

[61] Lisi E, Guida V, Blengino S, Pedrazzi E, Ossoli D, Parati G. Intra-aortic balloon pump for treatment of refractory ventricular tachycardia in Tako-Tsubo cardiomyopathy: a case report. Int J Cardiol 2014;174:135–6.

[62] van Zwet CJ, Rist A, Haeussler A, Graves K, Zollinger A, Blumenthal S. Extracorporeal membrane oxygenation for treatment of acute inverted takotsubo-like cardiomyopathy from hemorrhagic pheochromocytoma in late pregnancy. A A Case Rep 2016;7:196–9.

[63] Brunetti ND, Santoro F, De Gennaro L, Correale M, Gaglione A, Di Biase M. Drug treatment rates with beta-blockers and ACE-inhibitors/angiotensin receptor blockers and recurrences in takotsubo cardiomyopathy: a meta-regression analysis. Int J Cardiol 2016;214:340–2.

[64] Brunetti ND, Santoro F, De Gennaro L, Correale M, Gaglione A, Di Biase M, et al. Combined therapy with beta-blockers and ACE-inhibitors/angiotensin receptor blockers and recurrence of Takotsubo (stress) cardiomyopathy: a meta-regression study. Int J Cardiol 2017;230:281–3.

[65] Bonacchi M, Vannini A, Harmelin G, Batacchi S, Bugetti M, Sani G, et al. Inverted-Takotsubo cardiomyopathy: severe refractory heart failure in poly-trauma patients saved by emergency extracorporeal life support. Interact Cardiovasc Thorac Surg 2015;20:365–71.

[66] Santoro F, Ieva R, Musaico F, Ferraretti A, Triggiani G, Tarantino N, et al. Lack of efficacy of drug therapy in preventing takotsubo cardiomyopathy recurrence: a meta-analysis. Clin Cardiol 2014;37(7):434–9.

[67] Pelliccia F, Parodi G, Greco C, Antoniucci D, Brenner R, Bossone E, et al. Comorbidities frequency in Takotsubo syndrome: an international collaborative systematic review including 1109 patients. Am J Med 2015;128(6):654.e611–9.

[68] Salmoirago-Blotcher E, Rosman L, Wittstein IS, Dunsiger S, Swales HH, Aurigemma GP, et al. Psychiatric history, post-discharge distress, and personality characteristics among incident female cases of takotsubo cardiomyopathy: a case-control study. Heart Lung 2016;45(6):503–9.

[69] Neil CJA, Chong C-R, Nguyen TH, Horowitz JD. Occurrence of Tako-Tsubo cardiomyopathy in association with ingestion of serotonin/noradrenaline reuptake inhibitors. Heart Lung Circ 2012;21(4):203–5.

[70] Singh K, Carson K, Usmani Z, Sawhney G, Shah R, Horowitz J. Systematic review and meta-analysis of incidence and correlates of recurrence of takotsubo cardiomyopathy. Int J Cardiol 2014;174(3):696–701.

[71] Zalewska-Adamiec M, Bachorzewska-Gajewska H, Tomaszuk-Kazberuk A, Nowak K, Drozdowski P, Bychowski J, et al. Takotsubo cardiomyopathy: serious early complications and two-year mortality—a 101 case study. Neth Heart J 2016;24(9):511–9.

[72] Kato K, Di Vece D, Cammann VL, Micek J, Szawan KA, Bacchi B, et al. Takotsubo recurrence: morphological types and triggers and identification of risk factors. J Am Coll Cardiol 2019;73(8):982–4.

Editor's Summary: Takotsubo Cardiomyopathy

Sex Specific Features of Takotsubo Cardiomyopathy

Symptoms

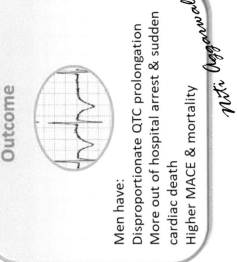

Women

Men

Chest pain is more common in women than men (73% vs 57%)

Outcome

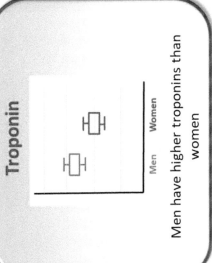

Men have:
Disproportionate QTC prolongation
More out of hospital arrest & sudden cardiac death
Higher MACE & mortality

Demographics

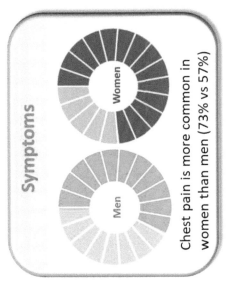

90% cases are in women
Mean age 67 years

Triggers

Physical triggers

Emotional Triggers

Majority (70–80%) have stressors.
Physical stress more common in men; emotional stress in women

Troponin

Men   Women

Men have higher troponins than women

Niti Aggarwal

Sex-Specific Features of Takotsubo Cardiomyopathy. Takotsubo cardiomyopathy is predominantly seen in women, and is often triggered by emotional stress. Men often express higher troponin levels and experience worse outcomes with higher incidence of sudden cardiac death and MACE (major adverse cardiovascular events) compared to women. *Data from [1,2]. Image courtesy of Niti R. Aggarwal.*

[1] Schneider B, Athanasiadis A, Sechtem U. Gender-related differences in takotsubo cardiomyopathy. Heart Fail Clin 2013;9(2):137–46.

[2] Medina de Chazal H, Del Buono MG, Keyser-Marcus L, Ma L, Gerard Moeller F, Berrocal D, et al. Stress cardiomyopathy diagnosis and treatment. J Am Coll Cardiol 2018;72(16):1955–71.

Chapter 14

Peripartum Cardiomyopathy

Jennifer Lewey and Zoltan Arany

Clinical Case

A 34year-old African American woman with a history of chronic hypertension and obesity presents with 1 week of dyspnea on exertion. She is a G3P2 and delivered a healthy baby girl at term gestation via cesarean section 4weeks prior. Her pregnancy was complicated by superimposed preeclampsia diagnosed just prior to delivery. Her review of systems is notable for a nonproductive cough and lower extremity edema and is negative for fever, chills, and orthopnea. She is taking nifedipine for hypertension. On exam, she is afebrile. She has mild resting tachypnea, her heart rate is 98 beats per minute, and her blood pressure is 130/78. Her jugular venous pressure (JVP) is difficult to assess. Lungs have diminished breath sounds at the bases bilaterally. Her heart rate is regular, she does not have a gallop, and a 2/6 systolic murmur is heard at the apex. Her abdomen is soft with a healing scar. She has 1+ pitting edema of her lower extremities. Her chest X-ray shows multifocal airspace opacities suspicious for pneumonia and a mildly enlarged cardiac silhouette (see **Figures 1**). A computed tomography (CT) pulmonary angiogram is performed and demonstrates extensive multifocal patchy groundglass opacities and no pulmonary embolism. Her laboratory analysis is notable for a normal comprehensive metabolic panel and complete blood count. An N-terminal pro-B-type natriuretic peptide (NT-proBNP) level is 3908pg/mL. A bedside echocardiogram shows severe dilatation of the left ventricle (LV) with an LV end-diastolic dimension of 7.3cm with normal LV wall thickness (**Figures 2**). The LV ejection fraction is 25%. She has moderate-to-severe mitral regurgitation, trace tricuspid regurgitation, and an estimated pulmonary artery systolic pressure of 46mmHg. What would be the next steps you would take in managing this patient's dyspnea and depressed left ventricular function?

Continued

Clinical Case—cont'd

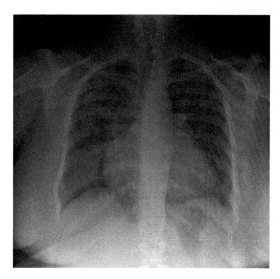

FIGURE 1 Chest X-Ray Showing Mild Cardio-
megaly and Multifocal Airspace Opacities.

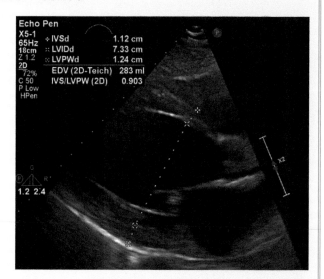

FIGURE 2 Parasternal Long-Axis View of the
Echocardiogram Showing Severe Dilation of the Left
Ventricle.

Abstract

Peripartum cardiomyopathy (PPCM) is an often severe
form of cardiomyopathy diagnosed in the peripartum period and is characterized by left ventricular systolic dysfunction and heart failure. The diagnosis is made in 1 in
every 1000–4000 live births in the United States and is more
commonly made in women of African ancestry. Other risk
factors include advanced maternal age, preeclampsia, and
twin gestations. The majority of women will recover cardiac function but short- and long-term complications are
common. Heart failure-specific therapies are the mainstay
of treatment and bromocriptine is currently under investigation as a disease-specific treatment in certain clinical cases.
Recent mechanistic work has highlighted the role of toxic
vascular hormones of late gestation and genetic predisposition in causing PPCM. In this chapter, we review the current literature on the epidemiology, pathophysiology, and
clinical management of PPCM, as well as future research
directions.

PPCM is a rare and often severe form of cardiomyopathy that occurs during or after pregnancy in the absence
of preexisting cardiac disease. PPCM is diagnosed in the
setting of new-onset heart failure occurring toward the
end of pregnancy or the months following delivery, accompanied by left ventricular (LV) systolic dysfunction,

operationally defined as an ejection fraction (EF) <45%
[1]. Patients diagnosed earlier in pregnancy may be clinically indistinguishable from those diagnosed during the traditional window [2–4]. Typically more than half of women
recover myocardial function, usually within 6 months of
diagnosis. However, many women go on to develop severe
persistent cardiomyopathy or heart failure, often requiring
the use of mechanical support or heart transplant.

The current prevalence of PPCM in the United States
is 1 in every 1000–4000 live births and may be rising.
Significant racial disparities exist: black women are more
likely to be diagnosed with PPCM and suffer from adverse outcomes. Although mortality rates are low, PPCM
is a significant contributor to pregnancy-associated deaths.
Maternal mortality is rising in the United States, in sharp
contrast to the rest of the developed world, and cardiomyopathy, usually PPCM, is the leading cause in the postpartum
period (**Figure 3**) [6–8]. Significant advances have been
made in recent years to understand the pathophysiology
and underlying mechanisms of PPCM. This chapter will
summarize our current understanding of the epidemiology,
clinical risk factors, proposed pathophysiology, management, and outcomes of PPCM.

Epidemiology

The incidence of PPCM in the United States varies based on
patient demographics and geography [9–12]. The disease

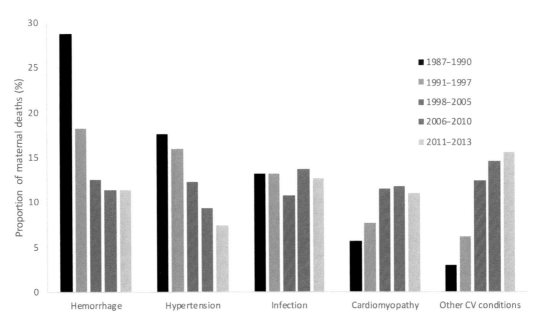

FIGURE 3 **Proportion of Maternal Deaths Attributed to Hemorrhage, Hypertension, Cardiomyopathy, and Other CV Conditions Over Time.** CV, cardiovascular. *Adapted from [5].*

prevalence may be underestimated due to underdiagnosis and varies due to differences in definitions used over time. In a statewide study in North Carolina, the overall prevalence was 1:2772 live births, with higher rates observed among black women (1:1087) compared to white women (1:4266) [10]. These prevalence rates are similar to those reported within the Kaiser healthcare system in northern and southern California [11, 13]. Epidemiological data in other countries are limited. The reported prevalence varies from 1 in every 10,000 births in Denmark [14] to 1 in every 4000 births in Taiwan [15], and 1 in every 1000 births in South Africa and Southeast Asia [16–18]. PPCM "hot spots" have been described in Haiti (1:300 births) and Nigeria (1:100 births), and may be related to environmental factors or higher prevalence of risk factors, such as preeclampsia [19, 20].

The incidence of PPCM appears to be increasing over time [12, 21]. Using administrative claims from a national inpatient database, the overall prevalence of PPCM increased from 1:1176 births in 2004 to 1:847 births in 2011 [21]. This temporal trend may be a result of the increasing proportion of women with risk factors for PPCM, such as older age and preeclampsia, in addition to an increasing familiarity with new billing practices, as the International Classification of Diseases billing code for PPCM was not introduced until 2003.

Risk Factors

Age

Maternal age ≥ 30 years is associated with 1.8 times greater odds of PPCM diagnosis compared to women aged 18–29,

after adjusting for other demographic and clinical risk factors [22]. Over half of all cases are observed in women ≥ 30 years, with the highest incidence observed among women aged 40–54 years, with an incidence of 1 in every 272 live births [21]. It is important to note, however, that PPCM can occur at any age.

Race

Women of African ancestry are at increased risk of developing PPCM but whether this association is caused by socioeconomic, environmental, or genetic risk is not clear [23]. The racial and ethnic composition of PPCM patients varies by region in the United States but nationwide studies suggest that 32% of all PPCM patients are black [12, 21]. The incidence of PPCM appears to be two to three times higher in black women compared to white women in the United States with a lower incidence described among Hispanic women [11, 21, 22]. Black women also present later in the postpartum period with more severely depression LV function and have lower rates of recovery, as described below.

Preeclampsia and Hypertensive Disorders

Hypertension is the most common medical complication of pregnancy, and preeclampsia affects up to 3–5% of pregnancies worldwide [24]. Preeclampsia is a leading risk factors for PPCM. In a meta-analysis of 22 cohort studies of PPCM, the overall prevalence of preeclampsia was 22%, and 37% of PPCM patients had any hypertensive disorder, including chronic hypertension, gestational hypertension, preeclampsia, or eclampsia [25]. Preeclampsia and PPCM share several risk factors, including black race, chronic hypertension, and older maternal age [26]. Controlling for these demographic and clinical characteristics, preeclampsia and eclampsia are associated with 2 and 5.9 times higher odds of PPCM, respectively [22]. The relationship between preeclampsia and PPCM does not appear to be mediated by race or region of the world [25, 27, 28], although chronic hypertension is more commonly reported among black women compared to white women with PPCM in the United States [29, 30]. The risk of PPCM appears related to the severity of preeclampsia, as gestational hypertension is associated with PPCM but to a lesser degree than preeclampsia [28].

Preeclampsia is also associated with subclinical cardiotoxicity, even in the absence of PPCM, as evidenced by diastolic dysfunction and/or abnormal myocardial strain on echocardiography, which can persist up to a year postpartum [31–33]. It is important to note that women with preeclampsia can present with pulmonary edema in the absence of depressed LV function. Only in the context of a reduced EF <45% should the diagnosis of PPCM should be made [34]. Despite the strong overlap between preeclampsia and PPCM and the potentially shared mechanistic pathway as discussed in later sections, these two diseases represent distinct entities, and over 90% of patients with preeclampsia will not develop PPCM [35].

Multiple Gestations

Women with multiple gestation pregnancy are at elevated risk of PPCM [22, 36]. The overall rate of twin or multiple gestation pregnancies across 16 cohort studies was 9%, significantly higher than the estimated prevalence in the United States of 3% [25, 34]. Case reports of triplet pregnancies have also been reported [37, 38].

Other conditions associated with PPCM have been described but less well substantiated, including substance abuse, anemia, asthma, thyroid disease, diabetes, autoimmune disease, diabetes and obesity, and prolonged tocolysis with terbutaline [13, 21, 36, 39–41].

Pathophysiology

Myocardial inflammation had long been postulated to play a possible role given the high prevalence of inflammatory infiltrates and viral genomes on right-sided endomyocardial biopsies in women with PPCM. However, when compared to populations of idiopathic dilated cardiomyopathy (DCM), the rates of inflammatory or infectious findings are not significantly different [42, 43]. Furthermore, in the prospective Investigations of Pregnancy-Associated Cardiomyopathy (IPAC) study of 100 women with PPCM, only 1 of the 40 patients undergoing CMR imaging demonstrated a pattern suggestive of myocarditis [35]. More recently, the IPAC investigators evaluated circulating immune cell profiles in PPCM and found that changes in immune cell subsets are not unique to PPCM but resemble those observed in other forms of cardiomyopathy [44]. The role of inflammation in PPCM thus remains unclear. Recent work, in contrast, has highlighted the two novel mechanisms: the role of toxic vascular hormones of late gestation and genetic predispositions (**Figure 4**). These are discussed below, with pathogenic factors listed in **Table 1**.

Prolactin Hypothesis

A landmark 2007 article provided the first evidence that PPCM is a vascular disease triggered by the hormonal changes of late pregnancy. In these experiments, the authors genetically deleted STAT3 (signal transducer and activator of transcription 3) in the cardiomyocytes of female mice, and noted that the mice developed PPCM [45]. STAT3 has been implicated in protecting the heart from ischemic injury, cardiotoxins, and chronic stress, and is critical for cardiac angiogenesis [46, 47]. Mechanistically, STAT3 acts in part by upregulating enzymes, such as manganese superoxide dismutase (MnSOD), that protect the heart by neutralizing reactive oxygen species (ROS) [48]. One of the outcomes of excess ROS in the absence of STAT3 in cardiomyocytes is the secretion of cathepsin D. Cathepsin D, in turn, cleaves multiple extracellular proteins. One of these, prolactin, is secreted by the pituitary gland in late pregnancy and during lactation. In the pregnant STAT3 cardiac knockout mice, inappropriate local cardiac secretion of cathepsin D cleaves prolactin into a 16 kD fragment, one of a host of vasculotoxic prolactin fragments known as vaso-inhibins. The 16-kDa prolactin fragment directly promotes apoptosis in endothelial cells, leading in the STAT3 mice to significant capillary dropout and consequent cardiomyopathy. Administration of bromocriptine, a dopamine agonist that suppresses pituitary secretion of prolactin, completely reversed the PPCM in these STAT3 cardiac knockout mice. STAT3 protein levels in cardiac tissue from five women with PPCM were reduced compared to normal human hearts, suggesting that these findings may extend to humans, although STAT3 suppression in cardiomyopathy appears not to be specific to PPCM [45].

In addition to direct vasculotoxicity, the 16 kD prolactin fragment also indirectly triggers cardiomyocyte damage through the expression of microribonucleic acid-146a (miR-146a) in endothelial cells. miR-146a is in turn packaged into vesicles, secreted by the endothelial cells, and internalized into cardiomyocytes, where it suppresses the neuregulin/ErbB pathway, ultimately promoting cardiomyocyte apoptosis [49]. The administration of antisense oligonucleotides to silence miR-146a in the STAT3 cardiac knockout mice

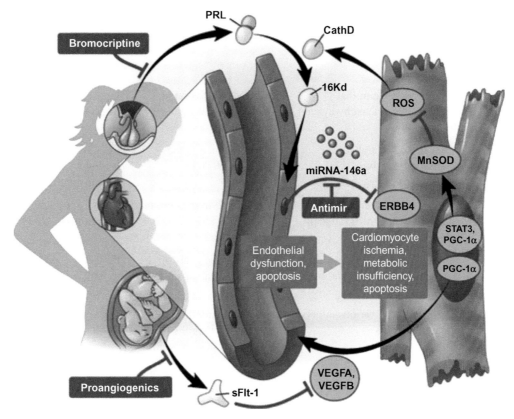

FIGURE 4   Mechanisms by Which Vasculotoxic Hormones and Prolactin Contribute to Peripartum Cardiomyopathy (PPCM). See text for details. CathD, cathepsin D; MnSOD, manganese superoxide dismutase; PGC-1α, peroxisomal proliferation activator receptor gamma coactivator-1α; PRL, prolactin; ROS, reactive oxygen species; sFlt-1, soluble fms-like tyrosine kinase-1; VEGF, vascular endothelial growth factor. *Reproduced with permission from [35].*

prevented the development of PPCM. Women with PPCM have markedly elevated levels of circulating miR-146a, compared to a group of patients with DCM. Thus, miR-146a may serve as both a biomarker and therapeutic target in PPCM [49].

Angiogenic Factors

The transcriptional coactivator PGC-1α (proliferator activated receptor gamma coactivator-1α), like STAT3, suppresses ROS via upregulation of MnSOD, and additionally promotes angiogenesis through the expression of vascular endothelial growth factor (VEGF) [50]. Genetic deletion of PGC-1α in mouse cardiomyocytes leads to PPCM, like the STAT3 mice [51]. However, PPCM in the PGC-1α knockout mice was only partially reversed with the administration of bromocriptine, and full rescue required additional treatment with VEGF, suggesting that that cardiotoxicity in this model is caused by both the activation of the antiangiogenic 16 kD prolactin-mediated pathway and inhibition of a pro-vascular VEGF-mediated pathway. This local VEGF pathway is

likely necessary to counteract a profound systemic VEGF inhibition that normally occurs in late gestation. In addition to the pituitary, the placenta secretes copious factors into the maternal circulation during late gestation. One of these, soluble fms-like tyrosine kinase-1 (sFlt-1) is a potent antiangiogenic protein that directly neutralizes VEGF. In the absence of PGC-1α, local secretion of VEGF cannot counteract systemic inhibition by sFlt-1, and vasculotoxicity and cardiomyopathy ensue. In fact, sFlt-1 administration was sufficient to cause cardiomyopathy in the PGC-1α mice, even in the absence of pregnancy. Higher sFlt-1 levels at the time of PPCM diagnosis in the IPAC cohort were associated with more severe functional limitation and major adverse clinical events, indicating that these findings extend to humans [52]. Women with recovered PPCM may continue to have elevated sFlt-1 levels compared to healthy controls [53].

Also supporting the importance of the sFlt-1 pathway in the development of PPCM is the observation that placental secretion of sFlt-1 is highly elevated in both preeclampsia

TABLE 1 Pathogenic Factors Associated With PPCM Mechanisms

	Physiological Role	Proposed Alteration in PPCM	Proposed Pathological Effect
STAT3	Transcription factor that regulates multiple pathways in the heart	Reduced expression in cardiomyocytes	Leads to lower expression of MnSOD, and higher ROS
MnSOD	Enzyme that neutralizes superoxide reactive oxygen species (ROS)	Downregulated in the absence of STAT3	Low MnSOD increases ROS, leading to increased secretion of cathepsin D
Cathepsin D	Lysosomal and secreted endo-protease, with a range of substrates	Upregulated in the absence of STAT3	Cleaves prolactin into vasculotoxic 16 kD fragment
Prolactin	Nursing hormone, secreted by the pituitary in late gestation and postpartum	Cleaved in the heart by presence of cathepsin D	Prolactin 16 kD fragment promotes endothelial cell and cardiomyocyte apoptosis
microRNA-146a	microRNA with various targets, including ErbB4	Induced in endothelial cells by 16 kD prolactin, and secreted via exosomes	146a-containing exosomes are taken up by cardiomyocyte and suppress ErbB4
ErbB4	Cell surface receptor on cardiomyocytes, promotes cellular viability and proliferation	Suppressed by microRNA-146a	Cardiomyocyte apoptosis
PGC-1α	Transcriptional coactivator that promotes expression of a wide range of metabolic genes	Reduced expression in cardiomyocytes	Leads to lower expression of MnSOD, as well as VEGF
VEGF	Growth factor vital for angiogenesis and to maintain vascular health	Downregulated in the absence of PGC-1α	Endothelial apoptosis, leading to cardiac ischemia
sFlt-1	Antiangiogenic protein that neutralizes VEGF, secreted by the placenta	Upregulated, in particular in context of preeclampsia	Neutralizes cardiac VEGF, leading to endothelial apoptosis and cardiac ischemia
TTN	Gene that encodes the protein titin, required for the integrity of the sarcomere	Genetic loss-of-function variants that lead to truncated protein	Contractile dysfunction that predisposes to "second hit," e.g., a vascular insult

microRNA, microribonucleic acid; microRNA-146a, microribonucleic acid-146a; MnSOD, manganese superoxide dismutase; PGC-1α, peroxisomal proliferation activator receptor gamma coactivator-1α; PPCM, peripartum cardiomyopathy; ROS, reactive oxygen species; sFlt-1, soluble fms-like tyrosine kinase-1; STAT3, signal transducer and activator of transcription 3; TTN, titin; VEGF, vascular endothelial growth factor.

and twin gestations, likely explaining the strong epidemiological link between these conditions and PPCM [25, 54, 55]. Increased levels of sFlt-1 in maternal circulation precede the development of preeclampsia, are associated with more severe preeclampsia, and presage worse outcomes [56]. Furthermore, as noted above, women with preeclampsia without evidence of PPCM experience subclinical cardiac changes, such as diastolic dysfunction and impairment of LV strain, and the severity of these abnormalities correlated with circulating sFlt-1 levels [51, 57].

In addition, PPCM likely represents a two-hit process, where vasculotoxic hormones secreted from the placenta (sFlt-1) and pituitary gland (16 kD prolactin fragment) provide a toxic challenge to the heart, and, in the absence of appropriate defenses (such as lack of PGC-1α or genetic predisposition), cardiotoxicity can ensue (**Figure 4**) [58].

Genetic Predisposition

Recent work has unveiled a significant genetic contribution to PPCM. Early suggestions to indicate that PPCM may have a genetic component included a small genome-wide association study in 79 patients that identified a single nucleotide polymorphism (SNP) near the parathyroid hormone like hormone (*PTHLH*) gene as associated with PPCM [59] and frequent reporting of familial clustering of PPCM [43, 60–63]. Evaluation of two large family pedigrees affected by both PPCM and idiopathic DCM identified a number of likely causal genetic variants in genes encoding myofibrillar proteins, including *TTN*, the gene encoding the large sarcomeric protein titin [64, 65]. Most recently, a study of 172 women with PPCM, who were not preselected for family history, revealed a striking 15%

prevalence in women with PPCM of high-impact nonsense mutation and splicing variants ($P = 1.3 \times 10^{-7}$), two-thirds of which were in *TTN* ($P = 2.7 \times 10^{-10}$) [66]. This profile of genetic variants is similar to that found in cohorts of idiopathic DCM, strongly suggesting a genetic overlap between these two diseases. In both cases, genetic variants have a predilection for affecting the A-band of TTN (**Figure 5**) [67, 68].

These findings demonstrate that PPCM often has genetic underpinnings but leave a number of unanswered questions. In post-hoc analyses, the subset of women enrolled in the clinically well-characterized IPAC cohort ($n = 83$) indicated that women with *TTN* variants had lower left ventricular ejection fraction (LVEF) at 1-year follow-up ($P = 0.005$), suggesting that the presence of *TTN* variants presages worse outcome, but this finding needs validation [29]. Damaging genetic variants were identified in only ~15% of women with PPCM, but the studies above were all focused on high-impact variants in myofibrillar genes, raising the question of whether other variants, or other genes, may contribute to the remaining 85%. In addition, the extent of genetic

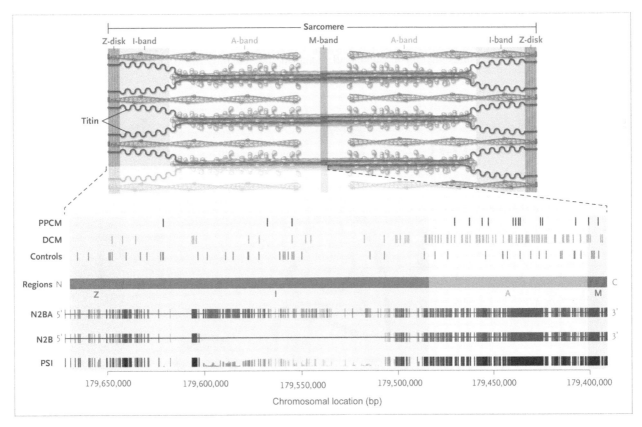

FIGURE 5   **Spatial Distribution of Truncating Variants Found in PPCM and DCM.** Schematic representation of cardiac sarcomere is shown at top. Titin spans the sarcomere from the Z-disk *(red)* through the I-band *(blue)* and A-band *(green)* and to the M-band *(purple)*. Spatial distributions of truncating variants in PPCM cases, in healthy volunteers, and in dilated cardiomyopathy (DCM) are indicated in the inset, as well as the two main cardiac transcripts N2BA and N2B, and exon usage in DCM hearts. DCM, dilated cardiomyopathy; PPCM, peripartum cardiomyopathy; PSI, proportion spliced in. *Reproduced with permission from [66].*

overlap between PPCM and DCM, which are not clinically identical, also requires further study. Finally, it will also be important to understand how truncations in *TTN* and other myofibrillar genes interact with the late gestational hormonal and anti-vascular context to ultimately cause PPCM.

Clinical Presentation and Diagnosis

The majority of women with PPCM present with new-onset heart failure, characterized by dyspnea, orthopnea, paroxysmal nocturnal dyspnea, and lower extremity edema (**Figure 6**). These symptoms may be mistaken for those associated with normal pregnancy and, as a result, the diagnosis may be missed or delayed. Less commonly, the initial presentation of PPCM may be ventricular tachycardia, acute ischemic stroke as a result of an LV thrombus, or other types of arterial thromboembolism [69–71]. Patients may also present with acute decompensated heart failure or cardiogenic shock, requiring urgent pharmacologic and mechanical support. PPCM is the leading cause of cardiogenic shock in women who have recently delivered a child [72].

Physical exam is usually notable for tachycardia, elevated jugular venous pressure, pulmonary rales, and peripheral edema. A third heart sound may be audible, although an S3 gallop may sometimes be present in normal pregnancy. The electrocardiogram shows sinus rhythm in almost all women. Sinus tachycardia, T-wave abnormalities, and ventricular hypertrophy are the most common abnormalities observed but are nonspecific [73, 74]. Chest radiography typically demonstrates an enlarged cardiac silhouette and pulmonary congestion. B-type natriuretic peptide (BNP) and N-terminal pro-BNP levels are usually elevated, while they are not in usual pregnancy and can be a helpful tool to distinguish cardiac vs. noncardiac causes of dyspnea in pregnancy [75]. Troponin T is elevated in the majority of patients [76]. It is important to note that women with severe preeclampsia but preserved LVEF can also present with all of the signs and symptoms above, including pulmonary edema, elevated BNP, and high-sensitivity troponin levels, underscoring the need for echocardiogram to confirm the diagnosis.

The echocardiographic criteria for PPCM includes systolic dysfunction with an EF < 45% in the absence of preexisting heart disease [1, 3]. Other echocardiographic features include LV dilatation, right ventricular (RV) dysfunction, elevated pulmonary pressures, biatrial enlargement, and mitral and tricuspid regurgitation [77, 78]. Cardiac magnetic resonance (CMR) imaging may be helpful to define LV function and rule out other causes of cardiomyopathy [79]. Magnetic resonance imaging (MRI) can be safely used during pregnancy but gadolinium crosses the placenta and should generally be avoided. MRI or echocardiography

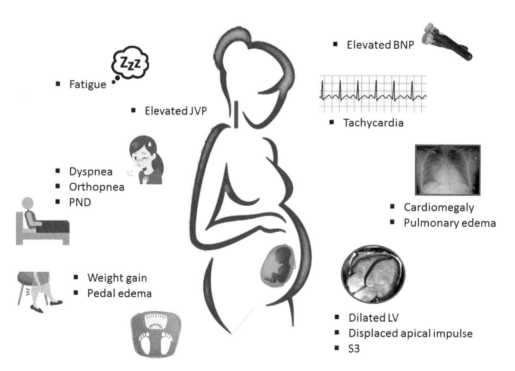

FIGURE 6 Signs and Symptoms of Peripartum Cardiomyopathy. Common symptoms, signs, and imaging results of peripartum cardiomyopathy. BNP, B-type natriuretic peptide; JVP, jugular venous pressure; LV, left ventricle; PND, postnatal depression. *Image courtesy of Niti R. Aggarwal.*

with contrast can assess for LV thrombus. An ischemic evaluation with left-heart catheterization or noninvasive imaging may be warranted for women at high risk for coronary artery disease or thromboembolic complications, but is usually not needed to make the diagnosis of PPCM [79]. Endomyocardial biopsy is generally not indicated unless there is suspicion for a competing diagnosis which would change management, such as giant cell myocarditis.

The majority of patients present postpartum, especially in the week after delivery, and less commonly in the later postpartum period [34]. Black women are more likely to present later postpartum compared to nonblack women, which may represent delayed time to diagnosis vs. a delayed onset of disease (**Figure 7**) [29, 30]. A small proportion of women present during pregnancy prior to the last gestational month and have similar baseline characteristics and long-term outcomes, suggesting that the timing of

diagnosis may represent a continuum rather than absolute cut-off, which is reflected in the updated European Society of Cardiology (ESC) guidelines [4].

The differential diagnosis includes pulmonary embolism, pneumonia, acute pulmonary edema associated with preeclampsia, and previously undiagnosed cardiac disease such as hypertrophic cardiomyopathy, cardiomyopathy from drugs or toxins, valvular disease, or congenital heart disease (**Table 2**). The risk of myocardial infarction from acute plaque rupture or spontaneous coronary artery dissection (SCAD) is elevated in the peripartum period, especially in the first few weeks postpartum, and often presents with chest pain and shortness of breath [83, 84]. Takotsubo cardiomyopathy is also on the differential of PPCM and is characterized on echocardiography by apical ballooning or variant forms and complete LV recovery, often within days [85, 86]. Takotsubo cardiomyopathy can also coexist

(A)

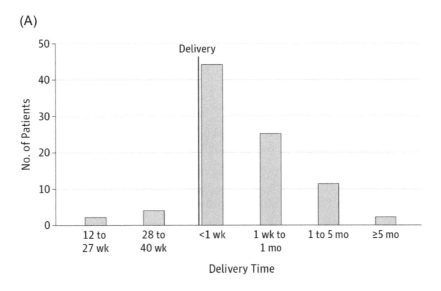

(B)

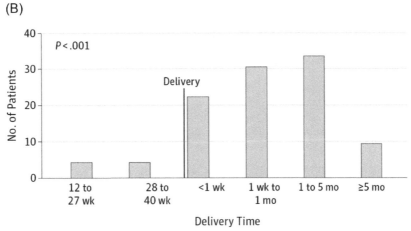

FIGURE 7 **Timing of Peripartum Cardiomyopathy (PPCM) Diagnosis by Timing of Delivery, Stratified by Race.**
(A) Non-African American women. (B) African American women. *Reproduced with permission from [30].*

TABLE 2 Differential Diagnosis of Dyspnea in Pregnancy

Benign Dyspnea of Pregnancy	Occurs in up to 76% of Women by 3rd Trimester [80]
Respiratory Pathologies	
Pulmonary embolism	Patients often present with acute-onset dyspnea and pleuritic chest pain. Increased risk during pregnancy and postpartum. Diagnosed on computed tomography (CT) pulmonary angiogram or ventilation perfusion (VQ) scan.
Pneumonia	Higher risk of infection during pregnancy and postpartum period due to anatomic/physiologic changes, increased risk of aspiration, and immune tolerance [81]. In rare cases can lead to acute respiratory distress syndrome (ARDS).
Iatrogenic fluid overload	Usually occurs when women have ≥ 3 L positive fluid balance over 48 h. Tocolytic agents can increase risk [82].
Asthma	Diagnosed in the setting of wheezing and history of asthma. Pulmonary function testing may be helpful if new diagnosis.
Placental Pathologies	
Preeclampsia	Diagnosed in setting of new-onset or worsening hypertension. Pulmonary edema may be present but echocardiogram shows normal LV function.
Amniotic fluid embolism	Acute-onset respiratory failure and/or circulatory collapse during or immediately after labor. Associated with disseminated intravascular coagulation (DIC).
Cardiomyopathies	
Peripartum cardiomyopathy	New-onset heart failure toward the end of pregnancy or the months following delivery. Echocardiogram shows EF <45%.
Preexisting cardiomyopathy	Diagnosis such as hypertrophic cardiomyopathy or familial dilated cardiomyopathy may not be known prior to pregnancy. Diagnosed by echocardiogram and family history. May become newly symptomatic following labor and delivery due to rapid fluid shifts.
Stress-induced cardiomyopathy	Echocardiogram most commonly shows LV systolic apical ballooning. Patients often present with acute chest pain and troponin elevation. May be triggered by cesarean section.
Other Cardiac Conditions	
Acute coronary syndrome (e.g., coronary dissection, atherosclerotic plaque rupture, vasospasm, or coronary embolism)	Patients often present with acute chest pain, ischemic electrocardiographic changes, and elevated troponin. Risk increases in 3rd trimester and postpartum period with coronary artery dissection as most common cause.
Cardiac arrhythmias	Patients often present with palpitations. Diagnosed with electrocardiogram, Holter monitor, or longer-term event monitoring. May be associated with development of heart failure (HF).
Valvular heart disease	Conditions may worsen in late 2nd and 3rd trimesters due to increase in plasma volume. Previously undiagnosed conditions include congenital bicuspid aortic valve or rheumatic valve disease, such as mitral stenosis.

EF, ejection fraction; LV, left ventricular.

with SCAD and therefore a careful examination of possible dissection, especially in the first week postpartum when pregnancy-associated SCAD is most common, may be warranted [87, 88].

Treatment and Complications

The management of patients with PPCM follows current heart failure treatment guidelines with specific consideration given to the safety of medications during pregnancy and lactation, when relevant. Loop diuretics should be used to treat pulmonary congestion and volume overload as needed, with caution to avoid overdiuresis during pregnancy which may lead to hypotension and uterine hypoperfusion. Neurohormonal blockade with angiotensin-converting enzyme (ACE) inhibitors, angiotensin receptor blockers (ARBs), and sacubitril/valsartan are contraindicated in pregnancy due to fetal toxic effects. The combination of hydralazine and nitrates can be safely used instead, especially when also treating hypertension. During lactation, enalapril and captopril can be safely used. Beta-blockade is generally safe and should be considered. Beta-1 selective agents are preferred during pregnancy due to the possible concern of nonselective agents facilitating uterine activity. Metoprolol tartrate is generally preferred during pregnancy due to more extensive clinical experience, especially with other cardiac conditions. Beta-blockers are secreted in breast milk at low concentration with little clinical consequence for the baby. Ivabradine can be considered in symptomatic patients who are not pregnant or breastfeeding and who have severely reduced EF and persistently elevated heart rates despite maximally tolerated doses of beta-blockade [89, 90]. Digoxin can be used in pregnancy for maternal and fetal arrhythmias and is compatible with breastfeeding. Mineralocorticoid receptor antagonists (MRAs) should be avoided during pregnancy but spironolactone can be considered during lactation. High resting heart rate and low systolic blood pressure are associated with adverse outcomes in PPCM, likely due to the inability to achieve therapeutic dosing of heart failure medications [91].

The optimal duration of heart failure therapies among patients with recovered LV function is not clear. In observational studies, withdrawal of heart failure medications in recovered PPCM has been associated with both good outcomes and increased risk of relapse [92–94]. Given the higher risk of relapse among patients with recovered DCM who discontinue heart failure therapies, continuing neurohormonal blockade and beta-blockers indefinitely may be preferred [95]. If limited by medication intolerance or patient preference, medical therapy should continue for at least 6 months after recovery followed by gradual tapering of one medication at a time [1]. Follow-up echocardiography after weaning and on an annual basis thereafter is reasonable to ensure lack of relapse.

Thromboembolism and Arrhythmia

The peripartum period is a hypercoagulable state and is associated with an increased risk of venous thromboembolism, especially in the first 6 weeks postpartum [96]. Consistent with this, PPCM is associated with a higher risk of thromboembolic complications compared to other forms of cardiomyopathy [34]. Numerous case reports have documented blood clots in both left- and right-sided cardiac chambers, and thromboembolic events, including stroke, can constitute the presenting symptom of PPCM [35, 97]. Thromboembolic events are frequently reported as the most common complications, affecting up to 6.6% of PPCM hospitalizations in the National (Nationwide) Inpatient Sample (NIS) and 6.8% of patients in the worldwide EURObservational registry [21, 27]. Anticoagulation is thus recommended in women with severely reduced EF and should be considered in all women with PPCM, especially if they are receiving bromocriptine, which is likely procoagulant [1, 79]. Heparin and low molecular weight heparin are preferred anticoagulants during pregnancy, whereas warfarin or low molecular weight heparin are compatible with breastfeeding.

Arrhythmias and Antiarrhythmic Therapies

The risk of ventricular arrhythmias and sudden cardiac death appears to be elevated in women with PPCM, especially among those presenting with severely reduced EF and delayed diagnosis [98–100]. For most women, however, the likelihood of myocardial recovery is high and may occur 12 months or more after diagnosis, thereby complicating the decision of when to place permanent implantable cardioverter-defibrillators (ICDs) for primary prevention. In women with an EF < 35% who are at high risk of poor outcomes, it may be reasonable to consider a wearable cardioverter/defibrillator (WCD) as a bridge to recovery or ICD implantation 3–6 months after diagnosis [1, 79].

A retrospective multicenter study in Germany found that 10% of women wearing a WCD, who had an EF ≤ 35%, experienced a shock for sustained tachyarrhythmias [101]. In contrast, a retrospective registry study of aftermarket WCDs in 107 women with PPCM in the United States (mean EF, 22%) revealed no shocks for ventricular tachycardia or fibrillation over an average follow-up of 4 months [102]. Although the average daily wear time was lower in this cohort compared to the German cohort above, the overall recovery rates were high and mortality was low.

The prevalence of other arrhythmias in PPCM is not well described but women with PPCM may be at increased risk of atrial fibrillation. In a Turkish cohort, 12% of women with PPCM experienced atrial fibrillation, a higher rate than reported in other series [94, 97]. Among women with

PPCM enrolled in a commercial health insurance plan, 1% had documented atrial fibrillation in the year after delivery [103].

Decompensated Heart Failure

Patients who develop low cardiac output with hypotension or evidence of end-organ perfusion may require vasopressors and/or inotropes for hemodynamic support. Levosimendan is a calcium sensitizing agent with phosphodiesterase (PDE)-3 inhibiting activity that improves cardiac filling pressures and cardiac output and, when compared to dobutamine, has similar or improved clinical outcomes [104–106]. Experimental evidence suggests that patients with PPCM may respond better to levosimendan rather than beta agonists, although a randomized trial of 24 women with PPCM showed no benefit of levosimendan on LV recovery or survival when added to standard treatment for heart failure [107]. Levosimendan is the recommended inotrope in European countries where it is available for clinical use [108, 109]. As in other cases of cardiogenic shock, norepinephrine is the preferred vasopressor [110, 111].

Advanced Mechanical Support

Patients with PPCM can present with severe depression of LV function and rapid hemodynamic deterioration. Given the high potential for recovery in this young patient population, aggressive treatment should be considered early for patients with circulatory shock despite maximal medical therapy [109]. Mechanical support with Impella device, extracorporeal membrane oxygenation (ECMO), and LV and biventricular assist devices can serve as a bridge to early recovery, durable support, or heart transplant. These strategies have been used successfully in varied contexts, but direct comparison between them is difficult because of differences in patient characteristics, severity of illness, and early management approaches [112–116].

Management of PPCM During Pregnancy

Among the minority of PPCM cases diagnosed during pregnancy, the management of heart failure during pregnancy requires a multidisciplinary approach by obstetricians, cardiologists, anesthesiologists, and neonatologists, with a focus on the health of both mother and fetus (**Figure 8**). Neonates born to mothers with PPCM are more likely to be premature, low birthweight, small for gestational age, and have lower Apgar scores compared to neonates born to mothers without PPCM [13]. Little evidence exists to guide the timing and mode of delivery in women with PPCM. Unless a cesarean section is indicated for other reasons, vaginal delivery with an epidural is preferred and can be performed with an assisted second stage to reduce maternal effort. In the setting of maternal deterioration, a cesarean section may be indicated to facilitate medical stabilization

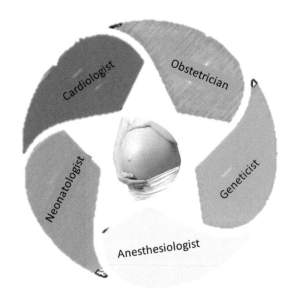

FIGURE 8   Multidisciplinary Heart Team Approach. *Image courtesy of Niti R. Aggarwal.*

of the mother. ECMO has been successfully used to stabilize women during pregnancy and cesarean delivery [113, 117]. Women with PPCM should undergo hemodynamic monitoring during labor and delivery and, in some cases, continuous monitoring of hemodynamics with a Swan-Ganz catheter may be necessary for medical optimization.

Bromocriptine

Bromocriptine is an ergot alkaloid and dopamine agonist that inhibits the secretion of prolactin in humans and is used for the treatment of prolactin-secreting adenomas and Parkinson's disease. Bromocriptine has also been shown to reverse the cardiomyopathy mimicking PPCM that develops in mice with a cardiomyocyte-specific deletion of STAT3 [45]. Bromocriptine was first tested, and compared to placebo, in a South African study of 20 women with PPCM and EF <35%. Fewer women receiving bromocriptine experienced the combined end point of poor outcome, defined by death, EF <35%, or New York Heart Association (NYHA) class III/IV heart failure at 6 months (10% vs. 80%, $P=0.006$) [118]. However, the small trial size and unusually high mortality rate in the control group (40%) has limited generalizability of these results into clinical practice. Prospective observational studies suggested improved outcomes in women receiving bromocriptine in Germany [63] and Quebec [119], but not among Danish women receiving a 2-day course of the prolactin inhibitor cabergoline [120].

The safety and therapeutic potential of bromocriptine were subsequently evaluated in an open-label multicenter randomized controlled trial in Germany, comparing two dosing regimens of bromocriptine (1 week vs. 8 weeks) along with prophylactic anticoagulation [121]. The dosing regimens compared 1 week of bromocriptine 2.5 mg twice daily to suppress lactation vs. 2 weeks of bromocriptine 2.5 mg twice daily followed by 6 weeks of 2.5 mg once daily. A placebo arm was not included due to ethical concerns of withholding treatment in light of the promising results observed by the authors in the studies described above [63, 118, 122]. No differences were observed in LVEF improvements between groups. No patients in the study died or required a left ventricular assist device (LVAD) or heart transplant, and bromocriptine was well tolerated, with no adverse events attributed to bromocriptine [121]. Comparing the pooled bromocriptine groups with EF <30% from this study with the subset of patients in the IPAC cohort with EF <30% (none of whom received bromocriptine) suggested that the clinical outcomes in women receiving bromocriptine in the German study appear quite favorable [121]. However, limiting the comparison to only white women in IPAC, or comparison to a Danish cohort that did not receive prolactin antagonists, reveals no differences in outcome between groups [120]. In summary, the clinical significance of these results is difficult to ascertain in the absence of a placebo control group in the German study, and generalization to racially diverse populations is limited.

Bromocriptine is generally well tolerated but can be associated with symptomatic hypotension, headache, and nausea. Despite its association with hypotension, it has been successfully used in women with cardiogenic shock treated with mechanical support devices [112, 114, 115]. Bromocriptine was previously used to prevent physiological lactation postpartum, but the Food and Drug Administration (FDA) withdrew approval for this indication in response to reported cases of hypertension, seizures, stroke, and myocardial infarction in the postpartum period [123–125]. However, a significant association between bromocriptine use and these adverse events has not been clearly demonstrated. When administered with anticoagulation, the safety profile of bromocriptine used in contemporary PPCM trials is reassuring. Finally, the potential clinical benefit of bromocriptine must also be weighed against the maternal and infant benefits of breastfeeding, especially in low- and middle-income countries [126].

Further study of bromocriptine in prospective, randomized, placebo-controlled trials is warranted to provide more information on the efficacy of bromocriptine therapy in PPCM in racially diverse populations. Safety data of bromocriptine when administered with anticoagulation in recent studies are reassuring, and a reasonable approach could include the administration of bromocriptine for patients with EF <35% and cardiopulmonary distress or other risk factors for poor prognosis after a detailed discussion with the patient regarding the potential benefits vs. risks. The European Society of Cardiology recommends consideration of bromocriptine in all PPCM patients (Class IIb recommendation), with the duration of treatment directed by severity of presentation [1].

Cessation of Breastfeeding

Preventing lactation suppresses prolactin secretion and may be predicted to be beneficial in PPCM. However, there are remarkably few published data that directly address this question, and most PPCM studies do not report on the issue. A small retrospective internet-recruited study in the United States suggested in fact that breastfeeding was associated with a better, rather than worse, maternal outcome [127]. Rates of breastfeeding in the prospective North American IPAC cohort were low (15%), and no difference was seen in mean change in LVEF from entry to 6 months or 12 months between women who breastfed and those who did not [128, 129]. Notably, rates of heart failure medications were similar among those who did and did not breastfeed. Most classes of heart failure medications can be safely used during lactation and should therefore not contraindicate breastfeeding. However, the metabolic demands of breastfeeding may be contraindicated for women with critical illness due to low output state or cardiogenic shock [130]. Finally, as noted above, the potential benefit of cessation of breastfeeding must also be weighed against infant benefits of breastfeeding, especially in low-income countries. Both the World Health Organization and the American Academy of Pediatrics recommend exclusive breastfeeding for 6 months, and continued breastfeeding for at least 1–2 years. In short, the indications for cessation of breastfeeding remain uncertain, and the risk vs. benefit of continued lactation should be discussed with women postpartum.

Prognosis

Myocardial recovery in contemporary cohorts in the United States is relatively high and usually occurs within 6 months postpartum. However, many women still experience severe complications, including death or severe heart failure requiring advanced mechanical support or heart transplantation. The long-term impact of PPCM on cardiac function, even in the setting of early recovery, has not been well-studied.

The IPAC cohort, as described above, enrolled 100 women diagnosed with PPCM across 30 medical centers in the United States and Canada [29]. Patients were followed clinically until 12 months postpartum and echocardiograms were reviewed by the core lab at 6 and 12 months postpartum. Survival free from LVAD and transplant at 12 months was 93%. Full recovery of LV function, defined as EF >50%, occurred in 72% of women by 12 months postpartum, with most patients demonstrating recovery by 6 months. Only 13% of women experienced a major adverse event or severe persistent cardiomyopathy (EF <35%) at 12 months postpartum. Several important associations with recovery were observed. First, no women with severe LV dilatation [left ventricular external end-diastolic diameter (LVEDD) ≥6.0 cm] and EF <30% at baseline experienced

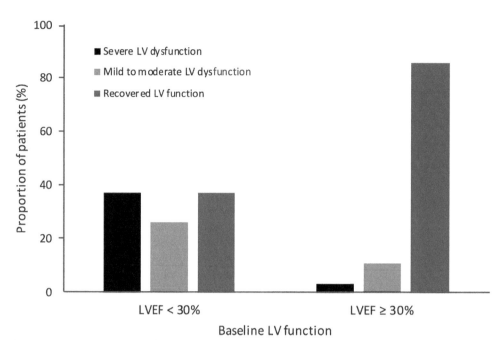

FIGURE 9   Recovery Rates at 1 Year Stratified by Baseline LV Function. Severe LV dysfunction is characterized by LVEF <35% or death, LVAD, or cardiac transplantation; mild-to-moderate LV dysfunction is EF 35–49%; and recovered LV function is LVEF ≥50%. EF, ejection fraction; LV, left ventricular; LVAD, left ventricular assist device; LVEF, left ventricular ejection fraction. *Adapted with permission from [29].*

full recovery at 12 months postpartum. Second, black women had more severely depressed LV function at baseline and had lower rates of recovery, even after adjusting for other clinical and echocardiographic features. Outcomes at 12 months stratified by baseline EF are presented in **Figure 9**.

Mortality at 1 year postpartum in the IPAC cohort was 4% [29]. In a statewide study in North Carolina utilizing medical record review and state pregnancy-related mortality files from 2002 to 2003, the mortality rate associated with PPCM was 11.8% at 1 year and 16.5% at 7 years of follow-up [10]. In this series, 7 of the 85 PPCM cases (8.2%) were diagnosed postmortem, representing half of all deaths. This suggests that mortality rates may be higher than reported in clinical registries, as a significant number of women will die before being properly diagnosed. Furthermore, the case fatality rate was fourfold higher for black compared to white women (24% vs. 6%).

Many women who recover myocardial function will do so within 6 months but several studies have demonstrated delayed recovery, in some cases occurring years after delivery. Among recovered patients in Haiti, 75% of patients took over 12 months to recover [131]. In Turkey, 48% women with PPCM had full recovery and average time to recovery was 19 months (3–42 months) [94]. A small PPCM cohort in Denmark demonstrated that one-third of women had further improvement in EF and two-thirds had stable EF after 12 months [132].

Women with recovered PPCM may continue to have subtle evidence of myocardial dysfunction in the setting of a normal EF, including lower E' septal and systolic S' septal velocities on tissue Doppler imaging and lower LV longitudinal and apical circumferential strain. Reports of long-term follow-up of women with PPCM are limited and may be confounded by attrition. The Danish PPCM cohort has been followed on average for 7 years (range 2.2–11.4 years) and demonstrates high rates of myocardial recovery. The majority of women with prior PPCM had normal EF but they also had subtle findings of diastolic dysfunction on CMR along with reduced peak VO_2 compared to women with a history of severe preeclampsia or healthy pregnancy [132].

Mechanical Support and Transplantation

A minority of women with PPCM will go on to develop refractory heart failure requiring heart transplantation and/or LVAD. Case reports of myocardial recovery with LVAD explantation are frequently published [133–135]. Among 99 women with PPCM treated with durable mechanical support in the Interagency Registry for Mechanically Assisted Circulatory Support (INTERMACS) database, 48% underwent transplant at 3 years and survival on device was 68% [136]. In nonischemic cardiomyopathy, acute-onset heart failure is associated with high rates of LV recovery and device explantation [137]. In contrast, only 6% of PPCM women experienced recovery in the INTERMACS database, which may be explained by the high proportion (59%) of women presenting with chronic heart failure, disease severity, or the lack of standardization in evaluating for recovery with ramp-down studies. Myocardial recovery on LVAD is an active area of investigation and future studies may further inform how women with PPCM differ from other etiologies of nonischemic cardiomyopathy. Among the 485 women transplanted for PPCM between 1987 and 2010, graft survival and age-adjusted survival were lower compared to survival for women transplanted for other indications, perhaps reflecting younger age and higher allosensitization [138].

Predictors of Recovery

Baseline LV Function and Other Cardiac Imaging Parameters

Baseline LV function has been consistently demonstrated to be a strong predictor of subsequent recovery [127, 139, 140]. In a study of 187 women with PPCM, mean EF at time of diagnosis was 31% in those with recovery at 6 months compared to 23% without recovery (P < 0.0001). A baseline EF > 30% was associated with a 6.4-fold higher odds of recovery compared to a baseline EF of 10–19% [141]. In a prospective German cohort of 115 women with PPCM, baseline EF was significantly higher in those who experienced improvement in LV function at 6 months compared to those without improvement (EF 28% vs. 17%, respectively, P < 0.0001) [63]. In both of these studies, LV dimensions were also strongly associated with outcomes. Because full and partial recovery are possible in women with severely depressed LV function, baseline EF alone is not an indication for premature device implantation or cardiac transplantation [34].

Additional markers on baseline imaging may offer insight into subsequent outcomes. Right ventricular (RV) function on echocardiogram or CMR imaging at time of diagnosis is increasingly emerging as an important predictor of subsequent myocardial recovery and clinical outcomes [78, 140, 142]. LV strain parameters were not associated with clinical outcomes in a small cohort of PPCM patients [143]. The presence of myocardial fibrosis as measured by late gadolinium enhancement (LGE) on CMR imaging is associated with adverse clinical outcomes among patients with nonischemic DCM [144, 145]. The role of LGE in PPCM is less clear, however. CMR was performed in a cohort of 36 German women with PPCM shortly after diagnosis and demonstrated LGE was present in a majority (71%) of cases, yet LGE was not associated with subsequent recovery [142]. In contrast, in the subset of women in the IPAC study who underwent serial CMR, LGE was observed in only 2 of 40 women (5%) at baseline. Neither

of the 2 women in this cohort who died had LGE [146]. Among women with PPCM undergoing CMR in Denmark on average 7 years after diagnosis, only 1 of 25 women had LGE (4%) [14]. Differences in LGE prevalence between these studies may reflect patient population or different algorithms for LGE detection.

Racial Differences

Black women with PPCM have a worse prognosis than nonblack women in the United States, and likely elsewhere. In a combined cohort of women with PPCM treated at the University of Southern California and the Louisiana State University Health Sciences Center, black women had significantly lower rates of complete recovery compared to white women (40% vs. 61%, respectively, P = 0.02) as well as higher combined rate of death or transplant (17.3% vs. 6.7%, P = 0.03) [147, 148]. In this cohort, baseline EF was similar between the two groups but black women had larger LV dimensions. In the IPAC cohort, black women achieved a lower EF at 12 months postpartum compared to nonblack women (EF 47% vs. 56%, P = 0.001). Compared to nonblacks, black women had lower EF and larger LV dimensions at baseline, and even when taking this into account, black race remained one of the strongest predictors of poor recovery [29]. In a cohort of 220 women treated at the University of Pennsylvania, outcomes in black women were significantly worse. Black women had lower rates of complete recovery compared to nonblacks (69% vs. 75%, P = 0.004), and among those who did recover, median time to recovery was significantly longer [8.8 months, interquartile range (IQR) 3–18.4 months vs. 4.2 months, IQR 1.7–9.5 months] [30]. Black women are more likely to present later in the postpartum period compared to nonblacks and this delayed diagnosis may be partially responsible for worse LV function at baseline and lower rates of recovery [149]. The causes for these racial differences in presentation and recovery are not clear and may reflect a combination of socioeconomic factors, including access to care, health literacy, institutionalized racism, environmental factors, and genetics.

Preeclampsia and Hypertension

Hypertensive disorders of pregnancy (HDP), such as preeclampsia and gestational hypertension, strongly predispose to the development of PPCM. As noted above, preeclampsia and PPCM share pathophysiological mechanisms, including the upregulation of antiangiogenic and vasculotoxic hormones secreted by the placenta. In patients who develop PPCM in the context of preeclampsia, the resolution of hypertension and normalization of angiogenic imbalance may theoretically accelerate resolution of PPCM. Indeed, cohorts in Germany and Japan have suggested improved rates of recovery in PPCM associated with hypertensive disorders [63, 150]. In contrast, racially diverse studies in the United States have shown no association with recovery of LV function, despite similar rates of HDP [29, 141]. A study of a small, predominately black cohort in St. Louis (39 patients, 77% black) found that women with preeclampsia had lower 1-year rates of event-free survival, although among the 26 survivors with complete echocardiographic follow-up, women with preeclampsia had significantly higher rates of LV recovery (80% vs. 25%, $P = 0.014$) [151]. In a large racially diverse PPCM cohort at the University of Pennsylvania (220 patients, 55% black), hypertensive disorders were not associated with improved recovery, although a trend toward improved recovery was observed among nonblack women (79% vs. 66%, $P = 0.285$), suggesting that different outcomes observed in international PPCM cohorts may be related to differences in racial composition, or other socioeconomic or biological variables [149]. Further complicating the study of this question is the grouping together of hypertensive disorders that share different degrees of interaction with PPCM and that may therefore influence outcomes differently.

Other baseline factors associated with subsequent recovery include normal electrocardiogram (ECG), higher relaxin-2 level, and lower levels of troponin T and NT-proBNP levels [52, 73, 76, 152].

Subsequent Pregnancy Recommendations

Many women diagnosed with PPCM will desire to become pregnant again [153]. Persistently reduced LV function prior to subsequent pregnancy is associated with a high risk of relapse and higher mortality in the postpartum period. Among women with a history of PPCM undergoing a subsequent pregnancy with persistent LV dysfunction, 48% experienced significant deterioration in LV function and 16% died [153]. In contrast, women with full LV recovery after PPCM who subsequently become pregnant have low mortality rates but remain at elevated risk of developing heart failure symptoms or deterioration in LV function. In a review of the literature published in 2014, 27% of women with recovered LV function experienced deterioration, compared to a deterioration rate of 14% among women with PPCM in Denmark, 21% at the Mayo Clinic, and 44% in an international combined cohort in Germany, Scotland, and South Africa [154–156]. Notably, none of the women with recovered LV function undergoing a subsequent pregnancy in these cohorts died.

Among women with persistent LV dysfunction after PPCM, subsequent pregnancy should generally be discouraged given the high risk of clinical deterioration and death [1]. Guidance for women with recovered function is less clear. Demonstrating that LV function remains normalized

after discontinuation of cardiac medications, especially ACE inhibitors or ARBs, is prudent, especially since these medications will need to be discontinued prior to conception. The presence of normal contractile reserve on stress echocardiogram has been suggested to have a positive prognosis but has not been rigorously evaluated [157].

Women with PPCM who subsequently become pregnant should be counseled about the potential risk of relapse during pregnancy and the postpartum period. Women should be followed closely with serial echocardiograms. Measuring BNP in the first trimester can provide a baseline for comparison if shortness of breath develops later on in pregnancy [153]. Teratogenic medications should be discontinued prior to attempting conception. Beta-blockers, especially metoprolol tartrate, can usually be continued safely during pregnancy without significant risk to the fetus [155]. Bromocriptine use immediately after delivery in conjunction with a heart failure medication regimen is well tolerated and may be associated with LV recovery; however, further data are needed before broadly recommending this treatment option [156].

Long-acting contraception should be offered to women with PPCM, especially if normalization of LV function has not been achieved. Tubal ligation and vasectomy are irreversible options for women who desire no further pregnancies. The safest and most effective forms of reversible contraception include progesterone-releasing intrauterine device or implant. Estrogen-containing contraception is discouraged given their associated with increased thromboembolic risk and hypertension [158].

Conclusion

PPCM is an uncommon but potentially severe form of cardiomyopathy that predominately affects women in the weeks to months after delivery. Black women are significantly more likely to develop PPCM, more likely to be diagnosed later in the postpartum period, less likely to have myocardial recovery, and, when they do recover, take longer to do so. PPCM is now a leading cause of maternal mortality in the postpartum period in the United States, especially among black women. Prompt recognition of heart failure symptoms in pregnant and postpartum women is key to the early diagnosis and initiation of treatment that supports myocardial recovery. NT-proBNP is a useful biomarker to differentiate cardiac from lung pathologies but is not specific to PPCM. Echocardiography is diagnostic. Treatment mirrors that of generic cardiomyopathy, focusing on neurohormonal blockade. Ongoing research in PPCM, including better understanding of the underlying disease mechanisms, genetic predisposition, and necessary factors for recovery, will help to address the large burden of PPCM in contributing to maternal morbidity and mortality in the United States, especially among black women.

Key Points

1. **Peripartum cardiomyopathy (PPCM) is a form of cardiomyopathy diagnosed in the peripartum period and is characterized by left ventricular systolic dysfunction and heart failure.**

2. **Factors associated with PPCM include African ancestry, advanced maternal age, preeclampsia, and twin gestations.**

3. **PPCM affects 1 in every 1000–4000 live births in the United States.**

4. **PPCM is now a leading cause of maternal mortality in the postpartum period in the United States, especially among black women.**

5. **Recovery of systolic dysfunction occurs in the majority of patients.**

6. **Management of PPCM includes guideline-directed medical therapy (GDMT) for LV systolic dysfunction and heart failure, and bromocriptine is currently under investigation as a possible therapeutic modality.**

7. **Underlying pathophysiologic mechanisms include the role of toxic vascular hormone of late gestation on cardiovascular function as well as possible underlying genetic predisposition.**

8. **Biochemical and genetic factors linked to the development of PPCM include prolactin, cathepsin D, MnSOD, PGC-1α, STAT3, microRNA-146a, VEGF, sFlt-1, and TTN.**

9. **Persistent left ventricular systolic dysfunction is associated with increased risk with subsequent pregnancy.**

Back to Clinical Case

Returning to our case at the beginning of the chapter, we recognize that our patient had multiple risk factors for developing PPCM, including her age (older than 30 years), black race, recent preeclampsia, and obesity. Her initial diagnostic work-up was suggestive of pneumonia, but the highly elevated NT-proBNP level provided a key clue to a cardiac cause of dyspnea, leading her physicians to perform an echocardiogram that provided the diagnosis of PPCM. The patient was admitted to the hospital, underwent significant diuresis, and was started on an optimal heart failure medication regimen. At 12 months after delivery, her EF had improved to 30–35%. She underwent placement of an ICD for primary prevention. She elected for placement of an intrauterine device to protect against future pregnancies.

CH
14

References

[1] Regitz-Zagrosek V, Roos-Hesselink JW, Bauersachs J, Blomström-Lundqvist C, Cífková R, De Bonis M, et al. 2018 ESC guidelines for the management of cardiovascular diseases during pregnancy. Eur Heart J 2018;39:3165–241.

[2] Demakis JG, Rahimtoola SH, Sutton GC, Meadows WR, Szanto PB, Tobin JR, et al. Natural course of peripartum cardiomyopathy. Circulation 1971;44:1053–61.

[3] Pearson GD, Veille JC, Rahimtoola S, Hsia J, Oakley CM, Hosenpud JD, et al. Peripartum cardiomyopathy: National Heart, Lung, and Blood Institute and Office of Rare Diseases (National Institutes of Health) workshop recommendations and review. JAMA 2000;283:1183–8.

[4] Elkayam U, Akhter MW, Singh H, Khan S, Bitar F, Hameed A, et al. Pregnancy-associated cardiomyopathy: clinical characteristics and a comparison between early and late presentation. Circulation 2005;111:2050–5.

[5] Creanga AA, Syverson C, Seed K, Callaghan WM. Pregnancy-related mortality in the United States, 2011–2013. Obstet Gynecol 2017;130:366–73.

[6] GBD 2015 Maternal Mortality Collaborators. Global, regional, and national levels of maternal mortality, 1990–2015: a systematic analysis for the Global Burden of Disease Study 2015. Lancet Lond Engl 2016;388:1775–812.

[7] Creanga AA, Berg CJ, Syverson C, Seed K, Bruce FC, Callaghan WM. Pregnancy-related mortality in the United States, 2006-2010. Obstet Gynecol 2015;125:5–12.

[8] Petersen EE, Davis NL, Goodman D, Cox S, Mayes N, Johnston E, et al. Vital signs: pregnancy-related deaths, United States, 2011–2015, and strategies for prevention, 13 states, 2013–2017. MMWR Morb Mortal Wkly Rep 2019;68. [Internet]; [cited 6 June 2019]. Available from: https://www.cdc.gov/mmwr/volumes/68/wr/mm6818e1.htm.

[9] Gunderson EP. Epidemiologic trends and maternal risk factors predicting postpartum weight retention. In: Obesity during pregnancy in clinical practice. London: Springer; 2014. p. 77–97. [Internet]; [cited 5 March 2018]. Available from: https://link.springer.com/chapter/10.1007/978-1-4471-2831-1_5.

[10] Harper MA, Meyer RE, Berg CJ. Peripartum cardiomyopathy: population-based birth prevalence and 7-year mortality. Obstet Gynecol 2012;120:1013–9.

[11] Brar SS, Khan SS, Sandhu GK, Jorgensen MB, Parikh N, Hsu J-WY, et al. Incidence, mortality, and racial differences in peripartum cardiomyopathy. Am J Cardiol 2007;100:302–4.

[12] Mielniczuk LM, Williams K, Davis DR, Tang ASL, Lemery R, Green MS, et al. Frequency of peripartum cardiomyopathy. Am J Cardiol 2006;97:1765–8.

[13] Gunderson EP, Croen LA, Chiang V, Yoshida CK, Walton D, Go AS. Epidemiology of peripartum cardiomyopathy: incidence, predictors, and outcomes. Obstet Gynecol 2011;118:583–91.

[14] Ersbøll AS, Johansen M, Damm P, Rasmussen S, Vejlstrup NG, Gustafsson F. Peripartum cardiomyopathy in Denmark: a retrospective, population-based study of incidence, management and outcome. Eur J Heart Fail 2017;19:1712–20.

[15] Wu VC-C, Chen T-H, Yeh J-K, Wu M, Lu C-H, Chen S-W, et al. Clinical outcomes of peripartum cardiomyopathy: a 15-year nationwide population-based study in Asia. Medicine (Baltimore) 2017;96:e8374.

[16] Sliwa K, Damasceno A, Mayosi BM. Epidemiology and etiology of cardiomyopathy in Africa. Circulation 2005;112:3577–83.

[17] Hasan JA, Qureshi A, Ramejo BB, Kamran A. Peripartum cardiomyopathy characteristics and outcome in a tertiary care hospital. JPMA J Pak Med Assoc 2010;60:377–80.

[18] Pandit V, Shetty S, Kumar A, Sagir A. Incidence and outcome of peripartum cardiomyopathy from a tertiary hospital in South India. Trop Dr 2009;39:168–9.

[19] Fett JD, Christie LG, Carraway RD, Murphy JG. Five-year prospective study of the incidence and prognosis of peripartum cardiomyopathy at a single institution. Mayo Clin Proc 2005;80:1602–6.

[20] Isezuo SA, Abubakar SA. Epidemiologic profile of peripartum cardiomyopathy in a tertiary care hospital. Ethn Dis 2007;17:228–33.

[21] Kolte D, Khera S, Aronow WS, Palaniswamy C, Mujib M, Ahn C, et al. Temporal trends in incidence and outcomes of peripartum cardiomyopathy in the United States: a nationwide population-based study. J Am Heart Assoc 2014;3, e001056.

[22] Kao DP, Hsich E, Lindenfeld J. Characteristics, adverse events, and racial differences among delivering mothers with peripartum cardiomyopathy. JACC Heart Fail 2013;1:409–16.

[23] Gentry MB, Dias JK, Luis A, Patel R, Thornton J, Reed GL. African-American women have a higher risk for developing peripartum cardiomyopathy. J Am Coll Cardiol 2010;55:654–9.

[24] Mol BWJ, Roberts CT, Thangaratinam S, Magee LA, de Groot CJM, Hofmeyr GJ. Pre-eclampsia. Lancet 2016;387:999–1011.

[25] Bello N, Hurtado Rendon IS, Arany Z. The relationship between preeclampsia and peripartum cardiomyopathy: a systematic review and meta-analysis. J Am Coll Cardiol 2013;62:1715–23.

[26] Anon. ACOG Practice Bulletin No. 202: gestational hypertension and preeclampsia. Obstet Gynecol 2019;133:e1–e25.

[27] Sliwa K, Mebazaa A, Hilfiker-Kleiner D, Petrie MC, Maggioni AP, Laroche C, et al. Clinical characteristics of patients from the worldwide registry on peripartum cardiomyopathy (PPCM): EURObservational Research Programme in conjunction with the Heart Failure Association of the European Society of Cardiology Study Group on PPCM. Eur J Heart Fail 2017;19:1131–41.

[28] Behrens I, Basit S, Lykke JA, Ranthe MF, Wohlfahrt J, Bundgaard H, et al. Hypertensive disorders of pregnancy and peripartum cardiomyopathy: a nationwide cohort study. PLoS ONE 2019;14. [Internet]; [cited 30 May 2019], Available from: https://www.ncbi.nlm.nih.gov/pmc/articles/PMC6382119/.

[29] McNamara DM, Elkayam U, Alharethi R, Damp J, Hsich E, Ewald G, et al. Clinical outcomes for peripartum cardiomyopathy in North America. J Am Coll Cardiol 2015;66:905–14.

[30] Irizarry OC, Levine LD, Lewey J, Boyer T, Riis V, Elovitz MA, et al. Comparison of clinical characteristics and outcomes of peripartum cardiomyopathy between African American and non-African American women. JAMA Cardiol 2017;2:1256–60.

[31] Melchiorre K, Sutherland GR, Baltabaeva A, Liberati M, Thilaganathan B. Maternal cardiac dysfunction and remodeling in women with preeclampsia at term. Hypertension 2011;57:85–93.

[32] Melchiorre K, Sutherland GR, Liberati M, Thilaganathan B. Preeclampsia is associated with persistent postpartum cardiovascular impairment. Hypertens Dallas Tex 1979 2011;58:709–15.

[33] Vaught AJ, Kovell LC, Szymanski LM, Mayer SA, Seifert SM, Vaidya D, et al. Acute cardiac effects of severe pre-eclampsia. J Am Coll Cardiol 2018;72:1–11.

[34] Elkayam U. Clinical characteristics of peripartum cardiomyopathy in the United States. J Am Coll Cardiol 2011;58:659–70.

[35] Arany Z, Elkayam U. Peripartum cardiomyopathy. Circulation 2016;133:1397–409.

[36] Afana M, Brinjikji W, Kao D, Jackson E, Maddox TM, Childers D, et al. Characteristics and in-hospital outcomes of peripartum cardiomyopathy diagnosed during delivery in the United States from the Nationwide Inpatient Sample (NIS) database. J Card Fail 2016;22:512–9.

[37] Kotlica BK, Cetković A, Plesinac S, Macut D, Asanin M. Peripartum cardiomyopathy: a case of patient with triplet pregnancy. Clin Exp Obstet Gynecol 2016;43:274–5.

[38] Altun İ, Akın F, Biteker M. Peripartum cardiomyopathy and triplet pregnancy. Anatol J Cardiol 2015;15:85–6.

[39] Barasa A, Rosengren A, Sandström TZ, Ladfors L, Schaufelberger M. Heart failure in late pregnancy and postpartum: incidence and long-term mortality in Sweden from 1997 to 2010. J Card Fail 2017;23:370–8.

[40] Dhesi S, Savu A, Ezekowitz JA, Kaul P. Association between diabetes during pregnancy and peripartum cardiomyopathy: a population-level analysis of 309,825 women. Can J Cardiol 2017;33:911–7.

[41] Lampert MB, Hibbard J, Weinert L, Briller J, Lindheimer M, Lang RM. Peripartum heart failure associated with prolonged tocolytic therapy. Am J Obstet Gynecol 1993;168:493–5.

[42] Bültmann BD, Klingel K, Näbauer M, Wallwiener D, Kandolf R. High prevalence of viral genomes and inflammation in peripartum cardiomyopathy. Am J Obstet Gynecol 2005;193:363–5.

[43] Ntusi NBA, Mayosi BM. Aetiology and risk factors of peripartum cardiomyopathy: a systematic review. Int J Cardiol 2009;131:168–79.

[44] McTIernan CF, Morel P, Cooper LT, Rajagopalan N, Thohan V, Zucker M, et al. Circulating T-cell subsets, monocytes, and natural killer cells in peripartum cardiomyopathy: results from the multicenter IPAC study. J Card Fail 2018;24:33–42.

[45] Hilfiker-Kleiner D, Kaminski K, Podewski E, Bonda T, Schaefer A, Sliwa K, et al. A cathepsin D-cleaved 16 kDa form of prolactin mediates postpartum cardiomyopathy. Cell 2007;128:589–600.

[46] Zouein FA, Altara R, Chen Q, Lesnefsky EJ, Kurdi M, Booz GW. Pivotal importance of STAT3 in protecting the heart from acute and chronic stress: new advancement and unresolved issues. Front Cardiovasc Med 2015;2:36.

[47] Hilfiker-Kleiner D, Limbourg A, Drexler H. STAT3-mediated activation of myocardial capillary growth. Trends Cardiovasc Med 2005;15:152–7.

[48] Negoro S, Kunisada K, Fujio Y, Funamoto M, Darville MI, Eizirik DL, et al. Activation of signal transducer and activator of transcription 3 protects cardiomyocytes from hypoxia/reoxygenation-induced oxidative stress through the upregulation of manganese superoxide dismutase. Circulation 2001;104:979–81.

[49] Halkein J, Tabruyn SP, Ricke-Hoch M, Haghikia A, Nguyen N-Q-N, Scherr M, et al. MicroRNA-146a is a therapeutic target and biomarker for peripartum cardiomyopathy. J Clin Invest 2013;123:2143–54.

[50] Arany Z, Foo S-Y, Ma Y, Ruas JL, Bommi-Reddy A, Girnun G, et al. HIF-independent regulation of VEGF and angiogenesis by the transcriptional coactivator PGC-1alpha. Nature 2008;451:1008–12.

[51] Patten IS, Rana S, Shahul S, Rowe GC, Jang C, Liu L, et al. Cardiac angiogenic imbalance leads to peripartum cardiomyopathy. Nature 2012;485:333–8.

[52] Damp J, Givertz MM, Semigran M, Alharethi R, Ewald G, Felker GM, et al. Relaxin-2 and soluble Flt1 levels in peripartum cardiomyopathy: results of the multicenter IPAC study. JACC Heart Fail 2016;4:380–8.

[53] Goland S, Weinstein JM, Zalik A, Kuperstein R, Zilberman L, Shimoni S, et al. Angiogenic imbalance and residual myocardial injury in recovered peripartum cardiomyopathy patients. Circ Heart Fail 2016. [Internet]; [cited 2018 Oct 10]. Available from: https://www.ahajournals.org/doi/10.1161/CIRCHEARTFAILURE.116.003349.

[54] Powe CE, Levine RJ, Karumanchi SA. Preeclampsia, a disease of the maternal endothelium: the role of anti-angiogenic factors and implications for later cardiovascular disease. Circulation 2011;123. [Internet]; [cited 10 September 2018]. Available from: https://www.ncbi.nlm.nih.gov/pmc/articles/PMC3148781/.

[55] Bdolah Y, Lam C, Rajakumar A, Shivalingappa V, Mutter W, Sachs BP, et al. Twin pregnancy and the risk of preeclampsia: bigger placenta or relative ischemia? Am J Obstet Gynecol 2008;198:428. e1–6.

[56] Levine RJ, Maynard SE, Qian C, Lim K-H, England LJ, Yu KF, et al. Circulating angiogenic factors and the risk of preeclampsia. N Engl J Med 2004;350:672–83.

[57] Ramadan H, Rana S, Mueller A, Bajracharya S, Zhang D, Salahuddin S, et al. Myocardial performance index in hypertensive disorders of pregnancy: the relationship between blood pressures and angiogenic factors. Hypertens Pregnancy 2017;36:161–7.

[58] Bello NA, Arany Z. Molecular mechanisms of peripartum cardiomyopathy: a vascular/hormonal hypothesis. Trends Cardiovasc Med 2015;25:499–504.

[59] Horne BD, Rasmusson KD, Alharethi R, Budge D, Brunisholz KD, Metz T, et al. Genome-wide significance and replication of the chromosome 12p11.22 locus near the PTHLH gene for peripartum cardiomyopathy. Circ Cardiovasc Genet 2011;4:359–66.

[60] Massad LS, Reiss CK, Mutch DG, Haskel EJ. Familial peripartum cardiomyopathy after molar pregnancy. Obstet Gynecol 1993;81:886–8.

[61] Pearl W. Familial occurrence of peripartum cardiomyopathy. Am Heart J 1995;129:421–2.

[62] Pierce JA, Price BO, Joyce JW. Familial occurrence of postpartal heart failure. Arch Intern Med 1963;111:651–5.

[63] Haghikia A, Podewski E, Libhaber E, Labidi S, Fischer D, Roentgen P, et al. Phenotyping and outcome on contemporary management in a German cohort of patients with peripartum cardiomyopathy. Basic Res Cardiol 2013;108. [Internet]; [cited 23 August 2018]. Available from: https://www.ncbi.nlm.nih.gov/pmc/articles/PMC3709080/.

[64] van Spaendonck-Zwarts KY, Posafalvi A, van den Berg MP, Hilfiker-Kleiner D, Bollen IAE, Sliwa K, et al. Titin gene mutations are common in families with both peripartum cardiomyopathy and dilated cardiomyopathy. Eur Heart J 2014;35:2165–73.

[65] Morales A, Painter T, Li R, Siegfried JD, Li D, Norton N, et al. Rare variant mutations in pregnancy-associated or peripartum cardiomyopathy. Circulation 2010;121:2176–82.

[66] Ware JS, Li J, Mazaika E, Yasso CM, DeSouza T, Cappola TP, et al. Shared genetic predisposition in peripartum and dilated cardiomyopathies. N Engl J Med 2016;374:233–41.

[67] Roberts AM, Ware JS, Herman DS, Schafer S, Baksi J, Bick AG, et al. Integrated allelic, transcriptional, and phenomic dissection of the cardiac effects of titin truncations in health and disease. Sci Transl Med 2015;7:270ra6.

[68] Herman DS, Lam L, Taylor MRG, Wang L, Teekakirikul P, Christodoulou D, et al. Truncations of titin causing dilated cardiomyopathy. N Engl J Med 2012;366:619–28.

[69] Puri A, Sethi R, Singh B, Dwivedi S, Narain V, Saran R, et al. Peripartum cardiomyopathy presenting with ventricular tachycardia: a rare presentation. Indian Pacing Electrophysiol J 2009;9:186–9.

[70] Manikkan A, Sanati M. Peripartum cardiomyopathy presenting as splenic infarct. J Hosp Med 2008;3:274–6.

[71] Zehir R, Karabay CY, Kocabay G, Kalayci A, Akgun T, Kirma C. An unusual presentation of peripartum cardiomyopathy: recurrent transient ischemic attacks. Rev Port Cardiol 2014;33:561. e1–3.

[72] Banayan J, Rana S, Mueller A, Tung A, Ramadan H, Arany Z, et al. Cardiogenic shock in pregnancy: analysis from the National Inpatient Sample. Hypertens Pregnancy 2017;36:117–23.

[73] Honigberg MC, Elkayam U, Rajagopalan N, Modi K, Briller JE, Drazner MH, et al. Electrocardiographic findings in peripartum cardiomyopathy. Clin Cardiol 2019;42:524–9.

[74] Tibazarwa K, Lee G, Mayosi B, Carrington M, Stewart S, Sliwa K. The 12-lead ECG in peripartum cardiomyopathy. Cardiovasc J Afr 2012;23:322–9.

[75] Tanous D, Siu SC, Mason J, Greutmann M, Wald RM, Parker JD, et al. B-type natriuretic peptide in pregnant women with heart disease. J Am Coll Cardiol 2010;56:1247–53.

[76] Hu CL, Li YB, Zou YG, Zhang JM, Chen JB, Liu J, et al. Troponin T measurement can predict persistent left ventricular dysfunction in peripartum cardiomyopathy. Heart 2007;93:488–90.

[77] Hibbard JU, Lindheimer M, Lang RM. A modified definition for peripartum cardiomyopathy and prognosis based on echocardiography. Obstet Gynecol 1999;94:311–6.

[78] Blauwet LA, Delgado-Montero A, Ryo K, Marek JJ, Alharethi R, Mather PJ, et al. Right ventricular function in peripartum cardiomyopathy at presentation is associated with subsequent left ventricular recovery and clinical outcomes. Circ Heart Fail 2016;9.

[79] Bozkurt B, Colvin M, Cook J, Cooper LT, Deswal A, Fonarow GC, et al. Current diagnostic and treatment strategies for specific dilated cardiomyopathies: a scientific statement from the American Heart Association. Circulation 2016;134:e579–646.

[80] Elkayam U, Gleicher N. Cardiac problems in pregnancy: diagnosis and management of maternal and fetal heart disease. John Wiley & Sons; 1998.

[81] Brito V, Niederman MS. Pneumonia complicating pregnancy. Clin Chest Med 2011;32:121–32.

[82] Arany ZP, Walker CM, Wang L. Case 22-2014—a 40-year-old woman with postpartum dyspnea and hypoxemia. N Engl J Med 2014;371:261–9.

[83] James AH, Jamison MG, Biswas MS, Brancazio LR, Swamy GK, Myers ER. Acute myocardial infarction in pregnancy: a United States population-based study. Circulation 2006;113:1564–71.

[84] Hayes SN, Kim ESH, Saw J, Adlam D, Arslanian-Engoren C, Economy KE, et al. Spontaneous coronary artery dissection: current state of the science: a scientific statement from the American Heart Association. Circulation 2018;137:e523–57.

[85] Citro R, Giudice R, Mirra M, Petta R, Baldi C, Bossone E, et al. Is Tako-tsubo syndrome in the postpartum period a clinical entity different from peripartum cardiomyopathy? J Cardiovasc Med (Hagerstown) 2013;14:568–75.

[86] Minatoguchi M, Itakura A, Takagi E, Nishibayashi M, Kikuchi M, Ishihara O. Takotsubo cardiomyopathy after cesarean: a case report and published work review of pregnancy-related cases. J Obstet Gynaecol Res 2014;40:1534–9.

[87] Buccheri D, Zambelli G. The link between spontaneous coronary artery dissection and takotsubo cardiomyopathy: analysis of the published cases. J Thorac Dis 2017;9:5489–92.

[88] Johnson SW, Hedgire SS, Scott NS, Natarajan P. Spontaneous coronary artery dissection masquerading as Takotsubo cardiomyopathy: a case report. Eur Heart J 2018;2. [Internet]; [cited 17 June 2019]. Available from: https://academic.oup.com/ehjcr/article/2/4/yty102/5127759.

[89] Haghikia A, Tongers J, Berliner D, König T, Schäfer A, Brehm M, et al. Early ivabradine treatment in patients with acute peripartum cardiomyopathy: subanalysis of the German PPCM registry. Int J Cardiol 2016;216:165–7.

[90] Demir S, Tufenk M, Karakaya Z, Akilli R, Kanadas M. The treatment of heart failure-related symptoms with ivabradine in a case with peripartum cardiomyopathy. Int Cardiovasc Res J 2013;7:33–6.

[91] Libhaber E, Sliwa K, Bachelier K, Lamont K, Böhm M. Low systolic blood pressure and high resting heart rate as predictors of outcome in patients with peripartum cardiomyopathy. Int J Cardiol 2015;190:376–82.

[92] Tahir U, Sam F. Withdrawal of heart failure medications in peripartum cardiomyopathy after myocardial recovery. Int J Cardiol 2015;190:212–3.

[93] Amos AM, Jaber WA, Russell SD. Improved outcomes in peripartum cardiomyopathy with contemporary. Am Heart J 2006;152:509–13.

[94] Biteker M, Ilhan E, Biteker G, Duman D, Bozkurt B. Delayed recovery in peripartum cardiomyopathy: an indication for long-term follow-up and sustained therapy. Eur J Heart Fail 2012;14:895–901.

[95] Halliday BP, Wassall R, Lota AS, Khalique Z, Gregson J, Newsome S, et al. Withdrawal of pharmacological treatment for heart failure in patients with recovered dilated cardiomyopathy (TRED-HF): an open-label, pilot, randomised trial. Lancet 2019;393:61–73.

[96] Kamel H, Navi BB, Sriram N, Hovsepian DA, Devereux RB, Elkind MSV. Risk of a thrombotic event after the 6-week postpartum period. N Engl J Med 2014;370:1307–15.

[97] Honigberg MC, Givertz MM. Arrhythmias in peripartum cardiomyopathy. Card Electrophysiol Clin 2015;7:309–17.

[98] Goland S, Modi K, Bitar F, Janmohamed M, Mirocha JM, Czer LSC, et al. Clinical profile and predictors of complications in peripartum cardiomyopathy. J Card Fail 2009;15:645–50.

[99] Sliwa K, Förster O, Libhaber E, Fett JD, Sundstrom JB, Hilfiker-Kleiner D, et al. Peripartum cardiomyopathy: inflammatory markers as predictors of outcome in 100 prospectively studied patients. Eur Heart J 2006;27:441–6.

[100] Mallikethi-Reddy S, Akintoye E, Trehan N, Sharma S, Briasoulis A, Jagadeesh K, et al. Burden of arrhythmias in peripartum cardiomyopathy: analysis of 9841 hospitalizations. Int J Cardiol 2017;235:114–7.

[101] Duncker D, Westenfeld R, Konrad T, Pfeffer T, Correia de Freitas CA, Pfister R, et al. Risk for life-threatening arrhythmia in newly diagnosed peripartum cardiomyopathy with low ejection fraction: a German multi-centre analysis. Clin Res Cardiol 2017;106:582–9.

[102] Saltzberg MT, Szymkiewicz S, Bianco NR. Characteristics and outcomes of peripartum versus nonperipartum cardiomyopathy in women using a wearable cardiac defibrillator. J Card Fail 2012;18:21–7.

[103] Dayoub EJ, Datwani H, Lewey J, Groeneveld PW. One-year cardiovascular outcomes in patients with peripartum cardiomyopathy. J Card Fail 2018;24:711–5.

[104] Nieminen MS, Akkila J, Hasenfuss G, Kleber FX, Lehtonen LA, Mitrovic V, et al. Hemodynamic and neurohumoral effects of continuous infusion of levosimendan in patients with congestive heart failure. J Am Coll Cardiol 2000;36:1903–12.

[105] Follath F, Cleland JGF, Just H, Papp JGY, Scholz H, Peuhkurinen K, et al. Efficacy and safety of intravenous levosimendan compared with dobutamine in severe low-output heart failure (the LIDO study): a randomised double-blind trial. Lancet Lond Engl 2002;360:196–202.

[106] Mebazaa A, Nieminen MS, Packer M, Cohen-Solal A, Kleber FX, Pocock SJ, et al. Levosimendan vs dobutamine for patients with acute decompensated heart failure: the SURVIVE randomized trial. JAMA 2007;297:1883–91.

[107] Biteker M, Duran NE, Kaya H, Gündüz S, Tanboğa Hİ, Gökdeniz T, et al. Effect of levosimendan and predictors of recovery in patients with peripartum cardiomyopathy, a randomized clinical trial. Clin Res Cardiol 2011;100:571–7.

[108] Stapel B, Kohlhaas M, Ricke-Hoch M, Haghikia A, Erschow S, Knuuti J, et al. Low STAT3 expression sensitizes to toxic effects of β-adrenergic receptor stimulation in peripartum cardiomyopathy. Eur Heart J 2017;38:349–61.

[109] Bauersachs J, Arrigo M, Hilfiker-Kleiner D, Veltmann C, Coats AJS, Crespo-Leiro MG, et al. Current management of patients with severe acute peripartum cardiomyopathy: practical guidance from the Heart Failure Association of the European Society of Cardiology Study Group on peripartum cardiomyopathy. Eur J Heart Fail 2016;18:1096–105.

[110] De Backer D, Biston P, Devriendt J, Madl C, Chochrad D, Aldecoa C, et al. Comparison of dopamine and norepinephrine in the treatment of shock. N Engl J Med 2010;362:779–89.

[111] Levy B, Clere-Jehl R, Legras A, Morichau-Beauchant T, Leone M, Frederique G, et al. Epinephrine versus norepinephrine for cardiogenic shock after acute myocardial infarction. J Am Coll Cardiol 2018;72:173–82.

[112] Sieweke J-T, Pfeffer TJ, Berliner D, König T, Hallbaum M, Napp LC, et al. Cardiogenic shock complicating peripartum cardiomyopathy: importance of early left ventricular unloading and bromocriptine therapy. Eur Heart J Acute Cardiovasc Care 2018. https://doi.org/10.1177/2048872618777876.

[113] Bouabdallaoui N, Demondion P, Leprince P, Lebreton G. Short-term mechanical circulatory support for cardiogenic shock in severe peripartum cardiomyopathy: La Pitié-Salpêtrière experience. Interact Cardiovasc Thorac Surg 2017;25:52–6.

[114] Horn P, Saeed D, Akhyari P, Hilfiker-Kleiner D, Kelm M, Westenfeld R. Complete recovery of fulminant peripartum cardiomyopathy on mechanical circulatory support combined with high-dose bromocriptine therapy. ESC Heart Fail 2017;4:641–4.

[115] Wiedemann D, Schlöglhofer T, Riebandt J, Neuner M, Tschernko E, Schima H, et al. Myocardial recovery in peripartum cardiomyopathy after hyperprolactinemia treatment on BIVAD. ASAIO J Am Soc Artif Intern Organs 1992 2017;63:109–11.

[116] Gevaert S, Van Belleghem Y, Bouchez S, Herck I, De Somer F, De Block Y, et al. Acute and critically ill peripartum cardiomyopathy and "bridge to" therapeutic options: a single center experience with intra-aortic balloon pump, extra corporeal membrane oxygenation and continuous-flow left ventricular assist devices. Crit Care Lond Engl 2011;15:R93.

[117] Park SH, Chin JY, Choi MS, Choi JH, Choi YJ, Jung KT. Extracorporeal membrane oxygenation saved a mother and her son from fulminant peripartum cardiomyopathy. J Obstet Gynaecol Res 2014;40:1940–3.

[118] Sliwa K, Blauwet L, Tibazarwa K, Libhaber E, Smedema J-P, Becker A, et al. Evaluation of bromocriptine in the treatment of acute severe peripartum cardiomyopathy: a proof-of-concept pilot study. Circulation 2010;121:1465–73.

[119] Tremblay-Gravel M, Marquis-Gravel G, Avram R, Desplantie O, Ducharme A, Bibas L, et al. The effect of bromocriptine on left ventricular functional recovery in peripartum cardiomyopathy: insights from the BRO-HF retrospective cohort study. ESC Heart Fail 2019;6:27–36.

[120] Ersbøll AS, Arany Z, Gustafsson F. Bromocriptine for the treatment of peripartum cardiomyopathy: comparison of outcome with a Danish cohort. Eur Heart J 2018;39:3476–7.

[121] Hilfiker-Kleiner D, Haghikia A, Berliner D, Vogel-Claussen J, Schwab J, Franke A, et al. Bromocriptine for the treatment of peripartum cardiomyopathy: a multicentre randomized study. Eur Heart J 2017;38:2671–9.

[122] Haghikia A, Podewski E, Berliner D, Sonnenschein K, Fischer D, Angermann CE, et al. Rationale and design of a randomized, controlled multi-centre clinical trial to evaluate the effect of bromocriptine on left ventricular function in women with peripartum cardiomyopathy. Clin Res Cardiol 2015;104:911–7.

[123] Hopp L, Haider B, Iffy L. Myocardial infarction postpartum in patients taking bromocriptine for the prevention of breast engorgement. Int J Cardiol 1996;57:227–32.

[124] Iffy L, Lindenthal J, Mcardle JJ, Ganesh V. Severe cerebral accidents postpartum in patients taking bromocriptine for milk suppression. Isr J Med Sci 1996;32:309–12.

[125] US Food and Drug Administration. 60 FR 3404; Sandoz Pharmaceuticals Corp.; Bromocriptine Mesylate (Parlodel); withdrawal of approval of the indication for the prevention of physiological lactation. Fed Regist 1995;60:3404–5.

[126] Victora CG, Bahl R, Barros AJD, França GVA, Horton S, Krasevec J, et al. Breastfeeding in the 21st century: epidemiology, mechanisms, and lifelong effect. Lancet Lond Engl 2016;387:475–90.

[127] Safirstein JG, Ro AS, Grandhi S, Wang L, Fett JD, Staniloae C. Predictors of left ventricular recovery in a cohort of peripartum cardiomyopathy patients recruited via the internet. Int J Cardiol 2012;154:27–31.

[128] Koczo A, Marino A, Jeyabalan A, Elkayam U, Cooper LT, Fett J, et al. Breastfeeding, cellular immune activation, and myocardial recovery in peripartum cardiomyopathy. JACC Basic Transl Sci 2019;4:291–300.

[129] Arany Z, Feldman AM. To breastfeed or not to breastfeed with peripartum cardiomyopathy. JACC Basic Transl Sci 2019;4:301–3.

[130] Kearney L, Wright P, Fhadil S, Thomas M. Postpartum cardiomyopathy and considerations for breastfeeding. Card Fail Rev 2018;4:112–8.

[131] Fett JD, Sannon H, Thélisma E, Sprunger T, Suresh V. Recovery from severe heart failure following peripartum cardiomyopathy. Int J Gynaecol Obstet 2009;104:125–7.

[132] Ersbøll AS, Bojer AS, Hauge MG, Johansen M, Damm P, Gustafsson F, et al. Long-term cardiac function after peripartum cardiomyopathy and preeclampsia: a Danish nationwide, clinical follow-up study using maximal exercise testing and cardiac magnetic resonance imaging. J Am Heart Assoc 2018;7. [Internet]; [cited 2 June 2019]. Available from: https://www.ncbi.nlm.nih.gov/pmc/articles/PMC6474952/.

[133] Oosterom L, de Jonge N, Kirkels J, Klöpping C, Lahpor J. Left ventricular assist device as a bridge to recovery in a young woman admitted with peripartum cardiomyopathy. Neth Heart J 2008;16:426–8.

[134] Emmert MY, Prêtre R, Ruschitzka F, Krähenmann F, Falk V, Wilhelm MJ. Peripartum cardiomyopathy with cardiogenic shock: recovery after prolactin inhibition and mechanical support. Ann Thorac Surg 2011;91:274–6.

[135] Lund LH, Grinnemo K-H, Svenarud P, van der Linden J, Eriksson MJ. Myocardial recovery in peri-partum cardiomyopathy after continuous flow left ventricular assist device. J Cardiothorac Surg 2011;6:150.

[136] Loyaga-Rendon RY, Pamboukian SV, Tallaj JA, Acharya D, Cantor R, Starling RC, et al. Outcomes of patients with peripartum cardiomyopathy who received mechanical circulatory support. Circ Heart Fail 2014;7:300–9.

[137] Boehmer JP, Starling RC, Cooper LT, Torre-Amione G, Wittstein I, Dec GW, et al. Left ventricular assist device support and myocardial recovery in recent onset cardiomyopathy. J Card Fail 2012;18:755–61.

[138] Rasmusson K, Brunisholz K, Budge D, Horne BD, Alharethi R, Folsom J, et al. Peripartum cardiomyopathy: post-transplant outcomes from the united network for organ sharing database. J Heart Lung Transplant Off Publ Int Soc Heart Transplant 2012;31:180–6.

[139] Cuenza LR, Manapat N, Jalique JRK. Clinical profile and predictors of outcomes of patients with peripartum cardiomyopathy: the Philippine Heart Center Experience. ASEAN Heart J 2016;24:9.

[140] Peters A, Caroline M, Zhao H, Baldwin MR, Forfia PR, Tsai EJ. Initial right ventricular dysfunction severity identifies severe peripartum cardio-myopathy phenotype with worse early and overall outcomes: a 24-year cohort study. J Am Heart Assoc 2018;7.

[141] Goland S, Bitar F, Modi K, Safirstein J, Ro A, Mirocha J, et al. Evaluation of the clinical relevance of baseline left ventricular ejection fraction as a predictor of recovery or persistence of severe dysfunction in women in the United States with peripartum cardiomyopathy. J Card Fail 2011;17:426–30.

[142] Haghikia A, Röntgen P, Vogel-Claussen J, Schwab J, Westenfeld R, Ehlermann P, et al. Prognostic implication of right ventricular involvement in peripartum cardiomyopathy: a cardiovascular magnetic resonance study. ESC Heart Fail 2015;2:139–49.

[143] Briasoulis A, Mocanu M, Marinescu K, Qaqi O, Palla M, Telila T, et al. Longitudinal systolic strain profiles and outcomes in peripartum cardiomy-opathy. Echocardiography 2016;33:1354–60.

[144] Pi S-H, Kim SM, Choi J-O, Kim EK, Chang S-A, Choe YH, et al. Prognostic value of myocardial strain and late gadolinium enhancement on car-diovascular magnetic resonance imaging in patients with idiopathic dilated cardiomyopathy with moderate to severely reduced ejection fraction. J Cardiovasc Magn Reson 2018;20:36.

[145] Gulati A, Jabbour A, Ismail TF, Guha K, Khwaja J, Raza S, et al. Association of fibrosis with mortality and sudden cardiac death in patients with nonischemic dilated cardiomyopathy. JAMA 2013;309:896–908.

[146] Schelbert EB, Elkayam U, Cooper LT, Givertz MM, Alexis JD, Briller J, et al. Myocardial damage detected by late gadolinium enhancement car-diac magnetic resonance is uncommon in peripartum cardiomyopathy. J Am Heart Assoc 2017;6.

[147] Goland S, Modi K, Hatamizadeh P, Elkayam U. Differences in clinical profile of African-American women with peripartum cardiomyopathy in the United States. J Card Fail 2013;19:214–8.

[148] Modi KA, Illum S, Jariatul K, Caldito G, Reddy PC. Poor outcome of indigent patients with peripartum cardiomyopathy in the United States. Am J Obstet Gynecol 2009;201:171.e1–5.

[149] Lewey J, Levine LD, Elovitz MA, Irizarry OC, Arany Z. Importance of early diagnosis in peripartum cardiomyopathy. Hypertension 2020;75:91–7.

[150] Kamiya CA, Kitakaze M, Ishibashi-Ueda H, Nakatani S, Murohara T, Tomoike H, et al. Different characteristics of peripartum cardiomyopathy between patients complicated with and without hypertensive disorders. Results from the Japanese Nationwide survey of peripartum cardiomyopathy. Circ J 2011;75:1975–81.

[151] Lindley KJ, Conner SN, Cahill AG, Novak E, Mann DL. Impact of pre-eclampsia on clinical and functional outcomes in women with peripartum cardiomyopathy. Circ Heart Fail 2017;10. [Internet]; [cited 20 August 2018]; Available from: https://www.ncbi.nlm.nih.gov/pmc/articles/PMC5520674/.

[152] Forster O, Hilfiker-Kleiner D, Ansari AA, Sundstrom JB, Libhaber E, Tshani W, et al. Reversal of IFN-γ, oxLDL and prolactin serum levels correlate with clinical improvement in patients with peripartum cardiomyopathy. Eur J Heart Fail 2008;10:861–8.

[153] Elkayam U. Risk of subsequent pregnancy in women with a history of peripartum cardiomyopathy. J Am Coll Cardiol 2014;64:1629–36.

[154] Guldbrandt Hauge M, Johansen M, Vejlstrup N, Gustafsson F, Damm P, Ersbøll AS. Subsequent reproductive outcome among women with peripartum cardiomyopathy: a nationwide study. BJOG 2018;125:1018–25.

[155] Codsi E, Rose CH, Blauwet LA. Subsequent pregnancy outcomes in patients with peripartum cardiomyopathy. Obstet Gynecol 2018;131:322–7.

[156] Hilfiker-Kleiner D, Haghikia A, Masuko D, Nonhoff J, Held D, Libhaber E, et al. Outcome of subsequent pregnancies in patients with a history of peripartum cardiomyopathy. Eur J Heart Fail 2017;19:1723–8.

[157] Fett JD, Shah TP, McNamara DM. Why do some recovered peripartum cardiomyopathy mothers experience heart failure with a subsequent pregnancy? Curr Treat Options Cardiovasc Med 2015;17. [Internet]; [cited 9 July 2019]. Available from: http://link.springer.com/10.1007/s11936-014-0354-x.

[158] Roos-Hesselink JW, Cornette J, Sliwa K, Pieper PG, Veldtman GR, Johnson MR. Contraception and cardiovascular disease. Eur Heart J 2015;36:1728–34.

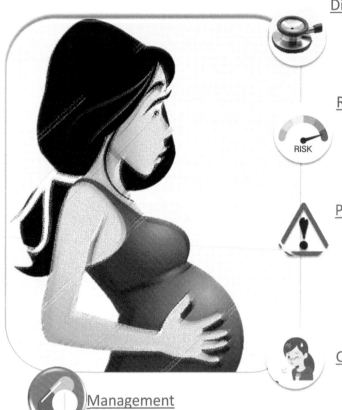

Diagnosis
- New onset HF, LVEF<45%
- Occurs in end pregnancy to 5 months postpartum
- Exclude other HF etiologies

Risk Factors
- Maternal age ≥ 30 years
- African race
- Preeclampsia, hypertensive d/o of pregnancy
- Multiple gestations

RISK

Poor Prognostic Markers
- LVEF < 30%, LV dilatation
- RV systolic dysfunction
- Obesity
- Black race
- Preeclampsia, hypertension
- Higher troponin and BNP levels
- Fibrosis on CMR

Clinical Presentation
- Dyspnea, orthopnea, PND
- Elevated JVP
- Weight gain, pedal edema
- Elevated BNP
- Pulmonary edema on CXR

Management

Pregnancy
- Avoid excess diuretics
- ACEI and ARBs are contraindicated in pregnancy
- Metoprolol Tartrate favored and safe

Delivery
- Favor vaginal delivery unless clinically unstable
- Multidisciplinary team

Post-Partum
- Contraception
- Avoid pregnancy with low LVEF
- Bromocriptine
- Prevent lactation

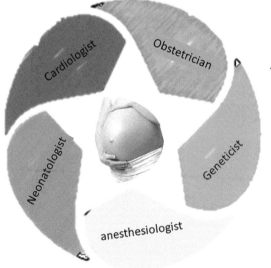

Cardiologist · Obstetrician · Geneticist · anesthesiologist · Neonatologist

Niti Aggarwal

Diagnostic criteria, risk factors, prognosis, presentation and management of peripartum cardiomyopathy. A multi disciplinary team is pivotal for success. ACEI, angiotensin converting enzyme inhibitor; ARB, angiotensin II receptor blockers; BNP, NT-pro-natriuretic peptide; CMR, cardiac magnetic resonance; CXR, chest X ray; d/o, disorder; HF, heart failure; JVP, jugular venous pressure; LV, left ventricle; LVEF, left ventricular ejection fraction; PND, paroxysmal nocturnal dyspnea. *Image courtesy of Niti R. Aggarwal.*

Valvular Heart Disease

Chapter 15

Valvular Heart Disease

Daniela Crousillat and Evin Yucel

Abstract

Valvular heart disease is an important cause of morbidity and mortality among women. Aortic stenosis (AS) is the most common valvular disease in the elderly and is associated with poor outcomes. There are significant sex-related differences in the pathophysiological response to severe AS, baseline characteristics, and outcomes after aortic valve replacement (AVR). Although female sex is a predictor of worse outcomes after surgical AVR (SAVR), women appear to have a mortality benefit after transcatheter AVR (TAVR) when compared to men. Data evaluating the sex-related differences among patients with aortic regurgitation (AR) are scarce and there is a lack of sex-specific guidelines for surgical intervention. Sex-related differences have been well described in the etiology, underlying pathology, and hemodynamic response to severe mitral regurgitation (MR). Mitral stenosis (MS) is most commonly secondary to rheumatic heart disease (RHD), although degenerative MS secondary to mitral annular calcification (MAC) is increasingly recognized and more common among women. Women have overall worse long-term outcomes after mitral valve (MV) surgery when compared with men, possibly owing to delayed referral. Tricuspid regurgitation (TR) is more prevalent among women and associated with increased mortality among both sexes. Further research is needed to establish guidelines for the most appropriate timing of tricuspid valve (TV) interventions. Women with left-sided stenotic valvular lesions carry a high risk of maternal and fetal adverse outcomes in the setting of pregnancy. Young women undergoing valvular replacement require special counseling regarding the risks and benefits inherent to each type of valvular prosthesis particularly if a future pregnancy may be desired.

Introduction

Valvular heart disease remains a growing and significant burden of disease nationwide with up to 2.5% prevalence in the United States. AS is the most common valvular lesion that requires replacement whereas MR is the leading cause of severe valvular heart disease in the United States [1]. Clinically significant TR is a growing problem and is disproportionately more prevalent among women. The current American Heart Association/American College of Cardiology (AHA/ACC) guidelines recommend AVR for patients with severe AS in the setting of symptoms or left ventricular (LV) dysfunction [2]. TAVR is now recommended as a reasonable alternative to surgical aortic valve replacement (SAVR) in patients with tricuspid aortic valve (AV) stenosis and indications are expected to expand to patients with bicuspid aortic valve (BAV) [3–5]. Similarly, MV surgery is recommended for patients with severe symptomatic MR or in the presence of LV dilation or LV dysfunction [2]. Transcatheter mitral valve repair (TMVr) is now approved for patients with symptomatic primary or secondary MR who are at high risk for surgery. Surgical treatment of TR is less well defined and only recommended as class I at the time of left-sided valve surgery. There is a growing body of evidence that sex-related differences in epidemiology, pathophysiology, treatment, and outcomes exist among patients with valvular heart disease. These differences may have significant clinical and therapeutic implications in the current era with respect to timing of referral for intervention, drug targets, and outcomes.

Aortic Stenosis

Clinical Case

A 78-year-old female with a history of breast cancer status post mastectomy and radiation therapy, hypertension, chronic kidney disease stage 2, and severe aortic stenosis (AS) presents with increased dyspnea on exertion. Her most recent transthoracic echocardiogram demonstrated a small and hypertrophied LV with a left ventricular ejection fraction (LVEF) of 75%. The left ventricular interventricular septum and posterior wall both measure 15 mm. There is AS with a peak gradient of 62 mmHg, mean gradient of 30 mmHg, and a calculated aortic valve area of $0.7\,cm^2$. The stroke volume index is $32\,cc/m^2$ and estimated pulmonary artery systolic pressure is 52 mmHg. Computed tomography shows severe calcification of her thoracic aorta. The aortic valve calcium score is 1378 Agatston units (AU). Her Society of Thoracic Surgery (STS) risk score is 2.9%. What is the appropriate next step in the management of her valve disease?

Epidemiology and Prevalence

More than half of all heart valve surgeries in the United States are for AVR, most of which are performed for severe AS. In developing countries, RHD is still the most common cause of AS whereas in the Western world, calcific degeneration of a tricuspid or bicuspid valve is the leading cause. BAV is the most common congenital valve lesion and constitutes about 1–2% of the general population [6, 7]. It is often associated with an aortopathy and can present with AS, AR, or both. There are noteworthy sex-specific differences in patients with BAV. BAV is three times more prevalent among men than women [7]. Although there are no sex

↑ incidence of AS in
patients with bicuspid AV

Normal LVEF
↑ concentric remodeling
↑ incidence of PLF-LG AS

Less myocardial fibrosis

Lower AV calcification
More AV fibrosis

Older, more frail, more
symptoms at the time of
AVR

Better long-term
outcomes after TAVR
compared to SAVR

↑ incidence of AR in
patients with bicuspid AV

Low LVEF
↑ concentric hypertrophy
↑ incidence of CLF-LG AS

More myocardial fibrosis

More AV calcification

Higher burden of
coronary and peripheral
artery disease at the time
of AVR

Lack of improvement in
LV remodeling and LVEF
after AVR

FIGURE 1 **Sex differences in aortic stenosis.** Summary of important sex differences in incidence, pathology, comorbidities and outcomes in patients with AS. AR, aortic regurgitation; AS, aortic stenosis; AV, aortic valve; AVR, aortic valve replacement; CLF-LG AS, classical low-flow low-gradient aortic stenosis; LV, left ventricular; LVEF, left ventricular ejection fraction; PLF-LG AS, paradoxical low-flow low-gradient aortic stenosis; SAVR, surgical aortic valve replacement; TAVR, transcatheter aortic valve replacement.

differences in BAV morphology and frequency of normal valve function, women are often referred for surgery at an older age, more commonly present with AS, and less often require aortic aneurysm repair than men. On the other hand, endocarditis and aortic dissections occur more frequently in men (**Figure 1**) [8–10].

In a large national database that used the Nationwide Inpatient Sample (NIS) to derive patient-relevant information between 2003 and 2014, out of a total of 166,809 patients who underwent SAVR, 63% were male and 37% were female [11]. A different study also utilizing the NIS between 2000 and 2012 demonstrated that among 113,847 patients admitted with an AV disorder diagnosis in the United States, 55.1% were women [12]. Similarly, in a large Scottish registry of 19,833 patients, men had a lower incidence of a new diagnosis of AS (46.8% in men vs. 53.2% in women) [13]. These findings point to a historical disparity

in referral for surgery among women with AS. In a single center study, women with severe, symptomatic AS with an AHA/ACC class I indication for AVR were less frequently referred to surgery compared with men (19% lower relative rate, $P=0.03$) despite similar outcomes after surgery [14]. Fortunately, development of TAVR has shifted the practice and women now account for 50% of patients undergoing TAVR in the United States [15].

Sex Differences in Ventricular Response to Severe AS and Reverse Remodeling After AVR

Severe AS, defined by a progressive decrease in AV opening, increases the pressure afterload on the LV and leads to LV hypertrophy (LVH), which initially represents adaptive changes of increased wall thickness to maintain normal wall stress and normal cardiac output. However, over time, it becomes maladaptive, leading to reduced myocardial perfusion [16], interstitial fibrosis [17], and impaired dia-stolic dysfunction and eventually systolic heart failure [18]. Interestingly, both echocardiographic and cardiac magnetic resonance (CMR) studies have demonstrated significant sex-related differences in the pathophysiological changes in response to AS. Although earlier echocardiography-based studies showed that women with severe AS have more concentric hypertrophy, contemporary data based on CMR demonstrate a higher LV mass-to-volume ratio [19] (0.68 ± 0.11 in men vs. $0.61 \pm 1\ 0.11$ in women, $P = 0.001$), lower systolic (LVEF, longitudinal and circumferential peak systolic strain) and diastolic (longitudinal and circumfer-ential peak early diastolic strain rate) function [19], and a higher LV mass index [19–21] in men compared to women. A recent CMR-based study showed clear sex-related differ-ences in patterns of LV remodeling in patients with severe AS who were referred for SAVR, where normal geometry (82% female) and concentric remodeling (60% female) were predominantly seen in women and concentric hypertrophy (71% male) and eccentric hypertrophy (76% male) domi-nated in men despite equal AS severity (**Figures 1 and 2**) [22]. These results suggest that men develop more advanced car-diac remodeling than women for a given degree of AS.

Following AVR, LVH regresses by 20–30% within 1 year [23, 24]. Capoulade et al. showed that concentric

hypertrophy, as compared with other remodeling patterns, is associated with a higher risk of mortality [hazard ra-tio (HR) = 1.27, 95% confidence interval (CI) 1.01–1.61, $P = 0.046$]; however, in their study, there was a significant interaction between sex and concentric hypertrophy with regards to impact on mortality. The authors demonstrated that concentric hypertrophy was associated with a near 60% increased risk of all-cause or cardiovascular (CV) mortality in women but not in men [25]. Consistent with these results, studies evaluating the response to SAVR demonstrated that residual LVH after SAVR is associated with increased mor-tality. Women with residual LVH have lower event-free survival (combined end point of all-cause death and CV hospitalization, 50% with no residual LVH vs. 93.2% with residual LVH, $P = 0.019$) and a trend toward lower survival free of all-cause death (67.8% vs. 96.4%, $P = 0.059$), and CV mortality (84.6% vs. 96.7%, $P = 0.086$), whereas there was no significant difference in event-free survival in men with or without residual LVH [26]. Whether these results reflect the consequence of delayed referral of women to SAVR at a later stage of disease compared with men is un-clear and further studies are needed to elucidate the factors responsible for the differential effects of LVH.

Focal, mid-wall myocardial fibrosis can occur in re-sponse to severe AS even in the absence of significant

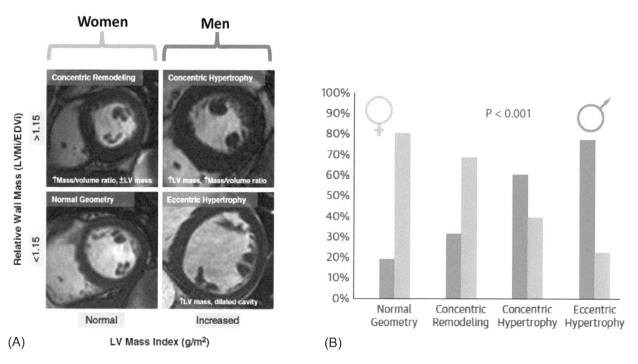

(A)

(B)

FIGURE 2 **Sex Differences in Left Ventricular Remodeling.** (A) Differentiation of LV remodeling patterns on cardiac magnetic resonance imaging based on LV mass index and relative wall thickness. (B) Sex-related differences in patterns of LV remodeling in patients with severe AS who were referred for SAVR based on CMR findings. Women have more normal geometry and concentric remodeling, whereas men have more concentric hypertrophy and eccentric hypertrophy. EDVi, end-diastolic volume index; LV, left ventricular; LVMi, left ventricular mass index. *Reproduced with permission from [22].*

coronary artery disease. Myocardial fibrosis demonstrated by late gadolinium enhancement on CMR represents irreversible scar formation and is differentiated from reactive diffuse myocardial fibrosis and extracellular volume expansion, which often regresses post-AVR [27]. In general, myocardial fibrosis is higher in men with AS with higher focal fibrosis and extracellular expansion (**Figure 1**) [19, 22]. Some authors postulate that estrogens decrease the gene expression of collagen I and III in women, thereby resulting in less fibrosis [28, 29]. Multiple studies highlight the negative prognostic impact of mid-wall myocardial fibrosis in patients with severe AS and its association with lack of improvement in LV remodeling and LVEF following SAVR [30–33]. However, despite a less favorable phenotype in men, women are more likely to develop symptoms for a similar severity of AS [19]. Larger studies evaluating the degree and impact of reverse remodeling on outcomes are required and may necessitate consideration of sex-specific thresholds for referral to AVR.

Evaluation of AS

With the advent of TAVR and the possibility of treating even nonagenarians, accurate diagnosis of severe AS in the elderly has become of utmost importance. Two-dimensional (2D) echocardiography is still the diagnostic cornerstone for the assessment of the hemodynamic severity of AS and its effects on the LV. Recently, "low-gradient" AS has emerged as a new entity which is of particular importance

in women. About 40% of patients with AS have low-gradient severe AS with discordant hemodynamic findings on Doppler echocardiography, where the calculated aortic valve area (AVA) is $\leq 1\,cm^2$, the mean trans-aortic gradient is $< 40\,mmHg$, and the maximal aortic velocity is $< 4\,m/s$. Low-gradient severe AS is further differentiated into three categories, as shown in **Figure 3**.

Classical low-flow low-gradient (CLF-LG) AS is often associated with preexisting coronary artery disease that leads to a dilated LV and depressed LVEF. The low-flow state is a result of a low cardiac output secondary to reduced LV systolic function and occasionally secondary MR. It is found in 5–10% of the AS population and is more prevalent in men (**Figure 4**). Paradoxical low-flow low-gradient (PLF-LG) AS comprises 10–15% of the AS population and is more common in elderly women. This AS phenotype consists of concentric remodeling of the LV leading to a small LV cavity and restrictive LV diastolic physiology resulting in a decrease in stroke volume (stroke volume index (SVi) $< 35\,mL/m^2$) despite a preserved LVEF (**Figure 4**). These patients have greater arterial stiffness, a higher level of global LV hemodynamic load reflected by higher ventriculo-arterial impedance, and pseudo-normalization of blood pressure due to a reduction in stroke volume [34]. Accurate diagnosis of PLF-LG AS is often challenging, and misdiagnosis can lead to inappropriate delay of AVR. Accurate prompt diagnosis is crucial in elderly women as PLF-LG AS has been associated with increased disease-related morbidity and mortality, and these patients have much

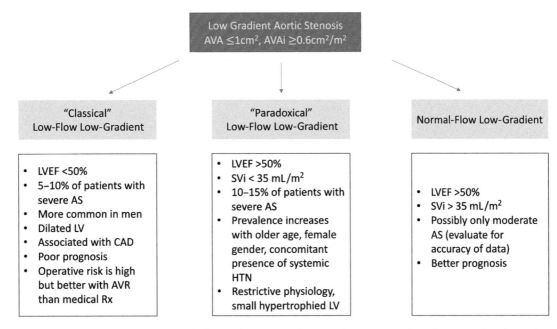

FIGURE 3 Low-Gradient Severe Aortic Stenosis. AS, aortic stenosis; AVA, aortic valve area; AVAi, aortic valve area index; AVR, aortic valve replacement; CAD, coronary artery disease; HTN, hypertension; LV, left ventricle; LVEF, left ventricular ejection fraction; Rx, therapy; SVi, stroke volume index.

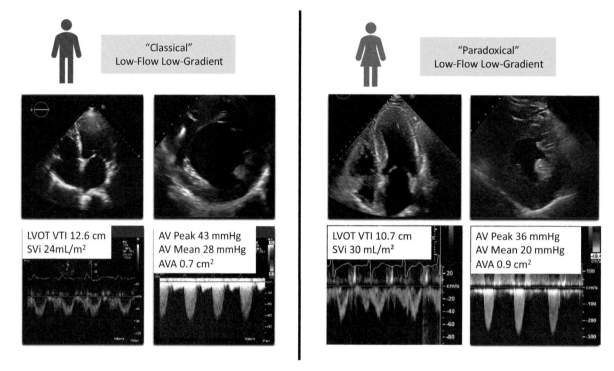

FIGURE 4 **Sex Differences in Low-Flow Low-Gradient Aortic Stenosis.** Classical low-flow low-gradient AS is more common among men. These patients have a dilated LV with low LVEF. The stroke volume index is low due to reduced LV systolic function with a corresponding low-mean transaortic valve gradient and a calculated aortic valve area less than 1 cm². Paradoxical low-flow low-gradient AS is more common among women. These patients have a thick ventricle with normal LVEF. The stroke volume index is low due to a small ventricle and often restrictive diastolic physiology. The mean transaortic gradient is less than 40 mmHg; however, the calculated aortic valve area is less than 1 cm². AS, aortic stenosis; AV, aortic valve; AVA, aortic valve area; LV, left ventricle; LVEF, left ventricular ejection fraction; LVOT, left ventricular outflow; SVi, stroke volume index; VTI, velocity time integral.

better prognosis when treated with AVR rather than medically despite increased procedural risks [35, 36]. Normal-flow low-gradient (NF-LG) AS, an entity that the current AHA/ACC Valvular Heart Disease management guidelines [5] do not address, is seen in 25% of patients with AS [37] and it can occur as a result of a normal stroke volume (SVi ≥ 35 mL/m²) but reduced mean transvalvular flow rate (defined as stroke volume/ejection time) and reduced arterial compliance and systolic hypertension. Therefore, assessment of AS severity at a normal blood pressure is essential for correct diagnosis. The European Association of Cardiovascular Imaging and the American Society of Echocardiography [38] also caution against misdiagnosing moderate AS as severe AS due to inaccurate measurement of the LV outflow tract diameter (LVOTd), which has been the source of inconsistencies in hemodynamic findings of a small valve area in the setting of low gradients.

In addition, current cutoffs for severe AS described in the AHA/ACC and European Society of Cardiology (ESC) guidelines pose a challenge to assessing the severity of AS in patients with smaller body size and smaller LVOTd, most of whom are women [39]. NF-LG AS is more prevalent in

this patient group [40] and some authors have challenged the guidelines in defining severe AS by an AVA ≤ 1 cm² for patients with smaller LVOTd (small defined as LVOTd < 1.9 cm) based on the finding that for patients with small LVOTd, AVA of 1 cm² corresponds to a mean gradient of 29 mmHg and decreasing the AVA cutoff to 0.8 cm² reduces this inconsistency [39].

Finally, quantification of AV calcium (AVC) by multidetector computed tomography (CT) has been recently accepted as an adjunct diagnostic tool in patients with low-gradient AS. Several studies have shown that AVC load correlates well with the hemodynamic severity of AS and clinical outcomes [41, 42]. Another notable sex-related difference is that women need less valvular calcium than men to develop severe AS even after accounting for body size [43]. Therefore, different thresholds for AVC score are proposed for men [≥ 2000 Agatston units or indexed to body surface area (BSA) 1067 AU/m²] vs. women (≥ 1275 Agatston units or indexed to BSA 637 AU/m²) to identify severe AS [44]. One potential explanation for the different thresholds is the observation that women have less valvular calcification but more fibrosis compared with men [45]. Multiple groups have demonstrated similar sex-specific AVC thresholds to accurately identify severe AS, but the use of CT-derived AVC is not yet included in the most recent AHA/ACC guidelines on the management of valvular heart disease [43, 46, 47]. These findings were replicated in a contemporary echocardiographic study, where moderate to severe AVC, assessed on 2D echocardiography using a validated visual score, was able to predict higher rates of major CV events during a median of 4.3 years follow-up in both sexes (HR 2.5, 95% CI 1.64–3.80; $P < 0.001$ in women; HR 2.2, 95% CI 1.54–3.17; $P < 0.001$ in men). Men with moderate-to-severe AVC at baseline exhibited 1.8-fold higher hazard rate of all-cause mortality independent of age and AS severity (95% CI 1.04–3.06, $P < 0.05$), but women did not [48].

The mechanisms leading to differential pathophysiology of AS progression in men and women are currently not well understood and require further research. Some authors hypothesize that sex-related differences in vitamin D receptors or growth factors may lead to more extensive fibrosis in women [43], whereas others emphasize the higher levels of androgen hormones in men, which has been shown to play a role in promoting vascular and aortic calcification in mice [45].

Baseline Characteristics of Women With Severe AS Presenting for AVR

Large registries have reported significant differences in preprocedural risk and characteristics between men and women. A study capturing data from the NIS consisting of 166,809 patients who underwent AVR between 2003 and 2014 found that compared to men, women were older (70 ± 13 vs. 67 ± 14 years, $P < 0.001$), had higher rates of nonatherosclerotic comorbid conditions such as hypertension, diabetes mellitus, obstructive pulmonary disease, atrial fibrillation, and anemia, but fewer incidences of coronary and peripheral arterial disease and prior sternotomies [11]. Correspondingly, men more often undergo concomitant coronary artery bypass grafting. In one large study on univariate analysis, women received bioprostheses more often than men at the time of AVR (63.6% vs. 54.7%; $P < 0.0001$); however after adjusting for age, there was only a trend toward more bioprostheses implanted in women at the time of AVR [odds ratio (OR) 1.2; CI 1.0–1.4; $P = 0.10$] [49]. Similarly, in the TAVR population, of 23,562 patients enrolled in the STS/ACC Transcatheter Valve Therapy (TVT) Registry, 49.9% were women, who were older, more frail, with higher STS risk scores, and New York Heart Association (NYHA) functional class III/IV symptoms despite a higher pre-procedural LVEF. Women also had a higher prevalence of porcelain aorta, at least moderate MR, and worse renal function at baseline, whereas men had a higher burden of CV risk factors and disease [50]. Pooled

data from five large TAVR registries demonstrated that compared to men, women were older, presented with more severe AS with a higher peak transaortic pressure gradient (61 ± 27 vs. 56 ± 24 mmHg), smaller AVA (0.7 ± 2 vs. 0.8 ± 2 cm^2) ($P < 0.001$ for both), and greater burden of severe pulmonary hypertension with a pulmonary artery systolic pressure > 60 mmHg (22.1 vs. 16.7%, $P < 0.001$) [51].

Complications and Outcomes After AVR

Mortality

Female sex is included as an independent risk factor for mortality after SAVR in the STS risk score model [52]. Among the patients who underwent isolated surgical AVR in the NIS database, after propensity matching, in-hospital mortality was higher in women than in men (3.3% in women vs. 2.9% in men, $P = 0.001$). However, after adjustment for age and comorbidities, women have significantly better long-term survival compared to men after bioprosthetic AVR (HR 0.5; CI 0.3–0.6). This difference is not apparent among patients who undergo mechanical AVR [49]. Vascular complications and blood transfusions were more frequent in women than in men, whereas rates of stroke, permanent pacemaker implantation, and acute kidney injury requiring dialysis were similar between sexes. Conversely to SAVR, there are emerging data demonstrating superior outcomes of TAVR in women compared to men with a higher magnitude of benefit of TAVR compared to surgical AVR (**Table 1**) [50, 51, 67, 68]. A meta-analysis of 17 studies (8 TAVR registries; 47,188 patients; 49.4% women) including the intermediate-risk cohort of the PARTNER II trial reported that female sex was associated with lower all-cause mortality at 1 year (relative risk (RR) 0.85; 95% CI 0.79–0.91; $P < 0.001$) and continued to be associated with lower all-cause mortality at a mid-term follow-up of 3.3 years ± 1 year (RR 0.86, 95% CI 0.81–0.92, $P < 0.001$) [67]. More favorable outcomes are possibly due to the lower comorbid risk profile at baseline with less coronary and peripheral artery disease, the lower rates of post-TAVR paravalvular regurgitation (PVR), and the greater reversal of cardiac remodeling after TAVR. Common imaging findings and outcomes after TAVR seen in women compared to men are summarized in **Figure 5**.

Procedural Complications

During TAVR, some commonly seen anatomical differences in women may predispose them to specific procedural complications, such as cardiac tamponade and ventricular rupture in patients with smaller LV cavities [50, 69], aortic annular rupture in patients with small annuli requiring TAVR-oversizing [70], and lastly, coronary obstruction in patients with lower height of coronary ostia and smaller sinuses of Valsalva dimensions (**Figure 5**) [71, 72]. Major

TABLE 1 TAVR Studies and Outcomes Among Men vs. Women

Study or First Author	Year	Type of Study	Total Population	Women, as % of Population	Follow-Up Duration in Months	All-Cause Mortality at 30-Days, RR (95% CI) for Men vs. Women	All-Cause Mortality at 1 Year, RR (95% CI) for Men vs. Women	All-Cause Mortality at >1-Year, RR (95% CI) for Men vs. Women
Ontario Cardiac Registry [53]	2017	Prospective observation	999	43.5%	15.5	1.37 (0.85, 2.22)	0.94 (0.73, 1.22)	0.94 (0.81, 1.09)
Levi et al. [54]	2017	Prospective observational	560	56.5%	24.7 (median)	1.31 (0.53, 3.29)	0.71 (0.42, 1.20)	0.94 (0.70, 1.26)
Brazilian TAVI Registry [55]	2016	Prospective observational	819	51%	16.8 ± 15.6 (mean)	1.77 (1.12, 2.80)	N/A	1.14 (0.92, 1.43)
STS/ACC TVT Registry [50]	2016	Prospective observational	23,562	49.9%	12	1.31 (1.17, 1.46)	0.94 (0.88, 1.01)	N/A
US CoreValve trial Registry [56]	2016	Retrospective from RCT	3687	46.3%	12	1.02 (0.78, 1.32)	0.90 (0.79, 1.03)	N/A
FRANCE 2 Registry [57]	2016	Prospective observational	3972	49.5%	12	1.04 (0.85, 1.26)	0.82 (0.72, 0.92)	0.84 (0.78, 0.91)
PARTNER Trial [58]	2016	Retrospective from RCT	2559	47.7%	12	1.10 (0.81, 1.48)	0.73 (0.63, 0.85)	N/A
PARTNER II Sapien 3 trial [59]	2016	Retrospective from RCT	1661	39.6%	12	1.66 (0.76, 3.61)	0.91 (0.67, 1.23)	N/A
UK TAVI Registry [60]	2015	Prospective observational	3980	52.7%	24	1.02 (0.79, 1.51)	1.20 (0.73, 1.96)	0.91 (0.81, 1.03)
Yakubov et al. [61]	2015	Prospective observational	489	52.1%	24	N/A	0.9 (0.78, 1.04)	0.86 (0.69, 1.08)
Italian CoreValve Registry [62]	2013	Prospective observational	659	55.8%	12 (median)	0.75 (0.39, 1.42)	0.85 (0.61, 1.18)	N/A
D'Ascenzo et al. [63]	2013	Prospective observational	377	57.2%	16.1 ± 8.3 (mean)	0.85 (0.43, 1.69)	N/A	0.73 (0.52, 1.02)
Hayashida et al. [64]	2012	Prospective observational	260	50.3%	7.1 (median)	0.69 (0.38, 1.24)	N/A	N/A
Humphries et al. [65]	2012	Prospective observational	584	51.3%	10 (median)	0.59 (0.34, 1.00)	0.63 (0.46, 0.85)	0.72 (0.57, 0.91)
Tamburino et al. [66]	2011	Prospective observational	663	56%	19 ± 6 (mean)	N/A	0.84 (0.60, 1.18)	N/A

ACC, American College of Cardiology; CI, confidence interval; N/A, not applicable; RCT, randomized control trial; RR, relative risk; STS, Society of Thoracic Surgery; TAVI, transcatheter aortic valve implantation; TVT, transcatheter valve therapy.

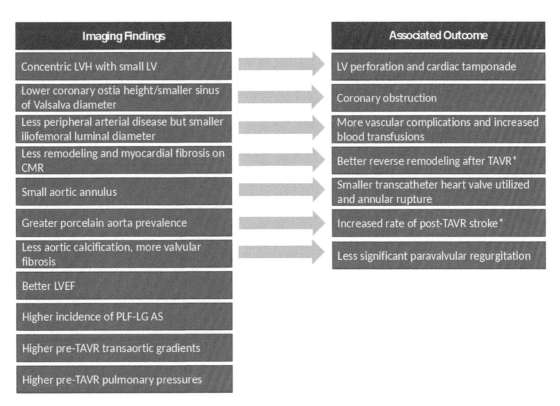

FIGURE 5 **Complications and Outcomes After Transcatheter Aortic Valve Replacement in Women Compared to Men.** CMR, cardiac magnetic resonance; LV, left ventricle; LVEF, left ventricular ejection fraction; LVH, left ventricular hypertrophy; PLF-LG AS, paradoxical low-flow low-gradient aortic stenosis; TAVR, transcatheter aortic valve replacement. * Possible association.

peripheral vascular complications occur more frequently in women [56, 63, 65, 73] owing to higher rates of non-transfemoral approaches, higher rates of porcelain aorta [50], and less favorable sheath-to-femoral artery ratio [64]. Parallel to vascular complications, major bleeding and need for blood transfusions post-TAVR have previously been reported to be higher in women compared to men. Fortunately, with the advent of new-generation, smaller-caliber devices, major vascular and bleeding complications were found to be lower in the Women's International Transcatheter Aortic Valve Implantation (WIN-TAVI) Registry [15].

Stroke

In a large observational study by Kulik et al., after adjusting for age and other comorbidities, women were more at risk for late stroke compared to men when an aortic mechanical prosthesis had been implanted (adjusted HR 1.7; CI 1.1, 2.7; *P*=0.02). [49]

There are contradicting data among studies regarding the incidence of stroke among women undergoing TAVR. Smaller studies [62, 65] along with the PARTNER II S3 trial [59], which included high-risk and intermediate-risk patients, did not demonstrate a significant difference in the rate of disabling stroke between sexes (30 days: 0.9% in women vs. 1.0% in men, *P*=0.87; 1 year: 2.4% in women vs. 2.3% in men; *P*=0.9).

Contrarily, in the STS/ACC TVT Registry [74], female sex was the only significant risk factor for stroke (HR 1.40; 95% CI 1.15–1.71) and a meta-analysis by O'Connor et al. reported a significantly increased rate of stroke at 30-days post-TAVR in women vs. men receiving a balloon-expandable valve (4.4% vs. 3.6%, *P*=0.029), but no difference in patients who received self-expandable valves [51]. The higher prevalence of porcelain aorta in women may be the major contributing factor to the observed higher rate of stroke, and future studies investigating the utility of cardioembolic protection devices are warranted. The WIN-TAVI Registry, which included data from experienced centers in which 43% of implanted devices were newer generation, reported a lower stroke rate than had been previously observed (1.3% at 30 days and 2.2% at 1 year) [15, 75].

Paravalvular Regurgitation

PVR portends a worse prognosis after both SAVR and TAVR. PVR post-SAVR is detected in 16% of patients whereas more than 90% of patients undergoing TAVR have some degree of PVR [76]. More than mild PVR is a known risk factor for adverse outcomes and mortality after TAVR, and is more commonly observed in men [51, 67, 77]. The "true cover index" defined by $100 \times$ ([prosthesis actual diameter at implantation depth-annulus diameter]/prosthesis actual diameter at implantation depth) is an objective measure of aortic annulus-TAVR prosthesis congruence, and strongly correlates with development of PVR after TAVR [78]. Owing to the smaller annular size and greater use of TAVR oversizing, women achieve a favorable true cover index $\geq 8\%$ more often than men, which along with lower AV calcification likely contributes to better long-term outcomes [50, 77, 79].

Patient-Prosthesis Mismatch

Patient-prosthesis mismatch (PPM) after SAVR has been attributed to smaller valve sizes and is associated with increased early and late morbidity and mortality. PPM is more common post-SAVR in women. Surgical valve manufacturers have attempted to design supra-annular stented and stentless valves to minimize the incidence of PPM. TAVR valves innately have a more favorable design with a minimal thin stented platform without a sewing ring occupying the annular space. As expected, data from the PARTNER A high-risk trial [80] and the CoreValve US High Risk Pivotal trial [81] both showed that PPM was less of a concern with TAVR as compared to SAVR, but PPM remained an independent predictor of mortality. Patients with a smaller aortic annulus (<20 mm in PARTNER Cohort A and <26 mm in the CoreValve trial) appeared to have more pronounced benefit from TAVR as compared to SAVR. Based on these results, TAVR may be more desirable over SAVR in patients who are more vulnerable to PPM, namely women with small annular sizes. The effect of PPM is also demonstrated in the VIVID (Valve-in-Valve International Data) Registry, where mortality after valve-in-valve (ViV) was strongly associated with small surgical valves and stenosis as the mechanism of failure, both of which occur more commonly in women. This effect was especially pronounced in large body-size women treated with surgical valves smaller than or equal to 21 mm [82]. Therefore, redo cardiac surgery with aortic root enlargement and larger valve size may be preferable in patients who are operative candidates.

Back to Clinical Case

This patient is a 78-year female who has symptomatic PLF-LG severe AS. Aortic valve replacement is recommended for optimal outcomes. Although her surgical risk is low by STS risk scoring, further imaging shows porcelain aorta which is a relative contraindication for surgery, therefore TAVR is the preferred method. It is expected for her 1-year post-TAVR follow-up echocardiogram to show regression of LVH.

This case highlights the key findings that are common among women, such as a small and hypertrophied LV with normal LVEF, low-mean-aortic valve gradient, pulmonary hypertension, and porcelain aorta. Women tend to have lower AVC scores compared with men, possibly due to more fibrosis rather than calcification. Although AVC scoring is not yet incorporated into the current guidelines, it can be helpful in differentiating severe from moderate AS, particularly among women with low-flow low-gradient (LF-LG) AS. Her surgical risk score is in the low-risk category, but porcelain aorta, commonly seen in women, adds a significant risk to surgery due to the requirement of aortic cannulation during surgery, therefore she would benefit from TAVR. In addition, women tend to have less myocardial fibrosis in response to pressure overload of AS and better regression in the degree of LVH at 1 year after AVR.

Aortic Regurgitation

The most common cause of chronic AR in the developing world is RHD, whereas in developed countries, aortic root dilation, congenital BAV, and calcific degeneration are the leading causes. Data evaluating sex-related differences in AR are scarce. There is a male predominance among patients with BAV; however, in one large multicenter international BAV registry, there were no significant sex differences regarding the morphology and frequency of normal valve function. Men who had moderate to severe aortic valve dysfunction were more likely to have AR than women (33.8% vs. 22.2%, $P < 0.001$) [8]. In a contemporary large database, these findings were replicated where any degree of AR was significantly higher in men compared to women (64% vs. 48%; $P = 0.003$). Correspondingly, the 20-year AVR rates were $41 \pm 4\%$ and $24 \pm 5\%$ for men and women, respectively ($P = 0.01$). At the time of AVR for AR, men have more severe AR (23% of men vs. 10% of women, $P < 0.001$), more likely to have aortic aneurysm (24% of men vs. 14% of women, $P < 0.0001$), whereas women are more likely to have NYHA III/IV class heart failure symptoms (46% of men vs. 58% of women, $P < 0.0001$). Among patients referred for AVR, 20-year survival was lower than that in the general population ($P < 0.0001$) and age-adjusted relative death risk was 1.16 (95% CI 1.05–1.29) for men vs. 1.67 (95% CI 1.38–2.03) for women ($P = 0.001$).

Based on the natural history studies which demonstrated high mortality in patients who develop symptoms, the AHA/ACC guidelines recommend AVR in patients with symptomatic severe chronic AR [47]. It is known that low LVEF and large LV dimensions are associated with increased late mortality; however, there are limited data on the sex-related differences in outcomes. In an earlier study that included patients who underwent surgery for isolated AR between 1980 and 1989, 10-year survival was worse for women than for men ($39 \pm 9\%$ vs. $72 \pm 4\%$, $P = 0.0002$). Independent predictors of late survival were different for men (age and ejection fraction) and women (age and concomitant coronary bypass grafting), and women rarely reached the unindexed LV dimension threshold for AVR, suggesting that women were referred for surgery late in the disease course when they developed symptoms [83]. A later study from the same institute, which included 306 patients who had AVR for AR between 1996 and 2006, demonstrated that sex was not a predictor of late survival. In this study, LV end-diastolic dimension (LVEDD) and LV end-systolic dimension (LVESD) were predictive of late mortality but only after indexing for BSA and the best discrimination in survival was seen at $20\,\mathrm{mm/m^2}$ for indexed LVESD and $30\,\mathrm{mm/m^2}$ for indexed LVEDD [84]. Therefore, for those who are asymptomatic, indications for AVR include an LVEF $< 50\%$ (class I) and LVESD $> 50\,\mathrm{mm}$ (class IIa). There are no sex-specific cutoffs for LV dilation, but a caveat is included to consider indexed end-systolic dimension (indexed LVESD $> 25\,\mathrm{mm/m^2}$).

Mitral Regurgitation

Clinical Case

A 72-year-old female with a history of anterior leaflet mitral valve prolapse (MVP) and chronic severe mitral regurgitation (MR) presents for routine cardiovascular evaluation. She has been followed expectantly for severe, asymptomatic MR with serial echocardiography. Her most recent transthoracic echocardiogram demonstrated LVEF of 62%, LVEDD of 55 mm (LVEDD indexed to BSA $34\,\mathrm{mm/m^2}$), LVESD of 37 mm (LVESD indexed to BSA $23\,\mathrm{mm/m^2}$), and a normal estimated pulmonary artery systolic pressure. She denies any shortness of breath or changes in her functional status. What is the most appropriate management of her mitral valve (MV) disease at this time?

Epidemiology and Prevalence

MR remains a growing and significant burden of disease nationwide with up to 2.5% prevalence in the United States. It is now the most common valvular heart disease among both men and women older than 75 years of age with prevalence rates of greater than 10% with increasing age [1]. MR can be classified as primary due to an abnormality of the MV apparatus or secondary due to LV dysfunction resulting in annular dilatation and/or restricted motion of the MV leaflets. Worldwide, the leading cause of primary MR is RHD. In the United States, the most common cause is myxomatous degeneration of the MV leaflets resulting in mitral valve prolapse (MVP). Despite the near-equal prevalence of MR among both sexes in the United States, there are distinct sex differences in the etiology of MV disease. A retrospective analysis of more than 23,000 Cardiovascular Information Registry patients (9712 women, 40%) from 1993 to 2016 assessed sex differences in the etiology of surgical MV disease over time and found a decrease in both ischemic heart disease (14.1–5.7%, $P < 0.001$) and RHD (19.9–8.3%, $P < 0.001$) as etiologies of MV disease in the United States. However, women maintained a significantly higher prevalence of rheumatic involvement than men ($P < 0.001$) and lower rates of degenerative MV disease ($P < 0.001$) [85] (**Figure 6A and B**). These trends highlight important differences in the etiology of MV disease, which account for significant sex-based differences in the treatment and outcomes of MV disease.

Sex-Based Differences in the Pathophysiology of Mitral Regurgitation

Mitral Valve Prolapse

Important sex differences with implications for the treatment of MV disease have been well documented for mitral valve prolapse (MVP). Anatomically, anterior leaflet prolapse is much more common among women. In addition, women have markedly higher rates than men of concomitant MV stenosis (13.9% vs. 2.7%, $P < 0.001$), and posterior MV leaflet calcification (20.1% vs. 6.5%) [85, 86]. In contrast, posterior leaflet prolapse (22% of women vs. 31% of men) and flail leaflet physiology (2% of women vs. 8% of men, all $P < 0.001$) are more common among men [87, 88]. These morphological differences have important implications in the selection of treatment for MV disease (**Figure 7**).

In addition, there are sex-based differences in chamber dimensions and regurgitation severity, which can be partially accounted for by women's smaller body surface area (BSA). Women with MVP have smaller absolute left atrium (LA) and LV diameters for the same degree of MR compared to men [87]. In a single institution study composed of 600 patients (217 women) with primary MR from

CH
15

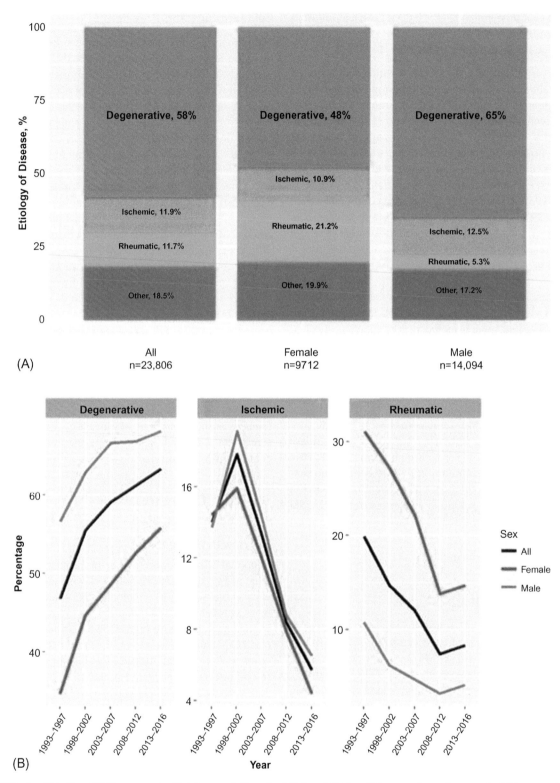

(A)

(B)

FIGURE 6 **(A) Sex Differences in Etiology of Surgical Mitral Valve Disease.** Prevalence of different etiologies of surgical mitral valve disease for the entire cohort and each sex. $P < 0.001$ for difference in etiologies between sexes. (B) Sex differences in etiology of surgical mitral valve disease. Sex-based differences for the most prevalent causes of surgical mitral valve disease from 1993 to 2016. $P < 0.001$ for degenerative and rheumatic disease. $P < 0.01$ for ischemic disease. *Reproduced with permission from [85].*

Rheumatic heart
disease more common

Anterior or bileaflet
prolapse

Delayed referral to MV
intervention

Higher rates of MV
replacement

Higher mortality after
MV interventions

Degenerative valve
disease predominates

Posterior leaflet
prolapse

Timely referral to MV
intervention

Higher rates of MV
repair

Restoration of life
expectancy after MV
repair

FIGURE 7 **Sex Differences in Mitral Valve Disease.** Summary of important sex differences in the pathology, access and referral to surgical interventions, and long-term outcomes between men and women with severe mitral regurgitation. MV, mitral valve.

1990 to 2000, preoperative echocardiographic analysis demonstrated smaller LA and LV dimensions, in addition to smaller regurgitant volume (RVol) and effective regurgitant orifice area (EROA). However, after indexing for BSA, cavity dilation and severity of regurgitation were similar to slightly more severe among women. The absence of sex-specific echocardiographic criteria for grading of MR severity [89] resulted in a smaller percentage of women meeting classification for severe MR compared to their male counterparts [90]. In a different study, evaluating the association between LVESD and survival among patients with primary MR, end-systolic dimension ≥ 40 mm (LVESD indexed to BSA 22 mm/m^2) independently determined mortality after diagnosis [91]. These findings highlight the potential for the underestimation of MR severity among women and its implications on survival [2, 92, 93].

Ischemic MR

Similar sex-based differences in cardiac dimensions and MR hemodynamics have been observed among women with ischemic MR. In a secondary analysis from the Cardiothoracic Surgical Trials Network (CTSN), among patients with severe ischemic MR undergoing MV surgery, women had smaller LV volumes, LV diameters, and EROA even after normalization for BSA. However, women had a larger EROA to left ventricular end-diastolic volume (LVEDV) ratio compared to men (0.24 ± 0.09 vs. 0.20 ± 0.09, $P = 0.002$), suggestive of a disproportionate degree of MR for the amount of LV dilation [94, 95]. Unlike primary MR, given the absence of data to support a mortality benefit from surgical MV interventions, treatment of secondary MR has traditionally focused on the diseased left ventricle as the primary driver of MR. However, these hemodynamic and morphological differences have important ramifications as women with secondary MR out of

proportion to LV dilatation may benefit from percutaneous MV interventions in addition to the current recommendations for guideline-directed medical therapy (GDMT) for treatment of LV dysfunction [92, 95].

Challenges to the Treatment of Mitral Regurgitation in Women

Referral to Surgical Intervention

Several studies have demonstrated reduced and delayed rates of surgical referrals among women with severe MR, which poses a significant challenge in the treatment of MV disease for women. In a large retrospective cohort of more than 8000 patients (4461 women), referral to surgical intervention was less likely among women [adjusted risk reduction (ARR) 0.79 (0.74–0.84), $P < 0.001$] and women with severe MR underwent surgical valve intervention less frequently ($52\% \pm 3$ vs. $60 \pm 2\%$, $P = 0.03$) compared to men after adjusting for age, LVEF, and regurgitation severity [87].

Lower rates of MV surgical referrals among women in the United States are likely related to the absence of sex-specific guidelines for surgical referral [92]. Current AHA/ACC guidelines for the management of chronic, severe primary MR emphasize worsening LV function defined by LVEF 30–60% and/or LV dilation (LVESD ≥ 40 mm) for surgical referral [2, 92]. In the absence of clinical guidelines that account for sex-based differences in women's chamber dimensions, women are at risk of failing to reach necessary thresholds for MV surgical interventions. For example, in a large retrospective cohort study of more than 7000 patients with MVP, only 5.7% of women met the surgical threshold for LV dilation vs. 9.6% of men ($P < 0.01$), and LVEF was slightly higher in women than in men ($63\% \pm 10$ women vs. $61\% \pm 9$ men, $P = 0.01$) [87].

Delayed referral of clinically significant MR is further reflected in women's worse preoperative clinical profiles at the time of MV surgery. A retrospective review of Centers for Medicare and Medicaid Services MV surgeries from 2000 to 2009 noted women's older age at presentation, greater burden of preoperative comorbidities, and greater need for additional concomitant surgical procedures compared to men [96]. At the time of MV repair, women had a higher prevalence of atrial fibrillation (43.8% vs. 36.9%, $P=0.0001$), a known perioperative marker of significant morbidity and decreased survival after MV repair [96, 97]. In addition, women had more symptomatic heart failure (HF) (41% vs. 19%, $P<0.01$), which more often triggered surgery [90] and was associated with increased risk of postoperative HF and worse clinical outcomes [94]. Similarly, a retrospective review of the NIS from 2005 to 2008 demonstrated higher rates of concomitant TV surgery among women undergoing MV surgery (21.7% vs. 15.4%, $P=0.0001$), suggestive of clinical presentations with more advanced disease compared to their male counterparts [96, 98]. Lastly, when women were referred for MV intervention, the indication for surgery/admission was more often urgent compared to elective in males with the same degree of MR [98, 99].

Choice of Mitral Valve Intervention

Surgical Mitral Valve Repair or Replacement

Over the last two decades, there has been marked improvement in operative risks associated with both MV repair and MV replacement. From 2000 to 2008, the STS Adult Cardiac Surgery Database (STS ACSD) demonstrated an overall increase in MV repairs and markedly lower mortality rates associated with MV repair (1.2%) compared with MV replacement (3.8%) in contemporary practice [99]. The durability of surgical MV interventions has been well described with improved long-term survival after MV repair compared to MV replacement in patients with degenerative MV disease [100]. For this reason, MV repair, when feasible, remains the primary treatment of choice for chronic, primary severe MR [2]. The current AHA/ACC guidelines recommend MV surgery for severe, primary MR (Class I indication) for symptomatic patients with LVEF > 30% and for asymptomatic patients with evidence of LV dysfunction (LVEF 30–60% and/or LVESD ≥ 40 mm). In contrast, surgical MV repair and replacement have failed to demonstrate any mortality benefit among patients with secondary MR [101, 102]. As a result, the treatment of secondary MR consists of GDMT for HF including cardiac resynchronization therapy and/or coronary revascularization when appropriate, with a focus on treatment of the diseased ventricle [92].

Despite the superiority of MV repair, multiple databases have documented sex-based disparities with decreased rates of MV repair among women [96, 98, 99] (**Figure 7**).

Vassileva et al. used Centers for Medicare and Medicaid Services data to demonstrate that women were less likely to receive MV repair (32% of women vs. 44% of men, $P<0.0001$) and more likely to undergo MV replacement, a similar trend that was highlighted using NIS data during a similar time period (37.9% of women vs. 55.9% of men, $P<0.001$). The basis of these sex disparities remains to be well defined; however, differences in MV morphology, perceived disease severity based on echocardiography criteria, preoperative comorbidities, and physician referral factors as well as patient preferences are also likely at play. For example, pure posterior leaflet prolapse and flail leaflets have demonstrated greater success with MV repair compared to other anatomic subsets (anterior and/or bileaflet prolapse) which are more common among women [88, 93]. Likewise, MS and rheumatic changes of the valve more often require MV replacement.

In addition to lower rates of MV repair among women, multiple large database studies have highlighted persistent sex disparities in both operative mortality and long-term outcomes among men and women undergoing MV surgeries (**Figure 7**). In the largest retrospective series among Medicare beneficiaries (> 28,000 women, 60.8% of the population) undergoing isolated MV operations in the United States between 2000 and 2009, women had increased in-hospital mortality (7.7% of women vs. 6.1% of men, $P<0.0001$) compared to men. In addition, although mortality rates were lower for MV repair among the entire population, women had increased mortality rates with both MV repair (4.2% of women vs. 3.5% of men, $P=0.0112$) and replacement (9.3% of women vs. 8.2% of men, $P=0.0018$) compared to men [96]. However, after adjustment for baseline comorbidities and severity of valve disease, survival after MV repair was similar among both sexes, suggestive of women's worse preoperative profiles and delayed interventions leading to worse outcomes. Interestingly, 5 years postrepair, MV repair restored normal life expectancy (compared to the age-matched general US population) for men but failed to do so for women (**Figure 8A**). There were no sex differences in survival after MV replacement (**Figure 8B**) [96].

In a secondary analysis from the CTSN trial comparing MV repair to replacement for ischemic MR, female sex was found to be an independent predictor of all-cause mortality (27.1% vs. 17.4%, adjusted HR 1.85, 95% CI: 1.05–3.26; $P=0.03$) (**Figure 9**) [94]. Women also had a trend toward a higher risk of treatment failure and higher rate of major adverse cardiac and cerebrovascular events (49% vs. 38.1%, adjusted HR 1.58, 95% CI: 1.06–2.37; $P=0.02$) over 2 years with MV surgery (**Figure 9**) [94]. After MV surgery, both men and women demonstrated improvements in health and quality of life (QoL) scores; however, women had higher Minnesota Living with Heart Failure (MLHF) questionnaire scores and a trend toward a higher prevalence of NYHA

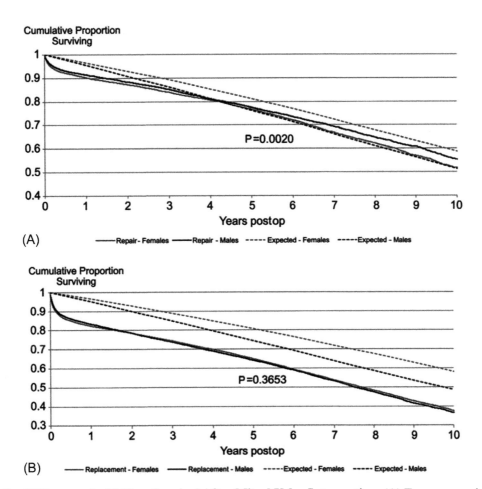

FIGURE 8   Sex Differences in 10-Year Survival After Mitral Valve Intervention. (A) Ten-year survival for male vs. female Medicare fee-for-service beneficiaries undergoing mitral valve *repair* from 2000 to 2009 compared with expected survival in the US population matched for age and sex. (B) Ten-year survival for male vs. female Medicare fee-for-service beneficiaries undergoing mitral valve *replacement* from 2000 to 2009 compared with expected survival in the US population matched for age and sex. *Reproduced with permission from [96].*

functional class III or IV (16.7% vs. 8.3% adjusted OR: 2.37; 95% CI: 0.87–6.41; $P=0.09$). Interestingly, these differences were not explained by differences in LV remodeling (**Figure 10**). In terms of long-term survival after MV surgery, a single center study conducted by Seeburger et al. in Germany of 3761 patients (1637 female, 43.5%) undergoing minimally invasive MV surgery between 1999 and 2011 demonstrated long-term survival rates among males of 96%, 89%, and 72% survival compared with 92%, 82%, and 58% survival in females after 1, 5, and 10 years, respectively ($P<0.0001$) [88].

Transcatheter Mitral Valve Repair

In the last decade, TMVr has emerged as an alternative treatment for patients with severe, symptomatic MR at prohibitive risk for MV surgery. The most well-studied device is the MitraClip, a leaflet coaptation clip that was studied in the landmark endovascular valve edge-to-edge

repair (EVEREST) II trial among patients with moderate to severe, symptomatic primary MR with excessive surgical risk for MV surgery. Compared to MV surgery (86% MV repair, 14% MV replacement), TMVr demonstrated superior safety outcomes at 30 days with major adverse cardiac events occurring in 15% with the MitraClip vs. 48% with surgery ($P<0.001$) [103]. Although there was no difference in efficacy (composite of death, MV dysfunction requiring surgery, or moderate to severe residual MR) at 12 months, TMVr was associated with higher rates of MV dysfunction requiring MV surgery (20% vs. 2%, $P<0.001$). Importantly, TMVr demonstrated sustained durability with similar 5-year long-term outcomes [104].

Given its promising results for primary MR, TMVr has been compared to medical therapy in two randomized clinical trials [105, 106] among patients with secondary MR, LV dysfunction, and symptomatic HF. The Cardiovascular Outcomes Assessment of the MitraClip Percutaneous

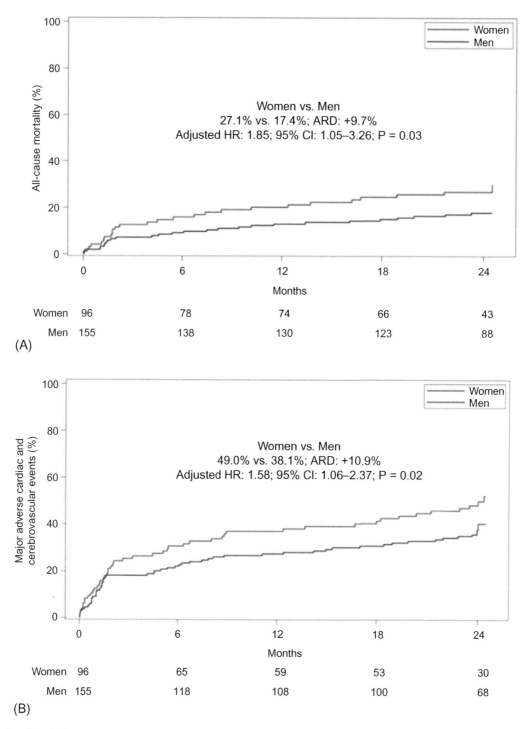

FIGURE 9 **Sex Differences in All-Cause Death and MACCE in Ischemic Mitral Regurgitation.** Kaplan-Meier curves are shown for (A) all-cause death and (B) MACCE in women and men. ARD, absolute risk difference; CI, confidence interval; HR, hazard ratio; MACCE, major adverse cardiac and cerebrovascular event(s). *Reproduced with permission from [94].*

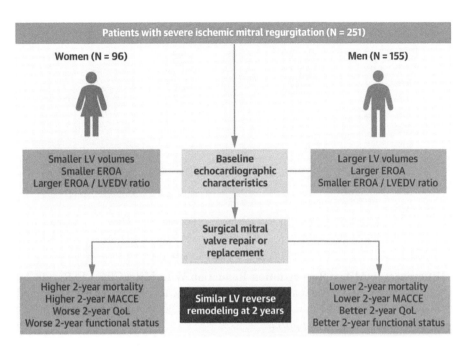

FIGURE 10 **Sex Differences in Severe Ischemic Mitral Regurgitation.** Women displayed features of more disproportionate mitral regurgitation despite smaller LV volumes and EROA. After mitral valve surgery, women were at significantly higher risk of mortality, adverse events, and worse quality of life and functional status over 2 years. EROA, effective regurgitant orifice area; LV, left ventricle; LVEDV, LV end-diastolic volume; MACCE, major adverse cardiac and cerebrovascular events; QoL, quality of life. *Reproduced with permission from [94].*

Therapy for Heart Failure Patients with Functional Mitral Regurgitation (COAPT) trial evaluated the efficacy of this device among patients with secondary MR and found clinically significant reductions in HF hospitalizations and mortality (29.1% vs. 46.1%, HR 0.62, 95% CI, 0.46–0.82, $P<0.001$) at 24 months compared to medical therapy [106]. In contrast, the MITRA-FR (Percutaneous Repair with MitraClip Device for Severe Functional/Secondary MR) trial failed to show any differences in death or hospitalizations at 12 months [105]. These disparate results have provided the impetus for better characterization of the heterogenous secondary MR populations studied in each trial. Importantly, the majority of patients in the COAPT trial who benefited from TMVr had MR disproportionate to the degree of LV chamber enlargement, a finding that is more common among females [94] and may represent a subset of patients who benefit preferentially from MV repair [95] compared to MR that is proportionate to the degree of LV dilation.

There are limited observational registries that have evaluated sex differences in the safety and efficacy of TMVr among men and women. A multicenter European registry study by Estevez-Loureiro et al. reported on the effect of sex (173 patients, 37% women) on short- and long-term outcomes after percutaneous edge-to-edge MV repair

and found no differences in the reduction of MR (MR grade $\leq 2 + 98.2\%$ of men, 96.8% of women, $P=0.586$) (**Figure 11**) or NYHA functional class (Class $<$ II 78.3% of men, 77% of women, $P=0.851$). Interestingly, preprocedure, women had smaller LV dimensions when indexed to BSA and higher LVEF. At a mean follow-up of 16 months, there were no sex differences in mortality or HF admissions ($P=0.798$) [107]. In addition, an Italian echocardiographic study by Attizaani et al. using the Getting Reduction of Mitral Insufficiency by Percutaneous Clip Implantation (GRASP) Registry (171 patients, 38% women) likewise found no significant sex differences in device success and no differences in rates of hospitalization for HF at 12 months (8.2% of men, 7.0% of women, $P=0.783$). There were additionally no differences between reduction in MR at 12 months ($P=0.197$); however, women more frequently experienced a reduction to $\leq 1 + MR$ post-procedure despite similar etiologies and severity of MR [108]. Importantly, in both observational studies, women were older, similar to trends observed among MV surgery recipients, but they had notably smaller LV dimensions even after indexing for BSA, which could partially account for the superior efficacy in some postprocedural outcomes. Given the limited data and small numbers, further studies are needed to further evaluate for the presence of any significant sex differences in the outcomes of percutaneous MV interventions.

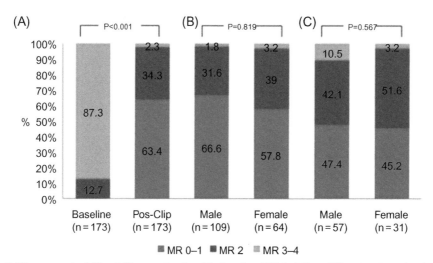

FIGURE 11 Sex Differences in Mitral Regurgitation Reduction With MitraClip. (A) Results for MR reduction before and after clip implantation. (B) MR reduction at 1 month. No significant differences were observed. (C) MR reduction at 6 months. A slight increase in severe MR was noted in the male group, but this difference was not significant. MR, mitral regurgitation. *Reproduced with permission from [107].*

Back to Clinical Case

The case represents a 72-year-old female who has a history of anterior leaflet MVP complicated by asymptomatic, chronic severe MR and has echocardiographic evidence of LV dilation after indexing for BSA (indexed LVESD 23 mm/m^2). Based on the 2017 update of the 2014 AHA/ACC Guidelines for the Management of Patients with Valvular Heart Disease, MV surgery is recommended (Class I, B) for asymptomatic patients with chronic severe primary MR with evidence of LV dysfunction (LVEF 30–60% and/or LVESD ≥ 40 mm) with preference for MV repair even in patients involving the anterior leaflet or bileaflet prolapse when durable repair can be accomplished. In addition, even in the absence of progressive LV dysfunction, in patients in whom the likelihood of a successful MV repair is greater than 95% with an expected mortality rate <1%, MV repair is reasonable (Class IIa, B) when performed at a Heart Valve Center of Excellence [92].

This case highlights the importance of indexing echocardiographic parameters for BSA to decrease underestimation of MR severity and hemodynamic and remodeling effects of MR, which are triggers for earlier referral for MV intervention. This case also highlights the importance of pursuing MV repair despite the less favorable MV leaflet morphology based on improved outcomes and survival compared with MV replacement even in patients with anterior leaflet prolapse.

Mitral Stenosis

Epidemiology and Prevalence

MS remains a significant cause of morbidity and mortality. Worldwide, rheumatic MS, a delayed complication of RHD, remains the primary cause of MS; however, in developed nations, degenerative (calcific) MS (DMS) is increasingly recognized as a growing cause of clinically significant MS. RHD has a female predilection and causes MS via commissural fusion. In contrast, DMS occurs secondary to pathological MAC caused by dystrophic calcification of the MV annulus, resulting in restricted annular dilation and anterior MV leaflet mobility [109]. Although the prevalence of DMS has not been well defined, small population studies estimate 6–8% of patients with severe MAC have associated hemodynamically significant MS [110] with an increasing prevalence with age, in contrast to rheumatic MS, which is most common in the third and fourth decades of life. Importantly, both rheumatic and calcific MS are more common among women. In a large retrospective study of more than 24,000 echocardiograms (53% female), MS was significantly more prevalent in women (1.6% vs. 0.4%, $P < 0.001$) [111]. A similar study demonstrated the incidence of MAC-associated DMS to be three times higher in women compared to men [112].

Challenges to the Treatment of Mitral Stenosis in Women

Degenerative (Calcific) Mitral Stenosis

The current mainstay of therapy in patients with DMS consists of diuretics and beta-blocker therapy to maximize diastolic filling time [2]. To date, there are no therapies known to prevent progression of annular calcification. MV replacement is considered in patients with refractory symptoms; however, the presence of MAC poses a significant challenge for surgical repair due to the need for partial or complete debridement of MAC for adequate anchoring of the prosthetic valve within the MV annulus. In addition, women more often have low-gradient very severe MS defined as a mean transmitral gradient < 10 mmHg with a planimetered MV area $\leq 1 \, cm^2$, which has been linked to less favorable left atrial remodeling and a greater risk of persistent symptoms post-MV replacement [113].

Tricuspid Regurgitation

Epidemiology and Prevalence

TR is the most common right-sided valvular heart lesion. Over two-thirds of the population have echocardiographic evidence of trace to mild TR [114, 115]; however, the community prevalence of at least moderate TR is estimated to be < 1% with a disproportionately higher prevalence among women ($P < 0.01$) [116]. TR is most commonly functional or secondary to a heterogenous group of etiologies that cause right atrial or right ventricular (RV) dilation and result in annular dilation and tethering of TV leaflets [115]. Functional TR is most commonly associated with left-sided valvular heart disease, pulmonary hypertension, and LV dysfunction with less than 10% of TR classified as isolated in origin [116].

Despite the higher prevalence of clinically significant TR among women, surgical correction of TR remains low [117, 118]. Among a community cohort (63% female) with at least moderate TR, only 2.6% of patients underwent surgical repair during 10 years of follow-up [116]. Based on the current ACC/AHA guidelines, TV repair or replacement for severe functional TR at the time of left-sided valve surgery has a class I indication [2].

Challenges to the Treatment of Tricuspid Regurgitation in Women

Concomitant Left-Sided Heart Disease

TR in the setting of left-sided heart disease is very common; however, current guidelines only recommend surgical correction in the setting of severe, functional TR (class I indication) [92] with controversial management of lesser degrees of TR. Women have higher rates of concomitant TR at the time of MV surgery [88, 98], and there is mounting evidence to suggest progression of preexisting TR years after correction of MR. In a single center retrospective study of 500 patients undergoing MV surgery for degenerative MVP, 40% of patients had preexisting concomitant trace to moderate TR, of whom more than a quarter developed progressive TR and right ventricular dysfunction post-MV surgery [119]. In addition, female gender was shown to be an independent risk factor (HR, 5.0; 95% CI 2.0–12.7; $P = 0.001$) for development of late significant TR (>1 grade increase or ≥ grade 3 or 4), which was associated with a lower 8-year clinical event-free survival rate (76% vs. 91%, $P < 0.001$) [120]. In patients with moderate TR undergoing MV surgery, concomitant tricuspid annuloplasty was predictive of recovery of RV function ($P = 0.02$) despite no difference in long-term survival [121]. Given the high prevalence of TR among women and the possible benefit from earlier surgical repair, additional studies are needed to better define the appropriate timing for maximal benefits from surgical intervention.

Isolated Tricuspid Regurgitation

Isolated TR is more common among women. From 2004 to 2013, using NIS data of TV surgeries, women represented 58% of the isolated TV surgeries in the United States [118]. Since 2004, the rate of isolated TV surgeries has increased without a change in operative mortality (8.8%) and a significantly higher mortality associated with TV replacement compared to TV repair (OR 1.91, 95% CI: 1.18–3.09; $P = 0.009$) [118]. Although female sex is a known risk factor for increased mortality following cardiac surgery, a sex-stratified propensity score matched analysis of patients undergoing isolated TV surgery demonstrated no differences in in-hospital mortality between men and women (7.9% vs. 7.7%, $P = 0.99$) [117]. However, the mortality rates were significantly higher compared to other valve surgeries setting a high threshold for the surgical correction of isolated TV disease [99, 100, 122]. In addition, in a large retrospective analysis of more than 3000 patients (54% female) with isolated severe TR who underwent surgery (5% surgery, 95% medical management), there was no difference in overall survival (HR 1.34 [0.78–2.30], $P = 0.288$) [123]. The current guidelines recommend TV repair or replacement in patients with isolated severe symptomatic primary TR but do not provide any guidance regarding the appropriate timing of surgical correction leading to heterogeneity in clinical practice. As a result, the absence of difference in mortality could reflect the delay in pursuit of TV surgical repair and the added morbidity associated with development of progressive HF. Given the disproportionate prevalence of TR among women, further research should aim to address the appropriate timing of TR interventions to maximize benefit and minimize detrimental effects of long-standing TR.

Valvular Heart Disease in Pregnancy

The normal physiological and hemodynamic changes associated with normal pregnancy pose a unique challenge to women with significant valvular heart disease. Pregnancy is associated with a significant increase in cardiac output and plasma volume resulting in a significant risk of adverse maternal and fetal outcomes in the setting of severe left-sided stenotic lesions. As a result, prepregnancy CV evaluation and counseling are recommended for any woman with known or suspected valvular heart disease [2]. The risk of complications with pregnancy depends on multiple factors including prior cardiac history, specific lesions (i.e., presence of a mechanical valve, left-sided stenotic lesions), and delivery of care [124] . However, in general, pregnancy is not advised (modified World Health Organization (WHO) Class IV) in women with severe MS and severe symptomatic AV stenosis [125, 126].

Aortic Stenosis

The most common cause of AS in women of childbearing age is bicuspid AV disease. This can be associated with an aortopathy as well as coarctation of the aorta which can further increase the risk of maternal complications during pregnancy. As a result, all women with bicuspid AV disease should undergo imaging of the ascending aorta [125]. In general, pregnancy is not advised in the setting of severe, symptomatic AS. Valvular intervention is recommended before pregnancy in patients with severe AS in the presence of symptoms, LV dysfunction (LVEF <50%), or symptoms during exercise testing [125]. For pregnant women with refractory symptoms, percutaneous balloon aortic valvuloplasty can be considered. In severe, symptomatic AS, cesarean delivery is preferred, but an individualized approach should always be undertaken [125].

Mitral Stenosis

Severe rheumatic MS remains a challenging clinical entity to manage, particularly among women of childbearing age, due to the poorly tolerated hemodynamic effects of pregnancy in the setting of relative LV inflow obstruction. All patients with significant MS should be counseled against pregnancy and considered for intervention prior to conception. The current ACC/AHA 2014 guidelines on the management of valvular heart disease recommend (Class I) percutaneous mitral balloon commissurotomy for symptomatic patients with severe MS (MV area $\leq 1.5\,cm^2$). For women with severe MS prior to conception, percutaneous commissurotomy is recommended before pregnancy even in the absence of symptoms [2]. Due to the normal physiologic changes of pregnancy, women with severe MS are at heightened risk for HF secondary to the large increases in cardiac output associated with labor and the autotransfusion of venous return from the lower extremities after delivery. Expert preconception counseling by a specialty CV team is essential for adequate risk stratification and

counseling of maternal and fetal risks associated with MS. In the international, prospective Registry of Pregnancy and Cardiac Disease (ROPAC), although mortality associated with rheumatic MS was <2%, about 50% of patients with severe rheumatic MS developed HF during pregnancy. Prepregnancy NYHA class > I was an independent predictor of maternal cardiac events. In addition, severe MS was independently associated with adverse fetal outcomes including preterm birth, small for gestational age fetus, and the need for an emergency caesarean section due to CV compromise [127]. Importantly, women with severe MS who undergo valvular intervention before pregnancy have a significant reduction in adverse cardiac events compared to those without prior intervention (14% vs. 66%, $P=0.014$) [127]. For patients in whom percutaneous commissurotomy is contraindicated (left atrial thrombus, more than moderate MR), the maternal and fetal risks of valve surgery including anticoagulation strategies, mechanical valve thrombosis, and bioprosthetic valve degeneration during pregnancy must be fully discussed. Both percutaneous and surgical valvular interventions during pregnancy are deferred if necessary, except in the presence of refractory symptoms and HF [125]. Vaginal delivery is favored in patients with mild MS; however, cesarean section is generally considered in patients with severe MS with symptomatic HF, pulmonary HTN, or in whom valvular intervention cannot be performed or fails [125].

Pulmonic Stenosis

Valvular pulmonic stenosis (PS) is a relatively common congenital defect occurring in 5–10% of children with congenital heart disease with equal occurrence in both sexes. The presence of symptoms secondary to PS or a mean Doppler gradient >40mmHg in the presence of a domed pulmonic valve is an indication for balloon valvotomy [128]. Severe PS is generally well tolerated even during pregnancy. For pregnant women with PS, close surveillance is warranted with serial echocardiography for assessment of right ventricular function with consideration of valvotomy for refractory symptoms [125].

Prosthetic Valves in Pregnancy

In young women with severe valvular heart disease requiring valvular intervention, special consideration is needed when choosing the type of bioprosthesis, particularly in women considering future pregnancies. In young women, bioprosthetic heart valves are generally recommended due to their lower risk of thromboembolism (compared to mechanical heart valves) and absent need for long-term anti-coagulation. However, in limited studies, bioprosthetic heart valves have been associated with a higher risk of structural valve deterioration and a high risk of valve reoperation (80% at 10 years) [129]. For young women with severe AV disease, the Ross procedure (autologous pulmonary valve in the aortic position and pulmonary homograft) provides an attractive alternative to a bioprosthesis although autograft failure and accelerated neoaortic dilatation have been recognized as complications during subsequent pregnancy. In a small retrospective study, there was no significant change in the aortic root diameter after first pregnancy although there was significant dilatation with subsequent pregnancies [2nd pregnancy $(4.3 \pm 0.7\,cm$, $P=0.009)$, 3rd pregnancy $(4.5 \pm 0.7\,cm$, $P=0.009)$] [130]. In contrast to a bioprosthesis, mechanical heart valves require meticulous anticoagulation throughout pregnancy to minimize teratogenicity to the fetus, valve thromboembolism, and hemorrhagic risks. The presence of a mechanical heart valve is associated with a high risk of pregnancy complications (modified WHO Class III) with a 58% chance of an event-free pregnancy with a live birth compared to 79% for women with a bioprosthetic heart valve [131]. The current AHA/ACC guidelines recommend the use of a low-dose vitamin K antagonist (VKA) ($<5\,mg/day$) or low-molecular weight heparin in the first trimester (if VKA dose $>5\,mg/day$) followed by VKA for the remainder of the pregnancy to help minimize the risk of the dose-dependent VKA teratogenicity in the first trimester and decrease the risks of valve thrombosis [2]. The advantages and disadvantages of different anticoagulation strategies should be extensively discussed with all women before pregnancy. Although VKAs are the most effective regimen for the prevention of valve thrombosis (compared to heparin), they are associated with risks of embryopathy, fetal loss, and hemorrhage [125]. Regardless of the anticoagulation strategy, planned delivery is necessary in order to ensure a safe transition to peripartum anticoagulation. All women on VKAs should be transitioned to intravenous heparin at least 36 h prior to planned delivery given the increased risk of fetal intracranial hemorrhage with VKAs. Anticoagulation during pregnancy is further discussed in Chapter 21.

Summary and Future Directions

Although there are no sex differences in the prevalence of valvular heart disease (VHD), there are important differences in diagnosis, treatment, and outcomes of VHD among men and women. Pathophysiology of AS in women appears to be significantly different from men, which may necessitate different targets for future drug development. Women more often present with more heart failure symptoms in the setting of VHD and are less likely to reach guideline-directed changes in chamber size, indicating the need for developing sex-specific clinical guidelines to help guide the timely diagnosis and referral for surgical and transcatheter interventions. Furthermore, pathophysiological changes during pregnancy pose a challenge to the care of the mother and the fetus. Given the complexity of VHD in pregnancy, women with AS, MS, and mechanical prosthetic valves should be managed by a multidisciplinary team consisting of a maternal-fetal specialist and a cardiologist specializing in the care of women through pregnancy.

CH
15

Key Points

1. Calcific degeneration is the most common cause of severe AS in developed countries and prevalence is roughly equal among both sexes.

2. Women with severe AS tend to have more concentric left ventricular (LV) remodeling, less myocardial focal fibrosis, more aortic valve (AV) fibrosis, and smaller aortic annuli compared to men. In contrast, men have more concentric LV hypertrophy, myocardial focal fibrosis, and aortic valve calcification. Paradoxical low-flow low-gradient AS is more common in women whereas classical low-flow low-gradient AS is more common among men.

3. Female sex is an independent risk factor for mortality post-SAVR. Conversely, post-TAVR, although women have higher vascular complications, more specific procedural complications, higher incidence of stroke, many studies have reported more favorable short-, mid-, and long-term outcomes in women compared with men.

4. The most common causes of aortic regurgitation (AR) in the developed world are aortic dilation, bicuspid aortic valve (BAV), and calcific degeneration. Among patients with BAV, men present more with AR, whereas women present with more heart failure symptoms. Women may be referred for surgery late in the disease course due to lack of sex-specific guidelines.

5. Degenerative mitral valve (MV) disease is the most common cause of mitral regurgitation (MR) between both sexes, but there remains a significantly higher burden of rheumatic heart disease (RHD) among women.

6. Women less frequently meet diagnosis for severe MR due to the absence of indexing morphological and hemodynamic parameters on echocardiography for body surface area (BSA).

7. Compared to males, women with severe MR are less often referred for MV surgery, more often require MV replacement, and have worse outcomes compared to men.

8. There are no significant sex differences in the efficacy and long-term outcomes of emerging transcatheter mitral valve repair (TMVr), but additional studies are needed.

9. Rheumatic mitral stenosis (MS) and degenerative MS secondary to mitral annular calcification (MAC) are more common among women.

10. Severe valvular heart disease in pregnancy, specifically left-sided stenotic lesions, is associated with adverse maternal and fetal outcomes.

11. Tricuspid regurgitation (TR) is more common among women and is associated with poor prognosis and long-term outcomes.

12. Women with severe rheumatic MS warrant special consideration for percutaneous balloon commissurotomy prior to seeking pregnancy.

References

[1] Nkomo VT, Gardin JM, Skelton TN, Gottdiener JS, Scott CG, Enriquez-Sarano M. Burden of valvular heart diseases: a population-based study. Lancet 2006;368:1005–11.

[2] Nishimura RA, Otto CM, Bonow RO, Carabello BA, Erwin JP, Guyton RA, et al. 2014 AHA/ACC guideline for the management of patients with valvular heart disease: a report of the American College of Cardiology/American Heart Association Task Force on Practice Guidelines. J Am Coll Cardiol 2014;63:e57–185.

[3] Popma JJ, Deeb GM, Yakubov SJ, Mumtaz M, Gada H, O'Hair D, et al. Transcatheter aortic-valve replacement with a self-expanding valve in low-risk patients. N Engl J Med 2019;380:1706–15.

[4] Mack MJ, Leon MB, Thourani VH, Makkar R, Kodali SK, Russo M, et al. Transcatheter aortic-valve replacement with a balloon-expandable valve in low-risk patients. N Engl J Med 2019;380:1695–705.

[5] Nishimura RA, Otto CM, Bonow RO, Carabello BA, Erwin JP, Fleisher LA, et al. 2017 AHA/ACC focused update of the 2014 AHA/ACC guideline for the Management of Patients with Valvular Heart Disease: a report of the American College of Cardiology/American Heart Association Task Force on Clinical Practice Guidelines. J Am Coll Cardiol 2017;70:252–89.

[6] Braverman AC. The bicuspid aortic valve and associated aortic disease. In: Otto CM, Bonow RO, editors. Valvular heart disease. Philadelphia: Saunders/Elsevier; 2013. p. 179.

[7] Siu SC, Silversides CK. Bicuspid aortic valve disease. J Am Coll Cardiol 2010;55:2789–800.

[8] Kong WKF, Regeer MV, Ng ACT, McCormack L, Poh KK, Yeo TC, et al. Sex differences in phenotypes of bicuspid aortic valve and aortopathy: insights from a large multicenter, international registry. Circ Cardiovasc Imaging 2017;10.

[9] Michelena HI, Suri RM, Katan O, Eleid MF, Clavel M-A, Maurer MJ, et al. Sex differences and survival in adults with bicuspid aortic valves: verification in 3 contemporary echocardiographic cohorts. J Am Heart Assoc 2016;5.

[10] Andrei AC, Yadlapati A, Malaisrie SC, Puthumana JJ, Li Z, Rigolin VH, et al. Comparison of outcomes and presentation in men-versus-women with bicuspid aortic valves undergoing aortic valve replacement. Am J Cardiol 2015;116:250–5.

[11] Chaker Z, Badhwar V, Alqahtani F, Aljohani S, Zack CJ, Holmes DR, et al. Sex differences in the utilization and outcomes of surgical aortic valve replacement for severe aortic stenosis. J Am Heart Assoc 2017;6.

[12] Badheka AO, Singh V, Patel NJ, Arora S, Patel N, Thakkar B, et al. Trends of hospitalizations in the United States from 2000 to 2012 of patients > 60 years with aortic valve disease. Am J Cardiol 2015;116:132–41.

[13] Berry C, Lloyd SM, Wang Y, Macdonald A, Ford I. The changing course of aortic valve disease in Scotland: temporal trends in hospitalizations and mortality and prognostic importance of aortic stenosis. Eur Heart J 2013;34:1538–47.

[14] Hartzell M, Malhotra R, Yared K, Rosenfield HR, Walker JD, Wood MJ. Effect of gender on treatment and outcomes in severe aortic stenosis. Am J Cardiol 2011;107:1681–6.

[15] Chieffo A, Petronio AS, Mehilli J, Chandrasekhar J, Sartori S, Lefèvre T, et al. Acute and 30-day outcomes in women after TAVR: results from the WIN-TAVI (Women's INternational Transcatheter Aortic Valve Implantation) real-world registry. JACC Cardiovasc Interv 2016;9:1589 600.

[16] Steadman CD, Jerosch-Herold M, Grundy B, Rafelt S, Ng LL, Squire IB, et al. Determinants and functional significance of myocardial perfusion reserve in severe aortic stenosis. JACC Cardiovasc Imaging 2012;5:182–9.

[17] Flett AS, Sado DM, Quarta G, Mirabel M, Pellerin D, Herrey AS, et al. Diffuse myocardial fibrosis in severe aortic stenosis: an equilibrium contrast cardiovascular magnetic resonance study. Eur Heart J Cardiovasc Imaging 2012;13:819–26.

[18] Dweck MR, Boon NA, Newby DE. Calcific aortic stenosis: a disease of the valve and the myocardium. J Am Coll Cardiol 2012;60:1854–63.

[19] Singh A, Chan DCS, Greenwood JP, Dawson DK, Sonecki P, Hogrefe K, et al. Symptom onset in aortic stenosis: relation to sex differences in left ventricular remodeling. JACC Cardiovasc Imaging 2019;12:96–105.

[20] Dobson LE, Fairbairn TA, Musa TA, Uddin A, Mundie CA, Swoboda PP, et al. Sex-related differences in left ventricular remodeling in severe aortic stenosis and reverse remodeling after aortic valve replacement: a cardiovascular magnetic resonance study. Am Heart J 2016;175:101–11.

[21] Lee JM, Park S-J, Lee S-P, Park E, Chang S-A, Kim H-K, et al. Gender difference in ventricular response to aortic stenosis: insight from cardiovascular magnetic resonance. PLoS One 2015;10, e0121684.

[22] Treibel TA, Kozor R, Fontana M, Torlasco C, Reant P, Badiani S, et al. Sex dimorphism in the myocardial response to aortic stenosis. JACC Cardiovasc Imaging 2018;11:962–73.

[23] Lim E, Ali A, Theodorou P, Sousa I, Ashrafian H, Chamageorgakis T, et al. Longitudinal study of the profile and predictors of left ventricular mass regression after stentless aortic valve replacement. Ann Thorac Surg 2008;85:2026–9.

[24] Beach JM, Mihaljevic T, Rajeswaran J, Marwick T, Edwards ST, Nowicki ER, et al. Ventricular hypertrophy and left atrial dilatation persist and are associated with reduced survival after valve replacement for aortic stenosis. J Thorac Cardiovasc Surg 2014;147:362–9. e8.

[25] Capoulade R, Clavel M-A, Le Ven F, Dahou A, Thébault C, Tastet L, et al. Impact of left ventricular remodelling patterns on outcomes in patients with aortic stenosis. Eur Heart J Cardiovasc Imaging 2017;18:1378–87.

[26] Gavina C, Falcao-Pires I, Pinho P, Manso M-C, Gonçalves A, Rocha-Gonçalves F, et al. Relevance of residual left ventricular hypertrophy after surgery for isolated aortic stenosis. Eur J Cardiothorac Surg 2016;49:952–9.

[27] Treibel TA, Kozor R, Schofield R, Benedetti G, Fontana M, Bhuva AN, et al. Reverse myocardial remodeling following valve replacement in patients with aortic stenosis. J Am Coll Cardiol 2018;71:860–71.

[28] Petrov G, Regitz-Zagrosek V, Lehmkuhl E, Krabatsch T, Dunkel A, Dandel M, et al. Regression of myocardial hypertrophy after aortic valve replacement: faster in women? Circulation 2010;122:S23–8.

[29] Medzikovic L, Aryan L, Eghbali M. Connecting sex differences, estrogen signaling, and microRNAs in cardiac fibrosis. J Mol Med (Berl) 2019;97:1385–98.

[30] Dweck MR, Joshi S, Murigu T, Alpendurada F, Jabbour A, Melina G, et al. Midwall fibrosis is an independent predictor of mortality in patients with aortic stenosis. J Am Coll Cardiol 2011;58:1271–9.

[31] Azevedo CF, Nigri M, Higuchi ML, Pomerantzeff PM, Spina GS, Sampaio RO, et al. Prognostic significance of myocardial fibrosis quantification by histopathology and magnetic resonance imaging in patients with severe aortic valve disease. J Am Coll Cardiol 2010;56:278–87.

[32] Barone-Rochette G, Pierard S, De Meester de Ravenstein C, Seldrum S, Melchior J, Maes F, et al. Prognostic significance of LGE by CMR in aortic stenosis patients undergoing valve replacement. J Am Coll Cardiol 2014;64:144–54.

[33] Milano AD, Faggian G, Dodonov M, Golia G, Tomezzoli A, Bortolotti U, et al. Prognostic value of myocardial fibrosis in patients with severe aortic valve stenosis. J Thorac Cardiovasc Surg 2012;144:830–7.

[34] Dumesnil JG, Pibarot P, Carabello B. Paradoxical low flow and/or low gradient severe aortic stenosis despite preserved left ventricular ejection fraction: implications for diagnosis and treatment. Eur Heart J 2010;31:281–9.

CH
15

[35] Hachicha Z, Dumesnil JG, Bogaty P, Pibarot P. Paradoxical low-flow, low-gradient severe aortic stenosis despite preserved ejection fraction is associated with higher afterload and reduced survival. Circulation 2007;115:2856–64.

[36] Lauten A, Figulla HR, Mollmann H, Holzhey D, Kötting J, Beckmann A, et al. TAVI for low-flow, low-gradient severe aortic stenosis with preserved or reduced ejection fraction: a subgroup analysis from the German Aortic Valve Registry (GARY). EuroIntervention 2014;10:850–9.

[37] Clavel MA, Burwash IG, Pibarot P. Cardiac imaging for assessing low-gradient severe aortic stenosis. JACC Cardiovasc Imaging 2017;10:185–202.

[38] Baumgartner H, Hung J, Bermejo J, Chambers JB, Evangelista A, Griffin BP, et al. Recommendations on the echocardiographic assessment of aortic valve stenosis: a focused update from the European Association of Cardiovascular Imaging and the American Society of Echocardiography. J Am Soc Echocardiogr 2017;30:372–92.

[39] Michelena HI, Margaryan E, Miller FA, Eleid M, Maalouf J, Suri R, et al. Inconsistent echocardiographic grading of aortic stenosis: is the left ventricular outflow tract important? Heart 2013;99:921–31.

[40] Eleid MF, Sorajja P, Michelena HI, Malouf JF, Scott CG, Pellikka PA. Flow-gradient patterns in severe aortic stenosis with preserved ejection fraction: clinical characteristics and predictors of survival. Circulation 2013;128:1781–9.

[41] Clavel MA, Pibarot P, Messika-Zeitoun D, Capoulade R, Malouf J, Aggarval S, et al. Impact of aortic valve calcification, as measured by MDCT, on survival in patients with aortic stenosis: results of an international registry study. J Am Coll Cardiol 2014;64:1202–13.

[42] Dulgheru R, Pibarot P, Sengupta PP, Piérard LA, Rosenhek R, Magne J, et al. Multimodality imaging strategies for the assessment of aortic stenosis: viewpoint of the Heart Valve Clinic International Database (HAVEC) Group. Circ Cardiovasc Imaging 2016;9, e004352.

[43] Aggarwal SR, Clavel M-A, Messika-Zeitoun D, Cueff C, Malouf J, Araoz PA, et al. Sex differences in aortic valve calcification measured by multidetector computed tomography in aortic stenosis. Circ Cardiovasc Imaging 2013;6:40–7.

[44] Clavel M-A, Messika-Zeitoun D, Pibarot P, Aggarwal SR, Malouf J, Araoz PA, et al. The complex nature of discordant severe calcified aortic valve disease grading: new insights from combined Doppler echocardiographic and computed tomographic study. J Am Coll Cardiol 2013;62:2329–38.

[45] Simard L, Cote N, Dagenais F, Mathieu P, Couture C, Trahan S, et al. Sex-related discordance between aortic valve calcification and hemodynamic severity of aortic stenosis: is valvular fibrosis the explanation? Circ Res 2017;120:681–91.

[46] Pawade T, Clavel M-A, Tribouilloy C, Dreyfus J, Mathieu T, Tastet L, et al. Computed tomography aortic valve calcium scoring in patients with aortic stenosis. Circ Cardiovasc Imaging 2018;11, e007146.

[47] Nishimura RA, Otto CM, Bonow RO, Carabello BA, Erwin JP, Guyton RA, et al. 2014 AHA/ACC guideline for the Management of Patients with Valvular Heart Disease: a report of the American College of Cardiology/American Heart Association Task Force on Practice Guidelines. Circulation 2014;129:e521–643.

[48] Thomassen HK, Cioffi G, Gerdts E, Einarsen E, Midtbø HB, Mancusi C, et al. Echocardiographic aortic valve calcification and outcomes in women and men with aortic stenosis. Heart 2017;103:1619–24.

[49] Kulik A, Lam B-K, Rubens FD, Hendry PJ, Masters RG, Goldstein W, et al. Gender differences in the long-term outcomes after valve replacement surgery. Heart 2009;95:318–26.

[50] Chandrasekhar J, Dangas G, Yu J, Vemulapalli S, Suchindran S, Vora AN, et al. Sex-based differences in outcomes with transcatheter aortic valve therapy: TVT registry from 2011 to 2014. J Am Coll Cardiol 2016;68:2733–44.

[51] O'Connor SA, Morice MC, Gilard M, Leon MB, Webb JG, Dvir D, et al. Revisiting sex equality with transcatheter aortic valve replacement outcomes: a collaborative, patient-level meta-analysis of 11,310 patients. J Am Coll Cardiol 2015;66:221–8.

[52] O'Brien SM, Shahian DM, Filardo G, Ferraris VA, Haan CK, Rich JB, et al. The Society of Thoracic Surgeons 2008 cardiac surgery risk models: part 2—isolated valve surgery. Ann Thorac Surg 2009;88:S23–42.

[53] Czarnecki A, Qiu F, Koh M, Prasad TJ, Cantor WJ, Cheema AN, et al. Clinical outcomes after trans-catheter aortic valve replacement in men and women in Ontario, Canada. Catheter Cardiovasc Interv 2017;90:486–94.

[54] Levi A, Landes U, Assali AR, Orvin K, Sharony R, Vaknin-Assa H, et al. Long-term outcomes of 560 consecutive patients treated with transcatheter aortic valve implantation and propensity score-matched analysis of early- versus new-generation valves. Am J Cardiol 2017;119:1821–31.

[55] Katz M, Carlos Bacelar Nunes Filho A, Caixeta A, Antonio Carvalho L, Sarmento-Leite R, Alves Lemos Neto P, et al. Gender-related differences on short- and long-term outcomes of patients undergoing transcatheter aortic valve implantation. Catheter Cardiovasc Interv 2017;89:429–36.

[56] Forrest JK, Adams DH, Popma JJ, Reardon MJ, Deeb GM, Yakubov SJ, et al. Transcatheter aortic valve replacement in women versus men (from the US CoreValve Trials). Am J Cardiol 2016;118:396–402.

[57] Biere L, Launay M, Pinaud F, Hamel JF, Eltchaninoff H, Iung B, et al. Influence of sex on mortality and perioperative outcomes in patients undergoing TAVR: insights from the FRANCE 2 registry. J Am Coll Cardiol 2015;65:755–7.

[58] Kodali S, Williams MR, Doshi D, Hahn RT, Humphries KH, Nkomo VT, et al. Sex-specific differences at presentation and outcomes among patients undergoing transcatheter aortic valve replacement: a cohort study. Ann Intern Med 2016;164:377–84.

[59] Szerlip M, Gualano S, Holper E, Squiers JJ, White JM, Doshi D, et al. Sex-specific outcomes of transcatheter aortic valve replacement with the SAPIEN 3 valve: insights from the PARTNER II S3 high-risk and intermediate-risk cohorts. JACC Cardiovasc Interv 2018;11:13–20.

[60] Ludman PF, Moat N, de Belder MA, Blackman DJ, Duncan A, Banya W, et al. Transcatheter aortic valve implantation in the United Kingdom: temporal trends, predictors of outcome, and 6-year follow-up: a report from the UK Transcatheter Aortic Valve Implantation (TAVI) Registry, 2007 to 2012. Circulation 2015;131:1181–90.

[61] Yakubov SJ, Adams DH, Watson DR, Reardon MJ, Kleiman NS, Heimansohn D, et al. 2-Year outcomes after iliofemoral self-expanding transcatheter aortic valve replacement in patients with severe aortic stenosis deemed extreme risk for surgery. J Am Coll Cardiol 2015;66:1327–34.

[62] Buja P, Napodano M, Tamburino C, Petronio AS, Ettori F, Santoro G, et al. Comparison of variables in men versus women undergoing transcatheter aortic valve implantation for severe aortic stenosis (from Italian Multicenter CoreValve registry). Am J Cardiol 2013;111:88–93.

[63] D'Ascenzo F, Gonella A, Moretti C, Omedè P, Salizzoni S, La Torre M, et al. Gender differences in patients undergoing TAVI: a multicentre study. EuroIntervention 2013;9:367–72.

[64] Hayashida K, Morice MC, Chevalier B, Hovasse T, Romano M, Garot P, et al. Sex-related differences in clinical presentation and outcome of transcatheter aortic valve implantation for severe aortic stenosis. J Am Coll Cardiol 2012;59:566–71.

[65] Humphries KH, Toggweiler S, Rodes-Cabau J, Nombela-Franco L, Dumont E, Wood DA, et al. Sex differences in mortality after transcatheter aortic valve replacement for severe aortic stenosis. J Am Coll Cardiol 2012;60:882–6.

[66] Tamburino C, Capodanno D, Ramondo A, Petronio AS, Ettori F, Santoro G, et al. Incidence and predictors of early and late mortality after transcatheter aortic valve implantation in 663 patients with severe aortic stenosis. Circulation 2011;123:299–308.

[67] Saad M, Nairooz R, Pothineni NVK, Almomani A, Kovelamudi S, Sardar P, et al. Long-term outcomes with transcatheter aortic valve replacement in women compared with men: evidence from a meta-analysis. JACC Cardiovasc Interv 2018;11:24–35.

[68] Williams M, Kodali SK, Hahn RT, Humphries KH, Nkomo VT, Cohen DJ, et al. Sex-related differences in outcomes after transcatheter or surgical aortic valve replacement in patients with severe aortic stenosis: insights from the PARTNER Trial (Placement of Aortic Transcatheter Valve). J Am Coll Cardiol 2014;63:1522–8.

[69] Gaglia Jr MA, Lipinski MJ, Torguson R, Gai J, Ben-Dor I, Bernardo NL, et al. Comparison in men versus women of co-morbidities, complications, and outcomes after transcatheter aortic valve implantation for severe aortic stenosis. Am J Cardiol 2016;118:1692–7.

[70] Barbanti M, Yang TH, Rodes Cabau J, Tamburino C, Wood DA, Jilaihawi H, et al. Anatomical and procedural features associated with aortic root rupture during balloon-expandable transcatheter aortic valve replacement. Circulation 2013;128:244–53.

[71] Naoum C, Blanke P, Dvir D, Pibarot P, Humphries K, Webb J, et al. Clinical outcomes and imaging findings in women undergoing TAVR. JACC Cardiovasc Imaging 2016;9:483–93.

[72] Ribeiro HB, Webb JG, Makkar RR, Cohen MG, Kapadia SR, Kodali S, et al. Predictive factors, management, and clinical outcomes of coronary obstruction following transcatheter aortic valve implantation: insights from a large multicenter registry. J Am Coll Cardiol 2013;62:1552–62.

[73] Onorati F, D'Errigo P, Barbanti M, Rosato S, Covello RD, Maraschini A, et al. Different impact of sex on baseline characteristics and major periprocedural outcomes of transcatheter and surgical aortic valve interventions: results of the multicenter Italian OBSERVANT Registry. J Thorac Cardiovasc Surg 2014;147:1529–39.

[74] Holmes Jr DR, Brennan JM, Rumsfeld JS, Dai D, O'Brien SM, Vemulapalli S, et al. Clinical outcomes at 1 year following transcatheter aortic valve replacement. JAMA 2015;313:1019–28.

[75] Chieffo A, Petronio AS, Mehilli J, Chandrasekhar J, Sartori S, Lefèvre T, et al. 1-Year clinical outcomes in women after transcatheter aortic valve replacement: results from the first WIN-TAVI registry. JACC Cardiovasc Interv 2018;11:1–12.

[76] Padang R, Ali M, Greason KL, Scott CG, Indrabhinduwat M, Rihal CS, et al. Comparative survival and role of STS score in aortic paravalvular leak after SAVR or TAVR: a retrospective study from the USA. BMJ Open 2018;8, e022437.

[77] Abdel-Wahab M, Zahn R, Horack M, Gerckens U, Schuler G, Sievert H, et al. Aortic regurgitation after transcatheter aortic valve implantation: incidence and early outcome. Results from the German transcatheter aortic valve interventions registry. Heart 2011;97:899–906.

[78] Vavuranakis M, Kalogeras K, Lavda M, Kolokathis MA, Papaioannou T, Oikonomou E, et al. Correlation of CoreValve implantation 'true cover index' with short and mid-term aortic regurgitation: a novel index. Int J Cardiol 2016;223:482–7.

[79] Rodes-Cabau J, Pibarot P, Suri RM, Kodali S, Thourani VH, Szeto WY, et al. Impact of aortic annulus size on valve hemodynamics and clinical outcomes after transcatheter and surgical aortic valve replacement: insights from the PARTNER trial. Circ Cardiovasc Interv 2014;7:701–11.

[80] Pibarot P, Weissman NJ, Stewart WJ, Hahn RT, Lindman BR, McAndrew T, et al. Incidence and sequelae of prosthesis-patient mismatch in transcatheter versus surgical valve replacement in high-risk patients with severe aortic stenosis: a PARTNER trial cohort—a analysis. J Am Coll Cardiol 2014;64:1323–34.

[81] Zorn 3rd GL, Little SH, Tadros P, Deeb GM, Gleason TG, Heiser J, et al. Prosthesis-patient mismatch in high-risk patients with severe aortic stenosis: a randomized trial of a self-expanding prosthesis. J Thorac Cardiovasc Surg 2016;151:1014–22. 1023. e1–3.

[82] Dvir D, Webb JG, Bleiziffer S, Pasic M, Waksman R, Kodali S, et al. Transcatheter aortic valve implantation in failed bioprosthetic surgical valves. JAMA 2014;312:162–70.

[83] Klodas E, Enriquez-Sarano M, Tajik AJ, Mullany CJ, Bailey KR, Seward JB. Surgery for aortic regurgitation in women. Contrasting indications and outcomes compared with men. Circulation 1996;94:2472–8.

[84] Brown ML, Schaff HV, Suri RM, Li Z, Sundt TM, Dearani JA, et al. Indexed left ventricular dimensions best predict survival after aortic valve replacement in patients with aortic valve regurgitation. Ann Thorac Surg 2009;87:1170–5 [discussion 1175-6].

[85] Vakamudi S, Jellis C, Mick S, Wu Y, Gillinov AM, Mihaljevic T, et al. Sex differences in the etiology of surgical mitral valve disease. Circulation 2018;138:1749–51.

[86] Turi ZG. Mitral valve disease. Circulation 2004;109:e38–41.

[87] Avierinos JF, Inamo J, Grigioni F, Gersh B, Shub C, Enriquez-Sarano M. Sex differences in morphology and outcomes of mitral valve prolapse. Ann Intern Med 2008;149:787–95.

[88] Seeburger J, Eifert S, Pfannmuller B, Garbade J, Vollroth M, Misfeld M, et al. Gender differences in mitral valve surgery. Thorac Cardiovasc Surg 2013;61:42–6.

[89] Zoghbi WA, Adams D, Bonow RO, Enriquez-Sarano M, Foster E, Grayburn PA, et al. Recommendations for noninvasive evaluation of native valvular regurgitation: a report from the American Society of Echocardiography developed in collaboration with the Society for Cardiovascular Magnetic Resonance. J Am Soc Echocardiogr 2017;30:303–71.

[90] Mantovani F, Clavel MA, Michelena HI, Suri RM, Schaff HV, Enriquez-Sarano M. Comprehensive imaging in women with organic mitral regurgitation: implications for clinical outcome. J Am Coll Cardiol Img 2016;9:388–96.

[91] Tribouilloy C, Grigioni F, Avierinos JF, Barbieri A, Rusinaru D, Szymanski C, et al. Survival implication of left ventricular end-systolic diameter in mitral regurgitation due to flail leaflets a long-term follow-up multicenter study. J Am Coll Cardiol 2009;54:1961–8.

[92] Nishimura RA, Otto CM, Bonow RO, Carabello BA, Erwin JP, Fleisher LA, et al. 2017 AHA/ACC focused update of the 2014 AHA/ACC guideline for the management of patients with valvular heart disease: a report of the American College of Cardiology/American Heart Association Task Force on Clinical Practice Guidelines. Circulation 2017;135:e1159–95.

[93] O'Gara PT, Grayburn PA, Badhwar V, Afonso LC, Carroll JD, Elmariah S, et al. 2017 ACC expert consensus decision pathway on the management of mitral regurgitation: a report of the American College of Cardiology Task Force on expert consensus decision pathways. J Am Coll Cardiol 2017;70:2421–49.

[94] Giustino G, Overbey J, Taylor D, Ailawadi G, Kirkwood K, DeRose J, et al. Sex-based differences in outcomes after mitral valve surgery for severe ischemic mitral regurgitation: from the cardiothoracic surgical trials network. JACC Heart Fail 2019;7:481–90.

[95] Grayburn PA, Sannino A, Packer M. Proportionate and disproportionate functional mitral regurgitation: a new conceptual framework that reconciles the results of the MITRA-FR and COAPT trials. J Am Coll Cardiol Img 2019;12:353–62.

[96] Vassileva CM, McNeely C, Mishkel G, Boley T, Markwell S, Hazelrigg S. Gender differences in long-term survival of Medicare beneficiaries undergoing mitral valve operations. Ann Thorac Surg 2013;96:1367–73.

[97] Eguchi K, Ohtaki E, Matsumura T, Tanaka K, Tohbaru T, Iguchi N, et al. Pre-operative atrial fibrillation as the key determinant of outcome of mitral valve repair for degenerative mitral regurgitation. Eur Heart J 2005;26:1866–72.

[98] Vassileva CM, Stelle LM, Markwell S, Boley T, Hazelrigg S. Sex differences in procedure selection and outcomes of patients undergoing mitral valve surgery. Heart Surg Forum 2011;14:E276–82.

[99] Gammie JS, Sheng S, Griffith BP, Peterson ED, Rankin JS, O'Brien SM, et al. Trends in mitral valve surgery in the United States: results from the Society of Thoracic Surgeons Adult Cardiac Surgery Database. Ann Thorac Surg 2009;87:1431–7 [discussion 1437-9].

[100] Badhwar V, Peterson ED, Jacobs JP, He X, Brennan JM, O'Brien SM, et al. Longitudinal outcome of isolated mitral repair in older patients: results from 14,604 procedures performed from 1991 to 2007. Ann Thorac Surg 2012;94:1870–7 [discussion 1877-9].

[101] Goldstein D, Moskowitz AJ, Gelijns AC, Ailawadi G, Parides MK, Perrault LP, et al. Two-year outcomes of surgical treatment of severe ischemic mitral regurgitation. N Engl J Med 2016;374:344–53.

[102] Acker MA, Parides MK, Perrault LP, Moskowitz AJ, Gelijns AC, Voisine P, et al. Mitral-valve repair versus replacement for severe ischemic mitral regurgitation. N Engl J Med 2014;370:23–32.

[103] Feldman T, Foster E, Glower DD, Kar S, Rinaldi MJ, Fail PS, et al. Percutaneous repair or surgery for mitral regurgitation. N Engl J Med 2011;364:1395–406.

[104] Feldman T, Kar S, Elmariah S, Trento A, Siegel RJ, Apruzzese P, et al. Randomized comparison of percutaneous repair and surgery for mitral regurgitation: 5-year results of EVEREST II. J Am Coll Cardiol 2015;66:2844–54.

[105] Obadia JF, Messika-Zeitoun D, Leurent G, Iung B, Bonnet G, Piriou N, et al. Percutaneous repair or medical treatment for secondary mitral regurgitation. N Engl J Med 2018;379:2297–306.

[106] Stone GW, Lindenfeld J, Abraham WT, Kar S, Lim DS, Mishell JM, et al. Transcatheter mitral-valve repair in patients with heart failure. N Engl J Med 2018;379:2307–18.

[107] Estevez-Loureiro R, Settergren M, Winter R, Jacobsen P, Dall'Ara G, Sondergaard L, et al. Effect of gender on results of percutaneous edge-to-edge mitral valve repair with MitraClip system. Am J Cardiol 2015;116:275–9.

[108] Attizzani GF, Ohno Y, Capodanno D, Cannata S, Dipasqua F, Immé S, et al. Gender-related clinical and echocardiographic outcomes at 30-day and 12-month follow up after MitraClip implantation in the GRASP registry. Catheter Cardiovasc Interv 2015;85:889–97.

[109] Sud K, Agarwal S, Parashar A, Raza MQ, Patel K, Min D, et al. Degenerative mitral stenosis: unmet need for percutaneous interventions. Circulation 2016;133:1594–604.

[110] Labovitz AJ, Nelson JG, Windhorst DM, Kennedy HL, Williams GA. Frequency of mitral valve dysfunction from mitral anular calcium as detected by Doppler echocardiography. Am J Cardiol 1985;55:133–7.

[111] Movahed MR, Ahmadi-Kashani M, Kasravi B, Saito Y. Increased prevalence of mitral stenosis in women. J Am Soc Echocardiogr 2006;19:911–3.

[112] Pasca I, Dang P, Tyagi G, Pai RG. Survival in patients with degenerative mitral stenosis: results from a large retrospective cohort study. J Am Soc Echocardiogr 2016;29:461–9.

[113] Cho I-J, Hong G-R, Lee SH, Lee S, Chang B-C, Shim CY, et al. Differences in characteristics, left atrial reverse remodeling, and functional outcomes after mitral valve replacement in patients with low-gradient very severe mitral stenosis. J Am Soc Echocardiogr 2016;29:759–67.

[114] Singh JP, Evans JC, Levy D, Larson MG, Freed LA, Fuller DL, et al. Prevalence and clinical determinants of mitral, tricuspid, and aortic regurgitation (the Framingham Heart Study). Am J Cardiol 1999;83:897–902.

[115] Arsalan M, Walther T, Smith 2nd RL, Grayburn PA. Tricuspid regurgitation diagnosis and treatment. Eur Heart J 2017;38:634–8.

[116] Topilsky Y, Maltais S, Medina Inojosa J, Oguz D, Michelena H, Maalouf J, et al. Burden of tricuspid regurgitation in patients diagnosed in the community setting. J Am Coll Cardiol Img 2019;12:433–42.

[117] Chandrashekar P, Fender EA, Zack CJ, Reddy YNV, Bennett CE, Prasad M, et al. Sex-stratified analysis of national trends and outcomes in isolated tricuspid valve surgery. Open Heart 2018;5, e000719.

[118] Zack CJ, Fender EA, Chandrashekar P, Reddy YNV, Bennett CE, Stulak JM, et al. National trends and outcomes in isolated tricuspid valve surgery. J Am Coll Cardiol 2017;70:2953–60.

[119] Goldstone AB, Howard JL, Cohen JE, MacArthur JW, Atluri P, Kirkpatrick JN, et al. Natural history of coexistent tricuspid regurgitation in patients with degenerative mitral valve disease: implications for future guidelines. J Thorac Cardiovasc Surg 2014;148:2802–9.

[120] Song H, Kim MJ, Chung CH, Choo SJ, Song MG, Song J-M, et al. Factors associated with development of late significant tricuspid regurgitation after successful left-sided valve surgery. Heart 2009;95:931–6.

[121] Chikwe J, Itagaki S, Anyanwu A, Adams DH. Impact of concomitant tricuspid annuloplasty on tricuspid regurgitation, right ventricular function, and pulmonary artery hypertension after repair of mitral valve prolapse. J Am Coll Cardiol 2015;65:1931–8.

[122] Kilic A, Saha-Chaudhuri P, Rankin JS, Conte JV. Trends and outcomes of tricuspid valve surgery in North America: an analysis of more than 50,000 patients from the Society of Thoracic Surgeons database. Ann Thorac Surg 2013;96:1546–52 [discussion 1552].

[123] Axtell AL, Bhambhani V, Moonsamy P, Healy EW, Picard MH, Sundt TM, et al. Surgery is not associated with improved survival compared to medical therapy in isolated severe tricuspid regurgitation. J Am Coll Cardiol 2019.

[124] Silversides CK, Grewal J, Mason J, Sermer M, Kiess M, Rychel V, et al. Pregnancy outcomes in women with heart disease: the CARPREG II Study. J Am Coll Cardiol 2018;71:2419–30.

[125] Regitz-Zagrosek V, Roos-Hesselink JW, Bauersachs J, Blomström-Lundqvist C, Cífková R, De Bonis M, et al. 2018 ESC guidelines for the management of cardiovascular diseases during pregnancy. Kardiol Pol 2019;77:245–326.

[126] Elkayam U, Goland S, Pieper PG, Silversides CK. High-risk cardiac disease in pregnancy. Part I. J Am Coll Cordiol 2016;68:396–410.

[127] van Hagen IM, Thorne SA, Taha N, Youssef G, Elnagar A, Gabriel H, et al. Pregnancy outcomes in women with rheumatic mitral valve disease: results from the registry of pregnancy and cardiac disease. Circulation 2018;137:806–16.

[128] Stout Karen K, Daniels Curt J, Aboulhosn Jamil A, Bozkurt B, Broberg CS, Colman JM, et al. 2018 AHA/ACC guideline for the management of adults with congenital heart disease: a report of the American College of Cardiology/American Heart Association Task Force on Clinical Practice Guidelines. Circulation 2019;139:e698–800.

[129] Elkayam U, Bitar F. Valvular heart disease and pregnancy: part II: prosthetic valves. J Am Coll Cardiol 2005;46:403–10.

[130] Carvajal HG, Lindley KJ, Shah T, Brar AK, Barger PM, Billadello JJ, et al. Impact of pregnancy on autograft dilatation and aortic valve function following the Ross procedure. Congenit Heart Dis 2018;13:217–21.

[131] van Hagen IM, Roos-Hesselink JW, Ruys TPE, Merz WM, Goland S, Gabriel H, et al. Pregnancy in women with a mechanical heart valve: data of the European Society of Cardiology Registry of Pregnancy and Cardiac Disease (ROPAC). Circulation 2015;132:132–42.

Editor's Summary: Aortic Stenosis

Incidence

Bicuspid AV is 3X more common in men

Presentation

Higher prevalence of CAD
Higher prevalence of PAD
Older
More symptomatic
More hypertension

Niti Aggarwal

Valvular Pathology

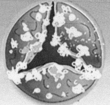 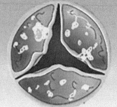

More calcification	Less calcification
Less fibrosis	More fibrosis
More inflammation	Less inflammation
Larger AV annulus	Smaller AV annulus

LV Response to Stress

More myocardial fibrosis	Less myocardial fibrosis
Higher LV mass index	Lower LV mass index
More hypertrophy	More concentric remodeling
LV dilatation	Smaller LV cavity
Greater stroke volume	Lower stroke volume
Lower ejection fraction	Preserved EF
Normal diastology	More diastolic dysfunction

Management

Women more often undergo TAVR, vs CABG and SAVR in men

Outcomes

Female sex is risk factor for ↑ mortality after SAVR
Superior outcomes in women with TAVR vs SAVR
Women less likely to get post TAVR paravalvular regurgitation
Women more likely to get post TAVR stroke and bleeding

Sex-Specific Incidence, Presentation, Pathophysiology and Management of Aortic Stenosis. Women with aortic stenosis are more likely to have CAD, and express less calcification and more fibrosis on AV compared to men. They also have a smaller LV cavity and more diastolic dysfunction. Women fare better after TAVR than SAVR. AV, aortic valve; CABG, coronary artery bypass grafting; CAD, coronary artery disease; EF, ejection fraction; LV, left ventricle; PAD, peripheral arterial disease; SAVR, surgical aortic valve replacement; TAVR, trans aortic valve replacement. *Image courtesy of Niti R. Aggarwal.*

Editor's Summary: Mitral Regurgitation

Rheumatic heart disease more common

Anterior or bileaflet prolapse

Delayed referral to MV intervention

Higher rates of MV replacement

Higher mortality after MV interventions

Degenerative valve disease predominates

Posterior leaflet prolapse

Timely referral to MV intervention

Higher rates of MV repair

Restoration of life expectancy after MV repair

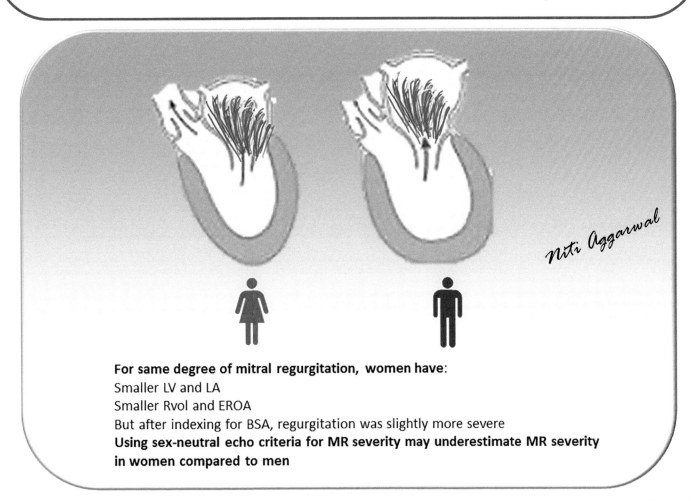

For same degree of mitral regurgitation, women have:
Smaller LV and LA
Smaller Rvol and EROA
But after indexing for BSA, regurgitation was slightly more severe
Using sex-neutral echo criteria for MR severity may underestimate MR severity in women compared to men

Sex-Differences in Mitral Regurgitation. Absence of sex-specific criteria for diagnosis of MR results in underestimation of MR severity in women. Resultantly women experience delays in presentation, and in turn higher post intervention mortality. BSA, body surface area; EROA, effective regurgitant orifice area; LA, left atrium; LV, left ventricle; MR, mitral regurgitation; MV, mitral valve; Rvol, regurgitant volume. *Image courtesy of Niti R. Aggarwal.*

Section VI

Arrhythmias

Chapter 16

Atrial Fibrillation

Christina K. Anderson, Anne B. Curtis, and Annabelle Santos Volgman

Clinical Case

A 76-year-old female is referred for newly diagnosed atrial fibrillation incidentally found on a routine physical examination. The patient has a history of hypertension, hypothyroidism, and hyperlipidemia. She complains of shortness of breath with moderate exertion but no chest pain or palpitations. Her blood pressure is 120/80 mmHg, heart rate of 70 bpm. Heart rate is irregularly irregular, but otherwise has a normal cardiovascular exam. Her medications included metoprolol 50 mg and losartan 100 mg. Her electrocardiogram showed atrial fibrillation at a rate of 70 bpm with left atrial enlargement. Laboratory studies including electrolyte panel and complete blood count are within normal limits. A stress test did not show any evidence of ischemia and her echocardiogram showed a moderately enlarged left atrium and mild left ventricular hypertrophy. How would you manage her next and counsel her on her risk for stroke? Would the risk be different if this was a 76-year-old man?

Abstract

Atrial fibrillation (AF) is the most common sustained arrhythmia and continues to increase in prevalence and incidence as the population ages. The diagnosis of AF confers a huge burden due to healthcare costs, patient symptoms, and impaired quality of life. This chapter examines sex differences in AF with respect to epidemiology, pathophysiology, presentation, access to treatment, prognosis, and risk of thromboembolism. While lifetime incidence of AF is the same between the sexes, women tend to be older at the time of presentation, lagging behind men by about 10 years. Compared to men, women are more likely to be

Sex Differences in Cardiac Disease. https://doi.org/10.1016/B978-0-12-819369-3.00008-3

symptomatic and experience more functional limitations. They have a higher mortality risk and are less likely to be referred for interventions such as catheter ablation or electrical cardioversion. Therapeutic anticoagulation is underused in elderly women, who are at a particularly high risk for debilitating stroke. Women are more likely to experience major bleeding while on anticoagulants, have more medication side effects, and have more procedural complications than men. This chapter highlights the unanswered questions in sex-specific differences in AF, and emphasizes the need for future studies. Recognizing and understanding key sex differences can improve treatment of all patients with AF. Given the vast prevalence and ever-growing incidence of AF, the global impact of such adaptations cannot be underemphasized.

Introduction

Atrial fibrillation (AF) is a pervasive and ever-growing problem in the United States affecting over 33 million people worldwide, or about 0.5% of the world's population. Because of increasing longevity worldwide and the general aging of the population, the overall prevalence and incidence of AF is predicted to increase further. The disease imposes a high burden on healthcare systems due to frequent hospital readmissions and increased risk of stroke, heart failure (HF), and death in affected patients. There is global evidence of progressive increases in overall burden, incidence, prevalence, and AF-associated mortality between 1990 and 2010, with significant public health implications [1]. The economic burden of AF cannot be overlooked, and estimates suggest that AF accounts for 1% of the National Health Service budget in the United Kingdom and $16–26 billion of annual US expenses [1]. The majority of this cost is driven by costs associated with inpatient care and hospitalizations but also with a significant contribution from outpatient care, cost of prescription medications and outpatient testing, as well as complications from treatments such as major bleeding [2].

While men are more predisposed to developing AF, the overall lifetime prevalence of AF is similar between the two sexes. Women with AF experience more frequent symptomatic episodes, worse quality of life, more drug-related adverse events, and have a higher adjusted risk of death [3–5]. This chapter examines sex differences in AF with respect to epidemiology, pathophysiology, presentation, access to treatment, prognosis, and risk of thromboembolism. It further highlights the unanswered questions pertaining to sex-specific differences in AF for future research and helps reduce disparities in AF outcomes.

Epidemiology

Prevalence

The prevalence and incidence of AF varies by sex, age, race/ethnicity, and geographic regions. High-income countries are experiencing a higher prevalence and incidence of AF [6]. Over a 20-year period, between 1990 and 2010, there was a modest increase in prevalence but a major increase in incidence of AF [1]. Globally, in 2010, the estimated number of individuals with AF was 33.5 million {20.9 million men [95% uncertainty interval (UI), 19.5–22.2 million] and 12.6 million women [95% UI, 12.0–13.7 million]} [1]. **Figure 1** shows the changes every 5 years from 1990 to 2010 globally in developed and developing countries of the estimated age-adjusted prevalence rates AF (per 100,000 population) stratified by sex [1]. Overall, there was a modest increase in the prevalence rates between 1990 and 2010 in both men and women. Although developed countries had higher prevalence rates compared with developing countries, the difference was more pronounced in men than in women. Men had a higher prevalence than women throughout the whole period.

The age-dependent nature of the disease has been well established. It has been predicted that as the population ages, over 6–12 million Americans will be affected with AF by 2050 and 17.9 million Europeans by 2060 [1]. In the Framingham Heart Study (FHS) between 1958–1967 and 1998–2007, in individuals 45–94 years old, the age-adjusted prevalence quadrupled in both women (13.7–49.4 cases per 1000 person-years) and men (from 20.4 to 96.2 cases per 1000 person-years) [7]. Because women on average live longer than their male counterparts, the absolute numbers of women and men with AF are similar in the United States. A pooled analysis of community cohorts in Europe found that the cumulative risk of developing AF was higher in men than women over most of the life span with women lagging about a decade behind men [8]. People of European ancestry have a higher prevalence of AF compared to other races/ethnicities [9].

Incidence

The rate of incidence of AF has also been increasing over time [7, 9]. The incidence rates per 1000 person-years based on optimal, borderline, or elevated levels of risk factors were analyzed in the Atherosclerosis Risk in Communities (ARIC) study [10]. Compared with those with no risk factors, the age-adjusted incidence rates were three times higher in those with one or more elevated risk factors: 2.19 vs. 6.59 per 1000 person-years, respectively [10]. Most of the literature reports a higher incidence of AF in men compared to women [7, 9–11]. However, in the Cohorts for Heart and

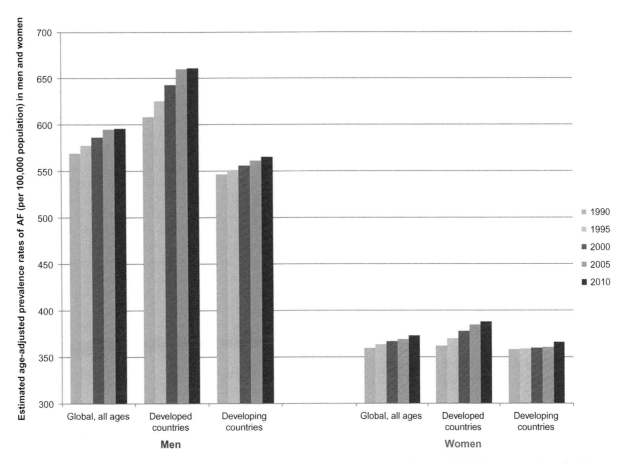

FIGURE 1 **Estimated Age-Adjusted Worldwide Prevalence Rates of AF (Per 100,000 Population) in Men and Women.** Men have higher prevalence of AF compared to women globally and prevalence is increasing over the years, mostly in developed countries, in both men and women. *Data from [1].*

Aging Research in Genomic Epidemiology (CHARGE-AF) consortium risk prediction model, based on the FHS, ARIC, and Cardiovascular Health Study, sex was not selected as a predictor in the model. The model suggested that sex differences in the distribution of AF predictors may account for this disparity [11].

Lifetime Risk

Table 1 shows that in the FHS, which comprised subjects mostly of European descent, the lifetime risk of AF varied by sex and the presence and burden of risk factors [12]. Men had a higher lifetime risk compared to women at all risk factor burden categories. In contrast, in the ARIC cohort, there were no sex differences in lifetime risk of AF in black population: 36% (white men), 30% (white women)/21% (black men), and 22% (black women) [13]. In the BiomarCaRE Consortium, the incidence of AF was very low overall before age 50 with the incidence increasing in men after age 50 and in women after age 60. However, the incidence

converged at the age of 90, demonstrating that lifetime risk among both sexes is very similar and again reinforcing the close relationship of AF with advancing age [8].

TABLE 1 Lifetime Risk of Atrial Fibrillation by Sex and Risk Factors

Risk Factor Burden and Lifetime Risk of Developing AF			
	Optimal (%)	Borderline (%)	Elevated (%)
Men	29.8	39.7	43.3
Women	20.5	28.0	34.6

The lifetime risk of atrial fibrillation by section risk factor burden in the visuals from Framingham Heart Study (European ancestry), at age 55 years is depicted. AF, atrial fibrillation. *Reproduced with permission from [12].*

Pathophysiology

Mechanism of Disease

The pathophysiology of AF is complex and requires both a trigger and an appropriate anatomic substrate. Despite multiple studies and resultant theories, the exact pathophysiology remains unclear. It is currently believed that AF is a triggered arrhythmia initiated by rapid repetitive discharges from the pulmonary veins. Atrial myocardial fibers around the pulmonary veins and the posterior left atrium are oriented in different directions, causing the triggered activity to spread in a disorganized fashion [14]. There is considerable anatomic variability among individuals causing differences in the ability to maintain sustained AF. Prior animal studies have been able to demonstrate sex-related differences in both pulmonary vein and left atrial electrical properties. For example, male rabbits have been found to have a higher incidence of burst firing and faster spontaneous activity in the pulmonary vein, which implies higher pulmonary vein arrhythmogenesis in males which may potentially contribute to a higher incidence of AF in males. Another interesting finding is that while it has been well established that the tissue surrounding the pulmonary veins is an important substrate for the initiation of AF, female gender has been associated with the presence of nonpulmonary vein triggers of AF such as ectopic beats from the superior vena cava [15]. Abnormal intracellular calcium handling may cause a diastolic calcium leak from the sarcoplasmic reticulum, which can trigger delayed afterdepolarizations leading to more disorganized atrial contractions [14].

The left atrium sustains both rapid focal firing and re-entry, allowing the maintenance of AF. In addition, various other systems play a role in triggering and maintaining AF, including the autonomic nervous system, atrial tachycardia remodeling, inflammation and oxidative stress, and the renin-angiotensin-aldosterone system [14]. There is an accumulating body of evidence regarding the relationship of inflammation and AF. There is a clear correlation between inflammatory states like myocarditis or postpericardiotomy and the onset of AF. However, the exact relationship (correlation vs. causation) is not clear. The most substantial support for this relationship is derived from studies that have related inflammatory biomarkers such as C-reactive protein (CRP) to AF [16]. In one study evaluating over 5800 subjects, baseline levels of CRP were higher in patients with AF and CRP level was a strong predictor of future AF, even after adjustment for multiple variables [17]. Another significant study evaluated 10,276 subjects from the Copenhagen City Heart Study for 12–15 years and again found that higher CRP levels were associated with a 2.19-fold increased risk of AF [18].

The pathophysiology of sex differences in the incidence, prevalence, and complications of AF is multifactorial and not completely understood [19, 20]. There are many current theories as to why these sex differences exist, including differences in electrical properties, structural properties, and sex hormones. Important structural differences include a larger left atrial dimension and volume in men, which are independently associated with an increased incidence of AF and could promote the maintenance of the arrhythmia [8, 11, 21]. It has also been shown that women with AF have a relatively greater burden of atrial fibrosis identified using delayed-enhanced magnetic resonance imaging (left atrial fibrosis of $23.0 \pm 7.9\%$ in men vs. $29.9 \pm 6.2\%$ in women, $P = 0.003$) [22]. The greater burden of atrial fibrosis seen in women has been associated with a higher prevalence of stroke and systemic embolism [22, 23]. Important differences in electrical properties have also been identified. For example, men have greater expression of repolarizing ion channel subunits, which could accelerate atrial repolarization and favor reentry [20]. Animal models have also demonstrated that female ventricular myocytes have longer duration of action potentials related to significantly lower delayed rectifier potassium current (IKr and IKl) densities [15]. The genetic contribution to these differences remains unclear as genetic variants have not identified sex-specific differences in AF predictability.

The lower incidence of AF in women may also be due to hormonal and variation in the electrophysiological substrate of atrial cells [19, 20]. Women have higher mean heart rates than men, about 2–6 beats/min [24], which are likely due to differences in intrinsic sinus node properties and persist despite total autonomic blockade [25]. It is not known whether this has any impact on the incidence or rate control of AF in women. Sex hormones may play a role in the higher heart rates in women since there are no differences in fetal heart rates and this difference is not seen until puberty [26]. This difference in heart rates gradually declines after the age of 50 and eventually disappears [26]. Importantly, women have longer QTc intervals than men and this has an impact on the risk of torsades de pointes when given antiarrhythmic drugs [27]. One study evaluating sex differences in cycle length-dependent QT and potassium currents in rabbits found that female rabbits have significant lower IKr and IKl outward current densities in their ventricular myocytes as compared to male cells which may explain the gender difference in QT intervals [28].

In addition, it has been well established that sex hormones affect the pharmacokinetics of drugs and may contribute to observed differences between men and women. Prior animal studies found that females experience dynamic changes in QT intervals and torsades de pointes risks during the menstrual cycle, which suggests these changes may be related to serum ovarian steroids. However, further studies would be needed to elucidate the molecular mechanisms of these observed differences [28]. Important differences in absorption, body composition, distribution, and excretion of drugs are responsible for a greater risk for adverse drug reactions in women [29]. Sex-specific differences in pharmacotherapy in women may be explained by the following factors: a lower body mass index; a higher proportion of body fat; lower creatinine clearance; and smaller organs [30]. The activity of drug-metabolizing enzymes modulates sex differences, particularly the cytochrome (CYP) P450 system [29]. CYP3A4 contributes to the first-pass metabolism of most cardiovascular drugs. Liver biopsies in women reveal a higher expression of CYP3A4 messenger ribonucleic acid (RNA) and twofold higher CYP3A4 levels [29].

Risk Factors

A US study showed that risk factors for AF shared by both sexes included advanced age, white race, increased height and weight, use of blood pressure-lowering medications, and history of cardiovascular disease [31]. Diabetes was a risk factor in women but not in men. Among the risk factors shared by both sexes, advanced age showed a stronger association in women than in men with an interaction *P*-value of 0.003 [31].

A European study showed that although blood pressure, smoking, alcohol consumption, N-terminal pro B-type natriuretic peptide (NT-proBNP), and prevalent cardiovascular disease are largely similar predictors of incident AF in both sexes, total cholesterol concentrations may show sex differences. Unlike the US study, the European study showed that a higher body mass index (BMI) and obesity are stronger risk factors in men [8].

Worldwide several major risk factors for AF shared by both sexes including advanced age, body mass index, blood pressure, hypertension treatment, diabetes mellitus, valvular heart disease, HF, and myocardial infarction. Women with AF had a high prevalence of hypertension and valvular heart disease and lower prevalence of coronary heart disease than men with AF. Recently, valvular heart disease has decreased except in countries with a higher incidence of rheumatic heart disease such as low- and middle-income countries.

Other risk factors that have not consistently shown sex differences include alcohol intake, physical activity, hyperthyroidism, and inflammatory pathways. Physical activity has a complicated association with AF. Vigorous exercise and sedentary lifestyle are associated with higher AF risk but moderate activity showed decreased association in men. In women, moderate exercise was also shown to decrease risk but contrary to men, vigorous exercise did not increase AF risk. Two studies in women showed that leisure time was associated with a lower risk of AF [19]. **Figure 2** presents the different risk factors for AF in men and women, showing the risk factors that are stronger in women vs. in men, or similar in men and women.

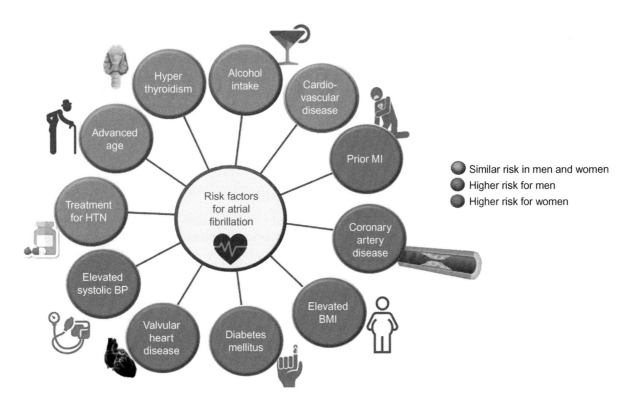

FIGURE 2   **Sex Differences in Risk Factors for Atrial Fibrillation.** BMI, body mass index; BP, blood pressure; HTN, hypertension; MI, myocardial infarction.

TABLE 2 Sex Differences in Presentation of Atrial Fibrillation			
Variable	Female	Male	Difference (95% CI)
Number	470	642	
Age (year) median (IQR)	79 (71–85)	71 (60–79)	8 (3–13)
Admitted to hospital	236 (50.2)	265 (41.3)	8.9 (2.9–14.9)
Weakness	183 (38.9)	136 (21.2)	17.8 (12.2–23.3)
Hypertension	334 (71.1)	405 (63.1)	8.0 (2.2–13.6)
Heart failure	120 (25.5)	109 (17.0)	8.6 (3.6–13.6)

Significant sex differences exist in demographics and risk profiles of patients presenting to the emergency room with atrial fibrillation. CI, confidence interval; IQR, interquartile range. *Data from [35].*

Genetics and Atrial Fibrillation

AF can be a heritable disease and genetic-wide association studies (GWAS) have extensively studied this disease. Despite the heterogeneous nature of the AF clinical phenotypes and the diverse mechanisms of AF, GWAS have contributed to the knowledge of the biological basis of AF and the roles of developmental pathways, electrophysiological signaling, and underlying cardiomyopathy involved in AF [32]. In a large meta-analysis, GWAS in AF identified multiple loci contributing to susceptibility, but one locus, near PITX2 at 4q25, has the strongest association with the heritability of AF [33].

Another large meta-analysis found that about 22% of the variance in AF risk was accounted for by additive genetic variation and similar for both men and women. Genetic variation contributed substantially to AF risk but does not fully account for genetic susceptibility to AF [34]. GWAS trials are underway to further study the genetic contribution in AF in various ethnic populations.

Clinical Presentation

Presenting Symptoms

Women presenting to the emergency room were older, had a higher admission rate, presented with weakness, and had more hypertension and HF compared to men, as seen in **Table 2** [35]. Presenting symptoms in AF in both sexes include palpitations, lightheadedness, fatigue, and dyspnea. Women tend to report more severe symptoms than men (potentially due to faster heart rates or smaller body habitus), but findings vary between studies [3, 5, 36, 37]. The reported sex differences are summarized in **Table 3**.

In the Outcomes Registry for Better Informed Treatment of AF (ORBIT-AF), women were more affected by AF than men; they reported more functional impairment, experienced more limitations in their daily activities, and had

lower quality of life scores than men [3]. This association remained even after accounting for AF therapies received and despite the fact that women had less persistent forms of AF compared to men [3]. This sex difference in duration of AF was not seen in the EURObservational Research Programme-Atrial Fibrillation General Registry Pilot Phase (EORP-AF) [38]. In addition, women with AF experienced higher levels of anxiety and depression compared to men [39].

Access to Treatment

Diagnostic Testing

Diagnostic testing such as noninvasive stress testing and invasive coronary angiography is less likely to be performed in women with AF [40]. Studies evaluating sex-based disparities in access to transthoracic or transesophageal echo (echocardiogram) have yielded disparate results [38, 40, 41].

Management and Prognosis

Mortality

Globally, the mortality risk in developed and developing countries is quite different with regard to sex. In developed countries, men with AF have a higher mortality. Mortality associated with AF hospitalizations decreased from 7.5% in 2006 to 4.3% in 2015 (relative decrease 42%), although hospital costs per year increased exponentially, by 468%, during this 10-year period [2]. In 2016, in a meta-analysis of 30 studies with over 4.3 million participants, women with AF had a 12% increased risk of death compared to men (95% CI, 1.07–1.17) [42]. The relative risk (RR) of mortality in women and men with AF was 1.69 (95% CI, 1.50–1.90) and 1.47 (95% CI, 1.32–1.65), respectively, compared to those without AF. In absolute terms, per 1000 patient-years, women, compared with men, had an 80% increased excess risk of death associated with AF (95% CI, 1.1–2.6).

TABLE 3 Sex Differences in Symptoms of Patients With Atrial Fibrillation

| Author (Year) | Number of Patients | | | Criteria | Symptoms | | Statistical Significance |
	Total	Women	Men		Women	Men	Odds Ratio or *P*-Value
Schnabel (2017) [36]	6412	2546	3866	EHRA > II	62.10%	49.60%	1.68 (1.51, 1.87)
Piccini (2016) [3]	10,135	4293	5842	EHRA I	32.10%	42.50%	*P* < 0.001
				EHRA II	49.10%	42.40%	*P* < 0.001
				EHRA III	16.30%	13.30%	*P* < 0.001
				EHRA IV	2.20%	1.50%	*P* < 0.001
				Palpitations	40.00%	27.00%	*P* < 0.001
				Lightheadedness	23.00%	19.00%	*P* < 0.001
				Fatigue	28.00%	25.00%	*P* < 0.001
				Dyspnea on exertion	29.00%	27.00%	*P* < 0.01
				Dyspnea at rest	11.00%	9.00%	*P* < 0.001
Lip (2015) [5]	3119	1260	1859	Currently symptomatic	65.60%	56.8%	*P* < 0.0001
				AF symptoms in the past	64.2%	54.6%	*P* = 0.0011
				Palpitations	80.20%	68.5%	*P* < 0.0001
				Fear/anxiety	14.60%	10.5%	*P* = 0.0007
				Dyspnea/shortness of breath	55.70%	52.1%	*P* = 0.1189
				Chest pain	23.1%	23.8%	*P* = 0.7237
				General nonwell-being	36.6%	33.6%	*P* = 0.1636
				Dizziness	25.4%	22.9%	*P* = 0.2160
				Fatigue	45.1%	47.9%	*P* = 0.2328
Blum (2017) [37]	1542	454	1088	Characteristic	Odds ratio[a] (95% CI)		Multivariate *P*-value
				Any symptoms	2.6 (2.1–3.4)		< 0.0001
				Palpitations	2.6 (2.1–3.2)		< 0.0001
				Dizziness	2.9 (2.1–3.9)		< 0.0001
				Dyspnea	2.1 (1.6–2.8)		< 0.0001
				Fatigue	1.6 (1.2–2.2)		0.0008
				Chest pain	1.8 (1.3–2.6)		0.001
				Effort intolerance	0.7 (0.5–1.0)		0.051

[a] *Odds ratio of women compared to men.*
AF, atrial fibrillation; CI, confidence interval; EHRA, European Heart Rhythm Association; EHRA I, no symptoms; EHRA II, mild symptoms daily activity not affected; EHRA III, severe symptoms normal activity affected; EHRA IV, disabling symptoms daily activity discontinued.

Heart Failure

AF is associated with a higher risk of HF [42]. Women and men with AF have similar incidence of HF (56.0 and 50.7 per 1000 person-years, respectively) [43]; however, women have a higher risk of incident HF [hazard ratio (HR) 1.16 (95% CI 1.07–1.27)] [42]. It has also been shown that women have a higher incidence of HF with preserved ejection fraction (HFpEF) (35.1 vs. 21.2 events per 1000 person-years) and lower incidence of HF with reduced ejection fraction (12.4 vs. 27.2 events per 1000 person-years) compared to men [43]. In the EORP-AF [38] and the Canadian Registry of Atrial Fibrillation [44], the higher incidence of HFpEF in women may be partially due to the higher prevalence of hypertension, thyroid disorders, and valvular disease in women [40, 45]. In an analysis from the Treatment of Preserved Cardiac Function Heart Failure With an Aldosterone Antagonist Trial (TOPCAT), women with AF and HFpEF had a stronger risk for hospitalization (HR = 1.63, 95% CI = 1.40, 1.91) than men

with AF and HFpEF (HR = 1.37, 95% CI = 1.18, 1.58; p-interaction = 0.032) [46].

Rate vs. Rhythm Control

Both rate and rhythm control are recommended as first-line therapies for management of AF [American College of Cardiology (ACC)/American Heart Association (AHA)/ Heart Rhythm Society (HRS) guidelines and European Society of Cardiology (ESC) guidelines] [47, 48]. Rate control has been shown to be noninferior to rhythm control [49, 50], and instrumental to improving arrhythmia-related symptoms and tachycardia-induced cardiomyopathy. However, many patients remain symptomatic despite adequate rate control and may need rhythm control strategy with antiarrhythmic drugs, electrical cardioversion, and ablative therapies. Sex-specific differences exist in utilization of rate vs. rhythm control strategy. **Figure 3** shows a summary of underrepresentation of women in major AF trials

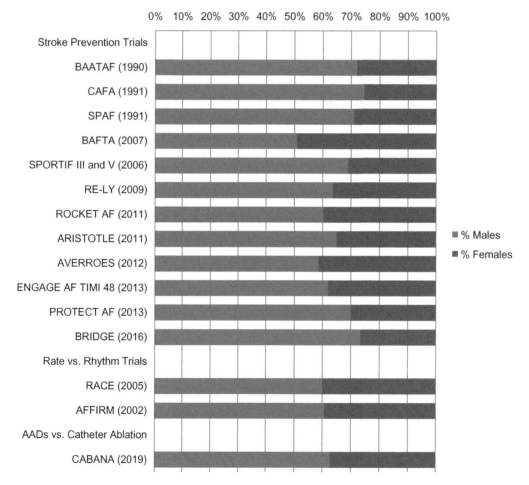

FIGURE 3   Percentages of Males vs. Females in Major Atrial Fibrillation Trials. Underrepresentation of women in major AF trials for stroke prevention, rate and rhythm control, and catheter ablation treatment. AADs, antiarrhythmic drugs.

including rate and rhythm control. There has been some increase in the percentage of women in clinical trials over the years, but more efforts can be made to improve this.

Utilization of Rate vs. Rhythm Control

Women tend to be more symptomatic than men, yet are more likely to receive rate control over rhythm control treatment than men [5]. Reviews of treatment disparities of AF patients found that compared with white men, women and blacks: (1) experienced longer-lasting and more frequent symptomatic AF episodes with worse quality of life [3, 5, 36, 37, 51, 52]; (2) had less stroke prevention treatment [53]; (3) had more drug-related adverse events [4]; (4) were treated less aggressively to maintain sinus rhythm [4, 42, 51, 54, 55]; and (5) had a higher adjusted mortality risk [42].

Within rate control strategy, utilization of β-blocker therapy was similar between men and women (72.5% and 70.0%, respectively), but women were more likely to be prescribed digoxin (25.0% vs. 19.8%) than men ($P = 0.0056$) [5].

In the ORBIT-AF study, women and men had similar rates of utilization of antiarrhythmic drugs (28.9% vs. 28.6%, respectively) [3]. However, women, especially if asymptomatic from AF, are less likely to undergo electrical cardioversion [5, 40, 56]. In the EORP-AF registry, the use of electrical cardioversion was much lower in women (18.9% women vs. 25.5% men, $P < 0.0001$) [38]. Women, compared to men, were more likely to undergo pharmacologic cardioversion compared to electrical cardioversion in the EORP-AF registry (28.2% vs. 22.4%, $P = 0.0002$) [5]. Similar results were seen in the substudy of the Prevention of Thromboembolic Events-European Registry in AF (PREFER in AF) [36]; of 7243 patients from seven European countries, women were less likely to undergo electrical cardioversion (14.9% vs. 20.6%, $P < 0.0001$) or ablation (3.3% vs. 6.3%, $P < 0.0001$) compared to men [36]. Similar trends were seen in the United States, which had inpatients with higher utilization electrical cardioversion in men compared to women [odds ratio (OR) 1.26, 95% CI 1.23–1.31; $P < 0.0001$] [57]. The use of electrical cardioversion was found to be associated with a significant lower rate of in-hospital stroke and mortality, decreased length of hospital stay, and hospitalization costs for both women and men [57].

Outcomes With Rate Control Medications

Beta-blockers and calcium channel blockers are equally effective in both men and women. However, there are sex differences in response to cardiac glycoside therapy. Even with lower doses of prescribed digoxin, women have higher serum levels of digoxin and a higher mortality rate with digoxin treatment compared to men [58]. In patients who were treated with digoxin, there was a significant sex interaction with its use for HF ($P = 0.014$). Digoxin therapy was associated with a small, nonsignificant reduction in mortality risk from any cause among men (HR 0.93; 95% CI 0.85–1.02) after multivariable adjustment. In contrast, women experienced a significant increase in all-cause mortality with digoxin therapy (HR 1.23; 95% CI 1.02–1.47) [58]. However, a post-hoc analysis of the Rivaroxaban vs. Vitamin K Antagonist for Prevention of Stroke and Embolism Trial (ROCKET) AF trial found a significant digoxin-sex interaction, with an increased risk of all-cause mortality and vascular death among men with AF compared with women [men—1.34 (1.17–1.55); women—1.01 (0.83–1.23); interaction $P = 0.0179$] [59]. In addition, two studies found an increased risk of breast cancer in women when given cardiac glycosides for AF [60, 61]. Postmenopausal women with AF enrolled in the Women's Health Initiative ($n = 93,676$) from 1994 to 98 and followed for 15 years were found to have a 5.7% incidence of invasive breast cancer [60]. A 19% excess risk of invasive breast cancer was associated with cardiac glycoside use independent of AF and other confounders (HR 1.68, 95% CI 1.33, 2.12). A meta-analysis of 29 studies found that cardiac glycoside use was associated with increase in estrogen receptor positive breast cancer (RR 1.330, 95% CI 1.247 ± 1.419) and increased all-cause mortality (HR 1.35, 95% CI 1.248 ± 1.46) [61]. This may be due to the estrogen-mimetic properties of cardiac glycosides, which may promote growth of breast cancer cells.

Outcomes With Rhythm Control

Significant sex-specific differences in outcome exist with rhythm control strategy, with women experiencing more adverse events than men. When randomized to a rhythm control strategy in the Rate Control vs. Electrical Cardioversion (RACE) trial, women were three times more likely to die compared to those in the rate arm (95% CI 1.5–6.3; $P = 0.002$) and a greater rate of CV events [i.e., cardiovascular (CV) death, HF, thromboembolic complications, bleeding, severe adverse effects of antiarrhythmic drugs, and need for pacemaker implantation (adjusted HR 4.8, 95% CI 1.2–18.8; $P = 0.02$)], due mostly from antiarrhythmic drugs [50]. In contrast, men had improved survival benefit when randomized to the rate control arm (adjusted HR 0.5, 95% CI 0.2–1.0; $P = 0.058$). In contrast, no sex-specific differences in mortality were seen in the subgroup analysis of the Atrial Fibrillation Follow-up Investigation of Rhythm Management (AFFIRM) trial [49].

Unlike prior studies, the AFFIRM trial [49] demonstrated female sex had a greater risk of torsade de pointes with the use of class III antiarrhythmic drugs [27]. This may have been due to the use of QT-prolonging drugs, including sotalol, dofetilide, and ibutilide, all of which carry a greater risk of torsade de pointes in women [27, 62].

Catheter Ablation

ACC/HRS/European guidelines recommend the use of pulmonary vein catheter ablation for patients who are unable to tolerate rate control medications or remain symptomatic despite rate control drugs. These guidelines have the same recommendation regardless of the sex of the patient.

Utilization of Catheter Ablation

In clinical practice, sex-specific differences also exist in utilization of AF ablation. Women are less likely to undergo AF ablation compared to men (4.9% vs. 5.9%, $P = 0.04$), as reported in a large observational cohort study in the United States that included 10,135 patients with AF (ORBIT-AF Registry) [4]. This is also seen in Western countries and Asian countries despite having more severe symptoms [42, 55]. In Canada, "first" AF ablation procedures have increased sevenfold over the past 10 years, but this increase was not seen in the relatively small proportion of women undergoing AF ablation [63]. Referral bias is most likely playing a role since women were three times less likely to be referred to a specialized outpatient arrhythmia clinic for management of AF [64]. However, following referral to an electrophysiologist, the difference in the proportion of women and men undergoing AF ablation was no longer seen [64, 65]. Sex differences and not referral bias may be operating here, since women were noted to have more comorbidities and increased age at the time of diagnosis of AF. In addition, the possibility that women might be more likely to decline subspecialty referral for consideration of invasive procedures cannot be excluded [66–73]. A study that analyzed the sex disparity with regards to referral for catheter ablation for AF found no difference in the number of failed antiarrhythmic drugs before undergoing catheter ablation (1.8 ± 1.3 vs. 1.8 ± 1.5 in females; $P = 0.67$). However, significantly fewer women than men underwent multiple catheter ablation procedures (41.5% vs. 32.0% in females, $P = 0.04$). Interestingly, the primary reason for the lower proportion of females undergoing multiple catheter ablation procedures was patient refusal (21.1% vs. 33.3% in females, $P = 0.01$) rather than medical reasons (2.8% vs. 1.9% in females, $P = 0.58$). Women in this study declined repeat catheter ablation because either: (1) they had marked improvement in their symptoms after the first ablation; or (2) antiarrhythmic drug use was effective in preventing recurrence of AF [68].

Outcomes With Catheter Ablation

Differences in utilization and referral to ablation procedure are also related to the sex-specific differences in clini-

cal characteristics, procedural success, and complications of AF catheter ablation. Compared to men who undergo catheter ablation for AF, women: (1) are referred less often and are older [66–70]; (2) have a longer history of AF [66, 70]; and (3) are more likely to have valvular heart disease [66, 68], hypertension [66], HFpEF [7], and have more complex comorbidities [74]. In addition to these sex differences, compared to men who undergo catheter ablation for AF, women: (1) are less likely to have "lone AF" [71]; (2) have a lower prevalence of paroxysmal AF [70]; (3) tend to be more symptomatic [64, 75]; and (4) have failed more antiarrhythmic agents prior to ablation [69, 70].

Reports of sex differences in complication rates can vary, but more studies report higher complication rates in women [67, 70–74, 76–78] than no sex difference [66, 68]. Adverse events such as bleeding or vascular complications, hematomas, and pseudoaneurysms have been reported to be higher in women than men [67, 70–73]. A higher rate of cardiac tamponade has also been noted in women compared to men [74, 79]. Women were found to have a twofold increased risk (1.24% in women vs. 0.67% in men, OR 1.83, $P < 0.001$) in a worldwide survey of almost 35,000 AF ablation procedures [79]. The reasons for these differences are not well understood but are thought to be potentially related to body habitus, inherent bleeding risk, and greater presence of comorbidities.

Stroke and Thromboembolism Risk

Higher Risk of Stroke in Women

Multiple AF studies have shown that women compared to men have a higher risk of stroke and thromboembolism. In some studies, women with AF were found to have about a 20–30% higher risk of stroke than men, even after adjusting for differences in stroke risk factors and stroke prevention treatment [42, 80–85]. A meta-analysis of 17 studies found that among AF patients, women had a 31% higher risk of stroke and thromboembolism than men (RR 1.31; 95% CI 1.18–1.46), with women aged ≥ 75 years at even higher risk [81]. An updated meta-analysis of 44 reports confirmed the higher risk of stroke and thromboembolism in women [RR 1.24 (1.14–1.36)], this time finding that the risk increased after age 65 years [82]. However, some studies found that sex influences the risk only in people over the age of 75 years [19]. Female sex was added as a risk factor in the CHA_2DS_2-VASc (1 point for congestive HF, hypertension, age > 65 years, diabetes, female sex, and vascular disease, and 2 points for stroke and age > 75 years) risk score for stroke from AF [86] and in 2014 the American and European guidelines on anticoagulant treatment for AF endorsed the use of this risk stratification [14, 87]. Whether female sex was actually an independent risk factor or risk modifier in the CHA_2DS_2-VASc score was further evaluated in an analysis of 240,000 newly diagnosed AF patients (48.7% females) [88, 89]. The excess female sex-related risk of stroke was seen only if there were other CHA_2DS_2-VASc risk factors, especially if they had ≥ 2 concomitant nongender CHA_2DS_2-VASc risk factors. The authors argued that the American guidelines should consider changing the recommendations to reflect that low-risk females should not be offered any oral anticoagulation or aspirin if their score was 1 or 2.

The 2019 AHA/ACC/HRS focused update of the 2014 AHA/ACC/HRS guidelines for the management of patients with AF stated that the excess risk for females was especially evident among those with two nonsex-related stroke risk factors and that female sex is a risk modifier and is age dependent [48]. The updated guidelines had a class I recommendation and level of evidence A for treating with oral anticoagulants those AF patients with a CHA_2DS_2-VASc score of 2 or greater in men or 3 or greater in women. In addition, a class IIa recommendation with level of evidence B was made to forgo anticoagulation in AF patients (except with moderate-to-severe mitral stenosis or a mechanical heart valve) with a CHA_2DS_2-VASc score of 0 in men or 1 in women. In other words, it is reasonable to omit anticoagulant or aspirin therapy in a woman whose only risk factor is being a woman. The North American recommendations are now similar to the European Society of Cardiology and the Canadian Cardiovascular Society (**Table 4**) [90].

TABLE 4 Comparison of Guideline Recommendations for Use of Oral Anticoagulants in Patients With Different CHA_2DS_2-VASc Scores

AHA/ACC/HRS 2019	European Society of Cardiology 2016	Canadian Cardiovascular Society 2018
CHA_2DS_2-VASc = 0 No OAC	CHA_2DS_2-VASc = 0 No OAC	CHADS$_2$ = 0 No OAC
CHA_2DS_2-VASc = 1 Aspirin may be considered CHA_2DS_2-VASc ≥ 2 OAC	CHA_2DS_2-VASc ≥ 1 OAC	Age > 65 years or CHADS$_2$ ≥ 1 OAC

ACC, American College of Cardiology; AHA, American Heart Association; HRS, Heart Rhythm Society; OAC, oral anticoagulant. *Reproduced with permission from [90].*

In high-risk patients, female sex was associated with a twofold increased risk of severe disabling or fatal ischemic stroke in consecutive AF patients with stroke according to the "Get With the Guidelines" stroke database, even after adjustment for possible confounders [91]. Moreover, compared with men, women are significantly more likely to be living alone or widowed before a stroke which highlights the significant public health burden of this increased risk [92].

Although some studies showed that women had a higher risk of stroke than men even after adjusting for differences in stroke risk factors and stroke prevention treatment [42, 80–85], other studies found that the higher risk of stroke in women compared to men with AF may be due to the distribution and effect of various cardiovascular risk factors [7, 8, 93, 94]. On average, women with AF are older than men with AF and more likely to have hypertension, valvular heart disease, HFpEF, and thyroid disease [19]. Other factors that may contribute are differences in body mass index, hormonal exposure, and cardiac hemodynamics [8, 22]. Interestingly, a population-based study from Quebec found that the 16% higher risk of stroke in women with AF was no longer seen when the men and women were matched by baseline risk factors at entry [95]. Other studies have questioned whether women are actually at higher risk of stroke than men when matched appropriately for risk factors of stroke.

Sex Differences in Stroke Prevention

Efficacy of Warfarin and Direct Oral Anticoagulant Use Compared to Antiplatelet Agents and Placebo

It was established in many randomized controlled trials (RCTs) that warfarin was superior to antiplatelet agents in the 1990s and early 2000s in patients with nonvalvular AF with a 40% greater reduction [96]. The AFFIRM study confirmed that patients had a higher risk of stroke when not anticoagulated regardless of the type of AF, i.e., paroxysmal, persistent, or chronic [49]. A multivariable analysis found that in AF patients, women had higher annual rates of thromboembolism off warfarin than did men [3.5% vs. 1.8%; adjusted rate ratio (RR), 1.6; 95% CI 1.3–1.9] [97]. In an analysis of five major antithrombotic trials, women treated with warfarin reduced the risk of stroke by 84% (95% CI 55–95), compared with 60% (95% CI 35–76) in men. Aspirin significantly reduced the risk of stroke in men (44%), but in women, there was only a 23% reduction which was not significant [98].

Differences in Warfarin Utilization, Efficacy, and Control of Adequate Anticoagulation

Women spend less time in the therapeutic range compared to men, further adding to the elevated stroke risk. Sex differences in the AFFIRM trial found that even when men

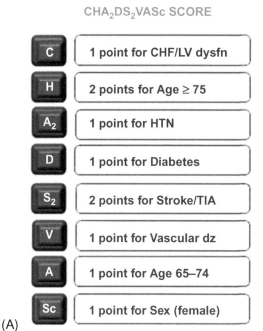

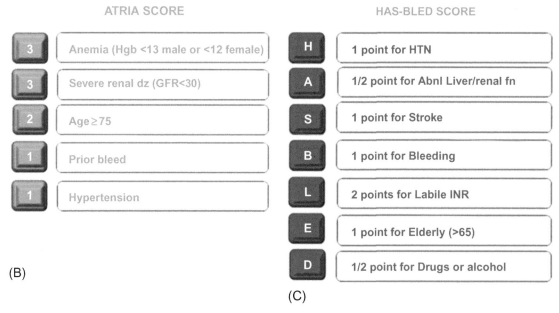

FIGURE 4   Common Risk Scores for Predicting Thromboembolic and Bleed Risk in Atrial Fibrillation. Common risk scores for predicting risk of (A) thromboembolic stroke risk and (B and C) bleeding risk with anticoagulation in patients with atrial fibrillation. Abnl, abnormal; CHF, congestive heart failure; dysfn, dysfunction; dz, disease; fn, function; GFR, glomerular filtration rate; Hgb, hemoglobin; HTN, hypertension; INR, international normalized ratio; LV, left ventricle; TIA, transient ischemic attack. *Image courtesy of Niti R. Aggarwal.*

and women had comparable time in the therapeutic range on warfarin, women were found to have a higher risk of systemic thromboembolism [99].

The PINNACLE National Cardiovascular Data Registry reported that women with AF were significantly less likely than men to receive oral anticoagulants (56.7% vs. 61.3%;

$P < 0.001$). At all levels of the CHA_2DS_2-VASc score, women received oral anticoagulants less often than men with adjusted risk ratios ranging from 9% to 33% lower ($P < 0.001$) [53]. However, in the global anticoagulant registry in the FIELD (GARFIELD)-AF, no significant difference in the overall rate of anticoagulant use was found between women (60.8%) and men (60.9%) [100].

In European studies, there has been an improvement in anticoagulant use for stroke prevention in AF. In 2011, women were more likely to receive aspirin only (35% women vs. 30% men) and fewer women received oral anticoagulants (53% men vs. 48% women), but by 2015, there were no longer sex differences in anticoagulant treatment except in women ≥ 80 years and patients with complicated comorbidities [100, 101].

A large study of AF patients (overall $N = 73,004$ from 35 countries) in the GARFIELD-AF and the ORBIT-AF I and II registries showed that oral anticoagulant use was 46% and 57% for patients with a CHA_2DS_2-VASc $= 0$ and 69% and 87% for CHA_2DS_2-VASc ≥ 2 in GARFIELD-AF and ORBIT-AF II, respectively [102]. There was substantial geographic heterogeneity in the use of oral anticoagulant [range: 31%–93% (GARFIELD-AF) and 66%–100% (ORBIT-AF II)]. It was also found that among patients with new-onset AF, direct oral anticoagulant (DOAC) use increased over time to 43% in 2016 for GARFIELD-AF and 71% for ORBIT-AF II, and the use of antiplatelet monotherapy decreased from 36% to 17% (GARFIELD-AF) and 18% to 8% (ORBIT-AF I and II). Although improvement in the appropriate use of anticoagulation was seen in AF patients, inappropriate use of anticoagulants was seen in low-risk patients with CHA_2DS_2-VASc scores of 0. The significant geographic variability in the use of oral anticoagulants highlights an opportunity for education and implementation of consistent guideline-based recommendations [102]. However, any changes in the guideline recommendations may cause confusion and require a lag time before actual implementation.

Safety and Bleeding With Warfarin

Bleeding is a major risk with the use of antiplatelet and anticoagulant therapy. Risk of bleeding during anticoagulant therapy is well studied and algorithms have been compared. Commonly used risk scores are: HAS-BLED [assigns 1 point each to hypertension, abnormal renal/liver function, stroke, bleeding history or predisposition, labile INR, elderly (>65 years), drug/alcohol concomitantly]; ORBIT-AF; AnTicoagulation and Risk factors In Atrial fibrillation (ATRIA); and the ABC (age, biomarkers, clinical history), which uses selected biomarkers, developed mainly in patients with AF (**Figure 4**). The accuracy of $CHADS_2$, CHA_2DS_2-VASc HAS-BLED, ORBIT, and ATRIA is similar in predicting DOAC-associated major and intracranial bleeding in patients with AF [103]. Sex differences using these risk models have not been published.

Data on Efficacy and Safety of Direct Anticoagulants

A meta-analysis of warfarin compared to DOACs found that women had better outcomes for stroke, thromboembolism,

and major bleeding with DOACs compared to warfarin with no sex differences [104]. However, another meta-analysis of slightly different DOAC trials found different outcomes; men treated with either warfarin or DOACs had a higher risk of systemic thromboembolism (OR: 1.21, 95% CI: 1.11–1.32, $P < 0.0001$). However, men were more protected from systemic thromboembolism, whereas women had a lower risk of major bleeding when treated with DOACs [105]. This meta-analysis included all major phase III DOAC trials of dabigatran, rivaroxaban, apixaban, and edoxaban. Some of these RCTs did not have enough women enrolled to evaluate sex-specific differences.

These two meta-analysis studies showed slightly different outcomes but are in agreement that women have better outcomes for ST-elevation (STE) and major bleeding with DOACs compared to warfarin. An indirect comparison study of all four DOACs revealed no significant difference with regard to their safety and efficacy in women compared with dose-adjusted warfarin, suggesting that they can be used interchangeably in women [106].

Table 5 summarizes sex-related differences in the risk and prevention of strokes in AF patients. The sex-related differences may be due to differences in age distribution, underrepresentation of females, and higher comorbid risk factors in women compared to men. These differences should be evaluated in larger-scale studies targeted toward real-life data to improve risk stratification, particularly with increased use of DOACs.

Percutaneous Left Atrial Appendage Closure Devices

The left atrial appendage (LAA) is the source of large embolic strokes and 90% of thrombi in patients with non-valvular AF [107]. Patients who are intolerant of oral anticoagulation due to bleeding or difficult anticoagulation management may benefit from LAA endocardial occlusion devices. These devices have been developed for clinical use for stroke risk reduction in AF patients, including: the Watchman, the Amplatzer Cardiac Plug and Amulet, the WaveCrest, and the Lariat.

The Watchman is the most extensively studied device approved by the US Food and Drug Administration [108]. There are currently no trials that specifically examine sex-specific systemic thromboembolism outcomes after percutaneous LAA occlusion. However, there were no significant associations with sex and a composite efficacy end point of stroke, systemic embolism, and CV death in a meta-analysis of the PROTECT-AF (Percutaneous Left Atrial Appendage Closure vs. Warfarin for Atrial Fibrillation) and PREVAIL (Prospective Randomized Evaluation of the Watchman Left Atrial Appendage Closure Device in Patients with Atrial Fibrillation Versus Long-Term Warfarin Therapy) trials [109]. Women, compared to men, were initially found to

TABLE 5 Differences in Prevention and Risk of Stroke

	Statistical Significance	Author (Year)
Stroke Prevention		
Women less likely to receive anticoagulation (United States)	56.7% vs. 61.3%; $P < 0.001$	Thompson (2017) [53]
No difference in receiving oral anticoagulation (global)	73.4% vs. 73.45; $P = 0.4456$	Lip (2015) [100]
Women spent more time outside the therapeutic range	40% vs. 37%; $P = 0.0001$	Sullivan (2012) [99]
Women spent more time below the therapeutic range	29% vs. 26%; $P = 0.0002$	Sullivan (2012) [99]
Strokes		
Women have higher risk of stroke than male atrial fibrillation patients	1.31 (1.18–1.46)	Wagstaff (2014) [81]
	1.24 (1.14–1.36)	Marzona (2018) [82]
Women with more debilitating strokes than men	1.99 (1.07–3.72)	Martin (2017) [91]

have more incidence of device-related thrombus (75% vs. 34%, $P = 0.094$) using the Watchman [110], but more recent data reported no significant sex difference in device-related thrombus with the same device [111]. However, female sex (OR 4.22; $P = 0.027$) and cigarette smoking (OR 5.79; $P = 0.017$) were independent predictors of device-related thrombus in univariate and multivariable logistic regression models with the use of the Amplatzer Cardiac Plug [112]. It is important to note that although the current clinical trials did not show a significant association between sex and various composite end points, they were likely underpowered to identify such sex differences due to the relatively small number of women that are historically included in cardiovascular clinical trials.

Future Directions

Compared to men, women with AF have more symptoms and have poorer quality of life. In order to improve symptoms in patients with AF, studies specifically addressing symptom improvement may improve the quality of life of AF patients. The use of cardiac glycosides is associated with increased breast cancer and mortality, and is more often used in women. More efforts are needed to educate healthcare professionals to avoid unnecessary or inappropriate use of cardiac glycosides. Women are not referred for catheter ablation or electrical cardioversions as frequently as men, and work need to be undertaken to understand this disparity. Lastly, more efforts need to be made to increase the percentage of women in clinical AF trials.

Conclusion

It has long been recognized that important sex differences exist between women and men with AF. Recognition of these differences and adaptation of clinical practices as a result is crucial to improving the outcomes for patients. **Figure 5** summarizes the important sex differences in women and men with AF. Multiple prior studies have demonstrated that while lifetime incidence of AF is the same between the sexes, women are more likely to be older at the time of presentation than men. Women are also more likely to be symptomatic, experience more functional limitations, and, as a result, seek medical attention more often than men. Interestingly, some studies have found an increased mortality risk for women compared to men, even when accounting for possible confounders like differences in therapies. Several studies have found that women are referred less for certain therapies such as catheter ablation or electrical cardioversion, but when such therapies are offered, women are more likely to refuse these therapies. Importantly, anticoagulation is underused in elderly women, despite a proven therapeutic benefit and the fact that these patients are at a particularly high risk for stroke. Women are also more likely to experience major bleeding in the setting of anticoagulation, have more medication side effects, and have more procedural complications than men. In conclusion, increasing recognition and understanding of key sex differences allows for a more tailored approach to the treatment of both men and women with AF. Given the vast prevalence and ever-growing incidence of AF, the global impact of such sex-specific adaptations cannot be overemphasized.

Epidemiology
- Incidence & prevalence are increasing
- Incidence in men occurs at a younger age than women
- Incidence/prevalence increases with age in both sexes
- Lifetime incidence of AF is similar between the sexes as women tend to live longer
- People of European ancestry & higher-income countries have a higher prevalence of AF

Pathophysiology
- Sex hormones play an important role in sex-related differences of AF mgmt
- Men have larger LA
- Women tend to have more atrial fibrosis associated with a higher prevalence of stroke and systemic embolism

Presentation
- Women have longer duration of AF symptoms
- Women experience more atypical symptoms
- Women experience more significant functional impairment and report worse quality of life
- Women have higher heart rates
- Women have greater proportion of AF recurrences

Access to Treatment
- Some diagnostic tests are less likely to be performed in women
- Women are more likely to receive rate control vs. rhythm control
- Women are equally likely to get AADs as men
- Women are less likely to undergo electrical cardioversion
- Women are less likely to be referred for catheter ablations
- Women have higher rates of complications with ablations

Prognosis
- Mortality risk is variable worldwide
- Women may have a higher mortality risk than men
- Women with AF and HFpEF have worse outcomes
- Women more likely to experience adverse drug reactions and side effects from AADs
- Women treated with rhythm control have worse outcome compared to men
- Women more likely to develop sick sinus syndrome requiring pacemaker with rate control strategy

Risk of Thromboembolism
- Female sex is associated with higher risk of stroke in AF
- Women suffer more from disabling stroke
- Women with AF are older than men & have more comorbidities which may explain increased stroke risk
- Women have more strokes on warfarin than men
- DOACs are equally safe and effective in men and women
- Therapeutic ant coagulation may be used less often in women despite a higher risk of thromboembolic events

FIGURE 5   Summary of Sex-Specific Differences in Patients With Atrial Fibrillation. Differences in epidemiology; pathophysiology, presentation, access to treatment, prognosis, and risk of thromboembolism are highlighted. AADs, antiarrhythmic drugs; AF, atrial fibrillation; DOACs, direct oral anticoagulants; HFpEF, heart failure with preserved ejection fraction; LA, left atria; mgmt, management.

Key Points

1 Substantial sex-specific differences exist in epidemiology, pathophysiology, presentation, outcomes, and response to therapy with AF.

2 Compared to men, women have a lower-age-adjusted incidence and prevalence of AF. However, as women tend to live longer, the lifetime incidence of AF is similar between the two sexes.

3 Women with AF are older than men with AF and more likely to have other medical comorbidities which may explain increased risk of disabling stroke.

4 Fibers in the left atrium around the pulmonary veins are often the origin of AF and women tend to have more atrial fibrosis which is associated with a higher prevalence of stroke and systemic embolism.

5 Women tend to have longer duration of symptoms, have more atypical symptoms, experience more significant functional impairment, and report worse quality of life.

6 Both men and women are equally treated with antiarrhythmic drugs.

7 Women are less likely to undergo electrical cardioversion and catheter ablations.

8 Some studies show women have higher rates of complications with ablations.

9 Women are more likely to experience adverse drug reactions and side effects from antiarrhythmic drugs.

10 Some studies suggest women have a higher mortality risk than men and women with AF and HFpEF have worse outcomes.

11 There is a need for a greater representation of women in clinical trials and future sex-specific research to improve gaps in our understanding of the pathophysiologic differences, address disparities, and improve sex-specific therapeutic approaches.

Back to Clinical Case

The clinical case describes a 76-year-old asymptomatic female with new-onset AF, who is rate controlled. Women often have more symptoms compared to men so electrical cardioversion may be preferred. However, compared to men, women have more adverse CV events with a rhythm control approach. Hence, in this asymptomatic patient, a rate control strategy may be preferred, especially since her heart rate is already controlled.

Among patients with atrial fibrillation, even after adjusting for risk factors, women have a nearly 20% higher risk of stroke compared to men. This patient's CHA_2DS_2VASc score was calculated to be 4 (2 points for age > 75; 1 for hypertension; 1 for female sex), giving her a 4.8% risk of stroke, and placing her in the high-risk category. In contrast, on the ATRIA score she scores 2 points (1 point each for age and hypertension). This calculates to a 0.76% (low) risk of hemorrhage with initiation of anticoagulation. The 2019 AHA/ACC/HRS guidelines for the management of patients with atrial fibrillation gives a class I recommendation for initiation of anticoagulation [48]. In addition, the use of direct oral anticoagulants resulted in improved outcomes for stroke, thromboembolism, and major bleeding compared to warfarin use with no sex differences. Given that her thromboembolic stroke risk outweighed her risk of bleeding, she was initiated on a direct anticoagulant, and educated on the signs of stroke and risk of bleeding. She was also told to continue her blood pressure medications. This case highlights the unique differences in sex-specific management of atrial fibrillation.

References

[1] Chugh SS, Havmoeller R, Narayanan K, Singh D, Rienstra M, Benjamin EJ, et al. Worldwide epidemiology of atrial fibrillation: a Global Burden of Disease 2010 Study. Circulation 2014;129:837–47.

[2] Ribeiro AL, Otto CM. Heartbeat: the worldwide burden of atrial fibrillation. Heart 2018;104:1987–8.

[3] Piccini JP, Simon DN, Steinberg BA, Thomas L, Allen LA, Fonarow GC, et al. Differences in clinical and functional outcomes of atrial fibrillation in women and men: two-year results from the ORBIT-AF registry. JAMA Cardiol 2016;1:282–91.

[4] Rienstra M, Van Veldhuisen DJ, Hagens VE, Ranchor AV, Veeger NJ, Crijns HJ, et al. Gender-related differences in rhythm control treatment in persistent atrial fibrillation: data of the Rate Control versus Electrical Cardioversion (RACE) study. J Am Coll Cardiol 2005;46:1298–306.

[5] Lip GYH, Laroche C, Boriani G, Cimaglia P, Dan GA, Santini M, et al. Sex-related differences in presentation, treatment, and outcome of patients with atrial fibrillation in Europe: a report from the Euro Observational Research Programme Pilot survey on atrial fibrillation. Europace 2015;17:24–31.

[6] Benjamin EJ, Muntner P, Alonso A, Bittencourt MS, Callaway CW, Carson AP, et al. Heart disease and stroke statistics—2019 update: a report from the American Heart Association. Circulation 2019;139:e56–66.

[7] Schnabel RB, Yin X, Gona P, Larson MG, Beiser AS, McManus DD, et al. 50 year trends in atrial fibrillation prevalence, incidence, risk factors, and mortality in the Framingham Heart Study: a cohort study. Lancet 2015;386:154–62.

[8] Magnussen C, Niiranen TJ, Ojeda FM, Gianfagna F, Blankenberg S, Njølstad I, et al. Sex differences and similarities in atrial fibrillation epidemiology, risk factors, and mortality in community cohorts: results from the BiomarCaRE consortium (biomarker for cardiovascular risk assessment in Europe). Circulation 2017;136:1588–97.

[9] Piccini JP, Hammill BG, Sinner MF, Jensen PN, Hernandez AF, Heckbert SR, et al. Incidence and prevalence of atrial fibrillation and associated mortality among Medicare beneficiaries, 1993-2007. Circ Cardiovasc Qual Outcomes 2012;5:85–93.

[10] Huxley RR, Lopez FL, Folsom AR, Agarwal SK, Loehr LR, Soliman EZ, et al. Absolute and attributable risks of atrial fibrillation in relation to optimal and borderline risk factors: the Atherosclerosis Risk in Communities (ARIC) study. Circulation 2011;123:1501–8.

[11] Alonso A, Krijthe BP, Aspelund T, Stepas KA, Pencina MJ, Moser CB, et al. Simple risk model predicts incidence of atrial fibrillation in a racially and geographically diverse population: the CHARGE-AF consortium. J Am Heart Assoc 2013;2:e000102.

[12] Staerk L, Wang B, Preis SR, Larson MG, Lubitz SA, Ellinor PT, et al. Lifetime risk of atrial fibrillation according to optimal, borderline, or elevated levels of risk factors: cohort study based on longitudinal data from the Framingham Heart Study. BMJ 2018;361:k1453.

[13] Mou L, Norby FL, Chen LY, O'Neal WT, Lewis TT, Loehr LR, et al. Lifetime risk of atrial fibrillation by race and socioeconomic status: ARIC study (Atherosclerosis Risk in Communities). Circ Arrhythm Electrophysiol 2018;11:e006350.

[14] January CT, Wann LS, Alpert JS, Calkins H, Cigarroa JE, Cleveland JC, et al. 2014 AHA/ACC/HRS guideline for the management of patients with atrial fibrillation: a report of the American College of Cardiology/American Heart Association task force on practice guidelines and the Heart Rhythm Society. J Am Coll Cardiol 2014;64:e1–e76.

[15] Tsai WC, Chen YC, Lin YK, Chen SA, Chen YJ. Sex differences in the electrophysiological characteristics of pulmonary veins and left atrium and their clinical implication in atrial fibrillation. Circ Arrhythm Electrophysiol 2011;4:550–9.

[16] Galea R, Cardillo MT, Caroli A, Marini MG, Sonnino C, Narducci ML, et al. Inflammation and C-reactive protein in atrial fibrillation: cause or effect? Tex Heart Inst J 2014;41:461–8.

[17] Aviles RJ, Martin DO, Apperson-Hansen C, Houghtaling PL, Rautaharju P, Kronmal RA, et al. Inflammation as a risk factor for atrial fibrillation. Circulation 2003;108:3006–10.

[18] Marott SCW, Nordestgaard BG, Zacho J, Friberg J, Jensen GB, Tybjaerg-Hansen A, et al. Does elevated C-reactive protein increase atrial fibrillation risk? A Mendelian randomization of 47,000 individuals from the general population. J Am Coll Cardiol 2010;56:789–95.

[19] Ko D, Rahman F, Schnabel RB, Yin X, Benjamin EJ, Christophersen IE. Atrial fibrillation in women: epidemiology, pathophysiology, presentation, and prognosis. Nat Rev Cardiol 2016;13:321–32.

[20] Andrade JG, Deyell MW, Lee AYK, Macle L. Sex differences in atrial fibrillation. Can J Cardiol 2018;34:429–36.

[21] Kishi S, Reis JP, Venkatesh BA, Gidding SS, Armstrong AC, Jacobs DR, et al. Race-ethnic and sex differences in left ventricular structure and function: the Coronary Artery Risk Development in Young Adults (CARDIA) Study. J Am Heart Assoc 2015;4:e001264.

[22] Cochet H, Mouries A, Nivet H, Sacher F, Derval N, Denis A, et al. Age, atrial fibrillation, and structural heart disease are the main determinants of left atrial fibrosis detected by delayed-enhanced magnetic resonance imaging in a general cardiology population. J Cardiovasc Electrophysiol 2015;26:484–92.

[23] Akoum N, Mahnkopf C, Kholmovski EG, Brachmann J, Marrouche NF. Age and sex differences in atrial fibrosis among patients with atrial fibrillation. Europace 2018;20:1086–92.

[24] Curtis AB, Narasimha D. Arrhythmias in women. Clin Cardiol 2012;35:166–71.

[25] Burke JH, Goldberger JJ, Ehlert FA, Kruse JT, Parker MA, Kadish AH. Gender differences in heart rate before and after autonomic blockade: evidence against an intrinsic gender effect. Am J Med 1996;100:537–43.

[26] Mason JW, Ramseth DJ, Chanter DO, Moon TE, Goodman DB, Mendzelevski B. Electrocardiographic reference ranges derived from 79,743 ambulatory subjects. J Electrocardiol 2007;40:228–34.

[27] Wolbrette DL. Risk of proarrhythmia with class III antiarrhythmic agents: sex-based differences and other issues. Am J Cardiol 2003;91:39d–44d.

[28] Liu XK, Katchman A, Drici MD, Ebert SN, Ducic I, Morad M, et al. Gender difference in the cycle length-dependent QT and potassium currents in rabbits. J Pharmacol Exp Ther 1998;285:672–9.

[29] Nicolson TJ, Mellor HR, Roberts RRA. Gender differences in drug toxicity. Trends Pharmacol Sci 2010;31:108–14.

[30] Kaiser J. Gender in the pharmacy: does it matter? Am Assoc Adv Sci 2005;308(5728):1572.

[31] Bose A, O'Neal WT, Wu C, McClure LA, Judd SE, Howard VJ, et al. Sex differences in risk factors for incident atrial fibrillation (from the Reasons for Geographic and Racial Differences in Stroke [REGARDS] Study). Am J Cardiol 2019;123:1453–7.

[32] Roden DM, Below JE. My cousin also has atrial fibrillation: family relationships in a genomic era. JACC Clin Electrophysiol 2019;5:501–3.

[33] Roselli C, Chaffin MD, Weng L-C, Aeschbacher S, Ahlberg G, Albert CM, et al. Multi-ethnic genome-wide association study for atrial fibrillation. Nat Genet 2018;50:1225–33.

[34] Weng L-C, Choi SH, Klarin D, Smith JG, Loh P-R, Chaffin M, et al. Heritability of atrial fibrillation. Circ Cardiovasc Genet 2017;10(6):e001838.

[35] Scheuermeyer FX, Mackay M, Christenson J, Grafstein E, Pourvali R, Heslop C, et al. There are sex differences in the demographics and risk profiles of emergency department (ED) patients with atrial fibrillation and flutter, but no apparent differences in ED management or outcomes. Acad Emerg Med 2015;22:1067–75.

[36] Schnabel RB, Pecen L, Ojeda FM, Lucerna M, Rzayeva N, Blankenberg S, et al. Gender differences in clinical presentation and 1-year outcomes in atrial fibrillation. Heart 2017;103:1024–30.

[37] Blum S, Muff C, Aeschbacher S, Ammann P, Erne P, Moschovitis G, et al. Prospective assessment of sex-related differences in symptom status and health perception among patients with atrial fibrillation. J Am Heart Assoc 2017;6(7):e005401.

[38] Lip GYH, Laroche C, Ioachim PM, Rasmussen LH, Vitali-Serdoz L, Petrescu L, et al. Prognosis and treatment of atrial fibrillation patients by European cardiologists: one year follow-up of the EURObservational Research Programme-Atrial Fibrillation General Registry Pilot Phase (EORP-AF Pilot registry). Eur Heart J 2014;35:3365–76.

[39] Gleason KT, Dennison Himmelfarb CR, Ford DE, Lehmann H, Samuel L, Han HR, et al. Association of sex, age and education level with patient reported outcomes in atrial fibrillation. BMC Cardiovasc Disord 2019;19:85.

[40] Dagres N, Nieuwlaat R, Vardas PE, Andresen D, Lévy S, Cobbe S, et al. Gender-related differences in presentation, treatment, and outcome of patients with atrial fibrillation in Europe: a report from the Euro Heart Survey on Atrial Fibrillation. J Am Coll Cardiol 2007;49:572–7.

[41] Sinner MF, Greiner MA, Mi X, Hernandez AF, Jensen PN, Piccini JP, et al. Completion of guideline-recommended initial evaluation of atrial fibrillation. Clin Cardiol 2012;35:585–93.

[42] Emdin CA, Wong CX, Hsiao AJ, Altman DG, Peters SA, Woodward M, et al. Atrial fibrillation as risk factor for cardiovascular disease and death in women compared with men: systematic review and meta-analysis of cohort studies. BMJ 2016;532:h7013.

[43] Santhanakrishnan R, Wang N, Larson MG, Magnani JW, McManus DD, Lubitz SA, et al. Atrial fibrillation begets heart failure and vice versa: temporal associations and differences in preserved versus reduced ejection fraction. Circulation 2016;133:484–92.

[44] Humphries KH, Kerr CR, Connolly SJ, Klein G, Boone JA, Green M, et al. New-onset atrial fibrillation: sex differences in presentation, treatment, and outcome. Circulation 2001;103:2365–70.

[45] Ball J, Løchen ML, Wilsgaard T, Schirmer H, Hopstock LA, Morseth B, et al. Sex differences in the impact of body mass index on the risk of future atrial fibrillation: insights from the longitudinal population-based Tromsø study. J Am Heart Assoc 2018;7:e008414.

[46] O'Neal WT, Sandesara P, Hammadah M, Venkatesh S, Samman-Tahhan A, Kelli HM, et al. Gender differences in the risk of adverse outcomes in patients with atrial fibrillation and heart failure with preserved ejection fraction. Am J Cardiol 2017;119:1785 90.

[47] Kirchhof P, Benussi S, Kotecha D, Ahlsson A, Atar D, Casadei B, et al. 2016 ESC guidelines for the management of atrial fibrillation developed in collaboration with EACTS. Eur Heart J 2016;37:2893–962.

[48] January CT, Wann LS, Calkins H, Chen LY, Cigarroa JE, Cleveland JC, et al. 2019 AHA/ACC/HRS focused update of the 2014 AHA/ACC/HRS guideline for the management of patients with atrial fibrillation: a report of the American College of Cardiology/American Heart Association Task Force on Clinical Practice Guidelines and the Heart Rhythm Society in Collaboration With the Society of Thoracic Surgeons. Circulation 2019;140:e125–51.

[49] Wyse DG, Waldo AL, DiMarco JP, Domanski MJ, Rosenberg Y, Schron EB, et al. A comparison of rate control and rhythm control in patients with atrial fibrillation. N Engl J Med 2002;347:1825–33.

[50] Van Gelder IC, Hagens VE, Bosker HA, Kingma JH, Kamp O, Kingma T, et al. Rate control versus electrical cardioversion for atrial fibrillation: a randomised comparison of two treatment strategies concerning morbidity, mortality, quality of life and cost-benefit—the RACE study design. Neth Heart J 2002;10:118–24.

[51] Bhave PD, Lu X, Girotra S, Kamel H, Vaughan Sarrazin MS. Race- and sex-related differences in care for patients newly diagnosed with atrial fibrillation. Heart Rhythm 2015;12:1406–12.

[52] Golwala H, Jackson 2nd LR, Simon DN, Piccini JP, Gersh B, Go AS, et al. Racial/ethnic differences in atrial fibrillation symptoms, treatment patterns, and outcomes: Insights from Outcomes Registry for Better Informed Treatment for Atrial Fibrillation Registry. Am Heart J 2016;174:29–36.

[53] Thompson LE, Maddox TM, Lei L, Grunwald GK, Bradley SM, Peterson PN, et al. Sex differences in the use of oral anticoagulants for atrial fibrillation: a report From the National Cardiovascular Data Registry (NCDR®) PINNACLE Registry. J Am Heart Assoc 2017;6(7), e005801.

[54] Kummer BR, Bhave PD, Merkler AE, Gialdini G, Okin PM, Kamel H. Demographic differences in catheter ablation after hospital presentation with symptomatic atrial fibrillation. J Am Heart Assoc 2015;4:e002097.

[55] Lee JM, Kim T-H, Cha M-J, Park J, Park J-K, Kang K-W, et al. Gender-related differences in management of nonvalvular atrial fibrillation in an Asian population. Korean Circ J 2018;48:519–28.

[56] Alegret JM, Vinolas X, Martinez-Rubio A, Pedrote A, Beiras X, García-Sacristán JF, et al. Gender differences in patients with atrial fibrillation undergoing electrical cardioversion. J Women's Health 2015;24:466–70.

[57] Rochlani YM, Shah NN, Pothineni NV, Paydak H. Utilization and predictors of electrical cardioversion in patients hospitalized for atrial fibrillation. Cardiol Res Pract 2016;2016:8956020.

[58] Rathore SS, Wang Y, Krumholz HM. Sex-based differences in the effect of digoxin for the treatment of heart failure. N Engl J Med 2002;347:1403–11.

[59] Washam JB, Stevens SR, Lokhnygina Y, Halperin JL, Breithardt G, Singer DE, et al. Digoxin use in patients with atrial fibrillation and adverse cardiovascular outcomes: a retrospective analysis of the Rivaroxaban Once Daily Oral Direct Factor Xa Inhibition Compared with Vitamin K Antagonism for Prevention of Stroke and Embolism Trial in Atrial Fibrillation (ROCKET AF). Lancet 2015;385:2363–70.

[60] Wassertheil-Smoller S, McGinn AP, Martin L, Rodriguez BL, Stefanick ML, Perez M. The associations of atrial fibrillation with the risks of incident invasive breast and colorectal cancer. Am J Epidemiol 2017;185:372–84.

[61] Osman MH, Farrag E, Selim M, Osman MS, Hasanine A, Selim A. Cardiac glycosides use and the risk and mortality of cancer; systematic review and meta-analysis of observational studies. PLoS One 2017;12:e0178611.

[62] Reisinger J, Gatterer E, Lang W, Vanicek T, Eisserer G, Bachleitner T, et al. Flecainide versus ibutilide for immediate cardioversion of atrial fibrillation of recent onset. Eur Heart J 2004;25:1318–24.

[63] Avgil Tsadok M, Gagnon J, Joza J, Behlouli H, Verma A, Essebag V, et al. Temporal trends and sex differences in pulmonary vein isolation for patients with atrial fibrillation. Heart Rhythm 2015;12:1979–86.

[64] Roten L, Rimoldi SF, Schwick N, Sakata T, Heimgartner C, Fuhrer J, et al. Gender differences in patients referred for atrial fibrillation management to a tertiary center. Pacing Clin Electrophysiol 2009;32:622–6.

[65] Gerstenfeld EP, Callans D, Dixit S, Lin D, Cooper J, Russo AM, et al. Characteristics of patients undergoing atrial fibrillation ablation: trends over a seven-year period 1999-2005. J Cardiovasc Electrophysiol 2007;18:23–8.

[66] Forleo GB, Tondo C, De Luca L, Dello Russo A, Casella M, De Sanctis V, et al. Gender-related differences in catheter ablation of atrial fibrillation. Europace 2007;9:613–20.

[67] Zylla MM, Brachmann J, Lewalter T, Hoffmann E, Kuck K-H, Andresen D, et al. Sex-related outcome of atrial fibrillation ablation: insights from the German Ablation Registry. Heart Rhythm 2016;13:1837–44.

[68] Takigawa M, Kuwahara T, Takahashi A, Watari Y, Okubo K, Takahashi Y, et al. Differences in catheter ablation of paroxysmal atrial fibrillation between males and females. Int J Cardiol 2013;168:1984–91.

[69] Winkle RA, Mead RH, Engel G, Patrawala R A. Long-term results of atrial fibrillation ablation: the importance of all initial ablation failures undergoing a repeat ablation. Am Heart J 2011;162:193–200.

[70] Patel D, Mohanty P, Di Biase L, Sanchez JE, Shaheen MH, Burkhardt JD, et al. Outcomes and complications of catheter ablation for atrial fibrillation in females. Heart Rhythm 2010;7:167–72.

[71] Zhang X-D, Tan H-W, Gu J, Jiang W-F, Zhao L, Wang Y-L, et al. Efficacy and safety of catheter ablation for long-standing persistent atrial fibrillation in women. Pacing Clin Electrophysiol 2013;36:1236–44.

[72] Kuck K-H, Brugada J, Furnkranz A, Chun KRJ, Metzner A, Ouyang F, et al. Impact of female sex on clinical outcomes in the FIRE AND ICE trial of catheter ablation for atrial fibrillation. Circ Arrhythm Electrophysiol 2018;11:e006204.

[73] De Greef Y, Ströker E, Schwagten B, Kupics K, De Cocker J, Chierchia GB, et al. Complications of pulmonary vein isolation in atrial fibrillation: predictors and comparison between four different ablation techniques: results from the MIddelheim PVI-registry. Europace 2018;20(8):1279–86.

[74] Kaiser DW, Fan J, Schmitt S, Than CT, Ullal AJ, Piccini JP, et al. Gender differences in clinical outcomes after catheter ablation of atrial fibrillation. JACC Clin Electrophysiol 2016;2:703–10.

[75] Reynolds MR, Lavelle T, Essebag V, Cohen DJ, Zimetbaum P. Influence of age, sex, and atrial fibrillation recurrence on quality of life outcomes in a population of patients with new-onset atrial fibrillation: the Fibrillation Registry Assessing Costs, Therapies, Adverse events and Lifestyle (FRACTAL) study. Am Heart J 2006;152:1097–103.

[76] Spragg DD, Dalal D, Cheema A, Scherr D, Chilukuri K, Cheng A, et al. Complications of catheter ablation for atrial fibrillation: incidence and predictors. J Cardiovasc Electrophysiol 2008;19:627–31.

[77] Shah RU, Freeman JV, Shilane D, Wang PJ, Go AS, Hlatky MA. Procedural complications, rehospitalizations, and repeat procedures after catheter ablation for atrial fibrillation. J Am Coll Cardiol 2012;59:143–9.

[78] Elayi CS, Darrat Y, Suffredini JM, Misumida N, Shah J, Morales G, et al. Sex differences in complications of catheter ablation for atrial fibrillation: results on 85,977 patients. J Interv Card Electrophysiol 2018;53:333–9.

[79] Michowitz Y, Rahkovich M, Oral H, Zado ES, Tilz R, John S, et al. Effects of sex on the incidence of cardiac tamponade after catheter ablation of atrial fibrillation: results from a worldwide survey in 34 943 atrial fibrillation ablation procedures. Circ Arrhythm Electrophysiol 2014;7:274–80.

[80] Cheng EY, Kong MH. Gender differences of thromboembolic events in atrial fibrillation. Am J Cardiol 2016;117:1021–7.

[81] Wagstaff AJ, Overvad TF, Lip GYH, Lane DA. Is female sex a risk factor for stroke and thromboembolism in patients with atrial fibrillation? A systematic review and meta-analysis. QJM 2014;107:955–67.

[82] Marzona I, Proietti M, Farcomeni A, Romiti GF, Romanazzi I, Raparelli V, et al. Sex differences in stroke and major adverse clinical events in patients with atrial fibrillation: a systematic review and meta-analysis of 993,600 patients. Int J Cardiol 2018;269:182–91.

[83] Raccah BH, Perlman A, Zwas DR, Hochberg-Klein S, Masarwa R, Muszkat M, et al. Gender differences in efficacy and safety of direct oral anticoagulants in atrial fibrillation: systematic review and network meta-analysis. Ann Pharmacother 2018. 1060028018771264.

[84] Bassand J-P, Accetta G, Al Mahmeed W, Corbalan R, Eikelboom J, Fitzmaurice DA, et al. Risk factors for death, stroke, and bleeding in 28,628 patients from the GARFIELD-AF registry: rationale for comprehensive management of atrial fibrillation. PLoS One 2018;13:e0191592.

[85] Camm AJ, Savelieva I. Female gender as a risk factor for stroke associated with atrial fibrillation. Eur Heart J 2017;38:1480–4.

[86] Lip GYH, Nieuwlaat R, Pisters R, Lane DA, Crijns HJGM. Refining clinical risk stratification for predicting stroke and thromboembolism in atrial fibrillation using a novel risk factor-based approach: the euro heart survey on atrial fibrillation. Chest 2010;137:263–72.

[87] Kirchhof P, Benussi S, Kotecha D, Ahlsson A, Atar D, Casadei B, et al. 2016 ESC guidelines for the management of atrial fibrillation developed in collaboration with EACTS. Kardiol Pol 2016;74:1359–469.

[88] Nielsen PB, Skjoth F, Overvad TF, Larsen TB, Lip GYH. Female sex is a risk modifier rather than a risk factor for stroke in atrial fibrillation: should we use a CHA2DS2-VA score rather than CHA2DS2-VASc? Circulation 2018;137:832–40.

[89] Bai CJ, Volgman AS. Editorial commentary: sex, strokes and atrial fibrillation. Trends Cardiovasc Med 2019;29:153–4.

[90] Mtwesi V, Amit G. Stroke prevention in atrial fibrillation: the role of oral anticoagulation. Med Clin North Am 2019;103:847–62.

[91] Martin RC, Burgin WS, Schabath MB, Kirby B, Chae SH, Fradley MG, et al. Gender-specific differences for risk of disability and death in atrial fibrillation-related stroke. Am J Cardiol 2017;119:256–61.

[92] Ball J, Carrington MJ, Stewart S, SAFETY Investigators. Mild cognitive impairment in high-risk patients with chronic atrial fibrillation: a forgotten component of clinical management? Heart 2013;99:542–7.

[93] Wei Y-C, George NI, Chang C-W, Hicks KA. Assessing sex differences in the risk of cardiovascular disease and mortality per increment in systolic blood pressure: a systematic review and meta-analysis of follow-up studies in the United States. PLoS One 2017;12:e0170218.

[94] Poorthuis MHP, Algra AM, Algra A, Kappelle LJ, Klijn CJM. Female- and male-specific risk factors for stroke: a systematic review and meta-analysis. JAMA Neurol 2017;74:75–81.

[95] Renoux C, Coulombe J, Suissa S. Revisiting sex differences in outcomes in non-valvular atrial fibrillation: a population-based cohort study. Eur Heart J 2017;38:1473–9.

[96] Hart RG, Pearce LA, Aguilar MI. Meta-analysis: antithrombotic therapy to prevent stroke in patients who have nonvalvular atrial fibrillation. Ann Intern Med 2007;146:857–67.

[97] Fang MC, Singer DE, Chang Y, Hylek EM, Henault LE, Jensvold NG, et al. Gender differences in the risk of ischemic stroke and peripheral embolism in atrial fibrillation: the anticoagulation and risk factors in atrial fibrillation (ATRIA) study. Circulation 2005;112:1687–91.

[98] Atrial Fibrillation Investigators. Risk factors for stroke and efficacy of antithrombotic therapy in atrial fibrillation. Analysis of pooled data from five randomized controlled trials. Arch Intern Med 1994;154:1449–57.

[99] Sullivan RM, Zhang J, Zamba G, Lip GYH, Olshansky B. Relation of gender-specific risk of ischemic stroke in patients with atrial fibrillation to differences in warfarin anticoagulation control (from AFFIRM). Am J Cardiol 2012;110:1799–802.

[100] Lip GYH, Rushton-Smith SK, Goldhaber SZ, Fitzmaurice DA, Mantovani LG, Goto S, et al. Does sex affect anticoagulant use for stroke prevention in nonvalvular atrial fibrillation? The prospective global anticoagulant registry in the FIELD-atrial fibrillation. Circ Cardiovasc Qual Outcomes 2015;8:S12–20.

[101] Loikas D, Forslund T, Wettermark B, Schenck-Gustafsson K, Hjemdahl P, von Euler M. Sex and gender differences in thromboprophylactic treatment of patients with atrial fibrillation after the introduction of non-vitamin K oral anticoagulants. Am J Cardiol 2017;120:1302–8.

[102] Steinberg BA, Gao H, Shrader P, Pieper K, Thomas L, Camm AJ, et al. International trends in clinical characteristics and oral anticoagulation treatment for patients with atrial fibrillation: results from the GARFIELD-AF, ORBIT-AF I, and ORBIT-AF II registries. Am Heart J 2017;194:132–40.

[103] Yao X, Gersh BJ, Sangaralingham LR, Kent DM, Shah ND, Abraham NS, et al. Comparison of the CHA2DS2-VASc, CHADS2, HAS-BLED, ORBIT, and ATRIA risk scores in predicting non-vitamin K antagonist oral anticoagulants-associated bleeding in patients with atrial fibrillation. Am J Cardiol 2017;120:1549–56.

[104] Pancholy SB, Sharma PS, Pancholy DS, Patel TM, Callans DJ, Marchlinski FE. Meta-analysis of gender differences in residual stroke risk and major bleeding in patients with nonvalvular atrial fibrillation treated with oral anticoagulants. Am J Cardiol 2014;113:485–90.

[105] Proietti M, Cheli P, Basili S, Mazurek M, Lip GYH. Balancing thromboembolic and bleeding risk with non-vitamin K antagonist oral anticoagulants (NOACs): a systematic review and meta-analysis on gender differences. Pharmacol Res 2017;117:274–82.

[106] Moseley A, Doukky R, Williams KA, Jaffer AK, Volgman AS. Indirect comparison of novel oral anticoagulants in women with nonvalvular atrial fibrillation. J Women's Health 2017;26:214–21.

[107] Blackshear JL, Odell JA. Appendage obliteration to reduce stroke in cardiac surgical patients with atrial fibrillation. Ann Thorac Surg 1996;61:755–9.

[108] Casu G, Gulizia MM, Molon G, Mazzone P, Audo A, Casolo G, et al. ANMCO/AIAC/SICI-GISE/SIC/SICCH consensus document: percutaneous occlusion of the left atrial appendage in non-valvular atrial fibrillation patients: indications, patient selection, staff skills, organisation, and training. Eur Heart J Suppl 2017;19:D333–53.

[109] Holmes Jr DR, Doshi SK, Kar S, Price MJ, Sanchez JM, Sievert H, et al. Left atrial appendage closure as an alternative to warfarin for stroke prevention in atrial fibrillation: a patient-level meta-analysis. J Am Coll Cardiol 2015;65:2614–23.

[110] Kaneko H, Neuss M, Weissenborn J, Butter C. Predictors of thrombus formation after percutaneous left atrial appendage closure using the WATCHMAN device. Heart Vessel 2017;32:1137–43.

[111] Dukkipati SR, Kar S, Holmes DR, Doshi SK, Swarup V, Gibson DN, et al. Device-related thrombus after left atrial appendage closure: incidence, predictors, and outcomes. Circulation 2018;138:874–85.

[112] Saw J, Tzikas A, Shakir S, Gafoor S, Omran H, Nielsen-Kudsk JE, et al. Incidence and clinical impact of device-associated thrombus and peri-device leak following left atrial appendage closure with the Amplatzer cardiac plug. J Am Coll Cardiol Intv 2017;10:391–9.

Editor's Summary: Atrial Fibrillation

Epidemiology

- Incidence & prevalence are increasing
- Incidence in men occurs at a younger age than women
- Incidence/prevalence increases with age in both sexes
- Lifetime incidence of AF is similar between the sexes as women tend to live longer
- People of European ancestry & higher-income countries have a higher prevalence of AF

Pathophysiology

- Sex hormones play an important role in sex-related differences of AF mgmt
- Men have larger LA
- Women tend to have more atrial fibrosis associated with a higher prevalence of stroke and systemic embolism

Presentation

- Women have longer duration of AF symptoms
- Women experience more atypical symptoms
- Women experience more significant functional impairment and report worse quality of life
- Women have higher heart rates
- Women have greater proportion of AF recurrences

Access to Treatment

- Some diagnostic tests are less likely to be performed in women
- Women are more likely to receive rate control vs. rhythm control
- Women are equally likely to get AADs as men
- Women are less likely to undergo electrical cardioversion
- Women are less likely to be referred for catheter ablations
- Women have higher rates of complications with ablations

Prognosis

- Mortality risk is variable worldwide
- Women may have a higher mortality risk than men
- Women with AF and HFpEF have worse outcomes
- Women more likely to experience adverse drug reactions and side effects from AADs
- Women treated with rhythm control have worse outcome compared to men
- Women more likely to develop sick sinus syndrome requiring pacemaker with rate control strategy

Risk of Thromboembolism

- Female sex is associated with higher risk of stroke in AF
- Women suffer more from disabling stroke
- Women with AF are older than men & have more comorbidities which may explain increased stroke risk
- Women have more strokes on warfarin than men
- DOACs are equally safe and effective in men and women
- Therapeutic anticoagulation may be used less often in women despite a higher risk of thromboembolic events

Sex Specific Differences in Atrial Fibrillation. Sex-differences in epidemiology, pathophysiology, presentation, access to treatment, prognosis and risk of thromboembolism are highlighted. AAD, anti-arrhythmic drugs; AF, atrial fibrillation; DOACs, direct oral anticoagulants; HFpEF, heart failure with preserved ejection fraction; LA, left atrium.

Chapter 17

Ventricular Arrhythmias

Niti R. Aggarwal and Andrea M. Russo

Clinical Case

A 52-year-old female with hypertension, treated with hydrochlorothiazide 25 mg daily and a 1-week history of intermittent palpitations, suddenly collapsed at home and experienced an out-of-hospital cardiac arrest, witnessed by her husband. He promptly started cardiopulmonary resuscitation (CPR) after calling 911. On arrival, the paramedic found her to be in an accelerated idioventricular rhythm without a palpable pulse. Following initial resuscitation, she was found to have ventricular fibrillation and was successfully resuscitated after two shocks and four rounds of CPR. On presentation to the hospital, electrocardiogram (ECG) demonstrated normal sinus rhythm with normal QRS duration and normal QTc, with nonspecific T-wave abnormalities. Echocardiogram revealed global hypokinesis with an ejection fraction of 45%. Laboratories revealed a low-normal serum K^+ of 3.5 and otherwise within normal limits. Serial cardiac troponins were unremarkable. Coronary angiography revealed mild nonobstructive coronary atherosclerosis. How would you recommend managing her next? What features of sex differences in arrhythmias and presentation of out-of-hospital cardiac arrest are highlighted in this situation?

Abstract

There are sex differences in epidemiology and presentation of ventricular arrhythmias and sudden cardiac arrest. Sudden cardiac arrest is less common in women than in men. Women are less likely to present with ventricular fibrillation or ventricular tachycardia and more likely to present with pulseless electrical activity or asystole than men at the time of out-of-hospital arrest. Women receiving an implantable cardioverter defibrillator (ICD) for primary prevention are less likely to experience ventricular arrhythmias during follow-up than men. The pathophysiological basis for sex differences in arrhythmias is poorly understood and requires further investigation.

Sex Differences in Cardiac Disease. https://doi.org/10.1016/B978-0-12-819369-3.00024-1

Sudden Cardiac Death

Sudden cardiac death (SCD) represents a major health problem and is thought to account for approximately 50% of all deaths due to cardiovascular disease [1]. There are wide variations in the reported incidence of SCD likely due to differences in definition, various data sources, year of study, as well as differences in treatment during and after the arrest. In the United States, the estimated annual incidence of sudden cardiac arrest (SCA) varies widely from 180,000 to >450,000 [2–4].

Between 2004 and 2016, the incidence of SCD in Multnomah County Oregon in adults demonstrated a U-shaped pattern, with a nadir in 2011 in both men and women [5].

Women represent 19–42% of SCA victims in various observational studies [6]. Sex differences in the epidemiology and presentation of SCD have been described.

Epidemiology and Predictors

Multiple studies have demonstrated that the incidence of SCD is considerably lower in women than in men [7–9]. The absolute incidence of SCD increases with age in both men and women and parallels the increase in incidence of ischemic heart disease [7]. Women have a lower lifetime risk of SCD than men at any given age group, with a 10–20 year lag in women (**Figure 1**) [7, 8]. In the Framingham Heart Study, lifetime risk of SCD indexed at age 45 was 10.9% for men and 2.8% for women [8].

Women who suffer a cardiac arrest are typically older [9–12], are less likely to have underlying coronary artery disease (CAD) [10, 13], and are more likely to have other forms of heart disease or structurally normal hearts [13] than men. Among 355 consecutive out-of-hospital cardiac arrest survivors, women were significantly less likely to have CAD compared to men (45% vs. 80%), and more likely to have nonischemic heart disease ($P<0.001$) (**Figure 2**) [13]. In the Oregon Sudden Unexpected Death Study, women were less likely to have a diagnosis of structural heart disease, including CAD or severe left ventricular (LV) dysfunction, recognized prior to cardiac arrest, than men [10]. In a study from Northern Finland that collected clinical and autopsy data from subjects who suffered SCD between 1998 and 2017, the most frequently identified cause of death was ischemic heart disease in both men and women, although men were still more likely than women to have CAD and women more often had nonischemic etiologies [9]. Women were more likely to have no macroscopic findings and normal histology compared to men (0.6% vs. 0.2%). In addition, women also had a lower prevalence of any degree of myocardial fibrosis compared to men (89.3% vs. 92.4%) (**Figure 3**).

The predictors of outcome differ between men and women survivors of cardiac arrest. CAD status was the most important predictor of mortality in women while impaired left ventricular function was the most important predictor in men [13]. Sex differences in risk factors for SCA may have important clinical implications. As SCD is more likely to occur without known prior CAD and with a higher

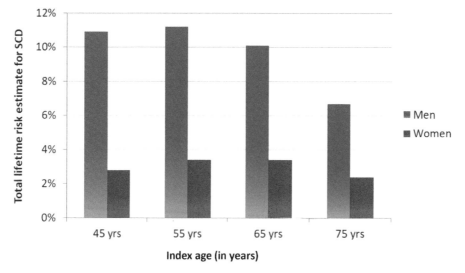

FIGURE 1   Lifetime Risk Estimate for Sudden Cardiac Death Stratified by Sex for Selected Index Ages. Women have a lower lifetime risk of SCD than men at any given age group. SCD, sudden cardiac death. *Data from [8].*

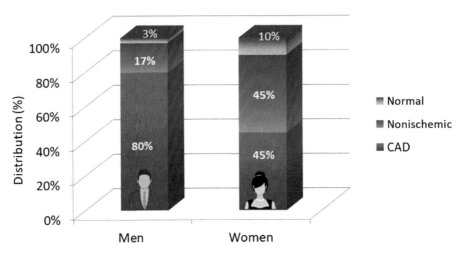

FIGURE 2   Underlying Pathology in Survivors of Sudden Cardiac Arrest. Women were significantly less likely to have CAD compared to men (45% vs. 80%) and more likely to have nonischemic heart disease ($P < 0.001$). CAD, coronary artery disease. *Data from [13].*

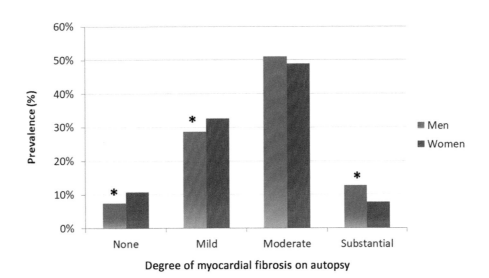

FIGURE 3   Degree of Fibrosis in Victims of SCD by Sex. On autopsy findings, women had a lower prevalence of myocardial fibrosis compared to men. *$P < 0.05$. *Data from [9].*

left ventricular ejection fraction (LVEF) in women than in men, primary prevention modalities to prevent SCA may be more difficult to identify in women based on current risk stratification criteria. While improved methods for risk stratification to prevent SCA are important in both men and women, this suggests it may have even greater importance for women.

In the Northern Finland study examining data from subjects with SCD between 1998 and 2017, in women were more likely to have a prior normal ECG than men (22.2% in

women vs. 15.3% in men, $P < 0.001$) [9]. However, women were more likely to have an increased marker for SCD risk based on ECG criteria for left ventricular hypertrophy with repolarization abnormalities (8.2% in women vs. 4.9% in men, $P = 0.036$) [9].

SCD may be the first manifestation of heart disease. This is true for 69% of women who suffered SCD without any previously diagnosed cardiac disease prior to death [14]. However, in women as in men, coronary heart disease risk factors predict risk of SCD [14]. At least one coronary heart disease risk factor was reported in 94% of women who died suddenly. Smoking, hypertension, and diabetes conferred a markedly elevated (2.5- to 4.0-fold) risk of SCD, while family history of myocardial infarction before 60 years old and obesity were associated with moderate (1.6-fold) elevations in risk [14]. This implies that the prevention of atherosclerosis may help to reduce the incidence of SCD in women.

Sex Differences in Presenting Arrhythmia

There are sex differences in presenting rhythm noted at the time of SCA. Women are less likely to present with ventricular fibrillation (VF)/ventricular tachycardia (VT) and more likely to present with pulseless electrical activity (PEA) or asystole than men at the time of out-of-hospital SCA [10–12, 15] (**Table 1** and **Figure 4**). Women are more likely to present with cardiac arrest at home, while men are more likely to present in a public location [6, 11, 16]. Several prognostic factors have been associated with successful resuscitation including initial shockable rhythm (VT/VF), witnessed arrest, and bystander CPR [17]. In a meta-analysis of SCA observational studies, women were more likely to present with SCA at home, were less likely to have a witnessed arrest, had a lower frequency of an initial shockable rhythm, and were more likely to receive bystander CPR than men [6]. Despite findings that women are less likely to present with VF/VT, are older, and less likely to present in a public location, more women were success-

TABLE 1 Presenting Rhythm at the Time of Cardiac Arrest in Men and Women

Study	Sex	VT/VF (%)	Asystole (%)	PEA (%)	Unknown (%)	P-Value (Men vs. Women)
Wigginton et al. [15]	Men	41	37	18	4	<0.05
	Women	30	42	24	4	
Chugh et al. [10]	Men	47	27	24	2	<0.0001
	Women	33	35	28	4	
Teodorescu et al. [11]	Men	55	21	24		<0.0001
	Women	38	31	31		
McLaughlin et al. [12][a]	Men	59			19	0.001
	Women	46			25	

[a]*Includes 56% in-hospital SCA.*
PEA, pulseless electrical activity; VF, ventricular fibrillation; VT ventricular tachycardia.

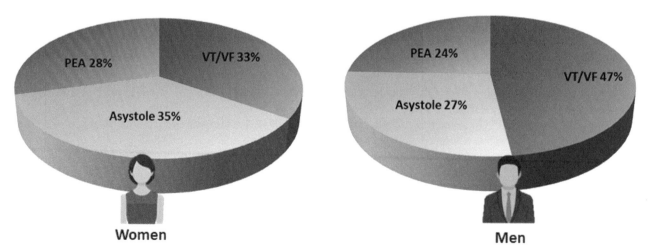

FIGURE 4   Sex Differences in Initial Presenting Rhythm During a Sudden Cardiac Arrest. Women are less likely to present with VF/VT and more likely to present with PEA or asystole than men at the time of out-of-hospital sudden cardiac arrest. $P<0.0001$. PEA, pulseless electrical activity; VF, ventricular fibrillation; VT ventricular tachycardia. *Data from [10].*

	OR (95% CI)	P
Johnson et al.	1.23 (1.09−1.38)	0.001
Teodorescu et al.	1.85 (1.12−3.05)	0.016
Adielsson et al.	1.43 (1.17−1.75)	<0.001
Akahane et al.	1.06 (1.02−1.10)	0.002
Bray et al.	1.11 (0.92−1.33)	0.27
Kitamura et al.	1.19 (1.03−1.37)	0.002
Ahn et al.	0.82 (0.64−1.05)	0.12
Cline et al.	0.29 (0.08−1.03)	0.06
Kim et al.	1.09 (0.93−1.27)	0.28
Arrich et al.	0.91 (0.60−1.38)	0.66
Mahapatra et al.	0.87 (0.43−1.75)	0.70
Pell et al.	1.04 (0.87−1.24)	0.66
Wissenberg et al.	1.10 (0.92−1.33)	0.31
Overall	**1.1 (1.03 − 1.20)**	**0.006**

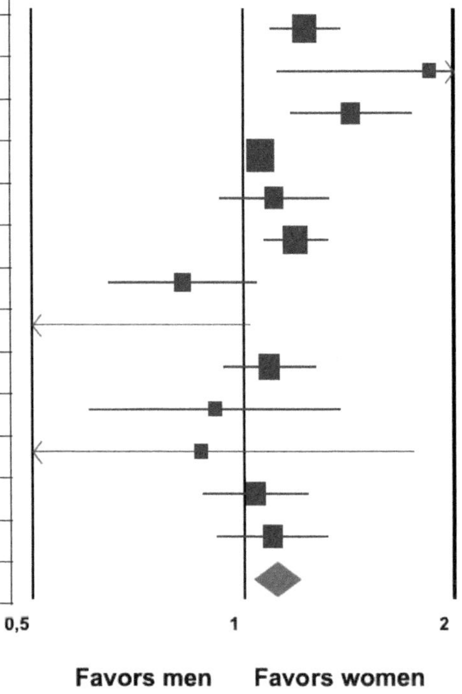

(I2 = 61%, Egger regression test for publication bias, P = .78)

Favors men   Favors women

FIGURE 5   Meta-Analysis of the Association Between Sex and Survival After SCA at or After Hospital Discharge (13 Observational Studies, N=409,323). CI, confidence interval; OR, odds ratio. *Reproduced with permission from [6].*

fully resuscitated and were more likely to survive to hospital discharge than men [6, 11, 15] (**Figure 5**). In a large consortium evaluating prehospital resuscitation in 15,584 patients, intervals from emergency medical services (EMS) dispatch to first rhythm capture and first EMS CPR were longer in women than in men, and women were less likely to receive successful intravenous or intraosseous access or certain drugs (including adrenaline, atropine, lidocaine, or amiodarone), even after adjusting for other factors such as location of arrest and whether or not it was witnessed [16].

Ventricular Arrhythmias Associated With Structural Heart Disease

Ischemic and Nonischemic Cardiomyopathy

The most common cause of ventricular arrhythmias associated with structural heart disease is related to scar in the setting of ischemic cardiomyopathy and prior myocardial infarction (MI). Patients with ischemic heart disease are more likely to have inducible ventricular arrhythmias at electrophysiological (EP) study. Studies have demonstrated that women survivors of cardiac arrest were less likely to have inducible sustained ventricular arrhythmias at EP study compared to men [18]. In fact, multivariate analysis revealed that only male sex and prior MI predicted inducibility in these survivors of cardiac arrest [18]. In MUSTT,

a primary prevention study in patients with ischemic cardiomyopathy and prior MI with LVEF <40%, women were also less likely to have inducible sustained ventricular VT than men (36% of men vs. 24% of women, P=0.001), and inducibility was a criterion for enrollment in the randomized arm of the trial [19]. This suggests the presence of sex differences in arrhythmia substrate, although reasons for lower susceptibility of women with structural heart disease to ventricular arrhythmias remain uncertain.

Women are underrepresented in randomized trials evaluating the role of catheter ablation for the treatment of VT post-MI, representing only 6–13% of those studied [20–22]. This is at least in part due to the prevalence of CAD in men vs. women. In another study that included patients with sustained VT and nonischemic cardiomyopathy, 22% of that cohort were women [23]. In a large international VT ablation collaborative group study that included 2062 patients with various types of structural heart disease, only 13% were women [24].

Results of studies evaluating sex differences in outcomes following VT ablation in the setting of structural heart disease show variable results. In a large international VT ablation collaborative study in patients with structural heart disease, women had higher rates of 1-year VT recurrence following ablation than men (30.5% vs. 25.3%, P=0.03) [24]. This difference in arrhythmia outcome was only partially explained by higher prevalence of nonischemic cardiomyopathy among women. In contrast, another large study showed no difference in acute success or rates of VT recurrence in men vs. women with structural heart disease [including ischemic cardiomyopathy, nonischemic cardiomyopathy, or arrhythmogenic right ventricular (RV) dysplasia] following VT ablation [23]. Complication rates did not differ between men and women [23, 24].

Arrhythmogenic Right Ventricular Cardiomyopathy

Arrhythmogenic right ventricular cardiomyopathy (ARVC) is an inherited cardiomyopathy that is characterized by fibrofatty replacement in the right ventricle. Studies examining sex-based differences in incidence of ARVC are conflicting, with some suggesting a male predominance [25, 26] and others showing no clear sex predominance [27–29]. Men have a higher incidence of sustained VT/VF or SCA as an initial manifestation and have a higher risk of developing VT/VF than women [27, 30]. Women have a higher risk of heart failure death or heart transplantation [30]. Ventricular arrhythmias were more often inducible in men with ARVC [29]. Electroanatomic mapping demonstrated that men with ARVC have a larger epicardial RV unipolar low-voltage zone, a larger area with late potentials, and longer duration of local ventricular electrograms [27].While there was no difference in acute procedural success (reported 100% acute success in men and women), men were more likely to develop recurrent ventricular arrhythmias after ablation than women [27]. Male sex and the presence of a larger area of abnormal electrograms independently predicted ventricular arrhythmia recurrence after radiofrequency catheter ablation [27].

Hypertrophic Cardiomyopathy

Despite the autosomal dominant inheritance, male predominance of hypertrophic cardiomyopathy (HCM) has been described, with men comprising approximately 55–72% of HCM published cohorts [31–35]. Women with HCM were older at presentation, more symptomatic, more likely to develop advanced heart failure, had more obstructive physiology, more mitral regurgitation, and may be referred to HCM centers later than men [31, 34, 35]. Women also had worse exercise performance than men [33, 35]. In patients with prophylactic ICDs, there was no difference in appropriate shock therapy between men and women [31]. Rates of SCD, out-of-hospital cardiac arrest, and appropriate ICD therapies for VT/VF, as well as overall mortality, were similar in men and women [31]. In contrast, other studies report a greater risk of HCM-related cardiovascular events and higher mortality in women, even after adjusting for other risk factors

[33, 35]. At the time of referral for myectomy, women were older than men and had more advanced diastolic dysfunction and had a larger degree of interstitial fibrosis than men [32]. This suggests the importance of increased recognition and earlier diagnosis of HCM in women to help provide therapy that may mitigate disease progression.

Ventricular Arrhythmias in the Absence of Underlying Structural Heart Disease

Idiopathic VT

Ventricular arrhythmias can occur in the absence of underlying structural heart disease, typically associated with a benign course and rarely associated with SCD. Idiopathic arrhythmias can arise in the right ventricular outflow tract (RVOT), left ventricular outflow tract (LVOT), LV septum, or other locations. In a review of 748 patients, RVOT-VT was more common in women than in men (male/female ratio 0.49) and LV septal VT was more common in men than in women (male/female ratio 3.37) [36]. LVOT-VT was slightly more common in men than in women (male/female ratio 1.38 or 1.37) [36, 37]. Sex differences in triggers for RVOT-VT have also been reported. The most common triggers for RVOT-VT in women were states of hormonal flux (premenstrual, gestational, or perimenopausal states, or administration of oral contraceptives), while men were more likely to report arrhythmia triggered by exercise, stress, or caffeine [38]. After adjustment for the site of ventricular arrhythmia origin and other baseline differences, female sex did not impact on success of radiofrequency catheter ablation of idiopathic VT [38].

Channelopathies

In congenital long QT syndrome (LQTS), there is a higher female prevalence of disease [39]. The risk of cardiac events (syncope, cardiac arrest, or SCD) is higher in males before puberty and higher in females during adulthood in LQTS [40]. Sex differences in timing of events are likely related to known shortening of QT intervals seen after puberty in men. Acquired LQTS, which can be seen with medications or electrolyte abnormalities, is also more common in women. Women are more likely to develop torsade de pointes following administration of drugs that prolong the QT interval [41].

Brugada syndrome occurs more often in men than in women [42]. Women are less likely to show a spontaneous Type 1 Brugada ECG pattern [42, 43] and are less likely to have arrhythmia identified at electrophysiology study [42, 43]. SCD or VF is more common in men than in women with Brugada syndrome [43]. An SCN5A mutation was more often found in women than in men [42].

Potential Mechanisms for Sex Differences in Sudden Cardiac Death and Ventricular Arrhythmias

Reasons for sex differences in the epidemiology of SCD or differences in ventricular arrhythmia occurrence are poorly understood. Sex differences in hormones, autonomic function, cardiac size, or arrhythmia substrate might contribute to a lower overall arrhythmic risk of sudden death in women. Female sex hormones appear protective in CAD, and the later occurrence of SCD in women seems to parallel the incidence of CAD with a 10–20-year lag in women. The role of sex hormones in ventricular arrhythmias remains unclear.

To investigate the relationship between sex hormones and idiopathic outflow tract ventricular arrhythmias, the effect of estrogen therapy in postmenopausal females was evaluated [44]. The concentration of estradiol in postmenopausal women with idiopathic outflow tract ventricular arrhythmias was significantly lower than control postmenopausal women. Following 3 months of estrogen replacement therapy, ventricular arrhythmia burden was significantly lower than before therapy, suggesting that hormonal replacement therapy may inhibit these idiopathic arrhythmias [44].

Sex hormones may play a role in arrhythmias related to LQTS. Estradiol prolongs repolarization and increases the QT interval, potentially increasing the risk for torsade de pointes in patients with congenital and acquired LQTS [45]. In contrast, testosterone and progesterone shorten QT and exert an antiarrhythmic effect with a reduced susceptibility to sympathetic stimuli [45].

Significant differences in sex hormones have also been identified in patients who suffered SCA compared to controls [46]. Higher testosterone levels were associated with a lower risk of cardiac events in men while higher estradiol levels were associated with an increased risk of cardiac arrest in both men and women (**Figure 6**) [46].

In patients with ARVC, serum levels of sex hormones have been associated with major arrhythmic cardiovascular events. While the disease is transmitted as an autosomal dominant trait and not sex linked, men with ARVC have been reported to develop ventricular arrhythmias at an earlier age than women, suggesting the possibility that sex hormones may influence the onset and progression of the disease. Testosterone levels were significantly higher in men with ARVC who experienced major arrhythmic cardiovascular events when compared to men who had favorable outcomes [47]. In women, estradiol was significantly lower in those who experience major arrhythmic cardiovascular events compared to women who had favorable outcomes [47].

Implantable Cardioverter Defibrillator Therapy and Ventricular Arrhythmias

Multiple randomized clinical trials have demonstrated the efficacy of the implantable cardioverter defibrillator (ICD) for the primary [48–52] and secondary prevention [53, 54] of SCD. However, women have been underrepresented in clinical trials and the benefit of ICD therapy has been questioned. Potential sex differences in ICD therapy, complications, and outcomes have been described (**Figure 7**).

Secondary Prevention

While population studies have demonstrated sex differences in the incidence of SCD, most data suggest no significant sex differences in arrhythmia occurrence or appropriate ICD therapy in patients presenting with sustained ventricular arrhythmias or syncope with inducible sustained ventricular arrhythmias [55–58]. In contrast, one study showed that female sex was an independent clinical predictor of shock therapy in cardiac arrest survivors [59]. In the antiarrhythmics versus implantable defibrillators (AVID) trial, women were younger, less often had CAD, more often had a nonischemic cardiomyopathy, and more often had VF rather than VT as the index arrhythmia compared with men [60]. Despite sex differences in baseline characteristics, there was no significant difference in ICD implantation rate with a similar 1-year mortality in women and men in this trial. In an analysis of a Medicare cohort, women were less likely to receive ICD therapy for a secondary prevention indication, but for those who did receive devices, the mortality benefit was significant for women as well as men [61].

Primary Prevention

Multiple randomized clinical trials have demonstrated the benefit of prophylactic ICD therapy in reducing mortality or SCD risk [48–52]. Women have been underrepresented in randomized primary prevention trials evaluating ICD therapy, representing only 8–28% of patients receiving devices in these trials (**Table 2**). Reasons for the underrepresentation of women enrolled in these trials are likely due to multiple factors, including sex differences in prevalence of ischemic heart disease, as some trials enrolled only subjects with CAD, as well as sex differences in arrhythmia inducibility in earlier trials that required EP testing and inducible ventricular arrhythmias for eligibility. Selection bias or patient refusal cannot be excluded.

In the multicenter clinical trials evaluating ICD therapy for primary prevention, sex differences in baseline characteristics have been identified [19, 62, 64]. In the Multicenter

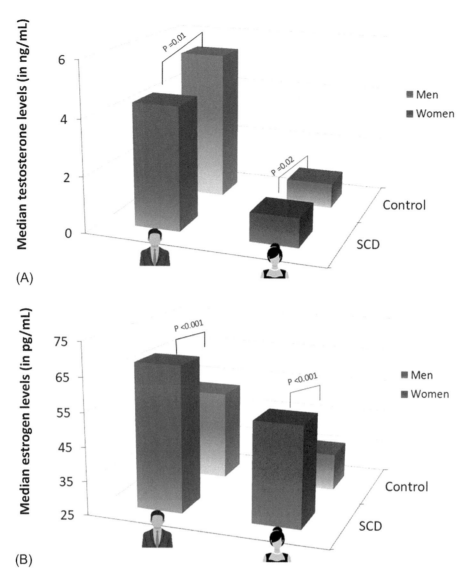

(A)

(B)

FIGURE 6 **Sex Differences in Estrogen and Testosterone Levels in Patients With Sudden Cardiac Arrest.** Higher testosterone levels were associated with a lower risk of cardiac events in men, with no difference in testosterone levels in women victims of SCA compared to controls (panel A). In contrast, higher estradiol levels were associated with an increased risk of cardiac arrest in both men and women compared to controls (panel B). SCD, sudden cardiac death. *Data from [46].*

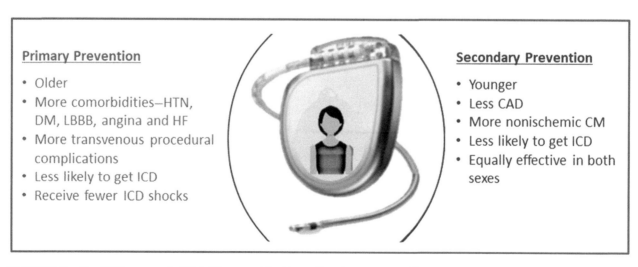

FIGURE 7 **Sex Differences in ICD Therapy.** The figure lists key sex-specific features of ICD therapy for primary and secondary prevention of sudden cardiac death in women, compared to men. CAD, coronary artery disease; CM, cardiomyopathy; DM, diabetes; HF, heart failure; HTN, hypertension; ICD, implantable cardioverter defibrillator; LBBB, left bundle branch block.

TABLE 2 Primary Prevention ICD Trials

Study	Year	Total Patients	Ischemic or Nonischemic	LVEF (%)	NYHA Class	% Women (Control)	% Women (ICD)	# Women (ICDs)	Follow-Up (Months)	HR for Death (95% CI)
MADIT [49]	1996	196	Ischemic	≤35	II–III	8	8	8	27	NA
MUSTT [19]	1999	704	Ischemic	≤40	I–III	10	10	18	39	Men vs. women: 1.51 (0.86, 2.64)
MADIT II [50,62]	2002	1232	Ischemic	≤30	I–III	15	16	119	20	Men: 0.66 (0.48–0.91)
										Women: 0.57 (0.28–1.18)
DEFINITE [52,63]	2004	458	Nonischemic	≤35	I–III	30	28	64	29	Men: 0.49 (0.27–0.90)
										Women: 1.14 (0.50–2.64)
SCD-HeFT [48,64]	2005	2521	Ischemic + nonischemic	≤35	II–III	23	23	185	45	Men: 0.71 (0.57, 0.88)
										Women: 0.90 (0.56, 1.43)

CI, confidence interval; HR, hazard ratio; ICD, implantable cardioverter defibrillator; LVEF, left ventricular ejection fraction; NA, not applicable; NYHA, New York Heart Association.

Unsustained Tachycardia Trial (MUSTT), women were older and more likely to have a history of heart failure, recent angina, or have suffered an infarction within 6 months than men [19]. In the Multicenter Automatic Defibrillator Implantation Trial II (MADIT II), women were more likely to have hypertension, diabetes, or LBBB, and less likely to have undergone prior CABG surgery [62]. In the Sudden Cardiac Death in Heart Failure Trial (SCD-HeFT), women were more likely to have class III heart failure and nonischemic heart disease [64]. Women were older and had more comorbidities than men when they presented for ICD implantation in a study that examined a Medicare population [61]. In this Medicare cohort, women were significantly less likely to receive ICD therapy for primary prevention of SCD than men [61].

The benefit of primary prevention ICD therapy has been questioned in women. **Table 2** outlines enrollment criteria, representation of women, and outcomes in the randomized trials [48–52]. In a subgroup analysis of the MUSTT, there were no significant sex influences on risk of arrhythmic death or cardiac arrest or overall mortality; however, very few women were enrolled [19]. In the MADIT II, there was no significant interaction between sex, mortality, and ICD therapy, suggesting similar effectiveness of ICD therapy in men and women [62]. In the Defibrillators In Nonischemic Cardiomyopathy Treatment Evaluation (DEFINITE), a reduction in all-cause mortality was seen in men but not in women [52, 63].

In the SCD-HeFT, which randomized therapy to ICD vs. amiodarone vs. placebo, overall mortality risk was lower in women than in men after adjusting for differences in baseline characteristics [48, 64] (**Figure 8**). When examining the different randomized groups, a sex difference in overall mortality was seen in the placebo group; there was no sex difference in overall mortality in the ICD group. While this suggests that women may have a smaller ICD benefit than men, the test for an interaction between sex and therapy was not significant. It is possible that lower overall mortality risk of women in the placebo group and the small number of women enrolled may make treatment differences in women more difficult to detect.

Caution is needed when interpreting post-hoc analyses, as trials enrolled small numbers of women and were not adequately powered to detect sex differences in outcomes. A meta-analysis evaluated the impact ICD therapy in 7229 patients from five primary prevention ICD trials, including MADIT II, MUSTT, SCD-HeFT, DEFINITE, and COMPANION [Comparison of Medical Therapy, Pacing, and Defibrillation in Heart Failure, which included cardiac resynchronization therapy (CRT)] trials, and women represented 22% of subjects in this analysis [65]. After adjusting for baseline characteristics, there was no significant

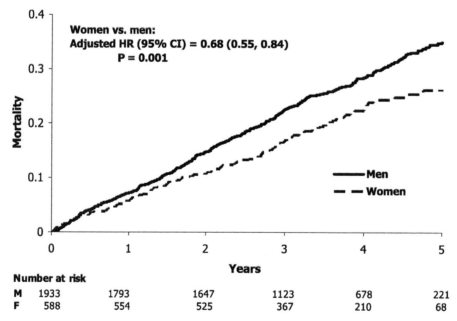

FIGURE 8 Mortality in SCD-HeFT. The figure includes subjects enrolled in all three arms of the trial, including placebo, amiodarone, and ICD therapy groups. Overall mortality was lower in women than in men ($P=0.001$) in SCD-HeFT. This lower mortality was seen in women in the placebo group with a trend in the amiodarone group. However, there was no difference in mortality between men and women in the ICD group. CI, confidence interval; HR, hazard ratio. *Reproduced with permission from [64].*

difference in overall mortality in women compared to men (**Figure 5**). While primary prevention ICD implantation significantly reduced mortality in men (HR 0.67, 95% CI 0.58–0.78, $P < 0.001$), there was no significant reduction in mortality in women (HR 0.78, 95% CI 0.57–1.05, $P = 0.1$).

It is possible that this meta-analysis may still have been underpowered to demonstrate a benefit of ICD therapy in women, based on calculations performed in the post-hoc analysis of SCD-HeFT [64]. Alternatively, it is possible that there may be less benefit of ICD therapy in women compared to men. Presentation of women at an older age, at a different stage of disease with more competing comorbidities, or presence of unmeasured confounders, might also impact on outcome and benefit of ICD therapy. Based on what is known from the epidemiology of SCD, it is also possible that sex differences in arrhythmia substrate, hormones, or autonomic differences might contribute to a lower overall arrhythmic risk in women. A study that examined five randomized trials or registries in heart failure patients revealed that women with heart failure have a lower mortality rate than men, and fewer of those deaths are sudden [66]. This supports the possibility that sex differences in benefit from ICD therapy may exist.

Procedural outcomes have been examined in men and women receiving ICD therapy. Data from a large national registry demonstrate that women who receive transvenous ICDs for primary prevention experience worse outcomes than men, including higher rates of device-related complications, all-cause readmissions, and heart failure admissions after adjusting for differences in baseline characteristics and device type [67] (**Table 3**). Procedural complications, including pneumothorax, cardiac tamponade, and mechanical complications requiring revision, were more common in women [67].

Following primary prevention ICD implantation, sex differences in appropriate ICD therapy for ventricular arrhythmias have been examined. Analysis of individual primary prevention studies and registries reveals conflicting findings, including a tendency for women to have lower rates [62, 63, 68–72] or similar rates [64, 73] of appropriate ICD therapies during follow-up. In SCD-HeFT, there was no significant difference in the risk of appropriate shock therapy for women vs. men [64]. However, a meta-analysis of five randomized primary prevention trials (including MADIT II, MUSTT, SCD-HeFT, DEFINITE, and COMPANION) demonstrated that women received fewer appropriate ICD therapies for ventricular arrhythmias than men [65] (**Figure 9**). Another meta-analysis of 20 studies revealed that women were significantly less likely to receive appropriate ICD shock therapy than men (HR 0.62; 95% CI 0.44–0.88, $P = 0.001$) [74]. A sex difference in arrhythmic risk supports the concept that SCD may have a smaller impact on total mortality in women than in men.

TABLE 3 Relationship of Sex With Complications, Mortality, and Hospitalizations in Patients Receiving Primary Prevention ICDs in the National Cardiovascular Data Registry (NCDR) ICD Registry

Description	Complications, 30 or 90 d		Mortality, 6 mo		All-Cause Readmission, 6 mo		HF Readmission, 6 mo	
	OR	P	OR	P	OR	P	OR	P
Unadjusted	1.53 (1.40–1.68)	<0.001	1.18 (1.07–1.29)	0.001	1.27 (1.21–1.33)	<0.001	1.45 (1.35–1.55)	<0.001
Adjusted[a]	1.39 (1.26–1.53)	<0.001	1.08 (0.98-1.20)	0.123	1.22 (1.16-1.28)	<0.001	1.32 (1.23–1.42)	<0.001

[a]*Adjusted for demographics (age and admission reason), medical history, and risk factors (family history of sudden death, chronic heart failure, NYHA class—current status, atrial fibrillation/atrial flutter, nonsustained VT, nonischemic dilated cardiomyopathy, ischemic heart disease, previous MI, previous coronary artery bypass grafting (CABG), previous percutaneous coronary intervention, previous valvular surgery, cerebrovascular disease, chronic lung disease, diabetes, hypertension, renal failure-dialysis), diagnostics (LVEF, QRS, blood urea nitrogen), ICD type, and discharge medications.*
OR, odds ratio. *Reproduced with permission from [67].*

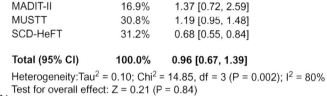
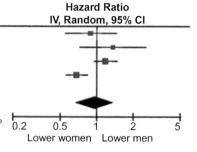

FIGURE 9 Meta-Analysis of Primary Prevention Trials. Hazard ratio of overall mortality (A) and appropriate ICD intervention (B) in men and women. CI, confidence interval; ICD, implantable cardioverter defibrillator. Square boxes denote hazard ratio; horizontal lines represent 95% confidence interval (CI). *Reproduced with permission from [65].*

Conclusion

There are sex differences in epidemiology and presentation of a variety of different ventricular arrhythmias and SCA. Studies suggest potentially less benefit of ICD therapy in women although even large multicenter primary prevention trials included only a relatively small number of women and were not powered to show the benefit of ICD therapy in women. In several studies, women experienced less appropriate ICD therapy for VT/VF than men, consistent with epidemiological findings of lower tachyarrhythmic mortality risk in women. The reasons for sex differences in ventricular arrhythmias and potential differences in the impact of device therapy on outcome are currently unclear and warrant further investigation.

Key Points

1 The incidence and prevalence of SCD are lower in women than in men.

2 Women are older at the time of SCD than men.

3 SCD in women is less likely to be associated with CAD and more likely to be associated with a nonischemic etiology than in men.

4 Women are less likely to present with ventricular fibrillation (VF)/ventricular tachycardia (VT) and more likely to present with pulseless electrical activity (PEA) or asystole than men at the time of out-of-hospital sudden cardiac arrest.

5 As SCD is more likely to occur without known prior CAD and with a higher LVEF in women than in men, primary prevention strategies to prevent SCA may be more difficult to identify in women based on current risk stratification criteria, and improved methods for risk stratification are needed.

6 Women receiving an ICD for primary prevention are less likely to experience ventricular arrhythmias than men.

7 Women are less likely to be referred for ICD therapy than men.

8 Women are more likely to experience a significant complication from transvenous ICD implantation than men.

9 Sex differences also exist in the incidence of idiopathic VT (i.e., ventricular arrhythmias in the absence of structural heart disease); RVOT-VT is twice as common in women than in men while LV septal VT is more common in men.

10 Women are underrepresented in studies evaluating catheter ablation for VT in the setting of structural heart disease.

11 Women are less likely to have inducible sustained ventricular tachycardia in the setting of CAD than men.

12 Variable outcomes with respect to VT recurrence after ablation have been reported, suggesting the possibility of higher recurrence rates or similar recurrence rates in women vs. men, with similar risks for complications in men and women.

13 The pathophysiological basis for sex differences in arrhythmias is poorly understood and requires further investigation.

Back to Clinical Case

The clinical case describes a 52-year-old female with witnessed out-of-hospital PEA cardiac arrest. As with this patient, women are more likely to experience PEA or asystole as the initial presenting rhythm, rather than ventricular tachycardia or fibrillation, which is more prevalent in men. They are also more likely to have a normal ECG and normal myocardial substrate without fibrosis, and less likely to have underlying coronary atherosclerosis. In this patient, she did undergo cardiac magnetic resonance imaging (MRI) to rule out structural heart disease and other causes of nonischemic cardiomyopathy. Cine images at the time of catheterization demonstrated mild left ventricular global hypokinesis with an ejection fraction of 47%, and normal right ventricular function. There was no evidence of myocardial fibrosis on late gadolinium enhancement imaging or T1 mapping sequences. She underwent ICD implantation for secondary prevention of sudden cardiac death. While women may experience more procedural complications with transvenous ICD implantation, ICD implantation for secondary prevention indications is supported by the American College of Cardiology (ACC)/American Heart Association (AHA)/Heart Rhythm Society (HRS) guidelines for ventricular arrhythmias regardless of sex [75].

References

[1] Huikuri HV, Castellanos A, Myerburg RJ. Sudden death due to cardiac arrhythmias. N Engl J Med 2001;345(20):1473–82.

[2] Kong MH, Fonarow GC, Peterson ED, Curtis AB, Hernandez AF, Sanders GD, et al. Systematic review of the incidence of sudden cardiac death in the United States. J Am Coll Cardiol 2011;57(7):794–801.

[3] Go AS, Mozaffarian D, Roger VL, Benjamin EJ, Berry JD, Blaha MJ, et al. Heart disease and stroke statistics—2014 update: a report from the American Heart Association. Circulation 2014;129(3):e28–e292.

[4] Rosamond W, Flegal K, Friday G, Furie K, Go A, Greenlund K, et al. Heart disease and stroke statistics—2007 update: a report from the American Heart Association Statistics Committee and Stroke Statistics Subcommittee. Circulation 2007;115(5):e69–e171.

[5] Reinier K, Stecker EC, Uy-Evanado A, Chugh HS, Binz A, Nakamura K, et al. Sudden cardiac death as first manifestation of heart disease in women: the Oregon Sudden Unexpected Death Study, 2004-2016. Circulation 2020;141(7):606–8.

[6] Bougouin W, Mustafic H, Marijon E, Murad MH, Dumas F, Barbouttis A, et al. Gender and survival after sudden cardiac arrest: a systematic review and meta-analysis. Resuscitation 2015;94:55–60.

[7] Kannel WB, Schatzkin A. Sudden death: lessons from subsets in population studies. J Am Coll Cardiol 1985;5(6 Suppl.):141B–9B.

[8] Bogle BM, Ning H, Mehrotra S, Goldberger JJ, Lloyd-Jones DM. Lifetime risk for sudden cardiac death in the community. J Am Heart Assoc 2016;5(7). pii: e002398.

[9] Haukilahti MAE, Holmström L, Vähätalo J, Kenttä T, Tikkanen J, Pakanen L, et al. Sudden cardiac death in women. Circulation 2019;139(8):1012–21.

[10] Chugh SS, Uy-Evanado A, Teodorescu C, Reinier K, Mariani R, Gunson K, et al. Women have a lower prevalence of structural heart disease as a precursor to sudden cardiac arrest: the Ore-SUDS (Oregon Sudden Unexpected Death Study). J Am Coll Cardiol 2009;54(22):2006–11.

[11] Teodorescu C, Reinier K, Uy-Evanado A, Ayala J, Mariani R, Wittwer L, et al. Survival advantage from ventricular fibrillation and pulseless electrical activity in women compared to men: the Oregon Sudden Unexpected Death Study. J Interv Card Electrophysiol 2012;34(3):219–25.

[12] McLaughlin TJ, Jain SK, Voigt AH, Wang NC, Saba S. Comparison of long-term survival following sudden cardiac arrest in men versus women. Am J Cardiol 2019;124(3):362–6.

[13] Albert CM, McGovern BA, Newell JB, Ruskin JN. Sex differences in cardiac arrest survivors. Circulation 1996;93(6):1170–6.

[14] Albert CM, Chae CU, Grodstein F, Rose LM, Rexrode KM, Ruskin JN, et al. Prospective study of sudden cardiac death among women in the United States. Circulation 2003;107(16):2096–101.

[15] Wigginton JG, Pepe PE, Bedolla JP, DeTamble LA, Atkins JM. Sex-related differences in the presentation and outcome of out-of-hospital cardiopulmonary arrest: a multiyear, prospective, population-based study. Crit Care Med 2002;30(4 Suppl.):S131–6.

[16] Mumma BE, Umarov T. Sex differences in the prehospital management of out-of-hospital cardiac arrest. Resuscitation 2016;105:161–4.

[17] Sasson C, Rogers MA, Dahl J, Kellermann AL. Predictors of survival from out-of-hospital cardiac arrest: a systematic review and meta-analysis. Circ Cardiovasc Qual Outcomes 2010;3(1):63–81.

[18] Freedman RA, Swerdlow CD, Soderholm-Difatte V, Mason JW. Clinical predictors of arrhythmia inducibility in survivors of cardiac arrest: importance of gender and prior myocardial infarction. J Am Coll Cardiol 1988;12:973–8.

[19] Russo AM, Stamato NJ, Lehmann MH, Hafley GE, Lee KL, Pieper K, et al. Influence of gender on arrhythmia characteristics and outcome in the Multicenter UnSustained Tachycardia Trial. J Cardiovasc Electrophysiol 2004;15(9):993–8.

[20] Reddy VY, Reynolds MR, Neuzil P, Richardson AW, Taborsky M, Jongnarangsin K, et al. Prophylactic catheter ablation for the prevention of defibrillator therapy. N Engl J Med 2007;357:2657–65.

[21] Kuck KH, Schaumann A, Eckardt L, Willems S, Ventura R, Delacrétaz E, et al. Catheter ablation of stable ventricular tachycardia before defibrillator implantation in patients with coronary heart disease (VTACH): a multicentre randomised controlled trial. Lancet 2010;375:31–40.

[22] Sapp JL, Wells GA, Parkash R, Stevenson WG, Blier L, Sarrazin JF, et al. Ventricular tachycardia ablation versus escalation of antiarrhythmic drugs. N Engl J Med 2016;375:111–21.

[23] Baldinger SH, Kumar S, Romero J, Fujii A, Epstein LM, Michaud GF, et al. A comparison of women and men undergoing catheter ablation for sustained monomorphic ventricular tachycardia. J Cardiovasc Electrophysiol 2017;28:201–7.

[24] Frankel DS, Tung R, Santangeli P, Tzou WS, Vaseghi M, Di Biase L, et al. Sex and catheter ablation for ventricular tachycardia: an international ventricular tachycardia ablation center collaborative group study. JAMA Cardiol 2016;1:938–44.

[25] Corrado D, Basso C, Thiene G, McKenna WJ, Davies MJ, Fontaliran F, et al. Spectrum of clinicopathologic manifestations of arrhythmogenic right ventricular cardiomyopathy/dysplasia: a multicenter study. J Am Coll Cardiol 1997;30:1512–20.

[26] Bauce B, Frigo G, Marcus FI, Basso C, Rampazzo A, Maddalena F, et al. Comparison of clinical features of arrhythmogenic right ventricular cardiomyopathy in men versus women. Am J Cardiol 2008;102:1252–7.

[27] Lin CY, Chung FP, Lin YJ, Chang SL, Lo LW, Hu YF, et al. Gender differences in patients with arrhythmogenic right ventricular dysplasia/cardiomyopathy: clinical manifestations, electrophysiological properties, substrate characteristics, and prognosis of radiofrequency catheter ablation. Int J Cardiol 2017;227:930–7.

[28] Bhonsale A, Groeneweg JA, James CA, Dooijes D, Tichnell C, Jongbloed JD, et al. Impact of genotype on clinical course in arrhythmogenic right ventricular dysplasia/cardiomyopathy-associated mutation carriers. Eur Heart J 2015;36:847–55.

[29] Choudhary N, Tompkins C, Polonsky B, McNitt S, Calkins H, Mark Estes 3rd NA, et al. Clinical presentation and outcomes by sex in arrhythmogenic right ventricular cardiomyopathy: findings from the North American ARVC registry. J Cardiovasc Electrophysiol 2016;27:555–62.

[30] Kimura Y, Noda T, Otsuka Y, Wada M, Nakajima I, Ishibashi K, et al. Potentially lethal ventricular arrhythmias and heart failure in arrhythmogenic right ventricular cardiomyopathy: what are the differences between men and women? JACC Clin Electrophysiol 2016;2:546–55.

[31] Rowin EJ, Maron MS, Wells S, Patel PP, Koethe BC, Maron BJ. Impact of sex on clinical course and survival in the contemporary treatment era for hypertrophic cardiomyopathy. J Am Heart Assoc 2019;8, e012041.

[32] Nijenkamp LLAM, Bollen IAE, van Velzen HG, Regan JA, van Slegtenhorst M, Niessen HWM, et al. Sex differences at the time of myectomy in hypertrophic cardiomyopathy. Circ Heart Fail 2018;11, e004133.

[33] Ghiselli L, Marchi A, Fumagalli C, Maurizi N, Oddo A, Pieri F, et al. Sex-related differences in exercise performance and outcome of patients with hypertrophic cardiomyopathy. Eur J Prev Cardiol 2019;7. 2047487319886961.

[34] Lu DY, Ventoulis I, Liu H, Kudchadkar SM, Greenland GV, Yalcin H, et al. Sex-specific cardiac phenotype and clinical outcomes in patients with hypertrophic cardiomyopathy. Am Heart J 2020;219:58–69.

[35] Geske JB, Ong KC, Siontis KC, Hebl VB, Ackerman MJ, Hodge DO, et al. Women with hypertrophic cardiomyopathy have worse survival. Eur Heart J 2017;38:3434–40.

[36] Nakagawa M, Takahashi N, Nobe S, Ichinose M, Ooie T, Yufu F, et al. Gender differences in various types of idiopathic ventricular tachycardia. J Cardiovasc Electrophysiol 2002;13:633–8.

[37] Tanaka Y, Tada H, Ito S, Naito S, Higuchi K, Kumagai K, et al. Gender and age differences in candidates for radiofrequency catheter ablation of idiopathic ventricular arrhythmias. Circ J 2011;75:1585–91.

[38] Marchlinski FE, Deely MP, Zado ES. Sex-specific triggers for right ventricular outflow tract tachycardia. Am Heart J 2000;139:1009–13.

[39] Moss AJ, Schwartz PJ, Crampton RS, Tzivoni D, Locati EH, MacCluer J, et al. The long QT syndrome. Prospective longitudinal study of 328 families. Circulation 1991;84:1136–44.

[40] Locati EH, Zareba W, Moss AJ, Schwartz PJ, Vincent GM, Lehmann MH, et al. Age- and sex-related differences in clinical manifestations in patients with congenital long-QT syndrome: findings from the International LQTS Registry. Circulation 1998;97:2237–44.

[41] Makkar RR, Fromm BS, Steinman RT, Meissner MD, Lehmann MH. Female gender as a risk factor for torsades de pointes associated with cardiovascular drugs. JAMA 1993;270:2590–7.

[42] Milman A, Gourraud JB, Andorin A, Postema PG, Sacher F, Mabo P, et al. Gender differences in patients with Brugada syndrome and arrhythmic events: data from a survey on arrhythmic events in 678 patients. Heart Rhythm 2018;15:1457–65.

[43] Benito B, Sarkozy A, Mont L, Henkens S, Berruezo A, Tamborero D, et al. Gender differences in clinical manifestations of Brugada syndrome. J Am Coll Cardiol 2008;52:1567–73.

[44] Hu X, Wang J, Xu C, He B, Lu Z, Jiang H. Effect of oestrogen replacement therapy on idiopathic outflow tract ventricular arrhythmias in postmenopausal women. Arch Cardiovasc Dis 2011;104:84–8.

[45] Odening KE, Koren G. How do sex hormones modify arrhythmogenesis in long QT syndrome? Sex hormone effects on arrhythmogenic substrate and triggered activity. Heart Rhythm 2014;11:2107–15.

[46] Narayanan K, Havmoeller R, Reinier K, Jerger K, Teodorescu C, Uy-Evanado A, et al. Sex hormone levels in patients with sudden cardiac arrest. Heart Rhythm 2014;11:2267–72.

[47] Akdis D, Saguner AM, Shah K, Wei C, Medeiros-Domingo A, von Eckardstein A, et al. Sex hormones affect outcome in arrhythmogenic right ventricular cardiomyopathy/dysplasia: from a stem cell derived cardiomyocyte-based model to clinical biomarkers of disease outcome. Eur Heart J 2017;38:1498–508.

[48] Bardy GH, Lee KL, Mark DB, Poole JE, Packer DL, Boineau R, et al. Ip JH for the Sudden Cardiac Death in Heart Failure Trial (SCD-HeFT) investigators: amiodarone or an implantable cardioverter-defibrillator for congestive heart failure. N Engl J Med 2005;352:225–37.

[49] Moss AJ, Hall WJ, Cannom DS, Daubert JP, Higgins SL, Klein H, et al. Improved survival with an implanted defibrillator in patients with coronary disease at high risk for ventricular arrhythmia. N Engl J Med 1996;335:1933–40.

[50] Moss AJ, Zareba W, Hall WJ, Klein H, Wilber DJ, Cannom DS, et al. Prophylactic implantation of a defibrillator in patients with myocardial infarction and reduced ejection fraction. N Engl J Med 2002;346:877–83.

[51] Buxton AE, Lee KL, Fisher JD, Josephson ME, Prystowsky EN, Hafley G, et al. A randomized study of the prevention of sudden death in patients with coronary artery disease. N Engl J Med 1999;341:1882–90.

[52] Kadish A, Dyer A, Daubert JP, Quigg R, Estes NAM, Anderson KP, et al. Prophylactic defibrillator implantation in patients with nonischemic dilated cardiomyopathy. N Engl J Med 2004;350:2151–8.

[53] Antiarrhythmics Versus Implantable Defibrillators (AVID) Investigators. A comparison of antiarrhythmic-drug therapy with implantable defibrillators in patients resuscitated from near-fatal ventricular arrhythmias. N Engl J Med 1997;337(22):1576–83.

[54] Connolly SJ, Gent M, Roberts RS, Dorian P, Roy D, Sheldon RS, et al. Canadian implantable defibrillator study (CIDS): a randomized trial of the implantable cardioverter defibrillator against amiodarone. Circulation 2000;101(11):1297–302.

[55] Pires LA, Sethuraman B, Guduguntla VD, Todd KM, Yamasaki H, Ravi S. Outcome of women versus men with ventricular tachyarrhythmias treated with the implantable cardioverter defibrillator. J Cardiovasc Electrophysiol 2002;563–8.

[56] Pacifico A, Ferlic LL, Cedillo-Salazar FR, Nasir N, Doyle TK, Henry PD. Shocks as predictors of survival in patients with implantable cardioverter-defibrillators. J Am Coll Cardiol 1999;34:204–10.

[57] Ruppel R, Schluter CA, Boczor S, Meinertz T, Schlüter M, Kuck K-H, et al. Ventricular tachycardia during follow-up in patients resuscitated from ventricular fibrillation: experience from stored electrograms of implantable cardioverter-defibrillators. J Am Coll Cardiol 1998;32:1724–30.

[58] Kies P, Boersma E, Bax JJ, van der Burg AE, Bootsma M, van EL, et al. Determinants of recurrent ventricular arrhythmia or death in 300 consecutive patients with ischemic heart disease who experienced aborted sudden death: data from the Leiden out-of-hospital cardiac arrest study. JCE 2005;16:1049–56.

[59] Dolack GL. Clinical predictors of implantable cardioverter-defibrillator shocks (results of the CASCADE trial). Cardiac Arrest in Seattle, Conventional versus Amiodarone Drug Evaluation. Am J Cardiol 1994;73:237–41.

[60] Engelstein ED, Friedman PL, Yao Q, Coromilas J, Beckman KJ, Buxton AE, et al. Gender differences in patients with life-threatening ventricular arrhythmias: impact on treatment and survival in the AVID trial. Circulation 1997;96(Suppl.):I-720.

[61] Curtis LH, Al-Khatib SM, Shea AM, Hammill BG, Hernandez AF, Schulman KA. Sex differences in the use of implantable cardioverter-defibrillators for primary and secondary prevention of sudden cardiac death. JAMA 2007;298(13):1517–24.

[62] Zareba W, Moss AJ, Jackson Hall W, Wilber DJ, Ruskin JN, McNitt S, et al. Clinical course and implantable cardioverter defibrillator therapy in postinfarction women with severe left ventricular dysfunction. J Cardiovasc Electrophysiol 2005;16:1265–70.

[63] Albert CM, Quigg R, Saba S, Estes 3rd NAM, Shaechter A, Subacius H, et al. Sex differences in outcome after implantable cardioverter defibrillator implantation in nonischemic cardiomyopathy. Am Heart J 2008;156:367–72.

[64] Russo AM, Poole JE, Mark DB, Anderson J, Hellkamp AS, Lee KL, et al. Primary prevention with defibrillator therapy in women: results from the Sudden Cardiac Death in Heart Failure Trial. J Cardiovasc Electrophysiol 2008;19(7):720–4.

[65] Santangeli P, Pelargonio G, Dello A, Casella M, Bisceglia C, Bartoletti S, et al. Gender differences in clinical outcome and primary prevention defibrillator benefit in patients with severe left ventricular dysfunction: a systemic review and meta-analysis. Heart Rhythm 2010;7:876–82.

[66] Rho RW, Patton KK, Poole JE, Cleland JG, Shadman R, Anand I, et al. Important differences in mode of death between men and women with heart failure who would qualify for a primary prevention implantable cardioverter-defibrillator. Circulation 2012;126:2402–7.

[67] Russo AM, Daugherty SL, Masoudi FA, Wang Y, Curtis J, Lampert R. Gender and outcomes after primary prevention implantable cardioverter-defibrillator implantation: findings from the National Cardiovascular Data Registry (NCDR). Am Heart J 2015;170:330–8.

[68] Lampert R, McPherson CA, Clancy JF, Caulin-Glaser TL, Rosenfeld LE, Batsford WP. Gender differences in ventricular arrhythmia recurrence in patients with coronary artery disease and implantable cardioverter-defibrillators. J Am Coll Cardiol 2004;43(12):2293–9.

[69] Santini M, Russo M, Botto G, Lunati M, Proclemer A, Schmidt B, et al. Clinical and arrhythmic outcomes after implantation of a defibrillator for primary prevention of sudden death in patients with post-myocardial infarction cardiomyopathy: the Survey to Evaluate Arrhythmia Rate in High-risk MI patients (SEARCH-MI). Europace 2009;11:476–82.

[70] MacFadden DR, Crystal E, Krahn AD, Mangat I, Healey JS, Dorian P, et al. Sex differences in implantable cardioverter-defibrillator outcomes: findings from a prospective defibrillator database. Ann Intern Med 2012;156:195–203.

[71] Providência R, Marijon E, Lambiase PD, Bouzeman A, Defaye P, Klug D, et al. Primary prevention implantable cardioverter defibrillator (ICD) therapy in women-data from a multicenter French registry. J Am Heart Assoc 2016;5(2). https://doi.org/10.1161/JAHA.115.002756. pii: e002756.

[72] Sticherling C, Arendacka B, Svendsen JH, Wijers S, Friede T, Stockinger J, et al. Sex differences in outcomes of primary prevention implantable cardioverter-defibrillator therapy: combined registry data from eleven European countries. Europace 2018;20:963–70.

[73] Chen HA, Hsia HH, Vagelos R, Fowler M, Wang P, Al-Ahmad A. The effect of gender on mortality or appropriate shock in patients with nonischemic cardiomyopathy who have implantable cardioverter-defibrillators. Pacing Clin Electrophysiol 2007;30:390–4.

[74] Conen D, Arendacká B, Röver C, Bergau L, Munoz P, Wijers S, et al. Gender differences in appropriate shocks and mortality among patients with primary prophylactic implantable cardioverter-defibrillators: systematic review and meta-analysis. PLoS One 2016;11(9). https://doi.org/10.1371/journal.pone.0162756.

[75] Al-Khatib SM, Stevenson WG, Ackerman MJ, Bryant WJ, Callans DJ, Curtis AB, et al. 2017 AHA/ACC/HRS guideline for management of patients with ventricular arrhythmias and the prevention of sudden cardiac death: a report of the American College of Cardiology/American Heart Association Task Force on Clinical Practice Guidelines and the Heart Rhythm Society. J Am Coll Cardiol 2018;72(14):e91–e220.

Pathology in Survivors of SCA

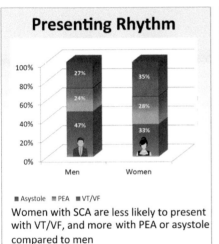

Women

Men

■ Ischemic
■ Non ischemic
■ Normal

Women are more likely to have non ischemic etiology for SCA compared to men

Lifetime risk of SCA by Age and Sex

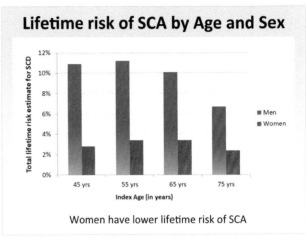

Women have lower lifetime risk of SCA

Presenting Rhythm

Men   Women

■ Asystole ■ PEA ■ VT/VF

Women with SCA are less likely to present with VT/VF, and more with PEA or asystole compared to men

ICD Therapy

Primary Prevention

- Older
- More comorbidities- HTN, DM, LBBB, angina and HF
- More transvenous procedural complications
- Less likely to get ICD
- Receive fewer ICD shocks

Secondary Prevention

- Younger
- Less CAD
- More non ischemic CM
- Less likely to get ICD
- Equally effective in both sexes

Characteristics, complications and outcomes of ICD therapy in women vs men

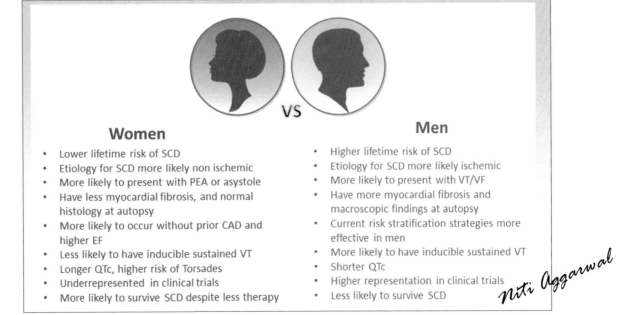

VS

Women

- Lower lifetime risk of SCD
- Etiology for SCD more likely non ischemic
- More likely to present with PEA or asystole
- Have less myocardial fibrosis, and normal histology at autopsy
- More likely to occur without prior CAD and higher EF
- Less likely to have inducible sustained VT
- Longer QTc, higher risk of Torsades
- Underrepresented in clinical trials
- More likely to survive SCD despite less therapy

Men

- Higher lifetime risk of SCD
- Etiology for SCD more likely ischemic
- More likely to present with VT/VF
- Have more myocardial fibrosis and macroscopic findings at autopsy
- Current risk stratification strategies more effective in men
- More likely to have inducible sustained VT
- Shorter QTc
- Higher representation in clinical trials
- Less likely to survive SCD

Niti Aggarwal

CAD, coronary artery disease; CM, cardiomyopathy; DM, diabetes; EF, ejection fraction; HF, heart failure; HTN, hypertension; ICD, implantable cardioverter device; LBBB, left bundle branch block; PEA, pulseless electric activity; SCA, sudden cardiac arrest; VF, ventricular fibrillation; VT, ventricular tachycardia. *Image courtesy of Niti R. Aggarwal.*

Chapter 18

Role of ICD and CRT

Theofanie Mela

Clinical Case

A 44-year-old female presented after a resuscitated out-of-hospital cardiac arrest and was diagnosed with a nonischemic cardiomyopathy, as her coronary angiogram showed clean coronary arteries and her echocardiogram showed a left ventricular ejection fraction (LVEF) of 40%, LV end-diastolic dimension (LVEDD) 58 mm, and moderate mitral regurgitation. A right ventricular biopsy did not show evidence of myocarditis. She had a left bundle branch block (LBBB) with a QRS duration of 138 ms. She had no family history of premature sudden death, cardiomyopathy, or heart failure symptoms.

She underwent placement of a single ventricular lead implantable cardioverter defibrillator. Ten years later, she subsequently had shocks for ventricular fibrillation (VF) with syncope. She was then initiated on mexiletine 200 mg orally twice daily, in addition to sotalol 80 mg orally twice daily. Her LVEF gradually diminished. Despite optimal medical therapy, her LVEF declined to 21%, LVEDD 59 mm and she developed moderate mitral regurgitation.

She also experienced progressive decline of her exercise tolerance. She used to be able to go up three flights of stairs to her office without problems. Now, her office was moved to the first floor and even after one flight of stairs she had a lot of shortness of breath. She also noticed a significant loss of energy, especially since the last VF episode. She required intermittent use of Lasix for the treatment of pedal edema. She denied any orthopnea or paroxysmal nocturnal dyspnea or significant weight loss. She continued to have a baseline LBBB with a QRS of 138–142 ms at different recordings. How would you manage this patient going forward?

Abstract

Implantable cardioverter defibrillators (ICD) and cardiac resynchronization therapy (CRT) devices have been proven to improve survival in patients with heart failure (HF) and reduced ejection fraction (EF). Sex-specific differences in the incidence and etiology of sudden cardiac death (SCD) and HF exist, but the indications for ICD and CRT implants are universal for both sexes. However, the studies from which the indications were obtained include a much smaller percentage of women than men, and they were not powered to examine sex-specific differences in outcome. This is a systematic review of the epidemiology and incidence of SCD, pivotal ICD and CRT trials, and exploration of the sex-specific differences in outcomes and utilization of these devices. Efforts should be made in the future for a higher enrollment of women in such trials, ultimately resulting in a decrease in sex-based disparities in the utilization of these devices.

Sex Differences in Cardiac Disease. https://doi.org/10.1016/B978-0-12-819369-3.00004-6

Incidence of Sudden Cardiac Death and HF

SCD and HF are two major health risks worldwide. There are more than 356,000 out-of-hospital cardiac arrests annually in the United States, nearly 90% of which are fatal, according to the American Heart Association's recently released Heart Disease and Stroke Statistics [1]. About 5.7 million adults in the United States have HF [1]. One in 9 deaths in 2009 included HF as a contributory cause and half of the people who develop HF die within 5 years of diagnosis [1].

There are sex-related differences in the epidemiology of these conditions. The annual rate of SCD in women has been estimated to be half of that in men [2, 3]. Large epidemiologic studies, such as the Nurses' Health Study, have illuminated the epidemic of SCD in women. The study included 121,701 nurses, aged 30–55, and enrollment started in 1976, with the main end point being SCD occurring before June 1, 2004 [1]. Most (69%) women who suffered SCD had no history of heart disease before their death. However, almost all of the women who died suddenly (94%) had reported at least one cardiac risk factor. Smoking, hypertension, and diabetes conferred markedly

elevated (2.5- to 4.0-fold) risk of SCD, similar to that conferred by a history of nonfatal myocardial infarction (MI; relative risk, 4.1; 95% confidence interval, 2.9–6.7). Family history of MI before age 60 and obesity were associated with moderate elevation in risk (1.6-fold). With regard to the mechanism, 88% of SCDs were classified as arrhythmic. In 76% of them the first rhythm documented was ventricular tachycardia (VT) or fibrillation. A small transient risk of SCD during moderate to vigorous exercise was minimized by regular exercise in women [4]. Among survivors of SCD, men and women have differences in the underlying cardiac structural abnormalities. In a single center study, 45% of women had underlying coronary artery disease (CAD), with dilated cardiomyopathy in another 19% and valvular heart disease in another 13%. For men, however, 80% had underlying CAD, 10% had dilated cardiomyopathy, and 5% had valvular heart disease (**Table 1**) [5].

In a more recent report, SCD occurring between 1998 and 2017 was studied in Finland [6], with special attention given to women's SCD. The cohort consisted of 5869 subjects with SCD, with available autopsy data. Women were older than men with an average age of 70 vs. 63 years, and CAD was still the predominant cause of SCD in both sexes (71.7% of women with SCD vs. 75.7% in men with SCD).

TABLE 1 Cardiac Disease by Sex in Survivors of Sudden Cardiac Death

Sex	Albert et al. 1996 [5] 355 Pts With Resuscitated Cardiac Arrest		Haukilahti et al. 2020 [6] 5869 Pts With SCD and Autopsy	
	Men	Women	Men	Women
n(%)	271 (76%)	84 (24%)	4631 (79%)	1238 (21%)
CAD (%)	80	45	75.7	71.7
DCM (%)	10	19	1.0	0.5
Valvular disease (%)	5	13	1.1	1.9
Normal (%)	3	10	0.2	0.6
Congenital (%)	0	2		
RV dysplasia (%)	0	2	0	0.4
Long QT (%)	0	2		
Spasm (%)	0	5		
Other (%)	2	2		
HTA CM (%)			6.7	7.3
Obesity CM (%)			5.7	6.1
Alcoholic CM (%)			5.3	4.0
Myocarditis (%)			0.9	1.6

TABLE 1 Cardiac Disease by Sex in Survivors of Sudden Cardiac Death—cont'd

Sex	Albert et al. 1996 [5] 355 Pts With Resuscitated Cardiac Arrest		Haukilahti et al. 2020 [6] 5869 Pts With SCD and Autopsy	
	Men	Women	Men	Women
PMF (%)			2.6	5.2
Anomalies (%)			0.1	0.1
HCM (%)			0.7	0.4
Missing info (%)			0.1	0.1

CAD, coronary artery disease; CM, cardiomyopathy; DCM, dilated cardiomyopathy; HCM, hypertrophic cardiomyopathy; HTA, hypertensive; PMF, primary myocardial fibrosis; Pts, patients; RV, right ventricular; SCD, sudden cardiac death.

Nonischemic causes were the second most common cause in both sexes, but higher in women at 28.3% compared to 24.3% for men (**Table 1**). As the authors had autopsy results available, they observed a higher incidence of primary myocardial fibrosis in women than men (5.2% vs. 2.3%). One-third of females with SCD had a normal electrocardiogram (ECG) prior to death, but women were more likely to have markers for left ventricular hypertrophy than men were (8.2% vs. 4.9%).

In addition, women were more likely to have no macroscopic findings at autopsy and normal histology compared to men (0.6% vs. 0.2%, $p < 0.001$). In addition, men were more likely to have fibrosis (92.4% vs. 89.3%). They were also likely to have a greater severity of fibrosis, with 12.7% of men having substantial fibrosis compared to only 7.8% of women ($p < 0.001$) [6].

Large quantities of basic science and clinical data have shown sex differences in arrhythmia susceptibility. There have been animal studies showing sex differences in potassium channel kinetics, calcium sensitivity and handling, autonomic modulation, and differences in Na-Ca exchanger that may lower susceptibility to triggered activity [7–10].

Pathophysiologic variations between men and women with CAD may explain the difference in propensity for the development of SCD. In earlier studies, although a smaller percentage of women were included, women had a lower incidence of angiographically significant epicardial CAD and a higher incidence of microvascular ischemia [11]. The presence of both myocardial fibrosis and microvascular ischemia can be visualized by magnetic resonance imaging, suggesting that development of other imaging techniques for women may be warranted for an improved understanding of sex differences in pathophysiology of arrhythmias and SCD. Interestingly, data from the Multicenter UnSustained Tachycardia Trial (MUSTT) study suggest that the induc-

ibility of VT/VF during the electrophysiologic study in post-MI women who met the enrollment criteria was lower than that of men. This indicates a potentially different arrhythmogenic mechanism for post-MI women than that for men [12].

Use of Implantable Cardioverter Defibrillators to Prevent Sudden Cardiac Death

Although women were underrepresented in the majority of studies on which the current guidelines for primary and secondary ICD and cardiac resynchronization therapy (CRT) utilization are based, these guidelines are recommended equally in both sexes [13].

Primary Prevention of SCD With ICD Trials

The primary SCD prevention recommendations have been based on a number of pivotal studies which demonstrated that the incidence of SCD is reduced by the use of ICD in patients with severely reduced systolic cardiac function, LVEF $\leq 35\%$, and congestive HF NYHA class II or more [Sudden Cardiac Death in Heart Failure Trial (SCD-HeFT)] [14], patients with prior MI and LVEF $\leq 30\%$ [Multicenter Automatic Defibrillator Implantation Trial II (MADIT-II)] [15], and patients with ischemic cardiomyopathy and inducible VT during an electrophysiologic study (MADIT-I, MUSTT) [16, 17].

Underrepresentation of women in cardiac device trials is a major impediment to understanding sex differences in outcomes, with women typically representing only a quarter of the enrolled subjects (**Table 2**). In MUSTT, only 9.7% of the study participants were women [17]. Similarly, 23% of the SCD-HeFT participants were women [14]. In SCD-HeFT, women were more likely to have nonischemic

TABLE 2 Women in ICD and CRT Trials

Trial-Publication Year	Mean or Median Follow-Up	Enrollment Population	Randomized Therapy	Total No.	Women (%)	Primary Outcome End Point	Primary Outcome	Interaction With End Point by Sex
AVID-1997 [18]	18.2 ± 12.2 mo	Cardiac arrest or syncopal VT or symptomatic VT if EF ≤40%	ICD or NYHA class III AADs	1106	21	All-cause mortality	3.32-fold increased survival with ICD vs. AADs ($p=0.02$)	Substudy showed no significant difference by sex
CIDS-2000 [19]	Amiodarone 2.9 y; ICD patients 3.0 y	Cardiac arrest or syncopal VT or symptomatic VT if EF <35%	ICD or amiodarone	659	15	All-cause mortality	Nonsignificant decrease with ICD; relative risk reduction, 19.7% (95% CI, 27.7–40; $p=0.142$)	No significant interaction by sex
MADIT-II-2002 [15]	21 mo	Prior remote MI; EF ≤30%	ICD or conventional medical therapy	1232	15	All-cause mortality	Lower mortality with ICD (HR, 0.69; 95% CI, 0.51–0.93; $p=0.016$)	No significant interaction by sex
DEFINITE-2004 [20]	29 ± 14.4 mo	Nonischemic CM, EF ≤35%, PVC or NSVT, NYHA class I–III	ICD or conventional medical therapy	916	29	All-cause mortality	ICD vs. optimal medical therapy (HR, 0.65; 95% CI, 0.40–1.06; $p=0.08$)	No significant interaction by sex
SCD-HeFT-2005 [14]	45.5 mo	Ischemic and nonischemic CM, EF <35%, NYHA class II or III	ICD or optimal medical therapy	1676	23	All-cause mortality	Lower mortality with ICD vs. placebo (HR, 0.77; 95% CI, 0.62–0.96; $p=0.007$)	No significant interaction by sex. Women: HR, 0.96 (95% CI, 0.58–1.61) Men: HR, 0.73 (95% CI, 0.57–0.93)

Trial	Duration	Inclusion criteria	Intervention/Comparison	N	N	Primary endpoint	Outcome	Sex interaction
COMPANION-2004 [21]	Range 11.9–16.2 mo	Ischemic or non-ischemic CM, NYHA class III or IV, EF ≤35%, QRS ≥120 ms	CRT-P, CRT-D, or optimal HF medical therapy	1520	33	All-cause mortality and hospitalization for any cause	Improved outcome with CRT-D vs. optimal HF medical therapy (HR, 0.80; $p=0.01$) and with CRT-P vs. optimal HF medical therapy (HR, 0.81; $p=0.014$)	No significant interaction by sex
CARE-HF-2005 [22]	29.4 mo	Ischemic CM or NIDCM, NYHA class III or IV, EF ≤35%, QRS ≥120 ms	CRT-P or optimal medical therapy	813	27	All-cause mortality and unplanned CV hospitalization	Improved outcome with CRT-P (HR, 0.63; 95% CI, 0.51–0.77; $p<0.001$)	No significant interaction by sex Men: HR, 0.62 (95% CI, 0.49–0.79) Women: HR, 0.64 (95% CI, 0.42–0.97)
MADIT-CRT-2009 [23]	2.4 yr	Ischemic or nonischemic CM, NYHA class I or II, EF ≤30%, QRS ≥130 ms	CRT-D or ICD	1820	26	All-cause mortality or nonfatal HF event	CRT-D associated with improved outcome (HR, 0.66; 95% CI, 0.52–0.84; $p=0.001$)	Significant interaction by sex Women: HR, 0.37 (95% CI, 0.22–0.61)

AAD, antiarrhythmic drugs; CI, confidence interval; CM, cardiomyopathy; CRT, cardiac resynchronization therapy; CRT-D, CRT-defibrillator; CRT-P, CRT-pacemaker; CV, cardiovascular; EF, ejection fraction; HF, heart failure; HR, hazard ratio; ICD, implantable cardioverter defibrillator; MI, myocardial infarction; mo, months; NIDCM, nonischemic dilated cardiomyopathy; NYHA, New York Heart Association; NSVT, nonsustained ventricular tachycardia; PVC, premature ventricular contractions; VT, ventricular tachycardia; y, years.

cardiomyopathy compared to men (66% vs. 43%) and were more likely to have NYHA class III HF than men (36% vs. 26%). Although there was no significant difference in the ICD mortality benefit by sex, women had a smaller benefit than that observed for men (women: HR, 0.96; 95% CI, 0.58–1.61 vs. men: HR, 0.73; 95% CI, 0.57–0.93). Women comprised 15% of MADIT-II study participants and experienced a similar benefit from ICD therapy to men (HR 0.57 vs. 0.66) [15].

The benefit of an ICD vs. conventional medical therapy alone for patients with nonischemic cardiomyopathy was evaluated in two landmark trials: Defibrillators in Non-Ischemic Cardiomyopathy Treatment Evaluation (DEFINITE) [20] and Danish ICD Study in Patients With Dilated Cardiomyopathy (DANISH) [24]. In DEFINITE, the ICD therapy significantly reduced the risk of SCD from arrhythmia, but was not statistically significant (HR 0.65, $p=0.08$) [20]. All-cause mortality was decreased among men ($p=0.02$) but not women ($p=0.75$). There was no arrhythmic death difference between men and women who comprised 29% of the study population. In contrast, in the DANISH trial prophylactic ICD implantation improved the risk of SCD (4.3% vs 8.2%, HR, 0.50; 95% CI 0.31–0.82; $p=0.005$) but no difference in all-cause mortality was seen (21.6% vs 23.4%, HR, 0.87; 95% CI 0.68–1.12, $p=0.87$) [24]. Similar results were seen in both men and women.

Secondary Prevention of SCD With ICD Trials

Secondary prevention of SCD was studied with the Antiarrhythmics Versus Implantable Defibrillators (AVID) [18] and Canadian Implantable Defibrillator Study (CIDS) [19] trials. The AVID trial included patients who had survived a cardiac arrest, or patients with hemodynamically unstable sustained VT and LVEF $\leq 40\%$. Patients were randomized to an ICD or antiarrhythmic medical therapy, primarily amiodarone. Only 21% of the patients in the study were women. Survival was highest in those who received an ICD. The CIDS trial included patients who had a cardiac arrest or syncopal VT or symptomatic VT with LVEF $\leq 35\%$. The patients were randomized to ICD vs. amiodarone. Only 15% of the patients were women. The study found a nonsignificant decrease in all-cause mortality with ICD and no significant interaction by sex. A sex-specific subgroup analysis was conducted for the AVID trial and found that the 1-year mortality was similar between men and women [25]. Women were younger, more likely to have nonischemic heart disease, and more likely to experience VF rather than VT as their index arrhythmia.

Use of Cardiac Resynchronization Therapy Devices

With the advent of CRT, multiple studies have demonstrated the significant decrease in mortality in patients with a wide QRS ≥ 120 ms, reduced LVEF, and advanced HF. The first of these studies was the Comparison of Medical Therapy, Pacing, and Defibrillation in Chronic Heart Failure (COMPANION) trial [21, 26], which included 33% women and enrolled patients with NYHA class III and IV HF. No sex difference in the primary end point of HF hospitalization and mortality was observed in this study.

The Multicenter Automatic Defibrillator Implantation Trial With Cardiac Resynchronization Therapy (MADIT-CRT) [23] demonstrated that CRT-D-treated patients with NYHA class I and II HF symptoms, left ventricular ejection fraction (LVEF) $\geq 30\%$, and QRS duration ≥ 130 ms had a 34% risk reduction in HF hospitalization or death, when compared to patients who received an ICD alone. The MADIT-CRT trial was comprised of 1820 subjects (26% women) who were followed for an average duration of 2.4 years. CRT was more beneficial in women compared to men, independent of the QRS duration (women: HR, 0.37; 95% CI, 0.22–0.61 vs. men: HR, 0.76; 95% CI, 0.59–0.97) ($p=0.01$ for the interaction). This difference was attributed to the fact that women were more likely to have nonischemic cardiomyopathy, which responds better to CRT than ischemic cardiomyopathy [27] (**Figure 1**). The benefit was more pronounced for women with QRS ≥ 150 ms and LBBB, resulting in all-cause mortality reductions of 82% and 78%, respectively. The beneficial effects of CRT-D in women were consistently associated with greater reverse cardiac remodeling by echocardiography compared to men [27, 28]. The benefit of CRT was attenuated in patients without an LBBB or wide QRS.

Several previous CRT trials have included patients with NYHA class III and IV HF. In the Multicenter InSync Randomized Clinical Evaluation (MIRACLE) study, 33% of the study participants were women [29]. The study showed that patients with CRT were less likely to experience HF hospitalizations or death compared to patients without CRT; this benefit was more pronounced in women than in men (HR, 0.157).

In the CARE-HF study, 27% of the study participants were women; both men and women benefited equally from CRT, with a 37% improvement in time to cardiac hospitalization or death (men: HR, 0.62 vs. women: HR, 0.64) [22]. Results from subgroup analyses must be interpreted with caution, since only a small proportion of women were enrolled and the trials were underpowered to show differences in smaller groups.

Meta-Analyses and Registries for ICD and CRT

A meta-analysis of the six major ICD trials, MADIT-II, MUSTT, DEFINITE, SCD-HeFT, COMPANION, and DANISH, was published in 2018 comprising of 6934 patients (23% of whom were women) [30]. The MADIT-II

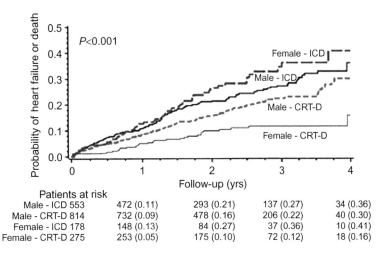

FIGURE 1 **Kaplan-Meir Estimates of Cumulative Probability of Heart Failure or Death Stratified by Sex and ICD or CRT-D Therapy in MADIT-CRT Study.** The four curves reflect the probability of heart failure or death, whichever comes first, over time in women and men having the device therapy to which they were randomized. Women randomly assigned to receive CRT-D had the best result and performed better than men assigned to CRT-D and women with ICD alone. CRT-D, cardiac resynchronization therapy-defibrillator; ICD, implantable cardioverter defibrillator; MADIT, Multicenter Automatic Defibrillator Implantation Trial; yrs, years. *Reproduced with permission from [27].*

and MUSTT trials studied patients with ischemic cardiomyopathy. The SCD-HeFT and COMPANION trials studied both ischemic and nonischemic cardiomyopathy, while the DEFINITE and DANISH trials studied patients with nonischemic cardiomyopathy. Results were analyzed separately for female and male patients. In 5356 male patients, the use of an ICD resulted in a 25% lower mortality rate compared with optimal medical therapy alone (HR, 0.75; 95% CI, 0.67–0.84, $p < 0.001$). In contrast, ICD was not associated with improved survival in women (HR, 0.73; 95% CI, 0.73–1.39, $p = 0.96$). CRT has proven to be beneficial in women. Given the inclusion of CRT trials (the COMPANION study), this meta-analysis may have overestimated the benefit of ICD therapy without concomitant CRT in women.

Similarly, an older meta-analysis of five primary prevention ICD trials (MADIT-II, MUSTT, SCD-HeFT, DEFINITE, and COMPANION) [31] evaluated the effect of prophylactic ICD therapy on the end points of total mortality, appropriate ICD therapies, and survival of women compared with men. This analysis included 7229 patients; women represented 22% of the study group and they presented with more comorbidities and more advanced HF. In addition, fewer women received renin-angiotensin blockers and underwent coronary revascularization procedures. Women received fewer appropriate ICD therapies for ventricular arrhythmias than men (HR, 0.63; 95% CI, 0.49–0.82; $p < 0.001$). After adjustment for baseline cofounders and covariates, there was no significant difference in overall mortality for women compared with men (HR, 0.96; 95% CI, 0.67–1.39; $p = 0.84$), suggesting a smaller impact of SCD on overall mortality in women. Prophylactic ICD implantation significantly reduced mortality in men

(HR, 0.67; 95% CI, 0.58–0.78; $p < 0.001$), while there was no significant reduction in mortality for women (HR, 0.78; 95% CI, 0.57–1.05; $p = 0.1$). These results were reproduced a year later by another group and a meta-analysis of the same studies with a similar outcome [32].

Similar sex-specific observations have also been noted in longitudinal registries. The Ontario ICD Database [33] provided longitudinal follow-up for complications, deaths, and device outcomes, studying prospectively 6021 patients (21% of whom were women) referred for ICD implantation. The total mortality after ICD implantation did not differ between men and women during 1-year follow-up (HR, 1.00; 95% CI, 0.64–1.55; $p = 0.99$). Women were less likely than men to receive appropriate ICD shock therapy (HR, 0.69; 95% CI, 0.51–0.93; $p = 0.02$) or appropriate shock or antitachycardia pacing therapy (HR, 0.73; 95% CI, 0.59–0.90; $p = 0.003$) during follow-up.

Analysis from the National Cardiovascular Data Registry (NCDR) ICD Registry [34], the Get With The Guidelines-HF (GWTG-HF) database, and the Centers for Medicare and Medicaid Services on women eligible for a primary prevention ICD compared 490 women with primary prevention ICD with matched 490 women without an ICD who were eligible for a primary prevention ICD. After a median follow-up of 4.6 years, women with ICD had better survival than women without ICD (HR, 0.79; 95% CI, 0.66–0.95; $p < 0.013$) and there was no difference by sex in the effectiveness of ICD therapy. A similar mortality benefit was seen in both men and women with HF and device use in IMPROVE HF—a registry to Improve the Use of Evidence-Based Heart Failure Therapies in the Outpatient Setting [35]. A retrospective European registry with 11 participating countries and 14 centers enrolled patients who received primary prevention ICD between 2002 and 2014 [36]. Of the 5033 patients, 19% were women. During a median follow-up of 33 months, death occurred in 13% of women and 20% of men (HR, 0.65; 95% CI, 0.53, 0.79; $p < 0.0001$). HR was adjusted for age, cause of HF, LVEF, and presence of CRT. An appropriate ICD shock occurred in 8% of women and 14% of men (HR, 0.61; 95% CI, 0.47–0.79; $p = 0.0002$). A large European registry, the French-UK-Sweden CRT Network, followed 5307 patients with ischemic and non-ischemic cardiomyopathy and cardiac resynchronization therapy pacemaker (CRT-P, $n = 1270$) and cardiac resynchronization therapy defibrillator (CRT-D, $n = 4037$) devices [37]. After a median follow-up of 34 months, no survival advantage of CRT-P vs. CRT-D was observed in neither men nor women after propensity score matching. With inverse probability weighting, a benefit of CRT-D was seen in male patients. The excess mortality of CRT-P vs. CRT-D was due to SCD in 7.4% in men and 2.2% in women. Similarly, men at high risk for competing noncardiac death did not reach a benefit from the ICD modality. A very interesting editorial in that same issue suggested that studies like this underscore the need for a personalized approach to the decision for an ICD implant and the assessment of multiple risk factors unique to each patient [38].

With the advent of remote monitoring, an observational cohort of consecutive US patients newly implanted with pacemakers (PPMs), ICD, and CRT devices between 2008 and 2011 was collected and comprised a significantly larger cohort than previously studied [39]. A total of 269,471 patients were assessed over a median of 2.9 years. Unadjusted mortality rates (deaths/100,000 patient-years) were similar between women and men receiving ICD ($n = 85,014$, 74% male, adjusted HR, 0.98; 95% CI, 0.93–1.02; $p = 0.244$). In contrast, survival was superior in women receiving CRT defibrillators ($n = 61,475$, 72% male; adjusted HR, 0.73; 95% CI, 0.70–0.76; $p < 0.001$) and also CRT pacemakers ($n = 7906$, 57% male; adjusted HR, 0.69; 95% CI, 0.61–0.78; $p < 0.001$). This relative difference increased with time. These results were unaffected by age or remote monitoring utilization. Similarly, in 31,892 CRT-D patients on the NCDR, ICD registry between 2006 and 2009, with a median of 2.9 years of follow-up, among patients with LBBB, women had 21% lower mortality risk than men (HR, 0.79; 95% CI, 0.74–0.84; $p < 0.001$); however, there was no sex difference in non-LBBB patients (HR, 0.95; 95% CI, 0.85–1.06; $p = 0.37$) [40]. In an even larger population of 75,079 NCDR patients with NYHA class III and IV HF, reduced LVEF, and prolonged QRS ≥ 120 ms receiving either CRT-D or an ICD, mortality comparisons were made according to the QRS morphology and 10-ms increments in QRS duration [41]. Women with LBBB who received a CRT-D had a lower mortality than women receiving an ICD

(absolute difference, 11%; HR, 0.74; 95% CI, 0.68–0.81). In men, the lower mortality with CRT-D vs. ICD was less pronounced (absolute difference 9%; HR, 0.84; 0.79–0.89; $p = 0.025$). In those without LBBB, the mortality difference was modest and did not differ between women and men.

Explaining the Differential Benefits of Primary ICD and CRT on Mortality in Women

While women are underrepresented in all large clinical trials, an earlier meta-analysis [42], published in 2012, showed that women with HF have a lower mortality than men and only a few of those deaths are sudden. In a cohort of 8337 patients with HF NYHA class II and III who were good candidates for a primary ICD implant, 20% of whom were women, women's age-adjusted all-cause mortality was 24% lower and the risk of SCD was 34% lower compared to men [42]. It is interesting to note that only 50% of patients were on a beta blocker. Treatment with beta blockers, CRT, and sacubitril/valsartan are modalities which improve survival further and potentially decrease the benefit of an ICD implant.

The potential explanation of women having less ischemic and more nonischemic cardiomyopathy has been postulated as an additional explanation of the lack of benefit from ICD implant in women. However, even in the pooled analysis of the placebo groups of five MI studies, Yap et al. demonstrated that the all-cause mortality was elevated for 2 years post MI. The incidence of sudden death exceeded other modes of death for 2 years in men, but only for 6 months in women [43].

In contrast to the primary prevention ICD, women derive a greater benefit than men from CRT [22, 23, 27–29]. It has been shown that improvements in clinical symptoms and echocardiography parameters are associated with a reduction in ventricular arrhythmias [44]. As women derive a greater benefit from CRT, their risk for arrhythmic events decreases and, therefore, there is no mortality benefit from CRT-D vs. CRT-P. In addition, women have shorter QRS duration than men and have LBBB at shorter QRS duration [45]. In the MADIT-CRT trial, there was evidence that women start benefiting from CRT at shorter QRS duration than men (QRS ≥ 130 ms vs. QRS ≥ 140 ms) [46]. There was even a suggestion that women should have different ECG criteria than men [47], but this idea has not been incorporated in the current guidelines. Moreover, heart size [48] and height [49] may play an important role in the increased benefit that women gain from CRT, although the evidence is only hypothesis-generating.

However, the non-LBBB women did not derive a benefit with CRT. The MADIT-CRT non-LBBB population included

59 females or 11% of the non-LBBB population. Only five women had a PR ≥ 230 ms. Patients with the longer PR were the only ones who derived a benefit from the implantation of a CRT-D, irrespective of the duration of the QRS. CRT-D was associated with a higher all-cause mortality risk in patients with a normal PR < 230 ms (HR = 2.27; p = 0.014). Therefore, women with a non-LBBB may have a harmful effect from CRT, as their PR interval is frequently normal [50].

In summary, women may have a lower all-cause mortality benefit compared with male recipients of an ICD and a lower rate of appropriate ICD therapy. However, none of the randomized ICD trials were powered to examine sex-specific differences. Women are more likely to benefit from CRT than men.

Therefore, women who meet guideline-directed indications for ICD therapy should receive an ICD. Observational sex-specific differences in the benefits of ICD therapy should not be considered for the stratification of patients who may be eligible for a primary prevention ICD, as clearly documented by the European Heart Rhythm Association consensus document [51]. Women with LBBB and QRS > 150 ms and LVEF < 35% despite optimal medical therapy should be referred for CRT. Women with LBBB and LVEF < 35% despite optimal medical therapy are highly likely to respond to CRT and should be referred for CRT [51].

Utilization of Cardiac Devices

A sex-based bias in the use of ICD and CRT devices has been present for nearly three decades. An analysis of data from the GWTG-HF program revealed that, out of 13,034 patients with HF who were eligible for ICD therapy, 44% eligible white men as opposed to only 28% of eligible women received an ICD [52]. When, as part of the IMPROVE HF study [53], clinical decision-making support tools and educational material to healthcare providers were given, after 2 years of follow-up, ICD use went up from 40%–50% to 75%–80% and CRT use from 35%–40% to 65%–75% in both men and women. A subsequent data analysis from the GWTG-HF database showed that there was an improved trend toward the overall use of ICD (30% in 2005 to 42% in 2007). Although the increase was greatest for African-American women (23.3% increase), a disparity between men and women remained [54]. Women with HF and LVEF < 35% received less predischarge ICD counseling compared to men with similar indication (19.3% vs. 24.6%, p < 0.001) during HF hospitalization in another GWTG-HF analysis. Among those counseled, women and men were similarly likely to receive an ICD [55].

In the French-UK-Sweden CRT Network [37], there were 42% women in the CRT-P group compared to 15% in the CRT-D group (p < 0.001) which by itself suggests a bias in the device selection (ICD vs. pacemaker) for women.

In the large contemporary cohort of the National Inpatient Sample database, 311,009 patients who underwent CRT implantation in the United States between 2006 and 2012 were analyzed. In this population, women were less likely to have ≥ 3 predictors of ICD efficacy (27% vs. 37%, p < 0.003), and more likely to have ≥ 3 predictors of CRT response (47% vs. 33%, p < 0.001). Despite this, men had a 10% higher likelihood of receiving a CRT-D compared to women (88.6% vs. 80.1%, p < 0.001); this disparity was more prevalent in patients who were older and had atrial fibrillation, renal insufficiency, and more comorbidities [56]. Sex differences in CRT-D implantation were inversely related to the predicated CRT benefit and have increased over time.

These results are not well understood, as women undergoing CRT were more likely to have a comorbidity profile associated with CRT and ICD benefit, in this study as well as in the earlier European CRT Survey [57]. Women were also significantly more likely to have clinical characteristics associated with a greater likelihood of reverse remodeling [41, 56], which is a predictor of improved outcomes. Of note, the utilization of CRT-D was similar between men and women who received the device for secondary prevention (VT or cardiac arrest).

Complications of Cardiac Devices

Using the NCDR, patients undergoing a primary prevention ICD implantation between January 2006 and December 2007 were evaluated for major and minor complications. Out of 161,470 patients, 27% were women. In multivariable models, women had a significantly higher risk for any adverse events [odds ratio (OR) 1.32; 95% CI, 1.24–1.39] and major adverse events (OR 1.71; 95% CI, 1.57–1.86) [58].

The overall complication rate for women was 4.4% with a major event rate of 2.0%, and these complications were mostly mechanical (cardiac perforation, lead dislodgment, coronary venous dissection, hemothorax, pneumothorax, pericardial tamponade), which were attributed to the smaller body size of women. The Ontario ICD database [33] had similar results. With a complication rate of 4.1%, female sex was one of the risk factors (HR, 1.49) for the development of complications. Similar outcomes were shown by the analysis of the MADIT-CRT trial and, therefore, individual risk-benefit assessment was suggested [59].

Acute and 6-month device-related complications from ICD implantation were identified among 38,912 Medicare patients (25% women) enrolled in the NCDR [60]. Women were found to have a much higher acute risk than men (7.2% vs. 4.8%, $p < 0.001$) but they also continued to have higher odds for complications up to 6 months (OR 1.22; 95% CI, 1.16–1.28; $p < 0.001$). The authors postulated that the higher complication rate for women may relate to their smaller size and smaller vessels, increasing the risk for pneumothorax, and thinner right ventricular wall thickness, making them higher risk for perforation.

One other significant aspect associated with the implantation procedure is pain, which results from the procedure and the size of the device placed in a subcutaneous pocket. Overall, women tend to have more severe postoperative pain [61] and give higher pain scores than men, although it is questionable whether this leads to less satisfaction of pain management in women [62].

Preoperative counseling for pregnancy and childbirth after an ICD implant can be useful in reducing anxiety and childbirth fear for younger recipients [63].

Women with cardiovascular disease are thought to be more susceptible to psychosocial distress due to a variety of reasons, including concerns about body image, shifts in role responsibilities, and changes in physical functioning. The FEMALE-ICD study [63] showed that ICD-specific cognitive behavioral therapy strategies and group social support provide improvements in shock anxiety and device acceptance at 1-month reassessment. Young women <50 years may experience a greater benefit from such an intervention.

Key Points

1. There are sex-specific differences in the epidemiology of SCD, with women constituting one-third of the individuals who suffer SCD.

2. Women are poorly represented in clinical trials examining the effect of ICD and CRT implants in their overall mortality.

3. The benefit of ICD for secondary prevention of SCD appears to be similar for men and women.

4. Women appear to have a smaller benefit from primary prevention ICD implants. However, due to the smaller number of women participating in all the primary prevention ICD trials, these results should be interpreted cautiously and, for the time being, the implantation guidelines should be applied similarly to men and women.

5. Women with CRT appear to derive a greater benefit than men, independent of the QRS width.

6. Trends of benefit or lack of it are intriguing and hypothesis-generating. Further studies to examine not only the actual ICD benefit but also the mechanism of SCD and development of HF in women are needed. Efforts should be made in the future for a higher enrollment of women in cardiovascular trials and to reduce bias in sex-based utilization of ICD and CRT devices.

Back to Clinical Case

The case represented a 44-year-old female with nonischemic cardiomyopathy who presented with sudden cardiac arrest, had an LBBB, and underwent ICD implant. Over time her LV function and exercise capacity declined. She then underwent an upgrade to a CRT-D (**Figures 2 and 3**). Following her upgrade, her first echo showed an LVEF of 41%, LVEDD 59 mm. Her symptoms of HF gradually improved. Her LVEF also gradually improved to 48%, with LVEDD 61 mm and mild mitral regurgitation. She has not had any further ICD shocks since her upgrade to CRT and in the 27 years since her cardiac arrest and first diagnosis of her cardiomyopathy. During her last clinic visit, she was only NYHA class I HF and had not required any hospitalization for HF.

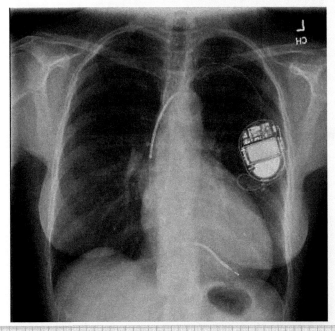

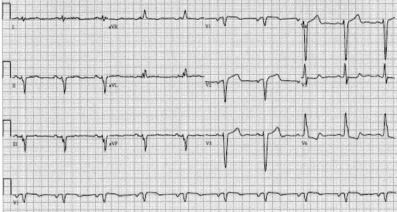

FIGURE 2   The Chest X-Ray and 12-Lead Electrocardiogram of the Patient Following the Single Chamber ICD.

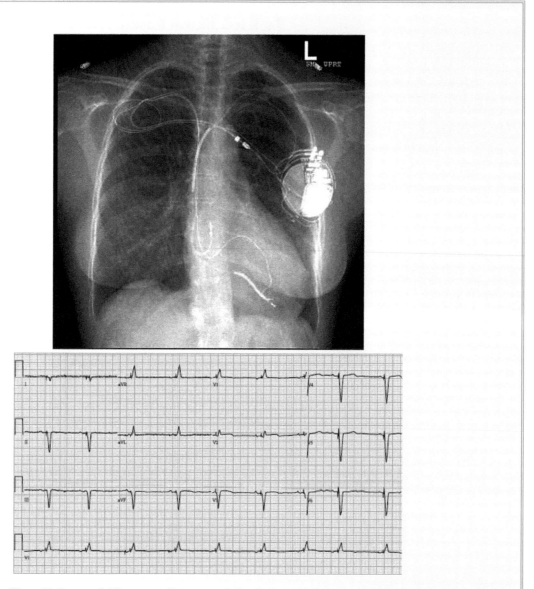

FIGURE 3   The Chest X-Ray and Electrocardiogram of the Patient Following the Upgrade of Her Single Chamber ICD to a CRT-D Device With the Addition of a Right Atrial and a Coronary Sinus Lead. Note that the two new leads had to be introduced through the right subclavian vein and tunneled to the left pectoral pocket, as the left subclavian vein was occluded.

References

[1] Mozaffarian D, Benjamin EJ, Go AS, Arnett DK, Blaha MJ, Cushman M, et al. Heart disease and stroke statistics-2016 update: a report from the American Heart Association. Circulation 2016;133:e38–e360.

[2] Albert CM, Chae CU, Goldstein F, Rose LM, Rexrode KM, Ruskin JN, et al. Prospective study of sudden cardiac death among women in the United States. Circulation 2003;107(16):2096–101.

[3] Schatzkin A, Cupples LA, Heeren T, Morelock S, Kannel WB. Sudden death in the Framingham heart study. Differences in incidence and risk factors by sex and coronary artery disease status. Am J Epidemiol 1984;120:888–99.

[4] Whang W, Manson JE, Hu FB, Chae CU, Rexrode KM, Willett WC, et al. Physical exertion, exercise, and sudden cardiac death in women. JAMA 2006;295:1399–403.

[5] Albert CM, McGovern BA, Newell JB, Ruskin JN. Sex differences in cardiac arrest survivors. Circulation 1996;93:1170–6.

[6] Haukilahti MAE, Holmstrom L, Vahatalo J, Kentta T, Tikkanen J, Pakanen L, et al. Sudden cardiac death in women. Causes of death, autopsy findings, and electrocardiographic risk markers. Circulation 2019;139:1012–21.

[7] Dash R, Frank KF, Carr AN, Moraves CS, Kranias EG. Gender influences on sarcoplasmic reticulum calcium handling in failing human myocardium. J Moll Cell Cardiol 2001;33:1345–53.

[8] Vizginda VM, Wahler GM, Sondgeroth KL, Ziolo MT, Schwertz DW. Mechanism of sex differences in rat cardiac myocyte response to beta-adrenergic stimulation. Am J Physiol Heart Circx Physiol 2002;282:H256–63.

[9] Nakagawa M, Ooie T, Ou B, Ichinose M, Takahashi N, Hara M, et al. Gender differences in autonomic modulation of ventricular repolarization in humans. J Cardiovasc Electrophysiol 2005;16:278–84.

[10] Wei S-K, McCurley JM, Hanlon SU, Haigney MCP. Gender differences in Na/Ca exchanger current and beta-adrenergic responsiveness in heart failure in pig myocytes. Ann N Y Acad Sci 2007;1099183–9.

[11] Poon S, Goodman SG, Yan RT, Bugiardini R, Biernan AS, Eagle KA, et al. Bridging the gender gap: insights from a contemporary analysis of sex-related in the treatment and outcomes of patients with acute coronary syndromes. Am Heart J 2012;163(1):66–73.

[12] Russo AM, Stamato NJ, Lehman MH, Hafley GE, Lee KL, Pieper K, et al. Influence of gender on arrhythmia characteristics and outcome in the Multicenter UnSustained tachycardia trial. J Cardiovasc Electrophysiol 2004;15(9):993–8.

[13] Al-Khatib SM, Stevenson WG, Ackerman MJ, Bryant WJ, Callans DJ, Curtis AB, et al. 2017 AHA/ACC/HRS guideline for management of patients with ventricular arrhythmias and the prevention of sudden cardiac death. Heart Rhythm 2018;15(10):e73–e189.

[14] Bardy GH, Lee KL, Mark DB, Poole JE, Packer DL, Boineau R, et al. Amiodarone or an implantable cardioverter-defibrillator for congestive heart failure. N Engl J Med 2005;352:225–37.

[15] Moss AJ, Zareba W, Hall WJ, Klein H, Wilber DJ, Cannom DS, et al. Prophylactic implantation of a defibrillator in patients with myocardial infarction and reduced ejection fraction. N Engl J Med 2002;346:877–83.

[16] Moss AJ, Hall WJ, Cannom DS, Daubert JP, Higgins SL, Klein H, et al. Improved survival with an implanted defibrillator in patients with coronary disease at high risk for ventricular arrhythmia. Multicenter Automatic Defibrillator Trial Investigators. N Engl J Med 1996;335:1933–40.

[17] Buxton AE, Lee KL, Fisher JD, Josephson ME, Prystowsky EN, Hafley G. A randomized study of the prevention of sudden death in patients with coronary artery disease. N Engl J Med 1999;341(25):1882–90.

[18] Antiarrhythmics versus implantable defibrillators (AVID) investigators. A comparison of antiarrhythmic-drug therapy with implantable defibrillators in patients resuscitated from near-fatal ventricular arrhythmias. N Engl J Med 1997;337:1576–83.

[19] Connolly SJ, Gent M, Roberts RS, Dorian P, Roy D, Sheldon RS, et al. Canadian implantable defibrillator study (CIDS): a randomized trial of the implantable cardioverter defibrillator against amiodarone. Circulation 2000;101:1297–302.

[20] Kadish A, Dyer A, Daubert JP, Quigg R, Estes NAM, Anderson KP, et al. Prophylactic defibrillator implantation in patients with nonischemic dilated cardiomyopathy. N Engl J Med 2004;350:2151–8.

[21] Bristow MR, Saxon LA, Boehmer J, Krueger S, Kass DA, De Marco T, et al. Cardiac resynchronization therapy with or without an implantable defibrillator in advanced chronic heart failure. N Engl J Med 2004;350:2140–50.

[22] Cleland JGF, Dauber J-C, Erdmann E, Freemantle N, Gras D, Kappenberger L, et al. The effect of cardiac resynchronization on morbidity and mortality in heart failure. N Engl J Med 2005;352:1539–49.

[23] Moss AJ, Hall WJ, Cannom DS, Klein H, Brown MW, Daubert JP, et al. Cardiac resynchronization therapy for the prevention of heart failure events. N Engl J Med 2009;361(14):1329–38.

[24] Kober L, Thune JJ, Nielsen JC, Harbo J, Videbæk L, Korup E, et al. Defibrillator implantation in patients with nonischemic csystolic heart failure. N Engl J Med 2016;375:1221–30.

[25] Engelstein ED, Friedman PL, Yao Q, Coromilas J, Beckman KJ, Buxton AE, et al. Gender differences in patients with life-threatening ventricular arrhythmias: impact on treatment and survival in the AVID trial. Circulation 1997;96:1–720.

[26] Saxon LA, Bristow MR, Boehmer J, Krueger S, Kass DA, De Marco T, et al. Predictors of sudden cardiac death and appropriate shock in the comparison of medical therapy, pacing, and defibrillation in heart failure (COMPANION) trail. Circulation 2006;114:2766–72.

[27] Arshad A, Moss AJ, Foster E, Padeletti L, Barsheshet A, Goldenberg I, et al. Cardiac resynchronization is more effective in women than in men in the MADIT-CRT trial. J Am Coll Cardiol 2011;57:813–20.

[28] Hsu JC, Solomon SD, Bourgoun M, McNitt S, Goldenberg I, Klein H, et al. Predictors of super-response to cardiac resynchronization therapy and associated improvement in clinical outcome: the MADIT-CRT. J Am Coll Cardiol 2012;59:2366–73.

[29] Woo GW, Petersen-Stejskal S, Johnson JW, Conti JB, Aranda JA, Curtis AB. Ventricular reverse remodeling and 6-month outcomes in patients receiving cardiac resynchronization therapy: analysis of the MIRACLE study. J Interv Card Electrophysiol 2005;12:107–13.

[30] Barra S, Providencia R, Boveda S, Narayanan K, Virdee M, Marijon E, et al. Do women benefit equally as men from the primary prevention implantable cardioverter-defibrillator? Europace 2018;20:897–901.

[31] Ghanbari H, Dalloul G, Hasan R, Dacarett M, Saba S, David S, et al. Effectiveness of implantable cardioverter-defibrillators for the primary prevention of sudden cardiac death in women with advanced heart failure: a meta-analysis of randomized controlled trials. Arch Intern Med 2009;169:1500–6.

[32] Santageli P, Pelargonio G, Dello Russo A, Casella M, Bisceglia C, Barloletti S, et al. Gender differences in clinical outcome and primary prevention defibrillator benefit in patients with severe left ventricular dysfunction. Heart Rhythm 2010;7:876–82.

[33] MacFadden DR, Crystal E, Krahn AD, Mangat I, Healy JS, Dorian P, et al. Sex differences in implantable cardioverter-defibrillator outcomes: findings from a prospective defibrillator database. Ann Intern Med 2012;156(3):195–203.

[34] Zeitler EP, Hellkamp AS, Fonarow GC, Hammill SC, Curtis LH, Hernandez AF, et al. Primary prevention implantable cardioverter-defibrillator and survival in older women. JACC Heart Fail 2015;3:159–67.

[35] Wilcox JE, Fonarow GC, Zhang Y, Albert NM, Curtis AB, Gheorghiade M, et al. Clinical effectiveness of cardiac resynchronization and implantable cardioverter-defibrillator therapy in men and women with heart failure: findings from IMPROVE HF. Circ Heart Fail 2014;7:146–53.

[36] Sticherling C, Arendacka B, Svendsen JH, Wijers S, Friede T, Stockinger J, et al. Sex differences in outcomes of primary prevention implantable cardioverter-defibrillator therapy: combined registry data from eleven European countries. Europace 2018;20:963–70.

[37] Barra S, Providencia R, Duehmke R, Boveda S, Marijon E, Reitan C, et al. Sex-specific outcomes with addition of defibrillation to resynchronization therapy in patients with heart failure. Heart 2017;103:753–60.

[38] Kutyifa V. Gender-specific outcomes of cardiac resynchronization therapy with and without defibrillator. Heart 2017;103(10):732–3.

[39] Varma N, Mittal S, Prillinger JB, Snell J, Dalal N, Piccini JP. Survival in women versus men following implantation of pacemakers, defibrillators, and cardiac resynchronization therapy devices in a large, nationwide cohort. J Am Heart Assoc 2017;6, e005031.

[40] Zusterzeel R, Curtis JP, Canos DA, Sanders WE, Selzman KA, Pina IL, et al. Sex-specific mortality risk by QRS morphology and duration in patients receiving CRT: results from the NCDR. J Am Coll Cardiol 2014;64:887–94.

[41] Zusterzeel R, Spatz ES, Curtis JP, Sanders WE, Selzman KA, Pina IL, et al. Cardiac resynchronization therapy in women versus men: observational comparative effectiveness study from the national cardiovascular data registry. Circ Cardiovasc Qual Outcomes 2015;8:S4–S11.

[42] Rho RW, Patton KK, Poole JE, Cleland JG, Shadman R, Anand I, et al. Important differences in mode of death between men and women with heart failure who would qualify for a primary prevention implantable cardioverter-defibrillator. Circulation 2012;126:2402–7.

[43] Yap YG, Duong T, Bland M, Malik M, Torp-Pedersen C, Køber L, et al. Temporal trends on the risk of arrhythmic versus non-arrhythmic deaths in high risk patients after myocardial infarction: a combined analysis from multicenter trials. Eur Heart J 2005;26:1385–93.

[44] Barsheshet A, Wang PJ, Moss AJ, Solomon SD, Al-Ahmad A, McNitt S, et al. Reverse remodeling and risk of ventricular tachyarrhythmias in the MADI-CRT. J Am Coll Cardiol 2011;57:2416–23.

[45] Linde C, Stahlberg M, Benson L, Braunschweig F, Edner M, Dahlstrom U, et al. Gender, underutilization of cardiac resynchronization therapy, and prognostic impact of QRS prolongation and left bundle branch block in heart failure. Europace 2015;17:424–31.

[46] Zareba W, Klein H, Cygankiewicz I, Hall WJ, McNitt S, Brown M, et al. Effectiveness of cardiac resynchronization therapy by QRS morphology in the Multicender Automatic Defibrillator Implantation Trail-Cardiac Resynchronization Therapy (MADIT-CRT). Circulation 2011;123:1061–72.

[47] Lee NS, Lin F, Birgersdotter-Green U. Should women have different ECG criteria for CRT for CRT than men? J Cardiol 2017;101:1800–6.

[48] Varma N, Lappe J, He J, Niebauer M, Manne M, Tchou P. Sex-specific response to cardiac resynchronization therapy: effect of left ventricular size and QRS duration in left bundle branch block. JACC Clin Electrophysiol 2017;3:844–53.

[49] Linde C, Cleland JGF, Gold MR, Daubert JC, Tang ASL, Young JB, et al. The interaction of sex, height, and QRS duration on the effects of cardiac resynchronization therapy on morbidity and mortality: an individual-patient data meta-analysis: results from a case-based meta-analysis. Eur J Heart Fail 2018;20(4):780–91. https://doi.org/10.1002/ejhf.113.

[50] Stockburger M, Moss AJ, Klein HU, Zareba W, Goldenberg I, Biton Y, et al. Sustained clinical benefit of cardiac resynchronization therapy in non-LBBB patients with prolonged PR-interval: MADIT-CRT long-term follow up. Clin Res Cardiol 2016;105:944–52.

[51] Linde C, Bongiorni MG, Birgedotter-Green U, Curtis AB, Deisenhofer I, Furokawa T, et al. Sex differences in cardiac arrhythmia: a consensus document of the European Rhythm Association, endorsed by the Heart Rhythm Society and Asia Pacific Heart Rhythm Society. Europace 2018;20:1565.

[52] Hernandez AF, Fonarow GC, Liang L, Al-Khatib SM, Curtis LH, LaBresh KA, et al. Sex and racial differences in the use of implantable cardioverter defibrillators among patients hospitalized with heart failure. JAMA 2007;298:1525–32.

[53] Walsh MN, Yancy CW, Albert NM, Curtis AB, Gheorghiade M, Heywood JT, et al. Equitable improvement for women and men in the use of guideline-recommended therapies for heart failure: findings from IMPROVE HF. J Card Fail 2010;16(12):940–9.

[54] Al-Khatib SM, Hellkamp AS, Hernandez AF, Fonarow GC, Thomas KL, Al-Khalidi HR, et al. Trends in use of implantable cardioverter defibrillator therapy among patients hospitalized for heart failure. Have the previously observed sex and racial disparities changed over time? Circulation 2012;125:1094–101.

[55] Hess PL, Hernandez AF, Bhatt DL, Hellkamp AS, Yancy CW, Schwamm LH, et al. Sex and race/ethnicity differences in implantable cardioverter defibrillator counseling and use among patients hospitalized with heart failure:findings from the Get With The Guidelines—Heart Failure Program. Circulation 2016;134(7):517–26.

[56] Chatterjee NA, Borgquist R, Chang Y, Lewey J, Jackson VA, Singh JP, et al. Increasing sex differences in the use of cardiac resynchronization therapy with or without implantable cardioverter-defibrillator. Eur Heart J 2017;38:1485–94.

[57] Bogale N, Priori S, Gitt A, Alings M, Linde C, Dickstein K, et al. The European cardiac resynchronization therapy survey: selection and implantation practice vary according to centre volume. Europace 2011;13:1445–53.

[58] Peterson PN, Daugherty SL, Wang Y, Vidaillet HJ, Heidenreich PA, Curtis JP, et al. Gender differences in procedure-related adverse events in patients receiving implantable cardioverter-defibrillator therapy. Circulation 2009;119:1078–84.

[59] Jamerson D, McNitt S, Polonsky S, Zareba W, Moss A, Tompkins C. Early procedure-related adverse events by gender in MADIT-CRT. J Am Coll Cardiol 2014;25:985–9.

[60] Russo AM, Daugherty SL, Masoudi FA, Wang Y, Curtis J, Lampert R. Gender and outcomes after primary prevention implantable cardioverter-defibrillator implantation: findings from the National Cardiovascular Data Registry (NCDR). Am Heart J 2015;170(2):330–8.

[61] Pereira MP, Pogatzki-Zahn E. Gender aspects in postoperative pain. Curr Opinion Anesthesiol 2015;28:546–58.

[62] Schwenkglenks M, Gerberhagen HJ, Taylor RS, Pogatzki-Zahn E, Komann M, Rothaug J, et al. Correlates of satisfaction with pain treatment in the acute postoperative period: results from the international PAIN OUT registry. Pain 2014;155:1401–11.

[63] Toohill J, Fenwick J, Gamble J, Creedy DK, Buist A, Turkstra E, et al. A randomized controlled trial of a psycho-education intervention by midwives in reducing childbirth fear in pregnant women. Birth 2014;41:384–94.

Vascular Disease

Chapter 19

Peripheral Arterial Disease

Kajenny Srivaratharajah and Beth L. Abramson

Clinical Case

A 58-year-old woman presents to your office with generalized fatigue and leg pain. The pain is worse in her right leg and described as mostly exertional but sometimes also at rest. Pain limits her from walking more than 200 m and can persist long after exertion. Her past medical history is significant for prediabetes and dyslipidemia. Her only medication is Crestor 10 mg daily. She has a 30-pack-year smoking history and continues to smoke two to three cigarettes per day. Further history reveals pregnancy-related vascular complications including gestational diabetes and severe preeclampsia with her last pregnancy 20 years ago. On examination, her body mass index (BMI) is 32, blood pressure is 148/85 mmHg, and peripheral pulses are diminished in the right leg. What are this woman's risk factors for vascular disease? What are some diagnostic tests to clarify the etiology of her leg pain? How would you assess and manage this woman's risk and symptoms?

Abstract

Peripheral arterial disease (PAD) carries a significant health and economic burden with an estimated prevalence of 8.5 million Americans over the age of 40 and >200 million adults worldwide. PAD is often associated with polyvascular (the presence of more than one affected vascular bed) disease and confers up to a threefold increase in all-cause and cardiovascular mortality. Despite this, PAD remains underrecognized and undertreated compared to coronary artery and cerebrovascular disease. Significant sex differences exist in epidemiology, clinical presentation, treatment, and prognosis of PAD. Women comprise more than half of the total population affected by PAD. However, studies suggest women receive less optimal medical therapy and fewer endovascular or surgical revascularizations compared to men. Women often present with atypical or asymptomatic disease, are older with more comorbidity, have polyvascular or more advanced disease, and suffer from more functional impairment and cardiovascular and procedural mortality. This may be due to a combination of biological differences, diagnostic delays, and undertreatment. Greater enrollment of women in large clinical trials, specific focus of cardiovascular research on sex and gender differences in PAD, and reinforcement of preventative strategies, especially in women, are warranted.

Atherosclerosis involving vascular beds other than coronary and cerebral vasculature defines peripheral vascular disease. This chapter will focus on peripheral vascular disease that involves arterial beds below the aortoiliac bifurcation, generally referred to as peripheral arterial disease or PAD throughout this chapter. The focus of this section will remain on sex-specific differences in PAD.

PAD is the third most common presentation of cardiovascular disease after coronary artery disease (CAD) and stroke [1]. It is also associated with all-cause, cardiovascular, and cerebrovascular mortality [1]. More than 6.5 million or 5.5% of Americans over the age of 40 have PAD, as defined by low ankle-brachial index (ABI; <0.9) [1]. This definition may grossly underestimate true prevalence. In fact, when accounting for false-negative ABI results and incorporation of individuals who may have already undergone revascularization, estimates of data from 2000 suggest this number may be closer to ~8.5 million or 7.2% of Americans ≥ 40 years [1]. The burden of PAD is not unique to North America, but the first publication to highlight

global prevalence of PAD came from the Global Peripheral Artery Disease Study of 2013 [2]. This suggests that worldwide, there are approximately 202 million individuals with ABI ≤ 0.9, with women outnumbering men [1]. In fact, in low- and middle-income countries, percentage of population prevalence curves for PAD are higher in women for all ages up until 85 years, after which men have higher PAD prevalence [2]. Updated estimates from 2015 show similar numbers with overall global prevalence being ~236 million individuals ≥ 25 years, among whom ~123 million or ~52% are women [3] (see **Figure 1**).

Risk Factors for PAD

Risk factors for the development of PAD are similar to those of CAD. However, some notable differences exist including the stronger correlation between PAD and smoking [odds ratio (OR) for heavy smoking of 3.94 for symptomatic PAD vs. 1.66 for coronary heart disease] [1]. Similar to other forms of cardiovascular disease, PAD risk generally increases with age in both sexes [1–3]. A recent

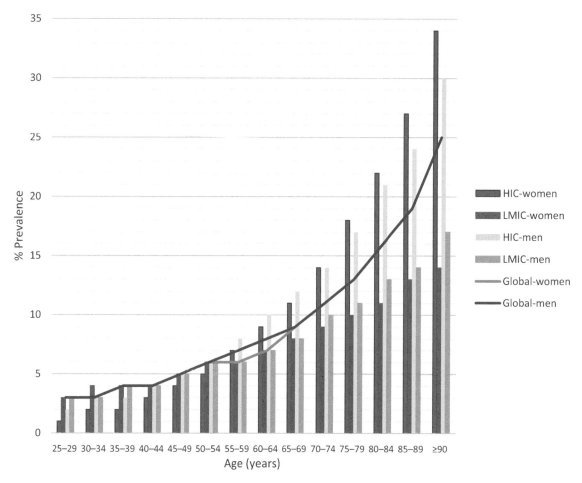

FIGURE 1 **Prevalence of Peripheral Artery Disease by Age in Men and Women in HIC and LMIC and Overall Global Prevalence in Men and Women.** HIC, high-income countries; LMIC, low- or middle-income countries. *Adapted from [3].*

systematic review and analysis of 118 articles found that smoking, diabetes, hypertension, and hypercholesterolemia were major risk factors for PAD (worldwide OR 1.70 for former smokers, 2.82 for current smokers, 1.89 for diabetes, 1.67 for hypertension, and 1.34 for hypercholesterolemia) [3]. Among traditional risk factors, diabetes and dysglycemia (or abnormal blood glucose including diabetes, impaired fasting glucose, and impaired glucose tolerance) are noted to increase the risk of intermittent claudication (IC) in women by fourfold [4]. Other risk factors include renal impairment [OR of 1.79 (1.03–3.12) in high-income countries only] and obesity (BMI $\geq 30 \text{kg/m}^2$; worldwide OR 1.55 [1.23–1.96]) [3]. In individuals with renal impairment under the age of 70, PAD is more prevalent in women than men [5]. Despite smoking, hypertension, and diabetes being more prevalent in men, global estimates suggest similar prevalence of PAD in women and men in high-income countries and higher prevalence of PAD in women than men in low- and middle-income countries [3]. This finding suggests other nontraditional, sex-specific risk factors may be involved.

There are limited data on sex-specific risk factors for PAD given the underenrollment of women in large clinical trials [6, 7]. In fact, women comprise only one-third of subjects in most large clinical trials to date [7]. **Table 1** outlines some recent randomized control trials in PAD and percentage of women enrolled. Sex-specific differences in clinical presentation have been observed, with PAD presenting on average 10–20 years later in women than men, suggesting a possible protective effect of estrogen which is lost following onset of menopause [8]. Estrogen's vascular effect is multifold and ranges from promotion of vasodilation, antioxidative effect, and reduction of vascular injury to reduced cytokine-mediated inflammation [8, 9]. However, the role of hormone replacement therapy in postmenopausal women and overall outcome benefits on PAD are not clear [8, 9]. In premenopausal women, the use of oral contraceptive pills (OCP) may in fact be associated with elevated vascular risk [10]. A case-control study of women between 18 and 49 years of age on OCP showed that all types of oral contraceptives were associated with increased PAD risk [10].

Sex-specific risk factors affecting women only including vascular complications in pregnancy such as preeclampsia and conditions with higher female prevalence such as hypothyroidism and osteoporosis warrant mention. Cardiovascular Health After Maternal Placental Syndromes (CHAMPS), a large retrospective cohort study, showed that the presence of maternal placental syndromes (defined as any of preeclampsia, gestational hypertension, placental abruption, and placental infarction) was associated with a threefold increase in PAD [11]. Underlying endothelial dysfunction may be a common driver for both maternal vascular complications and future risk of cardiovascular disease, including PAD [12] (see **Figure 2**). This suggests an earlier

window of opportunity to identify younger women at elevated risk of future and often premature vascular disease and implement preventative strategies. However, there are limited guidelines on this topic [12]. Evidence for the association of hypothyroidism and osteoporosis with PAD is less clear [8]. **Table 2** provides a summary of women-specific risk factors for PAD.

Role of Inflammatory Markers in PAD

Biomarkers for inflammation and the link to cardiovascular disease are a growing area of research interest. Inflammation plays a crucial role in atherosclerosis; therefore, markers of inflammation may function as early signals for PAD. Unfortunately, there are limited sex-specific data on inflammatory markers in PAD. However, in general, studies suggest that inflammatory biomarkers are more strongly correlated with PAD than CAD [13]. Of the various biomarkers of vascular wall biology studied to date, high-sensitivity C-reactive protein (CRP) (with a proposed threshold of 2 mg/L) seems to have the most evidence for independent association with both diagnosis and prognosis in PAD [14]. Therefore, this may be of clinical utility for risk stratification in PAD. A prospective cohort study of 27,935 US female health professionals ≥ 45 years of age without a diagnosis of vascular disease at baseline was conducted to determine nontraditional biomarkers that predict future risk of symptomatic PAD [15]. This study showed that among four other risk factors, high-sensitivity CRP was significantly associated with increased risk of symptomatic PAD [adjusted hazard ratio (HR) 2.1, 95% confidence interval (CI) 1.2–3.7]. Results from the Scottish Heart Health Extended Cohort (SHHEC) of 15,737 disease-free men and women between the ages of 30 and 75 suggest that after age, the second highest-ranked continuous variable based on HR for PAD is high-sensitivity C-reactive protein (hsCRP), whereas for CAD it is total cholesterol. In this study, the HR for PAD in those with elevated hsCRP was 1.51 (95% CI 1.27–1.80) in women vs. 1.37 (95% CI 1.19–1.58) in men [13]. A recent cross-sectional study of 156 hemodialysis patients suggests that higher CRP level is independently associated with PAD in this population as well [16].

Studies also suggest an association between markers of inflammation such as D-dimer and CRP, and PAD prognosis and/or functional impairment. For example, a systematic review and meta-analysis of 16 studies involving 5041 patients with PAD suggests that CRP may be a useful prognostic marker that is predictive of major cardiovascular events (HR 1.38, 95% CI 1.16–1.63, $p < 0.001$ for every increase in $\log_e$CRP) [17]. An older study suggests poor functional status among individuals with PAD and elevated D-dimer [18]. More recent studies call for a multimarker approach or biomarker panel to screen for PAD. The following panel

TABLE 1 Key Clinical Trials Related to PAD

Study	Year	Authors	Total Enrollment	% Women	Intervention	Primary Outcome	Secondary or PAD-Specific Outcomes
CAPRIE	1996	CAPRIE Steering Committee	19,185	28	Clopidogrel vs. aspirin	Composite outcome of ischemic stroke, MI, or vascular death occurred in 5.32% of intervention group and 5.83% of controls ($p = 0.043$)	In the PAD subgroup analysis, the primary outcome occurred in 3.71% of the intervention group and 4.86% of the control group ($p = 0.0028$)
Effects of an angiotensin-converting enzyme inhibitor, ramipril, on cardiovascular events in high-risk patients	2000	Heart Outcomes Prevention Evaluation Study Investigators	9297	27	Ramipril vs. placebo	22% reduction in risk of composite outcome of MI, stroke, or death from CV causes with ramipril use ($p < 0.001$)	Death from CV causes was 8.1% in placebo group vs. 6.1% in intervention group (RR 0.74; $p < 0.001$)
Telmisartan, ramipril, or both in patients at high risk for vascular events	2008	Yusuf S. et al. (ONTARGET investigators)	25,620	27	Ramipril vs. telmisartan vs. combination	Composite outcome of death from CV causes, MI, stroke, or hospitalization for heart failure was not different between groups	Composite of death from CV causes, MI, or stroke was also not significantly different between groups
Patients with peripheral arterial disease in the CHARISMA trial	2009	Cacoub et al.	15,603 (subgroup analysis of 3096 patients with PAD)	30	Clopidogrel plus low-dose aspirin vs. placebo plus low-dose aspirin	Composite of MI, stroke, or death from CV causes was not significantly different between the groups	Rate of MI and rate of hospitalization for ischemic events was lower in dual antiplatelet group ($p = 0.029$ and $p = 0.011$, respectively)
Statin therapy and long-term adverse limb outcomes in patients with peripheral artery disease: insights from the REACH registry	2014	Kumbhani et al.	5861	27	Statin use (not RCT, prospective registry data)	Statin use was associated with 18% lower adverse limb outcome rates	Statin use was associated with a reduction in composite of CV death, MI, and stroke (HR 0.83; $p = 0.01$)

Trial	Year	Author	N	Ref	Intervention	Primary outcome	Safety/secondary outcome
EUCLID	2017	Hiatt et al.	13,885	28	Ticagrelor vs. clopidogrel	Composite of CV death, MI, or ischemic stroke occurred in 10.8% of intervention group and 10.6% of controls ($p=0.65$)	Safety outcome of major bleeding was comparable in both groups (HR 1.1, 95% CI 0.84–1.43, $p=0.49$) as was acute limb ischemia (HR 1.03, 95% CI 0.79–1.33, $p=0.85$)
Low-density lipoprotein cholesterol lowering with evolocumab and outcomes in patients with peripheral artery disease	2018	Bonaca et al.	27,564	25	Evolocumab vs. placebo	Primary outcome (a composite of CV death, MI, stroke and hospital admission for UA, or coronary revascularization) was reduced in patients with PAD (HR 0.79, 95% CI 0.66–0.94; $p=0.0098$)	Secondary end point of composite CV death, MI, or stroke was reduced in those with PAD (HR 0.73, 95% CI 0.59–0.91; $p=0.0040$). Evolocumab reduced risk of MALE in all patients (HR 0.58, 95% CI 0.38–0.88, $p=0.0093$)
MACE and mortality in patients with peripheral artery disease: the COMPASS trial	2018	Anand et al.	6391	28	Rivaroxaban 2.5 mg twice daily + aspirin vs. rivaroxaban alone vs. aspirin alone	The intervention reduced the incidence of MALE by 43% ($p=0.01$)	The intervention also reduced total vascular amputations by 58% ($p=0.01$), peripheral vascular intervention by 24% ($p=0.03$), and all peripheral vascular outcomes by 24% ($p=0.02$)

CI, confidence interval; CV, cardiovascular; HR, hazard ratio; MALE, major adverse limb events; MI, myocardial infarction; PAD, peripheral arterial disease; RCT, randomized controlled trial; RR, relative risk; UA, unstable angina.

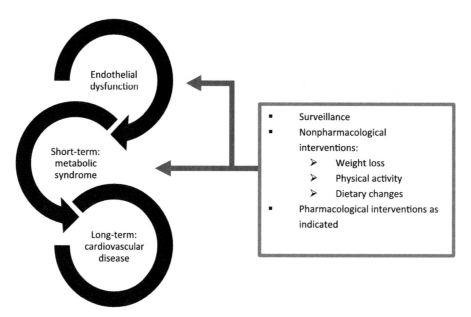

FIGURE 2 Cardiovascular Disease Cascade Relating to Nontraditional, Sex-Specific Risk Factors Such as Maternal Placental Syndromes (e.g., Preeclampsia). *Adapted from [12].*

Variable	Women-Specific Factors	Sex-Specific Data Not Available or No Differences Found
TABLE 2 Sex Differences in PAD Risk, Presentation, Screening, Treatment, and Outcomes		
Risk factors	Menopause	
	OCP	
	Elevated C-reactive protein	
	Vascular complications in pregnancy (e.g., preeclampsia)	
	Osteoporosis-conflicting evidence	
	Hypothyroidism-conflicting evidence	
Presentation	Often older (may be related to diagnostic delays)	
	Often asymptomatic or with atypical symptoms	
	More femoropopliteal lesions	
	More advanced disease and critical limb ischemia	
Screening and diagnosis	Presymptomatic screening may be warranted	No difference in ABI values in contemporary data. Further study warranted
	Screening symptomatic patients detects only ~10% of cases	
Treatment	Lower use of medical therapy, including antiplatelet agents, statins, and angiotensin-converting enzyme inhibitors	No difference in treatment response. Further study warranted
Cardiovascular outcomes	Two to fourfold increased risk of cardiovascular mortality and morbidity compared with women without PAD	
	Greater admission to hospital with acute myocardial infarction	

Variable	Women-Specific Factors	Sex-Specific Data Not Available or No Differences Found
Limb outcomes	Fewer minor amputations and arterial bypass operations; however, this may represent referral bias and under treatment	
Perioperative outcomes	Increased in-hospital mortality regardless of disease severity More perioperative bleeding complications and longer perioperative stay	

TABLE 2 Sex Differences in PAD Risk, Presentation, Screening, Treatment, and Outcomes—Cont'd

ABI, ankle-brachial index; OCP, oral contraceptive pills; PAD, peripheral arterial disease. *Adapted from [8].*

of markers was found to have increased association with PAD, independent of known traditional risk factors such as diabetes, hypertension, and smoking: beta2-microglobulin, cystatin C, hsCRP, and glucose [19]. Further study on the role of inflammatory biomarkers in screening, diagnosis, and outcomes in PAD is needed, with a specific focus on sex differences.

Presentation of PAD

The classic symptom of PAD is IC; however, less than 10% of PAD fits this description. Up to 40% of individuals with PAD may not complain of leg symptoms and the remainder often endorse atypical symptoms [1]. The majority of women diagnosed with PAD are asymptomatic at the time of diagnosis, yet present with an ABI < 0.9. However, when symptomatic, women often tend to be older, with more complex or multilevel and severe disease [8]. It is not clear whether this is the result of initial atypical or absent symptoms resulting in delays in recognition, diagnosis, and management of PAD in women.

ABI remains the method of choice for screening for PAD and is the basis for diagnosis of PAD in the vast majority of clinical trials. Although a cutoff of < 0.9 is used in most studies, the ABI Collaboration reports risk increase as the ABI falls below 1.10 [20]. Moreover, an ABI > 1.40 also poses risk in that this signifies arterial calcification, which in turn is frequently associated with occlusive PAD [21]. Data show a strong correlation between an abnormal ABI and coronary and cerebral vascular disease [21]. The Atherosclerosis Risk in Communities (ARIC) study showed similar association between low ABI and CAD among men and women but no association between low ABI and stroke in women [21]. This sex difference is not consistent in other studies [21]. Further study is needed on possible sex-specific differences in ABI cutoffs and correlation to clinical outcomes.

Despite a large portion of PAD presenting as asymptomatic disease, the United States Preventive Services Task Force (USPSTF) reports that current evidence is insufficient to assess net benefit or harms of screening for PAD and cardiovascular disease (CVD) risk with the use of ABI in asymptomatic adults [22]. On the contrary, data from the Viborg Vascular (VIVA) trial, an RCT in Denmark assessing population-based vascular screening, showed a 7% reduction in relative risk of mortality among Danish men and cost-effectiveness [23, 24]. Similar benefit is expected for women with PAD and studies looking at a comprehensive population-based screening program are warranted. Limited studies explore other risk stratification models in PAD and require further validation [25].

Treatment of Patients With PAD

Nonoperative management of PAD involves a combination of nonpharmacological and pharmacological strategies for aggressive risk factor modification including smoking cessation, lipid management, blood pressure and diabetes management, antiplatelet therapy, and exercise and rehabilitation. No clear differences in response to medical therapy exist between men and women [26, 27]. However, there is evidence that adherence to preventative therapy in all individuals, both men and women, is suboptimal. NHANES data from 1999 to 2004 show a 65% reduction in all-cause mortality with use of multiple preventative therapies in individuals with PAD (diagnosed based on ABI, without cardiovascular disease), but only 24–34% adherence to these therapies [28]. Several studies further suggest that pharmacological therapies are underused in women, with less than 40% of women on antiplatelet therapy compared to 58% of men and 63% of women on statin therapy compared to 87% in men [29]. Similarly, Canadian data suggest that the use of systemic vascular treatment for PAD remains suboptimal, particularly in older patients and women [30]. In this observational retrospective study including 43.8% women, fewer women were on all three preventative therapies [statin, antiplatelet therapy, and angiotensin-converting enzyme inhibitor (ACEI)] compared to men (18.2% vs. 22.4%, $p < 0.001$) [30]. Other pharmacotherapy used primarily for symptom management includes cilostazol, pentoxifylline, naftidrofuryl, and carnitine [29,31]. The theorized effects of these agents are multifold including vasodilation, inhibition of platelet aggregation, and correction of secondary metabolic deficiencies. Given that these agents are not consistently available internationally and are supported by variable evidence with no specific focus on the effects of these agents on women, further discussion on this topic is limited in this chapter [29, 31].

Antiplatelet and Antithrombotic Therapy

Earlier trials that established the role of antiplatelet therapy in PAD include CAPRIE and CHARISMA [32, 33]. The former suggested benefit of clopidogrel in patients with atherosclerosis via reduced CV events when compared to aspirin alone, particularly in the PAD subgroup [32]. CHARISMA assessed the role for dual antiplatelet therapy with clopidogrel and aspirin in those with symptomatic atherosclerosis, including PAD, and showed no improvement over clopidogrel alone, with excess bleeding complications [33]. EUCLID (examining use of ticagrelor in peripheral artery disease) is a multicenter, double-blind RCT examining whether ticagrelor is superior to clopidogrel in symptomatic PAD when comparing a composite of cardiovascular death, MI, or ischemic stroke [34]. The primary outcome in this trial was similar. Subgroup analysis did not suggest a sex-specific difference in outcomes but only 28% of the enrolled were women [34]. A priori sex differences were not studied in these trials.

A recent trial, although not a sex-specific analysis, COMPASS, has shown relevant outcomes with the combination of rivaroxaban low dose at 2.5 mg twice daily and aspirin 100 mg daily [35, 36]. This study included 6391 patients with PAD categorized on the basis of a history of claudication (46%), prior peripheral revascularization (27%), prior amputation (4%), ≥50% carotid stenosis or prior carotid revascularization (26%), and ABI of <0.90 but no clinical PAD (19%). Outcome data show 28% reduction in major adverse cardiovascular events, which include 46% reduction in major adverse limb events and 44% reduction in acute limb ischemia [35]. This benefit is offset by an increase in major bleeding (HR 1.61, $p = 0.0009$) [35]. However, net analysis still suggests benefit of this strategy over ASA alone for high-risk patients [35, 36]. Less than one-third of enrolled patients or approximately 1786 participants were women. For the primary outcome of cardiovascular death, myocardial infarction, or stroke, HR for rivaroxaban plus aspirin vs. aspirin alone was 0.76 for men

and 0.72 for women (p for interaction = 0.075) [36]. No sex-specific outcome analysis was completed for patients in the PAD subgroup [35].

Statin Therapy

Although there are limited sex-specific data, there is strong evidence for lipid lowering in PAD to reduce cardiovascular mortality. Guidelines including American Heart Association (AHA)/American College of Cardiology (ACC) make Grade A recommendations for utilization of lipid-lowering therapy in PAD [37]. A Cochrane review of lipid-lowering therapy in PAD of the lower limb examined 18 trials with a total of 10,049 participants and reported that overall pooled data did not show statistically significant effect on mortality or CV events [38]. However, with the exclusion of one specific trial that reported possible harm of lipid-lowering therapy, the subgroup analysis showed reduced risk of CV events (OR 0.74, CI 0.55–0.98). Some smaller trials included in this review also suggest improved walking distance and pain-free walking distance with lipid-lowering therapy [38]. The Reduction of Atherothrombosis for Continued Health (REACH) registry data suggest approximately 17% reduction in cardiovascular events (CV death, MI, and stroke) and 18% reduction in rate of adverse limb outcomes (worsening symptoms, peripheral revascularization, and ischemic amputations) with statin use [39]. Recent data also support the use of PCSK9 inhibitor therapy for lipid lowering in PAD [40]. Evolocumab showed not only a 21% reduction in a composite of CV death, MI, stroke, and hospital admission for UA or coronary revascularization, but also a 42% reduction in major adverse limb events [40].

Antihypertensive Therapy

Risk factor modification in PAD is important. Apart from lifestyle modification such as reduced dietary sodium, weight loss, and increased physical activity, pharmacological intervention to target blood pressure (BP) < 130/80 is recommended in the 2017 ACC/AHA multisociety guideline combined guideline for hypertension management in adults [41]. This guideline reports no evidence for any one class of antihypertensive medication on the basis of results from the International Veramapil-Trandolapril (INVEST) study which compared atenolol ± hydrocholorothiazide with verapamil ± perinodpril and ALLHAT, both reporting no difference in the classes of BP medications used on PAD. INVEST enrolled a majority of women, ~52% of 22,576 participants, with a mean age of 66 years, whereas 47% of the enrollment in ALLHAT were women with the majority (57.6%) being >65 years. Although not a blood pressure-lowering trial, the Heart Outcomes Prevention Trial (HOPE) and the Ongoing Telmisartan Alone and in Combination with Ramipril Global Endpoint Trial (ONTARGET) have both shown reduced CV events in patients with PAD treated with angiotensin-converting enzyme (ACE)/angiotensin receptor blocker (ARB) therapy [42, 43]. Less than one-third of the total 9297 participants in the HOPE trial and 25,620 participants in the ONTARGET trial were women. Sex-specific outcomes were reported in HOPE with RR 0.76 (0.5–0.88) in men vs. RR 0.79 (0.69–0.98) in women [44].

Exercise and Rehabilitation

There is a paucity of data on sex differences in outcomes following supervised exercise therapy. For both women and men, a meta-analysis of studies comparing endovascular and noninvasive therapy in treatment of IC showed that despite improved maximum walking distance with endovascular treatment compared to medical treatment, the addition of a supervised exercise program (SEP) for patients further improved outcomes [45]. A similar conclusion was reached in a more recent meta-analysis of seven RCTs with a total of 987 patients [46]. These trials enrolled approximately 40% women and showed that the addition of a SEP to endovascular treatment improved maximum walking distance, ABI, and risk of future revascularization or amputation in all individuals. Sex-specific analysis was not provided. On the basis of these results, general recommendations are for SEP at least three times per week for 30–60 min for at least 12 weeks in those with claudication [29]. A Cochrane review of 10 RCTs, with a total 1087 patients, provided further support for the synergistic effect of combined endovascular revascularization and supervised exercise [47]. A recent trial suggests sex- or gender-based IC treatment strategies need to be developed as women with IC appear to have less benefit from SEP in the first 3 months and an overall lower absolute walking distance after 1 year [48].

Prognosis and Outcomes After Revascularization

Global mortality from PAD, expressed as percent change rate from 2006 to 2016, shows 4.7% reduction in men vs. only 0.8% reduction in women [1]. American data from 2016 show 45% mortality among men of all ages vs. 54.9% in women with PAD [1]. This disparity may be explained by underrecognition and undertreatment of PAD in women. Similar differences are notable in morbidity, quality-of-life measures, and revascularization outcomes in women with PAD. Data from the PORTRAIT study, a multicenter, international prospective study, suggest that despite similar ABI, women were more symptomatic, had poor physical and social functioning, and higher rates of depression than men [49].

PAD increases the risk of CVD mortality and morbidity in both men and women, with studies suggestive of a three-fold increase for all-cause and CVD mortality in those with low ABI (<0.9) [20]. Comparing women with and without PAD, women with PAD have a two- to fourfold increased risk of CVD mortality and morbidity [50]. Women with PAD also seem to have fewer hospitalizations for vascular procedural interventions, but higher in-hospital mortality regardless of disease severity or type of revascularization [51]. Emergent admissions and postadmission complications including perioperative bleeding, longer length of stay and need for rehabilitation, and/or discharge to nursing homes are more frequent in women than men [52].

A large US nationwide sample of 1,797,885 patients with IC and critical limb ischemia showed that in-hospital mortality was higher in women than men (0.5% vs. 0.2% after percutaneous transluminal angioplasty and 1% vs. 0.7% after surgical intervention for those with IC; 2.3% vs. 1.6% after percutaneous transluminal angioplasty; and 2.7% vs. 2.2% after surgical intervention for critical limb ischemia) [51]. Observational data from a Korean nationwide, multicenter registry of 550 women with PAD suggest more comorbid disease, severe and complex target lesions, higher death rates, MI or major amputations and higher major adverse limb events, and procedural complications in women than men [53]. Overall, this difference may be explained by delays in diagnosis and, therefore, more advanced disease, comorbidity and frailty at time of intervention for women, and overall lower rates of preventative strategies as outlined above.

FMD and Other Forms of Nonatherosclerotic Vascular Disease

The most common cause of PAD is atherosclerotic disease. However, other less common causes such as vascular inflammatory and noninflammatory diseases warrant mention. Specifically, fibromuscular dysplasia (FMD), which is more prevalent in women, is an example of noninflammatory PAD [54–56]. The prevalence of FMD among the general population is not known but a 3–4% range is found based on data from kidney donors [55]. Moreover, there is a paucity of studies on FMD assessment and management. Data from the first 447 patients in a US national FMD registry show that 91% were women, with mean age at diagnosis 51.9 and common areas of involvement being renal arteries, extracranial carotid arteries, and vertebral arteries [55]. Claudication was the presenting symptom in only 5% of these FMD patients [55]. The etiology of FMD is unclear, but a genetic link as a subset with autosomal dominant inheritance has been suggested [56]. Currently there are no genetic tests specific for FMD.

FMD is diagnosed on the basis of characteristic findings on duplex ultrasound (US), computed tomography angiography (CTA), or magnetic resonance angiography (MRA). Findings include a string of beads appearance (or multifocal FMD) and focal lesions involving medium or small arteries, in addition to arterial aneurysm, dissection, and arterial tortuosity [56, 57]. Other overlapping arteriopathies should be excluded. Common clinical signs and symptoms of FMD are related to the most common presentations of FMD renal artery and cerebrovascular FMD. In the Cardiovascular Outcomes in Renal Atherosclerotic Lesions (CORAL) study, 5.8% of patients with renovascular hypertension had incidental renal FMD. Clinical signs of renal artery FMD include hypertension in individuals <30 years of age, especially women; drug-resistant hypertension (suboptimal control despite three drug therapies at optimal doses including a diuretic); unilateral small kidney without other urological etiology; abdominal bruit without atherosclerotic disease or risk factors; and renal artery dissection/infarction [55, 56]. The hallmark of cerebrovascular FMD includes severe or chronic headaches, pulsatile tinnitus, cervical bruit, stroke/transient ischemic attack (TIA)/amaurosis fugax, or focal neurological findings suggestive of cervical artery dissection [55, 56]. Much less common, FMD in the lower extremity arteries may present with symptoms such as claudication, foot/ toe ischemia, or atypical leg symptoms [56]. Despite the initial vascular territory involved, all patients with FMD are recommended to undergo an assessment for intracranial aneurysm with CTA or MRA [56, 57].

Recent guidelines on FMD suggest general management strategies including antiplatelet therapy to prevent thrombotic complications in FMD, use of ACEI and ARBs in the setting of renovascular hypertension management, and catheter-based angiography and revascularization where appropriate [56]. Nonpharmacological strategies such as smoking cessation also seem to have an effect on FMD prognosis such as reduced need for vascular procedures [56]. Despite questions being raised on exogenous estrogen use, currently there is no clear association between exogenous estrogen and FMD [56].

Spontaneous coronary artery dissection (SCAD), although a generally uncommon cause for acute myocardial infarction (AMI), may represent up to a quarter of AMI among women under age 50 [56]. Small studies have shown a prevalence of FMD ranging from 16% to 60% in those with SCAD [56]. Studies suggest multifocal extracoronary FMD in those with SCAD. Imaging all vessels from brain to pelvis with CTA or MRA is recommended, at least once in those patients diagnosed with SCAD, in order to assess for noncoronary involvement of FMD [56]. In contrast, <3% of patients in the US FMD Registry had SCAD. Therefore, routine screening of patient with FMD for SCAD, in the absence of symptoms such as angina, is not recommended.

Although rare, other forms of nonatherosclerotic disease warrant brief discussion, given higher prevalence among women. Takayasu's arteritis is a large vessel vasculitis affecting predominantly young women (median age at onset 25 years) and Asian populations [57]. It is exceedingly rare in the United States, with incidence rate estimated at 2.6 cases per million individuals per year [57]. Diagnosis is made on the basis of at least three of the following six findings: age at disease onset ≤40, claudication of extremities, diminished brachial artery pulse, difference of >10 mmHg in systolic pressure between arms, bruit over a subclavian artery or aorta, and arteriographic narrowing of entire aorta, its main branches or large arteries in the proximal extremities in the absence of features suggestive of other etiology such as atherosclerosis or FMD [57]. CTA or MRA may be used in the diagnosis of Takayasu's. Once the diagnosis is made, distinguishing active disease and appropriate treatment, which may involve biologic agents, requires subspecialist referral (rheumatologist) and a multidisciplinary approach. Comparatively common in the population, Raynaud's disease and chilblains affect smaller vessels and are driven by vasospastic and vasoconstrictive changes [57]. The latter results in tissue injury and subsequent swelling, blistering, and ulceration of extremities [57]. Although Raynaud's disease may be seen with FMD [56], the connection with chilblains is less clear. Treatment for both Raynaud's disease and chilblains includes nonpharmacological measures such as cold avoidance and pharmacological therapy including vasodilation using calcium channel blockers [57].

Conclusion

In summary, limited data are available on sex-specific differences in diagnosis and management of PAD. However, studies to date suggest PAD prevalence is similar or higher in women than men, but in general, women receive less optimal care. Greater enrollment of women and more focus on PAD in cardiovascular trials are warranted. See **Figure 3** for a summary of features of vascular disease in women discussed in this chapter.

FIGURE 3 **Features of Vascular Disease in Women.** PAD, peripheral arterial disease.

Key Points

1   Women often present with atypical or asymptomatic disease.

2   Sex-specific cutoffs for diagnosis and prognostication using ABI need further study.

3   Biomarkers such as hsCRP may be considered in early risk assessment and prognostication, especially in asymptomatic women.

4   Sex-specific management of PAD should focus on aggressive secondary prevention for women with PAD, and perhaps earlier consideration of revascularization therapies.

5   Women with PAD are often older with more comorbidity, have polyvascular or more advanced disease, and suffer from more functional impairment and cardiovascular and/or procedural mortality.

Back to Clinical Case

In this middle-aged woman with atypical leg symptoms, it is important to consider both atherosclerotic and nonathero-sclerotic etiologies. With regard to atherosclerotic risk factors, smoking and hyperlipidemia are notable in this case. However, one must also acknowledge sex-specific risk factors such as preeclampsia. In risk stratification and diagnosis, ABI with cutoff of < 0.9 is recommended. Further diagnostic testing may involve US doppler of lower extremities to quantify degree of obstruction. Computed tomography (CT) angiogram may be useful in diagnosis of FMD or other nonatherosclerotic arteriopathy. In the management of this patient, nonpharmacological strategies should include smoking cessation, weight loss, healthy diet, and physical exercise. Pharmacological strategies recommended include the introduction of an antiplatelet agent such as aspirin, consideration of low-dose rivaroxaban, optimization of statin therapy to target low-density lipoprotein (LDL) < 1.8 mmol/L or 50% reduction, and ACEI/ARB therapy for blood pressure management to target $< 130/80$.

References

[1] Benjamin EJ, Muntner P, Alonso A, Bittencourt MS, Callaway CW, Carson AP, et al. Heart disease and stroke statistics—2019 update: a report from the American Heart Association. Circulation 2019;139(10):e56–e528. https://doi.org/10.1161/CIR.0000000000000659.

[2] Fowkes FGR, Rudan D, Rudan I, Aboyans V, Denenberg JO, McDermott MM, et al. Comparison of global estimates of prevalence and risk factors for peripheral artery disease in 2000 and 2010: a systematic review and analysis. Lancet 2013;382:1329–40. https://doi.org/10.1016/S0140-6736(13)61249-0.

[3] Song P, Rudan D, Zhu Y, Fowkes FJI, Rahimi K, Fowkes FGR, et al. Global, regional, and national prevalence and risk factors for peripheral artery disease in 2015: an updated systematic review and analysis. Lancet Glob Health 2019;7(8):e1020–30. https://doi.org/10.1016/S2214-109X(19)30255-4.

[4] Kannel WB, McGee DL. Update on some epidemiologic features of intermittent claudication: the Framingham study. J Am Geriatr Soc 1985;33:13–8.

[5] Wang GJ, Shaw PA, Townsend RR, Anderson AH, Xie D, Wang X, et al. Sex differences in the incidence of peripheral artery disease in the chronic renal insufficiency cohort. Circ Cardiovasc Qual Outcomes 2016;9:S86–93.

[6] Vyas MV, Mrkobrada M, Donner A, Hackman DG. Under representation of peripheral artery disease in modern cardiovascular trials: systematic review and meta-analysis. Int J Cardiol 2013;168:4875–6.

[7] Jelani QU, Petrov M, Martinez SC, Holmvang L, Al-Shaibi K, Alasnag M. Peripheral arterial disease in women: an overview of risk factor profile, clinical features, and outcomes. Curr Atheroscler Rep 2018;20(8):40. https://doi.org/10.1007/s11883-018-0742-x.

[8] Srivaratharajah K, Abramson BL. Women and peripheral arterial disease: a review of sex differences in epidemiology, clinical manifestations, and outcomes. Can J Cardiol 2018;34(4):356–61. https://doi.org/10.1016/j.cjca.2018.01.009.

[9] Nguyen L, Liles DR, Lin PH, Bush RL. Hormone replacement therapy and peripheral vascular disease in women. Vasc Endovascular Surg 2004;38:547–56.

[10] Van Den Bosch MAAJ, Kemmeren JM, Tanis BC, Mali WPTM, Helmerhorst FM, Rosendaal FR, et al. The RATIO study: oral contraceptives and the risk of peripheral arterial disease in young women. J Thromb Haemost 2003;1(3):439–44.

[11] Ray JG, Vermeulen MJ, Schull MJ, Redelmeier DA. Cardiovascular health after maternal placental syndromes (CHAMPS): population-based retrospective cohort study. Lancet 2005;366(9499):1797–803. https://doi.org/10.1016/S0140-6736(05)67726-4.

[12] Srivaratharajah K, Abramson BL. Identifying and managing younger women at high risk of cardiovascular disease. Can Med Assoc J 2019;191(6):E159. https://doi.org/10.1503/cmaj.180053.

[13] Tunstall-Pedoe H, Peters SAE, Woodward M, Struthers AD, Belch JJF. Twenty-year predictors of peripheral arterial disease compared with coronary heart disease in the Scottish Heart Health Extended Cohort (SHHEC) [published correction appears in J Am Heart Assoc. 2017 Dec 23;6(12):e004215]. J Am Heart Assoc 2017;6(9). https://doi.org/10.1161/JAHA.117.005967.

[14] Vlachopoulos C, Xaplanteris P, Aboyans V, Brodmann M, Cífková R, Cosentino F, et al. The role of vascular biomarkers for primary and secondary prevention. A position paper from the European Society of Cardiology Working Group on peripheral circulation: endorsed by the Association for Research into arterial structure and physiology (ARTERY) society. Atherosclerosis 2015;241(2):507–32. https://doi.org/10.1016/j.atherosclerosis.2015.05.007.

[15] Pradhan AD, Shrivastava S, Cook NR, Rifai N, Creager MA, Ridker PM. Symptomatic peripheral arterial disease in women. Circulation 2008;117(6):823–31. https://doi.org/10.1161/CIRCULATIONAHA.107.719369.

[16] Ašćerić RR, Dimković NB, Trajković GŽ, Ristić BS, Janković AN, Đurić PS, et al. Prevalence, clinical characteristics, and predictors of peripheral arterial disease in hemodialysis patients: a cross-sectional study. BMC Nephrol 2019;20:281. https://doi.org/10.1186/s12882-019-1468-x.

[17] Singh TP, Morris DR, Smith S, Moxon JV, Golledge J. Systematic review and meta-analysis of the association between C-reactive protein and major cardiovascular events in patients with peripheral artery disease. Eur J Vasc Endovasc Surg 2017;54(2):220–33. https://doi.org/10.1016/j.ejvs.2017.05.009.

[18] McDermott MM, Greenland P, Green D, Guralnik JM, Criqui MH, Liu K, et al. D-dimer, inflammatory markers, and lower extremity functioning in patients with and without peripheral arterial disease. Circulation 2003;107(25):3191–8. https://doi.org/10.1161/01.CIR.0000074227.53616.CC.

[19] Cooke JP, Wilson AM. Biomarkers of peripheral arterial disease. J Am Coll Cardiol 2010;55(19):2017–23. https://doi.org/10.1016/j.jacc.2009.08.090.

[20] Ankle Brachial Index Collaboration, Fowkes FG, Murray GD, Butcher I, Heald CL, Lee RJ, et al. Ankle brachial index combined with Framingham risk score to predict cardiovascular events and mortality: a meta-analysis. JAMA 2008;300:197–208.

[21] Aboyans V, Criqui MH, Abraham R, Allison MA, Creager MA, Diehm C, et al. Measurement and interpretation of the ankle-brachial index: a scientific statement from the American Heart Association. Circulation 2012;126:2890–909.

[22] US Preventive Services Task Force. Screening for peripheral artery disease and cardiovascular disease risk assessment with the ankle-brachial index: US preventive services task force recommendation statement. JAMA 2018;320(2):177–83. https://doi.org/10.1001/jama.2018.8357.

[23] Lindholt JS, Søgaard R. Population screening and intervention for vascular disease in Danish men (VIVA): a randomised controlled trial. Lancet 2017;390(10109):2256–65. https://doi.org/10.1016/S0140-6736(17)32250-X.

[24] Søgaard R, Lindholt JS. Cost-effectiveness of population-based vascular disease screening and intervention in men from the Viborg vascular (VIVA) trial. Br J Surg 2018;105:1283–93. https://doi.org/10.1002/bjs.10872.

[25] Mansoor H, Elgendy IY, Williams RS, Joseph VW, Hong Y-R, Mainous AG. A risk score assessment tool for peripheral arterial disease in women: from the National Health and nutrition examination survey. Clin Cardiol 2018;41:1084–90. https://doi.org/10.1002/clc.23032.

[26] Abramson BL, Huckell V, Anand S, Forbes T, Gupta A, Harris K, et al. Canadian cardiovascular society consensus conference: peripheral arterial disease – executive summary. Can J Cardiol 2005;21:997–1006.

[27] Hirsch AT, Allison MA, Gomes AS, Corriere MA, Duval S, Ershow AG, et al. A call to action: women and peripheral artery disease: a scientific statement from the American Heart Association. Circulation 2012;125:1449–72.

[28] Pande RL, Perlstein TS, Beckerman JA, Creager MA. Secondary prevention and mortality in peripheral artery disease: National Health and nutrition examination study, 1999-2004. Circulation 2011;124:17–23.

[29] Saati A, AlHajri N, Ya'qoub L, Ahmed W, Alasnag M. Peripheral vascular disease in women: therapeutic options in 2019. Curr Treat Options Cardiovasc Med 2019;21:68. https://doi.org/10.1007/s11936-019-0769-5.

[30] Pâquet M, Pilon D, Tétrault JP, Carrier N. Protective vascular treatment of patients with peripheral arterial disease: guideline adherence according to year, age and gender. Can J Public Health 2010;101(1):96–100.

[31] Salhiyyah K, Forster R, Senanayake E, Abdel-Hadi M, Booth A, Michaels JA. Pentoxifylline for intermittent claudication. Cochrane Database Syst Rev 2015;9. https://doi.org/10.1002/14651858.CD005262.pub3.

[32] CAPRIE Steering Committee. A randomized, blinded, trial of clopidogrel versus aspirin in patients at risk of ischemic events (CAPRIE). Lancet 1996;348:1329–39.

[33] Cacoub PP, Bhatt DL, Steg PG, Topol EJ, Creager MA, CHARISMA Investigators. Patients with peripheral arterial disease in the CHARISMA trial. Eur Heart J 2009;30(2):192–201.

[34] Hiatt WR, Fowkes FGR, Heizer G, Berger JS, Baumgartner I, Held P, et al. Ticagrelor versus clopidogrel in symptomatic peripheral artery disease. N Engl J Med 2017;376(1):32–40. https://doi.org/10.1056/NEJMoa1611688.

[35] Anand SS, Caron F, Eikelboom JW, Bosch J, Dyal L, Aboyans V, et al. Major adverse limb events and mortality in patients with peripheral artery disease: the COMPASS trial. J Am Coll Cardiol 2018;71(20):2306–15. https://doi.org/10.1016/j.jacc.2018.03.008.

[36] Eikelboom JW, Connolly SJ, Bosch J, Dagenais GR, Hart RG, Shestakovska O, et al. Rivaroxaban with or without aspirin in stable cardiovascular disease. N Engl J Med 2017;377(14):1319–30. https://doi.org/10.1056/NEJMoa1709118.

[37] Gerhard-Herman MD, Gornik HL, Barrett C, Barshes NR, Corriere MA, Drachman DE, et al. 2016 AHA/ACC guideline on the Management of Patients with Lower Extremity Peripheral Artery Disease: executive summary: a report of the American College of Cardiology/American Heart Association task force on clinical practice guidelines. Circulation 2017;135:e686–725.

[38] Aung PP, Maxwell H, Jepson RG, Price J, Leng GC. Lipid-lowering for peripheral arterial disease of the lower limb. Cochrane Database Syst Rev 2007;4. https://doi.org/10.1002/14651858.CD000123.pub2.

[39] Kumbhani DJ, Steg PG, Cannon CP, Eagle KA, Smith Jr SC, Goto S, et al. Statin therapy and long-term adverse limb outcomes in patients with peripheral artery disease: insights from the REACH registry. Eur Heart J 2014;35:2864–72.

[40] Bonaca MP, Nault P, Giugliano RP, Keech AC, Pineda AL, Kanevsky E, et al. Low-density lipoprotein cholesterol lowering with evolocumab and outcomes in patients with peripheral artery disease: insights from the FOURIER trial (further cardiovascular outcomes research with PCSK9 inhibition in subjects with elevated risk). Circulation 2018;137:338–50. https://doi.org/10.1161/CIRCULATIONAHA.117.032235.

[41] Whelton PK, Carey RM, Aronow WS, Casey DE, Collins KJ, Himmelfarb CD, et al. 2017 ACC/AHA/AAPA/ABC/ACPM/AGS/APhA/ASH/ASPC/NMA/PCNA Guideline for the prevention, detection, evaluation, and management of high blood pressure in adults. J Am Coll Cardiol 2018;71(19):e127. https://doi.org/10.1016/j.jacc.2017.11.006.

[42] Yusuf S, Sleight P, Pogue J, Bosch J, Davies R, Dagenais G. Effects of an angiotensin-converting-enzyme inhibitor, ram-ipril, on cardiovascular events in high-risk patients. The Heart Outcomes Prevention Evaluation Study Investigators. N Engl J Med 2000;342:145–53.

[43] Yusuf S, Teo KK, Pogue J, Dyal L, Copland I, Schumacher H, et al. Telmisartan, ramipril, or both in patients at high risk for vascular events. N Engl J Med 2008;358:1547–59.

[44] Rabi DM, Khan N, Vallee M, Hladunewich MA, Tobe SW, Pilote L. Reporting on sex-based analysis in clinical trials on angiotensin-converting enzyme inhibitor and angiotensin receptor blocker efficacy. Can J Cardiol 2008;24(6):491–6.

[45] Ahimastos AA, Pappas EP, Buttner PG, Walker PJ, Kingwell BA, Golledge J. A meta-analysis of the outcome of endovascular and non-invasive therapies in the treatment of intermittent claudication. J Vasc Surg 2011;54(5):1511–21. https://doi.org/10.1016/j.jvs.2011.06.106.

CH
19

[46] Pandey A, Banerjee S, Ngo C, Mody P, Marso SP, Brilakis ES, et al. Comparative efficacy of endovascular revascularization versus supervised exercise training in patients with intermittent claudication: meta-analysis of randomized controlled trials. J Am Coll Cardiol Intv 2017;10(7):712–24. https://doi.org/10.1016/j.jcin.2017.01.027.

[47] Fakhry F, Fokkenrood HJP, Spronk S, Teijink JAW, Rouwet EV, Hunink MGM. Endovascular revascularisation versus conservative Management for Intermittent Claudication. Cochrane Database Syst Rev 2018;(3). https://doi.org/10.1002/14651858.CD010512.pub2.

[48] Gommans LNM, Scheltinga MRM, van Sambeek MRHM, Maas AHEM, Bendermacher BLW, Teijink JAW. Gender differences following supervised exercise therapy in patients with intermittent claudication. J Vasc Surg 2015;62(3):681–8. https://doi.org/10.1016/j.jvs.2015.03.076.

[49] Roumia M, Aronow HD, Soukas P, Gosch K, Smolderen KG, Spertus JA, et al. Sex differences in disease-specific health status measures in patients with symptomatic peripheral artery disease: data from the PORTRAIT study. Vasc Med 2017;22(2):103–9. https://doi.org/10.1177/1358863X16686408.

[50] Higgins JP, Higgins JA. Epidemiology of peripheral arterial disease in women. J Epidemiol 2003;13:1–14.

[51] Lo RC, Bensley RP, Dahlberg SE, Matyal R, Hamdan AD, Wyers M, et al. Presentation, treatment, and outcome differences between men and women undergoing revascularization or amputation for lower extremity peripheral arterial disease. J Vasc Surg 2014;59:409–18.

[52] Rieß HC, Debus ES, Heidemann F, Stoberock K, Grundmann RT, Behrendt CA. Gender differences in endo-vascular treatment of infra-inguinal peripheral artery disease. Vasa 2017;46:296–303.

[53] Choi KH, Park TK, Kim J, Ko Y-G, Yu CW, Yoon C-H, et al. Sex differences in outcomes following endovascular treatment for symptomatic peripheral artery disease: an analysis from the K-VIS ELLA registry. J Am Heart Assoc 2019;8:2. https://doi.org/10.1161/JAHA.118.010849.

[54] Ketha SS, Bjarnason H, Oderich GS, Misra S. Clinical features and endovascular management of iliac artery fibromuscular dysplasia. J Vasc Interv Radiol 2014;25(6):949–53. https://doi.org/10.1016/j.jvir.2014.03.002.

[55] Olin JW, Froehlich J, Gu X, Bacharach JM, Eagle K, Gray BH, et al. The United States registry for fibromuscular dysplasia: results in the first 447 patients. Circulation 2012;125(25):3182–90. https://doi.org/10.1161/CIRCULATIONAHA.112.091223.

[56] Gornik HL, Persu A, Adlam D, Aparicio LS, Azizi M, Boulanger M, et al. First international consensus on the diagnosis and management of fibromuscular dysplasia. Vasc Med 2019;24(2):164–89.

[57] Joseph L, Kim ESH. Non-atherosclerotic vascular disease in women. Curr Treat Options Cardio Med 2017;19:78. https://doi.org/10.1007/s11936-017-0579-6.

Risk Factors

Menopause
Osteoporosis
Pregnancy complications
Oral contraception
Inflammation

Treatment

Lower use of Rx
(statins, antiplatelet
agents, ACEI)
Fewer amputations
& bypass surgeries

Presentation

Asymptomatic or
atypical symptoms
Presents 10-20 years
later in women
More comorbidities
More advanced
disease

Prognosis

Higher morbidity &
CV mortality
More procedural
complications
Lower QOL scores

Diagnosis

Currently no sex-
specific cut off for
ABIs exist

Niti Aggarwal

Research

Women comprise
only 1/3 of all
patients in large
clinical trials

Risk Factors, Presentation, Diagnosis, Prognosis and Management of Peripheral Artery Disease in Women Compared to Men.
Women with PAD are often older, with more comorbidities, and underrepresented in trials. Presently guidelines recommend sex-neutral diagnostic criteria for PAD. Women with PAD are prescribed fewer pharmacotherapies and revascularization procedures, resulting in worse QOL. ABI, ankle brachial index; ACEI, angiotensin converting enzyme inhibitor; CV, cardiovascular; QOL, quality of life; RX, pharmacotherapy. *Image courtesy of Niti R. Aggarwal.*

Section VIII

Congenital Heart Disease

Chapter 20

Congenital Heart Disease

Yamini Krishnamurthy and Ami B. Bhatt

Clinical Case

A 26-year-old woman presents to her primary care clinic for management of hypertension and consideration of pregnancy. She had been taking lisinopril for several years for hypertension diagnosed in her early 20s, but self-discontinued this as she wanted to get pregnant. Her blood pressure is 150/90 in her right arm, 140/90 in her left arm, and 110/70 in her bilateral legs. Her exam shows a palpable radio-femoral delay, systolic click with a soft systolic murmur. Her workup includes a transthoracic echocardiogram which shows a bicuspid aortic valve, mild aortic insufficiency, and coarctation of the aorta with a gradient of 30 mmHg at the suprasternal notch. She was referred to adult congenital heart disease clinic. What does this new diagnosis of congenital heart disease mean for this female patient as she considers pregnancy? Does she warrant intervention? Medications? What are her long-term risks?

Continued

Sex Differences in Cardiac Disease. https://doi.org/10.1016/B978-0-12-819369-3.00006-X

Clinical Case—cont'd

An increasing number of patients with congenital heart disease (CHD) are surviving until adulthood because of advances in surgical and medical management in childhood. The impact of sex on risk stratification, counseling, management, outcomes, and mortality has received little attention in the CHD population compared to other cardiovascular diseases. As more women with CHD survive to reproductive age, special attention also needs to be directed toward contraception counseling, pregnancy risk stratification, and multidisciplinary management of pregnancy. A better understanding of the influence of sex on the adult congenital heart disease (ACHD) population is important to improve care of this growing population.

Abstract

The population of adults with congenital heart disease continues to grow as these individuals are surviving longer as a result of medical and surgical advances. Determining the role sex and gender plays on overall mortality, morbidity, transitions of care, reproductive health, and end-of-life decision-making is crucial to improving the comprehensive care of the adult with congenital heart disease. Preconception assessment is essential to understand the risk in pregnancy. Complications during pregnancy are predictable, and most women with congenital heart disease can have a successful pregnancy, labor, and delivery under the care of a multidisciplinary cardio-obstetric team.

Impact of Sex on Congenital Heart Disease: Incidence and Outcomes

The prevalence of congenital heart anomalies differs by sex (**Table 1**) [1]. Double outlet right ventricle, transposition of the great arteries, hypoplastic left heart syndrome, congenital aortic stenosis (AS), pulmonary and tricuspid atresia, coarctation of the aorta, and bicuspid atresia have a higher prevalence in male patients. Patent ductus arteriosus, Ebstein's anomaly, truncus arteriosus, atrioventricular septal defects (AVSDs), and Tetralogy of Fallot (TOF) have a higher prevalence in female patients. CHD risk factors and outcomes are also affected by sex (**Figure 1**).

Morbidity and Mortality

Verheugt et al. analyzed the Dutch CONCOR (CONgenital CORvitia) national registry to report on sex differences in outcomes of congenital heart disease (CHD). This database comprises of over 7000 adult patients with CHD. No sex difference in mortality was found (the mean age at the end of the study follow-up period was 35 years) [3]. This is not consistent across the literature. Engelfriet and Mulder, who utilized the European Heart Survey on adult congenital

TABLE 1 Prevalence of Congenital Heart Anomalies by Sex

Congenital Heart Anomaly	Range of Ratios (Male:Female)
Double outlet right ventricle	2.68:1 [1]
d-Loop transposition of the great arteries	2.11:1 [1]
Hypoplastic left heart syndrome	2.25:1 [1]
Congenital aortic stenosis	1.95:1 [1]
Pulmonary atresia	1.55:1 [1]
Tricuspid atresia	1.45:1 [1]
Coarctation of the aorta	1.30:1 [1]
l-Loop transposition of the great arteries	1.25:1 [1]
Patent ductus arteriosus	1:1.66 [1]
Ebstein's anomaly	1:1.57 [1]
Truncus arteriosus	1:1.22 [1]
Atrioventricular septal defect	1:1.17 [1]
Tetralogy of Fallot	1:1.12 [1]
Bicuspid aortic valve	2:51:1 [2]

heart disease (ACHD) found that cumulative mortality was greater in the male population [hazard ratio 1.63; 95% confidence interval (CI) 1.12–2.38; $P = 0.011$] [4]. Zomer et al. studied both the Dutch CONCOR registry and the Quebec CHD database to compare 30-day in-hospital mortality for all admissions between men and women [5]. Adjusted 30-day in-hospital mortality for all admissions was not significantly different in patients aged 18–65; mortality was higher in men than in women in patients aged 18–45 years (adjusted rate ratio 1.58, 95% CI 1.08–2.30). If pregnancy

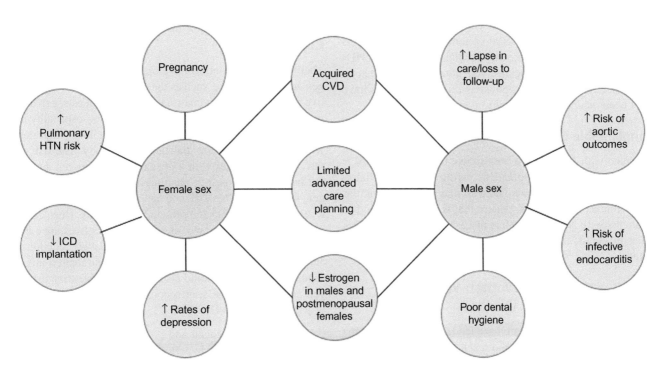

FIGURE 1   Sex-Specific Risk Factors for Morbidity and Mortality in Adult Congenital Heart Disease Patients.
Sex-specific risk factors affect male and female patients with ACHD. Underlying congenital heart defect and physiology in combination with these risk factors create a unique risk profile for each ACHD patient. CVD, cardiovascular disease; HTN, hypertension, ICD, implantable cardioverter-defibrillator.

admissions were excluded, the adjusted rate ratio for 30-day in-hospital mortality remained higher in men than in women (adjusted rate ratio 1.36, 95% CI 1.02–1.81). Importantly, a history of pregnancy was not found to be associated with an overall increase in medical encounters. This increased risk of mortality among men with ACHD is not well understood. Estrogen may play a cardioprotective role during the reproductive years, and increased health surveillance in women during these reproductive years may confer some degree of protection.

Pulmonary Hypertension

Women with ACHD have been found to have a 33% higher risk of pulmonary hypertension [odds ratio (OR) 1.33; 95% CI 1.07–1.65] [3]. Among Eisenmenger patients, gender distribution was roughly equal. Several other studies have shown a similar pattern [6–8]. The pathophysiology behind this increased risk is not clear and may reflect hormonal differences and the hemodynamic alterations associated with pregnancy. However, this may be too simplistic an approach. Of note, idiopathic pulmonary arterial hypertension is more common in women without CHD with an OR of 2.5:1; however, none of the various genetic mutations and polymorphisms associated with idiopathic pulmonary arterial hypertension appear to be sex linked [6].

Aortic Outcomes

Women with ACHD have been shown to have a 33% lower risk of unfavorable aortic outcomes, defined as aneurysm, dissection, and aortic surgery, and a 30% lower probability of aortic surgery [3]. This is likely because bicuspid aortic valve (BAV), AS, and coarctation of the aorta, all CHD associated with aortopathy, are more common in men (see **Table 1**). In those with BAVs, men were more likely to have dilation of the aortic root and ascending aorta as well as aortic dissection [2, 9]. Female patients were more likely to have AS [2, 9, 10]. Importantly, in the noncongenital population, men are more likely to undergo aortic surgery, though women with aortic aneurysms are more likely to have higher risk of rupture, and higher rates of complications following repair [11, 12]. Fukui et al. found no differences in early and long-term outcomes between men and women undergoing surgery for aortic dissection [13].

Infective Endocarditis

Women with ACHD have been shown to have a 47% lower risk of infective endocarditis [3]. This could be related to the sex-specific prevalence of vulnerable lesions [such as ventricular septal defects (VSDs), aortic valve disease, etc.; see **Table 1**]. Another explanation is that women may have better oral hygiene which then lowers their risk. Schmidt et al. demonstrated that female gender and higher infective endocarditis knowledge score (determined by an adapted version of the Leuven Knowledge Questionnaire for CHD) were independently associated with good oral hygiene [14]. Sex-specific antibiotic prophylaxis recommendations have not been incorporated into current guidelines.

Implantable Cardioverter-Defibrillator Use

Women with ACHD have been shown to have a 55% lower risk of receiving an implantable cardioverter-defibrillator, although men and women had a similar rate of ventricular arrhythmias [3]. This is a pattern that is well recognized in patients with acquired cardiac disease. The reason for this discrepancy is not clear, and should be the basis of further study.

Contraception in Women With Congenital Heart Disease

Advancements in CHD care have led to an increasing number of women with CHD who reach reproductive age. It is well documented that the physiologic changes inherent to pregnancy may increase the risk of adverse cardiovascular events for women with certain congenital heart defects. Thus, contraceptive counseling is an important component to the comprehensive care of women with CHD starting in the teenage years. Current counseling practices for contraception in women with CHD are inadequate. A study of approximately 500 women with CHD followed at ACHD specialty centers demonstrated that only 43% of the study population received contraceptive counseling from their ACHD provider [15]. Unplanned pregnancy was reported by 25% of the study population. While this risk is lower than the national average, this risk did not statistically differ from those with simple complexity CHD and those with complex CHD. Another study of women with CHD, only 51% reported receiving contraception advice; 50% of those who had a prohibitive risk for pregnancy were unaware of this risk [16]. ACHD providers must incorporate contraception counseling into their routine care of patients, and non-ACHD providers should refer patients if they do not feel comfortable or adequately trained to provide contraception counseling.

Barrier Method Contraception

Barrier method contraception (diaphragms, male and female condoms) are useful for preventing sexually transmitted infections (**Table 2**). However, the efficacy of barrier method contraception is highly user dependent, so barrier method contraception should be recommended as an adjunct to an alternative method of contraception for ACHD patients.

Combination Estrogen and Progesterone Contraception

There are several contraceptive options that combine estrogen and progesterone, such as the combined oral contraceptive pill, patch, and vaginal ring. Combination estrogen and progesterone increase the risk of thrombosis—especially in older individuals who have a history of smoking, migraine with aura, hypertension, diabetes, and obesity. The risk of thrombosis increases with higher doses of estrogen, and is attributed to increased production of prothrombin, fibrinogen, factor VII, factor VIII, and factor X, along with reduced levels of antithrombin and protein S levels [18]. Thus, combination estrogen and progesterone contraceptives are contraindicated for patients with prosthetic valves, conduits, and baffles [19]. ACHD patients with an unrepaired atrial septal defect (ASD) should also not use combination estrogen and progesterone contraceptives due to increased risk of paradoxical emboli [20].

Combination estrogen and progesterone contraceptives can induce or worsen hypertension by stimulation of the renin angiotensin aldosterone system [21]. Combination estrogen and progesterone contraceptives should be used with caution in ACHD patients with aortopathies, as worsening hypertension can result in aortic dilation and diastolic dysfunction among other complications.

Progesterone-Only Contraception

Progesterone-only contraceptives come in the form of a pill, injection, implant, or intrauterine device (IUD), and do not carry the risk of thrombosis associated with estrogen use. Progesterone-only contraceptives come with the risk of fluid retention, abnormal uterine bleeding, weight gain, and mood changes. The progesterone-only pill requires strict adherence as missed or delayed doses increase failure rate. Similarly, the progesterone-only injection must be administered every 90 days, and delayed injections increase failure rate. Progesterone-only implants or IUDs are the most efficacious hormonal options to prevent pregnancy, with failure rates estimated at 0.05% and 0.2%, respectively [22]. Progesterone-only IUDs are associated with a significant reduction of menstrual blood loss, and thus is an excellent option for ACHD patients who require anticoagulation—this benefit does not apply to nonhormonal IUDs which may increase menstrual blood loss [23]. With IUD placement, there is a small risk of infection; therefore, past guidelines recommended antibiotic prophylaxis for the insertion and removal of IUDs. However, there is no evidence to support this practice, and this recommendation was removed from the 2007 American College of Cardiology/American Heart Association guidelines on endocarditis prophylaxis [24].

TABLE 2 Contraception Methods for Adults With Congenital Heart Disease

Contraception Method	Pros	Cons	Efficacy With Perfect Use[a] [17]	Efficacy With Typical Use[a] [17]	Contraindications Within the ACHD Population
Barrier method contraception (diaphragms, female and male condoms)	Protects against sexually transmitted infections No hormone-related side effects	Efficacy is highly user dependent	Diaphragm: 94% Female condom: 95% Male condom: 98%	Diaphragm: 88% Female condom: 79% Male condom: 82%	Should be used with another method of contraception to prevent unintended pregnancy
Combined estrogen/ progesterone contraception (pill, patch, and ring)	Effective when used correctly	Increases risk of thrombosis Increases risk of hypertension	Pill: 99.7% Patch: 99.7% Vaginal ring: 99.7%	Pill: 91% Patch: 91% Vaginal ring: 91%	Contraindicated in those: – With prosthetic valves, conduits, and baffles – On anticoagulation – With atrial septal defects – With aortopathies – With hypertension – With pulmonary hypertension
Progesterone-only contraception (pill, depot injection, implant, and IUD)	Risk of thrombosis is not increased	Pill and depot injection require strict adherence Abnormal menstrual bleeding[b] Weight gain Fluid retention Theoretical risk of infective endocarditis with intrauterine device placement	Pill: 99.7% Injection: 99.8% Implant: 99.95% IUD: 99.8%	Pill: 91% Injection: 94% Implant: 99.95% IUD: 99.8%	
Nonhormonal IUD	No hormone-related side effects	Increases menstrual bleeding Theoretical risk of infective endocarditis with IUD placement	99.4%	99.2%	Contraindicated in those: – On anticoagulation
Nonreversible contraception	Highly effective	Requires invasive procedure and possible anesthesia	Tubal ligation: 99.5% Vasectomy: 99.9% Essure: 99.3%	Tubal ligation: 99.5% Vasectomy: 99.85% Essure: 99.3%	Contraindicated in those with a prohibitively high risk of anesthesia

[a] *Efficacy in the first year of use.*
[b] *Progesterone-only intrauterine devices typically reduce menstrual bleeding.*
ACHD, adult congenital heart disease; IUD, intrauterine device.

Nonreversible Contraception

Sterilization is a potential contraception option for women with ACHD who choose not to become pregnant or who have a high risk associated with pregnancy. Tubal ligation is a laparoscopic procedure where the fallopian tubes are transected or cauterized, and is typically performed under general anesthesia. The risk of general anesthesia must be considered when choosing tubal ligation, and may be prohibitive for some patients with ACHD, including those with severe pulmonary hypertension. Newer sterilization techniques that do not require general anesthesia are available. Partner vasectomy can be an alternative option if the patient does not have any other sexual partners. It is important to realize that the partner may outlive the ACHD patient, and may wish to have children in the future. Partner vasectomy then becomes a less ideal option.

Risk Assessment and Counseling of Adult Congenital Heart Disease Patients

The transition between pediatric and adult cardiology, mental health, acquired cardiovascular disease, and preconception counseling has become an important part of the comprehensive care of ACHD patients, and the influence of sex should be noted.

Transition Between Pediatric and Adult Cardiology

Patients with CHD require lifelong cardiac surveillance by specialized practitioners, and therefore are typically transitioned between pediatric and adult cardiology as an adolescent or young adult. However, several studies have demonstrated that as many as 63% of adolescents with CHD are either lost to follow-up or experience lapses in care after leaving pediatric cardiology; the lapse in care was found to be a predictor of mortality [25]. A study of 1828 ACHD patients in Germany found significant sex differences when studying pediatric to adult transitions of care. Women were more likely to be treated at a specific ACHD clinic (27% vs. 23.3%, $P < 0.05$). In addition, women were more likely to know the ACHD certifications status of their physician (29.7% vs. 25.4%, $P < 0.05\%$) [26]. In a study of 794 patients with CHD, a multivariable logistic regression revealed that male sex was associated with lack of follow-up after leaving pediatric care [27]. Ongoing research efforts are focusing on innovative strategies to reduce lapses in care as patient transition between pediatric and adult medicine.

Mental Health Risk Assessment and Counseling

Patients with CHD are at higher risk of psychiatric and neurodevelopmental disorders such as depression, anxiety, and attention deficit hyperactive disorder [28, 29]. The prevalence of depression in the ACHD patients ranges from 13% to 33%, which is higher than in the general population [30]. ACHD patients may have several characteristics that may predispose them to developing depression including functional limitations, recurrent hospitalizations or encounters with the healthcare system, and poor self-esteem from being different than their peers. Variables associated with increased depression prevalence include older age, female sex, poor perceived health status, and low quality of life [31–34]. Women with ACHD have been shown to have greater functional limitations than men despite lower mortality [4]. Depression was not associated with CHD lesion complexity. Depression has been shown to be an independent predictor of adverse clinical outcomes including nonadherence to care [30]. Given the adverse impact that psychiatric and neurodevelopmental disorders have on ACHD patients' overall morbidity, particular attention must be given to ensure that mental health comorbidities are recognized and treated accordingly.

Acquired Cardiovascular Disease Risk Assessment and Counseling

In the non-CHD literature, sex-specific differences for the pathophysiology, risk assessment, prevention, and management for acquired cardiovascular disease exist [35–37]. Traditional risk factors for atherosclerosis such as obesity, smoking, diabetes, hypertension, and family history apply to both men and women, though risk factors such as autoimmune disease tend to affect women disproportionally [38]. The development of atherosclerotic disease in ACHD patients is largely driven by traditional risk factors, though patients with coarctation of the aorta, aberrant coronary arteries, or those who have prior surgery which manipulated the coronary arteries may be at higher risk [39–42]. Acquired cardiovascular disease risk assessment and lifestyle modification counseling should at minimum follow guidelines for the general population starting at a young age [43]. Those with increased atherosclerotic risk may benefit from additional screening and preventative efforts.

Sex-specific differences regarding stroke risk in the ACHD population are important to consider as well. In a retrospective study of nearly 30,000 patients with ACHD, the incidence of embolic stroke was shown to be higher in the ACHD population than the general population [44]. Between ages 18 and 65, 8.9% of men and 6.8% of women experienced at least one stroke. Little is known about sex-specific differences for acquired cardiovascular and cerebrovascular disease in patients with ACHD, and further study is needed to best guide prevention and treatment in the ACHD population.

Pregnancy in Women With Congenital Heart Disease

As more women with CHD are surviving until childbearing age, there has been a significant increase in the number of pregnancies and deliveries among women with CHD in recent years [45]. Many ACHD patients are lost to follow-up during the pediatric to adult cardiology transition and often return to care when considering pregnancy or already pregnant. Preconception evaluation is crucial as the physiologic changes inherent to pregnancy can have deleterious effects on patients' hemodynamic stability. Pregnancy in the ACHD population requires a multidisciplinary team including the primary cardiologist, ACHD specialist, high-risk obstetrics, and anesthesia, and can be successful with close monitoring.

Maternal Risk

Women with CHD have been shown to tolerate pregnancy better than women with other heart disease, such as valvular heart disease or cardiomyopathies [46]. Maternal risk of pregnancy in women with CHD is largely driven by the patient's congenital heart defect (lesion-specific risk is described below). However, several studies have described the morbidity and mortality of pregnancy of the general ACHD population. Opotowsky et al. analyzed the Nationwide Inpatient Sample (NIS) from 1998 to 2007, and found that women with CHD had higher odds of mortality during pregnancy (OR 6.7; 95% CI 2.9–15.4) and higher odds of an adverse cardiovascular event when compared to women without CHD [47]. Arrhythmia was the most common cardiac event (OR 8.3; 95% CI 6.7–10.1), followed by heart failure (OR 7.0; 95% CI 4.6–10.7), cerebrovascular accident (OR 2.9; 95% CI 1.5–5.6), and embolism (OR 41.6; 95% CI 25.8–67.1). These findings were supported by more recent analyses of the NIS [48, 49]. In addition, pregnant women with CHD have significantly greater odds of hypertension in pregnancy (adjusted OR 1.4; 95% CI 1.3–1.6), preeclampsia (adjusted OR 1.5; 95% CI 1.3–1.7), placenta previa (adjusted OR 1.5; 95% CI 1.2–1.8), hemorrhage (adjusted OR 1.6; 95% CI 1.3–1.8), and placental abruption (adjusted OR 1.5; 95% CI 1.1–1.9). Women with CHD had greater odds of requiring assistance with delivery, including cesarean section, induction of labor, and operative vaginal delivery [49].

Preconception Counseling and Risk Assessment

To best guide risk assessment and preconception counseling, a thorough evaluation is necessary, and may include a review of past medical history, assessment of functional status, physical examination including vitals, imaging studies, and cardiopulmonary exercise testing. Cardiac catheterization can be considered when insufficient information is available on pulmonary hypertension or obstruction lesions. Risk scores which utilize these data have emerged to risk-stratify ACHD patients considering pregnancy (**Table 3**).

The CARPREG I (Cardiac Disease in Pregnancy) Risk Score assigns one point for each risk factor noted in **Table 3**. Zero points confers a 5% risk of cardiac complications, one point a 27% risk, and two or more points a 75% risk [50]. The ZAHARA I (Zwangerschap bij Aangeboren HARtAfwijkingen I) risk score assigns a risk factor-specific number of points. The risk of cardiac complications is 2.9% with <0.5 points, 7.5% with 0.5–1.5 points, 17.5% with 1.51–2.50 points, 43.1% with 2.51–3.5 points, and 70% with >3.5 points [51]. Studies have revealed limitations with these two risk scores, as they seem to overestimate cardiac risk and were unable to reliably predict outcomes in neonates [53, 54]. The modified World Health Organization classification, which is largely based on disease complexity and underlying congenital abnormality, has been shown to be the more reliable and comprehensive risk stratification method when compared to the CARPREG I and ZAHARA I risk scores [45]. The CARPREG II risk score, which includes five general factors, four lesion-specific predictors, and one delivery of care predictor, outperformed these three existing risk stratification models [52].

Cardiopulmonary exercise testing can be used for further risk stratification in pregnancy. A peak heart rate of 150 beats/min and/or a peak oxygen consumption greater than or equal to 25 mL/kg/min may confer a lower cardiac risk during pregnancy [55]. An abnormal chronotropic response correlated with higher cardiac risk during pregnancy among ACHD patients [56]. Cardiopulmonary exercise parameters correlated with neonatal birth weight [55].

Cardiac biomarkers may also have a role in pregnancy risk stratification in CHD. Kampman et al. studied 234 women with CHD, and performed N-terminal pro-B-type natriuretic peptide (NT-proBNP) measurements at 20-week gestation [57]. An NT-proBNP level greater than 128 ng/mL was associated with increased adverse events (OR 10.6, $P = 0.039$), and an NT-proBNP level less than 128 was associated with a negative predictive value for adverse events of 96.9%.

Fetal Risk and Neonatal Risk

Fetal and neonatal outcomes are closely correlated with maternal health. Ramage et al. studied 2214 women with ACHD, and found that neonatal morbidity and mortality were higher compared with those for women without ACHD (OR 1.8; 95% CI 1.6–2.1); however, substantial

Risk Score	Risk Factor	Risk Points
TABLE 3 Predictors of Maternal Cardiovascular Events		
CARPREG [50]	Prior cardiac event (heart failure, transient ischemic attack, stroke before pregnancy, or arrhythmia)	1
	Baseline NYHA functional class > II or cyanosis	1
	Left heart obstruction (mitral valve area < 2 cm², aortic valve area < 1.5 cm², peak left ventricular outflow tract gradient > 30 mmHg by echocardiography)	1
	Reduced systemic ventricular systolic function (EF < 40%)	1
Modified WHO classification [45]	WHO I: No significant risk elevation • Uncomplicated, small or mild pulmonary stenosis, patent ductus arteriosus, mitral valve prolapse • Isolated atrial or ventricular ectopic beats • Successfully repaired simple lesions Who II: Mildly elevated risk • Unoperated atrial or ventricular septal defect • Repaired tetralogy of Fallot • Most arrhythmias WHO II–III: Mildly to moderately elevated risk • Mild left ventricular impairment • Hypertrophic cardiomyopathy • Native or tissue valvular heart disease not considered WHO I or IV • Marfan syndrome without aortic dilatation • Bicuspid aortic valve with aorta < 45 mm • Repaired coarctation WHO III: Significantly elevated risk • Nonsevere systemic ventricular dysfunction • Mechanical valve • Systemic right ventricle • Fontan circulation • Cyanotic heart disease and other complex congenital heart disease • Aortic dilatation 40–45 mm in Marfan syndrome • Bicuspid aortic valve with aorta 45–50 mm WHO IV: High risk • Pulmonary arterial hypertension • Systemic severe ventricular dysfunction (EF < 30%, NYHA III–IV) • Previous peripartum cardiomyopathy with residual impairment of left ventricular function • Severe mitral stenosis, severe symptomatic aortic stenosis, severe coarctation of the aorta • Marfan syndrome with aorta dilated > 45 mm • Bicuspid aortic valve with aorta > 50 mm	
ZAHARA [51]	History of arrhythmias	1.50
	NYHA functional class III/IV	0.75
	Left heart obstruction (peak LVOT gradient > 50 mmHg or aortic valve area < 1.0 cm²)	2.50
	Mechanical valve prosthesis	4.25

TABLE 3 Predictors of Maternal Cardiovascular Events—cont'd

Risk Score	Risk Factor	Risk Points
	Systemic AV valve regurgitation (moderate/severe)	0.75
	Pulmonary AV valve regurgitation (moderate/severe)	0.75
	Cardiac medication before pregnancy	1.50
	Cyanotic heart disease (corrected and uncorrected)	1.00
CARPREG II [52]	Prior cardiac events or arrhythmias	3
	Baseline NYHA III–IV or cyanosis	3
	Mechanical valve	3
	Ventricular dysfunction	2
	High-risk left-sided valve disease/LVOT obstruction	2
	Pulmonary hypertension	2
	Coronary artery disease	2
	High-risk aortopathy	2
	No prior cardiac intervention	1
	Late pregnancy assessment	1

AV, atrioventricular; EF, ejection fraction; LVOT, left ventricular outflow tract; NYHA, New York Heart Association; WHO, World Health Organization.

variation was observed between women with different congenital defects [58]. Maternal anomalies of atrioventricular junctions and valves were associated with an adjusted OR of 0.9 (95% CI 0.7–1.1) of severe neonatal morbidity and mortality, while women with single-ventricle physiology were found to have an adjusted OR of 4.4 (95% CI 2.9–6.6). Women with CHD are at risk for early pregnancy loss or intrauterine fetal demise, intrauterine growth restriction, preterm delivery, and having small-for-gestational-age neonates as a result of maternal cardiac disease [49, 59, 60]. These adverse events are attributed to insufficient uteroplacental blood flow secondary to maternal cardiac anomalies. Subaortic ventricular outflow tract gradient >30 mmHg independently predicted adverse neonatal outcomes [59]. Given this increased fetal risk, serial growth ultrasounds are recommended.

Children born to mothers with CHD are at higher risk of CHD themselves. Perry et al. found that 829 women with CHD in the Registry of Pregnancy and Cardiac Disease (ROPAC) had successful pregnancies, and 3.5% of infants were found to have CHD [61]. Maternal to offspring transmission rates as high as 10% have been found in left-sided outflow tract lesions. Mothers affected by conotruncal lesions should be screened for 22q11 mutations [62]. Marfan and Noonan syndrome have an autosomal dominant inheritance pattern, thus 50% of offspring will be affected. Genetic counseling is crucial as a part of preconception counseling, and fetal echocardiography should be offered to women with CHD who become pregnant.

Congenital Heart Lesions in Pregnancy

Atrial Septal Defects

ASDs are a common congenital heart defect with pretricuspid shunt physiology. Initially, the shunt is left to right, and increases right atrial pressure due to increased blood volume, which can lead to right atrial and ventricle dilation and pulmonary hypertension. Pulmonary hypertension can lead to a shunt reversal (referred to as Eisenmenger syndrome). Pregnancy is well tolerated in women with ASDs without pulmonary hypertension or ventricular dysfunction. Compared with the general population, women with an unrepaired ASD are at increased risk of preeclampsia, small-for-gestational-age births, and fetal mortality. Arrhythmias are the most common complication in pregnancy, followed by thromboembolic disease (specifically paradoxical emboli) [63]. Thromboembolic disease is more common in women with an isolated ASD than in pregnant women with other CHD [49]. ASD closure may be pursued prior to pregnancy to reduce the risk of paradoxical emboli, and management during pregnancy consists of heparin prophylaxis when immobilized and early ambulation after delivery. In those with ASD and pulmonary hypertension or ventricular dysfunction, pregnancy poses significant risks to the mother and fetus; thus, pregnancy should be avoided.

Atrioventricular Septal Defects

AVSD is a defect in the endocardial cushion, and its physiology is distinct from a simple ASD or VSD. Complications in adulthood include residual ASD, atrioventricular valve insufficiency, arrhythmias, and heart failure. Drenthen et al. identified 29 women with AVSD in the Dutch CONCOR registry who underwent pregnancy [64]. They found that 60.4% of pregnancies were associated with temporary deterioration of NYHA class, and 22.9% of pregnancies were associated with perseverance of NYHA class after delivery. A total of 17% of pregnancies were associated with deterioration of preexisting left atrioventricular valve regurgitation. Offspring had high rates of congenital heart defects and mortality.

Ventricular Septal Defects

VSDs are a common congenital heart defect with posttricuspid shunt physiology. The shunt is initially left to right, though over time can lead to pulmonary hypertension and shunt reversal. Patients with an isolated VSD with successful closure tolerate pregnancy well. Those with an unrepaired VSD are at higher risk of preeclampsia. Interestingly, women with a repaired VSD are at higher risk of premature labor and small-for-gestational-age births compared to women with an unrepaired VSD [65]. Women with pulmonary hypertension or ventricular dysfunction should be advised to avoid pregnancy given the high maternal and fetal risk.

Congenital Mitral Stenosis

Congenital mitral stenosis is rare. Patients with moderate to severe mitral stenosis are at risk for heart failure symptoms and pulmonary edema especially during the second trimester when blood volume peaks.

Fuchs et al. compared outcomes between left heart obstructive lesions [mitral stenosis, left ventricular outflow tract obstruction (subvalvular, valvular, and supravalvular), coarctation of the aorta] and found that isolated mitral stenosis and serial left heart obstructive lesions carry the highest risk of maternal cardiovascular events [66]. Surgical correction should be considered before pregnancy to minimize pregnancy complications, though mitral stenosis may be amenable to balloon dilation during pregnancy.

Congenital Aortic Valve Disease

Congenital aortic valve disease, including congenital AS and BAV disease, is common. Orwat et al. showed that in patients with at least moderate AS, adverse maternal cardiac events were common [67]. A total of 21% of women were hospitalized for a cardiac indication during pregnancy; heart failure was the most common cardiac complication. Pregnant women with AS who develop heart failure symptoms are typically managed medically; however, percutaneous balloon dilation or percutaneous valvulopathy can be considered if heart failure symptoms are refractory to medical management. Women with left ventricular dysfunction should be advised to avoid pregnancy. Infants born to mothers with severe AS were more likely to be low birth weight and born preterm [67]. Women with aortopathies associated with congenital aortic valve disease are at risk of aortic dissection and infective endocarditis [68]. Those with significant aortic dilation should also be advised to avoid pregnancy [45].

Coarctation of the Aorta

Coarctation of the aorta (CoA) is a discrete narrowing of the proximal descending aorta, and this lesion can be associated with mitral stenosis, aortic valvulopathy, VSDs, aortic arch abnormalities, or aberrant subclavian arteries. Patients with repaired CoA who are normotensive typically do well during pregnancy (**Figure 2**). Krieger et al. compared pregnancy and delivery outcomes between women with and without CoA in the NIS [69]. They found that women with CoA are more likely to have hypertensive complications of pregnancy [defined as pregnancy-induced hypertension, preeclampsia, and eclampsia (multivariate OR 3.6; 95% CI 2.5–5.2)], delivery by cesarean section (multivariate OR 2.0; 95% CI 1.4–2.8), and have adverse combined cardiovascular outcomes (multivariate OR 16.7; 95% CI 6.7–41.5). Patients with repaired and unrepaired CoA carry a risk of aortic dissection and aortic rupture [70, 71]. Infants born to mothers with CoA were found to have lower birth weight [72]. Interestingly, right ventricular dysfunction before pregnancy and at 20 weeks gestation was associated with impaired uteroplacental Doppler flow. Of note, a study of 28 women with CoA who had cardiac magnetic resonance imaging within 2 years of delivery had a higher risk of hypertensive events in pregnancy if they were found to have a minimum aortic diameter less than 12 mm; there may be a role for prepregnancy cardiac magnetic resonance imaging for risk stratification of CoA patients [73].

Marfan Syndrome

Marfan syndrome is a connective tissue disorder with an autosomal dominant inheritance pattern caused by a mutation in the fibrillin-1 gene. A total of 80% of patients with

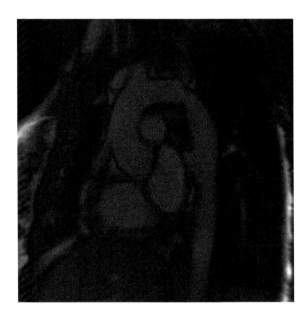

FIGURE 2 Cardiac Magnetic Resonance Imaging for a Pregnant Patient With Coarctation of the Aorta. Cardiac magnetic resonance imaging of a 33-year-old G2P1 at 35 weeks gestational age diagnosed with coarctation of the aorta during her second pregnancy. Imaging was performed to assess for aortic dissection, which was not identified. The patient reported a history of preeclampsia during her first pregnancy, and was found to have gestational hypertension during her second pregnancy. She underwent successful stent placement following the delivery of her second child.

Marfan syndrome will have cardiac involvement including aortic dilatation, aortic regurgitation, and mitral or tricuspid valve prolapse. Pregnancy in Marfan syndrome is associated with an increased risk of aortic dissection, both types A and B, particularly in the immediate last trimester or postpartum period [74, 75]. This risk of dissection increases with increasing aortic root diameter. Those with an aortic root diameter less than 4.0 cm have a 1% risk of cardiovascular event, while those with an aortic root diameter greater than 4.0 cm have a 10% risk of dissection (**Table 4**) [76, 77]. Thus, those with an aortic root measuring greater than

4.0 cm should consider surgical repair before pregnancy [45]. Cesarean section rates are higher in those with Marfan syndrome, and a small but significant average increase in aortic root diameter was noted after pregnancy [78]. Infants born to mothers with Marfan syndrome were more likely to be small for gestational age [79]. Given the autosomal dominant inheritance pattern of Marfan syndrome, preconception genetic counseling is crucial for all families in which either parent is affected.

Tetralogy of Fallot

TOF is the most common cyanotic heart disease. This anomaly comprises of a large conal VSD, an aorta that overrides the ventricular septum and right ventricular outflow tract obstruction that leads to right ventricular hypertrophy. TOF can also be associated with ASD, aortic arch abnormalities, and coronary artery abnormalities. Older patients with TOF typically underwent a palliative systemic to pulmonary artery shunt prior to complete repair. Younger patients typically undergo complete repair before the age of 6 months, and most do not require a palliative shunt. Repair consists of closure of the VSD and enlargement of the right ventricular outflow tract obstruction (which sometimes requires a transannular patch). Complications that may develop in adulthood include pulmonary regurgitation (those who underwent transannular patch placement are at particular risk) with subsequent right ventricular dilation and right heart failure, ventricular arrhythmias, and aortic dilation, as well as residual VSDs or right ventricular outflow tract obstruction.

Pregnancy is generally well tolerated in patients with repaired TOF. Predictors of adverse events are cardiac medication use before pregnancy, history of ablation, baseline cardiothoracic ratio on chest radiography, left ventricular dysfunction, severe pulmonary hypertension, and severe pulmonic regurgitation with right ventricular (RV) dysfunction [80–82]. Women with mild-to-moderate right ventricular dilation did not experience harmful effects of pregnancy on RV volume [83]. Peak plasma brain natriuretic peptide levels after the second trimester were found to be higher in

	Risk of Aortic Dissection During Pregnancy	Prepregnancy Considerations	Follow-Up During Pregnancy	Mode of Delivery
TABLE 4 Aortic Size in Marfan Syndrome and Pregnancy				
Aortic Size				
<4.0 cm	1%	Routine	Monthly	Vaginal
>4.0 cm	10%	Consider surgical repair	Monthly	Cesarean section

Refs. [76, 77].

patients with cardiac events [82]. Involvement of a multidisciplinary cardio-obstetric team during pregnancy is advised, and preconception valve revision should be considered in those with severe pulmonary regurgitation.

Ebstein Anomaly

Ebstein anomaly is a rare congenital heart defect that is caused by failure of delamination of the tricuspid valve leaflets resulting in apical displacement of the septal and posterior leaflets. This results in variable degrees of tricuspid regurgitation, proximal right ventricular "atrialization" (the proximal right ventricle becomes thin walled and poorly contractile), a small true right ventricle, and patients may develop progressive right heart failure. Associated anomalies include ASD, Wolff-Parkinson-White syndrome, and pulmonic stenosis. Women with Ebstein anomaly are more likely to suffer miscarriages [84]. Supraventricular tachycardias are a common complication during pregnancy—possibly related to the increased blood volume inherent in pregnancy along with the increased incidence of Wolff-Parkinson-White syndrome [85, 86]. Women with Ebstein anomaly and significant cyanosis or symptomatic heart failure should be advised against pregnancy, though women without cyanosis, arrhythmias, or heart failure symptoms tolerate pregnancy well. Of note, both maternal and paternal Ebstein anomaly conferred an increased risk of CHD in infants [84].

Congenitally Corrected Transposition of the Great Arteries

Congenitally corrected transposition of the great arteries (CC-TGA) or l-looped transposition of the great arteries involves systemic venous flow entering the right atrium, crossing the mitral valve to the left ventricle, then exiting to the pulmonary artery and lungs. The pulmonary veins enter the left atrium, but cross the tricuspid valve to the right ventricle, and then exit to the aorta. CC-TGA can be associated with VSD, pulmonic stenosis, anomalies of the atrioventricular valves, and complete heart block. As the morphologic right ventricle is the systemic ventricle, CC-TGA patients are at risk for systemic atrioventricular valve insufficiency and heart failure over time. Patients with CC-TGA typically tolerate pregnancy well. Connolly et al. studied 22 women with CC-TGA, and found that the rate of fetal loss is increased (17%) [87]. Two patients developed heart failure, though no pregnancy-related deaths occurred. Kowalik et al. studied 13 women with CC-TGA; two patients developed supraventricular arrhythmias and one patient developed heart failure [88].

d-Looped Transposition of the Great Arteries

In d-looped transposition of the great arteries (d-TGA), blood enters the right atrium via the systemic venous circulation,

traverses the tricuspid valve to the right ventricle, and then is ejected out into the aorta. Pulmonary venous return enters the left atrium, then the left ventricle, and then is ejected into the pulmonary arteries. Typically, an atrial or ventricular level shunt exists. Original surgical correction was an atrial switch operation, which resulted in a correction of the directionality of the blood flow, though the right ventricle was left in the systemic position. Long-term consequences include atrial arrhythmias, systemic atrioventricular valve regurgitation, and systemic (right) ventricle dysfunction. Miller et al. identified 56 pregnancies in 23 women with d-TGA who underwent an atrial switch repair; maternal cardiac complications included arrhythmias, hemoptysis (from pulmonary hypertension), and volume overload [89]. Infants born to a mother with d-TGA who underwent atrial switch repair have a higher risk of low birth weight [90]. Cataldo et al. compared cardiac events in pregnant and nonpregnant women with d-TGA who underwent atrial switch repair, and found that tricuspid regurgitation occurs more frequently in pregnant women, though cardiac events were common in both cohorts [91].

Since the mid-1980s, the preferred repair for d-TGA is the Jatene arterial switch surgery. This surgery consists of a two-stage repair; the first stage involves pulmonary artery banding supplemented by a systemic-pulmonary shunt, followed by complete repair. Long-term complications include aortic dilation, neo-aortic regurgitation, suprapulmonic and aortic stenosis, and coronary ostial narrowing. A study of 20 pregnancies in 10 women with d-TGA who underwent an arterial switch operation found that maternal cardiac events (heart failure and arrhythmias) occurred in four patients [92]. Risk factors for cardiac events were older age at arterial switch operation, older age at delivery, and higher brain natriuretic peptide in the first trimester. Patients were followed for 7–60 months after delivery, and did not have worsening ventricular function or functional status during this time span. Studies indicate that women who undergo an arterial switch operation tolerate pregnancy well [92, 93]; however, further study is needed as more women with d-TGA who underwent the arterial switch operation are reaching child-bearing age.

Fontan Circulation

The Fontan procedure is a common palliative procedure for individuals with single-ventricle physiology; systemic venous return is directed to the pulmonary artery allowing separation of the pulmonary and systemic venous flow. Many long-term complications occur including ventricular dysfunction, arrhythmias, thromboembolic events, and systemic to pulmonary collaterals (which leads to cyanosis). Preconception counseling is crucial for this patient population, and women with Fontan circulation who become pregnant should be managed by a multidisciplinary team including high-risk obstetrics and ACHD cardiologists.

Adverse cardiac events are high in patients with Fontan circulation, though normal and successful pregnancies are possible [45]. A systematic review of women with Fontan circulation demonstrated that the most common adverse cardiac events were supraventricular arrhythmias (8.4%; range 3–37%) and heart failure (3.9%; 3–11%) [94]. In patients who do experience NYHA class deterioration, most return to baseline following pregnancy [95]. Because of an increased rate of thromboembolic events during pregnancy, anticoagulation during pregnancy and postpartum should be considered [76]. Patients with a Fontan circulation have high rates of miscarriages and preterm birth [96]. In a study of 21 women with 55 pregnancies, median gestational age at delivery was 32 weeks [95].

Anesthesia and Delivery

The delivery and immediate postpartum period can be the most dangerous period for women with CHD, and thus multidisciplinary planning is important. Monitoring during delivery should be individualized, and can range from standard obstetric care to continuous pulse oximetry in those with cyanosis or right-to-left shunting, telemetry for those at risk for arrhythmias, arterial catheter for close blood pressure monitoring, and pulmonary artery catheter to monitor cardiac output and pulmonary pressures in those with or at risk for heart failure or cardiogenic shock [76]. Vaginal delivery is generally preferred unless there is an obstetric or fetal indication for cesarean section. Even for most NHYA class III and IV patients, a trial of labor is safe with close monitoring [97]. The second stage of labor is sometimes assisted to minimize the effects of "pushing" on maternal hemodynamics; however, this practice is largely based on expert opinion, and studies have shown that prolonged pushing is not associated with increased cardiac events [98]. Those with significant aortopathy or pulmonary hypertension are at significant risk with vaginal delivery, and thus cesarean section is recommended [45]. An epidural is typically preferred over general anesthesia though it may prolong the second stage of labor. Infective endocarditis prophylaxis should also be tailored to the individual patient.

Sex Differences in Advanced Care Planning in Patients With Adult Congenital Heart Disease

The majority of children born with CHD will reach adulthood. However, ACHD patients continue to have substantial morbidity and mortality, as well as increased healthcare utilization at the end of life [99]. However, there is a discrepancy between patient-reported interests in discussing advanced directives and physician-reported discussions, with more patients interested in discussions than recognized by providers. A meta-analysis of seven studies which focused on advanced care planning for ACHD patients found that only 1–28% of ACHD patients recalled participating in advanced care planning discussions with a physician, but 69–68% of patients had a strong interest and desire to participate [100]. Other single-center studies have found a similar willingness from patients to engage in advanced care planning discussions with their provider [101, 102]. Furthermore, while literature from non-ACHD patients has shown that sex influences a patient's advanced care planning preferences, the sex-specific differences in advanced care planning within the ACHD population are unclear [103]. This is an important area of future study that will help frame advanced care planning conversations with ACHD patients.

Conclusion

The 26-year-old patient with a new diagnosis of coarctation of the aorta, BAV, and mild aortic insufficiency described at the beginning of the chapter underwent stent placement of her coarctation before pregnancy. Following the procedure, her the gradient across her coarctation normalized. Despite this, she still developed gestational hypertension, which was managed with oral labetalol. She did not develop preeclampsia, and was able to carry her pregnancy to term. However, her risk of aortic dissection, re-coarctation, and premature coronary artery disease still exists and must be addressed as part of her long-term care.

A review of ACHD literature suggests that sex has a significant impact on the care of ACHD patients. Sex affects the incidence of congenital heart defects, and outcomes for ACHD patients including aortic outcomes and pulmonary hypertension. This suggests the need for a sex-specific clinical approach toward ACHD patient care. Sex should be considered for acquired cardiovascular risk assessment, transitions of care, mental health screening, and contraception counseling. Pregnancy is an opportunity to bring ACHD patients back into care, and preconception counseling is important to understand the risk to both the mother and the fetus. Care of pregnant ACHD patients includes a multidisciplinary team. Further research is needed to understand the relationship between sex and advance care planning. Understanding the influence of sex on ACHD is important in that it informs our ability to better counsel, risk-stratify, and improve the care that this growing population requires.

Key Points

1. Sex-specific differences within incidence of and outcomes for congenital heart disease occur, and suggest the need for a sex-specific clinical approach toward ACHD patients.

2. Acquired cardiovascular risk assessment, transitions of care, mental health screening, and contraception counseling should be a priority and begin at an early age.

3. Preconception counseling should be individualized according to the congenital anatomy, objective studies, and current physiology.

4. Care of pregnant patients with congenital heart disease includes a multidisciplinary cardiology/obstetric/anesthesiology team.

5. Further research is needed to identify the role of sex in advanced care planning.

References

[1] Samánek M. Boy:girl ratio in children born with different forms of cardiac malformation: a population-based study. Pediatr Cardiol 1994;15(2):53–7.

[2] Kong WKF, Regeer MV, Ng ACT, McCormack L, Poh KK, Yeo TC, et al. Sex differences in phenotypes of bicuspid aortic valve and aortopathy: insights from a large multicenter, international registry. Circ Cardiovasc Imaging 2017;10(3).

[3] Verheugt CL, Uiterwaal CSPM, van der Velde ET, Meijboom FJ, Pieper PG, Vliegen HW, et al. Gender and outcome in adult congenital heart disease. Circulation 2008;118(1):26–32.

[4] Engelfriet P, Mulder BJM. Gender differences in adult congenital heart disease. Neth Heart J 2009;17(11):414–7.

[5] Zomer AC, Ionescu-Ittu R, Vaartjes I, Pilote L, Mackie AS, Therrien J, et al. Sex differences in hospital mortality in adults with congenital heart disease: the impact of reproductive health. J Am Coll Cardiol 2013;62(1):58–67.

[6] Warnes CA. Sex differences in congenital heart disease: should a woman be more like a man? Circulation 2008;118(1):3–5.

[7] Weiss BM, Hess OM. Pulmonary vascular disease and pregnancy: current controversies, management strategies, and perspectives. Eur Heart J 2000;21(2):104–15.

[8] Engelfriet PM, Duffels MGJ, Möller T, Boersma E, Tijssen JG, Thaulow E, et al. Pulmonary arterial hypertension in adults born with a heart septal defect: the Euro Heart Survey on adult congenital heart disease. Heart 2007;93(6):682–7.

[9] Kong WKF, Delgado V, Bax JJ. Bicuspid aortic valve: what to image in patients considered for transcatheter aortic valve replacement? Circ Cardiovasc Imaging 2017;10(9).

[10] Ren X, Li F, Wang C, Hou Z, Gao Y, Yin W, et al. Age- and sex-related aortic valve dysfunction and aortopathy difference in patients with bicuspid aortic valve. Int Heart J 2019;60(3):637–42.

[11] Stoberock K, Kölbel T, Atlihan G, Debus ES, Tsilimparis N, Larena-Avellaneda A, et al. Gender differences in abdominal aortic aneurysm therapy—a systematic review. Vasa 2018;47(4):267–71.

[12] Stoberock K, Rieß HC, Debus ES, Schwaneberg T, Kölbel T, Behrendt CA. Gender differences in abdominal aortic aneurysms in Germany using health insurance claims data. Vasa 2018;47(1):36–42.

[13] Fukui T, Tabata M, Morita S, Takanashi S. Gender differences in patients undergoing surgery for acute type A aortic dissection. J Thorac Cardiovasc Surg 2015;150(3):581–7;e581.

[14] Schmidt S, Ramseier-Hadorn M, Thomet C, Wustmann K, Schwerzmann M. Gender-related differences in self-reported dental care in adults with congenital heart disease at increased risk of infective endocarditis. Open Heart 2017;4(1):e000575.

[15] Miner PD, Canobbio MM, Pearson DD, Schlater M, Balon Y, Junge KJ, et al. Contraceptive practices of women with complex congenital heart disease. Am J Cardiol 2017;119(6):911–5.

[16] Kovacs AH, Harrison JL, Colman JM, Sermer M, Siu SC, Silversides CK. Pregnancy and contraception in congenital heart disease: what women are not told. J Am Coll Cardiol 2008;52(7):577–8.

[17] Guttmacher Institute. United States contraception. Guttmacher Institute; 2019. [Accessed August 21, 2019].

[18] Tchaikovski SN, Rosing J. Mechanisms of estrogen-induced venous thromboembolism. Thromb Res 2010;126(1):5–11.

[19] Canobbio MM, Perloff JK, Rapkin AJ. Gynecological health of females with congenital heart disease. Int J Cardiol 2005;98(3):379–87.

[20] Thorne S, MacGregor A, Nelson-Piercy C. Risks of contraception and pregnancy in heart disease. Heart 2006;92(10):1520–5.

[21] Kang AK, Duncan JA, Cattran DC, Floras JS, Lai V, Scholey JW, et al. Effect of oral contraceptives on the renin angiotensin system and renal function. Am J Physiol Regul Integr Comp Physiol 2001;280(3):R807–13.

[22] Bradley SEK, Polis CB, Bankole A, Croft T. Global contraceptive failure rates: who is most at risk? Stud Fam Plann 2019;50(1):3–24.

[23] Shum KK, Gupta T, Canobbio MM, Durst J, Shah SB. Family planning and pregnancy management in adults with congenital heart disease. Prog Cardiovasc Dis 2018;61(3–4):336–46.

[24] Wilson W, Taubert KA, Gewitz M, Lockhart PB, Baddour LM, Levison M, et al. Prevention of infective endocarditis: guidelines from the American Heart Association: a guideline from the American Heart Association Rheumatic Fever, Endocarditis, and Kawasaki Disease Committee, Council on Cardiovascular Disease in the Young, and the Council on Clinical Cardiology, Council on Cardiovascular Surgery and Anesthesia, and the Quality of Care and Outcomes Research Interdisciplinary Working Group. Circulation 2007;116(15):1736–54.

[25] Yeung E, Kay J, Roosevelt GE, Brandon M, Yetman AT. Lapse of care as a predictor for morbidity in adults with congenital heart disease. Int J Cardiol 2008;125(1):62–5.

[26] Helm PC, Kaemmerer H, Breithardt G, Sticker EJ, Keuchen R, Neidenbach R, et al. Transition in patients with congenital heart disease in Germany: results of a nationwide patient survey. Front Pediatr 2017;5:115.

[27] Goossens E, Stephani I, Hilderson D, Gewillig M, Budts W, Van Deyk K, et al. Transfer of adolescents with congenital heart disease from pediatric cardiology to adult health care: an analysis of transfer destinations. J Am Coll Cardiol 2011;57(23):2368–74.

[28] Bellinger DC, Watson CG, Rivkin MJ, Robertson RL, Roberts AE, Stopp C, et al. Neuropsychological status and structural brain imaging in adolescents with single ventricle who underwent the fontan procedure. J Am Heart Assoc 2015;4(12).

[29] DeMaso DR, Calderon J, Taylor GA, Holland JE, Stopp C, White MT, et al. Psychiatric disorders in adolescents with single ventricle congenital heart disease. Pediatrics 2017;139(3).

[30] Ko JM, Cedars AM. Depression in adults with congenital heart disease: prevalence, prognosis, and intervention. Cardiovasc Innov Appl 2018;3(1):97–106.

[31] Ko JM, Tecson KM, Rashida VA, Sodhi S, Saef J, Mufti M, et al. Clinical and psychological drivers of perceived health status in adults with congenital heart disease. Am J Cardiol 2018;121(3):377–81.

[32] Amedro P, Basquin A, Gressin V, Clerson P, Jais X, Thambo J-B, et al. Health-related quality of life of patients with pulmonary arterial hypertension associated with CHD: the multicentre cross-sectional ACHILLE study. Cardiol Young 2016;26(7):1250–9

[33] White KS, Pardue C, Ludbrook P, Sodhi S, Esmaeeli A, Cedars A. Cardiac denial and psychological predictors of cardiac care adherence in adults with congenital heart disease. Behav Modif 2016;40(1–2):29–50.

[34] Eslami B, Sundin O, Macassa G, Khankeh HR, Soares JJF. Anxiety, depressive and somatic symptoms in adults with congenital heart disease. J Psychosom Res 2013;74(1):49–56.

[35] Rexrode K. Sex differences in sex hormones, carotid atherosclerosis, and stroke. Circ Res 2018;122(1):17–9.

[36] Yahagi K, Davis HR, Arbustini E, Virmani R. Sex differences in coronary artery disease: pathological observations. Atherosclerosis 2015;239(1):260–7.

[37] Mathur P, Ostadal B, Romeo F, Mehta JL. Gender-related differences in atherosclerosis. Cardiovasc Drugs Ther 2015;29(4):319–27.

[38] Fairweather D. Sex differences in inflammation during atherosclerosis. Clin Med Insights Cardiol 2014;8(Suppl. 3):49–59.

[39] Krishnamurthy Y, Stefanescu Schmidt AC, Bittner DO, Scholtz J-E, Bui A, Reddy R, et al. Subclinical burden of coronary artery calcium in patients with coarctation of the aorta. Am J Cardiol 2019;123(2):323–8.

[40] Awerbach JD, Krasuski RA, Camitta MGW. Coronary disease and modifying cardiovascular risk in adult congenital heart disease patients: should general guidelines apply? Prog Cardiovasc Dis 2018;61(3–4):300–7.

[41] Lui GK, Fernandes S, McElhinney DB. Management of cardiovascular risk factors in adults with congenital heart disease. J Am Heart Assoc 2014;3(6):e001076.

[42] Lui GK, Rogers IS, Ding VY, Hedlin HK, MacMillen K, Maron DJ, et al. Risk estimates for atherosclerotic cardiovascular disease in adults with congenital heart disease. Am J Cardiol 2017;119(1):112–8.

[43] Flannery LD, Fahed AC, DeFaria YD, Youniss MA, Barinsky GL, Stefanescu Schmidt AC, et al. Frequency of guideline-based statin therapy in adults with congenital heart disease. Am J Cardiol 2018;121(4):485–90.

[44] Lanz J, Brophy JM, Therrien J, Kaouache M, Guo L, Marelli AJ. Stroke in adults with congenital heart disease: incidence, cumulative risk, and predictors. Circulation 2015;132(25):2385–94.

[45] Regitz-Zagrosek V, Blomstrom Lundqvist C, Borghi C, Cifkova R, Ferreira R, Foidart JM, et al. ESC guidelines on the management of cardiovascular diseases during pregnancy: the Task Force on the Management of Cardiovascular Diseases during Pregnancy of the European Society of Cardiology (ESC). Eur Heart J 2011;32(24):3147–97.

[46] Roos-Hesselink JW, Ruys TPE, Stein JI, Thilén U, Webb GD, Niwa K, et al. Outcome of pregnancy in patients with structural or ischaemic heart disease: results of a registry of the European Society of Cardiology. Eur Heart J 2013;34(9):657–65.

[47] Opotowsky AR, Siddiqi OK, D'Souza B, Webb GD, Fernandes SM, Landzberg MJ. Maternal cardiovascular events during childbirth among women with congenital heart disease. Heart 2012;98(2):145–51.

[48] Thompson JL, Kuklina EV, Bateman BT, Callaghan WM, James AH, Grotegut CA. Medical and obstetric outcomes among pregnant women with congenital heart disease. Obstet Gynecol 2015;126(2):346–54.

[49] Schlichting LE, Insaf TZ, Zaidi AN, Lui GK, Van Zutphen AR. Maternal comorbidities and complications of delivery in pregnant women with congenital heart disease. J Am Coll Cardiol 2019;73(17):2181–91.

[50] Franklin WJ, Benton MK, Parekh DR. Cardiac disease in pregnancy. Tex Heart Inst J 2011;38(2):151–3.

[51] Drenthen W, Boersma E, Balci A, Moons P, Roos-Hesselink JW, Mulder BJM, et al. Predictors of pregnancy complications in women with congenital heart disease. Eur Heart J 2010;31(17):2124–32.

[52] Silversides CK, Grewal J, Mason J, Sermer M, Kiess M, Rychel V, et al. Pregnancy outcomes in women with heart disease: the CARPREG II study. J Am Coll Cardiol 2018;71(21):2419–30.

[53] Balci A, Sollie-Szarynska KM, van der Bijl AGL, Ruys TPE, Mulder BJM, Roos-Hesselink JW, et al. Prospective validation and assessment of cardiovascular and offspring risk models for pregnant women with congenital heart disease. Heart 2014;100(17):1373–81.

[54] Martins LC, Freire CMV, Capuruçu CAB, Nunes MCP, Rezende CAL. Risk prediction of cardiovascular complications in pregnant women with heart disease. Arq Bras Cardiol 2016;106(4):289–96.

[55] Ohuchi H, Tanabe Y, Kamiya C, Noritake K, Yasuda K, Miyazaki A, et al. Cardiopulmonary variables during exercise predict pregnancy outcome in women with congenital heart disease. Circ J 2013;77(2):470–6.

[56] Lui GK, Silversides CK, Khairy P, Fernandes SM, Valente AM, Nickolaus MJ, et al. Heart rate response during exercise and pregnancy outcome in women with congenital heart disease. Circulation 2011;123(3):242–8.

[57] Kampman MAM, Balci A, van Veldhuisen DJ, van Dijk APJ, Roos-Hesselink JW, Sollie-Szarynska KM, et al. N-terminal pro-B-type natriuretic peptide predicts cardiovascular complications in pregnant women with congenital heart disease. Eur Heart J 2014;35(11):708–15.

[58] Ramage K, Grabowska K, Silversides C, Quan H, Metcalfe A. Association of adult congenital heart disease with pregnancy, maternal, and neonatal outcomes. JAMA Netw Open 2019;2(5):e193667.

[59] Khairy P, Ouyang DW, Fernandes SM, Lee-Parritz A, Economy KE, Landzberg MJ. Pregnancy outcomes in women with congenital heart disease. Circulation 2006;113(4):517–24.

[60] Siu SC, Colman JM, Sorensen S, Smallhorn JF, Farine D, Amankwah KS, et al. Adverse neonatal and cardiac outcomes are more common in pregnant women with cardiac disease. Circulation 2002;105(18):2179–84.

[61] Perry DJ, Mullen CR, Carvajal HG, Brar AK, Eghtesady P. Familial screening for left-sided congenital heart disease: what is the evidence? What is the cost? Diseases 2017;5(4).

[62] Rao S, Ginns JN. Adult congenital heart disease and pregnancy. Semin Perinatol 2014;38(5):260–72.

[63] Bredy C, Mongeon F-P, Leduc L, Dore A, Khairy P. Pregnancy in adults with repaired/unrepaired atrial septal defect. J Thorac Dis 2018;10(Suppl. 24):S2945–52.

[64] Drenthen W, Pieper PG, van der Tuuk K, Roos-Hesselink JW, Voors AA, Mostert B, et al. Cardiac complications relating to pregnancy and recurrence of disease in the offspring of women with atrioventricular septal defects. Eur Heart J 2005;26(23):2581–7.

[65] Yap S-C, Drenthen W, Pieper PG, Moons P, Mulder BJM, Vliegen HW, et al. Pregnancy outcome in women with repaired versus unrepaired isolated ventricular septal defect. BJOG 2010;117(6):683–9.

[66] Fuchs M, Zaidi AN, Rose J, Sisk T, Daniels CJ, Bradley EA. Location matters: left heart obstruction in pregnancy. Eur J Obstet Gynecol Reprod Biol 2016;196:38–43.

[67] Orwat S, Diller G-P, van Hagen IM, Schmidt R, Tobler D, Greutmann M, et al. Risk of pregnancy in moderate and severe aortic stenosis: from the multinational ROPAC registry. J Am Coll Cardiol 2016;68(16):1727–37.

[68] Yuan S-M. Bicuspid aortic valve in pregnancy. Taiwan J Obstet Gynecol 2014;53(4):476–80.

[69] Krieger EV, Landzberg MJ, Economy KE, Webb GD, Opotowsky AR. Comparison of risk of hypertensive complications of pregnancy among women with versus without coarctation of the aorta. Am J Cardiol 2011;107(10):1529–34.

[70] Beauchesne LM, Connolly HM, Ammash NM, Warnes CA. Coarctation of the aorta: outcome of pregnancy. J Am Coll Cardiol 2001;38(6):1728–33.

[71] Vriend JWJ, Zwinderman AH, de Groot E, Kastelein JJP, Bouma BJ, Mulder BJM. Predictive value of mild, residual descending aortic narrowing for blood pressure and vascular damage in patients after repair of aortic coarctation. Eur Heart J 2005;26(1):84–90.

[72] Siegmund AS, Kampman MAM, Bilardo CM, Balci A, van Dijk APJ, Oudijk MA, et al. Pregnancy in women with corrected aortic coarctation: uteroplacental Doppler flow and pregnancy outcome. Int J Cardiol 2017;249:145–50.

[73] Jimenez-Juan L, Krieger EV, Valente AM, Geva T, Wintersperger BJ, Moshonov H, et al. Cardiovascular magnetic resonance imaging predictors of pregnancy outcomes in women with coarctation of the aorta. Eur Heart J Cardiovasc Imaging 2014;15(3):299–306.

[74] Roman MJ, Pugh NL, Hendershot TP, Devereux RB, Dietz H, Holmes K, et al. Aortic complications associated with pregnancy in Marfan syndrome: the NHLBI National Registry of Genetically Triggered Thoracic Aortic Aneurysms and Cardiovascular Conditions (GenTAC). J Am Heart Assoc 2016;5(8).

[75] Hassan N, Patenaude V, Oddy L, Abenhaim HA. Pregnancy outcomes in Marfan syndrome: a retrospective cohort study. Am J Perinatol 2015;32(2):123–30.

[76] Bhatt AB, DeFaria Yeh D. Pregnancy and adult congenital heart disease. Cardiol Clin 2015;33(4):611–23, ix.

[77] Goland S, Elkayam E. Pregnancy and Marfan syndrome. Ann Cardiothorac Surg 2017;6(6):642–53.

[78] Cauldwell M, Steer PJ, Curtis SL, Mohan A, Dockree S, Mackillop L, et al. Maternal and fetal outcomes in pregnancies complicated by Marfan syndrome. Heart 2019;105:1725–31.

[79] Curry RA, Gelson E, Swan L, Dob D, Babu-Narayan SV, Gatzoulis MA, et al. Marfan syndrome and pregnancy: maternal and neonatal outcomes. BJOG 2014;121(5):610–7.

[80] Balci A, Drenthen W, Mulder BJM, Roos-Hesselink JW, Voors AA, Vliegen HW, et al. Pregnancy in women with corrected tetralogy of Fallot: occurrence and predictors of adverse events. Am Heart J 2011;161(2):307–13.

[81] Veldtman GR, Connolly HM, Grogan M, Ammash NM, Warnes CA. Outcomes of pregnancy in women with tetralogy of Fallot. J Am Coll Cardiol 2004;44(1):174–80.

[82] Kamiya CA, Iwamiya T, Neki R, Katsuragi S, Kawasaki K, Miyoshi T, et al. Outcome of pregnancy and effects on the right heart in women with repaired tetralogy of fallot. Circ J 2012;76(4):957–63.

[83] Cauldwell M, Quail MA, Smith GS, Heng EL, Ghonim S, Uebing A, et al. Effect of pregnancy on ventricular and aortic dimensions in repaired tetralogy of Fallot. J Am Heart Assoc 2017;6(7).

[84] Connolly HM, Warnes CA. Ebstein's anomaly: outcome of pregnancy. J Am Coll Cardiol 1994;23(5):1194–8.

[85] Zhao W, Liu H, Feng R, Lin J. Pregnancy outcomes in women with Ebstein's anomaly. Arch Gynecol Obstet 2012;286(4):881–8.

[86] Katsuragi S, Kamiya C, Yamanaka K, Neki R, Miyoshi T, Iwanaga N, et al. Risk factors for maternal and fetal outcome in pregnancy complicated by Ebstein anomaly. Am J Obstet Gynecol 2013;209(5):452. e451–456.

[87] Connolly HM, Grogan M, Warnes CA. Pregnancy among women with congenitally corrected transposition of great arteries. J Am Coll Cardiol 1999;33(6):1692–5.

[88] Kowalik E, Klisiewicz A, Biernacka EK, Hoffman P. Pregnancy and long-term cardiovascular outcomes in women with congenitally corrected transposition of the great arteries. Int J Gynaecol Obstet 2014;125(2):154–7.

[89] Miller E, Cannobio M, Koos B. Pregnancy outcomes in women with D-transposition of the great arteries after atrial switch. J Am Coll Cardiol 2019;73(9).

[90] Lipczyńska M, Szymański P, Trojnarska O, Tomkiewicz-Pająk L, Pietrzak B, Klisiewicz A, et al. Pregnancy in women with complete transposition of the great arteries following the atrial switch procedure. A study from three of the largest Adult Congenital Heart Disease centers in Poland. J Matern Fetal Neonatal Med 2017;30(5):563–7.

[91] Cataldo S, Doohan M, Rice K, Trinder J, Stuart AG, Curtis SL. Pregnancy following Mustard or Senning correction of transposition of the great arteries: a retrospective study. BJOG 2016;123(5):807–13.

[92] Horiuchi C, Kamiya CA, Ohuchi H, Miyoshi T, Tsuritani M, Iwanaga N, et al. Pregnancy outcomes and mid-term prognosis in women after arterial switch operation for dextro-transposition of the great arteries—Tertiary hospital experiences and review of literature. J Cardiol 2019;73(3):247–54.

[93] Stoll VM, Drury NE, Thorne S, Selman T, Clift P, Chong H, et al. Pregnancy outcomes in women with transposition of the great arteries after an arterial switch operation. JAMA Cardiol 2018;3(11):1119–22.

[94] Garcia Ropero A, Baskar S, Roos Hesselink JW, Girnius A, Zentner D, Swan L, et al. Pregnancy in women with a Fontan circulation: a systematic review of the literature. Circ Cardiovasc Qual Outcomes 2018;11(5):e004575.

[95] Arif S, Chaudhary A, Clift PF, Katie Morris R, Selman TJ, Bowater SE, et al. Pregnancy outcomes in patients with a fontan circulation and proposal for a risk-scoring system: single centre experience. J Congen Cardiol 2017;1(10).

[96] Cauldwell M, Von Klemperer K, Uebing A, Swan L, Steer PJ, Babu-Narayan SV, et al. A cohort study of women with a Fontan circulation undergoing preconception counselling. Heart 2016;102(7):534–40.

[97] Asfour V, Murphy MO, Attia R. Is vaginal delivery or caesarean section the safer mode of delivery in patients with adult congenital heart disease? Interact Cardiovasc Thorac Surg 2013;17(1):144–50.

[98] Cauldwell M, Von Klemperer K, Uebing A, Swan L, Steer PJ, Gatzoulis M, et al. The management of the second stage of labour in women with cardiac: a mixed methods study. Int J Cardiol 2016;222:732–6.

[99] Steiner JM, Kirkpatrick JN, Heckbert SR, Sibley J, Fausto JA, Engelberg RA, et al. Hospital resource utilization and presence of advance directives at the end of life for adults with congenital heart disease. Congenit Heart Dis 2018;13(5):721–7.

[100] Ludmir J, Steiner JM, Wong HN, Kloosterboer A, Leong J, Aslakson RA. Palliative care opportunities among adults with congenital heart disease—a systematic review. J Pain Symptom Manage 2019;58:891–8.

[101] Steiner JM, Stout K, Soine L, Kirkpatrick JN, Curtis JR. Perspectives on advance care planning and palliative care among adults with congenital heart disease. Congenit Heart Dis 2019;14(3):403–9.

[102] Deng LX, Gleason LP, Khan AM, Drajpuch D, Fuller S, Goldberg LA, et al. Advance care planning in adults with congenital heart disease: a patient priority. Int J Cardiol 2017;231:105–9.

[103] Perkins HS, Cortez JD, Hazuda HP. Advance care planning: does patient gender make a difference? Am J Med Sci 2004;327(1):25–32.

Pregnancy-Related Heart Disease

Chapter 21

Pregnancy and Cardiovascular Disease

Emily Lau, Anna O'Kelly, and Nandita S. Scott

Clinical Case

A 34-year-old, G1P1, asymptomatic woman, who grew up in Ethiopia presents for preconception counseling. She has a history of rheumatic heart disease resulting in mitral stenosis as a child.

She has no cardiac symptoms, at rest or with exertion. On physical exam, she has a regular rate and rhythm, with a loud S1, and normal S2. There is an opening snap followed by a 2/6 diastolic murmur at the left ventricular apex. No other extracardiac sounds are present. Lungs are clear to auscultation. Her current echocardiogram demonstrates a mitral valve area by pressure half-time method of 1.2 cm^2, mean gradient across the valve of 7 mmHg at a heart of 70 beats per minute, and a mitral valve score of 8, with trace mitral regurgitation. What are her pregnancy-associated risks and how do we optimize her preconception?

CH
21

Abstract

Maternal mortality in the United States is rising and cardiovascular disease causes 26.5% of these pregnancy-related deaths. The rising trend appears to be secondary to acquired heart disease with many factors contributing, including increasing maternal age, rise in multifetal pregnancies, and the increased burden of preexisting cardiovascular risk factors in the pregnant population. Pregnancy itself causes a significant hemodynamic burden and may cause decompensation in a patient with preexisting cardiovascular disease. Risk stratification for pregnancy should ideally occur preconception and there are several risk assessment tools that can aid decision making.

The care of the pregnant woman with heart disease also implicates several stakeholders including obstetrics, obstetric anesthesia, and cardiology. Unlike many cardiovascular diseases, there are no randomized controlled data to support decision making, and guidelines are largely built on expert consensus. It is therefore vital for cardiologists to continue to build on their knowledge on the management of cardiovascular disease during pregnancy and work in a multidisciplinary fashion to improve care.

Introduction

Maternal mortality in the United States is rising and is presently the highest in the developed world at an estimated rate of 26.4 deaths per 100,000 live births [1]. Maternal mortality has considerable racial and ethnic disparities, with even higher mortality in black non-Hispanic and American Indian/Alaskan Native women (**Figure 1**).

Death from cardiovascular disease (CVD) represents 26.5% of all pregnancy-related deaths in the United States [2]. The rising trend appears to be secondary to many contributing factors including increasing maternal age, rise in multifetal pregnancies, and the increased burden of preexisting cardiovascular risk factors in the pregnant population. In addition, survivors of childhood cancer, who have been exposed to cardiotoxic therapies, and women with congenital heart disease (CHD) are now able to pursue pregnancy due to advances in care.

Unlike many CVD processes, there are no randomized controlled data to support decision making, and guidelines are largely built on expert consensus. The care of the pregnant woman with heart disease also engages several disciplines including obstetrics, obstetric anesthesiology, and cardiology. These teams have different perspectives but a common goal: the delivery of a healthy fetus without maternal cardiac complications. It is therefore vital for cardiologists to continue to build on their knowledge on the management of CVD during pregnancy and work in a multidisciplinary fashion to improve care.

Hemodynamics of Pregnancy

The cardiovascular adaptations to pregnancy are significant and generally well tolerated. In women with CVD, however, the physiologic hemodynamic burden can lead to maternal decompensation and increased morbidity and mortality for the mother and the fetus.

Changes in cardiovascular hemodynamics begin as early as the first trimester with a rise in cardiac output that ultimately increases to 30–50% of nonpregnancy values [3]. This increase in cardiac output is driven by an increase in

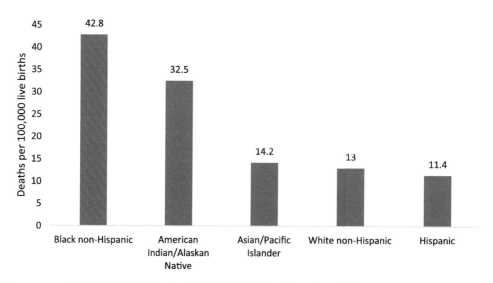

FIGURE 1 **Pregnancy-Related Mortality in the United States.** *Data from [2].*

1. There are marked cardiovascular hemodynamic adaptations that occur during pregnancy which begin as early as the first trimester. These include an increase in heart rate and plasma volume resulting in increased cardiac output as well as a fall in systemic vascular resistance.

2. Cardiac remodeling occurs secondary to these hemodynamic changes with the development of eccentric hypertrophy and chamber dilation.

3. The postpartum period remains an active period for the cardiovascular system and is when maternal decompensation often occurs.

then slowly rises to term. Importantly, the supine position can cause a substantial fall in cardiac output due to compression of the inferior vena cava by the gravid uterus.

Labor and delivery are a time of further change in cardiovascular hemodynamics. Cardiac output reaches its peak during labor [3, 5]. This is attributed to increases in both heart rate, due to sympathetic activation and pain, and stroke volume, due to autotransfusion from the contracting uterus [3, 5]. The postpartum phase remains an active period as well and often the most vulnerable period for maternal hemodynamic decompensation. There is further autotransfusion from uterine involution and mobilization of dependent edema as well as an increase in afterload due to the loss of the low vascular resistance placental unit. These adaptations are more marked in multifetal pregnancies [6].

Cardiac remodeling occurs as a result of these physiologic changes. The left ventricular mass increases with eccentric hypertrophy, and there is dilatation of all cardiac chambers (absolute values should remain in the normal range). Postpartum, hemodynamic parameters gradually return to their prepregnancy values [3].

The normal echocardiogram in pregnancy reflects these physiologic changes, with measurable increases in velocities across the valves, and physiologic valvular regurgitation across the tricuspid, pulmonary, and mitral valves but not the aortic valve. A small pericardial effusion can be seen in normal pregnancy. Diastolic function as measured by load-independent indices such as tissue Doppler imaging is unchanged during pregnancy despite the increase in preload and physiologic hypertrophy [7–9].

plasma volume and stroke volume, and, to a lesser extent, heart rate (**Figure 2**). Systemic vascular resistance (SVR) decreases during pregnancy, and is influenced by the changing hormonal milieu of pregnancy as well as flow into the low-resistance uteroplacental unit [3, 5]. Red blood cell mass increases because of increased erythropoietin levels. This rise, however, occurs to a lesser extent than that of plasma volume, and leads to a dilutional anemia. Blood pressure falls, reaches a nadir in the second trimester, and

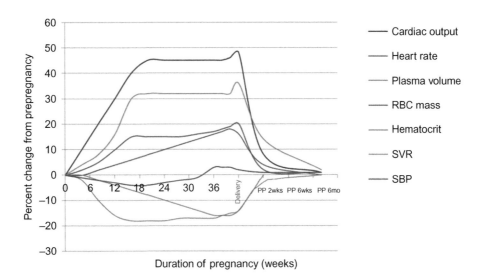

FIGURE 2   **Physiological Adaptations in Normal Pregnancy.** Increases in plasma volume and heart rate lead to a 30–50% increase in cardiac output. There is physiological anemia due to a larger rise in plasma volume than the red blood cell mass. PP, postpartum; RBC, red blood cell; SBP, systolic blood pressure; SVR, systemic vascular resistance. *Reproduced with permission from [4].*

Preconception Counseling and Risk Stratification Tools

> ### Key Point
>
> **1** For patients with existing cardiovascular disease desiring pregnancy, preconception counseling should include discussion of maternal and fetal risk, review of medications, preconception testing, and referral to maternal-fetal medicine. Risk scores including CARPREG, ZAHARA, and modified WHO assist in maternal risk stratification in patients with existing CVD.

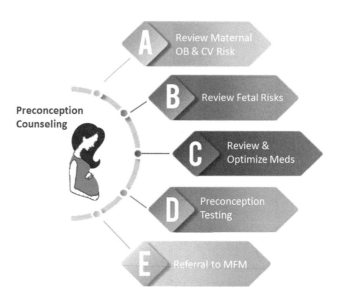

FIGURE 4 Overview of Preconception Counseling Considerations. CV, cardiovascular; MFM, maternal-fetal medicine; OB, obstetric. *Image courtesy of Niti R. Aggarwal.*

Given these hemodynamic changes, women with healthy pregnancies often experience palpitations, edema, orthostatic hypotension, and fatigue. Symptoms and physical exam findings of healthy pregnancy can at times be challenging to differentiate from those with cardiac pathology. **Figure 3** lists common signs and symptoms that should raise concern if experienced during pregnancy.

For patients with existing CVD, a thorough evaluation and discussion of pregnancy-specific risks should be conducted prior to conception. Preconception counseling can be broken down into the following components: (A) maternal risk, including optimization of current clinical status; (B) fetal risk, including genetic counseling if appropriate; (C) review of medications prior to conception; (D) preconception testing that may be needed to help risk assessment; and (E) referral to maternal-fetal medicine (MFM) for more nuanced obstetric assessment (**Figure 4**).

Maternal Risk

Several risk scores have been developed to help determine maternal risk for patients with existing CVD (**Figure 5**).

The most widely used and recognized clinical risk prediction tool for cardiac disease in pregnancy comes from the CARdiac disease in PREGnancy (CARPREG II) study (**Table 1**). The first iteration, CARPREG, was developed from data from 599 pregnancies, where the majority of women had CHD [10]. The maternal adverse outcome rate was 13%, and largely constituted heart failure and arrhythmia events, although rare events including maternal embolic stroke and sudden cardiac death were observed as well. In an effort to add complexity to the risk prediction tool and to capture cardiac risk more accurately, CARPREG II was developed [11]. The new risk index was derived from a study group of 1938 pregnancies including 289 pregnancies that were enrolled in the original CARPREG study. Adverse cardiac events occurred in 16% of the pregnancies, largely related to heart failure and arrhythmia. The CARPREG II risk score utilizes 10 predictors that incorporate general history and physical, specific lesions, imaging, and delivery of care factors: (1) prior cardiac event or arrhythmia; (2) baseline New York Heart Association (NYHA) functional class III/IV or cyanosis; (3) the presence of mechanical valve; (4) ventricular dysfunction with left ventricular ejection fraction (LVEF) less than 55%; (5) high-risk left-sided valve disease/left ventricular outflow tract (LVOT) (aortic valve area $< 1.5\,cm^2$, mitral valve area $< 2.0\,cm^2$, and subaortic gradient greater than 30 mmHg, moderate to severe mitral regurgitation); (6) pulmonary hypertension (right ventricular systolic pressure greater than 49 mmHg); (7) coronary artery disease (angiographically proven coronary artery obstruction or prior myocardial infarction); (8) high-risk aortopathy (Marfan syndrome, bicuspid aortopathy with aortic dimension greater than 45 mm, Loeys-Dietz syndrome, Ehlers-Danlos syndrome or prior aortic dissection or pseudoaneurysm); (9) no prior cardiac intervention; and (10) late

- Chest pain
- Dyspnea at rest
- Paroxysmal nocturnal dyspnea
- Syncope
- Sustained palpitations
- Symptoms starting and progressively worsening >20 weeks gestation

- Diastolic murmur
- Heart rate >100 bpm
- Cyanosis or clubbing
- Rales
- S4 or Gallop

FIGURE 3 Concerning Signs and Symptoms of CVD in Pregnancy. *Image courtesy of Niti R. Aggarwal.*

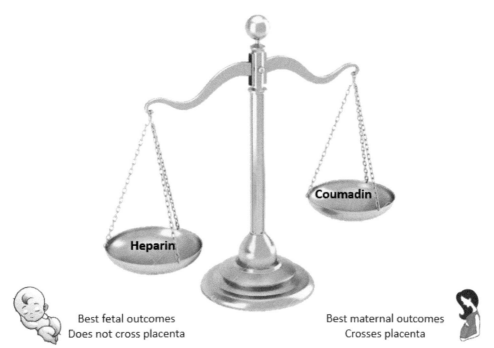

Best fetal outcomes
Does not cross placenta

Best maternal outcomes
Crosses placenta

Affected by change in volume of distribution & GFR
Ideal monitoring schedule not known
(peak/trough Xa levels and frequency)

Higher risk of fetal wastage & embryopathy
Obligate C section if anticoagulated in labor

FIGURE 5 **Vitamin K Antagonists vs.** Low Molecular Weight Heparin During Pregnancy. GFR, glomerular filtration rate. *Image courtesy of Niti R. Aggarwal.*

TABLE 1 CARPREG II Risk Predictors	
Predictor	Points
Prior cardiac events or arrhythmias	3
Baseline NYHA 3–4 or cyanosis	3
Mechanical valve	3
Systemic ventricular dysfunction LVEF < 55%	2
High-risk left-sided valvular disease or left ventricular outflow tract obstruction (aortic valve area < 1.5 cm², mitral valve area < 2 cm² subaortic gradient > 30, or moderate to severe mitral regurgitation)	2
Pulmonary hypertension, RVSP > 49 mmHg	2
High-risk aortopathy	2
Coronary artery disease	2
No prior cardiac intervention	1
Late pregnancy assessment	1

Primary cardiac event risk: score = 1, 5% risk, score = 2, 10% risk, score = 3, 15% risk, score = 4, 22% risk, and 41% risk if score greater than 4.
LVEF, left ventricular ejection fraction; RVSP, right ventricular systolic pressure. *Data from [11].*

pregnancy assessment (first antenatal visit after 20 weeks). The predicted risk for an adverse cardiac event was stratified by point estimate: 0–1 points (5%), 2 points (10%), 3 points (15%), 4 points (22%), and > 4 points (41%).

Beyond the CARPREG and CARPREG II scores, the modified World Health Organization (mWHO) (**Table 2**) [13] is another clinical tool for risk prediction in women with preexisting CVD. The mWHO score divides women into four risk categories, and recommendations regarding follow-up and treatment are based on risk category.

Risk in CHD

Women with preexisting CHD are a unique population for whom pregnancy is high risk. The ZAHARA [Zwangerschap bij Aangeboren HARtAfwijking (Pregnancy in Women with Congenital Heart Disease)] [14] risk score derives from 1302 pregnancies from 60 countries including 714 pregnancies in women with CHD (**Table 3**) and is another important tool in the risk stratification of pregnant women with CHD.

Fetal and Neonatal Risk

In women with preexisting CVD, neonatal complications are high: complications occur in 20–37% of cases, including a 1–4% neonatal mortality rate. Maternal morbidity is strongly associated with neonatal adverse events, so it is imperative that maternal health is optimized throughout

TABLE 2 Modified WHO Risk Categories

Risk Category	Predictors	Maternal Cardiac Event Rate	Recommendations
Class I No risk of maternal mortality No/mild increased maternal morbidity	• Mild PS • Small PDA • Mild MVP • Successfully repaired simple lesions (ASD or VSD, PDA, or anomalous pulmonary venous drainage) • Isolated atrial or ventricular ectopic beats	2.5–5%	Local hospital delivery, follow-up once or twice during pregnancy
Class II Small increased risk of maternal mortality, moderate increase in morbidity	• Unoperated ASD or VSD • Repaired tetralogy of Fallot • Most arrhythmias • Turner syndrome without aortic dilatation	5.7–10.5%	Local hospital delivery, follow-up every trimester
Classes II–III Intermediate increased risk of maternal mortality, moderate to severe increased risk morbidity	• Mild LV impairment • Hypertrophic cardiomyopathy • Native or tissue heart valve disease not considered WHO I or IV • Marfan syndrome without aortic dilatation and aorta <45 mm in aortic disease associated with bicuspid aortic valve • Repaired coarctation • Atrioventricular septal defect	10–19%	Referral hospital delivery, bimonthly follow-up
Class III Significantly increased risk of maternal mortality or severe morbidity	• Mechanical valves • Previous peripartum cardiomyopathy without any residual left ventricular impairment • Moderate left ventricular impairment • Systemic right ventricle • Fontan circulation • Unrepaired cyanotic heart disease • Complex congenital heart disease • Moderate mitral stenosis • Severe asymptomatic aortic stenosis • Aortic dilatation 40–45 mm in Marfan syndrome, 45–50 mm with bicuspid aortic valve • Ventricular tachycardia	19–27%	Expert center for pregnancy and cardiac disease, monthly or bi-monthly follow-up
Class IV Extremely high risk of maternal mortality and severe morbidity	• PAH of any cause • Severe systemic ventricular dysfunction (LVEF <30%, NYHA III–IV) • Previous peripartum cardiomyopathy with any residual impairment of LV function • Severe MS or severe symptomatic AS • Systemic right ventricle with moderate or severely decreased ventricular function • Marfan syndrome with dilated aorta >45 mm and aortic dilation >50 mm in aortic disease associated with bicuspid aortic valve • Vascular Ehlers-Danlos severe aortic coarctation • Fontan circulation with any complication	40–100%	Pregnancy contraindicated; termination should be discussed Expert center for pregnancy and cardiac disease If pregnant, monthly follow-up

AS, aortic stenosis; ASD, atrial septal defect; HF, heart failure; LV, left ventricle; LVEF, left ventricular ejection fraction; MS, mitral stenosis; MVP, mitral valve prolapse; NYHA, New York Heart Association; PAH, pulmonary arterial hypertension; PDA, patent ductus arteriosus; PS, pulmonic stenosis; VSD, ventricular septal defect; WHO, World Health Organization. *Data from [12].*

TABLE 3 ZAHARA Risk Predictors	
Predictor	Points
History of arrhythmias	1.5
> 2 NYHA functional class	0.75
LVOT obstruction with peak > 50 mmHg or AVA < 1 cm²	2.5
Mechanical valve prosthesis	4.25
Moderate/severe subpulmonic or systemic atrioventricular valve regurgitation	0.75
Use of cardiac medications prepregnancy	1.5
Repaired or unrepaired cyanotic heart disease	1

AVA, aortic valve area; LVOT, left ventricular outflow tract. *Data from [14].*

pregnancy for the health of both mother and fetus. Risk factors for fetal and neonatal risk include cyanosis, poor functional class, smoking, multiple gestations, use of anticoagulants during pregnancy, and mechanical valve prosthesis. The risk of CHD in the fetus must also be considered. In the largest series to date, the risk is estimated to be 2.7%. Some data suggest that the risk is higher for offspring of females with CHD vs. males. When it does occur, the fetal defect is not necessarily the same as that seen in the parent [15, 16].

Fetal Echocardiography

Performance of a fetal echocardiogram should be considered in certain situations because identification and management of fetal cardiac abnormalities can be initiated prior to birth, thus providing opportunities to improve and plan care. Indications for fetal echocardiogram include, but are not limited to, exposure to maternal cardiac teratogens, the presence of CHD in first-degree relative of fetus, fetal cardiac abnormality noted on obstetric ultrasound, first- or second-degree relative with disorder with Mendelian inheritance with CHD association, and fetal tachycardia or bradycardia [17]. The optimal gestational age for screening for fetal structural anomalies is between 18 and 22 weeks, although the structure of the fetal heart can be assessed as early as 10 weeks' gestation. The sensitivity and specificity of fetal echocardiography in fetuses at increased risk for CHD were 85% and 99%, respectively [18].

Preconception Testing

All women with existing CVD planning pregnancy should pursue preconception evaluation. The selection of testing

should be guided by cardiovascular pathology and patient history, and may include a transthoracic echocardiogram (TTE) and exercise testing [19]. Exercise testing provides an important assessment of baseline functional capacity and is especially important for women with existing asymptomatic severe valvular aortic stenosis [20]. Beyond exercise testing and TTE, measurement of brain natriuretic peptide (BNP) should also be part of the evaluation of all women. Normal range BNP in women with structural heart disease has a very strong negative predictive value for predicting adverse cardiac outcomes during pregnancy. However, BNP levels are often falsely decreased in obesity and gestational diabetes [8], so should be used with caution in these populations.

For special populations, additional specialized testing may be warranted. In women with CHD, cardiopulmonary exercise testing may provide additional information about functional capacity and oxygen utilization and transport beyond a traditional exercise test [21]. Stress echocardiography is useful to assess contractile reserve in patients with previous history of peripartum cardiomyopathy [22]. Finally, baseline imaging of the aorta is paramount in patients with aortopathy or aortic valve disease. Furthermore, in women with coarctation of the aorta specifically, evaluation for cerebral aneurysms preconception should also be performed.

Women choosing to avoid pregnancy should be counseled about contraception options (see Chapter 25 for additional details).

Mode of Delivery and Interdisciplinary Team

Key Points

 Vaginal delivery is the preferred mode of delivery for patients with existing CVD with few notable exceptions including obstetric indications, decompensated HF, Marfan syndrome with dilated aorta > 40–45 mm, history of aortic dissection, or therapeutic anticoagulation with VKA.

 A multidisciplinary team of cardiologists, maternal-fetal medicine specialists, obstetric anesthesiologists, and (when appropriate) cardiac surgeons is vitally important for the care of pregnant women with cardiovascular disease.

Despite a widely held perception that cesarean section (C-section) is safer than vaginal delivery in women with high-risk pregnancies, vaginal delivery with adequate analgesia is the recommended mode of delivery in most patients. Compared to C-section, vaginal deliveries have better outcomes including less hemorrhage requiring hysterectomy, fewer cardiac arrests, fewer major puerperal infections, and shorter in-hospital length of stays [23, 24]. Valsalva maneuvers performed during vaginal delivery result in decrease in preload, increased afterload, and decreased cardiac output [25]. Conversely, release of Valsalva maneuvers causes an overshoot in cardiac output. Avoidance of Valsalva maneuvers may be favored in women with pulmonary hypertension, fixed left-sided stenotic lesions, significant aortic dilatations, and decompensated heart failure [26]. Cardiac vaginal deliveries may be favored in these women, whereby after cervical dilation the mother does not push; instead the fetus descends down the birth canal solely by uterine contractions [27]. Delivery may be facilitated using low-outlet forceps or vacuum-assisted devices. While there is no consensus on absolute contraindications to vaginal delivery, C-section may be preferred in key populations including women who require urgent delivery due to decompensated heart failure refractory to medical therapy, women with Marfan syndrome with a dilated aorta > 40–45 mm, women who have a history of acute or chronic aortic dissection, or women who are therapeutically anticoagulated in labor with vitamin K antagonist (VKA) (as the fetus is anticoagulated as well) [28]. Timing and mode of delivery should be individualized depending on the clinical status of the mother and fetal maturity.

During vaginal delivery, adequate analgesia is important for the cardiopulmonary stability of the mother. Painful labor results in release of catecholamines, which then result in tachycardia, hypertension, increased cardiac output, and workload [27]. These changes may not be tolerated by pregnant women with high-risk cardiac lesions. Resultantly, most anesthesiologists recommend early administration of analgesia to avoid painful contractions.

Both spinal and epidural anesthesia may result in decreased venous tone and decrease in systemic venous resistance, resulting in hypotension. Spinal anesthetics have a more rapid onset, and hence a risk of more rapid decrease in SVR, and can cause cardiopulmonary perturbations [27]. In contrast, epidurals have a slower onset, but may not provide the same consistency of spinal anesthesia. Epidural anesthesia should be administered in small incremental doses rather than a single spinal dose to avoid hypotension [26]. General anesthesia may be indicated in patients who are anticoagulated, unable to lie flat, or need mechanical ventilation from pulmonary edema. The anesthetic should be carefully selected in these cases by an experienced anesthesiologist.

Intrapartum maternal monitoring may include continuous pulse oximetry to monitor maternal oxygen saturations, continuous telemetry for maternal electrocardiogram (ECG) monitoring, and arterial lines for monitoring fluids shifts [26, 27]. Large-volume boluses and excess blood loss should be avoided. Pregnancy can be a very vulnerable period of time for women with CVD. An interdisciplinary team of providers is vitally important for the care of this complex patient population in order to minimize maternal cardiac complications and to deliver a healthy baby. The team should include an MFM specialist, cardiologists with expertise in the care of pregnant women, an obstetric anesthesiologist, and/or a cardiac surgeon, should an intervention be required. The group should meet regularly to discuss, anticipate, and plan for issues that may arise for their patients and to formulate strategies for labor, delivery, and postpartum care. A multidisciplinary team should formulate a detailed delivery plan including mode of delivery, cardiac monitoring during labor and delivery, labor analgesia and anesthetics, and plan for obstetric or cardiac emergency. In centers with no expertise in pregnancy management, women can be referred to a specialized center for initial consultation. The specialists can then help to provide recommendations about where to deliver and continue care remotely through telemedicine if available.

Medication Use During Pregnancy

Key Points

 The first trimester is the highest risk period for teratogenicity.

 The physiologic changes of pregnancy, including increased plasma volume and altered hepatic and renal clearance, affect drug absorption and excretion.

Many factors need to be considered when using cardiovascular medications during pregnancy. Several drugs, for instance, cross the placenta during pregnancy and are shared via breast milk after delivery. The physiology of pregnancy also affects drug availability. Increased plasma volume expands the volume of distribution and as a result reduces drug concentration. The glomerular filtration rate (GFR) and liver metabolism increase, gastrointestinal motility and absorption fall, and there are reduced levels of plasma proteins; all of these alter drug concentration, absorption, and excretion. In general, it is advised during pregnancy to use only necessary drugs, and to use them at the lowest dose for

CH
21
Pregnancy and Cardiovascular Disease

the shortest duration of time necessary to achieve therapeutic goals [12, 29]. This is especially true in the first trimester, which is the highest risk period for teratogenicity. Adverse effects of medication can, however, occur throughout pregnancy. For further details regarding this topic, please refer to Chapter 22.

Cardiovascular medications that are contraindicated during pregnancy are HMG-Coa reductase inhibitors, angiotensin-converting enzyme inhibitors, angiotensin receptor blockers, aldosterone antagonists, and direct oral anticoagulants (DOACs). Atenolol is the least preferred beta-blocker. Additional details about medication use during pregnancy and lactation are summarized in **Table 4**.

TABLE 4 Medication Safety in Pregnancy and Lactation

Drug Name	Safety in Pregnancy	Safety in Lactation	Comments
Beta-Blockers			
Labetalol	Safe[a] [3, 29]	Use with caution given limited data, though likely safe [3, 29]	Preferred beta-blocker [30]
Metoprolol	Safe[a] [3, 29]	Use with caution given limited data, though likely safe [3, 29]	Preferred beta-blocker [30]
Carvedilol	Safe[a] [3, 29]	No available data, though likely safe [3, 29]	
Propranolol	Safe[a] [3, 29]	Use with caution given limited data, though likely safe [3, 29]	
Antiplatelets			
Aspirin	Use with caution given limited data [29]	Use with caution given limited data, though does not appear to increase risk of Reye's syndrome [29, 31]	Preferred antiplatelet agent, but can only be used at low dose given risk of PDA closure [30]
Clopidogrel	Use with caution given limited data [29]	Use with caution given limited data [29]	Limited data, must be stopped 7 days prior to spinal/epidural anesthesia [30, 32, 33]
Prasugrel	Use with caution given limited data [29]	Unknown [29]	
Ticagrelor	Use with caution given limited data [29]	Unknown [29]	
Anticoagulants			
DOACs	Contraindicated [3, 29]	Contraindicated [3, 29]	Limited data, drug also crosses the placenta [3, 7, 34]
Warfarin	Use with caution given limited data, though associated with risk of fetal hemorrhage and embryopathy [3, 29]	Safe [3, 29]	Drug crosses the placenta [3]
Heparin (unfractionated)	Safe [3, 29]	Safe [3, 29]	Drug does not cross the placenta [3]

Continued

TABLE 4 Medication Safety in Pregnancy and Lactation—cont'd

Drug Name	Safety in Pregnancy	Safety in Lactation	Comments
Heparin (fractionated)	Safe [3, 29]	Safe [3, 29]	Drug does not cross the placenta [29]
Diuretics			
Furosemide	Safe [3, 29]	Use with caution given limited data, though likely safe [3, 29]	
Hydrochlorothiazide	Use with caution given limited data [3, 29]	Safe [3, 29]	
Bumetanide	Safe [29]	Unknown [29]	
Torsemide	Use with caution given limited data [29]	Unknown [29]	
Metolazone	Use with caution given limited data [29]	Unknown [29]	

Antihypertensives

ACEi ARBs	Contraindicated [3, 29]	Use with caution given limited data, though likely safe [3, 29]	Strong evidence for teratogenicity during pregnancy [12, 35, 36]
Aldosterone antagonists	Contraindicated [29]	Contraindicated [29]	
Labetalol	Safe[a] [3, 29]	Use with caution given limited data, though likely safe [3, 29]	Preferred beta-blocker, can be used in hypertensive emergency [12, 26, 30]
Nifedipine	Safe [3, 29]	Safe [3, 29]	Preferred agent [12]
Amlodipine	Use with caution given limited data [3, 29]	Use with caution given limited data, though likely safe [3, 29]	
Clonidine	Use with caution given limited data [29]	Unknown [29]	
Hydralazine	Use with caution given limited data, though likely safe [3, 29]	Safe [3, 29]	Can be used in hypertensive emergency [26]
Methyldopa	Safe [29]	Safe [29]	
Nitroglycerin	Use with caution given limited data, though likely safe [3, 29]	Unknown [3, 29]	
Nitroprusside	Use with caution given limited data, must monitor for cyanide poisoning in fetus [3, 29]	Use with caution given limited data, though potentially dangerous [3, 29]	
Isosorbide dinitrate	Use with caution given limited data [29]	Unknown [29]	

TABLE 4 Medication Safety in Pregnancy and Lactation—cont'd			
Drug Name	Safety in Pregnancy	Safety in Lactation	Comments
Antiarrhythmics			
Amiodarone	Contraindicated [3, 29]	Contraindicated [3, 29]	Last-line agent for maternal arrhythmia, risk of neonatal thyroid dysfunction, and neurodevelopment abnormalities
Adenosine	Safe [3, 29]	Use with caution given limited data, though likely safe [3, 29]	
Digoxin	Safe [3, 29]	Safe [3, 29]	
Verapamil	Use with caution given limited data, though likely safe [3, 29]	Use with caution given limited data, though likely safe [3, 29]	
Diltiazem	Use with caution given limited data, though likely safe [3, 29]	Unknown, though likely safe [3, 29]	
Procainamide	Use with caution given limited data [3, 29]	Use with caution given limited data, though likely safe [3, 29]	
Sotalol	Use with caution given limited data [3, 29]	Unknown [3, 29]	
Flecainide	Use with caution given limited data [3, 29]	Use with caution given limited data, though likely safe [3, 29]	
Propafenone	Use with caution given limited data [29]	Use with caution given limited data [29]	

[a]Though generally considered safe, beta-blockers have been associated with small for gestational age, hypotension, bradycardia, and hypoglycemia in newborns. Data from [3, 7, 12, 26, 29–42].
ACEi, ACE inhibitors; ARBs, angiotensin receptor blockers; DOACs, direct oral anticoagulants; PDA, patent ductus arteriosus.

Safety and Utility of Imaging in Pregnancy

Key Points

1. It is important to weigh risks and benefits prior to imaging a pregnant woman.

2. Echocardiography is the test of choice in pregnancy given the absence of ionizing radiation.

3. The accepted maximum cumulative dose of ionizing radiation during pregnancy is 50 mSv.

Cardiovascular imaging is often essential to the diagnosis and management of women with CVD during pregnancy. Imaging modalities that are available for use including echocardiography, radiography, computed tomography (CT), nuclear imaging, and cardiac magnetic resonance imaging (MRI). Better understanding of the indications, limitations, and risks associated with each imaging modality is important to help determine the ideal test of choice. Prior to ordering an imaging study in a pregnant patient, consider the following questions:

- What is the gestational age of the fetus? Risk of teratogenicity is highest during the first trimester when organogenesis occurs.
- Will echocardiography be satisfactory for diagnosis?

- Is additional imaging required to address the clinical question and will it change management?
- Can imaging be delayed until later in pregnancy (second or third trimester) when organogenesis is complete, or after delivery?
- Is termination or early delivery being considered?
- Is imaging with a contrast agent and/or radiation essential for diagnosis and treatment?
- Are interventions appropriate to reduce fetal radiation exposure?

Echocardiography

Both the American College of Obstetricians Gynecologists (ACOG) and the Food and Drug Administration (FDA) recommend ultrasonography as the first-line imaging modality for pregnant patients [43]. Echocardiograms provide comprehensive assessments of ventricular function, valvular disease, and congenital anomalies, and can even visualize the ascending aorta. For this reason, conditions such as CHD, peripartum cardiomyopathy, valvular disorders, and ascending aortic disease are often evaluated with echocardiography. The safety profile of echocardiography is optimal as there have been no reports of adverse effects from diagnostic ultrasound. The only theoretical risk that has been proposed is tissue heating, which can easily be circumvented by minimizing scanning time and optimizing machine scanning parameters [43]. Additionally, agitated saline microbubble contrast or dobutamine stress studies are generally safe as well (FDA Category B) [28]. Stress echocardiography is rarely needed during pregnancy but exercise stress is safe in this population.

Transesophageal echocardiography (TEE) may be indicated, particularly for women being considered for advanced valve interventions such as percutaneous mitral balloon valvuloplasty or in suspected aortic dissection. TEE is considered safe in pregnancy but need for sedation does introduce risk to pregnant women, particularly of aspiration [28].

Ionizing Radiation

The potential for fetal radiation risk is an important consideration for the application of imaging modalities in the pregnant woman. The general consensus is that the risk to the embryo or fetus depends on the amount and type of radiation and gestational age at radiation [9]. However, there are no data linking diagnostic imaging to fetal harm as it is difficult to differentiate radiation-induced harm from spontaneous adverse fetal results [44]. Presently, the accepted maximum cumulative dose of ionizing radiation in pregnancy is 50 mGy or 50 mSv, below which there have not been any documented incidents of fetal anomalies, growth restriction, or abortion [44]. As a result, termination of pregnancy is not recommended unless there is

clear documentation of an estimated fetal radiation dose of > 150 mGy. The National Council on Radiation Protection and Measurements released a report on the potential health effects of fetal radiation doses > 100 mGy (**Table 5**).

Chest Radiography

Chest radiography is a simple and widely available diagnostic tool that is useful for the evaluation of CVD in the pregnant patient. Exposure of ionizing radiation is of concern and it is recommended that the gravid abdomen be shielded to minimize scattered radiation to the fetus. In general, ACOG recommends against the use of radiography for all pregnant patients unless the benefits far outweigh the risks [43]. Nevertheless, the overall dose exposure of chest radiography to the fetus is negligible and can be considered particularly in a pregnant woman presenting with new-onset dyspnea [46].

Computed Tomography

The most common scenario in which computed tomography angiography (CTA) imaging is required in pregnant women is for the assessment of suspected pulmonary embolism (PE). CTA is the preferred test of choice for pregnant women although ventilation-perfusion nuclear imaging scans are a reasonable alternative. According to the 2011 consensus guidelines from the American Thoracic Society (ATS), Society of Thoracic Radiology, Society of Nuclear Medicine and Molecular Imaging, and ACOG, suspected PE in the pregnant woman should first be evaluated with lower-extremity Doppler imaging and chest radiography followed by CTA if initial findings are abnormal or suspicion is high [47]. Minimizing the fetal radiation dose should always be considered when performing a CTA on a pregnant woman, and doses as low as 0.1–0.4 mSv can be achieved with dose reduction protocols [48]. Acquisition of good-quality CTA images in pregnant women is a particular challenge, with rates of diagnostically inadequate studies as high as 35.7% in the pregnant population (compared with 2.1% in the nonpregnant cohort) [49]. Specifically, hemodynamic changes of pregnancy including hemodilution, shunting of maternal blood to the fetus, and hyperdynamic circulation can lead to transient interruption of contrast material flow from the inferior vena cava, leading to inadequate enhancement of the pulmonary arterial tree.

Cardiac Magnetic Resonance Imaging

Cardiac MRI and magnetic resonance angiography (MRA) are important tools for the evaluation of ventricular function, tissue infiltration or scar, or the aorta. MRI and MRA without contrast have been used in the clinical assessment of pregnant women for several decades without reports of harm [43] and their long-term safety was recently

TABLE 5 Potential Health Effects (Other Than Cancer) of Prenatal Radiation Exposure

Acute Radiation Dose to the Embryo/Fetus	Time Postconception				
	Up to 2 Weeks	3–5 Weeks	6–13 Weeks	14–23 Weeks	24th Week to Term
< 100 mGy	Noncancer health effects NOT detectable				
100–500 mGy	Failure to implant Surviving embryos will probably have no health effects	Growth restriction possible	Growth restriction possible	Health effects unlikely	
> 500 mGy	Failure to implant will likely be high Surviving embryos will probably have no health effects	Probability of miscarriage may increase Probability of major malformations (neurological and motor deficiencies) increases Growth restriction likely	Probability of miscarriage increases Growth restriction likely	Probability of miscarriage increases Growth restriction possible Probability of major malformations may increase	Miscarriage and neonatal death may occur

Data from [45].

evaluated in a recent study of more than 1 million deliveries in Canada. There were no significant differences in risk of stillbirth or neonatal death between women who underwent first-trimester MRI and those who did not undergo MRI [relative risk (RR) 1.68, 95% confidence interval (CI) 0.97–2.90]. Gadolinium, however, was associated with an increased risk of stillbirth or neonatal death (adjusted RR 3.70, 95% CI 1.55–8.85) [50]. Use of MRI with gadolinium-based contrast agents is not recommended due to animal studies that have demonstrated teratogenicity. Gadolinium is known to cross the placenta and is filtered by the fetal kidney. Postpartum, gadolinium can be administered without interruption of breastfeeding [51].

Cardiac Catheterization

Cardiac catheterization can be safely performed in pregnant women and may be necessary in certain clinical conditions including high-risk acute coronary syndrome (ACS). Potential radiation exposure to the unborn fetus is due to scatter radiation to the lower pelvis. Therefore, it is important to take special measures to reduce radiation exposure to the lower pelvis during the procedure. Techniques that many catheterization labs have used to reduce radiation exposure to the pregnant mother include low-dose fluoroscopy, shortened fluoroscopy time, and abdominal shielding of the mother with lead aprons.

Pregnancy and Valvular Heart Disease

Key Points

 Rheumatic heart disease is the most common cause of valvular heart disease in pregnancy worldwide.

 Stenotic valvular lesions are poorly tolerated in pregnancy due to increases in cardiac output and heart rate.

3 **Regurgitant lesions are generally well tolerated in pregnancy as the fall in systemic vascular resistance reduces the regurgitant volume.**

Valvular heart disease (VHD) is one of the most formidable conditions facing pregnant women. Rheumatic heart disease is the most common cause of valvular disease in the developing world, while CHD now accounts for 35–50% of pregnancies complicated by CVD in the developed world [52]. In the European Registry of Pregnancy and Cardiac Disease (ROPAC) data, of 1321 patients, 25% of women had VHD [53]. Mitral valve disease (stenosis and regurgitation) was the most common cause of valvular lesions, followed by aortic valve disease.

TABLE 6 Summary of Valvular Lesions in Pregnancy

Valvular Lesion	Causes	Risk for Adverse Maternal Event	Management
Mitral stenosis	• Rheumatic heart disease • Congenital MS	High especially if moderate or severe MS	• Diuresis • Anticoagulation if AF • Beta-blockers • Activity restriction • Percutaneous mitral commissurotomy in severe cases
Aortic stenosis	• Rheumatic heart disease • Congenital bicuspid valve	High especially if severe symptomatic AS	• Diuresis • Beta-blockade • Activity restriction • Percutaneous valvuloplasty for intractable HF • AVR in rare and life-threatening cases
Mitral regurgitation	• Rheumatic heart disease • Congenital	Low unless severe or with concomitant LV dysfunction	• Diuresis • Vasodilator therapy • Immediate postpartum period most vulnerable • Surgery is reserved for high-risk cases
Aortic insufficiency	• Congenital bicuspid valve • Rheumatic heart disease • Endocarditis • Aortopathy	Low unless severe or with concomitant LV dilation or dysfunction	• Diuresis • Vasodilator therapy • Surgery for severe intractable cases
Right-sided obstructive lesions (PS, TS)	• Congenital	Well tolerated even in severe cases	• Diuresis • Balloon valvuloplasty if refractory to medical therapy

AF, atrial fibrillation; AS, aortic stenosis; AVR, aortic valve replacement; HF, heart failure; LV, left ventricle; MS, mitral stenosis; PS, pulmonic stenosis; TS, tricuspid stenosis.

The prognosis of VHD in pregnancy depends on the nature, acuity, and severity of the lesion (**Table 6**). In general, regurgitant valve disease is generally better tolerated during pregnancy due to the reduction in SVR. Stenotic lesions, however, fare worse in the face of the significant hemodynamic perturbations that accompany the pregnant state. The increased heart rate and plasma volume can increase the gradient across valves on echocardiography and result in overestimation of valvular disease severity.

Mitral Stenosis

Mitral stenosis (MS) is an important source of maternal morbidity and mortality. The predominant etiology of mitral stenosis during pregnancy is rheumatic, although congenital mitral stenosis is seen particularly in industrialized nations. More than 67% of women with severe MS develop a maternal cardiac event during pregnancy including pulmonary edema,

atrial tachyarrhythmias, thromboembolism, and death [54]. If unrecognized, MS in pregnancy can be destabilizing and even fatal with maternal mortality rates as high as 34% in older cohorts [55, 56]. Fortunately, contemporary registries of pregnant women with MS in North America and Europe have no maternal deaths related to MS [57]. Cardiac complications are related to the severity of MS. Compared to mild MS, women with severe MS are at significantly higher risk of developing HF (78% vs. 11%), arrhythmias (33% vs. 0%), and hospitalizations (43% vs. 11%) [57]. Adverse fetal outcomes are also increased in severe MS compared to mild MS including preterm delivery (23% vs. 6%), fetal growth restriction (21% vs. 0%), and stillbirth (4% vs. 0%).

The mainstays of medical management for MS during pregnancy include diuresis to reduce left atrial pressures, anticoagulation in cases of atrial fibrillation, or can be considered in sinus rhythm if spontaneous echocardiographic contrast is visualized in the left atrium, large left atrium $\geq 60 \, \text{mL/m}^2$ or HF, and rhythm control if atrial arrhythmias develop [12]. Beta-blockers are the first-choice agent for rate control but dose escalation is often required due to accelerated metabolism. Finally, activity should be restricted in symptomatic individuals or clinically significant pulmonary hypertension.

Percutaneous intervention should always be considered prepregnancy for patients with severe MS even if asymptomatic, especially if the valve area is < 1.0 cm [11, 12]. During pregnancy, percutaneous mitral commissurotomy is recommended after 20 weeks of gestation if the woman develops NYHA Class III or IV symptoms and/or elevation in PASP ≥ 50 mmHg despite optimal medical therapy. Due to the risk of fetal demise on cardiopulmonary bypass, open-heart surgery should only be considered in life-threatening cases.

Aortic Stenosis

Aortic stenosis (AS) in pregnancy is a relatively rare phenomenon but is associated with a high rate of adverse outcomes [58]. In the United States, the majority of cases are due to congenital bicuspid aortic valve, whereas rheumatic valve disease predominates outside the United States, particularly in the developing world. Isolated AS can be associated with adverse maternal (HF, tachyarrhythmias, and pulmonary edema), fetal (preterm birth and low birth weight), and obstetric (shorter median pregnancy durations and higher rates of C-section outcomes) outcomes, but fortunately, fatality rates are exceedingly low. Like MS, severity of AS and peak aortic gradient are independent predictors of maternal hospitalization, low birth weight, and small for gestational age neonates. In fact, women with moderate and severe AS are far more likely to be hospitalized during pregnancy, with rates up to 20% [53]. In a recent analysis of the ROPAC of 2966 pregnancies, 96 women with moderate to severe AS were identified [53]. In this cohort, 20.8% of women were hospitalized for cardiac reasons and patients with severe AS were more likely to experience cardiac complications compared with patients with moderate AS. However, no deaths were observed during pregnancy and during the first week after delivery [53]. In multivariable analyses, signs and symptoms of heart failure, the presence of severe AS, and peak aortic gradient were identified as independent predictors of adverse maternal outcomes.

Preconception evaluation of AS may include an exercise stress test to evaluate exercise capacity and blood pressure response to exercise, as well as a BNP. If these test results are abnormal, then pregnancy is not recommended [53].

Medical management for AS in pregnancy includes diuresis, beta-blockade, and restriction of activity to maintain normal intracardiac filling pressures [59]. For patients who develop intractable HF despite optimal medical therapy,

percutaneous valvuloplasty can be undertaken at an experienced center [60]. In rare and life-threatening cases, surgical aortic valve replacement (SAVR) may be recommended but should always be considered with caution as fetal mortality is as high as 30–40% due to cardiopulmonary bypass. Transcatheter aortic valve replacement (TAVR) may also be considered during pregnancy as an alternative to open surgery. Interestingly, rates of valve intervention following pregnancy are higher in women with previous pregnancy compared to nulliparous women. The mechanisms are not well described, but there is limited evidence suggesting that pregnancy may accelerate native heart valve disease [60]. In severe symptomatic AS, cesarean delivery with intubation may be considered, while vaginal delivery is favored in nonsevere cases [12, 26].

Regurgitant Lesions

Aortic Insufficiency

Aortic insufficiency during pregnancy is rare but associated with significant morbidity. The most common cause of aortic insufficiency in pregnancy is congenital bicuspid valve, although endocarditis, rheumatic heart disease, and aortopathy have all been associated with aortic insufficiency as well [52]. In general, regurgitant lesions are well tolerated in pregnancy because the reduction in SVR that accompanies pregnancy actually reduces the regurgitant volume.

Maternal risk depends on the severity of valvular regurgitation, symptoms, and LV function. Therefore, preconception evaluation includes an assessment of symptoms, echocardiographic evaluation of the regurgitant lesion (including severity, LV dimensions, and function) [61], and provocative testing including exercise testing in patients with moderate to severe aortic regurgitation. Ascending aortic dimensions should also be measured in women with aortic insufficiency, especially those with bicuspid valves. Women with asymptomatic mild to moderate aortic insufficiency and patients with severe aortic insufficiency but preserved LV function do well with pregnancy. However, pregnant women with severe aortic insufficiency accompanied by LV systolic dysfunction, severe LV dilation, or pulmonary hypertension often develop symptoms of heart failure during pregnancy. Data on aortic insufficiency in pregnancy are limited, but in a small study of 22 patients with aortic insufficiency, clinical deterioration (defined as receiving diuretic therapy) occurred in only 3 patients with enlarged LV size and depressed LVEF (< 55%) [61]. In this sample, 19 patients delivered spontaneously and 3 required cesarean section for obstetric indications. Based on these data, the European Society of Cardiology (ESC) Guidelines on the management of CVD in pregnancy recommend against pregnancy in women with severe aortic insufficiency (AI) with LV dysfunction (LVEF < 30%) [28].

For women with known severe aortic insufficiency with concomitant symptoms or LV systolic dysfunction, referral for prepregnancy valve repair or replacement should be considered. During pregnancy, medical management for aortic insufficiency largely relies on volume management with diuretics and blood pressure control with vasodilators such as nifedipine. In cases of acute severe regurgitation refractory to medical therapy, surgery during pregnancy may be unavoidable. In those cases, if the fetus is mature, delivery should be performed prior to cardiac surgery.

Mitral Regurgitation

Like aortic insufficiency, mitral regurgitation (MR) is well tolerated in the pregnant woman due to the reduction in SVR. MR in women of childbearing age is generally secondary to mitral valve prolapse, CHD, or rheumatic heart disease. The severity and acuity of the regurgitation along with LV function are important predictors of prognosis. In women with mild or moderate MR or severe MR without LV systolic dysfunction or pulmonary hypertension, the prognosis is favorable. However, women with symptomatic severe MR or MR with systolic dysfunction have a more guarded prognosis. In a study of 390 patients with rheumatic mitral valve disease from the ROPAC, 23.1% of patients with moderate to severe MR experienced heart failure and 1 patient with moderate to severe MR died [62]. For this reason, mitral valve repair or replacement should be considered prior to pregnancy in women with symptomatic moderate to severe MR. Women with severe MR and severe LV systolic dysfunction (LVEF < 30%) or significant pulmonary hypertension (HTN) should be advised to avoid pregnancy.

Medical therapy for pregnant women with MR involves diuresis and vasodilator therapy, similar to AI. Importantly, the immediate postpartum period is the most vulnerable period for women with MR because of autotransfusion of the placental circulation and sharp increase in SVR following the delivery of the placenta. Intravenous diuresis is often needed in that period and providers should be vigilant for signs and symptoms of volume overload after delivery. Again, surgery during pregnancy is reserved for the most high-risk cases given the high rate of fetal demise with cardiopulmonary bypass.

Right-Sided Obstructive Lesions

Right-sided obstructive lesions such as pulmonic stenosis or tricuspid stenosis are rare and usually congenital in etiology. Fortunately, they are well tolerated in pregnancy even when severe [63]. As cardiac output increases during pregnancy, valve gradients are expected to rise. Periodic monitoring with clinical exams and echocardiography are recommended in patients with severe right-sided obstructive lesions to monitor for the development of right-sided

HF. Balloon valvuloplasty may be considered in patients who are symptomatic despite medical management.

Management of Anticoagulation in Pregnant Women With Prosthetic Valves

Key Points

 Pregnancy is a hypercoagulable state.

For pregnant women with prosthetic heart valves, maternal risk is lowest with VKA therapy while fetal risk is lowest with low molecular weight heparin (LMWH).

Pregnancy is a period of marked hypercoagulability due to an increase in coagulation factors, decrease in natural anticoagulants, and inhibition of fibrinolysis. Populations most at risk for thrombosis during pregnancy include women with prosthetic heart valves, thromboembolism, or thrombophilias. Here, we highlight the management of anticoagulation in prosthetic heart valves. This population of women has an elevated risk of complications during pregnancy, with only a 58% chance of having an uncomplicated pregnancy with a live birth [53].

Pregnancy for women with prosthetic heart valves is complex and associated with high mortality (1.4% compared to 0.007–0.043% in normal population) [64]. The choice of anticoagulant involves a careful examination of both maternal and fetal risk (**Figure 5**). In general, maternal risk is lowest with VKA monotherapy, while fetal risk is lowest with low molecular weight heparin (LMWH) [65–67]. In a meta-analysis of 51 studies of 2113 pregnancies with mechanical heart valves, treatment with VKA therapy with strict international normalized ratio (INR) monitoring demonstrated the best maternal outcomes [maternal thromboembolic event: 2.79% in VKA only vs. 7.42% in unfractionated heparin (UFH) during the first trimester followed by VKA (H/VKA), 4.42% in LMWH only, and 29.9% in UFH only] [68]. However, the risk of fetal adverse events was highest in VKA monotherapy (2.13% VKA only vs. 0.74% in H/VKA and 0% in both LMWH and UFH monotherapy groups). In order to maximize benefit, strategies that reserve VKA after embryogenesis have been examined, but this does not mitigate fetal risk completely. Low-dose VKA (warfarin doses < 5 mg/day) is a potential therapeutic option, but there still seems to be residual risk of adverse fetal events [69]. If VKA therapy is ultimately chosen as the anticoagulant for the pregnant patient, strict INR dosing is important for prevention of valve thrombosis and targets are based on thrombogenicity of the prosthesis and patient-related factors (mitral or tricuspid valve replacement, previ-

TABLE 7 Target INR for Mechanical Prosthesis

Prosthesis Thrombogenicity	Patient-Related Risk Factors	
	None	≥1
Low Carbomedics Medtronic Hall ATS Medtronic Open-Pivot St Jude Medical On-X Sorin Bicarbon	2.5	3.0
Medium Other bileaflet valves with insufficient data	3.0	3.5
High Lillehei-Kaster Omniscience Starr-Edwards (ball cage) Björk-Shiley Other tilting-disc valves Any pulmonary valve prosthesis	3.5	4.0

INR, international normalized ratio. *Data from [12].*

ous thromboembolism, atrial fibrillation, mitral stenosis, or LVEF < 35%) (**Table 7**).

LMWH has generally been the preferred anticoagulant when considering fetal safety profile. Embryopathy risk associated with VKA therapy is 0.6–10% of cases. UFH and LMWH do not cross the placenta, and therefore have a near-zero risk of associated fetal embryopathy. However, maternal outcomes with LMWH therapy are inferior to those with VKA therapy. LMWH use throughout pregnancy with appropriate anti-Xa monitoring and dose adjustment is

associated with a valve thrombosis risk of 4.4–8.7%. Rates of valve thrombosis are even higher in cases of poor compliance and suboptimal target anti-Xa levels. However, cases of valve thrombosis have been described in patients where peak anti-Xa levels were within target levels. Careful dose adjustment is required because of the increase in GFR during pregnancy and changes in volume of distribution during pregnancy. Therefore, both peak and trough anti-Xa levels are recommended for routine monitoring [70].

Finally, there is increasing interest in the use of DOACs, but the experience in pregnancy is limited. Case reports have described rare birth abnormalities and it cannot be determined whether these rates actually exceed the prevalence of all major birth defects in the United States [71, 72]. However, due to incomplete data, DOAC use is currently not recommended in pregnancy. Ultimately, there needs to be a careful discussion with the interdisciplinary team to select the optimal anticoagulant for the patient [73].

The American College of Cardiology and European Society of Cardiology guidelines on management of anticoagulation for mechanical heart valves during pregnancy are shown in **Table 8**. For women with a mechanical valve taking low doses of VKA (warfarin < 5 mg/day), continuation of VKA therapy with INR monitoring throughout pregnancy should be considered given the favorable maternal profile and lower fetal risk profile. Though the fetal risk is lower, it is not negligible on lower-dose VKA; therefore, a switch to LMWH from weeks 6–12 with strict monitoring can also be considered. In women with mechanical valves taking higher doses of VKA therapy (warfarin > 5 mg/day), discontinuation of VKAs between weeks 6 and 12 and replacement with either UFH or LMWH should be considered given the higher risk of embryopathy, fetopathy, and fetal loss during the first trimester. During the second trimester, VKA therapy is reasonable although many women will remain on LMWH for the duration of their pregnancy and this is recommended as an option if the VKA dose is more than 5 mg [12]. After 36 weeks, anticoagulation options include continuing or transitioning to IV UFH or LMWH. The decision on

TABLE 8 American College of Cardiology and European Society of Cardiology Management Strategies for Anticoagulation During Pregnancy in Women With Mechanical Heart Valves

	First Trimester	Second and Third Trimesters	Peripartum
American College of Cardiology	Warfarin if dose less than 5 mg or Dose-adjusted LMWH or Dose-adjusted IV UFH	Warfarin plus aspirin	Dose-adjusted IV UFH
European Society of Cardiology	Warfarin if dose less than 5 mg or Dose-adjusted LMWH or Dose-adjusted IV UFH	Warfarin or LMWH if warfarin dose more than 5 mg	Dose-adjusted LMWH or IV UFH

IV, intravenous; LMWH, low-molecular-weight heparin; UFH, unfractionated heparin. *Data from [12, 74].*

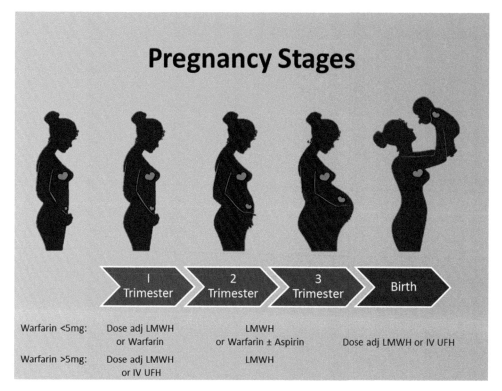

FIGURE 6   **Anticoagulation Management Strategies During Pregnancy.** adj, adjusted; IV, intravenous; LMWH, low-molecular-weight heparin; UFH, unfractionated heparin. *Image courtesy of Niti R. Aggarwal.*

which to use will depend on specific mechanical valve risk and position, the presence of atrial fibrillation or prior embolic event, and the guidelines that an epidural cannot be placed within 24 h of full-dose LMWH, or 4–6 h after discontinuation of IV UFH, with normal partial thromboplastin time (PTT). The American College of Cardiology also recommends aspirin during the second and third trimesters [74] (**Table 8** and **Figure 6**).

Ischemic Heart Disease and Pregnancy

Key Points

1   The fundamental principles of diagnosis of acute coronary syndrome in the pregnant patient are the same as in the nonpregnant patient.

2   The most common cause of acute MI in pregnancy is spontaneous coronary artery dissection (SCAD).

The presence of ischemic heart disease (IHD) is relatively uncommon in pregnancy, although the prevalence appears to be increasing as the number of preexisting comorbidities

and the average maternal age rise [75]. The rate of acute myocardial infarction (AMI) during pregnancy is also rising. In a large national US database, AMI occurred in 8.1 cases per 100,000 hospitalizations during pregnancy with an in-hospital mortality rate of 4.5% [76]. In a large contemporary review of AMI during pregnancy, fetal mortality was 5% [77].

There is increasing evidence that pregnancy itself is a possible risk factor for acute MI, as women who are pregnant have a three- to fourfold higher chance of having an acute MI compared to nonpregnant women of similar demographics [76, 78–80]. This is likely due to many of the physiologic changes that occur during pregnancy, including the inherent imbalance between oxygen demand and supply that occurs as a result of increased cardiac output as well as physiologic anemia and systemic vasodilation [76, 79, 81–83], the hormonal effects in pregnancy that predisposes to vulnerability of the vasculature [76, 83], and the inherently pro-thrombotic state of pregnancy [76, 84].

The most common cause of AMI in pregnancy is spontaneous coronary artery dissection (SCAD) [76, 85, 86], which occurred in 43% of women in a large contemporary review (as shown in **Figure 7**) [77].

There are a number of proposed mechanisms for the increased risk of SCAD during pregnancy, including elevated estrogen and progesterone levels of pregnancy lead-

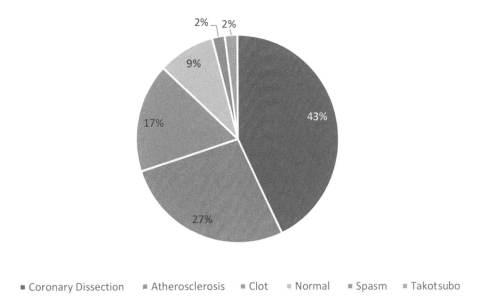

FIGURE 7   **Mechanism of Myocardial Infarction in Pregnancy.** *Data from [77].*

The Diagnosis of Acute MI

ing to changes in vascular wall integrity, which in turn are exacerbated by the hemodynamic changes of pregnancy and delivery [80, 87–89]. Pregnancy-associated SCAD is more likely to present with higher risk features including ST-segment elevation, left main or multivessel involvement, and a greater fall in left ventricular ejection fraction [77].

The Diagnosis of Acute MI

The clinical diagnosis of acute MI in pregnancy may be challenging as women frequently present with symptoms that may be considered a part of normal pregnancy including chest pain and dyspnea. The fundamental principles of diagnosis of acute coronary syndrome in the pregnant patient are the same as in the nonpregnant patient [3]. Pregnancy itself does affect the ECG including: left axis deviation due to diaphragmatic elevation [75]; T-wave inversions in leads III and aVF; a small Q wave in lead III; or an increased R/S ratio in leads V1 and V2 [32, 75, 90]. ST-segment elevation is never normal and when present should always prompt concern for myocardial infarction. Several small studies suggest that standard assays of both troponin T and I remain negative during normal pregnancy and delivery [91–93]. However, one study using a high-sensitivity troponin T assay reported a 4.3% incidence of concentrations in excess of the 99th percentile for a general, healthy population, 8–24h postdelivery [94]. The majority of these elevations were without clinical significance and did not differ between high- and low-risk women. More data are needed in this area, however. Although there are some data to suggest that high-sensitivity troponin may be intrinsically elevated in the hypertensive disorders of pregnancy

(HDP), an abnormal value should always be investigated further [3, 32, 94–97].

The Management of ACS

The management of ACS during pregnancy is guided by many of the same principles and interventions as in the nonpregnant patient, although consideration for fetal well-being is paramount and certain standard therapies must be avoided [75, 98]. As discussed in the "Medication use during pregnancy" section, low-dose aspirin (ASA) can be used safely with little risk of premature PDA closure. Beta-blockers should be used, especially in the SCAD population. Dual antiplatelet therapy (DAPT) can be used postpercutaneous coronary intervention; however, adenosine diphosphate inhibitors must be stopped 7 days prior to epidural analgesia.

In the setting of acute ST-segment elevation myocardial infarction, the same management principles should be used in pregnant patients as in nonpregnant patients, which include an invasive strategy with coronary angiography when indicated. Despite the appropriate role of percutaneous coronary intervention (PCI) in the early management of MI in pregnancy, there is evidence that coronary catheterization rates during pregnancy are lower than expected in the setting of ACS [32, 75, 78]. If the cause of infarction is SCAD, conservative management is preferred due to the risk of dissection propagation with catheter and balloon manipulation and a high rate of lesion recovery in 30 days [99]. Intervention should however be performed in patients with left main involvement, hemodynamic instability, and recurrent-ongoing ischemia [100].

Though there are limited data on using stents in pregnancy as these women were excluded from trials, both bare metal stents (BMS) and drug-eluting stents (DES) can be used during pregnancy [12, 101]. The choice depends on various factors including lesion characteristics and required length of DAPT course. As BMS require a shorter minimum duration of DAPT and that clopidogrel is held 7 days prior to epidural anesthesia, gestational age and timing for epidural anesthesia must be considered when choosing stent type. Given the shorter DAPT period, BMS may thus be preferred later in pregnancy. Epidural anesthesia can, however, be employed with aspirin use [30, 32, 33, 84]. In this setting, vaginal delivery remains the preferred mode of delivery, with C-section reserved for obstetric indications and to be considered in women with cardiogenic shock. The treatment for non-ST-segment elevation myocardial infarction (NSTEMI)/ST-elevation myocardial infarction (STEMI) takes precedence over the timing for delivery. If possible, delivery should be postponed for 2 weeks following an AMI [12].

When considering breastfeeding, high-dose aspirin is excreted into breastmilk, raising concerns for Reye's syndrome in the infant, though the true risk is unknown. Aspirin in low doses is not excreted into breastmilk and can therefore be considered as a potential antiplatelet agent in breastfeeding women [31]. There is no published information on the use of clopidogrel during breastfeeding, though the manufacturer reports that no adverse effects have been observed in a small number of postmarketing cases. Its use should therefore be considered on a case-by-case basis [31].

Hypertensive Diseases of Pregnancy

Key Points

 1 The hypertensive disorders of pregnancy are the most common cardiovascular complications during pregnancy.

2 The hypertensive disorders of pregnancy, especially preeclampsia, are strongly linked with the risk of developing cardiovascular disease long term.

The HDP represent a spectrum of disease ranging from uncomplicated preexisting hypertension to the development of hypertension complicated by end-organ function (**Table 9** and **Figure 8**). HDP are the most common cardiovascular complications of pregnancy and are thought to occur in 5–10% of pregnancies [103, 104]. As maternal age rises and

TABLE 9 Definitions of the Hypertensive Disorders of Pregnancy

Hypertensive Disorder of Pregnancy	Definition
Chronic hypertension	Hypertension (> 140/90 mmHg) that predates pregnancy, is diagnosed < 20 weeks' gestation, or persists > 12 weeks postpartum
Gestational hypertension	Hypertension (> 140/90 mmHg) that develops > 20 weeks' gestation without proteinuria or evidence of end-organ dysfunction
Preeclampsia	Hypertension (> 140/90 mmHg) that develops > 20 weeks' gestation and/or early in the postpartum period and is associated with proteinuria or evidence of end-organ dysfunction
Preeclampsia superimposed on chronic hypertension	The development of proteinuria or evidence of end-organ dysfunction, or increase in BP, in women with chronic hypertension

BP, blood pressure. *Data from [102].*

women present with a greater burden of preexisting medical comorbidities, this number is expected to rise [105]. The HDP are important to identify and manage, not only to optimize maternal and fetal well-being during pregnancy, but also because they are associated with increased risk of CVD after pregnancy.

Preeclampsia is one of the most concerning types of HDP as it is a multisystem disease that is the second most common cause of maternal mortality worldwide. It not only affects women during pregnancy, but also is strongly associated with future cardiovascular risk [105–110]. Preeclampsia is primarily attributed to placental pathology, and is thought to occur as a result of abnormal spiral artery development and placental ischemia leading to a number of sequelae including release of pro-inflammatory cytokines such as soluble fms-like tyrosine kinase 1 (sFlt-1) [106–113]. Women whose pregnancies are complicated by preeclampsia have higher postpartum rates of cardiovascular events. Preeclampsia is independently associated with increased risk of heart failure (RR, 4.19; 95% CI, 2.09–8.38), coronary heart disease (RR, 2.50; 95% CI, 1.43–4.37), CVD death (RR, 2.21; 95% CI, 1.83–2.66), and stroke (RR, 1.81;

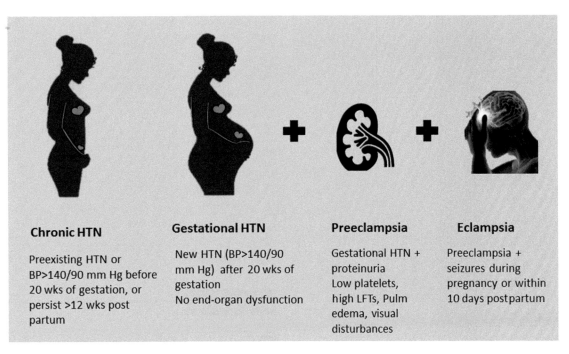

Chronic HTN

Preexisting HTN or BP>140/90 mm Hg before 20 wks of gestation, or persist >12 wks post partum

Gestational HTN

New HTN (BP>140/90 mm Hg) after 20 wks of gestation
No end-organ dysfunction

Preeclampsia

Gestational HTN + proteinuria
Low platelets, high LFTs, Pulm edema, visual disturbances

Eclampsia

Preeclampsia + seizures during pregnancy or within 10 days postpartum

FIGURE 8 **Hypertensive Disorders of Pregnancy.** BP, blood pressure; HTN, hypertension; LFTs, liver function tests; Pulm, pulmonary; wks, weeks. *Image courtesy of Niti R. Aggarwal.*

95% CI, 1.29–2.55) [109, 114]. Preeclampsia can also lead to eclampsia, which is the development of seizures in the context of preeclampsia. Though eclampsia is rare, it is associated with significant morbidity and mortality. Data are limited on the risk factors for the progression of preeclampsia to eclampsia, though what data are available suggest risk factors include preexisting comorbidities such as diabetes and heart disease, as well as pregnancy-specific factors such as multifetal pregnancies and nulliparity [115].

Delivery is the only cure for preeclampsia, and thus prevention is key. Initial screening should include a thorough medical history to evaluate for possible risk factors, and blood pressure screening should occur at every prenatal visit [102, 116, 117]. Data suggest that aspirin may have a role in preventing preeclampsia, reducing its frequency by approximately 10–20% [113, 118–120]. The US Preventative Services Task Force recommends low-dose aspirin (81 mg) after 12 weeks' gestation in women who are at high risk of preeclampsia (see **Table 10**) [121].

Preeclampsia is diagnosed when a measured blood pressure is ≥ 140/90 mmHg on two separate occasions at least 4h apart, and is accompanied by proteinuria (≥ 300 mg/dL in 24-h urine collection, protein to creatinine ratio ≥ 0.3, or urine protein dipstick reading > 1) or evidence of end-organ function such as cerebral or visual symptoms, pulmonary edema, thrombocytopenia, renal insufficiency, or liver dysfunction [117, 122].

The management of hypertension during pregnancy remains controversial, particularly in the mild to moderate level range. Antihypertensive guidelines in the nonpregnant population cannot be extrapolated to the pregnant population [123]. The concerns of treating blood pressure to lower values are the risks of hypotension, inhibiting fetal growth, and risks of fetal exposure to medications. However, to reduce the risks for maternal stroke and complications, it is recommended to use the threshold of blood pressure greater than or equal to 160/110 that is persistent for more than 15 min for treatment. In the setting of comorbidities or renal dysfunction, lower thresholds can be used. In a study evaluating less tight (target diastolic 100 mmHg) vs. tight (target DBP 85 mmHg) control of hypertension in pregnancy, there were no significant differences in the risk of pregnancy loss, need for higher level neonatal care, or maternal complications between the groups [124]. However, less tight control was associated with higher frequency of severe maternal hypertension.

In women with chronic hypertension, there are limited data to guide decision making on when to continue medications and what the goals of therapy are. The ACOG guidelines recommend treating women with chronic hypertension to systolic blood pressure goals of 120–159 mmHg, and diastolic 80–109 mmHg. Lower goals can be used in women with comorbid conditions. Preeclampsia with severe features (SBP greater than or equal to 160/110 on two

TABLE 10 Clinical Risk Assessment for Preeclampsia		
Risk Level	Risk Factors	Recommendation
High	History of preeclampsia Multifetal gestation Chronic hypertension Diabetes Renal disease Autoimmune disease	Low-dose aspirin if one or more of these risk factors
Moderate	Nulliparity Obesity Family history of preeclampsia Socioeconomic characteristics (African American race, low socioeconomic status) Age over 35 years Personal history (low birthweight or small for gestational age, prior adverse pregnancy outcome, greater than 10 years interpregnancy interval)	Consider low-dose aspirin if several of these risk factors
Low	Previous uncomplicated full-term delivery	Do not recommend low-dose aspirin

Data from [121].

occasions at least 4h apart, symptoms of central nervous system dysfunction, hepatic abnormality, thrombocytopenia, pulmonary edema, or renal abnormality) is considered an indication for delivery, to reduce the risk of maternal complications. Vaginal delivery is still the preferred mode of delivery [125]. Intravenous magnesium is given to reduce the risk of eclampsia. Magnesium does cross the placenta, and causes a decrease in fetal heart rate, though it gener-

ally remains within the normal range [126]. Based on meta-analysis, there are no clear fetal adverse outcomes [127].

The HDP are warnings for future cardiovascular risk. Obstetrical history is thus increasingly recognized as an important part of a woman's cardiovascular risk [128]. For pregnancies complicated by HDP, the American College of Obstetrics and Gynecology recommends follow-up cardio-vascular risk screening, including repeat blood pressure, focused cardiovascular history, and review of medications, within 3 months after delivery [3]. Women with HDP should be informed of their higher cardiovascular risk and coun-seled appropriately on lifestyle changes they can implement to mitigate these risks.

Aortopathies

Key Points

 The hemodynamic and hormonal changes of pregnancy are risk factors for aortic dissection in aortas of both normal and abnormal diameters.

 For women at risk of aortic dissection, frequent monitoring of aortic diameter throughout pregnancy is advised.

Aortic disease can be identified prior to or during pregnancy and can be exacerbated by the significant hemodynamic and hormonal changes of pregnancy. Even in normal pregnant aortas, pathologic data suggest there are changes that occur that result in less corrugation of the aortic elastic fibers in the aortic wall, fragmentation of the aortic reticulin fibers, and reduced amount of acid mucopolysaccharides [129]. Most pregnancy-related dissections occur in the third tri-mester (50%) and postpartum period (33%), and they tend to be in the ascending aorta [12, 130–132].

Women at especially high risk of dissection include those with known aortopathy or conditions that predispose them to aortic dissection such as Marfan syndrome, vascu-lar Ehlers-Danlos syndrome, Loeys-Dietz syndrome, bicus-pid aortic valve with aortic dilatation, Turner syndrome, and those with nonsyndromic heritable thoracic aortic disease. Advanced maternal age and poorly controlled hypertension are also risk factors [12]. Additionally, increasing parity is associated with increased aortic size [133]. In general, the risk of aortic complications is low when the ascending aor-tic diameter is < 4.0cm [12, 132, 134]; however, there is no level of aortic dilatation that can guarantee against an aortic complication in women with known aortopathy.

Preconception screening and counseling is important so that these women can understand the risks associated with pregnancy. This includes offering genetic counseling due to risk of disease transmission to offspring. Additionally, women should be considered for surgical repair prior to pregnancy when possible [132, 135], noting that surgical repair does not mitigate the risk of dissection in the non-intervened aortic segments. Pregnancy is felt to be contraindicated in women with: Marfan syndrome and aortic root diameters > 4.5 cm prior to pregnancy (> 5.0 cm if bicuspid aortic valve); history of prior aortic dissection; and vascular Ehlers-Danlos syndrome [136, 137]. Vascular Ehler-Danlos patients also carry the risk of uterine rupture, and their pregnancy-related mortality is estimated at 12% [138].

According to the American College of Cardiology (ACC)/American Heart Association (AHA)/American Association for Thoracic Surgery (AATS) guidelines, prophylactic aortic repair for thoracic aortic aneurysms in women contemplating pregnancy with Marfan syndrome should be performed when the maximal ascending aorta diameter exceeds 4.0 cm [132]. These recommendations differ from the 2011 European Society of Cardiology guidelines, which recommend surgical repair at aortic diameter ≥ 4.5 cm [28]. Repair is also considered for women with rapid growth of aortic diameter (> 0.5 cm/year), or family history of aortic dissection despite aortic diameter < 5 cm [131, 132, 139].

During pregnancy, the ACC/AHA/AATS guidelines recommend surveillance imaging monthly or bimonthly until delivery to assess for interval change in aortic diameter. TTE is the preferred imaging modality for the proximal aorta but may be limited by technical difficulty in image acquisition. If TTE is inadequate then MRI (without gadolinium) can be performed and is the preferred modality for imaging the aortic arch, or descending or abdominal aorta. CT is not the modality of choice due to ionizing radiation exposure to the mother and fetus.

During pregnancy, blood pressure control is recommended, with beta-blockers being the preferred antihypertensive due to their effect on reducing aortic wall stress. The patient should be managed in a multidisciplinary fashion with a cardio-obstetrics team to discuss mode of delivery, type of anesthesia, and monitoring during labor and delivery. Vaginal delivery remains the preferred mode of delivery but cesarean section should be considered in women with acute or chronic dissection and in those with maximal aortic diameters greater than 4.5 cm. An assisted second stage of labor is an option (forceps or suction) to reduce the duration of Valsalva maneuvers during labor [12]. Regional anesthesia is preferred to control pain and blunt sympathetic drive, and therefore minimize fluctuations or extremes of heart rate and blood pressure during labor.

Arrhythmias During Pregnancy

Key Points

 Arrhythmias are common during pregnancy, but most are benign.

 Many of the same interventions in the nonpregnant patient are possible in the pregnant patient, though careful attention must be paid to medication teratogenicity.

 In general, medications should be used at the lowest dose for the shortest period of time.

 Cardiac arrest during pregnancy follows the same general advanced cardiac life support (ACLS) principles with special considerations in the pregnant population, including left lateral uterine displacement when beyond 20 weeks' gestation and delivery of the fetus within 4 min.

Arrhythmias are frequently encountered during pregnancy, and the majority are benign [11, 140–142]. Both the anatomic changes, such as increased heart rate and chamber enlargement, as well as hormonal changes, such as the pro-arrhythmic effect of estrogen and progesterone, increase the risk of arrhythmia during normal pregnancy. This risk is even higher in women with CHD or preexisting arrhythmia [12, 141–143]. Nonsustained arrhythmias such as atrial and ventricular premature complexes occur in 50–60% of normal pregnancies [141, 143, 144], whereas new sustained arrhythmias such as atrial fibrillation occur in 24–68/100,000 of normal pregnancies [12, 142, 145, 146]. Sustained ventricular arrhythmias are fortunately rare, and are estimated to occur in 2 out of 100,000 hospitalized pregnant women [142]. Although most arrhythmias are benign, all are associated with increased risk of maternal and fetal complications, and death [12, 146]. Any concern for potential arrhythmia should be evaluated with electrocardiography, basic blood work including electrolytes and thyroid testing, and if possible echocardiography to evaluate for structural heart disease in the mother [143, 145].

The general approach to treating arrhythmia during pregnancy is similar to that outside of pregnancy, with the caveat that the evidence guiding the treatment of arrhythmia during pregnancy is largely based on case reports alone and the possible risk of therapeutic teratogenicity must be considered [141, 143, 147]. The risk of teratogenicity is especially high during fetal organogenesis. Despite this risk, treatment for arrhythmia should be initiated at any point during pregnancy if either the mother or the fetus becomes symptomatic or hemodynamically unstable, by using the lowest possible dose for the shortest period of time [141–143].

The majority of arrhythmias during pregnancy are supraventricular. For patients with supraventricular tachycardia (SVT), vagal maneuvers should be attempted as first-line intervention. For women who do not respond to vagal maneuvers, adenosine is the drug of choice, though depending on the nature of the SVT, beta-blockers (metoprolol and propranolol), calcium channel blockers (verapamil), or digoxin can be used, as well as sotalol or flecainide, if clinically indicated, if in the second trimester or later [141–143, 145]. For atrial fibrillation, beta-blockers are first-line agents for rate control, though expert consensus also recommends verapamil or digoxin if necessary [12, 142, 143, 145]. If a rate control strategy is unsuccessful, and rhythm-control strategies must be pursued, then sotalol or flecainide is recommended [141, 142, 145]. Amiodarone has significant fetal risk and should only be reserved for emergency situations [141, 142]. Given the intrinsic pro-thrombotic nature of pregnancy, appropriate anticoagulation in the setting of arrhythmia is especially important. Unfortunately, however, all anticoagulants carry risk during pregnancy. In non-valvular atrial fibrillation, LMWH and UFH are preferred because they, unlike warfarin, do not cross the placenta, and are thus considered less teratogenic [142, 145]. Warfarin can be used in the second or third trimester, with lower risk if the dose is less than 5 mg, though all women taking warfarin should have an alternate anticoagulant plan closer to delivery [142, 145]. Given the limited safety data in pregnancy, DOACs are not recommended.

Ventricular arrhythmias are rare in pregnancy. Expert consensus recommends electrical cardioversion for both hemodynamically stable and unstable sustained ventricular tachycardia (VT) [12]. In the structurally normal heart, stable monomorphic VT is often idiopathic and responds well to beta-blockers, though sotalol, flecainide, or verapamil can also be attempted if arrhythmia persists despite beta-blockade [142]. In the structurally abnormal heart, expert consensus also recommends acute management with beta-blockers, sotalol, or procainamide [12]. Lidocaine is an option as well [141–143]. Polymorphic VT, regardless of the substrate, is rarely stable, and most often progresses to hemodynamic instability; preparation for defibrillation should be initiated.

Bradyarrhythmias are generally well tolerated and should only be treated if patients are symptomatic. Treatment should follow the same guidelines as in nonpregnant patients [12, 141, 142].

Many of the same nonpharmacologic interventions for arrhythmia in the nonpregnant population are possible during pregnancy. Though data are limited, catheter ablation for medically unresponsive arrhythmia has been shown to be safe; the primary risk is fetal radiation, and thus delaying ablation until after the first trimester is recommended when possible [12, 142]. Similarly, both pacemakers and implantable cardioverter defibrillators can also be safely placed during pregnancy, but placement should occur after the first trimester when possible [12, 142]. For any unstable arrhythmia, cardioversion is considered safe in all trimesters of pregnancy [142, 143, 147].

Arrhythmias Warranting ACLS

Should medical management of arrhythmia be insufficient to prevent hemodynamic collapse, then advanced cardiac life support (ACLS) should be undertaken. Although many of the principles of ACLS that guide care in the nonpregnant and pregnant patients are similar, the unique maternal anatomy and physiology, as well as the growing fetus, warrant slight changes in approach. When undertaking ACLS in the pregnant patient, the team required for adequate resuscitation tends to be larger and more multidisciplinary (often including obstetricians, obstetric anesthesiologists, and pediatricians), and the focus should be on simultaneous intervention [3, 83–85].

Several key principles must be considered in the setting of maternal cardiac arrest [3] which include greater maternal oxygen demand and increased risk for aspiration, aortocaval compression beyond 20 weeks' gestation, and the planning of fetal delivery in parallel to resuscitative efforts. Standard ACLS algorithms should be used without concern for adverse fetal effects due to the gravity of the situation. Women should be placed supine, with attention to left lateral uterine displacement by a dedicated member of the resuscitative team.

Pulmonary Hypertension During Pregnancy

Women with pulmonary hypertension are considered to be at high risk for maternal mortality (25–56%), preterm delivery (85–100%), and low fetal birth weight (3–33%) with a high fetal loss (7–13%) [148–150]. As such, pregnancy is contraindicated in this patient population, and termination should be considered if pregnancy occurs. The physiologic changes in pregnancy that may be deleterious in this patient group include hypercoagulability, increased oxygen consumption, increased plasma volume, elevated heart rate, and fall in SVR which increases the risk for right to left shunting.

In a large contemporary series from the European Registry of Pregnancy and Cardiac Disease that included 151 women of different subtypes of pulmonary hypertension (idiopathic, secondary to CHD, or left-sided heart disease), there were no intrapartum maternal deaths [151]. However, 3.3% of the women died in the first week, and 2.6% by 6 months, with greater risk in those with greater severity pulmonary hypertension at baseline (RVSP > 70 mmHg). Cause of death included cardiogenic shock, sudden cardiac death, and mechanical valve thrombosis. These mortality values are lower than previously reported; however, mortality remained high in women with idiopathic pulmonary arterial hypertension at 43%. In this same study, premature delivery occurred in 21.7%, low birth weight in 19%, miscarriage in 6.6%, fetal mortality was 2.0%, and neonatal mortality was 0.7%.

In another large more recent review of 36 women with various subtypes of pulmonary hypertension, maternal mortality overall was 8.3% and occurred solely in women with WHO classified type 1 pulmonary arterial hypertension and in those with moderate or severe disease based on pulmonary artery systolic pressure at baseline [152]. Thus, while overall maternal mortality is lower than historically reported, risk remains especially high in this subgroup.

As pulmonary arterial hypertension predominantly occurs in women, effective birth control should be addressed as part of their general care. If pregnancy is pursued, then these women should be managed in a multidisciplinary expert care center that includes pulmonary hypertension expertise. There should be regular multidisciplinary discussion for intrapartum care as well as plans for labor and delivery and postpartum care. Serial echocardiography is the preferred method of monitoring, with the rare need for a right heart catheterization. The European Society of Cardiology/European Respiratory Society guidelines recommend women should be treated or should continue specific pulmonary arterial hypertension therapies (calcium channel blockers, phosphodiesterase inhibitors, and prostacyclins) except for endothelin receptor antagonists which are teratogenic [153].

While treatment is individualized, anticoagulation and diuretics should be considered. The optimal mode of delivery is unknown. Vaginal delivery is likely the preferred mode of delivery in most cases due to its lesser hemodynamic shifts, lower infection risk, lower thromboembolic risk, and lower expected blood loss; however, a long labor can be deleterious as well [154]. In all cases, women should be monitored carefully during labor, delivery, and postpartum. Lastly, women with a heritable cause of pulmonary hypertension should be counseled on the risks of fetal inheritance.

Future Directions

CVD is the leading cause of maternal mortality [2]. This rising incidence of cardiovascular disease during pregnancy and the rising maternal mortality in the United States should cause alarm. This should also serve as a call to action from all groups managing these high-risk patients. Specific challenges that need to be addressed include disparities in care. African American women have a greater than threefold higher rate of maternal mortality compared to white non-Hispanic women (**Figure 1**). As CVD is a major contributor to this elevated maternal risk, cardiologists need to be part of the solution. This will require addressing and optimizing cardiovascular risk factors preconception, evaluating maternal risk and counseling on this risk preconception, increasing patient and provider awareness of pregnancy risk, continued CVD risk attention postpartum, and engagement of a multidisciplinary cardio-obstetrics team for higher-risk women. Improved recognition and management of CVD risk factors in women with adverse pregnancy outcomes such as placental abruption, stillbirths, HDP, gestational diabetes, and preterm birth is needed. Improved awareness of increased maternal CVD mortality by both cardiologists and obstetricians is important, in addition to a collaborative approach to caring for these women.

A multidisciplinary team approach is optimal as each member of the team approaches the patient from a different perspective and with unique skills. Although there are increasing efforts to study CVD during pregnancy, it is not likely that this population of patients will be enrolled in large randomized clinical trials. As a result, it is increasingly clear that to improve care, there should be a combined multicenter collaboration across the United States to identify and resolve key gaps in patient care.

The Heart Outcomes in Pregnancy: Expectations (HOPE) for Mom and Baby Registry is currently being developed across the United States, with collaborations to date from more than 40 academic medical centers. By systematically and prospectively gathering data in this patient population, we will be able to identify optimal treatment strategies to improve both maternal and fetal outcomes by establishing future guidelines for this vulnerable patient population [155].

Back to Clinical Case

The case describes a young woman, now with severe mitral stenosis, presenting for preconception counseling. Mitral stenosis is considered a high-risk left-sided valvular lesion. Risks during pregnancy include death, heart failure, atrial arrhythmias, and thromboembolic events. Estimated risk for an adverse maternal event from the CARPREG II risk score is 10%. She should also be counseled about having an increased risk of adverse fetal outcomes, including preterm delivery, fetal growth restriction, and stillbirth. Her valve should be intervened upon preconception. If she had presented pregnant, medical management would include AV nodal agents to decrease heart rate and thus diastolic filling time, diuretics to manage volume overload, and anticoagulants for atrial fibrillation. This patient should be managed by a multidisciplinary team to plan intrapartum care, labor and delivery, and postpartum care. In this situation, this multidisciplinary team should also include an interventional cardiologist to consider percutaneous mitral valvuloplasty if medical management is ineffective during pregnancy.

References

[1] Kassebaum NJ, Barber RM, Bhutta ZA, Dandona L, Gething PW, Hay SI, et al. Global, regional, and national levels of maternal mortality, 1990–2015: a systematic analysis for the Global Burden of Disease Study 2015. Lancet 2016;388(10053):1775–812.

[2] Petersen EE, Davis NL, Goodman D, Cox S, Syverson C, Seed K, et al. Racial/ethnic disparities in pregnancy-related deaths—United States, 2007-2016. MMWR Morb Mortal Wkly Rep 2019;68(35):762–5.

[3] Hollier LM, Martin JN, Connolly H, Turrentine M, Hameed A, Arendt KW, et al. ACOG Practice Bulletin No. 212: pregnancy and heart disease. Obstet Gynecol 2019;133(5):e320–56.

[4] Yucel E, DeFaria Yeh D. Pregnancy in women with congenital heart disease. Curr Treat Options Cardiovasc Med 2017;19(9):73.

[5] Sanghavi M, Rutherford JD. Cardiovascular physiology of pregnancy. Circulation 2014;130(12):1003–8.

[6] Drenthen W, Pieper PG, Roos-Hesselink JW, van Lottum WA, Voors AA, Mulder BJM, et al. Outcome of pregnancy in women with congenital heart disease. J Am Coll Cardiol 2007;49(24):2303–11.

[7] Tello-Montoliu A, Seecheran NA, Angiolillo DJ. Successful pregnancy and delivery on prasugrel treatment: considerations for the use of dual antiplatelet therapy during pregnancy in clinical practice. J Thromb Thrombolysis 2013;36(3):348–51.

[8] Elkayam U, Goland S, Pieper PG, Silversides CK. High-risk cardiac disease in pregnancy: part I. J Am Coll Cardiol 2016;68(4):396–410.

[9] Colletti PM, Lee KH, Elkayam U. Cardiovascular imaging of the pregnant patient. Am J Roentgenol 2013;200(3):515–21.

[10] Siu SC, Samuel M, Harrison DA, Grigoriadis E, Liu G, Sorensen S, et al. Risk and predictors for pregnancy-related complications in women with heart disease. Circulation 1997;96(9):2789–94.

[11] Silversides CK, Grewal J, Mason J, Sermer M, Kiess M, Rychel V, et al. Pregnancy outcomes in women with Heart disease: the CARPREG II study. J Am Coll Cardiol 2018;71(21):2419–30.

[12] Regitz-Zagrosek V, Roos-Hesselink JW, Bauersachs J, Blomström-Lundqvist C, Cífková R, De Bonis M, et al. 2018 ESC guidelines for the management of cardiovascular diseases during pregnancy. Eur Heart J 2018;39(34):3165–241.

[13] Thorne S, MacGregor A, Nelson-Piercy C. Risks of contraception and pregnancy in heart disease. Heart 2006;92(10):1520–5.

[14] Drenthen W, Boersma E, Balci A, Moons P, Roos-Hesselink JW, Mulder BJM, et al. Predictors of pregnancy complications in women with congenital heart disease. Eur Heart J 2010;31(17):2124–32.

[15] Gill HK, Splitt M, Sharland GK, Simpson JM. Patterns of recurrence of congenital heart disease: an analysis of 6,640 consecutive pregnancies evaluated by detailed fetal echocardiography. J Am Coll Cardiol 2003;42(5):923–9.

[16] Romano-Zelekha O, Hirsh R, Blieden L, Green MS, Shohat T. The risk for congenital heart defects in offspring of individuals with congenital heart defects. Clin Genet 2001;59(5):325–9.

[17] Donofrio MT, Moon-Grady AJ, Hornberger LK, Copel JA, Sklansky MS, Abuhamad A, et al. Diagnosis and treatment of fetal cardiac disease: a scientific statement from the American Heart Association. Circulation 2014;129(21):2183–242.

[18] Rasiah SV, Publicover M, Ewer AK, Khan KS, Kilby MD, Zamora J. A systematic review of the accuracy of first-trimester ultrasound examination for detecting major congenital heart disease. Ultrasound Obstet Gynecol 2006;28(1):110–6.

[19] Tsiaras S, Poppas A. Cardiac disease in pregnancy: value of echocardiography. Curr Cardiol Rep 2010;12(3):250–6.

[20] Lui GK, Silversides CK, Khairy P, Fernandes SM, Valente AM, Nickolaus MJ, et al. Heart rate response during exercise and pregnancy outcome in women with congenital heart disease. Circulation 2011;123(3):242–8.

[21] Ohuchi H, Tanabe Y, Kamiya C, Noritake K, Yasuda K, Miyazaki A, et al. Cardiopulmonary variables during exercise predict pregnancy outcome in women with congenital heart disease. Circ J 2013;77(2):470–6.

[22] Betham R, Satish OS. Risk stratification of women with peripartum cardiomyopathy at initial presentation: a dobutamine stress echocardiography study. Indian Heart J 2018;70:S48–9.

[23] Liu S, Liston RM, Joseph KS, Heaman M, Sauve R, Kramer MS. Maternal mortality and severe morbidity associated with low-risk planned cesarean delivery versus planned vaginal delivery at term. Can Med Assoc J 2007;176(4):455–60.

[24] Ruys TPE, Roos-Hesselink JW, Pijuan-Domènech A, Vasario E, Gaisin IR, Iung B, et al. Is a planned caesarean section in women with cardiac disease beneficial? Heart 2015;101(7):530–6.

[25] Nishimura RA, Tajik AJ. The Valsalva maneuver—3 centuries later. Mayo Clin Proc 2004;79(4):577–8.

[26] Canobbio MM, Warnes CA, Aboulhosn J, Connolly HM, Khanna A, Koos BJ, et al. Management of pregnancy in patients with complex congenital heart disease: a scientific statement for healthcare professionals from the American Heart Association. Circulation 2017;135(8). [cited 2019 Feb 18] Available from: https://www.ahajournals.org/doi/10.1161/CIR.0000000000000458.

[27] Arendt KW, Lindley KJ. Obstetric anesthesia management of the patient with cardiac disease. Int J Obstet Anesth 2019;37:73–85.

[28] Regitz-Zagrosek V, Lundqvist CB, Borghi C, Cifkova R, Ferreira R, Foidart J-M, et al. ESC guidelines on the management of cardiovascular diseases during pregnancy The Task Force on the Management of Cardiovascular Diseases during Pregnancy of the European Society of Cardiology (ESC). Eur Heart J 2011;32(24):3147–97.

[29] Halpern DG, Weinberg CR, Pinnelas R, Mehta-Lee S, Economy KE, Valente AM. Use of medication for cardiovascular disease during pregnancy: JACC state-of-the-art review. J Am Coll Cardiol 2019;73(4):457–76.

[30] Pieper PG. Use of medication for cardiovascular disease during pregnancy. Nat Rev Cardiol 2015;12(12):718–29.

[31] Anon. Drugs and Lactation Database (LactMed). [cited 2019 May 28] Available from: https://toxnet.nlm.nih.gov/newtoxnet/lactmed.htm.

[32] Kealey AJ. Coronary artery disease and myocardial infarction in pregnancy: a review of epidemiology, diagnosis, and medical and surgical management. Can J Cardiol 2010;26(6):e185–9.

[33] Horlocker TT, Wedel DJ, Benzon H, Brown DL, Enneking FK, Heit JA, et al. Regional anesthesia in the anticoagulated patient: defining the risks (the second ASRA Consensus Conference on Neuraxial Anesthesia and Anticoagulation). Reg Anesth Pain Med 2003;28(3):172–97.

[34] Yarrington CD, Valente AM, Economy KE. Cardiovascular management in pregnancy: antithrombotic agents and antiplatelet agents. Circulation 2015;132(14):1354–64.

[35] Ruys TPE, Roos-Hesselink JW, Hall R, Subirana-Domènech MT, Grando-Ting J, Estensen M, et al. Heart failure in pregnant women with cardiac disease: data from the ROPAC. Heart 2014;100(3):231–8.

[36] Bowen ME, Ray WA, Arbogast PG, Ding H, Cooper WO. Increasing exposure to angiotensin-converting enzyme inhibitors in pregnancy. Am J Obstet Gynecol 2008;198(3):291.e1–5.

[37] Nakhai-Pour HR, Rey E, Bérard A. Antihypertensive medication use during pregnancy and the risk of major congenital malformations or small-for-gestational-age newborns. Birth Defects Res B Dev Reprod Toxicol 2010;89(2):147–54.

[38] Meidahl Petersen K, Jimenez-Solem E, Andersen JT, Petersen M, Brødbæk K, Køber L, et al. β-Blocker treatment during pregnancy and adverse pregnancy outcomes: a nationwide population-based cohort study. BMJ Open 2012;2(4), e001185.

[39] Ersbøll AS, Hedegaard M, Søndergaard L, Ersbøll M, Johansen M. Treatment with oral beta-blockers during pregnancy complicated by maternal heart disease increases the risk of fetal growth restriction. BJOG 2014;121(5):618–26.

[40] Chow T, Galvin J, McGovern B. Antiarrhythmic drug therapy in pregnancy and lactation. Am J Cardiol 1998;82(4, Supplement 1):58I–62I.

[41] Gladstone GR, Hordof A, Gersony WM. Propranolol administration during pregnancy: effects on the fetus. J Pediatr 1975;86(6):962–4.

[42] Klarr JM, Bhatt-Mehta V, Donn SM. Neonatal adrenergic blockade following single dose maternal labetalol administration. Am J Perinatol 1994;11(02):91–3.

[43] Anon. Guidelines for diagnostic imaging during pregnancy and lactation—ACOG. [cited 2019 Jan 30] Available from: https://www.acog.org/Clinical-Guidance-and-Publications/Committee-Opinions/Committee-on-Obstetric-Practice/Guidelines-for-Diagnostic-Imaging-During-Pregnancy-and-Lactation?IsMobileSet=false.

[44] Brent RL. Utilization of developmental basic science principles in the evaluation of reproductive risks from pre- and postconception environmental radiation exposures. Teratology 1999;59(4):182–204.

[45] Anon. National Council on Radiation Protection and Measurements. Report No. 174: Preconception and Prenatal Radiation Exposure: Health Effects and Protective Guidance; 2013.

[46] Morley CA, Lim BA. The risks of delay in diagnosis of breathlessness in pregnancy. BMJ 1995;311(7012):1083–4.

[47] Leung AN, Bull TM, Jaeschke R, Lockwood CJ, Boiselle PM, Hurwitz LM, et al. An Official American Thoracic Society/Society of thoracic radiology clinical practice guideline: evaluation of suspected pulmonary embolism in pregnancy. Am J Respir Crit Care Med 2011;184(10):1200–8.

[48] Winer-Muram HT, Boone JM, Brown HL, Jennings SG, Mabie WC, Lombardo GT. Pulmonary embolism in pregnant patients: fetal radiation dose with helical CT. Radiology 2002;224(2):487–92.

[49] Ridge CA, McDermott S, Freyne BJ, Brennan DJ, Collins CD, Skehan SJ. Pulmonary embolism in pregnancy: comparison of pulmonary CT angiography and lung scintigraphy. Am J Roentgenol 2009;193(5):1223–7.

[50] Ray JG, Vermeulen MJ, Bharatha A, Montanera WJ, Park AL. Association between MRI exposure during pregnancy and fetal and childhood outcomes. JAMA 2016;316(9):952–61.

[51] Chen MM, Coakley FV, Kaimal A, Laros RK. Guidelines for computed tomography and magnetic resonance imaging use during pregnancy and lactation. Obstet Gynecol 2008;112(2, Part 1):333–40.

[52] Soler-Soler J, Galve E. VALVE DISEASE: worldwide perspective of valve disease. Heart 2000;83(6):721–5.

[53] Orwat S, Diller G-P, van Hagen IM, Schmidt R, Tobler D, Greutmann M, et al. Risk of pregnancy in moderate and severe aortic stenosis: from the multinational ROPAC registry. J Am Coll Cardiol 2016;68(16):1727–37.

[54] Silversides CK, Colman JM, Sermer M, Siu SC. Cardiac risk in pregnant women with rheumatic mitral stenosis. Am J Cardiol 2003;91(11):1382–5.

[55] Avila WS, Rossi EG, Ramires JAF, Grinberg M, Bortolotto MRL, Zugaib M, et al. Pregnancy in patients with heart disease: experience with 1,000 cases. Clin Cardiol 2003;26(3):135–42.

[56] Roos-Hesselink JW, Ruys TPE, Stein JI, Thilén U, Webb GD, Niwa K, et al. Outcome of pregnancy in patients with structural or ischaemic heart disease: results of a registry of the European Society of Cardiology. Eur Heart J 2013;34(9):657–65.

[57] Hameed A, Karaalp IS, Tummala PP, Wani OR, Canetti M, Akhter MW, et al. The effect of valvular heart disease on maternal and fetal outcome of pregnancy. J Am Coll Cardiol 2001;37(3):893–9.

[58] Silversides CK, Colman JM, Sermer M, Farine D, Siu SC. Early and intermediate-term outcomes of pregnancy with congenital aortic stenosis. Am J Cardiol 2003;91(11):1386–9.

[59] Tzemos N, Silversides CK, Colman JM, Therrien J, Webb GD, Mason J, et al. Late cardiac outcomes after pregnancy in women with congenital aortic stenosis. Am Heart J 2009;157(3):474–80.

[60] Myerson SG, Mitchell ARJ, Ormerod OJM, Banning AP. What is the role of balloon dilatation for severe aortic stenosis during pregnancy? J Heart Valve Dis 2005;14(2):147–50.

[61] Leśniak-Sobelga A, Tracz W, KostKiewicz M, Podolec P, Pasowicz M. Clinical and echocardiographic assessment of pregnant women with valvular heart diseases—maternal and fetal outcome. Int J Cardiol 2004;94(1):15–23.

[62] van Hagen IM, Thorne SA, Taha N, Youssef G, Elnagar A, Gabriel H, et al. Pregnancy outcomes in women with rheumatic mitral valve disease: results from the registry of pregnancy and cardiac disease. Circulation 2018;137(8):806–16.

[63] Hameed AB, Goodwin TM, Elkayam U. Effect of pulmonary stenosis on pregnancy outcomes—a case-control study. Am Heart J 2007;154(5):852–4.

[64] van Hagen IM, Roos-Hesselink JW, Ruys TPE, Merz WM, Goland S, Gabriel H, et al. Pregnancy in women with a mechanical heart valve: data of the European Society of Cardiology Registry of Pregnancy and Cardiac Disease (ROPAC). Circulation 2015;132(2):132–42.

[65] Xu Z, Fan J, Luo X, Zhang W-B, Ma J, Lin Y-B, et al. Anticoagulation regimens during pregnancy in patients with mechanical heart valves: a systematic review and meta-analysis. Can J Cardiol 2016;32(10):1248.e1–9.

[66] D'Souza R, Ostro J, Shah PS, Silversides CK, Malinowski A, Murphy KE, et al. Anticoagulation for pregnant women with mechanical heart valves: a systematic review and meta-analysis. Eur Heart J 2017;38(19):1509–16.

[67] Steinberg ZL, Dominguez-Islas CP, Otto CM, Stout KK, Krieger EV. Maternal and fetal outcomes of anticoagulation in pregnant women with mechanical heart valves. J Am Coll Cardiol 2017;69(22):2681–91.

[68] Macle L, Cairns J, Leblanc K, Tsang T, Skanes A, Cox JL, et al. 2016 focused update of the Canadian Cardiovascular Society guidelines for the management of atrial fibrillation. Can J Cardiol 2016;32(10):1170–85.

[69] Chan WS, Anand S, Ginsberg JS. Anticoagulation of pregnant women with mechanical heart valves: a systematic review of the literature. Arch Intern Med 2000;160(2):191–6.

[70] Oran B, Lee-Parritz A, Ansell J. Low molecular weight heparin for the prophylaxis of thromboembolism in women with prosthetic mechanical heart valves during pregnancy. Thromb Haemost 2004;92(10):747–51.

[71] Hoeltzenbein M, Beck E, Meixner K, Schaefer C, Kreutz R. Pregnancy outcome after exposure to the novel oral anticoagulant rivaroxaban in women at suspected risk for thromboembolic events: a case series from the German Embryotox Pharmacovigilance Centre. Clin Res Cardiol 2016;105(2):117–26.

[72] Beyer-Westendorf J, Michalski F, Tittl L, Middeldorp S, Cohen H, Kadir RA, et al. Pregnancy outcome in patients exposed to direct oral anticoagulants—and the challenge of event reporting. Thromb Haemost 2016;116(10):651–8.

[73] Nishimura RA, Otto CM, Bonow RO, Carabello BA, Erwin JP, Guyton RA, et al. 2014 AHA/ACC guideline for the management of patients with valvular heart disease: a report of the American College of Cardiology/American Heart Association Task Force on Practice Guidelines. J Thorac Cardiovasc Surg 2014;148(1):e1–e132.

[74] Nishimura RA, Otto CM, Bonow RO, Carabello BA, Erwin JP, Guyton RA, et al. 2014 AHA/ACC guideline for the management of patients with valvular heart disease: executive summary: a report of the American College of Cardiology/American Heart Association Task Force on practice guidelines. J Am Coll Cardiol 2014;63(22):2438–88.

[75] Cauldwell M, Baris L, Roos-Hesselink JW, Johnson MR. Ischaemic heart disease and pregnancy. Heart 2019;105(3):189–95.

[76] Smilowitz NR, Gupta N, Guo Y, Zhong J, Weinberg CR, Reynolds HR, et al. Acute myocardial infarction during pregnancy and the puerperium in the United States. Mayo Clin Proc 2018;93(10):1404–14.

[77] Elkayam U, Jalnapurkar S, Barakkat MN, Khatri N, Kealey AJ, Mehra A, et al. Pregnancy-associated acute myocardial infarction: a review of contemporary experience in 150 cases between 2006 and 2011. Circulation 2014;129(16):1695–702.

[78] James AH, Jamison MG, Biswas MS, Brancazio LR, Swamy GK, Myers ER. Acute myocardial infarction in pregnancy. Circulation 2006;113(12):1564–71.

[79] Ladner HE, Danielsen B, Gilbert WM. Acute myocardial infarction in pregnancy and the puerperium: a population-based study. Obstet Gynecol 2005;105(3):480–4.

[80] Roth A, Elkayam U. Acute myocardial infarction associated with pregnancy. J Am Coll Cardiol 2008;52(3):171–80.

[81] Vinayagam D, Thilaganathan B, Stirrup O, Mantovani E, Khalil A. Maternal hemodynamics in normal pregnancy: reference ranges and role of maternal characteristics. Ultrasound Obstet Gynecol 2018;51(5):665–71.

[82] Melchiorre K, Sharma R, Thilaganathan B. Cardiovascular implications in preeclampsia: an overview. Circulation 2014;130(8):703–14.

[83] Ismail S, Wong C, Rajan P, Vidovich MI. ST-elevation acute myocardial infarction in pregnancy: 2016 update. Clin Cardiol 2017;40(6):399–406.

[84] Pacheco LD, Saade GR, Hankins GDV. Acute myocardial infarction during pregnancy. Clin Obstet Gynecol 2014;57(4):835–43.

[85] Elkayam U. Risk of subsequent pregnancy in women with a history of peripartum cardiomyopathy. J Am Coll Cardiol 2014;64(15):1629–36.

[86] Lameijer H, Kampman MAM, Oudijk MA, Pieper PG. Ischaemic heart disease during pregnancy or post-partum: systematic review and case series. Neth Hear J 2015;23(5):249–57.

[87] Saw J, Mancini GBJ, Humphries KH. Contemporary review on spontaneous coronary artery dissection. J Am Coll Cardiol 2016;68(3):297–312.

[88] Vijayaraghavan R, Verma S, Gupta N, Saw J. Pregnancy-related spontaneous coronary artery dissection. Circulation 2014;130(21):1915–20.

[89] Coulson CC, Kuller JA, Bowes WA. Myocardial infarction and coronary artery dissection in pregnancy. Am J Perinatol 1995;12(05):328–30.

[90] Soma-Pillay P, Catherine N-P, Tolppanen H, Mebazaa A, Tolppanen H, Mebazaa A. Physiological changes in pregnancy. Cardiovasc J Afr 2016;27(2):89–94.

[91] Shivvers SA, Wians FH, Keffer JH, Ramin SM. Maternal cardiac troponin I levels during normal labor and delivery. Am J Obstet Gynecol 1999;180(1):122–7.

[92] Hamad R, Larsson A, Pernow J, Bremme K, Eriksson MJ. Assessment of left ventricular structure and function in preeclampsia by echocardiography and cardiovascular biomarkers. J Hypertens 2009;27(11):2257–64.

[93] Dogan R, Birdane A, Bilir A, Ekemen S, Tanriverdi B. Frequency of electrocardiographic changes indicating myocardial ischemia during elective cesarean delivery with regional and general anesthesia: detection based on continuous Holter monitoring and serum markers of ischemia. J Clin Anesth 2008;20(5):347–51.

[94] Smith R, Silversides C, Downey K, Newton G, Macarthur A. Assessing the incidence of peripartum subclinical myocardial ischemia using the troponin T assay: an observational pilot study. Int J Obstet Anesth 2015;24(1):30–4.

[95] Joyal D, Leya F, Koh M, Besinger R, Ramana R, Kahn S, et al. Troponin I levels in patients with preeclampsia. Am J Med 2007;120(9):819.e13–4.

[96] Fleming SM, O'Gorman T, Finn J, Grimes H, Daly K, Morrison JJ. Cardiac troponin I in pre-eclampsia and gestational hypertension. BJOG 2000;107(11):1417–20.

[97] Ravichandran J, Woon SY, Quek YS, Lim YC, Noor EM, Suresh K, et al. High-sensitivity cardiac troponin I levels in normal and hypertensive pregnancy. Am J Med 2019;132(3):362–6.

[98] Silversides CK, Warnes CA. Pregnancy and heart disease. In: Braunwald's heart disease: a textbook of cardiovascular medicine. 11th ed. Philadelphia, PA: Elsevier; 2019. p. 1780–98. [cited 2019 Jan 30] Available from: http://www.clinicalkey.com/#!/content/book/3-s2.0-B9780323463423000906.

[99] Hassan S, Prakash R, Starovoytov A, Saw J. Natural history of spontaneous coronary artery dissection with spontaneous angiographic healing. JACC Cardiovasc Interv 2019;12(6):518–27.

[100] Hayes SNM, Kim ESH, Saw J, Adlam D, Arslanian-Engoren C, Economy KE, et al. Spontaneous coronary artery dissection: current state of the science: a scientific statement from the American Heart Association. Circulation 2018;137(19) [Miscellaneous Article].

[101] Ibanez B, James S, Agewall S, Antunes MJ, Bucciarelli-Ducci C, Bueno H, et al. 2017 ESC Guidelines for the management of acute myocardial infarction in patients presenting with ST-segment elevation. Eur Heart J 2018. [cited 2019 May 27] Available from: https://academic-oup-com.ezp-prod1.hul.harvard.edu/eurheartj/article/39/2/119/4095042.

[102] Roberts JM, August PA, Bakris G, Barton JR, Bernstein IM, Druzin M, et al. Hypertension in pregnancy: executive summary. Obstet Gynecol 2013;122(5):1122–31.

[103] Roberts JM, Pearson G, Cutler J, Lindheimer M. Summary of the NHLBI Working Group on research on hypertension during pregnancy. Hypertension 2003;41(3):437–45.

[104] Hutcheon JA, Lisonkova S, Joseph KS. Epidemiology of pre-eclampsia and the other hypertensive disorders of pregnancy. Best Pract Res Clin Obstet Gynaecol 2011;25(4):391–403.

[105] Ying W, Catov JM, Ouyang P. Hypertensive disorders of pregnancy and future maternal cardiovascular risk. J Am Heart Assoc Cardiovasc Cerebrovasc Dis 2018;7(17). [cited 2019 Feb 22] Available from: https://www.ncbi.nlm.nih.gov/pmc/articles/PMC6201430/.

[106] Ahmed R, Dunford J, Mehran R, Robson S, Kunadian V. Pre-eclampsia and future cardiovascular risk among women: a review. J Am Coll Cardiol 2014;63(18):1815–22.

[107] Powe CE, Levine RJ, Karumanchi SA. Preeclampsia, a disease of the maternal endothelium: the role of anti-angiogenic factors and implications for later cardiovascular disease. Circulation 2011;123(24). [cited 2019 Feb 22] Available from: https://www.ncbi.nlm.nih.gov/pmc/articles/PMC3148781/.

[108] Staff AC, Johnsen GM, Dechend R, Redman CWG. Preeclampsia and uteroplacental acute atherosis: immune and inflammatory factors. J Reprod Immunol 2014;101–102:120–6.

[109] Wu P, Haththotuwa R, Kwok CS, Babu A, Kotronias RA, Rushton C, et al. Preeclampsia and future cardiovascular health: a systematic review and meta-analysis. Circ Cardiovasc Qual Outcomes 2017;10(2). [cited 2019 May 26] Available from: https://www.ahajournals.org/doi/10.1161/CIRCOUTCOMES.116.003497.

[110] Lisowska M, Pietrucha T, Sakowicz A. Preeclampsia and related cardiovascular risk: common genetic background. Curr Hypertens Rep 2018;20(8). [cited 2019 May 26] Available from: https://www.ncbi.nlm.nih.gov/pmc/articles/PMC6028827/.

[111] Robertson SA. Preventing preeclampsia by silencing soluble Flt-1? N Engl J Med 2019;380(11):1080–2.

[112] McGinnis R, Steinthorsdottir V, Williams NO, Thorleifsson G, Shooter S, Hjartardottir S, et al. Variants in the fetal genome near *FLT1* are associated with risk of preeclampsia. Nat Genet 2017;49(8):1255–60.

[113] Shah S, Gupta A. Hypertensive disorders of pregnancy. Cardiol Clin 2019;37(3):345–54.

[114] Lin Y-S, Tang C-H, Yang C-YC, Wu L-S, Hung S-T, Hwa H-L, et al. Effect of pre-eclampsia–eclampsia on major cardiovascular events among peripartum women in Taiwan. Am J Cardiol 2011;107(2):325–30.

[115] Liu S, Joseph KS, Liston RM, Bartholomew S, Walker M, León JA, et al. Incidence, risk factors, and associated complications of eclampsia. Obstet Gynecol 2011;118(5):987–94.

[116] Sperling JD, Gossett DR. Screening for preeclampsia and the USPSTF recommendations. JAMA 2017;317(16):1629–30.

[117] Bibbins-Domingo K, Grossman DC, Curry SJ, Barry MJ, Davidson KW, Doubeni CA, et al. Screening for preeclampsia: US preventive services task force recommendation statement. JAMA 2017;317(16):1661–7.

[118] Askie LM, Duley L, Henderson-Smart DJ, Stewart LA. Antiplatelet agents for prevention of pre-eclampsia: a meta-analysis of individual patient data. Lancet 2007;369(9575):1791–8.

[119] Meher S, Duley L, Hunter K, Askie L. Antiplatelet therapy before or after 16 weeks' gestation for preventing preeclampsia: an individual participant data meta-analysis. Am J Obstet Gynecol 2017;216(2):121–8. e2.

[120] Shanmugalingam R, Hennessy A, Makris A. Aspirin in the prevention of preeclampsia: the conundrum of how, who and when. J Hum Hypertens 2019;33(1):1–9.

[121] Voelker R. USPSTF: low-dose aspirin may help reduce risk of preeclampsia. JAMA 2014;311(20):2055.

[122] Anon. ACOG Practice Bulletin No. 202: gestational hypertension and preeclampsia. Obstet Gynecol 2019;133(1):e1.

[123] Anon. ACOG Practice Bulletin No. 203: chronic hypertension in pregnancy. Obstet Gynecol 2019;133(1):e26–50.

[124] Magee LA, von Dadelszen P, Rey E, Ross S, Asztalos E, Murphy KE, et al. Less-tight versus tight control of hypertension in pregnancy. N Engl J Med 2015;372(5):407–17.

[125] Altman D, Carroli G, Duley L, Farrell B, Moodley J, Neilson J, et al. Do women with pre-eclampsia, and their babies, benefit from magnesium sulphate? The Magpie Trial: a randomised placebo-controlled trial. Lancet 2002;359(9321):1877–90.

[126] Duffy CR, Odibo AO, Roehl KA, Macones GA, Cahill AG. Effect of magnesium sulfate on fetal heart rate patterns in the second stage of labor. Obstet Gynecol 2012;119(6):1129–36.

[127] Shepherd E, Salam RA, Manhas D, Synnes A, Middleton P, Makrides M, et al. Antenatal magnesium sulphate and adverse neonatal outcomes: a systematic review and meta-analysis. PLoS Med 2019;16(12). [cited 2020 Feb 18] Available from: https://www.ncbi.nlm.nih.gov/pmc/articles/PMC6897495/.

[128] Arnett DK, Blumenthal RS, Albert MA, Buroker AB, Goldberger ZD, Hahn EJ, et al. 2019 ACC/AHA guideline on the primary prevention of cardiovascular disease: a report of the American College of Cardiology/American Heart Association Task Force on Clinical Practice Guidelines. J Am Coll Cardiol 2019;26029.

[129] Manalo-Estrella P, Barker AE. Histopathologic findings in human aortic media associated with pregnancy. Arch Pathol 1967;83(4):336–41.

[130] Kuperstein R, Cahan T, Yoeli-Ullman R, Ben Zekry S, Shinfeld A, Simchen MJ. Risk of aortic dissection in pregnant patients with the Marfan syndrome. Am J Cardiol 2017;119(1):132–7.

[131] Smok DA. Aortopathy in pregnancy. Semin Perinatol 2014;38(5):295–303.

[132] Hiratzka LF, Bakris GL, Beckman JA, Bersin RM, Carr VF, Casey DE, et al. 2010 ACCF/AHA/AATS/ACR/ASA/SCA/SCAI/SIR/STS/SVM guidelines for the diagnosis and management of patients with thoracic aortic disease. J Am Coll Cardiol 2010;55(14):e27–129.

[133] Pourafkari L, Ghaffari S, Ahmadi M, Salehi R, Mazani S, Parizad R, et al. Does multi-parity affect the size of the ascending thoracic aorta in women: a prospective cohort study. Curr Med Res Opin 2018;34(11):1907–12.

[134] Silversides CK, Beauchesne L, Bradley T, Connelly M, Niwa K, Mulder B, et al. Canadian Cardiovascular Society 2009 Consensus Conference on the management of adults with congenital heart disease: outflow tract obstruction, coarctation of the aorta, tetralogy of Fallot, Ebstein anomaly and Marfan's syndrome. Can J Cardiol 2010;26(3):e80–97.

[135] Boodhwani M, Andelfinger G, Leipsic J, Lindsay T, McMurtry MS, Therrien J, et al. Canadian Cardiovascular Society position statement on the management of thoracic aortic disease. Can J Cardiol 2014;30(6):577–89.

[136] Braverman AC, Moon MR, Geraghty P, Willing M, Bach C, Kouchoukos NT. Pregnancy after aortic root replacement in Loeys–Dietz syndrome: high risk of aortic dissection. Am J Med Genet A 2016;170(8):2177–80.

[137] Rao P, Isselbacher EM. Preconception counseling for patients with thoracic aortic aneurysms. Curr Treat Options Cardiovasc Med 2018;20(6):50.

[138] Chetty SP, Shaffer BL, Norton ME. Management of pregnancy in women with genetic disorders, part 1: disorders of the connective tissue, muscle, vascular, and skeletal systems. Obstet Gynecol Surv 2011;66(11):699–709.

[139] Goland S, Elkayam U. Pregnancy and Marfan syndrome. Ann Cardiothorac Surg 2017;6(6):642–53.

[140] Siu SC, Sermer M, Colman JM, Alvarez AN, Mercier LA, Morton BC, et al. Prospective multicenter study of pregnancy outcomes in women with heart disease. ACC Curr J Rev 2002;11(1):87.

[141] Joglar JA, Page RL. Management of arrhythmia syndromes during pregnancy. Curr Opin Cardiol 2014;29(1):36–44.

[142] Enriquez AD, Economy KE, Tedrow UB. Contemporary management of arrhythmias during pregnancy. Circ Arrhythm Electrophysiol 2014;7(5):961–7.

[143] Adamson DL, Nelson-Piercy C. Managing palpitations and arrhythmias during pregnancy. Heart 2007;93(12):1630–6.

[144] Shotan A, Ostrzega E, Mehra A, Johnson JV, Elkayam U. Incidence of arrhythmias in normal pregnancy and relation to palpitations, dizziness, and syncope. Am J Cardiol 1997;79(8):1061–4.

[145] MacIntyre C, Iwuala C, Parkash R. Cardiac arrhythmias and pregnancy. Curr Treat Options Cardiovasc Med 2018;20(8):63.

[146] Vaidya VR, Arora S, Patel N, Badheka AO, Patel N, Agnihotri K, et al. Burden of arrhythmia in pregnancy. Circulation 2017;135(6):619–21.

[147] Jeejeebhoy FM, Zelop CM, Lipman S, Carvalho B, Joglar J, Mhyre JM, et al. Cardiac arrest in pregnancy: a scientific statement from the American Heart Association. Circulation 2015;132(18):1747–73.

[148] Bedard E, Dimopoulos K, Gatzoulis MA. Has there been any progress made on pregnancy outcomes among women with pulmonary arterial hypertension? Eur Heart J 2008;30(3):256–65.

[149] Weiss BM, Zemp L, Seifert B, Hess OM. Outcome of pulmonary vascular disease in pregnancy: a systematic overview from 1978 through 1996. J Am Coll Cardiol 1998;31(7):1650–7.

[150] Yentis SM, Steer PJ, Plaat F. Eisenmenger's syndrome in pregnancy: maternal and fetal mortality in the 1990s. BJOG 1998;105(8):921–2.

[151] Sliwa K, van Hagen IM, Budts W, Swan L, Sinagra G, Caruana M, et al. Pulmonary hypertension and pregnancy outcomes: data from the Registry of Pregnancy and Cardiac Disease (ROPAC) of the European Society of Cardiology. Eur J Heart Fail 2016;18(9):1119–28.

[152] Sun X, Feng J, Shi J. Pregnancy and pulmonary hypertension. Medicine (Baltimore) 2018;97(44). [cited 2020 Feb 18] Available from: https://www.ncbi.nlm.nih.gov/pmc/articles/PMC6221755/.

[153] Galiè N, Humbert M, Vachiery J-L, Gibbs S, Lang I, Torbicki A, et al. 2015 ESC/ERS guidelines for the diagnosis and treatment of pulmonary hypertension The Joint Task Force for the Diagnosis and Treatment of Pulmonary Hypertension of the European Society of Cardiology (ESC) and the European Respiratory Society (ERS): endorsed by: Association for European Paediatric and Congenital Cardiology (AEPC), International Society for Heart and Lung Transplantation (ISHLT). Eur Heart J 2016;37(1):67–119.

[154] Olsson KM, Channick R. Pregnancy in pulmonary arterial hypertension. Eur Respir Rev 2016;25(142):431–7.

[155] Grodzinsky A, Florio K, Spertus JA, Daming T, Schmidt L, Lee J, et al. Maternal mortality in the United States and the HOPE Registry. Curr Treat Options Cardiovasc Med 2019;21(9):42.

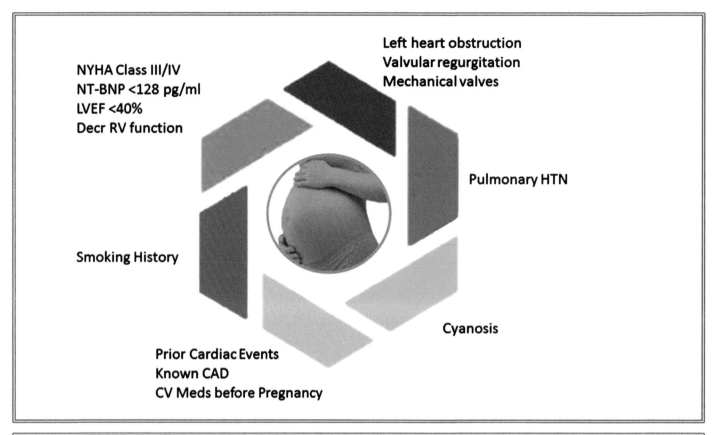

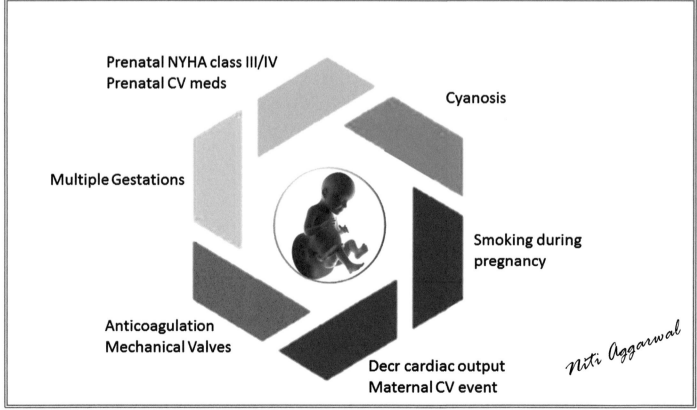

Predictors of Maternal and Neonatal Mortality. Women with preexisting CVD have a higher risk of maternal and neonatal mortality. Maternal morbidity is closely related to neonatal adverse events. BNP, basic patriuretic peptide; CAD, coronary artery disease; CV, cardiovascular; Decr, decreased; HTN, hypertension; LVEF, left ventricular ejection fraction; NYHA, New York Heart Association class; RV, right ventricular. *Image courtesy of Niti R. Aggarwal.*

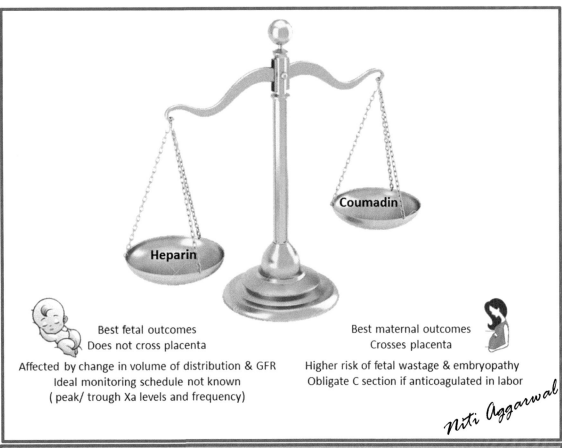

Best fetal outcomes
Does not cross placenta

Affected by change in volume of distribution & GFR
Ideal monitoring schedule not known
(peak/ trough Xa levels and frequency)

Best maternal outcomes
Crosses placenta

Higher risk of fetal wastage & embryopathy
Obligate C section if anticoagulated in labor

Niti Aggarwal

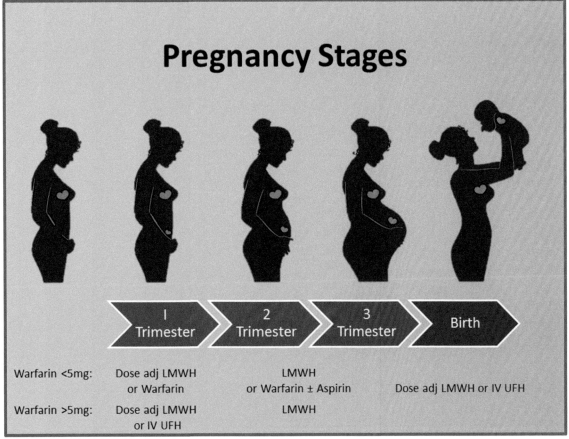

Pregnancy Stages

	1 Trimester	2 Trimester	3 Trimester	Birth
Warfarin <5mg:	Dose adj LMWH or Warfarin	LMWH or Warfarin ± Aspirin		Dose adj LMWH or IV UFH
Warfarin >5mg:	Dose adj LMWH or IV UFH	LMWH		

Recommended Anticoagulation Strategies During Pregnancy. GFR, glomerular filtration rate; IV intravenous; LMWH, low molecular weight heparin; UFH, unfractionated heparin. *Image courtesy of Niti R. Aggarwal.*

Chapter 22

Unique Features of Cardiovascular Pharmacology in Pregnancy and Lactation

Benjamin Laliberte and Debbie C. Yen

Sex Differences in Cardiac Disease. https://doi.org/10.1016/B978-0-12-819369-3.00005-8

Clinical Case

A 33-year-old female patient with chronic hypertension presents for prepregnancy planning. She was first diagnosed with benign essential hypertension at age 30. Her body mass index (BMI) is 35 kg/m². Her blood pressure is currently controlled with amlodipine/benazepril 5 mg/20 mg with typical readings in the 120–130/70–80 mmHg range. What initial workup and counseling should be done? How would you manage this patient?

At 24 weeks of gestation, you see the patient again in the emergency room with lower extremity edema and shortness of breath. She has developed worsening hypertension and new cardiomyopathy. Her blood pressure is 145/83, heart rate 90. An echocardiogram revealed a left ventricular ejection fraction of 35%. She had previously been transitioned to labetalol 400 mg *bis in die* (BID, twice a day) for blood pressure management. How might you alter her cardiac medications? Do these medications need to be changed following delivery?

Abstract

Cardiovascular disease in pregnancy has become more common due to increasing prevalence of cardiovascular risk factors, advanced maternal age, and improvements in the management of congenital heart disease. Approximately one-third of pregnant women use medication for the treatment of cardiovascular disease. The physiologic changes that occur during pregnancy may alter pharmacodynamic and pharmacokinetic properties of medications. Therefore, it is important for clinicians to be aware of both the safety and efficacy of medications in the developing fetus and breastfeeding infant. This chapter summarizes the published data and guidelines with regard to cardiovascular pharmacotherapy during pregnancy and lactation, and is intended to be a tool for clinicians to use in shared decision-making in this population.

Introduction

A wealth of high-quality studies and guidelines address the safe and effective use of medications in cardiovascular disease. Unfortunately, data in pregnancy and lactation are lacking. However, up to one-third of women with structural heart disease have used cardiac medications during pregnancy [1, 2].

When considering the use of medications in pregnancy and lactation, clinicians must understand the underlying disease, hemodynamic and pharmacokinetic changes during pregnancy, individual medication characteristics, and, most importantly, the risks to the mother and fetus.

Labeling Risks

From 1979 to 2015, a pregnancy letter category (A, B, C, D, X) was used in prescription labeling to describe individual medication risks [3, 4]. However, the system was vague and often misinterpreted, and in 2015, the US Food and Drug Administration (FDA) implemented new labeling requirements. Prescribing information now includes comprehensive summaries and clinical considerations for prescribing and monitoring of medications in pregnancy, lactation, and in females and males of reproductive potential [3].

Pharmacokinetics in Pregnancy

There is a complex interplay of changes in pharmacokinetic parameters throughout pregnancy. As depicted in **Figure 1**, physiologic changes that occur during pregnancy may significantly alter drug absorption, distribution, metabolism, and elimination [5–8]. A systematic review of 23 cardiovascular pharmacokinetic studies in pregnancy found consistently increased drug clearance and decreased half-life and peak drug concentrations [5]. However, there is a lack of data correlating the impact of these pharmacokinetic changes on clinical outcomes.

Absorption

Multiple factors alter absorption and bioavailability of orally administered medications during pregnancy [7]. Bioavailability is the fractional percent of drug that reaches systemic circulation. Increases in cardiac output during pregnancy improve intestinal blood flow and drug absorption. However, placental production of progesterone decreases small bowel motility and gastric and intestinal emptying time (~30–50%), thus decreasing the absorption of oral medications [2, 8]. Hormones also induce changes in the availability and activity of transport proteins, such as P-glycoprotein and organic anion transporters, potentially affecting drug serum levels. Furthermore, a reduction in gastric acid production and stimulation of mucous secretions increase gastric pH, theoretically decreasing absorption of weak acids (e.g., aspirin) [7, 8]. Finally, since nausea and vomiting are common during pregnancy, the extent of absorption of recently administered drugs may be inconsistent [2, 7]. Nevertheless, there is a lack of evidence that demonstrates clinically significant changes in oral bioavailability based on physiologic changes during pregnancy [6, 7].

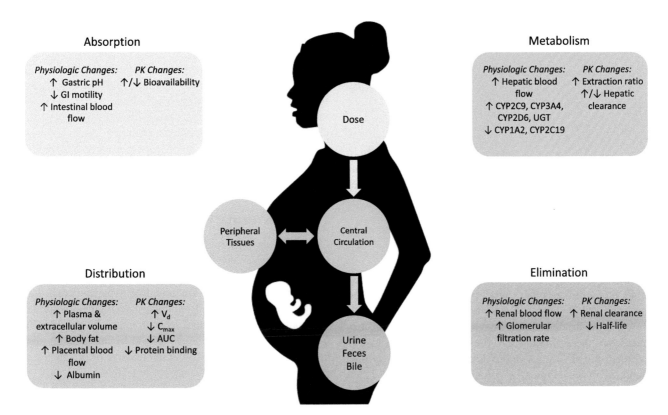

FIGURE 1   Physiologic Changes and Alterations in Pharmacokinetic Properties During Pregnancy. Pertinent pharmacokinetic properties that are affected by physiologic changes in pregnancy are depicted. The first step after administration of a drug is absorption *(yellow)*, followed by its distribution into the systemic circulation *(green)*, metabolism *(red)*, and elimination from the body *(blue)*. AUC, area under the curve; C_{max}, peak drug concentration; CYP, cytochrome P450; GI, gastrointestinal; pH, potential of hydrogen; PK, pharmacokinetic; UGT, uridine 5′-diphospho-glucuronosyltransferase; V_d, volume of distribution.

Distribution

After absorption, drugs transfer into systemic circulation in the distribution phase [5–8]. At steady state, drug concentrations in the central compartment (e.g., circulation) reach equilibrium with concentrations in peripheral tissues. Drugs may exist as bound (e.g., to receptor or albumin) or unbound (free). The most pertinent pharmacokinetic property in the distribution phase is the volume of distribution (V_d), which describes the extent of drug dispersal into tissues. V_d is influenced by various factors, including tissue perfusion, lipid solubility, and protein binding. For example, a highly lipophilic drug that disperses out of the water-based circulation into tissues will exhibit a higher V_d (e.g., amiodarone). Conversely, a highly hydrophilic drug (e.g., digoxin) or highly protein-bound drug (e.g., warfarin) typically exhibits a lower V_d. Various physiologic changes affect drug distribution during pregnancy, including a 40–50% plasma volume expansion, increased body fat, improved cardiac output to tissues, and additional compartments (e.g., feto-placental and amniotic fluid). Thus, both lipophilic and hydrophilic drugs have increased V_d in pregnancy, leading to decreased drug concentrations in the central compartment, which

could complicate monitoring for drugs with a narrow therapeutic index (e.g., digoxin). Finally, dilutional hypoalbuminemia and competitive protein binding from endogenous maternal hormones lead to more available free drug for distribution and pharmacodynamic activity.

Metabolism

Metabolism describes the chemical modification of drugs or toxins in order to facilitate safe elimination from the body [5–8]. This process is highly complex, with multiple factors (e.g., genetic polymorphisms, gender, and age) producing pharmacokinetic differences in enzyme metabolism. Two enzymatic systems are responsible for the majority of drug metabolism. In general, phase I metabolism via the cytochrome P450 (CYP) enzymatic family is most often affected by polymorphisms and drug interactions. However, the highly conserved phase II metabolism via glucuronidation or sulfation can also be affected in pregnancy. Progesterone and estrogen can also induce or inhibit specific enzyme activity [2, 7]. For example, enzymatic activity of CYP3A4, CYP2D6, and CYP2C9 increases during pregnancy, while CYP1A2 and CYP2C19 activity may decrease with advanced gestation. Therefore, drug levels may be lower for beta-blockers and calcium channel blockers (CCBs), which are primarily metabolized by CYP2D6 and CYP3A4, respectively, and may necessitate higher doses. However, many drugs are metabolized by multiple pathways and may assert their own pharmacodynamic effects on enzyme metabolism, making the net effect difficult to predict. Finally, improvements in hepatic blood flow during pregnancy increase the availability of drugs with high extraction ratios for metabolism (e.g., propranolol), thus improving their hepatic clearance. Drugs with low extraction ratios (e.g., warfarin) are unlikely to be affected by this physiologic change.

Elimination

Elimination is the removal of drugs from the body via urine, feces, bile, or other routes. For the fetus, elimination primarily occurs via diffusion back to the mother [8]. However, as the fetal kidney matures, more drugs are excreted into the amniotic fluid. Pregnancy improves maternal renal blood flow by approximately 50%, with similar increases in the glomerular filtration rate [2, 6–8]. Thus, renal clearance is improved and half-life decreased for renally cleared drugs. However, hormone-induced changes in transport proteins, as previously discussed, may limit the true magnitude of the effect.

General Approach to Prescription During Pregnancy and Lactation

When considering the use of medications during pregnancy and lactation, clinicians should employ a shared decision-making approach with patients to balance fetal and maternal risk. It is important to consider the necessity and urgency of the drug, as well as timing during gestation [2]. Congenital malformations can be caused by drug toxicity during organogenesis in the first trimester; therefore, drug administration should be avoided or minimized. In the second and third trimesters, drugs can potentially interfere with fetal growth and development. Patients considering pregnancy should be counseled on the risks and benefits of taking cardiovascular medications during pregnancy and postpartum during breastfeeding. Clinicians should also take measures to transition potentially teratogenic cardiovascular medications used in the prepregnancy period to safer alternatives. If clinical data are lacking, drug-specific properties should be considered, as these impact a drug's ability to pass into the placenta or breast milk [2, 7, 8]. Drugs with low oral bioavailability, high molecular weight, high protein binding, and short half-lives are preferable. Cardiovascular medications with a narrow therapeutic index (e.g., digoxin and warfarin) should be closely monitored for safety and efficacy.

Medications

In the following sections, medications are presented by therapeutic use and class. Each section includes available human data in pregnancy and lactation, incorporating guideline recommendations where appropriate. Comprehensive tables include labeling risks, animal data, pertinent pharmacokinetic properties, drug transfer to the placenta and breast milk, and compatibility in pregnancy and breastfeeding. **Figure 2** lists recommended and contraindicated medications for common cardiovascular disorders in pregnancy.

Lipid-Lowering Therapy

Lipids are an important component in fetal development [9]. However, many women have undiagnosed dyslipidemia prior to pregnancy, putting them at risk for obstetric and fetal complications. Treatment of hyperlipidemia during pregnancy is primarily nonpharmacological, as guidelines only consider bile acid sequestrants to be safe [9, 10]. Other lipid-lowering therapies should be immediately discontinued if a woman is trying to conceive or pregnancy is suspected.

Bile Acid Sequestrants

Bile acid sequestrants inhibit enterohepatic reuptake of intestinal bile salts and are considered third-line agents to lower low-density lipoprotein (LDL) in hypercholesterolemia [9, 11]. Cholestyramine has been studied in case reports for the treatment of cholestasis and inflammatory bowel disease in pregnancy, with no fetal adverse effects documented [12]. However, impaired vitamin absorption may have contributed to one fetal death. Bile acid sequestrants interfere with absorption of fat-soluble vitamins (i.e., A, D, E, K) and other drugs; therefore, vitamins should be taken at least 1 h before or at least 4–6 h after bile acid sequestrant administration. A case series of colesevelam in five pregnant women with familial hypercholesterolemia demonstrated uncomplicated pregnancies with healthy infants [12]. Guidelines recommend bile acid sequestrants as an alternative to statins in pregnant women who require pharmacotherapy for hypercholesterolemia or secondary atherosclerotic cardiovascular disease (ASCVD) prevention [9, 10].

Bile acid sequestrants have not been studied in lactation but guidelines consider them safe due to their lack of systemic absorption [9, 10, 12].

Ezetimibe

Ezetimibe inhibits the sterol transporter at the intestinal brush border, preventing cholesterol absorption. It is indicated as a second-line agent for patients with ASCVD [11].

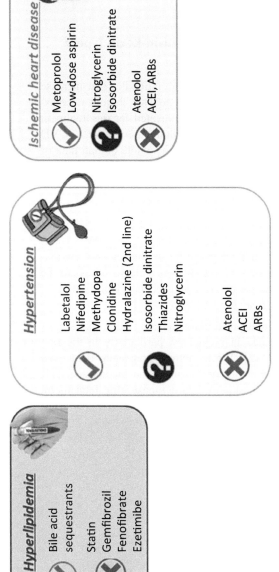
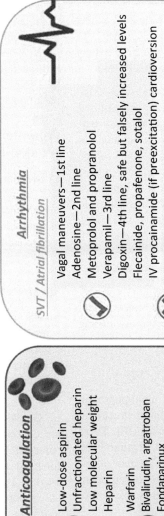
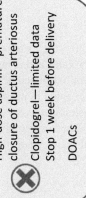
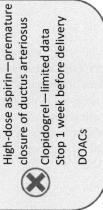

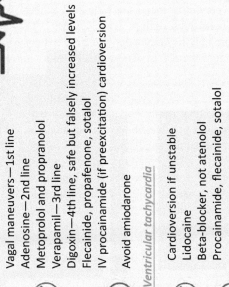

Hyperlipidemia

✓ Bile acid sequestrants

✗ Statin
Gemfibrozil
Fenofibrate
Ezetimibe

Hypertension

✓ Labetalol
Nifedipine
Methyldopa
Clonidine
Hydralazine (2nd line)

? Isosorbide dinitrate
Thiazides
Nitroglycerin

✗ Atenolol
ACEI
ARBs

Ischemic heart disease

✓ Metoprolol
Low-dose aspirin

? Nitroglycerin
Isosorbide dinitrate

✗ Atenolol
ACEI, ARBs

Heart failure

✓ Metoprolol
Carvedilol
Digoxin
Furosemide, bumetanide
(if pulmonary edema)
Hydralazine

? Isosorbide dinitrate
Dobutamine, milrinone,
dopamine

✗ ACEI, ARBs—fetal
malformations
Sacubitril/Valsartan
Spironolactone, eplerenone-
Feminization of fetus
Atenolol

Anticoagulation

✓ Low-dose aspirin
Unfractionated heparin
Low molecular weight
Heparin

? Warfarin
Bivalirudin, argatroban
Fondaparinux

✗ High-dose aspirin—premature
closure of ductus arteriosus
Clopidogrel—limited data
Stop 1 week before delivery
DOACs

Arrhythmia

SVT / Atrial fibrillation

✓ Vagal maneuvers—1st line
Adenosine—2nd line
Metoprolol and propranolol
Verapamil—3rd line
Digoxin—4th line, safe but falsely increased levels
Flecainide, propafenone, sotalol
IV procainamide (if preexcitation) cardioversion

✗ Avoid amiodarone

Ventricular tachycardia

✓ Cardioversion if unstable
Lidocaine
Beta-blocker, not atenolol
Procainamide, flecainide, sotalol

✗ Amiodarone, only if necessary

FIGURE 2 **Treatment of Common Cardiovascular Disorders During Pregnancy.** ACEI, angiotensin-converting enzyme inhibitor; ARBs, angiotensin receptor blockers; DOACs, direct oral anticoagulants; IV, intravenous; SVT, supraventricular tachycardia. ✓ Generally safe and recommended during pregnancy; ? Limited data/Use with caution during pregnancy; ✗ Avoid use/contraindicated during pregnancy. _Image courtesy of Niti R. Aggarwal._

Ezetimibe has not been studied independently in pregnancy or lactation [12].

Fibrates

Fibrates reduce triglycerides by increasing fatty acid oxidation and modifying gene transcription. They lack ASCVD outcomes data and are only recommended in severe hypertriglyceridemia to reduce the risk of pancreatitis [9, 11].

The use of fibrates in pregnancy is limited to case reports [12]. Two of three cases of fenofibrate use resulted in healthy infants. One fetal death was due to acute maternal pancreatitis secondary to fenofibrate discontinuation. Of five cases of gemfibrozil use in pregnancy, four women delivered healthy infants. The remaining infant required temporary intubation for respiratory distress. Finally, a Medicaid surveillance study identified 15 exposures to gemfibrozil in pregnancy, with one structural abnormality observed in an infant exposed during the first trimester [12]. Guidelines recommend fibrates for severe hypertriglyceridemia (triglycerides ≥ 500 mg/dL) in pregnancy in the second and third trimesters [9].

There are no available human data for fibrates in lactation, but guidelines consider them safe for use during breastfeeding [9, 12].

Lomitapide

Lomitapide inhibits the assembly of apolipoprotein B-containing lipoproteins and is indicated for the treatment of homozygous familial hypercholesterolemia [10]. There are no data for lomitapide in pregnancy and lactation [12]. Due to teratogenicity observed in animal studies, lomitapide is contraindicated in pregnancy; females of reproductive potential should have a negative pregnancy test before initiation [9, 10, 12].

Niacin

Niacin is a water-soluble B vitamin and a necessary component of enzymes involved in lipid metabolism. At therapeutic doses (e.g., 500–2000 mg daily), niacin decreases triglycerides and increases high-density lipoprotein concentrations, but its role is limited [9, 11]. While supplementation is recommended in pregnancy and lactation, there are no studies on medication dosage for hyperlipidemia [12].

Omega-3 Fatty Acids

There are several commercially available over-the-counter and prescription omega-3 fatty acids. They are used for the treatment of hypertriglyceridemia and recommended for supplementation in pregnancy and lactation [9, 11, 12]. A Cochrane review analyzed 70 randomized controlled trials (RCTs) of omega-3 fatty acids (supplements and food) in pregnancy, and found a statistically significant lower risk of preterm birth [11.9% vs. 13.4%; relative risk (RR) 0.89, 95% confidence interval (CI) 0.81–0.97], early preterm birth (2.7% vs. 4.6%; RR 0.58, 95% CI 0.44–0.77), and low birth weight (14% vs. 15.6%; RR 0.90, 95% CI 0.82–0.99) [13]. No other differences were observed in maternal or infant outcomes.

Proprotein Convertase Subtilisin-Kexin Type 9 (PCSK9) Inhibitors

PCSK9 inhibitors are monoclonal antibodies that increase LDL receptor availability on hepatocytes, thereby decreasing circulating LDL. They are indicated for the treatment of familial hyperlipidemia and adjunct therapy for ASCVD risk reduction [11]. There are no available human data of PCSK9 inhibitors in pregnancy and lactation [12].

Statins

Statins inhibit the rate-limiting enzyme in cholesterol synthesis, 3-hydroxy-3-methylglutaryl-CoA (HMG-CoA) reductase. Individual statins differ in potency and pharmacokinetic properties [11, 12, 14]. As a class, statins are the lipid cornerstone for primary and secondary prevention of ASCVD in the general population [9–11].

Guidelines and prescription labeling state that statins are contraindicated in pregnancy, particularly in the first trimester [1, 9–12]. While human pregnancy data exist for most statins, results are variable and inconsistent [12, 14–17]. Following animal data demonstrating fetal harm, case reports continued to highlight the risk of congenital abnormalities [12]. A case series of first-trimester statin exposures reported to the FDA from 1987 to 2001 found adverse outcomes in 29 of 52 cases (56%) [15]. All cases were associated with lipophilic statins (e.g., atorvastatin, lovastatin, simvastatin). The authors note that FDA reporting is limited by publication bias, with fewer reported incidents favoring more severe outcomes.

Higher-quality studies questioned the validity of the previous case reports. A systematic review identified 16 studies of statin exposure during pregnancy between 1996 and 2016, and found inconsistent results [16]. Five case series of first-trimester exposure described high rates of congenital abnormalities (3.8–100%). However, three cohort studies of 64–1152 women exposed during the first trimester found no increased risk. Three registry studies of 16–64 women and four other systematic reviews had similar conclusions. The authors stated that there was no clear relationship between statin use in pregnancy and adverse fetal outcomes, and negative outcomes were likely due to underlying risk factors. Due to the heterogeneity of the studies, no statistical analysis was conducted, making it difficult to draw direct comparisons.

The first RCTs evaluating statins in pregnancy were conducted in 20 pregnant women (12–16 weeks' gestation) deemed at high risk for preeclampsia, and randomized patients to receive pravastatin 10 mg daily or placebo until delivery [17]. No maternal or fetal deaths occurred, though two congenital anomalies were observed in each group. Four patients in the placebo group developed preeclampsia, compared to none in the pravastatin group ($P > 0.05$). No difference in infant birth weight was observed (2877 ± 630 g vs. 3018 ± 260 g, $P > 0.05$). Though not powered for clinical

outcomes, this study provided important safety information for pravastatin in pregnancy, which exhibits different pharmacokinetic properties from other statins [12, 14]. A systematic review of pravastatin in preeclampsia concluded that animal models and human case reports support this novel approach [14].

Since these publications, a retrospective cohort study analyzed 280 women exposed to statins in the first trimester [18]. The rate of congenital cardiac anomalies was significantly higher in this group compared to matched nonexposed pregnancies [5% vs. 1.4%; odds ratio (OR) 2.1, 95% CI 1. 2–3.6, $P = 0.009$], driven by a higher rate of ventricular septal defects (4.3% vs. 0.74%; OR 3.3, 95% CI 1.8–6, $P < 0.001$).

There are no clinical data for statins in breastfeeding [12]. Guidelines and prescription labeling state that statins are contraindicated in nursing mothers [1, 9–11] (**Table 1**).

TABLE 1 Lipid-Lowering Therapies in Pregnancy

Class	Drug	Prior FDA Category[a]	Prescription Labeling Risks	Pharmacokinetic Considerations and Drug Transfer	Compatibility[b]
Bile acid sequestrants	Cholestyramine Colestipol	C	May interfere with absorption of fat-soluble vitamins	Not absorbed systemically	
	Colesevelam	B	Studies of supratherapeutic doses in rats and rabbits revealed no fetal harm May interfere with absorption of fat-soluble vitamins	Placenta: Unknown; unlikely Breast milk: Unknown; unlikely	
Ezetimibe		C	Fetal skeletal abnormalities observed when given at supratherapeutic doses in rats and rabbits	> 90% Protein bound Half-life: 22 h Placenta: Unknown; unlikely Breast milk: Yes, in rats	
Fibrates	Fenofibrate (fenofibric acid) Gemfibrozil	C	Adverse effects and teratogenicity observed in rats and rabbits	99% Protein bound Placenta: Yes, in humans (gemfibrozil) Breast milk: Unknown; expected but limited	

Continued

CH 22

TABLE 1 Lipid-Lowering Therapy in Pregnancy—cont'd

Class	Drug	Prior FDA Category[a]	Prescription Labeling Risks	Pharmacokinetic Considerations and Drug Transfer	Compatibility[b]
Lomitapide		X	Teratogenicity and tumorigenicity observed in animals at a variety of doses Females of reproductive potential should have a negative pregnancy test before starting	7% Oral bioavailability 100% Protein bound Half-life: 40h Placenta: Unknown; expected but limited Breast milk: Unknown; expected but limited	
Mipomersen		B	No fetal harm in mice and rabbits at supratherapeutic doses Nursing pups of lactating rats receiving supratherapeutic doses had reduced weight gain	Very high (7kDa) molecular weight <10% Oral bioavailability >90% Protein bound Half-life: 1–2 months Placenta: Unknown; unlikely Breast milk: Yes, in rats	
Niacin		C	Animal and human studies with therapeutic doses for hyperlipidemia have not been conducted	Placenta: Yes, in humans Breast milk: Yes, in humans	
Omega-3 fatty acids		C	Embryocidal effects at supratherapeutic doses in rats and rabbits In milk of lactating rats, 6–15 times higher concentrations of EPA were observed compared to plasma	Placenta: Yes, in humans Breast milk: Yes, in humans	
PCSK9 inhibitors	Alirocumab Evolocumab	None	No adverse fetal effects in rats at supratherapeutic doses Suppression of the humoral immune response in monkeys at supratherapeutic doses (alirocumab) Human IgG is present in breast milk but does not substantially enter infant circulation	Not orally absorbed Very high (144–146kDa) molecular weight Placenta: Yes, in monkeys (unlikely in the first trimester) Breast milk: Unknown; unlikely	

TABLE 1	Lipid-Lowering Therapy in Pregnancy—cont'd				
Class	Drug	Prior FDA Category[a]	Prescription Labeling Risks	Pharmacokinetic Considerations and Drug Transfer	Compatibility[b]
Statins	Atorvastatin Fluvastatin Lovastatin Pitavastatin Pravastatin Rosuvastatin Simvastatin	X	Rare congenital abnormalities have been reported in humans; increased risk in the first trimester Animal fetal abnormalities observed at supratherapeutic doses in select agents (lova, pitava, and rosuva); no teratogenicity observed in others (atorva, fluva, prava, and simva)	5–51% Oral bioavailability 50–99% Protein bound Water-soluble statins: Fluvastatin, pravastatin, and rosuvastatin Long-acting statins: Atorvastatin (half-life: 14 h), pitavastatin (half-life: 12 h), and rosuvastatin (half-life: 19 h) Placenta: Yes, in humans (atorvastatin, lovastatin, pravastatin, simvastatin) Breast milk: Pravastatin and rosuvastatin excreted in small amounts in human breast milk	?

[a] *A = generally acceptable; controlled studies in pregnant women showed no evidence of fetal risk; B = may be acceptable; either animal studies showed no risk, or animal studies showed minor risks and human studies showed no risks; C = use with caution if benefits outweigh risks; animal studies showed risk and human studies unavailable, or no studies done; D = use in life-threatening emergencies when no safer drug available; positive evidence of human fetal risk; X = do not use in pregnancy; risks involved outweigh potential benefits, and safer alternatives exist.*
[b] ● *= Generally safe;* ○ *= Limited human data/use with caution;* ● *= Avoid use/contraindicated;* ○ *= No adequate human data;* 🤰 *= Pregnancy;* 🧍 *= Lactation.*

EPA, eicosapentaenoic acid; FDA, Food and Drug Administration; IgG, immunoglobulin G; PCSK9, proprotein convertase subtilisin-kexin type 9.

Antiplatelet and Antiinflammatory Agents

Pregnancy is associated with an increased risk of myocardial infarction, with ASCVD accounting for over 20% of maternal cardiac deaths [1]. Current guidelines recommend management of myocardial infarction in pregnancy similar to that of the general population. However, low-dose aspirin is the only antiplatelet agent deemed safe in pregnancy and lactation, as data for other therapies are lacking. Strategies to limit the duration of dual antiplatelet therapy, such as using newer-generation drug-eluting stents, are recommended.

Aspirin

Aspirin irreversibly inhibits cyclooxygenase-1 and prevents platelet aggregation; it is a cornerstone of ASCVD risk

reduction and prevention of thrombosis in valvular heart disease. Guidelines consider low-dose aspirin to be safe in pregnancy and recommend aspirin in pregnant women at moderate-high risk for preeclampsia from week 12 to 36–37 weeks [1]. The use of high-dose aspirin (> 150 mg) is not recommended, particularly in the third trimester due to risk of premature ductus arteriosus closure.

The safety and efficacy of low-dose aspirin in pregnancy has been extensively studied in RCTs and meta-analyses [1, 19–21]. In the Aspirin vs. Placebo in Pregnancies at High Risk for Preterm Preeclampsia (ASPRE) trial, 1776 women at high risk for preeclampsia were randomized to aspirin 150 mg or placebo from 11–14 to 36 weeks' gestation [19]. Aspirin significantly reduced the primary outcome of preterm preeclampsia (1.6% vs. 4.3%; OR 0.38, 95% CI 0.2–0.74). No differences in secondary outcomes or adverse maternal or neonatal events were observed. The authors concluded that initiation of aspirin 150 mg in high-risk pregnant women was effective in reducing pre-term preeclampsia.

In a meta-analysis of 45 RCTs, 20,909 pregnant women received 50–150 mg of aspirin [20]. Patients who received aspirin at ≤ 16 weeks' gestation demonstrated a dose-related decreased risk of preeclampsia (RR 0.57, 95% CI 0.43–0.5, $P < 0.001$) and fetal growth restriction (RR 0.56, 95% CI 0.44–0.7, $P < 0.001$). Aspirin use after 16 weeks reduced the risk of preeclampsia (RR 0.81, 95% CI 0.66–0.99, $P = 0.04$) but not fetal growth restriction (RR 0.85, 95% CI 0.64–1.14, $P = 0.28$).

Aspirin and other salicylates are excreted into breast milk in negligible amounts [12, 22]. Case reports described mixed results with high doses, as one infant had no detect-able serum concentrations after breastfeeding, while an-other developed salicylate intoxication [12, 22]. The latter exposure was questionable because direct administration of aspirin to the infant could not be ruled out. Guidelines con-sider low-dose aspirin generally safe in lactation [1].

Cilostazol

Cilostazol inhibits platelet aggregation and improves periph-eral vasodilation through inhibition of phosphodiesterase-3. There are no data for cilostazol in human pregnancy or lactation [12].

Colchicine

Colchicine inhibits leukocyte microtubule self-assembly and cytokine production, and has been used for centuries to treat inflammatory conditions. It is a first-line therapy for pericarditis and has also been studied in atrial fibrillation and prevention of stent restenosis [23].

The effects of colchicine on pregnancy have been evalu-ated in several observational studies, primarily in patients with familial Mediterranean fever [12, 24]. In case reports, no adverse fetal effects were observed [12]. A meta-analysis of four observational studies compared 554 pregnancies ex-posed to colchicine to 1575 controls [24]. Among women with familial Mediterranean fever, there were no significant differences in rates of miscarriage (10.9% vs. 16.7%; OR 0.6, 95% CI 0.43–1.11, $P = 0.12$), major malformations (1% vs. 1%; OR 0.59, 95% CI 0.12–2.93, $P = 0.52$), or birthweight (2956 ± 662 g vs. 3105 ± 452 g, $P = 0.12$) [24]. Similarly, in patients using colchicine for any indication, there was no difference in major congenital malformations (2.4% vs. 2.6%; OR 1.08, 95% CI 0.56–2.07, $P = 0.82$). Significantly lower rates of miscarriage (7.8% vs. 10%; OR 0.65, 95% CI 0.45–0.93, $P = 0.02$), higher rates of preterm birth (16.5% vs. 6.5%; OR 2.48, 1.65–3.71, $P < 0.001$), and lower birthweight (2985 ± 414 g vs. 3281 ± 187 g, $P = 0.02$)

were observed in the colchicine group. The authors con-cluded that colchicine is safe in pregnancy and should not be withheld in patients whose disease requires treatment.

In over 100 case reports of nursing infants exposed to colchicine up through 6 months, there were no adverse ef-fects observed [12, 25]. Similarly, a case-control study with 38 exposed infants compared to 75 controls breastfed for a mean of 9.1 months found no adverse events attributable to colchicine [25]. Pericarditis guidelines state that colchi-cine is contraindicated in pregnancy and lactation, but con-ceded that the data suggest otherwise [23]. The American Academy of Pediatrics (AAP) classifies colchicine as com-patible with breastfeeding [12, 22].

Dipyridamole

Dipyridamole is used in cardiac imaging and as an adjunct antiplatelet agent, inhibiting adenosine, phosphodiesterase, and thromboxane A2. Data for dipyridamole in pregnancy are limited due to coadministration of other antiplatelets and anticoagulants, but dipyridamole has not been associ-ated with fetal harm [12, 21].

Glycoprotein IIb/IIIa Inhibitors

Glycoprotein IIb/IIIa inhibitors, including eptifibatide and tirofiban, prevent cross-linking of fibrinogen between plate-lets and are indicated in acute coronary syndrome (ACS). They are typically utilized for short periods of time and thus data are limited to case reports [12]. Women given intrapro-cedural tirofiban (27 and 20 weeks' gestation) and eptifi-batide (8 weeks and 32 weeks' gestation) delivered healthy infants [12]. For the women who received eptifibatide, total durations of exposure were reported as 24 h and 7 days, re-spectively. No data are available in lactation.

P2Y$_{12}$ Inhibitors

P2Y$_{12}$ inhibitors (including clopidogrel, prasugrel, and ti-cagrelor) block adenosine diphosphate (ADP)-induced platelet activation and are indicated in patients with ACS or coronary stent placement. Guidelines favor clopidogrel in pregnancy and advocate for strategies to limit therapy du-ration [1]. P2Y$_{12}$ inhibitors should ideally be discontinued 1 week prior to delivery and prior to neuraxial anesthesia to decrease risk of epidural hematoma. Breastfeeding is not recommended.

Data on the use of P2Y$_{12}$ inhibitors in pregnancy and lactation are limited to case reports with oral agents [12]. In all, 11 of 12 cases of clopidogrel exposure resulted in healthy infants; one was born with benign anomalies. For prasugrel and ticagrelor, one case report exists for each [12, 26]. A 32-year-old woman received a drug-eluting stent early in gestation and was treated with prasugrel until 5 days before cesarean section [12]. Another 37-year-old woman became pregnant after ACS with a drug-eluting stent implantation. She continued ticagrelor for 8 months and discontinued therapy 1 week before planned delivery [26]. Healthy infants were delivered in both cases. Unpublished postmarketing reports with clopidogrel describe no adverse events during breastfeeding [25].

Vorapaxar

Vorapaxar antagonizes protease-activated receptor-1 and inhibits thrombin-induced platelet aggregation. It is indicated to reduce thrombotic events in patients with a history of ASCVD. There are no data in human pregnancy or lactation [12] (**Table 2**).

TABLE 2 Antiplatelet and Antiinflammatory Agents in Pregnancy

Class	Drug	Prior FDA Category[a]	Prescription Labeling Risks	Pharmacokinetic Considerations and Drug Transfer	Compatibility[b]
Aspirin (low-dose) Aspirin (high-dose)		C (D for full dose in third trimester)	Avoid high dose in the third trimester due to premature closure of the fetal ductus arteriosus	80–100% Bioavailability depending on route of administration (e.g., oral and rectal) Variable dose-dependent protein binding Placenta: Yes, in humans Breast milk: Yes, in humans (minimal)	
Cilostazol		C	Decreased fetal weights and increased cardiovascular, renal, and skeletal anomalies observed in rats and rabbits	95–98% Protein bound Placenta: Unknown; expected but limited Breast milk: Yes, in rats	
Colchicine		C	Data from limited number of studies found no increased risk of adverse fetal effects in pregnant mothers taking colchicine for FMF Published animal reproductive studies have demonstrated fetal harm	39% Protein bound Placenta: Yes, in humans Breast milk: Yes, in humans (<10%)	
Dipyridamole		B	No evidence of fetal harm in animals	91–99% Protein bound Placenta: Unknown; expected but limited Breast milk: Yes, in humans	
Glycoprotein IIb/IIIa inhibitors	Eptifibatide	B	No evidence of fetal harm at supratherapeutic doses in rabbits	Not orally bioavailable 25% Protein bound Half-life: 2.5h Placenta: Unknown; expected Breast milk: Unknown; expected	

Continued

CH
22

| | | | | Pharmacokinetic | |
Class	Drug	Prior FDA Category[a]	Prescription Labeling Risks	Considerations and Drug Transfer	Compatibility[b]
	Tirofiban	B	No evidence of fetal harm at supratherapeutic doses in rats and rabbits	65% Protein bound Half-life: 2 h Placenta: Yes, in rats and rabbits Breast milk: Yes, significant levels in rats	
P2Y$_{12}$ inhibitors	Cangrelor	C	No structural abnormalities observed in animals but fetal growth retardation in rats and rabbits	97–98% Protein bound Half-life: 3–6 min Placenta: Unknown; unlikely with short duration Breast milk: Unknown; unlikely with short duration	
	Clopidogrel	B	No evidence of fetal harm at supratherapeutic doses in rat	>50% Oral bioavailability High (98%) protein bound Placenta: Unknown; expected but limited Breast milk: Yes, in rats	
	Ticagrelor	C	Structural abnormalities observed at supratherapeutic doses in animals	36% Oral bioavailability ~100% Protein bound Placenta: Unknown; expected but limited Breast milk: Yes, in rats	
	Prasugrel	B	No evidence of fetal harm at supratherapeutic doses in rats	>79% Oral bioavailability 98% Protein bound Placenta: Unknown; expected but limited Breast milk: Yes, in rats	
Vorapaxar		B	No evidence of fetal harm at supratherapeutic doses in animals	100% Oral bioavailability 100% Protein bound Placenta: Unknown; expected but limited Breast milk: Yes, in rats	

TABLE 2 Antiplatelet and Antiinflammatory Agents in Pregnancy—cont'd

[a] *A = generally acceptable; controlled studies in pregnant women showed no evidence of fetal risk; B = may be acceptable; either animal studies showed no risk, or animal studies showed minor risks and human studies showed no risks; C = use with caution if benefits outweigh risks; animal studies showed risk and human studies unavailable, or no studies done; D = use in life-threatening emergencies when no safer drug available; positive evidence of human fetal risk; X = do not use in pregnancy; risks involved outweigh potential benefits, and safer alternatives exist.*

[b] ● *= Generally safe;* ○ *= Limited human data/use with caution;* ● *= Avoid use/contraindicated;* ○ *= No adequate human data;* 🤰 *= Pregnancy;* 🤱 *= Lactation.* FDA, Food and Drug Administration; FMF, familial Mediterranean fever.

Hypertension, Heart Failure, and Angina

Chronic hypertension occurs in approximately 5–10% of pregnancies and has been associated with maternal and fetal complications [2, 27]. Due to the physiologic fall in blood pressure that occurs during pregnancy, some pregnant women are able to withdraw antihypertensive medications in the first half of pregnancy [1]. If treatment is needed, methyldopa, labetalol, and nifedipine are recommended first-line drugs [1].

Pregnancy is typically poorly tolerated among women with preexisting dilated cardiomyopathy [1]. In heart failure in pregnancy, guidelines recommend continuation of beta-blockers in women who took them before pregnancy, and cautious initiation when clinically indicated [1]. Medications contraindicated in pregnancy, such as angiotensin-converting enzyme inhibitors (ACEIs) and angiotensin receptor blockers (ARBs), should be discontinued prior to conception and substituted with appropriate alternatives to reduce fetal harm. Diuretics should only be used in pregnant patients with acute decompensated heart failure. Patients should be monitored closely with fetal ultrasound and maternal echocardiography as needed.

Angiotensin-Converting Enzyme Inhibitors, Angiotensin Receptor Blockers, and Angiotensin Receptor-Neprilysin Inhibitor

ACEIs and ARBs are first-line options in the management of chronic hypertension in nonpregnant patients [28]. In conjunction with beta-blockers, ACEIs, ARBs, and angiotensin receptor-neprilysin inhibitor (ARNI) also form the foundation of guideline-directed medical therapy in heart failure with reduced ejection fraction (HFrEF) [29]. However, human studies have demonstrated that blockade of the renin-angiotensin-aldosterone system during pregnancy can be teratogenic, particularly in the second and third trimesters [1, 30]. Adverse effects include oligohydramnios, patent ductus arteriosus, intrauterine growth restriction (IUGR), and neonatal anuric renal failure leading to fetal demise [30].

Therefore, ACEIs and ARBs are contraindicated and should be discontinued once pregnancy is detected [12]. Due to the valsartan component, the combination sacubitril/valsartan should also be discontinued.

The effects of first-trimester exposure to ACEIs are not as well defined. One retrospective cohort study of 209 infants exposed to ACEIs found an increased risk of major congenital malformations compared to infants exposed to other antihypertensives (RR 2.71, 95% CI 1.72–4.27) [31]. A later retrospective cohort study of 4107 women exposed to ACEIs during the first trimester did not find a significant increase in the risk of major congenital malformations after adjustment for confounders [32]. The RR for overall malformations was 0.89 (95% CI 0.75–1.06), cardiac malformations 0.95 (95% CI 0.75–1.21), and central nervous system malformations 0.54 (95% CI 0.26–1.11) [32]. A smaller prospective study of 250 subjects found no differences in major congenital malformations with exposure to ACEIs and ARBs in the first trimester, but did find significantly lower birth weights and gestational ages than healthy controls [30].

During lactation, captopril and enalapril are preferred ACEIs due to minimal excretion into breast milk [4, 22]. For similar reasons, losartan is the preferred ARB in breastfeeding [33]. No adverse events were reported in 16 cases of breastfeeding infants whose mothers were taking captopril or enalapril [25]. There are no published data regarding breastfeeding for sacubitril.

Aldosterone Antagonists

Spironolactone and eplerenone are used in the treatment of hypertension and for the reduction of morbidity and mortality in patients with HFrEF [28, 29]. Neither agent is well studied in pregnancy. In 1 of 10 human case reports, spironolactone may have caused undervirilization of a male infant, resulting in ambiguous genitalia [34]. In contrast, eplerenone has not demonstrated antiandrogenic or teratogenic effects in human case reports [34]. When choosing an agent, it is important to consider the agent's antiandrogenic effects, particularly in the first trimester when sex differentiation of the fetus occurs. Similar to other renin-angiotensin-aldosterone system inhibitors, aldosterone antagonists can cause oligohydramnios and infants born small for gestational age. Aldosterone antagonists' diuretic effects can decrease maternal plasma volume, reducing placental blood flow and potentially causing IUGR [34].

The active metabolite of spironolactone, canrenone, is excreted into breast milk at low concentrations that are thought to be clinically inconsequential [22, 33]. Case reports have demonstrated no adverse events in breastfeeding infants [25]. There are no lactation data for eplerenone, and it is best to avoid use while breastfeeding [25].

Aliskiren

Since direct renin inhibition exerts similar physiologic effects to ACEIs and ARBs for hypertension treatment, it is recommended to discontinue aliskiren when pregnancy is detected [35]. There are no human studies available in pregnancy or lactation.

Beta-Blockers

Beta-blockers are used in the treatment of arrhythmias, hypertension, and HFrEF [28, 29, 36]. These agents are the most commonly used antihypertensives in the first trimester of pregnancy, with approximately 0.5% of pregnant women exposed [37]. Beta-blockers have not demonstrated teratogenicity, but have been associated with IUGR, low birth weight, fetal bradycardia, and respiratory depression.

An increased risk of birth defects has been observed with maternal use of beta-blockers, but studies are inconsistent in their results [37, 38]. One meta-analysis of beta-blockers in early pregnancy showed no increased odds of all or major congenital malformations (OR 1, 95% CI 0.91–1.10), but did find increased odds of cardiovascular defects (OR 2.01, 95% CI 1.18–3.42), cleft lip/palate (OR 3.11, 95% CI 1.79–5.43), and neural tube defects (OR 3.56, 95% CI 1.19–10.67) [37]. In a Medicaid study, late pregnancy exposure to beta-blockers was associated with increased risk of neonatal hypoglycemia (OR 1.68, 95% CI 1.50–1.89) and bradycardia (OR 1.29, 95% CI 1.07–1.55) [38].

A cohort study comparing perinatal outcomes in pregnant women receiving beta-blockers to those receiving methyldopa for chronic hypertension demonstrated an increased risk for small for gestational age [< 10th percentile (OR 1.95, 95% CI 1.21–3.15), < 3rd percentile (OR 2.10, 95% CI 1.06–4.44)], and hospitalization during infancy (OR 2.17, 95% CI 1.09–4.34) [39]. The rates for preterm birth, stillbirth, and infant death were similar between the two groups. Similarly, a systematic review of 29 trials found that beta-blockers were associated with an increase in small for gestational age infants (RR 1.36, 95% CI 1.02–1.82) [40]. In contrast, a separate systematic review of antihypertensive therapies in pregnant women found no additional risk for small for gestational age with any antihypertensive vs. beta-blockers (RR 1.19, 95% CI 0.76–1.84) [41].

Carvedilol, metoprolol, and bisoprolol are the three beta-blockers indicated for the reduction of morbidity and mortality in HFrEF [29]. In pregnant women, β_1-selective agents (i.e., bisoprolol and metoprolol) are favored because they avoid β_2-mediated uterine contraction [42]. For peripartum cardiomyopathy (PPCM), metoprolol succinate is the preferred beta-blocker due to extensive experience and demonstrated safety [43]. Of note, metoprolol oral clearance is significantly higher during mid- and late pregnancy compared to 3 months' postpartum ($P < 0.05$); aggressive doses or frequency adjustment may be required if clinical effect is not achieved [44]. A study of first-trimester use of bisoprolol did not increase the risk for spontaneous abortion or major congenital malformations, but there was a higher rate of reduced birthweight and preterm birth (OR 1.90, 95% CI 1.17–3.11) [45].

In pregnant women with acute and chronic hypertension, labetalol, a nonselective α-, β_1-, and β_2-blocker, is a first-line treatment option [27, 28, 46]. An RCT of 114 pregnant women found that labetalol and nifedipine achieved similar levels of blood pressure control with no significant differences in adverse fetal outcomes [47]. Like metoprolol, clearance of labetalol may be increased throughout pregnancy, requiring dose adjustments [38, 48].

Due to extensive experience and reports of safety in pregnant patients, beta-blockers are considered first-line agents for multiple arrhythmia types during pregnancy [36]. For the acute management of supraventricular tachycardia (SVT) in pregnant patients, intravenous metoprolol and propranolol are reasonable choices, with esmolol as a second-line option [36, 49]. Esmolol is an intravenous, short-acting β_1-blocker that has been used safely in conjunction with lidocaine to attenuate the adrenergic response associated with intubation prior to cesarean section [50].

Atenolol should be avoided in pregnancy because it has been associated with IUGR and low placental weight [51]. There are no human studies available for nebivolol.

Beta-blockers should be used with caution in breastfeeding women [22, 33]. Metoprolol and propranolol appear in small amounts in human breast milk but have not been associated with adverse effects [22, 33]. Bisoprolol was undetectable in breast milk in one case report [25]. Atenolol was associated with cyanosis and bradycardia in other reports, and is not recommended in breastfeeding [12, 22, 25].

Calcium Channel Blockers

CCBs are often prescribed in the third trimester to treat hypertension and labor tocolysis, with approximately 1–2% of pregnant patients exposed during this time [52]. CCBs have not demonstrated teratogenicity in humans [53]. A retrospective cohort study of 505 infants exposed to CCBs (primarily third trimester) found no increased risk of congenital malformations [53]. However, investigators found an increased risk of hematological disorders, perinatal jaundice, and neonatal convulsions (RR 3.61, 95% CI 1.26–10.37). Of note, the study did not adjust for differences between pregnant patients exposed to CCBs and those who were unexposed. A more recent retrospective cohort study of 22,988 pregnancy exposed to CCBs in the final month of pregnancy found no increase in risk of neonatal seizures after adjustment for confounders (OR 0.95, 95% CI 0.70–1.30) [54].

CCBs can prevent uterine contractions and help prolong pregnancy in patients with preterm labor [52]. However, CCBs may also prevent effective uterine contractions required to compress the uterine vasculature after delivery, resulting in postpartum hemorrhage [52]. A retrospective cohort study ($n = 9750$) compared pregnant patients exposed to CCBs ($n = 1226$) to controls exposed to methyldopa or labetalol, which have not been implicated in postpartum hemorrhage [55]. Postpartum hemorrhage occurred in 2.2%

in the CCB group and 2.7% in the methyldopa/labetalol group, with no meaningful association in the propensity score matched analysis (OR 0.77, 95% CI 0.50–1.18) [55].

Dihydropyridine CCBs

In practice guidelines, extended-release nifedipine is a first-line option for the treatment of chronic hypertension in pregnancy because of safety data from randomized trials [27, 28, 47, 56]. For acute-onset, severe hypertension, immediate-release oral nifedipine is first-line therapy, particularly when intravenous access is unavailable [46]. Nicardipine is an intravenous CCB option that effectively reduces blood pressure in pregnant patients with severe preeclampsia. In a prospective, observational trial, 10 pregnant women with preeclampsia (mean gestational age 28 weeks) were treated with nicardipine after insufficient blood pressure control with oral methyldopa, nifedipine, or labetalol [57]. The nicardipine infusion increased cardiac output with no effects on maternal diastolic function, fetal perfusion, or hemodynamics.

Data with amlodipine and felodipine are limited to case reports for the treatment of maternal hypertension [12]. Five cases with amlodipine resulted in two healthy infants, one with growth restriction likely unrelated to amlodipine, one intrauterine death at 12 weeks, and one who exhibited neonatal seizures and subcutaneous fat necrosis [12]. A case series of three infants exposed to felodipine throughout pregnancy demonstrated growth restriction that was attributed to maternal hypertension [12].

Nondihydropyridine CCBs

Verapamil and diltiazem are more commonly used in arrhythmic disorders. Verapamil is considered safe and well tolerated during pregnancy [8]. Intravenous verapamil is recommended as a second-line option for acute treatment of SVT and atrial fibrillation in pregnant patients after adenosine and beta-blockers [1, 36]. Two case reports demonstrated effective termination of paroxysmal SVT in pregnant patients who delivered healthy infants at term [8]. For long-term prevention or control of SVT or atrial fibrillation, verapamil is an option if beta-blockers fail, except in patients with Wolff-Parkinson-White syndrome [1]. One case-control study treated 90 pregnant women with verapamil and found no maternal, fetal, or newborn adverse effects [8].

Guidelines recommend avoidance of diltiazem in pregnancy due to limited human data [1]. One case report of diltiazem initiation in the first month of pregnancy to treat angina resulted in healthy twins [12]. Of note, diltiazem has been used as a tocolytic agent [2, 12].

CCBs are excreted into breast milk at low levels and are compatible with breastfeeding, as nifedipine, amlodipine, diltiazem, and verapamil have not demonstrated harm to breastfeeding infants [22, 25]. There are no lactation data for felodipine [25].

Digoxin

Digoxin is used for rate control of SVT and to reduce hospitalizations and symptoms in HFrEF [29, 36]. It is well tolerated in pregnancy and has a large pool of observational studies demonstrating safety [2, 58]. For long-term management of SVT and atrial fibrillation, digoxin is recommended if beta-blockers fail or are not tolerated [1, 36]. Digoxin transfers readily across the placenta, and at an accelerating rate as pregnancy progresses [8]. Similar serum digoxin concentrations have been detected in infants born at term and the mother at 6 h after the last maternal dose [8]. As pregnancy progresses, digoxin dose requirements can decrease due to decreased albumin concentrations, but increased renal clearance may prevail and some patients may require an increased digoxin dose [59]. Monitoring of digoxin levels is critical, as toxic levels have been associated with miscarriage and fetal demise [8]. However, digoxin levels may become difficult to interpret because of circulating digoxin-like fragments in pregnant women that can falsely elevate digoxin levels; therefore, dosing should also be based on clinical effect [59].

Digoxin is considered safe in lactation, as it is excreted at low levels into breast milk and has not been associated with adverse effects in case reports of breastfed infants [22, 33].

Diuretics

Loop and thiazide diuretics have been studied in pregnancy and are not teratogenic. However, their use in pregnancy is controversial because diuresis can reduce plasma volume, cardiac output, and placental blood flow [2, 59]. Diuretics are not recommended in the treatment of hypertension in pregnancy because of decreased plasma volume in preeclampsia; however, they can be used for volume management in acute decompensated heart failure [1].

One pilot study that compared furosemide 20 mg daily to other antihypertensives in pregnant women with chronic hypertension found no significant differences in birth weight, gestational age, maternal complications, or prematurity [60]. A Danish and Scottish cohort study comparing fetal growth in pregnant women exposed to loop diuretics to unexposed women found higher birth weights in the diuretic-exposed group, with a mean difference of 104.7 g (95% CI 2.6–206.9). However, a high prevalence of diabetes (10.3%) in the Danish cohort may have confounded the results [61]. Among women who received thiazide diuretics, the risk of having an infant weighing < 2500 g was increased in both cohorts (OR 2.6, 95% CI 1.4–5; OR 2.4, 95% CI 0.8–7.8, respectively) [61].

The use of thiazide diuretics during pregnancy has been associated with neonatal jaundice, thrombocytopenia, maternal pancreatitis, and hyponatremia [12]. However, a meta-analysis of 7000 patients exposed to hydrochlorothiazide, bendroflumethiazide, chlorothiazide, and chlorthalidone during pregnancy did not find significant differences in adverse effects compared to unexposed pregnant patients [62].

Furosemide is considered the loop diuretic of choice during breastfeeding due to poor oral bioavailability and history of use in neonates for volume management [33]. In general, diuresis by thiazide or loop diuretics may decrease lactation [22, 25].

No data are available for potassium-sparing diuretics such as amiloride or triamterene.

Inotropes

Dobutamine and milrinone are used for inotropic support in patients with acute decompensated heart failure. Typically, hemodynamic instability requiring inotropic support necessitates emergent delivery of the infant [4]. There are no human data available for the use of dobutamine and milrinone in pregnancy. In patients with PPCM, the use of dobutamine is less favorable because it increases myocardial oxygen demand [43].

There are no data on breastfeeding with dobutamine and milrinone [25].

Dopamine has no data in pregnant or lactating women [12, 25].

Ivabradine

Ivabradine should be avoided during pregnancy and breastfeeding due to teratogenicity in animal studies [1]. Available human data with ivabradine in pregnancy are limited to case reports [12, 63]. The most recent report described a patient in her 17th week of pregnancy admitted for persistent sinus tachycardia with a maternal heart rate of 156 beats per minute (BPM) despite multiple aortic valve (AV) nodal blockers. After ivabradine initiation, her heart rate decreased to 90 BPM, with fetal heart rate decreasing from 156 to 148 BPM at the end of 1 week [63]. During follow-up, there were no instances of maternal tachycardia or fetal bradycardia, and a healthy infant was delivered. The infant was breastfed and did not exhibit signs of bradycardia or cardiac dysfunction.

Peripartum Cardiomyopathy—Bromocriptine and Pentoxifylline

Bromocriptine and pentoxifylline have shown promise in RCTs for recovery of left ventricular ejection fraction in postpartum women with PPCM [4]. European guidelines recommend bromocriptine, a prolactin inhibitor, in the management of PPCM [43]. Due to reports of myocardial infarction and stroke with bromocriptine, concomitant prophylactic anticoagulation is recommended [1, 43].

Limited data are available in breastfeeding. Since bromocriptine suppresses lactation, it should be used with caution [12, 22]. In a case report describing bromocriptine use to reduce milk production, there were no detectable milk levels or adverse reactions in the breastfeeding infant [25]. Pentoxifylline is poorly excreted into breast milk and unlikely to cause adverse effects in breastfeeding infants [25].

Ranolazine

Ranolazine is indicated for the treatment of chronic angina. It inhibits the late sodium current in cardiac tissue, and is also postulated to modulate fatty acid oxidation in ischemic tissue. There are no data for ranolazine in human pregnancy and lactation [12].

Vasodilators

Alpha-2 Agonists

Methyldopa and clonidine are centrally acting α-2 agonists that decrease sympathetic outflow and blood pressure. There is extensive clinical experience and safety data with methyldopa [28, 41, 64]. In a prospective observational cohort study, use of methyldopa in the first trimester was not associated with significantly increased risk of major birth defects (3.7% vs. 2.5%; OR 1.24, 95% CI 0.4–4.0) [64]. However, there was a higher risk of preterm birth (OR 4.11, 95% CI 2.4–7.1) and a trend toward a higher rate of spontaneous abortions in the methyldopa group [64]. Though methyldopa is considered first-line in hypertension management in pregnancy, it is less effective than beta-blockers or CCBs in preventing an episode of severe hypertension (RR 0.70, 95% CI 0.56–0.88) [28, 41]. Of note, methyldopa has been associated with maternal hepatotoxicity in case reports [2, 27].

Like methyldopa, clonidine has not demonstrated teratogenicity in pregnancy [65]. However, it may exhibit a heterogeneous hemodynamic response. A retrospective study of clonidine in 66 pregnant women found that 52% of patients had reduced vascular resistance, while 22% had decreased cardiac output [65]. Infants born to mothers with reductions in cardiac output were less likely to have a higher birth weight than mothers with reductions in vascular resistance (RR 0.81, 95% CI 0.66–0.98). In addition, these infants had a higher rate of birth weight <10th percentile (41% vs. 8.8%, $P=0.008$).

In breastfed infants, methyldopa is excreted at low levels into breast milk and has not demonstrated adverse effects in case reports [22, 25]. Conversely, clonidine is found in high serum concentrations in breastfed infants [25]. It also exhibits dose-related effects on oxytocin and prolactin secretion. A group of nine infants who were exposed to clonidine through breast milk had no observed side effects [25]. However, a separate case described a 2-day-old infant who presented with drowsiness, hypotonia, potential generalized seizures, and apneic episodes that all resolved within 24 h of discontinuation of breastfeeding [25].

Alpha-Blockers

Alpha-blockers, including doxazosin and prazosin, competitively inhibit postsynaptic α-1 receptors, resulting in arterial vasodilation. They are considered second-line agents in the treatment of hypertension, with prazosin being the most studied in pregnancy [28, 35]. In an observational study of severe hypertension, 44 pregnant women received prazosin plus beta-blockers for up to 14 weeks without drug-related fetal adverse effects [12]. Two case reports describing third-trimester treatment of pheochromocytoma resulted in one healthy infant and one underweight infant delivered by cesarean section at 33 weeks [12]. Furthermore, a single-center study randomized 145 pregnant women taking methyldopa to receive prazosin or nifedipine to determine the best second-line antihypertensive [66]. Median gestational ages at recruitment were 28 ± 3 weeks and 27 ± 4 weeks, respectively. No differences in perinatal outcomes were observed, except for an increased risk of all-cause intrauterine death in patients randomized to prazosin (7 vs. 1; RR 0.14, 95% CI 0.02–1.09, $P=0.03$). Two case reports exist for doxazosin for pheochromocytoma at the end of the second trimester [12, 67]. One infant was prematurely delivered by cesarean section while the second pregnancy resulted in a healthy infant. There are no data on the effects of alpha-blockers on nursing infants [12, 25].

Hydralazine

Hydralazine can be used for hypertension or in combination with isosorbide dinitrate for HFrEF [1, 28]. In pregnancy, intravenous hydralazine is an option to treat acute-onset, severe hypertension, but has been associated with perinatal events such as maternal hypotension (SBP < 90 mmHg) and reflex tachycardia, as well as fetal tachyarrhythmias [1, 46]. Hydralazine can also cause a maternal lupus-like syndrome and neonatal thrombocytopenia, and is no longer considered a first-line treatment for hypertension in pregnancy [1, 12]. A meta-analysis of 35 trials of 3573 pregnant women with hypertension found that fewer women had persistent high blood pressure when treated with CCBs compared to hydralazine (8% vs. 22%, RR 0.37, 95% CI 0.21–0.66) [68]. However, in comparisons against labetalol, there were

insufficient data to make comparisons and determine the best antihypertensive choice in pregnancy.

Hydralazine is excreted minimally into milk, and is considered safe for breastfeeding based on case reports [22, 25, 33].

Minoxidil

Oral minoxidil is a vasodilator that can be used as adjunctive therapy for hypertension. There are no human data for oral minoxidil in pregnancy, but one case report of topical minoxidil for hair loss demonstrated significant fetal brain and cardiovascular malformations [69]. Minoxidil is excreted into breast milk at levels that follow maternal levels. One case report of a breastfed infant did not describe adverse effects associated with minoxidil exposure [25].

Nitrates

Long-acting nitrates are used in pregnant patients with preeclampsia and reduce maternal blood pressure without altering cerebral perfusion pressure [35]. However, oral administration of nitrates in pregnant patients has led to methemoglobinemia, alteration in embryonic cells, and malignant transformation [70]. Maternal nitrate exposure through drinking water has been linked to spontaneous abortions, IUGR, and other birth defects, although a direct relationship has not been confirmed [70]. A case-control study found that exposure to nitrates in drinking water above the 45 mg/L maximum contaminant level was associated with increased risk of anencephaly (OR 4.0, 95% 1.0–15.4) [71].

Intravenous nitroglycerin is recommended for hypertension with pulmonary congestion and preeclampsia [2]. Data on intravenous nitroglycerin use in pregnancy are limited to case reports, which have not demonstrated maternal, fetal, or neonatal harm [12]. Sublingual nitroglycerin has not been studied in pregnancy, but a study of 20 pregnant women treated with sublingual isosorbide dinitrate for hypertension found no significant differences in fetal heart rate (HR) or movements [72]. Sodium nitroprusside is reserved for refractory hypertension due to concerns for cyanide toxicity in the fetus [2]. However, case reports describe successful use with adverse events limited to transient fetal bradycardia [12].

Nitrates should be used with caution in breastfeeding, as there are limited human data outside of maternal exposure to contaminated water [4]. No adverse infant effects were reported with maternal use of topical nitroglycerin [25]. Sodium nitroprusside is unlikely to enter breast milk due to its short half-life, but its metabolites thiocyanate and cyanide enter breast milk and may cause toxicity [25] (**Table 3**).

TABLE 3 Medications for Hypertension, Heart Failure, and Angina in Pregnancy

Class	Drug	Prior FDA Category[a]	Prescription Labeling Risks	Pharmacokinetic Considerations and Drug Transfer	Compatibility[b]
ACEIs	Benazepril Lisinopril Enalapril Ramipril	C (first trimester) D (second and third trimester)	Increases risk of fetal and neonatal morbidity and mortality (benazepril and lisinopril) No teratogenicity in animal studies at supratherapeutic doses (enalapril and ramipril)	25–60% Oral bioavailability Protein binding: Lisinopril (0%), benazepril (96%), and ramipril (73%) Placenta: Yes, in humans Breast milk: Yes, in rats [benazepril (<0.1% of dose), lisinopril]; in humans (enalapril); no (ramipril)	
	Captopril	D	Increases risk of fetal and neonatal morbidity and mortality	75% Oral bioavailability 25–30% Protein bound Placenta: Yes, in humans Breast milk: Yes, in humans (1% of dose)	
ARBs	Candesartan Losartan Valsartan	D	Increases risk of fetal and neonatal morbidity and mortality	15–33% Oral bioavailability 95–99% Protein bound Placenta: Yes, in rats Breast milk: Yes, in rats	
ARNI	Sacubitril/valsartan	None	Increased embryo-fetal lethality in rats at supratherapeutic doses Fetal hydrocephaly at supratherapeutic doses	60% Oral bioavailability 94–97% Protein bound Placenta: Yes, in rabbits Breast milk: Yes, in rats	
Aldosterone antagonists	Eplerenone	B	No teratogenic effects at supratherapeutic doses in rats or rabbits Increased rabbit fetal resorption and postimplantation loss	69% Oral bioavailability 50% Protein bound Placenta: Yes, in rats and rabbits Breast milk: Yes, in rats	
	Spironolactone	C	No teratogenic or embryotoxic effects observed in mice, but supratherapeutic doses caused increased rate of resorption in rabbits Feminization of male rat fetuses Endocrine dysfunction in female rate fetuses	95% Oral bioavailability 90% Protein bound Placenta: Yes, in rats Breast milk: Yes, in rats	

Continued

		Pharmacokinetics	Fetal/Neonatal effects	Category
Direct renin inhibitor	Aliskiren	2.5% Oral bioavailability Placenta: Yes, in rabbits Breast milk: Unknown; unlikely	Increases risk of fetal and neonatal morbidity and mortality Decreased fetal birth weight at supratherapeutic doses in rabbits	D
Beta-blockers	Atenolol	50% Oral bioavailability 6–16% Protein bound Placenta: Yes, in humans Breast milk: Yes, in humans	Risk for hypoglycemia, bradycardia, infants born SGA Dose-related increase in embryo/fetal resorption at supratherapeutic doses in rats	D
	Bisoprolol	80% Oral bioavailability 30% Protein bound Placenta: Yes, in rats Breast milk: Yes, in rats	Increased fetotoxicity in rats and embryolethal in rabbits at supratherapeutic doses	C
	Carvedilol	25% Oral bioavailability 98% Protein bound Placenta: Yes, in rats Breast milk: Yes, in rats	Increased postimplantation loss, delayed skeletal development, and decreased fetal body weight at supratherapeutic doses in animal studies No adverse outcomes at clinically relevant doses in animals	C
	Esmolol	55% Protein bound Half-life: 9 min Placenta: Unknown: expected but limited Breast milk: Unknown; expected but limited	No teratogenicity at supratherapeutic doses in animal studies	C
	Labetalol	25% Oral bioavailability 50% Protein bound Increased oral clearance mid- to late pregnancy Placenta: Yes, in humans Breast milk: Yes, in humans (0.005% of dose)	Neonatal bradycardia, hypoglycemia, and respiratory depression No teratogenicity at supratherapeutic doses in animal studies	C

TABLE 3 Medications for Hypertension, Heart Failure, and Angina in Pregnancy—cont'd

Class	Drug	Prior FDA Category[a]	Prescription Labeling Risks	Pharmacokinetic Considerations and Drug Transfer	Compatibility[b]
	Metoprolol	C	Increase postimplantation loss and decreased neonatal survival at supratherapeutic doses in rats	77% Oral bioavailability 12% Protein bound Increased oral clearance mid- to late pregnancy Placenta: Yes, in rats Breast milk: Yes, in humans	
	Nebivolol	C	Decreased birth weight, prolonged gestation, dystocia, and increased fetal deaths and stillborn pups at supratherapeutic doses in rats No adverse effects at supratherapeutic doses in rabbits	98% Protein bound Genetic polymorphisms in cytochrome P450 CYP2D6 lead to unpredictable maternal plasma nebivolol levels Placenta: Yes, in rats Breast milk: Yes, in rats	
	Propranolol	C	Embryotoxicity and neonatal toxicity at supratherapeutic doses in rats	50% Oral bioavailability 90% Protein bound Placenta: Yes, in humans Breast milk: Yes, in humans	
Calcium channel blockers, dihydropyridine	Amlodipine	C	No evidence of teratogenicity or other embryo/fetal toxicity at supratherapeutic doses in animal studies; intrauterine deaths increased in rats	64–90% Oral bioavailability 95% Protein bound 90% Metabolized to inactive metabolites (hepatic) Placenta: Unknown; expected Breast milk: Unknown; expected but limited	
	Felodipine	C	Digital anomalies at supratherapeutic doses in rabbits and monkeys	20% Oral bioavailability >99% Protein bound Placenta: Yes, in animals Breast milk: Unknown; expected but limited	

Drug	Category	Effects	Pharmacokinetics
Nicardipine	C	No evidence of teratogenicity or embryolethality at supratherapeutic doses in rats or rabbits Reduced birth weight, neonatal survival, and reduced neonatal weight gain	35% Oral bioavailability >95% Protein bound Placenta: Unknown; expected Breast milk: Yes, in humans
Nifedipine	C	Increased fetotoxic effects and decreased neonatal survival at supratherapeutic doses in animal studies Placentotoxic effects in monkeys at equivalent or lower than human doses	Variable bioavailability: 40–77% for immediate-release capsule; 65–89% extended release 92–98% Protein bound Accelerated metabolism through CYP 3A4 during pregnancy Placenta: Yes, in animals Breast milk: Yes, in humans
Calcium channel blockers, non-dihydropyridine			
Diltiazem	C	Fetal lethality at supratherapeutic doses in animal studies	40% Oral bioavailability 70–80% Protein bound Placenta: Unknown Breast milk: Yes, in humans (0.9% of dose)
Verapamil	C	Embryocidal and retarded fetal growth at supratherapeutic doses in rats	20–35% Oral bioavailability 90% Protein bound Placenta: Yes, in humans Breast milk: Yes, in humans (<0.1% of dose)
Digoxin	C	No animal or human studies available	60–80% Oral bioavailability 25% Protein bound Digoxin renal clearance increases during pregnancy Placenta: Yes, in humans Breast milk: Yes, in humans

Continued

TABLE 3 Medications for Hypertension, Heart Failure, and Angina in Pregnancy—cont'd

Class	Drug	Prior FDA Category[a]	Prescription Labeling Risks	Pharmacokinetic Considerations and Drug Transfer	Compatibility[b]
Diuretics, thiazide, or thiazide-like	Chlorothiazide (IV)	C	No evidence of harm in animal studies at supratherapeutic doses Risk of fetal or neonatal jaundice, thrombocytopenia	Placenta: Yes, in humans Breast milk: Yes, in humans	
	Hydrochlorothiazide	B	No evidence of harm in animal studies at supratherapeutic doses Risk of fetal or neonatal jaundice, thrombocytopenia	65–75% Oral bioavailability 40–68% Protein bound Placenta: Yes, in humans Breast milk: Yes, in humans	
	Metolazone	B	No evidence of harm at supratherapeutic doses in animal studies Risk of fetal or neonatal jaundice, thrombocytopenia	Variable absorption in heart failure Placenta: Yes, in humans Breast milk: Yes, in humans	
Diuretics, loop, and potassium-sparing	Bumetanide	C	No evidence of teratogenicity or embryocide at supratherapeutic doses in mice Moderate growth retardation and delayed ossification in rabbits and rats One investigation in humans did not indicate adverse effects on fetus	~ 100% Oral bioavailability 94% Protein bound Placenta: Yes, in humans Breast milk: Unknown; likely	
	Furosemide	C	Unexplained maternal deaths and abortions in rabbits at supratherapeutic doses	60% Oral bioavailability 91–99% Protein bound Placenta: Unknown Breast milk: Yes, in humans	
	Torsemide	B	No evidence of fetotoxicity or teratogenicity in animal studies	80% Oral bioavailability > 99% Protein bound Placenta: Unknown Breast milk: Unknown; likely	

Drug	Category	Effects	Pharmacokinetics
Triamterene	C	No evidence of harm at supratherapeutic doses in animal studies. Risk of fetal or neonatal jaundice, thrombocytopenia	67% Protein bound. Placenta: Yes, in humans. Breast milk: Yes, in humans
Inotropes — Dobutamine	B	No evidence of fetal harm at supratherapeutic doses in animal studies	Half-life: 2 min. Placenta: Unknown. Breast milk: Unknown; expected but limited
Milrinone	C	No evidence of teratogenicity at supratherapeutic levels in animal studies	70% Protein bound. Half-life: 2.4h. Placenta: Unknown. Breast milk: Unknown; not orally absorbed
Dopamine	C	No teratogenic or embryotoxic effects at supratherapeutic doses in animal studies	Half-life: 2 min. Placenta: Unknown. Breast milk: Unknown; not orally absorbed
Ivabradine	None	Embryo-fetal toxicity and cardiac teratogenic effects observed in animal studies at 1–3 times human exposures. Females should use effective contraception	40% Oral bioavailability. 70% Protein bound. Placenta: Yes, in rats and rabbits. Breast milk: Yes, in rats
Bromocriptine	B	Dose-dependent harm at supratherapeutic doses in animal studies. No increased harm in human pregnancies	28% Oral bioavailability. 90–96% Protein bound. Placenta: Unknown; expected but limited. Breast milk: Yes, in humans
Pentoxifylline	C	No evidence of fetal malformation at supratherapeutic doses in animal studies	10–30% Oral bioavailability. Extensive first-pass metabolism with two active metabolites. Placenta: Unknown; expected. Breast milk: Yes, in humans

Continued

TABLE 3 Medications for Hypertension, Heart Failure, and Angina in Pregnancy—cont'd

Class	Drug	Prior FDA Category[a]	Prescription Labeling Risks	Pharmacokinetic Considerations and Drug Transfer	Compatibility[b]
	Ranolazine	C	Supratherapeutic doses in rats and rabbits produced offspring with reduced fetal weight and ossification. No harm was observed at standard doses. No adverse effects in nursing pups of lactating rats receiving maximum human doses	76% Oral bioavailability 62% Protein bound Placenta: Unknown; expected Breast milk: Yes, in rats	
Alpha-2-agonists	Methyldopa	B	No evidence of fetal harm at supratherapeutic doses in animal studies	70% Oral bioavailability Placenta: Yes, in humans Breast milk: Yes, in humans (<0.2% of maternal dose)	
	Clonidine	C	No teratogenic or embryotoxic effects at supratherapeutic doses in rabbits Increased resorptions at supratherapeutic in rats and mice	70–80% Oral bioavailability Placenta: Yes, in humans Breast milk: Yes, in humans (66% of maternal serum level)	
Alpha-blockers	Doxazosin Prazosin Terazosin	C	No teratogenicity or skeletal abnormalities were observed in rats and rabbits. Reduced fetal survival, delayed postnatal development, decreased fetal weight, and decreased litter sizes were found with supratherapeutic doses in these studies	90–99% Protein bound Placenta: Yes, in humans (umbilical cord concentrations: doxazosin 83% 12h after last dose and prazosin 9–23% 8–15h after last dose) Breast milk: Yes, in humans (doxazosin <1% and prazosin ≤3% of dose)	
Hydralazine		C	Teratogenic (cleft palate and malformations of facial and cranial bones) at supratherapeutic doses in animal studies	87% Protein bound Placenta: Yes, in humans Breast milk: Yes, in humans	

Drug	Category	Effects	Pharmacokinetics	
Minoxidil	C	No evidence of teratogenic effects in rats and rabbits Neonatal hypertrichosis in humans	90% Oral bioavailability 0% Protein bound Placenta: Unknown, expected Breast milk: Yes, in humans	
Nitrates				
Isosorbide mononitrate/dinitrate	B/C	Stillbirth and neonatal death at supratherapeutic doses in rats Dinitrate: Dose-related embryotoxicity in rabbits	Bioavailability: Mononitrate (100%), dinitrate (10–90%, undergoes extensive first-pass hepatic metabolism) Protein binding: Mononitrate (4%) Placenta: Unknown; expected Breast milk: Unknown; expected	
Nitroglycerin	C	No animal or human studies conducted No evidence of harm with topical nitroglycerin ointment at supratherapeutic doses in animal studies	40% Oral bioavailability 60% Protein bound Half-life: 3 min Placenta: Unknown; expected but limited Breast milk: Unknown; expected but limited	
Sodium nitroprusside	C	Evidence of fatal cyanide toxicity at supratherapeutic doses in fetuses of pregnant ewes	Not orally absorbed Half-life: SNP: 2 min; cyanide: 7.3 h; thiocyanate: 3 days Placenta: Yes, in sheep Breast milk: Yes, in humans (27–50% maternal serum levels)	

[a] A = generally acceptable; controlled studies in pregnant women showed no evidence of fetal risk; B = may be acceptable; either animal studies showed no risk, or animal studies showed minor risks and human studies showed no risks; C = use with caution if benefits outweigh risks; animal studies showed risk and human studies unavailable, or no studies done; D = use in life-threatening emergencies when no safer drug available; positive evidence of human fetal risk; X = do not use in pregnancy; risks involved outweigh potential benefits, and safer alternatives exist.

[b] ● = Generally safe; ◐ = Limited human data/use with caution; ● = Avoid use/contraindicated; ○ = No adequate human data; ⚭ = Pregnancy; = Lactation. ACEIs, angiotensin-converting enzyme inhibitors; ARBs, angiotensin receptor blockers; ARNI, angiotensin receptor-neprilysin inhibitor; CYP, cytochrome P450; FDA, Food and Drug Administration; IV, intravenous; SGA, small for gestational age.

Pulmonary Hypertension

Pulmonary hypertension and its subset, pulmonary arterial hypertension (PAH), are high-risk diseases, with guidelines recommending to avoid pregnancy due to high maternal mortality [1, 73]. Furthermore, many of the medications utilize for PAH are contraindicated in pregnancy based on animal studies. If a patient is to become pregnant, consultation with a multidisciplinary team, including a pulmonary hypertension expert, is strongly advised.

Endothelin Receptor Antagonists (ERAs)

ERAs (e.g., ambrisentan, bosentan, and macitentan) block the physiological effects of smooth muscle vasoconstriction in the pulmonary vasculature and are approved for use in PAH. In humans, data in pregnancy and lactation are limited to case reports with bosentan [12, 25]. One patient received bosentan throughout gestation, while two others discontinued therapy upon pregnancy confirmation. In all cases, mothers gave birth to healthy infants. The patient who continued bosentan also nursed her infant without adverse events.

Due to teratogenicity observed in animal studies, guidelines and prescription labeling contraindicate ERAs in pregnancy [1, 74, 75]. The FDA's Risk Evaluation and Mitigation Strategies (REMS) program requires females of reproductive potential to have negative pregnancy tests before initiation, during therapy, and 1 month after discontinuation, and to use reliable contraception during treatment and for 1 month following treatment discontinuation.

Phosphodiesterase-5 (PDE-5) Inhibitors

PDE-5 inhibitors (e.g., sildenafil and tadalafil) increase cyclic guanosine monophosphate (cGMP) levels in the nitric oxide pathway, promoting relaxation of pulmonary smooth muscle. The few case reports of sildenafil use during pregnancy resulted in healthy infants [12]. However, the extent and duration of exposure varied widely. PDE-5 inhibitors are recommended in pregnant patients with PAH and normal right ventricular function [1, 2, 75].

Studies indicate that sildenafil may provide a novel mechanism for treating fetal growth restriction and preeclampsia by increasing uteroplacental blood flow [76]. A meta-analysis of 22 animal studies and two human RCTs demonstrated significant improvements in fetal growth in patients with fetal growth restriction [ratio of means (ROM) 1.10, 95% CI 1.06–1.13] [76]. These effects were not observed in healthy pregnant groups (ROM 1.03, 95% CI 0.99–1.06, $P = 0.006$). In an RCT of 100 patients, patients who received sildenafil had significantly increased pregnancy duration (14.4 days vs. 10.4 days, $P = 0.008$) [77]. There were no differences in adverse effects or maternal or neonatal outcomes.

The larger STRIDER (Sildenafil TheRapy In Dismal prognosis Early-onset fetal growth Restriction) trials randomized patients to sildenafil or placebo for fetal growth restriction [78]. In the UK STRIDER trial, there was no difference observed in the primary end point of median time from randomization to delivery (17 vs. 18 days, $P = 0.23$) [78]. No differences in adverse effects or maternal or neonatal outcomes were observed. However, the Dutch STRIDER trial was stopped early due to an unexpected excess in neonatal mortality (27% vs. 14%; RR 1.87, 95% CI 0.91–3.84) and persistent pulmonary hypertension (27% vs. 5%) [79]. Pooled analyses of publicly presented neonatal mortality data from the United Kingdom, the Netherlands, and New Zealand/Australia were nonsignificantly higher for sildenafil compared to placebo (19.3% vs. 13.1%; RR 1.49, 95% CI 0.90–2.47).

One case report of maternal sildenafil use showed no adverse effects in the nursing infant [12, 25]. Of note, sildenafil is utilized in neonatal pulmonary hypertension.

Prostacyclins

Prostacyclins (e.g., epoprostenol, iloprost, and treprostinil) are naturally occurring hormones that vasodilate peripheral and pulmonary vasculature and inhibit platelet aggregation. Guidelines recommend inhaled or intravenous prostacyclins for World Health Organization (WHO) class III–IV patients with pulmonary hypertension and right ventricular dysfunction [1, 75].

Data regarding prostacyclins in pregnancy are limited to heterogeneous case reports [12]. No adverse effects or congenital abnormalities were observed in six reports with iloprost. Several reports with epoprostenol for the treatment of pulmonary hypertension in pregnancy and preeclampsia have demonstrated its safety and efficacy, with adverse fetal effects attributed to the mother's disease [12].

One case report exists for prostacyclins in lactation [25]. A woman was initiated on intravenous treprostinil at 32 weeks' gestation. A healthy infant was born and nursed for 1 year without adverse events. Of note, epoprostenol is used for the treatment of neonatal pulmonary hypertension.

Riociguat

Riociguat catalyzes synthesis of cGMP in the nitric oxide pathway, promoting pulmonary smooth muscle vasodilation. It is approved for use for PAH and chronic-thromboembolic pulmonary hypertension. No data exist in human pregnancy or lactation [12]. Due to teratogenicity observed in animal studies, guidelines and prescription labeling contraindicate riociguat in pregnancy, with a REMS program similar to that of ERAs [1, 74, 75] (**Table 4**).

TABLE 4 Medications for Pulmonary Hypertension in Pregnancy

Class	Drug	Prior FDA Category[a]	Prescription Labeling Risks	Pharmacokinetic Considerations and Drug Transfer	Compatibility[b]
Endothelin receptor antagonists	Ambrisentan Bosentan Macitentan	X	Teratogenic in animals at supratherapeutic doses Females of reproductive potential should have a negative pregnancy test before starting, monthly during treatment, and 1 month after discontinuation, in addition to contraception use	98–99% Protein bound Half-life: Bosentan (5–8h); macitentan (16–48h) Placenta: Unknown; expected but limited Breast milk: Unknown; expected but limited	
PDE-5 inhibitors	Sildenafil Tadalafil	B	No evidence of harm was observed with supratherapeutic doses in animals	94–96% Protein bound Placenta: Yes, in humans (sildenafil) Breast milk: Yes, in humans and rats (minimal)	
Prostacyclins	Epoprostenol	B	Limited published data in humans have not established an association with birth defects or adverse fetal outcomes No adverse fetal outcomes observed in animals receiving supratherapeutic doses	Not orally absorbed due to rapid degradation Half-life: 6 min Placenta: Unknown; expected but limited Breast milk: Unknown; expected but limited	
	Iloprost	C	Fetal structural abnormalities observed in rats when given intravenously but not orally; no adverse effects when given intravenously in rabbits or monkeys Higher mortality observed in pups of lactating rats	63% Oral bioavailability (via inhalation) 60% Protein bound Half-life: 20–30 min Placenta: Unknown; expected but limited Breast milk: Yes, in rats	
	Treprostinil	B	No adverse fetal outcomes in multiple rat studies with supratherapeutic doses of subcutaneous administration; increased incidence of fetal skeletal variations associated with maternal toxicity in rabbits No data by inhaled administration	64–72% Oral bioavailability via inhalation 91–96% Protein bound Half-life: 4h Placenta: Unknown; expected but limited Breast milk: Unknown; expected but limited	

Continued

TABLE 4 Medications for Pulmonary Hypertension in Pregnancy—cont'd

Class	Drug	Prior FDA Category[a]	Prescription Labeling Risks	Pharmacokinetic Considerations and Drug Transfer	Compatibility[b]
	Treprostinil diolamine	C	At supratherapeutic doses in rats and rabbits, a decrease in pregnancy rate and mean number of live births, along with increases in postimplantation loss and fetal skeletal malformations were observed	17% Oral bioavailability Placenta: Unknown; expected but limited Breast milk: Unknown; expected but limited	
Riociguat		X	Teratogenic in animals at supratherapeutic doses Females of reproductive potential should have a negative pregnancy test before starting, monthly during treatment, and 1 month after discontinuation, in addition to contraception use	94% Oral bioavailability 95% Protein bound Placenta: Unknown; expected but limited Breast milk: Yes, in rats	
Selexipag		None	No adverse fetal effects observed at supratherapeutic doses in animals	49% Oral bioavailability 99% Protein bound Placenta: Unknown; expected but limited Breast milk: Yes, in rats	

[a] *A = generally acceptable; controlled studies in pregnant women showed no evidence of fetal risk; B = may be acceptable; either animal studies showed no risk, or animal studies showed minor risks and human studies showed no risks; C = use with caution if benefits outweigh risks; animal studies showed risk and human studies unavailable, or no studies done; D = use in life-threatening emergencies when no safer drug available; positive evidence of human fetal risk; X = do not use in pregnancy; risks involved outweigh potential benefits, and safer alternatives exist.*

[b] ● *= Generally safe;* ○ *= Limited human data/use with caution;* ● *= Avoid use/contraindicated;* ○ *= No adequate human data;* ▲ *= Pregnancy;* ▲ *= Lactation. FDA, Food and Drug Administration; PDE-5, phosphodiesterase-5.*

Anticoagulants

Anticoagulation during pregnancy presents a unique balancing act, as pregnancy itself is a prothrombic state and anticoagulants carry increased risks of miscarriage and hemorrhagic complications [1, 2]. Currently recommended therapies are limited to low molecular weight heparin (LMWH), unfractionated heparin (UFH), and vitamin K antagonists (VKAs) such as warfarin. Important nuances, such as route of administration, pharmacokinetic properties, and fetal-maternal outcomes, exist between agents. Due to the high risk of maternal hemorrhage during delivery, careful planning is required. Please refer to Chapter 21 for specific management of anticoagulants during pregnancy and delivery.

Direct Oral Anticoagulants (DOACs)

DOACs are indicated for the treatment and prevention of venous thromboembolism and atrial fibrillation stroke prevention [80]. Four agents inhibit factor Xa (apixaban, betrixaban, edoxaban, and rivaroxaban) and one inhibits thrombin (dabigatran). The available human data in pregnancy are limited to case reports and pharmacovigilance analyses [12, 81, 82]. Per guidelines, DOACs are contraindicated during pregnancy [1]. No human data exist in lactation.

The largest description of DOACs in pregnancy is a review of 236 unique cases [82]. A majority of patients (84%) discontinued therapy within 2 months of initiation, with a maximum duration of 26 weeks. Only 140 pregnancies had sufficient outcome data (59.3%), with 69 live births (49.2%), 39 elective abortions (27.8%), and 31 miscarriages (22.1%). Most cases involved rivaroxaban ($n=105$), while others included apixaban ($n=12$), dabigatran ($n=13$), and edoxaban ($n=10$). Eight infants demonstrated fetal abnormalities, including facial dimorphism, mild hip dysplasia, abnormal limbs, and IUGR. All had been exposed to rivaroxaban in the first trimester. The authors labeled four of these abnormalities as possibly related to rivaroxaban exposure and the remaining as unlikely related. No fetal anomalies were reported with apixaban, dabigatran, or edoxaban. Of note, this review is limited by missing data and small sample sizes, making it difficult to draw clear conclusions for individual agents.

Direct Thrombin Inhibitors

Intravenous direct thrombin inhibitors (e.g., argatroban and bivalirudin) bind reversibly to the thrombin active site and are used in place of UFH or to treat heparin-induced thrombocytopenia. Data in pregnancy are limited to six case reports with argatroban, with exposures occurring in each trimester [12, 83]. In all cases, a healthy infant was delivered and no complications were attributable to the drug. No data exist in lactation.

Fondaparinux

Fondaparinux is a subcutaneously administered pentasaccharide that inhibits factor Xa. It is used for prophylaxis and treatment of venous thromboembolism and for heparin-induced thrombocytopenia. At least six case reports describe successful use of fondaparinux in pregnancy in all trimesters without fetal complications [12, 83]. No data are available in breastfeeding.

Heparin

UFH inhibits several activated clotting factors by binding antithrombin, and is widely used for the treatment and prevention of thrombotic disorders. Since UFH is not orally bioavailable and does not cross biological membranes, guidelines support its use in pregnancy [1, 12, 84].

However, a meta-analysis of anticoagulant regimens for mechanical heart valves in pregnancy found a high rate of fetal adverse events (7.6%, 95% CI 0.1–15) and valve thromboembolism (11.2%, 95% CI 2.8–19.6) with UFH monotherapy [85]. These results should be interpreted with caution as only four studies reported UFH monotherapy throughout pregnancy. It is important to note that extended durations of UFH therapy have been associated with maternal osteopenia [12].

Low Molecular Weight Heparins

LMWHs, including enoxaparin and dalteparin, inhibit factor Xa by binding antithrombin. They are also used for the treatment and prevention of thrombotic disorders. Guidelines recommend LMWHs with Xa monitoring for anticoagulation in pregnancy with venous thromboembolism or valvular disease [1, 80, 81, 86].

There is extensive experience with LMWHs in pregnancy [2, 12, 83]. A retrospective cohort study of 990 pregnancies found no increase in major congenital malformations due to enoxaparin exposure in the first trimester (OR 1.1, 95% CI 0.8–1.6), or increased risk of low birth weight in the third trimester (OR 1.1, 95% CI 0.8–1.4) [87]. A meta-analysis in mechanical heart valves found that LMWH regimens had the lowest risk of adverse fetal outcomes [ratio of average risk (RAR) 13.9%, 95% CI 3.7–29] compared to VKA (RAR 0.4, 95% CI 0.1–0.8) [88]. This effect was attenuated when compared to low-dose VKA regimens (RAR 0.9, 95% CI 0.3–2.1). However, maternal risk was higher with LMWH compared to VKA (16% vs. 5%; 3.2, 95% CI 1.5–7.5). A second analysis of 46 studies confirmed these results [85]. In patients who received LMWH in this study, no fetal or neonatal adverse events were identified. The estimated rate of live births was 92% (86.1–98), thromboembolism 8.7% (3.9–13.4), and maternal mortality 2.9% (0.2–5.7).

There are limited data available for LMWHs in lactation, as they are not orally absorbed [12]. Low amounts of anti-Xa activity were measured in the breast milk of lactating women receiving dalteparin.

Thrombolytics

Recombinant plasminogen activators produce local fibrinolysis in acute life-threatening conditions. Thus, thrombolytics in pregnancy should be reserved for patients with severe hypotension or shock [1, 2]. Guidelines state that in ACS, percutaneous intervention is favored and pregnancy is a contraindication to thrombolytic therapy [1, 2].

Several case reports for thrombolytics exist in pregnancy [12, 89]. One review identified 27 cases of intravenous and catheter-directed thrombolysis for massive pulmonary embolism in pregnancy [89]. Alteplase was the most common agent (52%), and gestational age at administration ranged from 8 to 35 weeks. Bleeding complications occurred in 37% of cases. Four fetal deaths and no maternal deaths were reported. Furthermore, one case report described the use of alteplase for ischemic stroke in a 36-year-old at 21 weeks' gestation [90]. A healthy infant was delivered at 38 weeks without complications.

No data are available in lactation [12, 89]. However, the likelihood of concomitant breastfeeding is nil.

Warfarin

Warfarin is a VKA that inhibits vitamin K epoxide reductase. This enzyme is responsible for reducing and activating vitamin K, an important coagulation cofactor. Warfarin crosses the placenta and has been associated with adverse fetal outcomes, including facial dysmorphism, limb hypoplasia, stippled epiphyses, cardiac defects, and central nervous system defects [1, 2, 81, 84]. Thus, women of reproductive potential should use effective contraception during treatment and for at least 1 month after the final dose.

However, some women require lifelong anticoagulation [1, 2, 81, 84]. For instance, in women with mechanical valves, warfarin is the preferred agent in pregnancy at a total daily dose ≤5 mg. Prior to a planned delivery, it is recommended to discontinue warfarin and initiate heparin. Women who take warfarin throughout pregnancy should be switched to UFH or LMWH at 36 weeks, or prior to planned delivery, to reduce the risk of fetal intracranial hemorrhage during vaginal delivery and reduce maternal delivery-related bleeding. A patient presenting for delivery on VKA must undergo C-section as there is no way to reverse VKA anticoagulation in the fetus. Please refer to Chapter 21 for specific management.

The correlation between warfarin dose and fetal complications was first observed in a retrospective cohort study of 43 women (58 pregnancies) with mechanical valves ($P < 0.0001$) [91]. Among patients taking a warfarin dose ≤5 mg (mean: 4 ± 0.8 mg), 5 of 33 (15.2%) pregnancies experienced a fetal complication (four spontaneous abortions and one growth retardation). In patients taking warfarin >5 mg (mean: 7.45 ± 0.9 mg), 22 of 25 (88%) pregnancies experienced a fetal complication (2 warfarin embryopathy, 18 spontaneous abortions, 1 stillbirth, and 1

ventricular septal defect). No maternal embolic or bleeding events were observed. A case series of four infants with fetal warfarin syndrome and mothers taking <5 mg highlights the continued risk in pregnancy [92].

A meta-analysis of 18 studies (eight prospective) observed the maternal and fetal outcomes of different anticoagulation strategies with mechanical valves in 800 pregnancies [88]. For the average risk of fetal outcomes, defined as spontaneous abortion, death, and presence of congenital defects, VKA regimens had the highest rates of adverse events (39.2%, 95% CI 27–52.1). However, maternal risk, defined as death, prosthetic valve failure, and thromboembolism, was lowest (5%, 95% CI 2.5–8.5). A second meta-analysis analyzing 46 articles and 2468 pregnancies found similar results [85]. While women receiving VKA had the lowest rate of live births (64.5%, 95% CI 48.8–80.2), they also demonstrated the lowest rates of thromboembolism (2.7%, 95% CI 1.4–4). Finally, in this study, the rate of fetal adverse events was not significantly elevated (2%, 95% CI 0.3–3.7).

Warfarin was studied in 15 nursing women and was not detected in breast milk or infant plasma [12]. The AAP classifies warfarin as compatible with breastfeeding [12, 22, 25] (**Table 5**).

Medications for Arrhythmia Management

In general, data on antiarrhythmic agent use in pregnancy tend to be limited to case reports and series, favoring older agents due to a greater availability of reports of safe use in pregnancy [36]. Antiarrhythmic drugs should be avoided in the first trimester; however, if benefits outweigh risks, clinicians should use the lowest recommended dose and monitor clinical response. For long-term management of highly symptomatic SVT, metoprolol, propranolol, and digoxin are recommended first-line options due to longer records of safety [36].

Adenosine

Adenosine is recommended for acute termination of SVT in pregnancy if vagal maneuvers are unsuccessful [36]. Similarly, adenosine is used to terminate narrow complex tachycardias in pregnant patients with Wolff-Parkinson-White syndrome, both before and during delivery [93]. Due to adenosine's ultrashort half-life, adverse fetal effects would not be expected. Dosages of up to 24 mg have been reported and well tolerated [94]. There are no data for adenosine in lactation [25].

TABLE 5 Anticoagulants in Pregnancy

Class	Drug	Prior FDA Category[a]	Prescription Labeling Risks	Pharmacokinetic Considerations and Drug Transfer	Compatibility[b]
Direct-acting oral anticoagulants	Apixaban	B	No increased risk of fetal malformations or toxicity observed in animals; increased incidence of maternal bleeding in rats	~50% Oral bioavailability 87% Protein bound Placenta: Yes, in humans Breast milk: Yes, in rats (12%)	
	Betrixaban	None	No adverse embryofetal or teratogenic effects observed in rats or rabbits at supratherapeutic doses, but maternal hemorrhage did occur	34% Oral bioavailability 60% Protein bound Placenta: Unknown; expected Breast milk: Unknown; expected	
	Dabigatran	C	At supratherapeutic doses in rats, increased incidence of delayed/ irregular bone ossification, dead offspring, and excess bleeding close to labor	3–7% Oral bioavailability 35% Protein bound Placenta: Unknown; expected Breast milk: Yes, in humans (minimal)	
	Edoxaban	C	No adverse developmental effects were observed at supratherapeutic doses in rats or rabbits; embryofetal toxicities did occur at maternally toxic doses	62% Oral bioavailability 55% Protein bound Placenta: Unknown; expected Breast milk: Yes, in rats	
	Rivaroxaban	C	Increased fetal toxicity, maternal bleeding, and maternal fetal death observed in animals at supratherapeutic doses	80–100% Oral bioavailability 92–95% Protein bound Placenta: Yes, in humans Breast milk: Yes, in humans (minimal)	
Direct thrombin inhibitors	Argatroban	B	No evidence of harm was observed with subtherapeutic doses in rats and rabbits	54% Protein bound Half-life: 39–51 min Placenta: Unknown; expected Breast milk: Yes, in rats	

Continued

CH
22

TABLE 5 Anticoagulants in Pregnancy—cont'd

Class	Drug	Prior FDA Category[a]	Prescription Labeling Risks	Pharmacokinetic Considerations and Drug Transfer	Compatibility[b]
	Bivalirudin	B	No evidence of harm was observed with supratherapeutic doses in rats and rabbits	Half-life: 25 min Placenta: Unknown; unlikely in normal renal function Breast milk: Unknown; unlikely in normal renal function	
Fondaparinux		B	No evidence of harm was observed with supratherapeutic doses in rats and rabbits	Very poor oral bioavailability Placenta: Yes, in humans (<10%) Breast milk: Yes, in rats	
Heparin		C	No adverse maternal or fetal outcomes observed at various doses and routes of administration in humans	Not orally absorbed Placenta: No Breast milk: No	
Low molecular weight heparin	Dalteparin Enoxaparin	B	No evidence of harm was observed with supratherapeutic doses in rats and rabbits Use preservative-free formulations without benzyl alcohol due to risk of fatal "gasping syndrome" in premature infants	Inactivated in the gut and not orally absorbed Placenta: No Breast milk: Yes, in humans and rats (minimal)	
Thrombolytics	Alteplase	C	Embryocidal in rabbits at human equivalent doses; no maternal or fetal toxicity at subtherapeutic doses	Not orally absorbed due to high molecular weights Half-life: 5–130 min Placenta: Unknown; unlikely Breast milk: Unknown, unlikely	
	Reteplase	C	Maternal and embryo toxicity in rabbits following multiple but not with single intravenous doses		
	Tenecteplase	C	Abortions observed in rabbits receiving supratherapeutic doses; no evidence of fetal abnormalities in rats		

TABLE 5 Anticoagulants in Pregnancy—cont'd					
Class	Drug	Prior FDA Category[a]	Prescription Labeling Risks	Pharmacokinetic Considerations and Drug Transfer	Compatibility[b]
Warfarin		X (D for mechanical valve)	Congenital malformations observed in 5% of pregnant women exposed during the first trimester; other adverse outcomes have been reported with second- and third-trimester exposure Females of reproductive potential should use effective contraception during treatment and for at least 1 month after final dose	79–100% Oral bioavailability 99% Protein bound Placenta: Yes, in humans Breast milk: Warfarin has not been detected in breast milk but other coumarins have	

[a] A = generally acceptable; controlled studies in pregnant women showed no evidence of fetal risk; B = may be acceptable; either animal studies showed no risk, or animal studies showed minor risks and human studies showed no risks; C = use with caution if benefits outweigh risks; animal studies showed risk and human studies unavailable, or no studies done; D = use in life-threatening emergencies when no safer drug available; positive evidence of human fetal risk; X = do not use in pregnancy; risks involved outweigh potential benefits, and safer alternatives exist.

[b] ● = Generally safe; ○ = Limited human data/use with caution; ● = Avoid use/contraindicated; ○ = No adequate human data; 🤰 = Pregnancy; 🤱 = Lactation. FDA, Food and Drug Administration.

Atropine

There are no human studies available for atropine in pregnancy and lactation [12]. It is noted that long-term use may decrease milk production [25].

Class Ia Antiarrhythmics

The class Ia antiarrhythmic agents block sodium channels and prolong the cardiac action potential, and are used in the treatment of supraventricular and ventricular arrhythmias, including Wolff-Parkinson-White syndrome [8].

Disopyramide

In addition to sodium channel-blocking properties, disopyramide has potent anticholinergic and negative inotropic effects [95]. While not teratogenic, it has been associated with low fetal weight at higher doses [8]. Several case reports describe premature uterine contractions caused by disopyramide and one case of abruptio placentae [8, 59]. A placebo-controlled study randomized 20 pregnant women to receive disopyramide or placebo for induction of labor [96]. At 48 h, 8 of 10 women in the disopyramide group

had delivered, compared to none in the control group ($P < 0.001$). The mean time to regular contractions was significantly shorter in the disopyramide group (4.15 ± 1.76 h vs. 56.13 ± 5.28 h, $P < 0.001$).

Disopyramide is excreted in breast milk, and has not been associated with adverse effects in breastfed infants [8].

Procainamide

There are few reports on the use of procainamide in pregnancy, but it is considered safe for the management of maternal and fetal arrhythmias [94]. Three case reports describe successful intrauterine cardioversion of fetal SVT with intravenous procainamide, with no observed adverse effects or teratogenicity [8]. There were inconsistent results for maternal and cord blood levels. Serum concentrations of procainamide and N-acetylprocainamide (NAPA) should be monitored for toxicity with long-term use [8, 94].

There are limited data for procainamide in breastfeeding. Based on one case report, a breastfed infant would receive 5–7% of the mother's weight-adjusted dosage of procainamide plus NAPA; these levels would be unlikely to cause adverse effects [8, 25].

Quinidine

Quinidine was among the first antiarrhythmic agents used to treat atrial fibrillation. It exhibits sodium channel-blocking effects at rapid heart rates and potassium channel-blocking effects at slower heart rates [95]. Quinidine has demonstrated fetal safety in numerous case reports over decades of use in the treatment of arrhythmias in pregnancy [94]. It has also been used successfully for transplacental treatment of fetal SVTs and atrial flutter [8]. Fetal side effects are rare, but there have been reports of fetal thrombocytopenia [8].

Quinidine is excreted into breast milk at low levels, and is considered safe in breastfeeding [12, 22, 25]. Monitoring of quinidine levels can rule out toxicity [25].

Class Ib Antiarrhythmics

Lidocaine and mexiletine are used for management of ventricular arrhythmias, and are generally well tolerated during pregnancy [94]. Lidocaine readily crosses the placenta within minutes of maternal intravenous administration, and is trapped in the more acidic fetal plasma [8]. A study of 57 pregnant women with lidocaine levels of 3 mg/L (therapeutic range: 1–5 mg/L) found that fetal lidocaine levels <2.5 mg/L did not cause central nervous system toxicity [97]. The authors concluded that maternal lidocaine levels <4 mg/L would likely avoid central nervous system toxicity in the fetus. Mexiletine is an oral agent with structural similarities to lidocaine. Individual case reports of mexiletine use in pregnancy have documented fetal bradycardia, small for gestational age, and neonatal hypoglycemia associated with its use [8].

Both lidocaine and mexiletine are excreted into breast milk, with negligible amounts of each drug being ingested by the breastfeeding infant [22, 25]. They are considered safe in lactation based on case reports of oral mexiletine and epidural lidocaine use in breastfeeding mothers [8, 12, 22].

Class Ic Antiarrhythmics

Flecainide and propafenone have been used effectively to treat both maternal and fetal tachycardias, including treatment and prevention of SVT in women with Wolff-Parkinson-White syndrome [36].

There are no randomized human studies of flecainide in pregnancy, but it has not demonstrated teratogenicity in observational data [58, 59, 94]. Flecainide has demonstrated greater efficacy than digoxin in the management of fetal arrhythmias, and is the treatment of choice in fetal SVT, especially in refractory cases or cases complicated by hydrops fetalis [8, 94]. In a meta-analysis of 10 studies evaluating antiarrhythmic agents for fetal tachyarrhythmia, digoxin achieved a lower rate of SVT termination compared to flecainide (OR 0.773, 95% CI 0.605–0.987), with no significant difference in maternal side effects between the two agents [58].

Two case reports of propafenone demonstrated no adverse effects or teratogenicity, with healthy infants delivered at term [8].

Limited data are available for flecainide and propafenone in breastfeeding. Since both are excreted at low levels into breast milk, they are unlikely to cause adverse effects in the infant [22, 25].

Class III Antiarrhythmics

Amiodarone

Amiodarone is an iodinated antiarrhythmic agent that exhibits properties of all Vaughn-Williams antiarrhythmic drug classes, and can be safely used in patients with structural heart disease [95]. Because of its association with fetal hypothyroidism and neurodevelopmental complications, amiodarone is contraindicated in pregnancy and should be avoided in women of reproductive age [2, 94]. Case reports of breastfed infants demonstrated no harm with maternal use; one separate report found transient signs of hypothyroidism in one breastfed infant [25].

Dronedarone

Dronedarone is structurally similar to amiodarone, but lacks iodine moieties. Due to adverse effects in animal studies, dronedarone is contraindicated in pregnancy and lactation [12, 25]. No data exist in humans.

Dofetilide and Ibutilide

Dofetilide and ibutilide are structurally similar to sotalol, but lack beta-blocking properties. The use of ibutilide in pregnancy is documented in case reports, in which pregnant women presenting with palpitations were successfully cardioverted to normal sinus rhythm, and delivered healthy infants at term [98]. There are no data available for dofetilide or ibutilide in lactation.

Sotalol

Sotalol displays beta-blocking and potassium channel-blocking properties, and is used in pregnancy to manage maternal and fetal arrhythmias [36]. It should be reserved for women without structural heart disease. One study utilized transplacental treatment of 14 fetuses with SVT (gestational age: 24–35 weeks), by first treating with digoxin and then oral sotalol 80–160 mg [99]. In all, 10 were successfully cardioverted. A meta-analysis evaluating antiarrhythmic treatment strategies for fetal SVT found no difference between sotalol and digoxin for termination of atrial fibrillation, but did find a higher incidence of maternal side effects with digoxin (OR 3.15, 95% 1.47–6.75, $I^2=0\%$) [58].

Sotalol is transferred into breast milk [22, 25]. However, two case reports reported no adverse events in nursing infants [25].

Isoproterenol

Isoproterenol is a β-agonist used for pharmacologic cardiac pacing to terminate electrical storm in patients with Brugada syndrome or torsades de pointes [100]. One case report documents the use of isoproterenol in a pregnant woman who had persistent premature ventricular contractions (PVCs)

when her HR was < 100 BPM. She failed defibrillation, intravenous lidocaine, metoprolol, and amiodarone, and progressed to ventricular fibrillation. Isoproterenol infusion was initiated to maintain her HR > 120 BPM, which successfully suppressed all PVCs and episodes of ventricular fibrillation. After a few days, she was weaned off without recurrence of PVCs. The patient received a dual-chamber implantable cardioverter-defibrillator, and delivered a healthy infant at term [100]. There are no data for isoproterenol in lactation [25] (**Table 6**).

TABLE 6 Medications for Arrhythmia Management in Pregnancy

Class	Drug	Prior FDA Category[a]	Prescription Labeling Risks	Pharmacokinetic Considerations and Drug Transfer	Compatibility[b]
Adenosine		C	No animal or human studies	Not orally absorbed Half-life: < 10 s Placenta: Unknown; unlikely Breast milk: Unknown; unlikely	
Atropine		B	No evidence of fetal harm in animal studies	Half-life: 3 h Placenta: Yes, in humans Breast milk: Yes, in humans	
Class Ia	Disopyramide	C	Decreased growth and survival of rat pups at supratherapeutic doses Can stimulate uterine contraction	Complete oral absorption 50–65% Protein bound Half-life increases with impaired renal function Placenta: Yes, in humans Breast milk: Yes, in humans	
	Procainamide	C	No animal or human studies	15–20% Protein bound Half-life: Procainamide: 2.5–4.7 h; NAPA: 6–8 h Placenta: Yes, in humans Breast milk: Yes, in humans	
	Quinidine	C	No animal or human studies	70–80% Oral bioavailability 80–88% Protein bound (decreased in pregnant women) Placenta: Yes, in humans Breast milk: Yes, in humans	
Class Ib	Lidocaine	B	No evidence of fetal harm at supratherapeutic doses in rat studies	60–80% Protein bound Half-life: 1.5–2 h Placenta: Yes, in humans Breast milk: Yes, in humans	

Continued

TABLE 6 Medications for Arrhythmia Management in Pregnancy—cont'd

Class	Drug	Prior FDA Category[a]	Prescription Labeling Risks	Pharmacokinetic Considerations and Drug Transfer	Compatibility[b]
	Mexiletine	C	No evidence of teratogenicity or impaired fertility at supratherapeutic doses in animal studies; increase in fetal resorption	90% Oral bioavailability 50–60% Protein bound Half-life: 10–12 h Placenta: Yes, in humans Breast milk: Yes, in humans (2% maternal dose)	
Class Ic	Flecainide	C	No teratogenic effects at supratherapeutic doses in rodent studies; delayed stenebral and vertebral ossification in rats	>90% Oral bioavailability 40% Protein bound Half-life: 3–5 days Placenta: Yes, in humans Breast milk: Yes, in humans (at 2.5 times plasma levels)	
	Propafenone	C	Case report data have not identified drug-associated risks of miscarriage, birth defects, or adverse maternal or fetal outcomes	3–40% Oral bioavailability >95% Protein bound Placenta: Yes, in humans Breast milk: Yes, in humans	
Class III	Amiodarone	D	Increased risk for cardiac thyroid, neurodevelopmental, neurological, and growth effects in neonates	50% Oral bioavailability 96% Protein bound Half-life: 60 days Placenta: Yes, in humans Breast milk: Yes, in humans	
	Dronedarone	X	Teratogenic in rats at maximum human dose, and in rabbits at half maximum human dose	4% Oral bioavailability >98% Protein bound Placenta: Yes, in rats Breast milk: Yes, in rats	?
	Dofetilide	C	Sternebral and vertebral abnormalities in rats and mice	>90% Oral bioavailability 60–70% Protein bound Placenta: Unknown; expected Breast milk: Unknown; expected	? ?
	Ibutilide	C	Teratogenic and embryocidal in rats	40% Protein bound Half-life: 6 h Placenta: Unknown; expected Breast milk: Unknown; expected	?
	Sotalol	B	Animal studies did not reveal teratogenic potential	90–100% Bioavailable Not protein bound Placenta: Yes, in humans Breast milk: Yes, in humans	

| | | | | Pharmacokinetic | |
| | | Prior FDA | Prescription Labeling | Considerations and Drug | |
Class	Drug	Category[a]	Risks	Transfer	Compatibility[b]
Beta-agonist	Isoproterenol	C	No animal or human studies	Half-life: 2–2.5 min Placenta: Unknown; expected but limited Breast milk: Unknown; expected but limited	

TABLE 6 Medications for Arrhythmia Management in Pregnancy—cont'd

[a] *A = generally acceptable; controlled studies in pregnant women showed no evidence of fetal risk; B = may be acceptable; either animal studies showed no risk, or animal studies showed minor risks and human studies showed no risks; C = use with caution if benefits outweigh risks; animal studies showed risk and human studies unavailable, or no studies done; D = use in life-threatening emergencies when no safer drug available; positive evidence of human fetal risk; X = do not use in pregnancy; risks involved outweigh potential benefits, and safer alternatives exist.*

[b] ◐ *= Generally safe;* ○ *= Limited human data/use with caution;* ● *= Avoid use/contraindicated;* ◑ *= No adequate human data;* 🤰 *= Pregnancy;* 🤱 *= Lactation.* FDA, Food and Drug Administration; NAPA, *N*-acetylprocainamide.

Conclusion and Future Directions

The safe and effective use of cardiovascular medications in pregnancy and lactation presents a unique challenge. Although several therapies have robust evidence, most are significantly lacking. Conflicting literature, guidelines, prescription labeling, and altered pharmacokinetics add further complications. This chapter has sought to present the available data to allow practitioners to make informed decisions when caring for women with cardiovascular disease during pregnancy and lactation.

During pregnancy, clinicians must balance appropriate treatment of the mother while limiting potential risks to the fetus. The AAP states that most drugs are safe in lactation, but clinicians and patients must carefully consider the benefits and risks of breastfeeding [22]. Breastfeeding infants should be monitored closely for adverse effects typically observed in adults. Future studies in pregnancy and lactation will provide necessary data to guide decision making in particular populations.

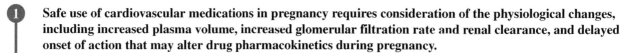

Key Points

1. Safe use of cardiovascular medications in pregnancy requires consideration of the physiological changes, including increased plasma volume, increased glomerular filtration rate and renal clearance, and delayed onset of action that may alter drug pharmacokinetics during pregnancy.

2. Prior prescription labeling and pregnancy categories did not fully describe the risks of medications in pregnancy and should no longer be used. Instead, a shared decision-making approach is essential, and patients should be counseled on the risks and benefits of using medications during pregnancy.

3. Data for medication use in pregnancy are often limited to case reports. However, registry studies and randomized trials are available for certain medications.

4. Figure 2 summarizes recommended and contraindicated medications for common cardiovascular disorders in pregnancy.

5. Most medications are considered safe in breastfeeding. However, limited evidence is available and infants should be monitored for side effects observed in adults.

6. Further studies are needed to improve our understanding of the safety and efficacy of pharmacotherapy in pregnancy and lactation.

Back to Clinical Case

The case presents a 33-year-old female with hypertension who is contemplating pregnancy. At the initial prepregnancy visit, it is important to counsel the patient on maternal and fetal risks of chronic hypertension during pregnancy. The clinician should also address maternal comorbidities and modifiable risk factors, including counseling on weight loss, diet, and exercise.

Prior to pregnancy, a woman with chronic hypertension should be evaluated for possible end-organ involvement. For baseline evaluation, it is recommended to check a basic metabolic panel (including serum creatinine and potassium levels), complete blood counts, transaminases, and spot urine protein/creatinine ratio. It is also reasonable to obtain a baseline electrocardiogram to assess for cardiac changes from long-standing hypertension. The goals of treatment for this patient would be: (1) decrease the maternal and fetal risks of severely elevated blood pressure during pregnancy; and (2) minimize the risk of teratogenicity with medications used to treat hypertension.

Since ACEIs are teratogenic and contraindicated in pregnancy, benazepril should be discontinued at this time to avoid fetal exposure. First-line options for the treatment of chronic hypertension in pregnancy include labetalol, nifedipine, and methyldopa. When managing blood pressure in pregnant patients, it is important to consider how the hemodynamic changes that can occur during pregnancy may affect management. For example, vasodilation of the systemic vasculature occurs in order to accommodate increases in plasma volume. In the first and second trimesters, systemic vascular resistance decreases by 30–50% from prepregnancy values. Low-dose aspirin therapy (< 150 mg) should be initiated at week 12 of gestation and continued until 36–37 weeks, since this patient is at risk for preeclampsia.

On follow-up, the patient developed postpartum cardiomyopathy. At this point, she should be transitioned from labetalol to a beta-blocker with a history of safety in pregnant patients, as well as a mortality benefit in HFrEF. She was transitioned from labetalol to long-acting metoprolol.

Diuretics, although one of the mainstays of heart failure therapy, can decrease maternal plasma volume and cardiac output, potentially compromising placenta blood flow. It is critical to be cautious, and reserve diuretics for patients with pulmonary congestion. Since the patient is exhibiting signs of pulmonary congestion and lower extremity edema, it is reasonable to use loop diuretics with careful monitoring of fluid status in the mother and fetus. Due to their outcomes data in HFrEF and general safety in pregnancy, a combination of hydralazine and isosorbide dinitrate would also be reasonable for afterload reduction and improved management of the patient's hypertension. As she progresses through her pregnancy, she should receive fetal monitoring and maternal echocardiography, and be monitored for signs and symptoms of worsening heart failure.

In the postpartum period, the patient should be counseled on the risks of breastfeeding while taking cardiovascular medications, as well as maternal risks associated with future pregnancies. Following delivery, the main goals of treatment should be optimization of heart failure therapies and recovery of cardiac function.

References

[1] Regitz-Zagrosek V, Roos-Hesselink JW, Bauersachs J, Blomström-Lundqvist C, Cífková R, De Bonis M, et al. 2018 ESC guidelines for the management of cardiovascular diseases during pregnancy. Eur Heart J 2018;39:3165–241.

[2] Halpern DG, Weinberg CR, Pinnelas R, Mehta-Lee S, Economy KE, Valente AM. Use of medication for cardiovascular disease during pregnancy. J Am Coll Cardiol 2019;73:457–76.

[3] Department of Health and Human Services. Content and format of labeling for human prescription drug and biological products. requirements for pregnancy and lactation labeling. Fed Regist 2014;79:72064–103.

[4] Laliberte B, Reed BN, Ather A, Devabhakthuni S, Watson K, Lardieri AB, et al. Safe and effective use of pharmacologic and device therapy for peripartum cardiomyopathy. Pharmacotherapy 2016;36:955–70.

[5] Pariente G, Leibson T, Carls A, Adams-webber T, Ito S, Koren G. Pregnancy-associated changes in pharmacokinetics: a systematic review. PLoS Med 2016;13, e1002160.

[6] Anderson GD. Pregnancy-induced changes in pharmacokinetics. Clin Pharmacokinet 2005;44:989–1008.

[7] Feghali M, Venkataramanan R, Caritis S. Pharmacokinetics of drugs in pregnancy. Semin Perinatol 2015;39:512–9.

[8] Qasqas SA, McPherson C, Frishman WH, Elkayam U. Cardiovascular pharmacotherapeutic considerations during pregnancy and lactation. Cardiol Rev 2004;12:240–61.

[9] Jacobson TA, Maki KC, Orringer CE, Jones PH, Kris-Etherton P, Sikand G, et al. National lipid association recommendations for patient-centered management of dyslipidemia: part 2. J Clin Lipidol 2015;9:S1–S122. e1.

[10] France M, Rees A, Datta D, Thompson G, Capps N, Ferns G, et al. HEART UK statement on the management of homozygous familial hypercholesterolaemia in the United Kingdom. Atherosclerosis 2016;255:128–39.

[11] Grundy SM, Stone NJ, Bailey AL, Beam C, Birtcher KK, Blumenthal RS, et al. AHA/ACC/AACVPR/AAPA/ABC/ACPM/ADA/AGS/APhA/ASPC/NLA/PCNA guideline on the management of blood cholesterol: executive summary. J Am Coll Cardiol 2018;2018, CIR0000000000000624.

[12] Briggs G, Freeman R, Towers C, Forinash A. Drugs in pregnancy and lactation. 11th ed. Lippincott Williams & Wilkins; 2017.

[13] Middleton P, Gomersall J, Gould J, Shepherd E, Olsen S, Makrides M. Omega-3 fatty acid addition during pregnancy. Cochrane Database Syst Rev 2018;11, CD003402.

[14] Esteve-Valverde E, Ferrer-Oliveras R, Gil-Aliberas N, Baraldès-Farré A, Llurba E, Alijotas-Reig J. Pravastatin for preventing and treating preeclampsia: a systematic review. Obstet Gynecol Surv 2018;73:40–55.

[15] Edison RJ, Muenke M. Central nervous system and limb anomalies in case reports of first-trimester statin exposure. N Engl J Med 2004;350:1579–82.

[16] Karalis DG, Hill AN, Clifton S, Wild RA. The risks of statin use in pregnancy: a systematic review. J Clin Lipidol 2016;10:1081–90.

[17] Costantine MM, Cleary K, Hebert MF, Ahmed MS, Brown LM, Ren Z, et al. Safety and pharmacokinetics of pravastatin used for the prevention of preeclampsia in high-risk pregnant women: a pilot randomized controlled trial. Am J Obstet Gynecol 2016;214:720.e1–720.e17.

[18] Lee MS, Hekimian A, Doctorian T, Duan L. Statin exposure during first trimester of pregnancy is associated with fetal ventricular septal defect. Int J Cardiol 2018;269:111–3.

[19] Rolnik DL, Wright D, Poon LC, O'Gorman N, Syngelaki A, de Paco Matallana C, et al. Aspirin versus placebo in pregnancies at high risk for preterm pre-eclampsia. N Engl J Med 2017;377:613–22.

[20] Roberge S, Nicolaides K, Demers S, Hyett J, Chaillet N, Bujold E. The role of aspirin dose on the prevention of preeclampsia and fetal growth restriction: systematic review and meta-analysis. Am J Obstet Gynecol 2017;216:110–20. e6.

[21] Askie LM, Duley L, Henderson-Smart DJ, Stewart LA. Antiplatelet agents for prevention of pre-eclampsia: a meta-analysis of individual patient data. Lancet 2007;369:1791–8.

[22] Committee on Drugs. The transfer of drugs and other chemicals into human milk. Pediatrics 2001;108:776–89.

[23] Adler Y, Charron P, Imazio M, Badano L, Barón-Esquivias G, Bogaert J, et al. The 2015 ESC guidelines on the diagnosis and management of pericardial diseases. Eur Heart J 2015;36:2921–64.

[24] Indraratna PL, Virk S, Gurram D, Day RO. Use of colchicine in pregnancy: a systematic review and meta-analysis. Rheumatology 2018;57:382–7.

[25] Anon. Drugs and lactation database (LactMed) [Internet]. Bethesda, MD: U.S. National Library of Medicine; 2006. [cited 20 April 2019]. Available from: http://toxnet.nlm.nih.gov/cgi-bin/sis/htmlgen. LACT Part of the Toxicology Data Network (TOXNET).

[26] Verbruggen M, Mannaerts D, Muys J, Jacquemyn Y. Use of ticagrelor in human pregnancy, the first experience. BMJ Case Rep 2015;2015:1–2.

[27] Anon. ACOG Practice Bulletin No. 203: chronic hypertension in pregnancy. Obstet Gynecol 2019;133:e26–50.

[28] Whelton PK, Carey RM, Aronow WS, Casey DE Jr, Collins KJ, Dennison Himmelfarb C, et al. 2017 ACC/AHA/AAPA/ABC/ACPM/AGS/APhA/ASH/ASPC/NMA/PCNA guideline for the prevention, detection, evaluation, and management of high blood pressure in adults: executive summary. J Am Coll Cardiol 2018;71:2199–269.

[29] Yancy CW, Jessup M, Bozkurt B, Butler J, Casey DE Jr, Drazner MH, et al. 2013 ACCF/AHA guideline for the management of heart failure. J Am Coll Cardiol 2013;62:e147–239.

[30] Moretti ME, Caprara D, Drehuta I, Yeung E, Cheung S, Federico L, et al. The fetal safety of angiotensin converting enzyme inhibitors and angiotensin II receptor blockers. Obstet Gynecol Int 2012;2012:658310.

[31] Cooper WO, Hernandez-Diaz S, Arbogast PG, Dudley JA, Dyer S, Gideon PS, et al. Major congenital malformations after first-trimester exposure to ACE inhibitors. N Engl J Med 2006;354:2443–51.

[32] Bateman BT, Patorno E, Desai RJ, Seely EW, Mogun H, Dejene SZ, et al. Angiotensin-converting enzyme inhibitors and the risk of congenital malformations. Obstet Gynecol 2017;129:174–84.

[33] Kearney L, Wright P, Fhadil S, Thomas M. Postpartum cardiomyopathy and considerations for breastfeeding. Card Fail Rev 2018;4:112–8.

[34] Riester A, Reincke M. Mineralocorticoid receptor antagonists and management of primary aldosteronism in pregnancy. Eur J Endocrinol 2015;172:R23–30.

[35] Podymow T, August P. Update on the use of antihypertensive drugs in pregnancy. Hypertension 2008;51:960–9.

[36] Page RL, Joglar JA, Caldwell MA, Calkins H, Conti JB, Deal BJ, et al. 2015 ACC/AHA/HRS guideline for the management of adult patients with supraventricular tachycardia: a report of the American College of Cardiology/American Heart Association Task Force on Clinical Practice Guidelines and the Heart Rhythm Society. J Am Coll Cardiol 2016;67:e27–115.

[37] Yakoob M, Bateman B, Ho E, Hernandez-Diaz S, Franklin JM, Goodman JE, et al. The risk of congenital malformations associated with exposure to β-blockers early in pregnancy: a meta-analysis. Hypertension 2013;62:375–81.

[38] Bateman B, Patorno E, Desai R, Seely EW, Mogun H, Maeda A, et al. Late pregnancy Beta blocker exposure and risks of neonatal hypoglycemia and bradycardia. Pediatrics 2016;138, e20160731.

[39] Xie RH, Guo Y, Krewski D, Mattison D, Walker MC, Nerenberg K, et al. Beta-blockers increase the risk of being born small for gestational age or of being institutionalised during infancy. BJOG 2014;121:1090–6.

[40] Magee L, Duley L. Oral beta-blockers for mild to moderate hypertension during pregnancy. Cochrane Database Syst Rev 2003;CD002863.

[41] Abalos E, Duley L, Steyn D, Gialdini C. Antihypertensive drug therapy for mild to moderate hypertension during pregnancy. Cochrane Database Syst Rev 2018;CD002252.

[42] Arany Z, Elkayam U. Peripartum cardiomyopathy. Circulation 2016;133:1397–409.

[43] Bauersachs J, Arrigo M, Hilfiker-Kleiner D, Veltmann C, Coats AJ, Crespo-Leiro MG, et al. Current management of patients with severe acute peripartum cardiomyopathy: practical guidance from the Heart Failure Association of the European Society of Cardiology Study Group on peripartum cardiomyopathy. Eur J Heart Fail 2016;18:1096–105.

[44] Ryu RJ, Eyal S, Easterling TR, Caritis SN, Venkataraman R, Hankins G, et al. Pharmacokinetics of metoprolol during pregnancy and lactation. J Clin Pharmacol 2016;56:581–9.

[45] Hoeltzenbein M, Fietz A-K, Kayser A, Zinke S, Meister R, Weber-Schoendorfer C, et al. Pregnancy outcome after first trimester exposure to bisoprolol. J Hypertens 2018;36:2109–17.

[46] Anon. ACOG Committee Opinion No. 767 Summary: emergent therapy for acute-onset, severe hypertension during pregnancy and the postpartum period. Obstet Gynecol 2019;133:409–12.

[47] Webster LM, Myers JE, Nelson-Piercy C, Harding K, Cruickshank JK, Watt-Coote I, et al. Labetalol versus nifedipine as antihypertensive treatment for chronic hypertension in pregnancy. Hypertension 2017;70:915–22.

[48] Fischer JH, Sarto GE, Hardman J, Endres L, Jenkins TM, Kilpatrick SJ, et al. Influence of gestational age and body weight on the pharmacokinetics of labetalol in pregnancy. Clin Pharmacokinet 2014;53:373–83.

[49] Too G, Hill J. Hypertensive crisis during pregnancy and postpartum period. Semin Perinatol 2013;37:280–7.

[50] Bansal S, Pawar M. Haemodynamic responses to laryngoscopy and intubation in patients with pregnancy-induced hypertension: effect of intravenous esmolol with or without lidocaine. Int J Obstet Anesth 2002;11:4–8.

[51] Brown CM, Garovic VD. Treatment of hypertension in pregnancy. Drugs 2014;74:283–96.

[52] Bateman B, Hernandez-Diaz S, Huybrechts K, Palmsten K, Mogun H, Ecker JL, et al. Outpatient calcium-channel blockers and the risk of postpartum haemorrhage: a cohort study. BJOG 2013;120:1668–76.

[53] Davis R, Eastman D, McPhillips H, Raebel MA, Andrade SE, Smith D, et al. Risk of congenital malformations and perinatal events among infants exposed to calcium channel and beta-blockers during pregnancy. Pharmacoepidemiol Drug Saf 2011;20:138–45.

[54] Bateman BT, Huybrechts KF, Maeda A, Desai R, Patorno E, Seely EW, et al. Calcium channel blocker exposure in late pregnancy and the risk of neonatal seizures. Obstet Gynecol 2015;126:271–8.

[55] Bateman B, Heide-Jørgensen U, Einarsdóttir K, Engeland A, Furu K, Gissler M, et al. β-Blocker use in pregnancy and the risk for congenital malformations. Ann Intern Med 2018;169:665–73.

[56] Clark SM, Dunn HE, Hankins GDV. A review of oral labetalol and nifedipine in mild to moderate hypertension in pregnancy. Semin Perinatol 2015;39:548–55.

[57] Cornette J, Buijs EAB, Duvekot JJ, Herzog E, Roos-Hesselink JW, Rizopoulos D, et al. Hemodynamic effects of intravenous nicardipine in severely pre-eclamptic women with a hypertensive crisis. Ultrasound Obstet Gynecol 2016;47:89–95.

[58] Alsaied T, Baskar S, Fares M, Alahdab F, Czosek RJ, Murad MH, et al. First-line antiarrhythmic transplacental treatment for fetal tachyarrhythmia: a systematic review and meta-analysis. J Am Heart Assoc 2017;6, e007164.

[59] Pieper PG. Use of medication for cardiovascular disease during pregnancy. Nat Rev Cardiol 2015;12:718–29.

[60] Vigil-De Gracia P, Dominguez L, Solis A. Management of chronic hypertension during pregnancy with furosemide, amlodipine or aspirin: a pilot clinical trial. J Matern Neonatal Med 2014;27:1291–4.

[61] Olesen C, de Vries C, Thrane N, MacDonald T, Larsen H, Sørensen H. Effects of diuretics on fetal growth: a drug effect or confounding indication? Pooled Danish and Scottish cohort data. Br J Clin Pharmacol 2001;51:153–7.

[62] Collins R, Yusuf S, Peto R. Overview of randomised trials of diuretics in pregnancy. Br Med J (Clin Res Ed) 1985;290:17–23.

[63] Sağ S, Çoşkun H, Baran I, Güllülü S, Aydınlar A. Inappropriate sinus tachycardia-induced cardiomyopathy during pregnancy and successful treatment with ivabradine. Anatol J Cardiol 2016;16:212–3.

[64] Hoeltzenbein M, Beck E, Fietz A-K, Wernicke J, Zinke S, Kayser A, et al. Pregnancy outcome after first trimester use of methyldopa: a propsective cohort study. Hypertension 2017;70:201–8.

[65] Rothenberger S, Carr D, Brateng D, Hebert M, Easterling TR. Pharmacodynamics of clonidine therapy in pregnancy: a heterogeneous maternal response impacts fetal growth. Am J Hypertens 2010;23:1234–40.

[66] Hall D, Odendaal H, Steyn D, Smith M. Nifedipine or prazosin as a second agent to control early severe hypertension in pregnancy: a randomized controlled trial. Br J Obstet Gynaecol 2000;107:759–65.

[67] Versmissen J, Koch BCP, Roofthooft DWE, Ten Bosch-Dijksman W, van den Meiracker AH, Hanff LM, et al. Doxazosin treatment of phaeochromocytoma during pregnancy: placental transfer and disposition in breast milk. Br J Clin Pharmacol 2016;82:568–9.

[68] Duley L, Meher S, Jones L. Drugs for treatment of very high blood pressure during pregnancy. Cochrane Database Syst Rev 2013;CD001449.

[69] Smorlesi C, Caldarella A, Caramelli L, Di Lollo S, Moroni F. Topically applied minoxidil may cause fetal malformation: a case report. Birth Defects Res Part A: Clin Mol Teratol 2003;67:997–1001.

[70] Bahadoran Z, Mirmiran P, Azizi F, Ghasemi A. Nitrate-rich dietary supplementation during pregnancy: the pros and cons. Pregnancy Hypertens 2018;11:44–6.

[71] Croen L, Todoroff K, Shaw G. Maternal exposure to nitrate from drinking water and diet and risk for neural tube defects. Am J Epidemiol 2001;153:325–31.

[72] Thaler I, Kahana H. The effect of a nitric oxide donor on Fetal heart rate patterns in patients with hypertension. Obstet Gynecol 2002;100:987–91.

[73] Galiè N, Humbert M, Vachiery J, Gibbs S, Lang I, Torbicki A, et al. 2015 ESC/ERS guidelines for the diagnosis and treatment of pulmonary hypertension. Eur Heart J 2016;37:67–119.

[74] Klinger JR, Elliott CG, Levine DJ, Bossone E, Duvall L, Fagan K, et al. Therapy for pulmonary arterial hypertension in adults. Chest 2019;155:565–86.

[75] Hemnes AR, Kiely DG, Cockrill BA, Safdar Z, Wilson VJ, Al Hazmi M, et al. Statement on pregnancy in pulmonary hypertension from the Pulmonary Vascular Research Institute. Pulm Circ 2015;5:435–65.

[76] Paauw ND, Terstappen F, Ganzevoort W, Joles JA, Gremmels H, Lely AT. Sildenafil during pregnancy: a preclinical meta-analysis on fetal growth and maternal blood pressure. Hypertension 2017;70:998–1006.

[77] Trapani A, Gonçalves LF, Trapani TF, Vieira S, Pires M, Pires MMDS. Perinatal and hemodynamic evaluation of sildenafil citrate for preeclampsia treatment. Obstet Gynecol 2016;128:253–9.

[78] Sharp A, Cornforth C, Jackson R, Harrold J, Turner MA, Kenny LC, et al. Maternal sildenafil for severe fetal growth restriction (STRIDER): a multicentre, randomised, placebo-controlled, double-blind trial. Lancet Child Adolesc Heal 2018;2:93–102.

[79] Sharp A, Cornforth C, Jackson R, Harrold J, Turner MA, Kenny L, et al. Mortality in the UK STRIDER trial of sildenafil therapy for the treatment of severe early-onset fetal growth restriction. Lancet Child Adolesc Heal 2019;3:e2–3.

[80] Kearon C, Akl EA, Ornelas J, Blaivas A, Jimenez D, Bounameaux H, et al. Antithrombotic therapy for VTE disease: CHEST guideline and expert panel report. Chest 2016;149:315–52.

[81] Lau E, DeFaria Yeh D. Management of high risk cardiac conditions in pregnancy: anticoagulation, severe stenotic valvular disease and cardiomyopathy. Trends Cardiovasc Med 2019;29:155–61.

[82] Lameijer H, Aalberts JJJ, van Veldhuisen DJ, Meijer K, Pieper PG. Efficacy and safety of direct oral anticoagulants during pregnancy: a systematic literature review. Thromb Res 2018;169:123–7.

[83] James AH. Prevention and management of thromboembolism in pregnancy when heparins are not an option. Clin Obstet Gynecol 2018;61:228–34.

[84] Nishimura RA, Otto CM, Bonow RO, Carabello BA, Erwin JP 3rd, Guyton RA, et al. 2014 AHA/ACC guideline for the management of patients with valvular heart disease: executive summary. Circulation 2014;129:2440–92.

[85] D'Souza R, Ostro J, Shah PS, Silversides CK, Malinowski A, Murphy KE, et al. Anticoagulation for pregnant women with mechanical heart valves: a systematic review and meta-analysis. Eur Heart J 2017;38:1509–16.

[86] Crousillat DR, Wood MJ. Valvular heart disease and heart failure in women. Heart Fail Clin 2019;15:77–85.

[87] Shlomo M, Gorodischer R, Daniel S, Wiznitzer A, Matok I, Fishman B, et al. The fetal safety of enoxaparin use during pregnancy: a population-based retrospective cohort study. Drug Saf 2017;40:1147–55.

[88] Steinberg ZL, Dominguez-Islas CP, Otto CM, Stout KK, Krieger EV. Maternal and fetal outcomes of anticoagulation in pregnant women with mechanical heart valves. J Am Coll Cardiol 2017;69:2681–91.

[89] Heavner MS, Zhang M, Bast CE, Parker L, Eyler RF. Thrombolysis for massive pulmonary embolism in pregnancy. Pharmacotherapy 2017;37:1449–57.

[90] Watanabe TT, Ichijo M, Kamata T. Uneventful pregnancy and delivery after thrombolysis plus thrombectomy for acute ischemic stroke: case study and literature review. J Stroke Cerebrovasc Dis 2019;28:70–5.

[91] Vitale N, De Feo M, De Santo LS, Pollice A, Tedesco N, Cotrufo M. Dose-dependent fetal complications of warfarin in pregnant women with mechanical heart valves. J Am Coll Cardiol 1999;33:1637–41.

[92] Basu S, Aggarwal P, Kakani N, Kumar A. Low-dose maternal warfarin intake resulting in fetal warfarin syndrome: in search for a safe anticoagulant regimen during pregnancy. Birth Defects Res Part A: Clin Mol Teratol 2016;106:142–7.

[93] Afridi I, Moise K, Rokey R. Termination of supraventricular tachycardia with intravenous adenosine in a pregnant woman with Wolff-Parkinson-White syndrome. Obstet Gynecol 1992;80:481–3.

[94] Joglar JA, Page RL. Management of arrhythmia syndromes in pregnancy. Curr Opin Cardiol 2014;29:36–44.

[95] January CT, Wann LS, Alpert JS. 2014 AHA/ACC/HRS guideline for the management of patients with atrial fibrillation. J Am Coll Cardiol 2014;64:e1–76.

[96] Tadmor O, Keren A, Rosenak D, Gal M, Shaia M, Hornstein E, et al. The effect of disopyramide on uterine contractions during pregnancy. Am J Obstet Gynecol 1990;162:482–6.

[97] Shnider S, Way E. Plasma levels of lidocaine (Xylocaine) in mother and newborn following obstetrical conduction anesthesia: clinical applications. Anesthesiology 1968;29:951–8.

[98] Kockova R, Kocka V, Kiernan T, Fahy GJ. Ibutilide-induced cardioversion of atrial fibrillation during pregnancy. J Cardiovasc Electrophysiol 2007;18:545–7.

[99] Sonesson S, Fouron J, Wesslen-Eriksson E, Jaeggi E, Winberg P. Foetal supraventricular tachycardia treated with sotalol. Acta Paediatr 1998;87:584–7.

[100] Mittadodla PS, Salen PN, Traub DM. Isoproterenol as an adjunct for treatment of idiopathic ventricular fibrillation storm in a pregnant woman. Am J Emerg Med 2012;30:251.e3–5.

Cardiac Disorders Unique or More Specific to Women

Chapter 23

CardioRheumatology

Rekha Mankad

Clinical Case

A 41-year-old female patient with a history of systemic lupus erythematosus presents to the emergency room with symptoms of chest pain. The pain is described as sharp and midsternal, with radiation to the left shoulder. The pain started while at rest and has been constant, although worsens with activity and deep breaths. There is associated shortness of breath but no diaphoresis, nausea, or lightheadedness. No additional symptoms of syncope or edema. The week prior she had experienced symptoms of a cough, malaise, and subjective fevers/chills. The patient was diagnosed with lupus in her late 20s when she presented with profound fatigue and joint pain as well as a malar rash. She is maintained on mycophenolate mofetil and prednisone 5 mg daily. Currently, she is experiencing some increase in the pain in her hands and feet, along with the aforementioned recent chest pain.

Upon presentation to the emergency room, her blood pressure was 112/78, heart rate 82. Her physical exam was remarkable for a malar rash, mild lower extremity edema, and a soft pericardial friction rub. An electrocardiogram was done in the emergency room (**Figure 1**) and did not show evidence of an acute myocardial infarction. Pertinent labs included a normal white blood cell count, mild anemia, a sedimentation rate of 95 mm/1 h (normal ≤ 29 mm/1 h), and C-reactive protein of 177 mg/L (normal ≤ 9 mg/L). Cardiac biomarkers revealed a troponin T of 0.17 ng/mL at 0 h, 0.19 ng/mL at 3 h, and 0.21 ng/mL at 6 h (normal ≤ 0.01 ng/mL). How would you manage this patient?

Continued

Sex Differences in Cardiac Disease. https://doi.org/10.1016/B978-0-12-819369-3.00012-5

Clinical Case—cont'd

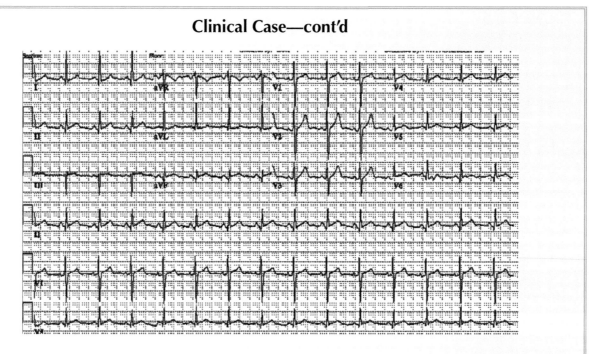

FIGURE 1   Electrocardiogram From Patient in the Clinical Case. Evidence of sinus rhythm with left axis deviation and some increased voltage suggestive of left ventricular hypertrophy. Small Q-waves inferiorly. Prominent T-waves in the V2-V3 leads.

Abstract

Women have an increased risk of certain chronic autoimmune inflammatory disorders (CIDs). All aspects of the heart's structure and function may be affected by the autoimmune disease. Many of the cardiovascular (CV) complications may be clinically silent and subclinical. However, there is still a clear burden of increased CV morbidity and mortality in these patients. Despite the growing recognition of the increased CV risk in the inflammatory autoimmune patient population, at present there are no specific guidelines that address the risk. A heightened awareness of the CV risk is important as one evaluates and treats the CID patient. Although the CID patient may not report CV symptomatology, their ongoing inflammatory condition may be resulting in damage to the entire cardiac system.

Introduction

It is becoming increasingly recognized that rheumatologic diseases can affect all aspects of cardiac structure and function (**Figure 2** and **Table 1**); however, many patients and providers remain unaware of this important link. In fact, cardiac events have been shown to be the culprit for the lower survival rates in patients with rheumatologic diseases compared to the general population. The patient with a rheumatologic condition may be plagued with noncardiac pain, deconditioning, and debility, which may impact the ability to fully ascertain the presence of cardiovascular (CV) symptomatology. In the general population, females are underdiagnosed and undertreated for heart disease because of atypical presentations along with a lack of recognition of individual risk [1]. As a few rheumatologic diseases occur predominantly in females, the understanding of the interaction between the inflammatory autoimmune condition and the CV system is critical to avoiding misinterpretation of signs and symptoms, which may indicate occult heart disease.

There are numerous rheumatologic diseases that tend to be united by their chronicity and the presence of inflammation, although to various degrees. These conditions can be arbitrarily divided into arthritides, connective tissue diseases, and vasculitides. The term "autoimmunity" is frequently attached to rheumatologic conditions and stems from the fact that many of these conditions occur when innate immunity has gone awry. Autoantibodies to self-antigens occur in these conditions, attacking various organs. Systemic inflammation brought about by the autoantibody destruction of tissues plays a critical role in the development and progression of cardiovascular diseases (CVDs) [2].

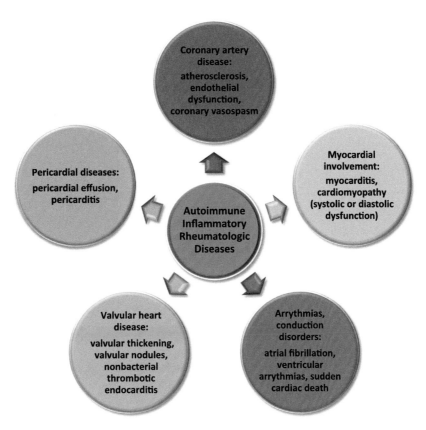

FIGURE 2 Various Pathologic Cardiovascular Conditions That May Manifest in Patients With Autoimmune Inflammatory Rheumatologic Diseases.

Sex-Specific Predominance of Rheumatologic Disorders

A few of the chronic autoimmune inflammatory disorders (CID) are more common in females. The lifetime risk for women to develop an autoimmune inflammatory condition is greater than the risk for men (8.4% compared to 5.1%, respectively) (**Table 1**) [3]. These include systemic lupus erythematosus (SLE), systemic sclerosis, antiphospholipid antibody syndrome (APS), and rheumatoid arthritis (RA). SLE is a rare condition in which autoantibodies and immune complexes cause widespread tissue damage; many organs can be affected such as the lungs, kidneys, or heart. Its high female predominance, occurring in a ratio of 9:1 in women to men, is largely unexplained [4]. In addition, African-American, Hispanic, Asian, and Native American women are two to three times more likely to develop SLE compared to white women [4]. The female predominance in lupus has been at least partially attributed to sex hormone influences, which may explain why the sex differences vary throughout the ages (less female predominance in prepubescence and postmenopausal ages) [5]. Systemic sclerosis (also commonly referred to as scleroderma) is a connective tissue disease that manifests as a small vessel vasculopathy

and fibrosis of the skin and organs. Due to the effects on the microvasculature, Raynaud's phenomena, pulmonary arterial hypertension, and renal crisis can develop. Systemic sclerosis affects women five times more than men [6]. APS is an autoimmune condition in which antibodies attack phospholipids. Having antiphospholipid antibodies does not equate with APS. The clinical entity of APS constitutes venous or arterial thrombosis and/or adverse pregnancy outcomes in the setting of persistent positive antiphospholipid antibodies. APS is frequently secondary to SLE but can occur independently as well. The prothrombotic state in APS results in venous and arterial thromboembolism. The incidence of APS as a primary disorder (that is, not associated with SLE) is equally prevalent in men and women [7]. RA is the most common CID and affects women three times more often than men [8]. RA is manifested as systemic joint inflammation due commonly to the autoantibodies, rheumatoid factor (RF), and citrullinated peptide.

Other CIDs that do not have a female predominance also carry an increased risk of CVD. In a meta-analysis, CVD and CV risk factors were more prevalent in psoriatic arthritis compared to the general population [9]. In a population-based study, untreated psoriasis and psoriatic

TABLE 1 Sex Differences in Prevalence, Cardiac and Noncardiac Features of Autoimmune Rheumatologic Conditions

Disease	Sex Difference in Prevalence; Female:Male	Noncardiac Features	Cardiac Features
Systemic lupus erythematosus (SLE)	9:1	Malar rash, nephritis, joint pain	Myocarditis, coronary artery disease, nonbacterial thrombotic endocarditis, heart block, pericarditis
Rheumatoid arthritis (RA)	3:1	Joint pain, swelling and stiffness (particularly in the mornings), fatigue, fever, loss of appetite	Coronary artery disease, pericarditis
Systemic sclerosis (SSc)	5:1	Hardening of patches of skin, Raynaud's disease, dysphagia, shortness of breath	Pulmonary hypertension, arrhythmias, cardiomyopathy, coronary artery disease
Antiphospholipid antibody syndrome (APS)	1:1	Arterial and venous thromboembolism, recurrent miscarriages	Nonbacterial thrombotic endocarditis
Psoriasis/psoriatic arthritis	1:1	Skin changes (red patches topped with silvery scales), joint pain, stiffness and swelling (foot pain due to Achilles tendinitis or plantar fasciitis), spondylitis	Coronary artery disease
Ankylosing spondylitis (the most common axial apondyloarthritis)	1:2–3	Stiffness and pain in the lower hips and lower back, fatigue, neck pain, uveitis	Aortic dilatation, aortic regurgitation, coronary artery disease
Relapsing polychondritis	1:1	Redness and pain of the ears and eyes. Saddle nose deformity, painful and swollen joints, shortness of breath	Aortic regurgitation, mitral regurgitation, pericarditis, heart block, coronary arteritis (causing myocardial infarction)

arthritis (which are equally prevalent in males and females) have an increased risk of CVD [10]. However, it is important to note that there may be a sex difference when it comes to addressing CV risk within the CID community. In a cross-sectional study, psoriatic women were noted to have a greater prevalence of CV risk factors compared to nonpsoriatic women; the opposite was true in men [11]. Ankylosing spondylitis has a greater prevalence in males, although the ratio is not as significant as it was decades prior [12]. Similar to patients with psoriatic arthritis, ankylosing spondylitis patients appear to have greater CV traditional risk factors and CV morbidity and mortality [13]; however, sex may play a role as ankylosing spondylitis men appear to be burdened with greater CV mortality compared to women with ankylosing spondylitis [standardized mortality rate for men 1.63 (95% confidence interval (CI) 1.29–1.97] vs. 1.38 [0.48–2.28] for women, $P<0.001$) [14]. Thus, there appears to be a complex interplay between traditional CV risk and sex differences, which have not been fully explored.

Most of the literature regarding CVD in the rheumatologic population has focused on SLE and RA (**Figures 3 and 4**). Studies in the 1970s–1980s described the increased incidence of CVD seen in both of these conditions [15, 16]. However, there does not appear to have been a general knowledge of this increased risk until relatively recently. This is highlighted by the fact that the discussion of the role of inflammatory conditions into a CV risk scoring system was only formally incorporated in the 2019 American College of Cardiology (ACC)/American Heart Association

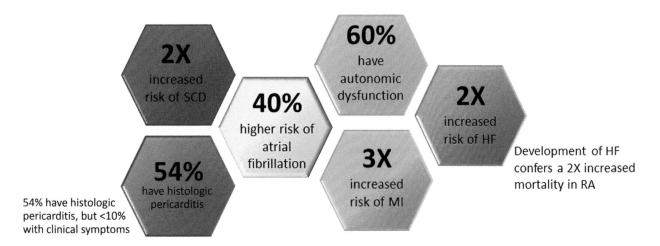

FIGURE 3   Cardiac Complications in Rheumatoid Arthritis. HF, heart failure; MI, myocardial infarction; RA, rheumatoid arthritis; SCD, sudden cardiac death. *Image courtesy of Niti R. Aggarwal.*

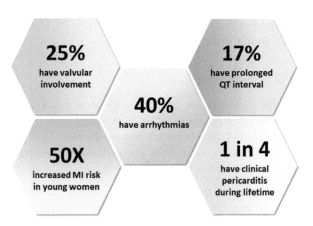

FIGURE 4   Cardiac Complications in Systemic Lupus Erythematosus. MI, myocardial infarction. *Image courtesy of Niti R. Aggarwal.*

(AHA) prevention guidelines [17]. Coronary artery disease (CAD) and atherosclerosis were the initial cardiac pathologies identified and discussed, but it is now evident that any part of the cardiac structure and function can be affected by the autoimmune inflammatory process. There do appear to be some differences between which CV conditions an individual patient with a specific rheumatologic condition is at risk for; not all of the potential CV complications are seen in all of the various autoimmune disorders.

The interplay between inflammation and an autoimmune process appears to be one of the drivers for the occurrence of CVD [18]. The process of atherosclerosis development within the coronary arteries has been shown to be due to more than just a cholesterol abnormality and the stages of inflammation are critical to the development and progression of the atherosclerotic plaque [19]. Thus, there

has been ongoing interest in the role of antiinflammatory treatments for CVD [20]. It is less well defined how the inflammatory burden affects the other components of the CV system.

The focus of this chapter will be on the various cardiac complications of common rheumatologic disorders, particularly those with female predominance.

Conduction Disease and Arrhythmias

Arrhythmias and conduction disturbances can occur in any of the rheumatologic diseases, during any stage of the course of the disease, or even be the first manifestation of the disease. These electrical abnormalities can be secondary to underlying cardiac abnormalities, such as CAD, pericarditis, or myocardial dysfunction. However, they can occur independent of structural heart disease.

Historically in SLE, arrhythmias and conduction defects occur in about 10% of patients [21]. Sinus tachycardia is the most common electrographic abnormality (seen in 18%), with atrial fibrillation (AF) being the most common (3%) tachyarrhythmia identified [22]. From a conduction standpoint, QT prolongation has been found to be the most common abnormality, present in 17% of patients with SLE [22]. Individuals with high titers of antismall cystoplamic ribonucleoprotein abnormalities appear to be at greater risk of QT lengthening [23]. Malignant arrhythmias are infrequently reported in SLE patients but can occur in the setting of acute myocarditis. The etiology for these electrical derangements may be due to small vessel vasculitis, infiltration of conduction tissue by fibrosis, or an interaction of autoantibodies to conduction tissue [24]. Bradyarrhythmias are less commonly reported but may be potentiated by antimalarial treatment for SLE [25].

In RA, most ventricular arrhythmias are secondary to CAD. Acute coronary syndromes leading to malignant arrhythmias are the leading cause of sudden cardiac death (SCD) [23]. In a study of over 600 RA patients studied for 15 years, SCD was twice as likely in RA compared to the general population (cumulative incidence in RA 7% vs. 4% in non-RA controls, $P=0.052$), but SCD risk was increased even after adjusting for the occurrence of an acute myocardial infarction (MI) [26]. The presence of rheumatoid nodules may also be a marker for greater cardiac involvement, which could explain a higher burden of arrhythmias, although that was not specifically found in Holter monitoring in a study of 35 nodular RA patients compared with 35 nonnodular RA patients [27]. However, this study did show that nodular RA patients had greater ST depression on Holter monitoring, which may be linked to greater CV morbidity and mortality [27]. Ventricular arrhythmias and SCD in RA may also be secondary to increased sympathetic activity, as measured by heart rate variability [28]. Reduction in heart rate variability suggests autonomic dysfunction, which may be a marker for malignant arrhythmias [29]. Autonomic dysfunction appears to be highly prevalent in RA, present in nearly 60% of RA patients, as suggested by a systematic review of the literature [30]. Disease activity in RA has been associated with heart rate variability in small studies [31]. Another marker of ventricular arrhythmia potential is QT dispersion/prolongation. Compared to healthy controls, RA patients have longer QT dispersion [32] and this dispersion may correlate with disease duration [33]. However, when QT prolongation is adjusted for C-reactive protein (CRP) levels, the association for CV mortality is lost [hazard ratio (HR) 1.73 compared to 2.17 when unadjusted]; this indicates that the QT interval is related to the level of systemic inflammation [34]. This relationship has been further corroborated by the finding that the antiinterleukin-6 receptor antibody, tocilizumab, which normalizes CRP levels, shortens QT lengths [35].

AF prevalence appears to be increased in RA as well. According to a Danish cohort study, RA subjects had a 40% higher incidence of AF compared to non-RA patients [36]. Although less common, heart block has been described in RA and has been attributed to the presence of rheumatoid nodules within the atrioventricular node or His bundles [31].

Systemic sclerosis patients have also been shown to have a higher burden of arrhythmias and conduction disturbances, although overall small numbers have been studied. The mortality related to an arrhythmia in systemic sclerosis is frequently related to the presence of pulmonary hypertension (PH) [37]. AF, atrial flutter, or paroxysmal supraventricular tachycardias are seen in 20–30% of systemic sclerosis patients [23]. Electrocardiographic abnormalities in systemic sclerosis are common (25–75% depending on the tool used) and can include left anterior fascicular block and first-degree atrioventricular blocks [38]. In systemic sclerosis patients without clinical evidence of heart disease, Holter monitoring has revealed a higher burden of premature supraventricular and ventricular contractions versus controls [39]. In an older, larger study of 183 systemic sclerosis patients, 67% had ventricular ectopy which strongly correlated with overall mortality and SCD [40]. These findings were associated with older age and greater evidence of other cardiac and pulmonary involvement of the systemic sclerosis [40]. The etiology for the electrical abnormalities in systemic sclerosis has been linked to fibrosis of the conduction system and/or myocardium.

The treatment for these arrhythmias in CIDs is no different than that used in the nonautoimmune inflammatory patient. A high index of suspicion for the presence of an arrhythmia is recommended given the prevalence of these abnormalities. In addition, the occurrence of an arrhythmia or conduction defect may be a signal of greater CV involvement than may be readily apparent. Treatment of the underlying systemic inflammation is important to curtail the CV events in the CID patient.

Pericardial Diseases

Pericardial involvement by CID is the most common CV manifestation. Asymptomatic, incidental finding of pericardial effusions or clinical pericarditis are the most likely presentations. Cardiac tamponade and pericardial constriction are much less common.

Pericarditis is the most common clinical cardiac abnormality, presenting in a large portion of patients with SLE [41]. The occurrence of pericarditis may be the first manifestation of SLE; thus, in a young woman presenting with pericarditis, the diagnostic testing for SLE should be entertained. Typically, the pericarditis is accompanied by a small pericardial effusion (**Figure 5**) and frequently occurs during SLE flares. However, the effusion may also be found incidentally. The prevalence of pericarditis with or without an accompanying effusion in SLE ranges from 20% to 50% [41]. Moderate to large effusions are infrequently encountered. Clinical tamponade is also uncommon, but when it occurs it is associated with low complement levels [42]. Recurrent episodes of pericarditis can occur in SLE but rarely does constrictive physiology develop [i.e., when the pericardium is thickened and leads to impaired ventricular filling and subsequently signs and symptoms of biventricular heart failure (HF)]. Diagnosis of pericarditis is made from the clinical presentation of pleuritic chest pain, although electrocardiography and transthoracic echocardiography (TTE) are frequently performed. Blood testing is usually done to assess the level of inflammation. Cardiac magnetic resonance imaging (MRI) (**Figure 6**) is infrequently needed for SLE pericarditis; however, it may be helpful to assess

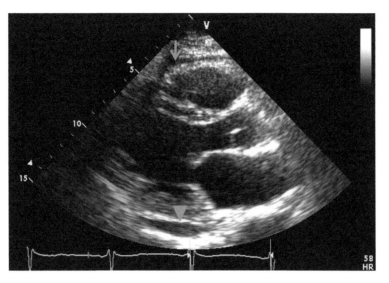

FIGURE 5 Echocardiogram With Pericarditis. The parasternal long-axis view still echocardiographic image demonstrates a trivial anterior pericardial effusion (*arrow*) and a slightly larger posterior pericardial effusion (*arrowhead*).

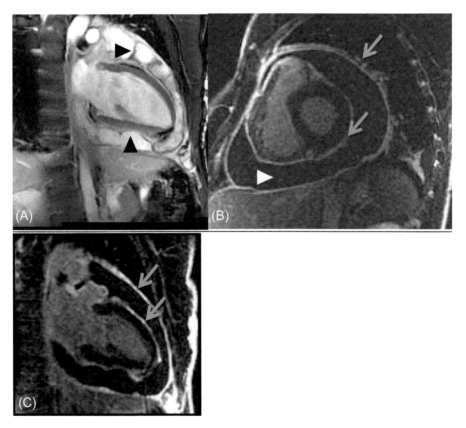

FIGURE 6 Magnetic Resonance Imaging in a Patient With SLE Who Presented With Acute Pericarditis. Panel A reveals the large pericardial effusion with evidence of septation within the pericardial space (*black arrowheads*). Panel B is an image of LGE in the same patient showing the large effusion (*white arrowhead*) as well as evidence of parietal and visceral pericardial inflammation (*arrows*). Panel C is another view of LGE in the same patient on long-axis imaging (*arrows* point to LGE in the visceral and parietal pericardium suggestive of inflammation). LGE, late gadolinium enhancement; SLE, systemic lupus erythematosus.

for inflammation in recurrent cases of pericarditis or to help establish the diagnosis of constrictive pericarditis. The treatments for pericarditis in SLE are agents that decrease the systemic inflammation, such as nonsteroidal antiinflammatory drugs (NSAIDs) and corticosteroids. If there are signs and symptoms of cardiac tamponade, then pericardiocentesis is indicated. Pericardectomy is reserved for refractory pericarditis or constrictive pericarditis.

Pericarditis can also occur in RA, and is seen in up to 50% of RA patients on autopsy studies [43]. However, it is far less common clinically than SLE. Less than 10% of RA patients present with pericarditis [43]. It is far more common to find pericardial effusions on echocardiography with an odds ratio of 10.7 and 95% CI 5.0–23.0 [44] or to discover pericardial involvement on autopsy studies [43–46]. Analysis of the pericardial fluid from an RA patient reveals immune complexes and RF. RA disease factors such as duration of disease or age do not appear to be associated with pericardial effusions, but more severe RA with greater inflammation and lower albumin levels do correlate with the presence of an effusion [46].

Pericardial involvement in systemic sclerosis is not common from a clinical perspective. A higher proportion of pericardial disease is identified on autopsy series (77.5% compared to controls) [47]. Pericardial effusions may be noted on TTEs done during the evaluation for PH, and in fact are seen more commonly in the presence of significant PH and portend a poorer prognosis [48]. Pericardial involvement can be seen in any of female-predominant CIDs, but overall the numbers studied have been small and not clinically relevant.

Treatment for the pericardial disease depends on the clinical manifestation. Larger pericardial effusions with evidence of tamponade physiology require urgent drainage. However, even asymptomatic moderate to large pericardial effusions may require pericardiocentesis for diagnostic purposes. Pericarditis is treated with antiinflammatory agents, which consist of NSAIDs, colchicine, and/or steroids. In the general population, avoidance of steroid treatment for pericarditis is typically recommended due to the increased risk of recurrences. However, this has not been studied particularly in a rheumatologic cohort. And given the marked inflammatory burden that may be present in the patient with CID (particularly the lupus patient), steroids do tend to be used more often as first-line therapy for pericarditis. In fact, most SLE patients may already be on steroids when they develop pericarditis.

Myocardial Dysfunction

Autoimmune inflammatory diseases can directly affect the myocardium. The clinical presentation of myocardial involvement could be an acute episode of myocarditis or a more chronic cardiomyopathic process. Subclinical disease has been identified on postmortem examinations. However, myocardial dysfunction is now recognized more often preclinically given our advanced imaging techniques. The reason for the myocardial involvement appears to be a direct result and damage related to inflammation and subsequent fibrosis [49].

In SLE, myocardial involvement of any type is reported to occur in 8–14% of patients [50]. Myocarditis appears to be relatively rare in lupus but may occur in times of severe flares with associated serositis (**Figure 7**) [41, 51]. Lupus myocarditis has a poor prognosis, with 40% of patients either dying or being left with residual myocardial dysfunction despite aggressive immunosuppressive treatment [51]. On echocardiography, lupus patients do have evidence of increased left ventricular (LV) mass and hypertrophy, even in the absence of hypertension [52]. Increases in LV mass may lead to stiffer ventricles, which can be characterized clinically by heart failure (HF) symptoms. In 41 SLE patients, cardiac magnetic resonance imaging (MRI) was performed to assess for the presence of myocardial fibrosis [53]. This study revealed that in SLE, there is a high propensity of mid-myocardial late gadolinium enhancement (LGE) in the left ventricle and this was associated with advanced age but not with SLE severity [53]. The presence of LGE can be associated with both systolic and diastolic dysfunction. A recent study showed that LV and right ventricular (RV) myocardial strain analysis (an echocardiographic measurement of myocardial deformation) in 50 SLE patients correlated with MRI evidence of myocardial fibrosis and may be a useful parameter to follow for CV disease progression [54]. It is important to recognize as well that one of the mainstays for treatment in SLE, hydroxychloroquine, has been implicated in the development of a cardiomyopathy [55]. This relationship requires further study given the high prevalence of use of this agent in SLE and other CIDs.

Like SLE, patients with RA may have myocardial disease related to inflammation and fibrosis. Myocardial involvement can subsequently lead to signs and symptoms of HF. This myocardial dysfunction is independent of ischemic heart disease [56]. In addition, HF in RA is less likely to be associated with obesity or hypertension and more likely to have higher ejection fractions compared to non-RA HF patients [57]. In fact, the diagnosis of HF may precede the diagnosis of RA. One year after the diagnosis of HF in an RA patient carries a higher mortality than in non-RA (35% vs. 19%; multivariable HR 1.89, 95% CI 1.26–2.84) [57]. The risk for HF in RA is 1.5–2 times the risk in the normal population [58] and is greater in RA patients with RF positivity (HR 2.59, 95% CI 1.95–3.43) [59]. In a recent Danish study, the relative risk for the development of HF was 1.3 (95% CI 1.17–1.45) in patients with RA compared to non-RA [60]. HF now appears to be driving the higher CV mortality in the RA population (greater than atherosclerotic heart disease) [57, 61, 62]. Elevations in markers of inflammation appear to precede the development of HF by several months, supporting the hypothesis that inflammatory

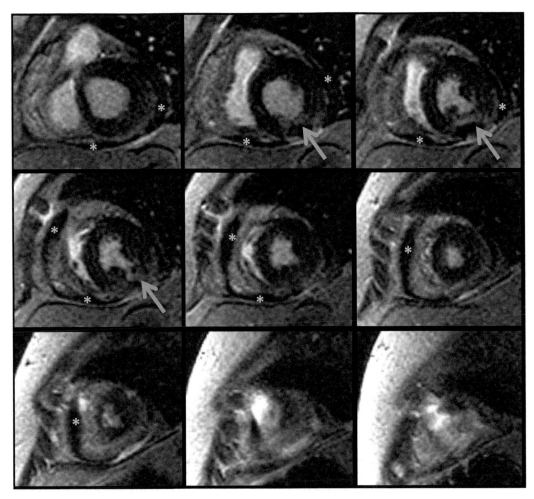

FIGURE 7   Cardiac MRI Showing Evidence of Edema and Inflammation in the Inferior and Inferolateral Walls (*Arrows*) Consistent With Myocarditis. In addition, there is evidence of a pericardial effusion (*).

stimuli drive the HF diagnosis [63]. However, the insults to the myocardium by inflammation may develop slowly over a period of time, during which there may not be any clinical evidence of HF (**Figure 8**). Changes in diastolic parameters on echocardiography seem to occur faster in RA patients compared to the general population; this may be one of the early markers of abnormal myocardial function leading subsequently to HF [64]. Specifically, RA patients had a faster increase in left atrial volume compared to a non-RA cohort as well as faster changes in mitral inflow, suggestive of diastolic filling abnormalities [64]. Compared to the general population, B-type natriuretic peptide (BNP) in the RA patient may not be as useful or helpful; RA patients are more likely to have elevations in BNP (16% vs. 9%; $P<0.001$) but which do no correlate with greater degrees of diastolic dysfunction [65]. Cardiac MRI is unique in its ability to image for both inflammation and fibrosis [66]. A study of 60 RA patients identified that MRI findings of inflammation/fibrosis correlated with RA disease activity as well as alterations in myocardial structure [67]. Myocardial strain assessment by TTE is another tool to assess for subclinical myocardial disease. RA patients have

worse global longitudinal strain, compared with normals, for the LV ($-15.7\pm3.2\%$ vs. $-18.1\pm2.4\%$, $P<0.001$) and the RV ($-17.9\pm4.7\%$ vs. $-20.7\pm2.4\%$, $P<0.001$), and this correlates with RA disease activity [68]. There has been some concern that the use of tumor necrosis factor alpha (TNF alpha) inhibitors increases the risk of developing HF when used as a treatment strategy in RA. That was not seen in a large study of RA patients who were treated with these agents; in fact, the benefit of reducing inflammation was greater than any risk [69]. It is uncertain at the present time if screening tools should be utilized in the RA population to predict those who may develop HF. In addition, there are no guidelines on approaches for prophylactic treatments to prevent HF in RA.

Pulmonary hypertension due to pathologic changes in the pulmonary vasculature leading to pulmonary fibrosis plays a large role in the CV mortality in systemic sclerosis. All forms of PH may be seen in systemic sclerosis, but the most common is Group 1 pulmonary arterial hypertension and Group 3 PH, which is due to chronic lung disease or chronic hypoxia [70]. There is a clear role for

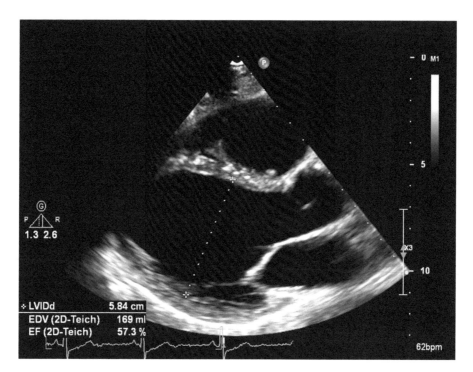

FIGURE 8 **Transthoracic Parasternal Long-Axis Image From an Asymptomatic 48-Year-Old Woman With Rheumatoid Arthritis.** Still image reveals a dilated left ventricular end-diastolic dimension (58 mm). Calculated ejection fraction was 50%. No identifiable cause for the dilated cardiomyopathy was determined beyond the presence of rheumatoid arthritis. Cardiac MRI did not reveal any active inflammation or fibrosis. MRI, magnetic resonance imaging.

echocardiography in systemic sclerosis patients during the initial diagnostic testing to assess for the presence of PH and also during any clinical changes in the patient (**Figure 9**) [71]. Pulmonary arterial hypertension is estimated to occur in 10–20% of systemic sclerosis patients [72, 73], but the overall prevalence is not fully known given the early asymptomatic nature of the condition [72, 74]. Transthoracic echocardiogram is a part of the current rheumatologic guidelines in systemic sclerosis as the presence of PH portends a poor prognosis [71, 75], with about 25% of SSc-related deaths being attributed to PH [76]. The presence of PH on echocardiography has been shown to be an independent risk for mortality (HR 2.02, 95% CI 1.44–2.83, $P < 0.001$) [76]. However, isolated CV involvement causing a cardiomyopathy accounts for over a third of the deaths in systemic sclerosis patients [77]. Myocardial fibrosis is seen in systemic sclerosis, and by MRI images, appears to be highly prevalent (45%) [78, 79]. Although the myocardial involvement in systemic sclerosis may be due to atherosclerosis, there is clearly a separate pathogenesis, which involves microvascular abnormalities. Hypoperfusion due to coronary vasospasm of the microvessels is the primary mode for repeated ischemia with subsequent myocardial damage [77, 78]. These changes in the microcirculation can then lead to ventricular diastolic and systolic dysfunction, which tend to be asymptomatic for a period of time. These early changes

may be identified on MRI or TTE with strain imaging [80]. The presence of a cardiomyopathy in a patient with SSc is associated with a lower survival compared with having an isolated "idiopathic" dilated cardiomyopathy (adjusted HR for death 1.75, 95% CI 0.93–3.29, $P = 0.081$) [81].

The treatment for myocardial involvement in CID is geared to standard HF medications for signs and symptoms. There is no consensus on prophylactic initiation of any therapies to decrease the onset of HF in these inflammatory disorders.

Valvular Heart Disease

There is paucity of available data on the occurrence of valvular heart disease (VHD) in the CID population. Most of the data on VHD is in patients with SLE or APS. The etiology of most of the valvular pathology identified in CID patients is not clear. It is likely that, inflammation and fibrosis play a role in the development of valvular dysfunction, but it is unclear which factors have the biggest impact in the initiation or progression of VHD in an individual patient. Most VHD is asymptomatic and typically discovered incidentally during echocardiography done for another indication. Thus, there are no specific guidelines for screening for the presence of valvular dysfunction in the CID population.

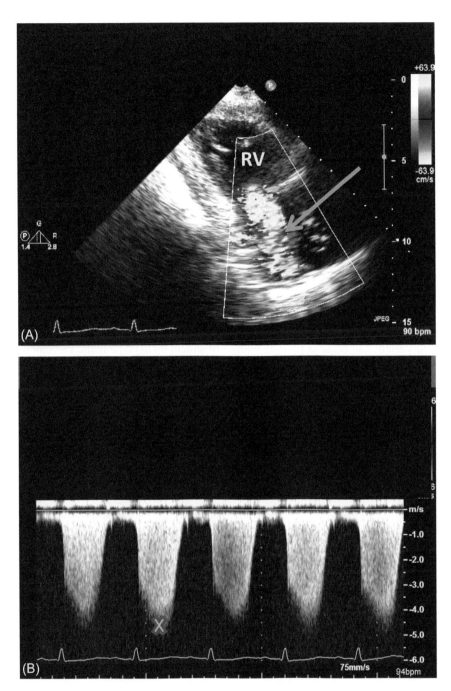

FIGURE 9   Transthoracic Echocardiographic Image From a Patient With Systemic Sclerosis. (A) Color Doppler reveals severe tricuspid regurgitation *(arrow)*. (B) Continuous wave Doppler recording through the tricuspid valve. The signal shows a tricuspid regurgitation velocity of approximately 4.5 m/s (X). This velocity corresponds to a pulmonary artery systolic pressure of at least 81 mmHg. RV, right ventricle.

Valve lesions have been commonly identified in the SLE population. Thickening of the valve leaflets with or without nodules have been described, more commonly on autopsy studies than is clinically significant. Valve regurgitation or stenosis can develop as well. VHD was seen in up to a quarter of SLE patients on TTE imaging in a 2016 study of 211 SLE patients [82]. Valve thickening was more

common than valve dysfunction; however, both were more likely with advanced age and with greater lupus activity [82]. Another study with an even larger population of SLE patients found that compared to age and sex-matched controls, SLE patients had a greater burden of all types of valve disease [aortic stenosis (AS), 1.08% vs. 0.35%, respectively, $P < 0.001$; aortic insufficiency, 1.32% vs. 0.29%, respectively, $P<0.001$; mitral stenosis, 0.74% vs. 0.21%, respectively, $P<0.001$; and mitral insufficiency, 1.91% vs. 0.39%, respectively, $P<0.001$] [83]. It is important to note that both of these studies found that SLE patients who also had antiphospholipid antibodies were at the highest risk of VHD. Thus, both SLE and APS may be discussed in concert when it comes to VHD risk and presentations.

Less than 20% of SLE patients have clinically significant (moderate or severe) valvular regurgitation [84], but this occurrence was seen more commonly in the setting of positive antiphospholipid antibodies [85]. Zuily et al. found a threefold higher risk of any valve lesion in SLE patients with antiphospholipid antibody based on echocardiographic studies [86]. An older study has suggested that valve lesions in SLE may either resolve or progress over time, occurring independently of treatment regimens or disease activity [87].

Nonbacterial thombotic endocarditis (NBTE) is another important consideration in SLE and APS patients. Although infectious endocarditis can occur in patients with CID given their immunocompromised state, the more "classic" abnormality is this noninfectious vegetation. NBTE is synonymous with verrucous endocarditis or Libman-Sachs endocarditis (named after the physicians who first described these valvular masses in lupus patients [88]). These lesions were initially described as large, mobile, verrucous masses on the valve leaflets. NBTE may be asymptomatic or can result in cerebral or coronary thromboemboli. In fact, the first manifestation of SLE or APS may be an embolic event related to NBTE. However, it is now recognized that often NBTE can be smaller masses, which most often occur on the left-sided valves (mitral valve > aortic valve) [89]. Careful imaging with echocardiography (transesophageal echocardiography being a better modality for discovering smaller lesions) can identify NBTE (**Figure 10**). Typically, these masses occur on the leaflet tips (so-called "kissing lesions") [89]. There may be mobility or multiple masses visualized. Pathologically, these lesions are bland thrombi which do not destroy leaflet tissue (which differentiates them from infective endocarditis). In primary or secondary APS, NBTE can occur. If these lesions are identified, first infectious endocarditis should be ruled out. A hypercoagulable work-up should be undertaken if the patient does not have known APS at the time of the valvular findings. In an SLE patient, if NBTE is identified, an evaluation for the concomitant diagnosis of APS should be undertaken. The treatment for asymptomatic valvular thickening is antiplatelet therapy. If there is evidence of vegetation or embolization, then anticoagulation is recommended, which should improve the valvular lesions. However, if there are ongoing embolic phenomena and/or incomplete resolution of the masses, surgical correction may need to be undertaken. However, in the setting of APS, ongoing anticoagulation is

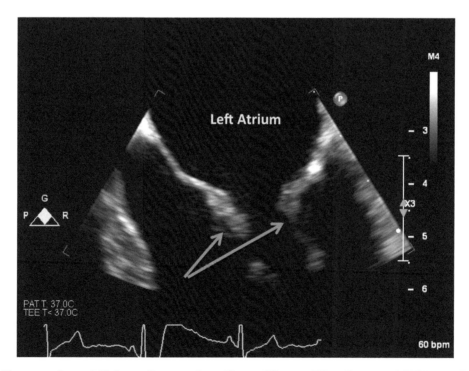

FIGURE 10 **Transesophageal Echocardiogram in a Young Woman Who Presented With an Embolic Stroke.** *Arrows* point to thickening at the leaflet tips. Small mobile elements were identified as well. Findings are consistent with nonbacterial thrombotic endocarditis. Patient was subsequently diagnosed with a lupus-like connective tissue disease.

necessary. Given the potential for VHD in the SLE and APS population, more routine echocardiography may be justified to identify early valvular abnormalities to allow for closer follow-up longitudinally.

In RA, clinically significant VHD appears to be uncommon. A small association has been found between RA and asymptomatic valvular abnormalities (15% compared to controls), related primarily to RA disease duration [90]. Transesophageal echocardiography has identified more subclinical and mild valve involvement in RA compared to controls; valve thickening and nodules along with regurgitant lesions are more common [91]. Interestingly, in one transesophageal echocardiography-based study of a small number of RA patients, 80% had some degree of mitral regurgitation, compared with 37% of matched controls [92]. A small study has looked at whether there is a difference in the progression rate of AS in the RA population compared to the general population. Overall, there was a greater rate of progression of AS in those RA patients with higher inflammation [as measures by erythrocyte sedimentation rate (ESR)]; however, the overall progression was actually slower than reported in the general population [93]. The etiology for this difference in RA compared to the general population was not clear, but it was hypothesized that medications to treat inflammation may have some effect on normal rates of valve progression [93].

Valvular abnormalities as a primary cardiac manifestation are relatively uncommon in systemic sclerosis [94]. Tricuspid regurgitation due to PH is not uncommon but is not due to a direct effect of systemic sclerosis on the tricuspid valve.

Although we are discussing female-predominant autoimmune conditions, it is important to mention other autoimmune conditions which also have valvular pathology as one of their cardiac complications. Axial spondyloarthritis (commonly ankylosing spondylitis which has a male predominance) is an inflammatory arthritis of the spine which may be associated with aortic root abnormalities or less commonly aortic valve pathology [95]. There is an inflammatory sclerosis that appears to affect all layers of the aorta, which can lead to aortic dilatation and secondary aortic regurgitation; however, there can also be an independent thickening of the aortic valve leaflets leading to an incompetent valve [95]. The risk for aortic pathology appears to be greater with longer duration of the spondyloarthropathy, with a 10% risk with disease duration of > 30 years [95, 96]. Relapsing polychondritis is a rare autoimmune condition (with equal sex predilection) in which inflammation primarily affects cartilage throughout the body; the eyes, ears, nose, and respiratory tract are involved, along with joints. Valvular involvement can occur in relapsing polychondritis, resulting in the need for surgery. In a review of over 400 cases of relapsing polychondritis, isolated aortic regurgi-

tation had a prevalence of 7.7%, isolated mitral regurgitation 1.8%, and combined valvular regurgitation 1.6% [97]. However, there is a high risk of reoperation required due to periprosthetic regurgitation, as well as a high mortality (38% died on follow-up in a small series) [98]. In addition, up to 10% of patients with relapsing polychondritis may have direct aortic involvement, resulting in dilatation and aneurysm formation as well as aortic regurgitation [95, 97].

Atherosclerosis/Coronary Artery Disease

Coronary artery involvement related to CID has been the cardiac abnormality that has been most often evaluated and studied in the literature. In fact, it was the finding of increased atherosclerotic disease with its associated mortality, which first garnered attention in the studies back in the 1980s–1990s. These data appear not to have been as readily appreciated by specialties outside of rheumatology.

Atherosclerotic cardiovascular disease (ASCVD) is itself a multifactorial condition in which many factors coexist to initiate the process of plaque deposition. Macrophages and activated lymphocytes have been identified within atherosclerotic plaque, supporting the notion that atherosclerosis at least partially develops because of an autoimmune, inflammatory process [2, 99–101]. However, inflammation also affects endothelial function and the microvasculature of the coronary circulation [102].

Although atherosclerosis as a risk in autoimmune conditions has been known for decades, the incorporation of this condition into risk scores did not occur until recently. The Framingham risk score (FRS) was evaluated for its ability to predict CVD in approximately 300 RA patients [103]. This study found that the FRS underestimated CV risk by 102% in women and 65% in men [103]. A European study evaluated multiple risk scores [FRS, systematic coronary risk evaluation (SCORE), and Reynolds risk score (RRS)] in RA and found that each of these scoring tools underestimated CV risk [104]. The American Heart Association/American College of Cardiology (AHA/ACC) 10-year pooled cohort score for individuals aged 40–75 was initiated in 2013 to better assess CV risk in the general population. When that calculator was used in a small RA cohort, it did not offer a greater advantage for accurate CV risk assessment [105]. In this study, RA patients underwent coronary artery calcium scoring on computed tomography; the study found that despite patients having coronary calcification (a marker of subclinical atherosclerosis), the FRS, RRS, and AHA/ACC risk scores only appropriately categorized 32%, 32%, and 41%, respectively, as high risk [105]. Specifically, this translated into 20 of the 34 patients with positive coronary calcium scores as "not high risk" [105]. There have been attempts to incorporate the presence of a CID into the scoring systems. The QRISK2 (a United Kingdom calculator for CV risk), the European League Against Rheumatism (EULAR), and the Expanded cardiovascular Risk Score for RA (ERAS-RA) have included RA in their algorithms for assessment of CV risk. The EULAR uses a multiplier of 1.5 for RA patients based on RA duration, seropositivity, and disease severity. The ERAS-RA was developed and internally validated using the Consortium of Rheumatology Researchers of North America (CORRONA) registry and incorporates RA-specific factors. However, when these three scoring systems were compared to current risk scores used in the general, non-RA, population, they did no better [106]. In this study, 100 patients developed CVD out of 1796 RA patients, over a mean of 6.9 years [106]. Discrimination was not improved for ERAS-RA (c-statistic=0.69), QRISK2, or EULAR multiplier applied to AHA/ACC compared with ACC/AHA (c-statistic=0.72 for all) [106]. Currently, the AHA/ACC pooled cohort risk score does include autoimmune, inflammatory conditions as "risk enhancers" [107]. However, there is no mathematical change to the score, merely a recognition that the score is higher due to the presence of a CID. Clearly, traditional risk factors, such as hypertension, hyperlipidemia, aging, and diabetes, play a role in the CID population, but these individuals carry the additional nontraditional risk of inflammation.

Atherosclerosis increases CV morbidity and mortality in men and women with autoimmune inflammatory conditions [108]. ASCVD is the leading cause of morbidity and mortality in the SLE population, particularly in younger women who have substantially higher rates of heart disease compared to matched controls [109]. Women with lupus aged 35–44 years had a 50-fold increased risk of MI compared to an age-matched cohort [109]. The ASCVD risk for SLE patients goes beyond the known traditional risk factors of hypertension (HTN), hyperlipidemia, obesity, and diabetes [110–113]. Although hyperlipidemia occurs with an equal frequency in SLE and the general population, the ASCVD risk of hyperlipidemia in SLE is greater than in a non-SLE patient [114]. There is also a proinflammatory high-density lipoprotein (HDL), which has been linked to increased ASCVD risk in the SLE patient [115]. A meta-analysis suggested a higher than expected burden of carotid plaque and increased carotid wall thickness (another subclinical marker of increased ASCVD risk) in SLE, which was linked to both traditional and nontraditional risk factors [116]. The process of atherosclerosis is independently associated with SLE-specific factors such as duration of disease, higher damage score (related to measures of end-organ damage related to SLE), and duration of steroid therapy, which has been best identified on studies assessing the carotid or coronary vasculature noninvasively [112, 117, 118].

This ASCVD risk beyond the traditional risk factors is why the current risk scoring systems for predicting CVD disease may not be as accurate in a patient with SLE [119]. Additionally, it has been identified that women with lupus may have angina chest pain without obstructive CAD, reflecting abnormalities in microcirculation [120]. Pharmacologic cardiac MRI and quantitative positron emission tomography (PET) flow imaging are other tools that may be helpful in identifying this endothelial dysfunction that is evident in the SLE population and which may be the etiology for ischemic heart disease symptomatology [120].

RA is the most common autoimmune inflammatory rheumatologic condition. Thus, most of the literature in regard to ASCVD in inflammatory disorders is concentrated on this patient population. Older epidemiologic data have shown lower survival for patients with RA compared to expected survival rates in the population, with women with RA having greater mortality rates compared to men [121]. For the RA group overall, the standardized mortality rate was 1.27 (95% CI 1.13–1.41) [121]. Women with RA had a standardized mortality rate of 1.41 (95% CI 1.21–1.61) and men with RA had a standardized mortality rate of 1.08 (95% CI 0.86–1.32), suggesting that women with RA had a lower life expectancy compared to non-RA women [121]. Data from the large Nurses' Health Study found that women with RA had an increased CVD-related mortality (HR 1.45, 95% CI 1.14–1.83) compared to women without RA, independent of traditional mortality risk factors, including smoking [122]. Accelerated atherosclerosis has been identified as the culprit for the increased CV mortality. A Swedish study comparing acute MI events in the RA population found that the standardized incidence ratio for acute MI was 2.9 in patients with RA compared with the general population ($P<0.05$). In addition, the RA patients experienced a higher case fatality rate compared to controls (HR 1.67, 95% CI 1.02–2.71), and this was despite similar care that was delivered to both groups [123]. Although the risk for ischemic heart disease is greater than controls, the reporting of angina is lower and thus silent MI is twice as likely [26]. The occurrence of a coronary event may actually predate the official diagnosis of RA by several years, indicating that subclinical inflammation may be occurring for a period of time before the official diagnosis of RA [26]. Subclinical atherosclerosis has also been identified on carotid imaging in RA patients [124, 125]. Specifically, greater carotid intima media thickness ($P=0.01$) and carotid plaque ($P<0.005$) at baseline imaging in RA patients were associated with more CVD events when compared to RA patients who did not have these abnormal parameters [126]. The presence of carotid plaque was seen in all the patients who sustained a CV event [126]. Endothelial dysfunction, which is a surrogate marker for increased ASCVD risk, when assessed by noninvasive parameters is abnormal in RA and associated with more severe inflammatory markers [127]. Similarly, severe RA inflammation is also associated with increased arterial stiffness, another surrogate marker for increased CVD risk [128]. As discussed for the other CIDs, the risk for ASCVD is not fully explained by traditional risk factors [26]. In fact, some traditional risk factors (smoking, male sex, and a personal history of cardiac history) actually impart a lower overall risk for ASCVD in the RA patient than the general population [129]. There are some risk factors that have a paradoxical association. One of these is related to lipid levels. High inflammatory burden is known to lower cholesterol levels [130]. Levels of low-density lipoprotein (LDL) and total cholesterol (TC) are lower during high inflammation [131]. When ESR is measured, the higher the ESR, the lower the TC/HDL ratio but with higher CV risk, which is deemed the "lipid paradox" [132]. The other paradoxical risk in RA is obesity. In the general population, obesity, through its link to hypertension, glucose intolerance, and dyslipidemia, is a risk factor for ASCVD. However, in RA, lower body mass index (BMI) is associated with higher CVD risk and mortality [129, 133]. This paradoxical relationship on survival appears to be mediated on the level of inflammation present (that is, lower BMI only confers a negative effect on survival in RA when ESR is high) [134].

In APS, there is clearly a risk of arterial thrombus as an etiology for an acute coronary event, given the highly prothrombotic state condition. However, the presence of antiphospholipid antibodies also appears to increase the risk for atherosclerosis [135]. Subclinical carotid artery abnormalities have been identified in APS and the presence of antiphospholipid antibodies has been discovered in patients with premature atherosclerosis [136].

In systemic sclerosis, the microvascular involvement of the disease increases the risk for ASCVD. Although there have been some conflicting data about whether systemic sclerosis truly accelerates the development of atherosclerosis, a couple of meta-analyses did confirm that coronary atherosclerosis, peripheral vascular disease, and cerebrovascular disease were all increased in systemic sclerosis patients compared to controls [137, 138]. One of these reviews assessed subclinical atherosclerosis with the use of carotid intima media thickness and found that systemic sclerosis patients had increased carotid intima media thickness (summary mean difference 0.11 mm, 95% CI 0.05 mm, 0.17 mm; $P = 0.0006$) compared to controls [137]. The other meta-analysis, using the studies with controls, showed an increased CAD prevalence (10–56%) or incidence (2.3%) in systemic sclerosis compared to controls (prevalence 2–44%; incidence 1.5%) [138]. Systemic sclerosis appears to be an independent risk over traditional risk factors for ASCVD, and the presence of renal involvement, duration of disease, and PH are associated with increased CAD [138]. In a large database study, 1344 systemic sclerosis patients were matched 10:1 to controls and found to have a substantially increased risk of acute MI (HR 2.45) [139]. In this study, there was no difference on the cardiac risk based on the use of immunosuppressive treatment or not [139]. A population-based study out of the United Kingdom also showed that the risk of MI, stroke, and peripheral vascu-

lar disease was higher in systemic sclerosis compared to comparison controls; the corresponding adjusted HRs were 1.80 (95% CI 1.07–3.05) for MI, 2.61 (95% CI 1.54–4.44) for stroke, and 4.35 (95% CI 2.74–6.93) for peripheral vascular disease [140]. Similar to patients with SLE, patients with systemic sclerosis also have abnormalities in coronary artery endothelial function. A small study using echocardiography to assess coronary flow reserve (CFR) found that asymptomatic patients with systemic sclerosis had lower CFR compared to controls and that this abnormality was not correlated to traditional risk factors or systemic sclerosis-specific factors [141]. In systemic sclerosis, there is frequently the concomitant presence of Raynaud's phenomena, which are related to vascular vasospasm. Therefore, it is reasonable to assume that some of the cardiac symptomatology in systemic sclerosis is related to macrovascular or microvascular coronary vasospastic disease [80]. The abnormalities of the microvascular may play a larger role in the development of CVD in systemic sclerosis given the fact that inflammation is less profound in this condition compared to SLE or RA.

The medications used in the treatment for CIDs may play a role in ASCVD risk. As inflammation is a driver of CVD risk, it is logical that measures to reduce inflammation will decrease that risk. NSAIDs are commonly used in the treatment armamentarium for a patient with CID. There are concerns about the risk of these drugs such that presently all NSAIDs have package labeling that warns of an increased CVD risk. This concern was brought about due to the increased CV mortality found in patients taking rofecoxib [a selective cyclo-oxygenase-2 (COX-2) inhibitor] [142]. However, newer data suggest that there is no increase in CVD risk with current COX-2 inhibitors and that COX-2 inhibition or NSAIDs have a relatively safe CV risk profile [143]. NSAIDs, however, do increase the risk of worsening HTN and kidney disease, which may be of particular concern in the SLE population. At the current time, NSAIDs and COX-2 inhibitors should be utilized very cautiously in those with established heart disease and in those without CVD, used intermittently and at low doses. Corticosteroids are utilized often in patients with CID, particularly when there are acute flares of the disease. Steroids have a profoundly beneficial effect on decreasing inflammation; however, they can also lead to hypertension, glucose intolerance, dyslipidemia, and weight gain. These adverse effects are more common with long-term steroid usage. Daily doses of greater than 7.5 mg of corticosteroids appear to incur the greatest risk [144]. The risk of any steroid use incurs the greatest risk in the seropositive patient compared to the seronegative patient [145]. Disease-modifying antirheumatic drugs (DMARDs) are typically the first-line drugs in the treatment of CIDs. Methotrexate (MTX) is a commonly used DMARD in RA and has been shown to decrease CV risk [146]. Anti-TNF alpha medications also appear to have beneficial effects on

CVD risk [147, 148]. As noted previously, there is a concern for the use of anti-TNF-alpha agents in HF patients [148], but more current data would suggest that the risk may not be as great as feared [69].

The treatment for ASCVD in the CID population is similar to that for the general population. Aggressive treatment of acute coronary syndromes with guideline-based measures pharmacologically and invasively should be the standard of care. Cholesterol treatments follow the current practices established for non-CID patients and have a clear benefit [149]. However, as inflammation plays such a profound role in the risk of ASCVD, treatment strategies addressing high inflammatory burden with repeated flares must be actively undertaken.

Conclusion

Cardiac involvement is common in autoimmune inflammatory rheumatologic conditions. Lupus, RA, systemic sclerosis, and APS occur more commonly in women and each of these conditions can have concomitant cardiac complications, which include arrhythmias, pericardial disease, myocardial dysfunction, VHD, and coronary atheroscle-

rosis. Although traditional risk factors play a role in the development of cardiac abnormalities, there is independent risk related to the autoimmune inflammatory process itself. The pathologic changes are related to the systemic inflammation and potential resultant fibrosis. Having a high index for suspicion of cardiac involvement will guide appropriate testing and evaluations. There are no specific guidelines on whom to test or which test to utilize. The conventional ASCVD risk assessment grossly underestimates the risk of ASCVD in patients with CID. It is reassuring that current prevention guidelines now recognize the increased risk of heart disease in the CID population. However, there is still no incorporation of these diseases into the calculation (or "score") for the 10-year or lifetime risk of CVD. Assessment for subclinical disease and early functional testing may help identify patients with CID who may be most prone to CVD. Additionally, data on sex-specific differences on risks, diagnosis, and management of CVD in CID patients are lacking, and there is a need for future research in this area. Treatment for cardiac complications does not differ between a CID patient or a non-CID patient. However, there may be a role for more aggressive treatment for underlying ongoing inflammation in patients with these cardiac manifestations.

Key Points

1. Women have a higher likelihood of certain autoimmune inflammatory conditions (rheumatoid arthritis, lupus, antiphospholipid antibody syndrome, and systemic sclerosis), all of which affect all aspects of cardiac structure and function.

2. Although conduction system disorders may be present in chronic inflammatory autoimmune conditions, clinically relevant arrhythmias are relatively rare but associated with greater burden of inflammation. Sudden death is increased in RA and systemic sclerosis.

3. Pericarditis is the most common clinical cardiovascular complication of SLE. Pericarditis may be the first manifestation of lupus.

4. Heart failure with preserved ejection fraction is now recognized as a greater cardiovascular burden in the RA population than atherosclerotic cardiovascular disease.

5. Nonbacterial thrombotic endocarditis occurs in APS and may be the cardiac manifestation that first occurs indicating the diagnosis of APS. SLE patients who develop NBTE are more likely to have antiphospholipid antibodies.

6. Women with lupus have a higher burden of atherosclerotic plaque (macrovascular disease), and also exhibit angina chest pain without obstructive CAD (microcirculation coronary disease).

7. Autoimmune inflammatory conditions are now included as "risk enhancers" in the ASCVD risk scoring evaluation. However, current ASCVD risk assessment underestimates the risk in patients with CID.

8. Traditional risk factors for ASCVD are often inadequate for risk assessment in the CID patient.

9. Traditional risk factors (smoking, male sex, and a personal history of cardiac history) actually impart a lower overall risk for ASCVD in the RA patient than in the general population. Additionally, LDL levels and degree of obesity are inversely related to the risk of ASCVD in RA patients.

10. Additionally, data on sex-specific differences on risks, diagnosis, and management of CVD in CID patients are lacking, and there is a need for future research in this area.

Back to Clinical Case

The case describes a 41-year-old female with history of SLE, on chronic immunosuppression, presenting with pleuritic chest pain, dyspnea, elevated erythrocyte sedimentation rate, and troponins, with an electrocardiogram (ECG) demonstrating left ventricular hypertrophy without obvious ischemia. Given this presentation, she underwent a coronary angiogram, which was negative for any obstructive coronary artery disease. An echocardiogram was performed (**Figure 5**) and revealed a left ventricular ejection fraction of 65%, without regional wall motion abnormalities, and a small pericardial effusion. Cardiac magnetic resonance imaging (**Figure 7**) demonstrated patchy mid-myocardial late gadolinium enhancement in the mid-inferior and inferolateral segments, along with a small pericardial effusion, consistent with myopericarditis. The patient was treated with an increase in her steroid and mycophenolate mofetil dosing along with the addition of colchicine. Her symptoms improved over a few days and she was discharged.

This case illustrates the complexities of cardiac presentation in a patient with systemic lupus erythematosus. This patient's symptoms did sound like pericarditis (the constant chest pain with the pleuritic component, occurring at rest). However, the elevated troponins would not be expected in that case. This suggests myocardial involvement. As lupus patients have an increased risk of coronary artery disease (with a much greater risk compared to premenopausal women without lupus), coronary angiography is not unreasonable. An echocardiogram did reveal the pericardial effusion, which would go along with the clinical story of symptoms of pericarditis. Cardiac MRI elucidated the etiology of the troponin elevation—myocarditis. This is a very important diagnosis to be made as the patient would be at risk for left ventricular systolic dysfunction and arrhythmias due to the myocardial inflammation. The treatment of choice is high-dose immunosuppression.

References

[1] Wenger NK. Women and coronary heart disease: a century after Herrick: understudied, underdiagnosed, and undertreated. Circulation 2012;126(5):604–11.

[2] Libby P, Ridker PM, Hansson GK. Inflammation in atherosclerosis: from pathophysiology to practice. J Am Coll Cardiol 2009;54(23):2129–38.

[3] Crowson CS, Matteson EL, Myasoedova E, Michet CJ, Ernste FC, Warrington KJ, et al. The lifetime risk of adult-onset rheumatoid arthritis and other inflammatory autoimmune rheumatic diseases. Arthritis Rheum 2011;63(3):633–9.

[4] Lau CS, Yin G, Mok MY. Ethnic and geographical differences in systemic lupus erythematosus: an overview. Lupus 2006;15(11):715–9.

[5] Lahita RG. The role of sex hormones in systemic lupus erythematosus. Curr Opin Rheumatol 1999;11(5):352–6.

[6] Mayes MD, Lacey Jr JV, Beebe-Dimmer J, Gillespie BW, Cooper B, Laing TJ, et al. Prevalence, incidence, survival, and disease characteristics of systemic sclerosis in a large US population. Arthritis Rheum 2003;48(8):2246–55.

[7] Duarte-García A, Pham MM, Crowson CS, Amin S, Moder KG, Pruthi RK, et al. The epidemiology of antiphospholipid syndrome: a population-based study. Arthritis Rheum 2019;71(9):1545–52.

[8] Gabriel SE. The epidemiology of rheumatoid arthritis. Rheum Dis Clin N Am 2001;27(2):269–81.

[9] Jamnitski A, Symmons D, Peters MJL, Sattar N, McInnes I, Nurmohamed MT. Cardiovascular comorbidities in patients with psoriatic arthritis: a systematic review. Ann Rheum Dis 2013;72(2):211–6.

[10] Ogdie A, Yu Y, Haynes K, Love TJ, Maliha S, Jiang Y, et al. Risk of major cardiovascular events in patients with psoriatic arthritis, psoriasis and rheumatoid arthritis: a population-based cohort study. Ann Rheum Dis 2015;74(2):326–32.

[11] Sondermann W, Djeudeu Deudjui DA, Korber A, Slomiany U, Brinker TJ, Erbel R, et al. Psoriasis, cardiovascular risk factors and metabolic disorders: sex-specific findings of a population-based study. J Eur Acad Dermatol Venereol 2020;34(4):779–86.

[12] Landi M, Maldonado-Ficco H, Perez-Alamino R, Maldonado-Cocco JA, Citera G, Arturi P, et al. Gender differences among patients with primary ankylosing spondylitis and spondylitis associated with psoriasis and inflammatory bowel disease in an iberoamerican spondyloarthritis cohort. Medicine 2016;95(51):e5652.

[13] Liew JW, Ramiro S, Gensler LS. Cardiovascular morbidity and mortality in ankylosing spondylitis and psoriatic arthritis. Best Pract Res Clin Rheumatol 2018;32(3):369–89.

[14] Bakland G, Gran JT, Nossent JC. Increased mortality in ankylosing spondylitis is related to disease activity. Ann Rheum Dis 2011;70(11):1921–5.

[15] Urowitz MB, Bookman AA, Koehler BE, Gordon DA, Smythe HA, Ogryzlo MA. The bimodal mortality pattern of systemic lupus erythematosus. Am J Med 1976;60(2):221–5.

[16] Linos A, Worthington JW, O'Fallon WM, Kurland LT. The epidemiology of rheumatoid arthritis in Rochester, Minnesota: a study of incidence, prevalence and mortality. Am J Epidemiol 1980;111:87–98.

[17] Arnett DK, Blumenthal RS, Albert MA, Buroker AB, Goldberger ZD, Hahn EJ, et al. 2019 ACC/AHA guideline on the primary prevention of cardiovascular disease: a report of the American College of Cardiology/American Heart Association task force on clinical practice guidelines. Circulation 2019;140(11):e596–646.

[18] Libby P. Inflammation in atherosclerosis. Nature 2002;420:868–74.

[19] Ross R. Atherosclerosis—an inflammatory disease. N Engl J Med 1999;340(2):115–26.

[20] Ridker PM, Everett BM, Thuren T, MacFadyen JG, Chang WH, Ballantyne C, et al. Antiinflammatory therapy with Canakinumab for atherosclerotic disease. N Engl J Med 2017;377(12):1119–31.

[21] Mandell BF. Cardiovascular involvement in systemic lupus erythematosus. Semin Arthritis Rheum 1987;17(2):126–41.

[22] Myung G, Forbess LJ, Ishimori ML, Chugh S, Wallace D, Weisman MH. Prevalence of resting-ECG abnormalities in systemic lupus erythematosus: a single-center experience. Clin Rheumatol 2017;36(6):1311–6.

[23] Seferovic PM, Ristic AD, Maksimovic R, Simeunovic DS, Ristic GG, Radovanovic G, et al. Cardiac arrhythmias and conduction disturbances in autoimmune rheumatic diseases. Rheumatology (Oxford) 2006;45(Suppl. 4):iv39–42.

[24] Natsheh A, Shimony D, Bogot N, Nesher G, Breuer GS. Complete heart block in lupus. Lupus 2019;28(13):1589–93.

[25] Tselios K, Gladman DD, Harvey P, Su J, Urowitz MB. Severe brady-arrhythmias in systemic lupus erythematosus: prevalence, etiology and associated factors. Lupus 2018;27(9):1415–23.

[26] Maradit-Kremers H, Crowson CS, Nicola PJ, Ballman KV, Roger VL, Jacobsen SJ, et al. Increased unrecognized coronary heart disease and sudden deaths in rheumatoid arthritis: a population-based cohort study. Arthritis Rheum 2005;52(2):402–11.

[27] Wislowska M, Sypula S, Kowalik I. Echocardiographic findings and 24-h electrocardiographic Holter monitoring in patients with nodular and non-nodular rheumatoid arthritis. Rheumatol Int 1999;18(5–6):163–9.

[28] Evrengul H, Dursunoglu D, Cobankara V, Polat B, Seleci D, Kabukcu S, et al. Heart rate variability in patients with rheumatoid arthritis. Rheumatol Int 2004;24(4):198–202.

[29] Schwemmer S, Beer P, Scholmerich J, Fleck M, Straub RH. Cardiovascular and pupillary autonomic nervous dysfunction in patients with rheumatoid arthritis—a cross-sectional and longitudinal study. Clin Exp Rheumatol 2006;24(6):683–9.

[30] Adlan AM, Lip GYH, Paton JFR, Kitas GD, Fisher JP. Autonomic function and rheumatoid arthritis: a systematic review. Semin Arthritis Rheum 2014;44(3):283–304.

[31] Eisen A, Arnson Y, Dovrish Z, Hadary R, Amital H. Arrhythmias and conduction defects in Rheumatological diseases—a comprehensive review. Semin Arthritis Rheum 2009;39(3):145–56.

[32] Goldeli O, Dursun E, Komsuoglu B. Dispersion of ventricular repolarization: a new marker of ventricular arrhythmias in patients with rheumatoid arthritis. J Rheumatol 1998;25(3):447–50.

[33] Voskuyl AE. The heart and cardiovascular manifestations in rheumatoid arthritis. Rheumatology (Oxford) 2006;45(Suppl. 4):iv4–7.

[34] Panoulas VF, Toms TE, Douglas KMJ, Sandoo A, Metsios GS, Stavropoulos-Kalinoglou A, et al. Prolonged QTc interval predicts all-cause mortality in patients with rheumatoid arthritis: an association driven by high inflammatory burden. Rheumatology (Oxford) 2014;53(1):131–7.

[35] Lazzerini PE, Acampa M, Capecchi PL, Fineschi I, Selvi E, Moscadelli V, et al. Antiarrhythmic potential of anticytokine therapy in rheumatoid arthritis: tocilizumab reduces corrected QT interval by controlling systemic inflammation. Arthritis Care Res 2015;67(3):332–9.

[36] Lindhardsen J, Ahlehoff O, Gislason GH, Madsen OR, Olesen JB, Svendsen JH, et al. Risk of atrial fibrillation and stroke in rheumatoid arthritis: Danish nationwide cohort study. BMJ 2012;344:e1257.

[37] Hugle T, Schuetz P, Daikeler T, Tyndall A, Matucci-Cerinic M, Walker UA, et al. Late-onset systemic sclerosis—a systematic survey of the EULAR scleroderma trials and research group database. Rheumatology (Oxford) 2011;50(1):161–5.

[38] Roberts NK, Cabeen Jr WR, Moss J, Clements PJ, Furst DE. The prevalence of conduction defects and cardiac arrhythmias in progressive systemic sclerosis. Ann Intern Med 1981;94(1):38–40.

[39] Bielous-Wilk A, Poreba M, Staniszewska-Marszalek E, Poreba R, Podgorski M, Kalka D, et al. Electrocardiographic evaluation in patients with systemic scleroderma and without clinically evident heart disease. Ann Noninvasive Electrocardiol 2009;14(3):251–7.

[40] Kostis JB, Seibold JR, Turkevich D, Masi AT, Grau RG, Medsger Jr TA, et al. Prognostic importance of cardiac arrhythmias in systemic sclerosis. Am J Med 1988;84(6):1007–15.

[41] Roman MJ, Salmon JE. Cardiovascular manifestations of rheumatologic diseases. Circulation 2007;116(20):2346–55.

[42] Rosenbaum E, Krebs E, Cohen M, Tiliakos A, Derk CT. The spectrum of clinical manifestations, outcome and treatment of pericardial tamponade in patients with systemic lupus erythematosus: a retrospective study and literature review. Lupus 2009;18(7):608–12.

[43] Kitas G, Banks MJ, Bacon PA. Cardiac involvement in rheumatoid disease. Clin Med (Lond) 2001;1(1):18–21.

[44] Corrao S, Messina S, Pistone G, Calvo L, Scaglione R, Licata G. Heart involvement in rheumatoid arthritis: systematic review and meta-analysis. Int J Cardiol 2013;167(5):2031–8.

[45] Amaya-Amaya J, Montoya-Sánchez L, Rojas-Villarraga A. Cardiovascular involvement in autoimmune diseases. Biomed Res Int 2014;2014:31.

[46] Sugiura T, Kumon Y, Kataoka H, Matsumura Y, Takeuchi H, Doi YL. Asymptomatic pericardial effusion in patients with rheumatoid arthritis. Cardiology 2008;110(2):87–91.

[47] Byers RJ, Marshall DA, Freemont AJ. Pericardial involvement in systemic sclerosis. Ann Rheum Dis 1997;56(6):393–4.

[48] Hosoya H, Derk CT. Clinically symptomatic pericardial effusions in hospitalized systemic sclerosis patients: demographics and management. Biomed Res Int 2018;2018:6812082.

[49] Ntusi NAB, Piechnik SK, Francis JM, Ferreira VM, Matthews PM, Robson MD, et al. Diffuse myocardial fibrosis and inflammation in rheumatoid arthritis: insights from CMR T1 mapping. JACC Cardiovasc Imaging 2015;8(5):526–36.

[50] Lee KS, Kronbichler A, Eisenhut M, Lee KH, Shin JI. Cardiovascular involvement in systemic rheumatic diseases: an integrated view for the treating physicians. Autoimmun Rev 2018;17(3):201–14.

[51] Tanwani J, Tselios K, Gladman DD, Su J, Urowitz MB. Lupus myocarditis: a single center experience and a comparative analysis of observational cohort studies. Lupus 2018;27(8):1296–302.

[52] Pieretti J, Roman MJ, Devereux RB, Lockshin MD, Crow MK, Paget SA, et al. Systemic lupus erythematosus predicts increased left ventricular mass. Circulation 2007;116(4):419–26.

[53] Seneviratne MG, Grieve SM, Figtree GA, Garsia R, Celermajer DS, Adelstein S, et al. Prevalence, distribution and clinical correlates of myocardial fibrosis in systemic lupus erythematosus: a cardiac magnetic resonance study. Lupus 2016;25(6):573–81.

[54] Tselios K, Gladman DD, Harvey P, Akhtari S, Su J, Urowitz MB. Abnormal cardiac biomarkers in patients with systemic lupus erythematosus and no prior heart disease: a consequence of antimalarials? J Rheumatol 2019;46(1):64–9.

[55] Tselios K, Deeb M, Gladman DD, Harvey P, Akhtari S, Mak S, et al. Antimalarial-induced cardiomyopathy in systemic lupus erythematosus: as rare as considered? J Rheumatol 2019;46(4):391–6.

[56] Crowson CS, Nicola PJ, Kremers HM, O'Fallon WM, Therneau TM, Jacobsen SJ, et al. How much of the increased incidence of heart failure in rheumatoid arthritis is attributable to traditional cardiovascular risk factors and ischemic heart disease? Arthritis Rheum 2005;52(10):3039–44.

[57] Davis 3rd JM, Roger VL, Crowson CS, Kremers HM, Therneau TM, Gabriel SE. The presentation and outcome of heart failure in patients with rheumatoid arthritis differs from that in the general population. Arthritis Rheum 2008;58(9):2603–11.

[58] Maradit-Kremers H, Nicola PJ, Crowson CS, Ballman KV, Gabriel SE. Cardiovascular death in rheumatoid arthritis: a population-based study. Arthritis Rheum 2005;52(3):722–32.

[59] Nicola PJ, Maradit-Kremers H, Roger VL, Jacobsen SJ, Crowson CS, Ballman KV, et al. The risk of congestive heart failure in rheumatoid arthritis: a population-based study over 46 years. Arthritis Rheum 2005;52(2):412–20.

[60] Khalid U, Egeberg A, Ahlehoff O, Lane D, Gislason GH, Lip GYH, et al. Incident heart failure in patients with rheumatoid arthritis: a nationwide cohort study. J Am Heart Assoc 2018;7(2).

[61] Myasoedova E, Gabriel SE, Matteson EL, Davis JM, Therneau TM, Crowson CS. Decreased cardiovascular mortality in patients with incident rheumatoid arthritis (RA) in recent years: dawn of a new era in cardiovascular disease in RA? J Rheumatol 2017;44(6):732–9.

[62] Nicola PJ, Crowson CS, Maradit-Kremers H, Ballman KV, Roger VL, Jacobsen SJ, et al. Contribution of congestive heart failure and ischemic heart disease to excess mortality in rheumatoid arthritis. Arthritis Rheum 2006;54(1):60–7.

[63] Maradit-Kremers H, Nicola PJ, Crowson CS, Ballman KV, Jacobsen SJ, Roger VL, et al. Raised erythrocyte sedimentation rate signals heart failure in patients with rheumatoid arthritis. Ann Rheum Dis 2007;66(1):76–80.

[64] Davis 3rd JM, Lin G, Oh JK, Crowson CS, Achenbach SJ, Therneau TM, et al. Five-year changes in cardiac structure and function in patients with rheumatoid arthritis compared with the general population. Int J Cardiol 2017;240:379–85.

[65] Crowson CS, Myasoedova E, Davis 3rd JM, Roger VL, Karon BL, Borgeson D, et al. Use of B-type natriuretic peptide as a screening tool for left ventricular diastolic dysfunction in rheumatoid arthritis patients without clinical cardiovascular disease. Arthritis Care Res 2011;63(5):729–34.

[66] Mavrogeni S, Markousis-Mavrogenis G, Koutsogeorgopoulou L, Kolovou G. Cardiovascular magnetic resonance imaging: clinical implications in the evaluation of connective tissue diseases. J Inflamm Res 2017;10:55–61.

[67] Kobayashi Y, Giles JT, Hirano M, Yokoe I, Nakajima Y, Bathon JM, et al. Assessment of myocardial abnormalities in rheumatoid arthritis using a comprehensive cardiac magnetic resonance approach: a pilot study. Arthritis Res Ther 2010;12(5):R171.

[68] Fine NM, Crowson CS, Lin G, Oh JK, Villarraga HR, Gabriel SE. Evaluation of myocardial function in patients with rheumatoid arthritis using strain imaging by speckle-tracking echocardiography. Ann Rheum Dis 2014;73(10):1833–9.

[69] Listing J, Strangfeld A, Kekow J, Schneider M, Kapelle A, Wassenberg S, et al. Does tumor necrosis factor alpha inhibition promote or prevent heart failure in patients with rheumatoid arthritis? Arthritis Rheum 2008;58(3):667–77.

[70] Simonneau G, Montani D, Celermajer DS, Denton CP, Gatzoulis MA, Krowka M, et al. Haemodynamic definitions and updated clinical classification of pulmonary hypertension. Eur Respir J 2019;53(1):1801913.

[71] McLaughlin VV, Archer SL, Badesch DB, Barst RJ, Farber HW, Lindner JR, et al. ACCF/AHA 2009 expert consensus document on pulmonary hypertension: a report of the American College of Cardiology Foundation Task Force on Expert Consensus Documents and the American Heart Association: developed in collaboration with the American College of Chest Physicians, American Thoracic Society, Inc., and the Pulmonary Hypertension Association. Circulation 2009;119(16):2250–94.

[72] Coghlan JG, Denton CP, Grunig E, Bonderman D, Distler O, Khanna D, et al. Evidence-based detection of pulmonary arterial hypertension in systemic sclerosis: the DETECT study. Ann Rheum Dis 2014;73(7):1340–9.

[73] Yang X, Mardekian J, Sanders KN, Mychaskiw MA, Thomas 3rd J. Prevalence of pulmonary arterial hypertension in patients with connective tissue diseases: a systematic review of the literature. Clin Rheumatol 2013;32(10):1519–31.

[74] Wigley FM, Lima JAC, Mayes M, McLain D, Chapin JL, Ward-Able C. The prevalence of undiagnosed pulmonary arterial hypertension in subjects with connective tissue disease at the secondary health care level of community-based rheumatologists (the UNCOVER study). Arthritis Rheum 2005;52(7):2125–32.

[75] Galie N, Humbert M, Vachiery J-L, Gibbs S, Lang I, Torbicki A, et al. 2015 ESC/ERS guidelines for the diagnosis and treatment of pulmonary hypertension: the joint task force for the diagnosis and treatment of pulmonary hypertension of the European Society of Cardiology (ESC) and the European Respiratory Society (ERS): endorsed by: Association for European Paediatric and Congenital Cardiology (AEPC), International Society for Heart and Lung Transplantation (ISHLT). Eur Heart J 2016;37(1):67–119.

[76] Tyndall AJ, Bannert B, Vonk M, Airo P, Cozzi F, Carreira PE, et al. Causes and risk factors for death in systemic sclerosis: a study from the EULAR Scleroderma Trials and Research (EUSTAR) database. Ann Rheum Dis 2010;69(10):1809–15.

[77] Allanore Y, Meune C. Primary myocardial involvement in systemic sclerosis: evidence for a microvascular origin. Clin Exp Rheumatol 2010;28(5 Suppl. 62):S48–53.

[78] Kahan A, Allanore Y. Primary myocardial involvement in systemic sclerosis. Rheumatology (Oxford) 2006;45(Suppl. 4):iv14–7.

[79] Rodriguez-Reyna TS, Morelos-Guzman M, Hernandez-Reyes P, Montero-Duarte K, Martinez-Reyes C, Reyes-Utrera C, et al. Assessment of myocardial fibrosis and microvascular damage in systemic sclerosis by magnetic resonance imaging and coronary angiotomography. Rheumatology (Oxford) 2015;54(4):647–54.

[80] Kahan A, Coghlan G, McLaughlin V. Cardiac complications of systemic sclerosis. Rheumatology (Oxford) 2009;48(Suppl. 3):iii45–8.

[81] Leyngold I, Baughman K, Kasper E, Ardehali H. Comparison of survival among patients with connective tissue disease and cardiomyopathy (systemic sclerosis, systemic lupus erythematosus, and undifferentiated disease). Am J Cardiol 2007;100(3):513–7.

[82] Vivero F, Gonzalez-Echavarri C, Ruiz-Estevez B, Maderuelo I, Ruiz-Irastorza G. Prevalence and predictors of valvular heart disease in patients with systemic lupus erythematosus. Autoimmun Rev 2016;15(12):1134–40.

[83] Watad A, Tiosano S, Grysman N, Comaneshter D, Cohen AD, Shoenfeld Y, et al. The association between systemic lupus erythematosus and valvular heart disease: an extensive data analysis. Eur J Clin Investig 2017;47(5):366–71.

[84] Galve E, Candell-Riera J, Pigrau C, Permanyer-Miralda G, Garcia-Del-Castillo H, Soler-Soler J. Prevalence, morphologic types, and evolution of cardiac valvular disease in systemic lupus erythematosus. N Engl J Med 1988;319(13):817–23.

[85] Perez-Villa F, Font J, Azqueta M, Espinosa G, Pare C, Cervera R, et al. Severe valvular regurgitation and antiphospholipid antibodies in systemic lupus erythematosus: a prospective, long-term, followup study. Arthritis Rheum 2005;53(3):460–7.

[86] Zuily S, Regnault V, Selton-Suty C, Eschwege V, Bruntz JF, Bode-Dotto E, et al. Increased risk for heart valve disease associated with antiphospholipid antibodies in patients with systemic lupus erythematosus: meta-analysis of echocardiographic studies. Circulation 2011;124(2):215–24.

[87] Roldan CA, Shively BK, Crawford MH. An echocardiographic study of valvular heart disease associated with systemic lupus erythematosus. N Engl J Med 1996;335(19):1424–30.

[88] Libman E, Sacks B. A hitherto undescribed form of valvular and mural endocarditis. Arch Intern Med 1924;33(6):701.

[89] Hojnik M, George J, Ziporen L, Shoenfeld Y. Heart valve involvement (Libman-Sacks endocarditis) in the antiphospholipid syndrome. Circulation 1996;93(8):1579–87.

[90] Beckhauser AP, Vallin L, Burkievcz CJ, Perreto S, Silva MB, Skare TL. Valvular involvement in patients with rheumatoid arthritis. Acta Reumatol Port 2009;34(1):52–6.

[91] Roldan CA, DeLong C, Qualls CR, Crawford MH. Characterization of valvular heart disease in rheumatoid arthritis by transesophageal echocardiography and clinical correlates. Am J Cardiol 2007;100(3):496–502.

[92] Guedes C, Bianchi-Fior P, Cormier B, Barthelemy B, Rat AC, Boissier MC. Cardiac manifestations of rheumatoid arthritis: a case-control transesophageal echocardiography study in 30 patients. Arthritis Rheum 2001;45(2):129–35.

[93] Bois JP, Crowson CS, Khullar T, Achenbach SJ, Krause ML, Mankad R. Progression rate of severity of aortic stenosis in patients with rheumatoid arthritis. Echocardiography 2017;34(10):1410–6.

[94] Lambova S. Cardiac manifestations in systemic sclerosis. World J Cardiol 2014;6(9):993–1005.

[95] Slobodin G, Naschitz JE, Zuckerman E, Zisman D, Rozenbaum M, Boulman N, et al. Aortic involvement in rheumatic diseases. Clin Exp Rheumatol 2006;24(2 Suppl. 41):S41–7.

[96] Eder L, Sadek M, McDonald-Blumer H, Gladman DD. Aortitis and Spondyloarthritis—an unusual presentation: case report and review of the literature. Semin Arthritis Rheum 2010;39(6):510–4.

[97] Lang-Lazdunski L, Hvass U, Paillole C, Pansard Y, Langlois J. Cardiac valve replacement in relapsing polychondritis. A review. J Heart Valve Dis 1995;4(3):227–35.

[98] Dib C, Moustafa SE, Mookadam M, Zehr KJ, Michet Jr CJ, Mookadam F. Surgical treatment of the cardiac manifestations of relapsing polychondritis: overview of 33 patients identified through literature review and the Mayo Clinic records. Mayo Clin Proc 2006;81(6):772–6.

[99] Hansson GK, Libby P, Schonbeck U, Yan ZQ. Innate and adaptive immunity in the pathogenesis of atherosclerosis. Circ Res 2002;91(4):281–91.

[100] Schaffner T, Taylor K, Bartucci EJ, Fischer-Dzoga K, Beeson JH, Glagov S, et al. Arterial foam cells with distinctive immunomorphologic and histochemical features of macrophages. Am J Pathol 1980;100(1):57–80.

[101] Hansson GK. Inflammation, atherosclerosis, and coronary artery disease. N Engl J Med 2005;352(16):1685–95.

[102] Foster W, Carruthers D, Lip GY, Blann AD. Inflammation and microvascular and macrovascular endothelial dysfunction in rheumatoid arthritis: effect of treatment. J Rheumatol 2010;37(4):711–6.

[103] Crowson CS, Matteson EL, Roger VL, Therneau TM, Gabriel SE. Usefulness of risk scores to estimate the risk of cardiovascular disease in patients with rheumatoid arthritis. Am J Cardiol 2012;110(3):420–4.

[104] Arts EEA, Popa C, Den Broeder AA, Semb AG, Toms T, Kitas GD, et al. Performance of four current risk algorithms in predicting cardiovascular events in patients with early rheumatoid arthritis. Ann Rheum Dis 2015;74:668–74.

[105] Kawai VK, Chung CP, Solus JF, Oeser A, Raggi P, Stein CM. The ability of the 2013 American College of Cardiology/American Heart Association cardiovascular risk score to identify rheumatoid arthritis patients with high coronary artery calcification scores. Arthritis Rheum 2015;67(2):381–5.

[106] Crowson CS, Gabriel SE, Semb AG, van Riel PLCM, Karpouzas G, Dessein PH, et al. Rheumatoid arthritis-specific cardiovascular risk scores are not superior to general risk scores: a validation analysis of patients from seven countries. Rheumatology (Oxford) 2017;56(7):1102–10.

[107] Arnett DK, Blumenthal RS, Albert MA, Buroker AB, Goldberger ZD, Hahn EJ, et al. 2019 ACC/AHA guideline on the primary prevention of cardiovascular disease: executive summary: a report of the American College of Cardiology/American Heart Association task force on clinical practice guidelines. J Am Coll Cardiol 2019;74(10):1376–414.

[108] Roifman I, Beck PL, Anderson TJ, Eisenberg MJ, Genest J. Chronic inflammatory diseases and cardiovascular risk: a systematic review. Can J Cardiol 2011;27(2):174–82.

[109] Manzi S, Meilahn EN, Rairie JE, Conte CG, Medsger TA, Jansen-McWilliams L, et al. Age-specific incidence rates of myocardial infarction and angina in women with systemic lupus Erythematosus: comparison with the Framingham study. Am J Epidemiol 1997;145(5):408–15.

[110] Petri M, Perez-Gutthann S, Spence D, Hochberg MC. Risk factors for coronary artery disease in patients with systemic lupus erythematosus. Am J Med 1992;93(5):513–9.

[111] Manzi S, Selzer F, Sutton-Tyrrell K, Fitzgerald SG, Rairie JE, Tracy RP, et al. Prevalence and risk factors of carotid plaque in women with systemic lupus erythematosus. Arthritis Rheum 1999;42(1):51–60.

[112] Roman MJ, Shanker BA, Davis A, Lockshin MD, Sammaritano L, Simantov R, et al. Prevalence and correlates of accelerated atherosclerosis in systemic lupus erythematosus. N Engl J Med 2003;349(25):2399–406.

[113] Bruce IN, Urowitz MB, Gladman DD, Ibañez D, Steiner G. Risk factors for coronary heart disease in women with systemic lupus erythematosus: the Toronto Risk Factor Study. Arthritis Rheum 2003;48(11):3159–67.

[114] Fischer LM, Schlienger RG, Matter C, Jick H, Meier CR. Effect of rheumatoid arthritis or systemic lupus erythematosus on the risk of first-time acute myocardial infarction. Am J Cardiol 2004;93(2):198–200.

[115] McMahon M, Grossman J, FitzGerald J, Dahlin-Lee E, Wallace DJ, Thong BY, et al. Proinflammatory high-density lipoprotein as a biomarker for atherosclerosis in patients with systemic lupus erythematosus and rheumatoid arthritis. Arthritis Rheum 2006;54(8):2541–9.

[116] Wu G-C, Liu H-R, Leng R-X, Li X-P, Li X-M, Pan H-F, et al. Subclinical atherosclerosis in patients with systemic lupus erythematosus: a systemic review and meta-analysis. Autoimmun Rev 2016;15(1):22–37.

[117] de Leeuw K, Freire B, Smit AJ, Bootsma H, Kallenberg CG, Bijl M. Traditional and non-traditional risk factors contribute to the development of accelerated atherosclerosis in patients with systemic lupus erythematosus. Lupus 2006;15(10):675–82.

[118] Asanuma Y, Oeser A, Shintani AK, Turner E, Olsen N, Fazio S, et al. Premature coronary-artery atherosclerosis in systemic lupus erythematosus. N Engl J Med 2003;349(25):2407–15.

[119] Jafri K, Ogdie A, Qasim A, Patterson SL, Gianfrancesco M, Izadi Z, et al. Discordance of the Framingham cardiovascular risk score and the 2013 American College of Cardiology/American Heart Association risk score in systemic lupus erythematosus and rheumatoid arthritis. Clin Rheumatol 2018;37(2):467–74.

[120] Recio-Mayoral A, Mason JC, Kaski JC, Rubens MB, Harari OA, Camici PG. Chronic inflammation and coronary microvascular dysfunction in patients without risk factors for coronary artery disease. Eur Heart J 2009;30(15):1837–43.

[121] Gabriel SE, Crowson CS, Kremers HM, Doran MF, Turesson C, O'Fallon WM, et al. Survival in rheumatoid arthritis: a population-based analysis of trends over 40 years. Arthritis Rheum 2003;48(1):54–8.

[122] Sparks JA, Chang S-C, Liao KP, Lu B, Fine AR, Solomon DH, et al. Rheumatoid arthritis and mortality among women during 36 years of prospective follow-up: results from the Nurses' health study. Arthritis Care Res 2016;68(6):753–62.

[123] Sodergren A, Stegmayr B, Lundberg V, Ohman ML, Wallberg-Jonsson S. Increased incidence of and impaired prognosis after acute myocardial infarction among patients with seropositive rheumatoid arthritis. Ann Rheum Dis 2007;66(2):263–6.

[124] Gonzalez-Juanatey C, Llorca J, Testa A, Revuelta J, Garcia-Porrua C, Gonzalez-Gay MA. Increased prevalence of severe subclinical atherosclerotic findings in long-term treated rheumatoid arthritis patients without clinically evident atherosclerotic disease. Medicine (Baltimore) 2003;82(6):407–13.

[125] Ambrosino P, Lupoli R, Di Minno A, Tasso M, Peluso R, Di Minno MND. Subclinical atherosclerosis in patients with rheumatoid arthritis. A meta-analysis of literature studies. Thromb Haemost 2015;113(5):916–30.

[126] Ikdahl E, Rollefstad S, Wibetoe G, Olsen IC, Berg I-J, Hisdal J, et al. Predictive value of arterial stiffness and subclinical carotid atherosclerosis for cardiovascular disease in patients with rheumatoid arthritis. J Rheumatol 2016;43(9):1622–30.

[127] Di Minno MND, Ambrosino P, Lupoli R, Di Minno A, Tasso M, Peluso R, et al. Clinical assessment of endothelial function in patients with rheumatoid arthritis: a meta-analysis of literature studies. Eur J Intern Med 2015;26(10):835–42.

[128] Ambrosino P, Tasso M, Lupoli R, Di Minno A, Baldassarre D, Tremoli E, et al. Non-invasive assessment of arterial stiffness in patients with rheumatoid arthritis: a systematic review and meta-analysis of literature studies. Ann Med 2015;47(6):457–67.

[129] Gonzalez A, Kremers HM, Crowson CS, Ballman KV, Roger VL, Jacobsen SJ, et al. Do cardiovascular risk factors confer the same risk for cardiovascular outcomes in rheumatoid arthritis patients as in non-rheumatoid arthritis patients? Ann Rheum Dis 2008;67(1):64–9.

[130] Choy E, Sattar N. Interpreting lipid levels in the context of high-grade inflammatory states with a focus on rheumatoid arthritis: a challenge to conventional cardiovascular risk actions. Ann Rheum Dis 2009;68(4):460–9.

[131] Hahn BH, Grossman J, Chen W, McMahon M. The pathogenesis of atherosclerosis in autoimmune rheumatic diseases: roles of inflammation and dyslipidemia. J Autoimmun 2007;28(2–3):69–75.

[132] Myasoedova E, Crowson CS, Kremers HM, Roger VL, Fitz-Gibbon PD, Therneau TM, et al. Lipid paradox in rheumatoid arthritis: the impact of serum lipid measures and systemic inflammation on the risk of cardiovascular disease. Ann Rheum Dis 2011;70(3):482–7.

[133] Kremers HM, Nicola PJ, Crowson CS, Ballman KV, Gabriel SE. Prognostic importance of low body mass index in relation to cardiovascular mortality in rheumatoid arthritis. Arthritis Rheum 2004;50(11):3450–7.

[134] Escalante A, Haas RW, del Rincon I. Paradoxical effect of body mass index on survival in rheumatoid arthritis: role of comorbidity and systemic inflammation. Arch Intern Med 2005;165(14):1624–9.

[135] Artenjak A, Lakota K, Frank M, Cucnik S, Rozman B, Bozic B, et al. Antiphospholipid antibodies as non-traditional risk factors in atherosclerosis based cardiovascular diseases without overt autoimmunity. A critical updated review. Autoimmun Rev 2012;11(12):873–82.

[136] Jara LJ, Medina G, Vera-Lastra O. Systemic antiphospholipid syndrome and atherosclerosis. Clin Rev Allergy Immunol 2007;32(2):172–7.

[137] Au K, Singh MK, Bodukam V, Bae S, Maranian P, Ogawa R, et al. Atherosclerosis in systemic sclerosis: a systematic review and meta-analysis. Arthritis Rheum 2011;63(7):2078–90.

[138] Ali H, Ng KR, Low AHL. A qualitative systematic review of the prevalence of coronary artery disease in systemic sclerosis. Int J Rheum Dis 2015;18(3):276–86.

[139] Chu S-Y, Chen Y-J, Liu C-J, Tseng W-C, Lin M-W, Hwang C-Y, et al. Increased risk of acute myocardial infarction in systemic sclerosis: a nation-wide population-based study. Am J Med 2013;126(11):982–8.

[140] Man A, Zhu Y, Zhang Y, Dubreuil M, Rho YH, Peloquin C, et al. The risk of cardiovascular disease in systemic sclerosis: a population-based cohort study. Ann Rheum Dis 2013;72(7):1188–93.

[141] Sulli A, Ghio M, Bezante GP, Deferrari L, Craviotto C, Sebastiani V, et al. Blunted coronary flow reserve in systemic sclerosis. Rheumatology (Oxford) 2004;43(4):505–9.

[142] Bombardier C, Laine L, Reicin A, Shapiro D, Burgos-Vargas R, Davis B, et al. Comparison of upper gastrointestinal toxicity of rofecoxib and naproxen in patients with rheumatoid arthritis. VIGOR Study Group. N Engl J Med 2000;343(21):1520–8. 2 p following 8.

[143] Nissen SE, Yeomans ND, Solomon DH, Luscher TF, Libby P, Husni ME, et al. Cardiovascular safety of celecoxib, naproxen, or ibuprofen for arthritis. N Engl J Med 2016;375(26):2519–29.

[144] Davis 3rd JM, Maradit-Kremers H, Gabriel SE. Use of low-dose glucocorticoids and the risk of cardiovascular morbidity and mortality in rheumatoid arthritis: what is the true direction of effect? J Rheumatol 2005;32(10):1856–62.

[145] Davis 3rd JM, Maradit Kremers H, Crowson CS, Nicola PJ, Ballman KV, Therneau TM, et al. Glucocorticoids and cardiovascular events in rheumatoid arthritis: a population-based cohort study. Arthritis Rheum 2007;56(3):820–30.

[146] Micha R, Imamura F, Wyler von Ballmoos M, Solomon DH, Hernan MA, Ridker PM, et al. Systematic review and meta-analysis of methotrexate use and risk of cardiovascular disease. Am J Cardiol 2011;108(9):1362–70.

[147] Barnabe C, Martin BJ, Ghali WA. Systematic review and meta-analysis: anti-tumor necrosis factor alpha therapy and cardiovascular events in rheumatoid arthritis. Arthritis Care Res 2011;63(4):522–9.

[148] Cacciapaglia F, Navarini L, Menna P, Salvatorelli E, Minotti G, Afeltra A. Cardiovascular safety of anti-TNF-alpha therapies: facts and unsettled issues. Autoimmun Rev 2011;10(10):631–5.

[149] Schoenfeld SR, Lu L, Rai SK, Seeger JD, Zhang Y, Choi HK. Statin use and mortality in rheumatoid arthritis: a general population-based cohort study. Ann Rheum Dis 2016;75(7):1315–20.

Editor's Summary: CardioRheumatology

Vasculature

AS associated with aortic dilatation
↑ arterial plaque and CIMT in pts with SLE & RA
RA pts have a 3X increased risk of carotid atherosclerosis compared to controls
Normal CIMT in pts with AS

Valvular Disease

1 in 4 pts with SLE have valvular lesions
SLE & APS pts often have NBTE
Mitral regurgitation is common in RA pts
AS pts often have AR, & may precede joint symptoms

Pericardium

Pericarditis seen in 20-50% pts with SLE, and up to 54% with RA, but often clinically silent
25% of SLE pts may have clinical pericarditis
Clinically significant pericarditis is rare in SSc

Pulmonary artery

10-20% of SSc pts have PHTN
25% of death in SSc are related to PHTN

Conduction System

Afib is the most common tachyarrhythmia in CID
17% of SLE pts have QT prolongation
2X increased risk of SCD in RA pts
60% of RA pts have autonomic dysfunction
Fibrosis of SA node & bundle branches seen on autopsy of SSc pts

Myocardium

SLE pts often have myocardial fibrosis by MRI
RA pts more likely to have HFpEF
Pts with SSc may have decreased RVEF even in absence of PHTN
Diastolic dysfunction is common in up to 20% pts with AS, & ↑ risk of HF

Coronary Arteries

Increased arterial stiffness with RA & SLE
Atherosclerosis progresses 2X faster in SLE pts than non SLE pts
SLE women have 50X increased risk of MI
SSc pts have MVD & 2.5X increased risk of MI
APS pts with ↑ arterial thrombosis, leading to MI

Niti Aggarwal

Cardiovascular Complications of Chronic Autoimmune Disorders. APS, antiphospholipid antibody syndrome; AR, aortic regurgitation; AS, ankylosing spondylitis; CID, chronic autoimmune disorders; CIMT, carotid intima media thickness; HF, heart failure; HFpEF, heart failure with preserved ejection fraction; MI, myocardial infarction; MRI, magnetic resonance imaging; MVD, microvascular disease; NBTE, nonbacterial thrombotic endocarditis; PHTN, pulmonary hypertension; RA, rheumatoid arthritis; SLE, systemic lupus erythematosus; SSc, systemic sclerosis. *Image courtesy of Niti R. Aggarwal.*

Chapter 24

CardioOncology

Iva Minga, Hena Patel, Tochi M. Okwuosa, and Niti R. Aggarwal

Sex Differences in Cardiac Disease. https://doi.org/10.1016/B978-0-12-819369-3.00028-9

Clinical Case

A 62-year-old female presented with a 1-month history of shortness of breath. She has medical history of postpartum cardiomyopathy 30 years prior. Her symptoms worsened with exertion, and she reported decreased exercise capacity. She denied any chest pain, nausea, or vomiting. Three months prior to her presentation, she had undergone partial mastectomy and received chemotherapy and radiation therapy (RT) for left-sided breast cancer (T3N1M0, stage IIIa). She underwent four cycles of doxorubicin chemotherapy (total dose ~240 mg/m²), followed by eight cycles of cyclophosphamide, methotrexate, and 5-fluorouracil. She was admitted to the hospital with orthopnea, and 10 kg weight gain. Physical examination revealed bibasilar crackles, elevated jugular vein distention (JVD) at 15 cm, S3 gallop, and 2+ bilateral lower extremity edema. Her chest X-ray showed pulmonary congestion with bilateral pleural effusions.

Echocardiogram before initiation of chemotherapy revealed a normal left ventricular size and left ventricular ejection fraction (LVEF) (5.3 cm in diastole and 68%, respectively). During the hospital admission, her initial electrocardiogram (ECG) revealed new T-wave inversions in the anterolateral leads. N-terminal protype natriuretic peptide was elevated to 6868 pg/mL (normal < 125 pg/mL). Serum troponins were negative. Her repeat echocardiogram revealed diffuse hypokinesis of the left ventricle (LV), an LV ejection fraction of 32%, and LV dimension in diastole of 6.5 cm. She had normal thyroid function test, and her cardiac catheterization revealed mild coronary artery disease (CAD) (20% nonobstructive atherosclerotic disease in the left circumflex artery), and elevated left ventricular end diastolic pressure of 20 mmHg. The patient was treated for an acute exacerbation of heart failure (HF) with diuretics. What is the most likely etiology for HF in this patient, and how would you manage her?

Abstract

Advances in cancer therapy have resulted in significant improvement in the long-term survival of cancer patients. As cancer patients live longer, cancer therapy-related cardiac dysfunction (CTRCD) is one of the major causes of morbidity and mortality among cancer survivors. Ultimately, the risk of death from cardiovascular disease (CVD) exceeds that of tumor recurrence for many cancers.

While CVD and cancer represent the main causes of morbidity and mortality among both men and women, there also exist gender-specific risk factors for the development of both cancer and CVD. Breast cancer is the most common cancer in women, while prostate cancer is the most common in men. This chapter highlights the sex differences in prevalence, risk factors, and management of CVD related to common cancer therapies, and identifies gaps in the literature for future research. In particular, the association of CVD with androgen deprivation therapy for prostate cancer and with breast cancer therapy is explored.

Introduction

Advances in cancer therapy have resulted in significant improvement in long-term survival of cancer patients. However, the side effects of cancer treatment often complicate further management of cancer and survivorship. Cancer therapy-related cardiac dysfunction (CTRCD) is one of the major causes of morbidity and mortality among patients undergoing cancer treatment and among cancer survivors. Ultimately, the risk of death from cardiovascular disease (CVD) exceeds that of tumor recurrence for many cancers. CVD is the leading cause of death among older female breast cancer survivors without preexisting diagnosis of CVD [1]. Heart failure (HF) as a result of cancer therapy has been linked to a 3.5-fold increase in mortality risk compared with idiopathic cardiomyopathy (CM) [2]. Cancer and CVD share many common risk factors including advanced age, unhealthy diet, obesity, and sedentary lifestyle. Even though cardiology and oncology are often considered separate medical subspecialties, they are frequently intertwined. Cardiovascular (CV) health and CVD can influence cancer outcomes. Cancer treatment can cause immediate or long-term CV toxicity [1].

While CVD and cancer represent a significant cause of morbidity and mortality among both men and women, there also exist gender-specific risk factors for both cancer and CVD. This chapter will focus on the most prevalent cancers in men and women, prostate and breast cancer, and discuss their treatment options, sex-specific risk factors, and management of CTRCD.

The Scope of the Problem and Epidemiology

CVD is the number one cause of mortality among both women and men in the United States [3]. Breast cancer and prostate cancer are two of the most common cancers affecting women and men, respectively, worldwide [4]. Cancer is associated with a high rate of morbidity and mortality among both men and women, and many chemotherapeutic agents used for breast cancer treatment have been associated with cardiotoxic side effects. Similarly, androgen deprivation therapy (ADT) used for prostate cancer treatment increases the risk of CV disease and myocardial infarction in men.

The lifetime risk of developing breast cancer in women is 12.4%, based on 2012–2016 data [1, 5]. The rate of new female breast cancer was 127.5 per 100,000 women per year, with a death rate of 20.6 per 100,000. Nearly 90% of breast cancer patients survive 5 years after initial diagnosis, resulting in 3 million breast cancer survivors in the United States [1, 5, 6]. Older survivors are more likely to die of diseases other than breast cancer, with CVD being the most frequent cause. The risk of CVD is higher in women with breast cancer history than in those without history of breast cancer [1, 7]. Cancer treatments unique to breast cancer can cause early or delayed cardiotoxicity, including severe hypertension (hypertension), arrhythmias, valvular heart disease, thromboembolic events, pericarditis, left ventricle (LV) dysfunction ischemia, and myocardial infarction [7].

In comparison, prostate cancer is the second most common cancer in men (followed by lung cancer), accounting for nearly 1.3 million new cases worldwide annually [4]. Its incidence increases with age, with an incidence rate of 60% in men over the age of 65 years. Despite its high prevalence, early prostate cancer often has an indolent course. The 5-year survival rate after diagnosis of prostate cancer is 98%. Despite this, it remains the second most common cause of cancer death in men. Intake of saturated fats and red meat, sedentary lifestyle, smoking, and obesity are all associated with increased risk of prostate cancer [4].

Sex Susceptibility to Cancer

The sex differences in cancer susceptibility are some of the most consistent findings in cancer epidemiology and can provide useful information to develop a causal hypothesis for disease, or define subgroups that are at the highest risk to guide preventive measures. Few cancers are more common in females, but overall males have higher susceptibility [8].

The latest analysis of the National Cancer Institute's Surveillance, Epidemiology and End Results (SEER) database showed that age-adjusted incidence rates were 501.9 for males and 417.9 for females per 100,000 persons, standardized to 2000 US standard population for all cancers from 2010 to 2014 [9]. The lifetime probability of developing cancer is 39.7% for males and 37.6% for females. The probability of developing cancers is higher for males despite a shorter life expectancy. The same analysis also showed that cancer mortality rates are 41% higher in males

vs. females (196.7 vs. 139.5) [9]. The common cancers that have the highest male-to-female predominance include: colorectal cancers: 1.35; lung and bronchus: 1.52; non-Hodgkin lymphoma (NHL): 1.44; and (urinary) bladder: 4.0 [8]. Other than breast cancer, which has < 1% incidence rate in males, only a few cancers are more common in females. For example, gallbladder, anal, and thyroid tumors consistently show a male-to-female ratio less than 1.0 [10].

The sex difference in cancer distribution and mortality (**Figures 1 and 2**) between both genders can be multifactorial due to immune surveillance, genome surveillance mechanism, hormonal differences, and number of X chromosomes [8].

Cancer and Cardiovascular Complications

Although primary cardiac tumors are extremely uncommon (postmortem studies report rates between 0.001% and 0.28%), secondary tumors are not [12]. Metastatic disease of the heart might range from 2.3% to 18.3%. The most common metastases are melanoma, lymphoma, leukemia, and carcinoma of the lung, breast, and esophagus [12]. Over 90% of cardiac involvement remains clinically silent, so its true incidence might be higher [13]. Indeed, malignant incidence at autopsy is around 10% in patients with known malignancies [14].

Pericardial metastasis, often leading to pericardial effusion, is the most common type of cardiac metastasis, followed by epicardial and myocardial metastasis [12]. Endocardial metastasis usually localized to the right heart is rare and associated with tumors with endovascular growth such as renal cell, liver, and uterine cancers [12]. Carcinoid heart disease is a rare cardiac manifestation occurring in patients with advanced neuroendocrine tumors, with carcinoid syndrome involving right-sided heart valves and leading to HF [15]. Carcinoid heart disease rarely involves the left side of the heart because of pulmonary metabolism and deactivation of hormonal substances by the lung [15].

Cancer can also predispose to arrhythmias [16]. One example is atrial fibrillation/flutter, with the manifestation of the arrhythmia sometimes preceding the diagnosis of malignancy. Chronic inflammation, metabolic change induced by cancer, and the presence of common risk factors such as obesity and alcohol use are the most plausible explanations for these associations.

Cancer is an independent risk factor for venous thromboembolism (VTE). It is estimated that approximately 4–20% of cancer patients will experience VTE at some stage of their course, with the highest incidence being at the initial period following diagnosis [17]. On the other hand, about 20% of new cases of VTE are associated with underlying cancers [18]. VTE in cancer is associated with a

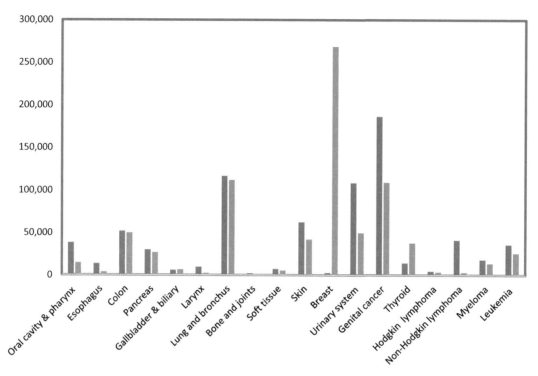

FIGURE 1 **Sex-Specific Incidence of Cancer.** Estimated new cases are based on 2001–2015 incidence data reported by the North American Association of Central Cancer Registries (NAACCR). The most common site of cancer diagnosis in women was breast compared to genital (prostate) in men. *Data from [11].*

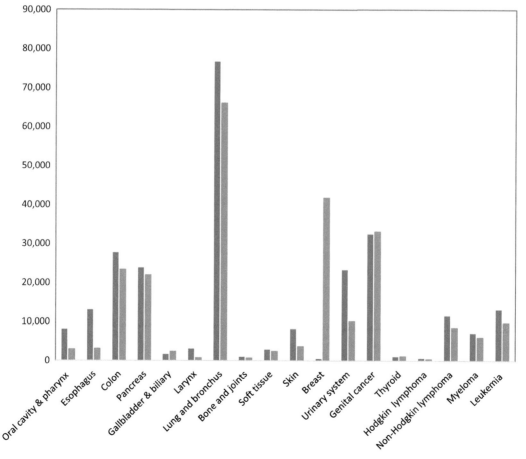

FIGURE 2 **Sex-Specific Death Rate Related to Cancer.** Estimated deaths are based on 2002–2016 US mortality data, National Center for Health Statistics, Centers for Disease Control and Prevention. Breast and prostate were the second most common sites of cancer in women and men, respectively, for cancer-related deaths, after lung and bronchus cancer. *Data from [11].*

21% annual risk of recurrent VTE, and 12% annual risk of bleeding complications [19]. The cancers associated with the highest incidence of VTE are pancreas, brain, stomach, and ovary [18]. The pathogenetic mechanisms of thrombosis involve complex interactions between tumor cells, the hemostatic systems, and characteristics of the patient [20]. Risk factors for thromboembolism and cancer include long-term immobilization (especially in-hospital), surgery, and chemotherapy. Prophylaxis in patients with cancers and VTE can be challenging due to the different complications these patients have [20].

Breast Cancer

Common Modifiable and Inherited Risk Factors That Predispose to CVD and Breast Cancer

Breast cancer and CVD share a number of common risk factors [1]. It is noted that approximately 80% of CVD can be prevented through risk factor modifications such as promoting a healthy diet, physical activity, healthy weight, abstinence from tobacco, blood pressure control, diabetes mellitus management, and optimal lipid control [21]. Adherence to a large number of ideal CV behaviors have been associated with a lower incidence of breast cancer [22].

Dietary habits: Dietary habits affect multiple established CV risk factors including blood pressure, lipid profile, and obesity [1]. Studies assessing the link between the Western diet and breast cancer have shown positive [23], negative [24], or no association with breast cancer [25].

Alcohol consumption: Alcohol consumption has been associated with both beneficial and negative effects in all-cause mortality in women [26]. Consuming ≥ 2 alcohol beverages per day for 5 years is associated with 82% increased breast cancer risk compared with no alcohol intake [27]. Alcohol might increase intracellular estrogen levels which can act through estrogen receptors (ERs) and promote breast cancer [1].

Physical inactivity: Physical inactivity is believed to be responsible for 12.2% of the burden of myocardial infarction after accounting for other CVD risk factors [28]. There is evidence that < 150 min/week of physical activity has implications for higher risk of both CVD and breast cancer [21, 29]. Studies have consistently shown that moderate to vigorous physical activity is associated with decreased risk of breast cancer between premenopausal and postmenopausal women [30]. Furthermore, a meta-analysis of 29 observational studies found significant reduction in breast cancer among the most physically active compared to the least active women [31]; another meta-analysis of 22 studies involving 123,574 participants found an inverse relationship between physical activity and breast cancer events and deaths [32].

Obesity: Similar to physical inactivity, overweight and obesity [body mass index (BMI) of ≥ 25 and $\geq 30 \, \text{kg/m}^2$, respectively] are risk factors for CVD [33]. Among women in the Women's Health Initiative ($n = 156,775$) with severe obesity compared with normal BMI, hazard ratios (HRs) for mortality were about 1.5 to 2.6 across black, Hispanic, and white women. For coronary heart disease, HRs were two- to nearly threefold higher [34]. Obesity is a complex risk factor, associated with higher risk of breast cancer in postmenopausal women, but lower risk in premenopausal women [1, 35, 36].

Smoking: Cigarette smoking is a major risk factor for CVD and stroke, but the risk of breast cancer with cigarette smoking remains inconclusive [1]. A large Canadian study found a nonsignificant trend toward increased breast cancer risk in premenopausal women who were active smokers [37]. Other studies have reported increased breast cancer risk among those who smoked for long durations or who were initiated to smoking early in life [38–40].

Age: Age is another factor affecting the risk of breast cancer and CVD. The incidence of breast cancer increases with advanced age, doubling approximately every 10 years until menopause, at which time the rate of increase slows [41]. On the other hand, the incidence of CVD increases steadily with advancing age but the rate becomes steeper at menopause [42].

Age of menarche and menopause: Age at menarche and menopause are additional risk factors for both breast cancer and CVD. Early menarche is associated with a two-fold increased risk of developing breast cancer [41] and higher risk of developing CVD, particularly in nonsmoking populations [43, 44]. In contrast, premature menopause confers a twofold increased risk of CVD even after adjusting for age, race, and other traditional CV risk factors, but 29% decreased risk of breast cancer for each year older at menopause [45, 46]. Premenopausal women experienced a 43% higher risk of breast cancer compared to age-matched postmenopausal women [46].

Hormone replacement therapy: The use of postmenopausal hormone replacement therapy (HRT) increases the risk of breast cancer. A National Health Service (NHS) study followed > 100,000 healthy women 30–50 years of age at the time of enrollment and found that the risk of developing breast cancer was 1.2–2 times greater for women who reported ≥ 5 years of current use of HRT than for those who never used HRT [47]. Similarly, HRT is also associated with an increased risk of CVD in older postmenopausal women and women with CHD [48, 49]. In the Women's Health Initiative, >16,000 postmenopausal women (average age 63.3 ± 7.1 years) with no history of coronary artery disease (CAD) were randomized to receive either HRT or placebo [49]. The HRT arm was prematurely terminated (after 5.2 years of follow-up) due to an increased risk of coronary heart disease with HRT (HR 1.29; nominal 95% CI 1.02–1.63) [49].

Cancer Therapy-Related Cardiac Dysfunction (CTRCD)

In simple terms, CTRCD refers to CM/LV dysfunction caused by cancer therapy. Several definitions of CTRCD have been proposed, posing a challenge for the development of uniformly accepted recommendations for diagnosis, surveillance, and treatment [2]. An expert consensus published by the American Society of Echocardiography (ASE) and the European Association of Cardiovascular Imaging (EACVI) defines CTRCD as a decrease in left ventricular ejection fraction (LVEF) of >10% to less than 53%, confirmed by repeat imaging 2–3 weeks after initial diagnosis [50]. The Breast Cancer International Research Group used >10% reductions from baseline LVEF to define asymptomatic LV dysfunction-associated with chemotherapy [51]. The Cancer Review and Evaluation Committee defined CTRCD as either: (1) a decrease in LVEF that is global or more severe in the septum with a decline in LVEF of at least 5% to <55% with accompanying signs of HF, or (2) or a decline of at least 10% to <55% without HF signs or symptoms [52]. The Food and Drug Administration defines anthracycline cardiac toxicity as >20% decrease in LVEF when baseline LVEF is normal, or >10% decrease when baseline LVEF is not normal [53]. The National Cancer Institute proposed the Common Terminology Criteria for Adverse Events that defines LV dysfunction and HF based on the severity of grades 1–5 [54]. Grade 1 is defined as an asymptomatic elevation in biomarkers and abnormalities in imaging. Grades 2 and 3 comprise symptoms on mild-to-moderate exertion. Grade 4 includes severe life-threatening symptoms and grade 5 involves death [54]. However, several trials have specified CTRCD with different parameters, making the estimation of the prevalence of CTRCD difficult [2].

Several attempts have been made to classify CTRCD. It has been defined as type I and type II based on structural abnormalities and reversibility. Type I is irreversible and dose related with myocyte injury. Type II includes reversibility, lack of dose relationship with myocyte injury, and the absence of structural abnormalities. Anthracyclines have been associated with type I toxicity and trastuzumab with type II. However, this categorization might be too simplistic to encompass the entire CTRCD spectrum, and it is therefore losing popularity among experts. For example, trastuzumab CTRCD is associated with irreversible troponin elevation. In clinical practice, CTRCD might represent a synergistic/combined action as often patients receive multiple drugs at the same time [2].

Breast Cancer Chemotherapy Agents

Cancer treatments commonly used in breast cancer often result in cardiotoxicity including conduction abnormalities, arrhythmias, LV dysfunction, and myocardial ischemia (**Table 1**).

Anthracyclines (Doxorubicin, Daunorubicin, Epirubicin, and Idarubicin)

Anthracyclines are the first chemotherapeutic agents to have been associated with CTRCD. Although the exact mechanism remains unknown, multiple mechanisms have been proposed (**Figure 3**) [55]. Anthracyclines induce the generation of reactive oxygen radicals that accelerate myocardial cell death. Another potential mechanism includes intercalation and inhibition of macromolecular biosynthesis and progression of the topoisomerase IIB, which preserves deoxyribonucleic acid (DNA) integrity within cardiac myocytes, disrupting replication and causing myocardial cell death [55, 56]. A recent computational study compared four anthracyclines (doxorubicin, epirubicin, idarubicin, and daunorubicin) [56]. Doxorubicin has the fastest rate of diffusion between the cell membrane bilayer, exhibits different orientation once incorporated into the cell membrane, and has a higher propensity to interact with the hydrocarbon tail within the lipids in the membrane bilayer—all characteristics thought to be contributing to higher cytotoxic effects [56].

A 9% incidence of cardiotoxicity events was observed in 2625 anthracycline-treated patients for various types of cancer, who underwent periodic scheduled surveillance after chemotherapy with a median follow-up of 5.2 years. Women composed 74% of this cohort with breast cancer making up 51% of all cases [57]. In a meta-analysis of 55 published randomized controlled trials for various types of cancer treated with doxorubicin, the risk of LVEF reduction was six times higher compared with a nonanthracycline containing regimens [odds ratio (OR) 6.25, 95% CI 2.58–15.13] [58]. Anthracycline-related cardiotoxicity represents a continuum that begins with subclinical myocardial cell injury followed by an early asymptomatic decline in LVEF that can progress to symptomatic HF if left untreated [59].

The myocardial toxicity of the anthracyclines can manifest either early (within 1 week) or late following exposure. The early manifestations are related to inflammation resulting in myocarditis and pericarditis. In contrast, late manifestations are related to actual myocyte damage leading to cardiac dysfunction, and resulting in HF [1, 60]. The risk of cardiotoxicity seems to be dose dependent: 5% risk at 400 mg/m^2, 26% risk at 550 mg/m^2, and 48% risk at a cumulative dose of 700 mg/m^2 [1]. However, there is no "safe dose." Even a lower dose of anthracyclines can lead to cardiac toxicity, especially in patients with preexisting CV

TABLE 1 Breast Cancer Treatments and Associated Cardiotoxicity

Treatment	Cardiotoxicity Manifestation
Anthracyclines (doxorubicin, daunorubicin, epirubicin, and idarubicin)	LV dysfunction Myocarditis/pericarditis Conduction abnormalities VT/Vfib, AF QTc prolongation
Alkylating Agents (cyclophosphamide, carboplatin)	LV dysfunction Hemorrhagic myocarditis Bradycardia, SVT, AF
Taxanes (docetaxel, paclitaxel)	Bradycardia, heart block VT Myocardial ischemia
HER2 Inhibitors (trastuzumab, pertuzumab, T-DM1, lapatinib)	LV dysfunction
Endocrine Therapy Tamoxifen Aromatase inhibitors Fulvestrant	**Tamoxifen** Favorable LDL profile but increases serum TG levels VTE (DVT, PE) Thrombophlebitis **Aromatase inhibitors** VTE less than tamoxifen Hypercholesterolemia Dysrhythmias Valvular dysfunction Pericarditis LV dysfunction **Fulvestrant** VTE
Antimetabolites (5-fluorouracil, capecitabine)	Coronary vasospasms Hypotension Myocardial ischemia
mTOR Inhibitors Everolimus	Hypertension Hyperglycemia Hyperlipidemia
CK4/6 Inhibitors (ribociclib, palbociclib, abemaciclib)	QTc prolongation VTE
Radiation Treatment	LV dysfunction Valvular heart disease Pericardial disease Arrhythmias Autonomic dysfunction

AF, atrial fibrillation; DVT, deep vein thrombosis; HER2, human epidermal growth factor receptor 2; LDL, low-density lipoprotein; LV, left ventricle; PE, pulmonary embolism; SVT, supraventricular tachycardia; T-DM1, trastuzumab emtansine; TG, triglyceride; Vfib, ventricular fibrillation; VT, ventricular tachycardia; VTE, venous thromboembolism.

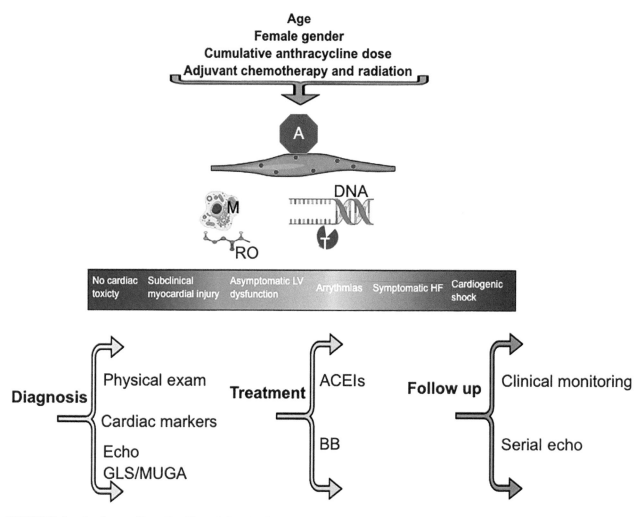

FIGURE 3 Anthracycline Cardiotoxicity. Anthracycline cardiotoxicity spans a spectrum from no cardiac toxicity to cardiogenic shock. A, anthracycline; ACEIs, angiotensin-converting enzyme inhibitors; BB, beta-blockers; DNA, deoxyribonucleic acid; echo, echocardiogram; GLS, global longitudinal strain using speckle-tracking echocardiography (STE); HF, heart failure; LV, left ventricle; M, mitochondria; MUGA, multigated radionuclide angiography; RO, radical oxygen species; T, topoisomerase.

risk factors [61–63]. On average, there is a persistent LVEF decline of ~4% over the follow-up duration (1-year change in LVEF −3.6%, 95% CI −4.4%, −2.8%; 3-year change −3.8%, 95% CI −5.1%, −2.5%) [63]. Though findings have been inconsistent, changes in diastolic parameters such as isovolumetric relaxation and deceleration time are present in patients as early as 3 months following doxorubicin, and precede systolic dysfunction [64]. The overall incidence of late-onset CM associated with anthracyclines is about 11%, with about 50% mortality [65]. Indeed, anthracycline-related CM confers higher mortality than idiopathic CM (HR 2.64, 95% CI 1.35–5.17, P = 0.005) [66].

The risk factors that predispose to anthracycline toxicity are summarized in **Table 2**, and include cumulative anthracycline dose and age (with a bimodal age distribution) as the strongest risk factors, concomitant CV risk factors, comorbid conditions such as arrhythmia, concomitant cancer therapies (such as trastuzumab and chest wall RT), female sex, and infusion route and rates. In addition, doxorubicin carries a higher risk than epirubicin or liposomal doxorubicin [69, 71, 73–76].

Anthracyclines have also been associated with conduction abnormalities and QTc prolongation, which can occur as early as 24 h after anthracycline infusion [1, 77]. Doxorubicin is related to arrhythmias and conduction abnormalities in 2.6% of patients as opposed to 1% of patients who did not receive doxorubicin, and atrial fibrillation in 2–10% of patients [1].

TABLE 2 Risk Factors for Anthracycline-Induced Cardiotoxicity

Risk Factor	Special Notes	Noted Increase in Risk
Cumulative dose	Significantly increased risk at doses over 500 mg/m²	7.5–26% at 550 mg/m² [67, 68] vs. 1.7% at 300 mg/m² [67], and 3.0–5.0% at 400 mg/m² [68]; 9% increased risk for every 50 mg/m² added [69]
Age	Bimodal age distribution: higher risk noted in age younger than 4 years or older than 65 years	32% for less than age 4 years [70] and 125% at age greater than 65 years [67]; 7% increased risk for each additional 5 years of age [69]
Duration of time postchemotherapy	Incidence increases progressively postchemotherapy	Median time for cardiotoxicity postchemotherapy is 3.5 months [69]
Comorbid cardiac risk factors	Particularly hypertension, but also diabetes, coronary disease, obesity, renal dysfunction, pulmonary disease, electrolyte abnormalities, infection, and pregnancy	45–58% for hypertension [66, 71], 27–74% for diabetes [66, 71], and 58% for coronary artery disease [71]
Concomitant cardiotoxic chemotherapy	Trastuzumab, taxanes, and cyclophosphamide; either potentiating risk or overlapping toxicity	Dose and drug dependent
Chest wall radiation	High-dose radiation for left-sided breast cancer, lung cancer, mediastinal lymphomas with concomitant or prior anthracycline therapy	23% for high-dose radiation vs. low-dose radiation [64]
Female sex	Particularly in the pediatric population	100% increased risk in females compared with males at a given dose in pediatric patients [65], 61% for all age females [69]
Echocardiographic evidence of toxicity	Drop in LVEF postchemotherapy from baseline prior to starting chemotherapy	37% increased risk of cardiotoxicity for each percent unit drop in LVEF at the end of chemotherapy [69]
Infusion rates	Continuous infusion is less cardiotoxic	9.5% with continuous infusion vs. 46.6% with bolus IV injection [72]

IV, intravenous; LVEF, left ventricular ejection fraction.

Alkylating Agents (Cyclophosphamide and Carboplatin)

Alkylating agents damage DNA by inhibiting transcription and affecting protein synthesis. They have been associated with the development of LV dysfunction in 7–28% of patients [78]. This effect for cyclophosphamide may be dose related (≥150 mg/kg and 1.5 g/m²d), occurring within 10 days after initial administration, and appears to be related to prior anthracycline therapy or mediastinal RT [1, 2]. Hemorrhagic myocarditis, supraventricular tachycardia, and atrial fibrillation have also been reported with use of alkylating agents [1, 79].

Taxanes (Docetaxel and Paclitaxel)

Taxanes block mitotic progression in the cell microtubules, resulting in apoptosis [80]. They are often administered sequentially or in combination with anthracyclines in early-stage breast cancer; they can affect metabolism and excretion of anthracyclines which may then potentiate the risk of LV dysfunction, particularly with use of high doses of anthracyclines. The incidence of HF with these agents has been reported at 1.6% vs. 0.7% in anthracycline sparing regimens [2]. Slower infusion of paclitaxel and doxorubicin decrease cardiac toxicity [2]. Furthermore, taxanes have been associated with conduction block, ventricular tachyarrhythmias, ischemia, and bradycardia in up to 29% of patients [1, 81].

Antimetabolites (5-Flouro Uracil)

Antimetabolites-related cardiotoxicity is incompletely understood and may arise from a combination of ischemia related to coronary vasospasm and direct myocardial cell toxicity [1]. Patients with preexisting CVD receiving continuous infusions of 5-Flouro Uracil (5-FU), as opposed to a bolus-based regimen, may be at increased risk. Management

of affected patients focuses on determining whether 5-FU can reasonably be attributed to the cardiotoxicity, identifying and treating other coexisting coronary disease, and determining if further 5-FU is required or whether acceptable alternative treatments can be safely used [82].

Targeted Therapies

The newer cancer therapies specifically target the hormone receptors on breast cancer cells and are summarized in **Figure 4**.

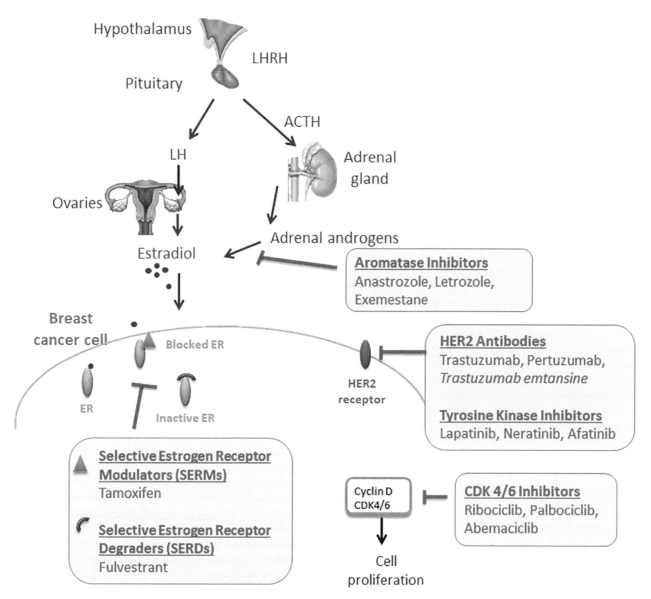

FIGURE 4   Mechanism of Action for Endocrine and Targeted Therapies for Breast Cancer. ACTH, adrenocorticotropic hormone; CDK, cycline-dependent kinase; ER, estrogen receptor; HER2, human epidermal growth factor receptor 2; LH, luteinizing hormone; LHRH, luteinizing hormone-releasing hormone.

HER2 Antibodies (Trastuzumab, Pertuzumab, and Trastuzumab Emtansine)

Breast cancer cells often express human epidermal growth factor receptor 2 (HER2). HER2 therapies (trastuzumab and pertuzumab) are monoclonal antibodies that bind to the extracellular domain of the ErbB2 used in the treatment of HER2-positive breast cancers. Trastuzumab-related cardiotoxicity is a well-studied phenomenon, and is of particular concern especially when patients have a history of previous or concomitant use of anthracyclines [1]. The mechanism leading to cardiotoxicity is thought to be related to myocyte dysfunction rather than myocyte death, as cardiac toxicity almost always occurs during treatment [83].

Risk factors for trastuzumab-induced cardiotoxicity include BMI $>25 \text{kg/m}^2$, low LVEF prior to treatment, age >50 years, and prior treatment with anthracyclines. In contrast to anthracycline-induced cardiotoxicity, trastuzumab-related cardiotoxicity is mostly reversible [1].

The use of trastuzumab has decreased breast cancer recurrence by 50% and mortality by 33%, but with incidence of cardiac events of up to 3.9% [1]. Population-based studies, especially in older individuals, have suggested higher rates of cardiac toxicity. In 9535 women with early-stage breast cancer, 23.1% received trastuzumab; the rate of HF was 29.4% compared to 18.9% in nonusers of trastuzumab ($P<0.001$) [84]. Furthermore, a meta-analysis of adjuvant trials with 1811 patients concluded that adjuvant therapy with trastuzumab is associated with a threefold increased risk of grade III and IV HF (95% CI 1.12–7.85; $P<0.01$) [85].

While single-agent therapy with HER2-targeted medications is well tolerated, the response even in the HER2 responsive cancers is suboptimal. As a result, these agents may be combined with more potent chemotherapeutic agents for greater cytotoxic effects. One such combination is **trastuzumab emtansine (T-DM1)**, where the transtuzumab is conjugated with a chemotherapy drug (DM1). Overall, while it has a favorable cardiotoxicity profile, CTRCD has been reported. In the EMILIA trial, 1.7% treated with T-DMI exhibited LVEF less than 50% [86], and grade III LV systolic dysfunction developed in 1 of 495 patients in the T-DM1 group. In the MARIANNE study, LVEF $\leq 50\%$ with ≥ 15 percentage point decrease from baseline was observed in 1.1% of patients in the T-DM1 group vs. 4.8% with trastuzumab plus taxane, and 3.0% with T-DM1 plus pertuzumab groups [87].

Tyrosine Kinase Inhibitors (Lapatinib, Neratinib, and Afatinib)

Tyrosine kinase inhibitors (TKIs) target the HER2 intracellular pathway and are used in women with HER2-positive breast cancers. They activate the AMP-kinase pathway and increase ATP reserves with subsequent induction of metabolic stress response in human cardiomyocytes, which might protect cardiomyocytes against cell death [83]. Lapatinib showed a favorable cardiac safety profile, but cardiac events have been reported [88]. In a landmark phase III trial, 12 cardiac events were reported. Decrease in LVEF was observed in 11 patients with 4 patients developing $\geq 20\%$ absolute decrease in LVEF from baseline values. The mean time to onset of decrease in LVEF was 63 days, and the mean nadir for absolute decline compared with baseline was 26% [89]. Neratinib and afatinib are a novel next generation of TKIs that are designed to overcome resistance by targeting multiple HER family members and irreversibly binding the targets [90]. They are under clinical development. Two trials in metastatic settings have become available, and grade 3 or higher cardiac events [defined as congestive heart failure (CHF), decreased LVEF, left ventricular systolic dysfunction (LVSD), and peripheral edema] were reported in 1.3% in the neratinib arm vs. 3.0% in the trastuzumab-paclitaxel arm [91].

Endocrine Therapy

Nearly 60–70% of breast cancers respond to estrogen and progesterone [92]. The presence of estrogen and progesterone receptors on breast cancer cells portends a favorable response to endocrine therapies. Strategies to inhibit estrogen effects include selective binding to the ER [selective estrogen receptor modulators (SERMs) and selective estrogen receptor degraders (SERDs)] and preventing intrinsic production of estrogen [aromatase inhibitors (AIs)]. Endocrine therapies are used to treat both early-stage and metastatic endocrine (estrogen and progesterone) receptor-positive breast cancers for long periods of time—up to 5–10 years for early-stage cancer—and are also used to treat endocrine receptor-positive metastatic breast cancers.

Selective Estrogen Receptor Modulators (Tamoxifen)

SERMs are the drug of choice in premenopausal women with ER-positive breast cancer. They antagonize the ER and bind to it with high affinity. In breast tissue, however, they act as an estrogen antagonist, and halt estrogen-dependent tumor growth. However, they often mimic effects of estrogen, and may result in vasomotor symptoms such as hot flashes. In contrast, in bone, uterine tissue, and the CV system, they portend an estrogen-agonistic effect. As a result, they have a black box warning for adverse effects on uterine lining and may result in uterine malignancies. *Tamoxifen* has also been associated with venous thromboembolism (HR 5.5; 95% CI 2.3–12.7) and has an unfavorable effect on serum triglycerides [1, 93]. However, despite an estrogen-agonist effect on the heart, tamoxifen does not confer a cardioprotective effect.

Selective Estrogen Receptor Degraders (Fulvestrant)

SERDs bind to ERs and prevent the binding of cancer cells. They act as an estrogen antagonist, and are particularly used as second-line agents in patients with ER-positive breast cancer, who have inadequate response to SERMs. With fulvestrant, ischemic CV disorder varies between 1.4% and 1.9%, and thromboembolic events range from 1.1% to 5.8%, but only limited data exist on the cardiac safety of SERDs [83].

Aromatase Inhibitors (Anastrozole, Exemestane, and Letrozole)

In the premenopausal state, most of the estrogen is synthesized in the ovaries and placenta. In contrast, following menopause, estrogen is predominantly synthesized in peripheral tissues including in subcutaneous fat, liver, and adrenal glands, and is facilitated by the aromatase enzyme. The aromatase inhibitors (AIs) block the transformation of androgens and testosterone into estrogen and estradiol. They are predominantly used in postmenopausal women with ER-positive breast cancer. Side effects of AIs are due to decreased effects of estrogen. They are associated with a 37% increased risk of myocardial infarction (HR, 1.37 [95% CI, 0.88–2.13]) and a 86% increased risk of heart failure (HR, 1.86 [95% CI, 1.14–3.03]) compared with the use of tamoxifen [94]. They also negatively impact the lipid profile with a 2.3-fold increase in risk of hypercholesterolemia compared with tamoxifen [95]. Patients receiving AIs have significantly higher risks of other forms of CVD including dysrhythmia, valvular dysfunction, pericarditis, or CM, but the clinical significance is yet to be described [96].

mTOR Inhibitor (Everolimus)

The mechanistic target of rapamycin (mTOR) inhibitor, *everolimus*, is an immunosuppressive drug that has been associated with hyperglycemia and dyslipidemia in 11% and 14% of patients, respectively [97].

Cycline-Dependent Kinase Inhibitors (Ribociclib, Palbociclib, and Abemaciclib)

The newer *cycline-dependent kinase (CDK) 4/6 inhibitors* (ribociclib, palbociclib, and abemaciclib) are being developed to overcome endocrine resistance, and are typically used in combination with other endocrine therapies for metastatic breast cancer. They inhibit the phosphorylation pathway which is mediated by estrogen, particularly in ER-positive cells. CDK 4/6 inhibitors have been associated with QTc prolongation, which seems to be dose dependent, as well as thromboembolic events ranging from 0.9% to 5% incidence [98].

Predicting Risk of CTRCD

There are multiple risk factors that may lead to the development of chemotherapy with cancer therapy (**Table 3** and **Figure 5**). Patient-related risk factors include preexisting LV dysfunction or HF, CAD, older age, female sex, and postmenopausal status. Therapy-related risk factors include combination cancer therapy (especially if administered simultaneously), type of chemotherapy, prior/current treatment with RT, and previous treatment with anthracyclines [2].

Several risk scores have been proposed to predict CTRCD. Herman et al. included patient-related risk factors and treatment-related risk factors [60]. Patient risk factors include CM/HF, CAD, hypertension, diabetes, prior anthracycline use, prior or concurrent RT, age < 15 years, age > 65 years, and female gender. Treatments were categorized

TABLE 3 Risk Factors Associated With Cancer Treatment-Related Cardiovascular Disease

Patient-Related Risk Factors	*Cardiovascular Risk Factors*
- Increased age	- Hypertension
- Race	- Diabetes
- Female sex	- Coronary artery disease
- Postmenopausal status	- Baseline reduced ejection fraction
Lifestyle	- Arrhythmias (atrial fibrillation)
- Increase alcohol consumption	- Renal failure
- Obesity	- Dyslipidemia
- Smoking	

Genetic Risk Factors	*Treatment-Related Risk Factors*
- Variant of amino acid 21,170 on the HER2 gene	- Cumulative doses of chemotherapy
- Ile655Val SNP	- Choice of anthracycline
	- Alternative delivery methods
	- Combination chemotherapy
	- Concomitant radiation therapy

HER2 gene, human epidermal growth factor receptor 2; SNP, single-nucleotide polymorphisms.

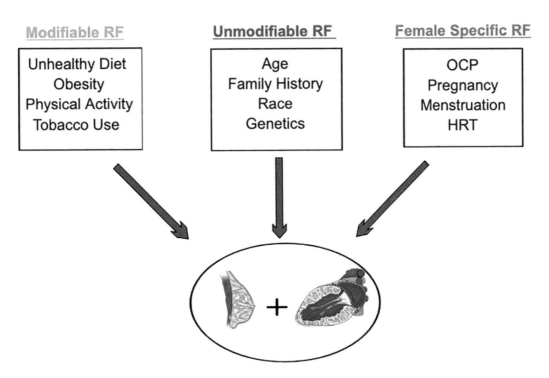

FIGURE 5 Risk Factors in Developing Breast Cancer and Cardiotoxicity. HRT, hormonal therapy; OCP, oral contraceptive pill; RF, risk factors.

into high risk, intermediate risk, low risk, and rare. A scoring system—cardiotoxicity risk score (CRS)—was developed with 0 indicating very low risk, 1–2 low, 3–4 intermediate, 5–6 high, and > 6 very high risk [60]. Ezaz et al. developed a seven-point risk factor score system including age, race, CAD, stroke/transient ischemic attack, atrial fibrillation, diabetes, hypertension, renal failure, and therapy type [99]. Similarly, Romond et al. included age and baseline LVEF in HER2 + breast cancer [100]. Dranitsaris et al. developed a risk assessment score in metastatic breast cancer patients treated with anthracyclines that includes age, weight, baseline LVEF, anthracycline exposure, and performance status for each chemotherapy cycle [101].

Female-Specific Risk Factors for Chemotherapy-Related Cardiovascular Dysfunction

Susceptibility in developing cardiac dysfunction during cancer therapy is multifactorial, including patient-specific risk factors, therapy, and environmental factors [2].

Genetics

Genetic factors play a significant role in developing breast cancer and CVD. BRCA1 and BRCA2 genes account for 5–10% of all breast cancer cases [1]. HER2 genetics is tar-geted therapeutically to prevent the progression of breast cancer, and genetic risk factors have also been identified in developing cardiotoxicity. Animal studies have shown an important role of the BRCA1 and BRCA2 genes in promoting DNA damage repair and regulating cardiomyocyte survival. Furthermore, mice that express BRCA (BReast CAncer gene) mutations are more susceptible to developing HF after anthracycline exposure [102]. However, a retrospective study with 102 breast cancer patients who were BRCA mutation carriers and were treated with anthracycline did not find increased risk of cardiotoxicity among the BRCA group vs. the comparison group [4.9% rate of HF in the BRCA mutation carriers vs. 5.2% in the matched controls (P = 0.99)] [103].

Age

Breast cancer patients make up the largest group of cancer survivors, accounting for 23% of all survivors, and breast cancer is more common in older women. With increasing life expectancy in the next 10 years, 70% of newly diagnosed patients with cancer will be older than 65 years. Those older than 65 years, account for more than half of CVD hospitalization and approximately 80% of CVD deaths [6, 104]. However, there exists a paucity of data on cardiooncology and older adults as they are often underrepresented in clinical trials [105].

The American Society of Clinical Oncology (ASCO) guidelines estimate 1.6- to 6.8-fold increased risk of cardiac dysfunction with anthracycline treatment in elderly patients (age ≥ 60 years), compared with younger patients [106]. In a population-based retrospective cohort study of 12,500 women with invasive breast cancer treated with anthracycline alone, trastuzumab alone, or combined chemotherapy, the 5-year cumulative incidence for CM was higher in older adults [107]. The incidence rate in the anthracycline group was higher among age 65–74 years, (cumulative incidence 6.2% [95% CI 3.9–8.5%]); and among age ≥ 75 years (cumulative incidence 10.6% [95% CI 3.9–16.9%]). Also, the risk of CM in the anthracycline plus trastuzumab group was higher among age 65–74 years (cumulative incidence 35.6% [95% CI 12.5–52.5%]); and among age ≥ 75 years (cumulative incidence 40.7% [95% CI 0.0–71.6%]) [107].

Female Sex

The role of sex in CTRCD is not well defined [108]. Female sex has been identified as an independent risk factor for anthracycline-induced cardiotoxicity in several pediatric studies [108]. In 1995, Lipshultz et al. reviewed echocardiograms from 120 children and adults who had received doxorubicin as part of their cancer therapy. Females had a significantly greater reduction in contractility than males ($P < 0.001$). This difference seems to be accentuated with a higher cumulative dose of anthracycline treatment [109]. Other studies have also reported female sex as an independent risk factor for cardiotoxicity, with a fourfold higher risk of HF in females than males [110–112]. However, there are mixed results, with some studies reporting female sex to be a protective factor for the development of cardiotoxicity [113, 114]. Hormonal status and body fat percentage found in girls can account for the difference in risk of CTRCD in this population [115].

In contrast to the many studies on sexual dimorphism in the pediatric cancer population, there is much less research documenting differences in adults. This might be due to the majority of CTRCD having been performed in breast cancer, predominantly a female pathology [108]. Female sex was protective against LV systolic function for subclinical CM in adult lymphoma patients (OR 0.324; 95% CI 0.106–0.989) [116]. Similarly, in a large cohort of patients receiving anthracyclines for different types of malignancies, major adverse cardiac events (symptomatic HF or cardiac death) were more frequent in men than in women ($P = 0.03$) [117].

The lack of consistency in study designs and the different definitions of cardiotoxicity preclude a definitive consensus regarding the role of sex as a risk factor for CTRCD. More clinical research using reliable techniques is needed to determine the role of sex in CTRCD [108].

Menopause

Postmenopausal status can be a risk factor for CTRCD; however, limited data exist [2]. In a retrospective study of 111 patients who received trastuzumab for their breast cancer treatment, postmenopausal status was associated with a decrease in LVEF ($P < 0.01$) [118]. However, in another retrospective study of 428 women who underwent trastuzumab treatment, postmenopausal status was not significantly associated with a decline in LVEF ($P = 0.13$) [119]. Postmenopausal status was associated with higher arterial stiffness in patients who had received treatment with anthracyclines or trastuzumab. Increased arterial stiffness has been described as an independent risk factor for subclinical cardiac damage ($P = 0.002$) [120]. As such, chemotherapy could indirectly contribute to coronary atherosclerosis, as premature menopause can occur in up to 40% of women under the age of 40, and in up to 80% of women older than 40, depending on the type of chemotherapy regimen [121].

Pregnancy

Pregnancy can pose a significant clinical challenge in women who had previously been exposed to cardiotoxic treatments, especially if they have experienced previous cardiotoxic complications. However, limited data exist on this topic. In a retrospective study, 78 cancer survivors with a total of 94 high-risk pregnancies were followed over a 10-year period [122]. All women had previously received cancer therapy; 55 women had received anthracycline-based chemotherapy, while 23 women had received either nonanthracycline chemotherapy or RT. The incidence of HF in women with prior history of cardiotoxicity was 31% vs. 0% in women without prior history of cardiotoxicity ($P < 0.001$) [122].

Hines et al. included 847 females who were childhood cancer survivors with 1556 live births [123]. Of all patients, 57% received anthracycline-based treatments, while 43% did not. Of the 847 survivors, 3 developed pregnancy-associated CM, 14 developed CM > 5 months postpartum, and 26 were diagnosed with CM prior to pregnancy. These data suggest the rate of cardiac dysfunction in pregnant childhood cancer survivors is approximately 1:500, which is much higher than the peripartum CM risk in the general population (1:3000 to 1:4000 women). The risk of worsening cardiac function was higher in patients with a prior decrease in cardiac function, with almost 30% of patients decompensating peripartum [123]. At MD Anderson, 58 women with subsequent pregnancies who were previously treated with anthracyclines and/or chest RT were compared with 80 nonpregnant women with similar treatment [124]. Approximately, 28% of these women had a decreased LVEF prior to, during, or after pregnancy. Younger age at the time of cancer diagnosis, longer time from cancer treatment to first pregnancy, and higher total anthracycline dose were associated with increased risk. Pregnancy was also identified as an independent risk factor with a 2.35-fold increase in cardiac risk [124].

Detection, Prevention, and Management of Cardiotoxicity Due to Chemotherapy

Detection of Cardiotoxicity Due to Chemotherapy

Imaging Biomarkers

Echocardiography

Echocardiography is currently the cornerstone of imaging before, during, and after potentially cardiotoxic therapy [125]. Advanced techniques using contrast echocardiography and three-dimensional (3D) echocardiography have significantly improved the accuracy of LVEF assessment [125]. Although there is no clear consensus, most studies use a decline in LVEF of 10% on serial studies as a marker of significant change in left ventricular function and a surrogate for cardiotoxicity. However, a major limitation to the use of LVEF for cardiotoxicity monitoring is that reduction in LVEF is usually a late manifestation of CTRCD when significant and irreversible myocardial damage has already occurred. Thus, efforts have focused on identifying more sensitive indicators of subclinical cardiotoxicity with the hypothesis that early detection of myocardial injury would translate into earlier treatment and higher chances of preventing long-term HF events.

Speckle-Tracking Echocardiography

Myocardial strain is a measure of regional myocardial deformation obtained by angle-independent two-dimensional (2D) speckle-tracking echocardiography (STE). Strain (the total deformation of the ventricular myocardium during a cardiac cycle expressed as a percentage) and strain rate (the rate of deformation) have been proposed as more sensitive tools for the detection of subclinical LV systolic dysfunction, compared to LVEF.

Global longitudinal strain (GLS) using STE has been shown to most consistently detect myocardial injury before the development of reductions in LVEF [63, 125–127]. In a prospective study of 81 patients receiving adjuvant anthracycline-based chemotherapy plus trastuzumab, a GLS of greater than − 19% after the completion of anthracycline treatment was predictive of later development of CTRCD [positive predictive value (PPV) 53%, negative predictive value (NPV) 87%], defined as a symptomatic reduction in LVEF of ≥5% to <55% or an asymptomatic reduction of LVEF ≤10% to <55% [126]. Others demonstrated that an 11% reduction in GLS after chemotherapy from baseline was the strongest predictor of CTRCD, defined by a reduction of LVEF ≥10% [128]. The EACVI/ASE and ASCO

guidelines recommend routine assessment of GLS in cancer patients at risk for CTRCD [128]. A relative percentage decrease in GLS >15% is indicative of subclinical LV dysfunction, whereas a relative decrease <8% is consistent with no evidence of subclinical LV dysfunction [50]. A relative decrease in GLS between 8% and 15% is a gray zone and closer follow-up should be considered to see if there is a trend toward further decrease in GLS at the next assessment. Strain assessment can also be used to help identify subclinical LV dysfunction at baseline prior to cancer treatment [129].

Cardiac MRI

Early detection of cardiotoxicity can allow for earlier treatment and decrease progression of CM. Cardiac magnetic resonance (CMR) is useful in the diagnosis of CTRCD and is the gold standard for assessing LV function. CMR can detect myocardial edema, which can be part of myocardial injury and fibrosis and can identify subclinical myocardial changes prior to clinical symptoms associated with LV dysfunction [130]. Anthracycline toxicity is associated with a decrease in LV mass, and CMR can measure LV mass in a reproducible manner [130]. CMR studies in women treated with anthracyclines have demonstrated abnormal LVEF even at 1 month after anthracycline therapy. The prognostic implications are yet to be known [131].

CMR can also identify an increase in afterload. Breast cancer and hematologic malignancies undergoing chemotherapy treatment with anthracyclines, cyclophosphamide, and/or trastuzumab have shown elevated measures of pulse-wave velocity in the aorta and decreased ascending thoracic aortic distensibility. In one study of 29 patients with breast cancer receiving sequential anthracycline and trastuzumab treatment, aortic pulse-wave velocity increased 4 months after therapy but resolved 1 year thereafter [130, 132].

CMR can also be used for the diagnosis of cancer treatment-related myocarditis, defects in myocardial perfusion, pericardial disease, and valvular disease [131]. Additional research is needed to integrate CMR measures into the guidelines for LV function monitoring in cancer survivors. It should be noted that the costliness and time-intensive nature of the scans and report generation, as well as the relatively lower availability of CMR, are all factors which limit its routine use for monitoring of CTRCD in clinical practice.

Multigated Radionuclide Angiography (MUGA)

MUGA scan is a noninvasive technique using ^{99m}Tc erythrocytes to visualize cardiac blood pool through a γ camera with gated acquisition [133]. It allows for highly reproducible quantification of LV volumes and LVEF during cancer therapy [134]. However, its use may be limited by soft tissue attenuation artifacts and may expose patients to ionizing radiation [135]. Nousianen et al. documented that a MUGA scan had 90% sensitivity and 72% specificity for predicting development of chronic HF [136]. However, the results of this study were not reproduced in other studies [137]. Serial MUGA scans are falling out of favor for LV function assessment. Radiation exposure is considered to be the main disadvantage of MUGA. Each MUGA scan is reported to be equivalent to 5–10 millisieverts of radiation [50].

Serum Biomarkers

Troponin

Cardiac biomarkers may play a complementary role to cardiac imaging in monitoring patients for cardiotoxicity. Elevations in cardiac troponins (Tn), a marker of cardiac injury, either early (checked with each cycle of chemotherapy) or late (1 month after completion of last cycle of chemotherapy), have been shown to be predictive of LVEF reduction and cardiac events [138]. Cardinale et al. performed serial Tn monitoring at multiple time points in 703 patients treated with high-dose chemotherapy and observed the highest incidence of cardiac events and LVEF reduction in patients with troponin ≥ 0.08 ng/mL, both early (within 3 days) and late (1 month) after chemotherapy. In contrast, patients with no chemotherapy-induced troponin rise had very low event rates (16% compared to 37% in the troponin-positive group) [138]. There was also a higher prevalence of women in the troponin-positive group compared to the normal troponin group (76% vs. 66%). A similar study performed in patients treated with trastuzumab demonstrated that elevated Tn was associated with lack of LVEF recovery despite HF therapy [139].

New high-sensitivity troponin assays have been under investigation for early detection of cardiotoxicity. A study by Sawaya et al. [140] demonstrated a significant increase in ultrasensitive Tn among patients with HER2-positive breast cancer treated with anthracyclines and trastuzumab. The mean ultrasensitive Tn concentration after completion of anthracycline chemotherapy was higher among women who developed cardiotoxicity (32 vs. 17 pg/mL), and an elevated ultrasensitive Tn > 30 pg/mL was predictive of subsequent cardiotoxicity ($P = 0.04$).

The greatest value of Tn may be the high NPV, such that patients who do not have an elevation in Tn have very low cardiac event rates and perhaps need less frequent surveillance imaging. The 2012 European Society for Medical Oncology (ESMO) Clinical Practice Guidelines for cardiotoxicity recommend troponin testing at baseline, during, and after cancer therapy (level of evidence III, grade B) [141]. However, this process may be complicated by variability in normal troponin values across laboratories and assays.

Brain-Type Natriuretic Peptides

Brain-type natriuretic peptide (BNP) and N-terminal protype BNP have also been studied as potential biomarkers for early detection of CTRCD. In the largest study of BNP cancer patients, made up of a heterogeneous cohort of 333 patients with different tumor types and cardiotoxic treatment exposures, Skovgaard et al. showed that BNP > 100 pg/mL was predictive of HF (HR 5.5, 95% CI 1.8–17.2, $P = 0.003$) [142]. However, conclusions regarding the utility of natriuretic peptides for the prediction and diagnosis of cardiac dysfunction remain conflicting and uncertain [143]. Challenges with BNP include variabilities with age, sex, weight, and renal function.

Combination and Novel Biomarkers in Risk Identification

Data are limited on associations between novel biomarkers and CTRCD. Recent HF biomarkers of interest which have been considered include serum galectin-3, ST-2, glycogen phosphorylase isoenzyme BB (GPBB), heart-type fatty acid binding protein (H-FABP), and high-sensitivity C-reactive protein (hsCRP) [144–146]. In their studies, Horacek et al. suggested some promise in GPBB [144–146], while another study has shown H-FASB and hsCRP as possible future markers for consideration [147]. Whether these markers will predict CTRCD or CV events in future is unclear. In one study, cardiac and inflammatory markers including N-terminal pro B-type natriuretic peptide (NT-proBNP), mid-regional pro-atrial natriuretic peptide (MR-proANP), mid-regional pro-adrenomedullin (MR-proADM), C-terminal proendothelin-1 (CT-proET-1), copeptin, high sensitivity troponin T (hs-troponin-T), interleukin-6 (IL-6), C-reactive protein (CRP), and cytokines serum amyloid A (SAA), haptoglobin, and fibronectin, were all associated with all-cause mortality in cancer patients [148]. Conversely, among a panel of biomarkers [troponin-I, CRP, proBNP, growth differentiation factor 15 (GDF-15), myeloperoxidase (MPO), placental growth factor (PlGF), soluble fms-like tyrosine kinase-1 (sFlt-1), and galectin-3 (gal-3)] in 78 breast cancer patients undergoing doxorubicin and trastuzumab chemotherapy, Ky and colleagues [149] found that troponin-I and MPO—but not the other biomarkers—improved risk prediction of chemotherapy-induced cardiotoxicity. In a longitudinal study of the same patients followed for up to 15 months, only MPO remained as a predictor [150]. Conversely, another biomarker panel of NT-proBNP, tumor necrosis factor alpha (TNF-α), gal-3, IL-6, troponin I, ST-2, and sFlt-1 showed only NT-proBNP as a predictor of detect subclinical cardiotoxicity after treatment with anthracyclines [151].

To date, controversy remains on the combinations of biomarkers that best predict cardiotoxicities associated with cancer therapy.

Screening Intervals for Cardiotoxicity Due to Chemotherapy

Over the last several years, many guidelines and position statements have been published to help guide the diagnosis and treatment of CTRCD, including ESMO in 2012 [141], EACVI/ASE in 2014 [50], the European Society of Cardiology in 2016 [152], and ASCO in 2017 [106]. Although not specific to cardiotoxicity, the 2013 American College of Cardiology/American Heart Association (ACC/AHA) guidelines for the management of HF also provide a relevant overview to the treatment of HF from stage A (at risk) to stage D (refractory HF) [153]. The ACC/AHA has also endorsed the more specific ASCO guidelines for care of patients with cancer. Though these recommendations for surveillance imaging vary, all algorithms generally agree that women undergoing breast cancer treatment should have baseline assessment of LVEF, with follow-up routine surveillance imaging during therapy, at intervals determined by the specific cancer therapeutic regimen. Serial LVEF evaluation is frequently performed with echocardiography given its widespread availability and the absence of RT exposure; however, CMR and gated radionucleotide angiography can also be used. Importantly, the same technique should be utilized for each patient's follow-up imaging. Unfortunately, there are no high-quality studies to help determine the optimal screening interval for imaging of patients receiving cardiotoxic medications. The ESMO guidelines recommend serial monitoring with echocardiogram for patients receiving anthracyclines and/or trastuzumab at baseline and at 3, 6, 9, 12, and 18 months [141]. The ASCO recommends that echocardiogram may be used for monitoring patients at increased risk, including those receiving trastuzumab, at unspecified intervals during cancer treatment and at 6 and 12 months posttreatment [106]. In general, we endorse individualized screening intervals tailored to the specific cancer agent of concern, and patient-related risks/factors.

Strategies to Prevent and Treat Cardiotoxicity Due to Chemotherapy

Baseline Risk Assessment

Patients at increased risk for LV dysfunction should be identified at initial cancer diagnosis by screening for traditional CV risk factors (hypertension, diabetes, dyslipidemia, obesity, and smoking) and treating them accordingly to minimize CV risks associated with cancer treatment. Any patient undergoing potentially cardiotoxic cancer therapy should be considered stage A HF. For these patients, the 2017 ACC/AHA HF guidelines give Class I recommendations for controlling hypertension, lipid disorders, and other CV risk factors [67].

Beyond the traditional CV risk factors, specific cancer therapy regimens that increase the long-term risk of cardiac dysfunction should also be reviewed, including high-dose anthracyclines (equivalent to $\geq 250\,mg/m^2$ of doxorubicin or $\geq 600\,mg/m^2$ of epirubicin), chest RT with the heart in the treatment field at a dose of 30 Gy or more, a combination of anthracyclines and RT even at lower dosages, and anthracyclines followed by trastuzumab [106].

In addition to a complete history and physical examination, a baseline LVEF assessment by echocardiography (with strain) is recommended to screen for asymptomatic LV dysfunction or other structural abnormalities, before initiating potentially cardiotoxic cancer therapy [106, 141].

Pharmacotherapy and Other Preventive Strategies

Beta-blockade, angiotensin-converting enzyme inhibitors (ACEIs), and angiotensin II receptor blockers (ARBs) have been evaluated in randomized controlled trials for prevention of anthracycline-induced cardiotoxicity. The OVERCOME trial (Prevention of Left Ventricular Dysfunction with Enalapril and Carvedilol in Patients Submitted to Intensive Chemotherapy for the Treatment of Malignant Hemopathies) found that cancer patients undergoing chemotherapy (82% anthracyclines) receiving combined treatment with carvedilol and enalapril had no reduction in LVEF compared with those who received placebo [68]. The PRADA (Prevention of Cardiac Dysfunction During Adjuvant Breast Cancer Therapy) trial showed that candesartan, but not metoprolol, helped to protect against LVEF decline in patients receiving anthracyclines with or without trastuzumab and RT [70]. This cardioprotective effect of candesartan was not seen in patients previously treated with anthracycline-based treatments receiving subsequent trastuzumab therapy [154]. The MANTICORE-101 (Multidisciplinary Approach to Novel Therapies in Cardiology Oncology Research) trial found that both perindopril and bisoprolol helped to protect against declines in LVEF during therapy with trastuzumab, though there was no difference in primary outcome for prevention of LV remodeling as assessed by CMR [155]. Lastly, CECCY (Carvedilol Effect in Preventing Chemotherapy-Induced Cardiotoxicity), a study of 192 patients undergoing anthracycline therapy for HER2-negative breast cancer, found that carvedilol did not protect against a drop in LVEF by more than 10%, though there was a decrease in troponin elevations and diastolic dysfunction [156]. It must be noted that these trials were limited by varying inclusion and exclusion criteria, short durations of follow-up, and low power. Thus, clinicians should apply guideline-directed medical therapy (GDMT) for stage A HF in this population, with an understanding of the limitations until further evidence is available.

Aldosterone antagonists may attenuate trastuzumab-induced myocardial dysfunction through inhibition of the epidermal growth factor receptor (EGFR) receptor, though further evaluation is needed [157]. Similarly, spironolactone with anthracycline therapy in breast cancer patients with preserved LV function may also attenuate LV dysfunction [158]. The role of other HF medications, such as hydralazine/nitrates and digoxin, has not been studied specifically in the cancer population but should be utilized in accordance with GDMT for stage A HF [153].

Statins, or 3-hydroxy-3-methylglutaryl-CoA (HMG-CoA) reductase inhibitors, in addition to their lipid lowering, may exert cardioprotective effects through pleotrophic mechanisms. A retrospective cohort study of over 600 cancer patients found that uninterrupted statin use resulted in reduced HF [159]. However, prospective data to address the role of statins in the prevention of CTRCD are warranted.

Dexrazoxane, an ethylenediaminetetraacetic acid analog, has been shown to attenuate anthracycline cardiotoxicity through iron chelation and decrease in production of free radicals. Dexrazoxane administration with anthracycline interferes with binding to topoisomerase 2β and reduces both cardiotoxicity and subsequent HF in high-risk patients [160, 161]. Multiple trials have demonstrated a reduction in decreased LVEF or development of HF with the addition of dexrazoxane [1]. Use of dexrazoxane has been associated with up to 82% reduction in HF [relative risk (RR) 0.18; 95% CI 0.1–0.32] [162]. However, because of uncertainty regarding the risk for secondary malignancy and possibly reduced tumor response rates, the use of dexrazoxane is narrow in breast cancer patients, and is limited only to patients with metastatic breast cancer who have received a cumulative doxorubicin dose of 300 mg/m^2 or more [1]. The ASCO guidelines recommend consideration of dexrazoxane for patients with planned high-dose anthracyclines (equivalent to ≥ 250 mg/m^2 of doxorubicin or ≥ 600 mg/m^2 of epirubicin) [106], and the ESMO guidelines recommend its use for all patients at high risk of cardiotoxicity without further specificity [141].

Other recommended strategies to reduce risk of cardiotoxicity in patients with planned high-dose anthracyclines include using a continuous anthracycline infusion (as opposed to bolus dosing), as well as the liposomal formulation of doxorubicin [141, 163]. For patients with planned mediastinal RT, techniques to reduce exposure to the heart including deep-inspiratory breath-hold and intensity-modulated radiotherapy are recommended [141].

More frequent monitoring and workup, along with referral to a cardiologist, are recommended if signs and/or symptoms of cardiac dysfunction occur. If a reduction in strain or LVEF is noted, appropriate interventions with cardioprotective medications (beta-blockers and/or ACEIs/ARBs) in accordance with the general ACC/AHA HF guidelines [67, 153] and/or adjustments to the cancer treatment regimen can be implemented. Management of cancer patients with HF may be hindered by relative hypotension during cancer treatment, thus reducing tolerance of guideline-directed medical therapy. There are no clear recommendations regarding the continuation or discontinuation of cancer therapy in patients with CTRCD. Any decision to discontinue cancer therapy should weigh the benefits and risks of the current treatment plan as well as alternative regimens, and be made within multidisciplinary cardiooncology teams that include oncologists, radiation oncologists, and cardiologists. In addition, other causes of LV dysfunction, including ischemia and thyroid disorders, should be considered.

Risk Factors, Detection, Prevention, and Management of Radiation Associated Cardiac Dysfunction (RACD)

Mechanism and Risk Factors

Radiation is often used in combination with chemotherapy to improve survival. However, when the heart is exposed to radiation field, it may incur deleterious effects on the pericardium, myocardium valve, conduction system, and coronary arteries. The mechanism of radiation associated cardiac dysfunction (RACD) remains incompletely understood and is most likely multifactorial. In rodent species, RT causes microvascular endothelial damage that leads to lymphocyte adhesion and extravasation, eventually leading to thrombus formation and capillary loss. Ongoing capillary damage eventually results in myocardial cell death and fibrosis [164]. RT also appears to have a significant effect on the macrovasculature, causing myocardial ischemia and injury. In large blood vessels, RT can cause inflammation and oxidative damage, which in the presence of high cholesterol leads to lipid peroxidation, and formation of foam cells, ultimately contributing to atherosclerosis. RT has been associated with accelerated atherosclerosis with thickening and fibrosis of the media and adventitia [165]. Accelerated atherosclerosis can be seen as early as 5 years after exposure among breast cancer survivors who receive thoracic RT, and risks can remain for up to 30 years after treatment [166]. The clinical implications of RT may be seen either acutely or chronically, and are listed in **Table 4**. **Figure 6** illustrates the different processes involved in cardiotoxicity from RT.

Individual and genetic differences between patients likely play a role in RT and CVD. RACD depends on the

TABLE 4 Radiation-Induced Cardiovascular Complications

Complication	Prevalence	Time of Onset From Radiation	Diagnostic Testing Modality
Pericardial disease [167, 168]	Most common form of radiation-induced radiation toxicity 6–30% of patients undergoing RT	Asymptomatic effusion or symptomatic pericarditis: immediately after RT Delayed pericarditis: months to years after treatment Constrictive pericarditis can be a complication of delayed pericarditis	ECG with diffuse ST elevations, clinical diagnosis Echocardiogram: delayed pericarditis often times presents with pericardial effusion CCTA/CMR: thickened calcified pericardium, pericardial effusion and tubular-shaped right ventricle (constrictive)
Coronary artery disease [60, 167]	Estimated in nearly 60% of patients with Hodgkin lymphoma Left chest RT has been associated with ~85% Left anterior descending artery disease	Hodgkin lymphoma survivors: 2–40 years Breast cancer survivors: typically stays latent until 10 year after exposure	Coronary angiography, stress testing, nuclear perfusion scan, or CCTA: early accelerated atherosclerosis
Valvular disease [169, 170]	Up to 42% asymptomatic valvular disease 6–15% of Hodgkin lymphoma survivors 0.5–4.2% in breast cancer	10–20 years after treatment Left-sided valvular disease is more common than right-sided. Incidence rate ratio: 1.54	Echocardiography shows leaflet thickening, fibrosis, and calcifications. Regurgitation is more common than stenosis
Restrictive cardiomyopathy [171, 172]	Asymptomatic diastolic dysfunction seen in 14% patients with mediastinal radiation.	Particularly common when patient receives concomitant cardiotoxic chemotherapy. Often after > 30 Gy radiation dose See asymptomatic changes in LV function within a few months of RT. Progression to restrictive cardiomyopathy at late stage (typically 10 years)	Sequelae of extensive fibrosis and diastolic dysfunction. Echo and MRI with systolic and diastolic dysfunction. LV strain may be a more sensitive measure of LV dysfunction
Radiation myocarditis [167, 172, 173]	7–15%	Increased risk 5 years after RT	ECG, echocardiogram with ST-T wave changes; regional wall motion abnormalities that do not fit a vascular territory. CMR: regional wall motion abnormalities and scarring with epicardial or mid-myocardial late gadolinium enhancement
Conduction abnormalities [172–174]	Up to 50% of patients reported to mediastinal irradiation	Nonspecific ECG abnormalities: within first year of treatment, asymptomatic and transient Serious arrhythmias/infranodal conduction block: late complication (>10 years after treatment)	ECG

CCTA, cardiac computed tomography angiography; CMR, cardiac magnetic resonance; ECG, electrocardiogram; echo, echocardiogram; LV, left ventricular; MRI, magnetic resonance imaging; RT, radiation therapy.

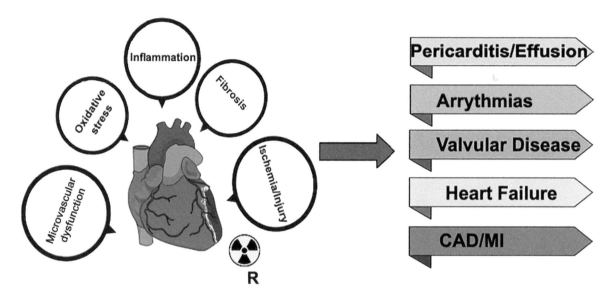

FIGURE 6 Radiation-Induced Cardiomyopathy. Radiation-induced cardiac dysfunction includes pericardial injury, conduction abnormalities, valvular disease, heart failure, and ischemic heart disease. CAD, coronary artery disease; MI, myocardial infarction; R, radiation.

dose of RT, baseline CV risk factors, and time from therapy [167]. LV dysfunction and HF can occur as acute RT myocarditis, but more commonly develop as long-term consequences of fibrosis leading to LV dysfunction or restrictive CM [2, 72].

The presence of other CVD risk factors, concomitant anthracycline use, and anterior or left chest irradiation increase the risk of RACD (**Figure 7**) [2]. Women with preexisting CV risk factors have four- to sixfold increased risk of cardiac events from RT compared to women without underlying CV risk factors [2, 61]. Cardiac perfusion scans before and after RT suggest early onset of RT-related CV damage. Onwudiwe et al. studied 90,000 elderly women ≥66 years with preexisting CVD and concluded that they were at greater risk of CV events in the first 6 months after RT compared to those who did not receive RT [175]. Patients were separated into three groups: low, intermediate, and high risk based on their baseline comorbidities. HF occurred in 71% of the high-risk group, 16% of the intermediate-risk group, and 14% of the low-risk group [175]. In a population-based study of 29,102 breast cancer patients who underwent RT 1 year after diagnosis, the 10-year cumulative incidence of HF was 41.3% (95% CI 40.2–42.3). However, no significant association was found with left-sided vs. right-sided breast cancer (subdistribution HR 0.97, 95% CI 0.92–1.01, $P=0.26$) [176]. In a meta-analysis, no significant association was found between left-sided vs. right-sided RT leading to HF (RR 1.00, 95% CI 0.91–1.12, $P=0.37$). However, patients with breast cancer who received RT were more likely to develop HF than those who did not receive RT (RR 1.17, 95% CI 0.90–1.51, $P<0.001$) [177].

While modern RT techniques have significantly lowered average RT of the heart over the past decades, there is still an excess risk of CV diseases following exposure to RT. A population-based case control study concluded that the relative risk of HF increased with increasing cardiac RT exposure during contemporary conformal RT for breast cancer treatment in older women (OR 9.1, 95% CI 3.4–24.4) [168]. Based on these findings, it is likely that contemporary RT techniques are associated with some but reduced risk of cardiac toxicity, compared with older techniques.

RT has also been associated with multiple other CV complications (**Table 4** [61, 72, 169–174, 178]). CV effects secondary to coronary atherosclerosis can be seen as early as 5 years in survivors who received left-sided thoracic RT, and could persist for up to 30 years posttreatment [166]. Patients can also develop microvascular dysfunction that results in impaired coronary flow reserve, myocardial ischemia, and myocardial fibrosis [1]. In a study of 2168 women with breast cancer, each 1-Gy increase in mean heart RT dose was associated with a 7.4% increase in coronary events [179]. Other cardiac complications reported with RT use include acute and chronic pericarditis, valvular regurgitation and stenosis, conduction abnormalities, and sudden cardiac death. Higher RT doses have been associated with autonomic dysfunction [1].

Detection of RACD

Diagnosis of RACD has some unique challenges. Symptoms are often heterogenous and typically present late into disease progression. The temporal delay in onset of symptoms

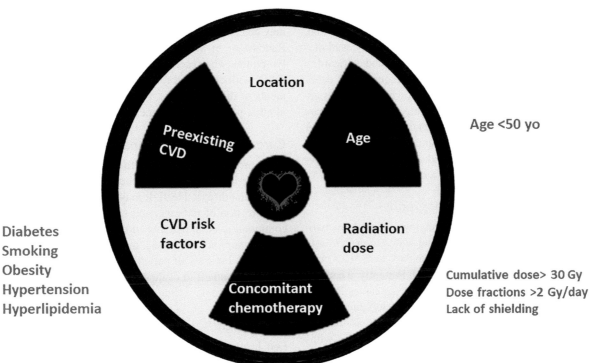

FIGURE 7   Risk Factors for Radiation Associated Cardiac Dysfunction. Risk factors for radiation associated cardiac dysfunction include location of the radiation, younger age, radiation dose, use of concomitant chemotherapy, preexisting CVD disease, and CVD risk factors. CVD, cardiovascular disease.

and exposure to RT requires a high index of suspicion and ongoing surveillance. The quest to identify subclinical disease in asymptomatic patients for early detection and treatment of RACD is ongoing.

Biomarkers

The utility of cardiac biomarkers acutely after RT demonstrates mixed results. One early study of 50 women with breast cancer did not find any change in serum troponin after a total dose of 45–46 Gy of RT [180]. Similarly, another study of 30 patients receiving RT did not show any significant elevations in creatine kinase MB (CK-MB), troponin, or NT-proBNP with RT. [181] Conversely, a more recent study found that both troponin and BNP increased significantly with chest RT, even though the absolute and mean values remained on a relatively low level [182]. Yet in another recent study of 58 patients receiving radiotherapy for left-sided breast cancer, hs-troponin-T increased during RT from baseline in 12 patients (21%). In this study, those with higher hs-troponin-T values had received significantly increased RT doses for the whole heart and LV than those with stable hs-troponin-T values [183]. Hence, cardiac bio-

markers might be useful for evaluation of RT-induced cardiotoxicity but require additional investigation. It is likely that normalized values of these biomarkers could prove more useful for monitoring of subclinical CVD associated with radiotherapy than absolute elevations in values.

Imaging Markers

The EACVI and ASE expert consensus statement recommends initiation of screening for CAD within 5 years of RT exposure, based on the timing of major coronary events in breast cancer survivors [184]. Screening for valvular heart disease generally starts 10 years after RT, with subsequent imaging performed at 5-year intervals [173]. Long-term follow-up after RT generally includes a yearly physical examination to assess for symptoms or signs of RACD. Additionally, a cardiac stress test is recommended after 5 years in high-risk individuals, and after 10 years in all others to assess for CAD related to RT vasculopathy [173].

Echocardiography

Echocardiography is the most common imaging modality used for detection and serial monitoring of RACD [173].

Lack of RT exposure, lower cost, and availability make it an easily accessible screening tool [185]. Echocardiographic features of RACD include biventricular systolic and diastolic dysfunction, multivalvular involvement with mixed valvular dysfunction, prominent calcification (pericardial, valvular, annular, aorto-mitral curtain, and aortic), wall motion abnormalities associated with CAD, and pericardial constriction [173]. The recommended frequency of echocardiographic screening in asymptomatic patients varies according to the individual but is typically performed 5 years after exposure in high-risk individuals and 10 years after exposure in others [173]. Subsequently, transthoracic echocardiography (TTE) is recommended for reassessment every 5 years (or more frequently if symptoms develop) [173].

Speckle-Tracking Echocardiography

As with chemotherapy, strain imaging assessed by two-dimensional (2D) STE may be better suited to detect early subclinical myocardial damage secondary to RACD prior to the development of symptoms [173]. Presently, a handful of studies have evaluated strain imaging to assess RACD. Lo et al. demonstrated lower strain parameters but normal LVEF after RT in women with left-sided breast cancer, early and up to 2 months after RT compared to those without exposure [186].

Stress Echocardiography

Stress echocardiography may be beneficial to evaluate for myocardial ischemia and dynamic assessment of RT-induced valvular heart disease. RT-associated CAD may manifest as resting or inducible regional wall motion abnormalities in typical coronary distributions or with global dysfunction and cavity enlargement at peak stress [173]. Patients who had received higher doses of mediastinal RT were more likely to have abnormal wall motion (34% vs. 15%; $P=0.001$), and more likely to have ischemia on stress imaging (23% vs. 12%; $P=0.02$) [187, 188]. Stress valvular assessment is usually reserved for symptomatic patients with mild or moderate disease at rest, whose symptoms appear proportionally worse than expected. Stress echo may also demonstrate stress-induced increase in valvular regurgitation, trans-valvular gradients, or pulmonary pressures, along with impaired ventricular contractile reserve [189].

Coronary Computed Tomographic Angiography

Coronary computed tomographic angiography (CCTA) is particularly useful in RT-associated CAD for its negative predictive value, as the absence of coronary calcification indicates a low risk for underlying CAD. CCTA can also evaluate for aortic, valvular, myocardial, and pericardial calcifications. Pericardial calcification, thickening, inferior vena cava enlargement, and ventricular conical deformity

are suggestive of pericardial constriction. Chest CT is also useful to evaluate extra-cardiac structures for surgical planning [189, 190]. Extensive mediastinal fibrosis or lack of a safety margin between the sternum and adjacent structures may be cause to reconsider a median sternotomy approach. Rademaker et al. followed nine patients with Hodgkin's lymphoma who received RT. Eight of the nine patients had CAD, with computed tomography (CT) showing long segments of diffuse disease. Calcium scores were higher than in other patients of the same age group [191].

Cardiac Magnetic Resonance

CMR provides simultaneous functional and structural data, enabling detection of RT-induced coronary, valvular, and pericardial disease [189]. It may serve as an invaluable tool for detection of ischemia and silent myocardial infarction, commonly seen in patients with RACD. In a CMR study of 20-year survivors of Hodgkin's disease, perfusion defects were found in 68% of patients [192]. CMR has emerged as the gold standard to evaluate myocardial infarction in both acute and chronic settings [173].

Cardiac Catheterization

Left heart catheterization allows assessment of the extent and severity of coronary stenosis. Right heart studies may be useful for calculation of intracardiac and pulmonary pressures, while simultaneous left and right heart measurements allow for evaluation of ventricular interdependence, with equalization of pressures confirming the presence of constrictive physiology. Proximal CAD may be underappreciated, thus there should be a low threshold for utilizing intravascular ultrasound [189].

Management of RACD

Radiation Associated Pericardial Disease

Pericardiectomy is reserved for constriction with fibrosed and extensively calcified pericardium or severe recurrent pericarditis despite medical therapy. However, outcomes from pericardiectomy are worse in RACD, likely reflecting that pericardial involvement is a marker of increased severity and extent of disease. Five-year survival rates post-pericardiectomy are only 11.0% for post-RT pericardial disease compared to 79.8% for idiopathic and 55.9% for postoperative pericardial disease [193].

Radiation Associated Coronary Artery Disease

Therapeutic options for management of RT-associated CAD are the same as those in nonirradiated patients, including medical therapy, percutaneous coronary intervention (PCI), and coronary artery bypass grafting (CABG). The decision between PCI and CABG depends on the anatomy of the lesion(s) and estimated surgical risk. Mediastinal fibrosis is common in patients treated with anterior irradiation, making surgical procedures more difficult [189, 194]. In addition, the internal mammary artery is often included in the RT field, rendering it unsuitable for harvesting [194]. Although PCI is generally preferable for reduced mortality, RT-induced CAD is typically diffused and extensively calcified, making stenting less amenable [189]. Although results are conflicting, there does not seem to be a significant difference in the rates of stent restenosis between patients with a history of RT and the general population [194, 195]. There are no studies directly comparing outcomes of PCI vs. CABG in patients with RT-induced CAD.

Radiation Associated Valvular Disease

Management of patients with RT-induced valvular disease is challenging, and associated with higher rates of perioperative morbidity and death compared to patients with valvular disease without chest RT. In one study, the long-term mortality rate was 45% in postradiotherapy patients undergoing single-valve surgery and 61% in those undergoing surgery on two or more valves, compared with 13% and 17%, respectively, in patients with no history of chest radiotherapy [196, 197]. Replacement is favored over repair because irradiated valve tissue is abnormal and tends to progressively fibrose and calcify, thereby increasing the risk of transforming a repaired regurgitant valve into a stenotic one. RT also results in extensive mediastinal fibrosis, again increasing the risk of complications with surgical therapy. Given increased risks of reoperation, mechanical prostheses are preferred, especially for younger patients [189]. Transcatheter aortic valve replacement has been successful in RT-induced valvular disease and may become the preferred method of aortic valve replacement in patients with radiation-induced aortic stenosis [198].

Strategies to Prevent Radiation Associated Cardiac Dysfunction

The most effective strategy to reduce RACD is to limit cardiac exposure to radiation dose. It is recommended that no more than 5% of the whole heart exceeds 20 Gy RT dose for left-sided breast cancers and that 0% of the heart should exceed 20 Gy for right-sided breast cancers [199]. Moreover, the normal tissue complication probability (NTCP) model-based estimates predict that 25 Gy below 10% (in 2 Gy per fraction RT dose plan) is associated with < 1% probability of cardiac mortality 15 years after RT [199].

In patients with breast cancer, the irradiated heart volume should be minimized without compromising the target coverage [200]. The main techniques employed to minimize RT exposure include cardiac shielding with lead blocks, reduced fraction size (below 2 Gy/day), reduced total dose of RT (below 30 Gy), using alteration in RT field or targeted RT, intensity-modulated RT (IMRT), and 3D conformal RT [201]. It should be noted that the use of a four-field IMRT technique can offer better sparing than the partial shielding technique, since the maximum heart depth is increased.

To further decrease cardiac toxicity among patients with breast cancer treated with RT, it is important to focus on the risk factors for radiation-induced heart disease (RIHD), in an attempt to optimize them. In particular, some cardiometabolic risk factors such as diabetes mellitus, hypertension, dyslipidemia, and obesity can be modified to some degree with pharmacotherapy and nonpharmacological approaches [185, 201].

Prostate Cancer Treatment-Related Cardiotoxicity

Prostate cancer is the most common cancer diagnosis made in men with more than 160,000 new cases diagnosed each year in the United States, adding to 3.3 million existing survivors. Although it often has an indolent course, it remains the third leading cause of death in men [202].

Androgens produced mostly in the testicles stimulate the growth of prostate cells. Lowering the levels of androgens can thereby slow the growth of prostate cancer cells. ADT is the mainstay for metastatic prostate cancer, either by surgical orchiectomy or by reversibly blocking the release of pituitary hormones with gonadotropin-releasing hormones (GnRH) agonists as demonstrated in **Figure 8** [202]. GnRH agonists (degarelix and abarelix) and GnRH antagonists (leuprolide, goserelin, and triptorelin) lead to lower testosterone levels. Testosterone binds to the androgen receptor to exert its biological effects on prostate cells. Antiandrogen drugs (bicalutamide, flutamide, and nilutamide) may work by inhibiting the androgen receptors and directly blocking the effect of androgens on prostate cancer cell growth.

Prostate cancer and CVD carry common risk factors including advanced age, smoking, and obesity [203]. In addition, ADT, by decreasing testosterone levels, may further increase the risk of CVD (**Figure 9**) [204, 205]. Overall, the data on ADT effects on CVD remain inconsistent. A meta-analysis that included randomized trials with follow-up ranging from 7.6 to 13.2 years found no link between ADT and increased CV death [206], whereas a retrospective analysis of the SEER database, examining 22,816 prostate cancer survivors 1 year after diagnosis, found that patients who received ADT had a 20% increased risk of CAD over a period of 5 years compared to patients who did not receive ADT (HR 1.20, 95% CI 1.15–1.26) [207]. However, no prospective trials have addressed the risks of CAD associated with ADT. ADT has also been associated with prolonged QTc and may contribute to higher risk of SCD [92]. Both GnRH agonists and antagonists have Food and Drug

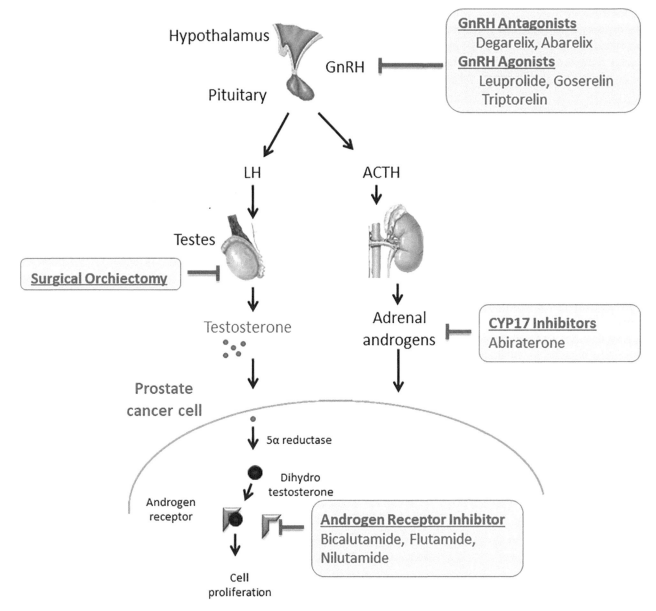

FIGURE 8 Mechanisms of Action for Prostate Cancer Drugs. ACTH, adrenocorticotropic hormone; GnRH, gonadotropin-releasing hormone; LH, luteinizing hormone.

Administration (FDA) black box warnings for QTc prolongation and can increase QTc by up to 10–20 ms [92].

Treatment with these agents also adversely affects body weight, visceral adiposity, insulin sensitivity, metabolic syndrome, and dyslipidemia [208]. Given these implications, the American Heart Association, the American Cancer Society, and the American Urological Association all recommend that patients receiving ADT should be evaluated annually with an examination of blood pressure, serum measurement of lipid profiles, and fasting glucose levels. They should also undergo primary and secondary preventive measures for CVD [209].

Androgen receptor inhibitors, bicalutamide, flutamide, and nilutamide, are FDA approved, and are often used in combination with GnRH agonists. Enzalutamide is a newer more potent androgen receptor inhibitor, and confers a greater incidence of hypertension [210].

While ADT is the cornerstone of prostate cancer treatment, new hormonal agents are able to prevent synthesis of testosterone by inhibition of the cytochrome $p450$ enzyme (CYP17), which is involved in the production of androgen in the adrenal glands (abiraterone) [211]. It is particularly indicated in patients with residual androgen synthesis despite castration. It may result in increased levels of aldosterone

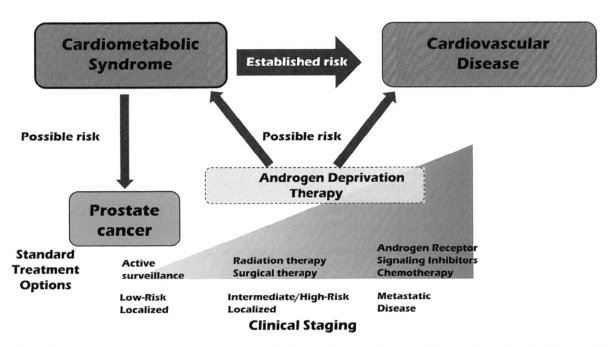

FIGURE 9 **Clinical Link Between Cardiometabolic Disorders and Prostate Cancer.** *Reproduced with permission from [204].*

precursors and hyperminearloidism. As a result, abiraterone is associated with higher incidence of hypokalemia and fluid retention.

It should be noted that the use of new hormonal agents was associated with an increased risk of all-grade (RR 1.36; 95% CI 1.13–1.64; $P=0.001$) and high-grade (RR 1.84; 95% CI 1.21–2.80; $P=0.004$) cardiac toxicity [211]. The use of new hormonal agents was also associated with an increased risk of all-grade (RR 1.98; 95% CI 1.62–2.43; $P=0.001$) and high-grade (RR 2.26; 95% CI 1.84–2.77; $P=0.004$) hypertension compared with controls. The incidence of all-grade and high-grade cardiotoxicity by abiraterone was 13.7% and 4.5%, respectively; this was significantly increased compared with placebo (RR 1.41; 95% CI 1.21–1.64; $P<0.001$ and RR 2.22; 95% CI 1.60–3.07; $P<0.001$) [211]. In a more contemporary study, patients with prior CVD conditions receiving either abiraterone or enzalutamide experienced a substantially higher mortality at 6 months compared to those without CVD [212].

Indeed, abiraterone conferred a new comorbid condition in 16% of patients on this drug, with obesity being the most

common, followed by hypertension. Among patients with at least one preexisting comorbid condition, 24% experienced worsening of that condition, with hypertension being the most common, followed by diabetes mellitus [211].

Conclusion

CTRCD is a significant cause of morbidity and mortality among cancer survivors, with noted differences based on type of cancer and type of therapy, age, and sex. Identification of those at risk is essential, in order to adequately implement monitoring and preventive strategies to help reduce the risk of CTRCD. Different biomarker and imaging modalities can be employed for monitoring, while prevention should focus on aggressive management of traditional CV risk factors, as well as those related to cancer and cancer therapy. As the number of cancer survivors continues to increase due to recent advances in cancer treatment, CTRCD will become more prevalent. Increasing awareness among patients and healthcare providers is necessary for early treatment and diagnosis of CTRCD.

Key Points

1. Sex differences in cancer susceptibility exist, with breast cancer being the most prevalent cancer in women, and prostate cancer in men. Treatment of both cancers is associated with significant cardiac implications, and requires close surveillance.

2. CVD is a leading cause of morbidity and mortality in breast cancer survivors or women currently being treated for breast cancer.

3. Breast cancer and CVD share several common risk factors, including unhealthy diet, sedentary lifestyle, smoking, age, early menarche, obesity, and hormonal therapy use.

4. Multiple breast cancer therapies have the potential to increase risk of CTRCD, with doxorubicin and trastuzumab being the classic breast cancer therapies associated with cardiotoxicity.

5. Endocrine therapies have become a cornerstone of breast and prostate cancer but can be associated with cardiovascular disease.

6. There are several manifestations of cardiac dysfunction from cancer-related therapy, but LV dysfunction is the most significant clinical manifestation associated with higher morbidity and mortality.

7. Chemotherapy often results in decreased cardiac function, and this has been classified as irreversible (anthracyclines) and reversible (trastuzumab). However, this classification does not encompass the full spectrum of cardiotoxicity and is falling out of favor.

8. Risk factors for CTRCD include age, female gender, previous chemotherapy, or RT. Baseline risk factors for LV dysfunction and specific cancer therapies contribute to the development of CTRCD.

9. Recommendations for surveillance vary, but all algorithms generally agree that for women undergoing breast cancer treatment, early detection and treatment of CTRCD are paramount. Baseline assessment of LVEF with follow-up routine surveillance imaging during therapy at intervals determined by the specific cancer therapeutic regimen is recommended.

10. Traditional heart failure medications including beta-blockers, ACEIs, ARBs, statins, and aldosterone antagonists hold promise to prevent and ameliorate severity of chemotherapy-induced cardiotoxicity, but larger randomized trials are required for further analysis.

11. There are multiple imaging modalities to evaluate CTRCD and RACD, including echocardiography (including speckle-tracking echo), CCTA, and CMR. MUGA, which used to be the main modality for LV function assessment during chemotherapy, has fallen out of favor.

12. Common complications of RT to the chest include CAD, valvular heart disease, tachy- or brady-arrhythmias, myocardial injury, and pericardial injury.

13. Identification of high-risk patients and close follow-up is essential for prevention, early diagnosis, and treatment of CTRCD.

Back to Clinical Case

The case describes a 62-year-old female with new-onset heart failure symptoms, after receiving chemotherapy and radiation therapy for breast cancer 3 months ago. She also underwent cardiac magnetic resonance imaging (MRI) without evidence for myocarditis or other infiltrative cardiomyopathy, and her other serum workups for secondary causes of cardiomyopathy were unrevealing. Following a negative coronary angiography and the above testing, she was diagnosed as having chemotherapy-induced cardiomyopathy. She gradually recovered and was discharged on day 10 after hospital admission, on guideline-directed medical therapy with metoprolol and lisinopril. Six months later, her LV size had normalized, and LVEF had recovered to 47% on TTE. She was deemed to have doxorubicin-induced cardiomyopathy, likely predisposed by her prior history of postpartum cardiomyopathy. She remains on guideline-directed medical therapy, and has been stable without further admissions.

References

[1] Mehta LS, Watson KE, Barac A, Beckie TM, Bittner V, Cruz-Flores S, et al. Cardiovascular disease and breast cancer: where these entities intersect: a scientific statement from the American Heart Association. Circulation 2018;137(8):e30–66.

[2] Bloom MW, Hamo CE, Cardinale D, Ky B, Nohria A, Baer L, et al. Cancer therapy-related cardiac dysfunction and heart failure: part 1: definitions, pathophysiology, risk factors, and imaging. Circ Heart Fail 2016;9(1), e002661.

[3] Centers for Disease Control and Prevention (CDC). Heart disease facts. 14 August; Available from: https://www.cdc.gov/heartdisease/facts. htm#:~:text=Heart%20disease%20is%20the%20leading,1%20in%20every%204%20deaths; 2020.

[4] Rawla P. Epidemiology of prostate cancer. World J Oncol 2019;10(2):63–89.

[5] National Cancer Institute. Cancer stat facts: female breast cancer. In: Surveillance, epidemiology, and end results program; 2020.

[6] de Moor JS, Mariotto AB, Parry C, Alfano CM, Padgett L, Kent EE, et al. Cancer survivors in the United States: prevalence across the survivorship trajectory and implications for care. Cancer Epidemiol Biomarkers Prev 2013;22(4):561–70.

[7] Bradshaw PT, Stevens J, Khankari N, Teitelbaum SL, Neugut AI, Gammon MD. Cardiovascular disease mortality among breast cancer survivors. Epidemiology 2016;27(1):6–13.

[8] Dorak MT, Karpuzoglu E. Gender differences in cancer susceptibility: an inadequately addressed issue. Front Genet 2012;3:268.

[9] Siegel RL, Miller KD, Jemal A. Cancer statistics, 2018. CA Cancer J Clin 2018;68(1):7–30.

[10] Cook MB, Dawsey SM, Freedman ND, Inskip PD, Wichner SM, Quraishi SM, et al. Sex disparities in cancer incidence by period and age. Cancer Epidemiol Biomarkers Prev 2009;18(4):1174–82.

[11] American Cancer Society Inc. Surveillance research. https://www.cancer.org/content/dam/cancer-org/research/cancer-facts-and-statistics/annual-cancer-facts-and-figures/2019/cancer-facts-and-figures-2019.pdf.

[12] Bussani R, De-Giorgio F, Abbate A, Silvestri F. Cardiac metastases. J Clin Pathol 2007;60(1):27–34.

[13] Burazor I, Aviel-Ronen S, Imazio M, Goitein O, Perelman M, Shelestovich N, et al. Metastatic cardiac tumors: from clinical presentation through diagnosis to treatment. BMC Cancer 2018;18(1):202.

[14] Hoffmeier A, Sindermann JR, Scheld HH, Martens S. Cardiac tumors—diagnosis and surgical treatment. Dtsch Arztebl Int 2014;111(12):205–11.

[15] Grozinsky-Glasberg S, Grossman AB, Gross DJ. Carcinoid heart disease: from pathophysiology to treatment—'something in the way it moves'. Neuroendocrinology 2015;101(4):263–73.

[16] Buza V, Rajagopalan B, Curtis AB. Cancer treatment-induced arrhythmias: focus on chemotherapy and targeted therapies. Circ Arrhythm Electrophysiol 2017;10(8).

[17] Abdol Razak NB, Jones G, Bhandari M, Berndt MC, Metharom P. Cancer-associated thrombosis: an overview of mechanisms, risk factors, and treatment. Cancers (Basel) 2018;10(10):380.

[18] Wun T, White RH. Epidemiology of cancer-related venous thromboembolism. Best Pract Res Clin Haematol 2009;22(1):9–23.

[19] Elyamany G, Alzahrani AM, Bukhary E. Cancer-associated thrombosis: an overview. Clin Med Insights Oncol 2014;8:129–37.

[20] Prandoni P, Falanga A, Piccioli A. Cancer and venous thromboembolism. Lancet Oncol 2005;6(6):401–10.

[21] Writing Group Members, Mozaffarian D, Benjamin EJ, Go AS, Arnett DK, Blaha MJ, et al. Heart disease and stroke statistics-2016 update: a report from the American Heart Association. Circulation 2016;133(4):e38–e360.

[22] Rasmussen-Torvik LJ, Shay CM, Abramson JG, Friedrich CA, Nettleton JA, Prizment AE, et al. Ideal cardiovascular health is inversely associated with incident cancer: the Atherosclerosis Risk in Communities study. Circulation 2013;127(12):1270–5.

[23] Edefonti V, Decarli A, La Vecchia C, Bosetti C, Randi G, Franceschi S, et al. Nutrient dietary patterns and the risk of breast and ovarian cancers. Int J Cancer 2008;122(3):609–13.

[24] Murtaugh MA, Sweeney C, Giuliano AR, Herrick JS, Hines L, Byers T, et al. Diet patterns and breast cancer risk in Hispanic and non-Hispanic white women: the Four-Corners Breast Cancer Study. Am J Clin Nutr 2008;87(4):978–84.

[25] Adebamowo CA, Hu FB, Cho E, Spiegelman D, Holmes MD, Willett WC. Dietary patterns and the risk of breast cancer. Ann Epidemiol 2005;15(10):789–95.

[26] Knott CS, Coombs N, Stamatakis E, Biddulph JP. All cause mortality and the case for age specific alcohol consumption guidelines: pooled analyses of up to 10 population based cohorts. BMJ 2015;350:h384.

[27] Berstad P, Ma H, Bernstein L, Ursin G. Alcohol intake and breast cancer risk among young women. Breast Cancer Res Treat 2008;108(1):113–20.

[28] Yusuf S, Hawken S, Ounpuu S, Dans T, Avezum A, Lanas F, et al. Effect of potentially modifiable risk factors associated with myocardial infarction in 52 countries (the INTERHEART study): case-control study. Lancet 2004;364(9438):937–52.

[29] Lee IM, Shiroma EJ, Lobelo F, Puska P, Blair SN, Katzmarzyk PT, et al. Effect of physical inactivity on major non-communicable diseases worldwide: an analysis of burden of disease and life expectancy. Lancet 2012;380(9838):219–29.

[30] Friedenreich CM. Physical activity and breast cancer: review of the epidemiologic evidence and biologic mechanisms. Recent Results Cancer Res 2011;188:125–39.

[31] Lynch BM, Neilson HK, Friedenreich CM. Physical activity and breast cancer prevention. Recent Results Cancer Res 2011;186:13–42.

[32] Lahart IM, Metsios GS, Nevill AM, Carmichael AR. Physical activity, risk of death and recurrence in breast cancer survivors: a systematic review and meta-analysis of epidemiological studies. Acta Oncol 2015;54(5):635–54.

[33] Poirier P, Giles TD, Bray GA, Hong Y, Stern JS, Pi-Sunyer FX, et al. Obesity and cardiovascular disease: pathophysiology, evaluation, and effect of weight loss: an update of the 1997 American Heart Association Scientific Statement on Obesity and Heart Disease from the Obesity Committee of the Council on Nutrition, Physical Activity, and Metabolism. Circulation 2006;113(6):898–918.

[34] McTigue KM, Chang YF, Eaton C, Garcia L, Johnson KC, Lewis CE, et al. Severe obesity, heart disease, and death among white, African American, and Hispanic postmenopausal women. Obesity (Silver Spring) 2014;22(3):801–10.

[35] Keum N, Greenwood DC, Lee DH, Kim R, Aune D, Ju W, et al. Adult weight gain and adiposity-related cancers: a dose-response meta-analysis of prospective observational studies. J Natl Cancer Inst 2015;107(2).

[36] Berstad P, Coates RJ, Bernstein L, Folger SG, Malone KE, Marchbanks PA, et al. A case-control study of body mass index and breast cancer risk in white and African-American women. Cancer Epidemiol Biomarkers Prev 2010;19(6):1532–44.

[37] Cotterchio M, Mirea L, Ozcelik H, Kreiger N. Active cigarette smoking, variants in carcinogen metabolism genes and breast cancer risk among pre- and postmenopausal women in Ontario, Canada. Breast J 2014;20(5):468–80.

[38] Xue F, Willett WC, Rosner BA, Hankinson SE, Michels KB. Cigarette smoking and the incidence of breast cancer. Arch Intern Med 2011;171(2):125–33.

[39] Gram IT, Braaten T, Terry PD, Sasco AJ, Adami HO, Lund E, et al. Breast cancer risk among women who start smoking as teenagers. Cancer Epidemiol Biomarkers Prev 2005;14(1):61–6.

[40] Reynolds P, Hurley S, Goldberg DE, Anton-Culver H, Bernstein L, Deapen D, et al. Active smoking, household passive smoking, and breast cancer: evidence from the California Teachers Study. J Natl Cancer Inst 2004;96(1):29–37.

[41] McPherson K, Steel CM, Dixon JM. ABC of breast diseases. Breast cancer-epidemiology, risk factors, and genetics. BMJ 2000;321(7261):624–8.

[42] Lerner DJ, Kannel WB. Patterns of coronary heart disease morbidity and mortality in the sexes: a 26-year follow-up of the Framingham population. Am Heart J 1986;111(2):383–90.

[43] Charalampopoulos D, McLoughlin A, Elks CE, Ong KK. Age at menarche and risks of all-cause and cardiovascular death: a systematic review and meta-analysis. Am J Epidemiol 2014;180(1):29–40.

[44] Mueller NT, Odegaard AO, Gross MD, Koh WP, Yuan JM, Pereira MA. Age at menarche and cardiovascular disease mortality in Singaporean Chinese women: the Singapore Chinese Health Study. Ann Epidemiol 2012;22(10):717–22.

[45] Wellons M, Ouyang P, Schreiner PJ, Herrington DM, Vaidya D. Early menopause predicts future coronary heart disease and stroke: the Multi-Ethnic Study of Atherosclerosis. Menopause 2012;19(10):1081–7.

[46] Collaborative Group on Hormonal Factors in Breast Cancer. Menarche, menopause, and breast cancer risk: individual participant meta-analysis, including 118 964 women with breast cancer from 117 epidemiological studies. Lancet Oncol 2012;13(11):1141–51.

[47] Colditz GA, Hankinson SE, Hunter DJ, Willett WC, Manson JE, Stampfer MJ, et al. The use of estrogens and progestins and the risk of breast cancer in postmenopausal women. N Engl J Med 1995;332(24):1589–93.

[48] Grady D, Herrington D, Bittner V, Blumenthal R, Davidson M, Hlatky M, et al. Cardiovascular disease outcomes during 6.8 years of hormone therapy: heart and Estrogen/progestin Replacement Study follow-up (HERS II). JAMA 2002;288(1):49–57.

[49] Rossouw JE, Anderson GL, Prentice RL, LaCroix AZ, Kooperberg C, Stefanick ML, et al. Risks and benefits of estrogen plus progestin in healthy postmenopausal women: principal results from the Women's Health Initiative randomized controlled trial. JAMA 2002;288(3):321–33.

[50] Plana JC, Galderisi M, Barac A, Ewer MS, Ky B, Scherrer-Crosbie M, et al. Expert consensus for multimodality imaging evaluation of adult patients during and after cancer therapy: a report from the American Society of Echocardiography and the European Association of Cardiovascular Imaging. Eur Heart J Cardiovasc Imaging 2014;15(10):1063–93.

[51] Slamon D, Eiermann W, Robert N, Pienkowski T, Martin M, Press M, et al. Adjuvant trastuzumab in HER2-positive breast cancer. N Engl J Med 2011;365(14):1273–83.

[52] Seidman A, Hudis C, Pierri MK, Shak S, Paton V, Ashby M, et al. Cardiac dysfunction in the trastuzumab clinical trials experience. J Clin Oncol 2002;20(5):1215–21.

[53] FDA. Drug label for DOXIL-doxorubicin hydrochloride liposome injection for intravenous use; 1995.

[54] NCI. Common terminology criteria for adverse events (CTCAE); 2017.

[55] McGowan JV, Chung R, Maulik A, Piotrowska I, Walker JM, Yellon DM. Anthracycline chemotherapy and cardiotoxicity. Cardiovasc Drugs Ther 2017;31(1):63–75.

[56] Toroz D, Gould IR. A computational study of anthracyclines interacting with lipid bilayers: correlation of membrane insertion rates, orientation effects and localisation with cytotoxicity. Sci Rep 2019;9(1):2155.

[57] Cardinale D, Colombo A, Bacchiani G, Tedeschi I, Meroni CA, Veglia F, et al. Early detection of anthracycline cardiotoxicity and improvement with heart failure therapy. Circulation 2015;131(22):1981–8.

[58] Smith LA, Cornelius VR, Plummer CJ, Levitt G, Verrill M, Canney P, et al. Cardiotoxicity of anthracycline agents for the treatment of cancer: systematic review and meta-analysis of randomised controlled trials. BMC Cancer 2010;10:337.

[59] Groarke JD, Nohria A. Anthracycline cardiotoxicity: a new paradigm for an old classic. Circulation 2015;131(22):1946–9.

[60] Herrmann J, Lerman A, Sandhu NP, Villarraga HR, Mulvagh SL, Kohli M. Evaluation and management of patients with heart disease and cancer: cardio-oncology. Mayo Clin Proc 2014;89(9):1287–306.

[61] van Nimwegen FA, Schaapveld M, Janus CP, Krol AD, Petersen EJ, Raemaekers JM, et al. Cardiovascular disease after Hodgkin lymphoma treatment: 40-year disease risk. JAMA Intern Med 2015;175(6):1007–17.

[62] Armenian SH, Hudson MM, Mulder RL, Chen MH, Constine LS, Dwyer M, et al. Recommendations for cardiomyopathy surveillance for survivors of childhood cancer: a report from the International Late Effects of Childhood Cancer Guideline Harmonization Group. Lancet Oncol 2015;16(3):e123–36.

[63] Narayan HK, Finkelman B, French B, Plappert T, Hyman D, Smith AM, et al. Detailed echocardiographic phenotyping in breast cancer patients: associations with ejection fraction decline, recovery, and heart failure symptoms over 3 years of follow-up. Circulation 2017;135(15):1397–412.

[64] Serrano JM, González I, Del Castillo S, Muñiz J, Morales LJ, Moreno F, et al. Diastolic dysfunction following anthracycline-based chemotherapy in breast cancer patients: incidence and predictors. Oncologist 2015;20(8):864–72.

[65] Chatterjee K, Zhang J, Honbo N, Karliner JS. Doxorubicin cardiomyopathy. Cardiology 2010;115(2):155–62.

[66] Felker GM, Thompson RE, Hare JM, Hruban RH, Clemetson DE, Howard DL, et al. Underlying causes and long-term survival in patients with initially unexplained cardiomyopathy. N Engl J Med 2000;342(15):1077–84.

[67] Yancy CW, Jessup M, Bozkurt B, Butler J, Casey DE Jr, Colvin MM, et al. 2017 ACC/AHA/HFSA focused update of the 2013 ACCF/AHA guideline for the management of heart failure: a report of the American College of Cardiology/American Heart Association task force on clinical practice guidelines and the Heart Failure Society of America. J Card Fail 2017;23(8):628–51.

[68] Bosch X, Rovira M, Sitges M, Domènech A, Ortiz-Pérez JT, de Caralt TM, et al. Enalapril and carvedilol for preventing chemotherapy-induced left ventricular systolic dysfunction in patients with malignant hemopathies: the OVERCOME trial (preventiOn of left Ventricular dysfunction with Enalapril and caRvedilol in patients submitted to intensive ChemOtherapy for the treatment of Malignant hEmopathies). J Am Coll Cardiol 2013;61(23):2355–62.

[69] Vasti C, Hertig CM. Neuregulin-1/erbB activities with focus on the susceptibility of the heart to anthracyclines. World J Cardiol 2014;6(7):653–62.

[70] Gulati G, Heck SL, Ree AH, Hoffmann P, Schulz-Menger J, Fagerland MW, et al. Prevention of cardiac dysfunction during adjuvant breast cancer therapy (PRADA): a 2 x 2 factorial, randomized, placebo-controlled, double-blind clinical trial of candesartan and metoprolol. Eur Heart J 2016;37(21):1671–80.

[71] Legha SS, Benjamin RS, Mackay B, Ewer M, Wallace S, Valdivieso M, et al. Reduction of doxorubicin cardiotoxicity by prolonged continuous intravenous infusion. Ann Intern Med 1982;96(2):133–9.

[72] Filopei J, Frishman W. Radiation-induced heart disease. Cardiol Rev 2012;20(4):184–8.

[73] Wang X, Liu W, Sun CL, Armenian SH, Hakonarson H, Hageman L, et al. Hyaluronan synthase 3 variant and anthracycline-related cardiomyopathy: a report from the children's oncology group. J Clin Oncol 2014;32(7):647–53.

[74] Visscher H, Ross CJ, Rassekh SR, Barhdadi A, Dubé MP, Al-Saloos H, et al. Pharmacogenomic prediction of anthracycline-induced cardiotoxicity in children. J Clin Oncol 2012;30(13):1422–8.

[75] Jensen BC, McLeod HL. Pharmacogenomics as a risk mitigation strategy for chemotherapeutic cardiotoxicity. Pharmacogenomics 2013;14(2):205–13.

[76] Shapiro CL, Hardenbergh PH, Gelman R, Blanks D, Hauptman P, Recht A, et al. Cardiac effects of adjuvant doxorubicin and radiation therapy in breast cancer patients. J Clin Oncol 1998;16(11):3493–501.

[77] Puppe J, van Ooyen D, Neise J, Thangarajah F, Eichler C, Krämer S, et al. Evaluation of QTc interval prolongation in breast cancer patients after treatment with epirubicin, cyclophosphamide, and docetaxel and the influence of interobserver variation. Breast Care (Basel) 2017;12(1):40–4.

[78] Braverman AC, Antin JH, Plappert MT, Cook EF, Lee RT. Cyclophosphamide cardiotoxicity in bone marrow transplantation: a prospective evaluation of new dosing regimens. J Clin Oncol 1991;9(7):1215–23.

[79] Dhesi S, Chu MP, Blevins G, Paterson I, Larratt L, Oudit GY, et al. Cyclophosphamide-induced cardiomyopathy: a case report, review, and recommendations for management. J Investig Med High Impact Case Rep 2013;1(1). https://doi.org/10.1177/2324709613480346.

[80] Mukhtar E, Adhami VM, Mukhtar H. Targeting microtubules by natural agents for cancer therapy. Mol Cancer Ther 2014;13(2):275–84.

[81] Rowinsky EK, Eisenhauer EA, Chaudhry V, Arbuck SG, Donehower RC. Clinical toxicities encountered with paclitaxel (Taxol). Semin Oncol 1993;20(4 Suppl 3):1–15.

[82] Sara JD, Kaur J, Khodadadi R, Rehman M, Lobo R, Chakrabarti S, et al. 5-fluorouracil and cardiotoxicity: a review. Ther Adv Med Oncol 2018;10. https://doi.org/10.1177/1758835918780140.

[83] Martel S, Maurer C, Lambertini M, Pondé N, De Azambuja E. Breast cancer treatment-induced cardiotoxicity. Expert Opin Drug Saf 2017;16(9):1021–38.

[84] Chavez-MacGregor M, Zhang N, Buchholz TA, Zhang Y, Niu J, Elting L, et al. Trastuzumab-related cardiotoxicity among older patients with breast cancer. J Clin Oncol 2013;31(33):4222–8.

[85] Long HD, Lin YE, Zhang JJ, Zhong WZ, Zheng RN. Risk of congestive heart failure in early breast cancer patients undergoing adjuvant treatment with trastuzumab: a meta-analysis. Oncologist 2016;21(5):547–54.

[86] Verma S, Miles D, Gianni L, Krop IE, Welslau M, Baselga J, et al. Trastuzumab emtansine for HER2-positive advanced breast cancer. N Engl J Med 2012;367(19):1783–91.

[87] Perez EA, Barrios C, Eiermann W, Toi M, Im YH, Conte P, et al. Trastuzumab emtansine with or without pertuzumab versus trastuzumab plus taxane for human epidermal growth factor receptor 2-positive, advanced breast cancer: primary results from the phase III MARIANNE study. J Clin Oncol 2017;35(2):141–8.

[88] Florido R, Smith KL, Cuomo KK, Russell SD. Cardiotoxicity from human epidermal growth factor receptor-2 (HER2) targeted therapies. J Am Heart Assoc 2017;6(9).

[89] Geyer CE, Forster J, Lindquist D, Chan S, Romieu CG, Pienkowski T, et al. Lapatinib plus capecitabine for HER2-positive advanced breast cancer. N Engl J Med 2006;355(26):2733–43.

[90] Zhang X, Munster PN. New protein kinase inhibitors in breast cancer: afatinib and neratinib. Expert Opin Pharmacother 2014;15(9):1277–88.

[91] Awada A, Colomer R, Inoue K, Bondarenko I, Badwe RA, Demetriou G, et al. Neratinib plus paclitaxel vs trastuzumab plus paclitaxel in previously untreated metastatic ERBB2-positive breast cancer: the NEfERT-T randomized clinical trial. JAMA Oncol 2016;2(12):1557–64.

[92] Barber M, Nguyen LS, Wassermann J, Spano JP, Funck-Brentano C, Salem JE. Cardiac arrhythmia considerations of hormone cancer therapies. Cardiovasc Res 2019;115(5):878–94.

CH
24

[93] Walker AJ, West J, Card TR, Crooks C, Kirwan CC, Grainge MJ. When are breast cancer patients at highest risk of venous thromboembolism? A cohort study using English health care data. Blood 2016;127(7):849–57. quiz 953.

[94] Khosrow-Khavar F, Filion KB, Bouganim N, Suissa S, Azoulay L. Aromatase inhibitors and the risk of cardiovascular outcomes in women with breast cancer: a population-based cohort study. Circulation 2020;141:549–59.

[95] Amir E, Seruga B, Niraula S, Carlsson L, Ocaña A. Toxicity of adjuvant endocrine therapy in postmenopausal breast cancer patients: a systematic review and meta-analysis. J Natl Cancer Inst 2011;103(17):1299–309.

[96] Haque R, UlcickasYood M, Xu X, Cassidy-Bushrow AE, Tsai HT, Keating NL, et al. Cardiovascular disease risk and androgen deprivation therapy in patients with localised prostate cancer: a prospective cohort study. Br J Cancer 2017;117(8):1233–40.

[97] Rugo HS, Pritchard KI, Gnant M, Noguchi S, Piccart M, Hortobagyi G, et al. Incidence and time course of everolimus-related adverse events in postmenopausal women with hormone receptor-positive advanced breast cancer: insights from BOLERO-2. Ann Oncol 2014;25(4):808–15.

[98] Finn RS, Crown JP, Lang I, Boer K, Bondarenko IM, Kulyk SO, et al. The cyclin-dependent kinase 4/6 inhibitor palbociclib in combination with letrozole versus letrozole alone as first-line treatment of oestrogen receptor-positive, HER2-negative, advanced breast cancer (PALOMA-1/TRIO-18): a randomised phase 2 study. Lancet Oncol 2015;16(1):25–35.

[99] Ezaz G, Long JB, Gross CP, Chen J. Risk prediction model for heart failure and cardiomyopathy after adjuvant trastuzumab therapy for breast cancer. J Am Heart Assoc 2014;3(1), e000472.

[100] Romond EH, Perez EA, Bryant J, Suman VJ, Geyer CE Jr, Davidson NE, et al. Trastuzumab plus adjuvant chemotherapy for operable HER2-positive breast cancer. N Engl J Med 2005;353(16):1673–84.

[101] Dranitsaris G, Rayson D, Vincent M, Chang J, Gelmon K, Sandor D, et al. The development of a predictive model to estimate cardiotoxic risk for patients with metastatic breast cancer receiving anthracyclines. Breast Cancer Res Treat 2008;107(3):443–50.

[102] Shukla PC, Singh KK, Quan A, Al-Omran M, Teoh H, Lovren F, et al. BRCA1 is an essential regulator of heart function and survival following myocardial infarction. Nat Commun 2011;2:593.

[103] Pearson EJ, Nair A, Daoud Y, Blum JL. The incidence of cardiomyopathy in BRCA1 and BRCA2 mutation carriers after anthracycline-based adjuvant chemotherapy. Breast Cancer Res Treat 2017;162(1):59–67.

[104] Weir HK, Anderson RN, King SM, Soman A, Thompson TD, Hong Y, et al. Heart disease and cancer deaths – trends and projections in the United States, 1969–2020. Prev Chronic Dis 2016;13:E157.

[105] Reddy P, Shenoy C, Blaes AH. Cardio-oncology in the older adult. J Geriatr Oncol 2017;8(4):308–14.

[106] Armenian SH, Lacchetti C, Lenihan D. Prevention and monitoring of cardiac dysfunction in survivors of adult cancers: American Society of Clinical Oncology clinical practice guideline summary. J Oncol Pract 2017;13(4):270–5.

[107] Bowles EJ, Wellman R, Feigelson HS, Onitilo AA, Freedman AN, Delate T, et al. Risk of heart failure in breast cancer patients after anthracycline and trastuzumab treatment: a retrospective cohort study. J Natl Cancer Inst 2012;104(17):1293–305.

[108] Meiners B, Shenoy C, Zordoky BN. Clinical and preclinical evidence of sex-related differences in anthracycline-induced cardiotoxicity. Biol Sex Differ 2018;9(1):38.

[109] Lipshultz SE, Lipsitz SR, Mone SM, Goorin AM, Sallan SE, Sanders SP, et al. Female sex and higher drug dose as risk factors for late cardiotoxic effects of doxorubicin therapy for childhood cancer. N Engl J Med 1995;332(26):1738–43.

[110] Ewer MS, Jaffe N, Ried H, Zietz HA, Benjamin RS. Doxorubicin cardiotoxicity in children: comparison of a consecutive divided daily dose administration schedule with single dose (rapid) infusion administration. Med Pediatr Oncol 1998;31(6):512–5.

[111] Krischer JP, Epstein S, Cuthbertson DD, Goorin AM, Epstein ML, Lipshultz SE. Clinical cardiotoxicity following anthracycline treatment for childhood cancer: the Pediatric Oncology Group experience. J Clin Oncol 1997;15(4):1544–52.

[112] Amigoni M, Giannattasio C, Fraschini D, Galbiati M, Capra AC, Madotto F, et al. Low anthracyclines doses-induced cardiotoxicity in acute lymphoblastic leukemia long-term female survivors. Pediatr Blood Cancer 2010;55(7):1343–7.

[113] Brouwer CA, Postma A, Vonk JM, Zwart N, van den Berg MP, Bink-Boelkens MT, et al. Systolic and diastolic dysfunction in long-term adult survivors of childhood cancer. Eur J Cancer 2011;47(16):2453–62.

[114] van der Pal HJ, van Dalen EC, Hauptmann M, Kok WE, Caron HN, van den Bos C, et al. Cardiac function in 5-year survivors of childhood cancer: a long-term follow-up study. Arch Intern Med 2010;170(14):1247–55.

[115] Zhang J, Knapton A, Lipshultz SE, Cochran TR, Hiraragi H, Herman EH. Sex-related differences in mast cell activity and doxorubicin toxicity: a study in spontaneously hypertensive rats. Toxicol Pathol 2014;42(2):361–75.

[116] Szmit S, Jurczak W, Zaucha JM, Drozd-Sokołowska J, Spychałowicz W, Joks M, et al. Pre-existing arterial hypertension as a risk factor for early left ventricular systolic dysfunction following (R)-CHOP chemotherapy in patients with lymphoma. J Am Soc Hypertens 2014;8(11):791–9.

[117] Wang L, Tan TC, Halpern EF, Neilan TG, Francis SA, Picard MH, et al. Major cardiac events and the value of echocardiographic evaluation in patients receiving anthracycline-based chemotherapy. Am J Cardiol 2015;116(3):442–6.

[118] Gunaldi M, Duman BB, Afsar CU, Paydas S, Erkisi M, Kara IO, et al. Risk factors for developing cardiotoxicity of trastuzumab in breast cancer patients: an observational single-centre study. J Oncol Pharm Pract 2016;22(2):242–7.

[119] Nowsheen S, Aziz K, Park JY, Lerman A, Villarraga HR, Ruddy KJ, et al. Trastuzumab in female breast cancer patients with reduced left ventricular ejection fraction. J Am Heart Assoc 2018;7(15), e008637.

[120] Yersal O, Eryilmaz U, Akdam H, Meydan N, Barutca S. Arterial stiffness in breast cancer patients treated with anthracycline and trastuzumab-based regimens. Cardiol Res Pract 2018;2018, 5352914.

[121] Rosenberg SM, Partridge AH. Premature menopause in young breast cancer: effects on quality of life and treatment interventions. J Thorac Dis 2013;5(Suppl 1):S55–61.

[122] Liu S, Aghel N, Belford L, Silversides CK, Nolan M, Amir E, et al. Cardiac outcomes in pregnant women with treated cancer. J Am Coll Cardiol 2018;72(17):2087–9.

[123] Hines MR, Mulrooney DA, Hudson MM, Ness KK, Green DM, Howard SC, et al. Pregnancy-associated cardiomyopathy in survivors of childhood cancer. J Cancer Surviv 2016;10(1):113–21.

[124] Thompson KA, Hildebrandt MA, Ater JL. Cardiac outcomes with pregnancy after cardiotoxic therapy for childhood cancer. J Am Coll Cardiol 2017;69(5):594–5.

[125] Thavendiranathan P, Grant AD, Negishi T, Plana JC, Popović ZB, Marwick TH. Reproducibility of echocardiographic techniques for sequential assessment of left ventricular ejection fraction and volumes: application to patients undergoing cancer chemotherapy. J Am Coll Cardiol 2013;61(1):77–84.

[126] Negishi K, Negishi T, Hare JL, Haluska BA, Plana JC, Marwick TH. Independent and incremental value of deformation indices for prediction of trastuzumab-induced cardiotoxicity. J Am Soc Echocardiogr 2013;26(5):493–8.

[127] Sawaya H, Sebag IA, Plana JC, Januzzi JL, Ky B, Cohen V, et al. Early detection and prediction of cardiotoxicity in chemotherapy-treated patients. Am J Cardiol 2011;107(9):1375–80.

[128] Negishi K, Negishi T, Haluska BA, Hare JL, Plana JC, Marwick TH. Use of speckle strain to assess left ventricular responses to cardiotoxic chemotherapy and cardioprotection. Eur Heart J Cardiovasc Imaging 2014;15(3):324–31.

[129] Larsen CM, Mulvagh SL. Cardio-oncology: what you need to know now for clinical practice and echocardiography. Echo Res Pract 2017;4(1):R33–41.

[130] Tan TC, Scherrer-Crosbie M. Cardiac complications of chemotherapy: role of imaging. Curr Treat Options Cardiovasc Med 2014;16(4):296.

[131] Jordan JH, Todd RM, Vasu S, Hundley WG. Cardiovascular magnetic resonance in the oncology patient. JACC Cardiovasc Imaging 2018;11(8):1150–72.

[132] Grover S, Lou PW, Bradbrook C, Cheong K, Kotasek D, Leong DP, et al. Early and late changes in markers of aortic stiffness with breast cancer therapy. Intern Med J 2015;45(2):140–7.

[133] Hesse B, Lindhardt TB, Acampa W, Anagnostopoulos C, Ballinger J, Bax JJ, et al. EANM/ESC guidelines for radionuclide imaging of cardiac function. Eur J Nucl Med Mol Imaging 2008;35(4):851–85.

[134] Altena R, Perik PJ, van Veldhuisen DJ, de Vries EG, Gietema JA. Cardiovascular toxicity caused by cancer treatment: strategies for early detection. Lancet Oncol 2009;10(4):391–9.

[135] D'Amore C, Gargiulo P, Paolillo S, Pellegrino AM, Formisano T, Mariniello A, et al. Nuclear imaging in detection and monitoring of cardiotoxicity. World J Radiol 2014;6(7):486–92.

[136] Nousiainen T, Jantunen E, Vanninen E, Hartikainen J. Early decline in left ventricular ejection fraction predicts doxorubicin cardiotoxicity in lymphoma patients. Br J Cancer 2002;86(11):1697–700.

[137] Swain SM, Whaley FS, Ewer MS. Congestive heart failure in patients treated with doxorubicin: a retrospective analysis of three trials. Cancer 2003;97(11):2869–79.

[138] Cardinale D, Sandri MT, Colombo A, Colombo N, Boeri M, Lamantia G, et al. Prognostic value of troponin I in cardiac risk stratification of cancer patients undergoing high-dose chemotherapy. Circulation 2004;109(22):2749–54.

[139] Cardinale D, Colombo A, Torrisi R, Sandri MT, Civelli M, Salvatici M, et al. Trastuzumab-induced cardiotoxicity: clinical and prognostic implications of troponin I evaluation. J Clin Oncol 2010;28(25):3910–6.

[140] Sawaya H, Sebag IA, Plana JC, Januzzi JL, Ky B, Tan TC, et al. Assessment of echocardiography and biomarkers for the extended prediction of cardiotoxicity in patients treated with anthracyclines, taxanes, and trastuzumab. Circ Cardiovasc Imaging 2012;5(5):596–603.

[141] Curigliano G, Cardinale D, Suter T, Plataniotis G, de Azambuja E, Sandri MT, et al. Cardiovascular toxicity induced by chemotherapy, targeted agents and radiotherapy: ESMO clinical practice guidelines. Ann Oncol 2012;23(Suppl 7):vii155–66.

[142] Skovgaard D, Hasbak P, Kjaer A. BNP predicts chemotherapy-related cardiotoxicity and death: comparison with gated equilibrium radionuclide ventriculography. PLoS One 2014;9(5):e96736.

[143] Yu AF, Ky B. Roadmap for biomarkers of cancer therapy cardiotoxicity. Heart 2016;102(6):425–30.

[144] Horacek JM, Tichy M, Jebavy L, Pudil R, Ulrychova M, Maly J. Use of multiple biomarkers for evaluation of anthracycline-induced cardiotoxicity in patients with acute myeloid leukemia. Exp Oncol 2008;30(2):157–9.

[145] Horacek JM, Tichy M, Pudil R, Jebavy L. Glycogen phosphorylase BB could be a new circulating biomarker for detection of anthracycline cardiotoxicity. Ann Oncol 2008;19(9):1656–7.

[146] Horacek JM, Vasatova M, Pudil R, Tichy M, Zak P, Jakl M, et al. Biomarkers for the early detection of anthracycline-induced cardiotoxicity: current status. Biomed Pap Med Fac Univ Palacky Olomouc Czech Repub 2014;158(4):511–7.

[147] Ozturk G, Tavil B, Ozguner M, Ginis Z, Erden G, Tunc B, et al. Evaluation of cardiac markers in children undergoing hematopoietic stem cell transplantation. J Clin Lab Anal 2015;29(4):259–62.

[148] Pavo N, Raderer M, Hülsmann M, Neuhold S, Adlbrecht C, Strunk G, et al. Cardiovascular biomarkers in patients with cancer and their association with all-cause mortality. Heart 2015;101(23):1874–80.

[149] Ky B, Putt M, Sawaya H, French B, Januzzi JL Jr, Sebag IA, et al. Early increases in multiple biomarkers predict subsequent cardiotoxicity in patients with breast cancer treated with doxorubicin, taxanes, and trastuzumab. J Am Coll Cardiol 2014;63(8):809–16.

[150] Putt M, Hahn VS, Januzzi JL, Sawaya H, Sebag IA, Plana JC, et al. Longitudinal changes in multiple biomarkers are associated with cardiotoxicity in breast cancer patients treated with doxorubicin, taxanes, and trastuzumab. Clin Chem 2015;61(9):1164–72.

[151] van Boxtel W, Bulten BF, Mavinkurve-Groothuis AM, Bellersen L, Mandigers CM, Joosten LA, et al. New biomarkers for early detection of cardiotoxicity after treatment with docetaxel, doxorubicin and cyclophosphamide. Biomarkers 2015;20(2):143–8.

[152] Zamorano JL, Lancellotti P, Rodriguez Muñoz D, Aboyans V, Asteggiano R, Galderisi M, et al. 2016 ESC position paper on cancer treatments and cardiovascular toxicity developed under the auspices of the ESC Committee for practice guidelines. Kardiol Pol 2016;74(11):1193–233.

[153] Yancy CW, Jessup M, Bozkurt B, Butler J, Casey DE Jr, Drazner MH, et al. 2013 ACCF/AHA guideline for the management of heart failure: executive summary: a report of the American College of Cardiology Foundation/American Heart Association Task Force on practice guidelines. Circulation 2013;128(16):1810–52.

[154] Boekhout AH, Gietema JA, Milojkovic Kerklaan B, van Werkhoven ED, Altena R, Honkoop A, et al. Angiotensin II-receptor inhibition with candesartan to prevent trastuzumab-related cardiotoxic effects in patients with early breast cancer: a randomized clinical trial. JAMA Oncol 2016;2(8):1030–7.

[155] Pituskin E, Mackey JR, Koshman S, Jassal D, Pitz M, Haykowsky MJ, et al. Multidisciplinary approach to novel therapies in cardio-oncology research (MANTICORE 101-breast): a randomized trial for the prevention of trastuzumab-associated cardiotoxicity. J Clin Oncol 2017;35(8):870–7.

[156] Avila MS, Ayub-Ferreira SM, de Barros Wanderley MR Jr, das Dores Cruz F, Gonçalves Brandão SM, Rigaud VOC, et al. Carvedilol for prevention of chemotherapy-related cardiotoxicity: the CECCY trial. J Am Coll Cardiol 2018;71(20):2281–90.

[157] Yavas G, Elsurer R, Yavas C, Elsurer C, Ata O. Does spironolactone ameliorate trastuzumab-induced cardiac toxicity? Med Hypotheses 2013;81(2):231–4.

[158] Akpek M, Ozdogru I, Sahin O, Inanc M, Dogan A, Yazici C, et al. Protective effects of spironolactone against anthracycline-induced cardiomyopathy. Eur J Heart Fail 2015;17(1):81–9.

[159] Seicean S, Seicean A, Plana JC, Budd GT, Marwick TH. Effect of statin therapy on the risk for incident heart failure in patients with breast cancer receiving anthracycline chemotherapy: an observational clinical cohort study. J Am Coll Cardiol 2012;60(23):2384–90.

[160] Hahn VS, Lenihan DJ, Ky B. Cancer therapy-induced cardiotoxicity: basic mechanisms and potential cardioprotective therapies. J Am Heart Assoc 2014;3(2), e000665.

[161] Kalam K, Marwick TH. Role of cardioprotective therapy for prevention of cardiotoxicity with chemotherapy: a systematic review and meta-analysis. Eur J Cancer 2013;49(13):2900–9.

[162] van Dalen EC, Caron HN, Dickinson HO, Kremer LC. Cardioprotective interventions for cancer patients receiving anthracyclines. Cochrane Database Syst Rev 2011;(6), CD003917.

[163] Lipshultz SE, Colan SD, Gelber RD, Perez-Atayde AR, Sallan SE, Sanders SP. Late cardiac effects of doxorubicin therapy for acute lymphoblastic leukemia in childhood. N Engl J Med 1991;324(12):808–15.

[164] Lee MO, Song SH, Jung S, Hur S, Asahara T, Kim H, et al. Effect of ionizing radiation induced damage of endothelial progenitor cells in vascular regeneration. Arterioscler Thromb Vasc Biol 2012;32(2):343–52.

[165] Berry GJ, Jorden M. Pathology of radiation and anthracycline cardiotoxicity. Pediatr Blood Cancer 2005;44(7):630–7.

[166] Iliescu CA, Grines CL, Herrmann J, Yang EH, Cilingiroglu M, Charitakis K, et al. SCAI expert consensus statement: evaluation, management, and special considerations of cardio-oncology patients in the cardiac catheterization laboratory (endorsed by the Cardiological Society of India, and sociedad Latino Americana de Cardiologia intervencionista). Catheter Cardiovasc Interv 2016;87(5):E202–23.

[167] Zhu Q, Kirova YM, Cao L, Arsene-Henry A, Chen J. Cardiotoxicity associated with radiotherapy in breast cancer: a question-based review with current literatures. Cancer Treat Rev 2018;68:9–15.

[168] Saiki H, Petersen IA, Scott CG, Bailey KR, Dunlay SM, Finley RR, et al. Risk of heart failure with preserved ejection fraction in older women after contemporary radiotherapy for breast cancer. Circulation 2017;135(15):1388–96.

[169] Chang HM, Okwuosa TM, Scarabelli T, Moudgil R, Yeh ETH. Cardiovascular complications of cancer therapy: best practices in diagnosis, prevention, and management: part 2. J Am Coll Cardiol 2017;70(20):2552–65.

[170] Gujral DM, Lloyd G, Bhattacharyya S. Radiation-induced valvular heart disease. Heart 2016;102(4):269–76.

[171] Handa N, McGregor CG, Danielson GK, Daly RC, Dearani JA, Mullany CJ, et al. Valvular heart operation in patients with previous mediastinal radiation therapy. Ann Thorac Surg 2001;71(6):1880–4.

[172] Heidenreich PA, Hancock SL, Vagelos RH, Lee BK, Schnittger I. Diastolic dysfunction after mediastinal irradiation. Am Heart J 2005;150(5):977–82.

[173] Lancellotti P, Nkomo VT, Badano LP, Bergler-Klein J, Bogaert J, Davin L, et al. Expert consensus for multi-modality imaging evaluation of cardiovascular complications of radiotherapy in adults: a report from the European Association of Cardiovascular Imaging and the American Society of Echocardiography. J Am Soc Echocardiogr 2013;26(9):1013–32.

[174] Gottdiener JS, Katin MJ, Borer JS, Bacharach SL, Green MV. Late cardiac effects of therapeutic mediastinal irradiation. Assessment by echocardiography and radionuclide angiography. N Engl J Med 1983;308(10):569–72.

[175] Onwudiwe NC, Kwok Y, Onukwugha E, Sorkin JD, Zuckerman IH, Shaya FT, et al. Cardiovascular event-free survival after adjuvant radiation therapy in breast cancer patients stratified by cardiovascular risk. Cancer Med 2014;3(5):1342–52.

[176] Boero IJ, Paravati AJ, Triplett DP, Hwang L, Matsuno RK, Gillespie EF, et al. Modern radiation therapy and cardiac outcomes in breast cancer. Int J Radiat Oncol Biol Phys 2016;94(4):700–8.

[177] Cheng YJ, Nie XY, Ji CC, Lin XX, Liu LJ, Chen XM, et al. Long-term cardiovascular risk after radiotherapy in women with breast cancer. J Am Heart Assoc 2017;6(5).

[178] Jaworski C, Mariani JA, Wheeler G, Kaye DM. Cardiac complications of thoracic irradiation. J Am Coll Cardiol 2013;61(23):2319–28.

[179] Darby SC, McGale P, Taylor CW, Peto R. Long-term mortality from heart disease and lung cancer after radiotherapy for early breast cancer: prospective cohort study of about 300,000 women in US SEER cancer registries. Lancet Oncol 2005;6(8):557–65.

[180] Hughes-Davies L, Sacks D, Rescigno J, Howard S, Harris J. Serum cardiac troponin T levels during treatment of early-stage breast cancer. J Clin Oncol 1995;13(10):2582–4.

[181] Kozak KR, Hong TS, Sluss PM, Lewandrowski EL, Aleryani SL, Macdonald SM, et al. Cardiac blood biomarkers in patients receiving thoracic (chemo)radiation. Lung Cancer 2008;62(3):351–5.

[182] Nellessen U, Zingel M, Hecker H, Bahnsen J, Borschke D. Effects of radiation therapy on myocardial cell integrity and pump function: which role for cardiac biomarkers? Chemotherapy 2010;56(2):147–52.

[183] Skytta T, Tuohinen S, Boman E, Virtanen V, Raatikainen P, Kellokumpu-Lehtinen PL. Troponin T-release associates with cardiac radiation doses during adjuvant left-sided breast cancer radiotherapy. Radiat Oncol 2015;10:141.

[184] Darby SC, Ewertz M, McGale P, Bennet AM, Blom-Goldman U, Brønnum D, et al. Risk of ischemic heart disease in women after radiotherapy for breast cancer. N Engl J Med 2013;368(11):987–98.

[185] Groarke JD, Nguyen PL, Nohria A, Ferrari R, Cheng S, Moslehi J. Cardiovascular complications of radiation therapy for thoracic malignancies: the role for non-invasive imaging for detection of cardiovascular disease. Eur Heart J 2014;35(10):612–23.

[186] Lo Q, Hee L, Batumalai V, Allman C, MacDonald P, Lonergan D, et al. Strain imaging detects dose-dependent segmental cardiac dysfunction in the acute phase after breast irradiation. Int J Radiat Oncol Biol Phys 2017;99(1):182–90.

[187] Heidenreich PA, Schnittger I, Strauss HW, Vagelos RH, Lee BK, Mariscal CS, et al. Screening for coronary artery disease after mediastinal irradiation for Hodgkin's disease. J Clin Oncol 2007;25(1):43–9.

[188] Yeh E, Ewer M. Cancer and the heart, 2. People's Medical Publishing House USA Ltd (PMPH); 2013.

[189] Desai MY. Radiation associated cardiac disease, expert analysis. Am Coll Cardiol 2019;74(7). https://doi.org/10.1016/j.jacc.2019.07.006.

[190] Kamdar AR, Meadows TA, Roselli EE, Gorodeski EZ, Curtin RJ, Sabik JF, et al. Multidetector computed tomographic angiography in planning of reoperative cardiothoracic surgery. Ann Thorac Surg 2008;85(4):1239–45.

[191] Rademaker J, Schöder H, Ariaratnam NS, Strauss HW, Yahalom J, Steingart R, et al. Coronary artery disease after radiation therapy for Hodgkin's lymphoma: coronary CT angiography findings and calcium scores in nine asymptomatic patients. AJR Am J Roentgenol 2008;191(1):32–7.

[192] Machann W, Beer M, Breunig M, Störk S, Angermann C, Seufert I, et al. Cardiac magnetic resonance imaging findings in 20-year survivors of mediastinal radiotherapy for Hodgkin's disease. Int J Radiat Oncol Biol Phys 2011;79(4):1117–23.

[193] George TJ, Arnaoutakis GJ, Beaty CA, Kilic A, Baumgartner WA, Conte JV. Contemporary etiologies, risk factors, and outcomes after pericardiectomy. Ann Thorac Surg 2012;94(2):445–51.

[194] Fender EA, Liang JJ, Sio TT, Stulak JM, Lennon RJ, Slusser JP, et al. Percutaneous revascularization in patients treated with thoracic radiation for cancer. Am Heart J 2017;187:98–103.

[195] Liang JJ, Sio TT, Slusser JP, Lennon RJ, Miller RC, Sandhu G, et al. Outcomes after percutaneous coronary intervention with stents in patients treated with thoracic external beam radiation for cancer. JACC Cardiovasc Interv 2014;7(12):1412–20.

[196] Gharagozloo F, Clements IP, Mullany CJ. Use of the internal mammary artery for myocardial revascularization in a patient with radiation-induced coronary artery disease. Mayo Clin Proc 1992;67(11):1081–4.

[197] Donnellan E, Phelan D, McCarthy CP, Collier P, Desai M, Griffin B. Radiation-induced heart disease: a practical guide to diagnosis and management. Cleve Clin J Med 2016;83(12):914–22.

[198] Makkar RR, Jilaihawi H, Mack M, Chakravarty T, Cohen DJ, Cheng W, et al. Stratification of outcomes after transcatheter aortic valve replacement according to surgical inoperability for technical versus clinical reasons. J Am Coll Cardiol 2014;63(9):901–11.

[199] Gagliardi G, Constine LS, Moiseenko V, Correa C, Pierce LJ, Allen AM, et al. Radiation dose-volume effects in the heart. Int J Radiat Oncol Biol Phys 2010;76(3 Suppl):S77–85.

[200] Rygiel K. Cardiotoxic effects of radiotherapy and strategies to reduce them in patients with breast cancer: an overview. J Cancer Res Ther 2017;13(2):186–92.

[201] Bovelli D, Plataniotis G, Roila F, ESMO Guidelines Working Group. Cardiotoxicity of chemotherapeutic agents and radiotherapy-related heart disease: ESMO clinical practice guidelines. Ann Oncol 2010;21(Suppl 5):v277–82.

[202] Litwin MS, Tan HJ. The diagnosis and treatment of prostate cancer: a review. JAMA 2017;317(24):2532–42.

[203] Yuan C, Cao Y, Chavarro J, Lindstrom S, Qiu W, Willett W, et al. Prediagnostic body-mass index, smoking and prostate cancer survival: a cohort consortium study of over 10,000 white men with prostate cancer. Cancer Epidemiol Biomarkers Prev 2015;24(4):759.

[204] Tanaka A, Node K. The emerging and promising role of care for cardiometabolic syndrome in prostate cancer. JACC: Cardiooncol 2019;1(2):307.

[205] Nguyen PL, Alibhai SM, Basaria S, D'Amico AV, Kantoff PW, Keating NL, et al. Adverse effects of androgen deprivation therapy and strategies to mitigate them. Eur Urol 2015;67(5):825–36.

[206] Nguyen PL, Je Y, Schutz FA, Hoffman KE, Hu JC, Parekh A, et al. Association of androgen deprivation therapy with cardiovascular death in patients with prostate cancer: a meta-analysis of randomized trials. JAMA 2011;306(21):2359–66.

[207] Saigal CS, Gore JL, Krupski TL, Hanley J, Schonlau M, Litwin MS, et al. Androgen deprivation therapy increases cardiovascular morbidity in men with prostate cancer. Cancer 2007;110(7):1493–500.

[208] Narayan V, Ky B. Common cardiovascular complications of cancer therapy: epidemiology, risk prediction, and prevention. Annu Rev Med 2018;69:97–111.

[209] Levine GN, D'Amico AV, Berger P, Clark PE, Eckel RH, Keating NL, et al. Androgen-deprivation therapy in prostate cancer and cardiovascular risk: a science advisory from the American Heart Association, American Cancer Society, and American Urological Association: endorsed by the American Society for Radiation Oncology. Circulation 2010;121(6):833–40.

[210] Zhu X, Wu S. Increased risk of hypertension with enzalutamide in prostate cancer: a meta-analysis. Cancer Invest 2019;37(9):478–88.

[211] Iacovelli R, Ciccarese C, Bria E, Romano M, Fantinel E, Bimbatti D, et al. The cardiovascular toxicity of abiraterone and enzalutamide in prostate cancer. Clin Genitourin Cancer 2018;16(3):e645–53.

[212] Dreicer R. Implications of cardiovascular disease for prostate cancer patients receiving androgen-signaling inhibitors. 6 February; Available from: https://www.jwatch.org/na50812/2020/02/06/implications-cardiovascular-disease-prostate-cancer; 2020. [Accessed 14 August 2020].

Chapter 25

Sex Hormones and Their Impact on Cardiovascular Health

Sasha De Jesus, Eugenia Gianos, Stephanie Trentacoste McNally, Dawn C. Scantlebury, and Stacey E. Rosen

Clinical Case

A 51-year-old woman postmenopausal woman with a medical history significant for polycystic ovarian syndrome (PCOS), diabetes, and borderline hypertension is concerned with her symptoms of increased frequency of hot flashes. Her last period was at age 45. Her family history is significant for early-onset atherosclerotic cardiovascular disease and breast cancer. She was referred to the gynecologist and would like to initiate menopausal hormone therapy (MHT). She is a physically active nonsmoker, with no symptoms on exertion. Her weight and blood pressure are optimal and her lipid panel reveals: total cholesterol 178 mg/dL; triglycerides 81 mg/dL; high-density lipoprotein cholesterol (HDL-c) 65 mg/dL; and low-density lipoprotein cholesterol (LDL-c) 97 mg/dL.

What are her underlying risk factors for MHT use?

What can you counsel her about her options and the safety measures associated with these treatment modalities?

What factors should be evaluated prior to initiation of MHT?

Sex Differences in Cardiac Disease. https://doi.org/10.1016/B978-0-12-819369-3.00029-0

Abstract

The sex steroids (hormones), the androgens, estrogens, and progestogens, are all present to variable levels in each sex. Aside from typically considered feminizing and masculinizing effects, through their interactions with the sex steroid receptors present in the heart and vasculature, and their metabolic effects, they play a vital role in cardiovascular (CV) health. These effects are complex, with both negative and positive effects being noted. Endogenous estrogen in women appears to infer protection from cardiovascular disease (CVD), but exogenous estrogen has been associated with increased CVD risk. Androgens appear to exert positive effects on the vasculature and myocardium and reduced testosterone levels are associated with higher CVD risk. However, exogenous testosterone shows mixed effects on CV outcomes. Therapies targeting hormones include menopause hormonal therapy, testosterone replacement therapy, hormonal antagonism in the treatment of cancer, contraception, and gender-affirming therapy in transgender individuals. The varying effects of hormonal manipulation on CV health mandate a methodical and integrated approach to their use, considering comorbidities and patient risks.

Overview of the Sex Hormones

General Concepts

The sex hormones (sex steroids), which include estrogens, androgens, and progestogens, are present in variable levels in each sex. Usually considered in relation to sexual development and reproduction, these hormones also have wide-ranging functions, including significant effects on cardiovascular (CV) health. Variations in hormonal levels across the life span, or in various pathologic states, impact CV risk in both men and women. Circulating estrogens protect women from atherosclerotic cardiovascular disease (CVD), a protection that declines following menopause. Elevated androgen levels in either sex, either from exogenous administration or in pathologic states such as polycystic ovarian syndrome (PCOS) in women, may be associated with elevated CV risk. Declining androgen levels in older men may also be associated with elevated CV risk. Understanding the impact of hormonal changes with aging, disease, hormone replacement, and suppression therapies therefore requires a complete appreciation of the complex roles of individual hormones in men, women, and the transgender population.

Gonadal production of sex hormones is regulated by circulating levels of the hypothalamic and pituitary peptides gonadotropin-releasing hormone (GnRH), luteinizing hormone (LH), and follicle-stimulating hormone (FSH). FSH stimulates the growth of a dominant ovarian follicle in females, and spermatogenesis in the Sertoli cells in the testes in males. LH promotes the production of androgens in the testes as well as the ovaries (as an estrogen precursor) (**Figure 1**). Ovulation is triggered by an LH surge, ultimately leading to progesterone production from the corpus luteum.

Sex hormones impact cellular processes by gene regulation and protein modification through their interaction with the sex steroid receptors which are all present in both sexes.

Androgens

Aside from pituitary LH-induced gonadal synthesis of androgens, extragonadal synthesis in sites such as adrenal gland, bone, and adipose tissue can occur, in both men and women (**Figure 1**), [1]. Testosterone, the most important naturally occurring androgen, is metabolized to the more potent dihydrotestosterone (DHT), by 5-α-reductase, as well as to estradiol via aromatase (**Figure 2**). Most testosterone circulates bound to sex hormone-binding globulin (SHBG) with only a fraction available to the androgen receptor as free testosterone, which is the biologically active form. Conditions that decrease SHBG levels, such as obesity and type 2 diabetes mellitus, and in women, PCOS, can result in elevated free testosterone levels relative to total testosterone [2].

In males, androgen levels are low in infancy, rising at puberty under the influence of the gonadotropins, and remain at high levels until late middle age, declining to levels 20–50% of peak levels by age 80 [3]. In women with hyperandrogenism, mean levels of total testosterone have been reported to be almost double normal levels [4]. Just as in older men, testosterone levels in women usually decrease in middle age, following menopause [5].

Most organs are responsive to the action of androgens. Physiologic effects of androgens include the development of "male" primary and secondary sexual characteristics such as voice, facial hair, and muscle development, as well as an increase in basal metabolic rate and red blood cell production [3].

Estrogens

There are several naturally occurring estrogens, with estrone (E1), 17β-estradiol, or simply, estradiol (E2), and estriol (E3) [6] being the only ones noted in significant quantities in the plasma of human females. In women, estrone and estradiol are produced in the granulosa cells of the follicles by the aromatase conversion of androstenedione and testosterone (**Figure 2**). E3 is a very weak estrogen produced from the oxidation of estradiol and estrone. Estetrol (E4) is produced by the fetal liver and is found only during pregnancy [7].

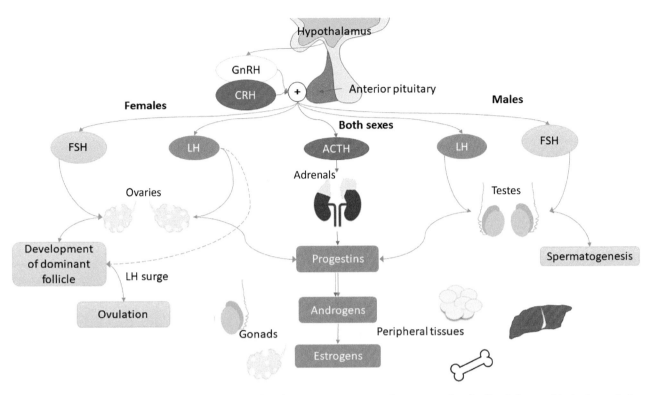

FIGURE 1   Male and Female Hormone Production. Hormones complete a negative feedback loop with the hypothalamus and pituitary. ACTH, adrenocorticotropic hormone; CRH, corticotropin-releasing hormone; FSH, follicle-stimulating hormone; GnRH, gonadotropin-releasing hormone; LH, luteinizing hormone.

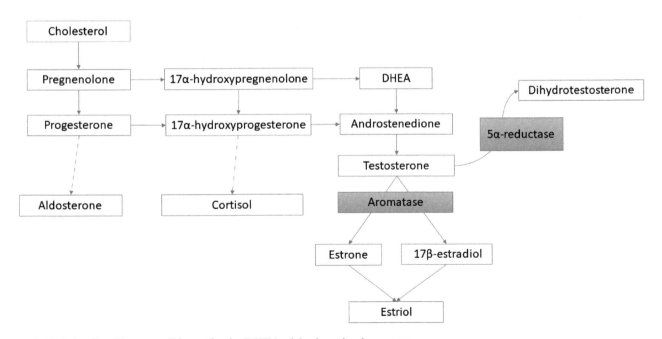

FIGURE 2   Sex Hormone Biosynthesis. DHEA, dehydroepiandrosterone.

E2 is the predominant estrogen produced during a woman's reproductive life. Levels increase during adolescence, ranging from 30 to 400 pg/mL during the reproductive years and decline in the premenopause period, to below 30 pg/mL after menopause [8]. During menopause, estrone (E1), produced primarily in adipose tissue, becomes the predominant estrogen [9]. In men, E2 is produced by aromatization of androgenic precursors from the testes and adrenal glands. Levels are typically significantly lower than in women but may be particularly elevated in obesity due to increased aromatase activity in adipose tissue. Elderly men may have higher levels of E2 than age-matched women. Aside from circulating hormone levels, a key factor influencing estrogen action is the activity of aromatase in tissues, including extragonadal tissues such as the brain, adipose tissue, and the heart and blood vessels of both sexes. Circulating levels may therefore not predict tissue levels of E2 [5, 10].

Estrogens exert their effects through interaction with varying isoforms of the estrogen receptor (ER) [10]. Effects include primary and secondary female characteristics [6] as well as effects on bone, skin, fat, and other subcutaneous tissues. Other actions include sodium and water retention (usually minimal but can be clinically significant during pregnancy).

Progesterone

The progestogens are pro-gestation, i.e., are essential for the maintenance of pregnancy. Progesterone is produced by the ovaries, placenta, and adrenal glands. During the normal menstrual cycle, progesterone is produced in the ovary in response to LH, and during pregnancy, placental production predominates after the 9th week of pregnancy. Progesterone has a wide range of effects aside from the reproductive system including effects on the immune system (T-lymphocytes), platelet aggregation, intestinal smooth muscle, and neurological and psychiatric effects [11].

Cardiovascular Effects of Sex Hormones: From Physiology to Epidemiology

Sex hormones, including androgens, estrogens, and progesterone, all have varied implications on cardiovascular health including effects on lipid profile, coronary artery reactivity, and response to insulin. These are summarized in **Figure 3**.

Androgens

Androgen receptors are present on cardiomyocytes and within the vasculature. In animal models, inhibition of androgen effect, either through inhibition of 5α-reductase by finasteride, resulting in reduced generation of the more potent DHT, or through orchiectomy, reduces adverse cardiac remodeling in response to various stressors [5]. On the other hand, exogenous testosterone reduces myocardial infarct size and demonstrates antiarrhythmic properties by decreasing the action potential duration, reducing early after depolarizations, and shortening the QT interval [12]. Testosterone has a vasodilatory effect on arteries that is independent of nitric oxide (NO), i.e., endothelium independent. It has a direct smooth muscle-relaxing effect, possibly via inhibition of L- and T-type calcium channels, and may also induce vasoconstriction under some circumstances [12].

With these varied (patho)physiologic effects, the clinical effect of androgens on the CV system is therefore a complex topic. Some studies show an increase in CV events associated with exogenous testosterone therapy [2]. Conversely, low testosterone levels in men are associated with increased risk of metabolic syndrome or its individual components. Araujo et al. demonstrated in a meta-analysis of 12 studies that lower endogenous testosterone levels were associated with increased risk of all-cause and CV mortality ($P < 0.001$ and 0.06, respectively) in community-based studies of men. There was significant between-study heterogeneity, however, related to study and subject characteristics [13]. Another meta-analysis by Ruije et al. showed a weak association of endogenous testosterone levels and CV risk, a finding which was modified by age, and the year of publication of the study [14]. Another meta-analysis of 70 studies showed that patients with CVD had lower testosterone and higher E2 levels in cross-sectional studies, but longitudinal studies did not show an increase in incident CVD in patients with low testosterone levels, although testosterone levels were lower in patients with incident all-cause and CVD mortality [15]. The authors of these meta-analyses suggest that these effects may be partly due to differences in overall health status, with low testosterone reflecting poor general health.

In women, higher androgen levels appear to be associated with CVD risk factors such as elevated blood pressures, C-reactive protein, and insulin resistance. Both high and low androgen levels have been associated with increased CV risk. The Multi-Ethnic Study of Atherosclerosis (MESA) is a prospective cohort study which enrolled men and women of four races/ethnic groups. The study evaluated the association of CV events with sex hormone levels among 2834 postmenopausal women and demonstrated that a higher testosterone/E2 ratio was associated with a higher risk for incident CVD, coronary heart disease, and heart failure events, in a 12-year follow-up period [16].

Estrogens

ER-α and β receptors have been found in the vascular endothelium and smooth muscle cells as well as

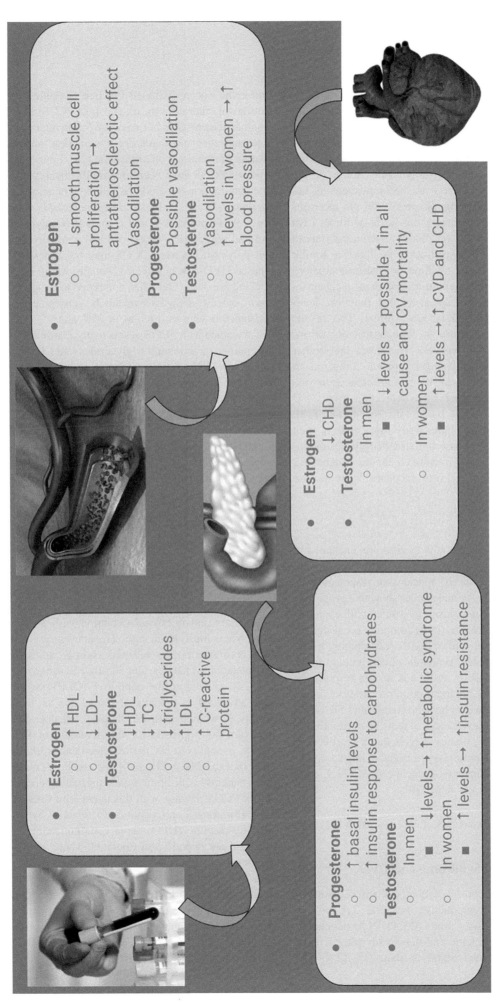

Estrogen
- ○ ↑ HDL
- ○ ↓ LDL
Testosterone
- ○ ↓ HDL
- ○ ↓ TC
- ○ ↓ triglycerides
- ○ ↑ LDL
- ○ ↑ C-reactive protein

Estrogen
- ○ ↓ smooth muscle cell proliferation → antiatherosclerotic effect
- ○ Vasodilation
Progesterone
- ○ Possible vasodilation
Testosterone
- ○ Vasodilation
- ○ ↑ levels in women → ↑ blood pressure

Progesterone
- ○ ↑ basal insulin levels
- ○ ↑ insulin response to carbohydrates
Testosterone
- ○ In men
  - ↓ levels → ↑ metabolic syndrome
- ○ In women
  - ↑ levels → ↑ insulin resistance

Estrogen
- ○ ↓ CHD
Testosterone
- ○ In men
  - ↓ levels → possible ↑ in all cause and CV mortality
- ○ In women
  - ↑ levels → ↑ CVD and CHD

FIGURE 3 **Cardiometabolic Effects of Endogenous Sex Hormones.** CHD, congenital heart disease; CV, cardiovascular; CVD, cardiovascular disease; HDL, high-density lipoprotein; LDL, low-density lipoprotein; TC, total cholesterol.

cardiomyocytes and cardiac fibroblasts. Estrogen effect on the CV system includes both metabolic and vascular effects. Levels can affect cholesterol production in the liver, increasing HDL and decreasing low-density lipoprotein (LDL). E2's vaso-protective effects are both endothelium dependent and endothelium independent. Some of the endothelium-dependent effects are accomplished via activation of the ERs, eliciting vasorelaxation via increases in endothelial NO and prostaglandin I2. In addition, ERα produces an antiatherosclerotic effect by decreasing smooth muscle proliferation. ERβ activation produces a similar antiproliferative effect in addition to antiinflammatory effects. The main endothelium-independent mechanism is increased K^+ channel expression and function, leading to diminished extracellular Ca^{2+} influx. This results in either decreased response to vasoconstrictors or increased vasodilation in females. Therefore, the higher circulating E2 levels in younger women compared to men are likely responsible for the sex differences in vascular function [17] and the relative protection from CVD that women enjoy in the premenopausal period [18]. CV events in women are significantly lower than in men during the reproductive phase but rise sharply postmenopause. In the MESA study mentioned above, a higher E2 level was associated with lower coronary heart disease events [16].

Progesterone

Progesterone effects on the CV system are unclear and likely mediated through its metabolic effects. The progesterone receptors (PR-A and B) appear to be responsible for smooth muscle growth. Animal studies show that progesterone reduces coronary vasospasm. In women with exercise-induced ischemia on stress testing, addition of progesterone increases exercise time, confirming the vasodilatory effect of progesterone in humans. Metabolic effects include an increase in both basal insulin levels and insulin response to carbohydrate intake. All preparations of progesterone lower total cholesterol and lower or have no change on LDL or triglyceride levels. HDL levels are also reduced or remain the same [19]. Despite this, studies suggest that a short course of oral progesterone does not have a negative impact on endothelial function or increased clinical risk of CVD, even with a lower HDL absolute value [20].

Hormone Excess or Deficiency in Men and Women

Polycystic Ovarian Syndrome and CVD

PCOS is one of the most common endocrine disorders affecting women of reproductive age [21, 22], with a global prevalence ranging from 4% to 21% depending on the population, ethnicity, and diagnostic criteria used. Prevalence in the United States is about 7% [23]. The Endocrine Society Clinical Practice Guideline prefers the Rotterdam criteria for the diagnosis of PCOS. These criteria mandate the presence of two of the following three findings: hyperandrogenism, ovulatory dysfunction, and polycystic ovarian morphology on ultrasound [24]. Four different subtypes of PCOS have been identified. Phenotypes A and B (the classic forms) include (A) all three characteristics used for diagnosis and (B) hyperandrogenism and ovulatory dysfunction. Phenotypes C and D include (C) hyperandrogenism and polycystic ovaries, and (D) ovulatory dysfunction and polycystic ovaries. PCOS may be associated with increased CVD risk in affected women. About half of those affected are obese and when compared to the general population, twice as many have metabolic syndrome [23]. Insulin resistance is seen as often as in 75% and 95% of lean and obese women with PCOS, respectively. Cardiometabolic abnormalities include high triglycerides and LDL-cholesterol levels, low HDL-cholesterol levels, impaired glucose tolerance, and hypertension. Some of these conditions, such as diabetes and hypertension, are present independent of body weight. C-reactive protein, homocysteine, carotid intima-media thickening, and coronary calcium score have been found to be higher in women with PCOS. It remains unclear whether CVD risk is conferred by PCOS itself or by traditional CVD risk factors; however, the presence of these cardiometabolic risk factors should prompt early recognition of risk and strategies to prevent and manage CVD. Classic forms of PCOS confer a higher CVD risk and risk of type 2 diabetes mellitus than nonclassic forms [22].

Nutritional interventions and exercise are foundational to the management of overweight and obese women with PCOS [24], with studies showing benefit in metabolic risk factors and improved ovulation with 5–10% weight loss. When lifestyle changes are not successful or if insulin resistance is present, metformin is indicated to control metabolic changes and improve hormone regulation [22, 24]. Glucagon-like peptide 1 (GLP1) agonists have been advanced as potential therapies for the treatment of obesity in women with PCOS [21], but data are limited. Although not approved for the treatment of PCOS, liraglutide is approved for the management of obesity [body mass index (BMI) $\geq 30\,kg/m^2$] and could be considered for these patients, with limited data suggesting greater weight loss in PCOS patients with this agent compared to placebo [25]. Hormonal contraceptives to regulate hypothalamic-pituitary-ovarian (HPO) axis communication are often used and are overall well tolerated. There are no data on the long-term effects of hormonal contraceptives in women with PCOS with and without diabetes. However, reports suggest that there are no adverse glycemic effects in healthy women or in those with history of gestational diabetes. Hormonal contraceptives can, therefore, be used in women with PCOS, insulin resistance, and diabetes without vascular complications [24].

Menopause and CVD

Hormone levels fluctuate throughout a woman's life span. The Stages of Reproductive Aging Workshop (STRAW) developed a system for staging ovarian aging, outlining menstrual and hormonal changes in three broad phases of female adult life: the reproductive, the menopausal transition, and postmenopausal periods, dividing into 10 stages centered on the final menstrual period (**Table 1**) [26]. In women with a uterus, menopause is defined as a complete loss of menses for 1 year. Menopause typically occurs during a woman's fifth decade of life and is a consequence of progressive decline in estrogen levels [27]. In women who have undergone hysterectomy but not oophorectomy, elevated levels of FSH and low estradiol concentrations (usually <20 pg/mL) support the diagnosis of menopause although they do not confirm it [28]. This loss of estrogen can cause vasomotor symptoms (VMS) such as hot flashes and sweating, which are a hallmark of menopause. Other symptoms frequently seen include vaginal dryness, insomnia, and mood changes. Most symptoms occur within the first year after the menstrual period stops but can last as long as 13 years. Some women experience no symptoms at all [28].

Hemodynamic sequelae due to loss of estrogen are related to changes in vascular endothelium, smooth muscle, and myocytes [29]. The term "vascular aging" refers to the arterial stiffness, vasodilatory ability, and vasoconstrictive response. Estrogen-mediated effects include NO bioavailability and attenuation of oxidative stress and inflammation [30]. Endothelial-dependent vasodilation is impaired in postmenopausal women, an impairment which accelerates more rapidly with aging than in men. Postmenopausal women with moderate obesity are even more susceptible to endothelial dysfunction [31]. Reduced synthesis of NO and enhanced inactivation of endothelial nitric oxide synthase (eNOS) is also seen with estrogen withdrawal [32], as are increased concentrations of proinflammatory markers [33]. Loss of estrogen results in cardiometabolic effects including alterations in lipid profile, blood pressure, glucose tolerance, and vascular reactivity, increasing a woman's risk for the development of ischemic heart disease [18].

Cardiovascular Impact of Premature Menopause

Primary ovarian insufficiency (POI) refers to the development of hypergonadotropic hypogonadism with dysfunction or depletion of ovarian follicles leading to the cessation of menses before the age of 40. Premature menopause has been defined as menopause occurring prior to the age of 40. "Early menopause" occurs before age 45. The term POI is now preferred over premature menopause as ovarian function in POI is intermittent and spontaneous conception can occur [34].

Women who experience POI or early menopause, regardless of the cause, have increased risk for morbidity and mortality [35]. A study evaluating a total of 1547 women,

---

TABLE 1 Terminology Related to Menopause

Stage		−5	−4	−3b	−3a	−2	−1	Final Menstrual Period (0)	+1a	+1b	+1c	2
Terminology	Menarche	Reproductive				Menopausal transition			Postmenopause			
		Early	Peak	Late		Early	Late		Early			Late
						Perimenopause						
Menstrual cycle		Variable → → regular → → subtle changes late				Variability in length of cycles, from 7 days to ≥60 days amenorrhea		1 year				
Symptoms							Vasomotor symptoms					Urogenital atrophy
FSH				Low → →		Increasing FSH levels → →				Stable (high) FSH levels		

Terminology related to menopause from the Stages of Reproductive Aging Workshop + 10 [26]. Criteria for these stages include menstrual cycle changes, endocrine changes, including FSH levels, ovarian antral follicle count, and menopausal symptoms. FSH, follicle-stimulating hormone.

with 659 having had bilateral oophorectomy, showed that after a 14-year follow-up, the group with ovarian resection, independent of the indication for oophorectomy, had an accelerated rate of accumulation of 18 chronic conditions including CV conditions such as hypertension, hyperlipidemia, diabetes, stroke, and coronary artery disease [36].

Several large cohorts have been examined to appreciate the effects of oophorectomy on long-term risk. The majority, including the Nurses' Health Study, the Mayo Clinic Cohort Study of Oophorectomy and Aging, the Swedish Health Care Registry, and the Women's Health Initiative (WHI), showed an increased risk of coronary heart disease that was most striking in younger women [37]. Analysis of a large cohort of women in the UK Biobank showed that both natural and surgically induced menopause before the age of 40 were associated with a small but significantly increased risk of a composite of CVD [38]. These associations held true even after adjustment for multiple CVD risk factors. The risk for the development of incident coronary artery disease, aortic stenosis, atrial fibrillation, venous thromboembolism (VTE), hyperlipidemia, and type 2 diabetes was progressively greater with younger age of menopause. The form of menopause was found to be associated with differences in CV risk. Surgical menopause was associated with higher hazard ratios (HRs) for CVD which may be explained by the differential development of risk factors since the association did not persist after adjustment for conventional CVD risk factors [38].

For incompletely understood reasons, hysterectomy appears to alter ovarian function over time. In observational studies, women posthysterectomy developed both symptoms and hormone profiles of menopausal women earlier than women who did not have a hysterectomy, even if the ovaries were retained. For example, a prospective cohort study showed that women with a hysterectomy reached menopause 3.7 years earlier than women without. In the hysterectomy group, women who had unilateral oophorectomy reached menopause 4.4 years earlier than those who retained both ovaries [39]. Interestingly, an analysis of data from the WHI showed a higher risk for CV events (HR 1.25) in women who underwent hysterectomy with or without simultaneous oophorectomy compared to women who did not have a hysterectomy. This risk was rendered nonsignificant after adjusting for demographic and CVD risk factors such as hypertension, diabetes, high cholesterol, obesity, and lower education, income, and physical activity, which were more prevalent in women with hysterectomies. The conclusion was that the higher CVD risk was due to the worse risk profile [40].

Menopausal Hormonal Therapy: Indications, Efficacy, Safety, Timing, and Duration

Menopausal hormone therapy (MHT) [consisting of estrogen therapy (ET) and estrogen-progestogen therapy (EPT)] is most effective when menopausal symptoms are affecting a woman's quality of life [41]. The foundation of MHT consists of using estrogen alone for women without a uterus, and estrogen in combination with progesterone for those with an intact uterus in order to prevent estrogen-associated hyperplasia [28].

Estrogen Therapy

The most frequently prescribed estrogens are conjugated equine estrone (CEE), synthetic conjugated estrogens, micronized 17β-estradiol (E2), and ethinyl estradiol, with no significant difference noted between CEE and estradiol in terms of effectiveness when treating VMS [41]. However, E2 has a better safety profile than CEE, with lower risk of thromboembolic disease, stroke, and coronary disease [17]. E2 has been found to provide a more anxiolytic and antidepressant effect compared to other types of estrogens [41]. The recommended dosing strategy is to use the lowest, most effective dose of systemic ET that will treat symptoms [27]. Routes of delivery include oral tablets, transdermal patches, sprays, gels, and vaginal creams.

Although the use of systemic estrogens is not advised in breast cancer survivors, low-dose vaginal ET can be considered for the treatment of genitourinary syndrome of menopause. This can be done once nonhormonal therapy has been attempted and after discussing it with the patient's oncologist, considering that even though minimal, nonoral routes of ET can be systemically absorbed [41]. Estetrol (E4), a low-potency estrogen, may play a role in hormone replacement in the future [42].

Progesterone Therapy

The main indication for progesterone use during menopause is to prevent the effects of unopposed estrogen, which lead to hyperplasia or endometrial cancer [27, 28]. Women with an intact uterus on estrogen replacement should be given either progesterone in the form of medroxyprogesterone acetate (MPA), norethindrone acetate, or micronized progesterone. However, when estrogens are used vaginally and at a low dose, progesterone replacement therapy is not recommended. In the WHI, the CEE + MPA postintervention group showed a decreased risk for endometrial cancer. A higher incidence of breast cancer was seen with CEE + MPA, followed by CEE alone, and less with placebo

[43]. The use of progestin+CEE, or CEE alone, may be a contributing factor for increased risk for CVD and memory loss when compared with the use of E2 [17]. An alternative to progesterone is bazedoxifene, a selective ER modulator that protects the uterus when CEE is used without the need for a progestogen [44].

Indications

In the absence of contraindications, the four US Food and Drug Administration (FDA)-approved indications for MHT are [41]:

(1) **Vasomotor symptoms**. MHT is first-line therapy for severe hot flashes impairing quality of life.
(2) **Prevention of bone loss and fractures** in postmenopausal women at high risk.
(3) **Premature hypoestrogenism**. In women who have POI, or premature surgical menopause, MHT is recommended until at least the median age of menopause (51 years).
(4) **Genitourinary syndrome of menopause/vulvovaginal atrophy**. Low-dose vaginal ET is recommended over systemic therapy for this indication.

Controversy Regarding Safety and Efficacy of Menopausal Hormone Therapy in the Prevention of Cardiovascular Disease

Over the years, many studies have been performed to address the potential risks and benefits of MHT in the prevention of CVD, with early studies demonstrating possible overall CV benefit (**Figure 4**). In a 10-year follow-up report from the Nurses' Health Study, published in 1991, investigators evaluated the association of postmenopausal ET and incident CVD. In all, 48,470 postmenopausal women, aged 30–63 years, who had no history of cancer or CVD at baseline were followed for 10 years. After adjustment for age and other risk factors, the relative risk for major coronary heart disease in women currently taking estrogens was 0.56 [95% confidence interval (CI) 0.40–0.80] and for CVD mortality in current and former estrogen users compared to those who had never used estrogen, it was 0.72 (95% CI 0.55–0.95). There was no difference in the risk of stroke [45].

Later randomized studies failed to confirm this protective effect of MHT on the CV system. The Heart and Estrogen/progestin Replacement Study (HERS), the first trial seeking efficacy and safety of postmenopausal hormone therapy to prevent CHD, was published in 1998. A total of 2763 postmenopausal women younger than 80 years (mean age 66.7 years) with CHD and intact uterus were randomized

to 0.625 mg/day of CEE plus 2.5 mg/day of MPA (1380 women) vs. placebo (1383). There was no significant difference between the groups in the primary outcome (nonfatal myocardial ischemia or CHD death). However, within this null effect, a statistically significant time trend was noted with more CHD events occurring in the first year of hormone treatment compared to fewer in years 4 and 5 [46].

The WHI, which concluded in 2002, enrolled 16,608 healthy postmenopausal women aged 50–79 years with an intact uterus. They were randomized to receive 0.625 mg/day of CEE plus 2.5 mg/day of MPA ($n=8506$), or placebo [8, 47]. The primary efficacy outcome was CHD [nonfatal myocardial infarction (MI) and CHD death], with the primary adverse outcome being invasive breast cancer. The planned duration of the study was 8.5 years; however, after an average follow-up of 5.2 years, the trial was stopped based on the observation of health risks exceeding benefits. A total of 196 cases of CHD in the CEE plus MPA group were observed, vs. 159 in the placebo group (HR 1.18, 95% CI 0.95–1.45). In addition, 206 cases of invasive breast cancer (HR 1.24, 95% CI 1.01–1.53) were observed, vs. 155 in the placebo group. A global index balancing risks and benefits, including the two primary outcomes, in addition to other conditions such as stroke, pulmonary embolism, colorectal cancer, and hip fracture, demonstrated an absolute excess risk of events of 19 per 10,000 person-years attributable to estrogen plus progestin. The benefit of MHT was only seen in the outcomes of colorectal cancer and hip fracture [48].

A parallel study of the WHI in women who had hysterectomy was continued beyond the 5.2 years. Women received 0.625 mg/day of CEE alone ($n=5310$) vs. placebo ($n=5429$) for 7.2 years. Here the risks and benefits were more balanced, with 204 CHD cases for the CEE-only group vs. 222 cases in the placebo group (HR 0.94, 95% CI 0.78–1.14). Additionally, only 104 cases of invasive breast cancer (HR 0.79, 95% CI 0.61–1.02) were observed vs. 135 in the placebo group. In the group of women taking CEE alone, younger women (aged 50–59 years) had more favorable results for all-cause mortality and MI [43].

In the KRONOS Early Estrogen Prevention Study (KEEPS), MHT (oral CEE and transdermal E2) did not affect the rate of increase in carotid intimal-medial thickness (CIMT) after 4 years when used in healthy postmenopausal women aged between 42 and 58 years [49]. However, the Early Versus Late Intervention Trial with Estradiol (ELITE) concluded that the use of estrogen with or without progesterone was associated with slowing in the progression of atherosclerosis as determined by CIMT [50]. This study

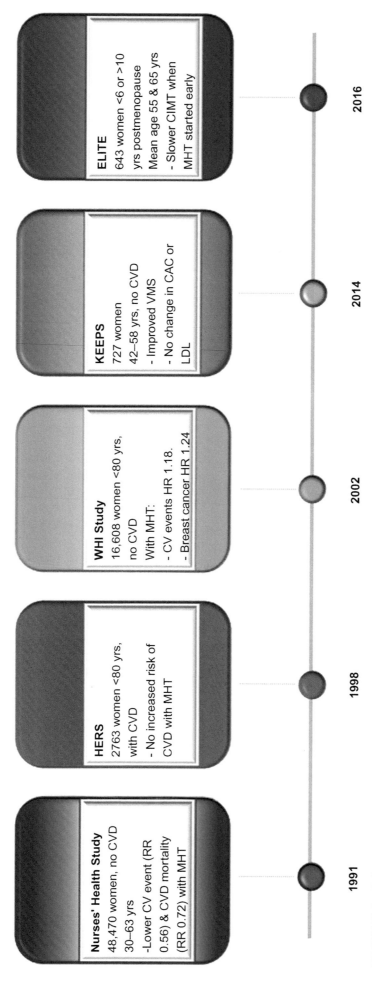

FIGURE 4 **Trials Evaluating the Effect of MHT on CVD.** CAC, coronary artery calcium score; CIMT, carotid intimal-medial thickness; CV, cardiovascular; CVD, cardiovascular disease; ELITE, Early Versus Late Intervention Trial with Estradiol; HERS, Heart and Estrogen/progestin Replacement Study; HR, hazard ratio; KEEPS, KRONOS Early Estrogen Prevention Study; LDL, low-density lipoprotein; MHT, menopause hormonal therapy; RR, relative risk; VMS, vasomotor symptoms; WHI, Women's Health Initiative.

comprised 643 healthy postmenopausal women who were stratified based on time since menopause [<6 years (early postmenopause) or ≥ 10 years (late postmenopause)]. They were randomly assigned to receive either 1 mg/day of E2 orally (plus 45 mg of vaginal gel progesterone for women with a uterus) or placebo for a median of 5 years. Slowing in atherosclerosis was seen when MHT was initiated early, but not in the late postmenopause group [50].

Timing

In general, MHT is a safe and effective way for treating menopause symptoms when initiated in healthy postmenopausal women who are younger than 60 years or who are within 10 years of menopause onset. Duration of therapy should be individualized [28, 41]. Even though MHT should not be used for prevention of CVD, the timing hypothesis, although untested, suggests that there may be a CV benefit when MHT is initiated before atherosclerotic CVD has developed [51]. In part, this timing impact could be due to age-dependent effects of E2. Studies suggest that a decrease in ER expression and function with age or menopause, loss of sex hormones, or both, could shift the vaso-protective properties of estradiol into a cytotoxic pathway [17].

Assessing Women Before Starting MHT

Prior to initiating MHT, a risk assessment should be performed to individualize therapy. The 2015 Clinical Practice Guidelines for the Treatment of Symptoms of the Menopause published by the Endocrine Society recommend assessing a woman's CVD risk using a population-based tool such as the American College of Cardiology/ American Heart Association 10-year ASCVD risk estimator, and breast cancer risk using tools such as the National Cancer Institute Breast Cancer Risk Assessment Tool, which calculates the 5-year risk of invasive breast cancer, or the International Breast Intervention Study calculator, which predicts 10-year and lifetime risk [28].

Nonhormonal therapy is recommended over MHT to alleviate VMS when a patient has a high risk for CVD and high or intermediate risk for breast cancer. The use of a drug with less effect on blood pressure, triglycerides and carbohydrate metabolism, such as transdermal estradiol alone, or transdermal E2 combined with micronized progesterone for women without and with a uterus, respectively, should be considered for women with moderate CVD risk [28]. MHT was previously considered to be contraindicated for hypertensive postmenopausal women because of the hypertensive effects seen with oral contraceptive pills [52]. MHT can be prescribed in hypertensive patients with transdermal preparations being preferred, given the neutral effect on blood pressure seen in hypertensive women. Blood pressure monitoring throughout the course of therapy should be encouraged [53].

Duration of Therapy

Given the reduced incidence of breast cancer seen in the WHI in women who were treated with CEE, the duration of ET use is more flexible. As there was a potential of increased risk of breast cancer after 3 years of EPT (standard-dose CEE + MPA) in the WHI, physicians are advised to have a clear discussion with the patient prior to its initiation [41]. The decision on abrupt discontinuation of MHT vs. a slow taper should also be a shared one. The patient's comorbidities and ongoing need for MHT should be reevaluated annually, and the possibility of lowering therapy should also be considered [28, 41].

Contraindications

The FDA advises against the use of MHT in women with a history of breast cancer [54]. Other contraindications for the use of MHT include estrogen-dependent endometrial cancer, CVD (coronary heart disease, stroke, and transient ischemic attack), venous thromboembolic event (deep venous thrombosis or pulmonary embolism), active liver disease, unexplained vaginal bleeding, thrombophilic disorders (protein C, protein S, or antithrombin deficiency), or known or suspected pregnancy [27, 41, 55]. MHT should be avoided or used with caution in women with gallbladder disease, hypertriglyceridemia, diabetes, migraine with aura, and intermediate or high risk for breast cancer [41] (see **Figure 5**).

Low Testosterone in Men

The term "male climacteric" was first used in 1944 by Heller to describe the complaints made by aging men that were reminiscent of women's menopausal symptoms, associated with decreased testosterone levels [56]. Male hypogonadism, the clinical and biochemical syndrome associated with low testosterone levels, may be classified as primary (testicular failure) or secondary (of either hypothalamic or pituitary origin). Hypogonadism may be congenital (e.g., Kallmann and Klinefelter syndromes), acquired (e.g., anorchia either due to trauma or orchiectomy, pituitary causes, or late-onset hypogonadism), or reversible, such as the hypogonadism that accompanies metabolic disorders, inflammatory diseases, or psychological conditions [57].

Clinical effects of male hypogonadism include VMS (hot flashes), decreased libido, and decreased bone mineral density. Although there is a considerable individual variation in serum total testosterone concentration, most laboratories report <300 ng/dL as low. Measurement of free testosterone is also recommended, as total testosterone may be elevated in situations where SHBG is elevated (aging) or reduced (obesity and diabetes) [2]. Testosterone levels decline by 1–1.4% per year [58]. Approximately one-quarter of men >65 years have low total testosterone levels.

- **Higher risk/avoid MHT**
  - Known ASCVD/ CAD/ PAD
  - Known venous thrombosis or pulmonary embolism
  - Known stroke/TIA or MI
  - Known clotting disorder
  - Known breast cancer
  - 10-year ASCVD risk ≥7.5%
- **Definite risk for CVD/caution with MHT**
  - Diabetes
  - Smoking
  - Uncontrolled HTN
  - Obesity/sedentary/limited mobility
  - SLE/RA/migraine with aura
  - High TG or uncontrolled cholesterol levels
  - 10-year ASCVD risk ≥5–7.4%
- **Lower risk/acceptable for MHT**
  - Recent menopause, normal weight, normal blood pressure, active female
  - 10-year ASCVD risk <5%

FIGURE 5   Guidance When Assessing Women for MHT Based on Their CVD Risk [55]. ASCVD, atherosclerotic cardiovascular disease; CAD, coronary artery disease; CVD, cardiovascular disease; HTN, hypertension; MHT, menopausal hormone therapy; MI, myocardial infarction; PAD, peripheral arterial disease; RA, rheumatoid arthritis; SLE, systemic lupus erythematosus; TG, triglyceride; TIA, transient ischemic attack.

Testosterone Replacement Therapy in Men

Testosterone replacement therapy (TRT) is indicated in boys and men with a clear diagnosis of hypogonadism due to a specific medical reason such as orchiectomy. In the mature male where aging is the sole reason for TRT, the benefit is less clear. In a review of four randomized controlled trials of TRT in diabetic men with hypogonadism, TRT showed reductions in fat mass, glycosylated hemoglobin, fasting glucose, and triglycerides, but produced no significant improvement in total and HDL cholesterol, blood pressure, or body mass index [59]. Another meta-analysis of 20 cross-sectional, longitudinal, and randomized studies showed that TRT was associated with both improved metabolic control and central obesity [60].

Studies attempting to translate the metabolic effects into CV outcomes have produced mixed data as are seen in the setting of MHT, likely related to underlying risk. Meta-analyses published between 2005 and 2010 suggested that TRT had a neutral effect on CV outcomes. However, subsequently, three large studies—the TOM trial (a randomized placebo-controlled trial of testosterone gel on physical function in older men [61]), a retrospective cohort trial of male veterans with low testosterone levels undergoing coronary angiography [62], and a study from a large healthcare database which evaluated the risk of developing a nonfatal MI after an initial prescription for testosterone [63]—all

reported adverse CV outcomes, such as increase in MI and stroke, with testosterone use. Conversely, other studies have reported no increase in events, or even benefits. A large cohort of Medicare beneficiaries who had received at least one dose of intramuscular testosterone showed no increase in MI risk, and indeed, in men at highest risk, TRT was associated with a reduced incidence of MI [64]. Both positive and negative trials had multiple criticisms, and different meta-analyses have yielded conflicting results. A large prospective randomized placebo-controlled trial is therefore necessary to clarify this issue [2].

Based on studies highlighting the adverse CV effects of TRT, the FDA recommended in its 2015 Drug Safety Communication that TRT should only be utilized in men with low testosterone due to medical disease. Conditions recognized by both the American College of Cardiology and the American Endocrine Society that should be avoided with TRT include poorly controlled heart failure, recent MI, revascularization, and stroke within the past 6 months [65]. Absolute contraindications to TRT include prostate or breast cancer. Prostate cancer risk should be assessed prior to initiating TRT using at least a prostate-specific antigen (PSA) measurement. Recommended surveillance during TRT includes prostate cancer screening, monitoring of hematocrit, and a bone density assessment in aging hypogonadal men [57].

Cardiovascular Impact of Therapeutic Hormones, Agonists and Antagonists

Contraception

As the field of "cardio-obstetrics" gains increasing importance, a multidisciplinary approach to optimal health for women requires cardiologists to participate in questions regarding contraception. According to Brown et al., of 101 adult and pediatric cardiology fellows questioned, only 69% felt that their fellowship adequately prepared them to counsel patients on contraception. Additionally, only 62% of responders felt that they were properly trained to guide patients during preconception counseling sessions [66]. This section aims to bridge that gap.

Maternal deaths have risen over the last 20 years, with CV causes being the predominant etiology [67]. Both congenital and acquired heart disease such as heart failure, MI, arrhythmia, and aortic dissection are on the rise [68, 69], contributing to increased maternal morbidity and mortality [70]. Certain CV conditions, such as severe pulmonary hypertension, severe valvular obstructive lesions, class III–IV heart failure, and a dilated aorta in women with Marfan syndrome impart a prohibitively high pregnancy risk to mother and fetus. In other conditions, pregnancy is relatively contraindicated [71]. For further details of risk stratification, see Chapter 21. Prevention of unintended pregnancy that could lead to a poor outcome is therefore an important part of the management of women with CV disease. Evaluation of a woman's risk for pregnancy is the first step in contraceptive counseling, followed by an appreciation of the safety profile of various contraceptive options in women with comorbidities.

There are various types of FDA-approved contraceptive methods in the United States (**Figure 6**). Hormonal, with or without estrogen, and nonhormonal contraceptive options are available. Combined estrogen/progesterone in pill or ring form are the most commonly used forms of hormonal contraception [73].

The estrogenic component of combined hormonal contraceptives (CHCs) increases the production of hepatic coagulation factors that increase the risks of VTE [74, 75]. Both estrogens and progestogens are associated with an elevation in blood pressures, in a dose-dependent manner [76]. This elevation is not usually clinically significant but may be marked in some women, and hypertensive emergencies have been reported [77]. CHCs may have a negative impact on lipid and carbohydrate metabolism [78], which is usually not clinically important, but may become so in women at higher risk, such as those with PCOS [79, 80].

Early studies evaluating the effect of CHCs on mortality, in general, did not demonstrate any significant effect. In the Oxford Family Planning Association study, the rate ratio for all-cause mortality was 0.89 (95% CI 0.77–1.02) except in smokers over the age of 35 [2.14 (95% CI 1.81–2.53) for women who smoked 15 or more cigarettes per day], driven by ischemic heart disease (IHD) deaths [81]. The Nurses' Health Study, reporting on data between 1976 and 2012, did not detect a difference in mortality between contraceptive users and nonusers [82].

These studies were performed using older formulations of CHCs which contained estrogen (ethinyl estradiol) doses of $\geq 50\,\mu g$, whereas many currently marketed contraceptive pills contain $10–20\,\mu g$ of ethinyl estradiol. The risks of both VTE and arterial thrombosis (MI and stroke) are higher in CHC users compared to nonusers, a risk which appears to have a dose-response relationship depending on the estradiol dose [83, 84]. Despite this higher thrombotic risk, it is important to note that the risk of VTE is lower with hormonal contraceptives than during pregnancy and the postpartum period, and therefore maintaining contraceptive benefits at a lower dose may pose the least long-term risk [85].

The Centers for Disease Control and prevention published the US Medical Eligibility Criteria for Contraceptive Use (USMEC) in 2010, updated 2015, to correlate medical risks with contraceptive use [86] (**Table 2**). This outline should help guide a prescriber's awareness of CV risk with estrogen-based hormonal options (**Table 3**). Table 4, adapted from the online document, focuses on CV conditions/risk factors.

Neither combined estrogen/progesterone formulations nor the less used estrogen/progesterone patch are recommended for women with underlying CV risk. Women with CVD (**Table 4**) may therefore be good candidates for progesterone-only or barrier forms of contraception. These include oral pills, intramuscular injections, implants, and four FDA-approved levonorgestrel intrauterine devices (IUDs). None of these described progesterone-only contraceptive options, whether they contain levonorgestrel, norethindrone, desogestrel, drospirenone, or etonogestrel increase CV risks for women with USMEC category 1–2 grades (**Table 4**) [74, 87, 88]. For USMEC category 3 women with a personal history of ischemic heart disease or stroke, Depo-Provera intramuscular (IM) injections suppress ovulation, which may cause an increase in total cholesterol levels that may lead to more complications for women in this category [89, 90].

According to the data collected by the Guttmacher Institute from 2014 [91], sterilization is one of the most frequently used nonhormonal forms of contraception. Vasectomy or tubal occlusion/removal (salpingectomy) are safe options for women with underlying CVD. If women have significant CVD and a high anesthesia risk, female sterilization may not be feasible. Most female sterilization

CH
25

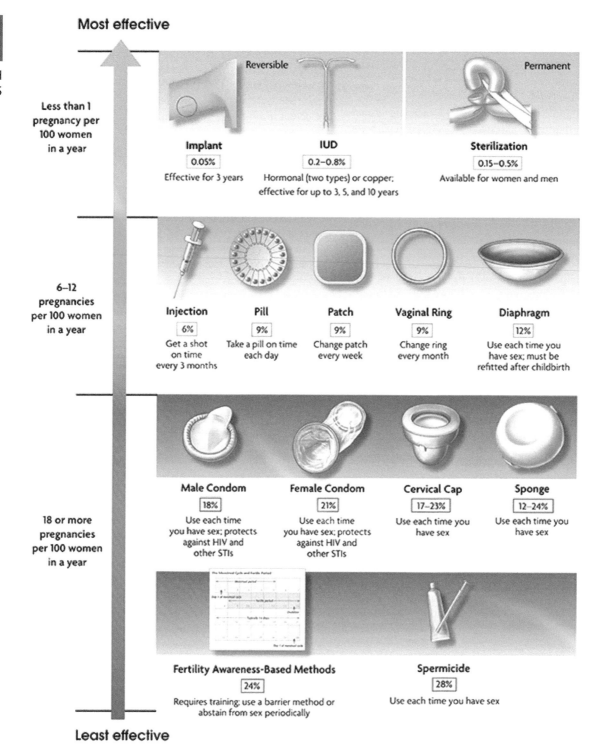

Most effective

Less than 1
pregnancy per
100 women
in a year

Reversible · Permanent

Implant
| 0.05% |
Effective for 3 years

IUD
| 0.2–0.8% |
Hormonal (two types) or copper;
effective for up to 3, 5, and 10 years

Sterilization
| 0.15–0.5% |
Available for women and men

6–12
pregnancies
per 100 women
in a year

Injection
| 6% |
Get a shot
on time
every 3 months

Pill
| 9% |
Take a pill on time
each day

Patch
| 9% |
Change patch
every week

Vaginal Ring
| 9% |
Change ring
every month

Diaphragm
| 12% |
Use each time you
have sex; must be
refitted after childbirth

18 or more
pregnancies
per 100 women
in a year

Male Condom
| 18% |
Use each time
you have sex; protects
against HIV and
other STIs

Female Condom
| 21% |
Use each time
you have sex; protects
against HIV and
other STIs

Cervical Cap
| 17–23% |
Use each time you
have sex

Sponge
| 12–24% |
Use each time you
have sex

Fertility Awareness-Based Methods
| 24% |
Requires training; use a barrier method or
abstain from sex periodically

Spermicide
| 28% |
Use each time you have sex

Least effective

Other methods of birth control
Lactational amenorrhea method: This is a temporary method of birth control that can be used for the first 6 months after giving birth by women who are exclusively breastfeeding.
Emergency contraception: Emergency contraceptive pills taken or a copper IUD inserted within 5 days of unprotected sex can reduce the risk of pregnancy.
Withdrawal: The man withdraws his penis from the vagina before ejaculating; 22 out of 100 women using this method will become pregnant in the first year.

FIGURE 6 FDA-Approved Contraceptive Methods [72]. HIV, human immunodeficiency virus; IUD, intrauterine device; STIs, sexually transmitted infections.

TABLE 2 US Medical Eligibility Criteria for Contraceptive Use (USMEC) 2016: Categories for Classifying Hormonal Contraception

Category	Associated Risk With Contraceptive Use
1	A condition for which there is no restriction
2	A condition for which the advantages of using the method generally outweigh the theoretical or proven risks
3	A condition for which the theoretical or proven risks usually outweigh the advantages of using the method
4	A condition that represents an unacceptable health risk if the contraceptive method is used

Data from [86].

TABLE 3 Risk Factors for VTE With Use of Combined Hormonal Contraceptives (USMEC Category 3 or 4)

Smoking AND > 35 years old

Major surgery with prolonged immobilization

History of deep vein thrombosis (DVT) or pulmonary embolism (PE)

Hereditary thrombophilia (category 4)

Inflammatory bowel disease with active or extensive disease, surgery, immobilization, corticosteroid use, vitamin deficiencies, or fluid depletion

Systemic lupus erythematosus (SLE) with positive or unknown antiphospholipid antibodies

Superficial venous thrombosis

Less than 21 days postpartum

21–42 days postpartum with the following factors: > 35 years old, previous VTE, thrombophilia, peripartum cardiomyopathy, immobility, transfusion at delivery, BMI at or greater than 30, postpartum hemorrhage, cesarean delivery, preeclampsia, and smoking

BMI, body mass index; VTE, venous thromboembolism.
Data from [86].

surgeries are done via a laparoscopic approach. Women with CV risks may need preoperative cardiac clearance prior to surgery and may not be candidates for this method of contraception. Women who opt for a sterilization procedure at the time of cesarean section may not have the same concerns during regional laparotomy. Transvaginal tubal occlusion procedures are no longer done.

Additional nonhormonal options are available for women with CVD. The rhythm method or natural family planning allows women to track their own cycle and to avoid intercourse during fertile windows. There are also barrier methods like condoms, diaphragms, or cervical caps that provide women with underlying CV risk hormone free options. The copper IUD (ParaGard) is another nonhormonal option. This type of IUD creates a sterile inflammation along the female reproductive track. In some cases, the ParaGard device can lead to menorrhagia, which can be significant for women on anticoagulation treatment [92]. Hemodynamic status and bleeding potential should be considered prior to its placement.

The decision to use hormonal or nonhormonal contraception should be a shared decision between the medical provider and the patient. The increase in cardiac output and decreased maternal systemic resistance that accompanies pregnancy and the reversal of these in the postpartum period may impose significant hemodynamic stress on women with some cardiac lesions [93]. It should therefore be determined if the risks from certain contraceptives are less than pregnancy risks (please refer to Chapter 21 for further details).

CVD Impact of Hormonal Antagonism in the Treatment of Cancer

Hormone Suppression for Gynecologic Cancer Treatment

The medications that are used to treat women with estrogen and/or progesterone receptor (PR)-positive tumors include selective ER blockers such as tamoxifen and aromatase inhibitors such as anastrozole and letrozole [94]. In addition, an effect similar to that achieved with surgical ovarian ablation in premenopausal and perimenopausal women with breast cancer is seen with the use of gonadotropin-releasing hormone (GnRH) agonists, in particular goserelin [95].

Research about the long-term CV effects of these medications in women is sparse, with no randomized controlled trials designed with a primary end point of CVD. However, a recent meta-analysis of 26 studies (both observational and randomized controlled data) assessed seven specific CVD outcomes including VTE, MI, stroke, angina, heart failure, and peripheral vascular disease. An increased risk of VTE was noted in tamoxifen users compared with both nonusers and aromatase inhibitor users. Higher risk of vascular disease, MI, and angina was noted in aromatase inhibitor users compared to tamoxifen users; however, this may be due to

TABLE 4 Summary of Contraceptive Methods for Cardiovascular Conditions

Condition	Cu-IUD	LNG-IUD	Implants	DMPA	POP	CHCs
			Contraceptive Method			
Personal Characteristics Increasing Risk for Cardiovascular Disease						
Smoking						
(a) Age <35 years	1	1	1	1	1	2
(b) Age ≥35 years						
(i) <15 cigarettes/day	1	1	1	1	1	3
(ii) ≥15 cigarettes/day	1	1	1	1	1	4
Obesity						
(a) BMI≥30 kg/m²	1	1	1	1	1	2
(b) Menarche to <18 years and BMI≥30 kg/m²	1	1	1	2	1	2
History of bariatric surgery *This condition is associated with increased risk for adverse health events as a result of pregnancy*						
(a) Restrictive procedures: decrease storage capacity of the stomach (vertical banded gastroplasty, laparoscopic adjustable gastric band, or laparoscopic sleeve gastrectomy)	1	1	1	1	1	1
(b) Malabsorptive procedures: decrease absorption of nutrients and calories by shortening the functional length of the small intestine (Roux-en-Y gastric bypass or biliopancreatic diversion)	1	1	1	1	3	COCs: 3 Patch and ring: 1
Cardiovascular Disease						
Multiple risk factors for atherosclerotic cardiovascular disease (e.g., older age, smoking, diabetes, hypertension, low HDL, high LDL, or high triglyceride levels)	1	2	2	3	2	3/4

Hypertension (systolic blood pressure ≥160 mmHg or diastolic blood pressure ≥100 mmHg are associated with increased risk for adverse health events as a result of pregnancy)

(a) Adequately controlled hypertension	1	1	1	2	3
(b) Elevated blood pressure levels (properly taken measurements)					
(i) Systolic 140–159 mmHg or diastolic 90–99 mmHg	1	1	1	2	3
(ii) Systolic ≥160 mmHg or diastolic ≥100 mmHg	1	2	2	3	4
(c) Vascular disease	1	2	2	3	4
History of high blood pressure during pregnancy (when current blood pressure is measurable and normal)	1	1	1	1	2
Deep venous thrombosis/pulmonary embolism					
(a) History of DVT/PE, not receiving anticoagulant therapy					
(i) Higher risk for recurrent DVT/PE (one or more risk factors) • History of estrogen-associated DVT/PE • Pregnancy-associated DVT/PE • Idiopathic DVT/PE • Known thrombophilia, including antiphospholipid syndrome • Active cancer (metastatic, receiving therapy, or within 6 months after clinical remission), excluding nonmelanoma skin cancer • History of recurrent DVT/PE	1	2	2	2	4
(ii) Lower risk for recurrent DVT/PE (no risk factors)	1	2	2	2	3
(b) Acute DVT/PE	2	2	2	2	4
(c) DVT/PE and established anticoagulant therapy for at least 3 months					

(Continued)

TABLE 4 Summary of Contraceptive Methods for Cardiovascular Conditions—cont'd

Condition	Cu-IUD	LNG-IUD	Implants	DMPA	POP	CHCs
(i) Higher risk for recurrent DVT/PE (one or more risk factors) • Known thrombophilia, including antiphospholipid syndrome • Active cancer (metastatic, receiving therapy, or within 6 months after clinical remission), excluding nonmelanoma skin cancer • History of recurrent DVT/PE	2	2	2	2	2	4
(ii) Lower risk for recurrent DVT/PE (no risk factors)	2	2	2	2	2	3
(d) Family history (first-degree relatives)	1	1	1	1	1	2
(e) Major surgery						
(i) With prolonged immobilization	1	2	2	2	2	4
(ii) Without prolonged immobilization	1	1	1	1	1	2
(f) Minor surgery without immobilization	1	1	1	1	1	1
Known thrombogenic mutations (e.g., factor V Leiden; prothrombin mutation; and protein S, protein C, and antithrombin deficiencies) ***This condition is associated with increased risk for adverse health events as a result of pregnancy.***	1	2	2	2	2	4
Superficial venous disorders						
(a) Varicose veins	1	1	1	1	1	1
(b) Superficial venous thrombosis (acute or history)	1	1	1	1	1	3

Condition	Cu-IUD	LNG-IUD Initiation	LNG-IUD Continuation	DMPA Initiation	DMPA Continuation	POP Initiation	POP Continuation	CHC Initiation	CHC Continuation
Current and history of ischemic heart disease **This condition is associated with increased risk for adverse health events as a result of pregnancy**	1	2	3	3	3	2	3	4	4
Stroke (history of cerebrovascular accident) **This condition is associated with increased risk for adverse health events as a result of pregnancy**	1	2	2	3	3	2	3	4	4
Valvular heart disease *Complicated valvular heart disease is associated with increased risk for adverse health events as a result of pregnancy*									
(a) Uncomplicated	1	1	1	1	1	1	1	2	2
(b) Complicated (pulmonary hypertension, risk for atrial fibrillation, or history of subacute bacterial endocarditis)	1	1	1	1	1	1	1	4	4
Peripartum cardiomyopathy *This condition is associated with increased risk for adverse health events as a result of pregnancy*									
(a) Normal or mildly impaired cardiac function (New York Heart Association Functional Class I or II: patients with no limitation of activities or patients with slight, mild limitation of activity)									
(i) <6 months	2	2	2	1	1	1	1	4	4
(ii) ≥6 months	2	2	2	1	1	1	1	3	3
(b) Moderately or severely impaired cardiac function (New York Heart Association Functional Class III or IV: patients with marked limitation of activity or patients who should be at complete rest)	2	2	2	2	2	2	2	4	4

Recommendations on safe use of contraceptive methods for women with various cardiovascular and related conditions are listed. Categories for classifying contraceptives are as follows: 1 = a condition for which there is no restriction for the use of contraceptive method. 2 = a condition for which the advantages of using the method generally outweigh the theoretical or proven risks. 3 = a condition for which the theoretical or proven risks usually outweigh the advantages of using the method. 4 = a condition that represents an unacceptable health risk if the contraceptive method is used. BMI, body mass index; CHCs, combined hormonal contraceptives; COCs, combined oral contraceptives; Cu-IUD, copper-containing IUD; DMPA, depot medroxyprogesterone acetate; DVT, deep vein thrombosis; HDL, high-density lipoprotein; IUD, intrauterine device; LDL, low-density lipoprotein; LNG-IUD, levonorgestrel-releasing IUD; PE, pulmonary embolism; POP, progestin-only pill. *Adapted from [86].*

the protective effect conferred from tamoxifen acting as an agonist [96].

Mixed effects on cardiometabolic risk are noted in patients undergoing adjuvant endocrine therapy. The variable effects of tamoxifen are attributed to the fact that it serves in both agonist and antagonist roles in different systems. It is noted to have adverse effects on body and hepatic fat accumulation with an increase in serum triglycerides but may reduce LDL cholesterol and lipoprotein (a) [Lp(a)] and may have beneficial effects on the arterial wall. These effects may explain the reduction in CV events noted with tamoxifen. Aromatase inhibitors on the other hand are associated with higher CV risks compared with tamoxifen, but whether this reflects aromatase inhibitor-related toxicity or a protective effect of tamoxifen remains uncertain [97].

Testosterone Suppression for Prostate Cancer Treatment

Testosterone suppression in men with prostate cancer, with agents such as leuprolide, results in several adverse cardiometabolic effects. However, despite well-known adverse effects on CV risk factors, and a possible association between androgen deprivation therapy (ADT) exposure and increased CV morbidity, no single prospective study has definitively established that ADT exposure increases the risk of CVD or CV mortality. Men with preexisting CVD, including a history of congestive heart failure or MI, appear to be at the highest risk for CV events with ADT exposure, particularly in the first 6 months of ADT initiation [98].

Cardiovascular Health in Transgender Individuals

Terminology

Table 5 summarizes some frequently used terms commonly used in the field of transgender medicine.

Role of Exogenous Hormone Therapy in Transgender Individuals

The number of individuals who identify as transgender has been estimated to range from just under 1 million to 1.4 million [100, 101], warranting dedicated research and registries to learn how to best care for this group. Drawing conclusions about exogenous hormones and CV health in transgender individuals, however, is challenging for several reasons. A higher prevalence of comorbidities, lifestyle factors and psychosocial issues, lower reported quality of life, and healthcare disparities may confound the effects of exogenous hormone therapy [47] as well as a variety of hormone regimens used and a relative brevity of the periods of observation [102]. A 56% prevalence of psychiatric disease has been noted in those seeking gender-affirming

TABLE 5 Terminology Used in Transgender Health	
Cisgender	A person whose self-identity matches the sex assigned at birth
Transgender	An individual whose gender identity differs from the sex assigned at birth based on anatomic characteristics
Transgender female	A person whose sex was assigned male at birth but who self-identifies as female
Transgender male	A person whose sex was assigned female at birth, but who self-identifies as male
Transition	The process during which a transgender person changes physical, social, and/or legal characteristics consistent with their self-identified gender
Transsexual	A diagnostic term used in the ICD-10 and in medical literature when discussing transgender identity. When referring to an individual, the term transgender should be used
Gender-affirming treatment	Hormonal or surgical treatment that some transgender people access to adapt physical characteristics to match their gender identity
Cross-sex hormone therapy	Hormonal therapy administered to a person of the opposite sex as part of gender-affirming treatment

ICD-10, International Classification of Diseases, 10th revision. *Adapted from [99].*

hormone therapy potentially influencing lifestyle and even comorbidities [103]. Separate from these confounding factors, the administration of sex hormones to the opposite sex may have different effects than in cisgender individuals. Lastly, the specific formulations and doses administered may be responsible for differences in outcomes.

With the potential risks of hormonal therapy in mind, specific goals of therapy should therefore include the patient's age, comorbidities, and risk factors, as well as the desired outcome. Some patients may wish to achieve results as close as possible to the opposite gender, while others may prefer to suppress any evidence of a specific gender [104]. An approach to modify risk at baseline is warranted prior to initiation of exogenous sex hormones or surgery that could be associated with additional risk.

A mainstay of gender-affirming hormone therapy (GHT) is not only hormonal administration, but also hormone suppression. For transgender males (TGMs), treatment consists of testosterone therapy, while for transgender females (TGFs), it includes estrogen administration often with either a GnRH analog such as goserelin, or an antiandrogenic agent such as cyproterone acetate [105]. In the younger subject who is prepubertal, GnRH agonists are started at the first signs of puberty to halt progression of puberty, and testosterone or estradiol are started in late adolescence [104].

The first observational study relating to GHT and CVD tracks back to 1989 and since then numerous studies have been conducted [105]. Studies assessing for CVD in TGFs on estrogen treatment show mixed data suggesting a mixed risk profile of increased and decreased risk for CVD which can be due to increased weight, increased body and visceral fat, prothrombotic blood changes, and higher triglycerides, all due to estrogens, while others suggest a decreased risk given estrogen-related homocysteine and LDL reduction [102]. On the other hand, a recent study suggests that there seems to be a higher risk for developing ischemic heart disease, and cerebrovascular accidents in TGFs compared with the scisgender population, especially when they have been using estrogens long term [105]. It is unclear whether this risk is due to the effects of GHT or if it is a legacy effect of natal sex.

GHT may increase blood pressure for both TGFs and TGMs, and increase triglycerides and LDL levels while reducing HDL levels in TGMs, without significantly affecting lipid profile in TGFs except for elevation in the triglyceride level. The relationship between GHT and glucose abnormalities remains uncertain. Despite the abnormalities seen in blood pressure and lipid profile in TGMs, there is no clear evidence that they have an increased risk for CV or cerebrovascular disease [105].

Effects of Testosterone Therapy in Transgender Males

A variety of formulations of testosterone therapy are used for masculinization in TGMs including transdermal, intramuscular, subcutaneous oral, and lozenge. While the increase of hemoglobin and hematocrit with testosterone therapy is a well-known side effect, it is unclear if TGMs are at an increased risk of DVT because there is a lack of evidence that testosterone treatment increases the incidence of DVT. However, untreated severe obstructive sleep apnea is a relative contraindication to testosterone therapy, as this increases the risk of DVT, as well as exacerbating the condition by increasing weight and muscle mass [104]. Testosterone use in transgender men has been associated with hypertension, increased LDL and triglycerides, reduced HDL, and reduced homocysteine levels. However, a recent observational study using a retrospective cohort of 2509 transwomen and 1893 transmen showed no significant associations between hormone therapy (HT) and hypertension [106]. Additionally, a recent prospective observational study from Europe also found no statistically significant relationship between HT and blood pressure in a cohort of transmen and transwomen after 1 year of HT [107]. Studies are unclear with respect to increased insulin resistance and type II diabetes [99].

Research to date does not support an increase in CV events and mortality; however, the findings are predominantly from smaller studies and in younger patients [108]. One large cross-sectional study showed a fourfold increased risk of MI in transgender men receiving cross-sex hormone therapy (CSHT) compared to cisgender women and a twofold increased risk compared to cisgender men [109]. Long-term smaller studies of transgender men show that 10-year risk for MI, stroke, and CVD appears to be similar to that of cisgender men; however, most of these studies include younger transgender men (average age 40) whose baseline risk is likely low and, therefore, longer-term studies and research in older transgender subjects are needed [109].

Estrogen Therapy in Transgender Females

Hormonal therapy for transgender women includes various formulations of estrogen (oral, transdermal, and intramuscular), but also a compound to reduce testosterone replicating the physiological hormone levels of premenopausal cisgender women. Transdermal varieties of estrogen appear to have the lowest risk of VTE, alterations in lipid profile, and markers of coagulation, but are also the costliest among preparations. The risk for thromboembolic events with estrogen is further increased with advancing age, cigarette smoking, and concomitant use of progestins. In cases where ET is not adequate to get to normal premenopausal female hormone levels, spironolactone, GnRH antagonists, 5-alpha reductase inhibitors, or orchiectomy may be needed to achieve optimal reductions in male pattern sex characteristics [105].

Potential effects of CSHT on lipids have also been reported, but data are limited. Variations in lipid metabolism include decreased LDL and increased triglycerides, with the alterations in HDL being even less clear. The role that estrogen plays in HDL may be affected by the route of administration, as transdermal application has been shown to increase HDL [105]. Some TGFs experience a reduction in blood pressure, but this is likely due to the reduction in testosterone rather than increased estrogen. Studies have shown that concurrent administration of estrogen and progesterone for trans females leads to increased blood pressure, as well as combined oral contraceptive use in cisgender females [99]. The study performed by Pyra et al., mentioned previously

[106], confirms that endogenous levels of testosterone were associated with hypertension, but only when adjusted for BMI, and the odds increased with age, showing significance only in the 45–70 years old age group of transwomen. Additionally, the study showed that transwomen who took progestin proved to be protected regarding hypertension, but there was a strong association with recent progestin use and VTE, increasing nearly threefold among transwomen when adjusted for other hormone therapies, race, insurance, HIV status, and BMI. There was no association found between VTE and either estradiol or testosterone in any of the models in the study, suggesting that perhaps safer regimens have become the norm, as oral ethinyl estradiol is no longer routinely prescribed [106].

Adverse CV effects of ET in transgender women include increased rates of MI and stroke, but there are no randomized controlled data in this realm. A large 2018 nationwide survey of transgender men and women noted an increased rate of MI in transgender women compared to cisgender women [odds ratio (OR) 2.9, 95% CI 1.6–5.3; $P < 0.001$], but not in comparison to cisgender men [109]. A nationwide survey of over 1.8 million adults by the Centers for Disease Control and Prevention (CDC) showed increased rates of MI in all transgender individuals compared to their cisgender counterparts. In transgender women, an increased twofold risk compared to cisgender women persisted even after adjusting for CVD risk factors, but there was no difference compared to cisgender men, suggesting that the natural protection that cisgender women have prior to menopause is not conveyed by CSHT [109]. A twofold increased risk of stroke has been described in some studies [110], and an increased prevalence of atrial fibrillation has also been reported in smaller studies [105].

Conclusion

Hormone levels fluctuate throughout an individual's life and may warrant treatment in the setting of menopause and andropause. An individual assessment of the patient, including CV risk among other risk factors, should be made to arrive at a shared decision on treatments. As with the case described at the start of the chapter, optimization of risk factors could then allow for the initiation of menopausal hormone therapy for symptom control. This is also confounded by a patient's comorbidities and careful assessment of their gender identification. A similar individualized approach is encouraged when deciding between contraceptive methods for the woman with underlying CVD. For these transgender individuals, there is evidence of increased CV events with estrogen treatment, but there is much to be learned with respect to optimization of CV risk. Long-term research studies and registries assessing all contributing factors should bring a greater understanding to the field. Based on the complexity of care for these individuals, clinicians should make efforts to understand the unique healthcare aspects and collaborate with experts in the field. Patient preference should also remain at the core of each healthcare decision.

Key Points

1. The sex hormones, present in variable levels in both sexes, have varying effects throughout the life span and play an important role in cardiovascular health.

2. CV effects of endogenous hormones include metabolic effects, e.g., lipid profile and insulin levels, vascular effects, e.g., smooth muscle proliferation and vasoreactivity and direct myocardial effects.

3. Women with PCOS, one of the most common endocrine disorders in women of reproductive age, are more likely to be obese and have metabolic syndrome and may have increased CVD risk. It is unclear whether this increased risk is conferred by PCOS itself or by the traditional risk factors seen more commonly with this condition.

4. Menopause typically occurs in the fifth decade. Declining estrogen levels result in a syndrome that may include vasomotor symptoms, mood and sleep disturbances, and genitourinary symptoms. Many women have no symptoms at all. CV effects include alterations in lipid profile, elevation in blood pressure, decreased glucose tolerance, and abnormal vascular reactivity.

5. Menopausal hormone therapy (MHT) is approved for vasomotor symptoms, prevention of bone loss and fracture, premature hypoestrogenism, and genitourinary (GU) symptoms of menopause/vaginal atrophy.

6. Early observational studies of MHT suggested a beneficial effect on CV risk but large, randomized studies revealed an increased risk of cardiovascular diseases and breast cancer. More recent data suggest that timing of initiation of HRT in relation to menopause onset and choice of medication may play an important role in enhancing the safety of this treatment. Individualized assessment of risk and benefit is critical.

7. Therapeutic use of sex hormones including contraception and cancer treatment requires careful evaluation of CV risk, clinical indication, and benefit.

8. CV issues are an important component of transgender health. Gender-affirming hormone therapy and the impact on long-term CV risk is an important area of ongoing research.

Back to Clinical Case

The case describes a 51-year-old woman postmenopausal woman with PCOS, as well as several traditional risk factors for CVD, who is 6 years after her last menstrual period and experiencing worsening vasomotor symptoms.

Her underlying risk factors for MHT use include PCOS, diabetes, hypertension, early menopause, and a family history of CVD. She should be counseled that she may be at higher risk for adverse CVD outcomes with these risk factors. MHT may not be fully contraindicated, but care should be taken if used, with close monitoring of risk factors after MHT initiation. She should understand that nonhormonal options are available.

Prior to initiation of MHT, personalized assessment of CVD and breast cancer risk should be performed, using tools such as the 10-year ASCVD risk estimator and the National Cancer Institute's breast cancer risk assessment tool or the International Breast Intervention Study calculator. Further diagnostic testing may be required based on these results. All risk factors for CVD and breast cancer should be evaluated and optimized. Strict adherence to guidelines, focused on treating hypertension, diabetes, and hyperlipidemia, should be a goal.

CH
25

References

[1] Yasui T, Matsui S, Tani A, Kunimi K, Yamamoto S, Irahara M. Androgen in postmenopausal women. J Med Investig 2012;59:12–27.

[2] Kloner RA, Carson 3rd C, Dobs A, Kopecky S, Mohler 3rd ER. Testosterone and cardiovascular disease. J Am Coll Cardiol 2016;67:545–57.

[3] Hall JE, Guyton AC. Guyton and hall textbook of medical physiology. Philadelphia, PA: Saunders; 2011.

[4] Skiba MA, Bell RJ, Islam RM, Handelsman DJ, Desai R, Davis SR. Androgens during the reproductive years: what is normal for women? J Clin Endocrinol Metab 2019;104:5382–92.

[5] Regitz-Zagrosek V, Kararigas G. Mechanistic pathways of sex differences in cardiovascular disease. Physiol Rev 2017;97:1–37.

[6] Norman AW, Henry HL. Hormones. 3rd ed. Elsevier; 2015.

[7] Holinka CF, Diczfalusy E, Coelingh Bennink HJ. Estetrol: a unique steroid in human pregnancy. Climacteric 2008;11(Suppl 1):1.

[8] Hoyt LT, Falconi AM. Puberty and perimenopause: reproductive transitions and their implications for women's health. Soc Sci Med 2015;132:103–12.

[9] Thurston RC, Bhasin S, Chang Y, Barinas-Mitchell E, Matthews KA, Jasuja R, et al. Reproductive hormones and subclinical cardiovascular disease in midlife women. J Clin Endocrinol Metab 2018;103:3070–7.

[10] Miller VM, Duckles SP. Vascular actions of estrogens: functional implications. Pharmacol Rev 2008;60:210–41.

[11] Taraborrelli S. Physiology, production and action of progesterone. Acta Obstet Gynecol Scand 2015;94(Suppl 161):8–16.

[12] Herring MJ, Oskui PM, Hale SL, Kloner RA. Testosterone and the cardiovascular system: a comprehensive review of the basic science literature. J Am Heart Assoc 2013;2, e000271.

[13] Araujo AB, Dixon JM, Suarez EA, Murad MH, Guey LT, Wittert GA. Endogenous testosterone and mortality in men: a systematic review and meta-analysis. J Clin Endocrinol Metab 2011;96:3007–19.

[14] Ruige JB, Mahmoud AM, De Bacquer D, Kaufman J-M. Endogenous testosterone and cardiovascular disease in healthy men: a meta-analysis. Heart 2011;97:870–5.

[15] Corona G, Rastrelli G, Monami M, Guay A, Buvat J, Sforza A, et al. Hypogonadism as a risk factor for cardiovascular mortality in men: a meta-analytic study. Eur J Endocrinol 2011;165:687–701.

[16] Zhao D, Guallar E, Ouyang P, Subramanya V, Vaidya D, Ndumele CE, et al. Endogenous sex hormones and incident cardiovascular disease in post-menopausal women. J Am Coll Cardiol 2018;71:2555–66.

[17] Pabbidi MR, Kuppusamy M, Didion SP, Sanapureddy P, Reed JT, Sontakke SP. Sex differences in the vascular function and related mechanisms: role of 17beta-estradiol. Am J Physiol Heart Circ Physiol 2018;315:H1499–518.

[18] Garcia M, Mulvagh SL, Merz CN, Buring JE, Manson JE. Cardiovascular disease in women: clinical perspectives. Circ Res 2016;118:1273–93.

[19] Bernstein P, Pohost G. Progesterone, progestins, and the heart. Rev Cardiovasc Med 2010;11:228–36.

[20] Prior JC, Elliott TG, Norman E, Stajic V, Hitchcock CL. Progesterone therapy, endothelial function and cardiovascular risk factors: a 3-month randomized, placebo-controlled trial in healthy early postmenopausal women. PLoS ONE 2014;9, e84698.

[21] Lamos EM, Malek R, Davis SN. GLP-1 receptor agonists in the treatment of polycystic ovary syndrome. Expert Rev Clin Pharmacol 2017;10:401–8.

[22] Osibogun O, Ogunmoroti O, Michos ED. Polycystic ovary syndrome and cardiometabolic risk: opportunities for cardiovascular disease prevention. Trends Cardiovasc Med 2020;30(7):399–404. https://doi.org/10.1016/j.tcm.2019.08.010. Epub 2019 Sep 4. PMID: 31519403.

[23] Williams T, Mortada R, Porter S. Diagnosis and treatment of polycystic ovary syndrome. Am Fam Physician 2016;94:106–13.

[24] Legro RS, Arslanian SA, Ehrmann DA, Hoeger KM, Murad MH, Pasquali R, et al. Diagnosis and treatment of polycystic ovary syndrome: an Endocrine Society clinical practice guideline. J Clin Endocrinol Metab 2013;98:4565–92.

[25] Nylander M, Frossing S, Clausen HV, Kistorp C, Faber J, Skouby SO. Effects of liraglutide on ovarian dysfunction in polycystic ovary syndrome: a randomized clinical trial. Reprod BioMed Online 2017;35:121–7.

[26] Harlow SD, Gass M, Hall JE, Lobo R, Maki P, Rebar RW, et al. Executive summary of the Stages of Reproductive Aging Workshop + 10: addressing the unfinished agenda of staging reproductive aging. J Clin Endocrinol Metab 2012;97:1159–68.

[27] Goodman NF, Cobin RH, Ginzburg SB, Katz IA, Woode DE, American Association of Clinical E. American Association of Clinical Endocrinologists Medical Guidelines for Clinical Practice for the diagnosis and treatment of menopause. Endocr Pract 2011;17(Suppl 6):1–25.

[28] Stuenkel CA, Davis SR, Gompel A, Lumsden MA, Murad MH, Pinkerton JV, et al. Treatment of symptoms of the menopause: an endocrine society clinical practice guideline. J Clin Endocrinol Metab 2015;100:3975–4011.

[29] Stice JP, Lee JS, Pechenino AS, Knowlton AA. Estrogen, aging and the cardiovascular system. Futur Cardiol 2009;5:93–103.

[30] Somani YB, Pawelczyk JA, De Souza MJ, Kris-Etherton PM, Proctor DN. Aging women and their endothelium: probing the relative role of estrogen on vasodilator function. Am J Physiol Heart Circ Physiol 2019;317:H395–404.

[31] Suboc TM, Dharmashankar K, Wang J, Ying R, Couillard A, Tanner MJ, et al. Moderate obesity and endothelial dysfunction in humans: influence of gender and systemic inflammation. Phys Rep 2013;1.

[32] Moreau KL. Modulatory influence of sex hormones on vascular aging. Am J Physiol Heart Circ Physiol 2019;316:H522–6.

[33] Nyberg M, Seidelin K, Andersen TR, Overby NN, Hellsten Y, Bangsbo J. Biomarkers of vascular function in premenopausal and recent postmenopausal women of similar age: effect of exercise training. Am J Phys Regul Integr Comp Phys 2014;306:R510–7.

[34] Anon. Committee opinion no. 605: primary ovarian insufficiency in adolescents and young women. Obstet Gynecol 2014;124:193–7.

[35] Shuster LT, Rhodes DJ, Gostout BS, Grossardt BR, Rocca WA. Premature menopause or early menopause: long-term health consequences. Maturitas 2010;65:161–6.

[36] Rocca WA, Gazzuola Rocca L, Smith CY, Grossardt BR, Faubion SS, Shuster LT, et al. Bilateral oophorectomy and accelerated aging: cause or effect? J Gerontol A Biol Sci Med Sci 2017;72:1213–7.

[37] Adelman MR, Sharp HT. Ovarian conservation vs removal at the time of benign hysterectomy. Am J Obstet Gynecol 2018;218:269–79.

[38] Honigberg MC, Zekavat SM, Aragam K, Finneran P, Klarin D, Bhatt DL, et al. Association of premature natural and surgical menopause with incident cardiovascular disease. JAMA 2019.

[39] Farquhar CM, Sadler L, Harvey SA, Stewart AW. The association of hysterectomy and menopause: a prospective cohort study. BJOG 2005;112:956–62.

[40] Howard BV, Kuller L, Langer R, Manson JE, Allen C, Assaf A, et al. Risk of cardiovascular disease by hysterectomy status, with and without oophorectomy: the Women's Health Initiative Observational Study. Circulation 2005;111:1462–70.

[41] NAMS Hormone Therapy Position Statement Advisory Panel T. The 2017 hormone therapy position statement of the North American Menopause Society. Menopause 2017;24:728–53.

[42] Coelingh Bennink HJ, Holinka CF, Diczfalusy E. Estetrol review: profile and potential clinical applications. Climacteric 2008;11(Suppl 1):47–58.

[43] Manson JE, Chlebowski RT, Stefanick ML, Aragaki AK, Rossouw JE, Prentice RL, et al. Menopausal hormone therapy and health outcomes during the intervention and extended poststopping phases of the Women's Health Initiative randomized trials. JAMA 2013;310:1353–68.

[44] Mirkin S, Komm BS, Pan K, Chines AA. Effects of bazedoxifene/conjugated estrogens on endometrial safety and bone in postmenopausal women. Climacteric 2013;16:338–46.

[45] Stampfer MJ, Colditz GA, Willett WC, Manson JE, Rosner B, Speizer FE, et al. Postmenopausal estrogen therapy and cardiovascular disease. Ten-year follow-up from the nurses' health study. N Engl J Med 1991;325:756–62.

[46] Hulley S, Grady D, Bush T, Furberg C, Herrington D, Riggs B, et al. Randomized trial of estrogen plus progestin for secondary prevention of coronary heart disease in postmenopausal women. Heart and Estrogen/progestin Replacement Study (HERS) Research Group. JAMA 1998;280:605–13.

[47] Schuster MA, Reisner SL, Onorato SE. Beyond bathrooms—meeting the health needs of transgender people. N Engl J Med 2016;375:101–3.

[48] Rossouw JE, Anderson GL, Prentice RL, LaCroix AZ, Kooperberg C, Stefanick ML, et al. Risks and benefits of estrogen plus progestin in healthy postmenopausal women: principal results from the Women's Health Initiative randomized controlled trial. JAMA 2002;288:321–33.

[49] Harman SM, Black DM, Naftolin F, Brinton EA, Budoff MJ, Cedars MI, et al. Arterial imaging outcomes and cardiovascular risk factors in recently menopausal women: a randomized trial. Ann Intern Med 2014;161:249–60.

[50] Hodis HN, Mack WJ, Henderson VW, Shoupe D, Budoff MJ, Hwang-Levine J, et al. Vascular effects of early versus late postmenopausal treatment with Estradiol. N Engl J Med 2016;374:1221–31.

[51] Mehta JM, Chester RC, Kling JM. The timing hypothesis: hormone therapy for treating symptomatic women during menopause and its relationship to cardiovascular disease. J Women's Health 2019;28:705–11.

[52] Felmeden DC, Lip GY. Hormone replacement therapy and hypertension. Blood Press 2000;9:246–9.

[53] Issa Z, Seely EW, Rahme M, El-Hajj FG. Effects of hormone therapy on blood pressure. Menopause 2015;22:456–68.

[54] U.S. Food & Drug Administration T. Menopause: Medicines to help you, https://www.fda.gov/consumers/free-publications-women/menopause-medicines-help-you; 2019.

[55] Lundberg GP, Wenger NK. Menopause hormone therapy: What a cardiologist needs to know. American College of Cardiology; 2019.

[56] Heller CG, Myers GB. The male climacteric, its symptomatology, diagnosis and treatment: use of urinary gonadotropins, therapeutic test with testosterone propionate and testicular biopsies in delineating the male climacteric from psychoneurosis and psychogenic impotence. JAMA 1944;126:472–7.

[57] Lunenfeld B, Mskhalaya G, Zitzmann M, Arver S, Kalinchenko S, Tishova Y, et al. Recommendations on the diagnosis, treatment and monitoring of hypogonadism in men. Aging Male 2015;18:5–15.

[58] Jakiel G, Makara-Studzinska M, Ciebiera M, Slabuszewska-Jozwiak A. Andropause—state of the art 2015 and review of selected aspects. Prz Menopauzalny 2015;14:1–6.

[59] Corona G, Monami M, Rastrelli G, Aversa A, Sforza A, Lenzi A, et al. Type 2 diabetes mellitus and testosterone: a meta-analysis study. Int J Androl 2011;34:528–40.

[60] Tanna MS, Schwartzbard A, Berger JS, Underberg J, Gianos E, Weintraub HS. Management of hypogonadism in cardiovascular patients: what are the implications of testosterone therapy on cardiovascular morbidity? Urol Clin North Am 2016;43:247–60.

[61] Basaria S, Coviello AD, Travison TG, Storer TW, Farwell WR, Jette AM, et al. Adverse events associated with testosterone administration. N Engl J Med 2010;363:109–22.

[62] Vigen R, O'Donnell CI, Baron AE, Grunwald GK, Maddox TM, Bradley SM, et al. Association of testosterone therapy with mortality, myocardial infarction, and stroke in men with low testosterone levels. JAMA 2013;310:1829–36.

[63] Finkle WD, Greenland S, Ridgeway GK, Adams JL, Frasco MA, Cook MB, et al. Increased risk of non-fatal myocardial infarction following testosterone therapy prescription in men. PLoS ONE 2014;9, e85805.

[64] Baillargeon J, Urban RJ, Kuo YF, Ottenbacher KJ, Raji MA, Du F, et al. Risk of myocardial infarction in older men receiving testosterone therapy. Ann Pharmacother 2014;48:1138–44.

[65] Ahmed T, Alattar M, Pantalone K, Haque R. Is testosterone replacement safe in men with cardiovascular disease? Cureus 2020;12, e7324.

[66] Brown AE, Bradbrook KE, Casey FE. A survey of adult and pediatric cardiology fellows on training received in family planning counseling. J Women's Health 2020;29:237–41.

[67] Hameed AB, Lawton ES, McCain CL, Morton CH, Mitchell C, Main EK, et al. Pregnancy-related cardiovascular deaths in California: beyond peripartum cardiomyopathy. Am J Obstet Gynecol 2015;213:379.e1–379.e10.

[68] Thompson JL, Kuklina EV, Bateman BT, Callaghan WM, James AH, Grotegut CA. Medical and obstetric outcomes among pregnant women with congenital heart disease. Obstet Gynecol 2015;126:346–54.

[69] Roos-Hesselink JW, Ruys TP, Stein JI, Thilén U, Webb GD, Niwa K, et al. Outcome of pregnancy in patients with structural or ischaemic heart disease: results of a registry of the European Society of Cardiology. Eur Heart J 2013;34:657–65.

[70] Callaghan WM, Creanga AA, Kuklina EV. Severe maternal morbidity among delivery and postpartum hospitalizations in the United States. Obstet Gynecol 2012;120:1029–36.

[71] Regitz-Zagrosek V, Roos-Hesselink JW, Bauersachs J, Blomström-Lundqvist C, Cífková R, De Bonis M, et al. 2018 ESC guidelines for the management of cardiovascular diseases during pregnancy. Eur Heart J 2018;39:3165–241.

[72] Committee on Gynecologic Practice Long-Acting Reversible Contraception Working G. Committee Opinion No. 642: increasing access to contraceptive implants and intrauterine devices to reduce unintended pregnancy. Obstet Gynecol 2015;126:e44–8.

[73] Kavanaugh ML, Jerman J. Contraceptive method use in the United States: trends and characteristics between 2008, 2012 and 2014. Contraception 2018;97:14–21.

[74] Dinger J, Bardenheuer K, Heinemann K. Cardiovascular and general safety of a 24-day regimen of drospirenone-containing combined oral contraceptives: final results from the International Active Surveillance Study of Women Taking Oral Contraceptives. Contraception 2014;89:253–63.

[75] Heit JA, Kobbervig CE, James AH, Petterson TM, Bailey KR, Melton 3rd LJ. Trends in the incidence of venous thromboembolism during pregnancy or postpartum: a 30-year population-based study. Ann Intern Med 2005;143:697–706.

[76] Woods JW. Oral contraceptives and hypertension. Hypertension 1988;11:II11–5.

[77] Lim KG, Isles CG, Hodsman GP, Lever AF, Robertson JW. Malignant hypertension in women of childbearing age and its relation to the contraceptive pill. Br Med J (Clin Res Ed) 1987;294:1057–9.

[78] Krauss RM, Burkman Jr RT. The metabolic impact of oral contraceptives. Am J Obstet Gynecol 1992;167:1177–84.

[79] Adeniji AA, Essah PA, Nestler JE, Cheang KI. Metabolic effects of a commonly used combined hormonal oral contraceptive in women with and without polycystic ovary syndrome. J Women's Health 2016;25:638–45.

[80] Cheang KI, Essah PA, Sharma S, Wickham 3rd EP, Nestler JE. Divergent effects of a combined hormonal oral contraceptive on insulin sensitivity in lean versus obese women. Fertil Steril 2011;96:353–9. e1.

[81] Vessey M, Painter R, Yeates D. Mortality in relation to oral contraceptive use and cigarette smoking. Lancet 2003;362:185–91.

[82] Charlton BM, Rich-Edwards JW, Colditz GA, Missmer SA, Rosner BA, Hankinson SE, et al. Oral contraceptive use and mortality after 36 years of follow-up in the Nurses' Health Study: prospective cohort study. BMJ 2014;349:g6356.

[83] Gerstman BB, Piper JM, Tomita DK, Ferguson WJ, Stadel BV, Lundin FE. Oral contraceptive estrogen dose and the risk of deep venous thromboembolic disease. Am J Epidemiol 1991;133:32–7.

[84] Lidegaard O, Lokkegaard E, Jensen A, Skovlund CW, Keiding N. Thrombotic stroke and myocardial infarction with hormonal contraception. N Engl J Med 2012;366:2257–66.

[85] van Vlijmen EF, Veeger NJ, Middeldorp S, Hamulyák K, Prins MH, Büller HR, et al. Thrombotic risk during oral contraceptive use and pregnancy in women with factor V Leiden or prothrombin mutation: a rational approach to contraception. Blood 2011;118:2055–61 [quiz 2375].

[86] Curtis KM, Tepper NK, Jatlaoui TC, Berry-Bibee E, Horton LG, Zapata LB, et al. U.S. Medical Eligibility Criteria for Contraceptive Use, 2016. MMWR Recomm Rep 2016;65:1–103.

[87] Dinger JC, Heinemann LA, Kuhl-Habich D. The safety of a drospirenone-containing oral contraceptive: final results from the European Active Surveillance Study on oral contraceptives based on 142,475 women-years of observation. Contraception 2007;75:344–54.

[88] Seeger JD, Loughlin J, Eng PM, Clifford CR, Cutone J, Walker AM. Risk of thromboembolism in women taking ethinylestradiol/drospirenone and other oral contraceptives. Obstet Gynecol 2007;110:587–93.

[89] Berenson AB, Rahman M, Wilkinson G. Effect of injectable and oral contraceptives on serum lipids. Obstet Gynecol 2009;114:786–94.

[90] Kongsayreepong R, Chutivongse S, George P, Joyce S, McCone JM, Garza-Flores J, et al. A multicentre comparative study of serum lipids and apolipoproteins in long-term users of DMPA and a control group of IUD users. World Health Organization. Task Force on Long-Acting Systemic Agents for Fertility Regulation Special Programme of Research, Development and Research Training in Human Reproduction. Contraception 1993;47:177–91.

[91] Guttmacher Institute. Contraceptive use in the United States. Guttmacher Institute; 2020. https://www.guttmacher.org/fact-sheet/contraceptive-use-united-states:.

[92] Burkman RT. Intrauterine devices. Curr Opin Obstet Gynecol 1991;3:482–5.

[93] Anon. ACOG Practice Bulletin No. 212: Pregnancy and Heart Disease. Obstet Gynecol 2019;133:e320–56.

[94] Fabian CJ. The what, why and how of aromatase inhibitors: hormonal agents for treatment and prevention of breast cancer. Int J Clin Pract 2007;61:2051–63.

[95] Robertson JF, Blamey RW. The use of gonadotrophin-releasing hormone (GnRH) agonists in early and advanced breast cancer in pre- and perimenopausal women. Eur J Cancer 2003;39:861–9.

[96] Matthews A, Stanway S, Farmer RE, Strongman H, Thomas S, Lyon AR, et al. Long term adjuvant endocrine therapy and risk of cardiovascular disease in female breast cancer survivors: systematic review. BMJ 2018;363:k3845.

[97] Cheung YM, Ramchand SK, Yeo B, Grossmann M. Cardiometabolic effects of endocrine treatment of estrogen receptor-positive early breast cancer. J Endocr Soc 2019;3:1283–301.

[98] Bhatia N, Santos M, Jones LW, Beckman JA, Penson DF, Morgans AK, et al. Cardiovascular effects of androgen deprivation therapy for the treatment of prostate cancer: ABCDE steps to reduce cardiovascular disease in patients with prostate cancer. Circulation 2016;133:537–41.

[99] T'Sjoen G, Arcelus J, Gooren L, Klink DT, Tangpricha V. Endocrinology of transgender medicine. Endocr Rev 2019;40:97–117.

[100] Flores AR, Herman JL, Gates GJ, Brown TNT. How many adults identify as transgender in the United States? The Williams Institute: Los Angeles, CA; 2016.

[101] Meerwijk EL, Sevelius JM. Transgender population size in the United States: a meta-regression of population-based probability samples. Am J Public Health 2017;107:e1–8.

[102] Seal LJ. Cardiovascular disease in transgendered people: a review of the literature and discussion of risk. JRSM Cardiovasc Dis 2019;8, 2048004019880745.

[103] Leinung MC, Urizar MF, Patel N, Sood SC. Endocrine treatment of transsexual persons: extensive personal experience. Endocr Pract 2013;19:644–50.

[104] Fishman SL, Paliou M, Poretsky L, Hembree WC. Endocrine care of transgender adults. In: Poretsky L, Hembree W, editors. Transgender medicine. Cham: Humana Press; 2019.

[105] Connelly PJ, Marie Freel E, Perry C, Ewan J, Touyz RM, Currie G, et al. Gender-affirming hormone therapy, vascular health and cardiovascular disease in transgender adults. Hypertension 2019;74:1266–74.

[106] Pyra M, Casimiro I, Rusie L, Ross N, Blum C, Keglovitz K, et al. An observational study of hypertension and thromboembolism among transgender patients using gender-affirming hormone therapy. Transgen Health 2020;5:1–9.

[107] van Velzen DM, Paldino A, Klaver M, Nota NM, Defreyne J, Hovingh GK, et al. Cardiometabolic effects of testosterone in transmen and estrogen plus cyproterone acetate in transwomen. J Clin Endocrinol Metab 2019;104:1937–47.

[108] Streed Jr CG, Harfouch O, Marvel F, Blumenthal RS, Martin SS, Mukherjee M. Cardiovascular disease among transgender adults receiving hormone therapy: a narrative review. Ann Intern Med 2017;167:256–67.

[109] Dutra E, Lee J, Torbati T, Garcia M, Merz CNB, Shufelt C. Cardiovascular implications of gender-affirming hormone treatment in the transgender population. Maturitas 2019;129:45–9.

[110] Nota NM, Wiepjes CM, de Blok CJM, Gooren LJG, Kreukels BPC, den Heijer M. Occurrence of acute cardiovascular events in transgender individuals receiving hormone therapy. Circulation 2019;139:1461–2.

Section XI

Psychological Health in CVD

Chapter 26

Psychosocial Issues in Cardiovascular Disease

Christina M. Luberto, Elyse R. Park, Jeff C. Huffman, and Gloria Y. Yeh

Clinical Case

"Katie" is a 38-year-old married non-Hispanic white woman with a history of spontaneous coronary artery dissection (SCAD), who was referred for psychotherapy by her cardiologist for the treatment of anxiety and stress symptoms; these symptoms were negatively impacting her quality of life (QoL), relationships, and health behaviors. Prior to the SCAD, Katie was otherwise physically healthy and had no history of mental health problems, but she had high stress from her work and family life. Katie owned her own consulting company and worked long hours, leaving her overwhelmed and irritable and taking away time from her relationships with her husband, young daughter, and friends.

Following the SCAD, Katie frequently worried about her health. She could not stop thinking about the possibility of having another SCAD and what it would mean for her health, her future, and her family. She reported frequent sadness at the thought of dying young and leaving her husband and young daughter. She was concerned that work stress was contributing to high blood pressure and increased her risk of another SCAD. When she was able to spend time with family and friends, Katie found herself short tempered and easily annoyed. She also began to overeat unhealthy foods when upset and stopped exercising. Katie expressed interest in psychotherapy to learn skills to reduce worries about her health, reduce work stress, and improve her relationship with her partner.

Abstract

Although cardiovascular disease (CVD) morbidity and mortality rates have decreased over the past 40–50 years,

CVD remains the number one cause of death worldwide [1]. Traditional risk factors for CVD include physical health conditions, such as diabetes and hypertension (HTN), and health behaviors such as cigarette smoking and

Sex Differences in Cardiac Disease. https://doi.org/10.1016/B978-0-12-819369-3.00025-3

decreased physical activity. A large body of research now demonstrates the significant role that psychosocial factors (i.e., psychological, emotional, and social experiences) play in the development, progression, and prognosis of CVD. The strength of the effect of psychosocial factors on CVD outcomes is often similar to the strength of traditional risk factors. In some cases, the negative effects of psychosocial risk factors are particularly salient and stronger for women as compared to men. In this chapter, we first present an overview of the literature establishing psychosocial factors [anxiety, depression, stress, hostility, socioeconomic status (SES), and social integration] as contributors to CVD outcomes. We then discuss the potential mechanisms linking psychosocial factors to CVD. Lastly, we review the efficacy of pharmacotherapy and behavioral treatments for improving health outcomes in patients with or at risk for CVD. In all instances, we note sex differences when relevant.

Epidemiology of Depression and Anxiety

In 2018, nearly 20% of adults in the United States were living with a mental illness (46.6 million people), most commonly major depressive disorder (MDD) or an anxiety disorder [2]. According to the Diagnostic and Statistical Manual-5 (DSM-5), MDD involves a 2-week period of depressed mood or loss of pleasure in previously enjoyed activities, which creates significant distress or functional impairment and is accompanied by additional symptoms (for a total of five or more symptoms): weight changes, sleep problems, psychomotor agitation or slowing, appetite changes, feelings of worthlessness or inappropriate guilt, impaired concentration, or recurrent suicidal ideation [3]. Depression is a leading cause of disability worldwide [4]. In a national survey of 36,309 adults in the United States, the annual prevalence rate of MDD was 10% and the lifetime prevalence rate was 21% [5]. Importantly, many people who do not meet the full diagnostic criteria for MDD still struggle with clinically significant elevations in depression symptoms, with up to 17% of patients in primary care settings demonstrating subthreshold elevations [6]. Therefore, throughout the chapter, we use the term "depression" as a general category to refer to both diagnoses of MDD and clinically significant elevations in depressive symptoms.

Anxiety disorders are the most common mental disorder in the United States, with an annual prevalence rate of 18% and lifetime prevalence rate of 31% [7]. As a category, "anxiety" encompasses generalized anxiety disorder (excessive worry across a number of life domains), social anxiety disorder (fear of being scrutinized by others), panic disorder (recurrent unexpected panic attacks), agoraphobia (fears of being in crowded spaces), specific phobia (fear of a specific object or situation such as flying or heights), trauma-related disorders (e.g., posttraumatic stress disorder), and

obsessive-compulsive disorders (e.g., repetitive distressing thoughts and behaviors [3]). Up to 50% of people with anxiety disorders experience severe functional impairment (e.g., missed days at work and missed social activities [8]). Similar to depression, even elevated symptoms that do not meet full criteria for an anxiety disorder can be significantly distressing and impairing [9]. For example, in terms of generalized anxiety disorder, the annual prevalence rate is 3% [7] but the rate of subthreshold symptoms is as high as 7%, and levels of distress and impairment are similar across diagnosed and subthreshold conditions [9]. Many people with elevated anxiety attempt to manage their symptoms by avoiding the sources of their fear, which creates a narrow and constricted way of living that decreases quality of life (QoL) and perpetuates anxiety symptoms over time. Again, throughout the chapter, we use "anxiety" to refer to anxiety disorder diagnoses and elevated anxiety symptoms.

Both depression and anxiety are overrepresented among women as compared to men. In a recent national survey, the past-year prevalence rate for depression was 13% for women vs. 7% for men, and the lifetime prevalence rate was 26% for women vs. 15% for men [5]. The annual prevalence rate of any anxiety disorder is 23% for women and 14% for men [10]. Women are also more likely to have multiple comorbid psychological disorders, which create more substantial functional disability for women [11]. Women tend to develop symptoms earlier, experience more severe symptoms, and have poorer long-term outcomes than men, with the increased risk for women beginning during puberty and persisting through adulthood [12].

There are several factors, including sociological, psychological, and biological, that contribute to sex differences in mental health. According to the World Health Organization (WHO), women occupy a disadvantaged social position as compared to men, affording them less power and control over social determinants of mental health, such as limited access to mental health resources [11]. Women may also experience harmful bias and discrimination within work settings and society more broadly as a result of their sex [11]. In addition, women are at a greater risk of certain adverse events during childhood that increase the risk of developing mental health disorders (e.g., sexual abuse), and they may also be more affected by these negative events as compared to men [12]. Sex differences in coping styles contribute to these differences: women tend to use more internalizing coping styles and experience greater rumination (i.e., unhelpful repetitive thinking), which can create greater suffering in the context of stressors [13]. Women may also be more likely to report mental health symptoms than men. Biological differences in sex hormones due to reproductive life events (e.g., puberty, menstruation, and pregnancy) are also thought to play a role. For example, hormonal fluctuations during reproductive periods may alter inflammation and immune pathways associated with depression [14]. The complexities of potential biological mechanisms, however, are not well understood and may be less significant than environmental factors [12, 13].

Relationship of Depression and Anxiety With CVD

As mentioned above, some studies focus on patients with clinical diagnoses of depression and anxiety, but many also take a broader perspective to include patients with clinically relevant symptom elevations that may not meet formal diagnostic criteria. Thus, throughout the chapter, we are often referring to patients who may not have diagnosed mental health disorders but who still struggle with emotional symptoms that nonetheless increase their risk of poor cardiovascular disease (CVD) outcomes.

Depression and CVD

Depression is common among CVD patients. Up to 45% of patients with CVD have clinically elevated depression symptoms and 15–20% meet the full diagnostic criteria for MDD, rates that are approximately 2–3× higher than those of the general population [15]. Among SCAD survivors, over 30% sought mental health treatment for depression [16]. Risk factors for depression in CVD patients include younger age, having a premorbid history of depression, and being a woman [15]. Depression is often chronic and recurrent in CVD patients, with the relationship between depression and CVD being bi-directional. For example, up to 70% of acute coronary syndrome (ACS) patients had ongoing depression symptoms prior to the cardiac event, suggesting that depression may lead to CVD, while at the same time, many patients experience their first onset of depression after a cardiac event, suggesting that CVD can lead to the development of depression (e.g., due to negative thoughts and feelings associated with new physical symptoms and limitations [15]).

The negative effects of depression on cardiovascular health are significant and well established. Depression has been identified as an independent risk factor for the development of CVD in healthy people, as well as CVD progression and mortality in people with CVD [15]. For example, in a meta-analysis of 22 longitudinal studies of over 111,000 medically healthy people with depression, MDD was significantly associated with the development of CVD and congestive heart failure (HF), as well as CVD-related death [17]. In terms of prognosis among patients with CVD, multiple meta-analyses of prospective studies have found that depression is an independent risk factor for cardiac events, hospitalizations, morbidity, and all-cause mortality among ACS patients, beyond other traditional risk factors [18, 19]. In 2014, the American Heart Association (AHA) declared depression an independent risk factor for

poor ACS prognosis and emphasized the need for safe and effective depression treatments [18]. A recent longitudinal study by Worcester and colleagues [20] found that depression significantly predicted death for up to 15 years after an ACS and that patients with mild depression symptoms actually had a greater mortality risk than those with moderate-severe depression. While other studies have found more of a dose-response relationship, these recent findings suggest that patients remain at risk for many years and even mild levels of depression symptoms pose a significant risk.

Anxiety and CVD

Anxiety is also common in patients with CVD. Approximately 20–30% of ACS patients, patients awaiting coronary artery bypass surgery, patients with HF, and patients with arrhythmia report elevated anxiety symptoms [21]. Up to 14% of coronary artery disease (CAD) patients are diagnosed with generalized anxiety disorder (as compared to up to 7% in the general population [21]). Anxiety is associated with CVD development, progression, and mortality, though the evidence is somewhat more limited as compared to depression. In a large meta-analysis of prospective studies of healthy adults, anxiety was associated with a 52% increased risk of developing CVD—a level of risk similar to that of traditional risk factors (i.e., well-established medical and behavioral CVD risk factors [22]). In another large meta-analysis of physically healthy adults, anxiety was significantly associated with a 41% increased risk of cardiovascular mortality, 41% increased risk of developing congenital heart disease (CHD), 71% increased risk of stroke, and 35% increased risk of HF [23]. Although some of the studies did not adjust for other important risk factors, the results did not change across adjusted and unadjusted analyses [23]. Among people with established CAD, a meta-analysis of over 30,000 patients found that anxiety in the setting of an acute cardiac event was less predictive of CVD outcomes, possibly because almost all patients are anxious at that time, but patients with chronic CAD and anxiety did have an increased risk of mortality [21]. These findings highlight the importance of measuring anxiety during a stable period when symptoms are not likely to be a transient response to an acute event. Overall, anxiety appears to significantly increase the risk of developing CVD among healthy people, though the effects for patients with existing CVD may not be as strong as for depression and may be explained by other factors.

A unique problem related to anxiety involves patients with noncardiac chest pain. This refers to patients who present for treatment in medical settings but do not have heart disease yet have high rates of anxiety disorders [21]. The lifetime prevalence of noncardiac chest pain may be as high as 33% [24]. In one study of patients presenting to an emergency department for acute chest pain, 64% were diagnosed as having noncardiac chest pain [25]. The primary reasons

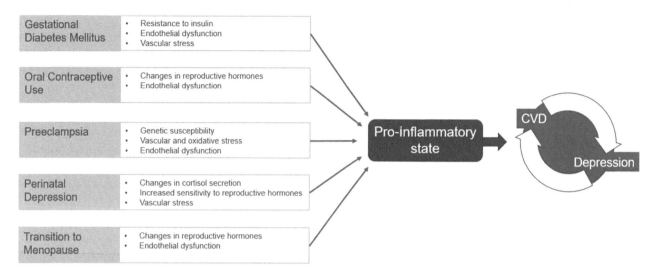

FIGURE 1   Sex-Specific Factors Contributing to Increased Inflammation and Depression in Women. Many hormonal changes and reproductive events in women result in increased inflammation and alter key biological processes, increasing vulnerability to depression. CVD, cardiovascular disease.

for seeking medical treatment were being worried about the physical symptoms (e.g., chest pain and heartburn) and the possibility of having a serious health problem [25]. Noncardiac chest pain is problematic because these patients still experience significant distress and functional impairment and incur significant healthcare costs as a result of seeking unnecessary medical treatment [21, 25]. QoL in patients with noncardiac chest pain is as low as in patients with CAD [24].

Similar to the general population, among patients with CVD, rates of depression and anxiety are higher among women as compared to men. In a meta-analysis of cohort and cross-sectional studies, 19% of women and 12% of men were diagnosed with depression during or after their CVD hospitalization [26]. In a cross-sectional study of postmyocardial infarction (MI) patients with no prior diagnosis of anxiety or depression, 71% of women and 60% of men reported depression or anxiety [27]. Mattina and colleagues [14] suggest that sex-related risk factors such as childhood trauma, hormonal fluctuations, and reproductive events (e.g., menstrual cycle, pregnancy, and menopause) lead to increased inflammation and alterations in key biological processes [e.g., hypothalamic-pituitary-adrenal (HPA) axis, renin-angiotensin-aldosterone system (RAAS), and serotonin/kynurenine pathway], which further increase inflammation to drive and maintain the increased vulnerability of comorbid depression and CVD among women [14]. **Figure 1** shows common sex-specific factors contributing to inflammation in women.

Relationship of Psychosocial Stressors With CVD

Beyond depression and anxiety as mental health disorders, there are additional psychosocial factors that contribute to

CVD. Stress and hostility are two significant psychological factors that contribute to poor CVD outcomes even though they are not diagnosable mental health problems. Social factors, such as low socioeconomic status (SES) and low social integration, also play a significant role. **Table 1** displays a summary of studies linking psychosocial factors to CVD.

Stress and CVD

Stress refers to feelings of being overwhelmed and the perception that environmental demands exceed coping resources [34]. Stress is highly prevalent in the United States and contributes to the development of psychological and physical health problems. Women report greater stress levels than men, greater increases in stress over a 5-year period, and greater physical and emotional symptoms due to stress (e.g., headache and crying [35]). Women and men also report different sources of stress, with women more likely to identify finances as a source of stress and men more likely to identify work as a source of stress [35].

A large body of research indicates that stress influences the development of CVD, conferring approximately a 40–60% increased risk of disease, and also worsens the prognosis for people with CVD [36]. Greater levels of stress are associated with episodes of myocardial ischemia and cardiac arrhythmia, and stress acutely increases the risk of ACS over a 24-h period [36]. A meta-analysis of over 166,000 healthy adults found that those with the highest levels of stress had a twofold higher rate of death due to HF, CVD, and stroke, and a fourfold increased risk of peripheral vascular disease, with the association between stress and mortality existing in a dose-dependent manner [37]. The negative health effects of stress extend from early life stressors including childhood sexual abuse, parental substance use, and low SES, as well as stressors in adulthood including work stress, marital

TABLE 1 Summary of Illustrative Meta-Analyses of Psychosocial Factors Influencing CVD

Reference	Psychosocial Factor	Number of Trials	Design of Included Trials	CVD Outcome	Results	Conclusion
Gan et al., 2014 [28]	Depression	$N = 39$ trials (893,850 participants)	Prospective cohort	CHD or MI incidence	RR = 1.30, 95% CI = 1.22–1.40, $p < 0.001$	People with depression have a 30% increased risk of CHD No differences based on sex
Correll et al., 2017 [17]	Depression	$N = 92$ trials, (3,211,768 participants)	Cross-sectional, retrospective, or prospective longitudinal	1. CVD incidence 2. CHD incidence 3. Cerebrovascular disease incidence 4. HF incidence 5. CVD-related death	1. HR = 1.72, 95% CI = 1.48–2.00 2. HR = 1.63, 95% CI = 1.33–2.00 3. HR = 2.04, 95% CI = 1.05–3.96 4. HR = 2.02, 95% CI = 1.48–2.75 5. HR = 1.63, 95% CI = 1.25–2.13	People with depression have an increased risk of cardiac disease and cardiac-related mortality
Batelaan et al., 2016 [22]	Anxiety	$N = 37$ trials (1,565,699 participants)	Prospective cohort	CVD incidence	HR = 1.52, 95% CI = 1.36–1.71, $p < 0.01$	Anxiety is associated with a 52% increased incidence of CVD
Kivimaki et al., 2015 (Lancet paper [29])	Stress (work stress)	$N = 25$ trials (603,838 participants)	Prospective cohort study	CVD incidence	RR = 1.13, 95% CI = 1.02–1.06, $p = 0.02$	People who work 55 h or more per week have a 1.13 times higher risk of incident CVD than those working standard hours
Chida and Steptoe, 2009 [30]	Hostility/anger	$N = 44$ trials (79,726 participants)	Prospective cohort study	1. Development of CHD in healthy people 2. Prognosis in people with CHD	1. HR = 1.19, 95% CI = 1.05–1.35, $p = 0.008$ 2. HR = 1.23, 95% CI = 1.08–1.42, $p = 0.002$	Anger/hostility are associated with increased CHD events in healthy adults and poor CHD prognosis in adults with existing CHD. This association is higher in men

Continued

TABLE 1 Summary of Illustrative Meta-Analyses of Psychosocial Factors Influencing CVD—cont'd

Reference	Psychosocial Factor	Number of Trials	Design of Included Trials	CVD Outcome	Results	Conclusion
Mostofsky et al., 2014 [31]	Hostility/anger	*Total:* N = 9 trials total (6119 participants total) *MI/ACS only:* N = 4 trials (5008 participants)	Case-crossover study	1. Incidence of MI or ACS in 2 h following an angry outburst 2. Incidence of stroke in 2 h following an angry outburst 3. Rate of ventricular arrhythmia in the 15 min following an angry outburst	1. RR = 4.74, 95% CI = 2.50–8.99, p < 0.001 2. RR = 3.62, 95% CI = 0.82–16.08, p = 0.09 3. RR = 1.83, 95% CI = 1.04–3.16	There is a higher risk of cardiovascular events shortly after outbursts of anger
Backholer et al., 2017 [32]	SES	N = 116 trials (22 million participants)	Cohort study	*Education* 1. CHD incidence among low- vs. high-education **men** 2. CHD incidence among low- vs. high-education **women** 3. CHD incidence among low-education **women vs. men** *Area deprivation* 4. CHD incidence among women as compared to men of highest vs. lowest area deprivation *Occupation* 5. CHD incidence among women as compared to men of manual vs. nonmanual occupation *Income* 6. CHD incidence among women as compared to men of lowest vs. highest income	*Education* 1. RR = 1.30, 95% CI = 1.15–1.48 2. RR = 1.66, 95% CI = 1.46–1.88 3. RRR = 1.34, 95% CI = 1.09–1.63 *Area deprivation* 4. RRR = 1.16, 95% CI = 0.98–1.37 *Occupation* 5. RRR = 0.99, 95% CI = 0.63–1.56 *Income* 6. RRR = 1.03, 95% CI = 0.74–1.42	Lower SES is associated with greater CHD risk, particularly for women
Valtorta et al., 2016 [33]	Social isolation/ integration	N = 23 trials (11 trials for CHD only, 4628 CHD participants)	Prospective longitudinal study	CHD risk	RR = 1.29, 95% CI = 1.04–1.59	There is a 29% increase in risk of CHD for people high in loneliness/social isolation

ACS, acute coronary syndrome; CHD, coronary heart disease; CI, confidence interval; CVD, cardiovascular disease; HF, heart failure; HR, hazard ratio; MI, myocardial infarction; RR, relative risk; RRR, ratio of the relative risks; SES, socioeconomic status.

problems, and the death of a child [36]. The negative health effects of work stress have received particular attention, with a meta-analysis of prospective cohort studies of over 600,000 healthy adults finding that working more than 55 h/week confers a 1.3-fold higher risk of incident stroke [29]. Long working hours also increased the risk of CHD, but this relationship was generally weaker and was moderated by SES, such that the negative health effects of long working hours were greater among those of lower SES [29].

Beyond the types of CVD discussed above, stress-induced cardiomyopathy, also called Takotsubo cardiomyopathy, is another specific type of CVD linked to stress. The pathophysiology of Takotsubo cardiomyopathy is poorly understood, but it is distinct from ACS and commonly occurs after a stressful event including physical triggers such as acute respiratory failure or infection (36%) and emotional triggers such as anxiety or interpersonal conflicts (28% [38]). Takotsubo cardiomyopathy is most common among postmenopausal, elderly women [38], although men with Takotsubo cardiomyopathy are more likely to experience serious in-hospital complications, major cardiac events shortly following discharge, and are more likely to die in the long term [38]. There are also sex differences in Takotsubo cardiomyopathy triggers, with women more commonly reporting emotional triggers and men more commonly reporting physical triggers [38]. More than half of patients have a history of neurologic or psychiatric disorder [38].

Hostility/Anger and CVD

Hostility and anger are specific emotional experiences that have been clearly linked to CVD. The terms hostility and anger are often used interchangeably. Hostility is considered a personality trait that involves intense and prolonged experiences of anger, ranging from annoyance and irritability to fury and rage [30]. There is a tendency to perceive external events as unfair and to negatively evaluate others, resulting in behavioral responses of verbal and/or physical aggression [39]. Physical and verbal aggression are more common in men, though indirect forms of hostility are more common in women (e.g., spreading gossip [40]).

The association between hostility/anger and CVD incidence, prognosis, progression, and mortality is well established [30]. In the seminal meta-analysis conducted to date, across 25 prospective cohort studies of healthy people and 19 prospective studies of patients with CHD, there was a significant positive association between hostility/anger and CHD outcomes in both groups. In healthy patients, hostility/anger increased the risk of developing CHD by 19% and, in patients with CHD at baseline, hostility/anger increased the risk of poor cardiac outcomes including death, ACS, and disease progression by 23% [30]. In the Heart and Soul Study, a prospective cohort study of patients with stable CHD, those in the highest quartile of hostility scores at baseline, as compared to those in the lowest

quartile, had more than twice the rate of mortality over a 5-year period [41]. Keith and colleagues [42] reported that self-reported hostility/anger significantly predicted cardiac-related hospitalizations and all-cause hospitalizations over 3 years [42]. In a small meta-analysis of nine case-crossover studies, Mostofsky et al. [31] demonstrated significantly increased risk of ACS, stroke, and arrhythmia in 2 h following an anger outburst [31].

Socioeconomic Status and CVD

A 2015 scientific statement from the AHA stated that addressing social determinants of CVD is currently the most significant way to reduce CVD death and disability in the United States [43]. SES, which encompasses income, education, employment, and neighborhood environment, is a particularly important social determinant of CVD.

The effect of SES on CVD outcomes is significant and well established, conferring a similar risk as traditional risk factors [43]. Low income and low education in particular increase the risk of developing CVD, as well as contribute to poorer CVD outcomes and mortality [43]. For example, increasing levels of family income correspond to a 40–50% decrease in cardiac mortality, and lower education is associated with greater incidence of CVD events, greater CVD traditional risk factors, and mortality, independent of other sociodemographic factors [43]. The relationship between occupation and CVD is not as widely studied, but there is some evidence that higher-status occupations and occupations with high control (e.g., managers and doctors) are protective of cardiac health and unemployment causes health problems [43]. Neighborhood environments, which refer to the structural, physical, and social aspects of one's living environment, have been associated with CVD outcomes in many studies. For example, decreased access to healthy food and physical activity are associated with increased rates of CVD, CVD mortality, and CVD risk factors such as increased blood pressure and body mass index [43]. There are also racial disparities in CVD that are related to SES, but are not entirely attributable to SES. For example, black Americans have higher rates of CVD risk factors and are two to three times more likely to die from heart disease than white Americans [43], and this increased risk exists even for those who are of high SES [44]. Systemic racism and discrimination, including within the healthcare system itself (e.g., implicit bias among healthcare providers), are thought to play a role in CVD racial disparities [43].

The SES and CVD association may be rooted in stress-related neurobiological mechanisms. In 2019, Tawakol and colleagues [45] conducted a longitudinal study using whole-body positron emission tomography (PET)/computed tomography (CT) imaging on 509 people without a history of CVD. The results indicated that lower SES predicted a greater risk of major adverse cardiac events over a 4-year period through a multitissue pathway involving

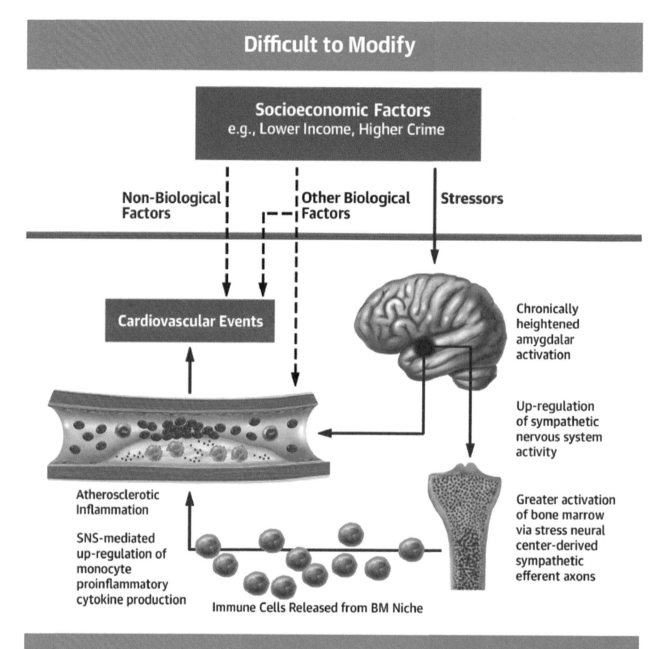

FIGURE 2 A Pathophysiologic Model of Lower Socioeconomic Status Leading to Adverse Coronary Events. Biological pathways included heightened activation of amygdala results in increased activation of the bone marrow (with release of inflammatory cells), which in turn leads to increased atherosclerotic inflammation and its atherothrombotic manifestations. Nonbiological (and likely other biological) paths also exist. While the social variables involved in this pathway are notoriously difficult to change, the biological factors are potentially more modifiable. BM, bone marrow; SNS, sympathetic nervous system. *Reproduced with permission from [45].*

arterial inflammation and elevated activity in the amygdala and bone marrow [45]. **Figure 2** includes a depiction of the pathways from this study [45].

Sex differences exist in the relationship between SES and CVD [32]. In a recent meta-analysis using data from

over 22 million people, lower SES was associated with increased risk of CVD and stroke in both men and women, but the relationships were stronger for women [32]. For example, women with the lowest level of education were at 24% excess risk of CHD compared to men with the lowest level of education [32]. Similarly, in the Jackson

Heart Study, a cross-sectional study of over 5000 black Americans, the inverse relationship between SES and CVD burden again appeared to be stronger for women than men [44]. Researchers speculate that the mechanisms underlying this increased risk for women include greater delays to presentation, diagnosis, and treatment, as well as a less typical CVD presentation that creates a lower awareness of CVD risk among women and interferes with their receiving appropriate care [32]. In addition, more women live in poverty as compared to men and thus are at greater risk for the negative health effects of low SES [32].

Social Integration and CVD

An individual's level of social integration versus social isolation also impacts CVD outcomes. Social integration refers to an individual's level of engagement in social relationships, including the social network that surrounds a person and how frequently they interact with that network [46]. Thus, social integration includes an individual's number of close relationships, community membership, and social roles (e.g., friend, parent, and partner). Social isolation is the opposite of social integration and refers to objectively having few social ties [47]. Social isolation often leads to loneliness, which refers to an individual's subjective feelings about their level of connection to others. However, individuals who are not socially isolated can still feel lonely; that is, while someone may objectively have a number of social ties, they may still feel alone, dissatisfied, or unhappy with their level of social connection or support. All of these dimensions of social relationships are associated with poorer CVD outcomes.

A large body of literature demonstrates that social integration protects against developing and dying from CVD. A recent meta-analysis [33] explored the prospective associations of social isolation and loneliness with the development of CHD and stroke, and found that there was approximately a 30% increased risk of CHD and stroke for those with high versus low loneliness or social isolation [33]. Thus, both objectively having few social ties and subjectively feeling alone are both important for preventing the development of CHD. There were no differences by sex [33]. Importantly, several of the studies were conducted among high-SES and healthy populations, potentially underestimating the negative health effects of social isolation and loneliness among low-SES groups with greater social problems [33]. In terms of mortality among patients with existing CVD, Barth and colleagues [48] conducted a meta-analysis of 32 prospective cohort studies and found a 71% increased risk of combined cardiac and all-cause mortality due to low social support, even when controlling for other risk factors [48].

The role of sex in the relationship between social integration and CVD is complex. A recent meta-analysis [49] found consistent evidence that having fewer social roles increases CVD incidence, short-term CVD progression, and

the risk of CVD mortality, but the relationships differ by sex. In this study, social integration was defined as the degree to which an individual participates in social relationships, forming a "graded hypothesis" where every increase in the number of social roles provides further protection. On the other hand, social isolation was defined as objectively having few social roles according to a "threshold hypothesis," where being below a certain number of social roles is problematic, but once that number is achieved, there is no greater impact of additional roles. The results of this meta-analysis indicated that, for women, both social integration and social isolation were relevant to CVD incidence, whereas for men, social isolation was most relevant [49]. These findings suggest that social integration may protect against CVD for men once a minimum number of social ties have been met, whereas women continue to benefit from additional relationships beyond that lower threshold. In terms of mortality risk, there was insufficient evidence to link social integration to CVD mortality in women, but for men, fewer social roles were clearly associated with greater mortality risk [49]. In a large study of healthy women in the Nurses' Health Study, followed over a 22-year period, Chang and colleagues [46] found that compared to women with the lowest level of social integration at baseline, those with the highest level had reduced risk of nonfatal CHD; however, this relationship was largely explained by smoking and, to a lesser extent, physical activity. In terms of fatal CVD events, women with the highest social integration had reduced risk, and this relationship was still significant when controlling for lifestyle [46]. Thus, the relationship between sex, social integration, and CVD outcomes may depend on types of social integration measurements and CVD outcomes.

Mechanisms Linking Psychosocial Factors to CVD

The range of psychosocial factors impacts CVD outcomes through complex interrelationships between psychological, social, behavioral, and biological pathways. Psychological factors and social factors are interrelated, and both impact CVD outcomes through behavioral and biological pathways [21, 50]. **Figure 3** depicts an overview of these potential mechanisms.

Patients with depression, anxiety, stress, and hostility/anger tend to smoke more, have more sedentary lifestyles, have poor diets, be less compliant with medications and medical visits, and are less likely to use healthy coping methods such as social support as a way to reduce stress [21]. Specific depression symptoms such as a lack of motivation and lack of interest can directly prevent patients from engaging in important health behaviors such as physical activity and healthy diet [51]. Anxiety may also lead patients to avoid physical activity because many people with anxiety also fear the physical symptoms of anxiety itself

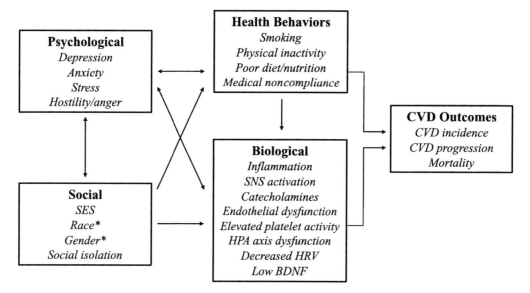

FIGURE 3 Depiction of Interrelationships Among Psychological, Social, Behavioral, and Biological Factors With CVD Outcomes. BDNF, brain-derived neurotrophic factor; CVD, cardiovascular disease; HPA, hypothalamic-pituitary-adrenal; HRV, heart rate variability; SES, socioeconomic status; SNS, sympathetic nervous system.
*The relationships of race and gender with psychological factors are considered to be unidirectional, where race and gender contribute to psychological factors but not the other way around.

(e.g., increased heart rate and sweating) and thus avoid exercise as a way to avoid these symptoms [52]. Stress is widely linked to poor health behaviors. A 10-year prospective trial of healthy men found that the association between hostility and mortality was largely accounted for by cigarette smoking, highlighting the role of health behaviors in the relationship between hostility and CVD [53].

There is also a link between psychological problems and social determinants of health. For example, psychological problems are more common and may also be more problematic among individuals from lower as compared to higher SES backgrounds. In the REGARDS study, a prospective observational cohort study of over 22,000 people, individuals earning <$35,000 per year who had both high stress and depression symptoms had a 48% increased risk of developing CVD and a 33% increased risk of all-cause mortality, adjusting for multiple potential confounders, but this increased risk of stress and depression did not exist for high-income individuals [54]. Given that psychological and social factors are interrelated, one way that social factors impact CVD outcomes is through psychological problems and, thereby, poor health behaviors [43]. At the same time, social factors can also directly impact health behaviors, as individuals from low-SES groups have decreased access to healthy food options, physical activity (e.g., due to unsafe neighborhood environments), and affordable health care [43]. Hostility [53] and social isolation are also linked to greater anxiety and depression symptoms and poor health behaviors [33, 46].

The biological mechanisms linking psychosocial factors to CVD are complex and involve the interactions between several physiological symptoms and processes. Processes that have been commonly noted to play a role include sympathetic nervous system activation and cardiovascular reactivity, endothelial dysfunction, increased inflammation [e.g., interleukin-6 (IL-6), C-reactive protein (CRP), and tumor necrosis factor alpha (TNF-a)], platelet activation, increased catecholamines, cortisol, HPA axis dysfunction, and low brain-derived neurotrophic factor [21, 36, 50]. For example, depression is widely linked to inflammatory biomarkers implicated in CVD [55] and anxiety also shows associations with inflammation [56]. Hostility/anger activates the sympathetic nervous system to create a physiological stress reaction leading to cardiovascular reactivity, activation of the HPA axis, and inflammation. Social determinants of health are also linked to these biological processes, but the data are mostly observational and not causal [43]. A recent prospective cohort imaging study demonstrated that greater activation of the amygdala, a key component of the brain's stress network, significantly predicted the development of CVD above and beyond traditional risk factors, and the association was mediated by increased arterial inflammation, which was due to upregulation of bone marrow activity [57]. This study was the first to link resting metabolic brain activity to the development of CVD and demonstrates the complexity of biological mechanisms underlying psychosocial effects on CVD.

A recent synthesis of available literature examining the association of inflammation with depression and CVD in women provides insight into potential biological differences that may account for sex differences in response to stress [14]. There is some evidence, both direct and indirect, from human and animal models, that suggests sex differences in stress physiology including alterations in the HPA axis, the RAAS, and the serotonin/kynurenine pathway. These differences result in women being more prone to pro-inflammatory states, thus contributing to the development of depression and CVD [14]. For example, changes to endothelial function in the postpartum period after preeclampsia suggest RAAS dysfunction during pregnancy and an increased sensitivity to angiotensin II contributing to persisting vasculature dysfunction [58]. Anti-depressant-like effects have been observed in mice treated with drugs that block the angiotensin-converting enzyme [59]. In the general population, women have higher levels of CRP and IL-6, and these are associated with risk of CVD as well as depression [14]. Interleukin-8 (IL-8) and interferon-gamma are also elevated in depressed women, with levels that correlate with depression severity [60]. In animal models, the ability to activate the HPA axis and a greater stress response with inflammatory cytokines is seen in female mice [61]. Similarly, in humans, women with higher depressive symptom severity had lower cortisol to high-sensitivity CRP ratios, suggesting heightened inflammation and dysregulation of these systems [62]. Many of the postulated differences in key pro-inflammatory cytokines relate to changes during

reproductive periods that make women uniquely vulnerable. It is noted that perinatal depression, gestational diabetes, preeclampsia, and the menopausal transition are all known to be pro-inflammatory states in women that have associations with mood symptoms—see **Figure 1** [14].

Management of Psychosocial Issues Among Patients With CVD

Depression and Anxiety Screening

Figure 4 shows a summary of psychosocial risk factors and clinical management approaches. Despite its important role, depression is underrecognized and undertreated in CVD. In 2008, the AHA recommended systematic depression screening of all CVD patients using a tiered approach beginning with the two-item Patient Health Questionnaire-2 (PHQ-2 [64]). These items ask: over the past 2 weeks, how often have you been bothered by (1) little interest or pleasure in doing things and (2) feeling down, depressed, or hopeless? If scores indicate possible depression (scores >3), then proceed with the full PHQ-9, which assesses additional symptoms (e.g., sleep and appetite changes [64]). Scores >10 on the PHQ-9 correspond with clinical diagnoses of MDD and

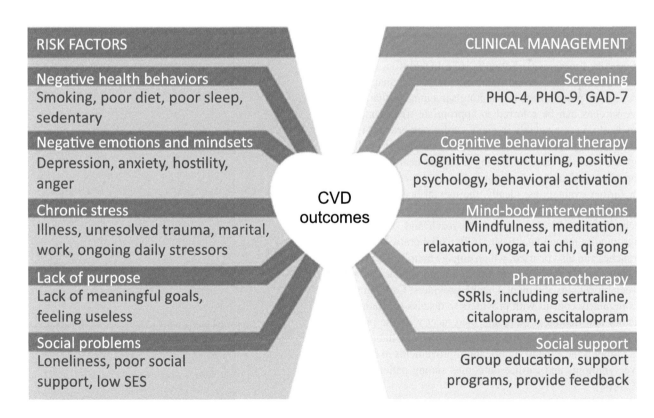

FIGURE 4 Summary of Psychosocial Risk Factors for CVD and Clinical Management. Both negative health behaviors and a variety of psychosocial risk factors increase the risk for heart disease, often in synergistic fashion. An increasing number of evidence-based techniques have been developed as management strategies for promoting healthy behaviors and the enhancement of psychosocial well-being. CVD, cardiovascular disease; GAD-7, Generalized Anxiety Disorder 7; PHQ, Patient Health Questionnaire; SES, socioeconomic status; SSRIs, selective serotonin reuptake inhibitors. *Adapted with permission from [63].*

scores >5 indicate clinically relevant symptom elevations [65]. Jha and colleagues [65] recommend that cardiologists follow this two-step screening approach and consider initiating anti-depression treatment when needed, by either prescribing medication, recommending psychotherapy, or encouraging physical activity. The PHQ-9 also assesses for suicidality, with positive responses requiring immediate arrangement for emergency care [65]. Although anxiety is also common and problematic for patients with CVD, anxiety screening is relatively uncommon in cardiac settings and has not been recommended by the AHA. Anxiety screening may be done efficiently and accurately using a similar two-step approach, starting with the GAD-2. The items ask about (1) feeling nervous, anxious, or on edge, and (2) being unable to stop or control worrying, in the past 2 weeks. If scores suggest elevated symptoms (scores > 3), then use the GAD-7 for verification [66]. Taken together, a four-item screening tool called the PHQ-4 that combines the PHQ-2 and GAD-2 could be used to screen for depression and anxiety, with the full versions of these measures used when the combined score is > 6 [66].

Routine screening requires minimal time and resources, and screening by cardiologists may reduce the stigma associated with mental health disorders [65]. However, there are concerns about misdiagnosis; many people who screen positive may not be able to access depression treatment; and assessing and managing depression can feel beyond the scope of practice for many cardiologists [67]. In addition, it is important that there is a clear treatment path and management protocol in place before conducting screening so that positive screens can be referred to appropriate treatment, but this is often not available in many cardiology settings [15]. Incorporating depression treatment into outpatient cardiology care can be done by having a mental health clinician consultant-liaison to whom cardiologists can refer patients for further evaluations; ideally, the mental health clinician would be colocated in the cardiology setting [67]. Using technology could also expand the reach and delivery of high-quality mental health care [67]. Collaborative care approaches can also be used, which engage nonphysician care managers to coordinate treatment recommendations between physicians and mental health providers. Care managers regularly consult with providers to discuss the patient's history and treatment plan, and then meet with the patient to relay information about treatment options. Collaborative care models have led to significant improvements in depressive symptoms and cardiac outcomes among patients with ACS [68].

Pharmacological Management

Medical management of depression and anxiety in patients with CVD is safe, generally effective for treating mood symptoms (except for patients with HF), and might but might not lead to cardiac benefits. Selective serotonin reuptake inhibitors (SSRIs) are particularly effective for mood management in CVD [69]. SSRIs are the most commonly prescribed class of antidepressants, which work by blocking the reabsorption of serotonin into neurons and thus increasing serotonin levels in the brain. For CVD patients, they are preferred over other classes of antidepressants, namely tricyclic antidepressants, because tricyclics have a greater risk of cardiotoxicity and adverse cardiac events [70]. There may be sex differences in the pharmacokinetics and responses to antidepressants due to differences in estrogen levels, but this research is unclear and limited by an under-representation of women in clinical trials and lack of sex-specific analyses [71]. **Table 2** provides a summary of key randomized controlled trials (RCTs) of medication trials.

There have been several randomized, placebo-controlled, double-blind trials of SSRIs for depression treatment in patients with various types of CVD, which support the efficacy of antidepressant medications. For example, the first such trial, the SADHART, examined sertraline for post-ACS patients with MDD and found greater improvements in psychological symptoms at 24 weeks [72]. Another trial examining citalopram for patients with stable CAD also found greater improvements in depression symptoms at 12 weeks [73]. However, many studies have not shown benefit of antidepressants for cardiac outcomes. In a meta-analysis of six randomized trials of patients with cooccurring depression and CHD, SSRIs had significant benefits for depression symptoms but no benefit on mortality or CHD readmission rates [77]. For example, in the SADHART trial among post-ACS patients, there was no difference in mortality over 7-year follow-up between sertraline or placebo [77]. Kim and colleagues [75] conducted an 8-year follow-up of escitalopram for post-ACS patients and found that escitalopram resulted in fewer major adverse cardiac events, but only in terms of depression remission and not based on escitalopram specifically (i.e., reduced depression symptoms were protective, whether those reductions were due to escitalopram or other factors [75]). Thus, antidepressants are safe and beneficial for improving mood symptoms in CVD patients, but whether these benefits extend to improved cardiac outcomes remains unclear.

Behavioral Treatment

Behavioral interventions focus on teaching patients how to develop and use practical skills to cope effectively with negative thoughts and emotions and promote positive behaviors. These approaches aim to increase a patient's self-efficacy, empowerment, and active engagement in their physical and mental health care by providing concrete tools, usually within the context of a supportive and nurturing therapeutic relationship with a therapist or teacher or other peers. Two types of behavioral approaches commonly used for patients with CVD include cognitive-behavioral therapy (CBT) and mind-body medicine. **Table 3** provides a summary of meta-analyses in this area.

TABLE 2 Summary of RCTs of Antidepressant Medications for Patients With CVD

Authors, Year	Sample Size	Medication	Outcome	Results	Conclusion
Glassman et al., 2002 [72]	$N = 369$	Sertraline (Zoloft)	1. Left ventricular ejection fraction 2. Severe cardiovascular adverse events 3. Depression (CGI-I) 4. Depression (HAM-D)	1. Sertraline: baseline 54% (10%); week 16, 54% (11%); placebo: baseline, 52% (13%); week 16, 53% (13%); $p > 0.05$ 2. Sertraline: 14.5%, placebo: 22.4% 3. $p = 0.049$ 4. $p = 0.14$	Sertraline is a safe medication to treat depression in patients with MI or unstable angina
Lesperance et al., 2007 [73]	$N = 284$	Citalopram (Celexa)	1. Depression (HAM-D) 2. Remission rates 3. Depression (BDI-II) 4. Depression in clinical management vs. interpersonal psychotherapy (HAM-D) 5. Depression in clinical management vs. interpersonal psychotherapy (BDI-II)	1. MD = 3.3, 96.7% CI = 0.80–5.85, $p = 0.005$, effect size = 0.33 2. 35.9% vs. 22.5%, $p = 0.1$ 3. MD = 3.6, 98.3% CI = 0.58–6.64, $p = 0.005$, effect size = .33 4. MD = −2.26, 96.7% CI = −4.78 to 0.27, $p = 0.06$, effect size = 0.23 5. MD = 1.13, 98.3% CI = −1.90 to 4.16, $p = 0.37$, effect size 0.11	Citalopram administered with clinical management is effective for treating CAD patients with major depression
Kim et al., 2015 [74]	$N = 217$	Escitalopram (Lexapro)	1. QoL improvement in physical, social, and environmental domains 2. Health outcomes after 1 year	1. Physical QoL: effect size = 0.45; social QoL: effect size = 0.34; environmental QoL: effect size = 0.33 2. $p < 0.17$	Escitalopram improved quality of life in ACS patients, and effects remained after 1 year
Kim et al., 2018 [75]	$N = 300$	Escitalopram (Lexapro)	1. Major adverse cardiac events 2. All-cause mortality 3. Cardiac death 4. MI frequency	1. HR = 0.69, 95% CI = 0.49–0.96, $p = 0.03$ 2. HR = 0.82, 95% CI = 0.51–1.33, $p = −0.43$ 3. HR = 0.79, 95% CI = 0.41–1.52, $p = 0.48$ 4. HR = 0.54, 95% CI = 0.27–0.96, $p = 0.04$	Escitalopram reduced the risk of major adverse cardiac events in depression patients with acute coronary syndrome
Hong et al., 2007 [70]	$N = 91$	Mirtazapine (Remeron)	1. Depression (HAM-D) 2. Depression (BDI) 3. Depression (dSCL-90) 4. Depression (CGI)	1. $F = 2.86$, df = 1, $p = 0.09$ 2. $F = 5.51$, df = 1, $p = 0.02$ 3. $F = 6.48$, df = 1, $p = 0.01$ 4. $F = 6.67$, df = 1, $p = 0.012$	Mirtazapine is safe and effective for treating depression in post-MI patients
Angermann et al., 2016 [76]	$N = 372$	Escitalopram (Lexapro)	1. Death or hospitalization 2. Depression	1. HR = 0.99, 95% CI = 0.76–1.27, $p = 0.92$ 2. MD = −0.9, 95% CI = −2.6 to 0.7, $p = 0.26$	Escitalopram is not an effective medication for patients with chronic systolic heart failure and depression

ACS, acute coronary syndrome; BDI-II, Beck Depression Inventory II; CAD, coronary artery disease; CGI, Clinical Global Impression Scale; CGI-I, Clinical Global Impression Scale Improvement Scale; CI, confidence interval; dSCL-90, Symptom Check List 90 Items; HAM-D, Hamilton Depression Rating Scale; HR, hazard ratio; MD, mean difference; MI, myocardial infarction; QoL, quality of life.

TABLE 3 Summary of Meta-Analyses of Behavioral Treatments (CBT and Mind-Body Medicine) for Patients With CVD

Reference	Number of Trials	Description of Included Trials	Psychosocial Outcome	Results	Conclusion
Abbott, 2014 [78]	8 trials (578 participants)	RCTs of MBSR and MBCT for patients with vascular disease. Two studies with active control groups	1. Stress 2. Depression 3. Anxiety 4. Systolic blood pressure 5. Diastolic blood pressure	1. SMD = −0.38, 95% CI = −0.67 to −0.09, $p = 0.01$ 2. SMD = −0.35, 95% CI = −0.53 to −0.16, $p < 0.001$ 3. SMD = −0.50, 95% = −0.70 to −0.29, $p < 0.001$ 4. SMD = −0.78, 95% = −1.46 to −0.09, $p = 0.03$ 5. SMD = −0.67, 95% = −1.26 to −0.08, $p = 0.03$	MBSR and MBCT may improve psychosocial health for patients with CVD
Cramer, 2014 [79]	44 trials (3168 participants)	RCTs of yoga for healthy people, patients with CVD risk, or type 2 diabetes. 18 studies with active control groups	1. Systolic blood pressure 2. Diastolic blood pressure 3. Heart rate 4. Respiratory rate	1. MD = −5.85 mmHg 2. MD = −4.12 mmHg 3. MD = −6.59 bpm 4. MD = −0.93 breaths/min	Yoga improves CVD risk factors
Jeyanantham, 2017 [80]	6 trials (320 participants)	RCTs and observational studies of CBT for cooccurring heart failure and depression. One study with active control group	1. Depression immediately following intervention 2. Depression at 3 months follow-up 3. Quality of life 4. All-cause mortality	1. SMD = −0.34, 95% CI = −0.60 to −0.08, $p = 0.01$, $I^2 = 0\%$ 2. SMD = −0.32, 95% CI = −0.59 to −0.04, $p = 0.03$, $I^2 = 0\%$ 3. SMD = −0.25, 95% CI = −0.68 to 0.18, $p = 0.26$, $I^2 = 52\%$ 4. RR = 0.99, 95% CI = 0.75–1.32, $p = 0.96$, $I^2 = 0\%$	CBT was associated with improvements in depression but not quality of life or mortality
Liu, 2018 [81]	13 trials (972 participants)	RCTs of tai chi for patients with CHD. Two studies with active control groups	1. Aerobic endurance 2. Anxiety 3. Depression 4. Quality of life 5. HDL-c	1. SMD = 1.12, 95% CI = 0.58–1.66, $p < 0.001$, $I^2 = 83\%$, $N = 434$ 2. MD = −9.28, 95% CI = −17.46 to −1.10, $p = 0.03$, $I^2 = 95\%$, $N = 168$ 3. MD = −9.42, 95% CI = −13.59 to −5.26, $p < 0.001$, $I^2 = 81\%$, $N = 168$ 4. SMD = 0.73, 95% CI = 0.39 to 1.08, $p < 0.001$, $I^2 = 56\%$, $n = 352$	Tai chi improves aerobic endurance and psychological well-being in CHD patients

Reavell, 2018 [82]	12 trials (2254 participants with depression, 605 patients with anxiety)	RCTs of CBT for patients with co-occurring depression or anxiety and CVD One study with active controls	1. Depression 2. Participants remaining depressed 3. Anxiety	1. SMD = −0.35, 95% CI = −0.52 to −0.17, $p < 0.001$, $I^2 = 59\%$ 2. OR = 0.29, 95% CI = 0.12 to 0.69, $p = 0.005$, $I^2 = 62\%$ 3. SMD = −0.34, 95% CI = −0.65 to −0.03, $p = 0.03$, $I^2 = 71\%$	CBT improves depression and anxiety in patients with CVD
Richards, 2017 [83]	35 trials (10,703 participants)	Cochrane review of RCTs of psychological therapies for patients with CHD No studies with active control	1. Total mortality 2. Depression 3. Anxiety 4. Stress	1. RR = 0.90, 95% CI = 0.77–1.05, $I^2 = 2\%$ 2. SMD = −0.27, 95% CI = −0.39 to −0.15, $I^2 = 69\%$ 3. SMD = −0.24, 95% CI = −0.38 to −0.09, $I^2 = 47\%$ 4. SMD = −0.56, 95% CI = −0.88 to −0.24, $I^2 = 86\%$	Psychological treatments are helpful for reducing depression and anxiety symptoms in CHD patients
Scott-Sheldon, 2020 [84]	16 trials (1476 participants)	RCTs and non-RCTs of mindfulness interventions for patients with CVD Five studies with active control	1. Systolic blood pressure 2. Diastolic blood pressure 3. Psychological outcomes (anxiety, depression, distress, and perceived stress)	1. $d_+ = 0.89$, 95% CI = 0.26–1.51, $p < 0.001$ 2. $d_+ = 0.07$, 95% CI = −0.47 to 0.60, $p = 0.601$ 3. $d = 0.49$–0.64	Mindfulness-based interventions improve physiological and psychological outcomes in patients with CVD
Younge, 2015 [85]	11 trials	RCTs of mind-body interventions for patients with CHD Seven studies with active control	1. Physical quality of life 2. Mental quality of life 3. Depression 4. Anxiety 5. Systolic blood pressure 6. Diastolic blood pressure 7. Resting heart rate	1. $d = 0.45$, 95% CI = 0.18–0.72 2. $d = 0.68$, 95% CI = 0.10–1.26 3. $d = 0.61$, 95% CI = 0.23–0.99 4. $d = 0.52$, 95% CI = 0.26–0.78 5. $d = 0.48$, 95% CI = 0.27–0.69 6. $d = 0.36$, 95% CI = 0.15–0.57 7. $d = 0.15$, 95% CI = −0.08 to 0.39	Mind-body practices can improve some psychosocial and physical symptoms

Note. Treatment as usual was not considered an active control in our summary of studies. CBT, cognitive-behavioral therapy; CHD, coronary heart disease; CI, confidence interval; CVD, cardiovascular disease; HDL-c, high-density lipoprotein cholesterol; MBCT, mindfulness-based cognitive therapy; MBSR, mindfulness-based stress reduction; MD, mean difference; OR, odds ratio; RCTs, randomized control trials; RR, risk ratio; SMD, standardized mean difference.

Cognitive-Behavioral Therapy

CBT is the gold-standard approach in treating depression and anxiety. CBT is a short-term treatment, usually anywhere from 8 to 24 weekly sessions (50 min each), during which patients actively work to learn and apply new coping skills for working with negative thoughts, emotions, and behaviors. Home practice in-between sessions is assigned to promote uptake of the skills. Patients are often first educated about the interconnections between thoughts, emotions, and behaviors. Interventions for working effectively with thoughts focus on learning to identify, challenge, and restructure negative thoughts to be more realistic. Patients are often asked to track their thoughts, consider the evidence for and against their thoughts, and replace negative thoughts with more positive ones. Behavioral interventions emphasize exposure exercises and behavioral activation. Exposure exercises are particularly relevant to anxiety disorders and involve teaching people how to slowly and systematically confront their fears while equipping them with skills to tolerate the temporary distress symptoms that arise. Behavioral activation is the gold-standard treatment for depression, which involves helping patients plan and engage in mood-enhancing pleasurable activities and activities that promote a sense of mastery (e.g., feelings of satisfaction that come from completing tasks). Multiple meta-analyses demonstrate the efficacy and superiority of CBT over other forms of psychotherapy for treating depression ($g = 0.87$ [86]) and anxiety ($g = 0.45–0.61$ [87, 88]).

CBT shows benefit for treating depression and anxiety in patients with CVD. The AHA has endorsed CBT as one option for depression treatment, alongside antidepressant medications and physical activity [64]. However, effect sizes of CBT are relatively small for patients with CVD as compared to medically healthy patients and the effects on cardiac outcomes are mixed.

A seminal RCT was the ENRICHD trial of CBT for almost 2500 post-MI patients with depression [89]. Patients were randomized to receive standard care or 11 sessions of CBT, with SSRIs for the subset of patients with particularly elevated symptoms or minimal improvement after 5 weeks of CBT. Results indicated greater improvement in depression symptoms following CBT but no significant difference in cardiac mortality [89].

A recent Cochrane review of 35 RCTs of psychosocial interventions for patients with CVD found statistically significant but small improvements in depressive symptoms [standardized mean difference (SMD = -0.27)] and anxiety symptoms (SMD = -0.24) and a 21% reduced risk of cardiac mortality, but no difference in all-cause mortality or revascularization [83]. Wells and colleagues [90] explored the subgroup of CBT studies included in that Cochrane review

[83] and found similarly small effect sizes (SMD = 0.29 for anxiety, 0.24 for depression). A separate meta-analysis focused on CBT for mixed CVD patients with anxiety or depression [i.e., CHD, ACS, and atrial fibrillation (AF)] and again found benefits of small-medium effect sizes (SMD = -0.35 for depression and SMD = -0.34 for anxiety [82]). Results were similar in a meta-analysis of six trials of CBT for patients with HF and depression (SMD = -0.35). In studies of patients with or at risk for CVD, CBT has shown benefits for improving health behaviors such as sleep [91], diet [92], physical activity [93], and smoking cessation [94]. These studies did not explore possible sex differences, though the study of CBT and diet was conducted among women.

Positive psychology interventions also take a cognitive-behavioral approach to improve emotional and behavioral health outcomes. Whereas CBT focuses on decreasing negative emotions, positive psychology interventions aim to increase positive emotions, such as happiness, optimism, and purpose in life [95]. Examples of positive psychology techniques include imagining and writing about a better future, recalling positive life events, practicing gratitude, using personal strengths, and performing acts of kindness. Although the research is limited, positive psychology interventions have shown promising benefits for psychological well-being, functional performance, and health behaviors in patients with CVD risk factors and existing CVD [95]. Further research is needed to establish the effects of positive psychological interventions on CVD outcomes [95].

Mind-Body Medicine

Another behavioral management approach involves mind-body medicine interventions. Mind-body medicine is an evidence-based healing approach that uses a variety of different techniques to unite the body and mind as a way to promote well-being. Women are more likely than men to use mind-body therapies (24% vs. 14%, respectively [96]). Mind-body interventions include relaxation, meditation, yoga, tai chi, and qi gong. Meditation interventions, particularly mindfulness meditation, have received significant attention over the past several years. Mindfulness meditation involves paying attention to the present moment in an open and nonjudgmental way, and thus incorporates elements of concentration, attention regulation, and emotional acceptance, as a way to promote self-regulation and healthy behavioral responses. Yoga, tai chi, and qi gong are often described as movement-based meditation practices, incorporating the cognitive and emotional components of mindfulness and combining them with various forms of physical activity. For example, different styles of yoga emphasize gentle stretching or invigorating physical exercises in combination with mindful awareness, and tai chi integrates slow and gentle flowing movements with mindful awareness. One common physiological pathway underlying mind-body interventions involves the elicitation of the relaxation response—that is, elicitation of a state of parasympathetic dominance and deactivation of the sympathetic nervous system, thus countering the stress response and resulting in physiological benefits including decreased heart rate, inflammation, and cortisol. A large and growing body of literature supports the potential for mind-body interventions to promote physical and emotional well-being across a variety of patient populations.

In 2002, 17% of patients with CVD in a nationally representative sample reported using mind-body therapies [97]. Since then, a systematic review of seven studies among patients with CVD found mind-body use rates ranging from 2% to 57%, with the most common techniques being meditation and deep breathing [98]. In 2018, the AHA released a scientific statement on the potential benefits of meditation for reducing CVD risk factors [99]. Focusing specifically on seated, nonmovement-based meditation practices to control for confounding effects of physical activity, the authors concluded that there is a possible benefit of meditation for CVD risk reduction including improvement of psychological risk factors (e.g., stress, anxiety, and depression), smoking cessation, and possibly atherosclerosis [99]. The authors noted that due to limited follow-up time, there are more data to support the reduction of risk factors and psychological factors than hard end points of cardiac incidents or cardiac death, and that more methodologically rigorous studies are needed [99]. However, given the generally low costs and low risks, the AHA statement suggested that meditation may be considered as an adjunct to existing American College of Cardiology/AHA risk reduction guidelines [99]. Interestingly, while the AHA focused on seated meditations, in fact, the integration of physical activity and meditation may be a unique advantage of movement-based practices such as yoga and tai chi. By using slow or gentle movements, these practices may be an attractive and accessible gateway to physical activity for patients with CVD, helping patients to get moving and increase self-efficacy for exercise while also providing the cognitive and emotional benefits of meditation.

Recent meta-analyses have summarized the effects of mind-body interventions for patients at risk for CVD. Abbott and colleagues [78] conducted a meta-analysis of two popular mindfulness-based interventions, mindfulness-based stress reduction and mindfulness-based cognitive therapy for patients with vascular diseases [e.g., hypertension (HTN) and diabetes]. These are both manualized mindfulness interventions consisting of 8 weekly group sessions focused on mindfulness training, but they differ in that mindfulness-based cognitive therapy also incorporates

an explicit CBT framework and techniques. In their meta-analysis of eight trials, they found small but significant benefits for stress and depression, and moderate effects for anxiety and blood pressure. In a meta-analysis of 44 RCTs of yoga for patients at risk for CVD (e.g., patients with HTN and type 2 diabetes), yoga was associated with greater improvements in blood pressure, heart rate, respiration rate, abdominal obesity, blood lipid levels, and measures of insulin resistance as compared to usual care, though studies were generally of low methodological quality [79]. Mind-body interventions have also been shown to improve health behaviors such as smoking cessation [100], physical activity [101], and diet [102]. Some studies explored safety and there were no serious adverse events reported [79].

Other meta-analyses have explored mind-body interventions for patients with existing CVD. A very recent meta-analysis looking specifically at mindfulness-based interventions for adults with CVD (primarily CHD, HTN, and HF) found medium-large improvements in psychological outcomes as well as in systolic but not diastolic blood pressure [84]. These effects were only found immediately postintervention (typically a 9-week group program) and were not found to persist at follow-up assessments [84]. Younge and colleagues [85] conducted a meta-analysis of a broader range of mind-body interventions, including mindfulness meditation, transcendental meditation, and relaxation techniques, for patients with cardiac disease, primarily HF, CAD, and MI. Across 13 RCTs, they observed statistically significant medium effect sizes for QoL, depression, anxiety, blood pressure, and exercise tolerance (VO_2 max and 6-min walk test), and small but significant effects for resting heart rate. However, the studies were overall of low quality. In a systematic review of nine trials (six RCTs) of relaxation interventions for patients with HF, most studies were of moderate quality and showed some benefit for reducing dyspnea and improving sleep as compared to both usual care and attention control [103]. In a meta-analysis among trials of tai chi ($N = 13$) for patients with CHD, tai chi showed a significant improvement for aerobic endurance (VO_2 max), anxiety symptoms, depression symptoms, and QoL as compared to nonactive controls, with some suggestion for greater benefit than walking and stretching exercises in terms of aerobic endurance [81].

Taken together, mind-body interventions show promising benefits for psychological and physical benefits in patients with or at risk for CVD, though in many areas the quality of the trials has been limited and more randomized trials are needed. Another area for future research is to explore the relative efficacy of CBT and mind-body approaches for patients with CVD, as well as the efficacy of approaches that combine CBT and mind-body techniques. For example, mindfulness-based cognitive therapy uses a

CBT framework and integrates mindfulness meditation training to treat depressive symptoms. A recent qualitative study in CVD patients suggests that an approach such as mindfulness-based cognitive therapy may be more relevant to CVD patients than a traditional CBT approach [90]. In a mindfulness-based approach, patients learn to notice and release unhelpful thoughts, rather than attempt to directly change what they are thinking as done in traditional CBT. This approach of noticing and releasing may be most useful to patients with CVD for whom negative thoughts may be fairly accurate and thus difficult to change (e.g., thoughts of having a recurrent cardiac event [90]). It also remains to be seen if mind-body interventions have any unique advantage for physical health outcomes in CVD as a result of directly targeting stress-related physiology (i.e., elicitation of the relaxation response and deactivation of the sympathetic nervous system) or impacting inflammatory or immune pathways. Lastly, CBT and mind-body interventions can both be applied using electronic health approaches, which can expand the reach and accessibility for patients, highlighting the importance of ongoing work in this area [104].

Key Points

1. Psychosocial problems including depression, anxiety, stress, hostility, low SES, and social isolation are common in patients with CVD and increase the risk of CVD mortality, often to a similar extent as traditional CVD risk factors (e.g., smoking and physical inactivity).

2. Psychological symptoms do not need to meet formal diagnostic criteria for a mental health disorder to substantially impact CVD. Effects of psychological symptoms on CVD are driven by both behavioral and biological mechanisms.

3. Women are more likely to experience elevated symptoms of depression, anxiety, and stress, and more likely to have low SES, as compared to men, placing them at greater risk for poor CVD outcomes. The negative effects of low SES on CVD, and the beneficial effects of social integration on CVD, may be stronger for women as compared to men. These sex differences should be considered in making treatment recommendations.

4. When there is a clear path to treatment referrals, screening for depression and anxiety can be done efficiently and accurately using the PHQ-4, a validated four-item measure that combines the PHQ-2 and GAD-2. Depression screening is recommended by the AHA.

5. Cardiologists should consider referrals to mental health providers for management of psychosocial issues and ideally may have mental health providers colocated in cardiology clinics, utilize consultant-liaison services, or consider collaborative care treatment approaches.

6. Cardiologists may consider a combination of SSRIs, cognitive-behavioral therapy, and mind-body medicine approaches such as meditation, yoga, and tai chi, which have evidence of efficacy for mood management in patients with CVD.

Back to Clinical Case

The case describes a 38-year-old female with spontaneous coronary artery dissection suffering from symptoms of anxiety and stress, which are significantly interfering with her day-to-day activities. She was referred for psychotherapy. She completed eight sessions of mindfulness-based cognitive therapy. Using psychoeducation from CBT theory, Katie learned how her worries about her health led to greater stress, sadness, and anxiety, and how these emotions contributed to unhealthy behaviors (e.g., emotional eating) as well as problems in her relationships (e.g., not being fully present in interactions with others, lashing out at her partner). Through mindfulness meditation practices, she developed the ability to notice her thoughts as they happened in real time, identify negative thoughts as being unhelpful, and disengage from the thoughts by redirecting her attention back to other aspects of the present moment.

Katie also benefited from mindfulness and CBT skills to improve behavioral and interpersonal treatment goals. She used short mindfulness practices to ground and calm herself during moments of stress so that she no longer lashed out at others. On her own, Katie realized she could integrate her blood pressure monitoring with her mindfulness practices. She enjoyed seeing that her blood pressure decreased after a mindfulness exercise, and seeing the numbers drop was reinforcing and motivated her to continue implementing these skills. Katie also scheduled time before work to take a mindful walk with her partner most days, which gave her a sense of comfort and self-efficacy about her physical health while also adding quality time into her relationship. Her improved health behaviors of reduced emotional eating and increased physical activity helped to further reduce her fear of another SCAD. In terms of work stress, Katie problem-solved and implemented strategies to set boundaries by learning to acknowledge when her calendar was full, say no when she was unavailable, and set appropriate timelines, which helped her to feel less overwhelmed.

At the end of treatment, Katie reported decreased anxiety and worry about her health, decreased work stress, and improvements in her relationships. She reported being more confident in her ability to take care of herself and that she planned to continue mindfulness practices and regular walks with her partner.

Acknowledgment

The authors would like to acknowledge and thank Ms Amy Wang for her assistance with this chapter.

References

[1] Anon. Cardiovascular diseases (CVDs). Available at: https://www.who.int/news-room/fact-sheets/detail/cardiovascular-diseases-(cvds; 2020. [Accessed 8 April 2020].

[2] Anon. Mental health by the numbers | NAMI: National alliance on mental illness. Available at: https://www.nami.org/learn-more/mental-health-by-the-numbers; 2020. [Accessed 8 April 2020].

[3] American Psychiatric Association. Diagnostic and statistical manual of mental disorders: diagnostic and statistical manual of mental disorders. 5th ed. Arlington, VA: American Psychiatric Association; 2013.

[4] Anon. Depression. Available at: https://www.who.int/news-room/fact-sheets/detail/depression; 2020. [Accessed 8 April 2020].

[5] Hasin DS, Sarvet AL, Meyers JL, Saha TD, Ruan WJ, Stohl M, et al. Epidemiology of adult DSM-5 major depressive disorder and its specifiers in the United States. JAMA Psychiatry 2018;75:336–46.

[6] Rodríguez MR, Nuevo R, Chatterji S, Ayuso-Mateos JL. Definitions and factors associated with subthreshold depressive conditions: a systematic review. BMC Psychiatry 2012;12:181.

[7] Anon. Facts & statistics | Anxiety and depression association of America, ADAA. Available at: https://adaa.org/about-adaa/press-room/facts-statistics; 2020. [Accessed 8 April 2020].

[8] Ruscio AM, Hallion LS, Lim CCW, Aguilar-Gaxiola S, Al-Hamzawi A, Alonso J, et al. Cross-sectional comparison of the epidemiology of DSM-5 generalized anxiety disorder across the globe. JAMA Psychiatry 2017;74:465–75.

[9] Haller H, Cramer H, Lauche R, Gass F, Dobos GJ. The prevalence and burden of subthreshold generalized anxiety disorder: a systematic review. BMC Psychiatry 2014;14:128.

[10] Anon. NIMH any anxiety disorder. Available at: https://www.nimh.nih.gov/health/statistics/any-anxiety-disorder.shtml; 2020. [Accessed 8 April 2020].

[11] Anon. WHO | Gender and women's mental health. WHO; 2020. Available at: https://www.who.int/mental_health/prevention/genderwomen/en/. [Accessed 8 April 2020].

[12] Piccinelli M, Wilkinson G. Gender differences in depression: critical review. Br J Psychiatry 2000;177:486–92.

[13] Altemus M, Sarvaiya N, Neill Epperson C. Sex differences in anxiety and depression clinical perspectives. Front Neuroendocrinol 2014;35:320–30.

[14] Mattina GF, Van Lieshout RJ, Steiner M. Inflammation, depression and cardiovascular disease in women: the role of the immune system across critical reproductive events. Ther Adv Cardiovasc Dis 2019;13. 1753944719851950.

[15] Huffman JC, Celano CM, Beach SR, Motiwala SR, Januzzi JL. Depression and cardiac disease: epidemiology, mechanisms, and diagnosis. Cardiovasc Psychiatry Neurol 2013;2013:1–14.

[16] Liang JJ, Tweet MS, Hayes SE, Gulati R, Hayes SN. Prevalence and predictors of depression and anxiety among survivors of myocardial infarction due to spontaneous coronary artery dissection. J Cardiopulm Rehabil Prev 2014;34:138–42.

[17] Correll CU, Solmi M, Veronese N, Bortolato B, Rosson S, Santonastaso P, et al. Prevalence, incidence and mortality from cardiovascular disease in patients with pooled and specific severe mental illness: a large-scale meta-analysis of 3,211,768 patients and 113,383,368 controls. World Psychiatry 2017;16:163–80.

[18] Lichtman JH, Froelicher ES, Blumenthal JA, Carney RM, Doering LV, Frasure-Smith N, et al. Depression as a risk factor for poor prognosis among patients with acute coronary syndrome: systematic review and recommendations: a scientific statement from the American heart association. Circulation 2014;129:1350–69.

[19] Meijer A, Conradi HJ, Bos EH, Anselmino M, Carney RM, Denollet J, et al. Adjusted prognostic association of depression following myocardial infarction with mortality and cardiovascular events: individual patient data meta-analysis. Br J Psychiatry 2013;203:90–102.

[20] Worcester MU, Goble AJ, Elliott PC, Froelicher ES, Murphy BM, Beauchamp AJ, et al. Mild depression predicts long-term mortality after acute myocardial infarction: a 25-year follow-up. Heart Lung Circ 2019;28:1812–8.

[21] Celano CM, Daunis DJ, Lokko HN, Campbell KA, Huffman JC. Anxiety disorders and cardiovascular disease. Curr Psychiatry Rep 2016;18:101.

[22] Batelaan NM, Seldenrijk A, Bot M, van Balkom AJLM, Penninx BWJH. Anxiety and new onset of cardiovascular disease: critical review and meta-analysis. Br J Psychiatry 2016;208:223–31.

[23] Emdin CA, Odutayo A, Wong CX, Tran J, Hsiao AJ, Hunn BHM. Meta-analysis of anxiety as a risk factor for cardiovascular disease. Am J Cardiol 2016;118:511–9.

[24] Campbell KA, Madva EN, Villegas AC, Beale EE, Beach SR, Wasfy JH, et al. Non-cardiac chest pain: a review for the consultation-liaison psychiatrist. Psychosomatics 2017;58:252–65.

[25] Eslick GD, Talley NJ. Non-cardiac chest pain: predictors of health care seeking, the types of health care professional consulted, work absenteeism and interruption of daily activities. Aliment Pharmacol Ther 2004;20:909–15.

[26] Shanmugasegaram S, Russell KL, Kovacs AH, Stewart DE, Grace SL. Gender and sex differences in prevalence of major depression in coronary artery disease patients: a meta-analysis. Maturitas 2012;73:305–11.

[27] Serpytis P, Navickas P, Lukaviciute L, Navickas A, Aranauskas R, Serpytis R, et al. Gender-based differences in anxiety and depression following acute myocardial infarction. Arq Bras Cardiol 2018;111:676–83.

[28] Gan Y, Gong Y, Tong X, Sun H, Cong Y, Dong X, et al. Depression and the risk of coronary heart disease: a meta-analysis of prospective cohort studies. BMC Psychiatry 2014;14:371.

[29] Kivimäki M, Jokela M, Nyberg ST, Singh-Manoux A, Fransson EI, Alfredsson L, et al. Long working hours and risk of coronary heart disease and stroke: a systematic review and meta-analysis of published and unpublished data for 603 838 individuals. Lancet 2015;386:1739–46.

[30] Chida Y, Steptoe A. The association of anger and hostility with future coronary heart disease: a meta-analytic review of prospective evidence. J Am Coll Cardiol 2009;53:936–46.

[31] Mostofsky E, Penner EA, Mittleman MA. Outbursts of anger as a trigger of acute cardiovascular events: a systematic review and meta-analysis. Eur Heart J 2014;7.

[32] Backholer K, Peters SAE, Bots SH, Peeters A, Huxley RR, Woodward M. Sex differences in the relationship between socioeconomic status and cardiovascular disease: a systematic review and meta-analysis. J Epidemiol Community Health 2017;71:550–7.

[33] Valtorta NK, Kanaan M, Gilbody S, Ronzi S, Hanratty B. Loneliness and social isolation as risk factors for coronary heart disease and stroke: systematic review and meta-analysis of longitudinal observational studies. Heart 2016;102:1009–16.

[34] Lazarus RS, Folkman S. Stress, appraisal, and coping. Springer Publishing Company; 1984.

[35] Anon. Gender and stress, https://www.apa.org. Available at: https://www.apa.org/news/press/releases/stress/2010/gender-stress; 2020. [Accessed 8 April 2020].

[36] Steptoe A, Kivimäki M. Stress and cardiovascular disease: an update on current knowledge. Annu Rev Public Health 2013;34:337–54.

[37] Batty GD, Russ TC, Stamatakis E, Kivimäki M. Psychological distress and risk of peripheral vascular disease, abdominal aortic aneurysm, and heart failure: pooling of sixteen cohort studies. Atherosclerosis 2014;236:385–8.

[38] Templin C, Ghadri JR, Diekmann J, Napp LC, Bataiosu DR, Jaguszewski M, et al. Clinical features and outcomes of Takotsubo (stress) cardiomyopathy. N Engl J Med 2015;373:929–38.

[39] Busch LY, Pössel P, Valentine JC. Meta-analyses of cardiovascular reactivity to rumination: a possible mechanism linking depression and hostility to cardiovascular disease. Psychol Bull 2017;143:1378–94.

[40] Archer J. Sex differences in aggression in real-world settings: a meta-analytic review. Rev Gen Psychol 2004;8:291–322.

[41] Wong JM, Sin NL, Whooley MA. A comparison of cook-medley hostility subscales and mortality in patients with coronary heart disease: data from the heart and soul study. Psychosom Med 2014;76:311–7.

[42] Keith F, Krantz DS, Chen R, Harris KM, Ware CM, Lee AK, et al. Anger, hostility, and hospitalizations in patients with heart failure. Health Psychol 2017;36:829–38.

[43] Havranek EP, Mujahid MS, Barr DA, Blair IV, Cohen MS, Cruz-Flores S, et al. Social determinants of risk and outcomes for cardiovascular disease: a scientific statement from the American heart association. Circulation 2015;132:873–98.

[44] Min Y-I, Anugu P, Butler KR, Hartley TA, Mwasongwe S, Norwood AF, et al. Cardiovascular disease burden and socioeconomic correlates: findings from the Jackson heart study. J Am Heart Assoc 2017;6:e004416.

[45] Tawakol A, Osborne MT, Wang Y, Hammed B, Hanje J, Durkalski V, et al. Stress-associated neurobiological pathway linking socioeconomic disparities to cardiovascular disease. J Am Coll Cardiol 2019;73:3243–55.

[46] Chang SC, Glymour M, Cornelis M, Walter S, Rimm EB, Tchetgen Tchetgen E, et al. Social integration and reduced risk of coronary heart disease in women. Circ Res 2017;120:1927–37.

[47] Townsend P. Isolation, desolation, and loneliness. In: Shanas E, Townsend P, Wedderburn D, Friis H, Milhoy P, Stehouwer J, editors. Old people in three industrial societies. New York, NY: Atherton Press; 1968. p. 258–87.

[48] Barth J, Schneider S, von Känel R. Lack of social support in the etiology and the prognosis of coronary heart disease: a systematic review and meta-analysis. Psychosom Med 2010;72:229–38.

[49] Chin B, Cohen S. Review of the association between number of social roles and cardiovascular disease: graded or threshold effect?. Psychosom Med 2020;82(5):471–86.

[50] Penninx BWJH. Depression and cardiovascular disease: epidemiological evidence on their linking mechanisms. Neurosci Biobehav Rev 2017;74:277–86.

[51] Chauvet-Gelinier J-C, Bonin B. Stress, anxiety and depression in heart disease patients: a major challenge for cardiac rehabilitation. Ann Phys Rehabil Med 2017;60:6–12.

[52] Muotri RW, Bernik MA. Panic disorder and exercise avoidance. Rev Bras Psiquiatr 2014;36:68–75.

[53] Appleton KM, Woodside JV, Arveiler D, Haas B, Amouyel P, Montaye M, et al. A role for behavior in the relationships between depression and hostility and cardiovascular disease incidence, mortality, and all-cause mortality: the prime study. Ann Behav Med 2016;50:582–91.

[54] Sumner JA, Khodneva Y, Muntner P, Redmond N, Lewis MW, Davidson KW, et al. Effects of concurrent depressive symptoms and perceived stress on cardiovascular risk in low- and high-income participants: findings from the reasons for geographical and racial differences in stroke (REGARDS) study. J Am Heart Ass 2016;5:e003930.

[55] Messay B, Lim A, Marsland AL. Current understanding of the bi-directional relationship of major depression with inflammation. Biol Mood Anxiety Disord 2012;2:4.

[56] Vogelzangs N, Beekman ATF, de Jonge P, Penninx BWJH. Anxiety disorders and inflammation in a large adult cohort. Transl Psychiatry 2013;3:e249.

[57] Tawakol A, Ishai A, Takx RA, Figueroa AL, Ali A, Kaiser Y, et al. Relation between resting amygdalar activity and cardiovascular events: a longitudinal and cohort study. Lancet 2017;389:834–45.

[58] Häupl T, Zimmermann M, Kalus U, Yürek S, Koscielny J, Hoppe B. Angiotensin converting enzyme intron 16 insertion/deletion genotype is associated with plasma C-reactive protein concentration in uteroplacental dysfunction. J Renin-Angiotensin-Aldosterone Syst 2015;16:422–7.

[59] Giardina WJ, Ebert DM. Positive effects of captopril in the behavioral despair swim test. Biol Psychiatry 1989;25:697–702.

[60] Birur B, Amrock EM, Shelton RC, Li L. Sex differences in the peripheral immune system in patients with depression. Front Psychiatry 2017;8:108.

[61] Bethin KE, Vogt SK, Muglia LJ. Interleukin-6 is an essential, corticotropin-releasing hormone-independent stimulator of the adrenal axis during immune system activation. PNAS 2000;97:9317–22.

[62] Suarez EC, Sundy JS, Erkanli A. Depressogenic vulnerability and gender-specific patterns of neuro-immune dysregulation: what the ratio of cortisol to C-reactive protein can tell us about loss of normal regulatory control. Brain Behav Immun 2015;44:137–47.

[63] Rozanski A. Behavioral cardiology: current advances and future directions. J Am Coll Cardiol 2014;64(1):100–10.

[64] Lichtman JH, Thomas Bigger J, Blumenthal James A, Frasure-Smith N, Kaufmann PG, Lespérance F, et al. Depression and coronary heart disease. Circulation 2008;118:1768–75.

[65] Jha MK, Qamar A, Vaduganathan M, Charney DS, Murrough JW. Screening and management of depression in patients with cardiovascular disease. J Am Coll Cardiol 2019;73:1827–45.

[66] Celano CM, Suarez L, Mastromauro C, Januzzi JL, Huffman JC. Feasibility and utility of screening for depression and anxiety disorders in patients with cardiovascular disease. Circ Cardiovasc Qual Outcomes 2013;6:498–504.

[67] Huffman JC, Celano CM. Depression in cardiovascular disease: from awareness to action. Trends Cardiovasc Med 2015;25:623–4.

[68] Huffman JC, Mastromauro CA, Beach SR, Celano CM, DuBois CM, Healy BC, et al. Collaborative care for depression and anxiety disorders in patients with recent cardiac events: the management of sadness and anxiety in cardiology (MOSAIC) randomized clinical trial. JAMA Intern Med 2014;174:927–35.

[69] Mavrides N, Nemeroff C. Treatment of depression in cardiovascular disease. Depress Anxiety 2013;30:328–41.

[70] Honig A, Kuyper AMG, Schene AH, van Melle JP, de Jonge P, Tulner DM, et al. Treatment of post-myocardial infarction depressive disorder: a randomized, placebo-controlled trial with mirtazapine. Psychosom Med 2007;69:606–13.

[71] Bigos KL, Pollock BG, Stankevich BA, Bies RR. Sex differences in the pharmacokinetics and pharmacodynamics of antidepressants: an updated review. Gender Med 2009;6:522–43.

[72] Glassman AH, O'Connor CM, Califf RM, Swedberg K, Schwartz P, Bigger JT, et al. Sertraline treatment of major depression in patients with acute MI or unstable angina. JAMA 2002;288:701–9.

[73] Lespérance F, Frasure-Smith N, Koszycki D, Laliberté M-A, van Zyl LT, Baker B, et al. Effects of citalopram and interpersonal psychotherapy on depression in patients with coronary artery disease: the Canadian cardiac randomized evaluation of antidepressant and psychotherapy efficacy (CREATE) trial. JAMA 2007;297:367–79.

[74] Kim J-M, Stewart R, Bae K-Y, Kang H-J, Kim S-W, Shin I-S, et al. Effects of depression co-morbidity and treatment on quality of life in patients with acute coronary syndrome: the Korean depression in ACS (K-DEPACS) and the escitalopram for depression in ACS (EsDEPACS) study. Psychol Med 2015;45:1641–52.

[75] Kim J-M, Stewart R, Yoon J-S. Cardiac outcomes after treatment for depression in patients with acute coronary syndrome—reply. JAMA 2018;320:2152–3.

[76] Angermann CE, Gelbrich G, Störk S, Gunold H, Edelmann F, Wachter R, et al. Effect of escitalopram on all-cause mortality and hospitalization in patients with heart failure and depression: the MOOD-HF randomized clinical trial. JAMA 2016;315:2683–93.

[77] Pizzi C, Rutjes AWS, Costa GM, Fontana F, Mezzetti A, Manzoli L. Meta-analysis of selective serotonin reuptake inhibitors in patients with depression and coronary heart disease. Am J Cardiol 2011;107:972–9.

[78] Abbott RA, Whear R, Rodgers LR, Bethel A, Thompson Coon J, Kuyken W, et al. Effectiveness of mindfulness-based stress reduction and mindfulness based cognitive therapy in vascular disease: a systematic review and meta-analysis of randomised controlled trials. J Psychosom Res 2014;76:341–51.

[79] Cramer H, Lauche R, Haller H, Steckhan N, Michalsen A, Dobos G. Effects of yoga on cardiovascular disease risk factors: a systematic review and meta-analysis. Int J Cardiol 2014;173:170–83.

[80] Jeyanantham K, Kotecha D, Thanki D, Dekker R, Lane DA. Effects of cognitive behavioural therapy for depression in heart failure patients: a systematic review and meta-analysis. Heart Fail Rev 2017;22:731–41.

[81] Liu T, Chan AW, Liu YH, Taylor-Piliae RE. Effects of Tai Chi-based cardiac rehabilitation on aerobic endurance, psychosocial well-being, and cardiovascular risk reduction among patients with coronary heart disease: a systematic review and meta-analysis. Eur J Cardiovasc Nurs 2018;17:368–83.

[82] Reavell J, Hopkinson M, Clarkesmith D, Lane DA. Effectiveness of cognitive behavioral therapy for depression and anxiety in patients with cardiovascular disease: a systematic review and meta-analysis. Psychosom Med 2018;80:742–53.

[83] Richards SH, Anderson L, Jenkinson CE, Whalley B, Rees K, Davies P, et al. Psychological interventions for coronary heart disease. Cochrane Database Syst Rev 2017;4.

[84] Scott-Sheldon LAJ, Gathright EC, Donahue ML, Balletto B, Feulner MM, DeCosta J, et al. Mindfulness-based interventions for adults with cardiovascular disease: a systematic review and meta-analysis. Ann Behav Med 2020;54:67–73.

[85] Younge JO, Gotink RA, Baena CP, Roos-Hesselink JW, Hunink MGM. Mind-body practices for patients with cardiac disease: a systematic review and meta-analysis. Eur J Prev Cardiol 2015;22:1385–98.

[86] Cuijpers P, van Straten A, Warmerdam L. Behavioral activation treatments of depression: a meta-analysis. Clin Psychol Rev 2007;27:318–26.

[87] Cuijpers P, Cristea IA, Karyotaki E, Reijnders M, Huibers MJH. How effective are cognitive behavior therapies for major depression and anxiety disorders? A meta-analytic update of the evidence. World Psychiatry 2016;15:245–58.

[88] Carpenter JK, Andrews LA, Witcraft SM, Powers MB, Smits JAJ, Hofmann SG. Cognitive behavioral therapy for anxiety and related disorders: a meta-analysis of randomized placebo-controlled trials. Depress Anxiety 2018;35:502–14.

[89] Berkman LF, Blumenthal J, Burg M, Carney RM, Catellier D, Cowan MJ, et al. Effects of treating depression and low perceived social support on clinical events after myocardial infarction: the enhancing recovery in coronary heart disease patients (ENRICHD) randomized trial. JAMA 2003;289:3106–16.

[90] McPhillips R, Salmon P, Wells A, Fisher P. Qualitative analysis of emotional distress in cardiac patients from the perspectives of cognitive behavioral and metacognitive theories: why might cognitive behavioral therapy have limited benefit, and might metacognitive therapy be more effective? Front Psychol 2019;9:2288.

[91] Conley S, Redeker NS. Cognitive behavioral therapy for insomnia in the context of cardiovascular conditions. Curr Sleep Med Rep 2015;1:157–65.

[92] Rapoport L, Clark M, Wardle J. Evaluation of a modified cognitive–behavioural programme for weight management. Int J Obes 2000;24:1726–37.

[93] Schneider JK, Mercer GT, Herning M, Smith CA, Prysak MD. Promoting exercise behavior in older adults: using a cognitive behavioral intervention. J Gerontol Nurs 2004;30:45–53.

[94] Stead LF, Lancaster T. Behavioural interventions as adjuncts to pharmacotherapy for smoking cessation. Cochrane Database Syst Rev 2012;12.

[95] Kubzansky LD, Huffman JC, Boehm JK, Hernandez R, Kim ES, Koga HK, et al. Positive psychological well-being and cardiovascular disease: JACC health promotion series. J Am Coll Cardiol 2018;72:1382–96.

[96] Barnes PM, Bloom B, Nahin RL. Complementary and alternative medicine use among adults and children: United States; 2007. National Health Statistics Reports 2008.

[97] Yeh GY, Davis RB, Phillips RS. Use of complementary therapies in patients with cardiovascular disease. Am J Cardiol 2006;98:673–80.

[98] Grant SJ, Bin YS, Kiat H, Chang DH-T. The use of complementary and alternative medicine by people with cardiovascular disease: a systematic review. BMC Public Health 2012;12:299.

[99] Levine GN, Lange RA, Bairey-Merz CN, Davidson RJ, Jamerson K, Mehta PK, et al. Meditation and cardiovascular risk reduction: a scientific statement from the American heart association. J Am Heart Assoc 2017;6.

[100] Brewer JA, Mallik S, Babuscio TA, Nich C, Johnson HE, Deleone CM, et al. Mindfulness training for smoking cessation: results from a randomized controlled trial. Drug Alcohol Depend 2011;119:72–80.

[101] Goodwin CL, Forman EM, Herbert JD, Butryn ML, Ledley GS. A pilot study examining the initial effectiveness of a brief acceptance-based behavior therapy for modifying diet and physical activity among cardiac patients. Behav Modif 2012;36:199–217.

[102] Katterman SN, Kleinman BM, Hood MM, Nackers LM, Corsica JA. Mindfulness meditation as an intervention for binge eating, emotional eating, and weight loss: a systematic review. Eat Behav 2014;15:197–204.

[103] Kwekkeboom KL, Bratzke LC. A systematic review of relaxation, meditation, and guided imagery strategies for symptom management in heart failure. J Cardiovasc Nurs 2016;31:457–68.

[104] Fish J, Brimson J, Lynch S. Mindfulness interventions delivered by technology without facilitator involvement: what research exists and what are the clinical outcomes? Mindfulness 2016;7:1011–23.

Editor's Summary: Psychosocial Determinants of CVD

Depression and anxiety more common in women
Annual prevalence
Depression: 26% women vs 15% men

Women face more bias & discrimination

Sex differences in coping styles

Social integration & isolation both important for women
Social integration may protect against CVD in men after a minimum number of social ties have been met. Women continue to benefit from additional relationship without a threshold.

Women report greater stress levels & more emotional symptoms compare to men

Women are at a socially disadvantaged position with less power and control

More women live in poverty & lower socio-economic status compared to men

Biological differences due to sex-hormones

Niti Aggarwal

Psychosocial Risk Factors for CVD in Women Compared to Men. Women are more often socioeconomically disadvantaged, experience more discrimination and psychological distress, and exhibit more emotional symptoms compared to men. CVD, cardiovascular disease. *Image courtesy of Niti R. Aggarwal.*

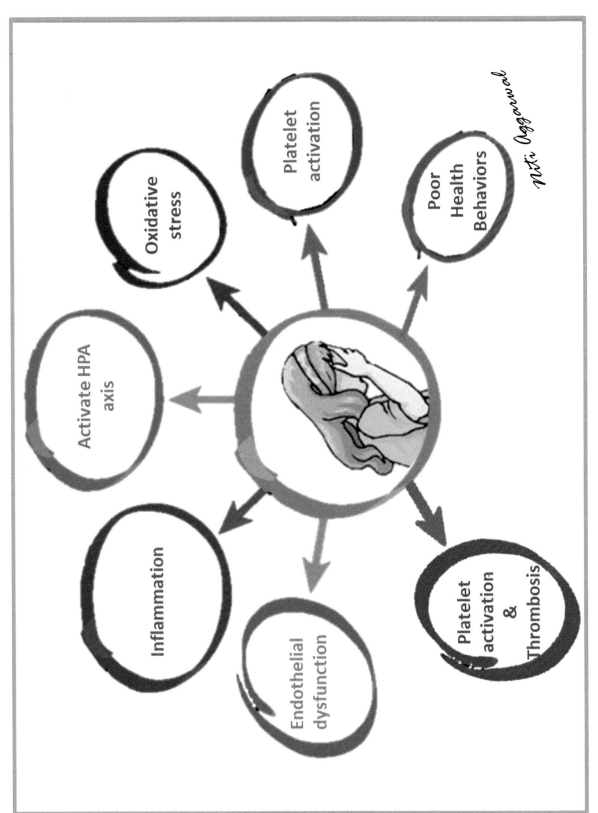

Mechanisms Linking Psychosocial Factors to CVD. Psychosocial distress results in biological changes such as increased inflammation and oxidative stress that ultimately result in increased risk of CVD (cardiovascular disease). HPA, hypothalamus pituitary adrenal. *Image courtesy of Niti R. Aggarwal.*

Section XII

Pharmacotherapy Considerations in Men and Women

Chapter 27

Cardiovascular Medications

Juan Tamargo, Niti R. Aggarwal, and María Tamargo

Clinical Case

A 69-year-old Caucasian woman was admitted in the emergency department after 1 week with fever, shortness of breath, malaise, dry cough, and gastroenteritis with 4-day history of diarrhea. Past medical history was significant for hypertension, dyslipidemia, paroxysmal atrial fibrillation, and depression. Her home medications included furosemide (20 mg), valsartan (80 mg), simvastatin (40 mg), sotalol (80 mg twice daily), warfarin, escitalopram (20 mg), and omeprazole (20 mg). She had also recently initiated treatment with solifenacin (10 mg) due to urinary incontinence. On admission, she was febrile (102 °F) and tachycardic (90 bpm), with oxygen saturations of 95% on room air. Chest X-ray confirmed an interstitial pneumonia and the physician prescribed ceftriaxone (1 g i.v.) and azithromycin (500 mg once daily for 3 days). An initial electrocardiogram (ECG) revealed a prolonged QTc interval of 560 ms. Serum potassium was 3.5 mmol/L and serum magnesium was 1.4 mg/dL. Renal, hepatic, and thyroid functions were normal. The echocardiogram showed a left ventricular hypertrophy consistent with hypertension, a left ventricular ejection fraction of 55%, a mildly dilated left atrium, and no significant valvular disease. About 24 h after initiating the antibiotic therapy, she experienced intermittent dizziness of short duration. A marked QT prolongation (QTc 650 ms) was observed on the telemetry, with premature ventricular complexes with prolonged compensatory pauses and an episode of torsades de pointes (TdP) (**Figure 1**). Azithromycin, citalopram, sotalol, and solifenacin were withdrawn. Magnesium sulfate and lidocaine were administered slowly intravenously and electrolyte disturbances were corrected. During the next 2 days, the QTc interval gradually returned to normal and no further arrhythmias were observed. She was discharged to home from hospital by day 4. Two weeks later she presented to the outpatient clinic for a follow-up. ECG demonstrated a QTc interval of 435 ms.

Continued

Sex Differences in Cardiac Disease. https://doi.org/10.1016/B978-0-12-819369-3.00020-4

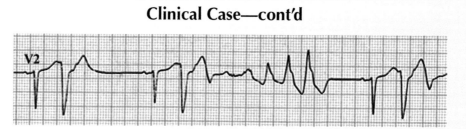

Clinical Case—cont'd

FIGURE 1   **Rhythm Strip in a Patient With Torsades de Pointes.** The rhythm strip of a 69-year-old female demonstrates two sinus beats with a prolonged QTc interval, the first one followed by a premature beat and the second one by several beats of torsades de pointes.

Abstract

Women and men exhibit significant differences in body composition and hormonal variations, and in drug pharmacokinetics and pharmacodynamics, such that they may respond differently to many cardiovascular drugs. However, the role of sex-related differences (SRDs) in clinical practice is not yet completely elucidated, probably because women are largely underrepresented in randomized clinical trials and sex-specific analysis usually is not included in the evaluation of clinical trials. Furthermore, women were prescribed more drugs, demonstrate less adherence to the prescribed medications, are less often treated with guideline-recommended drugs, and experienced more frequent and severe drug adverse reactions than men. But even when SRDs in cardiovascular pharmacology are well known, translation of these differences into clinical practice is slow. SRDs in drug efficacy and safety have to be considered at every step in the drug development to optimize medical therapy in both men and women. This chapter summarizes the most important SRDs in the pharmacokinetics, efficacy, and safety of the most frequently used cardiovascular drugs. A better understanding of SRDs in the efficacy and safety of cardiovascular drugs is the first step for developing a personalized drug therapy for the prevention and treatment of cardiovascular diseases, which represent the leading cause of morbidity and mortality among both men and women.

Introduction

Cardiovascular disease (CVD) represents the leading cause of morbidity and mortality among women in developed countries, and it kills twice as many women as all forms of cancer combined [1–6]. For decades, the risk of CVD has been underestimated in women because of the misperception that during the reproductive years they are "protected" against CVD. Therefore, CVD was commonly considered a public health problem solely for men. Consequently, women are less likely to receive the medication recommended in clinical guidelines for the prevention and treatment of CVD as compared to men [7]. However, the incidence of CVD increases rapidly after menopause, leaving women with untreated risk factors and vulnerable to development of CVD. Compared to men of the same age, women develop CVD 7–10 years later, but CVD represents the major cause of death in women older than 65 years of age [1, 2, 5, 6, 8, 9]. In addition, well-described differences between men and women, that is, SRDs, exist in the epidemiology, pathophysiology, clinical manifestations, diagnosis, drug prescription according to guideline recommendations, drug adherence, response (efficacy and safety) to therapy, and outcomes of CVD [1–7]. Women and men also differ in the anatomy and physiology of the cardiovascular system (body composition, hormonal changes during menstrual cycle, pregnancy, and menopause). There are also SRDs in the pharmacokinetics (i.e., drug absorption, distribution, metabolism, and excretion) and pharmacodynamics (i.e., the relationship between drug effect and drug concentration at the site of action) of some cardiovascular drugs [10–23]. Thus, it should not be a surprise that the efficacy and safety of some cardiovascular drugs can differ between men and women. For example, women experience adverse drug reactions (ADRs) more frequently and severely than men. Unfortunately, the clinical relevance of SRDs in the pharmacodynamics and pharmacokinetics of cardiovascular drugs is moderate or remains uncertain because women are largely underrepresented in randomized controlled trials (RCTs), which are not powered to ascertain such differences, and sex-specific analysis of drug efficacy and safety is usually not included in the evaluation of the results. As a result, the translation of

SRDs into daily clinical practice has been slow and clinical guidelines, mainly based on trials conducted predominantly on middle-aged men, generally provide sex-neutral recommendations for treatment.

Taking into account these issues, the aims of this chapter are to review the SRDs in the pharmacokinetics and pharmacodynamics of the commonly used cardiovascular drugs, elucidate the challenges for implementing sex-specific pharmacologic treatment, and propose recommendations to overcome these gaps.

Sex-Related Differences in Pharmacokinetics

SRDs in pharmacokinetics may arise from differences in body composition, physiological sex hormonal changes, and in drug absorption, distribution, metabolism, and excretion [14, 16–24] (**Table 1** and **Figure 2**). Women tend to have lower absorption rates than men, but SRDs in drug absorption, either orally or transdermally, have minor clinical relevance [10, 25]. Drug distribution depends on body composition, plasma volume, organ blood flow, and tissue and plasma protein binding [18, 20–22, 25]. Women have a higher percentage of body fat, but lower body weight/

size, muscle mass, organ sizes, plasma volume, cardiac output, organ blood flow, and volume of distribution (Vd) compared to men. This explains the faster onset, higher Vd, and greater duration of action of lipophilic drugs, while the Vd of hydrophilic drugs is smaller, reaching higher initial peak plasma levels (C_{max}) and having greater effects in women [11, 16, 17, 19, 22].

Drug elimination occurs via biotransformation and excretion. Hepatic biotransformation is reduced in women due to the lower cardiac output and hepatic blood flow and differences in drug-metabolizing enzymes and transporters [16, 17, 21, 22, 25, 67, 68]. Cytochrome P450 3A4 is responsible for the metabolism of more than 50% of therapeutic drugs and its activity is reduced in women [68–70]. CYP3A and the transporter P-glycoprotein (P-gp) present appreciable substrate overlap, so the increased clearance of CYP3A4 substrates in women might result from their lower P-gp activity [10, 17, 18, 69, 71]. Women have lower renal blood flow, glomerular filtration rate (GFR), and tubular secretion and/or reabsorption than men, and drugs primarily excreted unchanged in the urine are cleared more slowly in women [10, 13, 16, 17, 21, 22, 72, 73]. The main SRDs in the pharmacokinetics of cardiovascular drugs are summarized in **Table 2**.

Decreased drug absorption
- ⬇ Gastric acid secretion
- ⬇ GI motility and transit time

Pharmacodynamic differences in efficacy
& drug safety
⬆ Adverse drug reactions

Comorbidities
Women are older with more comorbidities,
so ⬆ sensitivity to drug effects

Poor adherence to prescribed medications

Polypharmacy
⬆ Consumption of drugs which increases
risk of drug interactions

Drug distribution
- ⬇ Plasma albumin and α1-acid glycoprotein
- Higher percentage of body fat
  ⬆ Vd for lipophillic drugs
  ⬇ Vd for hydrophillic drugs
- Smaller body size and plasma volume

Decreased drug metabolism and excretion
- ⬇ Activity of hepatic CYP450 enzymes & P-gp
- ⬇ Renal glomerular filtration rate

Variations in sex hormones affect drug pharmacokinetics

Underrepresented in clinical trials

Alignment with clinical guidelines
⬇ Likelihood to receive guideline-directed drugs

FIGURE 2 Sex-Related Differences in Drug Pharmacodynamics, Pharmacokinetics, and Drug Prescription. Figure depicts the drug pharmacodynamics, pharmacokinetics, and drug prescription effects seen in women compared to men. GI, gastrointestinal; P-gp, P-glycoprotein; Vd, volume of distribution.

TABLE 1 Differences in Pharmacokinetic Parameters in Women Compared to Men

Pharmacokinetic Parameters	Clinical Consequences in Women Compared to Men
Drug Bioavailability • ↓ Gastric acid secretion and gastric emptying • Slower gastrointestinal transit rate • No consistent changes in gut metabolism	• ↓ Oral availability of drugs requiring an acidic environment for absorption • Delayed/decreased absorption of enteric-coated aspirin, metoprolol, theophylline, and verapamil • Women need to wait longer after eating before taking drugs that must be administered on an empty stomach (ampicillin, captopril, cilostazol, and tetracyclines)
Drug Distribution	
1. Body composition • ↓ Body surface area, organ (heart) size, and blood flow • ↓ Total (intra-/extracellular) body water content and plasma volume • ↓ Cardiac output and functional lung capacity • ↑ Body fat content	• ↑ Vd and longer half-life of lipophilic drugs (anesthetics, benzodiazepines, opioids, and neuromuscular blockers) • ↓ Vd of hydrophilic drugs, reaching higher C_{max} and greater effects • Narrow therapeutic index drugs (amiodarone, lidocaine, procainamide, digoxin, heparins, and fibrinolytics) requiring loading dosages can reach higher C_{max} and produce more ADRs • ↓ Pulmonary exposure
2. Plasma protein binding to: • ↓ Albumin and α1-acid glycoprotein • ↑ Globulins	• Pregnancy and oral contraceptives ↓ plasma albumin and α1-acid glycoprotein and may ↑ free drug plasma levels of highly bound drugs (warfarin) • Exogenous estrogens ↑ serum-binding globulins
Drug Metabolism Phase I metabolism: • ↓ Activity of CYP1A2, 2C9, 2D6, 2C19, and 2E1 • ↑ Activity of CYP2A6, 2B6, 2D6, and 3A4 Phase II metabolism: • ↓ Activity of N-acetyl/sulfo/methyl/UDP-glucuronosyltransferases and alcohol dehydrogenase • ↑ Activity of xanthine oxidases Drug transporters: • ↓ P-glycoprotein, organic cationic transporter (OCT2), and anion-transporter polypeptides (OATP1B1-3)	• Oral contraceptives ↑ activity of CYP 1A2 and 2A6 • ↑ Activity of CYP 1A2, 2C9, 2D6, and 3A4 during pregnancy • ↓ Activity of CYP2C19 during pregnancy or use of oral contraceptives • ↑ Biotransformation of CYP3A4 substrates: calcium channel blockers and lidocaine • ↑ Biotransformation of CYP2D6 substrates: metoprolol, propafenone, and propranolol • ↓ Biotransformation of CYP2C19 substrates: warfarin • ↑ Drug exposure of P-glycoprotein substrates (colchicine, dabigatran, and digoxin)
Drug Excretion	
• ↓ Renal blood flow, GFR, and tubular secretion and/or reabsorption	• ↓ Renal Cl and longer elimination $t_{1/2}$ of renally excreted drugs (digoxin, flecainide, nadolol, procainamide, and sotalol) • Differences are reduced when doses are adjusted for body weight or creatinine clearance

ADRs, adverse drug reactions; Cl, clearance; CYP, cytochrome P(450); GFR, glomerular filtration rate; $t_{1/2}$, elimination half-life; UDP, uridine 5-diphosphate; Vd, volume of distribution. *Modified from [14, 16].*

TABLE 2 Differences in the Pharmacokinetic and Pharmacodynamic Characteristics of Cardiovascular Drugs in Women Compared to Men

Drug Class	Pharmacokinetics	Pharmacodynamics
Alcohol [26]	Faster absorption and higher plasma levels of alcohol due to a lower activity of the gastrointestinal alcohol dehydrogenase	Women cope worse with alcohol than men
Anesthetics [27–29]	Plasma levels declined more rapidly in women. Women wake up faster than men	Women require higher dosing for the same effect
Angiotensin-converting enzyme inhibitors [30]		No mortality benefit in women with asymptomatic LV systolic dysfunction
Antiarrhythmic drugs [31]	Plasma levels of procainamide are 30% higher due to the lower Vd	Higher risk of proarrhythmia
Aspirin [32–35]	Higher oral bioavailability, slower clearance, and longer half-life. Differences disappear with OCP	Better protective effect against stroke in women and against myocardial infarction in men
Beta-blockers: metoprolol and propranolol [36–39]	Higher exposure due to a lower Vd and reduced metabolism via CYP2D6. OCP increase drug exposure	Produce a greater reduction in systolic blood pressure and a smaller increase in heart rate while exercising
Calcium channel blockers [25, 40–45]	Faster clearance of verapamil, nifedipine, and amlodipine, probably due to the higher activity of CYP3A4 or lower activity of P-glycoprotein. No differences after i.v. administration of verapamil	More pronounced reduction in blood pressure in women
Digoxin [46, 47]	Higher serum digoxin concentrations due to reduced Vd and lower clearance. Clearance increases during pregnancy	Higher mortality in women with chronic HFrEF. Women require lower serum digoxin concentrations ($<0.9\,ng/mL$)
Glucocorticoids [48, 49]	Lower Vd, higher clearance, and shorter half-life	More effective in women than in men
Heparin [50]	Higher plasma levels due to slower clearance	Higher activated partial thromboplastin time values than men. Elderly women require lower doses
Iron [51]	Greater oral absorption	
Lidocaine [52]	Higher Vd and may require a higher intravenous bolus dose. Higher free plasma levels in women receiving OCP	
μ-Opioid receptor agonists (fentanyl and morphine) [53, 54]	Slower onset and offset of the analgesic effect	Greater analgesic response. Women require lower doses (30–40%) of opioids for the same pain relief
QT-prolonging drugs [55–58]		Higher risk of torsades de pointes

Continued

TABLE 2 Differences in the Pharmacokinetic and Pharmacodynamic Characteristics of Cardiovascular Drugs in Women Compared to Men—cont'd

Drug Class	Pharmacokinetics	Pharmacodynamics
Statins [10, 59, 60]	Higher plasma levels (15–50%)	Equally beneficial for reduction of LDL and decreasing risk of ASCVD in both men and women
Theophylline [61]	Faster drug metabolism and shorter half-life	
Torasemide [62]	Higher peak plasma drug concentration and lower clearance	
Thrombolytics [63]		Increased efficacy of rtPA in women with acute ischemic stroke compared to men
Warfarin [22, 25, 64–66]	Higher free plasma levels and slower clearance	Women need less warfarin per week than men. Doses should be modified to reduce the risk of excessive anticoagulation

ASCVD, atherosclerotic cardiovascular disease; HFrEF, heart failure with reduced ejection fraction; i.v, intravenous LDL, low-density lipoprotein; LV, left ventricle; OCP, oral contraceptives; P-gp, recombinant tissue plasminogen activator; rtPA, recombinant human tissue plasminogen activator; Vd, volume of distribution.

Because women present a lower body weight/size and creatinine clearance compared to men, the administration of fixed doses frequently results in higher drug exposure [C_{max} and area under the concentration-time curve (AUC)] leading to a higher incidence of ADRs, particularly when cardiovascular drugs with a narrow therapeutic index are administered (**Table 1**) [17, 20]. However, many SRDs in pharmacokinetics are not generally considered clinically relevant after normalizing the dose for body weight/size or renal function [10, 16, 17, 22, 72].

Sex-Related Differences in Pharmacodynamics

SRDs in pharmacodynamic properties have been reported with some cardiovascular drugs (**Table 2**) [11, 26–64, 72, 73]. However, differences in drug response are difficult to quantify because CVD is "different" in men and women, phase I and bioequivalence studies are performed mainly in healthy male volunteers, and women are historically underrepresented in RCTs [74–78]. Furthermore, RCTs often fail to evaluate and report SRDs in a systematic manner or sex-specific analysis is simply descriptive and the mechanisms responsible for the possible differences are not analyzed. In addition, SRDs can be partly modulated by endogenous sex hormones, oral contraceptives, and hormone replacement therapy (HRT), but the possible interactions between cardiovascular drugs and sex hormones are poorly understood [19, 21]. As a result, differences in clinical outcomes are uncertain for some cardiovascular drugs routinely used in clinical practice, are rarely translated into clinical practice, and clinical guidelines provide sex-neutral recommendations for treatment.

Next, the SRDs in the pharmacokinetics and pharmacodynamics (efficacy and safety) of several cardiovascular drugs will be reviewed.

Antithrombotic Drugs

Antithrombotic drugs, including anticoagulants, antiplatelets, and fibrinolytic agents, are the cornerstone of prevention and treatment of atherosclerosis (i.e., MI and stroke), venous thromboembolic disorders, and the complications of atrial fibrillation (AF) [79–81]. These agents reduce mortality from cardiovascular events in both men and women. However, women present a higher risk of major bleedings than men, possibly because: (a) their lower body weight and creatinine clearance increase the risk of overdosing; (b) women are older and present more comorbidities [diabetes, hypertension, heart failure (HF), and renal impairment] that increase the risk of bleeding; and (c) sex differences exist in response to antithrombotic agents [79–95].

Anticoagulants

Vitamin K Antagonists

Women require less doses of **warfarin** than men to maintain a therapeutic international normalized ratio (INR), with older women requiring the lowest doses [25, 96]. Thus, starting and maintenance doses should be modified to reduce the risk of inadequate therapy in young women and of excessive anticoagulation in elderly women [65]. Pooled data from five RCTs in patients (26% women) with AF revealed that warfarin decreased the risk of stroke (84% in women and 60% in men), with virtually no increase in the frequency of major bleeding [97]. In a post-hoc analysis of the AFFIRM trial, women with nonvalvular AF treated with warfarin were at greater risk of ischemic stroke compared to men despite similar anticoagulation patterns and after adjustment for differences in percent time in the therapeutic range [98]. This finding suggests that women with nonvalvular AF may require a higher therapeutic INR range or use direct oral anticoagulants (DOACs) to decrease their risk of stroke compared to men.

Warfarin was equally effective in lowering the risk of thromboembolism in men and women, but some studies reported a higher potential benefit in women [25, 96, 97, 99]. In some trials, women on warfarin had more minor bleeding than men [83, 100], but in a meta-analysis of 42 studies (94,293 patients), the risk of major bleeding was found to be similar in men and women with AF or venous thromboembolism (VTE) [101]. There is a possible interaction between warfarin and oral contraceptives and HRT; thus, frequent monitoring of INR is recommended when this combination is used [58].

Heparins

Unfractionated heparin (UFH) is distributed into plasma volume, which is proportional to body weight, and is eliminated less rapidly in women [102, 103]. Women who receive UFH after an acute myocardial infarction (AMI) achieved higher activated partial thromboplastin time (aPTT) than men, and aPTTs > 70 s increased the risk of death, stroke, bleeding, and reinfarction [104, 105].

In patients with unstable coronary artery disease (CAD), dalteparin was associated with a larger absolute (4.5% vs. 2.2%) and relative reduction (13.1% vs. 28.9%) of the primary end point of death and new myocardial infarction (MI) during the first 6 days in women than in men [106]. In this trial, minor bleeding was more frequent and anti-Xa activity was higher in women compared to men [107]. In patients with non-ST-segment elevation (NSTEMI) undergoing a percutaneous coronary intervention (PCI), enoxaparin showed similar pharmacodynamic/pharmacokinetic profiles in men and women [108] and achieved similar ef-

fectiveness to UFH but increased the risk of major bleeding in both sexes [109]. In a meta-analysis of the ESSENCE and TIMI 11B trials, enoxaparin was more effective than intravenous dose-adjusted UFH in reducing the risk of death, MI, or recurrent angina prompting urgent revascularization in patients unstable angina/non-ST-segment elevation myocardial infarction (UA/NSTEMI) in both sexes [110]. Furthermore, in the ExTRACT-TIMI 25 study, which randomized ST-segment elevation MI (STEMI) patients with planned fibrinolysis to enoxaparin or UFH, women and men had a similar relative benefit [111]. Given that women were older and at higher baseline risk, they derived a greater absolute benefit than men when treated with enoxaparin.

Factor Xa Inhibitors

In the OASIS-5 trial, **fondaparinux and enoxaparin** showed similar efficacy in reducing the risk of death and ischemic events (MI or refractory ischemia) at 9 days in patients with acute coronary syndrome (ACS) of both sexes, but the rate of major bleeding was significantly lower with fondaparinux [112]. Similarly, in STEMI patients not receiving reperfusion treatment, fondaparinux significantly reduced the risk of death or reinfarction at 30 days without increasing bleeding and strokes in men and women [113, 114].

Direct Thrombin Inhibitors

The pharmacodynamic/pharmacokinetic properties of **argatroban** are similar in both sexes [115]. Several RCTs compared the efficacy of bivalirudin and UFH, or low-molecular-weight heparins (LMWH) plus a glycoprotein (GP) IIb/IIIa inhibitor, in patients with STEMI undergoing urgent or elective PCI [116–120]. Bivalirudin conferred similar protection against ischemic end points compared with UFH/LMWH plus a GP IIb/IIIa inhibitor in both women and men. In the pooled analysis of the REPLACE-2, ACUITY, and HORIZONS-AMI trials, women (25% of the population) had a near twofold increase in bleeding complications compared to men after PCI [121]. **Bivalirudin** therapy in women significantly reduced 30-day PCI-related major bleeding (44%) and 1-year mortality compared to standard therapy. However, in a recent study in 7213 patients with ACS undergoing invasive management, the rates of major adverse cardiovascular events (MACE) and net adverse clinical events were not significantly lower with bivalirudin than with UFH in both sexes [122].

In these trials, women presented a higher risk of bleeding complications than men, probably because they were older and had more comorbidities that increased the long-term risk of ischemic events and were at increased risk of bleeding as compared to men [89, 121, 123–126].

Direct Oral Anticoagulants

There are no sex differences in the pharmacokinetics of rivaroxaban [127–129], apixaban [130], and edoxaban [131], and even when dabigatran exposure is ~30% higher in women, no dose adjustment is indicated [132]. A meta-analysis of major RCTs (RE-LY, ROCKET AF, ARISTOTLE, and ENGAGE AF-TIMI 48) including 71,683 patients (30–40% women) with nonvalvular AF showed no SRDs in the risk of stroke/systemic embolism or major bleeding risk among patients assigned to DOACs, but in these trials dose adjustments were made according to body weight and creatinine clearance, that is, for smaller female patients [133]. Nevertheless, these major trials were not designed or statistically powered to conduct sex-specific analyses. There was, however, a trend toward a reduction of stroke/systemic embolism for dabigatran 150 mg [134, 135] and of major bleedings for edoxaban 60 mg [136] and apixaban [137] in women compared with men. In a secondary analysis of the ARISTOTLE trial, women had a similar rate of stroke/systemic embolism but a lower risk of all-cause and cardiovascular mortality and of clinically relevant bleeding than men [138]. Furthermore, among patients with history of stroke or transient ischemic attack, women had a 30% lower risk of stroke/systemic embolism and of cardiovascular death compared with men [138]. In the ROCKET-AF trial, the bleeding risk of rivaroxaban was significantly higher in men than in women [139]. Moreover, in a post-hoc analysis of this trial, rivaroxaban increased the rates of both major and nonmajor clinically relevant gastrointestinal bleeding compared with warfarin, but the risk was significantly higher in men [140].

In a recent meta-analysis of five RCTs, women with nonvalvular AF treated with DOACs had a higher rate of stroke/systemic embolism compared with men, but a lower rate of major bleeding, suggesting differences between various DOACs in men and women [141]. In another meta-analysis of six trials (BAFTA, SPORTIF III/IV, RE-LY, ROCKET-AF, ARISTOTLE, and AVERROES), women with nonvalvular AF treated with warfarin have a significantly greater residual risk of stroke/systemic embolism, while no sex difference in residual risk was noted in patients receiving DOACs [142]. Because major bleeding was less frequent in women treated with DOACs, these results suggested an increased net benefit of DOACs compared to warfarin in women with nonvalvular AF.

Three population-based studies reported SRDs in the clinical outcomes for DOACs in patients with nonvalvular AF. In a Canadian study, women were more frequently treated with low-dose dabigatran. Women gained more stroke protection with a higher dabigatran dose (150 mg bid), while men using dabigatran benefit from lower bleeding rates compared with warfarin [135]. In an analysis of Medicare claims data for elderly patients, rivaroxaban use in men was associated with a reduced risk of MI admissions compared with either dabigatran or warfarin use, while the risk of MI was similar across all three anticoagulants in women [143]. Finally, in an Asian study, DOAC use was associated with a lower risk of intracerebral hemorrhage and all-cause mortality when compared with warfarin use in women only; the lower risk of intracerebral hemorrhage remained in women when compared to warfarin users with appropriate INR control [144].

Conversely, in several meta-analyses, no SRD was observed in the efficacy of DOACs in patients with nonvalvular AF or acute VTE, but contradictory results were reported on the risk of bleeding [93, 145–147]. In a sex-based meta-analysis of RCTs of DOACs for acute and extended VTE treatment, women suffered more bleeding complications than men without differences in treatment efficacy [93]. Similar results were reported in real-world VTE patients [148, 149]. Unfortunately, all-cause mortality was not reported by sex in patients with VTE.

Antiplatelet Drugs

Platelets play a key role in the pathogenesis of atherosclerosis, thrombosis, and acute thrombotic events [79–81]. Women have longer bleeding times, higher baseline platelet hyper reactivity, higher platelet aggregation in response to adenosine diphosphate (ADP), thrombin or collagen, and higher on-treatment (aspirin or clopidogrel) platelet reactivity compared to men [79, 80, 150, 151]. These SRDs in platelet function may result from the direct effect of sex hormones on platelets or, indirectly, on the vasculature [152–154]. Furthermore, platelets from postmenopausal women are more prothrombotic than platelets from men of similar age, probably because they had a higher prevalence of comorbidities (hypertension, hyperlipidemia, diabetes, chronic kidney disease, and HF) associated with an enhanced platelet reactivity [95, 155].

Acetylsalicylic Acid (Aspirin)

Low-dose aspirin is the cornerstone of treatment for patients with atherosclerotic cardiovascular disease (ASCVD) [80, 156]. Low-dose aspirin therapy produces complete and equal suppression of the cyclooxygenase-1 platelet activation pathway in both sexes, but women express increased baseline platelet reactivity and a higher prevalence of aspirin resistance as compared to men [150, 157–159]. However, pathways indirectly related to COX-1, that is, those stimulated by collagen, ADP, and epinephrine, were less inhibited by aspirin in women [150], and in ex vivo studies, aspirin was less effective at inhibiting platelet aggregation in women with a history of ischemic stroke or transient ischemic attack [160]. These findings suggest that women treated with low-dose aspirin retain more platelet reactivity compared with men and may benefit from the use of alternative antiplatelet drugs. There are also SRDs in aspirin pharmacokinetics. Aspirin presents faster absorption, higher AUC, slower clearance, and longer half-life in women, probably because men conjugate more aspirin with glycine and glucuronic acid [32, 33, 161]. These differences disappear in women taking hormonal contraceptives [32].

Several studies that analyzed the chronic use of aspirin in the primary prevention of CVD revealed SRDs in drug efficacy, but the risk of bleeding, mostly gastrointestinal, was increased to a similar degree in men and women. In the Hypertension Optimal Treatment (HOT) trial, aspirin (75 mg/day) significantly reduced the risk of MI in men (42%), but not in women [162]. In contrast, it did not reduce the risk of stroke in both sexes. In a sex-specific meta-analysis of six primary prevention trials, aspirin therapy reduced cardiovascular events (12%) in both sexes [163]. Among 51,342 women, aspirin reduced the risk of stroke (17%; 24% reduction in ischemic stroke without an increase in hemorrhagic stroke), but not of MI, all-cause, or cardiovascular mortality. Among 44,114 men, however, aspirin reduced the risk of MI (32%), but not of stroke (hemorrhagic strokes increased) or cardiovascular and all-cause mortality. The risk of bleeding significantly increased to a similar degree (~70%) in both sexes. In a collaborative meta-analysis of six primary prevention trials (95,000 individuals), aspirin significantly reduced many serious vascular events (12%) and major coronary events (18%), due mainly to a 23% reduction in nonfatal MI, but it did not significantly affect total stroke (14% risk reduction of ischemic stroke and 32% increase in hemorrhagic stroke) or vascular mortality [164]. In an update of the US Preventive Services Task Force that analyzed 11 primary prevention trials (118,445 participants), aspirin had little or no effect on CVD or all-cause mortality and provided a modest benefit in reducing nonfatal MI (22%) and nonfatal stroke (14%) [165]. Furthermore, this analysis showed that the SRDs previously described [163] relied on subanalyses with serious limitations. More recently, a systematic review of 15 RCTs (165,502 participants) revealed that, compared with control, aspirin did not modify all-cause, cardiovascular, and noncardiovascular death, but reduced the risk of nonfatal MI (18%), transient ischemic attacks (21%), and ischemic stroke (17%) and aspirin use was associated with an increased risk of major bleeding, intracranial bleeding, and major GI bleedings [166]. From these results, it can be concluded that there is insufficient evidence to recommend aspirin in the routine primary prevention of ASCVD because of the lack of net benefit [167, 168].

Conversely, in long-term **secondary prevention** trials, aspirin significantly reduced the risk of subsequent MI, stroke, and cardiovascular mortality [169, 170]. In the ISIS-2 trial, which evaluated the effects of aspirin in patients with AMI, aspirin significantly reduced the risk of vascular mortality, nonfatal MI, and nonfatal stroke in both women and men [171]. In a collaborative meta-analysis of 287 RCTs of antiplatelet therapy involving 135,000 patients, aspirin (75–325 mg) reduced the risk of serious vascular events (nonfatal MI, nonfatal stroke, or vascular death) by about one-quarter and the absolute benefits substantially outweighed the absolute risks of major extracranial bleeding in both sexes [169]. In another analysis of 11,265 patients with a history of MI and 6765 patients with a history of cerebrovascular disease (transient ischemic attack or stroke), aspirin reduced the risk of a major coronary events (nonfatal MI or coronary death) and stroke to a similar degree in men and women [170]. Furthermore, in a meta-regression analysis of 23 RCTs (113,494 participants), aspirin significantly reduced the risk of nonfatal MI (28%) [172]. Interestingly, trials recruiting predominantly men demonstrated the largest reduction in nonfatal MI (38%), while trials that recruited mostly women failed to demonstrate any benefit. Thus, sex accounts for ~27% of the variability in the efficacy of aspirin in reducing MI rates across these trials [172]. Finally, in 16 secondary RCTs (17,000 individuals), antiplatelet therapy (primarily aspirin) significantly reduced major vascular events (19%), ischemic stroke (22%), and major coronary events (20%), accompanied by a nonsignificant increase in hemorrhagic stroke [164]. No heterogeneity of treatment was observed by sex for any end point after controlling for multiple comparisons, suggesting that the absolute risk reduction mainly depends on the individual's absolute risk without treatment. In the setting of acute stroke, aspirin reduces the risk of death and recurrent ischemic strokes to a similar degree in men and women with no significant excess of hemorrhagic strokes [173, 174]. In conclusion, the benefits of aspirin clearly outweigh the risk of bleeding in both sexes and are widely recommended for the secondary prevention in patients with CVD [164, 167, 168].

Glycoprotein IIb/IIIa Receptor Inhibitors

These inhibitors reduce the risk of death or MI in patients with ACSs after PCI. No SRDs in the platelet response to GP IIb/IIIa inhibitors were observed in vitro [80]. In a pooled analysis of three trials (EPIC, EPILOG, and EPISTENT) in 6595 patients undergoing PCI, the reduction in death, MI, or urgent revascularization with abciximab at any time point was similar between men and women [86]. Similarly, in the ESPRIT trial, eptifibatide reduced the composite of death, MI, or urgent target vessel revascularization to a similar extent in both sexes [175]. Furthermore, in patients who received a GP IIb/IIIa inhibitor during PCI, despite several sex-specific differences in baseline characteristics, in-hospital and 1-month outcomes of women with

CAD treated with a GP IIb/IIIa inhibitor in the setting of PCI were similar to those of men [176]. However, in the PURSUIT trial eptifibatide reduced the risk of death and nonfatal MI in patients with NSTEMI-ACS in all major subgroups except women [177].

Two meta-analyses of GP IIb/IIIa inhibitors in patients with AMI did not provide a sex-specific analysis of outcomes [178, 179]. Another meta-analysis of six RCTs that enrolled 31,402 patients (35% women) with NSTEMI not routinely scheduled for early revascularization found a reduction in the rates of 30-day death or MI in men receiving GP IIb/IIIa inhibitors (OR 0.81) but not in women (OR 1.15) [180]. However, once patients were stratified by baseline troponin levels, a risk reduction was seen in troponin-positive, but not in troponin-negative, patients of either sex. In patients with NSTEMI, GP IIb/IIIa inhibitors increased the bleeding risk in men and women [177, 180, 181], but because of frequent excessive dosing in women, up to one-quarter of excess bleeding risk in women is avoidable by appropriate dose adjustment [181]. However, even after adaptation of dosage to body mass and serum creatinine levels, the bleeding risk is still higher in women, suggesting that the risk is the result of pharmacodynamic/pharmacokinetic differences.

Adenosine Diphosphate P2Y12 Receptor Antagonists

Clopidogrel is a prodrug that must be metabolized by CYP450-dependent enzymes (CYP2C19; minor contribution of CYP1A2, 2B6, and 3A4) to produce the active metabolite that selectively inhibits the binding of ADP to its platelet P2Y12 receptor [182]. Ex vivo studies found that women are more often hyporesponsive to clopidogrel. Given the lack of differences in the plasma levels of the active metabolite between sexes, differences in clopidogrel response are likely related to the higher baseline platelet reactivity in women [183, 184].

In patients with NSTEMI undergoing PCI, no SRDs were observed in the composite end point of cardiovascular death, nonfatal MI, or stroke in patients treated with clopidogrel plus aspirin compared to aspirin alone [185–187]. Similarly, in the CREDO trial enrolling patients undergoing elective PCI, the benefit of an early loading dose of clopidogrel before PCI followed by long-term treatment with clopidogrel significantly reduced the risk of death, MI, or stroke at 1 year in both sexes [188].

In a sex-specific meta-analysis of five RCTs (CURE, CREDO, CLARITY-TIMI 28, COMMIT, and CHARISMA) involving 79,613 patients (30% women), clopidogrel reduced the risk of major cardiovascular events in men and women. In women, the overall effect of clopidogrel was driven by a reduction of MI, while in men its effects on

MI, stroke, and all-cause mortality were separately significant [189]. However, there was no evidence of statistically significant heterogeneity in the efficacy and safety (risk of major bleeding), when compared between men and women. Also, in another meta-analysis of 20 RCTs (233,285 participants), the efficacy (defined as MI, stroke, or cardiovascular death) and safety (any bleeding) of clopidogrel did not differ by sex [190].

Systemic exposure of **prasugrel** and its active metabolite AR-C124910XX was not affected by sex [191, 192]. In patients with ACS with scheduled PCI, prasugrel produced a greater absolute (2.4% vs. 1.6%) and relative (21% vs. 12%) risk reduction in cardiovascular death, nonfatal MI, or nonfatal stroke in men than in women when compared with clopidogrel, but this difference disappeared after adjustment for baseline characteristics [193]. In women ≤55 years of age with ACS undergoing PCI, unadjusted 1-year MACE and bleeding were significantly higher compared with those in men, but this result was no longer significant after adjustment for baseline risk [194]. In these trials, being female was the strongest independent predictor of serious bleeding, possibly due to some extent to lower body weight [194, 195].

Systemic exposure to **ticagrelor** and its active metabolite AR-C124910XX was higher in women, but no dose adjustment is considered based on sex [196]. In patients with an STEMI-/NSTEMI-ACS, ticagrelor, as compared with clopidogrel, significantly reduced (16%) the rate of vascular death, MI, or stroke without an increase in the rate of overall major bleeding in both sexes [197, 198].

Because of its short half-life (2.6 min), no SRDs were reported with cangrelor. In a prespecified sex-specific analysis, compared with clopidogrel, **cangrelor** administered during PCI reduced the primary end point (death, MI, ischemia-driven revascularization, or stent thrombosis at 48 h) by 35% in women and by 14% in men and stent thrombosis by 61% and 16%, respectively. Severe bleeding was similar in both sexes, but moderate bleeding increased in women but not in men. Thus, the net clinical benefit favored cangrelor in both women and men [199].

In a sex-specific meta-analysis of six RCTs (TRITON-TIMI38, PLATO, CHAMPION PCI + PLATFORM, TRILOGY-ACS, CHAMPION PHENIX, and PEGASUS-TIMI 54) including 63,346 men and 24,494 women, the efficacy (reduced risk of MACE, MI, stent thrombosis, and cardiovascular mortality) and safety (major bleedings) of P2Y12 inhibitors (clopidogrel, prasugrel, ticagrelor, and cangrelor) were comparable in both sexes [155]. Thus, sex should not influence patient selection for the administration of P2Y12 inhibitors.

Selective Protease-Activated Receptor-1 (PAR-1) Antagonists

Vorapaxar exposure (C_{max} and AUC) is 30% higher in women, but no dose adjustment is necessary. In a meta-analysis of four RCTs, vorapaxar combined with dual-antiplatelet therapy reduced the risk of MACE, MI, and ischemic stroke in patients with ASCVD, and increased the risk of bleeding events in both sexes [200].

Fibrinolytic Drugs

Data are contradictory regarding the benefit and risk of morbidity and mortality in women compared to men receiving thrombolytic therapy for treatment of acute MI [201–203]. In the International Tissue Plasminogen Activator/Streptokinase Mortality Study, women had similar morbidity and mortality compared to men but presented a higher incidence of hemorrhagic stroke [201]. Similarly, there were no SRDs with regard to early infarct-related artery patency rates, reocclusion, or ventricular functional response to injury/reperfusion after fibrinolytic therapy [203]. In a pooled analysis of EPIC, EPILOG, and EPISTENT trials (6595 patients, 26.9% women) undergoing PCI, the benefit on 30-day and 6-month ischemic outcomes with abciximab was observed irrespective of sex [86]. Although women had higher rates of major and minor bleeding than men with abciximab, major bleeding in women was similar with and without abciximab. In patients with NSTEMI undergoing a PCI, the benefit with abciximab was greater in men than in women, but after adjustment for baseline clinical and angiographic characteristics, there was no significant interaction between sex and abciximab effect [204]. In another review of nine RCTs (58,600 patients), fibrinolytic therapy reduced mortality in patients with suspected AMI and the benefit was greater the earlier treatment began irrespective of sex, but women experienced more bleedings than men [84]. Of note, in patients with AMI the benefits of a single i.v. infusion of 1.5 million units of streptokinase on survival persisted up to 10 years in both men and women [205].

Conversely, in the GUSTO V study, death, reinfarction, stroke, and mechanical and bleeding complications (intracranial hemorrhage) were more common among fibrinolysis-treated women with MI [88]. Additionally, in a pooled analysis of RCTs, including 988 women with acute ischemic stroke, women were significantly more likely to benefit from recombinant tissue plasminogen activator (rtPA) compared to men and the usual sex difference in outcome favoring men was not observed in the fibrinolytic-treated therapy group [63]. Furthermore, among women with STEMI, early administration of abciximab improved patency of the infarct-related artery before and after primary PCI and decreased ischemic events, including 30-day mortality. In contrast, the reduction in 30-day and 1-year

mortality was not significant in men, but the bleeding events were similar in both men and women [206].

As already mentioned, women receiving fibrinolytic therapy have a greater risk of major bleeding, especially intracranial hemorrhage, but they were older and have more comorbidities—that is, they were at greater risk for both fatal and nonfatal complications than men [84, 86, 88, 201–203, 207–211]. The bleeding risk is only partly reduced by adjusting the dosage for body weight and renal function, which suggests an involvement of pharmacodynamic mechanisms.

Beta-Blockers

C_{max} and AUC (50–80%) of metoprolol and propranolol are higher in women due to higher oral bioavailability, lower Vd, and slower clearance via CYP2D6 as compared to men [10, 13, 16]. The increase in drug exposure persisted after normalization for body weight and was further increased by oral contraceptives [36–39, 212]. Consequently, women treated with metoprolol demonstrated a greater reduction in heart rate and systolic blood pressure during exercise than men [36, 212]. Thus, women may benefit from lower-than-standard doses to avoid ADRs [36–38, 213]. However, metoprolol might exert a significantly greater therapeutic effect on stress-induced angina pectoris in men than in women in spite of the latter having higher plasma levels [214].

Some early trials reported that β-blockers improved survival in males, but not in females, with hypertension [215], acute MI [216], or HF with reduced ejection fraction (HFrEF) [217]. These results can probably be related to the underrepresentation of women in these trials (<21%), which limited the number of deaths available for analysis, and to the act that women were older and sicker than men [12, 16]. However, in patients with HFrEF, the COPERNICUS and US-Carvedilol HF trials reported an equal benefit on all-cause death and HF hospitalizations in both sexes [218, 219]. Similarly, in a post-hoc sex-specific analysis of the MERIT-HF trial, metoprolol controlled release (CR)/extended release (XL) significantly reduced the primary combined end point of all-cause mortality/all-cause hospitalizations (21%) and in women with severe HF, it reduced all-cause (44%) and cardiovascular hospitalizations (57%) and hospitalization due to worsening HF (72%) [220]. Moreover, in a post-hoc analysis of the CIBIS II trial, bisoprolol significantly decreased all-cause, cardiovascular, and noncardiovascular mortality in women compared to men with HFrEF [221]. In a pooling of mortality data from the MERIT-HF, CIBIS-II, and COPERNICUS trials by sex (>8900 women), β-blocker therapy showed similar survival benefits in men and women [220]. In two other meta-analyses, β-blockers reduced all-cause mortality and HF admissions in patients with HFrEF regardless of sex [30, 222].

In patients with MI, β-blocker therapy reduced mortality, sudden death, and reinfarction with similar benefits in men and women [223]. In a retrospective cohort study in 115,015 elderly patients, after adjusting for potential confounders, β-blocker therapy reduced (14%) the risk of mortality at 1-year postdischarge in men and women [224]. Additionally, in five RCTs (5474 patients, 1121 women) metoprolol therapy after AMI significantly reduced mortality at 1 year after discharge in both men and women [225].

Calcium Channel Blockers

Sex-specific pharmacokinetic differences have been described for verapamil [40–42, 45], nifedipine [43], and amlodipine [44]. Women present with a faster clearance and lower plasma levels than men, probably due to the higher activity of CYP3A4 and/or lower activity of P-gp in women [16, 17, 45]. Drug clearance decreased with age in women, which explains why older women experienced a greater reduction in blood pressure and heart rate than younger women or men [22, 41, 226, 227]. In an open-label trial, even after dose adjustment for body weight, amlodipine produced a greater blood pressure reduction response in women [227]. In the HOT trial, where felodipine was administered alone or in combination with other antihypertensives, the rate of MI was significantly reduced in women with the lower diastolic BP target group (<80 or 85 mmHg), while a smaller (not significant trend) was found in men [162]. However, major hypertension trials with calcium channel blockers (ALLHAT, INSIGHT, STOP-Hypertension-2, and NORDIL) revealed no evidence for SRDs in cardiovascular outcomes [10, 16]. A higher incidence of peripheral edema, potentially resulting in decreased adherence and discontinuation of therapy, is more frequently observed in women than in men [227].

Digoxin

In patients with normal sinus rhythm and chronic HFrEF, digoxin did not reduce overall mortality, but reduced the rate of hospitalization, both overall and for worsening HF [228]. However, in a post-hoc analysis of this trial, women assigned to digoxin had a higher mortality risk than those assigned to placebo, while the mortality rate was similar in men assigned to digoxin or placebo [47]. Interestingly, in an analysis of the 1366 women enrolled in the DIG study, a bidirectional association was described between serum digoxin concentrations and all-cause mortality [229]. Compared with placebo, low serum digoxin concentrations (0.5–0.9 ng/mL) reduced mortality and all-cause and HF hospitalizations, while at higher serum digoxin concentrations, digoxin reduced HF hospitalizations, without any effect (or an increase) on mortality [46, 229]. Thus, it was hypothesized that increased mortality was related to higher

serum digoxin concentration due to the reduced Vd and renal clearance in women. However, sex-based differences in digoxin pharmacokinetics disappeared when doses were estimated with actual or ideal body weight [230]. Other explanations included: (a) differences in cellular sodium and calcium handling, because women have fewer Na$^+$ pumps than men, which might predispose HF women to fatal arrhythmias [231, 232]; and (b) because the mean age of women in the DIG trial was 64 years, a possible interaction between digoxin and HRT was suggested. In the HERS study, women under HRT treated with digitalis experienced an elevated incidence of coronary events in the first year [233]. It was hypothesized that progestin inhibits P-gp and increases serum digoxin concentration and digitalis toxicity.

On the contrary, in a post-hoc analysis of the SOLVD trials, there was no significant difference in the hazard ratios for men and women on digitalis with respect to all-cause, cardiovascular, or HF mortality [234]. However, digitalis therapy was not randomized in these trials and women on digitalis were sicker, which might explain the higher incidence of coronary events.

Diuretics

No SRDs have been described for thiazide diuretics on cardiovascular events and mortality in patients with hypertension of HF [235]. However, thiazide and loop diuretics produce more electrolyte disturbances in women, which may prolong the QT interval and increase the risk of long QT-associated arrhythmias [62, 236]. Torsemide clearance was reduced and drug exposure increased (40–50%) in women, which may explain why 66% of drug-induced hospitalizations occurred in women [62]. However, no sex-related dose adjustments are recommended for torsemide.

Drugs for Pulmonary Arterial Hypertension

It has been suggested that women with pulmonary arterial hypertension may benefit more from endothelin receptor antagonists and prostacyclin analogs than men, while men may benefit more from phosphodiesterase-5 inhibitors (PD5Is) than women [237–239]. However, most of the evidence came from retrospective studies and findings from RCTs with a direct comparison between endothelin receptor antagonists and PD5Is did not support this hypothesis [240, 241].

Glucose-Lowering Drugs

The pharmacodynamic/pharmacokinetic properties of most glucose-lowering drugs (metformin, pioglitazone, nateglinide, acarbose, and dipeptidyl peptidase-4 inhibitors) appear to be comparable in men and women with type 2 diabetes [242, 243]. In two recent meta-analyses, glucagon-like peptide-1 receptor agonists (GLP-1RAs) and sodium-glucose cotransporter 2 inhibitors (SGLT2Is) reduced the risk of MACE to a similar degree in patients with established ASCVD of both sexes [244, 245], and in an analysis of several RCTs (EMPA-REG OUTCOME, CANVAS Program, DECLARE TIMI-58, and CREDENCE), SGLT2Is provided similar protection against MACE, HF, vascular death, and total mortality in men and women [246]. A multinational, observational study, which included > 400,000 patients (48% women), reported that initiation of SGLT2Is reduced the risk of all-cause death, HF hospitalization, MI, and stroke in men and women with established CVD [247]. Furthermore, a subgroup analysis of the EMPA-REG OUTCOME trial concluded that empagliflozin reduced cardiovascular death, HF hospitalization, and incident or worsening nephropathy rate similarly in men and women [248]. However, a meta-regression analysis of 26 trials (22,256 patients, 58% women) suggested a progressive decrement in benefit in women, as the percentage of women included in the SGLT2 inhibitor arm was > 50% [249]. Additionally, a retrospective cohort study (167,254 type 2 diabetes mellitus metformin users, mostly without known established ASCVD; 46% women) reported that dipeptidyl peptidase-4 inhibitors, GLP-1RAs and SGLTIs, were associated with lower risk of major cardiovascular events than sulfonylureas, but the magnitude of the risk reduction was significantly greater in women users of GLP-1RAs than men ($P = 0.002$). Newer agents were also associated with a lower risk of ADRs, without differences between sexes [250]. Mycotic genital infections occurred four to six times more frequently with SGLT2Is than with other antidiabetic agents and more frequently in women [251]. Thus, further RCTs are required to confirm SRDs in the efficacy and safety of new glucose-lowering drugs.

Among diabetic patients at high risk of cardiovascular events, insulin degludec was not inferior to insulin glargine with respect to the incidence of MACE, with similar results in both sexes [252].

Ivabradine

No SRDs in efficacy and safety were observed in patients with stable CAD with or without LV systolic dysfunction [253, 254], or with symptomatic chronic HFrEF and a resting heart rate ≥ 70 bpm [255].

Inhibitors of the Renin-Angiotensin-Aldosterone System (RAAS)

Sexual hormones modulate the activity of the RAAS. Estrogens increase angiotensinogen and angiotensin II plasma levels but decrease, via negative feedback regulation, angiotensin-converting enzyme (ACE) and renin activity, AT1 receptor density, and aldosterone production, leading to a net inhibition of the RAAS. Conversely, androgens upregulate the RAAS [256, 257]. Thus, RAAS inhibitors can play a role in the cardioprotective effects of endogenous estrogens in premenopausal women [256, 257]. However, how sex hormones modulate the effectiveness of the RAAS inhibitors remains uncertain.

There are no SRDs in the pharmacokinetics of ACE inhibitors (ACEIs), angiotensin receptor blockers (ARBs), aliskiren, and eplerenone [16, 22, 258, 259]. RAAS inhibitors can cause fetotoxic and teratogenic effects and are not recommended in women during childbearing years, which limits the recruitment of young and middle-aged women in RCTs testing RAAS inhibitors [260, 261]. However, the lower use of ACEIs persists in women in all age groups, possibly because cough and angioedema are more frequent in women [262].

Angiotensin-Converting Enzyme Inhibitors

In a systematic review (98,496 patients, 25% women), ACEIs initiated during the acute phase of MI and continued for 4–6 weeks post-MI reduced 30-day mortality in both sexes [263]. Similarly, long-term ACEI therapy started early after MI complicated by LV dysfunction or HF reduced the rates of mortality, reinfarction, and hospital admissions for HF in both sexes [264].

In addition to improved outcomes in both sexes following MI, ACEIs are equally efficacious in both men and women with HFrEF. In early RCTs in patients with HFrEF (Consensus I, SAVE, and SOLVD), the reduction in mortality and HF hospitalization with enalapril and captopril were observed in men (30–40%), but not in women (< 5%) [265–267], probably due to the underrepresentation of women in these trials (20%), rather than to a reduced benefit of ACEIs in women [12, 16]. Indeed, the AIRE and HOPE trials reported a significant benefit of ACEIs in women, especially with regard to the secondary prevention of cardiovascular events in high-risk patients [268, 269]. In a review of 30 RCTs, in patients (5399 men and 1991 women) with symptomatic HFrEF, ACEIs (enalapril, captopril, ramipril, quinapril, and lisinopril) significantly reduced total mortality (23%) and the combined end point of mortality or hospitalization for congestive HF (35%) in men and women [270]. In another meta-analysis, RCTs were divided into those treating symptomatic HFrEF (CONSENSUS, SOLVD-Treatment, and TRACE) and those treating asymptomatic LV systolic dysfunction (SAVE, SOLVD-Prevention, and SMILE) [30]. Men and women with symptomatic HFrEF benefit from ACEIs, but women with asymptomatic LV systolic dysfunction may not achieve a mortality benefit when treated with ACEIs. In a post-hoc analysis of an Australian trial, the relative beneficial effects of ACEI treatment were evident in men [17% relative risk (RR) in first cardiovascular events and/or all-cause death] but not in women [hazard ratio (HR) 1.00 for both comparisons], despite similar reductions in blood pressure in both sexes [271]. However, this result has not been replicated.

Angiotensin AT1 Receptor Blockers

No sex-specific differences were found in RCTs investigating the effects of ARBs in the treatment of hypertension [272, 273], HFrEF [274–276], and post-MI [277–279]. However, women were underrepresented (<30%) in most RCTs with ACEIs and ARBs, and many trials were not designed to analyze SRDs [11, 16]. A retrospective analysis of 20,000 elderly patients discharged posthospitalization for HF suggested that women on ARBs had better survival than those on ACEIs, but there was no difference in survival in men prescribed ARBs or ACEIs [280]. However, these results contradict those obtained in RCTs [281].

Aldosterone Antagonists

No SRDs were described with spironolactone in patients with severe ischemic and nonischemic HFrEF [282] and with HF and preserved left ventricular ejection fraction (HFpEF) [283], or with eplerenone in patients with AMI complicated by LV dysfunction and HFrEF [284] or in patients with HFrEF and mild symptoms (EMPHASIS-HF trial) [285]. However, the EPHESUS trial showed a trend toward a greater benefit for 30-day all-cause mortality in women and a reduction in cardiovascular death or cardiovascular hospitalization in men only. However, because of the low enrolment of women (≤30%), no clear conclusions can be drawn [284].

Sacubitril-Valsartan

There are no SRDs in the pharmacokinetics of sacubitril-valsartan [286]. In the PARADIGM-HF trial, which enrolled 4187 patients with HFrEF (22% women), sacubitril-valsartan was superior to enalapril in reducing the risk of cardiovascular death and HF hospitalization in both sexes [287]. Furthermore, in the PARAGON-HF trial, which recruited 4600 patients with HFpEF (51.7% women), as compared to valsartan, sacubitril-valsartan significantly reduced the risk of HF hospitalization more in women than in men [288]. This is an interesting result, given that HFpEF is the predominant phenotype in women. However, the study did not provide a definite mechanistic basis for this finding.

Nitrates

In a pharmacokinetic study, women had significantly higher exposure to isosorbide-5-mononitrate than men, which was attributed to the difference in body mass index [289].

In African-American patients with HFrEF, the fixed-dose combination of isosorbide dinitrate and hydralazine had a more pronounced mortality benefit in women, but because of the low number of deaths, it is possible that this finding might be attributable to chance [290].

Lipid-Lowering Drugs

Statins

Dyslipidemia, especially high low-density lipoprotein cholesterol (LDL-C) levels, has the highest population-adjusted risk among women compared to all other known risk factors for ASCVD [59, 291]. After menopause, total cholesterol, LDL-C, and triglyceride levels increase, while high-density cholesterol levels decrease, so older women are at higher cardiovascular (CV) risk [292]. Statin therapy decreases the risk of CV events and mortality, and is equally effective in men and women for the primary and secondary prevention of ASCVD [167, 292–298]. Hence, the American College of Cardiology (ACC) and American Heart Association (AHA) prevention guidelines have a sex-neutral recommendation to use statins in both men and women for the prevention of ASCVD [299]. Despite this evidence, statins are prescribed less frequently in women for the primary and secondary prevention of ASCVD and fewer women receive high-intensity statin therapy and reached LDL-C levels <100 mg/dL compared to men [5, 59, 300–304]. This could reflect the perception that women have a lower risk of recurrent CV events despite the fact that they have a higher overall CV burden compared to men [1–6].

There are conflicting results on the role of statins in primary CVD prevention clinical trials, probably because women are underrepresented in RCTs and have a lower absolute risk of cardiovascular events [1, 295, 296]. In a Cochrane review of 18 primary prevention trials (56,934 participants), statin therapy reduced all-cause mortality (14%), fatal/nonfatal CVD events (27%), fatal/nonfatal stroke (22%), and revascularization rates (38%) in both sexes [296]. Additionally, in the USPSTF systematic review of 19 primary prevention RCTs (71,344 participants), statin therapy decreased the risk of all-cause (14%) and CV mortality (31%), stroke (29%), MI (36%), and composite cardiovascular outcomes (30%) in both sexes, the absolute benefits being higher in subgroups at higher baseline risk [298]. In both meta-analyses, statin therapy was not associated with increased risk of serious ADRs. In the JUPITER trial, which recruited apparently healthy enrolled men ≥50 years of age and women ≥60 years of age with high-sensitivity C-reactive protein values ≥2.0 mg/L and LDL-C <130 mg/dL, rosuvastatin reduced CVD events similarly in men and women. Absolute event rates were lower in women, with a significantly greater benefit for revascularization/unstable angina in women [305].

In the meta-analysis of 10 primary and secondary prevention trials (79,494 participants; 18.9% women), statin therapy reduced major coronary events, strokes, and all-cause mortality in both women and men [306]. Furthermore, in a sex-specific meta-analysis of 18 primary and secondary prevention trials (141,235 participants, 28.5% women), statin therapy produced a significant and similar benefit on cardiovascular events and all-cause mortality in men and women [307]. Moreover, in another sex-based meta-analysis of 11 secondary prevention trials (43,193 patients, 20.6% women), statins were equally effective in the secondary prevention of CV events in men and women, but they did not reduce all-cause mortality or stroke in women [308]; however, in a post-hoc-stratified analysis, the effect size of outcomes in women matches that in men [309]. The meta-analysis of 27 trials (174,149 participants, 27% women generally at lower cardiovascular risk) revealed that the proportional reductions per 1.0 mmol/L reduction in LDL-C in major vascular events, coronary revascularizations, stroke, and all-cause mortality did not differ significantly in men and women at an equivalent risk of CVD [297]. Thus, the absolute benefits of statin therapy are similar in both sexes and determined mainly by the absolute reduction in LDL-C and the underlying risk of vascular disease in the population treated [298].

The risk of ADRs (muscle symptoms, recent-onset diabetes, and central nervous system complaints) is greater in women, especially in older thin women, so women are more likely to discontinue the statin therapy than men [59, 310–314]. Interestingly, muscular adverse effects are mainly reported in observational studies but not RCTs [309, 315, 316]. Thus, in the JUPITER trial, which enrolled more women than any other statin trial or meta-analysis, no differences in the rate of myopathies between men and women were reported [315]. Nevertheless, the cardiovascular benefits appear to outweigh the risk for statin-related ADRs reported in adult women beyond childbearing years [310, 312].

Other Lipid-Lowering Drugs

A meta-analysis of 27 RCTs (22,231 patients, 47.7% women) suggested that men had slightly greater lipid responses to the ezetimibe-statin combination compared with women, but the clinical relevance of this finding is uncertain [317]. However, in other pooled analysis, ezetimibe produced a similar benefit in both sexes [318–320].

A prespecified analysis of the ACCORD study, which evaluated fenofibrate added to simvastatin therapy, suggested heterogeneity in treatment effect according to sex, with a benefit for men and possible harm for women ($P = 0.01$ for interaction) [321]. In a prespecified analysis of the FIELD trial, the largest study of fibrate use in women

with diabetes, fenofibrate usage was associated with a more improved lipoprotein profile in women compared to men, and reduced total cardiovascular events (cardiovascular death, fatal and nonfatal stroke, and carotid and coronary revascularization) by 30% in women and 13% in men, but these sex-specific treatment effects were not statistically different [322]. A similar benefit in both sexes was observed in patients with marked dyslipidemia, defined as triglyceride levels > 2.3 mmol/L and HDL-cholesterol levels ≤ 0.88 mmol/L in both sexes.

The pharmacokinetic and pharmacodynamic (LDL-C lowering) characteristics of alirocumab and evolocumab are similar in men and women with primary hypercholesterolemia and mixed dyslipidemia [323, 324]. No significant sex differences in ADRs were observed [325].

Table 3 summarizes the SRDs of commonly used cardiovascular drugs in main endpoint results described in RCTs as well as in meta-analysis/pooled analysis of clinical trials [326–331].

Adverse Drug Reactions

Women have a 1.5- to 1.7-fold greater risk of developing ADRs when drugs are prescribed at the same doses and ADRs tend to be more severe than in men, leading more often to hospital admissions [12, 13, 16, 348–351]. In fact, 8 of the 10 prescription drugs withdrawn from the US market between 1997 and 2000 caused more ADRs due to known pharmacodynamic differences in women than in men [352]. SRDs in ADRs produced by commonly prescribed CV drugs are summarized in **Table 4** [332–347]. This increased risk of ADRs may result from SRDs in organ physiology, pharmacodynamic/pharmacokinetic characteristics, and/or hormonal status, in the use of medications or the way that men and women report ADRs, a greater sensitivity to drug effects, polypharmacy, or simply because women present lower body weight and drug clearance than men, and if doses are fixed (not corrected for body weight/size or creatinine clearance), drug exposure increases leading to a higher incidence of ADRs [13, 22, 350, 353, 354] (**Table 5** and **Figure 3**). Therefore, when interpreting clinical trials, it is important to analyze whether doses were given on an mg/kg basis or the same dose was given to all subjects irrespective of body weight. Unfortunately, in most RCTs, ADRs are not reported by sex. The development of sex-specific pharmacological guidelines is the most effective strategy to minimize the higher incidence of ADR in women.

Cardiac arrhythmias. There are electrophysiologic differences between men and women. Women have a higher resting heart rate, shorter sinus node recovery time, and longer heart-rate corrected QT intervals (QTc) than men, which increases the risk of drug-induced torsades de

TABLE 3 Sex-Related Differences in Main End Point Results for Clinical Trials and Meta-Analysis of Common Cardiovascular Drugs

Study Acronym, Drug, CVD, [Reference]	Sample Size n (% Women)	End Point	Men (HR/RR, 95% CI) or RRR[a]	Women (HR/RR, 95% CI) or RRR[a]	P (Men vs. Women)[b]
ACCORD Study Group Simvastatin plus fenofibrate vs. simvastatin plus placebo, high-risk patients with type 2 DM [321]	5518 (30.7%)	1st occurrence of nonfatal MI, nonfatal stroke, or CV death	HR 0.83	HR 1.4	0.01
A-HeFT Investigators Isosorbide dinitrate + hydralazine, HFrEF [290]	1050 black patients (40%)	Mortality 1st hospitalization Event free survival	0.79 (0.46–1.35; P=0.38) 0.60 (0.42–0.89) 0.67 (0.49–0.92)	0.33 (0.16–0.71; P=0.003) 0.62 (0.41–0.96) 0.58 (0.39–0.86)	
ARISTOTLE Apixaban vs. warfarin Atrial fibrillation [137, 138]	18,201 (35.4%)	All-cause death CV death Major bleeding		0.63 (0.55–0.73) 0.62 (0.51–0.75) 0.86 (0.74–1.01)	<0.0001 <0.0001 0.066
	3538, prior history of stroke or transient ischemic attack	Stroke/systemic embolism CV death		0.70 (0.50–0.97) 0.85 (0.61–1.20)	0.036 0.027
CARE trial Pravastatin vs. placebo in patients with MI [331]	4159 (14%)	Death from CAD or nonfatal MI Fatal or nonfatal MI Stroke	21% (4–35; P=0.017) 20% (0–46; P=0.05) 25% (−9 to 49; P=0.14)	43% (4–66; P=0.035) 56% (18–76; P=0.009) 56% (−7 to 82; P=0.071)	0.048
CHAMPION PHOENIX Cangrelor vs. clopidogrel during PCI Patients undergoing PCI [199]	11,145 (27%)	Death, MI, ischemia-driving revascularization or stent thrombosis at 48 h Reduction in stent thrombosis	0.86 (0.70–1.05; P=0.14) 0.84 (0.53–1.33; P=0.44)	0.65 (0.48–0.89; P=0.01) 0.39 (0.20–0.77; P=0.01)	
CIBIS II Bisoprolol, HFrEF [221]	2467 (24%)	All-cause mortality CV mortality Non-CV mortality Pump failure death		0.64 (0.47–0.86) vs. men 0.64 (0.45–0.91) vs. men 0.11 (0.01–0.85) vs. men 0.30 (0.13–0.70) vs. men	0.003 0.013 0.034 0.005
DIG Digoxin, HFrEF [228]		All-cause death	0.93 (0.85–1.02)	1.23 (1.02–1.47)	0.014

Trial	N (%)	Endpoint			
FRISC Dalteparin plus aspirin. Unstable CAD [106]	1056 (36%)	Death or MI at 6 days	0.55 (0.28–1.11)	0.16 (0.05–0.56)	No statistical analysis
		Death or MI at 40 days	0.84 (0.58–1.23)	0.60 (0.33–1.07)	
GUSTO-I AMI treated with fibrinolytic therapy [202]	41,021 (25%)	Death (women vs. men)		1.15 (1.00–1.31)	0.04
		Stroke (women vs. men)		1.06 (0.88–1.28)	NS
		Reinfarction		1.30 (1.12–1.41)	–
		Congestive HF		1.32 (1.24–1.40)	–
GUSTO V Reteplase vs. abciximab and reteplase therapy, MI [88]	16,588 (24.6%)	30-day mortality		2.00 (1.59–2.53) vs. men	P<0.01
		Moderate or severe bleeding		1.31 (1.18–1.45) vs. men	P<0.001
HEAAL Losartan, HFrEF [326]	3834 (39%)	Death or HF admission	0.86 (0.88–0.96)	1.02 (0.85–1.23)	0.10
HOT Hypertensive patients Aspirin [162]	18,790 (47%)	Total MI	0.58 (0.41–0.81)	RR 0.81 (0.49–1.31)	Benefit only in men
JUPITER Primary prevention trial Rosuvastatin [305]	17,802 (38.2)	Revascularization/unstable angina	0.63 (0.46–0.85)	0.24 (0.11–0.51)	0.01
MERIT-HF • Metoprolol CR/XL in women with HFrEF NYHA II–IV, EF <40% [220]	3991 (29%)	Total mortality	0.90 (P=0.044)	0.63 (0.43–0.91) vs. men	0.015
		All-cause hospitalizations	0.87 (P=0.005)	0.81 (P=0.044)	
		CV hospitalization	0.70 (P=0.0001)	0.73 (P=0.013)	
		HF hospitalization		0.55 (P=0.021)	
• Metoprolol CR/XL in women with severe HFrEF-NYHA III–IV, EF <25% [220]	785 (23%)	All-cause hospitalizations	0.80 (P=0.044)	0.56 (P=0.016)	
		CV hospitalization	0.75 (P=0.01)	0.43 (P=0.005)	
		HF hospitalization	0.65 (P=0.01)	0.28 (P=0.0004)	
PARAGON-HF Sacubitril/valsartan, HFpEF [288]	4796 (51%)	First and recurrent HFH and CV death	1.03 (0.85–1.25)	0.73 (0.59–0.90)	0.017
		Total HF hospitalizations	1.06 (0.84–1.34)	0.67 (0.54–0.84)	0.004
		1st hospitalization for HF	1.09 (0.90–1.33)	0.72 (0.59–0.87)	0.002
PURSUIT Investigators Eptifibatide ACS-NSTEMI [177]	10,948 (35.5%)	Death and nonfatal MI occurring up to 30 days	0.80 (0.70–0.90)	1.10 (0.91–1.34)	Statistically significant

Continued

TABLE 3 Sex-Related Differences in Main End Point Results for Clinical Trials and Meta-Analysis of Common Cardiovascular Drugs—Cont'd

Study Acronym, Drug, CVD. [Reference]	Sample Size n (% Women)	End Point	Men (HR/RR, 95% CI) or RRR[a]	Women (HR/RR, 95% CI) or RRR[a]	P (Men vs. Women)[b]
ROCKET-AF Rivaroxaban vs. warfarin NVAF [139]	14,264 (39.7%)	Major and nonmajor clinically relevant bleeding	1.12 (1.02–1.22)	0.89 (0.79–1.01)	0.004
Second Australian National Blood Pressure Study ACEIs vs. diuretics in elderly hypertensives [271]	6083 (51%)	All-CV events or total death; 1st CV event or total death; All-cause death	0.83 (0.71–0.97; P=0.02); 0.83 (0.71–0.97; P=0.02); 0.83 (0.66–1.06)	1.00 (0.83–1.21); 1.00 (0.83–1.20); 1.01 (0.76–1.35)	NS
SENIORS Nebivolol, HFrEF, aged ≥70 years [327]	2128 (36.8%)	All-cause death or CV hospital admission	0.93 (0.78–1.11)	0.72 (0.55–0.93)	0.11
TOPCAT Spironolactone, HFpEF [328]	1767 (49.9%)	All-cause mortality	0.85 (0.59–1.22; P=0.37)	0.63 (0.42–0.96; P=0.026)	0.02
TRACE Trandolapril, HFrEF [329]	6676 (28%)	All-cause death	0.74 (0.62–0.89)	0.90 (0.69–1.18)	–
Meta-Analysis/Pooled Analysis					
ACEIs vs. ARBs in patients with HFrEF [280]	Men: 9475 (8484 on ACEIs, 991 on ARBs) Women: 10,223 (8627 on ACEIs, 1596 on ARBs)	Survival: ARBs vs. ACEIs; + Hypertension; – Hypertension	1.10 (0.95–1.30); 0.86 (0.64–1.15); 1.20 (1.01–1.43)	0.69 (0.59–0.80); 0.65 (0.52–0.80); 0.72 (0.59–0.90)	<0.0001; NS; <0.03
ACEIs Symptomatic/asymptomatic LV systolic dysfunction [30]	Treatment studies: CONSENSUS, SOLVD-T, TRACE 4497 (24%)	Mortality from HF	0.80 (0.68–0.93)	0.90 (0.78–1.05) RRR 1.15 (0.88–1.51)	0.07
	Prevention studies: SAVE, SOLVD-P, SMILE 8015 (16.1%)	Mortality from HF	0.83 (0.71–0.96)	0.96 (0.75–1.22) RRR 1.25 (0.94–1.65)	NS

Study	N	Outcome			P
Aspirin Primary prevention of CVD in women 3–5 RCTs [35]	39,876 healthy women ≥45 years	Major CV events Stroke Ischemic stroke Hemorrhagic stroke Myocardial infarction CV death		0.91 (0.80–1.03) 0.83 (0.69–0.99) 0.76 (0.63–0.93) 1.24 (0.82–1.87) 1.02 (0.84–1.25) 0.95 (0.74–1.22)	0.68 0.13 0.04 0.009 0.31 0.83
		MI Stroke	0.68 (0.54–0.86) 1.13 (0.96–1.33)	0.99 (0.83–1.19) 0.81 (0.69–0.96)	0.01 0.005
Aspirin Primary prevention of vascular disease 6 RCTs [164]	95,000	Major coronary events Ischemic stroke Serious vascular events		0.95 (0.77–1.17) 0.77 (0.59–0.99) 0.88 (0.76–1.01)	
Aspirin Secondary prevention of vascular disease 16 RCTs [164]	17,000	Major coronary events Ischemic stroke Serious vascular events		0.73 (0.51–1.03) 0.91 (0.52–1.57) 0.81 (0.64–1.02)	
Aspirin 6 primary prevention RCTs [163]	95,456 (53.8%)	Major CV events and MI MI Stroke Ischemic stroke Hemorrhagic stroke	0.86 (0.78–0.94; P=0.01) 0.68 (0.54–0.86; P=0.001) 1.13 (0.96–1.33; NS) 1.00 (0.72–1.41; NS) 1.69 (1.04–2.73; P=0.03)	0.88 (0.79–0.99; P=0.03) 1.01 (0.84–1.21; NS) 0.83 (0.70–0.97; P=0.02) 0.76 (0.63–0.93; P=0.008) 1.07 (0.42–2.69; NS)	Not reported
Aspirin for primary prevention 13 RCTs [330]	164,225 (43%)	MACE MI Stroke	0.89 (0.83–0.96; P=0.0008) 0.76 (0.57–1.01; P=0.06) 1.02 (0.72–1.44; P=0.93)	0.95 (0.88–1.22; P=0.16) 1.03 (0.84–1.25; P=0.26) 0.77 (0.63–0.94; P=0.01)	Not reported
Beta-blockers Pool data from CIBIS II, MERIT-HF, and COPERNICUS trials, HFrEF [220]	9829 (24%)	Total mortality	0.66 (0.58–0.75)	0.69 (0.51–0.93)	NS
Bivalirudin vs. heparin + Gp IIb/IIIa inhibitor 3 RCTs in women undergoing PCI [121]	7413 on bivalirudin (25%) 7371 on UFH + GP (26%)	30-day non-CABG-related major bleedings		1.80 (1.52–2.11) vs. men	<0.001
		1-year mortality bivalirudin vs. heparin + GP	0.90 (0.71–1.12; P=0.33)	0.66 (0.47–0.93; P=0.02)	NS

Continued

TABLE 3 Sex-Related Differences in Main End Point Results for Clinical Trials and Meta-Analysis of Common Cardiovascular Drugs—Cont'd

Study Acronym. Drug. CVD. [Reference]	Sample Size n (% Women)	End Point	Men (HR/RR, 95% CI) or RRR[a]	Women (HR/RR, 95% CI) or RRR[a]	P (Men vs. Women)[b]
Clopidogrel Sex-specific meta-analysis of 5 RCTs [189]	79,613 (30%)	Major CV events MI Stroke All-cause mortality	0.84 (0.78–0.91) 0.83 (0.76–0.92) 0.83 (0.71–0.96) 0.91 (0.84–0.97)	0.93 (0.86–1.01) 0.81 (0.70–0.93) 0.91 (0.69–1.21) 0.99 (0.90–1.08)	0.092 0.73 0.55 NS
Dabigatran Population-based cohort [135]	63,119 (50.4%)	Stroke Bleeding	150 mg: 0.98 (0.78–1.23) 150 mg: 0.73 (0.64–0.84)	150 mg: 0.79 (0.56–1.04) 150 mg: 0.85 (0.71–1.01)	0.008
Direct oral anticoagulants Atrial fibrillation [143]	146,869 (55.2%)	MI: rivaroxaban vs. warfarin MI: rivaroxaban vs. dabigatran	0.59 (0.38–0.91; P=0.03) 0.67 (0.44–1.01; P=0.06)	0.94 (0.65–1.37; P=0.76) 0.98 (0.67–1.43; P=0.92)	Not reported
Direct oral anticoagulants Atrial fibrillation [144]	9806 (49%)	Intracranial hemorrhage All-cause mortality	0.55 (0.27–1.10) 0.83 (0.59–1.16; P=0.27)	0.16 (0.06–0.40; P<0.001) 0.55 (0.39–0.77; P<0.001)	0.037 0.087
Direct oral anticoagulants Atrial fibrillation, 5 RCTs [141]	66,389 (37.8%)	Stroke and systemic embolism Major bleeding		1.19 (1.04–1.35) vs. men 0.86 (0.78–0.94) vs. men	<0.01 <0.0009
Direct oral anticoagulants VTE 8 RCTs [93]	9417 (43%)	Recurrent VTE (men vs. women) Bleeding (men vs. women)	1.02 (0.74–1.39) 0.79 (0.66–0.97)		NS 0.03
Glucose-lowering drugs DPP-4Is, GLP-1RAs and SGLT2Is added to metformin in T2DM [250]	167,254 (46%)	First nonfatal major CV: AMI, UA HF, or stroke (relative to sulfonylureas)	DPP-4Is: 0.85 (0.79–0.91) GLP-1RAs: 0.82 (0.71–0.95) SGLT2Is: 0.69 (0.57–0.83)	0.83 (0.77–0.89) 0.57 (0.40–0.68) 0.58 (0.46–0.74)	0.60 0.002 0.27
Glycoprotein IIb/IIIa inhibitors ACS 6 RCTs [180]	31,402 (35%)	Death or MI Death	0.81 (0.75–0.89) 0.83 (0.71–0.96)	1.15 (1.01–1.30) 1.08 (0.89–1.33)	<0.0001* 0.03*
Recombinant tissue plasminogen activator Acute ischemic stroke [63]	2178 (45%)	Modified Rankin score ≤1	RR 1.04, P=0.52	RR 1.34, P<0.0008	0.04

Statin Therapy Prevention of Recurrent CV Events Meta-analysis of 11 trials [308]	43,193 (20.6%)	Any CV event (cardiac and/or cerebrovascular)	0.82 (0.78–0.85; P<0.001)	0.81 (0.74–0.89; P<0.001)	NS
		All-cause mortality	0.79 (0.72–0.87; P<0.001)	0.92 (0.76–1.13; P=0.44)	No benefit in women
		Any stroke	0.81 (0.72–0.92; P<0.001)	0.92 (0.76–1.10; P=0.36)	No benefit in women

[a] HR are listed; when not available a risk ratio or relative risk reduction was listed. When not explicitly listed, these ratios were calculated from data presented in the manuscripts.

[b] P value between men and women, when provided.

* Differences disappeared when patients were stratified according to the troponin concentrations.

ACCORD, Action to Control Cardiovascular Risk in Diabetes; ACEI, angiotensin-converting enzyme inhibitor; ACS, acute coronary syndrome; ACS-NSTEMI, acute coronary syndromes without ST segment elevation; ACUITY, Acute Catheterization and Urgent Intervention Triage strategY; AMI, acute myocardial infarction; ARB, angiotensin-receptor blockers; ARISTOTLE, Apixaban for the Prevention of Stroke in Subjects With Atrial Fibrillation; CABG, coronary artery bypass graft; CAD, coronary artery disease; CARE, Cholesterol and Recurrent Events trial; CHAMPION PHOENIX, Clinical Trial Comparing Cangrelor to Clopidogrel Standard of Care Therapy in Subjects Who Require Percutaneous Coronary Intervention; CIBIS II, Cardiac Insufficiency BIsoprolol Study II; CONSENSUS, Cooperative North Scandinavian Enalapril Survival Study; COPERNICUS, Carvedilol Prospective Randomized Cumulative Survival Study; CR, controlled release; CV, cardiovascular; CVD, cardiovascular disease; DIG, Digitalis Investigation Group; DM, diabetes mellitus; DPP-4Is, dipeptidyl peptidase-4 inhibitors; EF, ejection fraction; FRISC, Fragmin and Fast Revascularization during In-Stability in Coronary artery disease; GLP-1RAs, glucagon-like peptide-1 receptor agonists; GP, glycoprotein; GUSTO-I, Global Utilization of Streptokinase and Tissue Plasminogen Activator for Occluded Coronary Arteries; GUSTO V, The Gusto V Randomised Trial; HEAAL, Heart failure Endpoint evaluation of Angiotensin II Antagonist Losartan; HF, heart failure; HFH, heart failure hospitalization; HFpEF, heart failure with preserved ejection fraction; HFrEF, heart failure with reduced ejection fraction; HORIZONS-AMI, Harmonizing Outcomes with Revascularization and Stents in Acute Myocardial Infarction; HOT, The Hypertension Optimal treatment trial; HR, hazard ratio; JUPITER, Justification for the Use of Statins in Primary Prevention: An Intervention Trial Evaluating Rosuvastatin; LV, left ventricle; MACE, major adverse cardiac events; MERIT-HF, Metoprolol Extended-release Randomized Intervention Trial in Heart Failure; MI, myocardial infarction; NS, not significant; NVAF, nonvalvular atrial fibrillation; NYHA, New York Heart Association; PARAGON-HF, Prospective Comparison of ARNI with ARB Global Outcomes in HF With Preserved Ejection Fraction; PCI, percutaneous coronary intervention; PROMETHEUS, The use of prasugrel versus Clopidogrel as used in clinical practice in patients with acute coronary syndromes undergoing percutaneous coronary intervention; PURSUIT, Platelet Glycoprotein IIb/IIIa In Unstable Angina:Receptor Suppression Using Integrilin Therapy; RCT, randomized clinical trials; REPLACE-2, Randomized Evaluation in PCI Linking Angiomax to Reduced Clinical Events-2; ROCKET-AF, Rivaroxaban Once Daily Oral Direct Factor Xa Inhibition Compared with Vitamin K Antagonism for Prevention of Stroke and Embolism Trial in Atrial Fibrillation; RR, relative risk; RRR, relative risk reduction; SAVE, Survival And Ventricular Enlargement; SENIORS, Study of the Effects of Nebivolol Intervention on Outcomes and Rehospitalisation in Seniors with Heart Failure; SGLT2Is, sodium-glucose cotransporter 2 inhibitors; SMILE, Survival of Myocardial Infarction Long-term Evaluation; SOLVD, Studies Of Left Ventricular Dysfunction; T2DM, type 2 diabetes mellitus; TOPCAT, Treatment of Preserved Cardiac Function Heart Failure With an Aldosterone Antagonist Trial; TRACE, Trandolapril Cardiac Evaluation; UA, unstable angina; UFH, unfractionated heparin; VTE, venous thromboembolism; XL, extended release.

TABLE 4 Sex-Related Differences in the Adverse Drug Reactions of Cardiovascular Drugs in Women Compared to Men

Drug Class	Outcome in Women
Alcohol [26]	Higher vulnerability to acute and chronic complications of alcoholism
Analgesic drugs [332]	More adverse effects to perioperative analgesic drugs
Anesthetic drugs [27–29]	More prone to the side effects of anesthetic drugs
Angiotensin-converting enzyme inhibitors [262, 333]	Cough occurs more frequently No sex differences in the occurrence of angioedema or urticaria
Anorexigens (dexfenfluramine and fenfluramine) [334]	Cardiac valvulopathy occurs more frequently
Antiarrhythmic drugs [55–58, 335]	Women have a longer QTc interval Increased risk of torsades de pointes on QT-prolonging drugs, and higher risk of torsades de pointes
Anticoagulants [89, 90, 121, 124–126]	Increased risk of minor/major bleedings
Antiplatelets [79–81, 86, 155, 180, 181]	Increased risk of minor/major bleedings
Aspirin [25, 35, 91, 158, 163, 164]	Increased risk of bleedings and gastrointestinal complications. Aspirin resistance tends to be more common among women
Beta-blockers [36–38, 213]	More adverse drug reactions when dosage is not adequately adjusted
Calcium channel blockers [227]	Higher risk of edema
Digoxin [46, 47, 336]	Higher mortality among women with HFrEF Digoxin plasma levels between 0.5 and 0.9 ng/mL are recommended in women, compared to men
Diuretics: thiazides and loop diuretics [62, 337, 338]	Higher rates of hospitalizations due to hypo-osmolarity and electrolyte disturbances (hypokalemia and hyponatremia). These electrolyte disturbances have the potential to cause cardiac arrhythmias
Drug-induced hepatotoxicity [339, 340]	Women present a higher risk of liver injury induced by amiodarone, antiepileptics, antimicrobials, immunosuppressants, NSAID (acetaminophen), or statins
Drug-induced lupus erythematosus [341, 342]	Higher risk of systemic lupus erythematosus secondary to hydralazine or procainamide
Fibrinolytic drugs [79, 80, 87, 88, 104, 201–203, 210, 211]	Higher risk of bleeding and intracranial hemorrhagic
Low-molecular-weight heparins [106–110]	Minor bleedings associated with the use of dalteparin were more frequent in women compared with men either with the weight adjusted and the fixed-dose treatment
Mineralocorticoid receptor antagonists [343]	Higher discontinuation of treatment because of hyperkalemia, and deterioration of renal function in women, and gynecomastia in men
Opioid receptor agonists (fentanyl and morphine) [27, 29, 53, 344]	More nausea, vomiting, and respiratory depression despite smaller dose requirements for pain control

TABLE 4 Sex-Related Differences in the Adverse Drug Reactions of Cardiovascular Drugs in Women Compared to Men—cont'd

Drug Class	Outcome in Women
Statins [59, 310–314]	Increased risk of statin-induced myalgias and diabetes
Thiazolidinediones [345–347]	Increased risk of fractures and decreased bone mineral density at the lumbar spine and hip in diabetic women, but not in men
Unfractionated heparin [79, 80, 101–103, 109]	More likely to achieve higher plasma levels, activated partial thromboplastin time (aPTT) and bleeding risk
Warfarin [25, 65, 66, 79–83, 96, 99, 100, 113]	Higher risk of bleeding, although in some studies the major bleeding risk associated with warfarin is similar in women and men and highlights a potentially higher benefit in women

aPTT, activated partial thromboplastin time; HFrEF, heart failure with reduced ejection fraction; NSAID, nonsteroidal antiinflammatory drugs.

TABLE 5 Potential Explanations for Sex-Related Differences in Adverse Drug Reactions

Sex Difference	Pharmacological Reasons
Women are more frequently overdosed	1. Pharmacokinetic factors (**Table 1**): • Differences in body composition • Smaller Vd for hydrophilic drugs • Higher Vd for lipophilic drugs • Higher fraction of protein-unbound drugs • Slower renal clearance of renally excreted drugs 2. Administration of fixed doses, not adapted to body weight or creatinine clearance, leading to higher plasma levels
Fluctuations in sex hormone levels	• Endogenous: during menstrual cycle or pregnancy • Exogenous hormones (oral contraceptives, hormone replacement therapy)
Women are more sensitive to the drug effects	1. Pharmacodynamic factors related to differences in: • Drug-receptor affinity • Receptor expression and/or binding • Signal transduction signaling pathways • Immunological and hormonal factors 2. Women are older and have more comorbidities than men • Drug exposure increases in elderly women due to reduced metabolism and excretion
Polypharmacy	• Women consume more drugs (prescribed, over-the-counter, herbal medicines, and supplements) than men • Polypharmacy increases the risk of drug interactions and adverse drug reactions • Differences in prescribing guideline-based drug therapy • Differences in the way that men and women report adverse drug reactions

Vd, volume of distribution. *Modified from [16].*

WOMEN

Antiarrhythmics
Increased risk of QT prolongation

Antithrombotics
Higher risk of bleeding

Aspirin
Greater protection for stroke

Beta-Blockers
Greater decrease in BP & HR
Lesser decrease in angina
Similar decrease in HF hosp. & mortality

Calcium Channel Blockers
More edema
Greater decrease in BP & HR
Similar decrease in CV outcomes

Diuretics
More electrolyte disturbances

Statins
Similar efficacy for prevention of ASCVD

Warfarin and Heparin
Increased sensitivity, lower dose needed

MEN

Antiarrhythmics
Decreased risk of QT prolongation

Antithrombotics
Lower risk of bleeding

Aspirin
Greater protection for MI

Beta-Blockers
Lesser decrease in BP & HR
Greater decrease in angina
Similar decrease in HF hosp. & mortality

Calcium Channel Blockers
Less edema
Lesser decrease in BP & HR
Similar decrease in CV outcomes

Diuretics
Fewer electrolyte disturbances

Mineralocorticoid Receptor Antagonists
More gynecomastia & hyperkalemia

Warfarin and Heparin
Decreased sensitivity, higher dose needed

FIGURE 3 Sex-Related Differences in Adverse Drug Reactions With Common Cardiovascular Drugs. ASCVD, atherosclerotic cardiovascular disease; BP, blood pressure; CV, cardiovascular; HF, heart failure; HR, heart rate; MI, myocardial infarction.

pointes (TdP) [55–58, 335, 355]. In fact, almost two-thirds of drug-induced TdP induced by cardiovascular (class IA and III antiarrhythmic drugs: amiodarone, d-sotalol, dofetilide, disopyramide, ibutilide, and quinidine; antianginal drugs: bepridil and prenylamine) or noncardiovascular drugs (some antibiotics, H1 antihistamines, tricyclic antidepressants, antipsychotics, and gastrokinetics) occurred in women [55–58, 335]. The prevalence of drug-induced QT prolongation and TdP remained higher in women than in men even at equivalent plasma concentrations, after careful dosing based on body weight and renal function [55, 356].

The longer QTc interval in women can be explained by fluctuations in circulating levels of sex hormones. SRDs in QTc interval are not present at birth, but at the onset of puberty, and the longest QT intervals occurred during menses and the ovulatory phase and the shorter intervals during the luteal phase, which correlated with higher serum concentrations of progesterone [357, 364, 365]. These observations suggest that sex hormones can modulate the expression and function of cardiac ion channels involved in ventricular repolarization (**Table 6**) [55–58, 329, 335, 357–363]. However, the finding that HRT did not modify the QTc interval in postmenopausal women [366] suggests that other nonhormonal mechanisms may be important in the differences in the QT interval between men and women.

Female Representation in Clinical Trials

Women have been underrepresented in RCTs of cardiovascular drugs, with the proportion of women in RCTs remaining much lower than the proportion of women with CVD [74, 75, 77]. A recent report confirmed that the proportion of women enrolled in 36 pivotal CV drug trials with Food and Drug Administration (FDA) approval between 2005 and 2015 ranged from 22% to 81% (mean 46%) across different cardiovascular areas [78]. Women were well represented in RCTs for arterial hypertension and AF, overrepresented for pulmonary arterial hypertension, and underrepresented (24%) in ASCVD and HF trials—that is, the most common CVD affecting women [1, 2, 6, 77, 78].

Little progress has been made in the reporting of sex in RCTs. Among 801,198 patients enrolled in 156 cardiovascular trials, only 30.6% were women and only one-third of the trials reported sex-specific results, which confirms that many RCTs are underpowered to examine sex-specific differences even if they do exist [9, 75, 367]. In a recent

TABLE 6 Mechanisms Involved in the Sex-Related Differences of the QT Interval

1. Sex hormones can modulate cardiac K^+ and Ca^{2+} ion channels involved in ventricular repolarization [55–58, 335]
   - Testosterone ↑ the rapid (I_{Kr}) and slow (I_{Ks}) components of the delayed rectifier and the inward rectifier (I_{K1}) K^+ currents and ↓ the L-type Ca^{2+} current (I_{CaL}), shortening the QT interval [357, 358]
   - Progesterone ↑ I_{Ks} and ↓ I_{CaL} shortening the QT interval [357, 358]
   - Estrogens ↓ I_{Kr}, I_{Ks}, and I_{CaL}, prolong the QT interval and facilitate bradycardia- and drug-induced QT prolongation and arrhythmias [57, 335, 358, 359]

2. Female hearts express fewer voltage-gated channels
   - ↓ Expression of K^+ channel subunits (Kv1.4, HERG, minK, KChiP2, SUR2, Kir2.3, and Kir6.2), L-type Ca^{2+} channels, connexin43, and phospholamban compared with male hearts [360–362]

3. Female ventricular myocytes present less I_{Kr} and I_{K1} density than male cells [360] and an increased transmural dispersion in I_{CaL} density (epicardium > endocardium) [363]
   - These differences prolong the repolarization and increase the transmural dispersion of repolarization, which are important for induction of TdP

TdP, torsades de pointes.

systematic analysis of 555 Cochrane reviews, only 25% reported analytic approaches for exploring sex, such as subgroup analyses, exploring heterogeneity, or presenting disaggregated data by sex [368]. Furthermore, a recent study concluded that the increasing trend in the number of studies reporting sex-differentiated results from 2008 to 2018 merely paralleled the increase in the number of papers published during this time period [369]. Therefore, important sex-related evidence, including findings of no differences in effectiveness based on sex, may be missed. Furthermore, despite the strong evidence that the incidence of ADRs is significantly higher in women, most phase III trials are not powered to ascertain ADRs and, therefore, ADRs were not analyzed stratified by sex. This is particularly relevant because ADRs reported during postmarketing surveillance are commonly reported in women, and could be responsible for the higher rates of drug discontinuation and nonadherence to cardiovascular drugs in women than in men [351]. In a recent systematic review of sex-specific ADR data for guideline-recommended HF drugs, only 7% of 155 studies reported ADR data for women and men separately [343]. Interestingly, this review suggested that despite the scarcity of sex-specific ADR, women may experience different ADRs than men when treated with the same HF drugs.

Because prespecified sex-specific analysis of drug efficacy and safety is not considered in the trial design, SRDs are retrieved from retrospective, observational nonrandomized studies, subgroup and post-hoc analysis, or from meta-analyses of multiple studies recruiting women with different baseline characteristics [11, 16, 74]. A retrospective analysis can at best generate plausible hypotheses and

although pooling studies might solve problems related to small underpowered trials, it is important to recognize that by increasing the numbers of women in the meta-analyses, the precision of a point estimate might increase, but the accuracy or validity of the treatment effect might not necessarily be improved. Indeed, biases of individual studies might be magnified, rather than mitigated, when pooled together [294]. On the other hand, subgroup analyses in systematic reviews are rarely credible and are not reproduced in subsequent studies [370, 371]. Therefore, at the present time, it seems unlikely to conclude that there are no SRDs in efficacy and safety of cardiovascular drugs. It is likely that the lack of differences between men and women is related to underpowered studies rather than a true absence of difference.

Many pharmacokinetic studies are performed mainly in healthy male volunteers, but the results are extrapolated to patients of both sexes with different CVDs, so the sex-specific pharmacokinetic properties of the CV drugs remain uncertain. Of note, the FDA reviewed 300 new drug applications between 1995 and 2000, and although 11 drugs showed a >40% difference in pharmacokinetics between men and women, no dosing recommendations were made based on sex [19].

Because clinical trials with cardiovascular drugs often failed to evaluate and report sex-related pharmacodynamic/pharmacokinetic differences systematically, the clinical relevance of many SRDs on clinical outcomes remains unproven. This is an important point because translation of evidence into clinical practice only occurs in populations adequately represented in RCTs. Therefore, current clinical guidelines for the prevention and treatment of CVD are based on RCTs conducted predominantly in men and even when the scarce information on SRDs in drug efficacy/safety might prevent the development of evidence-based guidelines, there are clear SRDs in drug safety (and sometimes in safety), which surprisingly are not incorporated in the guidelines [1]. Thus, current guidelines on CVD recommend to treat women like men, even when we lack proper evidence to support such a recommendation [16, 75]. As a result, treatment choices may be less optimal for women, leading to more frequent ADRs or suboptimal outcomes.

Medication Use, Adherence, and Prescribing Alignment With Clinical Guidelines

In general, women take more medications, but show lower adherence to the prescribed medications, and are less likely to receive guideline-recommended pharmacological therapies than men despite being at a similar cardiovascular risk [8, 23, 353, 372]. Nonadherence to life-saving cardiovascular drugs is a major problem in women as it worsens clinical outcomes and increases cardiovascular mortality and healthcare costs [353, 373, 374]. Therefore, increasing the adherence to cardiovascular drugs represents one of the most effective and efficient ways of improving health outcomes [373]. Women are also less likely to receive preventive treatment (i.e., statins and aspirin) than men at similar ASCVD risk. Similarly, when antihypertensive or cholesterol- or glucose-lowering drugs are prescribed, the treatment is less aggressive, contributing to increased challenges with achieving the recommended blood pressure, LDL-C, and HbA1c target levels, respectively [1, 5, 295, 302, 375].

Women are also less likely to receive the guideline-recommended pharmacological therapy for ASCVD, ACS, and HF [353]. A large Dutch database reported that younger women showed the lowest use of antithrombotics, statins, β-blockers, and other blood pressure-lowering drugs; they were also less likely than men to be on combination therapy [376]. In Sweden, women received significantly less antihypertensive drugs than men and women were more often treated with diuretics, ARBs, and β-blockers, while men received more ACEIs and calcium channel blockers [377]. In a review of three recent HFrEF trials, diuretics, aspirin, statins, ACEIs and ARBs, and anticoagulants were underutilized in women [378]. In a German study, women with HFrEF were less frequently treated with ACEIs, ARBs, or β-blockers and, when prescribed doses, these were lower than those for men [379]. Women with stable angina were significantly less likely to receive aspirin, β-blockers, and statins than men [380, 381]. Among patients with an ACS,

women were less likely than men to receive guideline-recommended secondary prevention drugs, including statins, aspirin, and ACEIs, potentially because MI has traditionally been considered a man's disease [9]. Despite having higher in-hospital risk, women with NSTEMI were less likely to receive UFH, ACEIs and GP IIb/IIIa inhibitors, and less commonly received aspirin, ACEIs, and statins at discharge [382]. Compared with men, women with STEMI were less likely to be prescribed β-blockers, ACEIs or ARBs, statins, and dual antiplatelet therapy at discharge, but were more likely to be prescribed nitrates (drugs with a less-established role in current guidelines) [383]. Furthermore, women with AF had a higher risk of stroke and thromboembolism than men, but sex differences in stroke risk disappeared with use of oral anticoagulation [384, 385]. However, women are significantly less likely than men to receive oral anticoagulants at all levels of the CHA_2DS_2-VASc score [386]. Additionally, in part related to underrecognition of their higher thromboembolic risk and their higher risk of major bleeding during hospitalization, women received less antiplatelets and anticoagulants than men.

Sometimes the intensity of guideline-recommended treatment of CVD is influenced by physicians' and patients' gender, physicians' interpretation of women's symptoms, or the reason is that women are older, forgetting that they outlive men [16]. In patients with chronic HF, male physicians used significantly less drugs and lower doses in female patients, and guideline-recommended drug use and achieved target doses compared to patients treated by female physicians [379]. A recent study found that following an AMI, women were less likely to survive when treated by male physicians and that male physicians with more exposure to female patients and female physicians have more success in treating female patients [387]. Physicians should be cognizant of this limitation in order to avoid bias in the treatment of their female patients.

Current Challenges and Next Steps

The present challenges in the analysis of SRDs in cardiovascular drug development and potential solutions to change the perception of the relevance of the SRDs of cardiovascular drugs are summarized in **Table 7** and **Figure 4**. Evaluation of the potential role of sex at each stage of drug development represents the first step to identify SRDs in dosing, efficacy, and safety of cardiovascular drugs. The enrollment of an adequate number of women at all stages of drug development with statistically sufficient power and the prespecified design, analysis, and report of SRDs in cardiovascular clinical end points that are important for women are key steps to better understand SRDs in drug efficacy and safety. It is also critical to reduce ADRs and improve adherence and compliance to cardiovascular pharmacotherapy

TABLE 7 Recommendations to Improve the Evidence-Based Clinical Practice in Women With Cardiovascular Diseases

Present Situation	Future Recommendations
There are sex-related differences in the efficacy and safety of cardiovascular drugs • There are SRDs in body composition, sex hormonal regulation, and in the PK and PD (efficacy and safety) of CV drugs • The interaction between CV drugs and sex hormones is poorly understood • SRDs in the efficacy/safety of some drugs have not been investigated • Women experience more ADRs and these were more severe, but most phase 3 trials are not powered to ascertain ADRs	• Evaluate the potential of sex and gender at each stage of drug development • Determine the role of sex on the changes in the PD/PK variations induced by CVD • Understand how endogenous sex hormonal changes, oral contraceptives, and hormonal replacement therapy can modify the PD/PK of CV drugs • Understand the mechanisms responsible for higher risk of ADRs in women • Identifying SRDs in the efficacy and safety of CV drugs is the first step for a personalized medicine • The lack of scientific evidence should stimulate basic and clinical research to better understand the mechanisms underlying SRDs in the efficacy and safety of CV drugs
Women are underrepresented in CV drug development • Inclusion of females in basic and clinical research is inadequate • Often RCTs are underpowered to reach sex-specific conclusions • This prevents optimization of therapy for women	• Enhance the inclusion of females at all stages of drug development • Identify and address barriers that limit participation of women in clinical trials • Develop strategies that enhance women recruitment in RCTs • Sample sizes must be adequately large to ensure a statistically sufficient power
The results of clinical trials are not analyzed to identify SRDs • RCTs often fail to report SRDs in a systematic manner • Sex-specific analysis is simply descriptive • Subgroup samples are often too small to provide reliable results • Post-hoc analyses and meta-analysis are frequently used • The mechanisms involved in SRDs are not analyzed	• Consider sex-based stratification analysis when designing and reporting clinical trials • RCTs should be powered to identify SRDs in the efficacy and safety of CV drugs • Outcomes important for women should be considered in the design and analysis of RCTs • Results of RCTs should analyze and report efficacy and safety separately for women and men • Scientific journals, advisory boards of drug companies and governmental regulatory agencies should require reporting sex-specific results
There are marked discrepancies in drug prescription and adherence • These discrepancies may reflect SRDs in the epidemiology, pathophysiology, and clinical manifestations of CVD • Women receive more drugs, but they are less adherent to medications • Women are less likely to receive evidence-based secondary prevention and treatments • Sex-differentiated dosage data are lacking for most CV drugs	• Clinicians need to keep in mind the SRDs in the epidemiology, pathophysiology and clinical manifestations of CVD • Women should be prescribed the CV drugs recommended in clinical guidelines for the prevention and treatment of CVD • Understand the reasons for SRDs in drug prescription and adherence to CV • An evidence-based pharmacotherapy in women is desirable to improve women's health • Necessary to know whether women need different doses of CV drugs • Improving drug adherence is one of the most efficient ways of improving clinical outcomes and decreasing total health-care costs

Continued

TABLE 7 Recommendations to Improve the Evidence-Based Clinical Practice in Women With Cardiovascular Diseases—cont'd

Present Situation	Future Recommendations
Poor translation of sex-related differences to clinical practice • Limited recognition of SRDs in prescription, adherence, and responses to cardiovascular drugs • Underrepresentation of women in RCTs • Evidence-based CV pharmacotherapy in women is still lacking • SRDs in drug efficacy and safety are generally not mentioned in clinical guidelines • Even for drugs that showed a >40% difference in PK between women and men, no dosing recommendations are made based on sex	• Pharmacological research should determine the influence of sex on CV drug effects • Further studies are needed to determine the clinical relevance of SRDs in the efficacy and safety of CV drugs with a narrow therapeutic margin • SRDs in drug efficacy and safety should be included in clinical guidelines • SRDs should be considered during the selection and dosages of CV medications • Sex-specific dosage recommendations for CV drugs should be included on labels • Improvement in women's cardiovascular health requires a multidisciplinary approach • Possible SRDs in ADRs should be studied and reported in phase 2 and phase 3 clinical trials • The development of sex-specific pharmacological guidelines is the most effective strategy to minimize the higher incidence of ADR in women
Education • Clinical awareness of the implications of SRDs on efficacy, dosing, and ADRs in daily clinical practice is inadequate	• SRDs of CV drugs should be incorporated in biomedical research and medical education • Develop educational programs to increase awareness of the SRDs of CV drugs • Disseminate results regarding any significant SRDs in CV drug efficacy and safety • Healthcare professionals need to incorporate SRDs in the efficacy and safety of CV drugs to ensure high-quality health care

ADRs, adverse drug reactions; CV, cardiovascular; CVD, cardiovascular diseases; PD, pharmacodynamic; PK, pharmacokinetic; RCTs, randomized clinical trials; SRDs, sex-related differences.

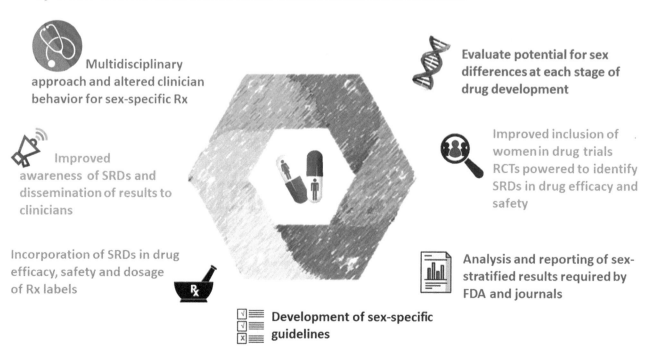

FIGURE 4 **Next Steps for Sex-Specific Practice of Drug Prescription.** FDA, Food and Drug Administration; RCTs, randomized clinical trials; Rx, prescription; SRDs, sex-related differences.

to improve clinical outcomes in women. SRDs in cardiovascular drug efficacy and safety should be part of medical education and must be incorporated in clinical guidelines.

Conclusion

Significant SRDs have been described in body composition and hormonal regulation as well as in the pharmacodynamics/pharmacokinetics of some cardiovascular drugs. There are also important SRDs in the prescription, adherence, and the response to many guideline-recommended cardiovascular drugs. Furthermore, women present a higher incidence of ADRs, leading to an increase in drug nonadherence to guideline-recommended drugs for the prevention and treatment of CVD, which ultimately increases CVD death and disability and total healthcare costs. However, the possible SRDs in the pharmacokinetics/pharmacodynamics of some cardiovascular drugs have not been investigated; the role of SRDs is still neglected in daily clinical practice and translation of these differences into clinical practice is slow. This gap of knowledge should stimulate basic and clinical research to better understand the role of SRDs in the efficacy and safety of cardiovascular drugs.

It is generally mentioned that the clinical impact of SRDs of cardiovascular drugs is moderate. This can be explained because historically women/females were underrepresented at all stages of drug development, leading to underpowered RCTs, and sex-specific analysis of drug efficacy and safety was not considered in the trial design, so it was difficult to identify SRDs even if they do exist. Because of the limited information on SRDs, current clinical guidelines generally recommend (with a few exceptions) to treat women like men, even in the absence of adequate evidence to support such recommendations. As a result, treatment choices may be less optimal for women, leading to more frequent ADRs or suboptimal outcomes.

Adequate inclusion of women at all stages of drug development, a priori sex-specific analysis, mandatory reporting of sex-specific results, improved sex-specific medical education, and development of sex-specific clinical guidelines will be important to better optimize medical therapy in both men and women. Ultimately, a better understanding of SRDs on cardiovascular drug efficacy and safety will help to design a more personalized and "tailored" therapy for the prevention and treatment of CVD and help reduce sex-specific disparities.

CH
27

Key Points

1. Women have a nearly 1.5-fold greater risk of experiencing an adverse drug reaction, when drugs are prescribed at the same dosage between men and women.

2. Women often experience more adverse drug reactions owing to having a smaller body size, slower renal clearance, differences in body composition, variations in sex hormone levels, more comorbidities, polypharmacy, and increased sensitivity to drug effects.

3. Aspirin offers better protective effect against stroke in women, and against myocardial infarction in men. It equally increases the risk of bleeding in both sexes. Women have a higher oral bioavailability, slower clearance, and longer half-life with aspirin compared to men. They also exhibit a higher prevalence of aspirin resistance compared to men.

4. Beta-blockers result in a similar decrease in CV outcome for ASCVD and heart failure in both men and women. However, women are more likely to experience a greater reduction in blood pressure and heart rate, and lesser decrease in angina compared to men.

5. ACEIs are equally efficacious in improving outcomes following a myocardial infarction and for heart failure with reduced ejection fraction in both men and women, but women with asymptomatic LV systolic dysfunction may not achieve a mortality benefit when treated with ACEIs.

6. Thiazide and loop diuretics result in greater electrolyte disturbances in women, compared to men. Women have a lower clearance for some diuretics, yet no sex-specific dose adjustments are recommended.

7. Statins decrease the risk of CV events and mortality equally in both men and women. Hence, contemporary guidelines have a sex-neutral recommendation to use statins for primary and secondary prevention of ASCVD.

8. Antithrombotic agents present a higher risk of bleeding in women, likely related to their (a) lower body weight, (b) lower creatinine clearance, (c) having more comorbidities, and (d) sex differences in response to antithrombotic agents.

9. Antiarrhythmic drugs are more likely to result in drug-induced QT prolongation and torsades de pointes in women, compared to men, even at equivalent plasma concentrations, and even after careful dosing based on weight and renal clearance. These differences may be related to variations in sex hormones which modulate the expression and activity of cardiac K^+ and Ca^{2+} ion channels involved in ventricular repolarization.

10. Women are underrepresented in clinical trials on CV drugs (30%), particularly for ASCVD and heart failure. Even when enrolled, only one in three trials reported sex-specific results.

11. Women take more medications, show lower adherence to prescribed medications, and are less likely to be prescribed guideline-recommended pharmacologic therapies despite being at a similar cardiovascular risk as men.

12. Sex-related differences in efficacy and safety of drugs exist, but are poorly appreciated at present. Improved inclusion of women in trials on CV drug development, sex-specific analysis, reporting improved basic science and clinical research focused on SRDs in CV drugs, and evaluating the role of sex hormones on pharmacodynamics and pharmacokinetics of drugs are pivotal for better understanding of these differences.

Back to Clinical Case

The clinical case represented a 69-year-old female with QTc interval of 560 ms on ECG, hypokalemia and hypomagnesemia at baseline, with worsening QTc prolongation, and development to torsades de pointes (TdP) following administration of azithromycin. The normal QTc interval in adults is 390–450 ms in men and 390–470 ms in women. Our patient exhibited a QTc > 500 ms, and represents a typical case of drug-induced (acquired) QT prolongation. Patients with drug-induced QT prolongation present one or more risk factors that can be "easily identified" in the medical history, including female sex, age over 65 years, bradycardia, electrolyte abnormalities (hypokalemia and hypomagnesemia), underlying heart disease, renal and hepatic dysfunction, drug interactions, excessive dosing, one or more QTc-prolonging drugs, and genetic predisposition. This patient was an older woman with electrolyte abnormalities possibly related to furosemide and augmented by the recent diarrhea. Extracellular potassium is a major factor modulating the rapidly activating cardiac delayed rectifier channel (hERG, Kv11.1), and there is an inverse relationship between QT interval and serum potassium so that ventricular action potentials are longer at low $[K^+]_o$. Other important factors are excessive dosing and drug interactions. This patient should have received only 10 mg/day of escitalopram because no additional benefits are seen at 20 mg/day. Additionally, she was receiving several drugs that prolong QT and/or cause TdP, particularly in patients with hypokalemia: azithromycin, escitalopram, furosemide, nadolol, and solifenacin (https://crediblemeds.org/pdftemp/pdf/CombinedList.pdf). Each of these drugs is contraindicated for concomitant use with other medicines known to prolong the QT interval. Since the QT interval became abnormally prolonged about 24 h after initiating the antibiotic therapy, it is highly likely that azithromycin precipitated the prolongation of QT interval and the TdP.

The Tisdale score can be used to assess the risk of our patient for developing TdP [388]. She was an elderly (≥ 68 years: 1 point) female (1 point), treated with diuretics (1 point), with serum potassium ≤ 3.5 mEq/L (2 points), a QTc ≥ 450 ms (2 points), receiving ≥ 2 QTc interval-prolonging drugs (6 points). This leads to a risk score of 13 points—that is, she was a high-risk (≥ 11) patient.

Drug interaction:

Azithromycin, escitalopram, furosemide, solifenacin, and sotalol prolong the QT interval and can induce TdP. The combination of these drugs should be avoided, particularly in women

Furosemide produces electrolyte disturbances (hypokalemia and hypomagnesemia) that increase the risk factors of development of drug-induced QT prolongation

Escitalopram is metabolized via CYP3A4 and 2C19 and its exposure increases (~50%) in women and in the elderly. CYP2C19 inhibitors (omeprazole) further increase its plasma levels of (~50%)

Omeprazole: hypomagnesemia has been reported in patients treated for > 3–6 months

Warfarin: omeprazole inhibits CYP2C19 and increases the exposure of warfarin. Women present higher plasma levels and need lower doses to reduce the risk of bleeding

Proposal for a new treatment:

- Replace furosemide by a calcium channel blocker
- Reduce the dose of escitalopram to 10 mg daily
- Replace sotalol with flecainide or propafenone if minimal signs for structural heart disease are present
- Administer omeprazole for short periods of time or replace it with H2 receptor blockers

Acknowledgment

We thank P. Vaquero for her invaluable technical assistance. This work was supported by grants from the Institute of Health Carlos III (PI16/00398 and CB16/11/00303).

References

[1] Mosca L, Benjamin EJ, Berra K, Bezanson JL, Dolor RJ, Lloyd-Jones DM, et al. Effectiveness-based guidelines for the prevention of cardiovascular disease in women—2011 update: a guideline from the American Heart Association. Circulation 2011;123:1243–62.

[2] Nichols M, Townsend N, Scarborough P, Rayner M. Cardiovascular disease in Europe: epidemiological update. Eur Heart J 2013;34:3028–34.

[3] Shaw LJ, Pepine CJ, Xie J, Mehta PK, Morris AA, Dickert NW, et al. Quality and equitable health care gaps for women: attributions to sex differences in cardiovascular medicine. J Am Coll Cardiol 2017;70:373–88.

[4] Timmis A, Townsend N, Gale C, Grobbee R, Maniadakis N, Flather M, et al. European Society of Cardiology: cardiovascular disease statistics 2017. Eur Heart J 2018;39:508–79.

[5] Garcia M, Mulvagh SL, Bairey Merz CN, Buring JE, Manson JA. Cardiovascular disease in women. Clin Perspect Circ Res 2016;118:1273–93.

[6] Benjamin EJ, Muntner P, Alonso A, Bittencourt MS, Callaway CW, Carson AP, et al. Heart disease and stroke statistics—2019 update: a report from the American Heart Association. Circulation 2019;139:e56–e528.

[7] Lundberg GP, Mehta LS, Sanghani RM, Patel HN, Aggarwal NR, Aggarwal NT, et al. Heart Centers for women. Circulation 2018;138:1155–65.

[8] Maas AHEM, van der Schouw YT, Regitz-Zagrosek V, Swahn E, Appelman YE, Pasterkamp G, et al. Red alert for women's heart: the urgent need for more research and knowledge on cardiovascular disease in women: proceedings of the workshop held in Brussels on gender differences in cardiovascular disease. Eur Heart J 2011;32:1362–8.

[9] Mehta LS, Beckie TM, DeVon HA, Grines CL, Krumholz HM, Johnson MN, et al. Acute myocardial infarction in women: a scientific statement from the American Heart Association. Circulation 2016;133:916–47.

[10] Jochmann N, Stangl K, Garbe E, Baumann G, Stangl V. Female-specific aspects in the pharmacotherapy of chronic cardiovascular diseases. Eur Heart J 2005;26:1585–95.

[11] Oertelt-Prigione S, Regitz-Zagrosek V. Gender aspects in cardiovascular pharmacology. J Cardiovasc Trans Res 2009;2:258–66.

[12] Seeland U, Regitz-Zagrosek V. Sex and gender differences in cardiovascular drug therapy. Handb Exp Pharmacol 2012;214:211–36.

[13] Franconi F, Campesi I. Pharmacogenomics, pharmacokinetics and pharmacodynamics: interaction with biological differences between men and women. Br J Pharmacol 2014;171:580–94.

[14] Rosano GMC, Lewis B, Agewall S, Wassmann S, Vitale C, Schmidt H, et al. Gender differences in the effect of cardiovascular drugs: a position document of the working group on pharmacology and drug therapy of the ESC. Eur Heart J 2015;36:2677–80.

[15] EUGenMed Cardiovascular Clinical Study Group, Regitz-Zagrosek V, Oertelt-Prigione S, Prescott E, Franconi F, Gerdts E, et al. Gender in cardiovascular diseases: impact on clinical manifestations, management, and outcomes. Eur Heart J 2016;37:24–34.

[16] Tamargo J, Rosano G, Walther T, Duarte J, Niessner A, Kaski JC, et al. Gender differences in the effects of cardiovascular drugs. Eur Heart J Cardiovasc Pharmacother 2017;3:163–82.

[17] Meibohm B, Beierle I, Derendorf H. How important are gender differences in pharmacokinetics? Clin Pharmacokinet 2002;41:329–42.

[18] Gandhi M, Aweeka F, Greenblatt RM, Blaschke TF. Sex differences in pharmacokinetics and pharmacodynamics. Annu Rev Pharmacol Toxicol 2004;44:499–523.

[19] Anderson GD. Sex and racial differences in pharmacological response: where is the evidence? Pharmacogenetics, pharmacokinetics, and pharmacodynamics. J Womens Health (Larchmt) 2005;14:19–29.

[20] Anderson GD. Gender differences in pharmacological response. Int Rev Neurobiol 2008;83:1–10.

[21] Nicolas J-M, Espie P, Molimard M. Gender and interindividual variability in pharmacokinetics. Drug Metab Rev 2009;408–21.

[22] Soldin OP, Chung SH, Mattison DR. Sex differences in drug disposition. J Biomed Biotechnol 2011;2011:187103.

[23] Stolarz AJ, Rusch NJ. Gender differences in cardiovascular drug. Cardiovasc Drug Ther 2015;29:403–10.

[24] Spoletini I, Vitale C, Malorni W, Rosano GMC. Sex differences in drug effects: interaction with sex hormones in adult life. Handb Exp Pharmacol 2012;214:91–105.

[25] Schwartz JB. The current state of knowledge on age, sex, and their interactions on clinical pharmacology. Clin Pharmacol Ther 2007;82:87–96.

[26] Baraona E, Abittan CS, Dohmen K, Moretti M, Pozzato G, Chayes ZW, et al. Gender differences in pharmacokinetics of alcohol. Alcohol Clin Exp Res 2001;25:502–7.

[27] Pleym H, Spigset O, Kharasch ED, Dale O. Gender differences in drug effects: implications for anesthesiologists. Acta Anaesthesiol Scand 2003;47:241–59.

[28] Hoymork SC, Raeder J. Why do women wake up faster than men from propofol anaesthesia? Br J Anaesth 2005;95:627–33.

[29] Ueno K. Gender differences in pharmacokinetics of anesthetics. Jpn J Anesthesiol 2009;58:51–8.

[30] Shekelle PG, Rich MW, Morton SC, Atkinson CSW, Tu W, Maglione M, et al. Efficacy of angiotensin-converting enzyme inhibitors and betablockers in the management of left ventricular systolic dysfunction according to race, gender, and diabetic status: a meta-analysis of major clinical trials. J Am Coll Cardiol 2003;41:1529–38.

[31] Koup JR, Abel RB, Smithers JA, Eldon MA, de Vries TM. Effect of age, gender and race on steady state procainamide pharmacokinetics after administration of procanbid sustained-release tablets. Ther Drug Monit 1998;20:73–7.

[32] Miners JO, Grugrinovich N, Whitehead AG, Robson RA, Birkett DJ. Influence of gender and oral contraceptive steroids on the metabolism of salicylic acid and acetylsalicylic acid. Br J Clin Pharmacol 1986;22:135–42.

[33] Ho PC, Triggs EJ, Bourne DW, Heazlewood VJ. The effect of age and sex on the disposition of acetylsalicylic acid and its metabolites. Br J Clin Pharmacol 1985;19:675–84.

[34] Berger JS, Brown DL, Becker RC. Low-dose aspirin in patients with stable cardiovascular disease: a meta-analysis. Am J Med 2008;121:43–9.

[35] Ridker PM, Cook NR, Lee I-M, Gordon D, Gaziano JM, Manson JE, et al. A randomized trial of low-dose aspirin in the primary prevention of cardiovascular disease in women. N Engl J Med 2005;352:1293–304.

[36] Luzier AB, Killian A, Wilton JH, Wilson MF, Forrest A, Kazierad DJ. Gender-related effects on metoprolol pharmacokinetics and pharmacodynamics in healthy volunteers. Clin Pharmacol Ther 1999;66:594–601.

[37] Walle T, Byington RP, Furberg CD, McIntyre KM, Vokonas PS. Biologic determinants of propranolol disposition: results from 1308 patients in the Beta-blocker heart attack Trial. Clin Pharmacol Ther 1985;38:509–18.

[38] Walle T, Walle UK, Cowart TD, Conradi EC. Pathway-selective sex differences in the metabolic clearance of propranolol in human subjects. Clin Pharmacol Ther 1989;46:257–63.

[39] Kendall MJ, Quarterman CP, Jack DB, Beeley L. Metoprolol pharmacokinetics and the oral contraceptive pill. Br J Clin Pharmacol 1982;14:120–2.

[40] Kang D, Verotta D, Krecic-Shepard ME, Modi NB, Gupta SK, Schwartz JB. Population analyses of sustained-release verapamil in patients: effects of sex, race, and smoking. Clin Pharmacol Ther 2003;73:31–40.

[41] Krecic-Shepard ME, Barnas CR, Slimko J, Jones M, Schwartz J. Gender-specific effects on verapamil pharmacokinetics and pharmacodynamics in humans. J Clin Pharmacol 2000;40:219–30.

[42] Krecic-Shepard ME, Barnas CR, Slimko J, Schwartz JB. Faster clearance of sustained release verapamil in men versus women: continuing observations on sex-specific differences after oral administration of verapamil. Clin Pharmacol Ther 2000;68:286–92.

[43] Krecic-Shepard ME, Park K, Barnas C, Slimko J, Kerwin DR, Schwartz JB. Race and sex influence clearance of nifedipine: results of a population study. Clin Pharmacol Ther 2000;68:130–42.

[44] Kang D, Verotta D, Schwartz JB. Population analyses of amlodipine in patients living in the community and patients living in nursing homes. Clin Pharmacol Ther 2006;79:114–24.

[45] Dadashzadeha SB, Javadiana B, Sadeghianb S. The effect of gender on the pharmacokinetics of verapamil and norverapamil in human. Biopharm Drug Dispos 2006;27:329–34.

[46] Rathore SS, Foody JM, Wang Y, Smith GL, Herrin J, Masoudi FA, et al. Race, quality of care, and outcomes of elderly patients hospitalized with heart failure. JAMA 2003;289:2517–24.

[47] Rathore SS, Wang Y, Krumholz HM. Sex-based differences in the effect of digoxin for the treatment of heart failure. N Engl J Med 2002;347:1403–11.

[48] Lew KH, Ludwig EA, Milad MA, Donovan K, Middleton E, Ferry JJ, et al. Gender-based effects on methylprednisolone pharmacokinetics and pharmacodynamics. Clin Pharmacol Ther 1993;54:402–14.

[49] Mattison D, Zajicek A. Gaps in knowledge in treating pregnant women. Gend Med 2006;3:169.

[50] Campbell NR, Hull RD, Brant R, Hogan DB, Pineo GF, Raskob GE. Different effects of heparin in males and females. Clin Invest Med 1998;21:71–8.

[51] Jacobs A. Sex differences in iron absorption. Proc Nutr Soc 1976;35:159–62.

[52] Routledge PA, Stargel WW, Kitchell BB, Barchowsky A, Shand DG. Sex-related differences in the plasma protein binding of lignocaine and diazepam. Br J Clin Pharmacol 1981;11:245–50.

[53] Bijur PE, Esses D, Birnbaum A, Chang AK, Schechter C, Gallagher EJ. Response to morphine in male and female patients: analgesia and adverse events. Clin J Pain 2008;24:192–8.

[54] Gear RW, Miaskowski C, Gordon NC, Paul SM, Heller PH, Levine JD. Kappa-opioids produce significantly greater analgesia in women than in men. Nat Med 1996;2:1248–50.

[55] Makkar RR, Fromm BS, Steinman RT, Meissner MD, Lehmann MH. Female gender as a risk factor for torsades de pointes associated with cardiovascular drugs. JAMA 1993;270:2590–7.

[56] Kannankeril P, Roden DM, Darbar D. Drug-induced long QT syndrome. Pharmacol Rev 2010;62:760–81.

[57] Drici MD, Clement N. Is gender a risk factor for adverse drug reactions? The example of drug-induced long QT syndrome. Drug Safe 2001;24:575–85.

[58] Delpón E, Tamargo J. Cardiovascular drugs—from A to Z. In: Kaski JC, Kjeldsen KP, editors. The ESC handbook on cardiovascular pharmacotherapy. 2nd ed. Oxford University Press; 2019. p. 813–96 [Chapter 9.1].

[59] Sirtori C. The pharmacology of statins. Pharmacol Res 2014;88:3–11.

[60] Cangemi R, Romiti GF, Campolongo G, Ruscio E, Sciomer S, Gianfrilli D, et al. Gender related differences in treatment and response to statins in primary and secondary cardiovascular prevention: the never-ending debate. Pharmacol Res 2017;117:148–55.

[61] Nafziger AN, Bertino Jr JS. Sex-related differences in theophylline pharmacokinetics. Eur J Clin Pharmacol 1989;37:97–100.

[62] Werner U, Werner D, Heinbüchner S, Graf BM, Ince H, Kische S, et al. Gender is an important determinant of the disposition of the loop diuretic torasemide. J Clin Pharmacol 2010;50:160–8.

[63] Kent DM, Price LL, Ringleb P, Hill MD, Selker HP. Sex-based differences in response to recombinant tissue plasminogen activator in acute ischemic stroke. A pooled analysis of randomized clinical trials. Stroke 2005;36:62–5.

[64] Yacobi A, Stoll RG, DiSanto AR, Levy G. Intersubject variation of warfarin binding to protein in serum of normal subjects. Res Commun Chem Pathol Pharmacol 1976;14:743–6.

[65] Garcia D, Regan S, Crowther M, Hughes RA, Hylek EM. Warfarin maintenance dosing patterns in clinical practice: implications for safer anticoagulation in the elderly population. Chest 2005;127:2049–56.

CH
27

[66] Whitley HP, Fermo JD, Chumney ECG, Brzezinski WA. Effect of patient-specific factors on weekly warfarin dose. Ther Clin Risk Manag 2007;3:499–504.

[67] Waxman DJ, Holloway MG. Sex differences in the expression of hepatic drug metabolizing enzymes. Mol Pharmacol 2009;76:215–28.

[68] Cotreau MM, von Moltke LL, Greenblatt DJ. The influence of age and sex on the clearance of cytochrome P450 3A substrates. Clin Pharmacokinet 2005;44:33–60.

[69] Nicolson TJ, Mellor HR, Roberts RRA. Gender differences in drug toxicity. Trends Pharmacol Sci 2010;31:108–14.

[70] Zanger UM, Schwab M. Cytochrome P450 enzymes in drug metabolism: regulation of gene expression, enzyme activities, and impact of genetic variation. Pharmacol Ther 2013;138:103–41.

[71] Cummins CL, Wu CY, Benet LZ. Sex-related differences in the clearance of cytochrome P450 3A4 substrates may be caused by p-glycoprotein. Clin Pharmacol Ther 2002;72:474–89.

[72] Berg UB. Differences in decline in GFR with age between males and females: reference data on clearances of inulin and PAH in potential kidney donors. Nephrol Dial Transplant 2006;21:2577–82.

[73] Gross JL, Friedman R, Azevedo MJ, Silveiro SP, Pecis M. Effect of age and sex on glomerular filtration rate measured by 51Cr-EDTA. Braz J Med Biol Res 1992;25:129–34.

[74] Stramba-Badiale M. Women and research on cardiovascular diseases in Europe: a report from the European heart health strategy (EuroHeart) project. Eur Heart J 2010;31:1677–81.

[75] Melloni C, Berger JS, Wang TY, Gunes F, Stebbins A, Pieper KS, et al. Representation of women in randomized clinical trials of cardiovascular disease prevention. Circ Cardiovasc Qual Outcomes 2010;3:135–42.

[76] Wallach JD, Sullivan PG, Trepanowski JF, Steyerberg EW, Ioannidis JPA. Sex based subgroup differences in randomized controlled trials: empirical evidence from Cochrane meta-analyses. BMJ 2016;355:i5826.

[77] Nguyen QD, Peters E, Wassef A, Desmarais P, Rémillard-Labrosse D, Tremblay-Gravel M. Evolution of age and female representation in the most-cited randomized controlled trials of cardiology of the last 20 years. Circ Cardiovasc Qual Outcomes 2018;11, e004713.

[78] Scott PE, Unger EF, Jenkins MR, Southworth MR, McDowell TY, Geller RJ, et al. Participation of women in clinical trials supporting FDA approval of cardiovascular drugs. J Am Coll Cardiol 2018;71:1960–9.

[79] Bailey AL, Scantlebury DC, Smyth SS. Thrombosis and antithrombotic therapy in women. Arterioscler Thromb Vasc Biol 2009;29:284–8.

[80] Capodanno D, Angiolillo DJ. Impact of race and gender on antithrombotic therapy. Thromb Haemost 2010;104:471–84.

[81] Basili S, Raparelli V, Proietti M, Tanzilli G, Franconi F. The impact of sex and gender on the efficacy of antiplatelet therapy: the female perspective. J Atheroscler Thromb 2015;22:109–25.

[82] Walker AM, Jick H. Predictors of bleeding during heparin therapy. JAMA 1980;244:1209–12.

[83] van der Meer FJ, Rosendaal FR, Vandenbroucke JP, Briët E. Bleeding complications in oral anticoagulant therapy. An analysis of risk factors. Arch Intern Med 1993;153:1557–62.

[84] Fibrinolytic Therapy Trialists' (FTT) Collaborative Group. Indications for fibrinolytic therapy in suspected acute myocardial infarction: collaborative overview of early mortality and major morbidity results from all randomised trials of more than 1000 patients. Lancet 1994;343:311–22.

[85] Dauerman HL, Andreou C, Perras MA, Spinner JS, Lessard D, Weiner BH. Predictors of bleeding complications after rescue coronary interventions. J Thromb Thrombolysis 2000;10:83–8.

[86] Cho L, Topol EJ, Balog C, Foody JM, Booth JE, Cabot C, et al. Clinical benefit of glycoprotein IIb/IIIa blockade with abciximab is independent of gender. Pooled analysis from EPIC, EPILOG and EPISTENT trials. J Am Coll Cardiol 2000;36:381–6.

[87] Van de Werf F, Barron HV, Armstrong PW, Granger CB, Berioli S, Barbash G, et al. Assessment of the safety and efficacy of a new thrombolytic. Incidence and predictors of bleeding events after fibrinolytic therapy with fibrin-specific agents: a comparison of TNK-tPA and rt-PA. Eur Heart J 2001;22:2253–61.

[88] Reynolds HR, Farkouh ME, Lincoff AM, Hsu A, Swahn E, Sadowski ZP, et al. Impact of female sex on death and bleeding after fibrinolytic treatment of myocardial infarction in GUSTO V. Arch Intern Med 2007;167:2054–60.

[89] Lansky AJ, Mehran R, Cristea E, Parise H, Feit F, Ohman EM, et al. Impact of gender and antithrombin strategy on early and late clinical outcomes in patients with non-ST-elevation acute coronary syndromes (from the ACUITY trial). Am J Cardiol 2009;103:1196–203.

[90] Steg PG, Huber K, Andreotti F, Arnesen H, Atar D, Badimon L, et al. Bleeding in acute coronary syndromes and percutaneous coronary interventions: position paper by the working group on thrombosis of the European Society of Cardiology. Eur Heart J 2011;32:1854–64.

[91] Mauer AC, Khazanov NA, Levenkova N, Tian S, Barbour EM, Khalida C, et al. Impact of sex, age, race, ethnicity and aspirin use on bleeding symptoms in healthy adults. J Thromb Haemost 2011;9:100–8.

[92] Daugherty SL, Thompson LE, Kim S, Rao SV, Subherwal S, Tsai TT, et al. Patterns of use and comparative effectiveness of bleeding avoidance strategies in men and women following percutaneous coronary interventions: an observational study from the National Cardiovascular Data Registry. J Am Coll Cardiol 2013;61:2070–8.

[93] Alotaibi GS, Almodaimegh H, McMurtry MS, Wu C. Do women bleed more than men when prescribed novel oral anticoagulants for venous thromboembolism? A sex-based meta-analysis. Thromb Res 2013;132:185–9.

[94] Yu J, Mehran R, Grinfeld L, Xu K, Nikolsky E, Brodie BR, et al. Sex-based differences in bleeding and long term adverse events after percutaneous coronary intervention for acute myocardial infarction: three year results from the HORIZONS-AMI trial. Catheter Cardiovasc Interv 2015;85:359–68.

[95] Wang WT, James SK, Wang TY. A review of sex-specific benefits and risks of antithrombotic therapy in acute coronary syndrome. Eur Heart J 2017;38:165–71.

[96] Fang MC, Singer DE, Chang Y, Hylek Em, Henault LE, Jensvold NG, et al. Gender differences in the risk of ischemic stroke and peripheral embolism in atrial fibrillation: the anTicoagulation and risk factors in atrial fibrillation (ATRIA) study. Circulation 2005;112:1687–91.

[97] Anon. Risk factors for stroke and efficacy of antithrombotic therapy in atrial fibrillation. Analysis of pooled data from five randomized controlled trials. Arch Intern Med 1994;154:1449–57.

[98] Sullivan RM, Zhang J, Zamba G, Lip GYH, Olshansky B. Relation of gender-specific risk of ischemic stroke in patients with atrial fibrillation to differences in warfarin anticoagulation control (from AFFIRM). Am J Cardiol 2012;110:1799–802.

[99] Gomberg-Maitland M, Wenger NK, Feyzi J, Lengyel M, Volgman AS, Petersen P, et al. Anticoagulation in women with non-valvular atrial fibrillation in the stroke prevention using an oral thrombin inhibitor (SPORTIF) trials. Eur Heart J 2006;27:1947–53.

[100] Pengo V, Legnani C, Noventa F, Palareti G, ISCOAT Study Group (Italian Study on Complications of Oral Anticoagulant Therapy). Oral anticoagulant therapy in patients with nonrheumatic atrial fibrillation and risk of bleeding. A Multicenter Inception Cohort Study. Thromb Haemost 2001;85:418–22.

[101] Lapner ST, Cohen N, Kearon C. Influence of sex on risk of bleeding in anticoagulated patients: a systematic review and meta-analysis. J Thromb Haemost 2014;12:595–605.

[102] Jick H, Slone D, Borda IT, Shapiro S. Efficacy and toxicity of heparin in relation to age and sex. N Engl J Med 1968;279:284–6.

[103] Cipolle RJ, Seifert RD, Neilan BA, Zaske DE, Haus E. Heparin kinetics: variables related to disposition and dosage. Clin Pharmacol Ther 1981;29:387–93.

[104] Granger CB, Hirsch J, Califf RM, Col J, White HD, Betriu A, et al. Activated partial thromboplastin time and outcome after thrombolytic therapy for acute myocardial infarction. Circulation 1996;93:870–8.

[105] Hirsh J, O'Donnell M, Eikelboom JW. Beyond unfractionated heparin and warfarin: current and future advances. Circulation 2007;116:552–60.

[106] Fragmin during Instability in Coronary Artery Disease (FRISC) Study Group. Low-molecular-weight heparin during instability in coronary artery disease. Lancet 1996;347:561–8.

[107] Toss H, Wallentin L, Siegbahn A. Influences of sex and smoking habits on anticoagulant activity in low-molecular-weight heparin treatment of unstable coronary artery disease. Am Heart J 1999;137:72–8.

[108] Becker RC, Spencer FA, Gibson M, Rush JE, Sanderink G, Murphy SA, et al. Influence of patient characteristics and renal function on factor Xa inhibition pharmacokinetics and pharmacodynamics after enoxaparin administration in non-ST-segment elevation acute coronary syndromes. Am Heart J 2002;143:753–9.

[109] White HD, Kleiman NS, Mahaffey KW, Lokhnygina Y, Pieper KS, Chiswell K, et al. Efficacy and safety of enoxaparin compared with unfractionated heparin in high-risk patients with non-ST-segment elevation acute coronary syndrome undergoing percutaneous coronary intervention in the superior yield of the new strategy of enoxaparin, revascularization and glycoprotein IIb/IIIa inhibitors (SYNERGY) trial. Am Heart J 2006;152:1042–50.

[110] Cohen M, Antman EM, Gurfinkel EP, Radley D, ESSENCE (Efficacy and Safety of Subcutaneous Enoxaparin in Non-Q-wave Coronary Events) and TIMI (Thrombolysis in Myocardial Infarction) 11B Investigators. Enoxaparin in unstable angina/non-ST-segment elevation myocardial infarction: treatment benefits in prespecified subgroups. J Thromb Thrombolysis 2001;12:199–206.

[111] Mega JL, Morrow DA, Ostör E, Dorobantu M, Qin J, Antman EM, et al. Outcomes and optimal antithrombotic therapy in women undergoing fibrinolysis for ST-elevation myocardial infarction. Circulation 2007;115:2822–8.

[112] Yusuf S, Mehta SR, Chrolavicius S, Fifth Organization to Assess Strategies in Acute Ischemic Syndromes Investigators, et al. Comparison of fondaparinux and enoxaparin in acute coronary syndromes. N Engl J Med 2006;354:1464–76.

[113] Yusuf S, Mehta SR, Chrolavicius S, Afzal R, Pogue J, Granger CB, et al. Effects of fondaparinux on mortality and reinfarction in patients with acute ST-segment elevation myocardial infarction: the OASIS-6 randomized trial. JAMA 2006;295:1519–30.

[114] Oldgren J, Wallentin L, Afzal R, Bassand J-P, Budaj A, Chrolavicius S, et al. Effects of fondaparinux in patients with ST-segment elevation acute myocardial infarction not receiving reperfusion treatment. Eur Heart J 2008;29:315–23.

[115] Swan SK, Hursting MJ. The pharmacokinetics and pharmacodynamics of argatroban: effects of age, gender, and hepatic or renal dysfunction. Pharmacotherapy 2000;20:318–29.

[116] Lincoff AM, Bittl JA, Harrington RA, Feit F, Kleiman NS, Jackman JD, et al. Bivalirudin and provisional glycoprotein IIb/IIIa blockade compared with heparin and planned glycoprotein IIb/IIIa blockade during percutaneous coronary intervention: REPLACE-2 randomized trial. JAMA 2003;289:853–63.

[117] Lincoff AM, Kleiman NS, Kereiakes DJ, Feit F, Bittl JA, Daniel Jackman J, et al. Long-term efficacy of bivalirudin and provisional glycoprotein IIb/IIIa blockade vs heparin and planned glycoprotein IIb/IIIa blockade during percutaneous coronary revascularization: REPLACE-2 randomized trial. JAMA 2004;292:696–703.

[118] Stone GW, Bertrand M, Colombo A, Dangas G, Farkouh ME, Feit F, et al. Bivalirudin in patients with acute coronary syndromes undergoing percutaneous coronary intervention: a subgroup analysis from the acute catheterization and urgent intervention triage strategy (ACUITY) trial. Lancet 2007;369:907–19.

[119] Stone GW, Witzenbichler B, Guagliumi G, Peruga JZ, Brodie BR, Dudek D, et al. Bivalirudin during primary PCI in acute myocardial infarction. N Engl J Med 2008;358:2218–30.

[120] Mehran R, Lansky AJ, Witzenbichler B, Guagliumi G, Peruga JZ, Brodie BR, et al. Bivalirudin in patients undergoing primary angioplasty for acute myocardial infarction (HORIZONSAMI): 1-year results of a randomised controlled trial. Lancet 2009;374:1149–59.

[121] Ng VG, Baumbach A, Grinfeld L, Lincoff AM, Mehran R, Stone GW, et al. Impact of bleeding and bivalirudin therapy on mortality risk in women undergoing percutaneous coronary intervention (from the REPLACE-2, ACUITY, and HORIZONS-AMI trials). Am J Cardiol 2016;117:186–91.

[122] Gargiulo G, da Costa BR, Frigoli E, Palmieri C, Nazzaro MS, Falcone C, et al. Impact of sex on comparative outcomes of bivalirudin versus unfractionated heparin in patients with acute coronary syndromes undergoing invasive management: a pre-specified analysis of the MATRIX trial. EuroIntervention 2019;15:e269–78.

[123] Shammas NW, Allie D, Hall P, Young J, Laird J, Safian R, et al. Predictors of in-hospital and 30-day complications of peripheral vascular interventions using bivalirudin as the primary anticoagulant: results from the APPROVE registry. J Invasive Cardiol 2005;17:356–9.

[124] Chacko M, Lincoff AM, Wolski KE, Cohen DJ, Bittl JA, Lansky AJ, et al. Ischemic and bleeding outcomes in women treated with bivalirudin during percutaneous coronary intervention: a subgroup analysis of the randomized evaluation in PCI linking Angiomax to reduced clinical events (REPLACE)-2 trial. Am Heart J 2006;151:1032.e1–7.

[125] Manoukian SV, Feit F, Mehran R, Voeltz MD, Ebrahimi R, Hamon M, et al. Impact of major bleeding on 30-day mortality and clinical outcomes in patients with acute coronary syndromes: an analysis from the ACUITY Trial. J Am Coll Cardiol 2007;49:1362–8.

[126] Madsen JK, Chevalier B, Darius H, Rutsch W, Wójcik J, Schneider S, et al. Ischaemic events and bleeding in patients undergoing percutaneous coronary intervention with concomitant bivalirudin treatment. Euro Interv 2008;3:610–6.

[127] Kubitza D, Becka M, Roth A, Mueck W. The influence of age and gender on the pharmacokinetics and pharmacodynamics of rivaroxaban—an oral, direct factor Xa inhibitor. J Clin Pharmacol 2013;53:249–55.

[128] Jiang J, Hu Y, Zhang J, Yang J, Mueck W, Kubitza D, et al. Safety, pharmacokinetics and pharmacodynamics of single doses of rivaroxaban—an oral, direct factor Xa inhibitor—in elderly Chinese subjects. Thromb Haemost 2010;103:234–41.

[129] Mega JL, Braunwald E, Mohanavelu S, Burton P, Poulter R, Misselwitz F, et al. Rivaroxaban versus placebo in patients with acute coronary syndromes (ATLAS ACS-TIMI 46): a randomised, double-blind, phase II trial. Lancet 2009;374:29–38.

[130] Frost CE, Song Y, Shenker A, Wang J, Barrett YC, Schuster A, et al. Effects of age and sex on the single dose pharmacokinetics and pharmacodynamics of apixaban. Clin Pharmacokinet 2015;54:651–62.

[131] Mendell J, Shi M. Safety, tolerability, pharmacokinetic (PK) and pharmacodynamic (PD) profiles of edoxaban in healthy post-menopausal or surgically sterile females, and healthy elderly males. Eur Heart J 2011;32:461.

[132] Stangier J. Clinical pharmacokinetics and pharmacodynamics of the oral direct thrombin inhibitor dabigatran etexilate. Clin Pharmacokinet 2008;47:285–95.

[133] Ruff CT, Giugliano RP, Braunwald E, Hoffman EB, Deenadayalu N, Ezekowitz MD, et al. Comparison of the efficacy and safety of new oral anticoagulants with warfarin in patients with atrial fibrillation: a meta-analysis of randomised trials. Lancet 2014;383:955–62.

[134] Connolly SJ, Ezekowitz MD, Yusuf S, Eikelboom J, Oldgren J, Parekh A, et al. Dabigatran versus warfarin in patients with atrial fibrillation. N Engl J Med 2009;361:1139–51.

[135] Avgil Tsadok M, Jackevicius CA, Rahme E, Humphries KH, Pilote L. Differences in dabigatran use, safety, and effectiveness in a population-based cohort of patients with atrial fibrillation. Circ Cardiovasc Qual Outcomes 2015;8:593–9.

[136] Giugliano RP, Ruff CT, Braunwald E, Murphy SA, Wiviott SD, Halperin JL, et al. Edoxaban versus warfarin in patients with atrial fibrillation. N Engl J Med 2013;369:2093–104.

[137] Granger CB, Alexander JH, McMurray JJV, Lopes RD, Hylek EM, Hanna M, et al. Apixaban versus warfarin in patients with atrial fibrillation. N Engl J Med 2011;365:981–92.

[138] Vinereanu D, Stevens SR, Alexander JH, Al-Khatib SM, Avezum A, Bahit MC, et al. Clinical outcomes in patients with atrial fibrillation according to sex during anticoagulation with apixaban or warfarin: a secondary analysis of a randomized controlled trial. Eur Heart J 2015;36:3268–75.

[139] Patel MR, Mahaffey KW, Garg J, Pan G, Singer DE, Hacke W, et al. Rivaroxaban versus warfarin in nonvalvular atrial fibrillation. N Engl J Med 2011;365:883–91.

[140] Sherwood MW, Nessel CC, Hellkamp AS, Mahaffey KW, Piccini JP, Suh EY, et al. Gastrointestinal bleeding in patients with atrial fibrillation treated with rivaroxaban or warfarin: ROCKET AF Trial. J Am Coll Cardiol 2015;66:2271–81.

[141] Raccah BH, Perlman A, Zwas DR, Hochberg-Klein S, Masarwa R, Muszkat M, et al. Gender differences in efficacy and safety of direct oral anticoagulants in atrial fibrillation: systematic review and network meta-analysis. Ann Pharmacother 2018;52:1135–42.

[142] Pancholy SB, Sharma PS, Pancholy DS, Patel TM, Callans DJ, Marchlinski FE. Meta-analysis of gender differences in residual stroke risk and major bleeding in patients with nonvalvular atrial fibrillation treated with oral anticoagulants. Am J Cardiol 2014;113:485–90.

[143] Palamaner Subash Shantha G, Mentias A, Inampudi C, Kumar AA, Chaikriangkrai K, Bhise V, et al. Sex-specific associations of oral anticoagulant use and cardiovascular outcomes in patients with atrial fibrillation. J Am Heart Assoc 2017;6, e006381.

[144] Law SWY, Lau WCY, Wong ICK, Lip GYH, Mok MT, Siu CW, et al. Sex-based differences in outcomes of oral anticoagulation in patients with atrial fibrillation. J Am Coll Cardiol 2018;72:271–82.

[145] Turpie AGG, Lassen MR, Eriksson BI, Gent M, Berkowitz SD, Misselwitz F, et al. Rivaroxaban for the prevention of venous thromboembolism after hip or knee arthroplasty. Pooled analysis of four studies. Thromb Haemost 2011;105:444–53.

[146] Dentali F, Sironi AP, Gianni M, Orlandini F, Guasti L, Grandi AM, et al. Gender difference in efficacy and safety of nonvitamin k antagonist oral anticoagulants in patients with nonvalvular atrial fibrillation or venous thromboembolism: a systematic review and a meta-analysis of the literature. Semin Thromb Hemost 2015;41:774–87.

[147] Loffredo L, Violi F, Perri L. Sex related differences in patients with acute venous thromboembolism treated with new oral anticoagulants. A meta-analysis of the interventional trials. Int J Cardiol 2016;212:255–8.

[148] Laporte S, Mismetti P, Décousus H, Uresandi F, Otero R, Lobo JL, et al. Clinical predictors for fatal pulmonary embolism in 15,520 patients with venous thromboembolism: findings from the Registro Informatizado de la Enfermedad TromboEmbolica venosa (RIETE) registry. Circulation 2008;117:1711–6.

[149] Yoshikawa Y, Yamashita Y, Morimoto T, Amano H, Takase T, Hiramori S, et al. Differences in clinical characteristics and outcomes of patients with venous thromboembolism—from the COMMAND VTE registry. Circ J 2019;83:1581–9.

[150] Becker DM, Segal J, Vaidya D, Yanek LR, Herrera-Galeano JE, Bray PF, et al. Sex differences in platelet reactivity and response to low-dose aspirin therapy. JAMA 2006;295:1420–7.

[151] Wang TY, Angiolillo DJ, Cushman M, Sabatine MS, Bray PF, Smyth SS, et al. Platelet biology and response to antiplatelet therapy in women: implications for the development and use of antiplatelet pharmacotherapies for cardiovascular disease. J Am Coll Cardiol 2012;59:891–900.

[152] Mikkola T, Turunen P, Avela K, Orpana A, Viinikka L, Ylikorkala O. 17 beta-estradiol stimulates prostacyclin, but not endothelin-1, production in human vascular endothelial cells. J Clin Endocrinol Metab 1995;80:1832–6.

[153] Caulin-Glaser T, García-Cardeña G, Sarrel P, Sessa WC, Bender JR. 17 beta-estradiol regulation of human endothelial cell basal nitric oxide release, independent of cytosolic Ca^{2+} mobilization. Circ Res 1997;81:885–92.

[154] Arora S, Veves A, Caballaro AE, Smakowski P, LoGerfo FW. Estrogen improves endothelial function. J Vasc Surg 1998;27:1141–6.

[155] Lau ES, Braunwald E, Murphy SA, Wiviott SD, Bonaca MP, Husted S, et al. Potent P2Y12 inhibitors in men versus women: a collaborative meta-analysis of randomized trials. J Am Coll Cardiol 2017;69:1549–59.

[156] Patrono C, García Rodríguez LA, Landolfi R, Baigent C. Low-dose aspirin for the prevention of atherothrombosis. N Engl J Med 2005;353:2373–83.

[157] Harrison MJG, Weisblatt E. A sex difference in the effect of aspirin on "spontaneous" platelet aggregation in whole blood. Thromb Haemost 1983;50:773–4.

[158] Gum PA, Kottke-Marchant K, Poggio ED, Gurm H, Welsh PA, Brooks L, et al. Profile and prevalence of aspirin resistance in patients with cardiovascular disease. Am J Cardiol 2001;88:230–5.

[159] Spranger M, Aspey BS, Harrison MJ. Sex difference in antithrombotic effect of aspirin. Stroke 1989;20:34–7.

[160] Cavallari LH, Helgason CM, Brace LD, Viana MA, Nutescu EA. Sex difference in the antiplatelet effect of aspirin in patients with stroke. Ann Pharmacother 2006;40:812–7.

[161] Buchanan MR, Rischke JA, Butt R, Turpie AGG, Hirsh J, Rosenfeld J. The sex-related differences in aspirin pharmacokinetics in rabbits and man and its relationship to antiplatelet effects. Thromb Res 1983;29:125–39.

[162] Kjeldsen SE, Kolloch RE, Leonetti G, Mallion JM, Zanchetti A, Elmfeldt D, et al. Influence of gender and age on preventing cardiovascular disease by antihypertensive treatment and acetylsalicylic acid. The hot study. Hypertension optimal treatment. J Hypertens 2000;18:629–42.

[163] Berger JS, Roncaglioni MC, Avanzini F, Pangrazzi I, Tognoni G, Brown DL. Aspirin for the primary prevention of cardiovascular events in women and men: a sex-specific meta-analysis of randomized controlled trials. JAMA 2006;295:306–13.

[164] Antithrombotic Trialists' (ATT) Collaboration, Baigent C, Blackwell L, Collins R, Emberson J, Godwin J, et al. Aspirin in the primary and secondary prevention of vascular disease: collaborative meta-analysis of individual participant data from randomised trials. Lancet 2009;373:1849–60.

[165] Guirguis-Blake JM, Evans CV, Senger CA, O'Connor EA, Withlock EP. Aspirin use for the primary prevention of cardiovascular disease and colorectal cancer: a systematic evidence review for the U.S. Preventive Services Task Force. Ann Intern Med 2016;164:804–13.

[166] Abdelaziz HK, Saad M, Pothineni NVK, Megaly M, Potluri R, Saleh M, et al. Aspirin for primary prevention of cardiovascular events. J Am Coll Cardiol 2019;73:2915–29.

[167] Piepoli MF, Hoes AW, Agewall S, Albus C, Brotons C, Catapano AL, et al. 2016 European Guidelines on cardiovascular disease prevention in clinical practice: the Sixth Joint Task Force of the European Society of Cardiology and Other Societies on Cardiovascular Disease Prevention in Clinical Practice (constituted by representatives of 10 societies and by invited experts) Developed with the special contribution of the European Association for Cardiovascular Prevention & Rehabilitation (EACPR). Eur Heart J 2016;37:2315–81.

[168] Arnett DK, Blumenthal RS, Albert MA, Buroker AB, Goldberger ZD, Hahn EJ, et al. 2019 ACC/AHA guideline on the primary prevention of cardiovascular disease: a report of the American College of Cardiology/American Heart Association task force on clinical practice guidelines. J Am Coll Cardiol 2019;74:e177–232.

[169] Antithrombotic Trialists' Collaboration. Collaborative meta-analysis of randomised trials of antiplatelet therapy for prevention of death, myocardial infarction, and stroke in high risk patients. BMJ 2002;324:71–86.

[170] Hennekens CH, Hollar D, Baigent C. Sex-related differences in response to aspirin in cardiovascular disease: an untested hypothesis. Nat Clin Pract Cardiovasc Med 2006;3:4–5.

[171] ISIS-2 (Second International Study of Infarct Survival) Collaborative Group. Randomised trial of intravenous streptokinase, oral aspirin, both, or neither among 17,187 cases of suspected acute myocardial infarction: ISIS-2. Lancet 1988;2:349–60.

[172] Yerman T, Gan WQ, Sin DD. The influence of gender on the effects of aspirin in preventing myocardial infarction. BMC Med 2007;5:29.

[173] International Stroke Trial Collaborative Group. The International Stroke Trial (IST): a randomised trial of aspirin, subcutaneous heparin, both, or neither among 19435 patients with acute ischaemic stroke. Lancet 1997;349:1569–81.

[174] CAST (Chinese Acute Stroke Trial) Collaborative Group. CAST: randomised placebo-controlled trial of early aspirin use in 20,000 patients with acute ischaemic stroke. Lancet 1997;349:1641–9.

[175] Fernandes LS, Tcheng JE, Weiner B, Lorenz TJ, Pacchiana C, et al. Is glycoprotein IIb/IIIa antagonism as effective in women as in men following percutaneous coronary intervention? Lessons from the ESPRIT study. J Am Coll Cardiol 2002;40:1085–91.

[176] Iakovou I, Dangas G, Mehran R, Lansky AJ, Kobayashi Y, Adamian M, et al. Gender difference in clinical outcome after coronary artery stenting with use of glycoprotein IIb/IIIa inhibitors. Am J Cardiol 2002;89:976–9.

[177] PURSUIT Trial Investigators. Inhibition of platelet glycoprotein IIb/IIIa with eptifibatide in patients with acute coronary syndromes. N Engl J Med 1998;339:436–43.

[178] Kandzari D, Hasselblad V, Tcheng JE, Stone GW, Califf RM, Kastrati A, et al. Improved clinical outcomes with abciximab therapy in acute myocardial infarction: a systematic overview of randomized clinical trials. Am Heart J 2004;147:457–62.

[179] De Luca G, Suryapranata H, Stone GW, Antoniucci D, Tcheng JE, Neumann F-J, et al. Abciximab as adjunctive therapy to reperfusion in acute ST-segment elevation myocardial infarction: a meta-analysis of randomized trials. JAMA 2005;293:1759–65.

[180] Boersma E, Harrington RA, Moliterno DJ, White H, Théroux P, Van de Werf F, et al. Platelet glycoprotein IIb/IIIa inhibitors in acute coronary syndromes: a meta-analysis of all major randomised clinical trials. Lancet 2002;359:189–98.

[181] Alexander KP, Chen AY, Newby LK, Schwartz JB, Redberg RF, Hochman JS, et al. Sex differences in major bleeding with glycoprotein IIb/IIIa inhibitors: results from the CRUSADE (can rapid risk stratification of unstable angina patients suppress ADverse outcomes with early implementation of the ACC/AHA guidelines) initiative. Circulation 2006;114:1380–7.

[182] Lenz T, Wilson A. Clinical pharmacokinetics of antiplatelet agents used in the secondary prevention of stroke. Clin Pharmacokinet 2003;42:909–20.

[183] Price MJ. Monitoring platelet function to reduce the risk of ischemic and bleeding complications. Am J Cardiol 2009;103(Suppl):35A–9A.

[184] Ferreiro JL, Angiolillo DJ. Clopidogrel response variability: current status and future directions. Thromb Haemost 2009;102:7–14.

[185] Yusuf S, Zhao F, Mehta SR, Chrolavicius S, Tognoni G, Fox KK, et al. Effects of clopidogrel in addition to aspirin in patients with acute coronary syndromes without ST-segment elevation. N Engl J Med 2001;345:494–502.

[186] Cannon CP, CAPRIE Investigators. Effectiveness of clopidogrel versus aspirin in preventing acute myocardial infarction in patients with symptomatic atherothrombosis (CAPRIE trial). Am J Cardiol 2002;90:760–2.

[187] Mehta SR, Yusuf S, Peters RJ, Bertrand ME, Lewis BS, Natarajan MK, et al. Effects of pretreatment with clopidogrel and aspirin followed by long-term therapy in patients undergoing percutaneous coronary intervention: the PCI-CURE study. Lancet 2001;358:527–33.

[188] Steinhubl SR, Berger PB, Mann 3rd JT, Fry ETA, DeLago A, Wilmer C, et al. Clopidogrel for the reduction of events during observation. Early and sustained dual oral antiplatelet therapy following percutaneous coronary intervention: a randomized controlled trial. JAMA 2002;288:2411–20.

[189] Berger JS, Bhatt DL, Cannon CP, Chen Z, Jiang L, Jones JB, et al. The relative efficacy and safety of clopidogrel in women and men: a sex-specific collaborative meta-analysis. J Am Coll Cardiol 2009;54:1935–45.

[190] Zaccardi F, Pitocco D, Willeit P, Laukkanen JA. Efficacy and safety of P2Y12 inhibitors according to diabetes, age, gender, body mass index and body weight: systematic review and meta-analyses of randomized clinical trials. Atherosclerosis 2015;240:439–45.

[191] Ernest 2nd CS, Small DS, Rohatagi S, Salazar DE, Wallentin L, Winters KJ, et al. Population pharmacokinetics and pharmacodynamics of prasugrel and clopidogrel in aspirin-treated patients with stable coronary artery disease. J Pharmacokinet Pharmacodyn 2008;35:593–618.

[192] Wrishko RE, Ernest 2nd CS, Small DS, Li YG, Weerakkody GJ, Riesmeyer JR, et al. Population pharmacokinetic analyses to evaluate the influence of intrinsic and extrinsic factors on exposure of prasugrel active metabolite in TRITON-TIMI 38. J Clin Pharmacol 2009;49:984–98.

[193] Wiviott SD, Braunwald E, McCabe CH, Montalescot G, Ruzyllo W, Gottlieb S, et al. Prasugrel versus clopidogrel in patients with acute coronary syndromes. N Engl J Med 2007;357:2001–15.

[194] Chandrasekhar J, Baber U, Sartori S, Faggioni M, Aquino M, Kini A, et al. Sex-related differences in outcomes among men and women under 55 years of age with acute coronary syndrome undergoing percutaneous coronary intervention: results from the PROMETHEUS Study. Catheter Cardiovasc Interv 2017;89:629–37.

[195] Hochholzer W, Wiviott SD, Antman EM, Contant CF, Guo J, Giugliano RP, et al. Predictors of bleeding and time dependence of association of bleeding with mortality: insights from the trial to assess improvement in therapeutic outcomes by optimizing platelet inhibition with prasugrel-thrombolysis in myocardial infarction 38 (TRITON-TIMI 38). Circulation 2011;123:2681-9.

[196] Teng R. Ticagrelor: pharmacokinetic, pharmacodynamic and pharmacogenetic profile: an update. Clin Pharmacokinet 2015;54:1125–38.

[197] Wallentin L, Becker RC, Budaj A, Cannon CP, Emanuelsson H, Held C, et al. Ticagrelor versus clopidogrel in patients with acute coronary syndromes. N Engl J Med 2009;361:1045–57.

[198] Husted S, James SK, Bach RG, Becker RC, Budaj A, Heras M, et al. The efficacy of ticagrelor is maintained in women with acute coronary syndromes participating in the prospective, randomized, PLATelet inhibition and patient outcomes (PLATO) trial. Eur Heart J 2014;35:1541–50.

[199] O'Donoghue ML, Bhatt DL, Stone GW, Steg PG, Gibson CM, Hamm CW, et al. Efficacy and safety of cangrelor in women versus men during percutaneous coronary intervention: insights from the cangrelor versus standard therapy to achieve optimal management of platelet inhibition (CHAMPION PHOENIX) Trial. Circulation 2016;133:248–55.

[200] Tan G, Chen J, Liu M, Yeh J, Tang W, Ke J, et al. Efficacy and safety of vorapaxar for the prevention of adverse cardiac events in patients with coronary artery disease: a meta-analysis. Cardiovasc Diagn Ther 2016;6:101–8.

[201] White HD, Barbash GI, Modan M, Simes J, Diaz R, Hampton JR, et al. After correcting for worse baseline characteristics, women treated with thrombolytic therapy for acute myocardial infarction have the same mortality and morbidity as men except for a higher incidence of hemorrhagic stroke. The Investigators of the International Tissue Plasminogen Activator/Streptokinase Mortality Study. Circulation 1993;88:2097–103.

[202] Weaver WD, White HD, Wilcox RG, Aylward PE, Morris D, Guerci A, et al. Comparisons of characteristics and outcomes among women and men with acute myocardial infarction treated with thrombolytic therapy GUSTO-I Investigators. JAMA 1996;275:777–82.

[203] Woodfield SL, Lundergan CF, Reiner JS, Thompson MA, Rohrbeck SC, Deychak Y, et al. Gender and acute myocardial infarction: is there a different response to thrombolysis? J Am Coll Cardiol 1997;29:35–42.

[204] Mehilli J, Ndrepepa G, Kastrati A, Neumann FJ, ten Berg J, Bruskina O, et al. Sex and effect of abciximab in patients with acute coronary syndromes treated with percutaneous coronary interventions: results from Intracoronary stenting and antithrombotic regimen: rapid early action for coronary treatment 2 trial. Am Heart J 2007;154:158.e151–7.

[205] Franzosi MG, Santoro E, De Vita C, Geraci E, Lotto A, Maggioni AP, et al. Ten-year follow-up of the first megatrial testing thrombolytic therapy in patients with acute myocardial infarction: results of the Gruppo Italiano per lo studio della Sopravvivenza nell'Infarto-1 study. The GISSI Investigators. Circulation 1998;98:2659–65.

[206] Dziewierz A, Siudak Z, Rakowski T, Birkemeyer R, Legutko J, Mielecki W, et al. Early administration of abciximab reduces mortality in female patients with ST-elevation myocardial infarction undergoing primary percutaneous coronary intervention (from the EUROTRANSFER registry). J Thromb Thrombolysis 2013;36:240–6.

[207] ISIS-2 (Second International Study of Infarct Survival) Collaborative Group. Randomised trial of intravenous streptokinase, oral aspirin, both, or neither among 17 187 cases of suspected acute myocardial infarction: ISIS-2. Lancet 1988;2:349–60.

[208] Lee KL, Woodlief LH, Topol EJ, Weaver WD, Betriu A, Col J, et al. Predictors of 30-day mortality in the era of reperfusion for acute myocardial infarction. Circulation 1995;91:1659–68.

[209] Moen EK, Asher CR, Miller DP, Weaver WD, White HD, Califf RM, et al. Long-term follow-up of gender-specific outcomes after thrombolytic therapy for acute myocardial infarction from the GUSTO-I trial. Global utilization of streptokinase and tissue plasminogen activator for occluded coronary arteries. J Womens Health 1997;6:285–93.

[210] Berkowitz SD, Granger CB, Pieper KS, Lee KL, Gore JM, Simoons M, et al. Incidence and predictors of bleeding after contemporary thrombolytic therapy for myocardial infarction. The global utilization of streptokinase and tissue plasminogen activator for occluded coronary arteries (GUSTO) I Investigators. Circulation 1997;95:2508–16.

[211] Gurwitz JH, Gore JM, Goldberg RJ, Barron HV, Breen T, Rundle AC, et al. Risk for intracranial haemorrhage after tissue plasminogen activator treatment for acute myocardial infarction. Participants in national registry of myocardial infarction 2. Ann Intern Med 1998;129:597–604.

[212] Gilmore DA, Gal J, Gerber JG, Nies AS. Age and gender influence the stereoselective pharmacokinetics of propranolol. J Pharmacol Exp Ther 1992;261:1181–6.

[213] Xie HG, Chen X. Sex differences in pharmacokinetics of oral propranolol in healthy Chinese volunteers. Zhongguo Yao Li Xue Bao 1995;16:468–70.

[214] Cocco G, Chu D. The anti-ischemic effect of metoprolol in patients with chronic angina pectoris is gender-specific. Cardiology 2006;106:147–53.

[215] Fletcher A, Beevers DG, Bulpitt C, Butler A, Coles EC, Hunt D, et al. Beta adrenoceptor blockade is associated with increased survival in male but not female hypertensive patients: a report from the DHSS hypertension care computing project (DHCCP). J Hum Hypertens 1988;2:219–27.

[216] Beta-Blocker Heart Attack Trial Research Group. A randomized trial of propranolol in patients with acute myocardial infarction. I. Mortality results. JAMA 1982;247:1707–14.

[217] MERIT-HF Study Group. Effect of metoprolol CR/XL in chronic heart failure: metoprolol CR/XL randomised intervention trial in congestive heart failure (MERIT-HF). Lancet 1999;353:2001–7.

[218] Packer M, Bristow MR, Cohn JN, Colucci WS, Fowler MB, Gilbert EM, et al. The effect of carvedilol on morbidity and mortality in patients with chronic heart failure. N Engl J Med 1996;334:1349–55.

[219] Packer M, Fowler MB, Roecker EB, Coats AJS, Katus HA, Krum H, et al. Effect of carvedilol on the morbidity of patients with severe chronic heart failure: results of the carvedilol prospective randomized cumulative survival (COPERNICUS) study. Circulation 2002;106:2194–9.

[220] Ghali JK, Pina IL, Gottlieb SS, Deedwania PC, Wikstrand JC, MERIT-HF Study Group. Metoprolol CR/XL in female patients with heart failure: analysis of the experience in metoprolol extended-release randomized intervention trial in heart failure (MERIT-HF). Circulation 2002;105:1585–91.

[221] Simon T, Mary-Krause M, Funck-Brentano C, Jaillon P. Sex differences in the prognosis of congestive heart failure: results from the cardiac insufficiency Bisoprolol Study (CIBIS II). Circulation 2001;103:375–80.

[222] Kotecha D, Manzano L, Krum H, Rosano G, Holmes J, Altman DG, et al. Effect of age and sex on efficacy and tolerability of beta blockers in patients with heart failure with reduced ejection fraction: individual patient data meta-analysis. BMJ 2016;353:i1855.

[223] López-Sendón J, Swedberg K, MacMurray J, Tamargo J, Maggioni AP, Dargie H, et al. Expert consensus document on β-adrenergic receptor blockers: the task force on beta-blockers of the European Society of Cardiology. Eur Heart J 2004;25:1341–62.

[224] Krumholz HM, Radford MJ, Wang Y, Chen J, Heiat A, Marciniak T. National use and effectiveness of beta-blockers for the treatment of elderly patients after acute myocardial infarction: National Cooperative Cardiovascular Project. JAMA 1998;280:623–9.

[225] Olsson G, Wikstrand J, Warnold I, Manger Cats V, McBoyle D, Herlitz J, et al. Metoprolol-induced reduction in postinfarction mortality: pooled results from five double-blind randomized trials. Eur Heart J 1992;13:28–32.

[226] Schwartz JB, Capili H, Daugherty J. Aging of women alters s-verapamil pharmacokinetics and pharmacodynamics. Clin Pharmacol Ther 1994;55:509–17.

[227] Kloner RA, Sowers JR, DiBona GF, Gaffney M, Wein M, Amlodipine Cardiovascular Community Trial Study Group. Sex- and age-related antihypertensive effects of amlodipine. Am J Cardiol 1996;77:713–22.

[228] Digitalis Investigation Group. The effect of digoxin on mortality and morbidity in patients with heart failure. N Engl J Med 1997;336:525–33.

[229] Ahmed A, Rich MW, Love TE, Lloyd-Jones DM, Aban IB, Colucci WS, et al. Digoxin and reduction in mortality and hospitalization in heart failure: a comprehensive post hoc analysis of the DIG trial. Eur Heart J 2006;27:178–86.

[230] Lee LS, Chan LN. Evaluation of a sex-based difference in the pharmacokinetics of digoxin. Pharmacotherapy 2006;26:44–50.

[231] Blaustein MP, Robinson SW, Gottlieb SS, Balke CW, Hamlyn JM. Sex, digitalis, and the sodium pump. Mol Interv 2003;3:68–72.

[232] Green HJ, Duscha BD, Sullivan MJ, Keteyian SJ, Kraus WE. Normal skeletal muscle Na(+)-K(+) pump concentration in patients with chronic heart failure. Muscle Nerve 2001;24:69–76.

[233] Furberg CD, Vittinghoff E, Davidson M, Herrington DM, Simon JA, Wenger NK, et al. Subgroup interactions in the heart and estrogen/progestin replacement study: lessons learned. Circulation 2002;105:917–22.

[234] Domanski M, Fleg J, Bristow M, Knox S. The effect of gender on outcome in digitalis-treated heart failure patients. J Card Fail 2005;11:83–6.

[235] Olde Engberink RHG, Frenkel WJ, van den Bogaard B, Brewster LM, Vogt L, van den Born B-JH. Effects of thiazide-type and thiazide-like diuretics on cardiovascular events and mortality: systematic review and meta-analysis. Hypertension 2015;65:1033–40.

[236] Chapman MD, Hanrahan R, McEwen J, Marley JE. Hyponatremia and hypokalemia due to indapamide. Med J Aust 2002;176:219–21.

CH
27

[237] Gabler NB, French B, Strom BL, Liu Z, Palevsky HI, Taichman DB. Race and sex differences in response to endothelin receptor antagonists for pulmonary arterial hypertension. Chest 2012;141:20–6.

[238] Marra AM, Benjamin N, Eichstaedt C, Salzano A, Arcopinto M, Gargani L, et al. Gender-related differences in pulmonary arterial hypertension targeted drugs administration. Pharmacol Res 2016;114:103–9.

[239] Mathai SC, Hassoun PM, Puhan MA, Zhou Y, Wise RA. Sex differences in response to tadalafil in pulmonary arterial hypertension. Chest 2015;147:188–97.

[240] Galiè N, Barberà JA, Frost AE, Ghofrani H-A, Hoeper MM, McLaughlin VV, et al. Initial use of Ambrisentan plus tadalafil in pulmonary arterial hypertension. N Engl J Med 2015;373:834–44.

[241] Wilkins MR, Paul GA, Strange JW, Tunariu N, Gin-Sing W, Banya WA, et al. Sildenafil versus endothelin receptor antagonist for pulmonary hypertension (SERAPH) study. Am J Respir Crit Care Med 2005;171:1292–7.

[242] Al-Shalamed A, Chanson P, Sucher S, Ringa V, Becquemont L. Cardiovascular disease in type 2 diabetes: a review of sex-related differences in predisposition and prevention. Mayo Clin Proc 2019;94:287–330.

[243] Vaccarino V, Parsons L, Peterson ED, Rogers WJ, Kiefe CI, Canto J. Sex differences in mortality after acute myocardial infarction: changes from 1994 to 2006. Arch Intern Med 2009;169:1767–74.

[244] Zelniker TA, Wiviott SD, Raz I, Im K, Goodrich EL, Furtado RHM, et al. Comparison of the effects of glucagon-like peptide receptor agonists and sodium-glucose cotransporter 2 inhibitors for prevention of major adverse cardiovascular and renal outcomes in type 2 diabetes mellitus. Circulation 2019;139:2022–31.

[245] Zelniker TA, Wiviott SD, Raz I, Im K, Goodrich EL, Bonaca MP, et al. SGLT2 inhibitors for primary and secondary prevention of cardiovascular and renal outcomes in type 2 diabetes: a systematic review and meta-analysis of cardiovascular outcome trials. Lancet 2019;393:31–9.

[246] Rådholm K, Zhou Z, Clemens K, Neal B, Woodward M. Effects of sodium-glucose co-transporter-2 inhibitors in type 2 diabetes in women versus men. Diabetes Obes Metab 2020;22:263–6.

[247] Kosiborod M, Lam CSP, Kohsaka S, Kim DJ, Karasik A, Shaw J, et al. Cardiovascular events associated with SGLT-2 inhibitors versus other glucose-lowering drugs: the CVD-REAL 2 Study. J Am Coll Cardiol 2018;71:2628–39.

[248] Zinman B, Inzucchi SE, Wanner C, Hehnke U, George JT, Johansen OE, et al. Empagliflozin in women with type 2 diabetes and cardiovascular disease – an analysis of EMPA-REG OUTCOME. Diabetologia 2018;61:1522–7.

[249] Mahmoud A, ELgendy IY, Saad M, Elgendy AY, Barakat AF, Mentias A, et al. Does gender influence the cardiovascular benefits observed with sodium glucose co-transporter-2 (SGLT-2) inhibitors? A meta-regression analysis. Cardiol Ther 2017;6:129–32.

[250] Raparelli V, Elharram M, Moura CS, Abrahamowicz M, Bernatsky S, Behlouli H, et al. Sex differences in cardiovascular effectiveness of newer glucose-lowering drugs added to metformin in type 2 diabetes mellitus. J Am Heart Assoc 2020;9, e012940.

[251] Tamargo J. Sodium-glucose cotransporter 2 inhibitors in heart failure: potential mechanisms of action, adverse effects and future developments. Eur Cardiol 2019;14:23–32.

[252] ORIGIN Trial Investigators, Gerstein HC, Bosch J, Dagenais GR, Díaz R, Jung H, et al. Basal insulin and cardiovascular and other outcomes in dysglycemia. N Engl J Med 2012;367:319–28.

[253] Tendera M, Borer JS, Tardif J-C. Efficacy of if inhibition with ivabradine in different subpopulations with stable angina pectoris. Cardiology 2009;114:116–25.

[254] Fox K, Ford I, Steg PG, Tendera M, Ferrari M, BEAUTIFUL Investigators. Ivabradine for patients with stable coronary artery disease and left-ventricular systolic dysfunction (BEAUTIFUL): a randomised, double-blind, placebo-controlled trial. Lancet 2008;372:807–16.

[255] Swedberg K, Komajda M, Böhm M, Borer JS, Ford I, Dubost-Brama A, et al. Ivabradine and outcomes in chronic heart failure (SHIFT): a randomised placebo-controlled study. Lancet 2010;376:875–85.

[256] Fischer M, Baessler A, Schunkert H. Renin angiotensin system and gender differences in the cardiovascular system. Cardiovasc Res 2002;53:672–7.

[257] Komukai K, Mochizuki S, Yoshimura M. Gender and the renin-angiotensin-aldosterone system. Fundam Clin Pharmacol 2010;24:687–98.

[258] Jarugula V, Yeh C-M, Howard D, Bush C, Keefe DL, Dole WP. Influence of body weight and gender on the pharmacokinetics, pharmacodynamics, and antihypertensive efficacy of aliskiren. J Clin Pharmacol 2010;50:1358–66.

[259] Israili ZH. Clinical pharmacokinetics of angiotensin II (AT1) receptor blockers in hypertension. J Hum Hypertens 2000;14:73–86.

[260] Walfisch A, Al-maawali A, Moretti ME, Nickel C, Koren G. Teratogenicity of angiotensin converting enzyme inhibitors or receptor blockers. J Obstet Gynaecol 2011;31:465–72.

[261] Pucci M, Sarween N, Knox E, Lipkin G, Martin U. Angiotensin-converting enzyme inhibitors and angiotensin receptor blockers in women of childbearing age: risks versus benefits. Expert Rev Clin Pharmacol 2015;8:221–31.

[262] Vukadinović D, Vukadinović AN, Lavall D, Laufs U, Wagenpfeil S, Böhm M. Rate of cough during treatment with angiotensin-converting enzyme inhibitors: a meta-analysis of randomized placebo-controlled trials. Clin Pharmacol Ther 2019;105:652–60.

[263] ACE Inhibitor Myocardial Infarction Collaborative Group. Indications for ACE inhibitors in the early treatment of acute myocardial infarction: systematic overview of individual data from 100 000 patients in randomized trials. Circulation 1998;97:2202–12.

[264] Flather MD, Yusuf S, Kober L, Pfeffer M, Hall A, Murray G, et al. Long-term ACE inhibitor therapy in patients with heart failure or left-ventricular dysfunction: a systematic overview of data from individual patients. Lancet 2000;355:1575–81.

[265] The Consensus Trial Study Group. Effects of enalapril on mortality in severe congestive heart failure. Results of the Cooperative North Scandinavian Enalapril Survival Study (CONSENSUS). N Engl J Med 1987;316:1429–35.

[266] Pfeffer MA, Braunwald E, Moye LA, Basta L, Brown EJ, Cuddy TE, et al. Effect of captopril on mortality and morbidity in patients with left ventricular dysfunction after myocardial infarction. Results of the survival and ventricular enlargement Trial. The save Investigators. N Engl J Med 1992;327:669–77.

[267] SOLVD Investigators, Yusuf S, Pitt B, Davis CE, Hood WB, Cohn JN. Effect of enalapril on survival in patients with reduced left ventricular ejection fractions and congestive heart failure. N Engl J Med 1991;325:293–302.

[268] Heart Outcomes Prevention Evaluation Study Investigators, Yusuf S, Sleight P, Pogue J, Bosch J, Davies R, et al. Effects of an angiotensin-converting-enzyme inhibitor, ramipril, on cardiovascular events in high-risk patients. The heart outcomes prevention evaluation study investigators. N Engl J Med 2000;342:145–53.

[269] The Acute Infarction Ramipril Efficacy (AIRE) Study Investigators. Effect of ramipril on mortality and morbidity of survivors of acute myocardial infarction with clinical evidence of heart failure. Lancet 1993;342:821–8.

[270] Garg R, Yusuf S. Overview of randomized trials of angiotensin-converting enzyme inhibitors on mortality and morbidity in patients with heart failure. Collaborative group on ACE inhibitor trials. JAMA 1995;273:1450–6.

[271] Wing LM, Reid CM, Ryan P, Beilin L, Brown MA, Jennings GLR, et al. A comparison of outcomes with angiotensin converting-enzyme inhibitors and diuretics for hypertension in the elderly. N Engl J Med 2003;348:583–92.

[272] Dahlöf B, Devereux RB, Kjeldsen SE, Julius S, Beevers G, de Faire U, et al. Cardiovascular morbidity and mortality in the losartan intervention for endpoint reduction in hypertension study (LIFE): a randomised trial against atenolol. Lancet 2002;359:995–1003.

[273] Julius S, Kjeldsen SE, Weber M, Brunner HR, Ekman S, Hansson L, et al. Outcomes in hypertensive patients at high cardiovascular risk treated with regimens based on valsartan or amlodipine: the VALUE randomised trial. Lancet 2004;363:2022–31.

[274] Pitt B, Segal R, Martinez FA, Meurers G, Cowley AJ, Thomas I, et al. Randomised Trial of losartan versus captopril in patients over 65 with heart failure (evaluation of losartan in the elderly Study, ELITE). Lancet 1997;349:747–52.

[275] Cohn JN, Tognoni G, Valsartan Heart Failure Trial Investigators. A randomized trial of the angiotensin-receptor blocker valsartan in chronic heart failure. N Engl J Med 2001;345:1667–75.

[276] Pfeffer MA, Swedberg K, Granger CB, Held P, McMurray JJV, Michelson EL, et al. Effects of candesartan on mortality and morbidity in patients with chronic heart failure: the Charm-overall programme. Lancet 2003;362:759–66.

[277] Pfeffer MA, McMurray JJV, Velazquez EJ, Rouleau JL, Køber L, Maggioni AP, et al. Valsartan, captopril, or both in myocardial infarction complicated by heart failure, left ventricular dysfunction, or both. N Engl J Med 2003;349:1893–906.

[278] Dickstein K, Kjekshus J. OPTIMAAL steering committee of the OPTIMAAL Study group effects of losartan and captopril on mortality and morbidity in high-risk patients after acute myocardial infarction: the OPTIMAAL randomised trial. Optimal trial in myocardial infarction with angiotensin II antagonist losartan. Lancet 2002;360:752–60.

[279] Carson P, Tognoni G, Cohn JN. Effect of valsartan on hospitalization: results from Val-Heft. J Card Fail 2003;9:164–71.

[280] Hudson M, Rahme E, Behlouli H, Sheppard R, Pilote L. Sex differences in the effectiveness of angiotensin receptor blockers and angiotensin converting enzyme inhibitors in patients with congestive heart failure—a population study. Eur J Heart Fail 2007;9:602–9.

[281] Strauss MH, Hall AS. Angiotensin receptor blockers may increase risk of myocardial infarction: unraveling the ARB-MI paradox. Circulation 2006;114:838–54.

[282] Pitt B, Zannad F, Remme WJ, Cody R, Castaigne A, Perez A, et al. The effect of spironolactone in morbidity and mortality in patients with severe heart failure. N Engl J Med 1999;341:709–17.

[283] Pitt B, Pfeffer MA, Assmann SF, Boineau R, Anand IS, Claggett B, et al. Spironolactone for heart failure with preserved ejection fraction. N Engl J Med 2014;370:1383–92.

[284] Pitt B, Remme W, Zannad F, Neaton J, Martinez F, Roniker B, et al. Eplerenone, a selective aldosterone blocker, in patients with left ventricular dysfunction after myocardial infarction. N Engl J Med 2003;348:1309–21.

[285] Collier TJ, Pocock SJ, McMurray JJV, Zannad F, Krum H, van Veldhuisen DJ, et al. The impact of eplerenone at different levels of risk in patients with systolic heart failure and mild symptoms: insight from a novel risk score for prognosis derived from the EMPHASIS-HF trial. Eur Heart J 2013;34:2823–9.

[286] Gan L, Langenickel T, Petruck J, Kode K, Rajman I, Chandra P, et al. Effects of age and sex on the pharmacokinetics of LCZ696, an angiotensin receptor neprilysin inhibitor. J Clin Pharmacol 2016;56:78–86.

[287] McMurray JJV, Packer M, Desai AS, Gong J, Lefkowitz MP, Rizkala AR, et al. Angiotensin-neprilysin inhibition versus enalapril in heart failure. N Engl J Med 2014;371:993–1004.

[288] McMurray JJV, Jackson AM, Lam CSP, Redfield MM, Anand IS, Ge J, et al. Effects of Sacubitril-valsartan, versus valsartan, in women compared to men with heart failure and preserved ejection fraction: insights from PARAGON-HF. Circulation 2020;141:338–51.

[289] Vree TB, Dammers E, Valducci R. Sex-related differences in the pharmacokinetics of isosorbide-5-mononitrate (60 mg) after repeated oral administration of two different original prolonged release formulations. Int J Clin Pharmacol Ther 2004;42:463–72.

[290] Taylor AL, Lindenfeld J, Ziesche S, Walsh MN, Mitchell JE, Adams K, et al. Outcomes by gender in the African-American heart failure trial. JACC 2006;48:2263–7.

[291] Yusuf S, Hawken S, Ounpuu S, Dans T, Avezum A, Lanas F, et al. Effect of potentially modifiable risk factors associated with myocardial infarction in 52 countries (the INTERHEART study): case-control study. Lancet 2004;364:937–52.

[292] Pilote L, Dasgupta K, Guru V, Humphries KH, McGrath J, Norris C, et al. A comprehensive view of sex-specific issues related to cardiovascular disease. CMAJ 2007;176:S1–S44.

[293] Truong QA, Murphy SA, McCabe CH, Armani A, Cannon CP, TIMI Study Group. Benefit of intensive statin therapy in women: results from PROVE IT-TIMI 22. Circ Cardiovasc Qual Outcomes 2011;4:328–36.

[294] Mosca L. Controversy and consensus about statin use. It is not about the sex. J Am Coll Cardiol 2012;59:583–4.

[295] Mosca L, Hammond G, Mochari-Greenberger H, Towfighi A, Albert MA. Fifteen-year trends in awareness of heart disease in women results of a 2012 American Heart Association National Survey. Circulation 2013;127:1254–63 [e1-29].

[296] Taylor F, Huffman MD, Macedo AF, Moore THM, Burke M, Davey Smith G, et al. Statins for the primary prevention of cardiovascular disease. Cochrane Database Syst Rev 2013;, CD00481.

[297] Cholesterol Treatment Trialists' (CTT) Collaboration, Fulcher J, O'Connell R, Voysey M, Emberson J, Blackwell L, et al. Efficacy and safety of LDL-lowering therapy among men and women: meta-analysis of individual data from 174,000 participants in 27 randomised trials. Lancet 2015;385:1397–405.

[298] Chou R, Dana T, Blazina I, Daeges M, Jeanne TL. Statins for prevention of cardiovascular disease in adults: systematic review for the U.S. preventive services task force. Evidence synthesis no. 139. AHRQ publication no. 14-05206-EF-2. Rockville, MD: Agency for Healthcare Research and Quality; 2016.

[299] Arnett DK, Blumenthal RS, Albert MA, Buroker AB, Goldberger ZD, Hahn EJ, et al. 2019 ACC/AHA guideline on the primary prevention of cardiovascular disease. A report of the American College of Cardiology/American Heart Association Task Force on Clinical Practice Guidelines. J Am Coll Cardiol 2019;74:e177–232.

[300] Butalia S, Lewin AM, Simpson SH, Dasgupta K, Khan N, Pilote L, et al. Sex-based disparities in cardioprotective medication use in adults with diabetes. Diabetol Metab Syndr 2014;6:117.

[301] Virani SS, Woodard LD, Ramsey DJ, Urech TH, Akeroyd JM, Shah T, et al. Gender disparities in evidence-based statin therapy in patients with cardiovascular disease. Am J Cardiol 2015;115:21–6.

[302] Magee MF, Tamis-Holland JE, Lu J, Bittner VA, Brooks MM, Lopes N, et al. Sex, prescribing practices and guideline recommended, blood pressure, and LDL cholesterol targets at baseline in the BARI 2D Trial. Int J Endocrinol 2015;2015:610239.

[303] Gamboa CM, Colantonio LD, Brown TM, Carson AP, Safford MM. Race-sex differences in statin use and low-density lipoprotein cholesterol control among people with diabetes mellitus in the reasons for geographic and racial differences in stroke study. J Am Heart Assoc 2017;6, e004264.

[304] Nanna MG, Wang TY, Xiang Q, Goldberg AC, Robinson JG, Roger VL, et al. Sex differences in the use of statins in community practice. Circ Cardiovasc Qual Outcomes 2019;12, e005562.

[305] Mora S, Glynn RJ, Hsia J, MacFadyen JG, Genest J, Ridker PM. Statins for the primary prevention of cardiovascular events in women with elevated high-sensitivity C-reactive protein or dyslipidemia: results from the justification for the Use of statins in prevention: an intervention trial evaluating Rosuvastatin (JUPITER) and meta-analysis of women from primary prevention trials. Circulation 2010;121:1069–77.

[306] Cheung BMY, Lauder IJ, Lau C-P, Kumana CR. Meta-analysis of large randomized controlled trials to evaluate the impact of statins on cardiovascular outcomes. Br J Clin Pharmacol 2004;57:640–51.

[307] Kostis WJ, Cheng JQ, Dobrzynski JM, Cabrera J, Kostis JB. Meta-analysis of statin effects in women versus men. J Am Coll Cardiol 2012;59:572–82.

[308] Gutierrez J, Ramirez G, Rundek T, Sacco RL. Statin therapy in the prevention of recurrent cardiovascular events: a sex-based meta-analysis. Arch Intern Med 2012;172:909–19.

[309] Taylor F, Ebrahim S. Statins work just as well in women as in men. Arch Intern Med 2012;172:919–20.

[310] Sattar N, Preiss D, Murray HM, Welsh P, Buckley BM, de Craen AJM, et al. Statins and risk of incident diabetes: a collaborative meta-analysis of randomised statin trials. Lancet 2010;375:735–42.

[311] Culver AL, Ockene IS, Balasubramanian R, Olendzki BC, Sepavich DM, Wactawski-Wende J, et al. Statin use and risk of diabetes mellitus in postmenopausal women in the Women's Health Initiative. Arch Intern Med 2012;172:144–52.

[312] Ma Y, Culver A, Rossouw J, Olendzki B, Merriam P, Lian B, et al. Statin therapy and the risk for diabetes among adult women: do the benefits outweigh the risk? Ther Adv Cardiovasc Dis 2012;7:41–4.

[313] Hsue PY, Bittner VA, Betteridge J, Fayyad R, Laskey R, Wenger NK, et al. Impact of female sex on lipid lowering, clinical outcomes, and adverse effects in atorvastatin trials. Am J Cardiol 2015;115:447–53.

[314] Thompson PD, Panza G, Zaleski A, Taylor B. Statin-associated side effects. J Am Coll Cardiol 2016;67:2395–410.

[315] Ridker PM, Danielson E, Fonseca FAH, Genest J, Gotto AM, Kastelein JJP, et al. Rosuvastatin to prevent vascular events in men and women with elevated C-reactive protein. N Engl J Med 2008;359:2195–207.

[316] Vonbank A, Drexel H, Agewall S, Lewis BS, Dopheide JF, Kjeldsen K, et al. Reasons for disparity in statin adherence rates between clinical trials and real-world observations: a review. Eur Heart J Cardiovasc Pharmacother 2018;4:230–6.

[317] Abramson BL, Benlian P, Hanson ME, Lin J, Shah A, Tershakovec AM. Response by sex to statin plus ezetimibe or statin monotherapy: a pooled analysis of 22,231 hyperlipidemic patients. Lipids Health Dis 2011;10:146.

[318] Bennett S, Sager P, Lipka L, Melani L, Suresh R, Veltri E, et al. Consistency in efficacy and safety of ezetimibe coadministered with statins for treatment of hypercholesterolemia in women and men. J Womens Health (Larchmt) 2004;13:1101–7.

[319] Ose L, Shah A, Davies MJ, Rotonda J, Meehan AG, Cutler DL. Consistency of lipid-altering effects of ezetimibe/simvastatin across gender, race, age, baseline low density lipoprotein cholesterol levels, and coronary heart disease status: results of a pooled retrospective analysis. Curr Med Res Opin 2006;22:823–35.

[320] Bays HE, Conard SE, Leiter LA, Bird SR, Lowe RS, Tershakovec AM. Influence of age, gender, and race on the efficacy of adding ezetimibe to atorvastatin vs. atorvastatin up-titration in patients at moderately high or high risk for coronary heart disease. Int J Cardiol 2011;153:141–7.

[321] ACCORD Study Group, Cushman WC, Evans GW, Byington RP, Goff DC, Grimm RH. Effects of combination lipid therapy in type 2 diabetes mellitus. N Engl J Med 2010;362:1563–74.

[322] d'Emden MC, Jenkins AJ, Li L, Zannino D, Mann KP, Best JD, et al. Favourable effects of fenofibrate on lipids and cardiovascular disease in women with type 2 diabetes: results from the Fenofibrate intervention and event lowering in diabetes (FIELD) study. Diabetologia 2014;57:2296–303.

[323] Gaudet D, Watts GF, Robinson JG, Minini P, Sasiela WJ, Edelberg J, et al. Effect of Alirocumab on lipoprotein(a) over ≥ 1.5 years (from the phase 3 ODYSSEY program). Am J Cardiol 2017;119:40–6.

[324] Sabatine MS, Giugliano RP, Keech AC, Honarpour N, Wiviott SD, Murphy SA, et al. Evolocumab and clinical outcomes in patients with cardiovascular disease. N Engl J Med 2017;376:1713–22.

[325] Gürgöze MT, Muller-Hansma AHG, Schreuder MM, Galema-Boers AMH, Boersma E, van Lennep JER. Adverse events associated with PCSK9 inhibitors: a real-world experience. Clin Pharmacol Ther 2019;105:496–504.

[326] Konstam MA, Neaton JD, Dickstein K, Drexler H, Komajda M, Martinez FA, et al. Effects of high-dose versus low-dose losartan on clinical outcomes in patients with heart failure (HEAAL study): a randomised, double-blind trial. Lancet 2009;374:1840–8.

[327] Flather MD, Shibata MC, Coats AJS, Van Veldhuisen DJ, Parkhomenko A, Borbola J, et al. Randomized trial to determine the effect of nebivolol on mortality and cardiovascular hospital admission in elderly patients with heart failure (SENIORS). Eur Heart J 2005;26:215–25.

[328] Merrill M, Sweitzer NK, Lindenfeld J, Kao DP. Sex differences in outcomes and responses to spironolactone in heart failure with preserved ejection fraction: a secondary analysis of TOPCAT trial. JACC Heart Fail 2019;7:228–38.

[329] Køber L, Torp-Pedersen C, Carlsen JE, Bagger H, Eliasen P, Lyngborg K, et al. A clinical trial of the angiotensin-converting-enzyme inhibitor trandolapril in patients with left ventricular dysfunction after myocardial infarction. Trandolapril cardiac evaluation (TRACE) Study Group. N Engl J Med 1995;333:1670–6.

[330] Gelbenegger G, Postula M, Pecen L, Halvorsen S, Lesiak M, Schoergenhofer C, et al. Aspirin for primary prevention of cardiovascular disease: a meta-analysis with a particular focus on subgroups. BMC Med 2019;17:198.

[331] Lewis SJ, Sacks FM, Mitchell JS, East C, Glasser S, Kell S, et al. Effect of pravastatin on cardiovascular events in women after myocardial infarction: the cholesterol and recurrent events (CARE) trial. J Am Coll Cardiol 1998;32:140–6.

[332] Richardson J, Holdcroft A. Results of forty years yellow card reporting for commonly used perioperative analgesic drugs. Pharmacoepidemiol Drug Saf 2007;16:687–94.

[333] Pillans PI, Coulter DM, Black P. Angiooedema and pharmacoge with angiotensin converting enzyme inhibitors. Eur J Clin Pharmacol 1996;51:123–6.

[334] Naqvi TZ, Gross SB. Anorexigen-induced cardiac valvulopathy and female gender. Curr Womens Health Rep 2003;3:116–25.

[335] Frommeyer G, Eckardt L. Drug-induced proarrhythmia: risk factors and electrophysiological mechanisms. Nat Rev Cardiol 2016;13:36–47.

[336] Yancy CW, Jessup M, Bozkurt B, Butler J, Casey DE, Drazner MH, et al. 2013 ACCF/AHA Guideline for the Management of Heart Failure: executive summary: a report of the American College of Cardiology Foundation/American Heart Association task force on practice guidelines. Circulation 2013;128:1810–52.

[337] Sharabi Y, Illan R, Kamari Y, Cohen H, Nadler M, Messerli FH, et al. Diuretic induced hyponatraemia in elderly hypertensive women. J Hum Hypertens 2002;16:631–5.

[338] Rodenburg EM, Stricker BHC, Visser LE. Sex-related differences in hospital admissions attributed to adverse drug reactions in the Netherlands. Br J Clin Pharmacol 2010;71:95–104.

[339] Reuben A, Koch DG, Lee WM, Acute Liver Failure Study Group. Drug-induced acute liver failure: results of a U.S. multicenter, prospective study. Hepatology 2010;52:2065–76.

[340] Licata A, Minissale MG, Calvaruso A, Craxì A. A focus on epidemiology of drug-induced liver injury: analysis of a prospective study. Eur Rev Med Pharmacol Sci 2017;21(1 Suppl):112–21.

[341] Marzano AV, Vezzoli P, Crosti C. Drug-induced lupus: an update on its dermatologic aspects. Lupus 2009;18:935–40.

[342] Vedove CD, Del Giglio M, Schena D, Girolomoni G. Drug-induced lupus erythematosus. Arch Dermatol Res 2009;301:99–105.

[343] Bots SH, Groepenhoff F, Eikendal ALM, Tannenbaum C, Rochon PA, Regitz-Zagrosek V, et al. Adverse drug reactions to guideline-recommended heart failure drugs in women. J Am Coll Cardiol HF 2019;7:258–66.

[344] Cepeda MS, Carr DB. Women experience more pain and require more morphine than men to achieve a similar degree of analgesia. Anesth Analg 2003;97:1464–8.

[345] Loke YK, Singh S, Furberg CD. Long-term use of thiazolidinediones and fractures in type 2 diabetes: a meta-analysis. CMAJ 2009;180:32–9.

[346] Bazelier MT, de Vries F, Vestergaard P, Herings RMC, Gallagher AM, Leufkens HGM, et al. Risk of fracture with thiazolidinediones: an individual patient data meta-analysis. Front Endocrinol (Lausanne) 2013;4:11.

[347] Home PD, Pocock SJ, Beck-Nielsen H, Curtis PS, Gomis R, Hanefeld M, et al. Rosiglitazone evaluated for cardiovascular outcomes in oral agent combination therapy for type 2 diabetes (RECORD): a multicentre, randomised, open-label trial. Lancet 2009;373:2125–35.

[348] Martin RM, Biswas PN, Freemantle SN, Pearce GL, Mann RD. Age and sex distribution of suspected adverse drug reactions to newly marketed drugs in general practice in England: analysis of 48 cohort studies. Br J Clin Pharmacol 1998;46:505–11.

[349] Rademaker M. Do women have more adverse drug reactions? Am J Clin Dermatol 2001;2:349–51.

[350] Zopf Y, Rabe C, Neubert A, Gassmann KG, Rascher W, Hahn EG, et al. Women encounter ADRs more often than do men. Eur J Clin Pharmacol 2008;64:999–1004.

[351] Parekh A, Fadiran EO, Uhl K, Throckmorton D. Adverse effects in women: implications for drug development and regulatory policies. Expert Rev Clin Pharmacol 2011;4:453–66.

[352] U.S. General Accounting Office. Better oversight needed to help ensure continued progress including women in health research. Washington, DC: U.S. General Accounting Office; 2015.

[353] Manteuffel M, Williams S, Chen W, Verbrugge RR, Pittman DG, Steinkellner A. Influence of patient sex and gender on medication use, adherence, and prescribing alignment with guidelines. J Womens Health (Larchmt) 2014;23:112–9.

[354] Watson S, Caster O, Rochon PA, den Ruijter H. Reported adverse drug reactions in women and men: aggregated evidence from globally collected individual case reports during half a century. Eclinicalmedicine 2019;17:100188.

[355] Curtis AB, Narasimha D. Arrhythmias in women. Clin Cardiol 2012;35:166–71.

[356] Torp-Pedersen C, Møller M, Bloch-Thomsen PE, Køber L, Sandøe E, Egstrup K, et al. Dofetilide in patients with congestive heart failure and left ventricular dysfunction. N Engl J Med 1999;341:857–65.

[357] Jonsson MK, Vos MA, Duker G, Demolombe S, van Veen TA. Gender disparity in cardiac electrophysiology: implications for cardiac safety pharmacology. Pharmacol Ther 2010;127:9–18.

[358] Arya A. Gender-related differences in ventricular repolarization: beyond gonadal steroids. J Cardiovasc Electrophysiol 2005;16:525–7.

[359] Hreiche R, Morissette P, Turgeon J. Drug induced long QT syndrome in women: review of current evidence and remaining gaps. Gend Med 2008;5:124–35.

[360] Liu XK, Katchman A, Drici MD, Ebert SN, Ducic I, Morad M, et al. Gender difference in the cycle length-dependent QT and potassium currents in rabbits. J Pharmacol Exp Ther 1998;285:672–9.

[361] Gaborit N, Varro A, Le Bouter S, Szuts V, Escande D, Nattel S, et al. Gender-related differences in ion-channel and transporter subunit expression in non-diseased human hearts. J Mol Cell Cardiol 2010;49:639–46.

[362] Kurokawa J, Kodama M, Clancy CE, Furukawa T. Sex hormonal regulation of cardiac ion channels in drug-induced QT syndromes. Pharmacol Ther 2016;168:23–8.

[363] Pham TV, Rosen MR. Sex, hormones, and repolarization. Cardiovasc Res 2002;53:740–51.

[364] Rodriguez I, Kilborn MJ, Liu XK, Pezzullo JC, Woosley RL. Drug-induced QT prolongation in women during the menstrual cycle. JAMA 2001;285:1322–6.

[365] Nakagawa M, Ooie T, Takahashi N, Taniguchi Y, Anan F, Yonemochi H, et al. Influence of menstrual cycle on QT interval dynamics. Pacing Clin Electrophysiol 2006;29:607–13.

[366] Larsen JA, Tung RH, Sadananda R, Goldberger JJ, Horvath G, Parker MA, et al. Effects of hormone replacement therapy on QT interval. Am J Cardiol 1998;82:993–5.

[367] Geller SE, Koch A, Pellettieri B, Carnes M. Inclusion, analysis, and reporting of sex and race/ethnicity in clinical trials: have we made progress? J Womens Health 2011;20:215–20.

[368] Petkovic J, Trawin J, Dewidar O, Yoganathan M, Tugwell P, Welch V. Sex/gender reporting and analysis in Campbell and Cochrane systematic reviews: a cross-sectional methods study. Syst Rev 2018;7:113.

[369] Carcel C, Woodward M, Balicki G, Koroneos GL, de Sousa DA, Cordonnier C, et al. Trends in recruitment of women and reporting of sex differences in large-scale published randomized controlled trials in stroke. Int J Stroke 2019;14:931–8.

[370] Sun X, Briel M, Busse JW, You JJ, Akl EA, Mejza F, et al. Credibility of claims of subgroup effects in randomised controlled trials: systematic review. BMJ 2012;344, e1553.

[371] Wallach JD, Sullivan PG, Trepanowski JF, Sainani KL, Steyerberg EW, Ioannidis JPA. Evaluation of evidence of statistical support and corroboration of subgroup claims in randomized clinical trials. JAMA Intern Med 2017;177:554–60.

[372] Chou AF, Scholle SH, Weisman CS, Bierman AS, Correa-de-Araujo R, Mosca L. Gender disparities in the quality of cardiovascular disease care in private managed care plans. Womens Health Issues 2007;17:120–30.

[373] Ferdinand KC, Senatore FF, Clayton-Jeter H, Cryer DR, Lewin JC, Nasser SA, et al. Improving medication adherence in cardiometabolic disease: practical and regulatory implications. J Am Coll Cardiol 2017;69:437–51.

[374] Kirkman MS, Rowan-Martin MT, Levin R, Fonseca VA, Schmittdiel JA, Herman WH, et al. Determinants of adherence to diabetes medications: findings from a large pharmacy claims database. Diabetes Care 2015;38:604–9.

[375] Kotseva K, Wood D, De Bacquer D, De Backer G, Rydén L, Jennings C, et al. EUROASPIRE IV: a European Society of Cardiology survey on the lifestyle, risk factor and therapeutic management of coronary patients from 24 European countries. Eur J Prev Cardiol 2016;23:636–48.

[376] Koopman C, Vaartjes I, Heintjes EM, Spiering W, van Dis I, Herings RMC, et al. Persisting gender differences and attenuating age differences in cardiovascular drug use for prevention and treatment of coronary heart disease, 1998-2010. Eur Heart J 2013;34:3198–205.

[377] Wallentin F, Wettermark B, Kahan T. Drug treatment of hypertension in Sweden in relation to sex, age, and comorbidity. J Clin Hypertens Greenwich 2018;20:106–14.

[378] Dewan P, Rørth R, Raparelli V, Campbell RT, Shen L, Jhund PS, et al. Sex-related differences in heart failure with preserved ejection fraction. Circ Heart Fail 2019;12, e006539.

[379] Baumhakel M, Muller U, Bohm M. Influence of gender of physicians and patients on guideline-recommended treatment of chronic heart failure in a cross-sectional study. Eur J Heart Fail 2009;11:299–303.

[380] Daly C, Clemens F, Sendon JLL, Tavazzi L, Boersma E, Danchin N, et al. Gender differences in the management and clinical outcome of stable angina. Circulation 2006;113:490–8.

[381] Enriquez JR, Pratap P, Zbilut JP, Calvin JE, Volgman AS. Women tolerate drug therapy for coronary artery disease as well as men do, but are treated less frequently with aspirin, beta-blockers, or statins. Gend Med 2008;5:53–61.

[382] Blomkalns AL, Chen AY, Hochman JS, Peterson ED, Trynosky K, Diercks DB, et al. Gender disparities in the diagnosis and treatment of non-ST-segment elevation acute coronary syndromes: largescale observations from the CRUSADE (can rapid risk stratification of unstable angina patients suppress Adverse outcomes with early implementation of the American College of Cardiology/American Heart Association guidelines) National Quality Improvement Initiative. J Am Coll Cardiol 2005;45:832–7.

[383] Redfors B, Angerås O, Råmunddal T, Petursson P, Haraldsson I, Dworeck C, et al. Trends in gender differences in cardiac care and outcome after acute myocardial infarction in Western Sweden: a report from the Swedish web system for enhancement of evidence-based Care in Heart Disease Evaluated According to recommended therapies (SWEDEHEART). J Am Heart Assoc 2015;4, e001995.

[384] Wagstaff AJ, Overvad TF, Lip GYH, Lane DA. Is female sex a risk factor for stroke and thromboembolism in patients with atrial fibrillation? A systematic review and meta-analysis. QJM 2014;107:955–67.

[385] Shantsila E, Wolff A, Lip GYH, Lane DA. Gender differences in stroke prevention in atrial fibrillation in general practice: using the GRASP-AF audit tool. Int J Clin Pract 2015;69:840–5.

[386] Thompson LE, Maddox TM, Lei L, Grunwald GK, Bradley SM, Peterson PN, et al. Sex differences in the use of oral anticoagulants for atrial fibrillation: a report from the National Cardiovascular Data Registry (NCDR) PINNACLE registry. J Am Heart Assoc 2017;6, e005801.

[387] Greenwood BN, Carnahanb S, Huang L. Patient-physician gender concordance and increased mortality among female heart attack patients. Proc Natl Acad Sci U S A 2018;115:8569–74.

[388] Tisdale JE. Drug-induced QT interval prolongation and torsades de pointes: role of the pharmacist in risk assessment, prevention, and management. Can Pharm J 2016;149:139–52.

Sex-related Differences in Drug Pharmacodynamics, Pharmacokinetics and Drug Prescription

Decreased Drug Absorption
- ↓ Gastric acid secretion
- ↓ GI motility and transit time

Pharmacodynamic differences in efficacy
& drug safety
↑ Adverse drug reactions

Comorbidites
Women are older with more comorbidities,
so ↑ sensitivity to drug effects

Poor adherence to prescribed medications

Polypharmacy
↑ Consumption of drugs which increases
risk of drug interactions

Alignment with clinical guidelines
↓ Likelihood to receive guideline-directed drugs

Drug Distribution
- ↓ Plasma albumin and α1-acid glycoprotein
- Higher percentage of body fat
  - ↑ Vd for lipophillic drugs
  - ↓ Vd for hydrophillic drugs
- Smaller body size and plasma volume

Decreased drug metabolism and excretion
- ↓ Activity of hepatic CYP450 enzymes & P-gp
- ↓ Renal glomerular filtration rate

Variations in sex hormones affect drug pharmacokinetics

Underrepresented in clinical trials

Niti Aggarwal

WOMEN	MEN
Anti-arrhythmics Increased risk of QT prolongation	**Anti-arrhythmics** Decreased risk of QT prolongation
Antithrombotics Higher risk of bleeding	**Antithrombotics** Lower risk of bleeding
Aspirin Greater protection for stroke	**Aspirin** Greater protection for MI
Beta Blockers Greater decrease in BP & HR Lesser decrease in angina Similar decrease in HF hosp & mortality	**Beta Blockers** Lesser decrease in BP & HR Greater decrease in angina Similar decrease in HF hosp & mortality
Calcium Channel Blockers More edema Greater decrease in BP & HR Similar decrease in CV outcomes	**Calcium Channel Blockers** Less edema Lesser decrease in BP & HR Similar decrease in CV outcomes
Diuretics More electrolyte disturbances	**Diuretics** Fewer electrolyte disturbances
Statins Similar efficacy for prevention of ASCVD	**Mineralocorticoid Receptor Antagonists** More gynecomastia & hyperkalemia
Warfarin and Heparin Increased sensitivity, lower dose needed	**Warfarin and Heparin** Decreased sensitivity, higher dose needed

Sex Differences in Cardiovascular Drugs in Women Compared to Men. Sex differences in CV (cardiovascular) pharmacodynamics, and pharmacokinetics results in differential effects in men and women. Women are less likely to be prescribed guideline-directed medications, more sensitive to their effect, experience more adverse effects and exhibit less adherence compared to men. ASCVD, atherosclerotic cardiovascular disease; BP, blood pressure; GI, gastrointestinal; HF, heart failure; HR, heart rate; MI, myocardial infarction; Vd, volume of distribution. *Image courtesy of Niti R. Aggarwal.*

Current Challenges, Opportunities, and Next Steps for Clinicians

Chapter 28

Disparity in Care Across the CVD Spectrum

Sonia A. Henry and Jennifer H. Mieres

Clinical Case

A 54-year-old immigrant South Asian woman presents with a strong family history of diabetes mellitus type II (DM) and premature coronary artery disease (CAD). She is a vegetarian and her diet consists of lentils, naan bread, rice, curries and stews, and occasional "fast food" when out of the home. She is a nonsmoker, body mass index (BMI) 27, abdominal circumference of 35 in., blood pressure 125/85, and heart rate 90. She is perimenopausal and her cholesterol panel reveals: high-density lipoprotein (HDL) 32, triglyceride (TG) 160, and low-density lipoprotein (LDL) 108. Her hemoglobin A1c (HbA1c) is 7.5 on oral medication. She does not exercise routinely. Recently, with walking uphill and stairs, she experiences exertional back pain, neck "tightness," leg cramps, dyspnea, and anxiety. Baseline electrocardiogram (ECG): nonspecific ST-T wave abnormality. An echocardiogram (echo) revealed normal left ventricular (LV) function with ejection fraction (EF) 60% with mildly reduced global longitudinal strain of -15%. She subsequently underwent exercise stress myocardial perfusion where she completed 5 min on the Bruce protocol. She experienced neck and back pain with peak exercise. There were 0.5 mm horizontal ST segment depressions

Continued

Sex Differences in Cardiac Disease. https://doi.org/10.1016/B978-0-12-819369-3.00009-5

Clinical Case—cont'd

in leads II, III, aVF, and V4–V6 starting at 2:00 min following exercise at heart rate (HR) 107 bpm and persisted 3:00 min in recovery. Her perfusion imaging revealed severe reversible defects in the mid-distal inferolateral wall with transient ischemic dilatation (**Figure 1**). Coronary angiography revealed severe three-vessel disease. [Proximal left anterior descending (LAD) 80%, mid-LAD 80%, OM1 99%, OM2 diffuse 99%; and small right coronary artery (RCA) 90%.] Despite small caliber of vessels with suboptimal bypass targets, she underwent coronary artery bypass grafting (CABG) × 5. On optimal medication, 1 week later two obtuse marginal (OM) grafts closed and she sustained a non-ST-elevation myocardial infarction.

What were this woman's cardiovascular risk factors and what is her pretest probability for developing coronary atherosclerosis?

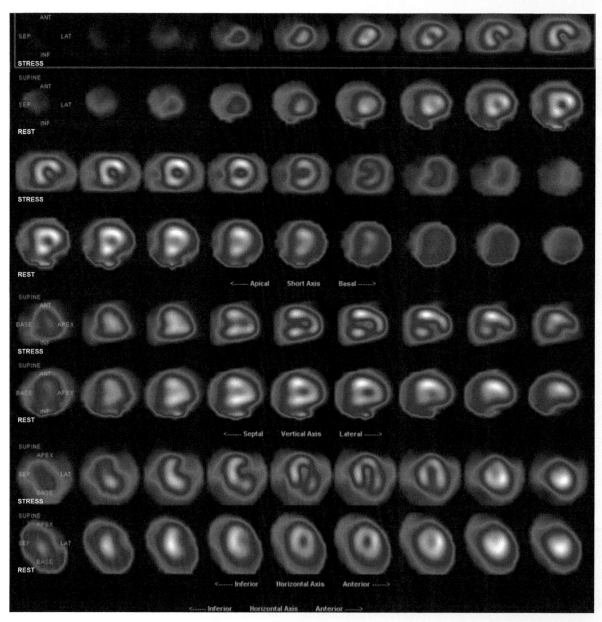

FIGURE 1   SPECT Myocardial Perfusion Imaging of 54-year-old South Asian Woman. Stress and rest images demonstrate severe reversible inferolateral perfusion defects and transient ischemic dilatation of the left ventricle.

Abstract

There has been an appropriate focus, since the turn of the 21st century, on sex- and gender-specific cardiovascular disease (CVD) as evidence suggests that there are substantial differences in the risk factor profile, social and environmental factors, clinical presentation, diagnosis, and treatment of CVD in women compared to men. As a result of increased awareness, detection, and treatment of CVD in women, there has been significant reduction (greater than 30%) in cardiovascular mortality in the United States [1–3]. Presently, more men than women die of CVD. Nevertheless, continued efforts are required as CVD remains the leading cause of cardiovascular morbidity and mortality of women in the Western world and in women younger than 55 there has been a rise in cardiovascular mortality [4]. The 2010 landmark Institute of Medicine (IOM) report, "Women's Health Research: Progress, Pitfalls, and Promise," highlighted that although major progress had been made in reducing cardiovascular mortality in women, there were disparities in disease burden among subgroups of women, particularly those women who are socially disadvantaged because of race, ethnicity, income level, and educational attainment [5]. The IOM recommended targeted research on these subpopulations with the highest risk and burden of disease. Causes of disparities are multifactorial and are related to differences in risk factor prevalence, access to care, use of evidence-based guidelines, and social and environmental factors. In this chapter, we review a few of the contributing factors to the disparities in CVD in women with a focus on the high-risk subgroups of women from black, Latino, and South Asian descent.

Introduction

During the last two decades, there has been an appropriate focus on women's cardiovascular health as increasing evidence suggests that there are substantial differences in the risk factor profile, presentation, diagnosis and treatment of ischemic heart disease (IHD), and the full spectrum of cardiovascular disease (CVD) in women compared to men. The landmark 2010 Institute of Medicine (IOM) publication, "Women's Health Research: Progress, Pitfalls, and Promise," highlights the fact that women's health involves two aspects: (1) sex differences from biological factors and (2) gender differences—those affected by broader social, environmental, and community factors [5]. The emphasis on sex- and gender-specific research and our enhanced understanding of sex-specific pathophysiology for coronary disease in women have resulted in the expanded spectrum of IHD in women to include obstructive and nonobstructive coronary artery disease (CAD) with supply-demand mismatch as well as dysfunction of the coronary

microvasculature and endothelium [4, 6]. Further, sex and gender CVD research in atrial fibrillation and heart failure is ongoing to broaden the understanding of female biology and physiology in these foci. Significant reduction (greater than 30%) in cardiovascular mortality in women has occurred for the first time this decade as a result of increased awareness and detection, sex- and gender-specific cardiovascular research, and the improved application of evidence-based treatments for IHD [1–3]. In 2013, for the first time, more US men than women died of CVD [2].

Despite progress, CVD remains the leading cause of morbidity and mortality of women in the Western world. Mortality from CVD is declining in women, but at a slower rate when compared to men [3]. Furthermore, women with IHD, atrial fibrillation, and advanced heart failure suffer worse outcomes as compared to men. Data from the VIRGO study and others demonstrate that female patients younger than 55 have seen a rise in CVD mortality [4]. The IOM report highlighted that although major progress had been made in reducing cardiovascular mortality in women, many subgroups of women defined by age, race, ethnicity, gender, socioeconomic status, and educational level still show striking disparities in cardiovascular health [5]. ARIC (Atherosclerosis Risk in Communities) 2005–2014 reports that the incidence of myocardial infarction (MI) or fatal CVD is higher in black women compared to white men at all ages [7]. Mexican American women have the highest prevalence of metabolic syndrome among females by race, according to NHANES, 1999–2010 [8]. South Asian women exhibit a greater prevalence of severe and extensive CAD compared with those of European and Chinese descent [9].

In this chapter, we review several contributing factors to the disparities in the full spectrum of CVD in women with a focus on the subgroups of women of black, Latino, and South Asian descent who are at high risk for morbidity and mortality from CVD. We review the disparities in prevalence, potency, and management of traditional risk factors. Flawed CVD risk scores are discussed. Disparities in sex-specific CVD diagnosis, treatment, management, and research are also reviewed. Lastly, we discuss a contemporary approach to focus on "well-women" visits and high-risk groups, and the requirements needed to implement this approach to prevent and reduce CVD in women.

Evidence of Sex and Gender Disparities

CVD in women has been misunderstood. Because of this and a myriad of factors, women have been mistreated, misdiagnosed, mismanaged, and misrepresented across the cardiovascular spectrum. This further gave rise to sex-based disparities in outcomes and quality of care (**Table 1**). The misfortune is worse for younger women (aged <55 years) and ethnic subgroups at higher risk.

TABLE 1 Sex-Based Disparities in Outcomes and Quality of Care

Factors	Setting	Sex-Specific Outcomes
Diagnostic testing	Stable IHD	1. Less than 1 in 10 women with angina and abnormal stress test had any change in pharmacotherapy or referral to diagnostic angiography.[38]
Delay in reperfusion	STEMI	1. Women had longer median first medical contact-to-device times compared with men (80 vs 75 min).[39]
		2. Women experienced a 30-minute prehospital delay from symptom onset to hospital presentation compared with men.[40]
		3. Young women were more likely to exceed door-to-needle time guidelines for PCI during STEMI compared with age-matched men (67% vs 32%). (OR, 2.62).[41]
Fewer revascularizations	ACS	1. Women were less likely to undergo revascularization after STEMI[41] and NSTEMI.[42]
		2. Women with documented 1-vessel disease were less likely to undergo PCI compared with men (OR, 0.78).[42]
		3. Women were less likely to be referred for surgical revascularization (OR, 0.81).[42]
		4. Young women were particularly less likely to have revascularization compared with men (28% vs 13%).[6]
		5. Despite known survival benefits of arterial grafts over vein grafts as conduits, women undergoing coronary artery bypass graft surgery (CABG) were less likely to receive arterial grafts compared with men.[43]
	Stable IHD	1. Despite higher angina class, women were less likely to undergo coronary angiography (31% vs 49% men).[44]
		2. Women with stable angina and confirmed CAD were less likely to undergo PCI compared with men (OR, 0.70).[44]
Less pharmacotherapy	Primary prevention	1. Women were less likely to have IHD risk factors measured, and young women (aged 35–54 y) were 37% less likely to be prescribed guideline-recommended medications.[45]
		2. Women were 65% less likely to have assessment of their smoking status, body habitus, blood pressure, and lipid profile.[46]
	ACS	1. In the CRUSADE study, women were less likely to received aspirin, ACE inhibitors, and statins on hospital discharge, even after adjustment for higher comorbidities in women.[47,48]
		2. Black women were significantly less likely to receive appropriate secondary prevention measures compared with age-matched White patients after MI, despite having ≥3 risk factors.[49]
		3. Medicare claims data demonstrated similar prescriptions patterns at hospital discharge but reported a 30% to 35% lower 12-month medication adherence after MI among Black and Hispanic women compared with White men.[50]
	Stable IHD	1. Women report a significantly lower use of statin and aspirin therapy compared with men.[44]
		2. Women were less likely to achieve guideline-directed secondary prevention targets for lipid (OR, 0.5), glucose (OR, 0.78), physical activity (OR, 0.74), or body mass index (OR, 0.82).[47] Similar findings were confirmed by the EUROASPIRE III and IV surveys in Europe.[51,52]
Cardiac rehabilitation		1. Despite higher event rates and worse outcomes after MI, women were less likely to access cardiac rehabilitation.
		2. Women were 32% less likely to be referred to cardiac rehabilitation (39.6% vs 49.4%).[53]
		3. Women were 36% less likely to enroll in cardiac rehabilitation (38.5% vs 45.0%).[54]
		4. Women adhered to median of 71.9% cardiac rehabilitation sessions compared with 75.6% men.[55]
Morbidity after MI	ACS	1. All women with acute MI had a 26% higher 1-year rate of rehospitalization even after adjusting for comorbidities, with even higher rehospitalizations in Black women.[56,57]
		2. Women with NSTEMIs had a higher in-hospital risk of recurrent MI (OR, 1.1) and heart failure (OR, 1.4).[48] Similar outcome disparities were noted in STEMI.[39]
		3. Black women were more likely to have angina at 1 y after MI treated with PCI (49% vs 31% in white men).[56]
		4. Young women with acute MI experienced more angina and depression, had worse quality of life, and were less likely to return to work within 12-months after their MI.[11,57]
	Stable IHD	1. Women with suspected angina, but angiographically normal vessels had more hospitalizations and repeat catheterizations for chest pain or ACS (OR 4.1).[58]
		2. These women exhibited a decreased quality of life.[58]
		3. 57% of women with confirmed CAD reported recurrent angina (compared with 47% men).[44]
Mortality	ACS	1. Women with STEMI had higher in-hospital mortality compared with men (10.2% vs 5.5%).[59]
		2. Women with ACS had higher in-hospital mortality after coronary angiography.[60]
		3. Men and women have similar 30-day post-MI mortality after adjusting for comorbidities and angiographic severity of disease.[2]
		4. Younger women (<55 y) had a 2-fold higher post-infarct in-hospital and 1-year mortality.[6]
	Stable IHD	1. Women with stable chest pain had higher in-hospital mortality at time of angiography (OR 1.25) and higher 1-year mortality compared with men.[60,61]
		2. Women with angiographic CAD had a 2-fold higher risk of 1-year death, even after adjusting for severity of disease and comorbidities.[44]

ACE, angiotensin-converting enzyme; ACS, acute coronary syndrome; CABG, coronary artery bypass graft; CAD, coronary artery disease; CRUSADE, Can Rapid risk stratification of Unstable angina patients Suppress Adverse outcomes with Early implementation of the American College of Cardiology/American Heart Association guidelines; EUROASPIRE, European Action on Secondary Prevention Through Intervention to Reduce Events; IHD, ischemic heart disease; MI, myocardial infarction; NSTEMI, non-ST-segment elevation MI; OR, odds ratio; PCI, percutaneous coronary intervention; STEMI, ST-segment elevation MI. *Reproduced with permission from [10].*

Misunderstood: *Moving the Needle to Improved Understanding of CVD in Women*

Historically, sex differences in pathophysiology, presentation, diagnosis, and management have been poorly recognized across the entire spectrum of CVD. For instance, previously, the diagnosis of IHD has focused on the identification of obstructive CAD as measured by degree of stenosis and candidacy for revascularization. Sex-specific research has resulted in an expanded understanding of the pathophysiology of IHD with a greater appreciation of the contributing roles of vascular endothelial dysfunction and microvascular disease as well as the prognostic importance and implications in the setting of nonobstructive CAD [1]. In women, nonobstructive CAD is more often found on angiography. Among symptomatic women with ACS and nonobstructive disease, especially among women ≤75, this has been shown to be associated with an elevated risk of adverse outcomes vs. men [adjusted hazard ratio (HR), 2.43, 95% confidence interval (CI), 1.08–5.49] [6, 11–13]. Furthermore, data from the Women's Ischemia Syndrome Evaluation (WISE) study showed 5-year annualized event rates for cardiovascular events of 16% and 7.9% in symptomatic women with nonobstructive IHD and normal coronary arteries, respectively [14]. In a recent Women's Health Alliance survey [15], although more than 90% of primary care physicians (PCPs) and cardiologists agreed that women with heart disease can present differently than men, only 49% of PCPs and 52% of cardiologists agreed that women's and men's hearts are physiologically different. Further work is needed to close the education gap about women's hearts among physicians.

Mistreated: *Suboptimal Awareness in Physician Leading to Under Treatment in Prevention*

The Women's Health Alliance survey reports [15], among PCPs, that heart disease was not their top priority in women. PCPs ranked heart disease after weight issues and breast health. Studies demonstrate that physicians are more likely to assign a lower CVD risk to female patients compared with risk-matched male patients and underestimate the probability of CVD in women [16], resulting in suboptimal cardiac preventative care [4, 16, 17]. Women were 65% less likely to have assessment of their smoking status, body habitus, blood pressure, and lipid profile [18]. In a large study of patients in a primary care setting, all women were less likely to be prescribed statins and antiplatelets, when indicated, compared with men. Furthermore, women in the youngest (35–54 years) age group were 37% less likely to be prescribed blood pressure (BP)-lowering medication, statins, and antiplatelets than men in the same age group [19]. Because CVD risk is suboptimally evaluated and symptoms are atypical, physicians were less likely to refer women and

ethnic minorities for cardiac testing and diagnostic cardiac catheterization [20]. Also, despite improved focus on sex-specific CVD care, women continue to demonstrate low use of ICDs for primary prevention of sudden cardiac death among patients hospitalized for heart failure [21].

Misdiagnosed: *Gender Disparities in the Presentation Leads to Worse Outcomes*

Disparities exist in the presentation of women and men with CHD. Women less often report chest pain and diaphoresis and more often complain of back or jaw pain, epigastric pain, palpitations, lightheadedness, or loss of appetite [22, 23]. Data from the National Registry of Myocardial Infarction (NRMI) demonstrated that 42% of men vs. 31% of women presented with chest pain in the setting of MI with in-hospital mortality of 14.6% vs. 10.3% for men. Younger women without chest pain in the NRMI were at highest risk [24]. This atypical presentation in women has been linked to delays in seeking care, delay in diagnosis, and delay in the delivery of life-saving treatment strategies, with worse outcomes [25, 26]. Women experienced a 30-min prehospital delay from symptom onset to hospital presentation compared with men [27]. Door-to-balloon times are increased for women, and young women (age 18–55) have the longest delays [28, 29]. Women were less likely to receive acute heparin, angiotensin-converting enzyme inhibitors, and glycoprotein IIb/IIIa inhibitors, and less commonly received aspirin, angiotensin-converting enzyme inhibitors, and statins at discharge [26].

With regards to atrial fibrillation, women were more symptomatic than men and had more atypical symptoms. The Outcomes Registry for Better Informed Treatment of Atrial Fibrillation found that women with atrial fibrillation had more functional impairment, more limitation in their daily activities, and worse quality of life than men [30]. Despite women having a higher risk for stroke, women are undertreated for stroke prevention [31].

The symptoms of heart failure are similar between men and women. However, women often present with more symptom burden, including more dyspnea, bronchitis-like symptoms, edema, fatigue, and worse quality of life. Provider perceptions of heart failure being a "man's syndrome" often lead to delay in diagnosis and treatment in women, instead treating a presumed upper respiratory syndrome without further investigating the cause of symptoms [32].

Mismanaged: *Inappropriate Management Resulting in Increased Mortality and Morbidity*

Women were less likely to get bystander cardiopulmonary resuscitation (CPR) than men. Bystander CPR improved survival to discharge, with greater survival among males compared to females [33]. Despite the proven efficacy of revascularization, women are less likely than men to be referred for revascularization for ST-segment elevation MI, non-ST-segment elevation MI, and stable angina [28, 34, 35]. Women also experience a higher incidence of cardiogenic shock (5.8% vs. 4.0%) and heart failure (5.8% vs. 3.4%) compared with men [36]. Despite known survival benefits of arterial grafts over vein grafts as conduits, women undergoing coronary artery bypass graft (CABG) surgery were less likely to receive arterial grafts compared with men [37]. Even with higher event rates and worse outcomes after MI, women were less likely to access cardiac rehabilitation [38].

Despite women being 52% of heart failure patients [39], they are poorly represented in heart failure in clinical trials [32], and as a result women receive fewer evidence-based therapies and likely incorrect doses of such therapies due to known differences in biology. Women can have different drug absorption, distribution, metabolism, excretion, and drug concentration of common heart failure drugs compared to men [32]. With advanced heart failure, in spite of proven efficacy of mechanical circulatory support for survival in both men and women, women were far less likely to receive these devices (22% of women compared to 78% of men) [40]. Similarly, sex differences persist in the rate of transplantation. Women receive fewer transplants (26%) compared to men (74%) [40]. Furthermore, women derive greater benefit from cardiac resynchronization therapy (CRT) than do men, including improved left ventricular reverse remodeling, decrease in left ventricular end-systolic volume, and improvement in hospitalizations and in mortality [41–43]. Despite these documented benefits, there is substantially less use of CRT in women [44].

Management in atrial fibrillation is also suboptimal for women. Women are less likely to receive anticoagulation and are referred less often for catheter ablation as rhythm control therapy [45, 46]. Furthermore, women are more often treated with cardiac glycosides compared to men (25.0% vs. 19.8%, $p=0.0056$) [47], which is associated with a higher risk of breast cancer and mortality [48]. The reasons for these differences are uncertain but certainly contribute to the higher adverse events and mortality in women [49].

Evidence supports a lack of recognition of CVD in women resulting in inappropriate management, leading to higher case-fatality rates and increased morbid complications among women [3, 50]. Early recognition of symptoms and accurate diagnosis of the full spectrum of CVD are critical to decreasing the mortality and morbidity in women.

Misfortune: *Worse Outcomes in Women vs. Men Across the CVD Spectrum*

Women have more complications after having a first acute myocardial infarction (AMI), such as increased bleeding risk after a first AMI treated with percutaneous coronary intervention (2.4% vs. 1.2% for men) [51]. Pooled data from the National Heart, Lung, and Blood Institute—sponsored cohort studies (1986–2007) indicate substantial disparities between men and women: 1 year after AMI, 19% of men and 26% of women ≥ 45 years of age will die; at 45–64 years of age, 15% of men and 22% of women have a recurrent AMI or fatal cardiovascular event within 5 years; within 5 years after a first AMI, 36% of men and 47% of women will die; and 8% of men and 18% of women develop heart failure within 5 years of a first AMI [52, 53]. Adverse outcomes continue over the long term for women in the WISE cohort with cardiovascular death or AMI at 10 years in 6.7%, 12.8%, and 25.9% of women with no, nonobstructive, and obstructive IHD ($p < 0.0001$), respectively [54]. With atrial fibrillation, studies continue to show that women have worse outcomes in terms of strokes and mortality compared to men. In healthy subjects 20–79 years old, the risk of stroke is 4.6-fold greater in women than men [55]. Furthermore, in a retrospective cohort study of 100,802 patients with atrial fibrillation, female gender was an independent risk factor for stroke (HR 1.18, 95% CI 1.12–1.24) even after adjusting for multiple confounding variables [56]. A meta-analysis of 30 studies published in 2016 with over 4.3 million participants found that women with atrial fibrillation had a 12% increased risk of death compared to men (95% CI 1.07–1.17) [49].

Misrepresented: *Disparities of CVD Research in Women*

It is affirmed that there are sex differences in presentation, pathophysiology, management, and outcomes in CVD. Despite advances in cardiovascular outcomes for women, most of the current treatment directing prevention and management of IHD are from studies of predominantly male subjects. Female representation in research was evaluated in clinical trials from 1997 to 2006 and was noted to be a meager 27% [57]. An analysis, in 2007, showed that the 156 IHD trials included only 25% women [58]. Furthermore, women comprised only about one out of three participants in 78 clinical trials of high-risk cardiovascular devices between 2000 and 2007, with the proportion of women in these trials unchanged over time [59]. Similarly, there is paucity of women being recruited in heart failure trials, with most landmark trials including on average 20–30% women [32]. Recognizing these inequalities, the IOM called for ongoing efforts to enhance inclusion of women in clinical trials [5]. In addition, the National Institute of Health (NIH) in 2015 announced that "sex" is a biological variable that should be included in basic sciences as well as clinical research design and analysis [60, 61]. The "Research for All Act" was passed from the 114th Congress 2015–2016 and supports equal inclusion of women in research and further subgroup analysis based on sex [62]. Continued responsibility and enforcement of sex- and gender-specific research are needed from government, industry, professional bodies, and publishing journals.

Sex and Gender Differences in CVD: Risk Factors and Disparities in Their Management

Although men and women share similar modifiable risk factors for CVD, there are differences in the incidence of these risk factors. Furthermore, several of these traditional risk factors are more potent in women with distinct impacts depending on whether a woman is premenopausal or postmenopausal. Data from the Framingham study highlighted the greater relative impact of diabetes mellitus (DM) on women compared with men [63], with diabetic women having a relative risk for fatal IHD 50% higher than men [64]. The INTERHEART study showed diabetic women have a 25% excess in stroke risk compared to diabetic men [14]. Hypertension is the leading cause of cardiovascular mortality worldwide, with an increased population-adjusted cardiovascular mortality for women compared with men, 29% vs. 15% [65]. Obesity increases coronary risk more for women than men, 64% vs. 46%. Central obesity with increase in visceral fat occurs with a higher presence of comorbid risk factors and components of the metabolic syndrome in women compared with aging men, including a higher prevalence of DM [66].

In addition to the disparities noted with traditional risk factors, there are female-specific and female-predominant risk factors. These risk factors such as early menopause, the effects of oral contraceptive pills, inflammatory diseases [systemic lupus erythematosus (SLE) and rheumatoid arthritis (RA)], complications of pregnancy (preeclampsia and gestational diabetes), and cardiac complications of breast cancer treatment (from cardiotoxic chemotherapy, targeted therapy, and left chest radiation) further widen the disparities for CVD in women. **Table 2** reviews the female-specific and female-predominant risk factors and their associated relative risk of causing CVD.

A total of 50% of reduction in CVD mortality is attributable to reducing risk factors [68]. Despite documented efficacy and guideline endorsement, women are less likely to receive guideline-based primary and secondary prevention treatment. Women were 65% less likely to have assessment of their smoking status, body habitus, blood pressure, and lipid profile [69] and young women (aged 35–54 years) were 37% less likely to be prescribed guideline-recommended therapies for primary prevention [19]. Despite the increased risk and potency of DM in women, they are less likely to be treated aggressively with regard to glucose control [70].

TABLE 2 Female-Specific and Female-Predominant Risks of CVD

Female-Specific and Female-Predominant Risks	CVD	Relative Risk
SLE	CVD	5–8
SLE (women aged 34–44 years)	Acute MI	50
Rheumatoid arthritis	CAD, HF, MI, CV death	2–3
OCP and smoking	CVD	7
OCP	CVA	1.5–2
Early menopause	CAD, CVA	2
Gestational HTN	HTN	4
Gestational DM	DM in 5–10 years	7–12
Preeclampsia/eclampsia	DM, HTN	4
Preeclampsia/eclampsia	IHD, VTE, CVA in 5–15 years	2
Pregnancy	Acute MI (i.e., SCAD)	2–3
Moderate-severe depression	Acute MI in 2 years	2
Breast cancer therapy: trastuzumab and anthracycline	HF	4.0; 1.5
Breast cancer therapy: anthracycline and trastuzumab	HF	7
Breast radiation therapy	CAD, CV death	1.3
Breast cancer survivor	CV death after 7 years	2

CAD, coronary artery disease; CV, cardiovascular; CVA, cardiovascular accident; CVD, cardiovascular disease; DM, diabetes mellitus; HF, heart failure; HTN, hypertension; IHD, ischemic heart disease; MI, myocardial infarction; OCP, oral contraceptive pill; SCAD, spontaneous coronary artery dissection; SLE, systemic lupus erythematosus; VTE, venous thromboembolism. *Adapted with permission from [67].*

From VIRGO, young women with DM were more likely to be told to lose weight and less likely to be treated with guideline-recommended statin therapy compared with similarly aged and more overweight men before an AMI [15]. The EUROASPIRE III and IV surveys demonstrated that despite the higher burden of comorbidities at the time of presentation of MI, women are less likely to receive recommended pharmacotherapies [18, 26] and less likely to achieve secondary prevention targets for hyperlipidemia (OR 0.5), hyperglycemia (OR 0.78), physical activity (OR 0.74), or body mass index (OR 0.82) [71, 72].

Race and Ethnic Disparities in CVD: Prevalence, Management, and Outcomes

Black and Hispanic Women

Disparities in CVD exist by race and ethnicity with significantly higher rates of disease prevalence (**Figure 2**), morbidity, and mortality in racial and ethnic minority groups compared to their white counterparts [74]. Causes of disparities are multifactorial and related to the entwinement of biologic and nonbiologic variables including genetics, risk factor prevalence, use of evidence-based guidelines, awareness, medication adherence, access to care, and social and environmental factors. In 2013–2016, the prevalence of IHD was higher among black women (57%) compared to non-Hispanic (NH) white women (43%) [7]. From ARIC surveillance 2005–2014 reports, the incidence of MI or fatal IHD was higher in black women compared to white men at all age ranges [7, 75]. In addition, when we look at pooled data (Framingham Heart Study, Atherosclerosis Risk in Communities Study, Cardiovascular Health Study, Multi-Ethnic Study of Atherosclerosis, Coronary Artery Risk Development in Young Adults, and Jackson Heart Study), black women aged 45–74 had the highest probability of death (20–22%) with recurrent stroke in 5 years after first stroke [7]. During 10 years of follow-up in REGARDS (Reasons for Geographic and Racial Differences in Stroke study), a higher percentage of black females (54%) developed hypertension than white males (38%) or females (27% for those 45–54 years of age and 40% for those ≥ 75 years old) [76, 77].

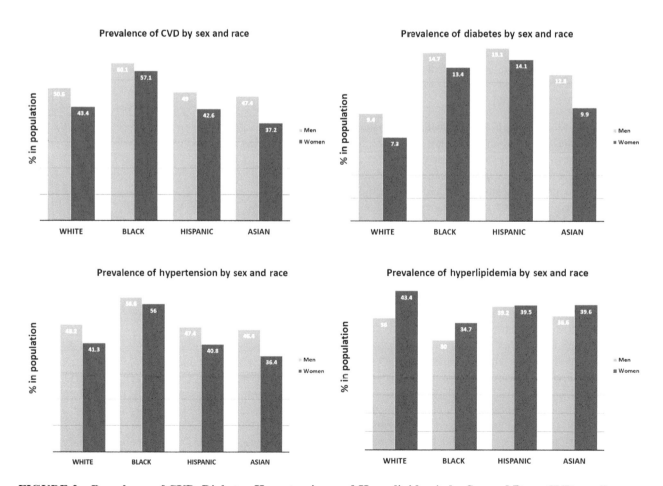

FIGURE 2   Prevalence of CVD, Diabetes, Hypertension, and Hyperlipidemia by Sex and Race. CVD, cardiovascular disease. *Data from [73].*

Higher systolic blood pressure (SBP) explains ≈50% of the excess stroke risk among blacks compared with whites [78].

Racial disparities also persist with regards to treatment of patients with atrial fibrillation. While blacks and Hispanics have a lower prevalence of atrial fibrillation compared to whites, they have a higher risk of stroke [79, 80]. In addition, they are at a greater risk of bleeding compared to white women [79]. Black patients with atrial fibrillation were three times less likely to be aware of having atrial fibrillation compared to white patients [81]. Among those aware, the odds of being treated with warfarin in blacks were only one-fourth as great as for whites. Furthermore, patients with atrial fibrillation and a CHADSVaSC of 5, only 50% of black patients compared to 75% of white patients were anticoagulated [81]. In addition, in this study, women were twice as likely to be aware of having atrial fibrillation, but less likely to be prescribed warfarin (OR 0.84). Similarly, data from the Ohio Medicaid population demonstrated that black patients with atrial fibrillation were 24% less likely to fill warfarin prescriptions for new incident atrial fibrillation [82]. Among patients undergoing catheter ablation for atrial fibrillation, both female sex (OR 0.83, 95% CI 0.79–0.87, $p < 0.001$), and black (OR 0.49, 95% CI 0.44–0.55, $p < 0.001$) and Hispanic race (OR 0.64, 95% CI 0.56–0.72, $p < 0.001$) were associated with lower likelihood of undergoing atrial fibrillation ablation [83].

Race and sex differences also exist in the prevalence and associations of heart failure risk factors. In a large low-income high-risk cohort of black and white adults, hypertension and diabetes were the only two risk factors significantly associated with heart failure risk in black men and women, whereas smoking and high BMI were also related to increased risk of heart failure in white participants [84]. Compared to whites, black patients are less likely to receive guideline-directed heart failure medications, less likely to be treated by a cardiologist, less likely to receive advanced heart failure therapies, and less likely to receive referral to transplantation [85, 86]. In addition, analysis of 13,034 patients admitted with heart failure and left ventricular ejection fraction (LVEF) ≤30% revealed that of those illegible for implantable cardioverter defibrillator (ICD) therapy for primary prevention of sudden cardiac death, odds of implantation of ICD were 0.73 for black men, 0.62 for white women, and 0.56 for black women, compared to white men [87].

Despite significant improvements in CVD quality of care and declines in cardiovascular mortality, racial/ethnic disparities in care and outcomes have persisted and the contributing factors to these disparities are complex and multifactorial [88]. The increased burden of cardiovascular risk factors among racial and ethnic minorities is definitely one significant contributor to the disparities: hypertension, diabetes, and obesity. From NHANES 2011–2016, hypertension is much more common among black women

(53%) compared to NH white women (38.8%) and Mexican American women (37.9%). Genetic variants may be associated with hypertension prevalence among black patients. NIH researchers recently identified 17 variants in the *ARMC5* gene that were associated with high blood pressure among black patients [89]. There are also known differences in medication response and pharmacogenetics in black vs. white patients with antihypertensive mediations. black patients have poor responses to beta-blocker and angiotensin-converting enzyme (ACE) inhibitors. Blacks responded to thiazide and calcium channel blocker antihypertensive agents, while whites responded similarly to all drug classes [90, 91].

Similarly, there is a higher prevalence of DM in ethnic minorities. According to the data from NHANES 2011–2016, DM is more prevalent in NH black (13.4%) and Mexican American women (14.1%) compared to NH white women (7.3%) [7]. Furthermore, African Americans are less likely to be aware of their DM and, when treated, are less likely to achieve adequate control according to common quality metrics defined by the Accountable Care Organization (hemoglobin $A_{1c} < 9\%$) [92]. Similar findings are noted for Hispanics [93]. Compared with their white counterparts, African Americans with DM are four times more likely to have visual impairment (caused by diabetic retinopathy) and 3.8 times more likely to have end-stage renal disease (resulting from diabetic nephropathy) [94, 95]. Chronic kidney disease and retinopathy from uncontrolled DM are also more prevalent in Hispanics compared to their white counterparts [96].

Of other modifiable risk factors, while smoking is higher among NH white women, other risk factors such as obesity (NH Asian 11.9%, NH white 35.2%, Hispanic 45.7%, black 56.9%) and physical inactivity (NH white 46.4%, Hispanic 58.2%, and black 60.9%) are much more common among black and Hispanic women [7]. The prevalence of metabolic syndrome was higher in Mexican American females than white and black females in the NHANES cycle 1999–2010 [8], while much of the data on CVD in the Hispanic/Latino population has focused on Mexican Americans; this is a very heterogeneous group with diverse backgrounds. A study of CVD risk factors in Hispanic/Latino groups in the United States showed variation of CVD risk factors by specific country of origin, with women of Puerto Rican background having the highest rates of obesity (31.7%), current smoking (51.4%), hypercholesterolemia (41%), and ≥3 risk factors compared to their Cuban, Dominican, Mexican, Central American, and South American counterparts [97].

With regards to risk factor management targets, in 2007–2010 (NHANES), 52.5% of adults with DM had a HbA1c <7.0%, 51.1% achieved a BP <130/80 mmHg, 56.2% had a low-density lipoprotein cholesterol (LDL-c) <100 mg/dL, and 18.8% had reached all three treatment targets. Compared with NH whites, Mexican Americans were less likely to meet HbA1c and LDL-c goals, and NH blacks were less likely to meet BP and LDL-c goals [98].

For various reasons minority groups, with a higher disease burden, are mishandled and have worse outcomes. The rate of premature CHD death is higher among blacks than their white counterparts [99]. Blacks with confirmed acute coronary syndrome (ACS) were younger, poorer, and less educated, and had a longer prehospital delay than whites [100]. From ACTION Registry-Get With The Guidelines (GWTG), Hispanics with ACS had longer time of symptom onset to hospital arrival, lower percentages of ambulance use, prehospital ECG, arrival to ECG in less than 10 min, and door-to-balloon time of less than 90 min [101].

Post MI, black women were significantly less likely to receive appropriate secondary prevention measures compared with age-matched white patients despite having ≥ 3 risk factors [102]. Black women were more likely to have angina at 1 year after MI treated with PCI, 49% vs. 31% in white men [103]. It has been recognized that the mortality rates and readmission rates post-CABG are worse in black patients than other races [104]. Black women with acute MI had the highest rate of 1-year rehospitalization even after adjusting for comorbidities. In addition, 12 months post MI, decreased medication adherence was present in the Medicare claims-based data with a 30–35% lower likelihood of medication adherence in black and Hispanic women. This decreased adherence was postulated to be secondary to lack of social support, community resources, cognitive deficiencies, and physician follow-up [105].

South Asian Women

Studies in migrant South Asian populations have shown they experience MI and IHD four times the rate of the general world's population with higher rates of premature IHD, up to a decade earlier, even with similar levels of risk factors [9]. Except for diabetes, based on current definitions, South Asians do not have a higher incidence of smoking, obesity, or hypertension; many are vegetarians. South Asian women have high overall heart disease rates, approaching those of South Asian men and their CAD mortality rate is much higher than that of white women [106]. The proportionate mortality burden from ischemic disease, as reflected by the proportional mortality rates, was highest in Asian Indian men (1.43) and women (1.12), compared with NH white men (1.08) and women (0.92) [107]. Angiographic studies in these women have shown high prevalence of severe and extensive disease, including three-vessel disease, despite the fact that majority of these women (56%) were premenopausal [108]. In a Canadian angiographic study, Asian Indian women were twice as likely to have left main or three-vessel CAD compared to white women [9].

In the Study of Health Assessment and Risk in Ethnic groups (SHARE), South Asians were found to have a higher prevalence of subclinical atherosclerosis. In fact, South Asian ethnicity was an independent predictor of CVD [106]. Unfortunately, conventional CVD screening and management do not address the high rates of premature MI in South Asians. Current defined risk thresholds likely underestimate IHD risk in South Asians and therefore undertreat the disease. The National Kidney Foundation states that Asian Americans are also at a higher risk for kidney disease and kidney failure compared with NH whites, and the high prevalence of DM and hypertension appears to be a contributing factor [109]. It has been suggested that the definitions of CVD risk cutoffs be redefined for South Asians. From the REGARDS cohort, the prevalence of metabolic syndrome varies by the definition used, with definitions such as that from the International Diabetes Federation suggesting lower thresholds for defining Asians (in particular, South Asians), Middle Easterners, sub-Saharan Africans, and Hispanics, which would result in higher prevalence estimates [110]. In addition, National Heart, Lung and Blood BMI cutoffs misclassify obesity of muscle mass on lower tails of the distribution and appear to underestimate risk in Asians [111]. For this reason, the American Diabetic Association and World Health Organization recommend lower BMI cutoffs in Asians to classify overweight and obesity (> 23 and $> 27.5 \,\mathrm{kg/m^2}$, respectively) [112, 113].

Genetics also play a role with CVD risk factors prevalence and atherosclerosis in South Asians. The MASALA study (Mediators of Atherosclerosis in South Asians Living in America), reports that South Asians have increased visceral fat in the abdominal area (truncal obesity), the liver, and around the heart [113]. This accumulation of visceral fat causes inflammation and activates biological pathways that contribute to atherosclerosis. Atherogenic dyslipidemia in South Asians is characterized by high triglycerides, low HDL, elevated apolipoprotein B (related to insulin resistance), and elevated lipoprotein (a) [114–116]. Although levels of low-density lipoprotein cholesterol (LDL-c) may not appear elevated, this population has a high incidence of qualitatively abnormal LDL-c particles characterized by smaller size and lower density [117]. The strongest contributor to atherosclerosis is high levels of lipoprotein (a) found in 35–40% of Asian Indians worldwide [116]. Current guidelines recommend statins in all ethnicities. In Asians, genetic differences in several genes influence rosuvastatin metabolism, including SLCO1B1, ABCG2, and CYP2C9 [118]. Further data are needed on efficacy and response to statins.

With DM type II, there is higher prevalence in South Asians (23%) compared to other ethnicities even after adjustment for age and adiposity (18% blacks, 17% Hispanics,

13% Chinese Americans, and 6% whites) [119]. Recent data from genome-wide association studies (GWAS) in South Asians found genetic variants associated with DM (GRB14, ST6GAL1, VPS26A, HMG20A, AP3S2, and HNF4A) which suggest potential genetic predisposition to DM II with a higher incidence of insulin resistance [120, 121].

Nonbiological Variables and Their Association With Increased CVD

Nonbiological factors play a significant role in CVD disease progression and worse outcomes for women; this is especially true for minority and ethnic groups. Some of these factors include bias, behavioral risk factors, social determinants of health, and psychosocial stress. **Figure 3** summarizes biological and nonbiological factors impacting CVD prevalence and outcomes.

Women are more objectified than men in society, which contributes to the implicit bias and belief inconsistent with evidence among both women and physicians that body weight is a CVD risk factor and weight management is an effective CVD risk reduction strategy [15]. Furthermore, embarrassment from elevated weight leads women to

decrease the conversations about risk factors and CVD risk to their family and social support. Social support and strong community or cultural infrastructure tends to be protective against extrinsic discrimination and lower stressful responses to bias, especially in minority communities [122, 123].

Implicit racial bias by the treating physician can introduce disparities in clinical decision making. In a recent study, Khadija et al. surveyed heart failure physicians and demonstrated that clinicians believed a black patient was sicker, less compliant with medications, had less social support, and recommended less heart transplant compared to a white patient despite similar medical and social history in both races [85].

Another important nonbiological factor is lower awareness of CVD as a threat to women. In surveys conducted by the American Heart Association (AHA), low levels of awareness and health literacy persisted with a minority of black and Hispanic women citing heart disease as the leading cause of death, levels similar to those of white women 15 years ago and almost half those of white women in 2012 [124]. Poor health literacy is linked to decreased awareness of the importance of physical activity and a healthy diet. The Southern, Latin, and Caribbean diets are disproportionally

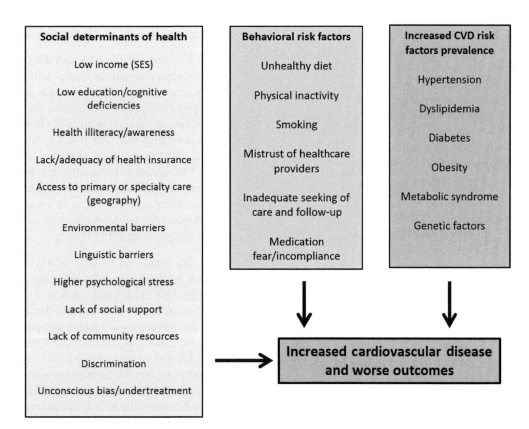

FIGURE 3 **Biological and Nonbiological Factors That Lead to CVD.** CVD, cardiovascular disease; SES, socioeconomic status.

high in sodium, carbohydrates, and saturated fats. Similarly, the South Asian diet is very rich in carbohydrates, refined grains, and saturated fats. With acculturation, any of these diets mixed with the "Western diet" (high intakes of red meat, fried foods, high-fat dairy products, high-fructose corn syrup, and high-sugar drinks) is a recipe for obesity, elevated CVD risks, and poor cardiovascular health.

Low socioeconomic status (SES) and social distress are fundamental causes of worse health outcomes. Low SES correlates with low income, low education, poorer living conditions, and inability to buy healthy food. There is also an inability to access primary or specialty care due to lack of insurance and cost-related issues [125]. Furthermore, geographical location of health providers is typically not located in underinsured and racially segregated communities, which leads minorities to turn to the emergency department (ED) for both acute and chronic health issues [126]. Use of ED care is significantly higher in blacks compared to whites and primary care physician (PCP) follow-up is only at a fraction of that of their white counterparts [127]. This process increases the burden of CVD risk factors and promotes unchecked development of overt CVD in minority populations.

Psychological and social effects also play crucial roles in predisposing minority populations to stress and to consequent pathology including CVD [125]. Racial discrimination is a source of stress and dissuades minorities from visiting PCPs for fear of inadequate treatment and prejudice. This further instigates a dangerous pattern of substandard medical attention and lack of continuity in care. In addition, differences in language, culture, and customs create potentials misunderstandings and obstacles in communication between a patient and a provider with different race/culture. Studies have shown that minorities prefer and trust race/ethnic concordant providers and rate better health because of higher-quality interpersonal interactions [128].

Another significant stressor is the increase in crime and violence associated with low-SES communities. Posttraumatic stress disorder (PTSD) secondary to violence or traumatic exposure occurs at higher rates in blacks as compared to whites and Hispanics in the United States. Alongside the devastating psychological and social ramifications of PTSD, Turner et al. [129] concluded that PTSD is an independent risk factor for cardiovascular consequences, namely myocardial ischemia.

Flawed Risk Scores Can Result in Risk Miscalculation for Women

Risk equations are a cornerstone of prevention. Several risk tools are currently available that predict cardiovascular outcome. The Framingham risk score (FRS) is the most commonly used global model for risk prediction [130]. The FRS utilizes age, gender, cholesterol profile, smoking, and blood pressure to predict one's chance of having a cardiac event in 10 years. The FRS was adopted in 2001 into the National Cholesterol Education Program/Adult Treatment Panel III (ATP III) cholesterol guidelines. One problem of this short-term risk estimation is that women traditionally have a lower short-term cardiovascular risk but a higher lifetime risk. Thus, the FRS characteristically underestimates cardiovascular risk for women. FRS underestimation of IHD in women was also recognized in the Multi-Ethnic Study of Atherosclerosis (MESA). In MESA, women in the highest quartile of coronary calcium scores were characterized as being at low risk by the FRS [131]. In addition, ATP III predicts future cardiovascular events but not angina or revascularization, which are two important end points for women [132].

The Reynolds risk score (RRS) was initially developed specifically for women and incorporates family history, inflammatory biomarkers (high sensitivity C-reactive protein), and hemoglobin A1C, in addition to the FRS risk factors [133]. The RRS reclassified 40–50% of women at intermediate risk into higher or lower categories in a cohort from the Women's Health Study [133]. The RSS also improves discrimination in multiethnic populations in comparison to the FRS. The RRS is a more accurate short-term predictor but the overall gains are modest [130]. As with the FRS, the RRS fails to include many of the risk factors unique to women.

A major limitation of the FRS and RRS is the narrow focus on short-term risk and the extrapolation of data from very high-risk or healthy patients—which do not apply to the general population [134]. To address these limitations, AHA guidelines on prevention of CVD in women classified women into "high risk," "at risk," or "ideal cardiovascular health," and was inclusive of novel CVD risk factors in women. This simplified classification emphasizes the concept of increased lifetime risk for women and is easily accessible to patients and providers. In a validation cohort, the model proposed by the AHA identified cardiac risk with an accuracy similar to the FRS [135].

In 2013, the American College of Cardiology (ACC) and the AHA published the "Guideline on Assessment of Cardiovascular Risk" [136] and a new tool for the estimation of cardiovascular events, namely the Pooled Cohort Equation. This tool is gender specific and provides specific information for Caucasians and African Americans. It provides both a 10-year atherosclerotic cardiovascular risk and a lifetime risk. The benefit for women is that their higher lifetime cardiovascular risk is included in this calculation. A concern is the overestimation of cardiovascular risk with advanced age [137, 138]. Another concern is that it was validated with a predominately white population and did not consider increased risk because of race and ethnicity.

The QRISK3 algorithm used in the United Kingdom astutely incorporate ethnicities at high CVD risk, i.e., South Asian (Indian, Pakistani, and Bangladeshi) and black (Caribbean and African). In addition, some female-predominant risks and other relevant diagnosis are included such as: premature CAD in family history, RA, SLE, mental illness, chronic kidney disease, and atrial fibrillation [139]. Pike et al. have validated the QRISK3 calculator and demonstrate that QRISK3 performed better that Framingham [140]. QRISK3 is recommended by the UK's National Institute of Health and is updated annually to reflect changes in the population.

History of National Efforts to Reduce CVD in Women

In an effort to increase awareness about heart disease in women, two major campaigns were established in 2003 and 2004, respectively. Sponsored by the National Heart, Lung, and Blood Institute, in 2003 the Heart Truth campaign was unveiled and in 2004, the AHA Go Red for Women campaign launched, with the release of the inaugural evidence-based guidelines focused on prevention of heart disease in women [141]. The AHA Go Red for Women campaign fervently continues its mission and has expanded to include the development of risk assessment tools and disease management guidelines. The adoption of these guidelines and insights from sex-specific research have contributed to the decrease in morbidity and mortality of CVD in women seen in the last 5 years [142]. Since the launch of the heart disease campaigns, awareness of heart disease as a leading cause of death among US women has increased from 30% in 1997 to 54% in 2012 [143], but overall remains suboptimal with lower awareness rates among women under 50 and those of ethnic minorities [15].

Additional heart health campaigns/programs, including AHA's Go Red for Women and "Make the Call. Don't Miss the Beat" initiative, have had success in increasing the percentage of women who reported a willingness to call 9-1-1 if they experienced signs of a heart attack, with improvement from 54% in 2009 [144] to 65% in 2012 [124]. The 2014 Women's Heart Alliance survey showed that an increase in awareness and a focus on personalizing the knowledge of heart disease is still needed as over two-thirds of women have never discussed heart health with their physicians despite having risk factors for heart disease [15].

In spite of the growing body of recommendations, guidelines, and legislation aimed at addressing gaps in scientific knowledge and reducing sex- and gender-based health disparities in CVD [5, 145–147], there is a gap in implementation. Obstacles to improving prevention, care of heart disease in women, and the improvement of health outcomes by the elimination of sex and gender differences

in CVD included ineffective policies, inconsistent practices and enforcement, and failure to adhere to and integrate sex- and gender-specific recommendations for healthcare delivery [148, 149].

In 2015, WomenHeart: The National Coalition for Women with Heart Disease convened the USA's first National Policy and Science Summit on Women's Cardiovascular Health. The summit's agenda included a review of the "10Q report" [150], and a review of the effects of existing policies, scientific advances, and current barriers, with the goal of providing recommendations for improving women's cardiovascular health [149]. The voices of women living with heart disease informed and added perspective to the recommendations. Key recommendations included: holding federal agencies and private research funders accountable for enforcing current policies regarding participation and enrollment of women in CVD research; incorporating sex and gender considerations in studies; importantly addressing sex- and gender-specific social determinants of health; and providing education, training, and support to medical professionals and women to foster the goal of improved CV health outcomes for women [149]. Also in 2015, the Food and Drug Administration (FDA) Strategy for Science and Innovation to Improve the Health of Women was published, which also highlighted the relevant priorities and strategies to address identified gaps in women's CV health [148]. **Figure 4** is a timeline of landmark initiatives and advances in the public awareness of heart disease in women [142]. There continues to be an urgent need for a collaborative approach among policy makers, scientists, researchers, clinicians, and patients, combining innovations in research design with policy solutions, to reduce women's burden of heart disease [149].

National and Other Significant Community Efforts to Reduce CVD in Ethnic Groups

A growing body of literature supports the racial/ethnic disparities in CV health, with a higher burden of chronic diseases, delivery of lower-quality care, and less access to medical care seen in members of minority groups compared to whites [125, 127]. Several components that contribute to the disparities in CV care include health illiteracy, language communication barriers, lack of medical team cultural competency, lack of awareness of CV risk factors, structural and psychosocial obstacles, and stress secondary to discrimination. Emerging national prevention campaigns have outlined specific goals to address health disparities in CVD through access to medical care, health promotion, and education via community-based approaches [125, 127].

The Patient Protection and Affordable Care Act (ACA) was passed in 2010, to guarantee coverage for all Americans

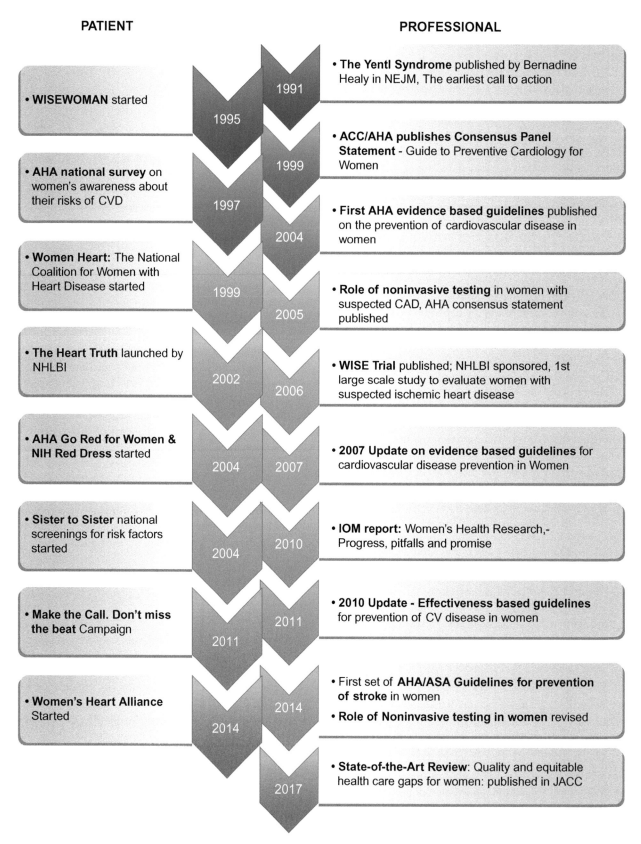

FIGURE 4 **Timeline of Landmark Initiatives and Advances in the Public Awareness of Heart Disease in Women.**
ACC, American College of Cardiology; AHA, American Heart Association; ASA, American Stroke Association; CAD, coronary artery disease; CVD, cardiovascular disease; IOM, Institute of Medicine; JACC, Journals of the American College of Cardiology; NEJM, New England Journal of Medicine; NHLBI, National Heart, Lung, and Blood Institute; WISE, Women's Ischemia Syndrome Evaluation. *Reproduced with permission from [142].*

and protect against insurer bias. With the ACA, it is estimated that approximately 16.9 million previously uninsured people have been able to access care. The Department of Health and Human Services estimates that the uninsured rate for African Americans has dropped by 9.2% (2.3 million) and by 12.3% (4.2 million) in Hispanics [151]. It is important to note that many state legislatures in conservative US states opted to decline ACA coverage expansion for economic/political reasons, limiting insurance coverage for lower-SES patients in these states [152].

Advances in improved awareness and efforts to reduce CVD in ethnic groups had been gained from nationally supported studies/surveys such as MESA, ARIC, MASALA, NHANES, REGARDS, and the Jackson Heart Study. Also critical to education and dissemination of data have been AHA scientific statements: "Cardiovascular Health in African Americans" [125], "Atherosclerotic Cardiovascular Disease in South Asians in the United States: Epidemiology, Risk Factors, and Treatments" [118], and "Status of Cardiovascular Disease and Stroke in Hispanics/Latinos in the United States" [153].

Community-based approaches have been a mainstay to aid in reducing CVD in ethnic groups. The largest multiracial ethnic community-based health initiative in the nation was Racial and Ethnic Approaches to Community Health Across the United States (REACH US). REACH US educated local minority groups about health and wellness, encouraged healthy lifestyle practices, promoted health screenings, and encouraged policy changes that would decrease health disparities [154]. Similarly, the Hispanic Community Health Study/Study of Latinos (HCHS/SOL), sponsored by the National Heart, Lung, and Blood Institute, have been correlated with lower diabetes prevalence [155].

Three additional, community-based, pivotal examples of successful cardiovascular health improvement initiatives are the Georgia Stroke and Heart Attack Prevention Program (SHAPP), the WISEWOMAN project, and the SAHELI study (South Asian Heart Lifestyle Intervention). SHAPP provided public clinics, health education, and hypertension medication at no or low cost. The overall cardiac complications were reduced by 46% [156]. As part of the national WISEWOMAN project, CVD risk screening was performed during routine breast and cervical cancer screenings in underinsured and uninsured women over 50. This intervention led to promotion of healthy diet and exercise, which led to improvements in hypertension of 7–9% [157]. The SAHELI study was conducted in a population of South Asian immigrants at risk for CVD. The intervention involved interactive classes focused on increased physical activity, healthful diet, and weight and stress management for a period of 6 months, which resulted in a significant weight loss and a greater decrease in hemoglobin A_{1c} [158].

Reducing disparities in CV health requires a multilevel, collaborative approach, to include a focus on identifying and building a partnership with individuals and communities at greatest risk to improve access to quality health care, increase cultural competence, and improve CV education [125, 127].

Current Challenges and Next Steps

Beyond the Traditional Risk Assessment: Contemporary Approach of Identifying Women At Risk of CVD

Despite forward progress in outcomes and mortality, CVD continues to be the leading cause of death in women, with recent data suggesting some reversal in the positive impact seen over the past 20 years. Significant disparities in diagnosis, treatment, and outcomes still exist compared to men. Women experience delays in reperfusion [27, 29], less revascularization [34], less mechanical circulatory support and less cardiac transplant [40], and less CRT [44]. They are less likely to receive anticoagulation and ablation for atrial fibrillation [45, 46], experience underutilization of pharmacotherapy for primary prevention [19], ACS [18, 26], and stable IHD [35], and experience worse morbidity [103, 159, 160] and mortality from CVD [35, 53, 161, 162]. At highest CV risk are black, Hispanic/Latino, and South Asian women [9, 163], those disadvantaged by income level and educational status, and young women. As mentioned, the death rate is rising in women younger than 55 years old [4]. In order to move the needle forward on outcomes and mortality, equity in care for women is needed and a novel approach for early identification and treatment of women at risk for CVD is crucial. **Figure 5** reviews components for a contemporary solution for equitable care and to improve cardiovascular outcomes in women.

Women's Healthcare Team

All clinicians who care for women have an urgent call to action to reduce CVD in women. A majority of women consider their obstetrician/gynecologist (OB/GYN) to be their sole physician, particularly during their childbearing years. The "healthcare team for women" approach, coined by Shaw et al. [164], is paramount in the care of women to reduce CVD, their number one health threat. The 2018, AHA/ American College of Obstetricians and Gynecologists (ACOG) Presidential Advisory statement [165] recognizes the importance of the "healthcare team." The annual OB/ GYN "well-woman visit" is a powerful opportunity to be comprehensive and counsel patients about maintaining a healthy lifestyle and minimizing health risks [166]. It should include family history and regularly screen for traditional, female-predominant, and female-specific cardiac risk factors. Paired with sex-specific CVD education, it is

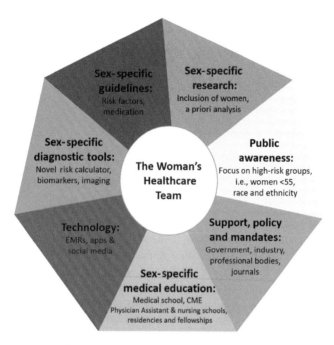

FIGURE 5 Contemporary Solution for Equitable Care and to Improve Cardiovascular Outcomes in Women. CME, continuing medical education; EMRs, electronic medical records. *Adapted with permission from [67].*

imperative to understand the woman's "healthcare team" approach in early training, furthered with continuing medical education (CME) and/or scientific sessions throughout a provider's career.

Sex-Specific Research, Diagnostic Tools, and Guidelines Required

The WISE (Women's Ischemia Syndrome Evaluation) Study and other sex-specific research have established that the pathophysiology of IHD in women extends beyond epicardial stenosis to include diffuse atherosclerosis and dysfunction of coronary microvasculature and endothelium. Similarly, heart failure with preserved ejection fraction is highly prevalent in older women [167]. Moreover, in patients with atrial fibrillation, female gender was an independent risk factor for stroke [56]. Despite this knowledge, there is a lack of proven guideline-directed diagnostic approaches and treatment recommendations for women with signs and symptoms of ischemia with nonobstructive CAD [168, 169]. There is also limited sex-specific guidance in atrial fibrillation and heart failure management. Landmark sex-specific clinical research is urgently needed. Advanced research is also needed on predictive and diagnostic tools such as novel circulating, genetic, and imaging biomarkers in women. Furthermore, a concise screening tool or risk calculator for women incorporating race/ethnicity, traditional, and contemporary (female-specific and female-predominant) risk factors is compulsory. These factors, collectively,

certainly impact risk stratification and should impact the aggressiveness of prevention treatment in women. Despite mandates on inclusion of women in research, women only represent a quarter of the participants in clinical trials [58]. Government, policy, industry, and professional bodies must mandate inclusion of women and focused recruitment of women in trials. It is only from research on women that accurate sex-specific guidelines can be generated.

Innovative Use of Technology to Impact CVD in Women

Electronic medical records (EMRs) could be used to improve cardiovascular health in women. Software algorithm should be designed and used to screen, calculate risk, and prompt patient education and referrals from EMRs. Social media and apps also have the wide potential to reach most populations, and are valuable resources to educate, empower, and motivate women, especially younger women.

Sustained Efforts in Education and Awareness

Public awareness campaigns have been successful to increase awareness in women that CVD is their number one killer, from 30% to 56% [170]. Continued action and novel strategies to further advance awareness are vital. A new focus to those at highest CV risk (women less than 55 years old and black/Hispanic/South Asian women) is desperately needed. Other current, potential barriers to health care in women include social stigma, i.e., "fat shaming," cultural/religious beliefs, caretaking responsibility, inadequate finances, and insufficient time [124]. Continued and new strategies to enhance public awareness should focus on these groups and utilize churches/cultural organizations, gyms, schools, workplaces, and community leaders to promote heart literacy and health behaviors that are culturally acceptable, easy, and affordable.

Conclusion

Substantial progress has been made to increase awareness and reduce CVD in women; however, we are still a long way from equitable care. CVD in women is multifaceted, in that it goes beyond traditional risk factors to include female-specific biological risk factors, unique pathophysiology and clinical presentation, and differences in diagnosis and treatments, and also incorporates race/ethnicity, social, and environmental factors. In order to improve risk prediction and risk stratification and decrease morbidity and mortality from CVD in women, we must move forward fervently to embrace the "healthcare team for women" approach. We are obligated to continue with new strategies in awareness, utilize technology, increase sex-specific research, and mandate clinicians, researchers, professional bodies, journals, and government to sustain these efforts to equitable care (**Figure 6**).

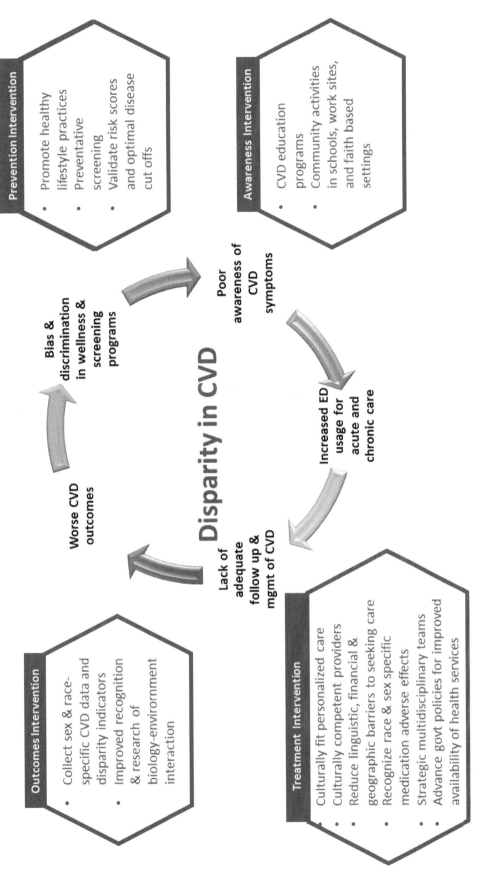

FIGURE 6 Summary of Factors That Promote and Are a Result of Disparity in Cardiovascular Disease. Also, a summary of suggested interventions to mitigate these factors and promote equitable care. CME, continuing medical education; CVD, cardiovascular disease; ED, emergency department; MD, medical doctor; NP, nurse practitioner; PA, physician assistant. *Image courtesy of Niti R. Aggarwal.*

Key Points

1. Although major progress has been made in reducing CV mortality in women, there are disparities in disease burden among subgroups of women, particularly those women who are socially disadvantaged because of race/ethnicity (black, Latino, and South Asian descent), income level, and educational attainment. Aggressive risk stratification and enhanced health literacy are needed in this group.

2. South Asian women have high overall heart disease rates, approaching those of South Asian men (four times the rate of the general world's population) and their CAD mortality rate is much higher than that of white women. Except for diabetes, South Asians do not have a higher incidence of smoking, obesity, or hypertension and many are vegetarians. Their cholesterol profile has low HDL and elevated TG. Current defined risk thresholds likely underestimate CHD risk and therefore undertreat disease.

3. The prevalence of CHD was more than 50% higher among black women compared to white women. Black women aged 45–74 had the highest probability of death (20–22%) with recurrent stroke in 5 years after first stroke. Hypertension is much more common among black women and blacks were less likely to meet BP and LDL-c goals.

4. Obesity and physical inactivity are much more common among black and Hispanic women, compared to white women.

5. The prevalence of CHD was 15% higher among Mexican American women compared to white women. The prevalence of metabolic syndrome was higher in Mexican American females and they were less likely to meet HbA1c and LDL-c goals.

6. Although men and women share similar modifiable typical risk factors for CHD (diabetes, smoking, elevated cholesterol, hypertension, sedentary lifestyle, and obesity), there are differences in the prevalence of disease. Furthermore, several of the traditional risk factors are more potent in women such as diabetes, smoking, and obesity.

7. Women are still misunderstood, mistreated, misdiagnosed, mismanaged, and misrepresented, and have greater misfortune across the CVD spectrum. We are obligated to continue with new strategies for increased awareness of CVD (IHD, atrial fibrillation, heart failure, etc.) in women, further sex and gender research, utilization of evidence-based guidelines, and establishment of accurate CV risk scores for women.

8. In order to move the needle forward on outcomes and mortality, equity in care for women is needed. A contemporary approach to focus on the "well-woman" for early identification and treatment of women at risk for CVD is crucial.

9. CVD in women is multifaceted, in that it goes beyond traditional risk factors to include female-specific biological risk factors. There continues to be an urgent need for a collaborative approach among policy makers, scientists, researchers, clinicians, and patients, combining innovations in research design with policy solutions, to reduce women's burden of heart disease.

Back to Clinical Case

The case demonstrated severe and extensive disease three-vessel disease in a 54-year-old South Asian woman despite risk factors not being "significantly uncontrolled" by traditional assessment. She was thought to be "borderline" to her primary care doctor. Of note, she had the typical profile of a South Asian woman with significant CAD risks: family history of premature CAD and DM II, South Asian diet mixed with Western diet, low HDL, elevated TG, and BMI 27. For South Asian women, this risk factor profile commonly falls under the radar for aggressive risk factor modification and is detrimentally deemed to be "borderline" and therefore undertreated. This case highlights the need to take a closer examination of subgroups of women at even higher risk and exemplifies why the need for aggressive risk stratification and treatment is paramount.

References

[1] Mieres JH, Gulati M, Bairey Merz N, Berman DS, Gerber TC, Hayes SN, et al. Role of noninvasive testing in the clinical evaluation of women with suspected ischemic heart disease: a consensus statement from the American Heart Association. Circulation 2014;130:350–79.

[2] Mozaffarian D, Benjamin EJ, Go AS, Arnett DK, Blaha MJ, Cushman M, et al. Executive summary: heart disease and stroke statistics—2016 update: a report from the American Heart Association. Circulation 2016;133:447–54.

[3] Wenger NK. Women and coronary heart disease: a century after Herrick: understudied, underdiagnosed, and undertreated. Circulation 2012;126:604–11.

[4] Lichtman JH, Leifheit EC, Safdar B, Bao H, Krumholz HM, Lorenze NP, et al. Sex differences in the presentation and perception of symptoms among Young patients with myocardial infarction: evidence from the VIRGO study (variation in recovery: role of gender on outcomes of Young AMI patients). Circulation 2018;137:781–90.

[5] Women's Health Research: Progress, pitfalls, and promise. Institute of Medicine; September 23, 2010. Report Brief.

[6] Shaw LJ, Bugiardini R, Merz CN. Women and ischemic heart disease: evolving knowledge. J Am Coll Cardiol 2009;54:1561–75.

[7] Benjamin EJ, Muntner P, Alonso A, Bittencourt MS, Callaway CW, Carson AP, et al. Heart disease and stroke Statistics-2019 update: a report from the American Heart Association. Circulation 2019;139:e56–e528.

[8] Beltran-Sanchez H, Harhay MO, Harhay MM, McElligott S. Prevalence and trends of metabolic syndrome in the adult U.S. population, 1999–2010. J Am Coll Cardiol 2013;62:697–703.

[9] Gupta M, Singh N, Verma S. South Asians and cardiovascular risk: what clinicians should know. Circulation 2006;113:e924–9.

[10] Aggarwal NR, Patel HN, Mehta LS, Sanghani RM, Lundberg GP, Lewis SJ, et al. Sex differences in ischemic heart disease: advances, obstacles, and next steps. Circ Cardiovasc Qual Outcomes 2018;11:e004437.

[11] Bairey Merz CN, Shaw LJ, Reis SE, Bittner V, Kelsey SF, Olson M, et al. Insights from the NHLBI-sponsored Women's ischemia syndrome evaluation (WISE) study. Part II. Gender differences in presentation, diagnosis, and outcome with regard to gender-based pathophysiology of atherosclerosis and macrovascular and microvascular coronary disease. J Am Coll Cardiol 2006;47:S21–9.

[12] Shaw LJ, Bairey Merz CN, Pepine CJ, Reis SE, Bittner V, Kelsey SF, et al. Insights from the NHLBI-sponsored Women's ischemia syndrome evaluation (WISE) study. Part I. Gender differences in traditional and novel risk factors, symptom evaluation, and gender-optimized diagnostic strategies. J Am Coll Cardiol 2006;47:S4–S20.

[13] Sedlak TL, Lee M, Izadnegahdar M, Merz CN, Gao M, Humphries KH. Sex differences in clinical outcomes in patients with stable angina and no obstructive coronary artery disease. Am Heart J 2013;166:38–44.

[14] Gulati M, Cooper-DeHoff RM, McClure C, Johnson BD, Shaw LJ, Handberg EM, et al. Adverse cardiovascular outcomes in women with nonobstructive coronary artery disease: a report from the Women's ischemia syndrome evaluation study and the St James women take heart project. Arch Intern Med 2009;169:843–50.

[15] Bairey Merz CN, Andersen H, Sprague E, Burns A, Keida M, Walsh MN, et al. Knowledge, attitudes, and beliefs regarding cardiovascular disease in women: the Women's Heart Alliance. J Am Coll Cardiol 2017;70:123–32.

[16] Mosca L, Linfante AH, Benjamin EJ, Berra K, Hayes SN, Walsh BW, et al. National study of physician awareness and adherence to cardiovascular disease prevention guidelines. Circulation 2005;111:499–510.

[17] Bairey Merz CN, Andersen HS, Shufelt CL. Gender, cardiovascular disease, and the sexism of obesity. J Am Coll Cardiol 2015;66:1958–60.

[18] Zhao M, Vaartjes I, Graham I, Grobbee D, Spiering W, Klipstein-Grobusch K, et al. Sex differences in risk factor management of coronary heart disease across three regions. Heart 2017;103:1587–94.

[19] Hyun KK, Redfern J, Patel A, Peiris D, Brieger D, Sullivan D, et al. Gender inequalities in cardiovascular risk factor assessment and management in primary healthcare. Heart 2017;103:492–8.

[20] Schulman KA, Berlin JA, Harless W, Kerner JF, Sistrunk S, Gersh BJ, et al. The effect of race and sex on physicians' recommendations for cardiac catheterization. N Engl J Med 1999;340:618–26.

[21] Al-Khatib SM, Hellkamp AS, Hernandez AF, Fonarow GC, Thomas KL, Al-Khalidi HR, et al. Trends in use of implantable cardioverter-defibrillator therapy among patients hospitalized for heart failure: have the previously observed sex and racial disparities changed over time? Circulation 2012;125:1094–101.

[22] DeVon HA, Zerwic JJ. Symptoms of acute coronary syndromes: are there gender differences? A review of the literature. Heart Lung 2002;31:235–45.

[23] Canto JG, Goldberg RJ, Hand MM, Bonow RO, Sopko G, Pepine CJ, et al. Symptom presentation of women with acute coronary syndromes: myth vs. reality. Arch Intern Med 2007;167:2405–13.

[24] Canto JG, Rogers WJ, Goldberg RJ, Peterson ED, Wenger NK, Vaccarino V, et al. Association of age and sex with myocardial infarction symptom presentation and in-hospital mortality. JAMA 2012;307:813–22.

[25] Akhter N, Milford-Beland S, Roe MT, Piana RN, Kao J, Shroff A. Gender differences among patients with acute coronary syndromes undergoing percutaneous coronary intervention in the American College of Cardiology-National Cardiovascular Data Registry (ACC-NCDR). Am Heart J 2009;157:141–8.

[26] Blomkalns AL, Chen AY, Hochman JS, Peterson ED, Trynosky K, Diercks DB, et al. Gender disparities in the diagnosis and treatment of non-ST-segment elevation acute coronary syndromes: large-scale observations from the CRUSADE (can rapid risk stratification of unstable angina patients suppress adverse outcomes with early implementation of the American College of Cardiology/American Heart Association guidelines) National Quality Improvement Initiative. J Am Coll Cardiol 2005;45:832–7.

[27] Bugiardini R, Ricci B, Cenko E, Vasiljevic Z, Kedev S, Davidovic G, et al. Delayed care and mortality among women and men with myocardial infarction. J Am Heart Assoc 2017;6.

[28] D'Onofrio G, Safdar B, Lichtman JH, Strait KM, Dreyer RP, Geda M, et al. Sex differences in reperfusion in young patients with ST-segment-elevation myocardial infarction: results from the VIRGO study. Circulation 2015;131:1324–32.

[29] Roswell RO, Kunkes J, Chen AY, Chiswell K, Iqbal S, Roe MT, et al. Impact of sex and contact-to-device time on clinical outcomes in acute ST-segment elevation myocardial infarction-findings from the National Cardiovascular Data Registry. J Am Heart Assoc 2017;6.

[30] Piccini JP, Simon DN, Steinberg BA, Thomas L, Allen LA, Fonarow GC, et al. Differences in clinical and functional outcomes of atrial fibrillation in women and men: two-year results from the ORBIT-AF registry. JAMA Cardiol 2016;1:282–91.

[31] Sullivan RM, Zhang J, Zamba G, Lip GY, Olshansky B. Relation of gender-specific risk of ischemic stroke in patients with atrial fibrillation to differences in warfarin anticoagulation control (from AFFIRM). Am J Cardiol 2012;110:1799–802.

[32] Eisenberg E, Di Palo KE, Pina IL. Sex differences in heart failure. Clin Cardiol 2018;41:211–6.

[33] Blewer AL, McGovern SK, Schmicker RH, May S, Morrison LJ, Aufderheide TP, et al. Gender disparities among adult recipients of bystander cardiopulmonary resuscitation in the public. Circ Cardiovasc Qual Outcomes 2018;11:e004710.

[34] Gudnadottir GS, Andersen K, Thrainsdottir IS, James SK, Lagerqvist B, Gudnason T. Gender differences in coronary angiography, subsequent interventions, and outcomes among patients with acute coronary syndromes. Am Heart J 2017;191:65–74.

[35] Daly C, Clemens F, Lopez Sendon JL, Tavazzi L, Boersma E, Danchin N, et al. Gender differences in the management and clinical outcome of stable angina. Circulation 2006;113:490–8.

[36] Redfors B, Angeras O, Ramunddal T, Petursson P, Haraldsson I, Dworeck C, et al. Trends in gender differences in cardiac care and outcome after acute myocardial infarction in Western Sweden: a report from the Swedish web system for enhancement of evidence-based Care in Heart Disease Evaluated According to recommended therapies (SWEDEHEART). J Am Heart Assoc 2015;4.

[37] Quyyumi AA. Women and ischemic heart disease: pathophysiologic implications from the Women's ischemia syndrome evaluation (WISE) study and future research steps. J Am Coll Cardiol 2006;47:S66–71.

[38] Colella TJ, Gravely S, Marzolini S, Grace SL, Francis JA, Oh P, et al. Sex bias in referral of women to outpatient cardiac rehabilitation? A meta-analysis. Eur J Prev Cardiol 2015;22:423–41.

[39] Adams Jr KF, Fonarow GC, Emerman CL, LeJemtel TH, Costanzo MR, Abraham WT, et al. Characteristics and outcomes of patients hospitalized for heart failure in the United States: rationale, design, and preliminary observations from the first 100,000 cases in the Acute Decompensated Heart Failure National Registry (ADHERE). Am Heart J 2005;149:209–16.

[40] Hsich EM. Sex differences in advanced heart failure therapies. Circulation 2019;139:1080–93.

[41] Cleland JG, Daubert JC, Erdmann E, Freemantle N, Gras D, Kappenberger L, et al. The effect of cardiac resynchronization on morbidity and mortality in heart failure. N Engl J Med 2005;352:1539–49.

[42] Varma N, Manne M, Nguyen D, He J, Niebauer M, Tchou P. Probability and magnitude of response to cardiac resynchronization therapy according to QRS duration and gender in nonischemic cardiomyopathy and LBBB. Heart Rhythm 2014;11:1139–47.

[43] Lilli A, Ricciardi G, Porciani MC, Perini AP, Pieragnoli P, Musilli N, et al. Cardiac resynchronization therapy: gender related differences in left ventricular reverse remodeling. Pacing Clin Electrophysiol 2007;30:1349–55.

[44] Yancy CW, Jessup M, Bozkurt B, Butler J, Casey Jr DE, Drazner MH, et al. 2013 ACCF/AHA guideline for the management of heart failure: executive summary: a report of the American College of Cardiology Foundation/American Heart Association Task Force on practice guidelines. Circulation 2013;128:1810–52.

[45] Thompson LE, Maddox TM, Lei L, Grunwald GK, Bradley SM, Peterson PN, et al. Sex differences in the use of Oral anticoagulants for atrial fibrillation: a report from the National Cardiovascular Data Registry (NCDR((R))) PINNACLE registry. J Am Heart Assoc 2017;6.

[46] Bai CJ, Madan N, Alshahrani S, Aggarwal NT, Volgman AS. Sex differences in atrial fibrillation-update on risk assessment, treatment, and long-term risk. Curr Treat Options Cardiovasc Med 2018;20:79.

[47] Lip GY, Laroche C, Boriani G, Cimaglia P, Dan GA, Santini M, et al. Sex-related differences in presentation, treatment, and outcome of patients with atrial fibrillation in Europe: a report from the Euro Observational Research Programme pilot survey on atrial fibrillation. Europace 2015;17:24–31.

[48] Wassertheil-Smoller S, McGinn AP, Martin L, Rodriguez BL, Stefanick ML, Perez M. The associations of atrial fibrillation with the risks of incident invasive breast and colorectal cancer. Am J Epidemiol 2017;185:372–84.

[49] Emdin CA, Wong CX, Hsiao AJ, Altman DG, Peters SA, Woodward M, et al. Atrial fibrillation as risk factor for cardiovascular disease and death in women compared with men: systematic review and meta-analysis of cohort studies. BMJ 2016;532:h7013.

[50] Vaccarino V. Angina and cardiac care: are there gender differences, and if so, why? Circulation 2006;113:467–9.

[51] Lichtman JH, Wang Y, Jones SB, Leifheit-Limson EC, Shaw LJ, Vaccarino V, et al. Age and sex differences in inhospital complication rates and mortality after percutaneous coronary intervention procedures: evidence from the NCDR((R)). Am Heart J 2014;167:376–83.

[52] Shehab A, Al-Dabbagh B, AlHabib KF, Alsheikh-Ali AA, Almahmeed W, Sulaiman K, et al. Gender disparities in the presentation, management and outcomes of acute coronary syndrome patients: data from the 2nd Gulf Registry of Acute Coronary Events (Gulf RACE-2). PLoS One 2013;8:e55508.

[53] Shaw LJ, Shaw RE, Merz CN, Brindis RG, Klein LW, Nallamothu B, et al. Impact of ethnicity and gender differences on angiographic coronary artery disease prevalence and in-hospital mortality in the American College of Cardiology-National Cardiovascular Data Registry. Circulation 2008;117:1787–801.

[54] Sharaf B, Wood T, Shaw L, Johnson BD, Kelsey S, Anderson RD, et al. Adverse outcomes among women presenting with signs and symptoms of ischemia and no obstructive coronary artery disease: findings from the National Heart, Lung, and Blood Institute-sponsored Women's ischemia syndrome evaluation (WISE) angiographic core laboratory. Am Heart J 2013;166:134–41.

[55] Schnohr P, Lange P, Scharling H, Jensen JS. Long-term physical activity in leisure time and mortality from coronary heart disease, stroke, respiratory diseases, and cancer. The Copenhagen City Heart Study. Eur J Cardiovasc Prev Rehabil 2006;13:173–9.

[56] Friberg L, Benson L, Rosenqvist M, Lip GY. Assessment of female sex as a risk factor in atrial fibrillation in Sweden: nationwide retrospective cohort study. BMJ 2012;344:e3522.

[57] Kim ES, Carrigan TP, Menon V. Enrollment of women in National Heart, Lung, and Blood Institute-funded cardiovascular randomized controlled trials fails to meet current federal mandates for inclusion. J Am Coll Cardiol 2008;52:672–3.

[58] Melloni C, Berger JS, Wang TY, Gunes F, Stebbins A, Pieper KS, et al. Representation of women in randomized clinical trials of cardiovascular disease prevention. Circ Cardiovasc Qual Outcomes 2010;3:135–42.

[59] Dhruva SS, Bero LA, Redberg RF. Gender bias in studies for Food and Drug Administration premarket approval of cardiovascular devices. Circ Cardiovasc Qual Outcomes 2011;4:165–71.

[60] Clayton JA, Collins FS. Policy: NIH to balance sex in cell and animal studies. Nature 2014;509:282–3.

[61] Collins FS, Tabak LA. Policy: NIH plans to enhance reproducibility. Nature 2014;505:612–3.

[62] Congress.gov. Research Act for All. 2015–2016.Congress.gov. Research Act for All. 2015–2016.

[63] Kannel WB, McGee DL. Diabetes and cardiovascular disease. The Framingham study. JAMA 1979;241:2035–8.

[64] Huxley R, Barzi F, Woodward M. Excess risk of fatal coronary heart disease associated with diabetes in men and women: meta-analysis of 37 prospective cohort studies. BMJ 2006;332:73–8.

[65] Wolf HK, Tuomilehto J, Kuulasmaa K, Domarkiene S, Cepaitis Z, Molarius A, et al. Blood pressure levels in the 41 populations of the WHO MONICA project. J Hum Hypertens 1997;11:733–42.

[66] Kip KE, Marroquin OC, Kelley DE, Johnson BD, Kelsey SF, Shaw LJ, et al. Clinical importance of obesity versus the metabolic syndrome in cardiovascular risk in women: a report from the Women's ischemia syndrome evaluation (WISE) study. Circulation 2004;109:706–13.

[67] Henry S, Bond R, Rosen S, Grines C, Mieres J. Challenges in cardiovascular risk prediction and stratification in women. Cardiovascular innovations and applications. vol. 3; 2019. p. 329–48 [20].

[68] Ford ES, Capewell S. Coronary heart disease mortality among young adults in the U.S. from 1980 through 2002: concealed leveling of mortality rates. J Am Coll Cardiol 2007;50:2128–32.

[69] Crilly M, Bundred P, Hu X, Leckey L, Johnstone F. Gender differences in the clinical management of patients with angina pectoris: a cross-sectional survey in primary care. BMC Health Serv Res 2007;7:142.

[70] Chou AF, Scholle SH, Weisman CS, Bierman AS, Correa-de-Araujo R, Mosca L. Gender disparities in the quality of cardiovascular disease care in private managed care plans. Womens Health Issues 2007;17:120–30.

[71] De Smedt D, De Bacquer D, De Sutter J, Dallongeville J, Gevaert S, De Backer G, et al. The gender gap in risk factor control: effects of age and education on the control of cardiovascular risk factors in male and female coronary patients. The EUROASPIRE IV study by the European Society of Cardiology. Int J Cardiol 2016;209:284–90.

[72] Dallongevillle J, De Bacquer D, Heidrich J, De Backer G, Prugger C, Kotseva K, et al. Gender differences in the implementation of cardiovascular prevention measures after an acute coronary event. Heart 2010;96:1744–9.

[73] Virani SS, Alonso A, Benjamin EJ, Bittencourt MS, Callaway CW, Carson AP, et al. Heart disease and stroke statistics—2020 update: a report from the American Heart Association. Circulation 2020.

[74] Mensah GA, Mokdad AH, Ford ES, Greenlund KJ, Croft JB. State of disparities in cardiovascular health in the United States. Circulation 2005;111:1233–41.

[75] Zhang ZM, Rautaharju PM, Prineas RJ, Rodriguez CJ, Loehr L, Rosamond WD, et al. Race and sex differences in the incidence and prognostic significance of silent myocardial infarction in the Atherosclerosis Risk in Communities (ARIC) study. Circulation 2016;133:2141–8.

[76] Thacker EL, Soliman EZ, Pulley L, Safford MM, Howard G, Howard VJ. Investigation of selection bias in the association of race with prevalent atrial fibrillation in a national cohort study: REasons for Geographic And Racial Differences in Stroke (REGARDS). Ann Epidemiol 2016;26:534–9.

[77] Carson AP, Howard G, Burke GL, Shea S, Levitan EB, Muntner P. Ethnic differences in hypertension incidence among middle-aged and older adults: the multi-ethnic study of atherosclerosis. Hypertension 2011;57:1101–7.

[78] Howard G, Safford MM, Moy CS, Howard VJ, Kleindorfer DO, Unverzagt FW, et al. Racial differences in the incidence of cardiovascular risk factors in older black and White adults. J Am Geriatr Soc 2017;65:83–90.

[79] Amponsah MK, Benjamin EJ, Magnani JW. Atrial fibrillation and race—a contemporary review. Curr Cardiovasc Risk Rep 2013;7.

[80] Mou L, Norby FL, Chen LY, O'Neal WT, Lewis TT, Loehr LR, et al. Lifetime risk of atrial fibrillation by race and socioeconomic status: ARIC study (Atherosclerosis Risk in Communities). Circ Arrhythm Electrophysiol 2018;11:e006350.

[81] Meschia JF, Merrill P, Soliman EZ, Howard VJ, Barrett KM, Zakai NA, et al. Racial disparities in awareness and treatment of atrial fibrillation: the REasons for Geographic and Racial Differences in Stroke (REGARDS) study. Stroke 2010;41:581–7.

[82] Schauer DP, Johnston JA, Moomaw CJ, Wess M, Eckman MH. Racial disparities in the filling of warfarin prescriptions for nonvalvular atrial fibrillation. Am J Med Sci 2007;333:67–73.

[83] Patel N, Deshmukh A, Thakkar B, Coffey JO, Agnihotri K, Patel A, et al. Gender, race, and health insurance status in patients undergoing catheter ablation for atrial fibrillation. Am J Cardiol 2016;117:1117–26.

[84] Kubicki DM, Xu M, Akwo EA, Dixon D, Muñoz D, Blot WJ, et al. Race and sex differences in modifiable risk factors and incident heart failure. JACC Heart Fail 2020;8:122–30.

[85] Breathett K, Yee E, Pool N, Hebdon M, Crist JD, Knapp S, et al. Does race influence decision making for advanced heart failure therapies? J Am Heart Assoc 2019;8:e013592.

[86] Breathett K, Liu WG, Allen LA, Daugherty SL, Blair IV, Jones J, et al. African Americans are less likely to receive care by a cardiologist during an intensive care unit admission for heart failure. JACC Heart Fail 2018;6:413–20.

[87] Hernandez AF, Fonarow GC, Liang L, Al-Khatib SM, Curtis LH, LaBresh KA, et al. Sex and racial differences in the use of implantable cardioverter-defibrillators among patients hospitalized with heart failure. JAMA 2007;298:1525–32.

[88] Lewey J, Choudhry NK. The current state of ethnic and racial disparities in cardiovascular care: lessons from the past and opportunities for the future. Curr Cardiol Rep 2014;16:530.

[89] Zilbermint M, Gaye A, Berthon A, Hannah-Shmouni F, Faucz FR, Lodish MB, et al. ARMC 5 variants and risk of hypertension in blacks: MH-GRID study. J Am Heart Assoc 2019;8:e012508.

[90] Brewster LM, van Montfrans GA, Kleijnen J. Systematic review: antihypertensive drug therapy in black patients. Ann Intern Med 2004;141:614–27.

[91] Johnson JA. Ethnic differences in cardiovascular drug response: potential contribution of pharmacogenetics. Circulation 2008;118:1383–93.

[92] Centers for Medicare & Medicaid Services. ACO shared savings program quality measures; 2017.

[93] Boltri JM, Okosun IS, Davis-Smith M, Vogel RL. Hemoglobin A1c levels in diagnosed and undiagnosed black, Hispanic, and white persons with diabetes: results from NHANES 1999–2000. Ethn Dis 2005;15:562–7.

[94] Nsiah-Kumi P, Ortmeier SR, Brown AE. Disparities in diabetic retinopathy screening and disease for racial and ethnic minority populations—a literature review. J Natl Med Assoc 2009;101:430–7.

[95] Narres M, Claessen H, Droste S, Kvitkina T, Koch M, Kuss O, et al. The incidence of end-stage renal disease in the diabetic (compared to the non-diabetic) population: a systematic review. PLoS One 2016;11:e0147329.

[96] Caballero AE. Type 2 diabetes in the Hispanic or Latino population: challenges and opportunities. Curr Opin Endocrinol Diabetes Obes 2007;14:151–7.

[97] Daviglus ML, Talavera GA, Aviles-Santa ML, Allison M, Cai J, Criqui MH, et al. Prevalence of major cardiovascular risk factors and cardiovascular diseases among Hispanic/Latino individuals of diverse backgrounds in the United States. JAMA 2012;308:1775–84.

[98] Stark Casagrande S, Fradkin JE, Saydah SH, Rust KF, Cowie CC. The prevalence of meeting A1C, blood pressure, and LDL goals among people with diabetes, 1988-2010. Diabetes Care 2013;36:2271–9.

[99] Gillespie CD, Wigington C, Hong Y, Centers for Disease Control and Prevention (CDC). Coronary heart disease and stroke deaths—United States, 2009. MMWR Suppl 2013;62:157–60.

[100] DeVon HA, Burke LA, Nelson H, Zerwic JJ, Riley B. Disparities in patients presenting to the emergency department with potential acute coronary syndrome: it matters if you are Black or White. Heart Lung 2014;43:270–7.

[101] Guzman LA, Li S, Wang TY, Daviglus ML, Exaire J, Rodriguez CJ, et al. Differences in treatment patterns and outcomes between Hispanics and non-Hispanic Whites treated for ST-segment elevation myocardial infarction: results from the NCDR ACTION Registry-GWTG. J Am Coll Cardiol 2012;59:630–1.

[102] Leifheit-Limson EC, D'Onofrio G, Daneshvar M, Geda M, Bueno H, Spertus JA, et al. Sex differences in cardiac risk factors, perceived risk, and health care provider discussion of risk and risk modification among Young patients with acute myocardial infarction: the VIRGO study. J Am Coll Cardiol 2015;66:1949–57.

[103] Hess CN, Kaltenbach LA, Doll JA, Cohen DJ, Peterson ED, Wang TY. Race and sex differences in post-myocardial infarction angina frequency and risk of 1-year unplanned rehospitalization. Circulation 2017;135:532–43.

[104] Rangrass G, Ghaferi AA, Dimick JB. Explaining racial disparities in outcomes after cardiac surgery: the role of hospital quality. JAMA Surg 2014;149:223–7.

[105] Lauffenburger JC, Robinson JG, Oramasionwu C, Fang G. Racial/ethnic and gender gaps in the use of and adherence to evidence-based preventive therapies among elderly Medicare part D beneficiaries after acute myocardial infarction. Circulation 2014;129:754–63.

[106] Palaniappan L, Wang Y, Fortmann SP. Coronary heart disease mortality for six ethnic groups in California, 1990-2000. Ann Epidemiol 2004;14:499–506.

[107] Jose PO, Frank AT, Kapphahn KI, Goldstein BA, Eggleston K, Hastings KG, et al. Cardiovascular disease mortality in Asian Americans. J Am Coll Cardiol 2014;64:2486–94.

[108] Dave TH, Wasir HS, Prabhakaran D, Dev V, Das G, Rajani M, et al. Profile of coronary artery disease in Indian women: correlation of clinical, non invasive and coronary angiographic findings. Indian Heart J 1991;43:25–9.

[109] National Kidney Foundation. Asian Americans and kidney disease; 2017.

[110] Brown TM, Voeks JH, Bittner V, Safford MM. Variations in prevalent cardiovascular disease and future risk by metabolic syndrome classification in the REasons for Geographic and Racial Differences in Stroke (REGARDS) study. Am Heart J 2010;159:385–91.

[111] Kanaya AM, Kandula NR, Ewing SK, Herrington D, Liu K, Blaha MJ, et al. Comparing coronary artery calcium among U.S. South Asians with four racial/ethnic groups: the MASALA and MESA studies. Atherosclerosis 2014;234:102–7.

[112] Consultation WHOE. Appropriate body-mass index for Asian populations and its implications for policy and intervention strategies. Lancet 2004;363:157–63.

[113] Shah AD, Kandula NR, Lin F, Allison MA, Carr J, Herrington D, et al. Less favorable body composition and adipokines in South Asians compared with other US ethnic groups: results from the MASALA and MESA studies. Int J Obes (Lond) 2016;40:639–45.

[114] Akeroyd JM, Chan WJ, Kamal AK, Palaniappan L, Virani SS. Adherence to cardiovascular medications in the South Asian population: a systematic review of current evidence and future directions. World J Cardiol 2015;7:938–47.

[115] Joshi P, Islam S, Pais P, Reddy S, Dorairaj P, Kazmi K, et al. Risk factors for early myocardial infarction in South Asians compared with individuals in other countries. JAMA 2007;297:286–94.

[116] Enas EA, Chacko V, Senthilkumar A, Puthumana N, Mohan V. Elevated lipoprotein(a)—a genetic risk factor for premature vascular disease in people with and without standard risk factors: a review. Dis Mon 2006;52:5–50.

[117] Kulkarni KR, Markovitz JH, Nanda NC, Segrest JP. Increased prevalence of smaller and denser LDL particles in Asian Indians. Arterioscler Thromb Vasc Biol 1999;19:2749–55.

[118] Volgman AS, Palaniappan LS, Aggarwal NT, Gupta M, Khandelwal A, Krishnan AV, et al. Atherosclerotic cardiovascular disease in south Asians in the United States: epidemiology, risk factors, and treatments: a scientific statement from the American Heart Association. Circulation 2018;138:e1–e34.

[119] Flowers E, Lin F, Kandula NR, Allison M, Carr JJ, Ding J, et al. Body composition and diabetes risk in South Asians: findings from the MASALA and MESA studies. Diabetes Care 2019;42:946–53.

[120] Chowdhury R, Narayan KM, Zabetian A, Raj S, Tabassum R. Genetic studies of type 2 diabetes in South Asians: a systematic overview. Curr Diabetes Rev 2014;10:258–74.

[121] Kooner JS, Saleheen D, Sim X, Sehmi J, Zhang W, Frossard P, et al. Genome-wide association study in individuals of South Asian ancestry identifies six new type 2 diabetes susceptibility loci. Nat Genet 2011;43:984–9.

[122] Noh S, Beiser M, Kaspar V, Hou F, Rummens J. Perceived racial discrimination, depression, and coping: a study of Southeast Asian refugees in Canada. J Health Soc Behav 1999;40:193–207.

[123] Mossakowski KN. Coping with perceived discrimination: does ethnic identity protect mental health? J Health Soc Behav 2003;44:318–31.

[124] Mosca L, Hammond G, Mochari-Greenberger H, Towfighi A, Albert MA, American Heart Association Cardiovascular Disease and Stroke in Women and Special Populations Committee of the Council on Clinical Cardiology, Council on Epidemiology and Prevention, Council on Cardiovascular Nursing, Council on High Bloo. Fifteen-year trends in awareness of heart disease in women: results of a 2012 American Heart Association national survey. Circulation 2013;127:1254–63 [e1-29].

[125] Carnethon MR, Pu J, Howard G, Albert MA, Anderson CAM, Bertoni AG, et al. Cardiovascular health in African Americans: a scientific statement from the American Heart Association. Circulation 2017;136:e393–423.

[126] Arnett MJ, Thorpe Jr RJ, Gaskin DJ, Bowie JV, LaVeist TA. Race, medical mistrust, and segregation in primary care as usual source of care: findings from the exploring health disparities in integrated communities study. J Urban Health 2016;93:456–67.

[127] Muncan B. Cardiovascular disease in racial/ethnic minority populations: illness burden and overview of community-based interventions. Public Health Rev 2018;39:32.

[128] Garcia JA, Paterniti DA, Romano PS, Kravitz RL. Patient preferences for physician characteristics in university-based primary care clinics. Ethn Dis 2003;13:259–67.

[129] Turner JH, Neylan TC, Schiller NB, Li Y, Cohen BE. Objective evidence of myocardial ischemia in patients with posttraumatic stress disorder. Biol Psychiatry 2013;74:861–6.

[130] Cook NR, Paynter NP, Eaton CB, Manson JE, Martin LW, Robinson JG, et al. Comparison of the Framingham and Reynolds risk scores for global cardiovascular risk prediction in the multiethnic Women's Health Initiative. Circulation 2012;125:1748–56 [S1-11].

[131] Lakoski SG, Greenland P, Wong ND, Schreiner PJ, Herrington DM, Kronmal RA, et al. Coronary artery calcium scores and risk for cardiovascular events in women classified as "low risk" based on Framingham risk score: the multi-ethnic study of atherosclerosis (MESA). Arch Intern Med 2007;167:2437–42.

[132] National Cholesterol Education Program (NCEP) Expert Panel on Detection, Evaluation, and Treatment of High Blood Cholesterol in Adults (Adult Treatment Panel III). Third report of the National Cholesterol Education Program (NCEP) Expert Panel on Detection, Evaluation, and Treatment of High Blood Cholesterol in Adults (Adult Treatment Panel III) final report. Circulation 2002;106:3143–421.

[133] Ridker PM, Buring JE, Rifai N, Cook NR. Development and validation of improved algorithms for the assessment of global cardiovascular risk in women: the Reynolds risk score. JAMA 2007;297:611–9.

[134] Mosca L, Benjamin EJ, Berra K, Bezanson JL, Dolor RJ, Lloyd-Jones DM, et al. Effectiveness-based guidelines for the prevention of cardiovascular disease in women—2011 update: a guideline from the American Heart Association. Circulation 2011;123:1243–62.

[135] Hsia J, Rodabough RJ, Manson JE, Liu S, Freiberg MS, Graettinger W, et al. Evaluation of the American Heart Association cardiovascular disease prevention guideline for women. Circ Cardiovasc Qual Outcomes 2010;3:128–34.

[136] Goff Jr DC, Lloyd-Jones DM, Bennett G, Coady S, D'Agostino RB, Gibbons R, et al. 2013 ACC/AHA guideline on the assessment of cardiovascular risk: a report of the American College of Cardiology/American Heart Association Task Force on Practice Guidelines. Circulation 2014;129:S49–73.

[137] DeFilippis AP, Young R, Carrubba CJ, McEvoy JW, Budoff MJ, Blumenthal RS, et al. An analysis of calibration and discrimination among multiple cardiovascular risk scores in a modern multiethnic cohort. Ann Intern Med 2015;162:266–75.

[138] Rana JS, Tabada GH, Solomon MD, Lo JC, Jaffe MG, Sung SH, et al. Accuracy of the atherosclerotic cardiovascular risk equation in a large contemporary, multiethnic population. J Am Coll Cardiol 2016;67:2118–30.

[139] Hippisley-Cox J, Coupland C, Brindle P. Development and validation of QRISK3 risk prediction algorithms to estimate future risk of cardiovascular disease: prospective cohort study. BMJ 2017;357:j2099.

[140] Pike MM, Decker PA, Larson NB, St Sauver JL, Takahashi PY, Roger VL, et al. Improvement in cardiovascular risk prediction with electronic health records. J Cardiovasc Transl Res 2016;9:214–22.

[141] Mosca L, Appel LJ, Benjamin EJ, Berra K, Chandra-Strobos N, Fabunmi RP, et al. Evidence-based guidelines for cardiovascular disease prevention in women. American Heart Association scientific statement. Arterioscler Thromb Vasc Biol 2004;24:e29–50.

[142] Lundberg GP, Mehta LS, Sanghani RM, Patel HN, Aggarwal NR, Aggarwal NT, et al. Heart centers for women. Circulation 2018;138:1155–65.

[143] Mosca L, Ferris A, Fabunmi R, Robertson RM, American Heart Association. Tracking women's awareness of heart disease: an American Heart Association national study. Circulation 2004;109:573–9.

[144] Mosca L, Mochari-Greenberger H, Dolor RJ, Newby LK, Robb KJ. Twelve-year follow-up of American women's awareness of cardiovascular disease risk and barriers to heart health. Circ Cardiovasc Qual Outcomes 2010;3:120–7.

[145] Mastroianni AC, Faden R, Federman D, editors. Women and health research: ethical and legal issues of including women in clinical studies, vol. I; 1994. Washington (DC): National Academies Press (US).

[146] Wizemann TM, Pardue ML, editors. Exploring the biological contributions to human health: does sex matter?; 2001. Washington (DC): National Academies Press (US).

[147] NIH Policy and Guidelines on The Inclusion of Women and Minorities as Subjects in Clinical Research NOT-OD-02-001; 2001.

[148] Women's Health Research Roadmap. A strategy for science and innovation to improve the health of women; 2015.

[149] Wood SF, Mieres JH, Campbell SM, Wenger NK, Hayes SN, Scientific Advisory Council of WomenHeart: The National Coalition for Women With Heart Disease. Advancing women's heart health through policy and science: highlights from the first National Policy and science summit on women's cardiovascular health. Womens Health Issues 2016;26:251–5.

[150] Friedewald VE, Hayes SN, Pepine CJ, Roberts WC, Wenger NK. The editor's roundtable: the 10Q report—advancing women's heart health through improved research, diagnosis, and treatment. Am J Cardiol 2013;112:1676–87.

[151] Health Insurance Coverage and the Affordable Care Act; 2015.

[152] Chen J, Vargas-Bustamante A, Mortensen K, Ortega AN. Racial and ethnic disparities in health care access and utilization under the affordable care act. Med Care 2016;54:140–6.

[153] Rodriguez CJ, Allison M, Daviglus ML, Isasi CR, Keller C, Leira EC, et al. Status of cardiovascular disease and stroke in Hispanics/Latinos in the United States: a science advisory from the American Heart Association. Circulation 2014;130:593–625.

[154] Liao Y, Siegel PZ, White S, Dulin R, Taylor A. Improving actions to control high blood pressure in Hispanic communities—racial and ethnic approaches to community health across the U.S. Project, 2009–2012. Prev Med 2016;83:11–5.

[155] Gallo LC, Fortmann AL, McCurley JL, Isasi CR, Penedo FJ, Daviglus ML, et al. Associations of structural and functional social support with diabetes prevalence in U.S. Hispanics/Latinos: results from the HCHS/SOL Sociocultural Ancillary Study. J Behav Med 2015;38:160–70.

[156] Georgia Stroke and Heart Attack Prevention Program (SHAPP). Million hearts; 2006.

[157] Stoddard AM, Palombo R, Troped PJ, Sorensen G, Will JC. Cardiovascular disease risk reduction: the Massachusetts WISEWOMAN project. J Womens Health (Larchmt) 2004;13:539–46.

[158] Kandula NR, Dave S, De Chavez PJ, Bharucha H, Patel Y, Seguil P, et al. Translating a heart disease lifestyle intervention into the community: the South Asian Heart Lifestyle Intervention (SAHELI) study; a randomized control trial. BMC Public Health 2015;15:1064.

[159] Dreyer RP, Dharmarajan K, Kennedy KF, Jones PG, Vaccarino V, Murugiah K, et al. Sex differences in 1-year all-cause rehospitalization in patients after acute myocardial infarction: a prospective observational study. Circulation 2017;135:521–31.

[160] Humphries KH, Pu A, Gao M, Carere RG, Pilote L. Angina with "normal" coronary arteries: sex differences in outcomes. Am Heart J 2008;155:375–81.

[161] Jneid H, Fonarow GC, Cannon CP, Hernandez AF, Palacios IF, Maree AO, et al. Sex differences in medical care and early death after acute myocardial infarction. Circulation 2008;118:2803–10.

[162] Kunadian V, Qiu W, Lagerqvist B, Johnston N, Sinclair H, Tan Y, et al. Gender differences in outcomes and predictors of all-cause mortality after percutaneous coronary intervention (data from United Kingdom and Sweden). Am J Cardiol 2017;119:210–6.

[163] Benjamin EJ, Blaha MJ, Chiuve SE, Cushman M, Das SR, Deo R, et al. Heart disease and stroke statistics-2017 update: a report from the American Heart Association. Circulation 2017;135:e146–603.

[164] Shaw LJ, Pepine CJ, Xie J, Mehta PK, Morris AA, Dickert NW, et al. Quality and equitable health care gaps for women: attributions to sex differences in cardiovascular medicine. J Am Coll Cardiol 2017;70:373–88.

[165] Brown HL, Warner JJ, Gianos E, Gulati M, Hill AJ, Hollier LM, et al. Promoting risk identification and reduction of cardiovascular disease in women through collaboration with obstetricians and gynecologists: a presidential advisory from the American Heart Association and the American College of Obstetricians and Gynecologists. Circulation 2018;137:e843–52.

[166] Committee on Gynecologic Practice. Committee Opinion No. 534. Well-woman visit. Obstet Gynecol 2012;120:421–4.

[167] Upadhya B, Kitzman DW. Heart failure with preserved ejection fraction in older adults. Heart Fail Clin 2017;13:485–502.

[168] Bairey Merz CN, Pepine CJ, Walsh MN, Fleg JL. Ischemia and no obstructive coronary artery disease (INOCA): developing evidence-based therapies and research agenda for the next decade. Circulation 2017;135:1075–92.

[169] Reynolds HR, Srichai MB, Iqbal SN, Slater JN, Mancini GB, Feit F, et al. Mechanisms of myocardial infarction in women without angiographically obstructive coronary artery disease. Circulation 2011;124:1414–25.

[170] Lansky AJ, Hochman JS, Ward PA, Mintz GS, Fabunmi R, Berger PB, et al. Percutaneous coronary intervention and adjunctive pharmacotherapy in women: a statement for healthcare professionals from the American Heart Association. Circulation 2005;111:940–53.

Misunderstood: Sex differences in CVD and risk factors are poorly recognized by clinicians

Mistreated: Suboptimal awareness leads to under treatment of women

Misdiagnosed: Sex disparities in presentation result in delayed diagnosis of CVD

Mismanaged: Inappropriate management results in increased morbidity and mortality

Misfortune: Experience worse outcome and QOL

Misrepresented: Inadequate inclusion in studies

6 common misfortunes contributing to sex-specific disparities in CVD care

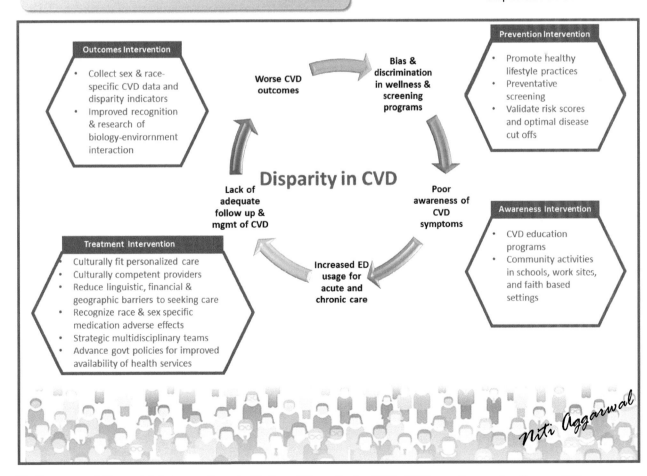

Challenges and Potential Interventions to Reduce Disparity in CVD. CVD, cardiovascular disease; ED, emergency department; mgmt, management; QOL, quality of life. *Image courtesy of Niti R. Aggarwal.*

Chapter 29

Women's Heart Programs

Niti R. Aggarwal and Sharon L. Mulvagh

Clinical Case

A 46-year-old female presented with a history of two acute coronary syndrome (ACS) episodes occurring over a 6-week interval, each confirmed by biomarker elevation and electrocardiogram (ECG) changes consistent with non-ST-elevation myocardial infarction (NSTEMI), and subsequent persistent intermittent chest pain, although of less severe degree than with the ACS episodes. She was perimenopausal, and experiencing significant emotional stress. A coronary angiogram at the time of the first ACS had shown nonobstructive coronary disease in the mid-left anterior descending artery (LAD); she was dismissed with reassurance, and initiation of aspirin and statin therapy, but because of persistent chest pain, and elevated troponin, a repeat angiogram was performed 6 weeks later. This showed resolution of the prior angiographic abnormality and no other abnormalities; she was started on a calcium channel blocker for suspected vasospasm and, since intermittent chest pain persisted, she was referred to the women's heart program. She had not been referred to cardiac rehabilitation. Additional historical details included preeclampsia and preterm birth with her second of two pregnancies; she subsequently had persistent hypertension, but no other traditional cardiovascular risk factors. Medications included: aspirin 81 mg qd, metoprolol 25 mg bid, rosuvastatin 10 mg qd, amlodipine 5 mg qd, and nitroglycerin spray sublingual as needed. On physical examination, the blood pressure was 130/73 on left and 135/75 on right, heart rate 80 bpm regular, and body mass index 34 kg/m². The only notable physical finding was of a left carotid bruit. Handheld echocardiogram showed normal left ventricle (LV) systolic function and no evidence of regional wall motion abnormality. How would you manage this patient?

Sex Differences in Cardiac Disease. https://doi.org/10.1016/B978-0-12-819369-3.00002-2

Abstract

Specialized cardiovascular care for women, delivered in "women's heart programs" (WHPs) focus on the cardiovascular needs of women. This chapter reviews the historical background leading to their development, their structure and specialized areas of focus including referral criteria for unique cardiovascular disease (CVD) risk factors in women, and CVD disorders resulting in acute coronary syndromes (ACS) with pathophysiologies seen exclusively, and/or more commonly in women including spontaneous coronary artery dissection (SCAD), myocardial infarction with nonobstructed coronary arteries (MINOCA), coronary microvascular dysfunction, and Takotsubo (stress) and pregnancy-related cardiomyopathies. The collaborative and multidisciplinary nature of WHPs is stressed, with key partners including those in obstetrics and gynecology, internal medicine, family medicine, maternal-fetal medicine, and cardiac rehabilitation specialists, specifically dieticians, physiotherapists, psychologists, and exercise physiologists. Finally, the roles of WHPs in education (clinical training programs and public awareness campaigns) and research in women with CVD are briefly discussed. The overarching goal of WHPs is to improve cardiovascular outcomes for women, through an enhanced awareness of sex-specific symptoms, cardiovascular risk factors, diagnoses, and treatments. Ultimately, successful achievement of reduction of glaring knowledge gaps in CVD care for women, with incorporation of evidence-based sex- and gender-specific cardiovascular guidelines into widespread and routine clinical practice, may reduce the need for these programs.

Introduction

Specialized cardiovascular care for women, delivered in "women's heart programs" (WHPs) also known in some institutions as "Heart Centers for Women" (HCW) began to emerge in the 1990s in response to the unmet cardiovascular needs of women and have become an integral component within the cardiovascular programs of many large healthcare centers and academic cardiology institutions around the world [1, 2]. A current estimate of the number of programs is challenging, as there is no centralized registration or credentialing, but search engines suggest that there are at least 50 in the United States [3], five in Canada [4], and a handful throughout the rest of the world [5, 6]. These programs have developed in response to increasing recognition that cardiovascular disease (CVD) is the most common cause of death for women worldwide; indeed, more women die from CVD than all cancers combined and at least five times more women die from CVD than breast cancer alone [7–9].

Subsequent to the early development of WHPs, national advocacy campaigns and grassroots organizations such as Go Red For Women, The Heart Truth, and WomenHeart were established in the early 2000s, promoting education and awareness of cardiovascular health in women, with steady reductions in CVD mortality rates to a lower level in females than in males for the first time since 1984 occurring in 2013. Thus, although these specialized programs might have initially been interpreted as a fad, or a marketing ploy on the part of larger multispecialty medical practices, it is concerning that the need only appears to have intensified with the recent observations of increasing rates of CVD mortality for both sexes, with an alarming plateau and increase in midlife women [10]. Of significant additional concern is that a surprisingly high number of physicians, both generalists and specialists, are not aware of the extent of CVD risks for women [11, 12], and moreover, they may be less confident in their knowledge, competence, and responsiveness to address CVD risk factors, diagnoses, and treatments in women, as compared with men [13, 14]. In addition, this knowledge gap does not seem to be improving; a recent survey reported that only 22% of primary care physicians and 42% of cardiologists felt "extremely prepared" to assess CVD risks in women; even fewer reported implementing guidelines for standard risk assessment in women [15].

Another consideration is the nuance of gendered communication. Although it is not a requirement, most WHPs are staffed primarily by female physicians, nurse practitioners, and nurses. Interestingly, it has been demonstrated that women physicians attending to female patients in the acute care setting achieve improved outcomes compared with care provided by their men colleagues. Although a similar comparative study in the outpatient clinical care setting has not been done, one might hypothesize that female patients receive improved cardiovascular care when provided in a setting predominantly staffed by women professionals who are motivated to provide optimal cardiovascular care for women.

Historical Perspective

As one of the first (1998) established WHPs, colleagues at the Mayo Clinic in Rochester, MN used their experience to address the broader need for focused cardiovascular care for women in a review article published in 2016 [1]. The purpose of the communication, initially conceived at a Mayo Clinic satellite symposium held in conjunction with the 2012 Scientific Sessions of the American Heart Association (AHA), "Women and Heart Disease: New Insights Across the Lifespan," was to provide guidance on CVD care for women and to encourage the establishment of focused centers for this care, staffed by providers familiar with sex- and gender-specific CVD issues, and promoting improved outcomes in women at risk of and living with CVD. The care model that was described was a team-based approach, consisting of cardiovascular subspecialty physicians and certified nurse practitioners, with available multidisciplinary and collaborative consultative resources to provide nutritional, physical activity, behavioral, rehabilitative, complementary, and integrative medical, interventional, and surgical support. The review compiled the evidence from MEDLINE and PubMed databases literature published from September 1, 1994 to September 1, 2015 using "CVD in women" plus the following search terms: adverse pregnancy outcomes (APOs), polycystic ovary syndrome, menopause, and menopausal hormone therapy (MHT), autoimmune disorders, and peripheral arterial disease; additional searches included the terms: Women's Heart Clinics, peripartum cardiomyopathy, coronary microvascular dysfunction (CMD), spontaneous coronary artery dissection (SCAD), apical ballooning syndrome, heart failure with preserved ejection fraction, and postural orthostatic tachycardia syndrome (POTS). The authors concluded there was emerging recognition of sex-based disparities in the treatment of, and survival from, heart disease which underscored a knowledge gap in the awareness, prevention, diagnosis, and treatment of heart disease in women and provided an overview of the understanding of cardiovascular physiology and pathophysiology across a woman's life span, identifying those CVD entities that are either uniquely or more often seen in women. Novel approaches to the evaluation and treatment of these disorders were discussed and the identification of newly recognized cardiovascular risk factors unique to women addressed. At that time, the concept of focused sex-specific cardiovascular care for women was new and the need for integrated multidisciplinary programs in women's heart health and disease was defined along with an introduction to unique approaches to primary and secondary preventive strategies, diagnostic testing, and treatment of ischemic heart disease in women.

Subsequently, the evidence has continued to mount. Most recently, a white paper on WHPs was published in Circulation [2]. This historical review (**Figure 1**) summarized that existing programs are present in both academic and private practice settings, are rapidly becoming integral components of most large healthcare centers and most academic cardiology programs, and are easily found through internet search engines. The authors noted that there is no single agency or organization that accredits or catalogs these programs, and some of the programs in academic institutions are active in research programs at the private, foundational, and federal levels. They also addressed the observation that even though awareness of CVD in women has increased among the public and healthcare providers, and both sex- and gender-specific research is currently required in all research trials, not all women have benefited equally in mortality reduction. Furthermore, they suggested that new strategies (**Figure 2** and **Table 2**) for these programs need to be developed to address these disparities and expand the current model with recommendations for the WHP care team to direct academic curricula on sex- and gender-specific research and care, expand to include other healthcare professionals and other subspecialties, provide new care models, address diversity, and include more male providers.

Structure and Specialized Areas of Focus of a Women's Heart Program

Most WHPs incorporate a team-based approach to care [1], comprised of cardiologists and advanced practice providers who form the core service, with referral access to additional resources to address nutritional, physical activity, cognitive, behavioral, rehabilitative, genetics, vascular, obstetric-gynecologic, complementary, and integrative medical, interventional, and surgical needs as identified.

Referral criteria and services offered in the WHPs usually include the following, and are summarized in **Figure 3**:

(1) CVD risk factor assessment [16], risk-reduction education, and treatment across the reproductive life span, including those associated with:
   a. differential impact of traditional CVD risks factors (diabetes, smoking, hypertension, hyperlipidemia, obesity/inactivity, and family history of premature CVD)
   b. sex-unique risk factors ("cardio-obstetric history") including:
      i. APOs [inclusive of hypertensive pregnancy disorders (HPD), gestational diabetes, preterm births, pregnancy loss, and fertility therapy]
      ii. ovarian disorder [polycystic ovarian syndrome (PCOS)]
      iii. menopausal (premature, early; vasomotor symptoms)
   c. CVD risk factors associated with autoimmune disorders more commonly seen in women:
      i. rheumatoid arthritis
      ii. systemic lupus erythematosus (SLE)
(2) CVD disorders resulting in ACS with unique pathophysiologies which are more common in women:
   a. SCAD and association of fibromuscular dysplasia (FMD)
   b. ACS associated with MINOCA
   c. CMD
   d. vasospasm
   e. Takotsubo (stress) cardiomyopathy (generally in postmenopausal women)
   f. premature ACS in young women with multiple CVD risk factors [e.g., familial hyperlipidemia (FH)]
   g. ACS more frequently caused by atherosclerotic plaque erosion than obstruction
(3) Evaluation, diagnostic testing, and treatment for symptomatic women with known or suspected ischemic heart disease [17]:
   a. obstructive CAD
   b. nonobstructive CAD/CMD

PATIENT

PROFESSIONAL

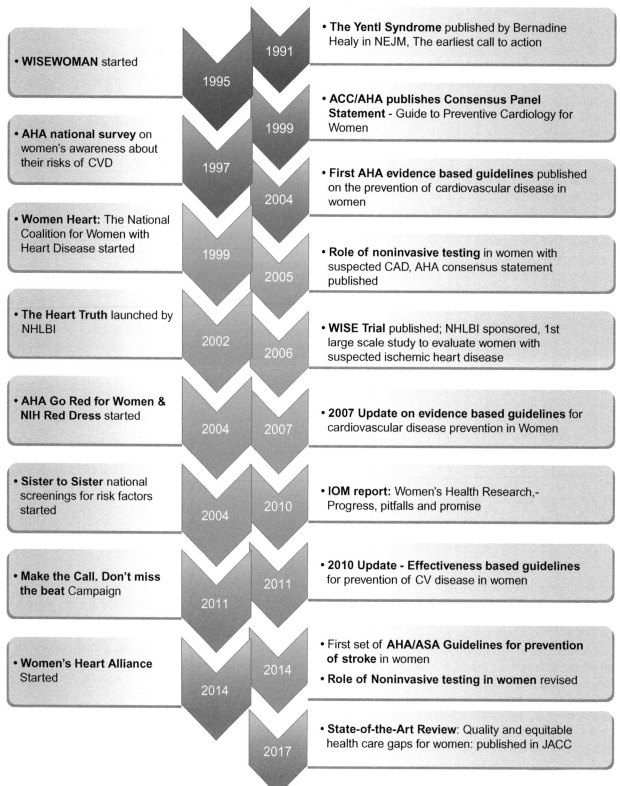

PATIENT

• **WISEWOMAN** started

• **AHA national survey** on women's awareness about their risks of CVD

• **Women Heart:** The National Coalition for Women with Heart Disease started

• **The Heart Truth** launched by NHLBI

• **AHA Go Red for Women & NIH Red Dress** started

• **Sister to Sister** national screenings for risk factors started

• **Make the Call. Don't miss the beat** Campaign

• **Women's Heart Alliance** Started

1995

1997

1999

2002

2004

2004

2011

2014

1991

1999

2004

2005

2006

2007

2010

2011

2014

2017

PROFESSIONAL

• **The Yentl Syndrome** published by Bernadine Healy in NEJM, The earliest call to action

• **ACC/AHA publishes Consensus Panel Statement** - Guide to Preventive Cardiology for Women

• **First AHA evidence based guidelines** published on the prevention of cardiovascular disease in women

• **Role of noninvasive testing** in women with suspected CAD, AHA consensus statement published

• **WISE Trial** published; NHLBI sponsored, 1st large scale study to evaluate women with suspected ischemic heart disease

• **2007 Update on evidence based guidelines** for cardiovascular disease prevention in Women

• **IOM report:** Women's Health Research,- Progress, pitfalls and promise

• **2010 Update - Effectiveness based guidelines** for prevention of CV disease in women

• First set of **AHA/ASA Guidelines for prevention of stroke** in women
• **Role of Noninvasive testing in women** revised

• **State-of-the-Art Review**: Quality and equitable health care gaps for women: published in JACC

FIGURE 1 Important Advances in Increasing Awareness and Research Regarding Cardiovascular Disease in Women. *Left,* Patient-targeted campaigns. *Right,* Healthcare professional-targeted research publications and significant initiatives. ACC, American College of Cardiology; AHA, American Heart Association; ASA, American Stroke Association; CAD, coronary artery disease; CVD, cardiovascular disease; IOM, Institute of Medicine; JACC, Journal of the American College of Cardiology; NEJM, New England Journal of Medicine; NHLBI, National Heart, Lung, and Blood Institute; WISE, Women's Ischemia Syndrome Evaluation. *Reproduced with permission from [2].*

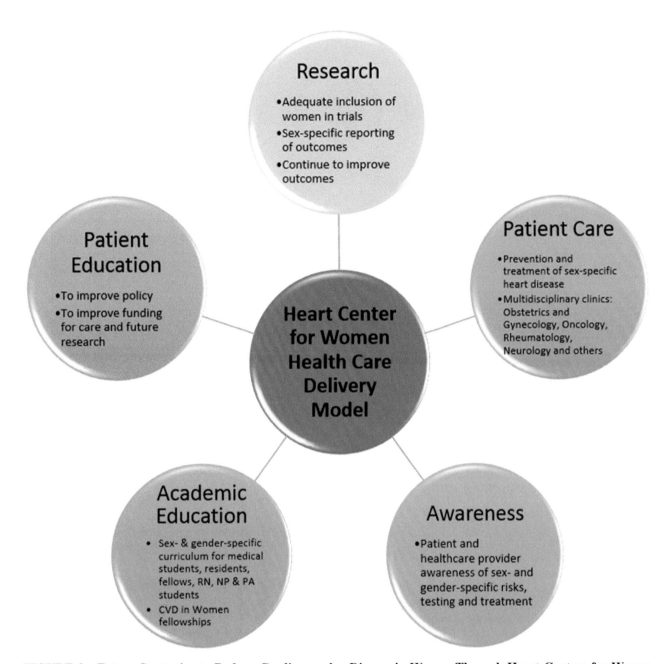

FIGURE 2 **Future Strategies to Reduce Cardiovascular Disease in Women Through Heart Centers for Women Healthcare Delivery Model.** CVD, cardiovascular disease; NP, nurse practitioner; PA, physician assistant; RN, registered nurse. *Reproduced with permission from [2].*

(4) Women with cancer-related treatment cardiac disease (CRTCD), unable to be addressed in a dedicated cardio-oncology clinic. Similarly, the shared risk factors for both breast cancer and CVD can be optimally addressed in WHPs [18]

Optional and emerging areas of care for women within WHPs include:

(1) Neurocognitive and geriatric CVD care; recent increasing awareness regarding the "connectedness" of heart-brain-mind conditions, including shared risk factors (hypertension, diabetes, smoking, hyperlipidemia, and obesity) and overlapping spectrum of cardiovascular disorders (myocardial infarction, heart failure, atrial fibrillation, and stroke) and vascular cognitive impairment (dementia), has particularly focused on the differential impact in women, who in most developed countries have longer life spans than men, and are more severely impacted [19].

(2) Some WHPs have developed a more comprehensive multidisciplinary pregnancy and CVD service that includes fertility and maternal-fetal medicine specialists,

Components of a Women's Heart Program

CardioRheumatology
Collaborate with
Rheumatology for CVD risk
assessment in pts with
autoimmune disorders

CVD & Menopause
Consultation for HRT in
women with
vasomotor sx

CVD Risk assessment
of a transgender pt

**Risk assessment &
education** of all women
with high CVD risk
**Mgmt of full spectrum
of CVD**--including SCAD,
micro & macro vascular
dz & Takotsubo CM

Pregnancy
Collaborate with Ob
to care for pregnant
pts with congenital
heart dz, prior
heart dz, or high-
risk pregnancies

CardioOncology
Collaborate with
oncology for CVD
prevention and
cardiotoxicity

**Neurocognitive &
Geriatric care**
Collaborate with
Neurology &
geriatrics for shared
risk factors of CVD &
dementia

FIGURE 3 Referral Criteria and Components of a Women's Heart Program. CM, cardiomyopathy; CVD, cardiovascular disease; dz, disease; HRT, hormone replacement therapy; Mgmt, management; Ob, obstetrics; Pts, patients; SCAD, spontaneous coronary artery dissection; Sx, symptoms.

TABLE 1 Suggested Ideal Elements for a Heart Center for Women
Ideal Elements for a Heart Center for Women
Accessible for all women regardless of geography, income, insurance, or time constraints
Ethnically, racially, and gender diverse multilingual healthcare teams committed to the mission of the center
Current, culturally appropriate and multilingual patient education materials
Personalized integrated care model utilizing existing guidelines or best practices when available
Multidisciplinary and integrated medical team approach (e.g., pregnancy and cardiovascular disease, cardiocognitive care, psychological support, nutrition, and exercise support)
Specialized patient intake forms with sex- and gender-specific detailed historical and screening questions pertinent to cardiovascular disease in women (reproductive history, migraines, and rheumatologic disorders) that include exercise, nutrition, lifestyle, sleep, stress, and cognitive information
Clinical database to track patient demographics, and patient volume, breadth of referral diagnosis, referral patterns and growth of the program, as well
Clinical database to track diagnoses with outcomes of lifestyle, behavior, and medical therapies
Addresses special needs of women throughout the life span from young adults to geriatric care, especially for women with genetic disorders such as adult congenital heart disease and familial hypercholesterolemia
Care models that address gender roles of women and social determinants of cardiovascular disease for women
Expertise in diseases unique to women or seen more often in women, such as spontaneous coronary artery dissection, Takotsubo cardiomyopathy, fibromuscular dysplasia, and coronary microvascular dysfunction

Reproduced with permission from [2].

obstetric anesthetists, general cardiologists, adult congenital heart disease specialists, and specialized nursing teams [2].

(3) An emerging area that is evolving within the domain of the WHP is CVD risk assessment and care for transgender patients, as there is a reported association between being transgender and myocardial infarction [20].

(4) An untapped area of potential collaboration is within radiology, where women routinely present for screening mammography; the presence of breast arterials calcifications and utilization of a deep learning algorithm may be utilized as a marker of atherosclerosis, and early detection of CVD [21]. Additionally, and/or alternatively, educational information on cardiovascular risk for women can be distributed within the waiting areas in the mammography suite and provide additional awareness regarding cardiovascular health and disease in women, and referral information.

Existing arrhythmia, heart failure, and valvular (structural heart disease) cardiology clinics generally serve women well, and while these areas are not necessarily considered core services within most WHPs, sex-specific issues are emerging, such as female predominance in heart failure with preserved ejection fraction (HFpEF) and POTS, poorer

outcomes of women with atrial fibrillation and stroke, differential sex-based outcomes associated with (i) timing of surgical intervention in valvular regurgitant lesions and (ii) novel procedural interventions, such as transcatheter aortic valve replacement (TAVR), of which not all cardiologists, and certainly few patients, may be aware, suggesting that the need for more specialized sex- and gender-specific care for women with CVD is becoming broader. Vascular clinics are increasingly recognizing the sex-specific aspects of vascular disorders [e.g., peripheral arterial disease (PAD)], and sex-specific aspects to physiological entities predisposing to them (e.g., hypertension), and although vascular disease assessments have generally remained in the domain of the vascular clinic, there has been a trend to provide more sex-specific care within these programs [22].

Suggested "Ideal Elements for a Heart Center for Women" were nicely summarized in the review article by Lundberg [2], and are included in **Table 1**.

Multidisciplinary Approach

A multidisciplinary approach has been successfully adopted by many WHPs, and direct communication and collaboration with internal medicine, family medicine,

TABLE 2 Future Goals of Heart Centers for Women Programs
Clinical Care
Develop evidence-based approaches to care incorporating existing sex- and gender-based guidelines
Provision of comprehensive patient-centered care customized to address cultural, ethnic, spiritual, and social determinants of the patient
Creation and implementation of a multidisciplinary healthcare team for women incorporating clinicians who care for women to improve the quality and equitable healthcare gaps in women including family physicians, primary care physicians, obstetricians and gynecologists, nurse practitioner, emergency department physicians, and nurses
Education and Health Literacy
Sex- and gender-specific cardiovascular education for all healthcare professionals
Patient education that integrates the tenets of health literacy
Development of educational modules and tool kits to facilitate the delivery of sex- and gender-specific cardiovascular care for women
Community Partnership
Creation of pathways to facilitate engagement with community groups and community health center staff to increase awareness of cardiovascular disease in women and provide information on navigating appointments with the medical team and health system for optimal cardiovascular care
Develop models of community-based participatory research to better assess unique cultural-, racial-, and community-based healthcare needs of women and to implement culturally sensitive approaches to meet the needs of underserved populations of women
Commitment to Research
Work with federal agencies, pharmaceutical, and device companies to continue to expand on current sex and gender registries and clinical trials and to ensure recruitment and participation of substantial numbers of women of diverse races and ethnicities
Develop novel research designed to investigate the impact of the social, racial, and ethnic determinants on cardiovascular health and disease in women and to gain insight into specific components that lead to high cardiovascular burden in black and South Asian women
Devise measures to assess the effectiveness of guidelines for the prevention, diagnosis, and treatment of women with cardiovascular disease

Reproduced with permission from [2].

maternal-fetal-medicine, obstetrics, and gynecology consultants caring for women seen in the WHP are key to successful outcomes (**Figure 4**). This multidisciplinary approach has been supported by collaborating specialties: the AHA and American College of Obstetrics and Gynecology developed a joint Presidential Advisory supporting improved collaboration between cardiology and obstetrics-gynecology to promote CVD risk identification and reduction during well-woman visits [23]. Similarly, supportive healthcare practitioners including dieticians, physiotherapists, psychologists, and exercise physiologists are key to a successful team approach in optimizing CVD care and prevention, both primary and secondary. Of increasing importance is the collaborative role of woman-to-woman patient peer support groups, and women's cardiovascular rehabilitation programs; this is particularly noted in the situation of younger women with ACS, including SCAD.

Unique Role of Women's Heart Programs: Enhancing Awareness of Sex-Specific Symptoms, Cardiovascular Risk Factors, Diagnoses, Treatments, and Improving Outcomes

Although recently it has been recognized that women who present with myocardial infarction (MI) usually have some type of chest pain syndrome [24], they may also have more

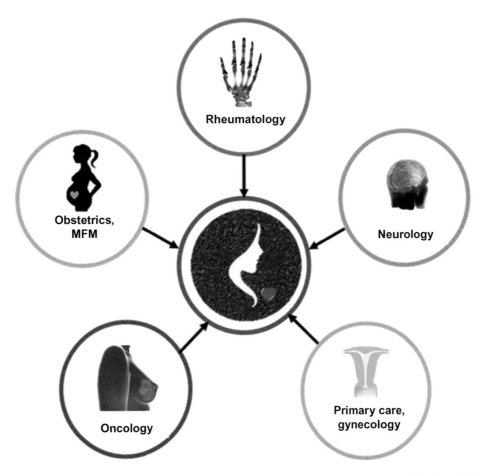

FIGURE 4 **Multidisciplinary Approach to a Women's Heart Program.** MFM, maternal-fetal medicine.

additional symptoms that may not be recognized as being consistent with ACS by the healthcare provider caring for them; moreover, they may have less commonly recognized etiologies such as SCAD, CMD, and coronary vasospasm, which result in further delay in time to diagnosis and appropriate care. Educating women and their healthcare providers about these variations in presentation and etiologies of ACS syndromes is imperative, in order to improve outcomes.

Thus, it has become apparent over the last several decades that angina symptoms in a woman have a broader differential diagnosis than in a man. The "female pattern" of ischemic heart disease is characterized by less classic obstructive coronary disease, and more nonobstructive, microvascular or nonatherogenic disease, which remains frequently unrecognized and demanding of specialized training on the part of the healthcare team and knowledge of appropriate testing.

Beyond the initial diagnosis, and acute treatment of ACS in women, which is associated with increased in-hospital and 30-day mortality compared to men [25], it is also important to recognize sex-specific differences in longer-term cardiovascular outcomes; women also fare worse than men with respect to 30-day and 1-year mortality after myocardial infarction [25].

WHPs can meet a need for much greater focus on secondary prevention of heart disease, and ensure that women are enrolled in cardiac rehabilitation (CR), which has been shown to be even more beneficial in women than men in improving cardiovascular outcomes [26]. It is a dismal fact that although referral to CR has proven benefits and has been deemed a performance measure of healthcare quality after acute myocardial infarction (AMI), over the last three decades more than 80% of eligible women have not had the benefit of CR [25]. CR for women includes CVD awareness, compliance with medications, mental fitness, and other forms of secondary prevention. Reported barriers to participation by women in CR include: lack of referral and/or lack of support by physicians, conflicts with time commitments due to work or family obligations, discomfort with the program, lack of transportation, insurance, and funding (**Figure 5**). Many HCW have coordinately developed CR programs exclusively for women. Although there

FIGURE 5 **Reported Barriers to Participation by Women in Cardiac Rehabilitation.** Rehab, rehabilitation.

have been a few conflicting reports regarding the efficacy of these programs, it seems logical that the encouragement and facilitation of involvement in CR provided by a WHP would impact positively on secondary prevention and on women's mental health, physical health, and general well-being.

A frequent concern heard from women at the conclusion of CR programs is that they do not yet feel confident to leave their safe place of the CR program. A novel solution has evolved from a Canadian Women's Heart program, which is based upon peer-to-peer learning and counseling, and is termed the "Women@Heart" program [27]. This program is a peer support program led *by* women with heart disease *for* women with heart disease that aims to create a caring environment for women to learn from each other. Women@Heart provides women who have heart disease, in every community, with access to emotional support, educational support, and a caring environment for a better recovery after a cardiac event. The program: (a) promotes coping to reduce risk of isolation associated with heart disease diagnosis in women, (b) empowers women to take charge of their heart health and to better understand their condition, (c) creates a caring environment for women to learn from each other by sharing their stories, and (d) is comprised of a volunteer force of women who become agents of change by being advocates for heart-healthy women in their communities. This concept of training women as grassroots patient advocates to support women with heart disease and educate the wider public about heart disease in women is

not new; WomenHeart.org (http://womenheart.org) was established over 20 years ago, and has partnered with Mayo Clinic Women's Heart Clinic to establish the WomenHeart Science and Leadership Symposium, an intense 5-day experience held in Rochester, MN. Each year since 2002, approximately 60–70 selected women with the lived experience of heart disease have participated in a comprehensive training program including a spectrum of educational offerings, from basic and applied medical science about cardiac disease in women, through media training, and have earned the designation of "WomenHeart Champions." These Champions have become community educators and support network coordinators for women heart patients as they tell their stories and share vital heart health information [28].

Another unique service that a WHP can provide includes counseling on MHT. Guidance for advising on MHT has recently been provided by an excellent commentary by Drs Lundberg and Wenger [29]. This topic is also covered in depth in Chapter 24. While the principal results of the Women's Health Initiative (WHI) [30] (in which the mean age of women was 63 years old) revealed no benefit and imparted increased risk for CVD in women on MHT, follow-up WHI studies, complemented by smaller subsequent studies, have identified no increased CVD risk if MHT is initiated earlier on in menopause with suggested improvement in imaging biomarkers that evaluate early detection of CVD [31–33]. Indeed, the effects of MHT on the risk of developing CVD appear to vary depending on when MHT is initiated in relation to a woman's age and/or her time since menopause, often termed the "timing hypothesis" [31]. Data suggest a reduced risk of CVD in women who start MHT when younger than 60 years and/or when within 10 years of menopause onset [34]. However, for women who initiate MHT > 10 or 20 years from menopause onset, risk of CVD, including CHD, stroke, and venous thromboembolism, appears increased [35]. Thus, counseling postmenopausal women for MHT requires an assessment of individual CVD risk, and history, as well as discussion on quality of life, and long-term follow-up availability; if indicated for symptoms, MHT should be administered in the lowest dose possible, preferably transdermal, and only for the necessary duration [36]. Postmenopausal women with known CVD or at high risk of CVD with significant postmenopausal symptoms and seeking MHT should be counseled on alternative nonhormonal therapies. In summary, MHT is currently not recommended for primary or secondary CVD prevention for women of any age [37], and its use as treatment for menopausal symptoms should be avoided in patients with established CVD [29].

The Role of Women's Heart Programs in Providing Trainee Guidance and Experience

Although there is no cohesive subspecialty training program for the development of expertise in women's cardiovascular health and disease, and no specific cardiology training requirements within the Accreditation Council for Graduate Medical Education (ACGME) accreditation process, many trainees at both the residency and fellowship levels rotate through WHPs to gain exposure to the unique aspects of female CVD diagnosis and care provided. Such programs can provide a concentrated experience of exposure to cases of women with or at risk for CVD, and unique sex-specific and/or CVD disorders more common in women such as SCAD, Takotsubo cardiomyopathy, FMD, familial hypercholesterolemia, and CMD. This focused approach to the care of women allows the trainees to develop clinical competencies in areas in which there are still few, if any, sex-specific guidelines and best practices. It behooves academic centers with WHPs to not only be leaders in clinical care and research but also lead efforts in undergraduate (medical school), postgraduate (residents and fellows), and continuing medical education (CME) (practicing physicians, and nurses) to raise awareness of CVD in women, and inform regarding the unique pathophysiologies, diagnosis, and treatment, in order to ultimately improve patient outcomes. Integration of sex- and gender-specific cardiovascular education throughout the training continuum, from medical school through postgraduate residency and fellowship training programs, must be introduced to develop clinical competencies.

An expansion of the diversity lens in healthcare education to also include sex- and gender-specific cardiovascular curricula is an important and emerging issue. This training similarly extends to nonphysician healthcare providers (nursing and other health science students). An example of a specialized training curriculum and dedicated clinical exposure to focused cardiovascular care of women to enhance the education of fellows has been proposed [38]. Similarly, the Canadian Women's Health Heart Alliance has compiled an "atlas"—a comprehensive review of the evidence focused on the sex and gender differences in epidemiology, diagnosis, and management of CVD—and hopes to update it annually [39]. Such publications will go a long way to teach and update the trainees and clinicians. Ongoing CME is essential for practicing physicians and healthcare providers to become aware of the unique aspects of CVD in women. Physicians and healthcare providers who staff WHPs are often tasked with providing and/or assisting in the delivery of these educational programs focused on sex- and gender-based differences in care, assessment of cardiovascular risk, treatment, and outcomes in women.

Role of Women's Heart Programs in Women-Centered CVD Registries and Research

The opportunity to be enrolled in clinical trials and registries related to unique aspects and disorders of women's CVD, and access to partnership with professional organizations aligned with women's CVD, are also important services and resources which may be organized and offered to patients within a WHP. This may serve as a mechanism to increase the proportion of women enrolled in CVD research studies, as despite their substantial CVD burden, women comprise only about 20–30% of enrolled patients in clinical trials, according to the AHA and the Heart and Stroke Foundation (Canada). WHPs provide opportunities to resolve the sex and gender imbalance and improve parity in clinical cardiovascular research in women by providing sites for the development of and recruitment into focused research protocols and develop evidence-based information for clinicians and patients. Although improvements have occurred since 1993, when the National Institutes of Health mandated that funded trials include female subjects and that sex-specific analyses of data be performed, actual implementation in research practice, publication, and translation into clinical practice has lagged. Although some of these metrics have been partially been achieved, those that have been most successful are disease specific, and studies of certain CVD conditions are lagging more in female participation than others [40]. The appropriateness of male:female distribution of study participants can be ascertained by the participation/prevalence ratio (PPR). For example, in a recent survey of participation of women in clinical trials supporting Food and Drug Administration (FDA) approval of cardiovascular drugs, women have been found to be over- or well represented in studies of pulmonary hypertension, hypertension, and atrial fibrillation, but underrepresented in studies of coronary artery disease, ACS, and heart failure [40]. There is a pressing need for further research into understanding sex and gender differences in cardiovascular conditions that disproportionately affect women such as MINOCA, CMD, HFpEF, and SCAD. Specialized women's heart centers should aim to encourage involvement of their patients in such research studies.

The establishment and maintenance of a robust clinical patient database is an essential component of a WHP, enabling interrogation and observations regarding local delivery of women's CVD care. External collaborations and development of specialized centers of excellence to facilitate the design and implementation of larger-scale research studies can subsequently evolve, in order to define and understand the unique features of CVD in women at the molecular, cellular, biological, patient, and population levels, with the ultimate goal of developing a strong evidence base for sex-specific guidelines in cardiovascular care.

Although both men and women investigators participate in cardiovascular research of both sexes, leadership by women in cardiovascular clinical trials has been lacking, in part due to the paucity of women within cardiovascular specialties. WHPs can play an important role in the training, support, mentorship, and sponsorship of junior and mid-career women in cardiology to become active participants and leaders in clinical trials, which is an essential foundational step to ensuring equity in cardiovascular research.

Role of Women's Heart Programs in Patient Education

A series of surveys from 1997 through 2012 [41] indicated a doubling in the improvement in awareness that heart disease was the number one killer of women from less than one-third of American women to approximately two-thirds; however, younger women and women of color were less aware. Disappointingly, a subsequent survey in 2014 conducted by the Women's Heart Alliance (WHA) revealed that 45% of women were unaware that heart disease is the leading killer of women, over 70% never raised the issue of heart health with their physician, and 26% described discussing CVD risk as embarrassing, assuming that risk was solely linked to weight [15]. Even in a medically focused community where 99% of participants in a longitudinal series study of Women's Heart Program-sponsored Go Red Awareness Luncheons and Healthcare Screening events were aware that CVD is the leading cause of death in women, less than half perceived themselves to be at personal risk, although 65% were "at risk" and 12% were "at high risk," demonstrating clear

discrepancies in theoretical knowledge and personal perception of CVD in women [42]. Beyond the imperative that all women "know their numbers" (cholesterol, blood pressure, blood sugar, and body mass index) and be knowledgeable about appropriate exercise, nutrition, and abstinence from smoking (the AHA's "Life's Simple 7") [43], there must be education regarding integration and personalization of this information, including discussion of CVD risk scores and "lifetime risk" with personalized implications. Importantly, review of sex-specific risk factors which are not included in conventional risk scores, with tailored lifestyle and medication intervention plans, is an important role served by WHPs in both the primary and secondary prevention of CVD in women. As mentioned above, educating both women *and their healthcare providers* in the sex- and gender-specific aspects of CVD in women is imperative in order to improve outcomes.

Similar to professional education, community education efforts regarding CVD in women are often undertaken by WHP personnel, at the local, regional, national, and international levels. Racial and ethnic differences in the prevalence of angina, MI, and congenital heart disease (CHD) mortality in women unfortunately exist and disproportionately affect black and South Asian women. Recognition and consideration of specific cultural, racial, age, and social determinants of health characteristics of the women in the community served can guide focused heart-healthy educational messaging that is customized and can promote health equity through elimination of cardiovascular healthcare disparities, with an overarching goal to improve cardiovascular outcomes in women [44].

Role of Women's Heart Programs in Guidelines and Policy Making

Professional organizations and health systems must develop sex-specific guidelines and introduce sex-specific best cardiovascular practices. A major challenge in the development of such sex-specific guidelines has been the glaring lack of sex-specific research analysis and reporting in the cardiovascular literature, hampering attempts to develop such guidelines due to lack of an adequate evidence base [45, 46]. It would seem that the professionals involved in WHPs are going to continue to be the responsible driving forces for the generation, acquisition, reporting, and publication of the needed sex- and gender-specific data analysis and interpretation. Perhaps, if one day these goals are achieved, there will no longer be a need for focused "women's heart programs," as all medical professionals and people will be aware of the sex-specific aspects of CVD, and there will no longer be a differential impact on healthcare equity and outcomes. Unfortunately, although efforts are coalescing and momentum is accelerating, through emerging enforcement of federal policies, and volunteer predominantly women-led professional-patient partnership organizations such as the Women Heart Association and Canadian Women's Heart Health Alliance (CWHHA), the timeline of such an achievement is likely to be fairly long, and until these goals are achieved, there will be continued and growing need for such centers of specialized cardiovascular care for women.

Summary and Conclusion

The cardiovascular health of women is strongly affected by sex-specific factors, including hormonal and metabolic disorders, APOs, menopausal status, and associated autoimmune diseases; women are predisposed to certain types of CVD which are more challenging to diagnose and treat, including SCAD, MINOCA, CMD, Takotsubo syndrome, and HFpEF, and women's outcomes are thus not surprisingly poorer [47]. Beyond sex-specific aspects of CVD, the impact of gender-specific factors and impact on CVD are only beginning to be understood [48], knowledge of gender assessment metrics is lacking, and implications on CVD care for women are only emerging. Awareness and recognition that women are at significant risk of CVD are crucial to provide appropriate care and avoid incorrect attribution of symptoms to noncardiac causes. Physicians, healthcare providers, and women themselves lack information on cardiovascular health and disease in women, making it less likely that they receive guidance on preventive strategies and re-

ferral for diagnostic testing, treatment, and CR. The public health cost of misdiagnosed or undiagnosed cardiac disease in women is significant. Moreover, although awareness of heart disease as the leading cause of death in women is improving, minority and younger women are often less aware, resulting in inadequate or nonexistent medical care and decreased likelihood of adopting necessary lifestyle changes, with an alarming recent increase in CVD mortality in young to midlife women.

Women's heart programs are currently essential not only for the recognition, diagnosis, and treatment of the CVD conditions and risk factors that are unique to, more common in, and/or uniquely expressed in women, but also for sex- and gender-specific education and research in CVD. The future goals of HCW programs are summarized in **Table 2** [2]. However, until education, awareness, and understanding of the unique sex and gender aspects of CVD are recognized, and existing knowledge gaps in pathophysiology, diagnosis, and treatment are alleviated with appropriate research, training, and achievement of competency by all healthcare providers caring for women, the need for these programs will continue. One could only hope that the success of these efforts will one day be that the very need for such programs will no longer exist, with elimination of these glaring knowledge gaps, and incorporation of evidence-based sex- and gender-specific cardiovascular guidelines into widespread and routine clinical practice. The ultimate success would then be achieved, and WHPs would then become obsolete.

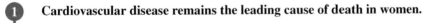

Key Points

1. **Cardiovascular disease remains the leading cause of death in women.**

2. **Women's heart programs have been developed to address sex differences in cardiovascular disease through improved education, clinical care, and sex-specific research.**

3. **Women's heart programs incorporate existing clinical evidence into sex-specific care pathways.**

4. **Multidisciplinary teams are an essential element of women's heart programs and lead to improved clinical care for women with sex-specific/female predominant conditions including pregnancy-associated heart disease, spontaneous coronary artery dissection, microvascular coronary artery disease, and stress cardiomyopathy.**

5. **Increased numbers of women's heart programs will most certainly result in improved awareness, not only of sex differences in heart disease in women, but also of the impact of socioeconomic disparities on health outcomes in women.**

6. **Women's heart programs provide an opportunity for focused basic, clinical, and translational research in women.**

Back to Clinical Case

The case represents a 46-year-old perimenopausal woman with two ACS episodes with intermittent chest pain, elevated troponins, and normal echocardiogram. Coronary angiography was unrevealing and demonstrated nonobstructive disease without an identifiable etiology to explain the episodes of ACS. Rereview of initial angiogram raised suspicion for possible spontaneous coronary artery dissection (SCAD) (type III) as the etiology of ACS; unfortunately, neither intracoronary nitroglycerin nor advanced intravascular imaging techniques (optical coherence tomography, or intravascular ultrasound) had been performed at the time of the initial angiogram. The possibility of a clinical SCAD diagnosis was discussed with the patient, the importance of optimizing blood pressure (BP) control was emphasized, and avoidance of extreme physical and/or emotional stress was emphasized. Computed tomographic angiography (CTA) of the entire vascular bed (head to pelvis imaging) was requested to assess for the presence of fibromuscular dypsplasia (FMD), given the reported association in approximately 50% of individuals with SCAD, and reinforced by the presence of the carotid bruit. A lipid profile was ordered. The patient was started on low-dose angiotensin-converting enzyme inhibitors (ACEis), and transdermal nitroglycerin, to further improve BP control and alleviate symptoms, and referred to the women's cardiac rehabilitation program for intensive lifestyle counseling and initiation of an exercise program.

Follow-up 1 month later: the CTA confirmed evidence of FMD in multiple vascular beds including cervical and renal. Lipid profile showed total cholesterol 96 mg/dL, low-density lipoprotein (LDL) 27 mg/dL, high-density lipoprotein (HDL) 42 mg/dL, triglycerides 126 mg/dL, and normal fasting glucose. A retrospective diagnosis of SCAD was thus presumed and further discussed with the patient. She had started the new medications, and was feeling better, with infrequent episodes of angina and rare need for nitroglycerin spray. She had started cardiac rehab and was enjoying this, especially the stress reduction guidance. On examination, BP was 110/82 mmHg, HR 65 bpm. Rosuvastatin was discontinued, and no new medications were advised. Inclusion in the SCAD registry, a referral to a vascular clinic (for further guidance on the presence of FMD and carotid bruit), and annual follow-up in the women's heart program were arranged.

References

[1] Garcia M, Miller VM, Gulati M, Hayes SN, Manson JE, Wenger NK, et al. Focused cardiovascular care for women: the need and role in clinical practice. Mayo Clin Proc 2016;91(2):226–40. https://doi.org/10.1016/j.mayocp.2015.11.001.

[2] Lundberg GP, Mehta LS, Sanghani RM, Patel HN, Aggarwal NR, Aggarwal NT, et al. Heart centers for women. Circulation 2018;138(11):1155–65. https://doi.org/10.1161/CIRCULATIONAHA.118.035351.

[3] Anon. Harvard heart letter. Harvard Health Publishing—Women's Heart Centers; 2009. Updated: 20 December 2016 https://www.health.harvard.edu/newsletterarticle/womens-heart-centers. [Accessed 11 January 2020].

[4] Anon. Canadian Women's Heart Health Alliance, https://cwhhc.ottawaheart.ca/tools-and-resources/womens-heart-health-programs-and-initiatives. [Accessed 20 February 2020].

[5] Anon. Women's Heart Center, Milan IT, https://www.cardiologicomonzino.it/en/operation-units/womens-heart-center/38/. [Accessed 20 February 2020].

[6] Low TT, Chan SP, Wai SH, Ang Z, Kyu K, Lee KY, et al. The women's heart health programme: a pilot trial of sex-specific cardiovascular management. BMC Womens Health 2018;18(1):56. https://doi.org/10.1186/s12905-018-0548-6.

[7] Institute of Medicine. (U.S.) committee on Women's Health Research. Women's Health Research: Progress, pitfalls, and promise. Washington, DC: National Academies Press; 2010.

[8] Anon. https://professional.heart.org/professional/ScienceNews/UCM_505440_Heart-Disease-and-Stroke-Statistics---2020-Update.jsp. [Accessed 20 February 2020].

[9] Norris CM, Yip CYY, Nerenberg KA, Clavel MA, Pacheco C, Foulds HJA, et al. State of the science in women's cardiovascular disease: a Canadian perspective on the influence of sex and gender. J Am Heart Assoc 2020;9(4). https://doi.org/10.1161/JAHA.119.015634 [Epub 2020 Feb 17].

[10] Shiels MS, Chernyavskiy P, Anderson WF, Best AF, Haozous EA, Hartge P, et al. Trends in premature mortality in the USA by sex, race, and ethnicity from 1999 to 2014: an analysis of death certificate data. Lancet 2017;389(10073):1043–54. https://doi.org/10.1016/S0140-6736(17)30187-3 [Epub 2017 Jan 26].

[11] Mosca L, Linfante AH, Benjamin EJ, Berra K, Hayes SN, Walsh BW, et al. National study of physician awareness and adherence to cardiovascular disease prevention guidelines. Circulation 2005;111(4):499–510.

[12] McDonnell LA, Turek M, Coutinho T, Nerenberg K, de Margerie M, Perron S, et al. Women's heart health: knowledge, beliefs, and practices of Canadian physicians. J Womens Health (Larchmt) 2018;27(1):72–82. https://doi.org/10.1089/jwh.2016.6240 [Epub 2017 Jun 12].

[13] Lichtman JH, Leifheit EC, Safdar B, Bao H, Krumholz HM, Lorenze NP, et al. Sex differences in the presentation and perception of symptoms among young patients with myocardial infarction: evidence from the VIRGO Study (Variation in Recovery: Role of Gender on Outcomes of Young AMI Patients). Circulation 2018;137(8):781–90. https://doi.org/10.1161/CIRCULATIONAHA.117.031650.

[14] Okunrintemi V, Valero-Elizondo J, Patrick B, Salami J, Tibuakuu M, Ahmad S, et al. Gender differences in patient-reported outcomes among adults with atherosclerotic cardiovascular disease. J Am Heart Assoc 2018;7(24). https://doi.org/10.1161/JAHA.118.010498.

[15] Bairey Merz CN, Andersen H, Sprague E, Burns A, Keida M, Walsh MN, et al. Knowledge, attitudes, and beliefs regarding cardiovascular disease in women: the Women's Heart Alliance. J Am Coll Cardiol 2017;70(2):123–32. https://doi.org/10.1016/j.jacc.2017.05.024 [Epub 2017 Jun 22. Erratum in: J Am Coll Cardiol. 2017 Aug 22;70(8):1106-1107].

[16] Agarwala A, Michos ED, Samad Z, Ballantyne CM, Virani SS. The use of sex-specific factors in the assessment of Women's cardiovascular risk. Circulation 2020;141(7):592–9. https://doi.org/10.1161/CIRCULATIONAHA.119.043429 [Epub 2020 Feb 17].

[17] Aggarwal NR, Patel HN, Mehta LS, Sanghani RM, Lundberg GP, Lewis SJ, et al. Sex differences in ischemic heart disease: advances, obstacles, and next steps. Circ Cardiovasc Qual Outcomes 2018;11(2). https://doi.org/10.1161/CIRCOUTCOMES.117.004437.

[18] Gulati M, Mulvagh SL. The connection between the breast and heart in a woman: breast cancer and cardiovascular disease. Clin Cardiol 2018;41(2):253–7. https://doi.org/10.1002/clc.22886.

[19] Anon. Heart-brain-mind connection, https://www.heartandstroke.ca/articles/report-2019. [Accessed 22 February 2020].

[20] Alzahrani T, Nguyen T, Ryan A, Dwairy A, McCaffrey J, Yunus R, et al. Cardiovascular disease risk factors and myocardial infarction in the transgender population. Circ Cardiovasc Qual Outcomes 2019;12. https://doi.org/10.1161/CIRCOUTCOMES.119.005597.

[21] Wang J, Ding H, Bidgoli F, Zhou B, Molloi S, Baldi P. Detecting cardiovascular disease from mammograms with deep learning. IEEE Trans Med Imaging 2017;36(5):1172–81. https://doi.org/10.1109/TMI.2017.2655486.

[22] Chung J, Coutinho T, Chu MWA, Ouzounian M. Sex differences in thoracic aortic disease: a review of the literature and a call to action. J Thorac Cardiovasc Surg 2020. https://doi.org/10.1016/j.jtcvs.2019.09.194. pii: S0022–5223(19)33123-X. [Epub ahead of print] No abstract available.

[23] Brown HL, Warner JJ, Gianos E, Gulati M, Hill AJ, Hollier LM, et al. Promoting risk identification and reduction of cardiovascular disease in women through collaboration with obstetricians and gynecologists: a Presidential Advisory From the American Heart Association and the American College of Obstetricians and Gynecologists. Circulation 2018;137:e843–52. https://doi.org/10.1161/CIR.0000000000000582.

[24] Ferry AV, Anand A, Strachan FE, Mooney L, Stewart SD, Marshall L, et al. Presenting symptoms in men and women diagnosed with myocardial infarction using sex-specific criteria. J Am Heart Assoc 2019;8(17). https://doi.org/10.1161/JAHA.119.012307 [Epub 2019 Aug 20].

[25] Mehta LS, Beckie TM, DeVon HA, Grines CL, Krumholz HM, Johnson MN, et al. Acute myocardial infarction in women: a scientific statement from the American Heart Association. Circulation 2016;133(9):916–47. https://doi.org/10.1161/CIR.0000000000000351 [Epub 2016 Jan 25].

[26] Thomas RJ, King M, Lui K, Oldridge N, Piña IL, Spertus J, et al. AACVPR/ACCF/AHA 2010 update: performance measures on cardiac rehabilitation for referral to cardiac rehabilitation/secondary prevention services: endorsed by the American College of Chest Physicians, the American College of Sports Medicine, the American Physical Therapy Association, the Canadian Association of Cardiac Rehabilitation, the clinical exercise physiology association, the European Association for Cardiovascular Prevention and Rehabilitation, the inter-American Heart Foundation, the National Association of clinical nurse specialists, the preventive cardiovascular nurses association, and the Society of Thoracic Surgeons. J Am Coll Cardiol 2010;56(14):1159–67. https://doi.org/10.1016/j.jacc.2010.06.006.

[27] Anon. https://cwhhc.ottawaheart.ca/programs-and-services/womenheart-program. [Accessed 27 April 2020].

[28] Anon. www.womenheart.org/about-us/become-a-womenheart-champion/. [Accessed 27 April 2020].

[29] Lundberg GP, Wenger NK. MHT menopause hormone therapy: what a cardiologist needs to know. In: Expert analysis; 2019. https://www.acc.org/latest-in-cardiology/articles/2019/07/17/11/56/menopause-hormone-therapy/. [Accessed 20 February 2022].

[30] Rossouw JE, Anderson GL, Prentice RL, LaCroix AZ, Kooperberg C, Stefanick ML, et al. Risks and benefits of estrogen plus progestin in healthy postmenopausal women: principal results from the Women's Health Initiative randomized controlled trial. JAMA 2002;288:321–33.

[31] Manson JE, Chlebowski RT, Stefanick ML, Aragaki AK, Rossouw JE, Prentice RL, et al. Menopausal hormone therapy and health outcomes during the intervention and extended poststopping phases of the Women's Health Initiative randomized trials. JAMA 2013;310:1353–68.

[32] Hodis HN, Mack WJ, Henderson VW, Shoupe D, Budoff MJ, Hwang-Levine J, et al. Vascular effects of early versus late postmenopausal treatment with estradiol. N Engl J Med 2016;374:1221–31.

[33] Harman SM, Black DM, Naftolin F, Brinton EA, Budoff MJ, Cedars MI, et al. Arterial imaging outcomes and cardiovascular risk factors in recently menopausal women: a randomized trial. Ann Intern Med 2014;161:249–60.

[34] Boardman HMP, Hartley L, Eisinga A, Main C, Roqué i Figuls M, Bonfill Cosp X, et al. Hormone therapy for preventing cardiovascular disease in post-menopausal women. Cochrane Database Syst Rev 2015. CD002229.

[35] Marjoribanks J, Farquhar C, Roberts H, Lethaby A, Lee J. Long-term hormone therapy for perimenopausal and postmenopausal women. Cochrane Database Syst Rev 2017. CD004143.

[36] Anon. The 2017 hormone therapy position statement of the North American Menopause Society. Menopause 2018;25(11):1362–87. https://doi.org/10.1097/GME.0000000000001241.

[37] Mosca L, Benjamin EJ, Berra K, Bezanson JL, Dolor RJ, Lloyd-Jones DM, et al. Effectiveness-based guidelines for the prevention of cardiovascular disease in women—2011 update: a guideline from the American Heart Association. Circulation 2011;123(11):1243–62. https://doi.org/10.1161/CIR.0b013e31820faaf8 [Epub 2011 Feb 14].

[38] Reza N, Adusumalli S, Saybolt MD, Silvestry FE, Sanghavi M, Lewey J, et al. Implementing a women's cardiovascular health training program in a cardiovascular disease fellowship: The MUCHACHA curriculum. Case Rep 2020;2(1). https://doi.org/10.1016/j.jaccas.2019.11.033. https://casereports.onlinejacc.org/content/2/1/164. [Accessed 20 February 2022].

[39] Norris NCM, Yip CYY, Nerenberg KA, Jaffer S, Grewal J, Levinsson ALE, et al. Introducing the Canadian Women's heart health Alliance ATLAS on the epidemiology, diagnosis, and Management of Cardiovascular Diseases in women. CJC Open 2020;2(3):145–50. https://doi.org/10.1016/j.cjco.2020.02.004.

[40] Scott PE, Unger EF, Jenkins MR, Southworth MR, McDowell TY, Geller RJ, et al. Participation of women in clinical trials supporting FDA approval of cardiovascular drugs. J Am Coll Cardiol 2018;71(18):1960–9. https://doi.org/10.1016/j.jacc.2018.02.070.

[41] Mosca L, Hammond G, Mochari-Greenberger H, Towfighi A, Albert MA. Fifteen-year trends in awareness of heart disease in women: results of a 2012 American Heart Association national survey. Circulation 2013;127(11):1254–63. e1-29 https://doi.org/10.1161/CIR.0b013e318287cf2f. [Epub 2013 Feb 19].

[42] Kling JM, Miller VM, Mankad R, Wilansky S, Wu Q, Zais TG, et al. Go red for women cardiovascular health-screening evaluation: the dichotomy between awareness and perception of cardiovascular risk in the community. J Womens Health (Larchmt) 2013;22(3):210–8. https://doi.org/10.1089/jwh.2012.3744.

[43] Anon. www.heart.org/en/healthy-living/healthy-lifestyle/my-life-check- -lifes-simple-7. [Accessed 27 April 2020].

[44] Shaw LJ, Pepine CJ, Xie J, Mehta PK, Morris AA, Dickert NW, et al. Quality and equitable health care gaps for women: attributions to sex differences in cardiovascular medicine. J Am Coll Cardiol 2017;70:373–88. https://doi.org/10.1016/j.jacc.2017.05.051.

[45] Norris CM, Tannenbaum C, Pilote L, Wong G, Cantor WJ, McMurtry MS. Systematic incorporation of sex-specific information into clinical practice guidelines for the management of ST-segment-elevation myocardial infarction: feasibility and outcomes. J Am Heart Assoc 2019;8(7). https://doi.org/10.1161/JAHA.118.011597.

[46] Tannenbaum C, Norris CM, McMurtry MS. Sex-specific considerations in guidelines generation and application. Can J Cardiol 2019;35:598–605. https://doi.org/10.1016/j.cjca.2018.11.011 [Epub 2018 Nov 24].

[47] Garcia M, Mulvagh SL, Merz CN, Buring JE, Manson JE. Cardiovascular disease in women: clinical perspectives. Circ Res 2016;118(8):1273–93. https://doi.org/10.1161/CIRCRESAHA.116.307547.

[48] Pelletier R, Khan NA, Cox J, Daskalopoulou SS, Eisenberg MJ, Bacon SL, et al. Sex versus gender-related characteristics: which predicts outcome after acute coronary syndrome in the young? J Am Coll Cardiol 2016;67(2):127–35. https://doi.org/10.1016/j.jacc.2015.10.067.

Editor's Summary: Women's Heart Programs

- Sex-specific history intake
- Integrate sex-specific guidelines into care
- Multilingual, ethnic and racially diverse workforce
- Accessible to all regardless of income, geography, insurance & geography
- Dedicated Women's heart programs

- Sex-specific CVD curriculum in medical schools, residencies and fellowships
- Sex-specific CVD educational modules and CME programs

- Improved inclusion of women in clinical trials
- Enforce sex- specific reporting of results

Niti Aggarwal

Clinical Care

Education

Research

Awareness

Community

Advocacy

- Multilingual educational material
- Improved awareness of sex- specific symptoms of CVD
- Public health awareness campaigns

- Identify underlying social, economic, and environmental forces that create health inequities in a community
- Recognize & resolve geographic and systemic barriers for widespread access to care, especially in underserved areas

- Address social determinants of health through policy enactment
- Improved funding for sex-specific research

Benefits of Women's Heart Programs. A women's Heart Program should encompass the tenets of clinical care, clinician education, public awareness, community partnership, research and advocacy to enhance CVD care of women. CME, continued medical education; CVD, cardiovascular disease. *Image courtesy of Niti R. Aggarwal.*

Index

Note: Page numbers followed by *f* indicate figures, *t* indicate tables, and *b* indicate boxes.

689

9780128193693